HANDBOOK OF
MANUFACTURING
ENGINEERING

MANUFACTURING ENGINEERING AND MATERIALS PROCESSING
A Series of Reference Books and Textbooks

FOUNDING EDITOR

Geoffrey Boothroyd
University of Rhode Island
Kingston, Rhode Island

Additional Volumes in Preparation

Metal Cutting Theory and Practice, *David A. Stephenson and John S. Agapiou*

HANDBOOK OF
MANUFACTURING ENGINEERING

edited by

JACK M. WALKER

Consultant, Manufacturing Engineering
Merritt Island, Florida

Marcel Dekker, Inc.   **New York•Basel•Hong Kong**

Library of Congress Cataloging-in-Publication Data

Handbook of manufacturing engineering / edited by Jack M. Walker.
    p. cm. — (Manufacturing engineering and materials processing ; 48)
    Includes bibliographical references and index.
    ISBN 0-8247-8962-8 (hardcover : alk. paper)
    1. Production engineering. I. Walker, Jack M.
  II. Series.
  TS176.H3365 1996
  658.5—dc20
                                                           96-18859
                                                           CIP

MARCEL DEKKER, INC.
270 Madison Avenue, New York, New York 10016

Current printing (last digit):
10 9 8 7 6 5 4 3 2 1

PRINTED IN THE UNITED STATES OF AMERICA

Foreword

In the mid-1940s, the United States was recognized worldwide as the leader in manufacturing. During World War II, we had pounded our plowshares into weapons to become the "arsenal of democracy." The outstanding success of the manufacturing buildup left the United States in a position to dominate many world markets that required these manufacturing skills. In the years since World War II, we have let our manufacturing capability atrophy; as a result, our share of the world's manufacturing capability has dropped drastically.

Many corporations have become aware of the fact that we have given away manufacturing markets to other countries. As a result, our workforce is becoming unbalanced between "soft" and "hard" products. Many say that we cannot compete in the manufacturing arena because of our wage structure. Others say that we can compete if we use all the productivity tools now available. The *Handbook of Manufacturing Engineering* makes a genuine contribution toward the goal of building market share in world manufacturing.

The editor of this book, Jack Walker, was active in manufacturing when I first met him in the early 1950s at McDonnell Aircraft Co. [now McDonnell Douglas Corp. (MDC)]. He was a manufacturing engineer in the Manufacturing Division, and I was a stress engineer in the Airplane Engineering Division. These two major divisions were usually at odds on who was "running the show," but Jack and I got along fine because we focused on solving the problem instead of defending our territory.

After moving up through most of the manufacturing disciplines, and spending a few years as head of Production Design Engineering, Jack became Director of Factory Operations for MDC's low-cost, high-rate missile plant. His knowledge and skills, combined with experience from several years in European manufacturing operations, make Jack well equipped to edit a book on manufacturing engineering. He has assembled a great team of contributing authors, all experts in their fields, who believe that we should make a turnaround to come back in the manufacturing world—and they tell the reader how this can be done. This is a modern handbook, primarily for the use of manufacturing engineers or other professionals who must perform the many different tasks in today's competitive manufacturing arena.

John F. Yardley
formerly
President, McDonnell Douglas Astronautics Co.
Senior Corporate Vice-President, McDonnell Douglas Corp.
NASA Associate Administrator for Manned Space Flight (Space Shuttle)

Preface

A handbook in its broadest sense, this volume is intended for use by the manufacturing engineer or other professional who must perform the many different tasks required in today's competitive manufacturing arena. Decision making by executives with a general surface knowledge of the manufacturing business, supported by a staff of narrow technical specialists, represents an outdated system—too costly and time consuming. Manufacturing professionals today are involved in management, product design, process design, process planning, tooling, equipment selection, facility planning and layout, plant construction, materials handling and storage, methods analysis, time standards, and production control—and all of the other working functions of a modern industrial plant.

This 1100-page handbook consists of 4 sections—Product and Factory Development, Factory Operations, Parts Fabrication, and Assembly Processes—each comprehensive enough to be a separate book! Today the manufacturing engineer is expected to understand each element of manufacturing in considerable depth and be able to work in teams that cross the old functional and departmental boundaries. The structure and organization of the handbook address the needs of the manufacturing engineer by permitting a logical progression through all elements of today's complex manufacturing processes.

One challenge we faced was to develop a definition of manufacturing engineering satisfactory to most of its practitioners. Since this initially seemed impossible, our approach was to provide "how to" information on the functions performed by most working manufacturing engineers and to include areas of related activities. But in today's tough manufacturing environment, the old functional lines have become blurred. The manufacture of a product demands consideration of not only the customer, who buys and uses the product, but also the manufacturing engineer's peers in product design, quality, accounting, and management. For this handbook, we define manufacturing engineering as the engineering approach to all factors involved in the production of goods or services.

To become broader in their knowledge of all the elements of the manufacturing process, manufacturing engineers must become more involved in understanding customers' needs and desires, and influencing product design to ensure the production of high-quality, low-cost, finished items. Section I covers modern techniques, such as Geoffrey Boothroyd's DFMA (design for manufacturing and assembly), that can be utilized by a multidisciplinary team. Here, the manufacturing engineer becomes the key player in establishing factory requirements (equipment required, material flow, product and process flow, and layout and arrangement). This information must then be converted into requirements for the architects, engineers, and contractors who will build or remodel the factory facilities. Tomorrow's factory will not be based on what worked in the past; rather it will be the result of thorough analysis and careful consideration of the complex infrastructure including computer networks required to support todays' sophisticated manufacturing processes and products.

The manufacturing engineer has to understand the human elements of the factory by learning and applying the principles of ergonomics, safety, and industrial hygiene in developing the factory processes. These directly influence workers' compensation and the other insurance premiums a company must pay—a measurable element of final product cost.

As Section II makes clear, the manufacturing engineer is heavily involved with the operation of the factory, starting with creating a detailed manufacturing plan for each product and product line. This continues with process planning to provide work instructions for the shop workers; estimating each operation and establishing standard hours and cycle times; and determining the machines, tools, and other factory facilities that will be used to produce the products. The control of quality is inherent in the manufacturing engineer's process planning. However, he or she must also work with the organization's quality professionals to be able to accurately control and measure the quality of the processes as well as the product. The control of production and materials is often the most vital part of running a factory. This involves plant capacity analysis, often using computer simulation of the production process to look for problems, bottlenecks, and inadequate work stations or tooling. Scheduling involves supplier deliveries as well as in-house operations. With the trend towards more "just-in-time" operations, shorter throughput cycle times, and lower inventory levels, this becomes a complex process, making use of computer programs that schedule the release of work to the shop and keep track of the production status and direct labor costs.

Detailed knowledge of parts fabrication operations is essential for engineers and other professionals in the manufacturing industry. Section III begins with a chapter on the principles of mechanics (forces, stresses, etc.) and an introduction to metallurgy and other materials characteristics. Products today are made from all different types of metals, plastics, and composites, and it is the material characteristics that help define, or limit, the fabrication processes selected. The processes utilized today, and in our factory of the future, include all the standard techniques of casting, forging, machining, pressing, molding, sheet metal fabrication, and others. However, the more nontraditional approaches also need to be understood, such as the thermal processes that employ lasers and electrical discharge machining, and the mechanical processes, including abrasive water-jet machining and burnishing.

All products are subjected to cleaning and finishing operations. This process starts with a thorough understanding of the corrosion mechanism, proceeds to cleaning, and concludes with the final protective and decorative finishes of plating or painting.

Assembly operations (Section IV), whether they are manual or involve some degree of mechanization or automation, contribute a major portion of the total cost of a product. The selection of the assembly process, and the influence of production rate and quantity, must be balanced against the equipment needed for the job. Materials handling plays an important role in assembly operations, which range from small, batch-type processing to continuous-flow assembly lines. In automation and robotics, the cost and flexibility of the system selected must be considered, another reason for the manufacturing engineer to be involved with the product design, as well as the production process. In addition to assembly, consideration is given to automated component testing and selection, which may be necessary to achieve a consistently assembled product.

This handbook is overdue in many respects. Although there are *design* reference handbooks that support the mechanical, electrical, and materials disciplines, and *process* references that are indispensable in the field of manufacturing engineering, there has not been a comprehensive design–process reference. The practicing professional, as well as the student, have had to collect a library of related information, much of it obtainable only from specialty books. There has been a need for a handbook to act as an overall umbrella for the many narrow-specialty technical books available.

We know many of the elements of the manufacturing engineer's role, and the kind of information he or she must have available in one volume, for use on a daily basis. Mechanical and electrical engineers remain the primary design groups for products to be manufactured, but many other people are involved in supporting the manufacturing process. They too need to be aware of these elements to varying degrees, depending on the particular product line, size of the plant, quantity and rate of production, and other factors. For example, architects who compete for the high-value contracts of designing and building new factory facilities need a good understanding of the manufacturing process. Plant engineers work to develop building specifications for new facilities, and remodel and maintain the present ones.

Tough, global competition is demanding better performance of all factories, wherever they are and whatever they build. This handbook will be valuable not only for those concerned with smaller plants, but also for those involved with manufacturing engineering in larger factories. Service organizations that supply manufacturers have an obvious need for information about manufacturing. Finally, colleges and universities offering engineering courses—particularly those with manufacturing-related subjects—will surely find this handbook useful.

The editor would like to thank the large group of professionals who contributed to this volume. The contributing authors' biographies are presented at the end of this book. In addition, William L. Walker was very helpful in preparing the overall outline and arrangement of the book, editing and revising sections to eliminate the overlap of material, and preparing computer-generated artwork. Finally, the editor would like to thank his wife Bernice, who typed manuscripts, prepared correspondence—and was supportive during the entire lengthy process of creating this handbook.

Jack M. Walker

Contents

Contents

III.   PARTS FABRICATION

Contributors

Frank Altmayer, M.S.   President, Scientific Control Laboratories, Chicago, Illinois

Shrikar Bhagath, M.S.   Project Engineer, Reliability, Parts and Materials Engineering Department, Delco Electronics Corporation, Kokomo, Indiana

Geoffrey Boothroyd, Ph.D., D.Sc.   President, Boothroyd Dewhurst, Inc., Wakefield, Rhode Island

Robert S. Busk, D.Eng.   President, International Magnesium Consultants, Inc., Hilton Head, South Carolina

Greg Chandler, M.B.A.   Manager, Manufacturing Engineering, Hubbell Premise Wiring, Inc., Marion, North Carolina

Stephen C. Cimorelli*   Manager, Manufacturing Systems, Learjet, Inc., Wichita, Kansas

Denise Burkus Harris, M.S.   Senior Electronics Packaging Engineer, Mechanical Design and Developmental Engineering Department, Westinghouse Corporation, Baltimore, Maryland

Alexander Houtzeel   Chairman, Houtzeel Manufacturing Systems Software, Inc., Waltham, Massachusetts

Robert L. Lints   Quality Consultant, Quality Assurance Systems, St. Louis, Missouri

John F. Maguire, Ph.D.   Program Director, Staff Scientist, Materials and Structures Division, Southwest Research Institute, San Antonio, Texas

Timothy L. Murphy   Group Manager, Human Resources, McDonnell Douglas Corporation, Titusville, Florida

Clyde S. Mutter   Manager of Tooling, CMS Defense Systems, Titusville, Florida

---

*Current affiliation: Manufacturing Systems Analyst, Square D Company, Asheville, North Carolina.

Michael Pecht, Ph.D. Professor and Director, CALCE Electronics Packaging Research Center (EPRC), University of Maryland, College Park, Maryland

Robert E. Persson Senior Corrosion Engineer, EG&G, Cape Canaveral, Florida

Allen E. Plogstedt, M.S.E.E.[†] Staff Director, MDC Fellow, McDonnell Douglas Corporation, Titusville, Florida

Marc Plogstedt Co-founder and Project Executive, ITEC Productions, Orlando, Florida

Lawrence J. Rhoades, M.B.A. President, Extrude Hone Corporation, Irwin, Pennsylvania

Paul R. Riedel Principal Industrial Hygienist, Dynacorp, Westminster, Colorado

Thomas J. Rose Technical Director, Advanced Processing Technology/Applied Polymer Technology, Inc., Norman, Oklahoma

Vijay S. Sheth, M.S. Principal Manufacturing Engineer, McDonnell Douglas Corporation, Titusville, Florida

John P. Tanner, M.B.A. President, Tanner and Associates, Orlando, Florida

V. M. Torbilo, Ph.D., D.Sc. Research Fellow, Mechanical Engineering Department, Ben-Gurion University of the Negev, Beer-Sheva, Israel

Jeffrey W. Vincoli, M.S., M.B.A. President, J. W. Vincoli and Associates, Titusville, Florida

Jack M. Walker Director of Factory Operations, McDonnell Douglas (Retired), and Consultant, Manufacturing Engineering, Merritt Island, Florida

William L. Walker Head of Facilities Engineering and Operations, National High Magnetic Field Laboratory, Florida State University, Tallahassee, Florida

Don Weed Staff Scientist, Southwest Research Institute, San Antonio, Texas

Bruce Wendle Senior Manufacturing Engineer, Advanced Input Devices, Coeur d'Alene, Idaho

Kjell Zandin Senior Vice President, International Business Development and Director, H. P. Maynard & Company, Inc., Pittsburgh, Pennsylvania

---

†Deceased.

1

Product Development

Jack M. Walker   *Consultant, Manufacturing Engineering, Merritt Island, Florida*

with

Geoffrey Boothroyd   *Boothroyd Dewhurst, Inc., Wakefield, Rhode Island*

The greatest durability contest in the world is trying to get a new idea into a factory.
Charles F. Kettering, CEO, General Motors

but

It's one heck of a lot easier to implement World Class Manufacturing principles than to go out of business.
William P. Conlin, President, Cal Comp, Inc.

and

You can't achieve World Class Manufacturing without first having a World Class Product Design!
Editor

1.0 INTRODUCTION TO PRODUCT DEVELOPMENT

This chapter on product development introduces the general field of concurrent engineering. In the first subchapter we discuss the need for teamwork among the various factory functions, starting with the customer's wants, and his early involvement in the product development cycle. Other subchapters describe in some detail two of the commercial systems (tools) available that may be used to formalize the satisfaction of the customer and in designing products that will make money for the company. These are Design for Assembly (DFA) and Quality Function Deployment (QFD). The final subchapter introduces some of the rapid prototyping techniques that allow design concepts to be produced quickly to assist in product development—and production start-up prior to availability of the planned "hard" tooling. Chapter 1 covers the first half of product development and design—the "systems" of determining who our real customers are and what they want, and the design of a product that meets these requirements at the lowest cost.

Chapter 2 introduces the detailed design parameters for parts fabricated by machining operations, by casting and forging, and of sheet metal; and a brief introduction of the materials and processes involved in each. All of the remaining chapters in this *Handbook of Manufacturing Engineering* provide additional information that is needed in order for the manufacturing engineer to participate properly in the product design process. Both Chapters 1 and 2 are part of the same process, but they look at the product from different perspectives.

1.0.1 World-Class Manufacturing

We hear a lot about *world-class manufacturing* today. Successful companies are very much involved; they consider it a requirement for survival in the competitive business environment. It is difficult to quantify the performance improvements required to become a world-class manufacturer. However, one significant study of several successful companies reveals the following features:

Costs to produce down 20–50%

Manufacturing lead time decreased by 50–90%

Overall cycle time decreased by 50%

Inventory down 50%+

Cost of quality reduced by 50%+ (Less than 5% of sales)

Factory floor space reduced by 30-70%

Purchasing costs down 5–10% every year

Inventory turns of raw material a minimum of 25

Manufacturing cycle times with less than 25% queue time

On-time delivery 98%

Quality defects from any cause less than 200 per million

While we can be certain that no one would agree that all of the above performance improvements are required, there is a strong message here that no one can dispute. We must make a paradigm shift in the way we operate to achieve all of these "future state" conditions—and must make some major changes to achieve any one of them! There is also general agreement that world-class manufacturing cannot be achieved without first having a *world-class product design.*

The product development and design process really starts by listening to the customer (the one who ultimately pays the bill!) and understanding his needs (and wants). The customer's requirements should be listed, prioritized, and answered completely during the product development and design phase. The other customer we must learn to listen to is our corporation—and especially our factory. We have heard messages such as:

Use the facilities we have.

Don't spend any more capital dollars beyond those needed to support our programs.

Watch your cash requirements. (This translates to any expenditure in the chain leading to product delivery—inventory, work in process, etc.)

Do the job with fewer people.

Continuous improvement is necessary, but we also need a paradigm shift in the way we do business!

Listen to our customers.

These are all great thoughts, but it is difficult to actually run a manufacturing operation with only these "high-level" goals. The bottom line (dollars of profit) was the *only* focus of many companies during the 1980s. The MBA mentality was micro-managing "profits" on a quarterly basis—to the detriment of the longer-term success of many companies. Today, we are all learning how this top-level focus has changed, and must "trickle down" to the working elements of the company, including the manufacturing engineering (ME) function. While we must do some rather uncomfortable things to keep our companies operating in the short term, the basic change in the way companies will operate in the future is our most important focus. The trained, experienced professional in today's manufacturing world—the person responsible for getting a "fantastic" product idea or dream or vision out the door of the factory—is facing a tremendous challenge. They will find that their training and experience has a lot of "holes" in it. It has nothing to do with how bright or hard-working the person is—the *job* has changed! This Manufacturing Engineer (ME) will be uncomfortable trying to do all elements of the job today, and their management will be even more uncomfortable! Management today is facing a worldwide struggle to keep their companies afloat. Many years ago, the owner of a struggling company understood this "new, recently discovered" requirement. One of my favorite stories may illustrate this.

Henry Ford defied all his "experts" by insisting that the wood shipping container for the Model A battery (the battery was just coming in to play as an addition to his latest model) be specified in a particular way. It must be made of a certain type of wood, be reinforced in places, and contain some rather peculiar "vent" holes in most of the pieces. He also insisted that this would not increase the cost of the battery from the supplier, since the quantities would be so large. (He was correct in this.) As the first Model A of the new series was coming down the line, he called all his department heads—engineers, buyers, accountants, and others—to the factory floor. The battery box was knocked down to get at the battery, and the battery was installed under the floor at the driver's feet. Henry then picked up the pieces of the box and fitted them above the battery, exactly forming the floor boards on the driver's side of the car. The holes in the boards were for the brake, starter switch, etc. The screws that held the box together were then used to bolt the floor boards down. Henry really understood the problems of total product design, cost of materials, just-in-time delivery, zero stock balance, low inventory cost, no material shortages, no waste disposal costs, good quality, and a host of others. He was a good ME and a one-man product design team!

In order to compete in today's marketplace, we must find new ways to increase productivity, reduce costs, shorten product cycles, improve quality, and be more responsive to customer needs. A good product design is probably the most important element to focus on.

Continuous Improvement is not a new idea. In the late 1980s, the author visited the IBM laptop computer assembly plant in Austin, Texas, and the Pro Printer line in Columbus, Georgia. These were excellent examples of the concurrent engineering process required for IBM to outperform their Japanese competitors. After studying these examples of computer integrated manufacturing (CIM), the McDonnell Douglas Missile Assembly Plant sponsored two CIM Application Transfer Studies with IBM's assistance: one concentrating on the production of low-lost, high-rate products, and the other on the production of high-cost, low-rate products.

Our strategy was for the studies to establish a "road map" for the transition to integrated manufacturing. This plan defined our existing state, what we wanted to become, projects and schedules that would be required, and the cost/benefits analyses. At that time, we had a somewhat fragmented approach to industrial automation, which included material requirements planning (MRP), manufacturing resource planning (MRPII), shop floor control, cost, process planning, bill of material, purchasing, stores, etc. Several of these were in independent functional areas with their own hardware, software, communications, files, databases, and so on. This was the situation that we were determined to improve. Our goal was to provide a factory that would support today's changing program requirements and that would be even more productive in the future. We believe that manufacturing systems and processes that simply modernize today's operations are not adequate for tomorrow's business environment. Rather, we need greater control over product cost and quality. We need to "outcompete" our competition—in both quality and cost.

Today, our company is well on the way to becoming the leading facility within the corporation in developing an architecture that ties our systems applications together and that fulfills the requirements of a truly integrated set of databases. CIM has evolved into the computer integrated enterprise (CIE), tieing the various plants in the corporation together—and now into an "extended enterprise" that includes our key suppliers. The goal of supporting the complete life cycle of product systems from design concept to development, to manufacturing, to customer support, and eventually to "disassembly" in a logical manner to perform maintenance or modifications, is still the focus.

1.0.2 Cost Analysis and Constraints

Although it may appear that many of us will build different products and systems, we see a network of common elements and challenges. All products have requirements for lower costs and improved quality. All can expect competition from producers of similar products, as well as from competing different products—both U.S. and foreign. All must accommodate rapid change in both product design and manufacturing processes.

The term "low-cost" is difficult to define. If your cost is about as low as your chief competitor, whatever the product or production rate is, you are probably a "low-cost" producer. Benchmarking against your most successful competitors is almost a necessity. In the manufacturing business, cost is partially attributable to the touch labor direct cost. This is modified by the direct support tasks performed by manufacturing engineering, production control, supervision, quality assurance and liaison engineering. On some programs this may equal or exceed the touch labor cost. Added to this is the more general overhead cost, which may double or triple the base cost. Also, since this in-house labor content may amount to a very small percentage of the overall product cost (which includes the cost paid to suppliers for material, parts, subassemblies, services, etc.), it may not be the only driving force in determining the price to the customer. Peter Drucker, in *The Practice of Management,* [1] states that "hourly employees' wages as a factor in product cost are down from 23% to 18%, and that productivity is on the rise." General Motors' hourly employees' costs are still in the 30% range, partially due to restrictive work rules in their labor contracts. Some Japanese car manufacturers who produce in the United States pay similar wages but operate at hourly employee costs of less than 20%. The trend is toward 15%. In selecting a design approach, it is perhaps more valuable to look at total cost in calculating earnings than to look at the amount of direct labor involved. One measurement of cost and earnings is return on investment (ROI).

ROI is a relationship between "bottom line" earnings, or profit, and the amount of money (assets) that must be spent (invested) to make these earnings. An example from the books of a medium size manufacturer shows the following:

(a)	Accounts receivable	$ 6.M
(b)	Inventory and work in process	14.M
(c)	Land	2.M
(d)	Buildings and equipment (net)	15.M
(e)	Total assets	$37.M

By looking closer at each item, we can improve the "bottom line."

(a) Submit billings sooner to reduce the amount that customers owe us.

(b) Reduce the amount of raw stock and hardware in inventory. Don't get the material in-house until it is needed. Reduce work in process by reducing the cycle time in the shop, which reduces the number of units in flow. Also, complete assemblies on schedule for delivery to permit billing the customer for final payment.

(c) Try not to buy any more land, since it cannot be depreciated: it stays on the books forever at its acquisition cost.

(d) Evaluate the payoff of any additional buildings and capital equipment. Of course, some new buildings and equipment may be needed in order to perform contract requirements. Additional investment may also be wise if it contributes to a lower product cost by reducing direct labor, support labor, scrap, units in process, cost of parts and assemblies, etc.

(e) The "bottom line" in investment may not be to reduce investment, but to achieve a greater return (earnings). In this simple example, we could add $10 million in equipment—and reduce accounts receivable, inventory, and work in process by $10 million—and have the same net assets. If this additional equipment investment could save $4 million in costs, we would increase our earnings and double our ROI.

In today's real world, we need to consider one more factor. Items (b), (c) and (d) above require company money to be spent. There is a limit to borrowing, however, and a point where cash flow becomes the main driver. It is therefore essential to reduce overall costs by utilizing existing buildings and capital equipment and by doing the job with fewer people and less inventory. The sharing of facilities and equipment between products becomes very attractive to all programs. This would reduce our need for additional capital and reduce the depreciation expense portion of our overhead. The design engineer and the manufacturing engineer are certainly key members of the product development and design team, but the input from all other factory functions becomes more important as we look toward the future.

1.0.3 Project Design Teams

In an increasing number of companies across the country, both the designer and the manufacturing engineer are climbing over the wall that used to separate the two functions. In addition, a full team consisting of all the development-related functions should participate in the design of a new product—or the improvement of an existing product. The team may include the following:

Design engineering

Process engineering

Manufacturing engineering

Manufacturing

Quality assurance

Marketing

Sales

Profitability, and even survival, depend on working together to come up with a product design that can be made easily and inexpensively into a quality product. Other appropriate team members may be representatives from:

Purchasing

Distribution

Accounting

Human Resources

Suppliers

Customers

An excellent example of quality improvement and cost reduction is Ford Motor Company. Ford has adopted "design for assembly" (and "design for manufacture and assembly") as one of their concurrent engineering approaches. The company acknowledges the importance of product improvement and process development before going into production. Assembly is a small part of overall product cost, about 10 to 15%. However, reducing this cost by a small amount results in big money saved in materials, work in process, warehousing, and floor space. Figure 1.1 shows how a small investment in good design has the greatest leverage on final product cost.

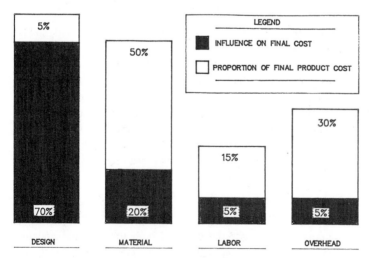

Figure 1.1 The cost of design in manufacturing at Ford. (Courtesy of Ford Motor Company.)

World-class manufacturers know that they cannot dedicate a single factory or production line to one model or product. The generic assembly line must have the flexibility to produce different models of similar products rapidly—and entirely new products with a minimum change-over time. There are exceptions where the production quantities of a single product are sufficiently large, and the projected product life is great enough, that a production factory dedicated to a single product is the best choice.

1.1 CONCURRENT ENGINEERING

1.1.0 Introduction to Concurrent Engineering

What is concurrent engineering? Why do we care about concurrent engineering? To remain competitive in industry, we must produce high-quality products and services the first time. We can accomplish this by impelmenting concurrent engineering (CE) principles and tools in our programs.

It does not matter whether we call the concept concurrent engineering, integrated product definition, or simultaneous engineering, as long as we consistently produce high-quality products on time for the best price. Agreeing on a common definition helps communication and understanding. We use the term concurrent engineering and the Department of Defense/Institute of Defense Analysis (DoD/IDA) definition because it has wide acceptance. IDA Report R-338, gives the following definition:

> Concurrent Engineering is a systematic approach to the integrated, concurrent design of products and their related processes, including manufacture and support. This approach is intended to cause the developers, from the outset, to consider all elements of the product life cycle from conception through disposal, including quality, cost, schedule and user requirements.

Four main elements have emerged as most important in implementing CE: the voice of the customer, multidisciplinary teams, automated tools, and process management. The underlying concept is not a new one: teamwork. The secret is to involve all the right people at the right time. We must increase our understanding of product requirements by being more effective in capturing the "voice of the customer." We must increase our emphasis on product producibility and supportability. This requires the involvement of all related disciplines on our CE teams. We must acquire and use the best tools to permit the efficient development of "build-to" technical data packages. Finally, we must increase our emphasis on developing production processes in parallel with development of the product.

Industry is continuing to refine and improve the elements of concurrent engineering. All of us must contribute to this process. More than that, all of us must be willing to change—and concurrent engineering requires change.

1.1.1 Why Concurrent Engineering?

Concurrent Engineering is a common-sense approach to product design, development, production, and support. By collecting and understanding all requirements that the product must satisfy throughout its life cycle at the start of concept definition, we can reduce costs, avoid costly redesign and rework, and shorten the development process. We do this by capturing all customer requirements and expectations and involving all related disciplines from the start. Working as teams on all product-related processes, we can provide for a smooth transition from development to production. Experience shows that concurrent engineering results in:

Well understood user requirements

Reduced cycle times

First time quality producible designs

Lower costs

Shorter development spans

A smoother transition to production

A new respect for other teammates

Highly satisfied customers

It is not surprising that some CE practices, namely voice of the customer and process management, are the least practiced. We need to leave our "comfort zones" in order to implement these practices effectively. Concurrent engineering pays off in:

Product development cycle time reduced 40–60%

Manufacturing costs reduced 30–40%

Engineering change orders reduced more than 50%

Scrap and rework reduced by as much as 75%

The primary elements of CE are:

Voice of the customer

Multidisciplinary teams

Automation tools and techniques

Process management

The voice of the customer includes the needs and expectations of the customer community, including end users. Concurrent engineering can best be characterized by the conviction that product quality can be achieved only by listening and responding to the voice of the customer. The ideal way is to capture, at the outset, all requirements and expectations in the product specifications.

The most effective way we have found to accomplish this is to conduct focus group sessions with different elements of the customer's organization. Properly staffed multi-disciplinary teams provide the means to enable all requirements, including producibility and supportability, to be an integral part of product design from the outset. CE teams are broadly based, including representatives from production, customer support, subcontract management and procurement, quality assurance, business systems, new business and marketing, and suppliers. Broadly based CE teams succeed because they can foresee downstream needs and build them into our products and processes.

Not all team members are necessarily full-time. In many cases, part-time participation is all that is needed and all that can be afforded.

Automation tools and techniques provide effective and efficient means of developing products and services. Computer-based tools such as Unigraphics can be used as an electronic development fixture (EDF) during prototyping, in lieu of mockups, to verify clearances and mechanism operations before hardware is fabricated.

There are a variety of home-grown and purchased CE tools in use on programs. Their importance to CE is an increased ability to communicate and transfer data readily among team members, customers, and suppliers. With reliable information sharing, we are able to review and comment (or import and use) product and process data more rapidly, while eliminating sources of error.

Process management is the final key to controlling and improving the organization as well as the processes used to develop, build, and support a product. This is probably the newest and least practiced element of CE. Big gains can be made by defining the program work flows and processes and then improving them. Processes define the relationships of tasks and link the organization's mission with the detailed steps needed to accomplish it. Process management is an effective way of managing in a team environment. Program-wide processes provide a means of identifying the players who need to be involved and indicate team interrelationships. Product processes are also a part of process management, requiring the definition and development of production processes in parallel with the definition and development of the product design.

1.1.2 Concurrent Engineering Throughout the Acquisition Process

Concurrent engineering practices are applicable to all programs, old or new, regardless of program type or acquisition phase. The biggest payoff is made by implementing CE at the very beginning of a program. Approximately 80% of a product's cost is determined during the concept phase, so it is very important to have manufacturing, producibility, support-ability, and suppliers involved then. There are benefits of CE to be realized during the later program phases, including reducing production costs for dual-source competitions or defining product improvements.

1.1.3 The Voice of the Customer

The voice of the customer (VOC) represents the needs and expectations by which the customer or user will perceive the quality of products and services. It is the program manager's responsibility to make sure that all CE participants understand the voice of the customer and that all program requirements can be traced back to those needs and expectations.

Quality is achieved by satisfying all customer needs and expectations—technical and legal documentation will not overcome bad impressions. Meeting the minimum contractual requirements will often not be enough—especially in a highly competitive environment. Customer quality expectations invariably increase based on exposure to "best-in-class" products and services. Nevertheless, the products and services must be provided consistent with contractual requirements and within the allotted budgets. This is not an easy task—it requires continuous attention and good judgement to satisfy the customer, while staying within program constraints.

Requirements definition begins at the program's inception and continues throughout the product life cycle. It is essential that the right disciplines, including suppliers, are involved at all times in order to avoid an incomplete or biased outcome. Methodologies such as Quality Function Deployment (QFD) can be used to enable the team to analyze customer requirements and to specify the appropriate product or service.

1.1.4 Capturing the Voice of the Customer

The concurrent engineering team must maintain a careful balance between satisfying customer needs and expectations and maintaining a reasonably scoped program. Early in the product life cycle, written requirements will be sparse and general. The team will use customer needs and expectations to interpret and expand the written requirements. It is at this time that the CE team has the greatest opportunity to influence the product life-cycle cost. Studies indicate that 80% of cumulative costs are already set by the end of the concept phase. Consequently, it is vital that all available information is used to choose the best product concept based on costs and user requirements.

Later in the program, the written requirements will become more detailed. The voice of the customer will then be used primarily to help clarify ambiguous requirements. In either case, the team will have to listen unceasingly to the voice of the customer in order to capture the needed requirements. The program manager will need to provide continuous support and encouragement—especially in light of the many obstacles the team might encounter. There are many real and perceived barriers to gathering the voice of the customer:

Restricted access to customers during competitions

User perception of VOC as a sales tactic

Confusion between higher quality and higher cost

Failure to identify all users

Lack of skills in analyzing the voice of the customer

Rush to design

Tradition and arrogance

1.1.5 Customer Focus Group Sessions

User focus group interview sessions can be very effective in gathering inputs from the users of a system. While there are many sources of customer needs and expectations, focus group sessions can uncover and amplify a broad range of requirements that might otherwise be overlooked. These requirements can range from the positive—"I want this!"—to the negative—"Don't do that again!" These inputs can provide a better understanding of user needs and even a competitive edge for your program. See Figure 1.2.

The use of focus group sessions is especially recommended at the beginning of

Figure 1.2 Translating customer requirements into requirements for the designer. (Courtesy of TECHNICOMP, Inc., Cleveland, OH.)

product conceptualization and during the support phase of a deployed system. The latter sessions identify product improvement opportunities.

A well-conducted session uses group dynamics to encourage individuals to provide inputs in a synergistic manner. A moderator leads the discussion in order to maintain focus on the critical issues and to enable each panel member to participate as much as possible. Post-It Notes (© 3M) should be used to capture, sort, and prioritize the panel members' comments using a disciplined form of brainstorming. A limited number of concurrent engineering team members attend the session as nonparticipating observers to help capture and understand the voice of the customer.

The moderator should prepare open-ended questions to stimulate and focus the session:

Why is a new product or system needed?

What do you like about the existing system?

What do you dislike about the existing system?

What makes your current job difficult?

The results of the focus group sessions can be analyzed using one of the formal process management tools or design for manufacturability tools such as DFA or QFD.

1.2 QUALITY FUNCTION DEPLOYMENT

1.2.0 Introduction to Quality Function Deployment

Quality Function Deployment (QFD) is a methodology used by teams to develop products and supporting processes based on identifying the customer needs and wants, and comparing how your product and the competition meet these customer requirements.

Quality: What the customer/user needs or expects

Function: How those needs or expectations are to be satisfied

Deployment: Making it happen throughout the organization

QFD starts with the voice of the customer. Using all practical means, the team gathers customer needs and expectations. The team has to analyze these inputs carefully to avoid getting caught up in providing what they *expect* or *believe* that the customer needs. Conceptual models, such as the Kano model shown in Figure 1.3, can be used to help each team member understand better how the customer will evaluate the quality of the product. The team can then use QFD as a planning tool to derive product and process specifications that will satisfy the voice of the customer while remaining within business and technical guidelines and constraints. The program utilizes a series of matrices, starting with the customer "wants" on one axis and the company's technical interpretation of how to accomplish these customer expectations on the other axis. These elements are weighted and quantified, to permit focus on the most important issues—or on the areas where the company product is most deficient.

1.2.1 Quality Function Deployment

QFD is defined as "a *discipline* for product planning and development in which key customer wants and needs are deployed throughout an organization." It provides a structure for ensuring that customers' wants and needs are carefully heard, then trans-

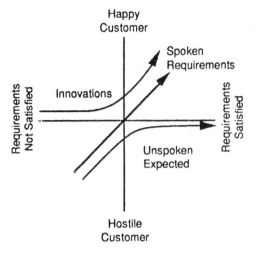

Figure 1.3   Kano model showing how the customer evaluates quality. (Courtesy of TECHNICOMP, Inc., Cleveland, OH.)

lated directly into a company's internal technical requirements—from component design through final assembly.

Strength of QFD

Helps minimize the effects of:

Communication problems

Differences in interpretation about product features, process requirements or other aspects of development

Long development cycles and frequent design changes

Personnel changes

Provides a systematic means of evaluating how well you and your competitors meet customer needs, thus helping to identify opportunities for gaining a competitive edge (sort of mini-benchmarking)

Offers a great deal of flexibility, so it may be easily tailored to individual situations

Brings together a multifunctional group very early in the development process, when a product or service is only an idea

Helps a company focus time and effort on several key areas which can provide a competitive edge

Applications of QFD

QFD is best applied to specific needs: areas in which significant improvements or "breakthrough" achievements are needed or desired. It can be used for virtually any type of product or service, including those from such areas as all types of discrete manufacturing, continuous and batch processes, software development, construction projects, and customer service activities in the airline, hotel, or other industries.

The Voice of the Customer

The first step in the QFD process is to determine what customers want and need from a company's product or service. This is best heard directly from customers themselves, and stated in their words as much as possible. This forms the basis for all design and development activities, to ensure that products or services are not developed only from the "voice of the engineer," or are technology-driven.

Background of QFD

QFD was developed by Technicomp, Inc., of Cleveland, Ohio. The first application as a structured discipline is generally credited to the Kobe Shipyard of Mitsubishi Heavy Industries Ltd. in 1972. It was introduced to the United States in a 1983 article in *Quality Progress,* a publication of the American Society of Quality Control (ASQC). Leaders in developing and applying QFD in Japan include Akao, Macabe, and Fukahara. U.S. companies that have utilized the Technicomp Quality Function Deployment program include Alcoa, Allen-Bradley, Bethlehem Steel, Boeing, Caterpillar, Chrysler, Dow Chemical, General Motors, Hexel, Lockheed, Magnavox, and others. The program consists of a series of videotapes, team member application guides, instructor guides, and other course materials.

Phases of QFD

QFD involves a series of phases (see Fig. 1.4) in which customer requirements are translated into several levels of technical requirements. Phases are often documented by a series of linked matrices.

> *Phase 1: Product Planning*—customer requirements are translated into technical requirements or design specifications in the company's internal technical language.
>
> *Phase 2: Product Design*—technical requirements are translated into part characteristics.
>
> *Phase 3: Process Planning*—part characteristics are translated into process characteristics.
>
> *Phase 4: Process Control Planning*—process characteristics are assigned specific control methods.

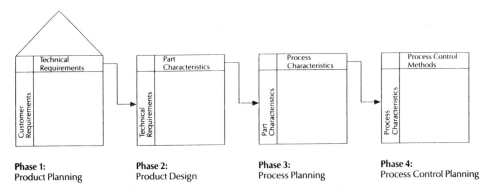

Figure 1.4   The four phases of quality function deployment (QFD). (Courtesy of TECHNICOMP, Inc., Cleveland, OH.)

Potential Benefits

Some of the results achieved by companies which have implemented QFD include:

30–50% reduction in engineering changes

30–50% shorter design cycles

20–60% lower start-up costs

20–50% fewer warranty claims

Other results include:

Better, more systematic documentation of engineering knowledge, which can be more easily applied to future designs

Easier identification of specific competitive advantages

More competitive pricing of products or services, due to lower development and start-up costs

More satisfied customers

Requirements for Success

Management commitment to QFD is the minimum requirement; support by the entire organization is ideal. Participation on a project team is required by individuals who support QFD and represent all development-related functions, such as:

Design engineering

Process engineering

Manufacturing engineering

Manufacturing

Quality assurance

Marketing

Sales

Other appropriate members may be representatives from:

Purchasing

Distribution

Accounting

Human resources

Suppliers

Customers

1.2.2 The House of Quality

The QFD program introduces a chart in Phase 1 that is commonly called "The House of Quality." The illustration shown in Figure 1.5 is a simplified example for large rolls of paper stock used in commercial printing. The following is a brief summary of the completed chart shown in Figure 1.5.

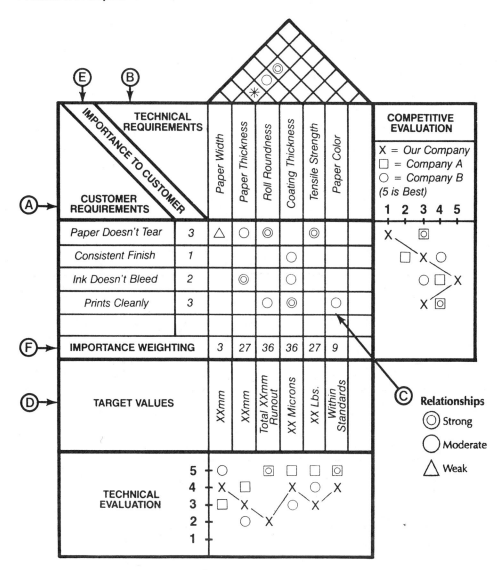

Figure 1.5 QFDs house of quality chart for rolls of paper stock. (Courtesy of TECHNICOMP, Inc., Cleveland, OH.)

(A) *Customer requirements*—Customers' wants and needs, expressed in their own words.

(B) *Technical requirements*—design specifications through which customers' needs may be met, expressed in the company's internal language.

(C) *Relationship matrix*—indicates with symbols where relationships exist between customer and technical requirements, and the strength of those relationships.

(D) *Target values*—show the quantifiable goals for each technical requirement.

(E) *Importance to customer*—indicates which requirements are most important
       to customers.
(F) *Importance weighting*—identifies which technical requirements are most im-
       portant to achieve. In this chart, each weighting is calculated by multiplying the
       "importance to customer" rating times the value assigned to a relationship, then
       totaling the column.

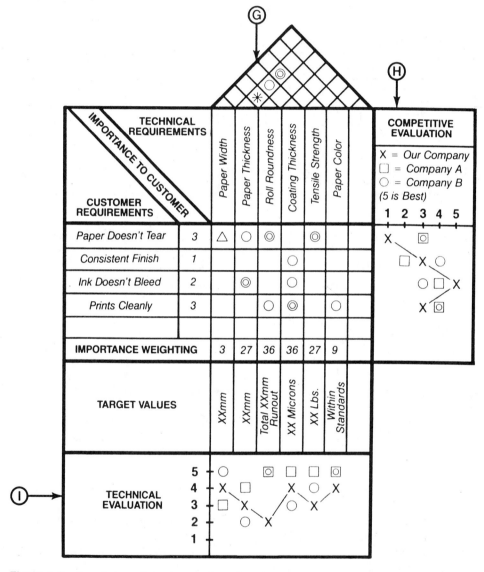

Figure 1.6 Completion of the house of quality chart for rolls of paper stock. (Courtesy of
TECHNICOMP, Inc., Cleveland, OH.)

The following are shown in Figure 1.6.

(G) *Correlation matrix*—indicates with symbols where relationships exist between pairs of technical requirements, and the strength of those relationships.

(H) *Competitive evaluation*—shows how well a company and its competitors meet customer requirements, according to customers.

(I) *Technical evaluation*—shows how well a company and its competitors meet technical requirements.

Product Planning (Phase 1 of QFD)

Most companies begin their QFD studies with a product planning phase. Briefly, this involves several broad activities including collecting and organizing customer wants and needs, evaluating how well your product and competitive products meet those needs, and translating those needs into your company's specific technical language.

The House of Quality

In very simple terms, the house of quality can be thought of as a matrix of *what* and *how*:

What do customers want and need from your product or service (customer requirements)

How will your company achieve the *what* (technical requirements)

The matrix shows where relationships exist between *what* and *how,* and the strength of those relationships. Before starting a chart, the scope and goals of the study must be clearly identified. A typical chart enables you to:

Learn which requirements are most important to customers

Spot customer requirements that are not being met by technical requirements, and vice versa

Compare your product or service to the competition

Analyze potential sales points

Develop an initial product or service plan

Constructing a House of Quality

Suggested elements and a recommended sequence of construction for a house of quality are given below. Also keep in mind that the chart is constructed by the entire project team, unless noted otherwise. (See Figures 1.7, 1.8, and 1.9.)

(A) Customer Requirements

Every chart must begin with the voice of the customer.

Identify all customer groups. Various groups will probably have some common needs, as well as some conflicting needs.

Collect accurate information from customers about their wants and needs.

Brainstorm to identify any additional customer requirements.

Use an affinity diagram to help group raw customer data into logical categories.

Transfer the list of customer requirements to the house of quality.

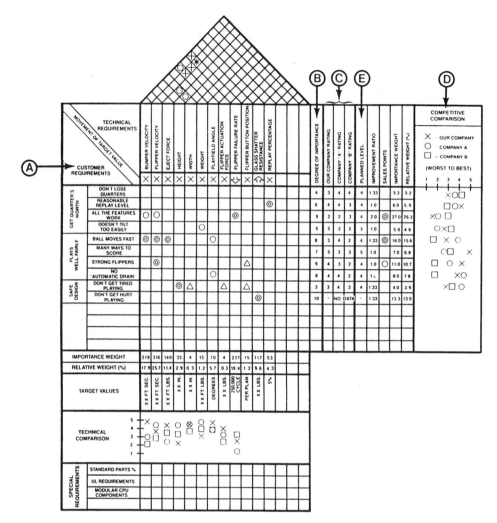

Figure 1.7 Construction a house of quality: (A) customer requirements; (B) degree of importance; (C) competitive comparison; (D) competitors' ratings; (E) planned level to be achieved. (Courtesy of TECHNICOMP, Inc., Cleveland, OH.)

Do not assume that your company already knows everything about its customers. QFD teams have been astounded at the results of focused efforts to listen to the voice of the customer. Primary benefits of QFD often include clearing up misconceptions and gaining an accurate understanding of customers' demands.

(B) Degree of Importance

Identify the relative priority of each customer requirement.

Use customer input as the basis for determining values, whenever possible.

Use a scale of 1 to 10, with 10 indicating very important items.

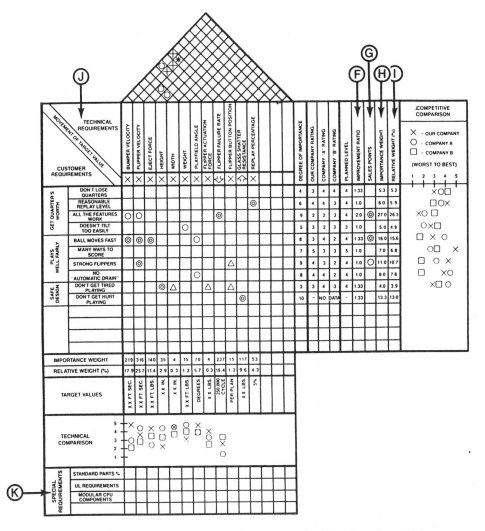

Figure 1.8 The house of quality (*cont.*): (F) improvement ratio; (G) sales points; (H) importance weight of customer requirements; (I) relative weight of customer requirements; (J) technical requirements; (K) special requirements. (Courtesy of TECHNICOMP, Inc., Cleveland, OH.)

A more rigorous statistical technique also can be effective for determining degrees of importance.

(C) Competitive Comparison

Identify how well your company and your competitors fulfill each of the customer requirements.

Use customer input as the basis for determining numeric ratings.

Use a scale of 1 to 5, with 5 being the best.

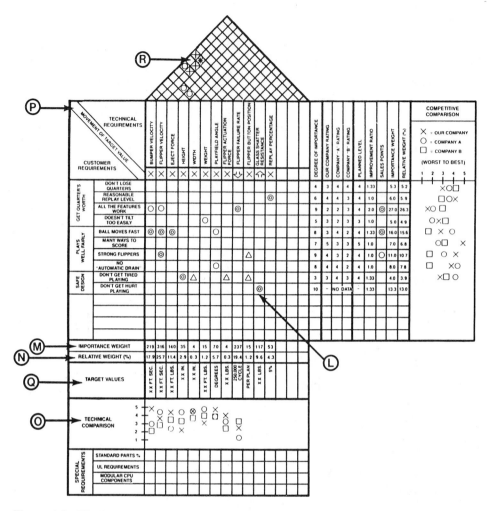

Figure 1.9 The house of quality (*cont.*): (L) relationships between technical requirements and customer requirements; (M) importance weight of technical requirements; (N) relative weight of technical requirements; (O) technical comparison; (P) movement of target values; (Q) target values. (Courtesy of TECHNICOMP, Inc., Cleveland, OH.)

(D) Competitive Comparison

Use symbols to depict each company's rating, so that you can easily see how well, in your customers' view, your company compares to the competition.

(E) Planned Level

Determine what level you plan to achieve for each customer requirement.

Base the rating on the competitors comparison, and use the same rating scale.

Focus on matching or surpassing the competition on items which will give your

product or service a competitive edge, or which are very important to customers.
(It is not necessary to outdo the competition on every item.)

(F) Improvement Ratio

Quantify the amount of improvement planned for each customer requirement.

To calculate, divide the value of the planned level by the value of the current
company rating:

Improvement ratio = planned level ÷ current company rating

(G) Sales Points

Identify major and minor sales points.

Limit the team to only a few points.

To indicate the importance of sales points, use symbols to which values are assigned:

⊙ = major point = 1.5

○ = minor point = 1.2

Major and minor sales points are often assigned values of 1.5 and 1.2, respectively.

Note that items with high degrees of importance are often logical sales points. Also,
remember that any items which will be new and "exciting" to customers are likely
sales points, although they probably will not have high degrees of importance.

(H) Importance Weight (Customer Requirements)

Quantify the importance of each customer requirement to your company.

To calculate, multiply the degree of importance times the improvement ratio times the
sales point value (if applicable):

$$\begin{matrix} \text{Importance} \\ \text{Weight} \end{matrix} = \left(\begin{matrix} \text{Degree of} \\ \text{Importance} \end{matrix} \right) \times \left(\begin{matrix} \text{Improvement} \\ \text{Ratio} \end{matrix} \right) \times \left(\begin{matrix} \text{Sales Point} \\ \text{Value} \end{matrix} \right)$$

(I) Relative Weight (%) (Customer Requirements)

Quantify the relative importance of each customer requirement by expressing it as
a percentage.

To calculate:

Total the importance weights,

Divide the importance weight of an item by the total,

Multiply by 100.

Use the relative weights as a guide for selecting the key customer requirements on
which to concentrate time and resources.

(J) Technical Requirements

Develop this list internally, using existing data and the combined experience of team members.

Begin by collecting available data.

Brainstorm to identify any additional requirements.

Follow general guidelines while developing the technical requirements:

Address global requirements of the product or service, not lower-level performance specifications,

Identify performance parameters, and avoid simply restating the components of the existing product or service,

Try not to include parts of mechanisms,

Establish definitions that are understood and agreed upon by all team members,

Use terminology which is understood internally.

Transfer the technical requirements to the house of quality chart.

(K) Special Requirements

List any unusual needs or demands, such as government standards, certification requirements, or special company objectives.

Note that these items must be considered during design and development, but normally do not appear in a list of customer demands.

(L) Relationship Matrix

Identify any technical requirements that bear on satisfying a customer requirement.

Evaluate each pair of requirements by asking if the technical requirement in any way affects the customer requirement.

For large charts, consider dividing the matrix into "strips," assigning them to groups of only a few team members for evaluation, and then have the full team review the combined results.

To indicate the strengths of relationships, use symbols to which values are assigned:

$\odot$ = strong relationship = 9

$\circ$ = moderate relationship = 3

$\triangle$ = weak relationship = 1

(M) Importance Weight (Technical Requirements)

Quantify the importance of each technical requirement.

To calculate this value, only those customer requirements which are related to a technical requirement are factored into the calculation:

Multiply the value of any relationships shown in the column of the technical requirement times the relative weight of the customer requirement,

Test the results.

(N) Relative Weight (%) (Technical Requirements)

Quantify the relative importance of each technical requirement by expressing it as a percentage.

To calculate:

Total the importance weights,

Divide the importance weight of an item by the total,

Multiply by 100.

At this point, the house of quality contains enough information for you to choose several key technical requirements on which to focus additional planning and definition. As a general rule, especially for pilot QFD projects, limit the number of key requirements to three to five. Those with the highest importance or relative weights are good candidates. They should have a strong bearing on satisfying the highest-priority customer requirements, or many of the customer requirements. However, don't look at just the numbers when choosing the key technical requirements. Also consider:

The potential difficulty of meeting a requirement

Any technical breakthroughs that may be needed

Requirements that are unfamiliar to your company

These factors can be particularly important if you must choose key requirements from several that have similar weightings.

(O) Technical Comparison

Identify how well your company and your competitors fulfill each of the technical requirements.

Use internal expertise from sources such as engineers, technicians, and field personnel to develop the comparisons.

Consider evaluation techniques such as bench testing, laboratory analysis, product teardowns, field observations, and reviews of data from outside testing labs or agencies.

Do not expect to be able to evaluate every technical requirement.

Convert the test data into values that are appropriate for a rating scale of 1 to 5, with 5 being the best.

"Plot" each company's performance using symbols, so that comparisons can be seen easily.

(P) Movement of Target Value

Use symbols to indicate the desired direction for each target value:

↑ = increase the value

↓ = decrease the value

x = meet a specified nominal value

(Q) Target Values

Assign specific target values for as many technical requirements as possible, in order to:

Establish concrete goals for designers and engineers

Help define further actions to ensure that customers' demands are carried through design and development

Compare your current actual values against those of the competition:

For each technical requirement, look at competitive ratings for the related customer requirements,

If you rate lower than the competition in key areas, assign target values which equal or better their values.

Check historical test results or operating data to help assign target values which reflect desired improvements to key items.

Consider using the results of designed experiments, Taguchi experiments, or other developmental work to achieve "optimal" values.

Try working backward from broader, less specific requirements to develop specific target values.

Be sure that each target value is measurable; if not, the team should develop an alternative that is measurable.

Make sure that the team agrees on exactly how target values will be measured.

Keep in mind that some of the values may not be feasible to achieve later in development; for now, they represent important starting points.

The vast majority of U.S. companies have completed only the first phase of QFD in their studies. A QFD study should proceed beyond phase 1 if any of the following are true:

Product or service requires more or better definition,

Additional detail is needed to make significant improvements,

Technical requirements shown on the house of quality cannot now be executed consistently.

1.2.3 Product Design (Phase 2)

The overall goal of phase 2 is to translate key technical requirements from the house of quality into specific parts characteristics. "Parts" are tangible items that eventually

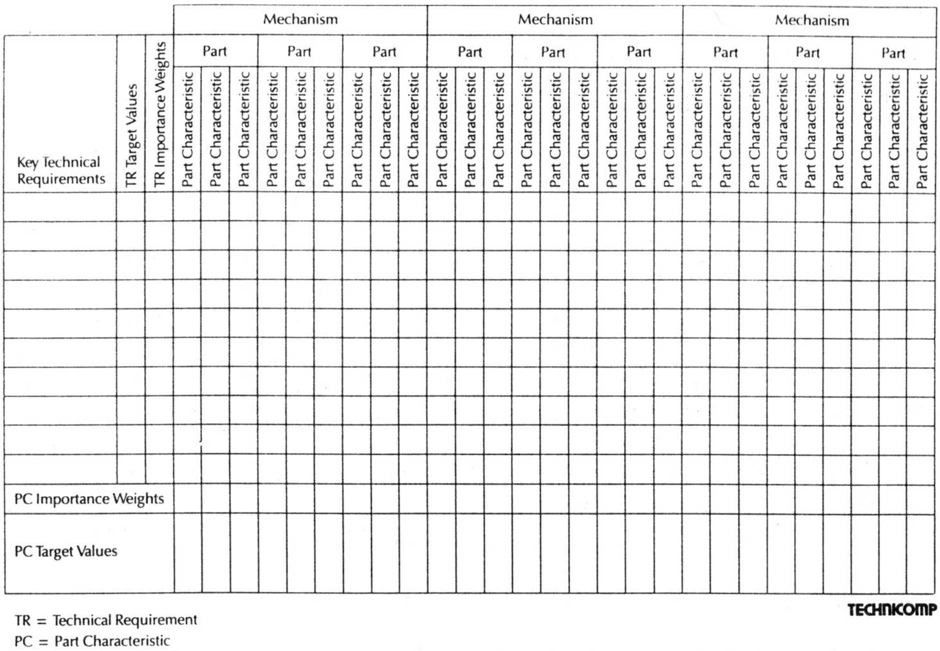

Key Technical Requirements	TR Target Values	TR Importance Weights	Mechanism									Mechanism									Mechanism								
			Part			Part			Part			Part			Part			Part			Part			Part			Part		
			Part Characteristic	Part Characteristic	Part Characteristic	Part Characteristic	Part Characteristic	Part Characteristic	Part Characteristic	Part Characteristic	Part Characteristic	Part Characteristic	Part Characteristic	Part Characteristic	Part Characteristic	Part Characteristic	Part Characteristic	Part Characteristic	Part Characteristic	Part Characteristic	Part Characteristic	Part Characteristic	Part Characteristic	Part Characteristic	Part Characteristic	Part Characteristic	Part Characteristic	Part Characteristic
PC Importance Weights																													
PC Target Values																													

TR = Technical Requirement
PC = Part Characteristic

TECHNICOMP

Figure 1.10 Product design matrix used to identify key part characteristics and to select best design concepts. (Courtesy of TECHNICOMP, Inc., Cleveland, OH.)

compose the product or service; its elements, raw materials, or subsystems. Inputs to this phase are:

Highest priority technical requirements from phase 1

Target values and importance weights for those requirements

Functions of the product or service

Parts and mechanisms required

The outcome of the product design phase is the identification of key part characteristics, and the selection of the new or best design concepts. Figure 1.10 shows a typical product design matrix chart as utilized in phase 2.

1.2.4 Other Elements of The QFD Program

The Technicomp Corporation goes on to explain the other two phases shown in Figure 1.4. In addition to QFD, the company has a large number of additional training programs.

Note: The editor wishes to thank Technicomp, Inc., for allowing us to utilize part of their QFD system in this Handbook, since we consider it one of the important tools that should be in the manufacturing engineer's "tool box."

1.3 DESIGN FOR ASSEMBLY*

Geoffrey Boothroyd and Jack M. Walker

1.3.0 Introduction to Design for Assembly

One example of the effectiveness of the DFA concept is the Proprinter line of computer printers launched by IBM in 1985. Before then, IBM got its personal printers from Japan's Seiko Epson Corp. By applying DFA analysis, IBM turned the tables on Japan, slashing assembly time from Epson's 30 minutes to only 3 minutes. Ford Motor Company trimmed more than $1.2 billion and in one year helped Ford edge out General Motors as Detroit's most profitable auto maker. GM later started using the software which has a database of assembly times established as the industry standard.

DFA is systematic in its approach and is a formalized step-by-step process. The techniques are concerned with simplifying the design in order to minimize the cost of assembly and the cost of the parts. The best way to achieve this minimization is first to reduce the number of parts to be assembled, and then to ensure that the remaining parts are easy to manufacture and assemble. It is important to have a measure of how efficient the design is in terms of assembly. The DFA process shows how to quantify this factor.

1.3.1 Choice of Manual or Machine Assembly System

It is important to decide at an early stage in design which type of assembly system is likely to be adopted, based on the system yielding the lowest costs. In product design requirements, manual assembly differs widely from automatic assembly due to the differences in ability between human operators and any mechanical method of assembly. An operation that is easy for an assembly worker to perform might be impossible for a robot or special-purpose workhead.

The cost of assembling a product is related to both the design of the product and the assembly system used for its production. The lowest assembly cost can be achieved by designing the product so that it can be economically assembled by the most appropriate assembly system. The three basic methods of assembly as shown in Figure 1.11 are:

1. Manual assembly,
2. Special-purpose machine assembly,

Editor's note: The "design for assembly" (DFA) process was developed by Geoffrey Boothroyd at the University of Massachusetts in the mid-1970s. He envisioned a method that would stress the economic implications of design decisions. This is crucial, because while design is usually a minor factor in the total cost of a product, the design process fixes between 70% and 95% of all costs. The National Science Foundation got things rolling in 1977 with a $400,000 research grant, with further grants during the 1980s.

Private companies began supporting Boothroyd's research in 1978, led by Xerox and AMP Inc. Then came Digital Equipment, General Electric, Westinghouse, and IBM. After Peter Dewhurst joined Boothroyd in 1981 at the University of Massachusetts, the two professors formed Boothroyd Dewhurst Inc. (BDI) to commercialize the concept in the form of personal computer software. After they moved to the University of Rhode Island, Ford and IBM quickly became the biggest supporters, investing $660,000 into research that has since moved beyond just assembly to include process manufacturing. Ford embraced the concept with great fervor and has trained roughly 10,000 people in DFMA (Design for Manufacture and Assembly). [DFMA is the registered trade mark of Boothroyd Dewhurst, Inc.]

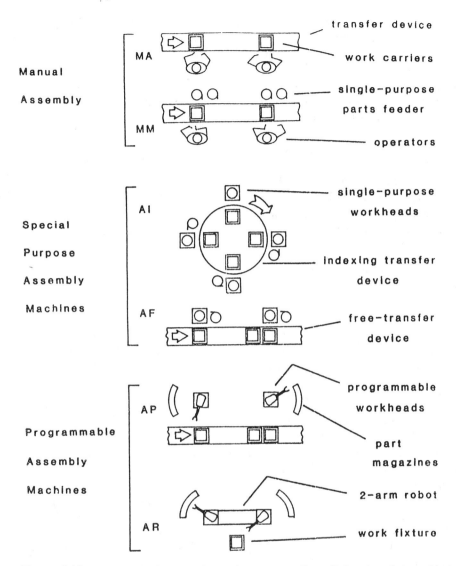

Figure 1.11 The three basic types of assembly systems. (From G. Boothroyd, *Assembly Automation and Product Design,* Marcel Dekker, New York, 1992.)

3. Programmable machine assembly.

In manual assembly (MA), the tools required are generally simpler and less expensive than those employed on automatic assembly machines, and the downtime due to defective parts is usually negligible. Manual assembly costs remain relatively constant and somewhat independent of the production volume; also, manual assembly systems have considerable flexibility and adaptability. Sometimes it will be economical to provide the assembly operator with mechanical assistance (MM) in order to reduce assembly time.

Special-purpose assembly machines are those machines that have been built to assemble a specific product; they consist of a transfer device with single-purpose workheads and parts feeders at the various workstations. The transfer device can operate on an indexing (synchronous) principle (AI) or on a free-transfer (nonsynchronous) principle (AF). They often require considerable prove-in time, and defective parts can be a serious problem unless the parts are of relatively high quality. (See references in Chapter 8, Tooling, and Chapter 21 Assembly Automation)

Programmable machines are similar to nonsynchronous special-purpose machines, except that some or all of the special-purpose workheads are replaced by programmable robots. This arrangement (AP) allows for more than one assembly operation to be performed at each workstation and provides for considerable flexibility in production volume and greater adaptability to design changes. For lower production volumes, a single robot workstation may be preferable (robotic assembly). Sometimes two robot arms will be working interactively at the same workfixture (AR). When considering the manufacture of a product, a company must take into account the many factors that affect the choice of assembly method. For a new product, the following considerations are generally important:

Suitability of the product design;

Production rate required;

Availability of labor; and

Market life of the product.

Figure 1.12 shows for a particular set of conditions how the range of production volume and number of parts in the assembly affects the choice of system.

1.3.2 Design for Manual Assembly

Although there are many ways to increase manufacturing productivity (utilizing improved materials, tools, processes, plant layout, etc.), consideration of manufacturing and assembly *during product design* holds the greatest potential for significant reduction in production costs and increased productivity. Robert W. Militzer, one-time president of the Society of Manufacturing Engineers, has stated: "as manufacturing engineers, we could do far more to improve productivity if we had greater input to the design of the product itself—it is the product designer who establishes the production (manufacture and assembly) process. For the fact is that much of the production process is implicit in product design."

In other words, if the product is poorly designed for manufacture and assembly, techniques can be applied only to reduce to a minimum the impact of the poor design. Improving the design itself may not be worth considering at this late stage; usually too much time and money have already been expended in justifying the design to consider a completely new design or even major changes. Only when manufacture and assembly techniques are incorporated early in the design stage (i.e., product design for ease of manufacture and assembly) will productivity be significantly affected.

Product costs are largely determined at the design stage. The designer should be aware of the nature of assembly processes and should always have sound reasons for requiring separate parts, and hence longer assembly time, rather than combining several parts into one manufactured item. The designer should always keep in mind that each combination of two parts into one will eliminate an operation and a workstation in manual assembly, or usually an entire workstation on an automatic assembly machine. It will also tend to reduce the part costs. The following discussion on manual assembly will be used to introduce the basic DFA principles. This is always a necessary step, even if the decision is

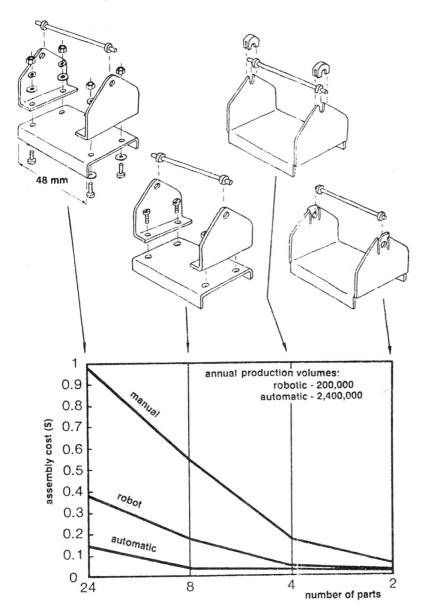

Figure 1.12 Example of the effect of design for assembly (DFA). (From G. Boothroyd, *Assembly Automation and Product Design,* Marcel Dekker, New York, 1992.)

ultimately to employ some degree of automation, in order to compare the results against manual assembly as a basis for comparison. In addition, when automation is being seriously considered, some operations may have to be carried out manually, and it is necessary to include the cost of these operations in the analysis.

In the DFA technique embodied in the widely used DFA software, developed by Boothroyd Dewhurst, Inc., features of the design are examined in a systematic way and a *design efficiency* or *DFA index* is calculated. This index can be used to compare different designs. There are two important steps for each part in the assembly:

1. A decision as to whether each part can be considered a candidate for elimination or combination with other parts in the assembly in order to obtain a theoretical minimum number of parts for the product
2. An estimation of the time taken to grasp, manipulate, and insert each part

Having obtained this information, it is then possible to obtain the total assembly time and to compare this figure with the assembly time for an "ideal" design. First you obtain the best information about the product or assembly. Useful items are:

Engineering drawings and sketches

Exploded three dimensional views

An existing version of the product

A prototype

Next we list the parts in assembly sequence. If the product contains sub-assemblies, treat these at first as "parts", and analyze them later. In the example of the controller assembly, Figure 1.13, the tube with nuts at each end is treated as one item and named "tube assembly." If a prototype of the product is available it is usually best to name the items as they are removed from the product. However, the software always lists them in a viable assembly order.

Finally, during the assembly (or reassembly) of the product, the minimum number of parts theoretically required is determined by answering the following questions:

1. Is the part or subassembly used only for fastening or securing other items?
2. Is the part or subassembly used only for connecting other items?

If the answer is Yes to either question, then the part or subassembly is not considered theoretically necessary. If the answer is No to both questions, the following criteria questions are considered:

1. During operation of the product, does the part move relative to all other parts already assembled? Only gross motion should be considered—small motions that can be accommodated by elastic hinges, for example, are not sufficient for a positive answer.
2. Must the part be of a different material than, or be isolated from, all other parts already assembled? Only fundamental reasons concerned with material properties are acceptable.
3. Must the part be separate from all other parts already assembled because otherwise necessary assembly or disassembly of other separate parts would be impossible?

If the answer to all three criteria questions is No, the part cannot be considered theoretically necessary.

When these questions have been answered for all parts, a theoretical minimum part count for the product is calculated. It should be emphasized, however, that this theoretical minimum does not take into account practical considerations or cost considerations but

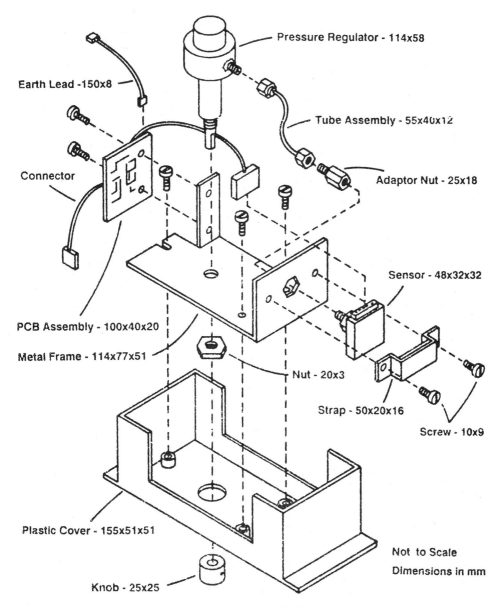

Earth Lead -150x8

Pressure Regulator - 114x58

Tube Assembly - 55x40x12

Connector

Adaptor Nut - 25x18

PCB Assembly - 100x40x20

Sensor - 48x32x32

Metal Frame - 114x77x51

Nut - 20x3

Strap - 50x20x16

Screw - 10x9

Plastic Cover - 155x51x51

Not to Scale

Dimensions in mm

Knob - 25x25

Figure 1.13 Exploded view of original controller assembly. (From G. Boothroyd, *Assembly Automation and Product Design,* Marcel Dekker, New York, 1992.)

simply provides a basis for an independent measure of the quality of the design from an assembly viewpoint.

Also, during reassembly of the product, you answer questions that allow handling and insertion times to be determined for each part or subassembly. These times are obtained from time standard databases developed specifically for the purpose. For estimation of the handling times (Figure 1.14) you must specify the dimensions of the item, its thickness, whether it nests or tangles when in bulk, whether it is fragile, flexible,

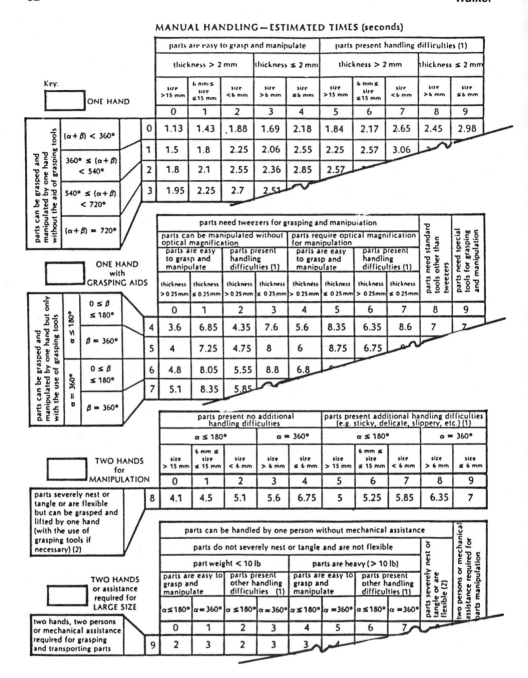

Figure 1.14 Classification, coding, and database for part features affecting manual assembly time. (Boothroyd Dewhurst, Inc., reproduced with permission.)

slippery, sticky, and whether it needs two hands, grasping tools, optical magnification, or mechanical assistance.

For estimation of insertion and fastening as shown in Figure 1.15, it is important to know whether the assembly worker's vision or access is restricted, whether the item is difficult to align and position, whether there is resistance to insertion, and

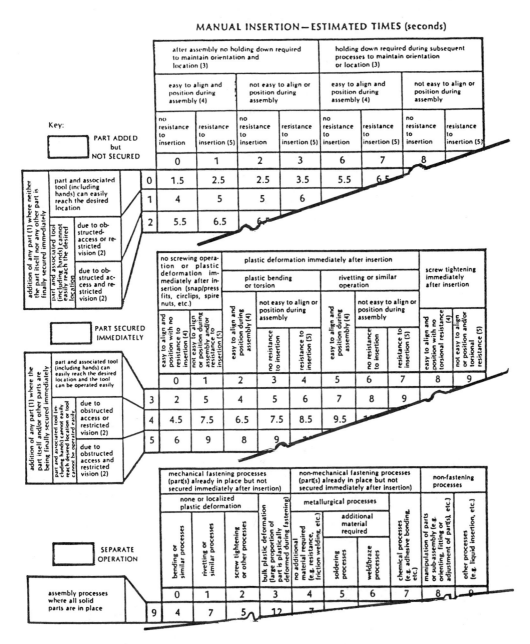

Figure 1.15 Classification, coding, and database for part features affecting insertion and fastening. (Boothroyd Dewhurst, Inc., Wakefield, Rhode Island, reproduced with permission.)

whether it requires holding down in order to maintain its position for subsequent assembly operations.

For fastening operations, further questions may be required. For example, a portion of the database for threaded fastening is shown in Figure 1.16. Here, the type of tool used and the number of revolutions required are important in determination of the assembly time.

A further consideration relates to the location of the items that must be acquired. The database for handling shown in Figure 1.14 is valid only when the item is within easy reach of the assembly worker. If turning, bending, or walking are needed to acquire the item, a different database (Figure 1.17) is used.

It can be seen in Figures 1.14 through 1.17 that for each classification, an average acquisition and handling or insertion and fastening time is given. Thus, we have a set of time standards that can be used to estimate manual assembly times. These time standards were obtained from numerous experiments, some of which disagree with some of the commonly used predetermined time standards such as methods time measurement (MTM) and work factor (WF). For example, it was found that the effect of part symmetry on handling time depends on both alpha and beta rotational symmetries illustrated in Figure 1.18. In the MTM system, the "maximum possible orientation" is employed, which is one-half the beta rotational symmetry of a part. The effect of alpha symmetry is not considered in MTM. In the WF system, the symmetry of a part is classified by the ratio of the number of ways the part can be inserted to the number of ways the part can be grasped preparatory to insertion. In this system, account is taken of alpha symmetry, and some account is taken of beta symmetry. Unfortunately,

Insertion and fastening times (seconds)
for threaded items using power tools

			easy to align and position		not easy to align or position	
			one part	several parts	one part	several parts
			0	1	2	3
autofeed	no vision or obstruction restrictions	0	1.9	2.3	3.6	4
screw or nut inserted into tool or assembly	no vision or obstruction restrictions	1	3.6	4	5.3	
	restricted vision only	2	6.3	6.7		
	obstructed access	3				

- Times do not include fastening tool acquisition and replacement.
- Times in column 0 - 5 assume screw or nut fastening requiring 5 revolutions. For operations requiring more than 5 revolutions, times for additional revolutions in column 6 should be added.

Figure 1.16 Classification, coding, and database for threaded fastening. (Boothroyd Dewhurst, Inc., Wakefield, Rhode Island, reproduced with permission.)

Acquisition and handling times (2) (seconds)
for one large part not within easy reach

average distance to location of parts (ft.)		easy to grasp	difficult to grasp (1)	requires two persons (3)	fixed swing crane
		0	1	2	3
14" to 4	0	2.54	4.54	8.82	19
4 to 7	1	4.25	6.25		
7 to 10	2	5.54			

1) For large items, no features to allow easy grasping (eg. no finger hold).
2) Times are for part acquisition only. Multiply by 2 if replacement time for a fixture is to be included. When cranes are used, the time to return the crane is included.
3) Times are equivalent times for one person.

Figure 1.17 Classification, coding, and database for turning, bending, and walking. (Boothroyd Dewhurst, Inc., Wakefield, Rhode Island, reproduced with permission.)

these effects are combined in such a way that the classification can be applied to only a limited range of part shapes.

Similar experimentation for DFA was carried out for part thickness, part size, weight, and many other factors which give a more precise basis for comparing one type of part design to another in the detailed analysis required by the DFA method.

1.3.3 Results of the Analysis

Once the analysis is complete, the totals can be obtained as shown in Figure 1.19. For the example of the controller, the total number of parts and subassemblies is 19, and there are

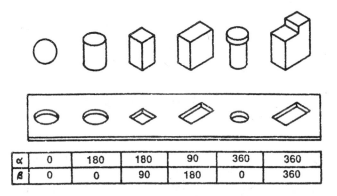

α	0	180	180	90	360	360
β	0	0	90	180	0	360

Figure 1.18 Alpha and beta rotational symmetries for various parts. (From G. Boothroyd, *Assembly Automation and Product Design,* Marcel Dekker, New York, 1992.)

MANUAL - BENCH ASSEMBLY	Manual handling code		Manual Insertion code		Total oper'n time RP*(TH+TI)	Figures for min. parts		Operator rate OP: 30.00 $/hr 0.83 c/s
Name of Assembly - $MAIN SUB	No. of items	Handling time per item (s)	Insertion time per item (s)		Total oper'n cost-cents TA*OP			
Item Name: Part, Sub or Pcb assembly No. or Operation	RP	HC	TH	IC	TI	TA	CA	NM	Description
1 $pressure regulator	1	30	1.95	00	1.5	3.5	2.9	1	place in fixture
2 metal frame	1	30	1.95	06	5.5	7.4	6.2	1	add
3 nut	1	00	1.13	39	8.0	9.1	7.6	0	add & screw fasten
4 Reorientation	1	-	-	98	9.0	9.0	7.5	-	reorient & adjust
5 $sensor	1	30	1.95	08	6.5	8.4	7.0	1	add
6 strap	1	20	1.80	08	6.5	8.3	6.9	0	add & hold down
7 Screw	2	11	1.80	39	8.0	19.6	16.3	0	add & screw fasten
8 Apply tape	1	-	-	99	12.0	12.0	10.0	-	special operation
9 adaptor nut	1	10	1.50	49	10.5	12.0	10.0	0	add & screw fasten
10 tube assembly	1	91	3.00	10	4.0	7.0	5.8	0	add & screw fasten
11 Screw fastening	1	-	-	92	5.0	5.0	4.2	-	standard operation
12 &PCB ASSEMBLY	1	83	5.60	08	6.5	12.1	10.1	1	add & hold down
13 Screw	2	11	1.80	39	8.0	19.6	16.3	0	add & screw fasten
14 connector	1	30	1.95	31	5.0	6.9	5.8	0	add & snap fit
15 earth lead	1	83	5.60	31	5.0	10.6	8.8	0	add & snap fit
16 Reorientation	1	-	-	98	9.0	9.0	7.5	-	reorient & adjust
17 $knob assembly	1	30	1.95	08	6.5	8.4	7.0	1	add & screw fasten
18 Screw fastening	1	-	-	92	5.0	5.0	4.2	-	standard operation
19 plastic cover	1	30	1.95	08	6.5	8.4	7.0	0	add & hold down
20 reorientation	1	-	-	98	9.0	9.0	7.5	-	reorient & adjust
21 screw	3	11	1.80	49	10.5	36.9	30.8	0	add & screw fasten

Figure 1.19 Completed design for assembly (DFA) worksheet for original controller design. (From G. Boothroyd, *Assembly Automation and Product Design,* Marcel Dekker, New York, 1992.)

6 additional operations. The total assembly time is 227 seconds, the corresponding assembly cost is $1.90, and the theoretical minimum number of items is 5.

If the product could be designed with only 5 items and each could be designed so that no assembly difficulties would be encountered, the average assembly time per item would be about 3 seconds, giving a theoretical minimum assembly time of 15 seconds. This theoretical minimum is divided by the estimated time of 227 seconds to give the design efficiency or DFA index of 6.6%. This index can be independently obtained and provides a basis for measuring design improvements.

The major problem areas can now be identified, especially those associated with the installation of parts that do not meet any of the criteria for separate parts. From the tabulation of results, it can be seen that attention should clearly be paid to combining the plastic cover with the metal frame. This would eliminate the assembly operation for the cover, the three screws, and the reorientation operation, representing as total time saving of 54.3 seconds, which constitutes 24% of the total assembly time. Of course, the designer must check that the cost of the combined plastic cover and frame is less than the total cost of the individual piece parts. A summary of the items that can be identified for elimination or combination and the appropriate assembly time savings are presented in Figure 1.20(a).

Design changes now identified could result in saving of at least 149.4 seconds of assembly time, which constitutes 66% of the total. In addition, several items of hardware would be eliminated, resulting in reduced part costs. Figure 1.20(b) shows a conceptual redesign of the controller in which all the proposed design changes have been made, and Figure 1.21 presents the revised tabulation of results. The total assembly time is now 84 seconds, and the assembly efficiency is increased 18%, a fairly respectable figure for this type of assembly. Of course the design team must now consider the technical and economic consequences of the proposed designs. Some of the rapid prototyping techniques discussed in subchapter 1.4 for rapidly producing a prototype part in a very short lead time may be able to furnish a part to the new design for evaluation or testing.

First, there is the effect on the cost of the parts. However, experience shows, and this example should be no exception, that the saving from parts cost reduction would be greater than the savings in assembly costs which, in this case is $1.20. Chapter 2 will assist in evaluating a design made by casting or forging, machining, or sheet metal processes. In this handbook, Chapter 10 on work measurement, Chapter 9 on process planning, and Chapter 7 on cost estimating for manufacturing, will be valuable aids for estimating parts costs.

It should be realized that the documented savings in materials, parts fabrication, and assembly represent direct costs. To obtain a true picture, overheads must be added, and these can often amount to 200% or more. Total cost per shop man hour could be assumed to be $30 to $50 per hour in the mid 1990's. In addition, there are other savings more difficult to quantify. (See subchapter 1.0, introduction to product design) For example, when a part such as the metal frame is eliminated, all associated documentation, including part drawings, is also eliminated. Also, the part cannot be misassembled or fail in service, factors that lead to improved reliability, maintainability, and quality of the product. It is not surprising, therefore, that many United States companies have been able to report annual savings measured in millions of dollars as a result of the application of the DFA analysis method described here.

The DFA system is much more comprehensive than outlined here. There could be wide variations in the results, depending on the degree of automation eventually selected, and the results of detail analysis of the parts fabrication of the final design selected. However, the example of the basic system approach outlined in this handbook should serve as an introduction to the DFA system.

Design change	Items	Time saving, sec
1. Combine plastic cover with frame and eliminate 3 screws and reorientation	19,20,21	54.3
2. Eliminate strap and 2 screws (snaps in plastic frame to hold sensor, if necessary)	6,7	27.9
3. Eliminate screws holding PCB assembly (provide snaps in plastic frame)	13	19 6
4. Eliminate 2 reorientations	4, 16	18.0
5.. Eliminate tube assembly and 2 screwing operations (screw adapter nut and sensor direct to the pressure regulator)	10, 11	12.0
6. Eliminate earth lead (not necessary with plastic frame)	15	10.6
7. Eliminate connector (Plug sensor into PCB)	14	7.0

(a)

Figure 1.20 (a) Summary of items to be considered for elimination or combination on the controller assembly. (b) Conceptual design changes of the controller assembly. (From G. Boothroyd, *Assembly Automation and Product Design,* Marcel Dekker, New York, 1992.)

1.3.4 Bibliography

Boothroyd, G., *Assembly Automation and Product Design,* Marcel Dekker, New York, 1992.
Boothroyd, G., *Product Design for Manufacture and Assembly,* Marcel Dekker, New York, 1994.

1.4 RAPID PROTOTYPING

1.4.0 Introduction to Rapid Prototyping

Rapid prototyping is one of the names given to a new group of technologies for converting designs from computer representations directly into solid objects without human intervention. The technologies are sometimes known collectively as solid free-form fabrication or computer automated fabrication. No single one of the many technologies has yet proven that it can meet all market requirements, so those intending to be involved in this industry must know the fundamental processes, limits, and potentials of the competing machines.

One of the goals of all successful companies is to reduce the time between initial concept of a product and production deliveries. The other vital concept is the early involvement of manufacturing expertise with the product design team. This includes outside suppliers along with in-house experts. The outside suppliers that must be involved are those trades and skills involving parts that your company may "buy" rather than build in-house. This often includes castings, forgings, plastic moldings, specialty sheet metal, and machining. Their knowledge can assist the original equipment manufacturer (OEM) to produce a better product, in a shorter time, at a lower cost.

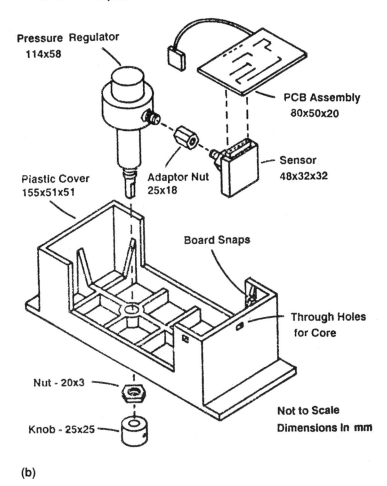

Pressure Regulator
114x58

PCB Assembly
80x50x20

Sensor
48x32x32

Plastic Cover
155x51x51

Adaptor Nut
25x18

Board Snaps

Through Holes
for Core

Nut - 20x3

Not to Scale
Dimensions in mm

Knob - 25x25

(b)

Other firms that may be of great assistance are firms that concentrate on producing prototype parts on an accelerated schedule to aid in the design and process selection. They may also produce the early production parts while you are waiting for your "hard tooling" to be completed. They may even be your eventual production supplier, or a back-up source if your production supplier has problems.

As we shall see in Chapter 2 on product design, often the small details of a part or process can greatly influence the design. In the product design and development phase of a project, it is often valuable to have a prototype part made of a proposed design—or of alternative designs—to permit members of your team to get a better "feel" for the design concept. In the automobile industry, for example, the appearance and "feel" of door handles, controls, etc., are important to the ultimate customer. We can look at drawings or computer screens of a concept design—but few of us can assess what a Shore Durometer hardness number means on paper as well as we can by handling several copies of a part with different flexibility, color, or made from different materials. Subchapter 5.1 on ergonomics, discusses the importance of considering the human capabilities of the user in the design of a product. The design of Logitec's new ergonomic mouse for computers is a recent success story. Most good designs are an iterative process, and most companies do

MANUAL - BENCH ASSEMBLY Name of Assembly - $MAIN SUB Item Name: Part, Sub or Pcb assembly No. or Operation	Manual handling code		Manual Insertion code	Total oper'n time RP*(TH+TI)		Figures for min. parts	Operator rate OP: 30.00 $/hr 0.83 c/s Description	SUB ASSEMBLY OR PART COSTS	
	No. of items RP	Handling time per item (s) HC TH	Insertion time per item (s) IC TI	Total oper'n cost-cents TA*OP TA CA		NM		Total item cost $ CT	Total tooling cost k$ CC
1 $pressure regulator	1 30	1.95 00	1.5 3.5	2.9	1	place in fixture	10.46	0.0
2 plastic cover	1 30	1.95 06	5.5 7.4	6.2	1	add & hold down	0.00	0.0
3 nut	1 00	1.13 39	8.0 9.1	7.6	0	add & screw fasten	0.20	0.0
4 $knob assembly	1 30	1.95 08	6.5 8.4	7.0	1	add & screw fasten	-	-
5 Screw fastening	1 -	- 92	5.0 5.0	4.2	-	standard operation	-	-
6 Reorientation	1 -	- 98	9.0 9.0	7.5	-	reorient & adjust	-	-
7 Apply tape	1 -	- 99	12.0 12.0	10.0	-	special operation	-	-
8 adaptor nut	1 10	1.50 49	10.5 12.0	10.0	0	add & screw fasten	0.30	0.0
9 $SENSOR	1 30	1.95 39	8.0 9.9	8.3	1	add & screw fasten	1.50	0.0
10 &PCB ASSEMBLY	1 83	5.60 30	2.0 7.6	6.3	1	add & snap fit	0.00	0.0

Figure 1.21 Completed (DFA) analysis for the redesigned controller assembly. (From G. Boothroyd, *Assembly Automation and Product Design,* Marcel Dekker, New York, 1992.)

not have the in-house capability to quickly make accurate prototypes in their production shops. A prototype part is more than a "model." It is as much like the eventual production part as possible—in a short period of time!

There are a few specialty companies is the United States that have the ability to produce models and patterns using the latest rapid prototyping techniques, including stereolithography, laminated object manufacturing (LOM), CAM hog-outs from solid stock, as well as traditional pattern making and craft skills for producing prototype plaster-mold metal castings and plastic parts. Some specialize in photo etching metal blanks, others in quickly producing sheet metal parts with temporary, low-cost tooling. Complex investment castings can be made without tooling. With these skills and methods, they can create premium-quality, economical parts in the shortest time possible. While 2 or 3 weeks is fairly common lead time, they sometimes can deliver prototype parts in less than a week.

In this subchapter we will introduce some of the techniques available, and the capabilities of some typical firms specializing in this field. Some of these companies build only prototype quantities, but most can produce the initial start-up quantities of parts while your "hard tooling" is being built. Many of these companies can produce parts directly from CAD files, as well as from drawings or sketches. This allows us to send data via modem, rather than relying on other communication systems. For example, Armstrong Mold Company, in East Syracuse, New York, uses IGES, DFX, CADL, and STL file formats directly, or can work from floppy disk or DC 2000 mini data tape cartridges to produce prototype parts with short lead times.

Advantages of Rapid Prototyping

Paperless manufacturing—work directly from CAD files or sketches

Functional parts as quickly as 1 week

Produce models which can be used as masters for making metal castings and plastic parts in small or large quantities

Fit and function models used to detect design flaws early in development

Allows prototype part to be used during qualification testing of a product

1.4.1 Prototype Models and Patterns

Stereolithography

Stereolithography is used to create three dimensional objects of any complexity from CAD data. Solid or surfaced CAD data is sliced into cross sections. A laser generating an ultraviolet beam is moved across the top of a vat of photo-sensitive liquid polymer by a computer-controlled scanning system. The laser draws each crosssection, changing the liquid polymer to a solid. An elevator lowers the newly formed layer to recoat, and establishes the next layer thickness. Successive cross sections are built layer by layer one on top of another, to form the a three-dimensional plastic model. Armstrong, along with a few other firms, takes advantage of this process.

Laminated Object Manufacturing

Laminated object manufacturing (LOM) is also used to create three dimensional objects of any complexity from CAD data. Solid or surfaced CAD data is sliced into cross sections. A single laser beam cuts the section outline of each specific layer out of paper from 0.002 to 0.020 in. thick coated with heat-seal adhesives. A second layer is then bonded to the first, and the trimming process is repeated. This process continues until all layers are cut and laminated, creating a three-dimensional, multilayered wood-like model. One LOM machine observed had a 32 in. × 22 in. × 20 in. working envelope.

Machined Prototypes

Three-dimensional objects can be cut from a variety of solid stocks (metal, resin, plastic, wood) directly from two- or three-dimensional CAD data using multiaxis CNC machining centers. Applicable machines can be utilized for making models from drawings or sketches.

1.4.2 Casting Metal Parts

Plaster Molds

Aluminum or zinc castings can be produced by pouring liquid metal into plaster (gypsum) molds. Typical applications at Armstrong include:

Castings for business machines, medical equipment, computers, automotive, aerospace,electronics, engineering

Molds for the plastics industry: rotational, vacuum form, expanded polystyrene, kirksite, and injection molds

The first step in the process is to make a model or master pattern. This can be done from a drawing or CAD file. The material can be wood, plastic, or machined from brass or other metal. The model is either hand-crafted or machined, using multiaxis CNC or tool room machines, depending on the complexity and availability of CAD data. The model will have shrinkage factors builtin, as well as draft, and machining allowance if desired. A customer-furnished model or sample part can by adapted to serve as the master pattern (See Figure 1.22)

The second step is to make a negative mold, and core plugs if required, from the master model. A positive resin cope and drag pattern is now made from the negative molds. Core boxes are builtup, and the gating and runnering system, with flasks as necessary, are added.

Next, a liquid plaster slurry is poured around the cope and drag pattern and into the core boxes. After this sets, the plaster mold is removed from the cope and drag patterns. The plaster mold and cores are then baked to remove moisture.

Molten metal is prepared by degassing, and spectrographic analysis of a sample checks the chemical composition of the material. The molten metal is then poured into the assembled plaster mold. The plaster is removed by mechanical knock-out and high-pressure water jet. When the casting has cooled, the gates and risers are removed.

The raw castings are then inspected and serialized. Any flash and excess metal is removed (snagged). Castings may then require heat treatment, X-ray, or penetrant inspec-

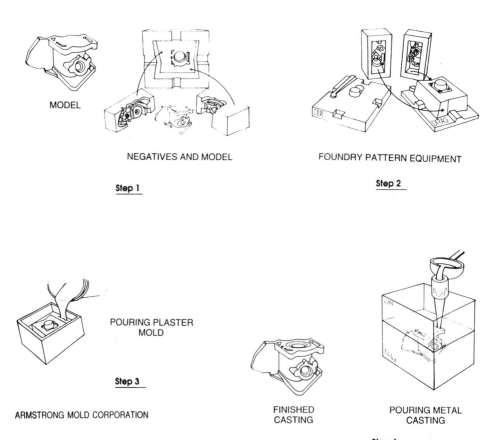

Figure 1.22   Plaster molding process. (Courtesy of Armstrong Mold Corp., East Syracuse, NY.)

tion as requested by the customer. After inspection, the casting is ready for machining, chemical film, chromate conversion, paint or special finishes, and assembly if required.

The object of this process is:

1. To produce precision and premium-quality aluminum and zinc castings
2. For esthetic applications where appearance is critical
3. Prototyping—to assist engineering in debugging design before committing to hard tooling
4. For low-volume applications of complex or unusual shapes
5. For high-volume applications of complex or unusual shapes
6. For castings with thin walls or where weight is critical
7. To simulate die castings for prototype and pilot production
8. Tooling is low cost and allows ease of modification
9. To reduce "time to market" on new programs and evaluate market potential
10. To reduce time for machining and secondary operations
11. To fill the gap if dies are lost, damaged, or delayed
12. To assist and reduce time for UL approval

Design and Technical Information

Size: No limitation, but best range is within 2 in. to 36 in. cube

Finish: Can hold 63 μin. but normally 90 μin.

Shape: Considerable design freedom for unusual and complex shapes

Wall thickness:

Thin wall,	0.030–0.060 in.
Average,	0.080–0.120 in.
Thick,	0.180–0.500 in.

General tolerances:

0–2 ± 0.010 in.	2–3 ± 0.012 in.
3–6 ± 0.015 in.	6–12 ± 0.020 in.
12–18 ± 0.030 in.	1–30 ± 0.040 in.

Tighter tolerances can be negotiated

The process is limited to nonferrous metals with pouring temperatures below 2000°F. This includes all aluminum and zinc casting alloys and some copper-based alloys. All aluminum and zinc casting alloys are per commercial and military specifications (see Figure 1.23). This includes the appropriate mechanical properties. Small holes, ¼ in. or less, are not economical to cast unless they are an odd shape or inaccessible for machining. Zero draft is possible in specified areas, but is typically ½ to 2°. Corner radii and fillets are as required, but are typically ¹⁄₁₆ in. R.

A variety of models and patterns can be made, depending on the customer's request:

Wood—to expedite for up to 20 pieces

Epoxy resin—usually up to 500 pieces

Metal (aluminum or brass)—used to obtain best tolerances and quality

Aluminum Alloys

Alloy	Heat Treating	Ultimate Tensile Strength KSI	Min. Yield Strength Set @ 2% KSI	Elongation % in 2"	Hardness Rockwell E Scale 1/16" Ball	Pressure Tightness	Machin-ability
319.0	T6	36	24	2.0	85.5	B	A
355.0	T51	28	23	1.5	76.0	A	C
355.0	T6	35	25	3.0	85.5	A	B
C355.0	T6	39	29	5.0	89	A	B
356.0	T51	25	20	2.0	71	A	C
356.0	T6	33	24	3.5	79	A	A
A356.0	T51	26	18	3.0	71	A	C
A356.0	T6	40	30	6.0	82.5	A	A
357.0	T51	26	17	3.0	80	B	C
357.0	T6	45	35	3.0	91	B	B
A357.0	T6	45	35	3.0	89	B	B
380.0	NONE	46	23	3.5	85.5	B	C
712.0	*	35*	25*	5.0*	82.5	C	A

712.0 also known as D712, D612, and 40E

Zinc Alloys

ZA3	NONE	41	-	10.0	87	C	C
ZA8	NONE	34	30	1.5	95.5	C	C
ZA12	NONE	43	30	2.0	93	C	B
ZA27	NONE	61	46	9.5	100	C	B

These values are for separately cast test bars, and are typical values.

* Test 30 days after casting

A= Excellent   B= Good   C= Fair

Applicable Military Specifications

- Mechanical/Chemical Inspection to MIL-A-21180 QQ-A-601, AMS-4217
- System Control to MIL-1-45208
- Gage Control to MIL-STD-45662
- Penetrant Inspection to MIL-STD-6866
- Radiographic Inspection to MIL-STD-453, MIL-STD-2175
- Heat Treat to MIL-H-6088
- N.D.T. to MIL-STD-410

Figure 1.23 Aluminum alloy casting mechanical properties. (Courtesy of Armstrong Mold Corp., East Syracuse, NY.)

Rubber—for quantities up to 1000 pieces; tooling can be duplicated easily from master tooling to expedite delivery or for higher volumes

A rule of thumb for costs of complex shapes in 15 in. cubes is:

Tooling: 10% of die cost tools

Piece price: 10 times die casting price

Normal delivery time is 1 to 2 weeks for simple parts, and 6 to 8 weeks for complex parts.

Plaster-and Sand-Molding Combination

A combination of plaster and sand molding is used for castings that require high metal-lurgical integrity as verified by radiographic or fluorescent penetrant inspection. When used in combination with no-bake sand molds, the properties of plaster mold castings can be enhanced considerably by taking advantage of the faster cooling rates inherent in sand molds in combination with the insulating aspect of plaster molds. (Material still pours and fills, maintaining its liquidity—but then cools quickly.)

Investment Casting

Investment casting prototypes can be produced from plastic patterns without permanent tooling. In brief terms, the supplier replaces the wax pattern that would have been produced from an injection die with a low-melting-point plastic pattern assembled from multiple pieces. The assembly is then hand radiused with wax, and from this point on, the process continues in the same manner as the typical investment casting process. This gives the OEM the opportunity to qualify a casting as part of the initial testing program, eliminating the need for requalification had a machined part or a weldment been used as a substitute in qualification test hardware. Depending on the complexity of the design, the supplier can deliver a prototype investment casting in 2 weeks or so, at less cost than a machining. Uni-Cast, Inc., in Nashua, New Hampshire, will provide engineering assistance to build a large, thinwalled initial plastic pattern in your model shop under the supervision of your design team. This pattern can then be taken back to their shop for the casting process (see Figure 1.24).

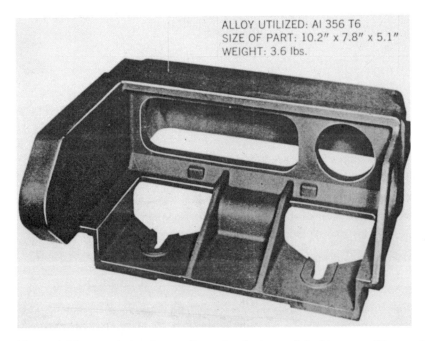

ALLOY UTILIZED: Al 356 T6
SIZE OF PART: 10.2″ x 7.8″ x 5.1″
WEIGHT: 3.6 lbs.

Figure 1.24 Part designed as a die casting because of the large quantities required. Smaller repeat orders at a later date were more economical as an investment casting. (Courtesy of Uni-Cast, Inc., Nashua, NH.)

1.4.3 Manufacture of Plastic Parts

Castable Urethane

Castable urethane is a liquid, two-component, unfilled urethane resin system that is poured into a closed mold. The result after solidification is a tough, high-impact-resistant, dimensionally stable part. This process is also used by Armstrong.

This process is often used:

1. To produce prototype and low-volume plastic parts.
2. To simulate injection-molded parts to check fit and function, and to debug design.
3. As in injection molding there is considerable design freedom.
4. Molds can be made from room temperature cure silicone (RTV) resin, sprayed metal, or, for precision parts, machined aluminum.
5. Tooling is low in cost and allows ease of modification.
6. A variety of inserts can be molded or postinstalled.
7. Stereolithographic molds can be duplicated.
8. Delivery depends on size and complexity; can be 1 to 7 weeks after receipt of order.
9. U.L. traceability.

Design and Technical Information

Size: Up to 24 in.

Finish: Molded color white (no-bake primer and paint available if required).

Wall thickness: Can vary but should be kept less than $1/4$ in.

General tolerances:

0–3 ± 0.010 in.
3–6 ± 0.015 in.
6–12 ± 0.020 in.
12–24 ± 0.030 in.

None: Tighter tolerances can be negotiated.

Holes: All holes molded to size, if precision required can be machined.

Draft: Recommended 1–2°, but no draft is possible if required.

Corner and filet radii: Can be sharp, but some radius ($1/16$ in.) preferred.

See Figure 1.25 for more information on properties of castable urethane. A variety of materials are available for this process which offer different properties.

Reaction Injection Molding

In the reaction injection molding (RIM) process, two liquid components—component A, a formulated polymeric isocyanate, and component B, a formulated polyol blend— are mixed in a highpressure head and then pumped into a mold cavity. A reaction then occurs in the mold, resulting in a high-density polyurethane structural foam part.

Typical Properties

		Test Method
Color	Dark Amber	Visual
Hardness, Shore D	83 ± 5	ASTM D2240
Tensile Strength, psi	7966	ASTM D412
		CONAP
Linear Shrinkage (in/in)	0.0001	CONAP
Heat Distortion Temperature	127.4°F	ASTM D648 (Surface Stress 66 psi)
Elongation	13%	ASTM D412
Flexural Modulus (psi)	174,593	ASTM D790
Flexural Strength (psi)	8,095	ASTM D790
Specific Gravity	1.23	
Flammability 1/16" Sample	94V-0	UL-94

The above properties are intended as a guide only and may vary depending on thickness and shape.

Figure 1.25 Typical properties of castable urethane. (Courtesy of Armstrong Mold Corp., East Syracuse, NY.)

Paul Armstrong of Armstrong Mold says that this process has the following features:

1. Can be used to produce covers, bases, keyboards, bezels, and housings for computers, business machines, and medical applications.
2. For prototype through medium volume applications (1 to 2000 pieces.)
3. Considerable design freedom to produce complex shapes.
4. Molds can be resin, sprayed metal, cast aluminum, or machined aluminum
5. Tooling is low in cost and allows for ease of modification
6. A variety of castings, sheet metal, and threaded inserts can be molded in place.
7. Alternative foams are available which offer a range of insulating properties.
8. UL traceability.

Design and Technical Information

Size: Up to 48 in.

Weight: Up to 18 lb.

Finish (if required):

> No-bake prime and paint
> RF shielding
> Silk screen

Wall thickness: Can vary, but should not be less than ¼ in., although local thicknesses of ⅛ in. can be produced.

General tolerances:

0–3 ± 0.010 in.
3–6 ± 0.015 in.
6–12 ± 0.020 in.
12–18 ± 0.030 in.
18–48 ± 0.040 in.

Note: Tighter tolerances can be negotiated.

Holes: All holes molded to size.

Draft: Recommended 1–2°, but no draft is possible when specified.

Radii and fillets: Should be as liberal as possible—⅛ to ¼ in.

Density: Can be varied depending on foam used to offer a variety of insulation properties.

1.4.4 Photochemical Machining

Blanking

Photochemical machining (PCM) allows burr-free, stress-free blanking on virtually all ferrous and nonferrous metals with no chemical or physical alteration to the materials. With precise multiple imaging, multiple parts blanks can be produced on sheets up to 30 in. wide. The process can work with metals of any temper with thickness ranging from 0.001 in. to heavy gauge. The tolerance range is ±0.001 to ± 0.010 in., depending on material type and thickness. See Figure 1.26 for some examples of parts made by this process.

The process starts with CAD drawings, either furnished by your company or made by the supplier. This is then copied multiple times to make a master negative containing the total parts required in the initial order, and checked for accurate registration. A metal sheet is sheared to size, cleaned, and coated with a photoresist. A contact print is made from the master sheet, and developed. The entire sheet is then immersed in an acid bath and the unwanted metal is "eaten" away, leaving the finished part. The photoresist is then removed. This can then be followed by machining, forming, or whatever subsequent process is required.

Forming and Fabrication

Many companies offer complete photochemical engineering services, using state-of-the-art computer-aided design. Precision plotting and photographic equipment allow for the design of complex and intricate parts with economical photo tooling costs. In general, these companies are very willing to work with your design team to bring your concept to production quickly and accurately. Because tooling is produced photographically, lead time can be reduced from weeks to hours, and design changes can be made quickly, without scrapping expensive dies. Typical lead time is 1 to 2 weeks using readily available materials, but this time can be shortened if the schedule demands it. An example of a firm with a complete in-house tooling, forming, and fabrication department, Microphoto, Inc. (Detroit, Michigan), can economically produce quality finished parts. Using both mechanical and hydraulic presses, they produce precision parts to most print tolerances after PCM blanking.

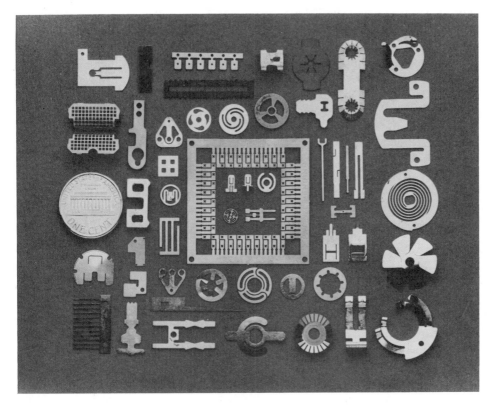

Figure 1.26 Examples of some parts made by photochemical machining. (Courtesy of Microphoto, Inc., Roseville, MI.)

1.4.5 Short-Run Stampings

Tooling costs for short-run stampings are kept to a minimum. You share the tools and holders with all of the suppliers' customers. Standard round and square punches meet many demands. Standard forming tools make a wide variety of shapes. Other standard tools permit many companies to build parts quickly at a minimum cost. Design changes are most easily incorporated with short-run tooling. This minimizes costly scrap and retooling. Tooling techniques for the short-run process at Dayton Rogers (Minneapolis, Minnesota) are unique. Their large tool rooms use the Andrew Linemaster, Electrical Discharge Machine, and other special tools to speed the production of dies. Short runs do not create unnecessary penalties in either quality or cost.

The quality of stamped parts is usually more reliable than from most other methods. The part-to-part repeatability is unusually good. A good firm's quality control department will be well equipped to assure that your parts meet print/specifications.

Delivery of sheet metal prototype parts is usually assured. Dayton Rogers, for example, has over 150 presses up to 300 tons. Their large metal inventory makes 2 million pounds of stock immediately available. Production control methods include a daily check of order status against schedule. Their "scoreboard performance" indicates that over 99% of all orders are shipped "on time" or ahead of schedule.

How to Use Short-Run Stampings

Short-run stampings offer an economical way to produce parts in quantities from prototype to 100,000 pieces with short lead time. It is an ideal method for checking the design, assembly process, and market acceptance of a new product,—all with minimum investment. The investment in short-run tooling is always a secure one. Even if the product exceeds sales forecasts, the tooling can be a major asset as a second source in the event of high-rate production die breakdown. Subcontractors review thousands of prints each year, searching for ways to save customers money. Their suggestions involve material specifications, dimensions, and tolerances. Use their expertise!

Standard warehouse materials save both time and money. Price and delivery are both much more favorable when the supplier has a large metals inventory. For small lots, you can save on part price by using metal bought in large lots. If thickness tolerance is critical, double disk grinding is an option. If a part is extremely thick for its shape, sometimes two pieces laminated together is a solution.

Tolerances

Consider the necessity of specific tempers or closer material tolerances. If a part requires a specific temper or closer material tolerances, strip steel is desirable. However, if temper or closer material tolerances are not a factor, sheet steel is available at a lesser cost. Tolerances should be no tighter than necessary to make a part functional. Close or tight tolerances result in high cost to build tools, high tool maintenance, higher run cost, and costly in-process inspection to assure specifications.

1.4.6 Bibliography

Armstrong Mold Company, 6910 Manlius Road, East Syracuse, NY 13057.
Dayton Rogers Mfg. Company, 2824 13th Avenue South, Minneapolis, MN 55407.
Microphoto, Inc., 6130 Casmere, Detroit, MI 48212.
Uni-Cast, Inc., 45 East Hollis Street, Nashua, NH 03060.

2

Product Design

Jack M. Walker   *Consultant, Manufacturing Engineering, Merritt Island, Florida*

2.0 INTRODUCTION TO PRODUCT DESIGN

There is no "magic formula" for guiding the design team of a new or redesigned product, or for support by the manufacturing engineering member of the team representing the factory. The considerations and techniques in Chapter 1 are valuable guides in (1) working together as an integrated team for product development and design, (2) examining the customer's wants and needs, (3) identifying the problems with assembly of the final product, and (4) developing the product cost. Chapter 10, on work measurement will aid in making estimates of the work content, as will Chapter 7, on practical cost estimating for manufacturing.

This chapter is devoted to discussion of some of the major manufacturing processes used in making detail parts. A discussion of each of the processes is presented in order to better understand the tolerances, lead times, costs, and the "do's and don'ts" of designing parts using each process. Subchapter 2.1 covers design for machined parts; Subchapter 2.2 goes into designing for castings and forgings; Subchapter 2.3 discusses designing parts from sheet metal; and 2.4 is a summary to aid in selecting the best process for a particular design application.

Brief discussions of material selection are included, although more complete coverage will be found in Chapter 14. More information on conventional detail fabrication processes and equipment is found in Chapter 15. The more nontraditional processes used in the fabrication of parts are discussed in Chapter 16, fabrication of plastic parts in Chapter 17, and composite materials in Chapter 18. Chapter 3 may be valuable when new equipment selection and purchase must be considered for some of the product requirements.

Experience tells us that it is necessary to involve production process expertise early in the design and development phases of all new programs. The best product design for manufacturing will be one that:

Considers all elements of the fabrication, subassembly, and assembly operations leading to a completed product;

Is able to accept the manufacturing process tolerances;

Uses straightforward, understandable manufacturing processes that lead to "first- time quality" during production start-up;

Can be built with automated process equipment using programmable process controllers (when appropriate);

Does not require separate inspection operations, but empowers the operator with the
responsibility for his work;

Allows statistical process control (SPC) to monitor the production operations and
permits operator adjustment of the process to ensure 100% quality every time;

Contributes to a shorter manufacturing cycle with minimum work in process; and

Can be built for the lowest cost.

2.1 DESIGN FOR MACHINING

2.1.0 Introduction to Design for Machining

An early comment on designing parts for machining should be to question the necessity
of machining the part. With the options available for other processes to consider, this is no
simple task. Leo Alting, *Manufacturing Engineering Processes* (Marcel Dekker, New
York, 1982) discusses "mass conserving" processes such as rolling, extrusion, forming,
swaging, cold-heading, and the like. The idea is that a piece of metal may be modified to
produce a finished part, without cutting chips. Many times, a part using "mass reducing"
techniques, such as cutting chips on some type of machining center, may retain only 20%
of the original metal, while 80% winds up as chips—or scrap. In many competitive job
shops that use production screw machines or other high rate machines, the profit comes
from the sale of the scrap chips! During the concurrent engineering process, whether you
utilize design for assembly (DFA) or some other formal technique to assist in design
analysis, all the various production options must be explored. Quantity of parts to be made,
the required quality, and the overall design requirements will influence the decision.
Figure 2.1 shows two examples where conventional "machining" is not the correct choice.
In Figure 2.1a, the part was changed to an extrusion, with a simple cutoff operation; and
in Figure 2.1b, the large quantities permitted use of cold heading plus standard parts. New
casting alloys and techniques, increased pressures available for forging and pressing, and

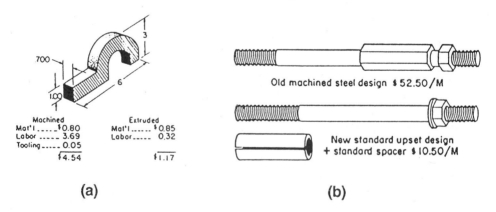

(a)   **(b)**

Figure 2.1   (a) Change from machined plate to extruded material produced at 74% savings on
quantities of 2000 pieces. (b) Stud used by General Electric required 200,000 pieces per year.
Automatic upset and rolled, threaded stud with a standard roll-formed spacer offered worthwhile
savings. (From R. A. Bolz, *Production Processes*, Industrial Press, New York, 1963.)

the progress made in powder metal and engineering plastics make "near net shape" worth strong consideration.

Machine Fundamentals

In order to acquire a sound working knowledge of the design of parts to be made by machining, it is essential to begin by becoming familiar with a few of the fundamentals. There are several advantages to studying the elements of a subject as complex as today's machine tools. In the first place, it provides us a chance to get acquainted with the main points before having to cope with the details. Once learned, these main points will serve as a filing system for mentally sorting and cataloging the details as they are brought up and discussed. It is then easier to study the subject in a logical fashion—progressing step by step from yesterday's equipment and processes to tomorrow's.

To begin, we should agree upon just what we mean by "machine tools." A machine is a device to transmit and modify force and motion to some desired kind of work. Therefore, we could say that a machine tool is a mechanism for applying force and motion to a cutting instrument called a tool. However, as we get into the subject a little further, we shall see that there are several qualifications to this broad definition. First, a machine tool does not always apply force and motion to the tool, but sometimes applies them to the material to be cut, while it is in contact with the fixed tool.

A further limitation of the term "machine tool" is usually observed in practice by not including such power tools as cutoff saws, shears, punch presses, routers, and the like. Although such machines do apply mechanical force to a cutting tool, they do not machine metals by "cutting chips" in the same way as does a lathe or milling machine.

Besides employing power and a cutting tool, true machine tools also provide means for accurately controlling the tool to take as deep or as shallow a cut as desired, as well as a means for applying this action consecutively in the direction needed to obtain the desired finished shape. Further, a true machine tool must provide means for interrupting this action whenever necessary, to allow the operator to check the progress of the work—or for some other function to occur.

Since all machine tool operations must be confined to the surface of the workpiece, so far as we are concerned here, the shape of any workpiece is determined by the nature of its surfaces. Shapes can thus be divided into three broad classifications:

1. Those composed primarily of flat surfaces
2. Those composed primarily of surfaces curving in one direction
3. Those composed primarily of surfaces curving in more than one direction

The cube is a typical example of the first class, the cylinder of the second class, and the sphere of the third class. Of course, we can machine the "interior" surface of a part, as well as the exterior surface.

All machine tools are basically alike in that their purpose is to change the shape, finish, or size of a piece of material. We find that no matter what kind of a machine it is, it accomplishes this objective by bringing a cutting edge or edges into contact with the workpiece when one or both of them are in motion. All the variations in machine design, tool design, operation, and procedure may be understood more readily if you know that the major function of a machine tool is to move a cutting edge or a piece of material while they are being held in contact with each other in such a way that one cut or a series of cuts can be made for the purpose of changing its size, shape, or finish.

Machine Motion

Fundamentally, the motion of either the material or the tool can occur in two forms (recip-rocating or rotating), but this motion can take place in two directions in the vertical plane (up or down), and four directions in the horizontal plane (left, right, forward, or backward). See Figure 2.2. (We will see later that there are additional "axes" for moving the material or the cutter, but we should start with the above definition.) Refer to Chapters 15 and 16 in this handbook for more information on the actual machines and machining processes.

2.1.1 The Lathe

In a lathe, the material is rotated and the tool is reciprocated back and forth along the length of the surface of the cylindrical part, with each pass of the cutting tool moved toward the centerline of the part to increase the depth of cut (See Figure 2.3a). Therefore a lathe must have some means of holding the material and rotating it—and some means of holding the cutting tool and controlling its movement, as shown in the schematic of a simple lathe in Figure 2.3b. Most lathes are provided with tool holders of various types and styles for holding turning tools, drills, reamers, boring bars, etc. They are also provided with collets, chucks, centers, and face plates for holding and driving the material from the headstock spindle. When a casting or forging requires machining, specially designed holding fixtures are needed to grasp the material and hold it in position to be presented to the cutting tools.

2.1.2 The Milling Machine

Much of the previous discussion on lathes applies equally to the milling machine process. In a sense, the milling machine is the opposite of the lathe, in that it provides cutting action by rotating the tool while the sequence of cuts is achieved by reciprocat-ing the workpiece. When the sequence of consecutive cuts is produced by moving the workpiece in a straight line, the surface produced by a milling machine will nor-

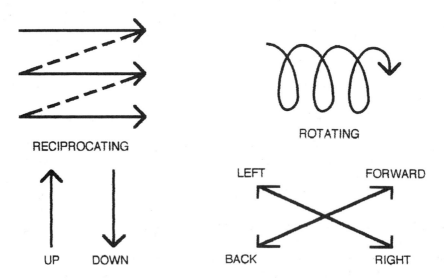

Figure 2.2 Fundamentals of motion.

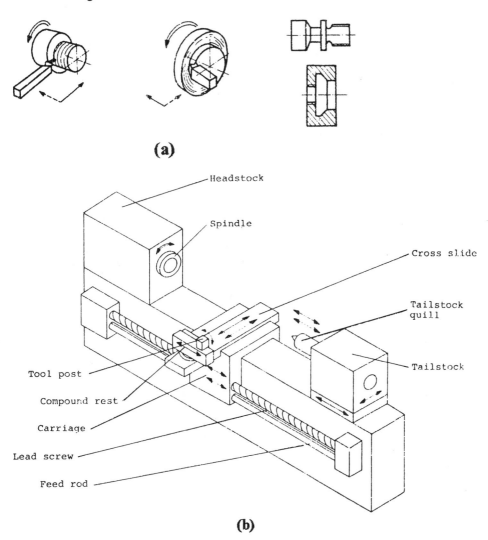

(a)

Headstock

Spindle

Cross slide

Tailstock quill

Tailstock

Tool post

Compound rest

Carriage

Lead screw

Feed rod

(b)

Figure 2.3 (a) Cutting action on a lathe. (b) Pattern of motions on a lathe. (From Leo Alting, *Production Processes*, Marcel Dekker, New York, 1982.)

mally be straight—in one direction at least. A milling machine, however, uses a multiple-edged tool, and the surface produced by a multiple-edged tool conforms to the contour of the cutting edges. Figure 2.4a shows the horizontal milling cutter on the left, a vertical milling cutter on the right, with additional types of milling cutters shown in Figure 2.4b. If the milling cutter has a straight cutting edge, a flat surface can be produced in both directions. The workpiece is usually held securely on the table of the machine, or in a fixture clamped to the table. It is fed to the cutter or cutters by the motion of the table. Multiple cutters can be arranged on the spindle, separated by precision spacers, permitting several parallel cuts to be made simultaneously. Figure 2.5 shows a schematic of a horizontal milling machine.

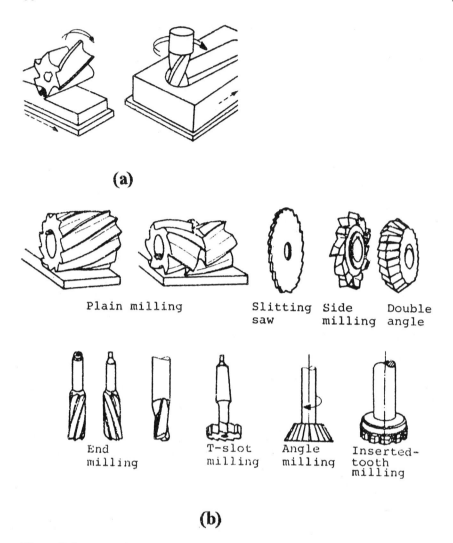

(a)

Plain milling   Slitting Side   Double
                            saw   milling angle

End   T-slot   Angle   Inserted-
milling   milling   milling   tooth
                                           milling

(b)

Figure 2.4   (a) Horizontal (left) and vertical (right) milling patterns. (b) Typical milling cutters. (From Leo Alting, *Production Processes*, Marcel Dekker, New York, 1982.)

2.1.3 Drilling

In the construction of practically all products, a great many holes must be made, owing to the extensive use of bolts, machine screws, and studs for holding the various parts together.

In drilling, the material is fixed and the tool is rotated and advanced to complete a given sequence of cuts. (An exception be drilling on a lathe, with the drill held fixed in the tailstock while the work is being rotated.)

Drilling Machines

Drilling machines or "drill presses" as they are often called, which are used for drilling these holes, are made in many different types designed for handling different classes of

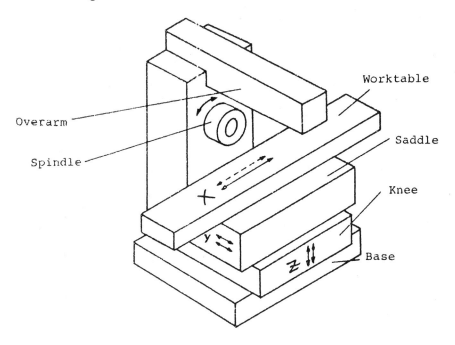

Figure 2.5 Plain column-and-knee type milling machine, showing motion patterns. (From Leo Alting, *Production Processes*, Marcel Dekker, New York, 1982.)

work to the best advantage. The various types are also built in a great variety of sizes, as the most efficient results can be obtained with a machine that is neither too small nor too large and unwieldy for the work which it performs. The upright drill press is one type that is used extensively. As the name indicates, the general design of the machine is vertical, and the drill spindle is in a vertical position. Figure 2.6 shows a schematic design of a single-spindle drill press. The heavy-duty machine is a powerful machine for heavy drilling. It has an adjustable knee in front of the column, and is supported by an adjusting screw, somewhat like a vertical milling machine. See Figure 2.7. The more common drill presses, for medium or lighter work, where the spindle can be moved up and down and the table adjusted to any height, can be arranged with multiple drill heads. Special-purpose drill presses include a sensitive drilling machine in which the spindle rotates very quickly, making small-diameter holes, as in printed circuit boards. There are multiple-head machines which may have from 4 to 48 spindles, driven off the same spindle drive gear in the same head.

Drills

The twist drill (originally made from twisted bar stock) is a tool generally formed by milling, or forging followed by milling, two equal and diametrically opposite spiral grooves in a cylindrical piece of steel. The spiral grooves, or flutes, have the correct shape to form suitable cutting edges on the cone-shaped end, to provide channels for the free egress of chips and to permit the lubricant to get down to the cutting edges when the drill is working in a hole. It is probably the most efficient tool used, for in no other tool is the

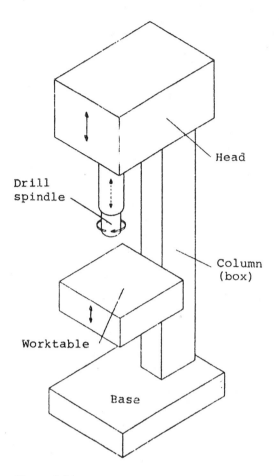

Figure 2.6 Vertical drilling machine. (From Leo Alting, *Production Processes*, Marcel Dekker, New York, 1982.)

cutting surface so large in proportion to the cross-sectional area of the body that is its real support. See Figure 2.8.

Drill sizes are designated under four headings:

1. Numerical: no. 80 to no. 1 (0.0135 to 0.228 in.)
2. Alphabetical: A to Z (0.234 to 0.413 in.)
3. Fractional: 1/64 in. to 4 in. and over, by 64ths
4. Millimeter: 0.5 to 10.0 mm by 0.1 mm increments, and larger than 10.0 mm by 0.5 mm increments

Odd sizes may be ground to any specified diameter, but are only available on special order—at a higher cost.

Speeds and Feeds

When we refer to the "speed" of a drill, we mean the speed at the circumference—called *peripheral speed.* Speed refers to the distance the drill would travel if, for example, it rolled 30 feet a minute. Speed usually does not refer to revolutions per minute unless specifically stated.

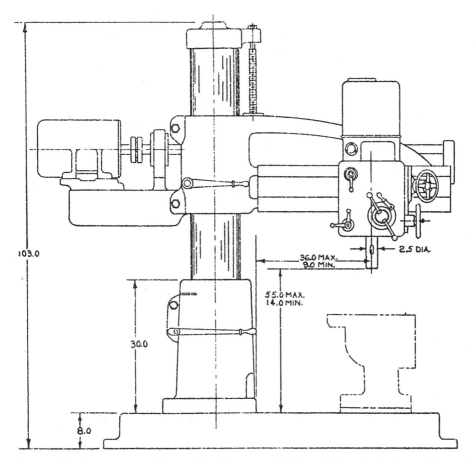

CARLTON RADIAL DRILLS
3-FOOT ARM ON 11-INCH COLUMN

Figure 2.7 Heavy-duty radial-arm drill. (Courtesy of McDonnell Douglas, St. Louis, Missouri.)

Except under certain rather rare conditions, a drill does not pull itself into the work like a corkscrew—it requires constant pressure to advance it. This advance (measured in fractions or decimals of an inch per revolution) is called *feed*. Feed pressure, however, is the pressure required to maintain the feed. Some typical feeds and speeds are shown in Figure 2.9.

Tolerances

Under average conditions drilled holes, preferably not over about four times the diameter in depth, can be held within the following tolerances:

#80 to #71	+0.002	−0.001
#70 to #52	+0.003	−0.001
#51 to #31	+0.004	−0.001
⅛″ to #3	+0.005	−0.001

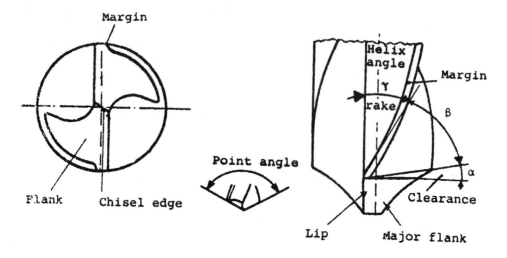

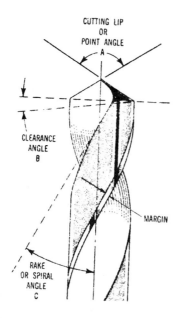

Figure 2.8 Cutting angles and margins on a twist drill. (From Leo Alting, *Production Processes*, Marcel Dekker, New York, 1982.)

1/32″ to R	+0.006	−0.001
11/32″ to 1/2″	+0.007	−0.001
33/64″ to 23/32″	+0.008	−0.001
47/64″ to 63/64″	+0.009	−0.001
1″ to 2″	+0.010	−0.001

TYPICAL SPEEDS FOR HIGH-SPEED STEEL DRILLS

Drill Diameter Inches	Peripheral Speed Feet Per Minute
Less than 1	600
1 to 1½	550
Over 1½	450

TYPICAL DRILL FEEDS FOR FIRST DRILL ENTRY①

Drill Diameter Inch	Tolerance Inch ±	Feed for 2011-T3 Inch/Rev	Feed for Other Alloys Inch/Rev
.0625	.0015	.004	.004
.125	.002	.012	.010
.187	.002	.014	.012
.250	.002	.017	.014
.375	.0025	.020	.017
.500	.0025	.020	.017
.750	.003	.020	.017

①For multiple drill entries these feeds should be reduced about 15 percent for each succeeding entry. Lower feeds should also be used for thin-wall parts.

Figure 2.9 Typical feeds and speeds for drilling aluminum. (Courtesy of Reynolds Metals Company, Richmond, Virginia.)

These tolerances are sufficiently liberal for general production, and although a definite set of limits suitable for all cases is out of the question, will provide a good conservative working base. Where more exacting specifications are desirable, redrilling or reaming will usually be necessary. Tolerances on drilled holes are normally plus, owing to the fact that drills invariably cut oversize from 1- to 5-thousandths, depending on the accuracy of sharpening. For this reason, the above tolerances are often specified as all minus 0.000 in. Straightness is dependent on the hole depth and the homogeneity of the material.

A tolerance of ±0.001 in. can be held on ordinary reaming operations. With additional boring operations preceding, reaming tolerances can be held to ±0.0005 in. Depending on the type material being cut, the surface quality left in turret lathe operations will be about 60 μin., rms, or less. Surface patterns depend on the type of tooling and cuts employed.

2.1.4 Design for Machining

In manufacturing, the product produced is always a compromise between acceptable quality and acceptable cost. The seldom-reached goal is always one of maximum attainable quality for minimum cost. Close tolerances and fine finishes may have very little influence on the functional acceptance by the ultimate user—who is the only one who can really judge "quality." With these few general guidelines, we should examine some of the "do's and don'ts" of design features for parts that will be machined by cutting chips.

Machinability

When an economical machining operation is to be established, the interaction among geometry, the material, and the process must be appreciated. Figure 2.10 shows the natural

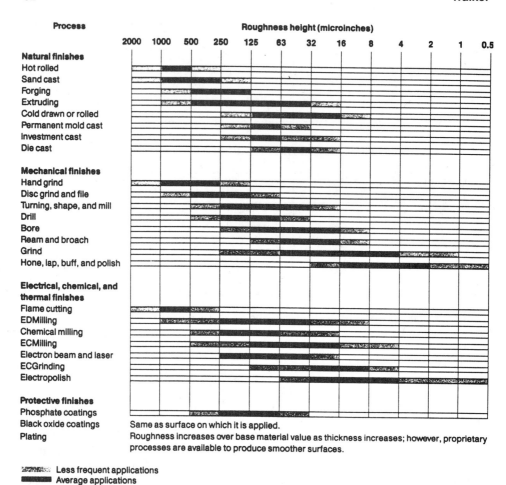

Figure 2.10 Surface finishes versus manufacturing process. (From John P. Tanner, *Manufacturing Engineering,* Marcel Dekker, New York, 1991.)

finish of several materials. The lowest cost would obviously be not to machine at all. It is not sufficient to choose a material for a part that fulfills the required functional properties; its technological properties describing the suitability of the material for a particular process must also be considered. In mass conserving processes, the material must possess a certain ductility (formability); and in mass reducing processes, such as machining, it must have properties permitting machining to take place in a reasonable way. The technological properties describing the suitability of a material for machining processes are collectively called its *machinability.*

Machinability cannot be completely described by a single number, as it depends on a complex combination of properties which can be found only by studying the machining process in detail. The term *machinablilty* describes, in general, how the material performs when cutting is taking place. This performance can be measured by the wear on the tool, the surface quality of the product, the cutting forces, and the types of chip produced. In many cases, tool wear is considered the most important factor, which means that a machinability index can be defined as the cutting speed giving a specified tool life. When a component is to be machined by several processes, the machinability index corresponding to the process most used is chosen.

The machinability of a material greatly influences the production costs for a given component Poor machinability results in high costs, and vice versa. In Figure 2.11 the machinability for the different materials groups is expressed as the removal rate per millimeter depth of cut when turning with carbide cutters. The chart can be used only as a general comparative guideline; in actual situations, accurate values must be obtained for the particular material chosen for the product design. The machinability of a particular material is affected primarily by its hardness, composition, and heat treatment. For most steels, the hardness has a major influence of machinability. A hardness range of HB 170 to 200 is generally optimal. Low hardnesses tend to lead to built-up edge formation on the cutters at low speeds. High hardnesses above HB 200, lead to increased tool wear. The heat treatment of the work material can have a significant influence on its machinability. A coarse-grained structure generally has a better machinability than does a fine-grained structure. Hardened, plain carbon steels ($> 0.35\%$ C) with a martensitic structure are very difficult to machine. Inclusions, hard constituents, scale, oxides, and so on, have a deteriorating effect on the machinability, as the abrasive wear on the cutting tool is increased.

2.1.5 Cost Versus Accuracy

Surface roughness is also a factor in design of machined parts. Typical roughness values (arithmetical mean value, R_a) for different processes is shown in Figure 2.10. It can be said that the roughness decreases when feed is decreased, the nose radius is increased, and the major cutting-edge angle and the minor cutting-edge angles are reduced on the cutter. Furthermore, increasing cutting speeds and effective cutting lubrications can improve surface quality.

It is difficult to equate minor design features with their proper influence on ultimate product cost. Figure 2.12 shows charts of general cost relationship on various degrees of accuracy. Also shown are some examples of tolerance versus cost. These

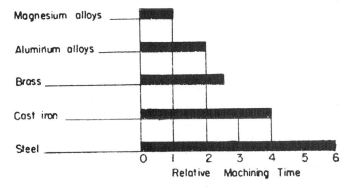

Figure 2.11 General comparative machining rates of various materials. (From R. A. Bolz, *Production Processes*, Industrial Press, New York, 1963.)

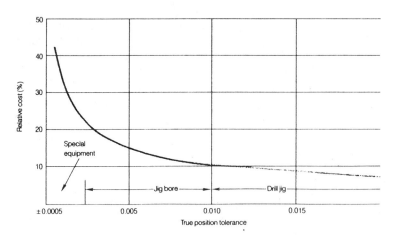

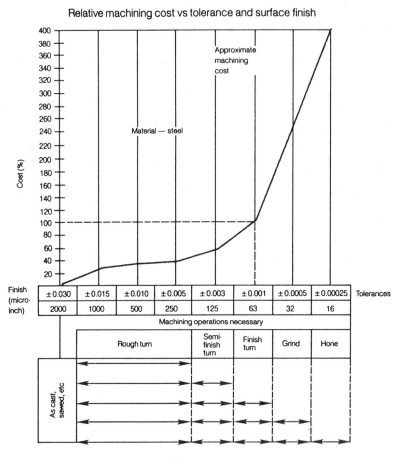

Figure 2.12 General cost relationship of various degrees of accuracy and surface finish. (From John P. Tanner, *Manufacturing Engineering,* Marcel Dekker, New York, 1991.)

data are plotted from a variety of sources, and the trend is only indicative of the cost of refinement and does not really take into account the possibilities offered by judicious designing for production. The profound effect of tolerances on cost can be seen by the examples in Figures 2.13, 2.14, and 2.15, which show a variety of turned and milled parts.

One trap that is easy to fall into today is the new machines and their ability to achieve very tight tolerances. Do not permit a tight tolerance to appear on the product drawings— just because it is possible to achieve it. As machines and tools wear, the "new machine" tolerances may not be held so easily—and the cost will be higher than necessary. Also, the looser tolerances will permit greater choice of machine utilization in the shop, making the new "tight" machines available for work that must have the closer tolerances.

2.1.6 General Rules for Design

John P. Tanner, president of Tanner and Associates, Orlando, Florida, suggests that when addressing surface tolerances and tolerances for machined parts, the following cautionary steps should be followed:

Avoid surface finishes that cannot be achieved in the first operation of fabrication (i.e., a forging requiring a 32 μin. finish will need a subsequent machining operation)
Generally, surface finishes of 63 μin. or less will require additional machine operations to meet specifications and should be avoided.
Avoid tolerances on machined parts that require several operations to meet specifications. Use the most economical design tolerances indicated for the type of precess selected.
Avoid hole depths which are three times greater than the hole diameter.
Generally, avoid tolerances tighter that ±0.004 in. in drilling operations.
Generally, avoid tolerances tighter than ±0.0025 in. in milling operations.
Generally, avoid tolerances tighter than ±0.0015 in. in turning operations.

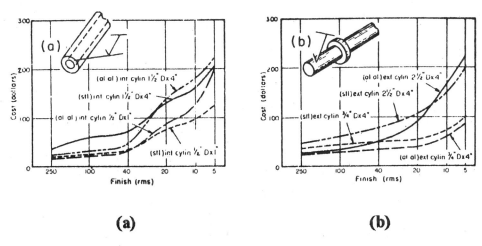

(a) **(b)**

Figure 2.13 (a) Comparative costs for internal holes with varying degrees of surface requirement. (b) Plot of surface refinement versus cost for some turned and ground surfaces. (From R. A. Bolz, *Production Processes*, Industrial Press, New York, 1963.)

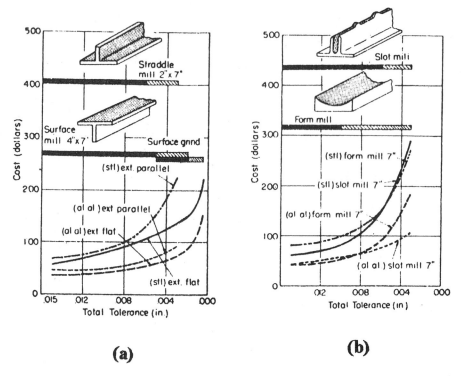

Figure 2.14 (a) Straddle and face milling costs plotted against tolerance requirements. (b) Tolerance effect of slotting and form milling costs. (From R. A. Bolz, *Production Processes*, Industrial Press, New York, 1963.)

An additional checklist for machined parts design follows:

1. Ensure maximum simplicity in overall design. Develop product for utmost simplicity in physical and functional characteristics. Production problems and costs have a direct relation to complexity of design, and any efforts expended in reducing a product to its "lowest common denominator" will result in real savings. Always ask: "Is it absolutely required?" The most obvious way to cut costs is to eliminate or combine parts to obviate the expense of attaching or assembling. Refinements of design and analysis are the key to sharp reductions in costs—not revolutionary improvements in production techniques.

2. Analyze carefully all materials to be used. Select materials for suitability as well as lowest cost and availability. For instance, though aluminum alloys and aluminum alloy forgings cost somewhat more than steel for parts of similar design, time required for machining is often sufficiently lower (about one-third that required for similar steel parts) to offset the original additional material cost.

3. Use the widest possible tolerances and finishes on parts. The manufacturing engineer should have a set of process capability charts available on every machine in the shop—including any contemplated new equipment purchases. Our goal should be 6 sigma quality, which says that statistical process control (SPC) will eliminate any inspection operations—we won't make any bad parts. However, the machine capability and tolerances produced must be tighter than any of the part

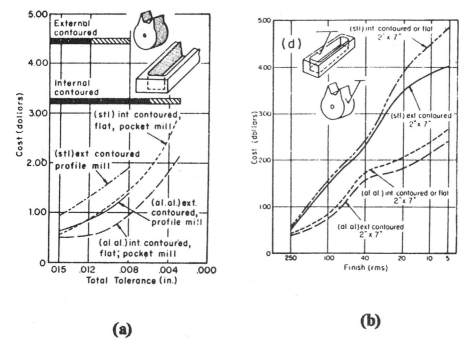

(a)

(b)

Figure 2.15 (a) Plot of tolerances versus costs for profiled and pocket milled parts. (b) Surface finish versus costs for profiled and milled parts. (From R. A. Bolz, *Production Processes*, Industrial Press, New York, 1963.)

tolerances. Be sure that surface roughness and accuracies specified are reasonable and in keeping with the product and its function. The roughest acceptable finish should be contemplated and so specified. There is a direct relationship between surface finish and dimensional tolerancing. To expect a tolerance of ±0.0001 in. to be held on a part turned or milled to an average of 500 μin. is rather foolish. Likewise, to specify a finish of 10 to 15 μin. for a surface which is merely intended to provide proper size for locating subsequent operations is also needless. A 40- to 60-μin. finish would be satisfactory and the cost will be at least 50 to 60% less.

4. Standardize to the greatest extent possible. Use standard available parts and standardize basic parts for one product as well as across a maximum number of products. Evolving the basic idea for a product constitutes about 5% of the overall effort in bringing it onto the market; some 20% of the effort goes toward general design of the product and about 50% of the effort is necessary to bring about economic production of the product, while the remainder goes toward removal of bugs from design and production, including servicing. The big job is in making the product economically—through design for more economic production.

5. Determine the best production or fabricating process. Tailor the product to specifically suit the production method or methods selected. It is possible to tailor the details of a product to assure economic production, and it is possible to have a basically similar part which is excessively expensive to produce. Small but extremely important details can create this difference. Figure 2.16 shows these difference in a design with a tangent radius. A concave cutter usually creates a nick on one or both sides of the piece shown in (a). The nontangent design shown

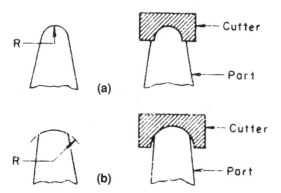

Figure 2.16   Tolerances have a profound effect on cost: (a) requires turning and finish grinding; (b) requires turning only. (From R. A. Bolz, *Production Processes*, Industrial Press. New York, 1963.)

in (b) offers considerable production economy. Every production method has a well-established level of precision which can be maintained in continuous production without exceeding normal costs. The practice of specifying tolerances according to empirical rules should be abandoned. Practical design for production demands careful observances of the "natural" tolerances available with specific processes. See Figure 2.17.

6. Minimize production steps. Plan for the smallest possible number or lowest-cost combination of separate operations in manufacturing the product. The primary aim here is to ensure the elimination of needless processing steps. Even though the processing may be automatic, useless steps require time and equipment. Reducing the number of separate processing operations required to complete a part will usually yield the greatest savings. Next in importance is reduction in the number of parts used, and following that comes savings effected through the use of stock parts or interchangeable parts.

7. Eliminate handling problems. Ensure ease in locating, setting up, orienting, feeding, chuting, holding, and transferring parts. Where parts must be held or located in a jig for a processing operation, consideration must be given to the means by which the operation will be accomplished. Lugs may be added, contours changed, or sections varied slightly to make handling a simple rather than a costly operation. The difficulty of handling certain parts in production is often overlooked. Pads or locating positions are almost invariably required in processing, regardless of quantity. Use of false lugs for locating and clamping often may be a necessity. While ingenious handling methods and mechanical production aids are continuously being developed, the designer who can eliminate the need for such equipment simplifies production problems and effectively reduces overall costs.

Drawing Notes

Each part machined must have a starting point in its process. The old story in process engineering was that the first operation on any machined part started with "Slab mill the base." This is still true of most parts. The part is then turned over, and all subsequent

Range of sizes		Tolerances								
From	Through									
0.000	0.599	0.00015	0.0002	0.0003	0.0005	0.0008	0.0012	0.002	0.003	0.005
0.600	0.999	0.00015	0.00025	0.0004	0.0006	0.001	0.0015	0.0025	0.004	0.006
1.000	1.499	0.0002	0.0003	0.0005	0.0008	0.0012	0.002	0.003	0.005	0.008
1.500	2.799	0.00025	0.0004	0.0006	0.001	0.0015	0.0025	0.004	0.006	0.010
2.800	4.499	0.0003	0.0005	0.0008	0.0012	0.002	0.003	0.005	0.008	0.012
4.500	7.799	0.0004	0.0006	0.001	0.0015	0.0025	0.004	0.006	0.010	0.015
7.800	13.599	0.0005	0.0008	0.0012	0.002	0.003	0.005	0.008	0.012	0.020
13.600	20.999	0.0006	0.001	0.0015	0.0025	0.004	0.006	0.010	0.015	0.025

Lapping & honing
Grinding, diamond, turning
  boring
Broaching
Reaming
Turning, boring, slotting
  & planing
Milling
Drilling

Figure 2.17 "Natural" tolerances available with various processes. (From John P. Tanner, *Manufacturing Engineering,* Marcel Dekker, New York, 1991.)

operations are taken from the base as the common reference. Three points define a plane, therefore there must be some reference on the blueprints as to the location of these points. These tooling points are used by the pattern maker, the die maker, and the casting or forging fabricator. Any subsequent holding fixtures required for machining use these same reference points. Final inspection (if required) will also start with these common reference points.

2.1.7 Bibliography

Alting, Leo, *Manufacturing Engineering Processes*, Marcel Dekker, New York, 1982.
Tanner, John P., *Manufacturing Engineering*, Marcel Dekker, New York, 1991.

2.2 DESIGN FOR CASTING AND FORGING

2.2.0 Introduction to Design for Casting and Forging

In casting, the liquid material is poured into a cavity (die or mold) corresponding to the desired geometry. The shape obtained in the liquid material is now stabilized by solidification (cooling) and removed from the cavity as a solid component. Casting is the oldest known process for producing metallic components. The main stages are producing a suitable mold cavity; melting the material; pouring the liquid material into the cavity; stabilizing the shape by solidification; removing or extracting the solid component from the mold; cutting off sprues and risers; and cleaning the component.

Forging, or its close cousin, pressing, consists of preheating a solid piece of wrought metal and placing it in a lower die. The upper die is then closed, squeezing the material into the closed die cavity and forming the desired shape of the part. The pressure may be applied slowly, as in pressing, or rapidly, with one or more hammer actions of the upper die, as in forging. Any excess material is squeezed out between the die halves as flash. The die is opened and the part ejected. The main stages are cutting the blank to the correct volume or shape; preheating the blank; forging the part; opening the die and ejecting the part; trimming off the flash (usually in a punch press die); and cleaning the part.

In both processes, depending on the properties of the metal and the desired requirements of the product, the finishing from this point on is similar. Both may require heat treatment, straightening, machining, plating or painting, etc. Both aluminum and steel, as well as most other metals, may be cast or forged. Different alloying ingredients are added to improve the processing characteristics, making the precise alloy selection important in both processes. Some alloys can only be cast and others can only be forged. In general, forging materials require a high ductility at the elevated temperatures required for forging. Forgings also have better mechanical properties in the finished part, since the grain structure of the material becomes somewhat flattened or stretched, as in other wrought alloys. Castings, on the other hand, have a more nonoriented grain structure, with the size of the grains being a function of the rate of cooling. The casting alloying ingredients usually have more effect on the liquidity at the pouring temperature. After heat treatment and the accompanying changes in the internal structure and properties of the metal, castings are generally of lower strength and ductility than forgings.

The design problems of casting and forging involve comparing the cost of one process against another—coupled with the limitations of web thicknesses, radii, strength, and the like. The quantity of parts to be produced is also an important factor. Lead times for various processes differ greatly. For example, the tooling lead time could vary by several months for a sand casting versus a die casting. Subchapter 1.4 discusses some short-time options that are available for use until the final production tooling is available. In nearly all cases, a part that will ultimately be cast or forged can be hogged-out of solid stock,—although the cost may be very high. The designer has more leeway in the physical shape of a part by utilizing one of the casting processes rather than a forging. Castings are usually lower in cost than forgings. Forging are superior in strength, ductility, and fatigue resistance, while castings may be better for absorbing vibration and are generally more rigid.

The introduction to casting and forging processes in this section are not meant to make the manufacturing engineer or product designer an expert in the field. The processes are described in enough detail to provide a reference as to the limitations of the various processes available, in order for the design team to consider the strength, tolerances, surface finish, machinability, lead time for tooling, and cost. Chapter 15 discusses the casting and forging processes in greater detail.

2.2.1 Sand Casting*

Sand casting is the oldest method of casting. The process is basically a simple one, with a few universal requirements. A mold cavity is formed around the pattern in a two-part box of sand; the halves are called the *cope* (upper) and the *drag* (see Figure 2.18). Fitted together, these form the complete mold. To withdraw the pattern from the mold, the pattern must be tapered and have no undercuts. This taper is called *draft* and usually has an angle of 3°. The most common molding material used in sand casting is green sand, which is a mixture, usually of ordinary silica sand, clay, water, and other binding materials. The molding sand is placed in the lower half of a two-part form and is packed around half the

---

*The material in this section is reproduced with permission from C. W. Ammen, *Complete Handbook of Sandcasting,* Tab Books, Blue Ridge Summit, PA, 1981.

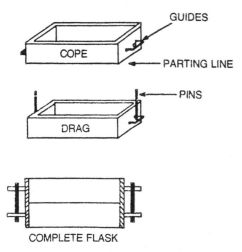

Figure 2.18 Basic flask for sand casting. (From C. W. Ammen, *Complete Handbook of Sand Casting,* Tab Books, Blue Ridge Summit, PA, 1981.)

pattern (up to the parting line). Then a dry molding powder is sprinkled over the pattern and the sand in the cope as a release agent, the upper half of the form is put in place on top of the drag, and the remaining sand is packed into place. When the sand is compacted sufficiently, the cope and drag are separated and the pattern is removed (see Figure 2.19). The two halves are then rejoined, and the metal is poured in through an opening (*sprue*) into the cavity. The size of the pattern must be adjusted to allow for the shrinkage of the metal as it cools. Aluminum and steel shrinks about 1/4 in./foot, depending on the thickness of the part—the thicker parts shrinking less. Molten metal may flow from the sprue to various entry ports (*gates*) in the cavity. Risers at each gate act as reservoirs of molten metal that feed the casting during solidification. A sprue and riser can be seen in Figure 2.20. Aluminum castings need larger gates and risers than other metal castings, thus the average weight of metal poured is about two or three times the weight of the finished casting. Figure 2.21 shows the weight comparison of equal volumes of various metals. The importance of proper gating has been demonstrated by the fact that redesign and relocation on the gates feeding a previously improperly gated casting have increased the strength of the casting as much as 50 to 100%.

Cores

A hollow casting is made by inserting a core of sand into the cavity before pouring. Figure 2.22 shows the maksing and use of cores in a sand casting mold. Any leakage between the halves of the mold cavity forms thin ridges of flash. Protuberances left from risers and sprues must be removed from the part, along with the flash, after it has solidified and the sand has been cleaned off.

Patterns

When more than one part is to be made, the pattern is split along the parting line and each half is mounted on a pattern board or match plate as shown in Figure 2.23. One half of the impression is made in the cope, the other half in the drag. You can see the locating holes for the core in both the cope and drag. The mold is then opened up and the pattern board removed, the core installed, and the mold reclosed for pouring.

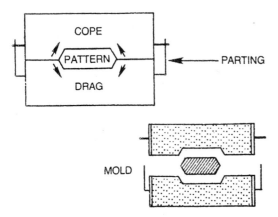

Figure 2.19 Development of split pattern mold. (From C. W. Ammen, *Complete Handbook of Sand Casting*, Tab Books, Blue Ridge Summit, PA, 1981.)

General

Sand casting is the most common casting process used with aluminum alloys. Casting has the lowest initial cost and permits flexibility in design, choice of alloy, and design changes during production. For these reasons, sand casting is used to make a small number of cast pieces, or to make a moderate number that require fast delivery with the likelihood of repeat production. A sand-cast part will not be as uniform in dimensions as one produced by other casting methods, so greater machining allowances must be made. Surface finish can be controlled somewhat by varying the quality of sand in contact with the metal and by applying special sealers to the sand surfaces. Hot shortness of some aluminum alloys is also important. Certain alloys have low strength at temperatures just below solidification. The casting may crack if these alloys are cast in

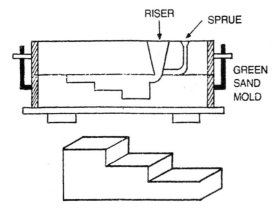

Figure 2.20 Casting a step pattern. (From C. W. Ammen, *Complete Handbook of Sand Casting*, Tab Books, Blue Ridge Summit, PA, 1981.)

Metal	Relative Weight Factor
Magnesium	.64
Aluminum	1.00
Titanium	1.68
Cast Iron	2.63
Zinc	2.64
Tin	2.70
Cast Steel	2.90
Cast Brass (35% Zinc)	3.14
Monel Metal	3.25
Cast Bronze (5% Tin)	3.28
Copper	3.32
Lead	4.20
Uranium	6.93

Figure 2.21 Comparison of weight of aluminum with equal volumes of other metals. (From Reynolds Metals Company, Richmond, VA, 1965.)

a mold that offers resistance to contraction as the metal solidifies. This is called *hot cracking*. Hot shortness varies with the alloy used. Aluminum-silicon alloys show considerably less hot shortness than aluminum-copper alloys. The wide range of aluminum alloys available enables the designer to choose an aluminum alloy and avoid hot cracking when this factor is important.

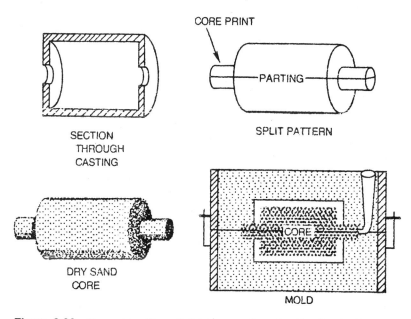

Figure 2.22 Core prints. (From C. W. Ammen, *Complete Handbook of Sand Casting*, Tab Books, Blue Ridge Summit, PA, 1981.)

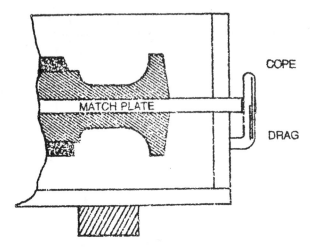

Figure 2.23   Split pattern mounted on pattern board. (From C. W. Ammen, *Complete Handbook of Sand Casting,* Tab Books, Blue Ridge Summit, PA, 1981.)

Basic Design Considerations

Often a casting is designed by someone who does not have the experience in how the casting is to be produced, and the foundry will then inherit problems caused by poor design. Someone who has no knowledge of the foundry and its requirements should talk with a local casting expert or the foundry before starting design. Some basic considerations are the following.

1. On castings that must present a cosmetic appearance along with function, the exterior should be designed to follow simple, flowing lines with a minimum of projections.
2. Avoid irregular or complicated parting lines whenever possible. Design for partings to be in one plane.
3. Use ample but not excessive draft. Avoid any no-draft vertical surfaces unless there is no other way out.
4. Avoid long, slender cores through heavy metal sections or long spans. When unavoidable, they should be straight and well anchored to the mold.
5. Avoid the use of pattern letters on any surface other than one parallel to the parting.
6. Avoid sudden changes in section thicknesses that will create hot spots.
7. Use ribs to stiffen or strengthen castings, thus reducing weight.
8. Avoid all sharp corners and fillet all junctions.
9. Stagger crossing ribs so that the junction will not create a hot spot which could shrink.

Figure 2.24 shows some examples of good and bad sand casting designs.

Volume of Parts

Since pattern equipment for sand casting is much less expensive than dies for permanent-mold or pressure die casting, sand casting is indicated where small quantities are wanted.

As larger quantities are called for, the point where it becomes economical to go to permanent-mold or die casting depends on the size and complexity of the casting and other factors.

2.2.2 Investment Casting

Investment castings have some of the advantages of the sand casting and some of the advantages of the closed-mold processes. The process begins by making a metal die (mold), often of aluminum, that is the shape of the finished part—plus normal shrinkage factors for the material. The mold is injected with wax or low-melting-point plastic. Several of these "wax" parts are attached to a central shaft, or sprue, forming a cluster of parts attached to the sprue. Each wax pattern has one or more gates, which are used to attach it to the sprue. The entire cluster of parts is then coated with a ceramic slurry or placed in a metal flask and the flask filled with a mold slurry. After the mold material has set and cured, the mold is heated, allowing the wax pattern to run out, leaving a hollow core. This is the origin of the term "lost wax" method. Molten aluminum (or other metal) is then poured into the cluster through the sprue and all cavities are filled. Upon

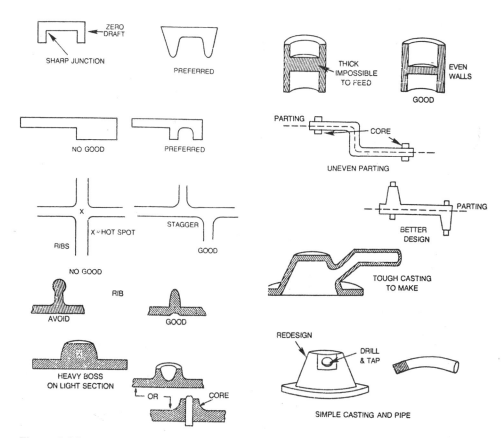

Figure 2.24 Examples of good and bad castings. (From C. W. Ammen, *Complete Handbook of Sand Casting,* Tab Books, Blue Ridge Summit, PA, 1981.)

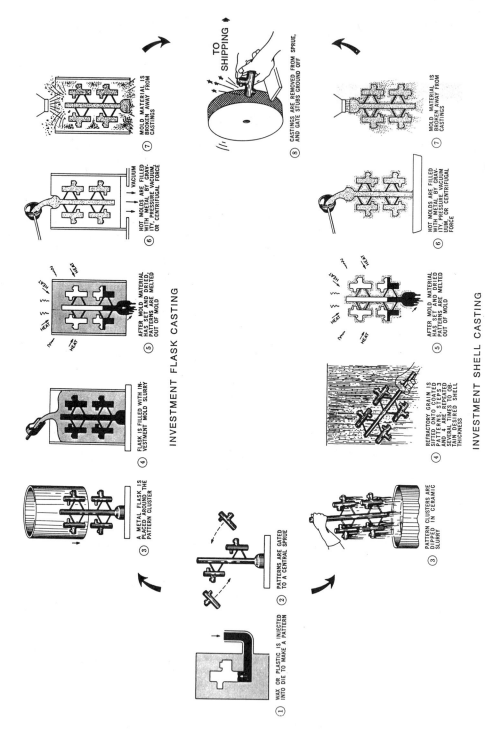

Figure 2.25 Basic production techniques for investment castings. (From *Investment Casting Handbook*, Investment Casting Institute, Chicago, 1968.)

solidification, the mold material is broken away and the parts broken off the sprue. Figure 2.25 shows both the investment flask and investment shell processes.

Design for Investment Casting

Wax or plastic temperature, pressure, die temperature, mold or shell composition, back-up sand, firing temperature, rate of cooling, position of the part on the tree, and heat treatment temperature all bear directly on tolerances required in the investment casting process. The amount of tolerance required to cover each process step is dependent, basically, on the size and shape of thes casting and will vary from foundry to foundry. This is because one foundry may specialize in thin-walled, highly sophisticated castings, another in mass-production requirements, and yet another in high-integrity aerospace or aircraft applications. One factor, however, is constant. The cost of any casting increases in proportion to the preciseness of specifications, whether on chemistry, nondestructive testing, or tolerance bands.

Tolerances

As a general rule, normal linear tolerance on investment castings can be up to 1 ± 0.010 in. For each additional inch thereafter, allow ± 0.003 in. Figure 2.26a shows a chart with expected normal and premium tolerances. Normal tolerances are tolerances that can be expected for production repeatability of all casting dimensions. Premium tolerances are those which require added operations at extra cost, and which provide for closer tolerances on selected dimensions.

Flatness, or the effect of dishing, tolerances cannot be quoted, as they vary with configuration and alloy used. Figure 2.26b is a rough guide in areas under 6 in.^{2}. The amount of dishing allowed is in addition to the basic tolerance. Thus, on a block 1 ± 0.005 in. thick, the tolerances as shown in Figure 2.26c would apply.

Roundness, or "out of round," is defined as the radial difference between a true circle and a given circumference. Tolerances are shown in Figure 2.27a. Figure 2.27b shows the relationship between roundness and concentricity. Two cylindrical surfaces sharing a common point or axis as their center are concentric. Any dimensional difference in the location of one center with respect to the other is the extent of eccentricity. When the length of a bar or tube does not exceed its component diameters by a factor of more than 2, the component diameters will be concentric within 0.005 in. per inch of separation. The roundness of a cast hole as shown in Figure 2.27c is affected by the mass of surrounding metal. If an uneven mass is adjacent, the hole will be pulled out of round. If the surrounding metal is symmetrical, holes up to $\frac{1}{2}$ in. in diameter can be held to ± 0.003 in. when checked with a plug gauge. Larger holes may be affected by interior shrinkage or pulling, and the foundry should be consulted. The longer the hole or the more mass of the section around it, the more pronounced will be the effect. Although the above notes are mostly taken from the *Investment Casting Handbook*, which is published by the Investment Casting Institute, it is strongly recommended that the design team consult with the supplier for a particular part if there might be a question regarding a particular feature not described above. The general castability rating of investment casting alloys is shown in Figure 2.28.

LINEAR TOLERANCE		
DIMENSIONS	NORMAL	PREMIUM
up to ½″	± .007″	± .003″
up to 1″	± .010″	± .005″
up to 2″	± .013″	± .008″
up to 3″	± .016″	± .010″
up to 4″	± .019″	± .012″
up to 5″	± .022″	± .014″
up to 6″	± .025″	± .015″
up to 7″	± .028″	± .016″
up to 8″	± .031″	± .0.17″
up to 9″	± .034″	± .018″
up to 10″	± .037″	± .019″
maximum variation	± .040″	

An exception to the Standard Linear Tolerance exists on thin wall thickness where the tolerance must be a minimum of ± .020″.

(a)

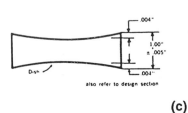

SECTION THICKNESS	POSSIBLE DISH PER FACE OF CASTING
up to ¼″	not significant
¼″ to ½″	0.002″
½″ to 1	0.004″
over 1″	0.006″

(b)

(c)

Figure 2.26 Investment casting tolerances. (From *Investment Casting Handbook,* Investment Casting Institute, Chicago, 1968.)

Drawing Notes

Linear tolerances, unless otherwise specified, ± 0.010 in. for first inch plus ± 0.003 in. per inch for every inch thereafter.
Corner radii 0.030 in. max., fillet radii 0.060 in. max. unless otherwise specified.
Wall thickness 0.080, +0.015 –0.000 in.
Surface finish 125 rms.
Add 0.040 to 0.060 in. stock to surfaces identified by machining mark.
Material: aluminum alloy, MIL-C-11866-T6 Comp. 2 (356-T6).

2.2.3 Die Casting

Permanent-Mold Casting

Permanent-mold casting employs split molds made from cast iron or steel. These molds are usually constructed with thick walls which provide sufficient thermal capacity to assure rapid chilling of cast metal. In semipermanent-mold casting, sand or plaster cores are used in a cast iron or steel mold to produce more intricate and complex shapes. The molten metal is poured by gravity into heated metal molds. Sand casting is essentially a batch process, while permanent-mold casting is suitable for quantity production of a continuous nature. This production-line approach requires a different arrangement of foundry equipment, metal handling methods, and production procedures compared to those used with sand casting.

With permanent-mold casting, a carefully established and rigidly maintained sequence of operations is essential. Every step in the foundry, from charging the furnace to removal of the cast piece from the mold, must be systematized. If any of the factors (pouring

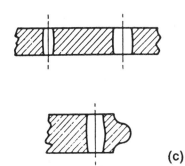

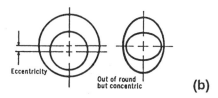

OUT OF ROUNDNESS	
Diameter	TIR or ½ difference between diameters
½"	.010"
1"	.015"
1½"	.020"
2"	.025"
On larger diameters, linear tolerances apply.	

(a)

Eccentricity

Out of round but concentric

(b)

(c)

Figure 2.27 Roundness and concentricity tolerances of investment castings. (From *Investment Casting Handbook,* Investment Casting Institute, Chicago, 1968.)

temperature, mold temperature, pouring rate, solidification rate) are thrown out of balance, the resultant castings may end up as scrap.

General

The improved mechanical properties of permanent-mold castings are the result of the advantageous crystalline structure that develops in the cast alloy as it cools and solidifies rapidly in the metal mold. Permanent-mold castings have closer dimensional tolerances and smoother surfaces than can be obtained in sand castings, so less machining and finishing is needed on permanent-mold castings. Thinner sections can be used, due to the higher mechanical properties.

Pressure Die Casting

Pressure die casting was developed because of the need to provide large numbers of cheap parts for assembly-line production methods. An automobile may have as many as 250 parts in it which are die cast, from door handles to aluminum-alloy engine blocks. Some simple die cast items can be produced on automatic machinery at the rate of hundreds per minute. Pressure die casting machinery is expensive and complicated to design, because it has three functions: to inject the molten metal into the die, to clamp the two halves of the die together at forces which may be as high as several thousand tons, and to eject the finished casting when it has solidified. The costs of such machinery can only be recouped over long

ALLOY	CASTABILITY	RATING	FLUIDITY SHRINKAGE	RESISTANCE TO HOT TEARING
Silicon Irons (Electrical alloys of pure iron and silicon)				
0.5% Si.	75	3	3	2
1.2% Si.	80	3	3	2
1.5% Si.	80	3	3	2
1.8% Si.	75	3	3	2
2.5% Si.	70	3	2	2
Carbon & Sulphur Steels (A.I.S.I. designations)				
1015	80	3	3	3
1018	80	3	3	3
1020	80	3	3	3
1025	85	3	3	2
1030	85	3	2	2
1035	85	3	2	2
1040	85	2	2	2
1045	85	2	2	2
1050	85	2	2	2
1060	85	2	1	2
1117	75	3	3	3
1140	80	2	2	3
Low Alloy Steel (A.I.S.I. designations)				
2345	90	2	2	2
3120	85	2	3	2
4130	90	2	2	2
4140	90	2	2	2
4150	90	2	2	2
4340	90	2	2	2
4615	85	2	3	2
4620	85	2	3	2
4640	90	2	2	2
5130	85	2	3	2
6150	90	2	2	2
8620	85	2	3	2
8630	85	2	3	2
8640	90	2	2	2
8645	90	2	2	2
8730	85	2	3	2
8740	90	2	3	2
52100	80	1	2	2
Nitralloy	75	2	3	2
400 Series Stainless (A.I.S.I. designations)				
405	90	2	3	2
410	95	1	3	2
416	85	1	3	2
420	90	1	3	2
430	90	1	3	2
430 F	90	1	3	2
431	90	1	3	2
440 A	85	1	3	2
440 C	85	1	3	2
440 F	85	1	3	3
AMS 5355 (Armco 17-4 PH)	85	1	3	2
AMS 5354	85	1	3	2
300 Series Stainless (A.I.S.I. designations)				
302	100	1	1	1
303	95	1	1	2
304	100	1	1	1
310	90	1	1	1
312	90	1	1	1
316	100	1	1	1

ALLOY	CASTABILITY	RATING	FLUIDITY SHRINKAGE	RESISTANCE TO HOT TEARING
347	95	1	1	1
CF-8M (ACI)	100	1	1	1
CN-7M (ACI)	95	1	1	1
High Nickel Alloys				
Monel (QQ-N-288-A)	85	1	2	2
Monel, R. H.	75	1	2	2
Monel, S (QQ-N-288-C)	75	1	2	3
Inconel (AMS-5665)	85	1	2	2
47-50 (47% Ni-50% Fe)	80	3	2	2
Invar	75	1	2	2
Cobalt Alloys				
Cobalt J	80	1	1	3
Cobalt 3	80	1	1	3
Cobalt 6 (AMS-5387)	80	1	1	2
Cobalt 19	85	1	1	2
Cobalt 21 (AMS-5385C)	90	1	1	2
Cobalt 31 (AMS-5382B)	90	1	1	3
Cobalt 93	70	1	1	3
N-155 (AMS-5531)	80	1	1	3
Tool Steels (A.I.S.I. designations)				
A-2	85	2	2	2
A-6	80	3	2	2
D-2	85	3	2	2
D-3	85	3	2	2
D-6	80	3	2	2
D-7 (BR-4)	80	3	2	3
BR-4 FM	80	3	2	2
F-2	75	3	2	2
H-13	85	2	2	2
L-6	80	2	2	2
M-2	80	2	2	3
M-4	75	2	2	3
O-1	80	2	1	2
O-2	80	2	1	2
O-7	80	2	1	2
S-1	90	2	2	2
S-2	90	2	2	2
S-4	90	2	2	2
S-5	90	2	2	2
T-1	80	2	1	3
Aluminum				
13	85	2	2	2
40E	75	3	3	3
43	90	2	2	2
356-A356	100	1	1	1
355-C355	95	1	1	1
B-195	85	1	2	2
Copper Base				
Al. Bronze Gr. C	80	1	3	1
Al. Bronze Gr. D	80	1	3	1
88-10-2(G Br. & Gun Metal)	85	1	3	1
Mn Bronze	80	2	3	1
Hi Tensile Mn Bronze	80	2	3	1
Naval Brass (Yellow Bronze)	85	2	2	1
Navy "M"	85	2	2	1
Navy "G"	85	2	2	1
Phosphor Bronze (SAE 65)	85	2	2	1
85-5-5-5 (Red Bronze)	90	1	1	1
Silicon Brass	100	1	1	1
Be Cu 10C	90	1	1	1
Be Cu 20C	100	1	1	1
Be Cu 275C	90	1	1	1

Castability ratings are based on a casting of relatively simple configuration and upon comparisons with three alloys having excellent foundry characteristics and assigned castability rating of 100. These are 302 stainless steel (ferrous), 20 C beryllium copper (non-ferrous) and aluminum alloy 356. The fluidity, shrinkage and hot tear ratings of each alloy are based on 1—best, 2—good and 3—poor.

Figure 2.28 Castability rating of investment casting alloys. (From *Casting Aluminum*, Reynolds Aluminum Company, Richmond, VA, 1965.)

production runs. Since a small die of only moderate complexity may cost several thousand dollars, the process should be considered only for a large production run.

The molds used in automated die casting methods are designed with water circulation tubing in them to cool the castings rapidly. The machine opens the mold so that the half of the casting which cools quickest is uncovered first, then ejecting pins push the casting out of the other half of the mold. The two halves of the mold are coated with a lubricant to facilitate the ejection of the finished casting. A typical pressure die casting die is shown in Figure 2.29. The fixed cover half of the die on the left shows the gate where the liquid shot of molten metal is injected. The ejector half of the die, on the right, shows the large alignment pins, the sprue pin, the cores for blind holes, and the ejector pins.

Die casting processes are divided into two types: cold chamber and hot chamber methods. In the cold chamber method, the molten metal is ladled into a cylinder and then shot into the mold by a hydraulic or pneumatic plunger. In the hot chamber method, shown in Figure 2.30, the injection cylinder itself is immersed in the molten alloy and successive shots are made either by a plunger or compressed air. There are also high-pressure and low-pressure techniques. The low-pressure system is newer and was developed to extend the versatility of die casting to alloys with higher melting points. Another development is the vacuum mold, in which a vacuum is produced in the mold, enabling the pressure shot to fill the mold faster.

Most die casting alloys have a low melting point. Almost the entire production of zinc-based alloys is used in the die casting industry. The higher the melting point, the shorter will be the useful life of the mold. A tool steel die may be able to pro-

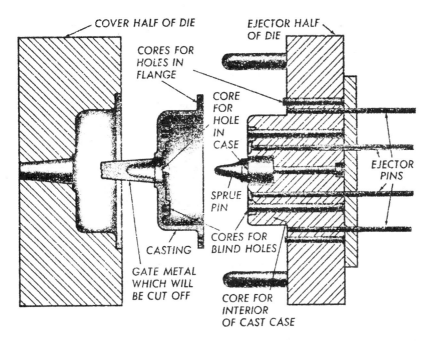

Figure 2.29 Typical pressure die casting die. All the fixed cores are aligned to permit ejection of the casting. (From *Designing for Alcoa Die Castings,* Aluminum Corporation of America, Pittsburgh, PA, 1949.)

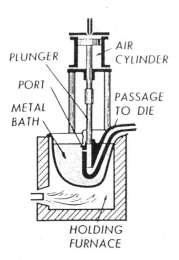

PLUNGER — AIR CYLINDER

PORT

PASSAGE TO DIE

METAL BATH

HOLDING FURNACE

Figure 2.30   The submerged type of injection system is suitable for use with low-melting-temperature alloys. (From *Designing for Alcoa Die Castings,* Aluminum Corporation of America, Pittsburgh, PA, 1949.)

duce up to a half-million castings in zinc, but the total for brass may be only a few thousand. Typical mechanical properties for an aluminum die cast part are shown in Figure 2.31.

Design and Tolerances

The cast surfaces that slide on the die cavity during die opening or ejection must have draft or taper to free the casting without binding. Insufficient draft causes sticking in the die, distortion of the castings, and galling of cast surfaces. Ample draft assures smoothness of cast surfaces. In general, outer walls or surfaces should have a minimum draft of 0.5° (0.009 in. per linear inch of draw). Inner walls which exert higher shrinkage forces on the die as the casting cools require greater draft allowance, as much as 2 or 3°, depending on the design. Surfaces which are on the outer part of the casting but which tend to bind against the die during contraction of the casting are treated as inner walls in determining draft.

Where the greatest degree of smoothness is required on die cast surfaces, draft should be increased to several times the allowable minimum. This will ensure rapid production and make it possible to maintain excellent surface smoothness on the castings. Generous fillets and rounded corners can be used to advantage as means to reduce the areas subject to sliding on the die.

The practical minimum wall thickness is usually governed by the necessity of filling the die properly. Under the most difficult conditions, a wall 9/64 to 5/32 in. thick may be needed for metal flow in the die. Where the casting is small (less than about 6 in. by 6 in. overall) and its form promotes good casting conditions, walls may be thin as 0.050 to 0.065 in.

Draft in a cored hole allows the core to be withdrawn from the casting, or the casting to be ejected from the core. The amount of draft needed is so slight that many cored holes perform the functions of straight cylindrical holes without the necessity of reaming. They

	Alcoa Alloy	13	43	A360	360	85	A380	380	384	218
Room Tem- perature 68°F.[3]	Tensile Strength, psi[1]	39,000	30,000	41,000	44,000	40,000	46,000	45,000	46,000	45,000
	Tensile Yield Strength, psi[2]	21,000	16,000	23,000	27,000	24,000	25,000	26,000	27,000	27,000
	Elongation in 2 Inches, %	2.0	9.0	5.0	3.0	5.0	3.0	2.0	1.0	8.0
	Shear Strength, psi	25,000	19,000	26,000	28,000	26,000	29,000	29,000	29,000	27,000
	Endurance Limit, psi[4]	19,000	17,000	18,000	19,000	22,000	19,000	20,000	21,000	23,000
212°F.[5]	Tensile Strength, psi	34,000	26,000	36,000	39,000	38,000	44,000	43,000	44,000	40,000
	Tensile Yield Strength, psi	21,000	16,000	23,000	27,000	24,000	26,000	27,000	28,000	25,000
	Elongation in 2 Inches, %	5.0	13.0	6.0	3.0	5.0	5.0	3.0	2.0	12.0
300°F.	Tensile Strength, psi	30,000	23,000	32,000	34,000	33,000	38,000	37,000	38,000	34,000
	Tensile Yield Strength, psi	19,000	15,000	21,000	25,000	22,000	23,000	24,000	25,000	23,000
	Elongation in 2 Inches, %	7.0	16.0	8.0	4.0	7.0	7.0	4.0	5.0	15.0
400°F.	Tensile Strength, psi	25,000	19,000	27,000	29,000	24,000	28,000	28,000	28,000	25,000
	Tensile Yield Strength, psi	18,000	14,000	20,000	24,000	19,000	20,000	21,000	22,000	19,000
	Elongation in 2 Inches, %	12.0	19.0	11.0	5.0	10.0	12.0	6.0	8.0	18.0
500°F.	Tensile Strength, psi	19,000	14,000	20,000	21,000	17,000	20,000	20,000	20,000	16,000
	Tensile Yield Strength, psi	13,000	10,000	15,000	17,000	13,000	14,000	15,000	16,000	12,000
	Elongation in 2 Inches, %	20.0	23.0	15.0	9.0	16.0	15.0	10.0	12.0	22.0

Figure 2.31 Typical mechanical properties of Alcoa aluminum die casting alloys. (From *Designing for Alcoa Die Castings,* Aluminum Corporation of America, Pittsburgh, PA, 1949.)

often serve as pivot bearings, screw holes, tap holes, or seats for mating parts. In such cases, where a close approach to parallel sides is desired, it is well to specify the minimum allowable draft. The minimum draft for cored holes is shown in Figure 2.32a, and the recommended depth of blind cored holes is shown in Figure 2.32b.

In addition to the commonly employed processes mentioned above, others include centrifugal casting, plaster-mold, and shell-mold casting. Figure 2.33 compares design and cost features of basic casting methods.

2.2.4 Forging and Pressing

This section provides a brief introduction to the forging and pressing industry. In earlier days, forging was the process of shaping metal by heating and hammering. Today, metal is not always heated for forging, and the work may be performed by several types of heavy machines which apply impact or squeeze pressure with swift precision. In today's forging industry the skill and seasoned judgment of the forgeman is enhanced by machines of modern technology that produce metal parts of unparalleled strength and utility.

Forging permits the structure of metals to be refined and controlled to provide improved mechanical properties. Also, forging produces a continuous "grain flow" in the metal which can be oriented to follow the shape of the part and result in maximum strength efficiency of the material. Since virtually all metals, from aluminum to zirconium, can be forged, extensive combinations of mechanical and physical properties are available to meet demanding industrial applications. Figure 2.34 shows the dollar sales by commercial forging industry to major end-use markets.

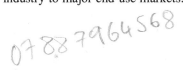

MINIMUM DRAFT FOR CORED HOLES

Diameter or Narrowest Section of Hole (Inches)	Draft on Diameter or Cross-Sectional Dimension per Inch of Depth (Inches)	
	Aluminum	Magnesium
$\frac{1}{10}$ to $\frac{1}{8}$	.020	.010
$\frac{1}{8}$ to $\frac{1}{4}$	.016	.008
$\frac{1}{4}$ to 1	.012	.006
Over 1	†.012 + .002 per inch of diameter over 1 inch.	†.006 + .001 per inch of diameter over 1 inch.

†As an example, a 3-inch deep cored hole of 1½ inches diameter in aluminum would require .012 plus ½ times .002, or .013 inch per inch of depth. The draft would be three times .013, or .039 inch.

(a)

DEPTHS OF CORED HOLES

Diameter or Narrowest Section of Hole (Inches)	Maximum Recommended Depth (Inches)	Diameter or Narrowest Section of Hole (Inches)	Maximum Recommended Depth (Inches)
$\frac{1}{10}$ (min.)	$\frac{3}{8}$	$\frac{1}{4}$	2
$\frac{1}{8}$	$\frac{1}{2}$	$\frac{1}{2}$	4
$\frac{5}{32}$	$\frac{3}{4}$	$\frac{3}{4}$	6
$\frac{3}{16}$	$1\frac{1}{8}$	1	8

(b)

Figure 2.32 Recommended minimum draft and depth for cored holes. (From *Designing for Alcoa Die Castings,* Aluminum Corporation of America, Pittsburgh, PA, 1949.)

Materials

In most cases the stock to be forged has been preworked by the material supplier to refine the dendritic structure of the ingot, remove defects inherent in the casting process, and further improve the structural quality. This is usually accomplished in successive rolling operations which reduce the cross section of the material under pressure, eliminating possible porosity, refining the crystalline structure of the base metal, and orienting any nonmetallics and alloy segregation in the direction of working. This directional alignment is called *grain flow*. The forging process employs this directionality to provide a unique and important advantage by orienting grain flow within the component so that it lies in the direction requiring maximum strength.

Casting Method	Cost		Produc-tion Quan-tities	Mechan-ical Proper-ties	Surface Finish	Limitations		Dimen-sional Accu-racy	Minimum Section Thick-ness inch
	Produc-tion Equip-ment	Unit Casting				Casting Size	Alloy		
Green Sand	Low	High	Medium	Medium	Fair	None	None	Fair	⅛
Baked Sand	Medium	High	Medium	Medium	Fair	None	None	Fair	⅛
Semi-Permanent Mold	High	Low	Medium Large	High	Fair to Good	To Medium	Medium to High Fluidity	Good	¹⁄₁₆
Permanent Mold	High	Low	Medium Large	High	Good	To Medium	Medium to High Fluidity	Very Good	¹⁄₁₆
Die Casting	Very High	Lowest	Very Large	High	Very Good	To Medium	High Fluidity	Excellent	¹⁄₃₂
Centrifugal	High	Low	Medium	Medium High	Good	Limited by Shape	None	Good	⅛
Investment	Medium	High	Small	Medium	Excellent	Small	High Fluidity	Excellent	⅛-¹⁄₃₂
Plaster	Medium	High	Small	Low	Excellent	Small	High Fluidity	Excellent	⅛-¹⁄₃₂

Figure 2.33 Design and cost features of basic casting methods. (From *Designing for Alcoa Die Castings,* Aluminum Corporation of America, Pittsburgh, PA, 1949.)

Properly developed grain flow in forgings closely follows the outline of the component. In contrast, bar stock and plate have grain flow in only one direction, and changes in contour require that flow lines be cut, exposing grain ends and rendering the material more liable to fatigue and more sensitive to stress corrosion. Figure 2.35 compares the grain structure in casting, bar stock machining, and forging. The forging process (through proper orientation of grain flow) develops maximum impact strength and fatigue resistance in a material—greater values in these properties than are obtainable by any other metal-working process. Thus forgings provide greater life expectancy than is possible with other metal products.

Consistency

The consistence of material from one forging to the next, and between separate quantities of forgings produced months or even years apart, is extremely high. Forge plants can usually rely on acceptance testing for the quality control procedures and certified reports of major metal producers, assuring close control of metal composition.

Dimensional Uniformity

The dimensional characteristics of impression die forgings are remarkably stable from piece to piece throughout an entire production run. This exceptional continuity of shape from the first to the last piece is due primarily to the fact that impressions in which forgings are shaped are contained in permanent dies of special steel. Successive forgings are produced from the same impression.

Materials

Since virtually all metals can be forged, the range of physical and mechanical properties available in forged products spans the entire spectrum of ferrous and nonferrous metalurgy.

Aluminum alloys are readily forged, primarily for structural and engine applications

MAJOR END-USE INDUSTRIES

PERCENT OF FORGING INDUSTRY SALES

Aerospace

        Aircraft Engines and Engine Parts   15.4%

        Airframes, Aircraft Parts & Aux. Equip.   13.2%

        Missiles and Missile Parts   3.0%   31.6%

Automotive and Truck   20.5%

Off-Highway Equipment

        Construction, Mining & Materials Handling   12.2%

Ordnance (except Missiles)   8.4%

Agricultural   4.2%

Plumbing Fixtures, Valves & Fittings   3.1%

Railroad   2.8%

Petrochemical   1.7%

Mechanical Power Transmission Equip. incl. Bearings   1.4%

Internal Combustion Engines (Stationary)   1.1%

Metalworking & Special Industry Machinery   1.0%

Pumps & Compressors   0.9%

Steam Engines & Turbines (except Locomotives)   0.8%

Refrigeration & Air-Conditioning   0.6%

Motors and Generators   0.5%

Motorcycles, Bicycles & Misc. Equipment   0.4%

Other   <u>8.8%</u>

                        TOTAL   100.0%

Figure 2.34 Major end-use industries: percent of forging industry sales. (From *Forging Industry Handbook,* Forging Industry Association, Cleveland, OH, 1970.)

GRAIN

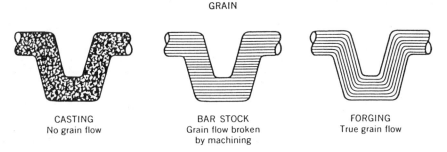

CASTING	BAR STOCK	FORGING
No grain flow	Grain flow broken by machining	True grain flow

Figure 2.35 Grain structure of casting, bar stock, and forgings. (From *Forging Industry Handbook,* Forging Industry Association, Cleveland, OH, 1970.)

in the aircraft and transportation industries, where temperature environments do not exceed approximately 400° F. Forged aluminum combines low density with good strength-to-weight ratio.

Magnesium forgings are usually employed for applications at service temperatures lower than 500° F, although certain alloys (magnesium-thorium, for example) are employed for short-time service up to 700° F. Magnesium forgings are used efficiently in lightweight structures and have the lowest density of any commercial metal.

Low-carbon and low-alloy steels comprise the greatest volume of forgings produced for service applications up to 800 to 900° F. These materials provide advantages of relatively low material cost, ease of processing, and good mechanical properties. The design flexibility of steel is due in great part to its varied response to heat treatment, giving the designer a wide choice of properties in the finished forging.

Sizes and Shapes

Forgings are produced in an extremely broad range of sizes, from parts measuring less than an inch to those well over 23 ft. in length. Accordingly, the designer has considerable freedom in developing mechanical components in sizes required for most applications in machines and conveyances.

Metallurgy

The cast state of metal and conversion to the wrought state is shown for aluminum in Figure 2.36. This shows the striking instability of various microstructures at elevated temperatures and the eventual disappearance of all the modes of hardening at sufficiently high temperatures. The solidification process (transformation from the liquid state to the solid state) produces a metallurgical structure which can be characterized by (1) grain size and shape, (2) segregation, (3) porosity, and (4) nonmetallic inclusions. In the cast state, the grain size in shape castings and ingots is relatively coarse and may be more or less spherical in shape, or cigar-shaped. Mechanical properties are diminished, in general, by coarse grain size. Segregation, as also seen in Figure 2.36, leads to variability in properties and can result in embrittlement by low-melting constituents in the grain boundaries (hot shortness). There is no clear-cut point at which the cast state is converted into the wrought state by the action of plastic deformation. It seems reasonable to mark this point at the degree of deformation which seals up the microporosity and eliminates the internal notch

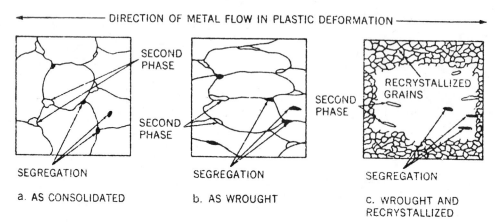

DIRECTION OF METAL FLOW IN PLASTIC DEFORMATION

SECOND PHASE

SECOND PHASE

SECOND PHASE

RECRYSTALLIZED GRAINS

SEGREGATION

SEGREGATION

SEGREGATION

a. AS CONSOLIDATED

b. AS WROUGHT

c. WROUGHT AND RECRYSTALLIZED

Figure 2.36 Typical development of grain flow in metal. (From *Forging Industry Handbook,* Forging Industry Association, Cleveland, OH, 1970.)

effect arising form this source. The typical transformation from the wrought condition to the final wrought and recrystallized grain development is shown in Figure 2.36c, and is the result of deformation of the wrought alloy.

Basic Forging Processes

For purposes of understanding forging design, it is helpful to consider several of the various forging operations in terms of the resulting characteristic metal flow. This allows breaking down complex operations into basic components and permits better understanding of the forging process.

Compression between flat dies is also called "upsetting" when an oblong workpiece is placed on end on a lower die, and its height is reduced by downward movement of the top die. Friction between the end faces of the workpiece and the dies in unavoidable. It prevents the free lateral spread of the ends of the workpiece and results in a barrel shape as shown in Figure 2.37. This operation is the basic type of deformation employed in flat die forging. The nature of local deformation can be followed in prepared test specimens. Bisected specimens show that, as deformation proceeds, material adjoining the dies remains almost stationary (area I in Figure 2.37). Material near the outer surface of the cylinder is deformed as a result of the center material moving radially outward (area III), and the bulk of the deformation is concentrated in the remaining area (area II). Within this latter zone, heaviest deformation occurs at the center of the test piece, forcing material out of the center and developing circumferential tensile stresses. Material nearest the dies behaves somewhat like a rigid cone penetrating the rest of the specimen. Slip occurs at the faces of this "cone."

It should be noted that while barreling is usually attributed to friction effects, the cooling of the specimen has much to do with it. Contact with the cool die surface chills the end faces of the specimen, increasing its resistance to deformation and enhancing the barreling. In this type of upsetting, the ratio of workpiece length to diameter (or square) is subject to limitations if one is to avoid buckling of the workpiece. For successful upsetting of unsupported material, ratios of 2.5:1 to 3:1 are generally considered appropriate, although some work requires a smaller ratio.

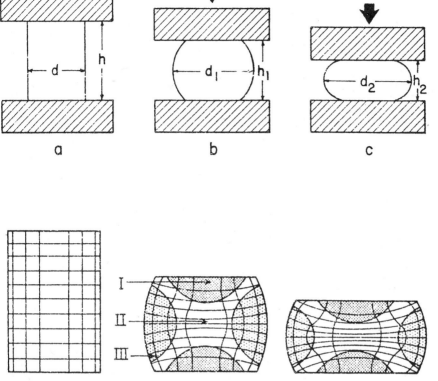

Figure 2.37 Compression of a cylindrical workpiece between flat dies. (From *Forging Industry Handbook,* Forging Industry Association, Cleveland, OH, 1970.)

Compression Between Narrow Dies

While upsetting between parallel flat dies is one of the most important of the *basic* forging operations, it is limited to deformation that is always symmetrical around a vertical axis. If preferential elongation is desired, the dies must be made relatively narrow, as shown in Figure 2.38. When compressing a rectangular (or round) bar, frictional forces in the axial direction of the bar are smaller than in the direction perpendicular to it. Therefore, most of the material flow is axial as shown in Figures 2.38b and 2.38c. Since the width of the tool still presents frictional restraint, spread is unavoidable. The proportion of elongation to spread depends on the width of the die. A narrow die will elongate better, but die width cannot be reduced indefinitely because cutting instead of elongation will result (see Figure 2.38c).

Fullering and Edging

The direction of material flow can be more positively influenced by using dies with specially shaped surfaces. As shown in Figure 2.39, material can be made to flow away from the center of the dies or to gather there. In *fullering*, the vertical force of deformation always possesses (except for the exact center) a component that tends to displace material away from the center as seen in Figure 2.39a. An *edging* or *roller* die has concave surfaces

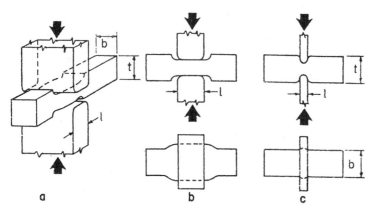

Figure 2.38 Compression between flat, narrow dies. (From *Forging Industry Handbook,* Forging Industry Association, Cleveland, OH, 1970.)

and forces material to flow toward the center, where the thickness can be increased above that of the original bar dimensions as seen in Figure 2.39b.

Impression Die Forming

In the cases discussed thus far, the material was free to move in at least one direction. Thus, all dimensions of the workpiece could not be closely controlled. Three-dimensional control requires impressions in the dies, and thus the term "impression die" forging is derived. In the simplest example of impression die forging, a cylindrical (or rectangular) workpiece is placed in the bottom die, as seen in Figure 2.40a. The dies contain no provision for controlling the flow of excess material. As the two dies are brought together, the workpiece undergoes plastic deformation until its enlarged sides touch the side walls of the die impressions as in Figure 2.40b. At this point, a small amount of material begins to flow

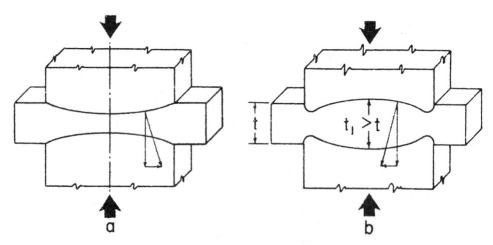

Figure 2.39 Compression between (a) convex and (b) concave die surfaces (fullering and edging or rolling, respectively). (From *Forging Industry Handbook,* Forging Industry Association, Cleveland, OH, 1970.)

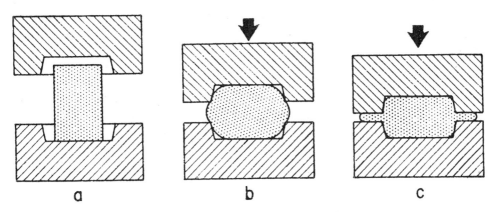

Figure 2.40 Compression in simple impression dies without special provision for flash formation. (From *Forging Industry Handbook,* Forging Industry Association, Cleveland, OH, 1970.)

outside the die impressions, forming what is known as *flash*. In the further course of die approach, this flash is gradually thinned. As a consequence, it cools rapidly and presents increased resistance to deformation. In this sense, then, the flash becomes a part of the tool and helps to build up high pressure inside the bulk of the workpiece. This pressure can aid material flow into parts of the impressions hitherto unfilled so that, at the end of the stroke, the die impressions are nearly filled with the workpiece material, as in Figure 2.40c.

The process of flash formation is also illustrated in Figure 2.41. Flow conditions are complex even in the case of the very simple shape shown. During compression, the material is first forced into the impression that will later form the hub, then the material begins to spread and the excess begins to form the flash as seen in Figure 2.41a. At this stage the cavity is not completely filled, and success depends on whether a high enough stress can be built up in the flash to force the material (by now gradually cooling) into the more intricate details of the impressions. Dimensional control in the vertical direction is generally achieved by bringing opposing die faces together as shown in Figure 2.41b. The

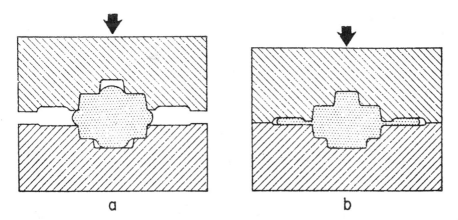

Figure 2.41 Formation of flash in a conventional flash gutter in impression dies. (From *Forging Industry Handbook,* Forging Industry Association, Cleveland, OH, 1970.)

flash presents a formidable barrier to flow on several counts. First, the gradual thinning of the flash causes an increase in the pressure required for its further deformation. Second, the relatively thin flash cools rapidly in intimate contact with the die faces and its yield stress rises correspondingly. Third, the further deformation of the flash entails a widening of the already formed ringlike flash, which in itself requires great forces.

Thus the flash can ensure complete filling even under unfavorable conditions, but only at the expense of extremely high die pressures in the flash area. Such dependence on excessive die pressures to achieve filling is usually undesirable, because it shortens die life and creates high power requirements. Forging design aims at easing metal flow by various means, making extremely high pressures in the flash area superfluous. In order to reduce the pressure exerted on the die faces, it is usual to form a flash gutter as shown in Figure 2.41b, which accommodates the flash and also permits the dies to come together to control the vertical (or thickness) dimension. The flash will be removed later in the trimming operation and thus represents a loss of material. While flash is usually a by-product that is necessary for proper filling of the die cavity, too large a flash is not only wasteful, it imposes very high stresses on the dies. Impression die forging, as just described, accounts for the vast majority of commercial forging production.

Extrusion

In extrusion, the workpiece is placed in a container and compressed by the movement of the ram until pressure inside the workpiece reaches the flow stress. At this point the workpiece is upset and completely fills the container. As the pressure is further increased, material begins to leave through an orfice and forms the extruded product. In principle, the orfice can be anywhere in the container or the ram as seen in Figure 2.42. Depending of the relative direction of motion between ram and extruded product, it is usual to speak of forward or direct (Figure 2.42a) and reverse or backward (Figure 2.42c) extrusion.

The extruded product may be solid or hollow. In the latter case, the outer diameter of the workpiece can be reduced as in Figure 2.43a, or allowed to remain at its original dimension as seen in Figure 2.43b. Tube extrusion is typical of forward extrusion of hollow shapes (a), whereas reverse or back extrusion is used for the mass production of containers as seen in (b).

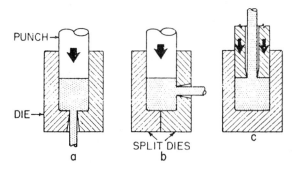

Figure 2.42 Extrusion of solid shapes: (a) forward (direct); (b) radial; (c) backward (reverse). (From *Forging Industry Handbook,* Forging Industry Association, Cleveland, OH, 1970.)

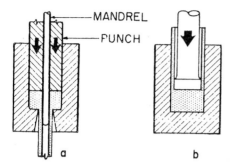

Figure 2.43 Extrusion of hollow shapes: (a) forward; (b) backward. (From *Forging Industry Handbook,* Forging Industry Association, Cleveland, OH, 1970.)

A common feature of all processes discussed thus far is that the material is purely plastically deformed and no part of it is removed in any form. In punching and trimming as seen in Figure 2.44, the mechanism is entirely different. The material is subjected to shear stresses between the punch and die until the stress exceeds the yield stress of the material in shear and cracks penetrate the material. Proper clearance between punch and trimming blade is important in obtaining best quality of the cut surface.

Figure 2.45 shows the four step forging of a gear blank. It starts with a cut billet figure 2.45a, which is preheated. The first preform is shown in Figure 2.45b, the second preform in Figure 2.45c, and the finished forging in Figure 2.45d, ready for the punching operation. This part required three sets of dies for forming, with subsequent heating and forming, plus a punch die to remove the excess in the center of the part.

Forging Design Principles

Establishing the location and shape of the parting line (or flash line) is one of the most important tasks of forging design. Aside from the obvious need to remove the forging from the dies, the parting line influences many other factors, including: selection of the forging equipment; the design, construction, and cost of dies; grain flow; the trimming procedure; material utilization; and the position of locating surfaces for subsequent machining. Figure 2.46 illustrates a variety of simple shapes showing undesirable and desirable parting lines for hammer and press forgings. Reasons for the preferred locations are as follows.

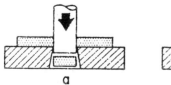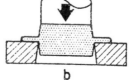

Figure 2.44 Shearing operations: (a) punching; (b) trimming. (From *Forging Industry Handbook,* Forging Industry Association, Cleveland, OH, 1970.)

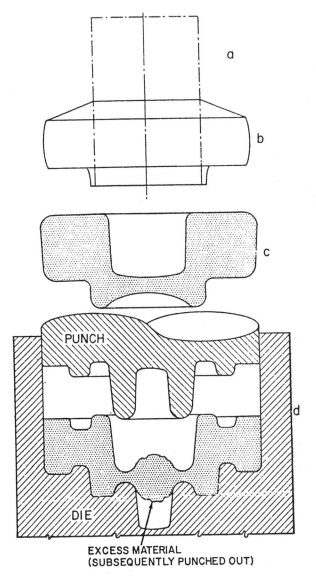

Figure 2.45 Forging of a gear blank with internal flash: (a) cut billet; (b) first preform; (c) second preform; (d) finished forging, ready for punching operation. (From *Forging Industry Handbook,* Forging Industry Association, Cleveland. OH, 1970.)

    Case 1. The preferred choice avoids deep impressions that might otherwise promote die breakage.

    Cases 2 and 3. The preferred choices avoid side thrust, which could cause the dies to shift sideways.

    Case 4. Preference is based on grainflow considerations. The "satisfactory" location provides the least expensive method of parting, since only one impression die is

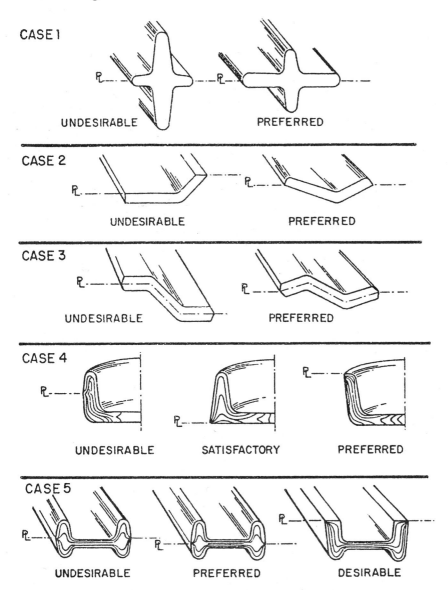

Figure 2.46 Forging shapes illustrating both undesirable and preferred parting line locations. (From *Forging Industry Handbook,* Forging Industry Association, Cleveland, OH, 1970.)

needed. The preferred location, however, produces the most desirable grainflow pattern.

Case 5. The choice in this case is also based on grainflow considerations. However, the "desirable" location usually introduces manufacturing problems and is used only when grainflow is an extremely critical factor in design. This, in turn, depends on the directional properties and cleanliness of the material being forged.

The most common draft angles are 3° for materials such as aluminum and magnesium and 5 to 7° for steels and titanium. Draft angles less that these can be used, but the cost will increase.

Webs and Ribs

A *web* is a thin part of the forging lying in or parallel to the forging plane, while a *rib* is perpendicular to it. Since virtually all designs can be broken down into component sections formed either parallel or perpendicular to the fundamental forging plane, a discussion of webs and ribs has meaning for a great variety of forgings. Figure 2.47 shows examples of four basic types of rib designs.

> *Type 1.* Centrally located ribs are formed by extruding metal from the body of a forging. As a general rule, the web thickness should be equal to or greater than the rib thickness to avoid defects.
>
> *Type 2.* Ribs at the edges of forgings with the parting line at the top are formed by

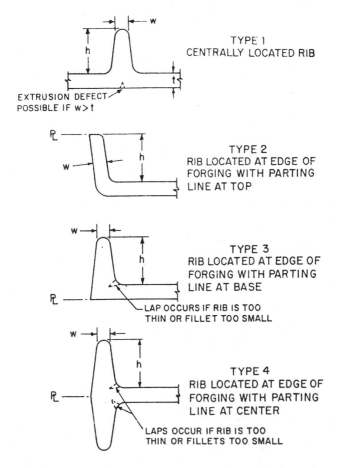

Figure 2.47   Four basic types of rib designs. (From *Forging Industry Handbook,* Forging Industry Association, Cleveland, OH, 1970.)

metal flow similar to that in reverse extrusion. Ribs thinner than the other three types are possible with this design.

Type 3. Ribs at the edges of forgings with the parting line at the base are also formed by an extrusion type of flow. Section thickness limits are similar to type 1.

Type 4. Ribs and bosses at the edge of a forging with a central web are the most difficult to forge. Such rib designs almost always require preliminary tooling to gather the volume of metal necessary for filling the final die. Minimum thicknesses for type 4 ribs are generally larger than are those for the other three types.

There are no hard-and-fast rules that apply to the dimensions of ribs. In general, the rib height should not exceed eight times the rib width. Most forging companies prefer to work with rib height-to-width ratios between 4:1 and 6:1. Figure 2.48 summarizes, for several alloys, suggested minimum section size limits for ribs and webs with conventional fillet and corner radii and draft angles. Fillet and corner radii are shown in Figure 2.49. In Figure 2.49a, the influence of impression depth on corner radii of ribs, bosses, and other edges for steel and aluminum forgings of average proportions is seen; in Figure 2.49b, the influence of rib height on minimum filet radius for steel and aluminum is shown; in Figure 2.49c, are representative fillet and corner radii for forgings of several materials with 1-in.-high ribs; and in Figure 2.49d, representative minimum fillet and corner radii of steel forgings on a weight basis are given. Figure 2.50a shows various die fillets for corner radii for steel and aluminum forgings, showing order of preference for 1-in.-high bosses and ribs. The schematic diagram (Figure 2.50b) shows the influence of fillet radius on metal flow during progressive stages of die closure.

Many forgings are used with little or no machining, where the normal forging variations and the surface condition typical for the material are acceptable. Finish allowance for machining varies with cost considerations, size of part, the way the

	Minimum Rib Thickness, in., for Forgings of Given Plan Area, sq. in.		Minimum Web Thickness, in., for Forgings of Given Plan Area, sq. in.			
Materials	Up to 10	10-100	Up to 10	10-100	100-300	Over 300
2014 (aluminum)	0.12	0.19	0.10	0.19	0.31	0.50
AISI 4340 (alloy steel)	0.12	0.19	0.19	0.31	0.50	0.75
H-11 (hot-work die steel)	0.12	0.19	0.20	0.38	0.60	1.00
17-7PH (stainless steel)		0.25		0.38	0.60	1.00
A-286 (super alloy)		0.25		0.38	0.75	1.25
Ti-6Al-4V (titanium alloy)		0.25		0.31	0.55	0.75
Unalloyed Mo		0.38		0.38		

Note: This table is based on data provided by several forging companies. In many cases, the companies did not agree on minimum values for rib widths and web thicknesses. The values presented here indicate the most advanced state of commercial forging practice. There was general agreement on the need for more liberal dimensions with the alloys requiring the greater forging pressures.

Figure 2.48 Representative minimum section thickness for rib-and-web forgings of several materials. (From *Forging Industry Handbook,* Forging Industry Association, Cleveland, OH, 1970.)

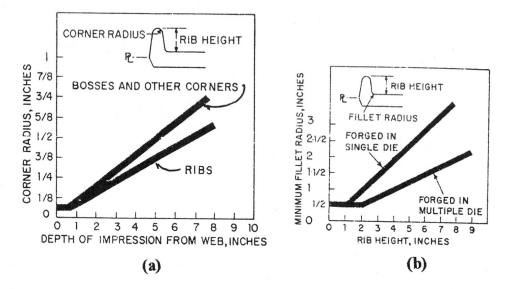

	Fillet Radius, inch		Corner Radius, inch	
Material	Preferred	Minimum	Preferred	Minimum
2014	1/4	3/16	1/8	1/16
AISI 4340	3/8-1/2	1/4	1/8	1/16
H-11	3/8-1/2	1/4	3/16	1/16
17-7PH	1/4-1/2	3/16	3/16	1/32
A-286	1/2-3/4	1/4-3/8	1/4	1/8
Ti-6Al-4V	1/2-5/8	3/8	1/4	1/8
Unalloyed Mo	1/2	—	Full radius up to 1/2 in.	3/8

(c)

Forging Weight, lb	Fillet Radius, inch	Corner Radius, inch
1	3/64-1/8	3/64-1/8
2	1/16-1/8	1/16-1/8
5	1/8-1/4	1/8
10	1/8-1/4	3/32-1/8
30	1/4-1/2	1/8-1/4
100	1/2	1/4

(d)

Figure 2.49 Representative minimum fillet and corner radii for forgings. (From *Forging Industry Handbook,* Forging Industry Association, Cleveland, OH, 1970.)

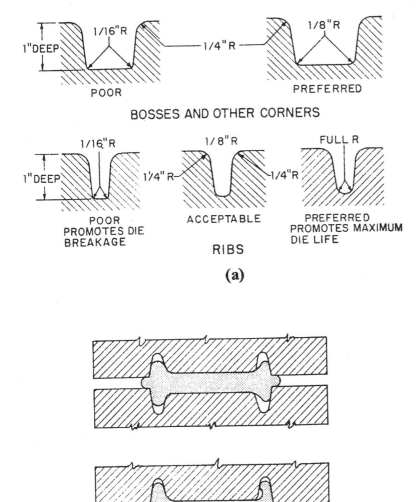

Figure 2.50 (a) Influence of fillet radius on metal flow during progressive stages of die closure; (b) Various die fillet designs for corner radii for steel and aluminum forgings, showing the order of preference for 1-in.-high bosses and ribs. (From *Forging Industry Handbook,* Forging Industry Association, Cleveland, OH, 1970.)

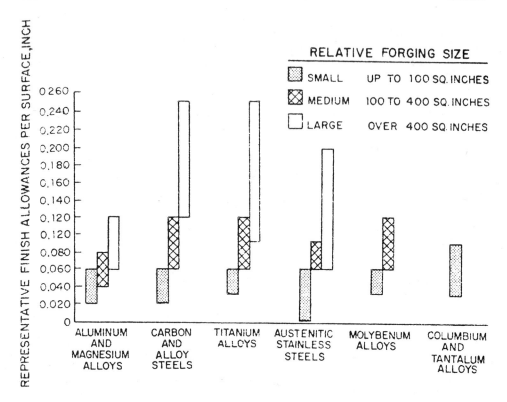

Figure 2.51 Representative finish allowances for impression die forgings of various materials. (Values are in addition to dimensional tolerances.) Finish allowance at corners is usually greater than values shown. (From *Forging Industry Handbook,* Forging Industry Association, Cleveland, OH, 1970.)

machinist sets it up, and the oxidation and forgeability characteristics of the metal. Figure 2.51 shows representative finish allowances for impression die forgings of various materials. (Values are in additional to dimensional tolerances.) Finish allowance at corners is usually greater than the values shown in the figure.

Tolerances

Tolerances, on the whole, represent a compromise: the desired accuracy achievable within reason. Therefore, tolerances cannot be regarded as rigid standards. Their values change with the development of technology and design concepts. The tolerances expressed in this handbook, the *Forging Industry Handbook,* and others represent reasonable average performance that can be expected from a forge plant. Tolerances closer than this obviously require greater care, skill, more expensive equipment, or a more expensive technology and, therefore, carry a premium. When tolerances are excessively tight, cost of production may become disproportionately high. This makes the cost of any additional machining important to the design of the product. It may be worthwhile to pay additional costs for a forging that has a "near-net-shape"—requiring little if any machining. On the other hand,

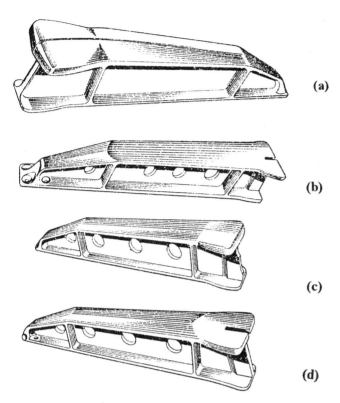

(a)

(b)

(c)

(d)

Figure 2.52 Standard aircraft forging (a), weighing 21 lb, and the same part as a precision forging (c), weighing 10.8 lb. See text. (From *Forging Product Information,* Kaiser Aluminum and Chemical Sales, Cleveland, OH, 1959.)

a lower-cost forging with greater tolerances in some dimensions and requiring more machining may be a better design. These are the trade-offs that a manufacturing engineer must consider during the product design and development phases of a project. Figure 2.52a shows the standard aluminum forging of this aircraft part, 26 in. long and weighing over 21 lb. It required 11.8 lb of metal machined off to obtain the machined fitting (Figure 2.52b), weighing 9.8 lb. The precision forging (Figure 12.52c), weighing 10.8 lb, is shown as it was received by the contractor; with flash and knockout lugs removed by the forging vendor with a saw. No weight-removal machining was required, though some hand finishing was necessary to smooth the rough surface finish at the parting line left by the saw cut. The finished forging, weighing 9.1 lb, after 1.7 lb had been removed, is shown in Figure 2.52d.

Flash and Mismatch

Conventional forging dies are designed with a cavity for flash, as shown in Figure 2.53. This sketch of a completed forging in the closed die shows a typical gutter design. Flash extension tolerances are based on the weight of the forging after trimming and relate to the amounts of flash extension. Flash is measured from the body of the forging to the

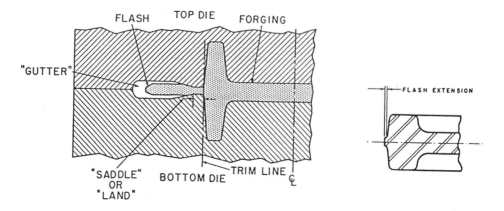

Tabulated figures are ranges of flash extension, expressed in inches.

Weights of Forgings After Trimming, in Pounds	Materials				
	Carbon and Low Alloy	Stain-less	Superalloys and Titanium	Aluminum and Magnesium	Refractory Alloys
10 and under	0 to $\frac{1}{32}$	0 to $\frac{1}{16}$	0 to $\frac{1}{16}$	0 to $\frac{1}{32}$	0 to $\frac{1}{8}$
Over 10 to 25 incl.	0 to $\frac{1}{16}$	0 to $\frac{3}{32}$	0 to $\frac{3}{32}$	0 to $\frac{1}{16}$	0 to $\frac{3}{16}$
Over 25 to 50 incl.	0 to $\frac{3}{32}$	0 to $\frac{1}{8}$	0 to $\frac{1}{8}$	0 to $\frac{3}{32}$	0 to $\frac{1}{4}$
Over 50 to 100 incl.	0 to $\frac{1}{8}$	0 to $\frac{3}{16}$	0 to $\frac{3}{16}$	0 to $\frac{1}{8}$	0 to $\frac{5}{16}$
Over 100 to 200 incl.	0 to $\frac{3}{16}$	0 to $\frac{1}{4}$	0 to $\frac{1}{4}$	0 to $\frac{3}{16}$	0 to $\frac{3}{8}$
Over 200 to 500 incl.	0 to $\frac{1}{4}$	0 to $\frac{5}{16}$	0 to $\frac{5}{16}$	0 to $\frac{1}{4}$	0 to $\frac{1}{2}$
Over 500 to 1000 incl.	0 to $\frac{5}{16}$	0 to $\frac{3}{8}$	0 to $\frac{3}{8}$	0 to $\frac{5}{16}$	0 to $\frac{5}{8}$
Over 1000	0 to $\frac{3}{8}$	0 to $\frac{1}{2}$	0 to $\frac{1}{2}$	0 to $\frac{3}{8}$	0 to $\frac{3}{4}$

Figure 2.53 Conventional flash gutter design. (From *Forging Industry Handbook,* Forging Industry Association, Cleveland, OH, 1970.)

trimmed edge of the flash. Flash tolerances are expressed in fractions of an inch as shown in the table. Mismatch, often the result of die wear, and the expected tolerances are shown in Figure 2.54.

Radii Tolerances

Radii tolerances relate to variation from purchaser's radii specifications on fillet radii, and on corner radii where draft is not subsequently removed by trimming, broaching, or punching. Radii tolerances are plus or minus one-half the specified radii, except where corner radii are affected by subsequent removal of draft.

Straightness

Straightness tolerances allow for slight and gradual deviations of surfaces and centerlines from the specified contour, which may result from postforging operations such as trimming, cooling from forging, and heat treating. Long, thin forgings are more susceptible to warpage than massive shapes; similarly, thin disks warp more than thick disks of comparable diameter. See Figure 2.55.

Figure 2.56 shows the influence of forgeability and flow strength on die filling. Ease of filling decreases from upper left to lower right.

Materials	Weights of Forgings After Trimming, in pounds								
	Less than 2 lbs.	Over 2 to 5 incl.	Over 5 to 25 Incl.	Over 25 to 50 incl.	Over 50 to 100 incl.	Over 100 to 200 incl.	Over 200 to 500 incl.	Over 500 to 1000 incl.	Over 1000
Carbon, Low Alloys	①	1/64	1/32	3/64	1/16	3/32	1/8	5/32	3/16
Stainless Steels	①	1/32	3/64	1/16	3/32	1/8	5/32	3/16	1/4
Super Alloys, Titanium	①	1/32	3/64	1/16	3/32	1/8	5/32	3/16	1/4
Aluminum, Magnesium	①	1/64	1/32	3/64	1/16	3/32	1/8	5/32	3/16
Refractory Alloys	①	1/16	3/32	1/8	5/32	3/16	1/4	5/16	3/8

① = Customarily negotiated with purchaser

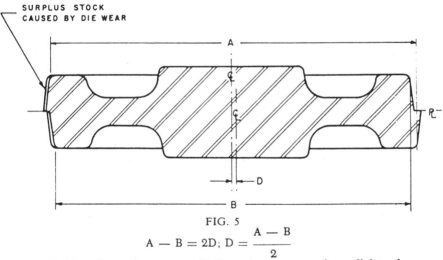

FIG. 5

$$A - B = 2D; \quad D = \frac{A - B}{2}$$

A = Projected maximum over-all dimensions measured parallel to the main parting line of the dies.

B = Projected minimum over-all dimensions measured parallel to the main parting line of the dies.

D = Displacement.

Figure 2.54 Forging die mismatch. (From *Forging Industry Handbook*, Forging Industry Association, Cleveland, OH, 1970.)

2.3 DESIGN FOR SHEET METAL

2.3.0 Introduction to Design for Sheet Metal

More parts are made from sheet metal than from any other material in the world. Most of the processes are well established, and design limits are well known and accepted. Many finishes are available, and customer acceptance and product life are well proven. More steel is produced than any other metal, with aluminum second in volume

Materials	Area at the Trim Line, expressed in sq. in.—Flash not included						
	10 and under	Over 10 to 30 incl.	Over 30 to 50 incl.	Over 50 to 100 incl.	Over 100 to 500 incl.	Over 500 to 1000 incl.	1000 Over
Carbon, Low Alloy	$\frac{1}{32}$	$\frac{1}{16}$	$\frac{3}{32}$	$\frac{1}{8}$	$\frac{5}{32}$	$\frac{3}{16}$	$\frac{1}{4}$
400 Series Stainless	$\frac{1}{32}$	$\frac{1}{16}$	$\frac{3}{32}$	$\frac{1}{8}$	$\frac{3}{16}$	$\frac{1}{4}$	$\frac{5}{16}$
300 Series Stainless	$\frac{1}{16}$	$\frac{3}{32}$	$\frac{1}{8}$	$\frac{5}{32}$	$\frac{3}{16}$	$\frac{1}{4}$	$\frac{5}{16}$
Aluminum, Magnesium	$\frac{1}{32}$	$\frac{1}{32}$	$\frac{1}{16}$	$\frac{3}{32}$	$\frac{1}{8}$	$\frac{3}{16}$	$\frac{1}{4}$

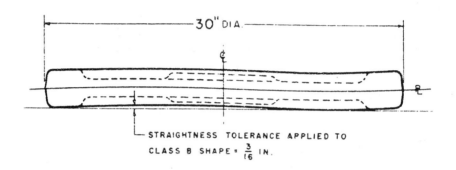

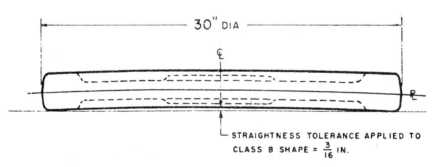

Figure 2.55 Straightness tolerances, class B shapes. (From *Forging Industry Handbook,* Forging Industry Association, Cleveland, OH, 1970.)

and in use in fabricating products. The weight of aluminum is approximately one third the weight of steel. In this subchapter we will concentrate our discussions on parts fabricated from these two materials. Although precision cold-rolled sheet is more expensive per pound that billets or bars, this can seldom offset the total part cost since sheet metal stampings are generally the lowest-cost parts to produce. Both the machinery and labor are relatively low in cost, and the production rates can be quite high.

One of the problems the author had in preparing this subchapter was separating design "do's and don'ts" from the "nuts and bolts" of the materials and fabrication processes. For the most part, the equipment and processes are described in Subchapter 15.1. Active participation by a manufacturing engineer (or other factory representative) in the design

INCREASING FLOW STRENGTH OR FORGING PRESSURE

	LOW GROUP	MODERATE GROUP	HIGH GROUP
GOOD	1030 (CARBON STEEL) 4340 (ALLOY STEEL) H-11 (TOOL STEEL) 6061 (Al ALLOY)	TYPE 304 (STAINLESS) Ti-6Al-4V	MOLYBDENUM 16-25-6 (STAINLESS)
MODERATE	AZ80 (Mg ALLOY) 7075 (Al ALLOY)	A-286 (STAINLESS) INCO 901 (Ni ALLOY) Ti-5Al-2.5Sn	WASPALOY (Ni ALLOY) Ti-13V-11Cr-3Al N 155 (Ni-Cr-Co ALLOY)
FAIR	RESULPHURIZED STEELS (e. g. 1130)	TYPE 321 (STAINLESS) 15-7Mo (STAINLESS)	RENE 41 (Ni ALLOY) HASTELLOY C (Ni ALLOY) HASTELLOY B (Ni ALLOY)

DECREASING FORGEABILITY

Figure 2.56 Influence of forgeability and flow strength on die filling. Ease of filling decreases from upper left to lower right. (From *Forging Industry Handbook,* Forging Industry Association, Cleveland, OH, 1970.)

of a new product requires a knowledge of materials, processes, and costs as their most important contribution to the design team. This subchapter therefore will primarily provide guidance to the design features of sheet metal parts as seen from the manufacturing point of view. However, we must introduce sufficient material and process information to understand why the design comments are applicable.

Stampings

Stamping parts from sheet metal is a relatively straightforward process in which the metal is shaped through deformation by shearing, punching, drawing, stretching, bending, or coining. Production rates are fairly high, and little if any secondary machining is required to produce finished parts within tolerances. In most cases such parts are fairly simple in shape and do not tolerate extreme loads. However, stamped parts can be produced in complex shapes and sizes, and with load-carrying capabilities that rival parts made by other production processes. In some cases, unusual techniques have been developed to deform the metal to its required shape without exceeding forming limits. In other instances, ingenuity combined with good design practice have produced sheet metal parts that would be impossible to fabricate by other methods. Comparing cost impact of the several design

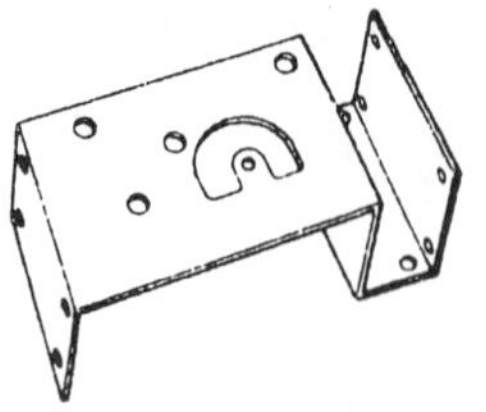

LOW COST DESIGN

Tolerance	± .010 on centers
	± .005 on hole diameters
Material	16 ga. 1010 sheet
	Std. tolerance .059 ± .006
Bends	Radius inside 1/4 thickness
Hole distortion	Print should note "distortion permissible" if hole is too close to a form
Outside corners	1/16" radius allowed

HIGH COST DESIGN

Tolerance	± .003 on centers
	± .001 on hole diameters
Material	1010 half-hard
	.055 ± .003 thickness
Bends	Inside .005 radius
Hole distortion	± .001 all holes
Outside corners	.001 radius allowed (sharp corners required)

PRICE COMPARISON
BASED ON TOOLING & RUN OF 1000 PIECES

	LOW COST	HIGH COST	DIFF
MATERIAL COST	$.19	$.38	+$.19
TOOLING	$300	$600	+$300
TOOLING/PIECE (1000 PCS)	.30	.60	.30
RUN LABOR (1000 PCS)	.20/Piece	.35/Piece	+ .15
TOTAL	$.69 each	$1.33	$.64

Figure 2.57 "Minor" differences in sheet metal design can double part cost. (Courtesy of Dayton Rogers Mfg. Company, Minneapolis, MN.)

variables is difficult. Figure 2.57 is an example of "minor" differences in the design of a sheet metal part which can double the part cost.

Quantity Considerations

The simple example shown in Figure 2.58 shows the influence of quantity production on the design and the manufacturing processes for producing a large tubular shape of large diameter and thin wall. Lowest unit cost is practical where substantial quantities are required for each production run. Welding may not be required in either the low or the high production design if locking tabs can be utilized.

2.3.1 Design for Blanking

The first step in forming a product from sheet, plate, or foil is usually the cutting of a blank. A blank may be any size or pattern, if it represents the most suitable initial shape for the

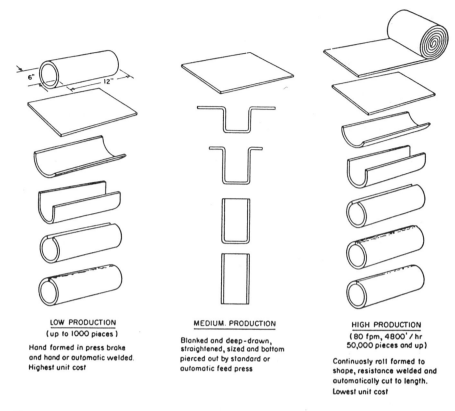

LOW PRODUCTION
(up to 1000 pieces)

Hand formed in press brake
and hand or automatic welded.
Highest unit cost

MEDIUM. PRODUCTION

Blanked and deep-drawn,
straightened, sized and bottom
pierced out by standard or
automatic feed press

HIGH PRODUCTION
(80 fpm, 4800' / hr
50,000 pieces and up)

Continuosly roll formed to
shape, resistance welded and
automatically cut to length.
Lowest unit cost

Figure 2.58 A simple example showing the influence of quantity on process selection for producing a tubular shape of large diameter and thin wall. (From R. A. Bolz, *Production Processes,* Industrial Press, New York, 1963.)

manufacture of the end product. Circular blanks are employed for roundware and most cylindrical-end products. Blanks with straight or curved edges and rounded corners are used for rectangular products. Irregularly shaped shells usually require blanks of irregular shape to provide material where needed. Blanks are made in a number of ways, depending on such factors as gauge, properties of the material, quantity, and size. Sheet metal can be furnished in sheets or in large rolls.

Blanks with straight edges can be produced from sheet, plate, and foil by employing either power- or foot-operated guillotine shears. Plate blanks which require good edge condition can be rough sheared and then machined or routed to final size.

Blanking with conventional punch and die is usually employed for sheet and thin plate where large quantities are required. Its application can be limited by such factors as press capacity, blank size, and material thickness. Therefore, some blanks may be more economically cut by special shears or other means.

Blanking action can be summarized as follows. As the blanking load is applied, the cutting edges of the punch and die enter the stock producing a smooth, sheared surface. As penetration increases, cracks in the stock form at the sharp cutting edges. The remainder of the blanking operation is accomplished by fracturing of the material rather than shearing. This produces the characteristic blanked edge, which is partly shear and partly fracture. A blanked edge produced with the correct tool clearance will exhibit a sheared surface for one-third to one-half of the metal thickness. Insufficient clearance will increase the amount of sheared surface, and will generally produce a blank with a higher burr. This reduced clearance gives increased blanking load and tool wear. The latter can lead to more rapid deterioration in blank edge quality. Excessive clearance will tend to increase the amount of fractured surface, and the blank will be "broken out" rather than sheared, producing a rough edge.

Figure 2.59, Example #1, illustrates results that can be expected from material such as cold rolled sheet steel, commercial quality, that is less than ¼ hard temper; or with a shear strength of approximately 45,000 psi. A slight pulldown (A) of the blanked edge and a straight sheared section (B) for about 25 to 30% of the material thickness will occur on the die side. The opposite will occur on piercing, as indicated by (C) and (D) on the punch side. The balance will have breakage. More pulldown and a greater sheared edge will occur for softer material. Example #2 indicates the results of less than the usual clearance. Increased blanking pressure will be required, and double breakage on the blanked edge will occur. At times, double breakage on thick parts might be more desirable than 8 to 10% taper. Example #3 indicates that on very hard materials, the pulldown is negligible and the sheared edge might amount to only 10% of the thickness of the material.

A tolerance of ±0.005 in. is normal in the blanking and piercing of aluminum alloy parts in a punch press. Using a power shear and press brake, it is possible to blank and pierce to a location tolerance of ±0.010 in. or less, although tolerances for general press-brake operations usually range from ±0.020 to ±0.030 in., since the individual punches are usually located manually on the brake. For economy in tool cost, specified tolerance should be no less than is actually necessary for the particular part. A tolerance of ±0.005 in. would probably require that the punch and die be jig-ground, adding 30 to 40% to the cost over a "normal" punch press part.

Poor blanks may be produced if there is too little stock material between the blank and the edge of the strip. Typical scrap allowances between the blank and strip edge, and between the blanks, are shown in Figures 2.60 and 2.61.

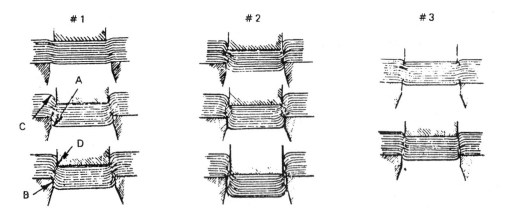

Figure 2.59 Typical blanking or punching operation on commercial cold-rolled sheet steel. (Courtesy of Dayton Rogers Mfg. Company, Minneapolis, MN.)

Thickness, inch	Blank diameter, inches	Scrap on the side, inch	Scrap between blanks, inch
0.013 to 0.057	0 to 9 10 to 19 20 and over	$\frac{1}{8}$ $\frac{3}{16}$ $\frac{1}{4}$	$\frac{1}{16}$ $\frac{3}{32}$ $\frac{1}{8}$
0.057 to 0.081	0 to 9 10 to 19 20 and over	$\frac{5}{32}$ $\frac{7}{32}$ $\frac{5}{16}$	$\frac{3}{32}$ $\frac{1}{8}$ $\frac{5}{32}$
0.081 to 0.128	0 to 9 10 to 19 20 and over	$\frac{3}{16}$ $\frac{1}{4}$ $\frac{3}{8}$	$\frac{1}{8}$ $\frac{5}{32}$ $\frac{1}{16}$

Figure 2.60 Scrap allowances for blanking. (From *Forming Alcoa Aluminum,* Aluminum Company of America, Pittsburgh, PA, 1981.)

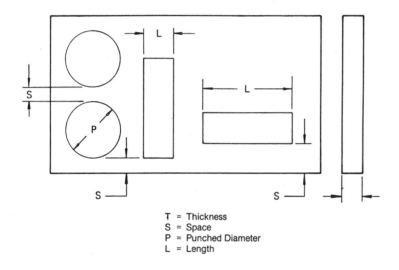

T = Thickness
S = Space
P = Punched Diameter
L = Length

Avoid Minimum Values of "S" Where Possible.

Length "L" is less Than	Punch Dia. "P" is less Than	Space "S" is Minimum
	5T	1.5T
	10T	2T
10T		2T
25T		4T

Figure 2.61 Minimum distance between cutouts and sheet edge. (Courtesy of Dayton Rogers Mfg. Company, Minneapolis, MN.)

Piercing

The piercing of sheet, thin plate and foil is similar to blanking except that the cut blank becomes the scrap and the stock material is kept. When a large number of closely spaced small holes are pierced, the process is described as *perforating*. Most techniques used for blanking also apply for piercing. Finish blanking becomes finish piercing, with a radius or chamfer on the punch rather than the die.

Piercing and perforating may either be preparatory operations in the blank development, as in the case of ceiling panels where perforated blanks are made for further forming, or they may be finish operations, as in the case of strainers and colanders, which are formed prior to being perforated.

Piercing is usually performed on a single-action press. The tooling is the same as for the blanking process, except that the die for piercing must match the contour of the part. The punch may be ground at an angle to reduce the load on the press. The dimensions of the pierced hole correspond to the dimensions of the punch.

Figure 2.62 shows typical punched holes. Rollover and/or bulge are natural consequences of the punching process, the mechanical properties of the material being punched, and the die application of techniques employed.

A common practice exists in dimensioning punched holes. For general-purpose holes:

Dimension feature size limits only.
Measure feature size only in the burnished land area.
Shape deviations within the feature size limits are permissible.

For special-purpose holes:

Dimension only those elements that affect part function.
When dimensioned, specify the burnished land as minimum only.
Specify burr height as maximum only.

To pierce holes with economical tools and operations, the hole diameter must not be less than the stock thickness. If the hole diameter is less than the material thickness (or less than 0.040in.), it usually must be drilled and deburred, and each of these operations

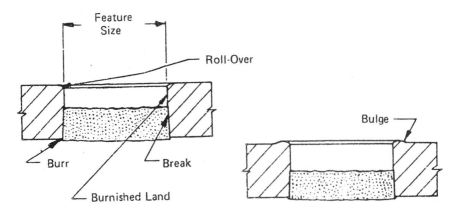

Figure 2.62 Typical punched holes showing roll-over and bulge. (Courtesy of Dayton Rogers Mfg. Company, Minneapolis, MN.)

A

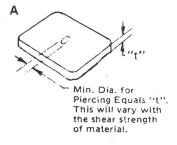

Min. Dia. for
Piercing Equals ''t''.
This will vary with
the shear strength
of material.

B

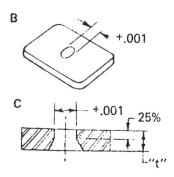

+.001

C

+.001

25%

"t"

D

+.001

E

+.001

Figure 2.63 Tolerances on punched holes in sheet metal. (Courtesy of Dayton Rogers Mfg. Company, Minneapolis, MN.)

is slower than punching. Figure 2.63B, indicates a hole diameter with a tolerance of ±0.001 in. A hole can be pierced within these limits on the punch side for approximately 25 to 30% of the material thickness, as indicated in Figure 2.63C. The percent of thickness varies with the shear strength of the materials. On machined holes punched undersize, redrilling and reaming are usually required where the finished hole size is somewhat larger than the stock thickness, as indicated in Figure 2.63D. If a finished hole is required through the stock thickness, as indicated in Figure 2.63E, where the hole size is equal to or less than the material thickness, more costly drilling is required in addition to deburring and reaming, since the part probably cannot be punched.

Figure 2.64 shows that when the web is a minimum of the stock thickness, a hole can be punched, which is less expensive than drilling and deburring (see example E). A web that is less than the stock thickness will result in a bulge on the blank. Bulge conditions increase progressively as the web decreases, until there is a complete breakthrough. However, the bulge is hardly visible until the web is reduced to less that one-half the stock thickness. These examples also apply to a web between holes. If a measurable bulge is not permitted, a drilling and deburring operation may be necessary. If the web is too narrow, the profile of the blank might be changed by adding an ear of sufficient dimensions and shape to eliminate the problem, as seen in example G in Figure 2.64. Another suggestion might be to change the contour of the blank to include the hole as a notch. The notch could be either be pierced, or be made wide enough in the blank without a piercing or notching operation.

When piercing adjacent to bends, Figure 2.65A indicates that the minimum inside distance required from the edge of a hole to a bend is 1½ times the material thickness (T) plus the bend radius (R). Otherwise, distortion will occur as indicated in Figure 2.65B, or

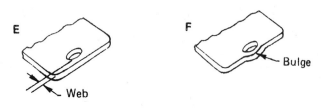

Figure 2.64   Bulge conditions when hole is pierced too close to edge. (Courtesy of Dayton Rogers Mfg. Company, Minneapolis, MN.)

piercing after forming must be considered. Figure 2.65C indicates a similar condition, except for openings with an edge parallel to bend. In this case the following requirements apply for economical tooling and production:

> When L is up to 1 in., the distance should be $2T + R$ (minimum).
> When L is 1 to 2 in., the distance should be $2\frac{1}{2}T + R$ (minimum).
> When L is 2 in. or more, the distance should be $3T$ to $3\frac{1}{2}T + R$ (minimum).

Burrs are ragged, usually sharp, protrusions of edges of metal stampings as seen in Figure 2.66. Commonly specified permissible burr conditions in order of increasing cost are:

1. Unless otherwise specified, normal acceptable burr can be 10% of stock thickness.
2. A note, "Conditioned for Handling", is interpreted to mean that normal stamping burrs are to be refined as necessary for average handling.
3. Remove burrs on specified edges.
4. Remove all burrs.
5. Break sharp edges or corners where specified.
6. Break all sharp edges and corners.

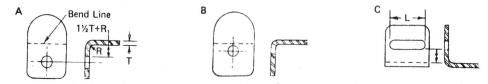

Figure 2.65   Problems when piercing too close to bends in sheet metal. (Courtesy of Dayton Rogers Mfg. Company, Minneapolis, MN.)

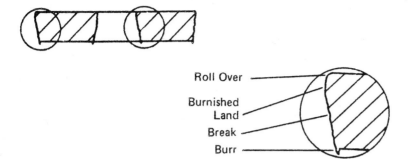

Figure 2.66 Section through a pierced hole showing roll-over, the burnished land, the break, and burr formation. (Courtesy of Dayton Rogers Mfg. Company, Minneapolis, MN.)

2.3.2 Design for Bending

Bending, one of the most common operations in sheet metal working, consists of forming flat metal about a straight axis. Press-brake bending and roll-bending are the usual methods, although punch-press tools are often built to produce multiple bends, or bends in conjunction with other operations, such as blanking. The same equipment and tools employed for bending steel and other metals can be used for aluminum, although appropriate changes in bend radii and technique should be considered. Most principles of bending also apply to such complementary operations as flanging, beading, curling, crimping, and lock seaming. Figure 2.67 shows some typical press-brake forming dies.

Factors Influencing Bending

It is important to consider carefully in the design stage the relationships among mechanical properties, dimensions of the part after bending, metal thickness, and "formability" of the metal. When a specific alloy is selected for strength or other reasons, the part should be designed in accordance with recommended bend radii to assure producible results (see

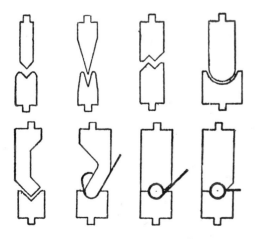

Figure 2.67 Typical press-brake forming dies. (Courtesy of Rafter Machine Company, Inc., Belleville, NJ.)

Alloys	Tempers	Thickness of Sheet—Inches									
		.016	.025	.032	.040	.050	.063	.090	.125	.190	.250
		Bend Radii in 32nds of an Inch									
1100	-0	0	0	0	0	0	0	0	0	0	0
	-H12	0	0	0	0	0	0	0	0	3	6
	-H14 ①	0	0	0	0	0	0	0	0	3	6
	-H16	0	0	0	0	1	2	3	4	—	—
	-H18	1	1	2	2	3	4	6	8	—	—
3003 & Alclad 3003 5005 5457 5357 5657	-0	0	0	0	0	0	0	0	0	0	0
	-H12 or -H32	0	0	0	0	0	0	0	0	3	6
	-H14 or -H34 ②	0	0	0	0	0	0	1	2	4	8
	-H16 or -H36	0	0	1	2	2	3	5	6	—	—
	-H18 or -H38	1	2	3	4	5	6	9	12	—	—
3004 Alclad 3004 5154 5454 5254	-0	0	0	0	0	0	0	0	0	2	4
	-H32	0	0	0	1	1	2	3	4	8	16
	-H34	1	1	1	2	2	3	5	6	12	24
	-H36	1	2	3	4	5	6	9	12	—	—
	-H38	2	3	4	5	7	8	12	16	—	—
5050	-0	0	0	0	0	0	0	0	0	0	0
	-H32	0	0	0	0	0	0	1	2	4	8
	-H34 ③	0	0	0	1	1	2	3	4	8	12
	-H36	1	1	2	2	3	4	6	8	—	—
	-H38	1	2	3	4	5	6	9	12	—	—
5052 5652	-0	0	0	0	0	0	0	0	0	0	0
	-H32	0	0	0	0	0	0	2	3	6	12
	-H34	0	0	0	1	1	2	3	4	8	16
	-H36	1	2	2	2	3	4	6	8	—	—
	-H38	1	2	3	4	5	6	9	12	—	—
5083 ①	-0	—	—	—	—	3	4	6	8	12	16
	-H32	—	—	—	—	6	7	10	14	20	28
	-H34	—	—	—	—	8	10	14	20	28	36
	-H113	—	—	—	—	—	—	—	14	20	28
5086 5155	-0	0	0	0	0	0	0	2	3	6	12
	-H32	1	1	1	2	2	3	5	6	12	24
	-H34	1	2	3	4	5	6	9	12	24	32
	-H36	3	4	5	6	7	9	12	18	—	—
5456 ①	-0	—	—	—	—	—	—	—	8	12	16
	-H24	—	—	—	—	8	10	14	20	32	40
	-H321	—	—	—	—	—	—	—	—	24	32
2024 Alcald 2024	-0	0	0	0	0	0	0	0	0	2	4
	-T3	1	3	4	5	7	8	12	16	32	48
6061 Alclad 6061	-0 ④	0	0	0	0	0	0	0	0	2	4
	-T6	1	2	2	2	3	4	6	8	18	32
2014	-0	0	0	0	0	0	0	0	0	2	4
	-T6	2	4	6	8	10	12	18	24	40	64
7075 Alclad 7075	-0	0	0	0	1	1	1	3	4	8	16
	-T6	2	4	6	8	10	12	18	24	40	64
7178 Alclad 7178	-0 .	0	0	0	1	1	1	3	4	8	16
	-T6	3	6	7	9	12	16	24	32	48	96

①Use this data for #1 reflector sheet.

②Use this data for #1, #2, #11 and #12 brazing sheet, #2 reflector sheet and RF10, RF15, RF20 finishing sheet.

③Use this data for porcelain enameling sheet.

④Use this data for #21 and #22 brazing sheet.

⑤These materials may require annealing or stress-relieving after cold forming. If formed at a metal temperature of 425-475°F, a 50 per cent smaller bend radii may be used.

Figure 2.68 Minimum recommended inside radii for 90° cold-bending aluminum sheet. (From *Forming Alcoa Aluminum*, Aluminum Company of America, Pittsburgh, PA, 1981.)

Figures 2.68 and 2.69). Conversely, when part dimensions are the prime consideration, the material selected should be readily formable to those dimensions. While it is often possible to form sheet to tighter radii than is shown in the tables, it usually costs more due to heating, tool maintenance, greater incidence of failure, etc.

The elongation values of a metal are not valid indicators of bendability, as they are taken over a 2-in. gauge length. The elongation in a very short gauge length controls bending characteristics, and this is frequently not related to the 2-in. value. The empirical data of Figures 2.68 and 2.69 represent the best measure of bendability available. Some other recommendations are given in Figure 2.70. The height of the form of the part has an influence on the bend. Diagram A in Figure 2.70 illustrates a 90° bend with insufficient height (h) to form properly. Consequently, stock must be added so the form is high enough (H), stock is then cut off, which means additional tooling and an additional operation. If h is not high enough, the cutoff die may not have sufficient strength to stand up for a particular material or thickness. This may result in an even higher cost secondary operation, such as milling. Diagram B shows how to determine the minimum inside height H, which in this case equals 2½ times the material thickness (T) plus the required bend radius (R). The concept illustrated by diagram B is shown in chart form below the diagram. These recommended minimum formed height dimensions are general to cover most variables of design, size, material types, tempers, and thicknesses but permit economical tooling and production. Proper design, small parts, and easily formed materials, such as

WORKING FROM THE INSIDE, THE FIGURES SHOWN ARE THE CORRECT BEND ALLOWANCES. SEE EXAMPLE FOR CLARITY

.75   .06 R. INSIDE   DEVELOPED LENGTH
.063   .750 + .880 + .015 = 1.645
.88

MATERIAL THICKNESS	INSIDE RADIUS OF BEND													
	.016	.030	.048	.063	.094	.125	.156	.188	.219	.250	.312	.375	.437	.500
.016	.004	.003	.010	.016	.029	.043	.056	.070	.083	.097	.124	.150	.177	.204
.020	.007	.000	.007	.014	.027	.041	.054	.067	.080	.094	.121	.148	.175	.201
.025	.010	.004	.004	.010	.023	.037	.051	.064	.077	.091	.118	.144	.171	.198
.030	.013	.007	.001	.007	.020	.034	.047	.061	.074	.088	.115	.141	.168	.195
.032	.014	.008	.000	.006	.019	.033	.046	.059	.073	.086	.113	.140	.167	.194
.036	.017	.010	.003	.003	.016	.030	.044	.057	.070	.084	.111	.137	.164	.191
.040	.020	.013	.006	.001	.014	.027	.040	.054	.067	.081	.108	.135	.161	.188
.045	.022	.015	.009	.002	.011	.025	.038	.052	.065	.078	.105	.132	.159	.186
.048	.025	.018	.011	.004	.008	.022	.035	.049	.062	.076	.103	.130	.156	.183
.050	.027	.020	.013	.007	.007	.020	.033	.047	.060	.074	.101	.127	.154	.181
.060	.032	.026	.019	.013	.001	.014	.028	.041	.054	.068	.095	.122	.148	.175
.063	.036	.029	.022	.015	.003	.011	.025	.038	.052	.065	.092	.119	.145	.172
.075	-	.036	.029	.022	.009	.005	.018	.031	.045	.058	.085	.112	.139	.165
.080	-	.039	.033	.026	.012	.000	.014	.028	.041	.055	.081	.108	.135	.162
.090	-	-	.040	.033	.020	.006	.007	.021	.034	.047	.074	.101	.128	.155
.125	-	-	-	.056	.042	.029	.015	.002	.011	.025	.051	.078	.105	.132
.160	-	-	-	.076	.063	.049	.036	.022	.009	.004	.031	.058	.085	.112
.190	-	-	-	-	.083	.070	.056	.043	.030	.016	.011	.038	.064	.091

NOTE: FIGURES TO THE LEFT OF THE HEAVY LINE ARE TO BE ADDED
FIGURES TO THE RIGHT OF THE HEAVY LINE ARE TO BE SUBTRACTED

Figure 2.69 Table of metal-forming bend allowances. (From John P. Tanner, *Manufacturing Engineering*, Marcel Dekker, New York, 1991.)

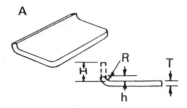

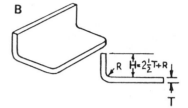

Stock Thickness		INSIDE BEND RADIUS									
		FRAC.	DEC.	FRAC.	DEC.	FRAC.	DEC.	FRAC.	DEC.	FRAC.	DEC.
Frac.	Dec.	Sharp	Sharp	1/32	.031	1/16	.062	3/32	.093	1/8	.125
1/32	.031	5/64	.078	7/64	.109	9/64	.140	11/64	.171	13/64	.203
1/16	.062	5/32	.156	3/16	.187	7/32	.218	1/4	.250	9/32	.281
3/32	.093	15/64	.234	17/64	.265	19/64	.296	21/64	.328	23/64	.359
1/8	.125	5/16	.312	11/32	.343	3/8	.375	13/32	.406	7/16	.437
5/32	.156	25/64	.390	27/64	.421	29/64	.453	31/64	.484	33/64	.515
3/16	.187	15/32	.468	1/2	.500	17/32	.531	9/16	.562	19/32	.593

Figure 2.70 Minimum height of form depending on material thickness and inside bend radius. (Courtesy of Dayton Rogers Mfg. Company, Minneapolis, MN.)

aluminum, brass, copper, or mild steel may be formed with slightly lower inside formed height (roughly 20% less).

Figure 2.71A illustrates a design that is not desirable in terms of either quality or economy. When the bend is inside the blank profile, as shown, the material must be torn through the stock thickness and the bend radius. If the part is under stress, this tear will likely cause fatigue failures. In addition, stock tooling cannot be used, because the flat area adjacent to the bend must be held in position during forming, which means extra tooling expense. Diagram B shows a similar condition, but with the bend just outside the blank profile. In this case, the tear proceeds to the tangent of the required bend radius. Figures 2.71C and 2.71D illustrate a possible solution by changing the blank profile to provide relief for the bend. Besides eliminating the chance of fatigue under stress, there is the possibility of using stock 90° V punches and dies in a press brake. The results are better quality and a less expensive tool price. If the relief notches shown in D are wide enough compared to the material thickness and shear strength, or are designed like the relief in E, they can be included in the blanking operation at very little extra cost for tools and no extra operation.

Distortion and interference of forming is another condition to consider. Figure 2.72A shows a distortion condition that occurs in forming. It is a noticeable distortion when heavy metal is bent with a sharp inside bend radius. On material thicknesses less than $\frac{1}{16}$ in., or when the inside forming radius is large in comparison to the material thickness, distortion is barely noticeable. The material on the inside of the bend is under compression, which results in this bulge condition on the edges. In addition, the edges on the outside of the bend are under tension and tend to pull inside. This bulge or distorted condition is usually of no concern and is accepted as standard practice unless bulging will cause interference

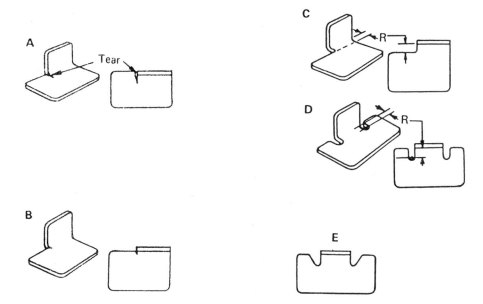

Figure 2.71 Examples of poor design for forming, and some alternatives. (Courtesy of Dayton Rogers Mfg. Company, Minneapolis, MN.)

with a mating part. This interference should be referred to on the part drawing so that a secondary operation can be considered to remove it. The extra operation may not require tooling, but it will add to the cost of production. Figure 2.72B illustrates a blank developed to prevent interference resulting from bulge without extra production cost. The upper left-hand form (enlarged section) in illustration C indicates a fracture condition that occurs when the burr side of the blank is on the outside of the bend. This fracture condition occurs because the burr side of the blank on the outside of the bend is under tension and causes the minute fracture on the sharp edge to open up and in extreme cases to become visible. Blanks should be produced so the burr side will be on the inside of the bend, which is under compression, as in the lower form. However, when blueprint requirements prevent this, or when a bend is in an opposite direction, as in the upper bend, fractures may occur.

Tumbling or deburring well before forming can minimize fracture in most cases. On extra heavy material with a very sharp inside bend radius, or on materials that are difficult to form, such as SAE 4130 steel, tumbling well before forming may not be adequate. It may be necessary to hand file or disk sand a radius on the sharp edges. Such secondary operations will add to the cost of production. Therefore, for the most economical production and if design will permit, ample inside bend radii should be permitted for heavy and difficult forming material when the burr side of the blank must be on the outsid of the bend. If slight fractures are permissible, this should be indicated on the part drawing.

Springback

Springback is the partial return of the metal to its original shape when bending forces are removed. This is encountered in all bending operations. Springback cannot be eliminated entirely; it must be compensated for by overbending. The proper amount of overbending is easily determined by varying the equipment settings when using V dies or rolls. When

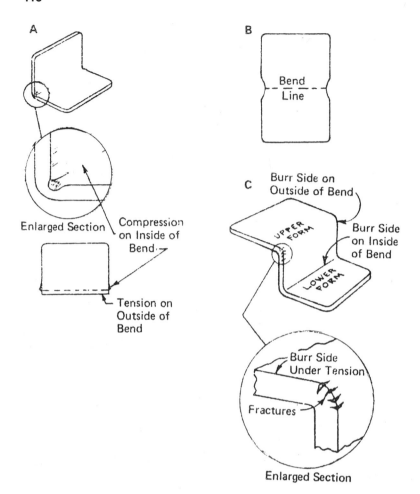

Figure 2.72 Examples of distortion and interference in forming. (Courtesy of Dayton Rogers Mfg. Company, Minneapolis, MN.)

bending on bottoming dies, the springback must be estimated before the dies are built and compensated for in the design. Analytical methods are available for calculating springback; their accuracy depends on the specific mechanical and physical metal properties, which usually can be obtained only by experiment. Samples should be made to determine the springback of specific bends prior to tool design, and the tools should be build with sufficient stock to allow for rework if close angular tolerances are required.

As a general rule, the springback characteristics of different alloys and tempers will be proportional to their yield strengths, and thin sheet will spring back more than thick sheet in an inverse ratio to the cube of the thickness. Springback is greater for large-radius bends than for tight bends.

Low-Cost Forming

Forming or bending is an easy way to convert flat, sheetmetal stampings into parts that are useful in three dimensions. However, each forming operation is an added cost. The

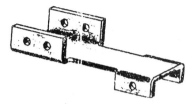 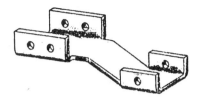

Figure 2.73 Redesigning to permit low-cost forming. (Courtesy of Dayton Rogers Mfg. Company, Minneapolis, MN.)

following suggestions will help to keep forming operations (and costs) to a minimum (see Figure 2.73). Important for this part are the locations of the hole centerlines. Two forming operations, one for each direction of bending, are required to produce this part. An alternative design, which preserves the relationships between the holes, requires only one forming step. The offset (or elongated Z bend) is easy to include in the forming die and is inexpensive.

The hole and boss in the center of the formed part in Figure 2.74 must be extruded. For the first part design, extruding is an additional operation, because the forming press must move in an opposite direction. The alternative design can be formed while at the same time the center hole and boss are extruded in one movement of the forming press.

In the assembly shown in Figure 2.75, two formed parts are nested together. Because R1 = R2, manufacturing variations in the two bend radii may prevent complete contact between the two parts. Either the need for selective assembly or high rejection rates can increase the cost of these parts. When R1 is smaller then R2, the parts do not touch in the bending zone. This design ensures that any two parts, selected at random for assembly, will fit together.

The part shown in Figure 2.76 is made by a process called *channel forming*. Severe angles—near 90°—make die removal difficult, create high stresses in the metal, and require high bending force. The alternative part design, also a formed channel, has slightly open angles and therefore is easy and inexpensive to form.

Slots and holes near the bending zone of a formed part distort with the metal during bending. Often, distortion can be compensated for by stamping a deformed slot in the blank as seen in Figure 2.77. In this example, the deformed slot (dashed lines) is transformed into the proper shape after forming. Usually, trial-and-error testing is required to determine the exact shape of the slot prior to bending. Twisting is neither a well-known nor a commonly used forming operation, even though it is a simple and inexpensive way to form parts. Twisting can be performed easily with special press tools, jigs, or even by hand. This

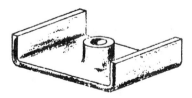 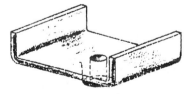

Figure 2.74 Redesigning to reduce the cost of the extruded hole. (Courtesy of Dayton Rogers Mfg. Company, Minneapolis, MN.)

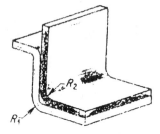

Figure 2.75 Assembly problem caused by nesting when R1 (inside) is larger than R2 (outside). (Courtesy of Dayton Rogers Mfg. Company, Minneapolis, MN.)

part, for example, requires less material than an equivalent part bent into an L shape. Twist angle is limited only by the ductility of the material; in some cases, angles over 360° are possible.

Optimizing Bracket Shape

Although a bracket is a rather prosaic component—basically a formed sheetmetal part designed to support or mount one component to another—a wide variety of designs are possible. In some cases, unconventional shapes can save material, reduce weight, or add strength. Here are a number of such improvements in bracket design (see Figure 2.78).

Dimples, beads, and other forms of compound curvature improve the strength of a bracket. The simplest application of this technique is shown by a bracket with two 45° bends at the corners. Ribs and beads provide more strength, however, and can be used as shown to stiffen brackets made from thin-gauge metal. Several of these designs reduce bracket weight by as much as 50% without sacrificing strength.

In many cases, ribs are welded or riveted to brackets for strength. All but two of the examples ahown in Figure 2.78 can be made from a single piece of sheet metal and then welded or riveted easily in a second operation. The two- and three-piece assemblies require jigs and fixtures for assembly prior to welding.

"Cutting corners" saves material. For relatively low loading, a bent plate with diagonally cut edges can reduce material requirements by 50%. The example calling for compound bending also saves material but can support higher loads. Shear forming or lancing creates a tab for fastening.

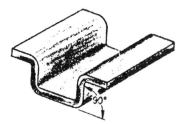
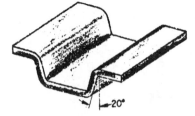

Figure 2.76 Angles near 90° make this channel difficult to form. (Courtesy of Dayton Rogers Mfg. Company, Minneapolis, MN.)

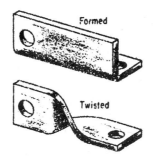

Figure 2.77 Example of a distorted slot near a bend, and an example of twisting. (Reprinted with permission from *MACHINE DESIGN*, a Penton publication.)

2.3.3 Design for Drawn Parts

Drawing is a process which forms a flat piece of metal into a seamless hollow shape. In a drawing operation, a punch applies force to the center portion of a blank to pull or draw the metal into a die cavity. A blankholder applies a restraining force to the edges of the blank to eliminate wrinkling and to control the movement of metal into the die. Closely related to the process are redrawing, which reduces the cross-section dimensions and

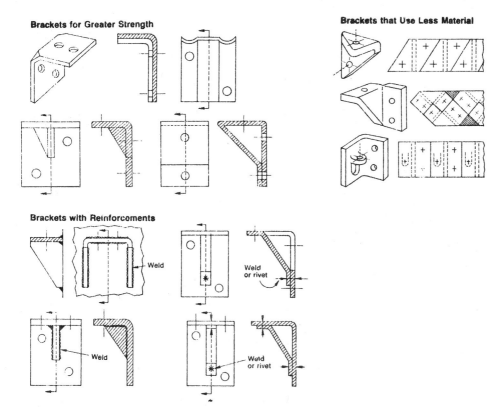

Figure 2.78 Some improvements in common bracket design. (Reprinted with permission from *MACHINE DESIGN*, a Penton publication.)

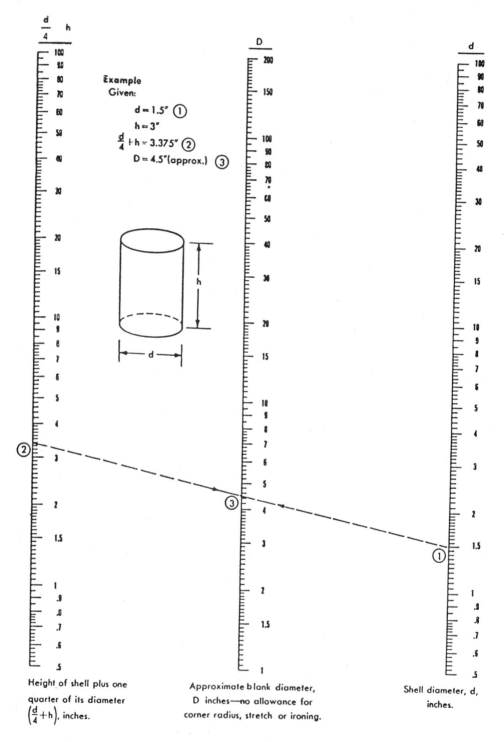

Figure 2.79 Nomograph for calculating blank diameter for a cylinder drawn from sheet. (From *Forming Alcoa Aluminum,* Aluminum Company of America, Pittsburgh, PA, 1981.)

increases the depth of a drawn part, and reverse drawing, where a draw and a redraw operation are combined in one press stroke. Some drawing operations, such as the forming of shallow shapes, are referred to as pressing or stamping, but the distinction is not consistent.

A blank is the starting form for practically all drawing operations. The surface area of the blank should be only slightly greater than that of the finished product, allowing sufficient metal to hold the blank during drawing and final trimming. Excessive blank metal, aside from incurring unnecessary cost, requires increased drawing pressures that can result in fracture of the metal. See Figure 2.79 for a nomograph for calculating blank diameter.

The word *drawing* is generic and includes the operation in which metal is drawn over a die radius with no attempt to reduce the gauge (pure drawing), the process in which the edges of the blank are gripped tightly by the blankholder and die so the stock cannot slip (pure stretching), and all draw operations that combine drawing and stretching. The metal requirements vary depending on the type of operation. When a part is to be formed primarily by stretching, high elongation in the metal is required. The elongation value, established in a tensile test, is a measure of the relative formability of metals (see Figure 2.80).

When the blank is allowed to move between the blankholder and the die as in the case of pure drawing, elongation is less important. Here, the prime factors are tensile strength and the rate of strain hardening. It is not uncommon in producing a round, flat-bottom, straight-sidewall part to take a greater draw reduction with high-strength, low-elongation metal than with annealed stock. The operations are not necessarily accomplished with the same set of tools, however, since higher-strength alloys normally require larger tool radii and greater clearance.

The force exerted by the punch varies with the percent of reduction, the rate of strain hardening, and the depth of draw. Aluminum, like other metals, strain-hardens during draw operations and develops a harder temper with a corresponding increase in tensile and yield strength. If the part is to be drawn successfully, the force exerted by the face of the punch must always be greater than the total loads imposed on that portion of the blank between the blankholder and the die. Also, the metal between the edges of the punch and die must be strong enough to transmit the maximum load without fracturing. This set of conditions establishes a relationship between the blank size and punch size for each alloy-temper combination that dictates the minimum punch size that can be employed in both drawing and redraw operations.

The edge condition of drawn parts can be seen in Figure 2.81, and the surfaces and characteristics of drawn shapes in Figure 2.82.

2.3.4 Design Without Fasteners

Tabs

One of the easiest and most economical methods of making permanent and semipermanent, moderate-strength joints in sheetmetal parts is by folding or bending of tabs. No screws, rivets, or other loose pieces are required, preparation is simple, and assembly is rapid. Limitations of fastening with tabs are few: The thickness of the deformed member should be between 0.012 and 0.080 in.; the method is not recommended for high-strength joints; and the metals should have sufficient ductility so they do not spring back after being bent. Soft steel, aluminum, copper, and brass are

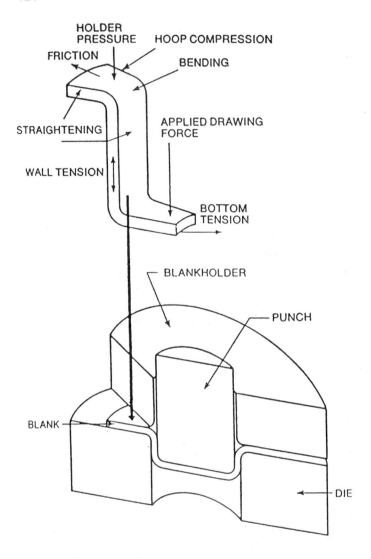

Figure 2.80 Cross section of a typical draw die and the forces in the drawn part. (From *Forming Alcoa Aluminum,* Aluminum Company of America, Pittsburgh, PA, 1981.)

the metals most commonly joined by this method. For best results, parts should be designed so that the bending is perpendicular to the direction of sheet rolling, or at least 45° from that direction. Also, the stamping burr should be at the inside of the bend when possible, and the final bend should be made in the same direction as the preliminary bend.

The working principle is relative straightforward as shown in Figure 2.83. In Figure 2.83a, one or more tabs (also called legs, ears, or tongues) are formed in one of the workpieces, and matching slots slightly longer than the tab widths are punched in the mating part. Both features are made in the press operations that fabricate the parts.

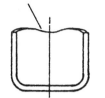

Edge of blank with earing

1. Cupped from the blank.

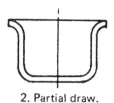

2. Partial draw.

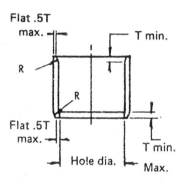

Flat .5T max.

T min.

R

R

Flat .5T max.

T min.

Hole dia.

Max.

3. Redrawn, struck, punched hole & pinch trimmed

4. Flat-edge trimmed after drawing.

Figure 2.81 Edge conditions of drawn parts. (Courtesy of Dayton Rogers Mfg. Company, Minneapolis, MN.)

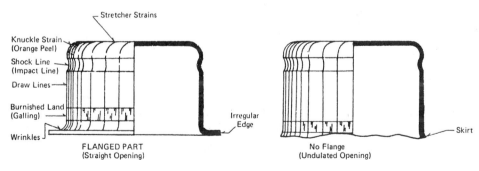

Figure 2.82 Surfaces and characteristics of drawn shapes. (Courtesy of Dayton Rogers Mfg. Company, Minneapolis, MN.)

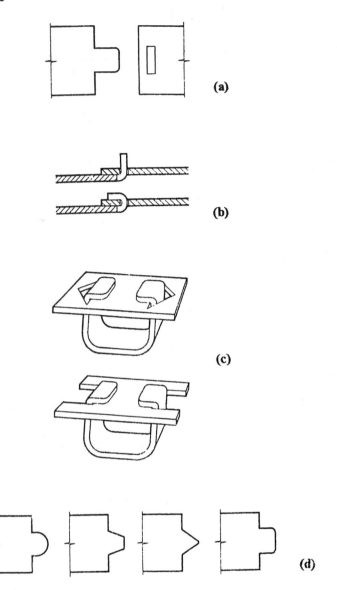

Figure 2.83 Use of tabs in joining parts. (Reprinted with permission from *MACHINE DESIGN*, a Penton publication.)

In Figure 2.83b, the tab is bent up at an angle that suits the assembly (usually 90°), inserted into the slot, and folded over to form the joint. Where the slot shape need not be a narrow rectangle, preferred shapes are triangular, round, or simple notches as seen in Figure 2.83c. Punches for these shapes are stronger and last longer. The shape and size of tabs can be varied to suit assembly requirements and appearance as shown in Figure 2.83d.

Joint strength is increased by using two or more tabs, folded in opposite directions as

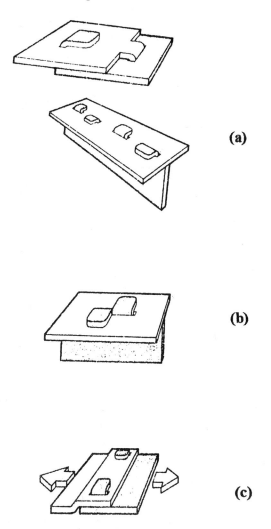

(a)

(b)

(c)

Figure 2.84 Joint strength is increased by using two or more tabs. (Reprinted with permission from *MACHINE DESIGN*, a Penton publication.)

seen in Figure 2.84a. Where space is limited, a single tab can be split so that it functions as two tabs, shown in Figure 2.84b. Figure 2.84c shows that joint strength is higher when the tabs are stressed in shear rather than in bending.

Assembly using tabs need not involve two separate parts; two ends of the same part can be joined by folded tabs, as shown in Figure 2.85. In some cases, three or more parts can be joined by tabs, in sandwich fashion. Figure 2.86 shows twisting replacing folding. For joints involving relatively thick metal, locking action can be improved by twisting the tabs instead of folding them. One drawback to the twisted tabs in Figure 2.86a is the increased height required; another is that the sharp edges can be a source of injury to personnel. The height of twisted tabs can be reduced using the designs in Figure 2.86b, which are more efficient but also more expensive to stamp. Bulky or non-sheet-metal parts

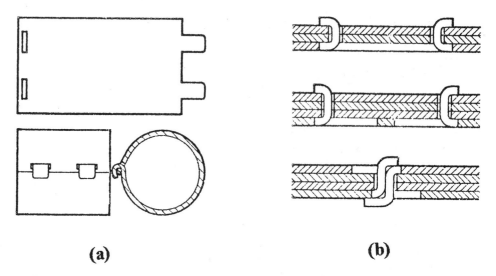

(a)   **(b)**

Figure 2.85 Joining of a single part or multiple parts with tabs. (Reprinted with permission from *MACHINE DESIGN*, a Penton publication.)

as shown in Figure 2.86c can be attached to a sheetmetal base with bent tabs or with a metal strap whose ends are twisted to secure the parts together.

Staking

Staking is a simple way to assemble parts in which one part is plastically deformed or "spread out" around the other. Although staking is most frequently used to fasten lugs, bosses, and covers to sheet metal and plastic parts, the process can be applied to nearly all assemblies where one component is a ductile metal. Here is a review of the many variations of staking, along with several suggestions that can make the procedure more effective.

Figure 2.87a shows that parts can be staked together in one of two ways. To stake this lug into a flat sheet, for example, either the sheet is forced inward, left, or the lug is deformed outward, right. The tool shown here is a circular impact punch with a wedge around the periphery. When a mating part has a small shoulder, as in Figure 2.87b, a washer should be used to distribute the staking impact. In the case of delicate stampings, washers may be required on both sides of the assembly. Figure 2.87c shows that a hollow die can be used when the assembled component must protrude at both sides. For heavy-duty applications, a flat punch shears the metal to produce a larger deformation. Examples so far have shown stepped or shouldered components that require staking from one side only. It is possible, however, to stake inserts from both sides, as shown in Figure 2.87d.

Parts are not always staked all around their periphery. As shown in Figure 2.88a, parts can be "prick punched" in specified locations, especially for thin or weak materials and when low joint strengths can be tolerated. Staking points are selected according to strength or impact requirements, but usually they are symmetrical around the insert.

The staked assemblies illustrated so far will not resist rotational load on the insert. One way to prevent rotation under light loads is to notch the surface of the plate, and stake the insert into the notches (Figure 2.88b). For high rotatational loading, however, non-circular inserts and matching holes should be specified. Two stampings, such as a cover and a base plate, can also be staked together. In Figure 2.88c, for example, an ear on one

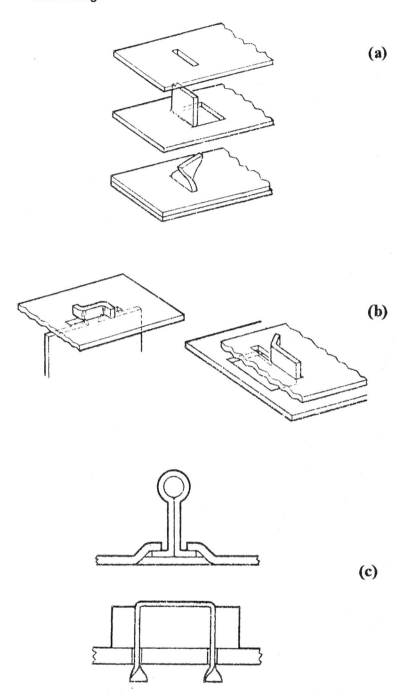

Figure 2.86 Use of assembly tabs by twisting rather than folding. (Reprinted with permission from *MACHINE DESIGN*, a Penton publication.)

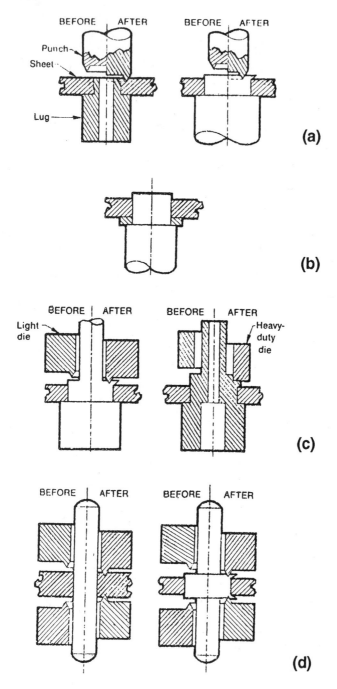

Figure 2.87   Recipe for well-done staking. (Reprinted with permission from *MACHINE DESIGN*, a Penton publication.)

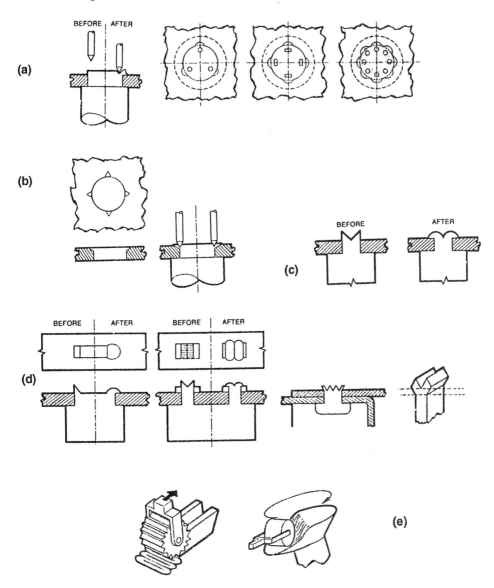

Figure 2.88 Additional staking methods. (Reprinted with permission from *MACHINE DESIGN*, a Penton publication.)

stamping is assembled and staked at right angles to the plate. Figure 2.88d shows two variations for heavy-duty assembly by staking using rivets (left) and longitudinal spreading of a sheet metal tab (right). Long assemblies may require several staking points or ears for rigidity. In addition, washers (right) can be used to prevent damage while staking.

"Progressive" staking, shown in Figure 2.88e, eliminates the sudden blow of a punch or die and is often used for light load-bearing assemblies. A toothed wheel (left) pushed along an ear or tab, and a roller or wheel forced against a rotating insert or rivet (right) are two progressive staking methods.

Lock Seams

Seams provide a simple way to join sheetmetal parts such as panels, tubes, housings, containers, and ducts. Basically, a seam is a shape formed in the two mating edges of sheetmetal parts, so that when the parts are assembled, they lock together by pressure or mechanical "flattening." Seams are economical and are often used instead of mechanical fasteners, especially when holes for fasteners would create stress concentrations, or when welding or brazing heat would deform thin metal. Seams are often stronger than other fastening alternatives and are as heatresistant as the sheet metal. In addition, the curved, overlapping column of a metal seam can help stiffen the assembled part. See Figure 2.89.

Seams require a great amount of plastic flow in a metal. Generally, seams are limited to 0.011- to 0.050-in.-thick sheet; the formability of the alloy determines whether the sheet is too thick to be seamed without cracking. A rule of thumb for seams in carbon steel is to design the seamwidth between $3/16$ and $1/2$ in. for sheets up to $1/32$ in. thick. Thicker sheets require wider seams.

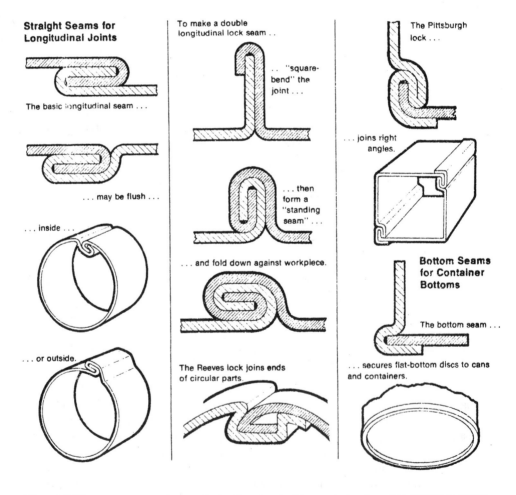

Figure 2.89 Additional staking methods. (Reprinted with permission from *MACHINE DESIGN*, a Penton publication.)

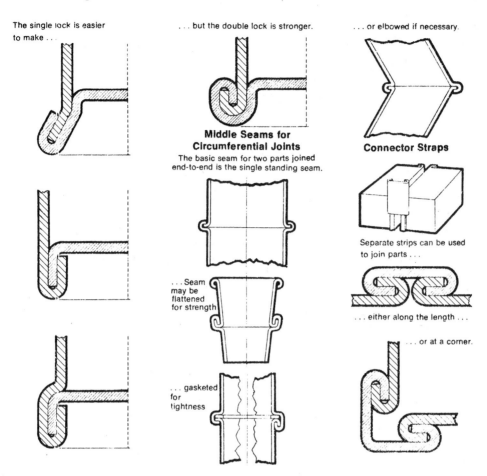

The single lock is easier to make . . .

. . . but the double lock is stronger.

. . . or elbowed if necessary.

Middle Seams for Circumferential Joints

The basic seam for two parts joined end-to-end is the single standing seam.

Connector Straps

Separate strips can be used to join parts . . .

. . . Seam may be flattened for strength

. . . either along the length . . .

. . . or at a corner.

. . . gasketed for tightness

Figure 2.90 Variations of lock seams in sheet metal fabrication. (Reprinted with permission from *MACHINE DESIGN*, a Penton publication.)

There are four types of seams. Straight seams are used to make tubelike parts out of flat sheets or to join parts such as formed sheetmetal panels. Closed-bottom seams can join a tubular body to a flat top or bottom (for example, sealing a top to a food can). Closed middle seams connect circular or rectangular tubing or similar parts end to end (as in connecting ductwork). A fourth type of seam is locked with a strap, clip, or band between the mating parts. Of course, this seam requires a third component for each joint. The illustrations here review most of the common variations of the four types of seams. See Figure 2.90.

2.3.5 Bibliography

Forming Alcoa Aluminum, Aluminum Company of America, Pittsburgh, PA., 1981.
Tanner, John. P., *Manufacturing Engineering*, Marcel Dekker, New York, 1991.
MACHINE DESIGN, Penton Publications.
Alting, Leo, *Production Processes*, Marcel Dekker, New York, 1982.

2.4 PROCESS SELECTION

In Chapters 1 and 2 we have covered a great deal of material in a relatively few pages. The selection of the optimum product design and production process is not a simple task; there is no "magic formula" to make the decisions for us. However, there are some good guidelines and techniques that will assist us in separating out the very bad designs and processes and help us in selecting the better ones.

Worldwide competition is real, and the market in general (and the customer in particular) is very demanding. It will take a team effort with a degree of cooperation that was unheard of a few years ago to design, produce, and market high-quality products at a low price.

We must understand who the customers are; what they want; what they expect the life of the product to be; whether or not they expect to be able to repair it, or have it repaired; when they want it; how much they are willing to pay for it; and what their actions will be if the product does not meet all their stated and unstated requirements.

Some customers want a product at the lowest possible price, that will just meet their minimum requirements for a short period of time, and then they plan to dispose of it. Other customers will want a good-quality product, that will always perform its intended function, and that will have a reasonable life, for which they expect to pay a "fair" price.

The best possible product at the lowest possible price should be our goal in every case. Good design does not automatically translate to a high price. Although the design effort is not without cost, it is usually only a very small percentage of the overall product price—however, it does have a major influence on product manufacturing cost and quality. Poor-quality workmanship in the shop is unforgivable, as is scrap and rework. A lot of our poor quality is due to a poor design—the product is not "producible." This is the kind of reasoning that leads us to believe that the functions of design, manufacturing, quality, marketing, purchasing, and others, must work together during the design and development phases of any product to develop the best product configuration, production processes, and production sources. Perhaps we should not try to build every part of every product in our factory—we should do the things we must, and the things that we can manage to do well, and work in cooperation with qualified suppliers during this early development phase and on into production.

We should never attempt to put a product into production that has not had sufficient environmental testing, in addition to functional testing. This applies to children's toys as surely as to aerospace products. The author had the opportunity to work with the head of R&D of a major toy manufacturer as a consultant while we were in the development phase of a sophisticated weapon system for the U.S. military. We were both surprised to learn that the toys were tested in about the same harsh environments as the military weapon— and that the toy "customer" was more demanding of the product. It was not enough that a toy function the day the child took it home—it was expected to lay out in the snow all winter and still work when the child picked it up the next spring.

We must insist that any manufacturing process continuously produce a good-quality product. I was discussing "quality" a few years ago with a manufacturer of camera flash bulbs. Their goal was never to produce a flash bulb that did not function every time—even after rough handling, and under low voltage conditions. The company once recalled 20 million flash bulbs and replaced them—at no charge to the dealers—because they thought that "maybe" a few of them might be defective due to a variation discovered in the manufacturing process. Their reasoning was that if someone used their flash bulbs to take a "once-in-a-lifetime" photo (and aren't they all?), and the bulb did not fire, the customer would never again buy their flash bulbs—and probably none of their other name-brand products!

We are capable of producing very high quality products at very low cost. Many of us have become a little careless—or short sighted or whatever—in the last couple of decades. We have designed and built products and then tried to sell them. For a long time, we did very well. Now we need to totally reverse the design process: First, what does the customer want; second, how can we make it; and finally, how shall we design this product so it can be built with high quality at low cost? Don't forget the worldwide market (of both competitors and customers).

Perhaps a very short summary of Chapter 1 is in order here to help us select the optimum product and process design. Subchapter 1.1 talks about the concurrent engineering process—where we need a change in our thinking, and perhaps in our company management thinking, in order to be able to combine our collective talents and experience and put them to better use. The "tribal knowledge" that all manufacturing companies possess, is usually a manufacturer's most valuable asset. The important "do's and don'ts" as well as the "how to's" are usually not written—they are scattered in the brains of people throughout the company. Concurrent engineering (by whatever name you call it) is a recognition of this, and an attempt to change from after the fact "I knew that wouldn't work," negative attitudes, to positive attitudes and activities before the fact and during the design and development process. The use of this "tribal knowledge" in a positive, cooperative environment is a *must* for future success in our business. There are some excellent "tools" available, as well as talented people that can help our companies in this effort. The author believes that an approach such as QFD (see Subchapter 1.2), where the customer's wants are reviewed and prioritized, and the technical requirements to achieve them defined in a matrix, is a pretty good way to start. We can see what the product requirements are—as a target, at least. We can also see how our competition does in comparison with our approach to customer satisfaction.

Once we have looked at this "big picture" of product design, we can start looking at some concepts of actual hardware design that will meet the technical requirements we think will provide what the customer wants in a product. The author feels that at this point, assistance from a process such as DFA or DFMA may be necessary (see Subchapter 1.3). We do not want the ultimate product design to be the result of "the loudest voice," who may be very persuasive in supporting his or her favorite design concept. We need to look very closely at the processes that are inherent in each design concept, and analyze the cost elements in a formal manner. This takes considerable skill, since seemingly "minor" or "unimportant" details may become significant costdrivers. However, until they are brought out during the review and analysis process, we really do not know (quantitatively) what the impact of each element of the design will be on the product production. The quality of the final product is also largely inherent in the detail design. We must make certain that the demonstrated manufacturing tolerances are smaller than the product demands.

The author's experience has shown the importance of trying to do things correctly the first time. If there is any question about some part of the product design, consider models, or preferably a prototype part that is truly representative of the final product. By utilizing some of today's rapid prototyping techniques (see Subchapter 1.4), and the companies that specialize in short-lead-time deliveries, we can have a lot more confidence in our proposed design. We should insist on sufficient functional and environmental testing, to prove that the design works under the anticipated conditions. While this sometimes may seem like a waste of time and money in getting a product to market, it is usually not. It may save the commitment of funds for the wrong production tooling, requiring expensive redesign and retooling after start of production—the worst of all situations!

Chapter 2 has introduced some of the more common manufacturing processes. There are advantages and disadvantages to all of them. Again, a rather formal estimating technique is the only approach that is "quantitative" rather than "intuitive." Checklists are valuable triggers, since no one can remember everything all the time. The remaining chapters in this handbook will also be of assistance in influencing the process selection, and the influence of the process on the ultimate product design. They include plastics, composites, nontraditional processing, assembly, factory requirements, factory design, and the human factors of the people involved. Things such as ergonomics, anthropometry, safety, industrial hygiene, and training are important to consider today and tomorrow.

Good luck! (or is it?)

3

Factory Requirements

Vijay S. Sheth   *McDonnell Douglas Corporation, Titusville, Florida*

3.0 INTRODUCTION TO FACILITIES REQUIREMENTS

A factory is one of the largest fixed assets that a manufacturing firm owns. According to the Internal Revenue Service, a building has a normal life of 40 years, and therefore $\frac{1}{40}$th of its cost becomes part of your depreciation expense every year. The equipment that is inside the building probably costs more than the building itself, and although it is depreciated over a shorter period of time, it is also a big part of your overhead cost. Buildings and equipment must be maintained in good repair. You could have the best product design—and the best production line—in the world, and still not be successful in business. The facility itself is a large part of the bottom-line equation. This might indicate that the smaller the facility, and the less expensive and simpler it could be—the better for the total enterprise!

Figure 3.1 shows such an "efficient" factory in Poland in the early 1900s. There was little concern for the employees; there were so few jobs that people would work hard under the worst of conditions for whatever wages were offered. The competition for jobs was such that workers developed very good individual skills in order to keep their jobs. The quality of the product produced was not really much of a concern, since it was probably as good as anyone else around was producing. In fact, the products were usually consumed locally, and sold at whatever price was necessary in order for the factory to make money. There was little in the way of competition for the product, and not much concern about inventory levels. The raw materials were quite simple, and low in cost. There was no need to be too concerned about reducing costs, as long as sufficient product came out the door to satisfy the market demand. Since labor was so cheap, there was no need to spend a lot of money for exotic tooling, and inspection as such was unheard of. One supervisor, usually the largest and "most positive" of the experienced workers, was sufficient to make sure that everyone was working. There was almost no government regulation to worry about. Waste was dumped out back, or in the river, and forgotten. Cost control and accounting procedures were minimal. Even then, however, there were two kinds of costs: those that added direct value to the product, and those that were necessary but did not add directly to the product value.

Now consider the world we live in today—and the one we will live in tomorrow. Life is more complicated, and running a successful factory today—and tomorrow—is a lot

137

Figure 3.1 An "efficient" factory in Poland in the early 1900s.

more complex. There is worldwide competition, both for products just like ours and for similar competing products. The customer holds the reins, and drives us to produce products of very high quality, with short delivery times, utilizing exotic materials and processes, all at competitive prices. In this environment, we need to improve our existing products constantly, and to develop and produce new products with a short lead time. We have rules and regulations on every imaginable subject, with fines and other penalties hanging over our heads. The majority of our workforce is not highly skilled, and in many cases not especially motivated or loyal to the company. Employees demand and expect benefits such as high wages, excellent working conditions, good tools to work with, training and instructions, health insurance, retirement benefits, and so on.

So—what is our plan? We must have a good product line, and a well-designed product that is capable of being produced with high quality and at low cost. Let us assume for the moment that we have accomplished this important step (see Chapters 1 and 2 on product development and design). We now need the facilities to produce this product effectively, at low cost and with high quality. We must furnish our workers with a safe, environmentally correct workplace, and furnish them the best tools and equipment that is available (within reason, after some accurate trade-off studies). We need to establish goals and standards for all operations on the product production floor, and have real time feedback on our status—and especially any problems that might come up. We need to know what our material requirements are, and where the material is. It may be in the warehouse, in stores, in transit, at a supplier, on the production floor waiting—or actually in the factory being processed as planned. We need a way to move the material, and the product in process, through the factory. We must keep our inventories of raw materials, work in process, and completed product to the very minimum. This says that we must have a good factory layout, the correct equipment and tooling, and a good flow through the factory for both the product and the materials. Our information and control systems must provide up-to-the-minute information about our total enterprise operation. Our systems and procedures for operating the factory must be correct and disciplined. The same goes for all the other "non-value-added" effort and costs.

What are the factory requirements that establish this "world-class factory"?

What kind of buildings do we need? (Or can we use the ones we have?)
How much floor space do we need for the production of our products?
What special features are required in the production buildings?
    Floor space
    Ceiling height
    Structure
    Floor strength and rigidity
    Enviriomental controls (ventilation, temperature, lighting, etc.)
    Safety considerations

What do we need to consider for material?
    Receiving and shipping (single point, or deliveries direct to various places in the production line?)
    Space for storage (warehouses, stores, on the production line, or at suppliers?)
    Transportation of materials to the work place (conveyors, AGVs, hand trucks, fork lifts?)
    Movement of material in process (hand pass, conveyors, handtrucks, forklifts, cranes?)
    Total accountability for all material (what, where, how much—versus need.)

What space do we need for other supporting functions?
    Offices (how many, where located, how fancy?)
    Meeting rooms
    Communications support (central mainframe computer, network distribution, telephones, video conferencing, etc.)
    Laboratories (quality, materials and processes, testing, etc.)

Fortunately, this complex task can be broken down into manageable pieces, and tools and techniques are available to assist us in the analysis of each of the elements of the factory. We will ensure that the facility and equipment is not only adequate to compete in the marketplace today, but will be flexible enough to permit us to continue in the future. This is the subject of Chapter 3. The emphasis is on the planning approach to facility requirements, layouts, and implementation in general, not on the mechanics of site selection, plant design, contracting, and installations, which are covered in Chapter 4.

3.1 GENERAL FACTORY REQUIREMENTS

Many people think that a manufacturing building is sortof like a warehouse. It has a floor, walls, and a roof. If we add some lighting, and perhaps a restroom or two, we will be ready for production. This makes one visualize a rectangular steel frame sitting on a concrete slab floor. Across the top are standard steel bar joists and a simple roof. The walls are prefabricated steel or aluminum panels, attached to stringers or purlins on the standard post-and-beam wall construction. This may not be a bad way to start our thinking. After all, we just want to do some work and make money—with the lowest possible fixed investment and operating costs!

Now back to the real world of today—and tomorrow. In our original thinking about the factory, we would probably need to do some site preparation (after having selected the site), including roads, drainage, parking lots, a loading dock, perhaps some security fencing, regardless of any other special conditions known at the time. We could continue this study for a while, and specify some landscaping out front, an office for the boss, water, telephone lines, and power lines coming in, and a sewer going out. However, we have only set the stage to ask some pretty basic general questions.

How strong should the floor be, and where should it be stronger or have a better base prepared underneath it?

How high should the walls be (what ceiling height is needed) and where?

What kind of loads should the roof support? Perhaps you had in mind an overhead crane system, or air-conditioning units mounted on the roof.

How about the walls—will we need insulation, or a finished interior?

Will there be a mezzanine, either now or later?

How much square footage, and cubic volume, are needed?

How many people will work here? How many in the shop, and how many in offices?

What kind of lighting levels are needed in the shop, and in the offices?

Does it look like a long, narrow building, or is it more square in shape?

Will we need multiple loading and unloading docks? How many need adjustable height provisions, or dock levelers of some sort?

Is there a limit on how close together the internal supporting columns for the roof can be?

How fancy should the offices be? How many will have full-height walls and ceilings, and how many can just have privacy partitions?

You can see that there are going to be a great number of "general specifications" needed, whether we remodel an existing building or build a new one. In any case, there will need to be some architectural and engineering (A&E) work, even at this early planning stage. The type of questions asked above, although very basic, will have a great influence on the cost of construction, and the time required to do the work. For example, how much are we willing to spend to develop an image for our company? This includes the facade and landscaping, among other factors.

3.1.1 Special Requirements

In addition to the "simple" questions above, there will be some more specific general requirements that must also be defined. The question of exhaust vents into the atmosphere is becoming a big design problem today. The exhaust may come from a paint shop, or a chemical processing line, or dust from a grinding operation, or whatever. The same goes for all types of waste disposal, especially anything considered "hazardous" by one of the environmental agencies. Even such things as a chemical sink in the materials laboratory needs consideration. If there are any types of cryogenics involved—and the use of CO_2 for cooling is quite common—they require special lines, etc. The problem of how clean your compressed air needs to be—or how dry—can make a difference in the plumbing and piping, as well as the compressors, dryers, etc. Your communication systems may require a "backbone" down the entire factory, with specified drops or closets which are much more cost effective if installed or provided for in the initial building plans. Some existing plants have actually installed radiofrequency transmitters at each computer workstation, as being more cost-effective than installing hard lines in an existing building. Fire protection may turn out to be a problem, or at least worth consideration in the general planning of the factory. Storm sewers and retention ponds on site are a requirement in some localities.

These types of decisions are part of the overall facility requirements, in addition to providing for the production process and material handling, which are usually thought of as the main drivers in factory planning. In todays' world they are not the only drivers.

3.2 REQUIREMENTS OF THE MANUFACTURING PROCESS

The manufacturing process is the primary basis for the factory requirements. It is essential that the process be thoroughly understood and revised if necessary for ease of material handling and manufacturability. For instance, it may be possible to combine operations, to modify the process to suit some existing equipment, or to alter the sequences so that a common line can be set up for multiple products.

The material going through the manufacturing facility is either waiting, in motion, or being worked on. The only productive time is when it is being worked on, and this may be only a fraction of the total time it spends in the plant. See Figure 3.2.

The process can be visualized by constructing a process flow chart which shows each operation, move, and any delays in the process. By analyzing various elements of the process flow chart, it is possible to layout a plant to minimize unproductive elements. Once requirements of the plant layout are clearly understood, whether for a new facility or for a modification of an existing plant, it is necessary to work on the details which help in defining the factory requirements, including any buildings and the preliminary layout of the factory.

In many instances, the process starts by understanding the manufacturing process for the product manufactured. Sometimes, depending on the existing plant and equipment, it may be beneficial to change the process before further work is done on the facility

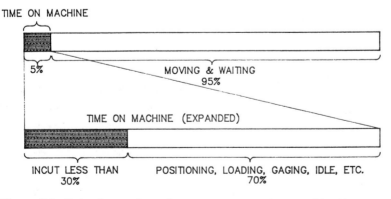

Figure 3.2 In batch-type shops, the average part spends most of its life in the shop just waiting to be cut, or waiting to be moved. The results: great work-in-process inventories, very slow through-put rates. (Courtesy of Cincinnati Milacron, Cincinnati, OH.)

planning. Many of these decisions involve capital investment considerations and a commitment to the process which is economical in the long range. This activity, the manufacturability of product, is often accomplished concurrently with the product design to save time and cost (see Chapters 1 and 2). However, it is essential that the manufacturing engineer understand the importance of the process required by the product as designed, and the available resources. Many decisions, such as regarding storage, handling of materials, equipment availability, manpower and support services are necessary during the process of laying out the facility.

3.2.1 Block Plan

Imagine designing and setting up a 50,000 ft^2 plant involving over 200 pieces of different equipment linked to produce an optimum flow. This would be a horrifying task if one were to start at the beginning with a piece-by-piece placement of equipment. However, there is a simpler and more logical approach to this problem in combining tasks or equipment into major functional groups. Simplification can also be achieved by dividing plant activities into smaller functional groups. In order to do this one must have a knowledge of process details, material flow, and the amount of support activities required. Once this information is obtained, it is possible to locate the different groups in the layout with their relationship to each other.

Block Layout Approach

One approach to layout which shows the outlines of activities or departments is known as a block layout, or block plan. For instance, a block layout for a department store has products assigned to different areas such as Mens, Ladies, Cosmetics, Kitchenware, Appliances, and Toys. It is usual practice to prepare a preliminary *block layout* before the detailed factory layout is prepared. See Figure 3.3 for an example of an early block layout.

Block layouts are used to provide preliminary information to other groups in the factory and to the A&E's involved in the construction of a new facility. A preliminary assembly flow chart as shown in Figure 3.4 may be about the level of detail available at the time the preliminary block plan is prepared. It will be based heavily on previous experience, and will be the baseline for preliminary work by the A&E. A block layout showing special requirements such as processed steam, chemical processing, storage, and overhead equipment will be required to address the key issues of facility planning. The preliminary block layout will change as more detailed analyses are performed. Early considerations must include the following:

> Size of the facilities: Minimum or maximum practical size for operating cost, organizational effectiveness, and risk exposure.
> Lease, buy, or build: A review of the alternatives available from architectural designs based on the block layout provided. This will also provide cost estimates for management review and perhaps a major shift in direction from original goals, based on financial commitment.
> Operating requirements: This consists of details regarding flexibility and expansion possibilities. It provides alternatives for material handling and storage requirements.

Requirements for preparing a more complete block layout include the following:

Square feet required by each block
Relationship of blocks to each other

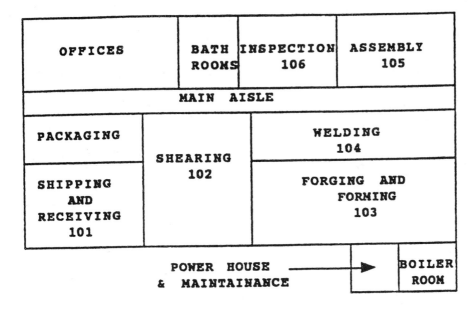

Figure 3.3 Preliminary block plan based on experience with similar product production.

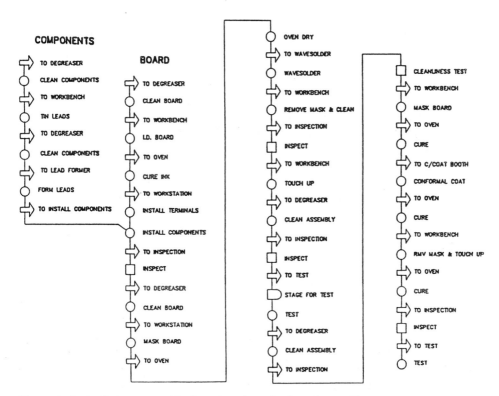

Figure 3.4 Preliminary assembly flow chart for a circuit card assembly.

Shipping and receiving dock locations
Department requiring the use of hazardous material
Special utility requirements
Plant service requirements, e.g., toolcribs, stores, maintenance, engineering support
Emergency exits
Main aisle locations

Block Layout Elements

It is important that a balance be achieved between the total square footage required by blocks, service areas and aisles, and the available square footage in the building. This is the first step in determining that all the activities to be housed in the building are provided for. The following are typical details that are summarized early in the facility planning process:

Department/function	Square footage required
Shipping and receiving	3,600
Storage	10,800
Square shear	9,800
Roll and swage	22,000
Flame cut and weld	15,000
Assemble	22,900
Clean and pickle	17,800
Paint and plating	15,000
Packaging	6,000
Support offices	20,500
Main aisles—10% assumed	14,400
Total	157,800 Square feet

If the required square footage is more than is available in the existing facility, a decision must be made as to how to provide the additional square footage. This might mean an extension to the existing building for some of the services, construction of a mezzanine to take advantage of the building cube, or a new building or buildings.

Notice that the square footage values are rounded to the nearest hundred square feet and the aisle requirement is established as an approximation based on experience. Questions still remain as to how to establish a more accurate square footage for each department and how to arrange the blocks. For this, we must have more knowledge of the process and production requirements. Some of the specific elements needed are shown below, followed by a more detailed description:

Process routing sheet
Flow process chart
Relationship chart
Trip frequency chart

Process Routing Sheet

One important document used in the plant by production workers, supervisors, and schedulers is the process routing document. Figure 3.5 shows a basic form of routing as an assembly process flow. This document, prepared by manufacturing engineers, depicts the best and the most practical method of processing incoming material to obtain the desired product. In the case of a new product manufactured in an existing facility, it is

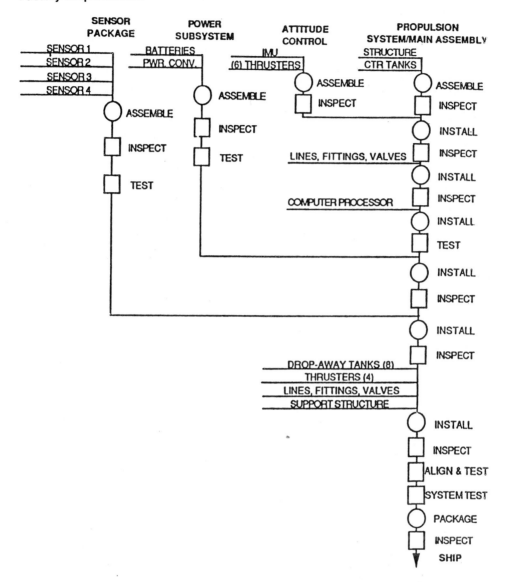

Figure 3.5 Assembly flow chart.

generally prepared to take advantage of existing resources which include equipment, the material handling system, storage, and the skill of the personnel. In the case of new facilities, the manufacturing process and layout require a thoughtful approach and heavy involvement of management. For instance, a part can be machined on a low-cost mill or on a high-speed, computerized numerical-controlled machine with an automatic tool changer, costing several hundred thousand dollars.

Flow Process Chart

A flow process chart (Figure 3.6) is developed for each component and subassembly in the manufacturing of the final product. It is a descriptive method in chart format to show

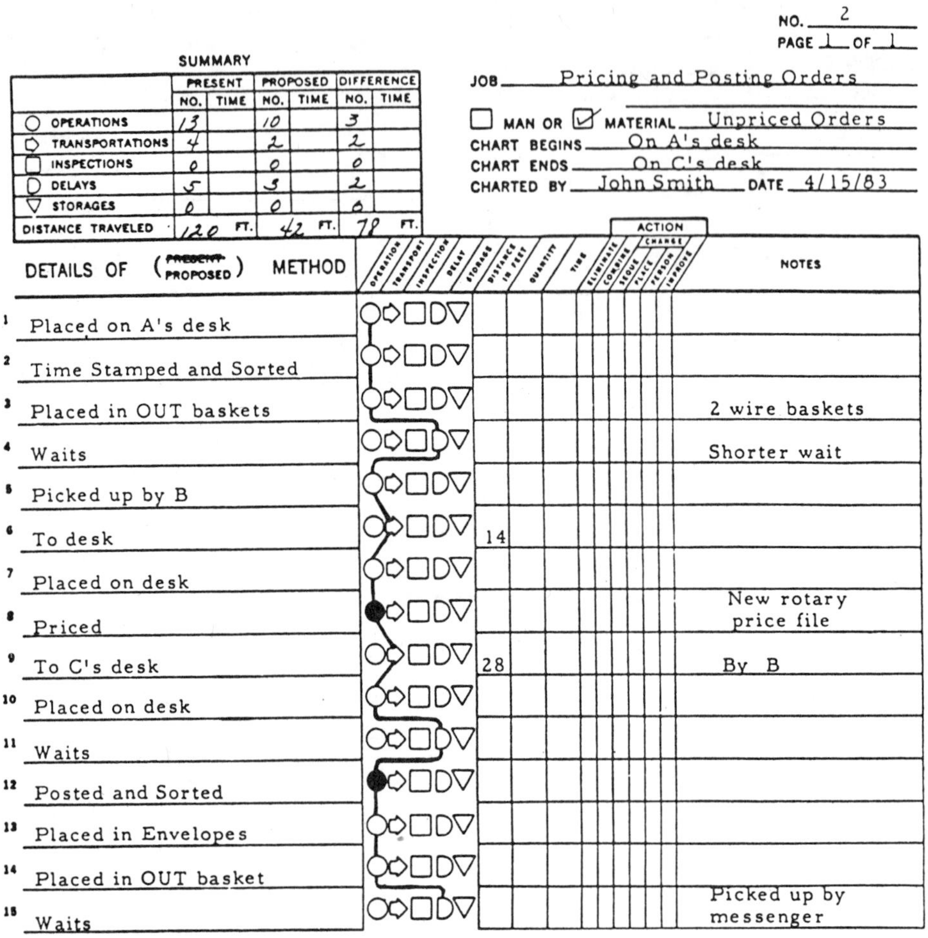

NO. 2
PAGE 1 OF 1

SUMMARY	PRESENT		PROPOSED		DIFFERENCE	
	NO.	TIME	NO.	TIME	NO.	TIME
◯ OPERATIONS	13		10		3	
⇨ TRANSPORTATIONS	4		2		2	
☐ INSPECTIONS	0		0		0	
◖ DELAYS	5		3		2	
▽ STORAGES	0		0		0	
DISTANCE TRAVELED	120 FT.		42 FT.		78 FT.	

JOB   Pricing and Posting Orders

☐ MAN OR ☑ MATERIAL   Unpriced Orders

CHART BEGINS   On A's desk

CHART ENDS   On C's desk

CHARTED BY   John Smith   DATE   4/15/83

DETAILS OF (PRESENT/PROPOSED) METHOD	ACTION	DISTANCE IN FEET	QUANTITY	TIME	NOTES
1 Placed on A's desk	◯⇨☐◖▽				
2 Time Stamped and Sorted	◯⇨☐◖▽				
3 Placed in OUT baskets	◯⇨☐◖▽				2 wire baskets
4 Waits	◯⇨☐◖▽				Shorter wait
5 Picked up by B	◯⇨☐◖▽				
6 To desk	◯⇨☐◖▽	14			
7 Placed on desk	◯⇨☐◖▽				
8 Priced	●⇨☐◖▽				New rotary price file
9 To C's desk	◯⇨☐◖▽	28			By B
10 Placed on desk	◯⇨☐◖▽				
11 Waits	◯⇨☐◖▽				
12 Posted and Sorted	●⇨☐◖▽				
13 Placed in Envelopes	◯⇨☐◖▽				
14 Placed in OUT basket	◯⇨☐◖▽				
15 Waits	◯⇨☐◖▽				Picked up by messenger

Figure 3.6 Flow process chart.

the flow of material through various process activities. These activities have been coded by the American Society of Mechanical Engineers (ASME). An explanation of these activities, which are also represented by symbols in the flow chart, follows:

O *Operation* (to change form): An operation is performed when material is intentionally changed in its physical or chemical form to obtain a desired result. This is preceded and followed by one or more of the other four classifications as described below.

⇒ *Transportation* (to move material): When the purpose is to move an object for a specific reason, transportation occurs. Many times transportation is combined with operation for savings in time and space. In this case, it is considered to be an operation activity. Transportation is identified separately in the flow process since it signifies: (1) the need for handling and storage methods and equipment and (2), a non-value-added function which should be minimized or preferably eliminated.

☐ *Inspection* (to verify): An inspection occurs when the objective is to verify quality or quantity separately from operation activities. Here, the only objective is to inspect. Frequent inspections performed by the operator during machining operations are not separate inspection classifications, but rather are a part of the operation.

D *Delays* (to wait): A delay occurs to an object when it is not possible to perform the operation, move, inspection, or store function intentended. Here the object is simply waiting for the next action to process it further toward its completion.

∇ *Storage* (to hold): Here an object is protected and held until further action is warranted. The material is kept ready for the move condition.

The productive time usually occurs only under the operation and inspection classifications. The others are contributors to material handling and storage costs, which should be kept to a minimum. This method is a very useful tool in determining handling requirements, process equipment, and possible process improvements based on the ratio of productive time to total time required to complete the process. The column marked "notes" can be used to estimate the time required to perform the corresponding activity (except storage), and also to identify any special requirements such as equipment, tooling, storage, etc. This information is vital to the layout, which in turn will help establish the size of the building requirements.

Reasons for Flow Process Chart

To understand the overall nature of the system being studied
To eliminate costly errors by properly analyzing the material flow
To allow adequate space to avoid safety problems
To eliminate flow patterns that are not suitable, resulting in additional handling
To allow storage space adequate to support the production rate
To locate and size aisles appropriate for product handled
To show the possibility of combining operations by grouping machines
or operations to avoid extra handling, storage, and delays
To avoid cross traffic and backtracking of the material in process
To decide whether product flow or process flow concepts of factory flow will be adopted

A flow process chart can be constructed by using a process routing sheet (see Figure 3.7) which provides details of processing the product. If the factory requirements must be established prior to the final design of the product, as often is necessary, preliminary process routing sheets can be prepared during the product design process. They are usually needed in any case during design evolution, in order to help select the optimum product design.

Relationship Chart

A relationship chart is extremely effective where product or traffic patterns cannot be established easily. Figure 3.8 shows a relationship chart using the following steps.

1. List all departments or activity centers in the blocks shown on the left.
2. Using appropriate closeness ratings from a value box, assign closeness ratings in the top half of the blocks intersecting two departments or activity centers.

ROUTING

Part Name	Blade Piston	Models	AD7562-A1	Part No.	6510777

Material	Bronze Forging A29	Date		No. of Sheets	1	Sheet No.	1	Issue No.	9

OPERATION	EQUIPMENT	Oper. No.	Dept. & Group	STANDARD TIME			No. Men	Group Standard Hours
				Regis. No.	Minutes	Hours		
Face, turn, chamfer & form groove in small O. D. (#5 W & S)		27A	44A3	2776		.1530		.2202
Face & turn large end (#5 W & S)		27B	44A3	2806		.0672		
Inspect & credit Group 44A3		24	33	Ind. Labor				
Drill (2) 21/64 holes & (2) 7/32" holes & drill burr (4) holes (4 Spdl. L. G.)		13	44B1	3539		.0730		.0730
Inspect & credit Group 44B1		24	33	Ind. Labor				
Mill slot 3/32 x 7/64" deep (Kent Owens)		31A	44B2	2883		.0237		.0237
Inspect & credit Group 44B2		24	33	Ind. Labor				
Mill threads (Lees Bradner)		31B	44B5	2897		.0721 (1 Oper. 2 Mach.)		.0721
Inspect & credit Group 44B5		24	33	Ind. Labor				
Burr slots, groove, & threads (Burr Room)		7A	44D	3255		.0330		.0467
Scratch brush slotted end (Burr Room)		7B	44D	3389		.0090		
Wash (Burr Room)		10	44D	3390		.0047		
Inspect & credit Group 44D		24	30	Ind. Labor				
MOVE TO BOND ROOM								
(Used on Piston Assembly - 6500386)								
	TOTALS							

Figure 3.7 Process routing sheet, showing standard times, that can be used to prepare a flow process chart.

3. Assign an appropriate code from the reason for the closeness block in the bottom half of the block corresponding to the closeness rating.

It is important to understand the functions of all departments and appropriately assign frequencies and closeness values between any two departments. Many times, inconsistency arises when two departments rate their relationship with each other differently. One way to avoid this inconsistency is to have one person who is knowledgeable about functions provide ratings for each pair of departments. Additional charts can be prepared between the factory floor departments, or within a department.

Trip Frequency Chart

The travel pattern between departments or activity centers can be shown by a chart such as the one in Figure 3.9, which can help in providing valuable data for location of the blocks. (Refer back to the original block plan in Figure 3.3.) The following steps will aid in constructing this chart.

1. Construct a matrix showing departments on both the left and the top of the matrix. Each department represents a block to be shown on the block plan.
2. Estimate the number of moves required by the material handlers to support production in an average shift or day between the two departments. e.g., eight trips are required to deliver material from Department 103 to Department 104.

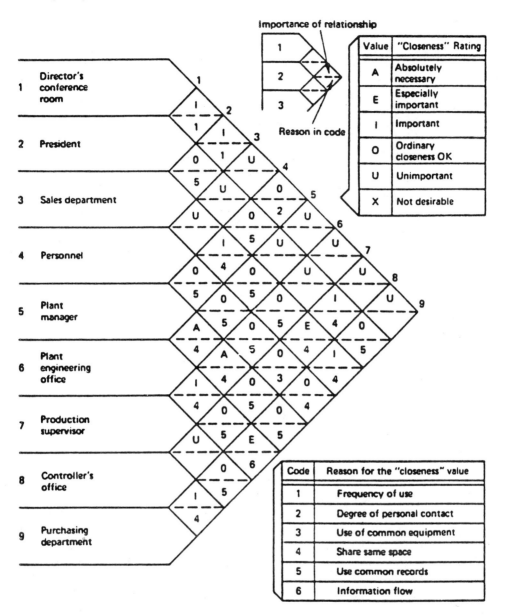

Figure 3.8 Relationship chart.

3. The higher the number, the closer the departments (blocks) should be located to reduce traveling distance, time, and downtime. The other advantage is that the material traffic is not going through unrelated areas, thereby eliminating safety problems and congestion.

The numbers shown in each block represents the relationship of one specific department to another for the movement of goods. The blocks are filled by analyzing product flow and tallying the number of times. Material movement is required over a period of one

To From	101	102	103	104	105	106
101	–	15	4	3	4	2
102	9	–	12	3	4	7
103	0	7	–	8	1	0
104	0	2	1	–	12	2
105	0	0	1	9	–	10
106	1	0	0	2	8	–

Figure 3.9 Trip frequency chart used to establish travel patterns between departments.

shift or one day. The greater the frequency, the more significant is the requirement to locate them closer to each other. A carefully planned block plan can save operation costs in handling equipment and overhead support labor. There are also other advantages—for example, savings in aisle space by providing wide aisles only between blocks where necessary.

Other Elements in a Block Layout

Flexibility in facility planning refers to allowances required for uncertainties. For instance, a specific department cannot be laid out due to unresolved questions on the method of material handling or not knowing the stage of production that the material is received from the supplier. Sometimes, there is flexibility in routing a product in more than one way between departments. If these variables are unanswered during the block layout, it is necessary to take precautions and allow for these uncertainties. Product rework should be allowed for.

The expansion is always difficult to plan unless one has a definite idea of what, how much, and when it is required. For an existing building which has more area available than the sum total of all blocks, it is possible to pro-rate additional square footage to the blocks which are apt to be affected by the expansion. A later version of the block plan is shown in Figure 3.10.

3.2.2 Requirements for Preparing a Plant Layout

Plant layout is one of the most important phases in any facility planning and material handling project related to either changes in existing products or new product manufacturing. It depicts the thoughtful, well-planned process of integrating equipment, material, and

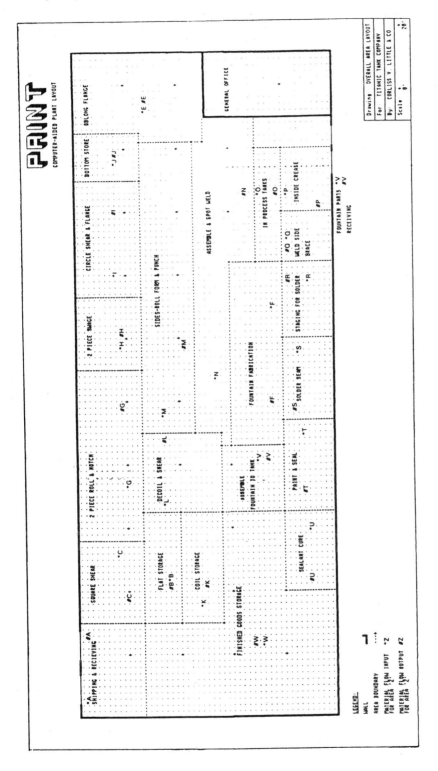

Figure 3.10 Later version of a block layout with pick-up (*) and set-down (#) points, developed using the cursor on an Exidy Sorcerer computer. Distances are computed automatically and provided to the scoring routine. (Courtesy of Corliss V. Little & Co., Raytown, MO.)

manpower for processing the product through the plant in an efficient manner. This means that the material moves from Receiving (as raw materials and parts) to Shipping (as finished product) in the shortest time and with the least amount of handling. This is important, since the more time the material spends in the plant, the more costs it collects in terms of inventory, obsolescence, overhead, and labor charges.

Plant layout is the beginning of a footprint of the actual plant arrangement to obtain the most efficient flow of products. See Figure 3.11 for some examples of desirable and undesirable flow patterns. Plant layout is also the basis for determining the cost of the facility in the case of new construction or alterations. Sometimes it is useful to work the problem of facility requirements in reverse. For example, if the final output of the studies is a layout of the factory (new or modified existing), how do we collect the information needed in order to complete the layout? The following tools, techniques, and other elements will help in our analysis and factory requirements definition.

1. Assembly flow chart
2. Flow pattern of product
3. Production rate and number of shifts
4. Equipment configuration
5. Flooring preparation
6. Utilities
7. Store arrangement
8. Shipping and Receiving
9. Material handling methodology
10. Support functions
11. Work-in-process storage
12. Safety and emergency evacuation

A: DESIRABLE FLOW PATTERNS

B: UNDESIRABLE FLOW PATTERNS

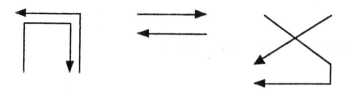

Figure 3.11 Examples of desirable and undesirable flow patterns in a factory.

Following are brief descriptions of some of these factors.

Flow Pattern of Product

Plant layout is not simply an arrangement of machines, departments, and services best suited to the physical dimensions of the plant. It is a carefully thought-out plan for installing equipment as the product smoothly follows the process as determined by the process flow chart. It is often difficult to obtain an ideal condition in an existing plant in which a new product is introduced because of several reasons. Some of them are:

Cost to rearrange equipment is prohibitive
Plant configuration does not allow for installation of the most effective arrangement
More than one product line shares common equipment
Incremental addition of machinery is not strategically located due to lack of space
Change in process sequence after the installation is completed

Process Flow or Product Flow?

One important consideration in layout is whether the plant arrangement will be based on a process or a product flow. This decision is based mainly on the characteristics of the product and its production rate. The basic difference is in how the product flows through the plant during manufacturing. It has implications for the initial investment in equipment, the size of the facility and future requirements. Figure 3.12 shows some of the differences in these arrangements. In process flow, similar functions to be performed on the product are grouped together. The functions can be by either machines or tasks. For example, all drill presses may be grouped together, or all manual packaging performed at one location. With this type of arrangement, various products manufactured in the facility travel haphazardly depending on which function is to be performed on it next.

In product flow, one product is processed independently of another in the same plant. In this case, all functions required to be performed on the product are so arranged as to obtain a smooth flow of material with the least amount of downtime and handling.

Process or product flow can be arranged either for machines or for departments. Figure 3.13 shows the department arrangements. In process flow, all similar machines are arranged together, which is not the case in product flow. It is not uncommon in a job shop to find all drill presses standing side by side, and then all the lathes, the milling machines, the grinders, and so on. Process flow plants usually have single-function departments, such as packaging, inspection, painting, or chemical processing, for all products.

In product flow plants, where more than one product may be manufactured, there will be some duplications of functions. This type of arrangement is most effective when the product is mass produced. Common examples are manufacturing facilities for appliances, automobiles, garden equipment, and housewares.

Advantages of Product Flow Arrangement

Material moves progressively in a definite pattern from machine to machine or from one department to another, causing less damage to material.
Workers involved with the handling of the product experience less confusion, since only one product is involved. They attain a degree of expertise in processing the product.
Material is handled efficiently. There is less possibility of a mixup with parts from other products.

OPERATION SEQUENCE

Part	Oper. 1	Oper. 2	Oper. 3
A	lathe	drill	lathe
B	drill	mill	
C	lathe	mill	drill

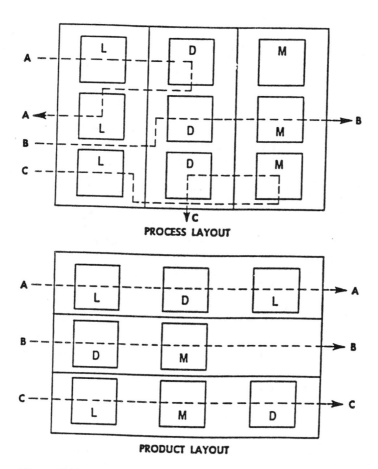

Figure 3.12 Influence on plant layout based on process flow or product flow through the plant.

Processing time is greatly reduced, since equipment and workers are assigned to a particular product.

Less work-in-process inventory is required.

Inspection can be reduced, since workers become proficient in one part or a function. Further, the part is frequently tried out on the very next operation, thereby eliminating the need for inspection.

Delays are reduced, since there is no schedule conflict with other products.

Production control tasks become simpler, since only one product is made in each

PRODUCT	FIRST SEQUENCE	SECOND SEQUENCE	THIRD SEQUENCE	FOURTH SEQUENCE
X	Forming (F)	Machining (M)	Chem. Process (CP)	Assembly (A)
Y	Machining	Assembly	Chem. Process	Forming
Z	Forming	Machining	Assembly	Chem. Process

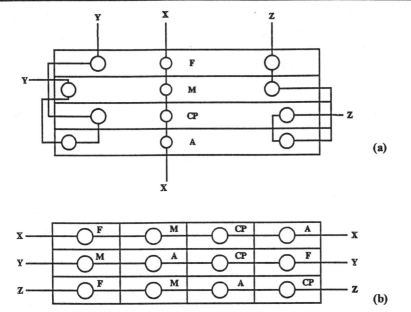

Figure 3.13 Flow patterns within a department for the same product: (a) Process flow; (b) product flow.

production area. This greatly reduces paperwork such as move tickets at every stage of production.

Material staging areas between machines can be reduced, since there are fewer delays and the product can be moved as a continuous flow instead of a batch.

Investment in material handling equipment can be reduced, since the product does not have to be moved in batches over a long distance, thereby eliminating the need for bigger or faster equipment.

It is easy to train workers, since only one product is involved. Supervisors can be trained easily for the processing of just one product.

Disadvantages of Product Flow Arrangement

Some duplication of machines is necessary, which requires high capital outlays.

It is difficult to justify high-speed, automatic machines if product volume is low.

Machines are not used to their fullest potential—in terms of both capabilities and utilization.

If specialized machines are used, they cost more and become obsolete if production of the product ceases

Balance of the production line is difficult. As a result, labor utilization may drop.

Any changes in processing of a part may cause major changes in the line layout, balancing, and equipment.

Major breakdown of a machine may idle the entire line.

Specialized material handling methods are used. They are costly and of little value if the product is discontinued.

Workers exhibit special skills and have to be retrained for different assignments.

Manufacturing cost is higher if the product line is not moving at full planned capacity.

Advantages of Process Flow Arrangement

Flexibility in scheduling is available, since a number of machines are available in the same group.

Product flows in batches. This simplifies handling and reduces the frequency of moves.

Machines are of a generic type and are better utilized. This reduces initial capital investment.

A single machine breakdown does not affect the schedule much, since the product can be routed to other, similar machines in the same area.

If specialized handling equipment is required, it can be shared with other products with some modification.

Workers are skilled to run any equipment within the group. This provides flexibility if an assigned worker is not available or quits. This eliminates delays waiting for a worker.

Supervisors become proficient in the process functions they supervise. They can help troubleshoot problems related to setup and quality without much outside help.

When a new product is introduced in the plant, it can be processed more efficiently initially, since most of the equipment is already in use.

Overtime schedules can be avoided, since flexibility exists with similar equipment.

Breakdown of one machine does not necessarily hold up successive operations if part can be routed differently.

Layout is less susceptible to changes in product mix over the long term.

Generally suitable for small production runs.

Disadvantages of Process Flow Arrangement

Material takes a long route and may be subjected to loss or damage.

Production control has to closely follow up and coordinate the required date of the product at each function.

Main aisles and aisles between the machines have to be wide for transportation of material in bulk.

A large number of rejection of parts results if the error is not caught until a subsequent assembly operation.

Delays can occur when parts obtained from previous operations have to be scrapped, and replaced by new ones from the very beginning of the line.

Production Rate and Number of Shifts

It is usually not possible to plan facilities without knowing the required rate of production. This requirement is established initially by management in order to facilitate not only the planning of the facility but also the commitment to capital outlays.

Equipment Configuration

Before plotting a layout, it is necessary to have all the physical characteristics available for the equipment to be housed in the plant. The following are the details required:

1. Size of the equipment: length, width, and in some cases height, to make sure that equipment can be placed under a low ceiling height or under the mezzanine.
2. Operator working area in relation to the equipment.
3. Access for material handling and storage equipment, such as the forklift, containers, and racks.
4. Storage for parts which are completed and waiting to be processed.
5. Storage for fixtures and tooling which are stored near the equipment for ease of handling.
6. Inspection table by the equipment if required.
7. Maintenance access from the sides and from behind the equipment.
8. Special characteristics of machines. A large press may require a foundation separated from the rest of the flooring to make sure that the vibration generated while operating the machine does not affect other areas where precision operations such as inspection, spot welding, or automated electronic assemblies are performed. If this is not taken into consideration, then quality of production may be affected in other areas when such tasks are performed simultaneously with press operations.
9. Pit requirements for equipment or for services such as an underground chip conveyor. It becomes expensive to construct a pit after the flooring is poured, because rerouting of underground utilities and drains may be necessary. In the case of existing facilities, if the details of underground utilities are known, it is advisable to consider alternative plant arrangements which do not require costly rerouting of the utilities.
10. Location of the controller and machine controls in case of CNC equipment or other special equipment when a separate operator control panel is provided with a flexible conduit. This provides flexibility in locating the panel to obtain smooth material flow from and to the machine. See Figure 3.14 for an example of the actual floor space required for a lathe and the supporting elements in a cell arrangement.
11. The material handling method is of concern in the case of heavy product or tooling. If a bridge crane is necessary, then obviously the equipment must be located in the area covered by the bridge crane. If a floor-mounted jib crane is required, then it is necessary to locate a jib crane column in the right relationship to the equipment and parts containers. Think of the situation in which a jib crane requirement is not considered in the initial layout. If there is not enough room, then it becomes expensive to relocate the equipment once the requirement surfaces. In the case of new facilities, it is also necessary to specify the load requirement of the bridge crane as part of facility planning so that the building

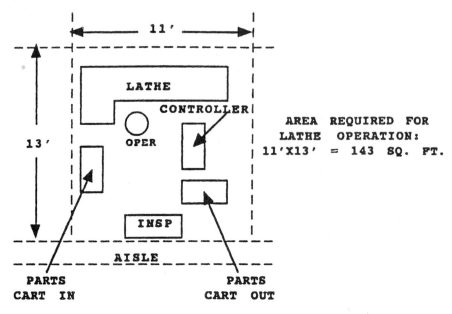

AREA REQUIRED FOR
LATHE OPERATION:
11'X13' = 143 SQ. FT.

Figure 3.14 Computation of functional area requirement for a cell or equipment which includes materials staging and secondary operations.

structure is designed to support the bridge crane load in addition to the building and the roof.

12. The upgrading plan for the equipment and material handling system or the production rate increase in the foreseeable future must also be considered. This is usually the case when a new product is in development, financial resources are limited, and there is uncertainty in the business. Many government contractors in the defense industry delay large investments in productivity improvements until a new product is developed and tried out. In many instances, the layout is revised when it is decided to go from the developmental phase of a new product to the production phase. However, a good estimate must be made when designing the new building.

13. Weight of equipment. Some processing equipment must be massive in order to handle heavy products such as steel fabrication. Not only must the flooring be able to support the equipment load without buckling or cracking, but also the other floor should be able to support the weight of equipment being moved inside the plant for replacement or relocation.

Machine Capabilities

The production rate desired in a plant is a function of the total output required from the plant. Suppose that the production schedule is set for 320 units per day. In order to determine the floor space requirement, one must know whether the plant is scheduled for one shift or two shifts. Further, what is the capacity of the equipment through which the product is processed? If a machine's capacity is 25 units per hour, then only 200 units can be produced in a given shift. There is a choice of either scheduling a second shift or installing a second, similar machine to obtain the same output.

On the surface, this situation may seem to have a minor effect since the labor cost is not affected. However, there are other concerns:

Investment in the second machine
Floor space occupied by the second machine
Maintenance of the second machine
Supervision for the second shift
Support services for the second shift
Future capacity requirements

Floor Requirements

In a new facility, it is necessary that some of the known requirements of the utilities be planned to be underground to avoid overhead exposure and high maintenance costs. For example, if the process water usage is known, underground sewer lines can be provided that can eliminate expenses for holding tanks and pump stations. Similarly, if the departmental process layout shows heavy usage of electrical power in a certain area, it is possible to lay cables underground. Well-planned utilities underground may result in savings in maintenance cost, valuable floor space, and expensive service equipment requirements, although the A&E design work will assist in making the final decisions.

Flooring requirements also include ramps for ingress and egress to the facility with forktrucks. The shipping and receiving dock must also be planned with cavities for a dock leveler to be installed.

Sometimes it is a challenge to accommodate new requirements in an existing plant. Simply knowing the size of the equipment without supporting details can lead to disaster when the actual facility work begins. The new equipment requirement must be fully supported by the layout and the various details to help the facility engineer in designing and planning the installation work.

Utilities

The basic utilities, such as water and electric power, are usually available in plentiful quantities. However, there are processes which demand high usage of both. For example, a totally airconditioned facility in Florida dealing in missile production requires the heavy use of electric power for heat treatment and curing ovens. Another example is a chemical treatment plant that uses large amounts of water. It is necessary to estimate requirements at the planning stage so that the A&E firm and contractors are aware of it, and the additional cost to bring the requirements to the plant is also known. A facility may require an electrical substation or a holding tank for water. It may also require a separate distribution center and a waste disposal system. All these requirements are shown on the plant layout or accompanying documents.

Other utilities of concern are compressed air for air-powered tools, fiber optic data lines for computers, telephone lines, and process steam requirements. A knowledge of these requirements is useful since the activities involving these services can be located strategically in order to control the overhead expense and to maintain safety in the plant.

Cellular Flow

When one studies a process flow chart, one will find that the process moves smoothly along flow lines. On the other hand, with a variety of products, each varying a little bit, there is another approach to processing in a plant that is as cellular flow (Figure 3.15). In this method, different products which take same processing equipment are grouped as a family. A cell consists of a group of machines arranged such that a product can be processed progressively from one workstation to another without having to wait for a batch to be completed or requiring additional handling between the operations. In essence, this arrangement is a small production flow line within a plant.

Cellular flow is not an entirely new concept, but has received wide exposure recently in large companies manufacturing a high volume of a variety of products. One automobile exhaust system manufacturing plant has set up its equipment in a pattern so that an exhaust pipe can be bent on a pipe bender, the pipe ends faced on a facing machine, one end sized to the required diameter and another one formed to accept a flange. All four operations are performed on four different machines by the same operator without laying the pipes down between operations. Further, equipment is built on casters so it can be wheeled in and out to suit cell requirements. The immediate benefits realized with this arrangement are reduction in the material handling, elimination of in-process inventory between operations, and reduction in floor space.

Essentials of a Cellular Layout

Multiple products can be grouped into families with similar process requirements.
Parts do not require storage between operations. They are transferred to the next

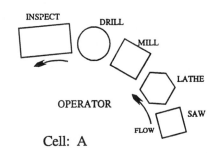

Cell: A

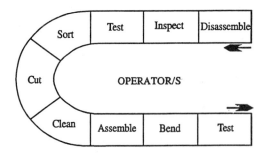

Cell: B.

Figure 3.15 Cellular flow consisting of a group of machines arranged such that a product can be processed from one workstation to another progressively.

operation either manually or via a simple transfer device such as a chute or a small conveyor section.

The majority of parts through the same cell are processed in the same way. No special setup is required, although some parts may skip a machine or two in the setup.

The operator is capable of operating all machines in a cell.

Equipment is arranged close together and there is no staging between equipment.

Equipment in cells are usually single-function, basic machines.

Concerns of a Cellular Layout

Equipment utilization is low.

A breakdown in one piece of equipment may paralyze the cell operation.

Tight quality control is necessary in order to reduce scrap.

In a block layout or plant layout, one must weigh each alternative and then decide between cellular and flow approaches. Based on each method, the areas required are calculated to provide the total requirement for that block. However, it must be noted that the length of a block is critical to a straight-line flow within the block.

Progressive Assembly Operations

Many assembly operations cannot be economically automated, and are performed by setting up progressive assembly lines as shown in Figure 3.16. Subassemblies and components requiring heavy equipment are set up away from the assembly lines, and the parts are generally fed to the line as required. This method provides the advantage of minimum movement of the material in a steady flow. Operations are also synchronized on the line, which gives the benefit of steady output provided all operations are performed as planned.

It is necessary to study in detail the operations affected by this approach as to space and equipment requirements before making a layout. This approach is most suitable for light assembly work requiring bench operations, such as assembly of pressure gauges, small kitchen appliances, electrical switches, and controls. However, mass production of automobiles uses a similar approach whereby there is a central facility setup with progressive parallel assembly lines for assembly of automobiles. The major subassemblies such as frame, body, transmission, engine assembly, axles, and doors are merged into these lines with the help of automated and material handling equipment. This equipment may consist of overhead, underground, and floor conveyors, AGVs, and industrial trucks. The feeder lines are either connected to separate facilities nearby or are serviced through receiving areas for parts received from outside the plant.

Line Balancing for Detail Layout

Line balancing refers to the requirements of equal time for operations performed on a progressive assembly line. In order to achieve a smooth flow of product through the line and gain high procuctivity, it is necessary that the manufacturing engineer distribute the total work in equal amounts to various stations on the line

Operations that are well suited for this type of application are manual and bench type, using small equipment. With large and expensive equipment it is necessary to perform detailed analyses to make sure that the balancing requirements do not end up with too much idle time on the equipment, or else result in a cumbersome line layout. In many situations, it is preferred that operations requiring the use of heavy equipment are performed separately and away from the line. The parts are then introduced to the

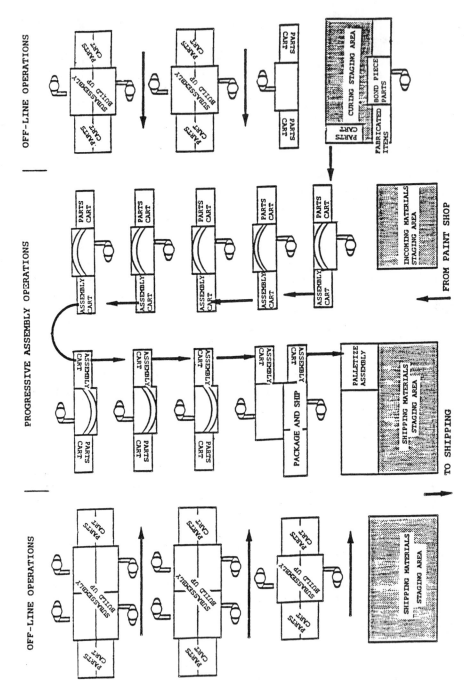

Figure 3.16 Progressive assembly lines supplemented by off-line operations.

progressive assembly line at the workstation on the line where they are assembled to the product moving line. This is also referred to as a feeder line concept, in which parts are either brought into the line as batches, or via a conveyor system which is synchronized with the main conveyor speed or operation time of the main line.

Even with manual operations, it is sometimes difficult to split the operations to even out the time. This is particularly true with light assembly-type operations. In these cases, it is necessary to introduce such activities as general cleanup, servicing other stations with supplies, inspection, testing, documentation, and labeling at the stations which have idle time left from the assigned task.

In the progressive line concept, the major elements which determine the layout are the production rate (pieces per hour), the clean breakdown of the entire process into separate operations, and the time required to perform each operation (also referred to as cycle time). Following are the steps in balancing a production line.

1. Determine whether a manufacturing process is conducive to forming a progressive assembly line.
2. Follow the process and itemize operations in chronological order.
3. Determine the time required to perform each operation by an average worker. Use stopwatch study or predetermined time standards.
4. Determine the production rate desired (units per hour) based on daily output expected.
5. Determine the total time required to manufacture one unit (labor hours per unit) by adding each operational time.
6. Determine the number of workstations to meet the production requirement by multiplying the production rate by labor hours per unit.
7. Determine the time allowed to perform the task on each unit (hours per unit) at each workstation (cycle time) by finding the reciprocal of units per hour.
8. Arrange the tasks on each workstation so as to progressively assemble the unit by either combining the operations, dividing into subassemblies, or duplicating the operations on two or more workstations so there is an even distribution of work. In cases where the operations are to be divided into more than one workstation, it may be necessary to review the time studies of operations to determine the break in the operation. Essentially, the product moves progressively from one workstation to the next at the end of the cycle time.

In process type operations, which take longer than the cycle time determined, it may be necessary to duplicate workstations for that operation. Again, there may be some idle time left over. This may be filled by splitting other operations and redistributing the task. However, line balancing requires a paper exercise in balancing the time for each workstation and often a rough layout to make sure that the area assigned to each workstation is adequate for smooth flow, safety to the worker, and efficient materials handling.

One method of overcoming the limitations of line balancing manufacturing is by utilizing a transporter as a material handling medium. A transporter does not require balancing of operations and yet allows flexibility to the extent that even unrelated products can be run on the workstations served by the line. In essence, the line can be run as a progressive assembly line just like line balancing manufacturing, or as a multiproduct operations line served by common powered material handling equipment. With the transporter it is necessary to deliver parts to the workstations served in small batches and to return the completed product to the dispatcher at the head of the transporter for rerouting to the next operation workstation.

Batch Processing Versus Continuous Flow

As discussed above, there will be some operations that are not a part of a fabrication center, nor part of an assembly line. The example in Figure 3.17 shows parts processed through a curing oven in batches. This type of process flow is easily adaptable to existing equipment or standard equipment. Although it is easy to set up and requires the minimum in equipment purchase, it will require more units in flow due to the batch process (perhaps 36 units cured at one time). It is less productive, and requires handling between operations. Figure 3.18 shows the same task setup with a continuous flow arrangement. The higher production rate capability must be weighed against the required production rate, to make certain that the higher cost of the conveyor system and flow-through oven, the longer lead time for the equipment, and the larger floor space requirement will be offset by the lower product production cost savings in labor. While there are advantages and disadvantages to both of these processes, one should be selected prior to making the final layout or ordering any capital equipment.

3.3 REQUIREMENTS FOR THE FLOW OF MATERIALS

3.3.1 Introduction

One of the three main ingredients of plant activities is material. The other two are labor and machines. It is necessary that raw materials and parts be available when required. Since delays are unproductive and expensive, they must be eliminated whenever possible.

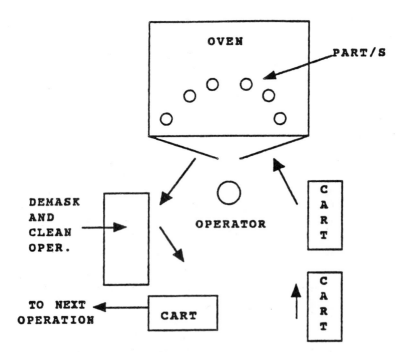

Figure 3.17 Parts being processed through a curing oven in batches.

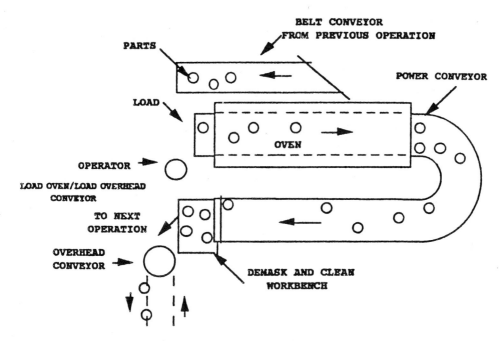

Figure 3.18 Continuous flow arrangement showing parts loaded from one end and flowing continuously on a conveyor which carries them in and out of the oven for processing.

For the plant layout to be effective, it must consider the movement and the handling of the production materials, supplies, and tooling. Many times the layout is prepared around the existing material handling equipment and methods presently in use in the plant. The manufacturing work centers must be adequately supplied with the material so there is a smooth flow of production and no downtime waiting for the materials, handling equipment, toteboxes, containers, and carts. An effective layout must take into consideration the production planning and control requirements, such as batch sizes, mode of transportation, storage space by the workstation, and the follow-through with shop work orders for effective control of the work-in-process material.

On the other hand, an improper layout which does not allow for the systematic flow of the product may result in parts shortages, quality problems, missing operations on some parts, and damage to the product due to handling.

With the modern concepts of minimum inventory and just-in-time manufacturing, the emphasis has been on making the material available when needed. It makes sense to carefully plan out the store requirements, since the cost of production in most industries includes over 50% material costs. In paper-converting facilities, the material cost may be as high as 80%. Large companies emphasizing low overhead costs have realized the value of subcontracting unskilled and semiskilled operations to smaller companies with less overhead, lower wages, and smaller benefit packages. This in turn increases material costs while reducing in-plant labor content.

On the other hand, companies have realized the value of high inventory turnover to keep inventory shrinkage and costs to the minimum. This has become more important than ever before since the advances in technology and global competition have made product life cycles shorter. Inventory shrinkage includes material obsolescence, damage, and loss. For continuous flow production, it is necessary that material flow be smooth. The material handling department must be well equipped to serve all production activities in the plant efficiently.

3.3.2 Materials Handling in Support of Production

In a manufacturing plant, the overall process is broken into various small operations connected by numerous transportation and staging steps. Materials handling should be considered an integral part of the total manufacturing operation. It is a catalyst which provides the reactions of people, materials, and machine. The material inside the plant is in one of two physical states in support of production: (1) properly stored and ready for its next move, and (2) properly transported. The goal is to keep the cost of these functions to a minimum, since neither of these add value to the product but both are an essential part of the total product cost. The material handling system in the factory operation must integrate the following functions:

> Transportation of raw material and parts to the factory
> Receiving, storage, and retrieval
> Movement of material to production centers
> Transportation between operations and staging
> Packaging and shipping

A proper flow pattern of material will help the manufacturing process by moving materials in the shortest possible time through the plant. It will also help in reducing rework, congestion, and safety problems. In order to obtain proper flow, one should consider the factors affecting the flow:

> Plan to minimize storage of material at the workstation.
> Minimize frequency of material handling.
> Move material the shortest distance and in a straight line wherever possible.
> Standardize containers and material handling equipment.
> Minimize paperwork accompanying material movement.
> Decide on the type of flow process best suited for manufacturing—batch, continuous, or cell.
> Standardize production equipment.
> Combine operations.
> Reduce distances between departments.
> Remove bulky outer packing from incoming materials.
> Avoid using specialized handling equipment.
> Avoid large variations in production output.
> Plan for common use of special requirements.
> Familiarize workers with system requirements.
> Locate support services nearby to reduce production downtime.

Figure 3.19 shows various flow patterns in manufacturing plants. The beginning and the end of the flows to a certain extent on the locations of both the receiving and shipping docks. Other factors which determine the flow are the type of the products manufactured and the locations of the key equipment in the plant. The key equipment here is defined as the equipment which must be placed only at certain locations inside the plant. In this case, the flow pattern is worked around the equipment.

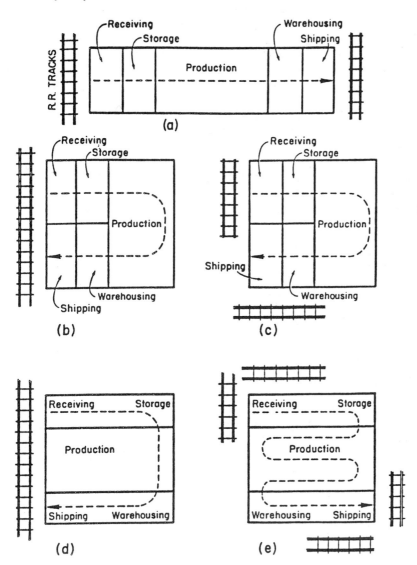

Figure 3.19 Material flow patterns in support of production.

Figures 3.20a–3.20c and 3.20f show parallel manufacturing line flows usually adaptable to a high-volume, light to medium products manufacturing plant, e.g., electrical motors of different sizes, lawn mowers, small kitchen appliances. Figures 3.20d and 3.20e show a looped flow for a processing plant with little variations in the product, e.g., corrugated containers, automobile bumpers, dishwashers, clothes washers and dryers. Figures 3.20g and 3.20h are for setups where more than one product is manufactured with subassemblies during processing, e.g., building hardware such as aluminum doors and windows.

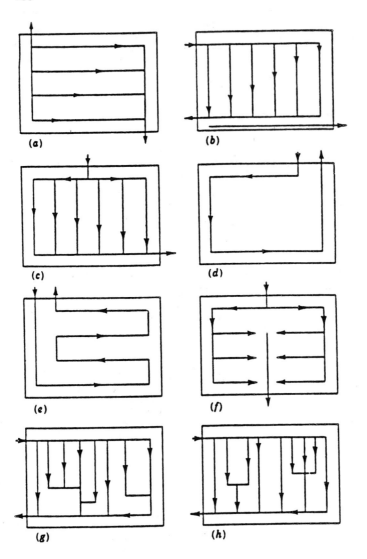

Figure 3.20 Types of manufacturing flow patterns: (a–c, f) parallel line flows; (d, e) a looped flow; (g,h) more than a single product.

3.3.3 Stores

Centralized Stores

Referring to Figure 3.21, method 1, the material is received at the receiving area, then inspected and routed for storage in a central store. From there it is retrieved for delivery to the workstations. However, provision has to be made to provide enough storage area for material received from the store. Frequently the workstations may end up being mini stock rooms for items which are brought in as unit loads containing a large quantity of items whose usage is minimal.

METHOD 1   CENTRALIZED STORES

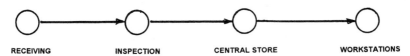

METHOD 2   DE-CENTRALIZED STORES

METHOD 3   CONTINUOUS FLOW MANUFACTURING (CFM)

METHOD 4   CENTRALIZED DISTRIBUTION STORE

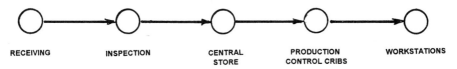

METHOD 5   LIMITED CONTINUOUS FLOW MANUFACTURING

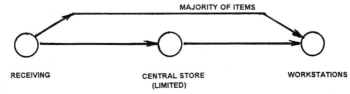

Figure 3.21   Material storage arrangements.

Benefits:

Materials can be bought in bulk for use on more than one product.
Avoids duplications of storage functions such as inventory control, supervision, and
    space in the storeroom.

Increased utilization of storage equipment such as automatic storage and retrieval systems (AS/RS), and vertical and horizontal carrousels.

Increased utilization of material handling equipment.

Material handling system can be automated, since there is economic justification due to the large volume of products handled.

Better control of material, since access to the stores can be limited to only a few authorized employees.

Flexibility due to cross-training of personnel and greater familiarity with a larger variety of parts.

Less prone to parts shortages for common materials, since they can be borrowed from materials allocated to other product lines.

Less investment in inventory, since duplicate storage is avoided of common items such as hardware, chemicals, and shop supply items.

Drawbacks:

Not all materials can be centralized, such as hazardous materials like oil, chemicals, explosives, and some packaging.

Excessive control on low-cost items such as nuts, bolts, or rivets can add cost burden and may cause production delays if not obtained in time from central store.

Breakdown of storage and handling equipment may paralyze entire plant operation.

Conflict in priority of servicing the needs of each production center by stores personnel.

Kitting function must be able to make materials available at proper workstation.

Production control has less flexibility and must properly coordinate production schedule and material requirement to avoid disruption and material jamups.

Emergency replacement and last-minute schedule changes may disrupt stores routine and affect other areas served.

Decentralized Stores

As shown in Figure 3.21, method 2, material is received at one receiving center, inspected, and routed to one of the many decentralized stores serving the corresponding production lines. These stores are naturally smaller in size than the centralized store, since each of them deals only with the materials required by the production line it serves. For instance, a central store might be storing 6000 items, whereas distributive stores might keep only 800 items. It is necessary that the receiving department have proper information on which store to route the material to when it is received. This can be done in a number of ways, such as coding receiving papers, identifying packaging material, or cross-referencing part numbers to store location on computer. However, with this method, the material is segregated at a common receiving center and distributed to properly identified decentralized stores.

Benefits:

Store personnel work in close relationship with the production line they support and with material handlers to avoid delays.

More familiarity with products, therefore less susceptibility to error in filling orders.

Less investment is required in automatic storage and handling equipment.

Store location is usually closer to the point of use, therefore avoiding delays in filling emergency requirements.

Store can easily be moved or rearranged in case of product change or reconfiguration of product line.

It is easier to implement in-store kitting requirements.

There is less damage to material since smaller quantities are handled.

There is better control of reordering, due to familiarity with parts and rate of usage.

Less storage space is required by workstations due to reduced lead time and improved response time.

Provides better support to continuous flow manufacturing concept.

Drawbacks:

Duplication of storage for common parts results in higher overall inventory cost.

Labor utilization may suffer due to low store activity level.

Duplication of clerical and accounting functions and computer equipment.

Unauthorized and undocumented withdrawal of materials by unauthorized personnel.

Coordination and control by central authority on all decentralized stores to maintain uniformity can be a challenge to management.

3.3.4 Continuous Flow Manufacturing

What is continuous flow manufacturing, and how does it affect material supply?

In recent years, with so much emphasis on the Japanese production technique of just-in-time, American industries have started seeing the benefits of reducing inventory cost. A way of achieving this is to maintain a continuous flow of products through the manufacturing floor by eliminating batch-type production orders. Batches create a stop-and-go environment with increased setup, inspection, and delays. Often operators have to be retrained to perform the same job, due to the time between batches. The tooling condition, material readiness, and support services have to be lined up when a new production batch is begun. All these create inefficiency in the operation and increase the overall cost of manufacturing.

Automobile assembly is one example of continuous flow production which has been in use for decades. However, over the years improvements have been made to allow for more flexibility and productivity.

In continuous flow, the batch size is reduced almost to one unit, with the flexibility of allowing necessary changes to be made in a particular production unit. This allows for production of a multitude of models or variations every day. The material is stored directly at the point of use. See Figure 3.21, method 3. Receiving inspection is also eliminated. It is the responsibility of the supplier to assure that 100% acceptable parts are supplied and stored at the workstation at the right time.

However, this system requires total control over labor, material, and machines. Labor must be skilled to adapt to new variations. The right material in the exact quantity must be available at the proper work center. Every operator is his own inspector and should be able to perform the job requiring no delays or further inspection. Equipment should be well maintained and thoroughly checked periodically to avoid downtime, since the effectiveness of a continuous production line depends on achieving and maintaining high reliability of all its components.

Continuous production is not a new concept. Chrysler, and others, have extended it by involving major suppliers in delivering products on dock as required. It may be necessary for the supplier to deliver parts and materials two or three times daily, properly inspected and arranged so that they can be brought right on to the production line for use the same day. This may call for several loading ramps to be built into the building, along the production line.

Benefits:

Quality awareness reduces product cost in the long run.

Inventory cost is greatly reduced.

Better service from suppliers, as they are part of the continuous production team.

Materials handling can be simplified and automated due to steady demand in small volumes.

Operators are involved in keeping production line running for maximum output.

Drawbacks:

Needs a great deal of planning and effort for start-up.

Many variables make it difficult to keep the line running smoothly.

May initially discourage management due to too many breakdowns, stoppages, and high costs.

As flow is depicted in Figure 3.21, method 4, the materials are received at the dock, inspected, and put away in a central store for future use. From there, they are retrieved by various production control cribs when required to support the production floor. In this case, the production control cribs usually act as temporary staging areas, so that materials are readily made available to production. The is usually the case in a large plant manufacturing several products. Production control cribs are usually situated near the production line to facilitate effective service. The production control crib functions to bring material from the store, unpack, and sort it out in kit form for deliveries directly to workstations.

Figure 3.21, method 5, shows yet another variation of a continuous flow system which is either adapted as a step toward continuous flow or used as a compromise in case of a product which is not fully adaptable to continuous flow. However, a properly designed system can yield some benefits to improve the efficiency for other systems used concurrently.

3.3.5 Establishing Stores Requirements

In arriving at the facility requirements for a manufacturing plant, the material handling and control system plays a large part. This system must be selected early in the process of developing a production block plan and layout. The number of parts, the quantity of each expected to be in plant at one time, the location, weight, and cube should be estimated. See Figure 3.22 for a materials storage analysis. Adequate floor space and volume can then be established, the material handling system defined, and equipment placed on order. Some of the equipment may be included in the cost of construction, and be provided by the building contractor. Other equipment may be handled by the company, and installed upon building completion. Some of the equipment that has "roots" is often assigned to the building contractor. This gives single-point responsibility for the coordination of structure, piping, and electrical requirements, and it is accepted upon demonstrated performance of the equipment. This often includes the air-conditioning equipment, compressed air, special cryogenic equipment, controls, and lines, communication and data lines, pumps, and other items that have complex interfaces or that could delay construction if they are not properly coordinated and available on schedule.

The completion of the material handling system requirements then feed in to a revised block plan and factory flow and layout.

MATERIALS STORAGE ANALYSIS

Approved By: GLP

By: VS
Date: 4-5-89

Page 2 of 9

Unit Load Characteristics

No.	Item	Cont. Type	Size (Ft)	Weight (Lbs)	Quant. per Unit Load	Space Req'd (CuFt)	Handl Equpt Req'd	Handl Storg. Limit.	Prefer Storg. Equpt	Storg. Method	Pick Method	Storg. Capac. Req'd Unit Load	Total Space Req'd (CuFt)	Storg. Category	Comments
1	VINYL FLOORING	WOOD PALLET	2X2 X13	1,100	4 ROLLS	52	SIDE-LOADER OR FORK	WIDE AISLE & FORK CANT. TRUCK RACKS	SIDE RACKS	FORK TRUCK & SIDE LOADER	FORK	12 LOADS	624	A	
2	LAWN MOWER	CARTON	1½ X 2½X4	450	4 PIECES	15	FORK TRUCK	/	FORK PALLET RACKS	TRUCK RACKS	FORK TRUCK	8 CARTON	120	B	
3	CAMPING FUEL (LIQUID)	CANS CARTON	½ X 2X4	15	10 GAL	12	PALLET JACK	HAZ. SUB.	FORK TRUCK	MARB AREAS	MANUAL	10	120	D	HAZARDOUS SUBSTANCE.
4	ANCHOR BOLTS-A	FIBERO BOX	0.8X 1X1	60	50 PIECES	0.8	HAND CART	/	FORK TRUCK-L	CAROLE	MANUAL	5	4	C	
5	FLOOR SEALANT KIT	CAN CARTON	1.5X 2X3	125	12 QUARTS	9	FORK TRUCK	/	/	/	MANUAL	4	36	D,E	SHELF LIFE CONTROL KEEP IN COOL PLACE

CATAGORIES: A - Extra Long Items, B - Bulk Storage, C - Medium Shelf Storage, D - Special Storage, E - Rotate Stock.

Figure 3.22 Material storage analysis worksheet.

3.4 ANALYSIS OF FACTORY EQUIPMENT NEEDS

3.4.0 Introduction and Overview

Equipment engineering encompasses many functions. Large companies have a person or department assigned to equipment engineering. Responsibilities may vary from company to company, but essentially the function is to select and procure the right equipment for the given set of data. It is essential that the responsible person or group involved in equipment engineering be business oriented and have an engineering background.

Methodology of equipment engineering is an important and integral part of an organization, since it often represents one of the most vital functions of facilities planning. Proper selection of equipment provides smooth operation and lowest operating costs. On the other hand, improper selection results in cost overruns, downtime, quality defects, and management headaches. Equipment engineering is also involved in environmental and safety issues, which may have a large effect on the corporate bottom line.

3.4.1 Business Considerations of Equipment Acquisition

"Equipment" is a piece of machinery to support production in the plant. Equipment is distinguished generally from tooling in the sense that equipment is general purpose and can be used for more than one particular application. It is a standard product for manufacturers. In contrast, tooling is especially made for a particular application, and is useless for other applications. Equipment is usually depreciated over its lifetime under Internal Revenue Service guidelines, except in the case of certain research and development projects where it could be expensed earlier. However, tooling is considered an up-front expense to produce an item. The yearly depreciation value provided for equipment is a bookkeeping entry which adds to the overhead expenses of doing business, but provides cash flow in terms of tax credits. The equipment in this category is commonly known as *capital equipment*, since its book value at any time is considered to be a part of the capital assets of a company.

There is a rather large "gray area" between tooling and capital equipment. For example, an off-the-shelf machine can have some special features added to make it unique to the support of only one product. This may be done by the equipment manufacturer, or it may be designed and added to the machine in-house after receipt. Depending on your accounting department interpretation (subject to agreement with the IRS), the above example could all be classified as capital equipment, since the modification is considered minor and perhaps usable on other products at some time in the future. Or it could all be considered tooling, since the equipment is made special purpose by your design, and the machine is only a part of the special-purpose tool. Or perhaps the most conservative approach would be to classify the standard part of the machine as capital, subject to depreciation, with the special features classified as tooling charged to the product and expensed at the outset. This "color of the money" is worthy of some serious discussion at times, and should be well understood by the manufacturing engineer. The other choice is to make the equipment a part of the basic building, and treat it totally differently. One of the reasons that this discussion becomes important is that you may not want the equipment to become a part of your continuing overhead rate, which you would have to consider in your product cost every year. It might be wise to lease the equipment, which would be part of your annual cost, but you could terminate the lease, or exchange the equipment and lease a more modern machine as one becomes available in the future. This would reduce the amount of cash expenditure by your company required to set up a new facility.

3.4.2 The Process of Equipment Acquisition

The following sequence of events describes in general the process of acquiring equipment:

1. Plant layout showing the process flow and equipment in general
2. Flow charts, process flows, and any other documents that describe in greater detail the production process that requires the piece of equipment
3. Review of equipment literature from various suppliers
4. Preparation of equipment specifications
5. Solicitation of quotes for qualified sources
6. Selection of equipment
7. Justification and procurement package
8. Installation, start-up, training, and acceptance
9. Preventive maintenance, record keeping, guarantees, and follow-up

Plant Layout

The function of the plant layout in acquiring equipment is to help in preparing specifications and in visualizing how the proposed equipment fits in the product flow of the plant. The selection of supply voltage and current, location of the breaker box, location of control panels in relation to building walls, access for maintenance, and other factors may be influenced by the equipment selection. It would be embarrassing to discover that the power drive for a machine cannot be replaced without moving the machine away from a wall, since there are no removable panels to provide access for maintenance from the front. It may be valuable for the equipment supplier to review the plant layout with you in terms of performance and accessibility of his equipment.

Certain equipment requires a special environment. A jig borer requires a clean, vibration-free, temperature-controlled room with a special foundation. Welding equipment requires protection from excessive air movement, which would otherwise blow off the arc or shield. An ammunition assembly machine may require a separation wall of reinforced concrete with a blow-off roof.

Manufacturers' Literature

Once the requirement for a particular type of machine is known, it is necessary to contact various manufacturers of this equipment and obtain their descriptive literature. A phone call, after checking the *Thomas Register* for prospective suppliers, will usually save correspondence time. Narrow down your selection to four or five manufacturers to consider asking for formal bids to your specification.

Equipment Specification

Equipment specification may be the most important part of the equipment engineering function. The preparation of equipment specifications is often a long, tedious process requiring help from other disciplines in the plant. It is the single common document to which all bidders respond. It spells out sufficient detail to ensure that the installed equipment will perform as planned. It establishes the responsibilities of the seller and the buyer during acquisition and after installation. It becomes a legal document binding both parties if there is a disagreement between buyer and seller in the future, since it becomes a part of the contract.

The suppliers' specifications and product descriptions will aid in preparation of the specification that you create, since you would like to buy "off the shelf" if possible, and not create an increased price by forcing the supplier to make a special model just to meet your request. However, in certain situations it is necessary for the supplier to modify his product to meet specification. One example is a forklift truck with a 3500-lb capacity requirement. Since models are available in 3000- or 4000-lb ratings, the manufacturer quotes on the 4000-lb model which uses a beefed-up chassis and frame assembly. The other alternative is to modify a 3000-lb truck by adding the appropriate counter balance to support a 3500-lb weight. This alternative is sometimes less costly, and is acceptable in many situations.

Some of the important features to be defined in the specification include:

Intended use

Performance requirement (size, capacity, etc.)

Design and construction (rigidity, tolerances, hook-up requirements, controls, etc.)

Data (maintenance procedures, schematics, etc.)

Installation (can vary from turnkey, to technical assistance, to none)

Warranty

Training (could be at your plant, at supplier's plant, may be included in price or additional charge)

Acceptance test (important in high-speed or special purpose equipment—may be witnessed at supplier's plant, or at yours)

Maintenance service (may be needed if you do not have in-house capability)

Spare parts (should obtain the critical ones, at supplier's recommendation, as part of equipment purchase)

Selection of Equipment

Prior to placing a purchase order, it is necessary to make sure the supplier is capable, both financially and technologically, of providing the product to your satisfaction. Most common problems are

Lack of working capital

Creditworthiness of the supplier

Unavailability of technical support

Uncontrollable growth at supplier's plant

Work delays due to safety and labor problems

Problems in meeting technical performance specification during construction

In any case it is necessary that the supplier selected be not only economical, but capable of supporting the quoted commitment to the equipment. A customer should pursue the following before placing the purchase order:

Obtain a credit reference from the financial institution with which the supplier is connected.

Call past customers who have obtained similar equipment regarding their experience.

Check sales, service, and warrantee support by the supplier.

Visit the supplier's facility to review the manufacturing process and safety records.

Obtain a surety bond for any down payment made with the purchase order.

When delivery of key equipment in a system installation is critical, one approach is to attach a penalty clause to the purchase order for late delivery by the vendor.

Once the bids are received, the review process begins. The first step is to review the bidders' responses to the specification to make sure that the offers meet all the requirements. In certain cases, the bidder may have proposed an alternative solution, which should also be reviewed. Conversations between the purchasing agent and the bidders must be documented to avoid any possible future misunderstanding. (See Figure 3.23 for an example of a telephone log.) If a compromise is given to one bidder, then the other bidders should have an opportunity to revise their bids as well. Once all the bids have been reviewed, and contacts made with the bidders' customers, the next step is to select the supplier. Award of the purchase order should not be based solely on the lowest bid received, but should also include reliability, experience of others, delivery lead time, installation, and operating costs. An example of a bid review sheet for four suppliers of a vertical carousel is shown in Figure 3.24. It provides salient features and other details of the equipment offered by the bidders. It may bring out irregularities which may trigger further inquiries and uncover hidden traits. For instance, one brand may be 30% less costly than the others. On checking, you may find that the base casting is machined in a foreign country by an unproven source. A table travel may be manual in one machine but powered in another. One supplier may have just switched to using ball bearings from heavy-duty roller bearings in order to reduce costs. Many construction details are generally not mentioned in the manufacturers' literature and are made available only by questioning the suppliers and their customers.

	By: ___MPN___
TELECON LOG	
SUPPLIER/EQUIPMENT :	ACME #3 CNC MILL

DATE	DETAILS
5-6-92	JW called to find out if we can accept single phase motor at a savings of $ 750. I agreed to accept it.
5-15-92	Called to discuss with S. Roberts (Chief Engineer, ACME) regarding reliability of Mars Computer System. He had no problem in five years and is user friendly. The equipment has 11/4" diameter drive shaft instead of 1" - industry standard.
5-18-92	Called to find out users in this area: They are: 1. Reliance Technology 407-268-9532. 2. Rocket Science Corp., Miami 305-212-7359.
5-20-92	Called Reliance Mr. Moxham, Chief Engineer. He experienced no major breakdown with the similar equipment - just a few oil leaks. Prompt service and parts inventory well kept at regional warehouse.

Figure 3.23 Telephone log used in the equipment selection process.

VERTICAL CAROUSEL FOR TOOL CRIB (DEPT. F203)

VENDOR:	1	2	3	4
MANUFACTURER:				
MODEL:	2412-8/28	140-1100-15	501 31/14	2400
DATA BASE	MMS-27640O-II INTERFACE	CIM/PL	F/P 2000	ACS10
DIMENSIONS: (HXWXD)	22.2 X 10'11" X 6' 2"	21.6' X 10.1'X 4-1/2'	21.1' X 10' X 5.6'	23' X 10' X 5 1/2'
PAINT:	TAN	BEIGE	BEIGE	PLANTINUM
SAFETY FEATURES:	SAFETY BARS & INFARED PHTOEYE	TOUCH BAR & PHOTOCELL	MICROSWITCH	INFRARED
UNEVEN LOAD:	20% IMBALANCE	20% IMBALANCE	IMBALANCE INDICATOR	UP TO 1800#
MOTOR HP:	7.5	6.5	5.9	5
CARRIER: PITCH (IN.)	14"	18"	14"	13 1/2"
DIMENSIONS (HXWXD):	14" X 24 3/8" X 102"	16.5" X 15.9" X 96.5"	12" X 24.4" X 98.4"	12" X 24" X 98"
NUMBER:	28	26	31	30
WEIGHT CAPACITY LBS	1,200	1,100	1,100	1,150
CU. FT. STORAGE (TOTAL)	564	381	517	490
WARRANTY:	1 YEAR (ON SITE P+L)	1 YEAR (PARTS ONLY)	1 YEAR (PARTS ONLY)	1 YEAR
DELIVERY:	12 WEEKS	16 WEEKS	16 WEEKS	12 WEEKS
FOB	WISCONSIN	BOSTON	BOSTON	N.J.
TURNKEY INST.	INCLUDED	FORKTRUCK & DRIVER REQUIRED	FORKTRUCK & DRIVER REQUIRED	INCLUDED
NOISE LEVEL: DbA.	N/A	65	60	-
SPEED IN/SECONDS:	5.1	6.0	5.3	-
PRICE:	$33,760.25	$28,839.00	$33,080.00	$48,468.00
INSTALLATION	2,400.00	2,500.00	3,576.00	3,400.00
BARCODE CAPABILITY:	3,622	INCLUDED	INCLUDED	INCLUDED
FIFO ACCESS:	872	INCLUDED	INCLUDED	INCLUDED
DATA BASE:	4,284.60	INCLUDED	4,495.00	4,555.00
MAIN FRAME INTERFACE:	1,850.00	INCLUDED	INCL. BASE PRICE	INCLUDED
TRAINING SOFTWARE:	$400/DAY+T&E ($1,400.)	$450/DAY+T&E ($1,500.)	1,275.00	1,960.00
INTERFACE BOX (NETWORK):	INCLUDED	INCLUDED	6,495 (NOT REQ'D)	INCLUDED
OVERLIGHTS:	INCLUDED	INCLUDED	150	316
POSITION LIGHTS:	INCLUDED	825	500	741
KIT PROCESSING:	INCLUDED	INCLUDED	995	-
19 DIGIT CAPACITY:	INCLUDED	INCLUDED	$1,877.	INCLUDED
EMERGENCY OPERATION:	HAND CRANK	HAND CRANK	HAND CRANK	HAND CRANK
MANUFACTURED IN:	USA	WEST GERMANY	WEST GERMANY	USA
SLOTTED CARRIERS:	INCLUDED	390 (STYLE e)	INCLUDED	INCLUDED
TOTAL COST:	$46,789.	$34,054.	$45,942.	$59,440.

Figure 3.24 Equipment bid review summary sheet.

3.4.3 Justification of Equipment Purchases

Even though the piece of equipment finally selected is shown on the preliminary plant layout, it will probably require separate purchasing approval at the management level. An important part of factory equipment engineering is to quantitatively justify the purchase of each piece of equipment. Rather than making a long list of advantages and disadvantages, it is better to present to the decision makers your reasoning in monetary terms. Also, any alternative solutions to the purchase of this specific piece of equipment should be presented. The following is an example cost justification for the installation of a bridge crane for a Cincinnati machining center:

A. Savings Calculation:

1. Savings in space and overhead expense because wide aisles and large staging area will not be necessary which otherwise is required for present use of fork truck for handling.

    Estimated square footage required with fork truck handling 1,627 sf
    Estimated square footage required with bridge crane 1,147 sf
    Yearly Savings: 480 sf @ $10/ sf/yr = $4,800/yr

2. Elimination of production delays as operator will not have to wait for fork truck which is also used in other areas of the plant.

Estimated at 1 hr/shift × $12/hr × 2 shifts/day × 240 days/yr = $5,760/yr

3. Savings from elimination of damage due to improper handling by fork trucks. Machined parts and fixtures are costly and if not handled properly while positioning, the machine may cause damage and dents which require expensive rework and replacement. Based on a minimum of one incident per year:

Material and Controls	$ 8,430
Labor (190 hrs. × $12/hr)	2,280
	$10,710 /yr

4. Delays and rework may require overtime operation to meet schedule requirements.

10 Hr/Wk × $6/Hr (OT prem.) × 50 wks./yr = $6,000

5. Handling of fixtures and parts without bridge crane may require additional time of 1 hr/shift.

1 hr/shift × 2 shifts/day × 240 days/yr × $12/hr = $5,760/yr.

6. Total cost savings estimated:

$4,800 + 5,760 + 10,710 + 6,000 + 5,760 = $33,030/yr

B. Equipment cost (capital investment):

Two-ton bridge crane with controls		$62,500
Installation, testing, & certification		9,000
Freight and taxes		8,500
	Total	$80,000

C. Prospective rate of return on investment:

Original cost	$80,000
Salvage value	$ 2,000
Expected life	20 years
Straight line depreciation	$3,900 /yr

Return on Investment: ($33,030 – $3,900 ÷ $80,000) × 100 = 36.4%

This method of calculating the prospective rate of return is also referred to as the "original book" method. The rate of return on the investment is the interest rate at which the present value of the net cash flow is zero. In this case, 36.4% is considered as the interest rate of money invested and shows the relative attractiveness of buying the bridge crane in comparison with other projects showing different returns on the investment.

If the investment money is borrowed from a financial institution at 12% interest to purchase the bridge crane, then the purchase in effect provides additional cash flow (profit) of $19,520 per year to the facility. (36.4 - 12. = 24.4% × 80,000 = $19,520)

However, if the total savings generated is only $12,000 per year instead of $33,030, then the return on investment would be ten percent. In this case, the manufacturing engineer would not recommend the purchase of the bridge crane since it creates a negative cash flow of $1,600 per year—meaning it would cost more to operate the bridge crane than continue the present method.

3.4.4 Equipment Purchase Priority

Earlier in this chapter, we discussed the possibility (likelihood!) of not being able to get everything in the way of a factory and all the equipment to optimize our operations in a "world class manufacturing" mode—at one time. We therefore suggested that we build the "optimum" building required, even if we could not afford all the equipment desired prior to starting production. The decision then becomes one of deciding which equipment, machine, or system to proceed with first, and which ones could come on line as funding was available.

Visualize a complete list of equipment prepared by reviewing the entire production and materials handling processes, and adding the equipment items that were omitted from the building program, and finally adding the other capital expenditure items from the remainder of the factory. These could include computer systems, networks, software packages, additional terminals, and a host of other pieces of equipment needed to properly outfit the new product facility.

Priority Assessment

The facility floor, walls, roof, lighting, etc. plus the equipment that is absolutely necessary to be able to operate the plant comes first.

There will be some machines and other pieces of equipment that are absolutely required in order to start production. Without a certain minimum amount of the necessary machine tools, etc., we could not produce the product. These pieces of equipment are next on our priority list.

From this point on, the decision as to which expenditure to make next will be based on its value to the operation by making financial analyses like we addressed in Subchapter 3.4.3. Although all companies seem to have different criteria for justifying equipment purchases, the bottom line is always the same—which one makes the most money?

3.5 IMPLEMENTATION PLAN FOR FACTORY AND EQUIPMENT

3.5.0 Introduction and Overview

We talked in Chapters 1 and 2 about the requirements for a good product design, paying close attention to the needs and wants of our customers. We analyzed and revised our existing or proposed product design, with a team of experts from all functions in the factory. We introduced concurrent engineering, QFD, and DFA as tools to guide us to assure that we did not overlook anything, that we had an organized approach to our analyses, and that our product was the best that we could develop. We discussed the advantage of rapid prototyping techniques to obtain some of the parts. This allows visual and "feel" evaluation, and permits testing in some cases to prove to ourselves that our product is as we had visualized it. We looked at the dimensions, limits, and tolerances of various processes, to help establish the details of the product design that could be produced. We considered the tolerances that the various processes and machines are capable of holding. We made cost studies of the various design and process alternatives, and understood that we could meet the cost and quality requirements of the product in production.

Now, in this chapter, we started thinking about the factory required to produce our product. In Subchapter 3.1 we looked at some of the general factory requirements, and outlined information that would be needed before we could define the physical building arrangement that would best suit our product production. In Subchapter 3.2 we looked at specific factory requirements from the product point of view: what would be needed to support the direct manufacturing operations that we had outlined? In Subchapter 3.3 we looked at the factory requirements from the materials and materials handling point of view. How should we receive, store, and transport the materials to the work stations when needed, and move the product progressively through the production process? More tools and techniques were explored that would help us evaluate our factory requirements from the product and materials point of view. Subchapter 3.4 talked about the selection and justification of capital quipment needed to support both production and material flow, since they are both part of the factory production operation.

The final "missing link" in the factory requirements process is to determine how we want to operate the plant. What does our top management have in mind for our future "world class factory" goals? What systems will we set up, and just how will we actually run the plant? This must be done prior to the final factory requirements definition, or selection of equipment, or making a final layout, or designing a factory (Chapter 4). There are a lot of options at this point, and perhaps we should look again at the "world-class" manufacturer characteristics that we introduced in Chapter 1. They are as follows:

Costs to produce down 20–50%
Manufacturing lead time decreased by 50–90%
Overall cycle time decreased by 50%
Inventory down 50%+
Cost of quality reduced by 50%+ (less than 5% of sales)
Factory floor space reduced by 30–70%
Purchasing costs down 5–10% every year

Inventory turns of raw material a minimum of 25
Manufacturing cycle times with less than 25% queue time
On-time delivery 98%
Quality defects from any cause less than 200 per million

While we can be certain that no one would agree that *all* of the above performance improvements are required, there is a strong message that no one can dispute. We must make a paradigm shift in the way we operate to achieve these "future state" conditions, and perhaps make some major changes to achieve any one of them!

If we believe (as your author does!) that the above are not just "goals", but rather essential elements for success in Manufacturing, perhaps we should let these elements guide us in establishing the final requirements for the factory required to produce our product.

It would be hard to believe that we could achieve *all* the above goals, and implement *all* the requirements in setting up our factory (existing or new construction) to produce our new product that we have so carefully developed—at least not prior to producing the first product for delivery to our customers. However, we must *consider* all of them in developing the requirements for our facility layout to build the new product—and future products. We need a plan to get there! Most companies will have yearly goals, and 5-year or 10-year operating plans. We can assume that the success of our new product will play a role in these future plans. The questions that come to mind include the following:

How much money for capital expenditures are available this year for the new factory (or new arrangement)?
How much money is available for capital expenditures on this product (or product line) in future years?
What do the cash flow analyses justify in the way of company investment for this product?
What will be the projected cost for the design and build of the new facility?
What will be the cost for the optimum machines and equipment to support production?
What would be the minimum investment required to support starting production— even if our direct labor cost goals could not be met in this configuration?
What would be the cost of the optimum material storage and handling system and equipment?
What would be the cost of the minimum equipment for storage and handling to support production—even if the support cost was not the lowest?
What are the risks involved in the success of this project?

These kinds of questions (and others) will be asked prior to the start of the project, and during the design development phase of the product. They will be asked again after the studies are more definitized, prior to giving approval to proceed with the optimum factory plan. The bottom line in factory requirements is going to be a compromise. The maximum expenditures required for buildings, production equipment, and material handling equipment to obtain the optimum product and material flow which gives us the highest quality and lowest cost product is one extreme. As manufacturing engineers we should define this configuration. However, we should be prepared to offer alternative plans, and plans for progressive implementation.

There are many unique circumstances in all companies that will give different answers to the above questions, however we must proceed to support the product production:

We should establish the factory building design for the optimum system—even though it may not have *all* the desired features and equipment in place the day we start production.

It will be a progressive activity, as the product production process matures—as the market is proven—and more funding is available.

We must never forget that the objective of running a factory is to make money for the investors, and this "real world" can only stand so much negative cash flow—or only so much time before the investment begins to actually pay off.

Having established this as a background check on the cost and feasibility of our factory, we can proceed with a summary of the requirements for its design and construction.

3.5.1 Implementing a Plan for Revisions to an Existing Factory

If rearrangement of the factory to meet the production needs is fairly minor, perhaps the work can be done with your in-house staff. This usually requires some sort of factory order, usually initiated by the manufacturing engineer, to start the actual layout and rearrangement process. After review by the appropriate group responsible for performing the work, a cost estimate should be prepared and approved. The actual layout can now be prepared in the form that is common with your plant procedures, and the necessary engineering calculations for power and other utilities made. This is followed by making more detailed drawings of the modifications to power, water, compressed air, air conditioning, lighting, and other systems. A master schedule of all these activities should then be prepared, and a detailed move schedule for all the major and minor pieces of equipment prepared. In some cases, it may be necessary to contact local trades for proposals to perform some of this work, in conjunction with your in-house maintenance workforce.

The coordination of this "minor" rearrangement is usually more difficult and time-consuming than originally visualized. People get very emotional about moving their offices, making certain that they have the required computer hookups, that the floor machines are accessible for maintenance, and, most important, that production of other products does not suffer. Sometimes part of the work may have to be scheduled at night or weekends. An example of such a problem would be in cleaning and sealing a concrete floor. This requires the use of strong chemicals, which could cause damage to electronic products and creates strong odors. Another example would be welding of brackets, etc., to the overhead structure, where the danger of fire, or damage to work being performed under the overhead activities, must be considered.

In other words, the process is about the same as in designing and constructing a new building, only on a much smaller scale. Quite often consultants will be needed, and in many cases an A&E firm will be hired to do some of the design and coordination. In many cases, a new building was lower in cost than the same area available by adding a mezzanine to an existing facility. This is one of the reasons for making in-house plans and estimates, followed by the appropriate management approvals.

3.5.2 Implementing a Plan for a New Factory Building

When it has been decided that a major add-on to an existing building, or an entirely new building, is required, a more complex set of circumstances is created. It becomes the task of the manufacturing company to provide to the A&E firm, or the construction management firm, sufficient information to assure that the new facility will meet your requirements. Chapter 4 outlines this program is considerable detail. The manufacturing company should set-up a building committee and appoint a project manager. The PM becomes the "funnel" for all information transfer in both directions. The building committee is his sounding board, and in most cases approval authority, during all phases from this point on. All of the flow charts, analyses, preliminary block plans, layouts, and studies that have been prepared by the ME, and others, should be provided to the design agent, through the PM. The company activities do not stop at this point, since there will be continual discussions and joint decisions made during design and construction. The company will want to check most of the work done as it progresses. They will help in obtaining permits, coordination, and approvals from the local regulatory agencies, and in general making certain that all factors internal and external to the construction proceed smoothly. This usually involves some local political contacts, and working with other influential and interested parties in the community and state. There may be state road funds available to help finance the project, or special agreements made that will ease some of the regulatory problems.

4

Factory Design

William L. Walker *National High Magnetic Field Laboratory, Florida State University, Tallahassee, Florida*

4.0 INTRODUCTION TO FACILITY DESIGN AND CONSTRUCTION

This chapter introduces the processes of facility design and construction in general, and some of the specific options which can significantly effect the outcome. We will start with a look at opportunities for the management of an existing manufacturing firm to shape the resulting designs. Then we will address the two principal design methodologies and explore some of their differences. Earlier chapters provided many of the technical considerations related to manufacturing. This chapter is intended to build upon that information in order to best support the real needs of the factory.

A successful factory design is much more than the sum of the electrical and mechanical requirements of the equipment nameplates. Design is more intuitive, based on experience with codes and with people. There are thousands of decisions which must be made during design and construction. One way to deal with the complexities is to break the process of design into smaller, more manageable groups. A good building program will help define and reinforce the expectations of management. One important benefit is that it can lead the architect or engineer to ask the right questions and make the right decisions during the process of design.

While every element of the new factory is important, some elements are more important than others. That hierarchy must be determined through careful evaluation. The building committee is responsible for determining such issues for the designers. The *building program* as discussed in this chapter is the best opportunity for the manufacturing company to define the requirements and expectations of the new factory.

You might think of the building program as a recipe for a $100-million, five-course meal. If you are going to pay that kind of money for something, you want to be confident of success before you put it in the oven. It must provide enough specific information to guide a complete stranger through a very complicated series of decisions. There are high

---

Editor's Note: The term "program" is not just a general plan for building the new facility. Used by A&E firms, it becomes the specific document which ultimately guides each architect, engineer, and designer in the preparation of construction drawings, specifications, and contracts for work to be done. It translates the users' requirements into the language of the building trades.

expectations and high stakes. Plans and specifications developed from a good program will yield superior results through cooperative teamwork. Construction costs are notoriously volatile, and time is always an important factor. Construction may become one of the highest-risk challenges many companies will face in the next century. Furthermore, in many cases factory refurbishing, additions, and new construction may be a requirement for survival in the near future.

The last section of this chapter deals with construction. In order to survive recent economic changes, many building construction companies have modified the way they do business. There are many advantages to each type of company, whether it be a general contractor or construction management firm. Subchapter 4.5 discusses some of their benefits and strengths. The language of the building designer is used on occasion to discuss some terms and processes.

4.1   THE TWENTY-FIRST-CENTURY MANUFACTURING FACILITY

Designing the twenty-first-century manufacturing facility will be more challenging than ever before. No longer can we expect to draw all of our inferences from what has worked in the past. We must now have a process of design which draws from the solutions used in past decades and develops appropriate responses for the future. Changes in management and society have mandated new performance criteria as well as new functions which must be planned for. Changes in materials and equipment are only beginning to affect the design and construction industries. With the current rapid market and technology changes we are faced with different evaluation criteria than in the past. The profit potentials of tomorrow, coupled with modern corporate philosophies, have created new facilities challenges.

One flawed decision can result in millions of dollars lost and potential human and business disasters. With this added pressure comes another series of changes in the way businesses make decisions. Many large companies distribute responsibility to a group of area managers such as fiscal, safety, engineering, manufacturing, support staff, legal, etc. This can make for a lot of lengthy meetings, difficulty in reaching decisions, and even major disagreements. While it does provide many ways of looking at the outcome of the decisions, it can also create delays in making one.

While there are always work-arounds to obstacles, some of them create their own negative results and can snowball into big problems. What, then, is the best way to deal with the challenge? Is it better to have one senior-level manager in charge of all the decisions, or maybe hire an outside guru with 40 years' experience? The decisions must be made for every firm based on its specific needs, resources, and situations. The method for determining future decisions comes from management's goals and objectives.

Accounting Systems

Years ago most firms utilized annual profit statements in evaluating the performance of the directors responsible for factory performance. As an example, accounting was usually based on a fairly narrow scope, seldom accounting for waste removal or treatment. Wastes could be dumped or stored in drums for indeterminate periods to avoid quarterly accounting for storage, disposal, and by-products in product cost analysis. This practice resulted in skewed data and large waste dumps.

Responsible firms are moving toward full-cycle cost accounting, from source to disposal, including energy and transportation costs. Some firms have begun to carry all expense and income figures as they are incurred, tracked against each product to which they apply. In this way management is provided a more complete and accurate indication of the real costs of doing business. Regardless of the way a firm tracks expenses, it is imperative that the decision makers understand the complete impact of their decisions in the case of new construction. Understanding the real meaning of the data available helps their decisions to yield more consistent success.

4.1.1 The Building Committee

The first thing which must be done by the manufacturing firm is to form a building committee. Upper-level management must be represented, preferably with full decision-making and budget authority. This representative may not be the company's eventual project manager (PM) for design and construction, but it must be a person who is strong willed, experienced, and capable of maintaining focus on the big picture. His or her primary responsibilities are to ensure that the design feedback supplied to the design firm is the best possible, and that the overall project cost targets are met. All area representatives offer their insights and desires. It is up to this senior manager to resolve conflicts and guide the committee to clear, enforceable agreements and directives. Often this person is responsible for the ultimate building program, whether an outside consultant is hired for the actual preparation of the document or it is done in-house. The size and makeup of the committee will depend on the structure of the firm, but experience indicates that it should not exceed nine people.

4.1.2 The Project Manager

The project manager is the single most important player throughout the design and construction process. This person must have experience in similar situations, common sense, and commitment to the project goals. Whether the goal is the cheapest initial-cost space or the most vibration-free work platform achievable, the day-to-day control of the project is the responsibility of the PM. Once the committee decides the issues and objectives and assigns them significance factors (hierarchy), the PM will be the primary point of interface. It may be the PM's responsibility to originate and then implement the hierarchy chart with help from the building committee. The PM will also be responsible for paperwork throughout the project. From the selection of the design firm through construction, this person will represent the firm in meetings with outside agents and agencies. See Figure 4.1 for a diagram showing the relationships among all parties.

Meetings may be held as often as six times a day for two years, in which decisions are made relative to the project. The PM will be called upon to evaluate options or to make recommendations for changes in the plans and specifications. At each phase of design, the PM will be responsible for review and approval of design documents. One of the most important duties is to make sure that other members of the team perform. Milestones must be met in order to complete things according to schedule. During construction there will be everyday opportunities to influence the process. The electrical and mechanical drawings are schematic at best; there will need to be field interpretation of the documents in many instances. A good PM can find answers to problems before anybody else even becomes aware of them. The PM must repeatedly evaluate the needs of the end users, in order to assure the best results. All the way through the project, the PM will evaluate materials on site and work in place for payments to the contractors. Cost and schedule savings are likely through good management.

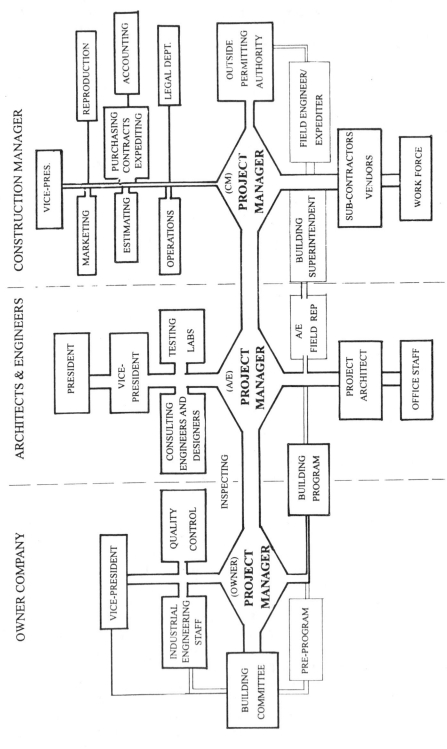

Figure 4.1 Relationships among participants in a typical construction program.

4.1.3 Hierarchy of Decisions

Designed by many, managed by few! One of the most frequently overlooked aspects of decision making as related to factory design is that of relative importance to the big picture, or *hierarchy*. There needs to be a fixed set of criteria which will help the designers through the process of design. It is usually up to the manufacturing firm to establish, and later to enforce, this critical list. With a clear understanding of the hierarchy of factors, the design team will keep in mind the values of the user group.

By assembling the proper representatives from each group within the manufacturing firm to serve on a building committee, management can achieve many things. The first is representation of all those impacted by design decisions while they have a chance to contribute, and at the right forum. One might think of applying the "Golden Mean" as ascribed to Aristotle: *Everything in the right proportion creates the best results.* We will take that two steps further by adding that it must be at the right time as well as at the right cost. Construction, like manufacturing, is a five-dimensional animal. There are the regular three dimensions (height, width, and length) which must be considered—along with time and money. A project that comes in on time and on budget is considered more successful than one that comes in exactly the right size. Therefore it is the right proportions of all components which will lead to the best results. Whether we apply that to building the best facility or the best product, it still applies.

Chart Of Issues

A typical project might have the following chart of issues, as arranged in decreasing level of relative importance (*hierarchy*):

1. Completed at a cost not to exceed $125.00 per square foot
2. Completed within 18 months
3. Must utilize local labor and 25% minority businesses
4. Highest possible reliability of operation
5. Lowest 10-year planning horizon for equipment costs
6. Cash flow during construction not to exceed interest from investments
7. Lowest operating and maintenance costs on equipment
8. Good relationship with local politicians and influential VIPs
9. All by-products must be treated and tracked—no dumping

You can see how this would result in a different end result than the following hierarchy chart:

1. Finish as fast as possible; 10 months is best
2. Construction budget not an issue
3. Lowest 5-year planning horizon for equipment and operating costs
4. No permits required; no building inspections required
5. Local labor may be utilized
6. Waste by-products can be dumped into river

It is very important to establish the hierarchy prior to beginning design, even though it may change during the process. This places a lot of importance on the building committee meetings. It is helpful to distribute copies of all correspondence with the design and build professionals to the building committee. No single person can be expected to make 10,000 right decisions under stress. This simple precaution can prevent a lot of problems later. Often the PM cannot know the impact of all his or her decisions on the ultimate specialized user. Using E-mail to a preprogrammed distribution group can save a lot of time, and much wasted paper.

4.1.4 Importance of Time and Money

This subsection is intended to provide information related to evaluating the importance of time and money for each specific project. Time and money tend to vary more widely than any other construction variable, and certainly they have the greatest impact on the factory design and construction process. The old adage, "time is money," does not provide the total story. There is always a trade-off between time and money. Seldom do we have a finite date for the completion and operation of a factory when we begin the design process. It should be understood that there is a point at which it is no longer cost-effective to try to shorten the delivery process—the point of diminishing return. A good goal would be to design and build the most cost-effective facility which meets the performance criteria set by the committee as spelled out in the building program. The best result is to finish ahead of time, under budget, and still meet or exceed performance expectations.

The Influence of Time

Factors which may have an impact on time as a determining factor might include the market for the end product, its sensitivity to time, the penalties and benefits for delayed delivery, and also the impact on the cost of construction. Construction costs over the long term have increased along with the inflation rate, often influencing the cost of living more than other factors. We have seen the cost of construction components fluctuate more than 100% in less than 90 days. These swings are most often driven by economic changes such as are caused by wars, by legislation, or by naturally destructive forces such as hurricanes. Labor costs must also be considered. Adding more people to perform tasks does not necessarily improve their performance. There is a maximum number of people who can work on welding one section of pipe, for example, so we must schedule work tasks during construction in such a manner as to maximize their performance. Overtime and additional shifts can be used to shorten construction time more successfully than overstaffing on one shift, but long hours tend to reduce worker yield as well. An experienced construction manager or project manager should know the benefit-cost ratio which is optimal. Major time components in a typical 3-year project schedule for a large, complex facility include:

	Percent	Weeks
Contract negotiation	6	9
Design development	11	14
Review by owner	4	6
Completion of contract documents	14	22
Permitting	7	11
Bidding and award	5	8
Construction	52	82
Total	100	156

Often projects can be delayed more than 14 weeks due to lengthy review periods or lack of communications during these planning stages. Delays from lengthy owner review periods, or from negotiating insignificant details on a construction contract, can cost time and money. Less obvious but equally important are decisions as to building type, materials, systems, connections, and finishes. Equipment can have more than 10 months lead time for delivery, then require extensive fit-up. The labor element of construction is one which can also affect project delivery time. Brick exterior walls, for example, take more labor hours per square foot than metal panel walls, bolted connections can be quicker than welded ones, skid mounted equipment is quicker to connect than loose components. Figure 4.2 shows a project planning timetable for a minor project and a major project, with average and expedited times shown.

It can be imprudent to try to expedite the *schematic* design phase very much. We must make efficient use of the available key personnel in the A&E teams' offices. There can really be only a few designers producing effectively during the preliminary/schematic design stage. This is usually a short period in the project schedule anyway, but it has significant implications for the end result. Trying to cut time here is usually counterproductive. Owner review time can be trimmed, especially if there is good interface during the design development. Weekly, or at most biweekly, team meetings combined with continuous design reviews can prevent problems.

The best opportunities for time savings are during the design development and contract documents phases. A large firm can assign more than 30 production staffers full time, and if there are consulting firms under a prime there could be more than 50 or 60 employees on the project at one time. The cost of this workforce may not be immediately visible, but it should be clear once proposals are evaluated. Providing two alternative schedules for document completion will provide a means for evaluating actual costs for any savings in time. The potential is for cutting more than 2 months from the design process.

The permitting period may be time-consuming or it may not even be an issue, depending on the project. Consultants can often submit early plan sets for review by code officials to prevent lengthy delays. Sometimes nothing can be done toward this end. Local A&E or design/build firms should have the best estimate for the time involved here.

Permitting may also be a good period for final owner review and approval of the design, and for contractors to secure bids from their subcontractors and suppliers. Ideally there will be no major redesign after approval of plans. Usually one addendum can be issued to the plans and specs after bids are in, prior to construction award. An addendum allows changes in the documents to be clarified to all bidders prior to commitment to one contractor. This is really the last chance for a competitive bid situation on known changes. Occasionally a short list of the top proposals will have another opportunity to rebid after clarifications of issues encountered during bid negotiations.

The cost of time is difficult to calculate. There are inflation rates, interest rates, and lost opportunity costs to consider. Usually the fiscal officers for the firm can closely predict short-term trends and long-term levels. Marketing personnel may provide input as to present and future demands for products.

The Cost of Money and Time

The cost of money continues to change, in terms of both interest rates and initiation costs to acquire loans. Similar factors must be weighed if the money is coming from internal sources. What do we have to give up in order to use this money for building? It costs nothing to expedite contract negotiations, unless you make a mistake in your haste. Therefore the issues of time and money are *interdependent* and are extremely important to a successful project. During the design phase, most decisions must be made with complete knowledge of their impact on time and money. The biggest thing to look out for here is

DESCRIPTION	MINOR PROJECTS		MAJOR PROJECTS	
	Min.	Max.	Min.	Max.
Preliminary				
Pre-Program	2	5	4	8
Hire Design Team	3	5	4	14
Schematic Design	2	6	4	10
Review by Owner	2	4	4	6
Budget Development	1	4	2	6
Site Analysis	2	6	2	12
Master Plan	3	5	4	8
Analysis & Review	1	2	2	4
Financing	2	6	2	10
Design Development				
50% Design Development	3	5	4	12
Review by Owner & CM	1	3	2	4
100% Design Development	6	8	6	18
Cost Estimating	1	4	2	5
Review by Owner & CM	1	2	1	4
Pre-permitting review	0	8	1	16
Construction Documents				
50 % CD's	4	10	4	14
Review by Owner & CM	1	2	2	6
95 % CD's	3	8	5	10
Final Cost Estimate check	2	5	2	6
Final Owner & CM Review	1	2	2	6
Plans and Specs complete	2	4	2	6
Permitting	2	6	4	14
Bidding				
Advertising for Bids	2	4	2	6
Selection of Firm	1	4	2	8
Bid Preparation	3	6	3	8
Evaluation and Review of Bids	1	2	2	6
Award Contracts	0	4	1	8
Shop Drawings Prep	2	6	4	16
Review, resubmit, approve	0	2	1	8
Building				
Site Prep	1	8	4	14
Utilities	1	6	3	12
Foundations	2	5	2	8
Construction	18	32	48	76
Start-up, Test & Balance	1	4	3	7
Punch List	1	2	1	4
Final Completion	2	3	2	10
Project Closeout	1	2	2	8
TOTAL IN WEEKS	81	200	145	398

	MINOR PROJECTS		MAJOR PROJECTS	
	Min.	Max.	Min.	Max.
Total Number of Years	1.6	3.8	2.8	7.7
Number of weeks for Reviews	8	27	17	60

Figure 4.2 Project planning timetable showing the minimum and maximum times from the preprogram phase through project completion.

the omission of clauses which would cover contract scope under *basic services*. Always take the time to think ahead for *services* such as field representation and inspections, interiors design, landscape and civil design, testing, multiple bid packages, and phasing. Include as many services in the original agreement as possible. Additional services contracts generally cost more than would the same work under a basic service agreement. To prevent contractors from using inflated estimates for loosely defined scopes of work, allowance figures can be developed with not-to-exceed amounts. In addition, agree to terms and conditions for potential future billing rates for out-of-scope work in advance of signing a design contract.

The cost of money is much easier to calculate than the cost of time. This cost can be developed by taking into account loan origination costs, interest rates, duration, and amounts. Another factor might be what other uses could be made with that money. If it is self-financed, the options would include reducing other debt, lost income from the money if invested or working, purchasing revenue-producing equipment, or many other things. For these reasons and more, it is important to have a qualified representative acting as PM to make these decisions. He or she should at least understand them well enough to present the major items to the proper management representatives for decision. Whether you choose to create a simple or complex equation will depend on your firm's specific situation.

Another issue is whether or not to include early equipment purchases in the general contractor (GC) or construction manager (CM) contract scope. If equipment is purchased by the owner, there will be one less markup on the equipment cost and orders can be placed prior to awarding a contract for construction. There are advantages to the subs purchasing the equipment as well, such as sole source responsibility and coordination of installation, which may make up for the markup. A good rule of thumb is to combine all long-lead items which would interfere with project critical path into early bid packages. Whether or not to buy them direct, or buy them through the CM or GC should be a simple decision based on the financial package. Unless your company buys a lot of large equipment direct, the CM can usually get more competitive pricing from major suppliers. It can certainly exceed the 5 to 10% you would pay the CM, and the CM would then be responsible for scheduling and expediting arrivals. This gives the CM additional leverage in negotiations with their subcontractors as well. One good way to assure performance by designers and constructors is to pay them performance bonuses for accelerated delivery, and to penalize them for late delivery. Most professionals will respond to this enticement better than other management techniques such as assessing liquidated damages for late delivery.

The last thing to define is the project budget. Early in the planning stages, it is important to develop accurate project cost estimates. Gross area-based system cost projections can be made from schematic drawings and current prices. It is important that all significant potential cost elements be identified. This should be done early and should be checked all along the way. There will be two or three times when the documents are developed with sufficient detail for the next level of cost estimating. The first budget needs to be general, with projected expenses and revenues. The example in Figure 4.3 shows a one page summary budget.

The five major issues we have addressed so far are: (1) Appoint a senior manager to head up a building committee. (2) Create a project manager position—hire a good one from outside or assign someone qualified from within. This person can work as head of facilities or plant engineering before, during, or after construction. Expect both managers to work more than 50 hours a week for as long as it takes. (3) Form a building committee

Description	Units	Cost / unit	Price
Site Costs			
Land Purchase, Acres	20	$21,650	$433,000
Utilities, incl El, Water, Sewer, Storm & FP	20	$3,500	$70,000
Clearing, leave trees on 4 acres	16	$650	$10,400
Development, includes fill & compaction	16	$1,512	$24,192
Taxes, Permits, Fees	20	$6,062	$121,240
	Subtotal	$33,374	$658,832
Other Costs	(sq. ft.)		
Professional Soft Costs (below)	Percentage	of Constr. Cost	$965,750
Construction Manager Fees	Percentage	of Constr. Cost	$756,480
Buildings, Type 1, Typical conditioned	170000	$84	$14,280,000
Buildings, Type 2, Storage, unconditioned	63000	$21	$1,338,750
Buildings, Type 3, Special Construction	17000	$145	$2,465,000
Early Equipment Purchase	bid package		$715,475
Parking, Paving and Planting	6570	$6	$37,121
Insurance	Percentage	of Constr. Cost	$126,000
Contingency, at 7 %	Percentage	of Constr. Cost	$1,050,000
	Estimated Project Cost		**$22,393,408**

Construction Soft Costs
Professional Fees, breakdown

A/E	$285,752.50
Mech, Elect, Structural Eng's	$460,087.50
Inspection Services	$38,393.75
Surveys and Tests	$38,630.00
Systems consultant	$48,287.50
Vibration consultant	$28,972.50
Civil Eng. & Irrig. design	$51,146.25
Materials Handling, Removals	$14,480.00
Sum	**$965,750.00**

Figure 4.3 Preliminary budget.

of representatives from diverse areas and work together to forge objectives and goals. (4) Establish the hierarchy. (5) Determine preliminary schedule and cost projections.

4.2 SITE SELECTION

We will proceed as though it is the responsibility of the factory facilities staff and plant management to determine the site for the new facility. An alternative procedure would be to select the A&E team prior to site determination, and to pay them for their assistance in site evaluation.

Firms with master planning expertise can be helpful in evaluating potential sites. They can often provide the personnel to perform labor-intensive comparisons when the in-house staff does not have the time. If you are adding on to an existing facility site, the fee will be minimal or perhaps the work can be done at no extra cost. Do not assume that it is cheaper to modify the existing facility than to build new, nor assume the opposite. In every case, perform an evaluation of actual cost in time and money prior to determining a course of action. You may be surprised at the outcome.

4.2.1 Expansion of the Existing Site

The first task is to determine the feasibility of adding a new building to an existing site. One must consider circulation, structure, utility needs, and support spaces, as well as the impact on the existing operation. Frequently there is an exterior wall common to the new space which would mean no obvious impact on plant operations. If the expansion is not too large, this may be the most cost-effective approach.

In some locations, site coverage may be limited because of impervious areas for percolation of rainfall, or one may be prevented from paving or roofing over green areas. Other regulations must be reviewed in relation to parking spaces, green spaces, numbers and sizes of existing trees, etc. There may be agreements in place which limit the allowable square footage for any building site, such as a "Development for Regional Impact" or similar document. Always take the time to research all applicable rules and regulations before deciding to expand on the existing site. Always look at alternatives, and research available rental space.

Attached or Detached Construction

Once it has been determined that the addition will be made on site, consider the two obvious options: attached or detached. If previous designs had planned for future attached expansions, there may be planned logical locations and utilities capped for connection to the new spaces. There may be vibration concerns or safety reasons for physical separation. All these issues and more should be considered prior to starting site design.

For tomorrow's manufacturing facility there will be new issues driving the decisions. It is possible that space which was once utilized for storage will be used for staging in the production process, and that overhead conveyors will take the place of pallets and shelving. We must be prepared to look for new means of solving old problems, challenge the traditional obvious solutions, and be ready to address the questions of cost, quality, or other factors.

Existing site utilities must be one of the first considerations. A rough estimate of existing and planned new electrical and mechanical capacities will provide good information. Total area requirements may also dictate the addition of a new building or buildings. It may be that the site cannot accommodate the addition of sufficient built area to be considered. One must look for a better utilization of existing space before thinking of a new building.

4.2.2 New Site Evaluation

The remainder of this subchapter on site selection will proceed as though the decision has been made to choose a new site, and that it will be accomplished without hiring outside consultants. One of the first things to resolve in site selection relates to geographic location. There are a myriad of issues, each of which may be factored into the equation.

First we must look at the materials which will be used in the manufacturing process, and see if there is a cost advantage to locating in a specific area. If a majority of materials originate in one geographic area, or the majority of the market is located in a certain area, these should receive strong consideration. If one city, state, or county is offering tax incentives, that may be another. It is important to look at the big picture when making such a decision. Issues such as being close to the home office may improve the buying power of both plants through combined volume. The answer will depend on the product and the process.

Look at the total cost, net profit, and net-to-gross ratios for mathematical evaluation for a projected planning horizon. Look at the biggest contributors to overall costs over a given period of time. For some factories these will be labor, for others, taxes. For others it could be the cost of operations, water, electrical power, or waste removal.

4.2.3 Analysis of Potential Sites

After looking at the general considerations, it is good to look in depth at some important factors one at a time. In this manner it is possible to evaluate the pros and cons with more depth. The order of evaluation should start with regional and contextual evaluations, then work toward the site specific.

Regional Stability

Regional stability can be measured by a number of factors. Some cities and counties have shown their commitment to growth by offering tax advantages, municipal bonds, and other benefits to large potential employers. Donations of land, interest-free loans, and cut-rate utilities are among some of the enticements offered. The most welcome factories of the future will be those offering a stable employment base for a wide variety of skills with a process yielding little or no hazardous waste.

The recent national trend has been for municipalities to charge more each year for every traceable waste product, including condensation water from air-conditioning systems which discharge into the sanitary sewer system. These costs can run upwards of $4000 per month for pure water being processed as waste. Solid and liquid waste costs are escalating across the country, in some regions much faster than others. Among the things to be considered, include these:

1. What is the economic base of the region? Is the economy based solely on tourism or defense industries? Look back 10 to 100 years for the whole picture.
2. What are the demographics? Can we hire 80% local workers, or will we have to recruit 100% from outside the area? Are work ethics, education, service and support industries in place?
3. What local restrictions apply? Are governmental entities currently considering property tax changes or growth management plans that will significantly change doing business in the future?
4. How are the other businesses in the region performing?

Political Stability

Before committing to a particular site, it is beneficial to find out the political stability of the locale. If today's governor loses a reelection bid, could the entire enticement package be put in jeopardy? Is there a track record of cleaning house every election year, or does the secretary of state serve for decades and usually control such issues? These questions and more can be answered through relatively brief communications with knowledgeable sources. Is new legislation about to be enacted which would affect the site operation? Are there old laws on the books which may?

If you are thinking of hiring a local design consultant, or even interviewing a number of potential ones, there is reason to find out as much as you can from each of them. They may have intimate knowledge that few people have access to. Sound them out for their overall savvy and business acumen. If the project faces extensive permitting and review processes, find out which firms have a reputation for submitting packages which make it through approval with the fewest delays. They are often the best-managed firms, with experienced personnel who are able to produce accurate, complete documentation.

Infrastructure

Now we come to an often overlooked aspect, infrastructure. Tomorrow's factories will require energy, but it is important to survey the major elements of infrastructure which affect any significant project. They include electricity, water, sanitary sewer, storm sewer, steam, gas, roads, and recreation facilities. Can the local utility company provide enough power with current capacities? Is the distribution system large enough? Who pays for the new roads and lighting? The cost of putting up a new electrical substation can run in the millions. See Figure 4.4 for examples of site electrical components. Roadways can cost more than $4 million a mile. If the local municipality will pay the initial cost of providing new services, you can expect them to attempt to recover that cost plus some extra through their billings. There are distribution costs, engineering costs, possibly even land costs to be factored in. Forty megawatts delivered to your site from just a mile away could cost more than you expected. It is better to know these things going into the decision. Work up a projected cost sheet based on current and future estimated utility costs—be certain to include all utilities.

Water has traditionally been one of the determining factors for site selection and site design. At the start of the twentieth century water was important from a transportation point of view. Today the quality and quantity of water delivered to the machines can be more important than port access. Machine, process, or system water is usually highly treated, to the point where the quality at the source is not a factor. Quantities are not often the limiting

Figure 4.4 Examples of the site electrical power components that may have a major impact on the total construction cost.

factor in site design, with storage tanks and wells capable of augmenting that which is available from the water lines. Therefore, the regional water supply and the distribution system in place may be a secondary concern.

Sewer capacities are often more limited than water. Some cities will impose a moratorium on sewer taps for several years. They may require large payments up front just to put your project on the list for future consideration. There are alternatives to every problem encountered, of course, but often the alternative is worse than the original option. Sewer treatment plants may be so regulated as to be cost prohibitive. Package treatment plants maybe the lowest initial-cost solution, yet cost more in the long term or even be prohibited by local law. Most areas have impact fees for every possible system which may be affected by the new facility.

New road systems could be required of your plant, extending beyond the property boundaries. Some entities may mandate miles of roads, all the way to the closest major artery. There may be an acceptable compromise which can benefit both the region and the employer and which can be reached without deceit or dishonesty. Some kind of phasing plan can be developed which plans for growth and pays for it as the real need materializes.

Another type of agreement would be to allow use of existing two-lane roads with the stipulation that the starting and ending times of the workforce be staggered to lessen the impact on existing intersections. Whatever the results of such negotiations in verbal form, always try to get them in writing before you act. The best-meaning committee of city council representatives can drag their feet for years or change their minds overnight upon public scrutiny of a pending agreement.

Recreational Facilities

The issue of recreational facilities should not be overlooked in today's factory site selection. If there are not enough in place to satisfy the need, the new site should contain sufficient area with the right topography to provide adequate space. Statistics have proven that individuals and their families benefit from such activities, and projections indicate that recreational opportunities will become even more important in the future. Hard-working employees need recreation, and even more sedentary people enjoy sitting around a nicely wooded park. A parcourse or walking path will improve the workers' stamina, strength, and health. This benefits everyone. One factor many families consider when relocating is the quality and quantity of recreational facilities. Golf courses, tennis courts, or baseball and softball fields are but a few of tomorrow's recreation activities. Fishing, sailing, and skiing are enjoying growth. Bicycling, climbing, and hiking have always been popular activities. Some of the bigger cities have professional sports teams to add to the entertainment side of recreation. Museums, plays, and more passive activities attract another segment of the population. Some people just like to get away and be alone. The best site allows for many types of activities for years to come.

4.2.4 Site Selection

Example of Geographical Site Selection

As a simplified example, an imaginary company chooses to build a new factory. There is a market for 20,000 widgets per day, which will sell in California only for $28.00 apiece. The expected material cost per piece is $12.23, and half of all materials come from Pennsylvania. The rest are common materials found everywhere in the continental United States. The factory will consist of 250,000 gross square feet of floor area. With truck docks, parking, and roads, this will require approximately 40 acres. Transportation costs per 1000 widgets are 35 cents per mile, and for raw materials the cost is 20 cents per mile for 1000 widgets.

Let us assume that three sites are being considered after preliminary evaluation of more than 10 potential locations across the nation. We can create a spreadsheet comparing the potential for net profits as determined by overall long-term expenses. If sales forecasts are reliable, actual projections can be used. Selecting the right planning horizon for cost accounting is critical and will depend on changing laws. For this example we will use 10-year, straight-line depreciation on equipment and buildings, and focus more on the direct-cost side. We will also assume that the list of potential sites has been narrowed to the best three. The paragraphs to follow give brief descriptions of the features of the three. They are in Arizona, California, and West Virginia. The information attributed to each is provided for the sake of illustrating the methodology for site selection. Actual data are available only through research at the time. The information is summarized in spreadsheet form in Figure 4.5, where the three sites are compared for initial and long-term costs. The bottom line is net profits after 10 years.

Phoenix, Arizona, where land is reasonable, climate is great, and utilities costs are minimal. It is 200 miles to the distribution network in California and 850 from the material suppliers. Forty acres would cost $4000 per acre and annual taxes would be constant at $125,000 per year. Labor costs would be $2.00 per widget, and the workforce is well trained and has good work ethics. These workers can each produce 18 widgets per day. Electrical supply and waste disposal would cost 65 cents per widget. Construction costs are running $68.00 per square foot.

Sacramento, California, where land is expensive and utilities costs are higher than average. The site is 5 miles from the sales network and 1200 miles from the Pennsylvania materials source. Forty acres would cost $320,000, and taxes would be waived in order to acquire such a prominent employer. Labor costs would be $2.80 per widget, and worker productivity will produce 15 widgets per day. Electrical supply and waste disposal would cost 85 cents per widget. Construction costs would run $74 per square foot.

Wheeling, West Virginia, which has very affordable real estate. Land costs there are $800 per acre, and taxes will be $80,000 per year. Labor is not highly trained but works cheap. Labor costs would be $1.50 per widget, and each worker can produce 12 widgets per day, average productivity for local industries. Electrical and waste costs would be 50 cents per widget. Construction costs would be $50 per square foot. This site is located 60 miles from the major materials source and 1150 miles from the sales network.

Next it helps to look at the sites on a local level. The key aspects here are the available workforce, economic stability, and available infrastructure. What has been the leading employment source in the past will reveal something about the potential employee base available. If food stores are the largest employer, you may not expect too many technically experienced workers.

It will also be important to look at the educational system, the numbers of technical and vocational schools, community colleges or universities. Furthermore, you can learn about the courses offered, numbers of students and degrees earned. This will help provide an indication of the population's level of readiness or commitment to learning. Look at other factors also, especially the demographics. What are the ages and sex of the

40 acres
250,000 s. f. building
$0.20 per mile raw transport/1000
$0.35 per mile fin transport/1000
500,000 widgets per yr.

10 planning horizon, years

Description	Site A Arizona	Site B California	Site C West Virginia
Cost per acre	$4,000	$25,000	$800
Total cost	$400,000	$2,500,000	$80,000
Annual taxes	$125,000	$0	$80,000
Total taxes	$1,250,000	$0	$800,000
El. & disposal per widget	$0.65	$0.85	$0.50
El. & disp. per year	$325,000	$425,000	$250,000
El. & disp. total	$3,250,000	$4,250,000	$2,500,000
Building initial cost/sf	$68	$74	$50
Building cost	$17,000,000	$18,500,000	$12,500,000
Operating cost/ year	$312,500	$686,000	$454,860
Total oper. costs/10 yrs	$3,125,000	$6,860,000	$4,548,600
Transport, raw miles	850	1200	60
Trans for 500,000 widgets, raw/yr	$85,000	$120,000	$6,000
Total raw trans cost/10 yrs	$850,000	$1,200,000	$60,000
Transport finished in miles	200	5	1150
Trans for 500,000 widgets, fin/yr	$35,000	$875	$201,250
Total transp fin/10yrs, flat	$350,000	$8,750	$2,012,500
Labor cost/widget	$2.00	$2.40	$1.50
Total labor/ yr	$1,000,000	$1,200,000	$750,000
Labor for 10 yrs	$10,000,000	$12,000,000	$7,500,000
Mat'l cost/widget	$14.40	$14.40	$14.40
Total mat'l for 10 yrs	$72,000,000	$72,000,000	$72,000,000
Widgets per day/ person	18	16	14
Total employees	111	125	143
Insurance rate/ person/ yr	$1,200	$1,440	$890
Ins. cost per yr.	$133,333	$180,000	$127,143
Ins cost/ 10 yrs, flat rate	$1,333,333	$1,800,000	$1,271,429
TOTAL EXPENSES PROJECTED	$109,558,333	$119,118,750	$103,272,529
TOTAL INCOME PROJECTED	$140,000,000	$140,000,000	$140,000,000
EXPECTED 10 YEAR NET PROFITS	$30,441,667	$20,881,250	$36,727,471

Figure 4.5 Site comparison spreadsheet used in the site selection process.

workforce, by percentage? By preparing a customized version of the site selection formula one can look at the most important determinants of all three sites on one sheet of paper. The difficulty is in determining which factors should be most significant. The program can provide such information. Once the site is chosen, final plans can be solidified.

4.3 THE BUILDING PROGRAM

Each new facility will require a unique *Building Program*. While the processes and products will be different, the major elements will be similar. A brief excerpt listing the spaces from one such prototype building program is provided in Figure 4.6. A complete

#	DESCRIPTION	AREA(GROSS S.F.)
1.	ACCOUNTING	1200
2.	ADMINISTRATION	1600
3.	ASSEMBLY AREAS	24000
4.	BREAK ROOMS	600
5.	CASTING	2000
6.	CENTRAL PLANT	4500
7.	CUSTODIAL	800
8.	ENGINEERING	2000
9.	FABRICATION	8000
10.	FIBERGLASS WORK	1450
11.	HAZARDOUS MATERIALS	2400
12.	JANITOR/ MECH. EQUIP.	1200
13.	WORK IN PROCESS STORAGE	1000
14.	MACHINING AREAS	2800
15.	MISCELLANEOUS, CIRC. ETC.	1000
16.	PAINT SHOP, AND OVENS	2000
17.	PAINT PREP	3500
18.	QUALITY ASSURANCE	600
19.	RECEIVING	800
20.	SECURITY	350
21.	SHIPPING	400
22.	SHORT TERM STORAGE	2150
23.	TESTING, MECH. & ELECT.	1400
24.	TOOL CRIB	600
25.	WELDING	1000

Figure 4.6   Elements of a program for a twenty-first-century manufacturing facility.

building program may exceed 200 pages in length and therefore cannot be included in this book. In the example are 25 areas of the facility which would require in-depth, comprehensive coverage. That information would include gross square footage needs along with performance expectations, lighting levels, utilities, and pertinent information on preferred relationships to other parts of the factory. These 25 areas become the major elements in building design. Space planning information sheets similar to Figure 4.7 can be utilized to provide this information to the design agency. Detailed information relative to each machine in the manufacturing facility is perhaps the most important information in the building program. Accurate plans and all four elevations must be included in the program if the information is known. If only block diagrams are available, it can lead to wasted space, time and money. Using "basis of design" information can tie down some

SPACE PLANNING INFORMATION SHEET

DATE / INITIALS WLW 4/3/92	GROUP NAME OPERATIONS	ROOM NUMBER OP 101	DESCRIPTION
REVIEWED, APPROVED DATE/ INITIALS 5/1/92 JRG	SECURITY LEVEL BLUE	ROOM NAME CASTING ROOM	

HAZARDS
PRESSURE PIPES
HIGH HEAT CURE
MOLTEN METAL

CHEMICAL NONE	FUELS, SOLVENTS 5 GAL. ACETONE
GASEOUS NONE	POTENTIAL EXPLOSION NOT REALLY
OTHER, DESCRIBE LEAKS, BURST FITTINGS	WASTE DISPOSAL TECHNIQUE RECYCLE TO SMELTER

VIBRATION

SOURCE, WAVELENGTH, FREQUENCY, NOT AN ISSUE	SOUND LEVELS DESIRED 55 DbA
SENSITIVITY TO: N. A. I.	ACOUSTIC TREATMENT RECOMMENDED NONE

ARCHITECTURAL

FLOOR SEALED CONCRETE	SPAN REQUIREMENTS MIN. 25 FOOT BAYS CLEAR
WALLS EPOXY PAINT	FUTURE CONNECTIONS, MODIFICATIONS NONE
CEILING EXPOSED STRUCTURE ABOVE	WIND LOADS EXTERIOR WALLS ONLY
WINDOWS FIXED GLASS, ALUM. FRAMES	COLUMN ENCLOSURES PIPE BOLLARDS
DOORS HOLLOW METAL, S. S. KICK PLATES 1/2	OPEN WEB JOISTS PIPE IN TRAPEZE HANGERS
SPECIAL CONSIDERATIONS	

STRUCTURAL

UNIFORM LOADS 100 PSF	VIBRATION CRITERIA NOT AN ISSUE
CONCENTRATED LOADS 800 PSI	ISOLATION? N. A. I.
CRANES, ETC. OVERHEAD	TYPE CRANE & CAPACITY GANTRY, 10 TON
ALLOWABLE DEFLECTION MINIMAL, LESS THAN 3/4" IN 10'	NATURAL FREQUENCY LOW APPROX. 2 HZ.
EXPLOSION TREATMENT N.A.I.	OTHER CONSIDERATIONS
REMARKS LOCATE ON GROUND FLOOR	

HEATING, VENTILATION & AIR CONDITIONING

DESIGN ROOM TEMP RANGE 72 TO 78 DEGREEF. F.	HUMIDITY RANGE 35 TO 60% RH	
ACCEPTABLE FLUCTUATIONS PER HR. 3 DEG.	SPECIAL CLIMATE NONE	
CLASS FOR CLEAN ROOMS 1000	CFM ALL HOODS 1500	%AGE USE FACTOR , HOODS .25
EST. EQUIP. HEAT GAIN 9800 WATTS, NIC CURE	EQUIPMENT USE FACTOR .40	
SPECIAL FILTRATION FINAL FILTERS IN MOLD PREP AREA	AIR CHANGES PER HR. 8	
OTHER CONSIDERATIONS CURING AREA NOT CLIMATE CONTROLLED		

PIPING, PLUMBING

FULLY SPRINKLED YES	WET PIPE YES	DRY PIPE CURING NO	DELUGE	STANDPIPE YES	CHEM. EXTINGUISHMENT NONE	FIRE EXTING. CLASS ABC
WATER REQUIREMENTS: DOMESTIC C. W. 6GPM MAX. 2HRS./DAY	140 DEG. H.W. SAME	180 DEG. H.W. YES 2GPD		DE-IONIZED NO	CHILLED 45 DEG. NO	EQUIP. COOLING YES, 12 GPM
GASES: COMPRESSED AIR 120 PSI	STEAM NO	HELIUM NO	HYDROGEN NO	NITROGEN NO	PHOSGENE NO	NATURAL GAS YES	OTHER NO
WASTE 4"	VENT YES	FLOOR DRAINS 2	ACID WASTE/ VENT NO	INDIRECT DRAIN YES 1-1/2"	REMARKS	CLASSIFICATION OF HAZARD ORDINARY	

ELECTRICAL, SYSTEMS

LIGHTING LEVEL 65 FC	FIXTURE TYPE PREFERRED H. P.SODIUM	VOLTAGE 277/480	EMER. POWER EXIT	DIVERSITY .8
POWER 480V. 2000	AMPS 208V. 1500	AMPS 120V, 3Ph. 200	110V. SINGLE Ph. ω	LOAD FACTOR .35
SYSTEMS: VOICE YES	DATA YES	FIBER YES	ETHERNET NO	LAN YES	SPECIAL SHIELDING NO	GROUNDING YES	U. P. S. NO

MISCELLANEOUS COMMENTS

PROXIMITY/ RELATIONSHIPS MATERIALS FLOW QUANTITIES, METHODS, ETC. NEAR TO MACHINE SHOP, GRINDING AREA AND CONNECTED TO CRANES

Figure 4.7 Space planning information sheet.

dimensions, but may result in changes later. (See Chapter 3 for further discussion of requirements planning.)

4.3.1 Elements

From the information in the building program, one can develop an understanding of the elements which are most crucial to the manufacturing process. All future design decisions should be made relative to their ability to support these major functions. Rather than attempt many simultaneous equations with too many variables, it is good to group similar elements. Based on product and process flows, there will be dependent elements which require a certain relationship. First the designer tries to arrange all elements into fewer than five groups. At this stage of design it is better to use abstract geometric shapes to represent the groups of elements. Simple circles and squares are most commonly used, with square elements encircled to form the groups.

Materials flow analysis will often determine ajacencies of these groups. The sample bubble diagram in Figure 4.8 shows how the 25 boxes can be divided into three groups by function. Within each of the groups, the designer will later transform the squares into more meaningful shapes and determine which of them are on exterior walls, which are closest to the utility plant, etc. At this point, the boxes need only indicate the number of the program element which they represent. Next, more information will be added in order to make other decisions. Each element can be considered a stand-alone component, even if it is made up of a number of spaces. For example, the administrative area may contain as many private offices as there are staff, yet for elemental design it should be treated as one entity since the general performance criteria are the same. On the other hand, the

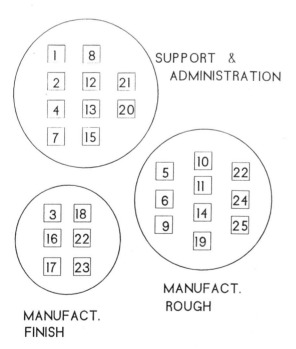

Figure 4.8 Bubble diagram showing the 25 major elements divided into three functional groups.

manufacturing shop may have many areas which have grossly differing needs, therefore we treat them as different space elements.

These abstract representations of the manufacturing components may be initially combined into two or three groups with similar activities. This will not necessarily dictate the final location for the elements; it provides only a means for representing a complex group of activities. The designers will then prepare as many alternative designs as it takes to exhaust potential solutions. Then all these alternative plans will be evaluated for the best fit for the program performance requirements, considering the advantages and disadvantages of different designs. For example, long linear arrangements can be more costly, while centralized ones may cause material flow problems.

Adjacencies

Another useful study in the programmatic stage is that of preferred adjacencies. One should arrange elements according to what are next to them. Such studies answer questions, such as: Which connections are required, and are they related to product flow or are they simply driven by stops along a conveyor? Are there safety reasons to be considered? Will we also consider human safety factors? Perhaps there is one element such as a pyrotechnics station where one would construct blast-proof walls and blow-out panels. This should be shown as removed from the location of the other building elements. A bubble diagram as shown in Figure 4.9 is an example of the effect of considering potential adjacencies.

Next, consider the mechanical and electrical adjacencies which would best support the manufacturing process. What would be the best arrangement to minimize expensive piping or cable runs? Which of the functions require "clean" conditioned air? Is it better to have them flanking a common HVAC unit somewhere, or to have noncentralized clusters for improved redundancy? The designer must consider all of these during the design phase, and try to determine the best options.

Next, one must evaluate choices for locating these systems in the completed building. Will they all be hanging from the roof structure, or under the floor? Is it more cost-effective to build three levels with laminar flow from top to bottom? The addition of pumps would permit us to run all waste and vent piping above the floor. This would enable future modifications to be made without cutting the floor. Traditionally, waste utilities are located under the slabs, while vents and piped gasses are run above the equipment. That relationship must be evaluated as well as plan (two-dimensional, planar) adjacencies. It may prove more cost-effective if all piping systems are located on a second level with branches above and below, as in some laboratories. In some process piping systems the cost of special plastic, glass, or stainless steel piping can escalate the building costs to more than $300 per square foot. The use of utility floors, pipe splines, or chases should be a conscious one. Factors such as access for future maintenance and reconfiguration for new process systems should be weighed in the decision along with initial cost. A photograph of an example of piping and cable trays mounted overhead is seen in Figure 4.10.

4.3.2 Relationships

Previous sections have alluded to the relationships required for optimum functional arrangements. There are other issues to be considered as well, such as minimizing expensive piping runs or maintaining adequate displacement for health and safety reasons. With more factory operations being automated, computer centers are controlling machines

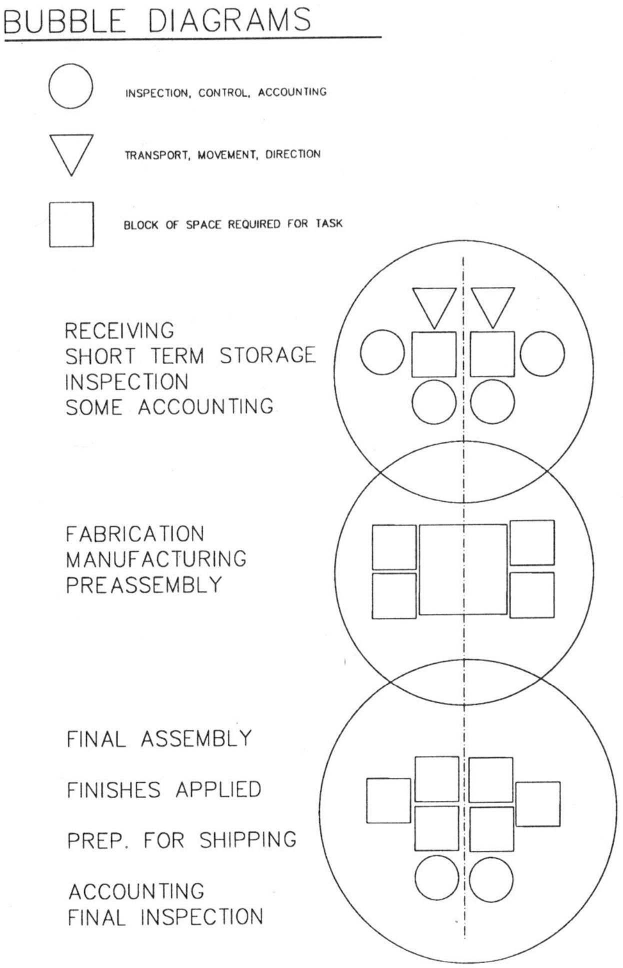

BUBBLE DIAGRAMS

Figure 4.9 Bubble diagram revised to consider the effects of adjacencies.

from centralized locations rather than having personnel in potentially hazardous situations. The designers must consider these factors and still provide for flexible operations in the future. Look first at the most critical operations in the process and their preferred adjacencies. In many facilities these will be fabrication and assembly areas. Next look at all the required connections for supply-side materials. Working backward from the final assembly process will often provide the most accurate and complete picture of the important relationships. A matrix which lists major spaces and their preferred adjacencies will provide a graphic illustration (see Chapter 3).

The next level of design includes actual requirements such as area (in gross square feet or meters), orientation, utilization, and other information found on the detailed space plan information sheets. From this second level of information, relationships can be evaluated for their ability to support the primary functional roles in the finished facility. The issues of safety must be factored into the arrangement of elements in the design. This is where the third dimension begins to play a more important role in the equation. *Elevation*

Figure 4.10   Example of piping and cable trays mounted overhead.

of elements and their support systems must be considered during design development. Section and elevation views must be developed concurrent with the plan.

4.3.3 Order

Once the best three-dimensional relationships are determined, the important issue of *order* should be evaluated for potential impact on the design. Similar to site plan design, the building element order can effect the initial price and long-term operating cost of the facility. Order can be defined for our use here as the arrangement of the elements in a structured system. There are two primary considerations in determining the order for a manufacturing facility. The first is how it will support the necessary functions of the manufacturing process, and the second is cost. Working from the information in Chapter 3, the sequence of manufacturing processes and machine operations is usually quite clear. That sequence will dictate to some degree the general order of the building elements which relate to the process. But what about the other building components? What determines the final layout of the building footprint?

A good place to start is by developing a module. In small areas, less than 100 ft^2 (9m^2) the module usually ranges from 3 to 4 ft, or around 1 m. In contrast, for large areas it may be 100 ft by 100 ft. In many cases it is approximately 4-ft for residential-scale buildings, and a minimum of 5-ft for factories. Some designers will use the width of their conveyance devices, such as AGVs or fork trucks, to assure that there is sufficient space for circulation. In any case, the module will be used as a design tool in the transition from bubble diagram to schematic plan drawing. This allows the designer to focus more on the big picture without being burdened by pragmatics.

If the manufacturing facility will contain mainly punch presses, each approximately 10-ft by 20-ft (including material work room), then that could lead to a 20-ft by 40-ft grid. If another area in the shop is mostly drill presses requiring about 5-ft by 7.5-ft, then a group of four could lead to a 20-ft by 30-ft module. Some designers create designs on quadrille pads, then assign some scaling to the grids to achieve the overall area targets. These tend to lead to grids of say 25-ft square. The economical spans of concrete and steel will usually range between 25 and 40 ft. These dimensions would be evenly divisible into 5-ft modules. Once that is determined, the process of factory ordering becomes easier. The block elements can be arranged so as to function best in sequential operation and still fit within a gridwork of 5-ft modules.

The configuration of these parts is the issue which affects the building footprint. As mentioned earlier, a square building has less exterior wall surface area than an elongated rectangle of the same total area. Therefore a relatively square building has a lower initial and recurring cost for maintenance. A ratio of about 1.6:1 has been proposed as having the best aesthetic appeal, and therefore the best subliminal acceptance by workers. Long, narrow spaces are perceived as boring, overwhelming, and monotonous. Large, square ones are impersonal and scaleless. The proportions between 1.3:1 and 1.6:1 often yield the best compromise of low cost and high yield. Once these decisions are made, the overall building form is taking shape. Rather than jump ahead to the finished plan and elevation views now, (and then try to make the rest work out,) it is best to consider the next issues. Studying all aspects first it can prevent inefficient use of space and potential problems.

When the designer combines those criteria with site-specific forces and employee needs, there is usually one ordering system which fits better than the rest. Most common in manufacturing facilities are centralized, clustered, radial, and nodal systems. Axial and linear systems are least common because of the increased surface areas and cost associated with their geometries. A successful designer will have the education and experience to make good decisions in this regard. See Figure 4.11.

Once the ordering system has been determined, the program yields other significant information such as shared needs for utilities. This leads to groupings of functions. Those groups are arranged by preferred adjacencies, or by methods covered in Chapter 3 on product and process flows. Other support functions, such as storage and administration, are then located for best circulation. See Figure 4.12, which is the natural result of applying adjacencies to the bubble diagram shown in Figure 4.9.

Figure 4.13 shows the natural bridge between this step and the beginning of the actual plant footprint.

Materials Flow

Starting from truck, rail, or waterway access routes, materials receiving and transportation becomes one of the most critical factors to be determined. Quantities of delivered goods will determine the total number of docks required. With just-in-time delivery becoming more prevalent, the number of docks has risen while the area for storage has decreased. Mechanized material transport devices such as automated guided vehicles (AGVs) and conveyors help keep both operating and labor costs low. The circulation area required for fork trucks and AGVs must be planned for during design. This can be done using a minimum-width module for the different types of vehicular passageways. The main aisles would permit the biggest fork trucks, of say 5 tons and 10 ft of maximum load width and a 12-ft height. Secondary paths would be designed for an 8-ft or 5-ft module. Each process will have its own conveyance requirements.

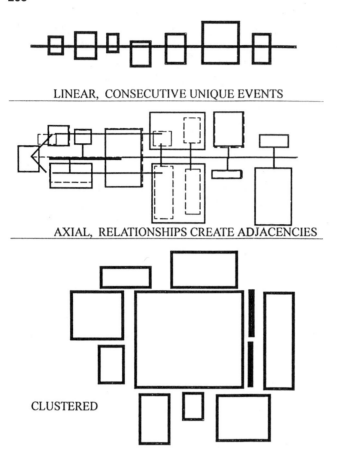

LINEAR, CONSECUTIVE UNIQUE EVENTS

AXIAL, RELATIONSHIPS CREATE ADJACENCIES

CLUSTERED

Figure 4.11   Linear, axial, and clustered ordering systems for factory design.

4.3.4   Circulation

A study focusing on circulation can yield some interesting results. It must look at pedestrian, vehicular, and product movement across the entire site and especially as related to support of the product. Simple diagrams can again be used, with symbols to represent the different groups, their trip frequencies, weights, directions, times, etc. (see Chapter 3). This information can then be added to the general block layout already developed. It may show bottlenecks in the flow, areas which need to be widened, or even reveal flaws in earlier design schemes. Areas where pedestrian paths cross vehicular paths must be treated in such a way as to avoid conflict or personal injury while not impeding product flows. Often pedestrian paths are elevated above machines routes to maintain safety. That also implies human's dominance over machines, which can help worker morale. Issues which seem unimportant to some may prove critical to others.

Elevated walkways can also be utilized to provide observation decks for tours without endangering personnel or reducing plant productivity. Employee break areas, cafeterias, restrooms, etc., can share the elevated areas with overhead conveyors and mechanical systems. This is another reason why three-dimensional studies are important in the design process. Coordination of the design disciplines is critical to successful project completion.

ADJACENCIES DIAGRAM

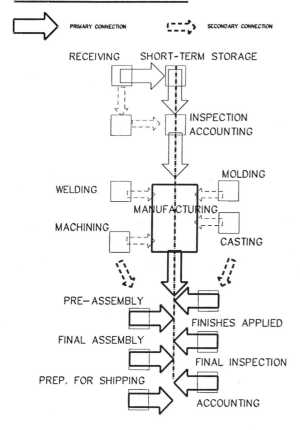

Figure 4.12 Results of applying adjacencies to the preliminary design evolution using the bubble diagram in Figure 4.9.

4.3.5 Employee Evaluations

Next is the very important aspect of listing information relative to the people who will be working in the facility. It should include the age, education, sex, religious, philosophical, mental, and physical makeup of the potential workforce. This list can then be combined into summary workforce groups. This can simplify the design response to the various workforce needs and preferences. The makeup of the workforce can be important in determining the space requirements beyond the production line. A workforce that is predominantly highly religious and well educated will have different needs than a workforce made up of uneducated agnostics. If there are mostly 6-ft-tall, 250-lb American males, the aisles will need to be wide. They could be narrower for a predominantly female Asian workforce, for whom the average is about 5 ft 3 in. tall and slight build. The section in this book on human factors (Subchapter 5.1) will provide useful data on dimensions, strengths, and spatial needs of human bodies in action and at rest.

Physically Challenged

Physically challenged individuals must also be accommodated in the design of new facilities. Ramps can be used for access rather than stairs. Winding, gradual ramps are

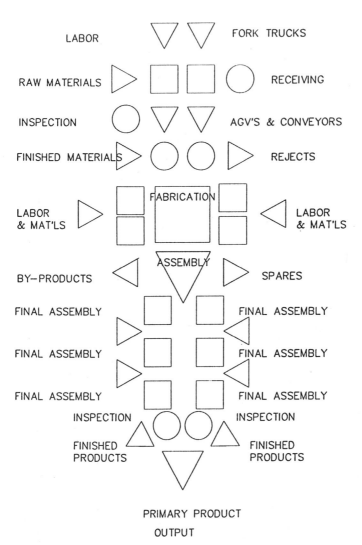

Figure 4.13 Example showing the natural bridge between Figure 4.12, and the actual factory layout.

preferred rather than short, steep ones. A good rule of thumb is that ramps should not exceed 1 ft of elevation change for every 13 ft of horizontal distance. Long ramps should not be used without intermediate flat landings of at least 80 in. Passages require a minimum width of 44 in. for one and 60 in. for two persons.

Fire exits must be designed to provide special considerations for those individuals who might need assistance in exiting a burning or smoking building. All elevators are no longer required to proceed immediately to the ground level for fire department use. Some emergency egress is permitted using elevators for assisted exit by challenged individuals. Many rated stairwells are designed utilizing safe zones for assisted egress in cases of emergency. In this way, mobility-challenged persons can wait safely in predetermined areas which do not obstruct others from using the stairwells. These spaces can define where emergency personnel can report to assist them. Be sure and review all safety plans with appropriate fire and police agencies.

4.3.6 Code Review

There are many rules and regulations which will apply to master plan decisions. Prior to initiating design development, the owner and some of the owners' chosen professionals will be required to work together to determine all codes which apply. Some of the codes may change during design and construction, which can affect the operation of the completed facility. It should be the responsibility of all concerned to make continuous efforts to see that all codes are met. We will not attempt to list all of them at this time. Generally speaking, however, they will include:

Uniform, Standard, or Southern building codes
National Electrical and Plumbing codes
ASHRAE codes
National Fire Protection Association codes
Life Safety codes
District, county, and municipal building, fire, and zoning codes
Handicap access codes

Other agencies which will also have potential jurisdiction might include the local fire marshals, building inspectors, traffic engineers, the Occupational Safety and Health Administration, water management districts, departments of environmental regulation, departments of natural resources, and many others too numerous to list. Failure to comply with any of them can have significant results. It should be the shared responsibility of the owner and the design team to research applicable codes. (Subchapters 5.2 and 5.3 may offer additional insight here.)

4.4 PROJECT DESIGN

There are many different ways to design and build a facility. The owner can use in-house engineering staff to design and the facilities department to construct. Proper administration of even modest construction projects can require more than five experienced full-time employees. Design of small projects can be accomplished by one registered professional in the firm with the help of specialized staff. Most manufacturing firms will not have an appropriate staff to cope with the time constraints and will therefore hire design professionals. Experienced A&E firms can provide the expertise and personnel to complete hundreds of design documents in a relatively short time. One aspect of their contract requirements should be for them to provide accurate documentation of the existing facility (if not already available), and for them to turn over to the owner a complete set of hard copies and software copies of the design documents when completed. It is common for firms to generate schedules and estimates on Lotus or Excel software, word processing copies of the specifications and key correspondence on WordPerfect or Word software, and drawing files as CAD drawing files (Autocad, Datacad, Versacad, etc.). The owner is entitled to copies of all of this at the end of the project. The form and content of the documents may affect future operation and maintenance of the plant and therefore become very important to the client.

The fees for A&E firms range from 4 to 20% depending on the range and scope of design services. The contracts can vary from minimal schematic design or master planning to complete design and contract administration. For complex projects or tight schedules, A&E firms can provide on-site field representatives to guide construction. In most areas, licensed design professionals are required to seal the construction documents prior to permitting. The choices are made easier by looking at project schedule and cost information developed in the preprogramming phase. Each unique project will create new challenges to the designer, to improve upon what has been done before. With the complexities inherent in preparing hundreds of pages of contract documents, it is often safest to rely on professionals who are in the mainstream of the specialty you need.

4.4.1 Design Firms

Many of today's A&E teams are comprised of specialists who are expert in a very narrow field. A list of specialized consultants for a factory project might include:

Architect
Civil engineer
Landscape architect
Irrigation designer
Structural engineer
Mechanical engineer
Systems specialist, communications, voice, fiber optics, etc.
Process controls engineer
Conveyance systems specialist
Environmental and permitting specialist
Interior design
Soils testing, materials testing and reports

It will usually be the architect's responsibility to guide the project through design, interfacing with all the consultants and often making recommendations for contract award. The architectural firm will also function as the main point of distribution for questions and answers between the owner and the constructors through the client's PM. The PM for the architect's office will be singularly responsible for implementing the wishes of the building committee. This person must be confident enough to exert authority and experienced enough to understand the implications of his or her decisions.

The project architect and project manager in the design firm will be the most important people during document preparation. For these reasons it is important to choose a firm with whom you can communicate completely. The importance of each role in communications can be seen in the earlier diagram (see Figure 4.1). The three PMs are the real points of contact, so choose them with that in mind.

Some manufacturing firms have established strong ties with design teams over time through successful projects. If there is no such relationship, it will be necessary to select the best A&E team or firm for your specific project. This is usually achieved through advertising a fixed scope of work at a fixed fee. The advertisement must include enough pertinent information for the interested firms to understand the expectations of the owner, the scope of the project, and any special experience which would be helpful in designing this project. If the time constraints are atypical, they must be expressed as well. From all applicants, a short list can be developed which contains only the highest-ranked firms based on certain evaluation criteria. A typical list of evaluation criteria would include:

1. Are they technically competent? Can they understand the project requirements?
2. Are their designs cost-effective? Can they be built for the target budget?
3. Can they meet the demands for production? How many employees are available?
4. How many years has the firm been practicing?
5. What is the experience and motivation of their key personnel?
6. List of consultants, past affiliations and competed projects, names and addresses.

Once the firms have submitted their preliminary proposals, one is usually able to create a short list of the highest-rated responding firms. Firms on that list will be invited to present their "pitch" in a scheduled interview format. Intangibles such as fit and attitude can be evaluated from these person-to-person exchanges. Next one should follow up referrals personally, by visiting completed projects if possible. By discussing performance with other clients of the design firm over the past few years, one can get a better picture of the performance of the design firm.

4.4.2 Final Site Analysis

After determining a preferred site, there are a number of site-specific issues which may contribute to the final design decisions. One should start with total area, orientation to transportation, and boundary information (length of each property line and bearings). Building setbacks and easements are critical to development of a site master plan. From these one can determine the net and gross areas of the site and these will lead to the developable area. Topographic information such as elevation above sea level and that of the surrounding areas are important to consider. Flood plain information may apply. There are zoning requirements, utility surveys—a whole host of issues which should be evaluated in depth. The best approach breaks them into subaspects and then studies them as if they were independent entities. Upon completion, the essential important kernels from these analyses are combined, weighting them according to their importance (hierarchy). There are some lowcost tests which may determine the needed foundation system and therefore construction cost. They provide a good starting point for analysis. Following are a few of the most common subaspects for site analysis.

Soil Tests

Soil borings located at 100-ft intervals in the footprint of the building can provide sufficient data for general subsoil evaluation. The depth should usually exceed two times the maximum projected building height. The testing agency will include a report as to conditions encountered, including ground water and bearing capacities of the different soils encountered. A minimum bearing capacity of 3000 lb/ft^2 at the surface is required for modest floor loading. Significantly higher values will be preferred. It is better to see a consistent profile, rather than levels of high and low bearing capacities in a stratified or mixed manner. If there are voids or very-low-density layers in the subsoil, there could be potential problems. If the preliminary tests are inconclusive, testing at 20-ft intervals will provide better information. If the manufacturing process will be extremely sensitive to vibration, the top 2 or 3 ft of soil should contain nothing but dry, coarse sand. Clay, rock, or wet soils can increase vibration transmission beneath the slabs. The consultant's report should also contain recommendations as to the type of foundation systems best suited for your specific performance criteria.

Water Quantity and Quality

A number of other site forces need to be taken into consideration. Water quality and quantity in both ground water and piped supplies should be quantified. Look to see if there is gray water piping, or separate irrigation water systems. Availability of fire protection (FP) water to the site can affect fire pump requirements. If sufficient water is available, a project can save perhaps $100,000 in pumps and related costs. When sufficient pressure exists, pumps may not be required. Generally, water pressure in excess of 120 psi will be sufficient. Ponds or lakes may function as fire water reserves in some remote areas. Low-cost elevated tanks can provide pressurized delivery systems without expensive highcapacity FP supply pumps, but they can add to the structural cost. Good designers will be able to determine what effect anticipated growth in the area will have on the water pressure. A factor of 0.75 is generally applied to the test results from fire department flow tests, in order to anticipate future conditions.

Depending on the particular factory processes, water may be a critical requirement. A stream on site may provide a heat exchange source, or a means of transporting treated waste to the properly classified river at its end. Many industrial facilities can operate "closed cycle," with minimal make-up water required. Cooling towers can require more than 50,000 gal per month to offset evaporation at the basin, while cooling coils in air handlers throughout the project can distill a similar quantity from the inside air of the factory. Presses, ovens, transformers, milling machines, and other machinery can require equipment cooling water. While the quality is not as critical here, some processes require very specific water. Industrial lasers, cleaning machines, and chemical processes have very specific requirements. A great water treatment systems can cost in the millions. For some factory processes, the elements present in the water supply can affect initial and operating cost of the treatment system. Commonly we find calcium, iron, and sulfur, but some trace elements can cause the biggest problems.

Other Factors

Some of the other regional criteria to be considered should include storm water, topography, vegetation, and wildlife. If there are endangered or protected wildlife habitats present, there will be more issues to consider. Similarly, historical artifacts found during excavation have been known to delay projects for years. Indian burial grounds, tar pits, or other animal remains on sites further complicate proceedings. The easiest way to protect against these difficult-to-anticipate occurrences might be to include provisions in the land purchase offer. These escape clauses require return of deposits and expenses when such delays take more than 30 or 90 days. In summary, there are mathematical formulas which can provide decision makers with quantitative data. Ultimately, however, no picture is complete, and management must make an educated compromise.

4.4.3 Site Master Plan Development

A *site master plan* is the design document which locates all major elements of the project and deals with gross area calculations. It addresses parking, paving, planting, and all building footprints. It can be thought of as the framework on which all the components are arranged. A good master plan will work for decades, regardless of changes in time or factory processes. It will be a flexible *ordering scheme* which combines the best natural features of the site with the necessary physical components of the manufacturing facility. It should be a skillful manipulation of the *built masses* which creates an effective *sense of place,* where the workers want to be and wherein they all work together toward common goals.

If this seems like too much to ask of the simple task of placing buildings on a site, then one has not grasped the power of design. Design is not simple, and it has a powerful effect on our subconscious minds. Having said this, let us see if we can conjure up one place in our memories which was a joyful place to be. It might be a childhood playground or a religious setting such as Saint Peter's square in Rome. Whatever the place, the arrangement of the components is what stirs the senses. The same can be done in arranging the building elements on a 10-acre site or in a 250-acre industrial park. Both require skillful master planning. Other good examples are Walt Disney World and Golden Gate Park. All were designed to create a *sense of place* where sensory experiences are pleasurable. The stimuli are controlled to heighten the senses of sight, smell, sound, touch, and imagination. Memories of such places are vivid because of the sheer number and intensity of sensory stimuli coupled with the cumulative effect of repetitive elements.

The challenge to create a similar atmosphere in a factory environment has been accomplished successfully in the past. The most important aspect is to consider the senses involved. Sight, smell, and sound are the three which we can easily control and must consider in design (see Subchapter 5.1, on human factors).

Through skillful manipulation we can create a master plan which will keep worker morale and performance high, while keeping illness and turnover low. It takes a careful, thorough examination of the workers and the work to be performed. Only after all this can we expect to make good decisions. We must define the needs of the workers and make conscious decisions to achieve fulfillment of their needs. If you believe that it is not important to consider these things, then get ready for the twenty-first century! Humans and machines work better and longer in the proper environment. Most successful manufacturing firms have been trying to improve their employee performance through improvements in the workplace. The master plan is a great vehicle to start to achieve that improvement.

Physical Studies of the Selected Site

The first study is to evaluate the physical attributes of the area to be built upon. Some of these attributes are the same as we looked at on a regional level, but with more attention to specifics and details. Others were considered less important then, but take on more meaning now. Among those specific site studies are:

Soil types
Topography
Vegetation
Circulation
Prevailing winds
Views
Utilities

Each component should be studied independently and evaluated in terms of its impact for determining the building zones. Building zones are located where there is best subsoil for bearing the anticipated building loads, and where the other factors support cost-effective construction. Excavation and replacement of 3-ft of topsoil or clay can cost hundreds of thousands of dollars for an average factory footprint. See the example in Figure 4.14 for an introduction to topography and other site forces.

Soil Types

Depending on the type of manufacturing facility, the soil types may be the determining factor for selecting a building zone or zones within the site. Bearing capacity of the subsoil should far exceed the future loading potential. For slab-on-grade foundations it can be as low as 2500 lb/ft^2, but for point-loading structural systems you may need more than 5000 lb/ft^2. Other considerations include the permeability of the soil, the amount of clay or rock encountered in the soil tests, and ground water levels. Each can affect the cost of construction, and should be considered prior to final siting of the buildings. Copies of soil information maps which illustrate the classification and capacities of soils on and around

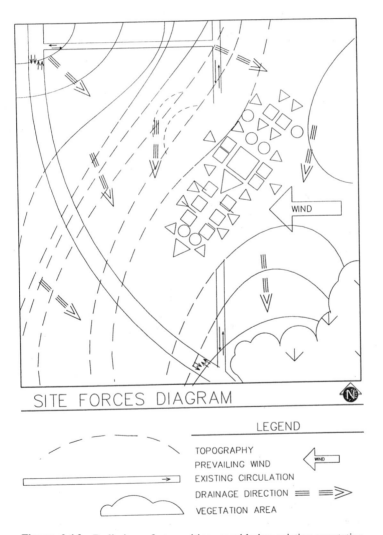

Figure 4.14 Preliminary factory siting, considering existing vegetation.

the projected site can usually be obtained from city or county records. More detailed or specific information can be provided through geotechnical consultants.

Topography

Positive natural drainage can usually be achieved through proper topographic considerations. A good master plan design often results in minimal cut and fill operations. The material cut can be used to backfill around the construction. Foundation fill may be obtained from other required on-site operations such as retention pond excavation. On some sites, there is no suitable fill available, and all must be trucked in from off-site at a cost of more than $5.00 per cubic yard. For a large project this may add up to more than $140,000 for 2 ft of fill under 250,000 ft^2 of floor area. Under-drain systems can be designed to remove surface water from building perimeters also. These systems often cost more than fill, and will require maintenance or replacement over time. The best site plan will avoid costly improvements. Parking lots can be terraced around the building perimeter to retain soil and maintain existing elevations and vegetation.

Vegetation

Vegetation studies usually reveal where the soils are richest and where bearing capacities are lowest. They also indicate where animal life is sustained, which may be a significant issue for people's comfort. There is a subliminal reassuring quality about seeing animal habitat, especially near a manufacturing environment. It eases people's concerns about their personal safety, and indirectly informs them of the company's values toward life outside the factory line. The once common tradition of simply clearing the entire site prior to design or construction is being replaced by thoughtful and selective clearing concurrent with design. A tree survey of all trees over 4-in. in diameter provides valuable information for use in the process of master plan design.

Twenty-first-century manufacturing facilities should be designed so as to cost-effectively work in harmony with the natural topography. Through careful site planning, amenities such as recreation facilities, fitness trails, parks, and more can be a natural extension of the site. Today's and tomorrow's best employees will weigh such factors in determining future employers. The addition of new plants after site construction is completed can be kept to a minimum, thereby reducing landscape purchase and maintenance costs. See Figure 4.14 for an example of vegetation zone identification.

Circulation

Circulation studies can include vehicular circulation on and around the site, as well as human and animal paths. Even if they are to be consciously altered later, they can provide useful information for the designer. Roads and walks can be designed to allow easy access to loading docks. Waste receptacles, dumpsters, and parking lots can be sited for best access, or for desired low visibility. Through good design, bottlenecks and dangerous intersections can be avoided, comfortable turning radius corners for tractor trailers can be accommodated, and all users will be happier. A complete study will show more information needed: minimum road width, existing road capacities, parking, sidewalks, etc. For preliminary parking designs, use 10 ft by 20 ft for automobile spaces and 25-ft roadways for two lanes. This does not include a factor for aprons, turning lanes, or circulation. A rule of thumb which is acceptable for preliminary estimates is to use 350 ft^2 per vehicle. Double-loaded 90° car parking lots with some walks and retention areas can be designed on 100-ft centers. Then the preliminary design studies can derive more accurate information, complete with minimum turning radii for different vehicle types, which should take into account trailer parking areas for just-in-time delivery.

Winds

Prevailing winds need to be determined so that fresh air intakes are not located downwind from noxious exhausts. A good rule of thumb is to try to locate all fresh air intakes upwind and on the side of buildings. All exhaust fans should be on downwind rooftops, extended 10-ft above the roof surface in order to assure mixing. All exterior employee functions should also be located away from the prevailing direction of exhaust discharge stacks.

Views

Views should be diagrammed to determine which areas should be visible from the inside. No longer do we avoid any windows (for theorized production enhancement), as we tried to do in the 1960s and 1970s. We learned that a windowless building *does not* create the best performance environment. Limited access to selected views is a more successful approach for the future. Allowing people to acquire their own information about their outside world is calming and allows them to focus on their work. Humans need to sense if it is dark or light, and if they are facing north or west. These needs are thousands of years old and cannot be ignored without negative results. Visual connection with the outside world is reassuring.

Utilities

Early site studies will have revealed where the existing utility connections must be made. These drawings are often crudely schematic and should always be field verified prior to design or construction. The quality and quantity of the utilities will affect the eventual project initial and operating costs. At this point in the process of design, detailed information will permit the most cost-effective design solutions. Site planners should always consider the cost of utility extensions along with all the other factors in design. Water, sewer and electric combined could cost more than $200 per linear foot. Asphalt roadways, by comparison, can cost around $50 per linear foot. Valving, metering, and directions of flow should be shown for water systems, along with size and specification of the conveying materials. Finally all carrying capacities must be determined. Then the designers can make the most economical design decisions. Figure 4.15 shows a possible schematic design which could result from the individual evaluations in the previous sections.

Summary

There is no good reason to avoid any of the simple aspects of site analysis listed above. Cost-effective initial and long-term results depend on it. If you won't do it, hire someone who will. A good master plan starts with a listing of the components of present needs, and should look 10 years or more into the future at potential developments. This will allow for the best arrangement today without erecting permanent structures or infrastructure in areas to be used later for other purposes. It will develop guidelines for an overall land use plan, general arrangement of buildings and spaces, and their connections to vehicular and pedestrian circulation. Net and gross square footage figures are calculated for the site, for total parking, and for each building type. From some of these early numbers, budgets can be accurately developed for economic forecasting. They are also useful in obtaining early review comments from governing regulatory agencies.

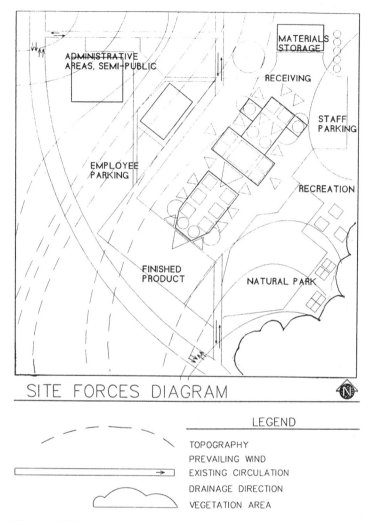

SITE FORCES DIAGRAM

LEGEND

TOPOGRAPHY
PREVAILING WIND
EXISTING CIRCULATION
DRAINAGE DIRECTION
VEGETATION AREA

Figure 4.15 Possible overall project design showing the factory outline taking form, and the related support functions.

Total area of the site will be the first and most limiting factor in a master plan. Zoning, height, setbacks, and minimum parking area will be of secondary importance. Next one must develop the gross area of each type of building space and specify the general activities to take place within. Then the issues of adjacency, sound levels, safety, physical limitations of equipment, and operational requirements of the facility must be factored in. For some facilities, such as those dealing with explosives or potentially dangerous materials, adjacencies will be the most significant determinants of the master plan. This will generate a widely clustered group of small assembly buildings located hundreds of feet apart.

Facilities which share expensive conveyance devices or which utilize cryogenic systems will normally be rather centralized in order to minimize waste and cost. Each site will generate a unique set of forces which should effect the master plan. Figure 4.16

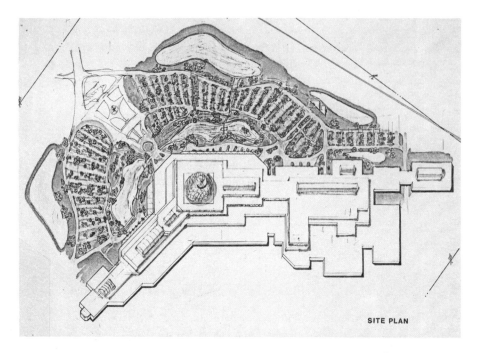

SITE PLAN

Figure 4.16 Example of the master plan for a manufacturing plant which includes employee housing and recreation facilities.

illustrates the possible impact of topography on a master plan for a manufacturing center which includes employee housing and recreation facilities to complement the factory.

4.4.4 Facility Design

Architecture

The most visible of all decisions are made during the actual facility design. Massing, form, and finish all work together to create an image of the completed facility, but how it fits with the needs of the manufacturing process is what counts most in factory design. This is when the designer take the programmatic information provided (from Chapter 3, as modified during the programming phase by the A&E firm and the client) and turns it into plans and specifications which yield the final buildings. Decisions made during design affect not only initial cost but how the buildings perform in the manufacturing mode over time. A thorough building program will provide such detailed information relative to the performance requirements that there should be only a few viable organizational schemes that support the optimum manufacturing process.

The traditional manufacturing building resembles a warehouse. The architect is challenged to create a cost-effective method of expressing the corporate image. Obviously, the building must support the specific manufacturing process as described previously. So the length and shape of the product line may be the primary factor in determining the ultimate building form. Steel column and beam construction has been used for hundreds of years for a reason—because of its simplicity. This leads to rectangular buildings which are limited in height by money; the higher they are the more they cost to build and operate.

Simple geometry tells us that the closer to square we shape our buildings, the less surface area will be required of exterior walls. Other factors include:

Cost-effective span for joists
Areas which require exterior walls
Height of machines, products, work in process
Material flow
Utilities

The first tool the designer has is the site. First impressions of a place are lasting. Are there visual or physical forces which will affect our perception? If not natural, are there site development requirements which could lead to new features such as retainage ponds, berms, or tanks which will create visual interest? The manner in which we deal with parked cars will have a significant visual impact. The next significant visual aspect of design is the massing of the buildings. Landscaping becomes the finishing touch, softening hard angles. Finally, finishes, color, texture, and details make the last impressions.

There are a few opportunities for the designer to infuse visual interest into a simple box without spending more for it. Through careful evaluation of the necessities of design one can keep the "beast" from growing either too large or too expensive.

Generally, a building using metal wall panels and steel post-and-beam construction will utilize spans of between 25 and 50 ft, with wall height between 8 and 16 ft. Typically the columns are within the outside skin and are not expressed on the exterior. Girts, purlins, and intermediate framing support the skin system. For conditioned buildings there is usually a liner panel with insulation between the two skins to reduce heat loss or gain. Open-web steel joists are most commonly used for secondary structural supports because of their low cost. Mechanical and electrical components are then supported from the bottom chords of the joists, so there is no natural beauty in the materials. It will be up to the skilled professional to create a complex which is more dynamic than the sum of the parts. One of the often overlooked decisions relates to the need for conditioning many of the factory spaces. The range of allowed tolerances for temperature and humidity will dramatically affect the mechanical and electrical systems' initial and long-term costs.

Depending on the particular site, building zones may influence the shape of the building. If land costs are high, it may be more economical to build vertically than horizontally. The shape of the roof is often a powerful image maker in the designer's toolbox. The span will determine the most cost-effective roof structure, and the climate will determine the material. Low-slope metal roofing is cheaper by far than flat single ply, so if the least dimension of the building is not too great, it will be cheapest to design a low-slope roof. If the span exceeds 100 ft, it is common to use internal roof drains and to segment the roof area with sloping insulation around a series of roof drains. The intermediate method is for a low slope roof with a center ridge, and roof gutters and leaders just inside the exterior wall parapet. This has the aesthetic of a higher-cost building without the cost of interior roof drains and leaders. One of the few other elements which has been used historically for decoration is the overflow scupper. Others are exterior lights and lamps, stair rails, and bolted connections.

Visual Impact

One of the low-cost choices with greatest visual impact is the column. Columns and beams which are internal of the skin are not revealed or expressed on the outside, resulting in smoother, planar massing. Columns expressed on the outside of a building can create rhythm, visual interest, and will break the large planes into smaller components. Primary and secondary steel can be painted in such a way as to reinforce the hierarchy of the parts. This will also provide greater visual interest than having all exposed materials the same color.

The choice of skin materials will have to be made with climate and programmatic considerations first. Pragmatics usually cause factory exterior walls to have low initial and maintenance costs. This leaves about four choices, each of which has a different aesthetic. Metal panels can be flat or can have ribs for lateral load resistance. For a planar facade, designers choose flat panels, usually with factory-applied finishes. Metal panels with raised ribs can provide additional resistance to lateral or wind loads. Second choice is a precast concrete panel system or site-cast "tilt-slab" concrete construction. A third choice is a relatively new system known in the industry as EIFS (for exterior insulation and finish system). It is composed of rigid foam panels with a mesh reinforcing layer and synthetic plaster finish. If the wall height is not too great, concrete block exterior walls are a common solution to low-maintenance construction. They can be specified with textures and color from the supplier. "Slump blocks" or split-faced blocks are rough-textured, with the appearance of jagged concrete. These are also available in a variety of colors. Other concrete block choices include smooth, stucco, and scored block. Each will yield a different texture and result in a different contrast in light, shade, or shadow. The skilled designer can combine these basic choices into an effective image-achieving solution. Massing, contrast, and color affect the way we perceive the composition of buildings, while axes and siting affect the power of the image.

One issue which controls how much information we use to form our image of a project is the speed at which we view it. There is not much time for observation at 65 mph. We first see the form and try to compare it with those we have seen before—what is it like? Is it like a barn, or a church, or does it resemble a "blah" warehouse? The three things which give us those lasting visual impressions are the manner in which the building meets the sky (roof shape, slope, scale), the way it breaks up the massing, and the way it meets the ground. Does it appear to be "plopped" on the site, or is it integral with the site? As human society changes with time, so does our perception. While once it was sufficient to spend a few bucks at the entrance in order to give a favorable impression, people today look more at the total picture than before. Few factories have the luxury of being able to spend much of the budget on applied decoration. This places more importance on the few areas where there is little or no additional cost for architectural treatment. It is best to manipulate the necessary items of the building and not rely on applied decoration. Rather than try to hide the systems components behind bulkheads, it is more cost-effective to treat them as elements to be exposed and to make them into the decorations. Spiral duct and cable trays, for example, reinforce the high-tech aesthetic. Leaving them accessible will also reduce maintenance costs and future remodeling costs.

Mechanical Design

Once the programmatic information is understood, it is time to determine the best way to provide them. It is also necessary to ask questions. Are the requirements real or perceived? The designer must decide how best to achieve the real needs, while still exceeding all applicable code minimums. The best designers will make use of low-tech, passive systems where possible. They will be able to provide for flexibility of future installations during initial design, without adding substantially to the initial cost. They should also look at many optional solutions for each problem before deciding upon one. This type of brainstorming often results in as many breakthroughs as it does rejects.

For example, rather than conditioning a space 14 ft high, which is 20 by 40-ft of floor space around a machine, it may be more cost-effective to construct an enclosure around the machine space, which can be conditioned for $\frac{1}{10}$th the initial, and $\frac{1}{100}$th the 10-year operating cost. In many instances it is possible to cap pipes for possible future installations, and to size them for future anticipated loads. This may result in slightly larger branch piping for about 5% of the distribution network, which will have minimal impact on initial cost. The use of zone isolation valves is another often-overlooked aspect of mechanical design. Their initial cost is low compared to the future cost of maintenance, and to draining the entire systems to install a new branch for a new piece of equipment.

The number of air changes per hour, for example, is usually determined by the classification of the use of the space (e.g., office versus industrial use). This affects the fire protection system design requirements, and the type of construction materials allowed. As a rule of thumb, between three and five air changes per hour is a basis for rough estimating. Areas with chemical processes may need more than 10 air changes per hour. Another method is to figure 1 cfm supply for every 10 ft^2 of building area. Of course, further design studies would take into account calculated heat losses, heat gains, insulation, humidity, and allowable fluctuations over time. The psychometric chart is a good way of understanding the relationships among creature comfort, humidity, and air velocity. Unrealistic design criteria can result in thousands of dollars in mechanical equipment initial and operating cost. This is a critical area in design, when the decision to use one system over another may affect air quality and cost more than at any other time in design or construction. These decisions must be evaluated by the ownership team in the schematic phase, prior to proceeding to construction contract documents.

An example is a requirement to maintain $40 \pm 2\%$ relative humidity. This could entail the use of expensive humidifiers in the winter and heating coils during the summer months to dry the conditioned air. If the real criteria is that the humidity must stay below 60% year round, that would be much easier and cheaper to achieve. In some instances it is better to increase the velocity of air rather then try to keep the temperature of the air introduced into the room lower. The resultant comfort level for the worker will be the same, and the initial and operating costs will be lower. Similarly, for a positive-pressured building, it may be cheaper to design less exhaust ducting and fans, and to allow perimeter losses through dampered grilles. A comprehensive evaluation of many alternatives should be included in the schematic design phase. Insulation of walls and roof surfaces should then be designed to minimize operational costs. The biggest challenge in mechanical systems design is to get to the essential information. Only through understanding of the intended goal can one design the best solution.

In general terms, the mechanical system should be designed to provide productivity, reliability, and safety. Profit is the goal, and predictable cost forecasting is a good way to get there. For a factory, the project cost of using superior mechanical products may be less than 3% higher than the low-cost competitor. Variable-speed pumps may cost more to purchase, but they can support a wider range of operating conditions. If the demand is relatively constant, fixed-speed pumps should be selected which most closely match the design criteria. There are many decisions which will affect the performance of the plant. Selection of the right mechanical components, and cost-effective layout of the utilities, are critical. See Figure 4.17 for an example of some mechanical installations.

Figure 4.17 Example of typical mechanical installations in a complex project.

Mechanical system design is far too complex to cover in a few pages of a book on manufacturing engineering. Materials selection alone would require a chapter, piping alone could barely be covered in 50 pages. Specifications and details affect construction cost nearly as much as area and scope. Many excellent references provide more comprehensive information on the subject. An overview of both mechanical and electrical equipment systems is presented in McGuiness, Stein, and Reynolds, *Mechanical and Electrical Equipment for Buildings*, 6th ed. (John Wiley, New York, 1980).

Electrical Systems

Electrical systems design is more objective than mechanical design, but there is the same challenge for cost-effective decision making. Starting from the big picture, one should design to utilize natural lighting where practical to augment artificial lighting. Power and lighting distribution should be done at the highest practical voltage. Since electrical systems are the most frequently changed, it is good to use cable trays where possible. This reduces initial and remodeling costs. Some facilities use a lot of 480V bus duct in lieu of cable in conduit. With recent changes in some manufacturers marketing schemes, bus is available in standard lengths in a matter of weeks. It is very flexible compared to modifying cable in conduit runs. Bus plugs can be added wherever needed to power step-down transformers at the point of use. This also provides a means of disconnect in lieu of mounting a distribution panel and breaker or main disconnects.

Usually, all large motors will be specified to operate at the highest available voltage. Interruptable or curtailable power can be placed on separate circuits, in order to negotiate a lower annual cost for power. All noncritical systems should be powered on the interruptable circuits. Power factor correction is another important factor in electrical systems design. Capacitor banks can be utilized to offset fluctuations in primary current from the local utility. After distribution and step-down transformers, line conditioners or surge supressors can be added.

Lighting control systems are available to optimize the efficiency of lighting systems. Experienced designers will choose the most cost effective fixtures for each space, looking first at the amount of direct and indirect light from the shape of the fixture. Next one would look at the bulbs, striving to achieve the best initial and operating cost for the owner's needs. Lighting consultants can also recommend special designed task lighting in areas for which it would be costly to provide general illumination fixtures.

With the continuing development of ultra-low-energy light sources such as we have seen in the last few years, lighting design is about to be revolutionized. Imagine telephone wire providing enough current to light a small office at 40 foot-candles. The material cost of the installation is going to be low due to the amount of copper in cross section, and there will be no need for conduit since it is all low-voltage. Furthermore the operating cost will be low due to the current draw, and switches should be cheaper and should last longer. It has been reported that bulbs will last more than 20 times as long as traditional bulbs. Inventory and replacement costs should be negligible. A valuable reference on this subject is Albert Thumann, *Plant Engineers and Managers Guide to Energy Conservation* (Van Nostrand Reinhold, New York, 1977).

Communications Systems

The reliance of business today on information systems will change the role of the manufacturing engineer dramatically. As part of the new team approach to product development, the ME will need skills in computers and information flow. Networks consist of the computers to be connected, the communication media for the physical connection between the computers, and the software for the network communications and management (see Subchapter 6.4).

As the name implies, the *local area network* (LAN) is geographically the smallest of the network topologies. In general, the LAN is limited to 1 to 2 miles in displacement. The *wide area network* (WAN) has the capability to allow connecting a LAN to other LANs within the company or with LANs of vendors and customers. The speed of the WAN is determined by the communication media used for the long-distance transmission. At the low end of the speed regime is the modem using regular phone lines. In most cases, however, if you have more than one computer, a modem will not suffice. Most of the data transmitted over long distances is sent over T1/E1 lines.

One of the most important factors to consider in the design of manufacturing support systems is the amount of data to be transmitted over the network and the response time of the data transfer. The speed of the network will depend on the type of connection media (copper wire, coax, or fiber optic), the distance to be covered, and the network software. The actual cabling that connects the various components on the network comes in many varieties. The three major ones are:

Coaxial cable, where one or more concentric conductors are used
Twin twisted-pair cable with two pairs of wires, each pair twisted together
Fiber optic cable, used for FDDI and can be used with Ethernet

The coax and twisted-pair cables have been used for some time. The fiber optic cable is fairly new, but this technology is expanding at a very rapid rate. With all of the phone lines being changed over, the price of this transmission medium is dropping to the point where the material is very competitive with wire. The cost of installation of a fiber optic system is still higher than for the other types, due in large part to the unit cost of converting the signals. The expanded capabilities and the immunity of a fiber optic system to electrical interference makes it a good choice in many applications. Once all the elements (i.e. mainframe, mini, PC, and programmable controllers) and all the necessary hardware and software parts for the network are determined, the *computer integrated enterprise* (CIE) becomes possible. The CIE consists of a combination of various functional units to perform the desired tasks to operate the business. These include financial, purchasing, material control, manufacturing process control, and reporting. To perform these tasks, the CIE must have certain capabilities. Processing (central processing unit, CPU), input and output terminals, printers, and an interconnecting network among the units make up the minimum capabilities. A typical architecture for a CIE system is shown in Chapter 6, Figure 6.12. This example assumes that your organization has used computer systems for some time, and you will have a mixture of types of hardware and legacy software programs in a multisite system. Medium to small, or new start-up, manufacturers would have the luxury of designing a system from the ground up. See Figure 6.9 for an example of a new start-up system.

With respect to the multi-site system, Figure 6.10, the mainframe is at one location while other elements are at remote locations. The data are transmitted from location to location via hard-line (T1 lines) or in some cases by satellite. For the system to be a true CIE, and to be as productive as possible, the user should input data into the system once. It is then up to the program logic to route and install the data into the correct tables and databases for processing and record keeping. Some of the processing will be handled at the local terminals and servers, while other data will be transmitted to the mainframe for processing. One thing to remember is that in a well-designed system, all applications should be available to the users at local terminals. Timely information at the lowest level of the organization will allow workers to make decisions and improve productivity. Figure 4.18 shows some overhead cable trays containing signal and data cables.

To be first in class, your CIE must provide data to and from your customers, vendors, and across your total factory. For you to accomplish your goals of meeting and beating your competition, you must look at all the factors that will affect your price (and profit). Inventory and work in process must be reduced to an absolute minimum. This can be accomplished by judicious design of your computer interfaces (electronic data interchange, EDI) with your vendors. With a *material requirements planning (MRP)* or a *manufacturing resources planning (MRP-ll)* system, you can effect a just-in-time delivery system to reduce your inventory, and work in process (see Chapter 11).

The definition of the communication systems for your factory must be transmitted to the A&E or other design agency as part of developing the factory requirements. For example, if you plan an Ethernet backbone using fiber optics down the center of the plant, you must describe it. If you plan on distribution closets from the backbone at strategic locations throughout the plant, with twisted-pair hardwire into all offices, receiving, stores, shipping, and most of the workstations on the manufacturing floor, you should specify the criteria the building designer will need. Telephones, FAX machines, and other systems should also be defined. If there are distributed numerical control (DNC) machines on the floor, you need to define the linkage to the programmers, minicomputers, etc.

Figure 4.18 Typical overhead cable trays.

4.4.5 Costs

Now that we have discussed the programmatic issues and started the real factory design, it is possible to make an accurate cost estimate based on the actual materials, equipment, and other items specified. The earlier estimates were partly parametric calculations, based on experience with other building elements and projects. The cost-estimating process has been an ongoing activity, as we have progressed through the process leading up to a preliminary design.

Once design documents are past the schematic approval stage, it is time to develop them to the point where specific materials selections and systems components are selected. Cost estimates need to be developed from the entire set of drawings including scale floor plans and elevations. We can now use actual area information along with lists of major equipment in order to determine how the decisions made during design have influenced costs, and to check current construction cost information relative to our earlier projections. For example, if the costs appear to be 7% over budget at the 50% design development point, reductions can be designed into the final documents. With computer stretch and auto-dimensioning functions, one can easily make area changes. Each subcomponent shown on the plans and specs should be accounted for in this estimate. Figure 4.19 is an example of a "warehouse"-type factory building estimate, which is representative of the "low end" of technology and cost. Figure 4.20 is a detailed, multipage example of an estimate for a prototype plant, which is representative of a cost breakdown for a more major "high-tech" factory building.

Description	Quantity	Units	Unit Price	Price
TYPICAL WAREHOUSE BUILDINGS				
CONSTRUCTION COST BREAKDOWN				
21,000 SF OF FLOOR SPACE, UNCONDITIONED				
600 SF OF CONDITIONED OFFICE				
Building Personnel				
Superintendent	24	wks	$650.00	$15,600.00
Travel	24	wks	$100.00	$2,400.00
Project Engineer, partial	24	wks	$200.00	$4,800.00
Clerical support	24	wks	$25.00	$600.00
Permits & Fees				
Electrical Tap fee	1	lump sum (ls)	$1,200.00	$1,200.00
Environmental Permit	1	ls	$240.00	$240.00
Water Tap fee	1	ls	$900.00	$900.00
Sewer Tap fee	1	ls	$2,450.00	$2,450.00
General Conditions				
General Labor	21600	sf	$0.06	$1,296.00
Cleanup	6	mos	$250.00	$1,500.00
Rubbish Removal	6	mos	$350.00	$2,100.00
Dump charges	6	ea	$150.00	$900.00
Temp. toilet	6	mos	$65.00	$390.00
Water & Ice	6	mos	$25.00	$150.00
Pager, mobile ph., partial	6	mos	$65.00	$390.00
Temp. Telephone	6	mos	$120.00	$720.00
Temp. Electric	6	mos	$125.00	$750.00
Temp. water, sewer	6	mos	$45.00	$270.00
Layout	1	ls	$250.00	$250.00
Reproduction, etc.	1	ls	$450.00	$450.00
Testing	1	ls	$550.00	$550.00
Equipment rental	1	ls	$1,000.00	$1,000.00
Signage	1	ls	$450.00	$450.00
Office Trailer	6	mos	$275.00	$1,650.00
Supplies	6	mos	$25.00	$150.00
Security	3	mos	$500.00	$1,500.00
Site Construction				
Site Clearing	1	ls	$13,500.00	$13,500.00
Rough Grading	1.5	acres	$5,000.00	$7,500.00
Finish Grading	22000	sf	$0.25	$5,500.00
Fill & Compaction, 18 inches	2000	cy	$4.50	$9,000.00
Paving	12000	sf	$1.15	$13,800.00
Erosion controls	1	ls	$650.00	$650.00
Drainage	1	ls	$4,500.00	$4,500.00
Landscaping	1	ls	$6,500.00	$6,500.00
Seed & Mulch	2	acres	$1,000.00	$2,000.00
Utilities Improvements	1	ls	$6,800.00	$6,800.00
Fencing	1	ls	$3,400.00	$3,400.00
Building Construction				
Slab edge forming	1100	lf	$0.35	$385.00
Concrete	680	cy	$55.00	$37,400.00
Steel	23000	sf	$0.80	$18,400.00
Concrete finishing	21500	sf	$0.65	$13,975.00
Masonry, block, w/pilasters	1400	ea	$1.35	$1,890.00
Ornamental Iron	1	ls	$1,200.00	$1,200.00
Metal doors & frames	6	ea	$400.00	$2,400.00
Overhead doors	3	ea	$2,875.00	$8,625.00
Windows	6	ea	$250.00	$1,500.00
Hardware	1	ls	$4,250.00	$4,250.00
Glass & Glazing	2	ea	$120.00	$240.00
Metal studs & Drywall	2000	sfsf	$4.35	$8,700.00
Flooring, office	320	sf	$1.15	$368.00
Painting	2000	sf	$1.25	$2,500.00
Louvers, vents	1	ls	$4,250.00	$4,250.00
Accessories	1	ls	$450.00	$450.00
Loading dock, ret. walls	1	ea	$3,125.00	$3,125.00
Pre-engineered Buildings	21500	sf	$5.25	$112,876.00
Plumbing	1	ls	$2,500.00	$2,500.00
HVAC, window unit for office	1	ls	$650.00	$650.00
Operable shutters & Fans	1	ls	$6,500.00	$6,500.00
Electrical	21500	sf	$1.56	$33,540.00
Labor	12	manweeks	$320.00	$3,840.00
Misc'l Material	1	allowance	$6,500.00	$6,500.00
Insurance	1	ls	$27,650.00	$27,650.00
Bonds	1	ls	$2,520.00	$2,520.00
Overhead & Profit @15% of hard	1	ea	$55,960.00	$55,960.00
TOTAL COST				**$478,000.00**

Figure 4.19 Typical "warehouse"-type factory building construction cost breakdown.

PROJECT NAME: PROTOWIDGET				
Prepared by : W. L. Widget				
Date prepared: 10/12/94				
Based upon Plans _____ and Specs_____ dated				
DESCRIPTION				
General Conditions	Quantity	Units	Cost per Unit	Cost
Office, trailer etc. cost	10	mos	$650.00	$6,500
Utilities cost, Temp. & connect	10	mos	$14,000.00	$140,000
Submittals, copying, blueprinting	10	mos	$250.00	$2,500
Special Conditions (vary greatly)				
Bid Alternates,deductive alternates , etc.			(be specific)	
Protection, erosion, sound, dust	4	acres	$1,500.00	$6,000
Parking, storage, security	10	mos	$800.00	$8,000
Lay-down space, staging			$0.00	$0
Staff; direct cost plus Insurance	10	mos	$15,000.00	$150,000
				Initial
Site Work	Quantity	Units	Cost per Unit	Cost
Clear and grub	1.8	acres	$2,200.00	$3,960
Misc'l demolition	14,500	sf	$0.05	$725
Cut	21,300	cy	$2.00	$42,600
Fill	5,000	cy	$6.00	$30,000
Haul	16,500	cy	$2.00	$33,000
Rough grade	196,000	sf	$0.05	$9,800
Compact	8,000	cy	$1.50	$12,000
Finish grade	125,000	sf	$0.02	$2,500
Soil treatment	100,000	sf	$0.02	$2,000
Vapor barrier	100,000	sf	$0.05	$5,000
Excavation for footings, piers, etc.	1,200	cy	$4.85	$5,820
Pilings, driven	1,220	lf	$13.50	$16,470
Pilings, sheet	250	lf	$7.50	$1,875
Dewatering	100,000	sf	$0.05	$5,000
Gravel under slab	3,500	cy	$8.00	$28,000
Expansion joints	1,800	lf	$0.28	$504
Waterproofing, sheet	1,200	sf	$4.50	$5,400
Waterproofing, coatings	3,000	sf	$2.90	$8,700
Hand excavation, miscl	45	cy	$45.00	$2,025
Road base, 6" crushed rock, stone	1,600	sf	$0.39	$624
Labor, edge work	2,500	lf	$0.25	$625
1 1-2" Asphaltic Topping	15,000	sy	$3.70	$55,500
Concrete curbs	4,750	lf	$8.24	$39,140
12" reinforced concrete drive, 4000 psi	432	sf	$9.50	$4,104
6" reinforced concrete drive, 4000	40000	sf	$5.80	$232,000
4" reinforced concrete drive, 4000	10000	sf	$3.20	$32,000
4" sidewalks, 3500 w/ mesh	3800	sf	$2.35	$8,930
Pavement markings	350	ea	$4.00	$1,400
Signage	6	ea	$125.00	$750
Utilities				
Fiber optics system, backbone	1250	lf	$0.78	$975
Telephone cable, vault and conn.	350	lf	$0.25	$88
Data cable, 16 UTP, from comp's	10250	lf	$2.80	$28,700
Power from property line	125	lf	$3.50	$438
Water line, domestic, 4"	1550	lf	$6.00	$9,300
Water line, domestic, 2"	225	lf	$5.40	$1,215
Fire [rotection line 6"	500	lf	$15.00	$7,500
Fire prot. 8"	1400	lf	$18.00	$25,200
Post indicator valve	6	ea	$1,325.00	$7,950
Siamese connections	2	ea	$1,500.00	$3,000
Hydrants	5	ea	$1,400.00	$7,000

Figure 4.20 **(a)** Detailed, multipage example of an estimate for a prototype plant. This is representative of a major "high-tech" factory building.

Thrust blocks	24 ea	$45.00	$1,080
Metering	3 ea	$3,200.00	$9,600
Backflow prevention devices	1 ea	$2,100.00	$2,100
Gas piping, 2" steel	125 lf	$9.50	$1,188
Sanitary sewer, 8" PVC	1280 lf	$12.00	$15,360
Sanitary sewer, 6" PVC	200 lf	$8.00	$1,600
Sanitary sewer, 4" PVC	124 lf	$5.00	$620
Storm sewer, culvert	2 ea	$1,500.00	$3,000
Storm sewer, swale	4 ea	$400.00	$1,600
Storm sewer, 15" RCP	1300 lf	$17.50	$22,750
Storm sewer, 12" PVC	250 lf	$15.00	$3,750
Storm sewer, 10" PVC	100 lf	$12.50	$1,250
Storm sewer, 8" PET	300 lf	$10.50	$3,150
Storm sewer, 6" PET	1300 lf	$9.50	$12,350
Trench drain	150 lf	$112.00	$16,800
Interceptor	2 ea	$1,250.00	$2,500
Neutralizing tank	1 ea	$2,500.00	$2,500
Treatment tank	3 ea	$2,130.00	$6,390
Liquid sto. 400,000 gal. tank	1 ea	$232,000.00	$232,000
Gas storage tanks	3 ea	$27,500.00	$82,500
Fences, gates, vinyl coated	3000 lf	$10.28	$30,840
Site Lighting, poles & wire	20 ea	$1,050.00	$21,000
Well pointing	0 ea	$0.00	$0
Rock excavation	0 tons	$200.00	$0
Landscaping, allowance	10000 sf	$12.00	$120,000
Rebar	74 tons	$789.00	$58,386
Mesh	9 tons	$800.00	$7,200
Concrete, column footers, caps	53 cy	$58.00	$3,074
Continuous footings	165 cy	$53.00	$8,745
Pilasters, piers	11 cy	$61.00	$671
Isolation mats, slabs	105 cy	$64.00	$6,720
Pits, elevator etc.	50 cy	$69.00	$3,450
Cooling tower sump	125 cy	$72.00	$9,000
Special reinforced pads	23 cy	$112.00	$2,576
Concrete floor slabs, 4000 PSI	1928 cy	$46.50	$89,652
Concrete floor slabs, 3500 PSI	35 cy	$45.50	$1,593

Structure	Quantity	Units	Cost per Unit	Price
Structural steel, primary	618	tons, m	$1,173.00	$724,914
Steel, secondary bracing	56	tons	$1,210.00	$67,760
Steel, small members	37	tons	$1,379.00	$51,023
Special framing	120	sf	$4.90	$588
Roof openings	12	ea	$1,000.00	$12,000
Bar joists	0	tons	$1,000.00	$0
Floor decking, 2" 20 ga.	14000	sf	$1.50	$21,000
Roof decking, 1 1/2" 20 ga.	13000	sf	$1.05	$13,650
Acoustical metal decking	800	sf	$1.85	$1,480
Misc'l metal supports, etc.	100000	sf	$0.33	$33,000
Catwalks, etc.	4	ea	$750.00	$3,000
Metal wall panels, steel, painted	33700	sf	14.5	$488,650
Metal wall panels, alum. painted	4000	sf	$16.00	$64,000
Explosion relief wall system	200	sf	$18.00	$3,600

Figure 4.20 **(b)** Continuation of multipage example of an estimate for a prototype plant.

Masonry wall, 8 " CMU	7000 sf	$8.25	$57,750
Masonry 2 hr. wall	2641 sf	$5.50	$14,526
Metal studs & 1/2" gyp., not rated	84500 sf	$2.00	$169,000
Metal studs & 1/2" gyp, 2 hr. rated	6600 sf	$3.00	$19,800
Exterior insul. & finish system	5000 sf	$10.75	$53,750
Glass block	240 sf	$15.00	$3,600
Brick veneer	0 sf	$0.00	$0
Cut stone veneer	500 sf	$6.00	$3,000
Glazed facing tile on block	500 sf	$12.80	$6,400
Miscellaneous			
Roof blocking	3500 lf	$14.00	$49,000
Cant and transition strips	800 lf	$20.00	$16,000
Roof penetrations	200 lf	$10.00	$2,000
Frames at roof drains	64 ea	$125.00	$8,000
Wall frame blocking, deadwood	54 ea	$100.00	$5,400
Telephone backboards	12 ea	$125.00	$1,500
Window sills	2800 lf	$8.00	$22,400
Rough carpentry	100000 sf	$0.50	$50,000
Roof insulation boards, 2 " thk	112000 sf	$0.45	$50,400
Roof insul. lt wt concrete, sloped	4800 sf	$2.85	$13,680
Sealants, caulking	100000 sf	$0.45	$45,000
Finish carpentry	100000 sf	$0.80	$80,000
Casework	15000 sf	$12.00	$180,000
Furnishings, desks, chairs, cab's	25000 sf	$1.75	$43,750
Shelving, metal, heavy	450 lf	$18.00	$8,100
Shelving, metal, light	220 lf	$9.00	$1,980
Fume, flow hoods, thru roof	8 ea	$6,750.00	$54,000
Special hardware for connections	8 ea	$1,000.00	$8,000
Dock levelors, w/bumpers	6 ea	$5,250.00	$31,500
Lockers	60 ea	$180.00	$10,800
Benches	12 ea	$140.00	$1,680
Marker boards	14 ea	$240.00	$3,360
Fire extinguishers & cab's	36 ea	$125.00	$4,500
Wall corner guards	2000 lf	$15.00	$30,000
Access flooring, raised	1200 sf	$10.50	$12,600
Ramps, steps	325 sf	$20.00	$6,500
Crane, 20 ton X-Y, remote, 500 sf	2 ea	$110,000.00	$220,000
Crane, 10 ton X-Y per 500 sf	2 ea	$80,000.00	$160,000
Telephone enclosures	0 ea	$1,000.00	$0
Pallet storage system	2000 sf	$45.00	$90,000
Safe storage cabinets, containm't	600 sf	$115.00	$69,000
Suspended ceiling systems	22000 sf	$1.45	$31,900
Acoustical treatment	1200 sf	$8.00	$9,600
Curtain wall glazing system, ext	2000 sf	$31.00	$62,000
Glass panels, interior	1000 sf	$12.00	$12,000
Epoxy floor coatings, f'gl reinf	800 sf	$3.50	$2,800
Epoxy floor paint, 2 coat	60000 sf	$2.50	$150,000
Concrete stain, seal	15000 sf	$1.75	$26,250
Ceramic tile , incl. cove & base	2400 sf	$10.00	$24,000
Carpet	18000 sf	$1.50	$27,000
Acid resist surfaces, w/backer bd.	4000 sf	$14.00	$56,000
Painting, ext. masonry	1200 sf	$0.88	$1,056
Paint, int. masonry	120000 sf	$0.78	$93,600
Block filler	120000 sf	$0.54	$64,800
Epoxy on walls	50000 sf	$0.75	$37,500
Latex walls	20000 sf	$0.50	$10,000
Ceilings	8000 sf	$0.75	$6,000

Figure 4.20 (c) Continuation of multipage example of an estimate for a prototype plant.

Accent paint colors	4000	sf	$0.85	$3,400
Color coded piping, safety striping	75000	sf	$0.33	$24,750
Miscl iron & stairs, rails	100000	sf	$0.17	$17,000
Elevator	0	ea	$50,000	$0
Mechanical Equipment Allowances				
Description	Quantity	Units	Cost per Unit	Price
Wells & pumps, 1000 GPM cap.	1	system	$30,000	$30,000
Cooling towers, per 30000 gpm	2	ea	$125,000	$250,000
Chillers, per 2000 ton unit	2	ea	$925,000	$1,850,000
Primary water pumps, 500 gpm	2	ea	$27,000	$54,000
Secondary pumps	6	ea	$21,000	$126,000
Strainers & filtration, varies greatly	1	allowance	$50,000	$50,000
CUP condenser water piping	1	allowance	$200,000	$200,000
Water treatment system	1	allowance	$50,000	$50,000
DI system, 5 megohm, polish, demin.	1	allowance	$350,000	$350,000
Dual air compress sys., complete	1	allowance	$50,000	$50,000
Heat exchanger, dual @ 1000gpm	1	allowance	$200,000	$200,000
Process fluid pumping, meter'g etc.	1	allowance	$60,000	$60,000
Process gases, bottles, tanks etc.	1	allowance	$50,000	$50,000
Reclamation system,scrubbers, etc.	1	allowance	$75,000	$75,000
Boiler, 6,000 MBH, gas, steam gen.	1	allowance	$60,000	$60,000
Boiler, gas, hot water, 1,500 MBH	1	allowance	$25,000	$25,000
Furnace, elect. 30 MBH, w/ controls	1	allowance	$750	$750
Oven, 5' horiz. gas fired, @ 450 C	1	allowance	$12,500	$12,500
Vibration isolation for process piping system	1	allowance	$35,000	$35,000
Fire pump system, 2 500 GPM	1	allowance	$55,000	$55,000
Circ'g pumps, 1/12 to 1 hp, total 10	1	allowance	$5,000	$5,000
Vertical turbines, 250 psi, 50' head	1	allowance	$25,000	$25,000
Fume hood, 100 cfm, thru roof	1	ea	$5,000	$5,000
Fan, exhaust, roof curb mt, 500 cfm	1	ea	$2,500	$2,500
Fan, exhaust, roof curb mt, 5000 cfm	1	ea	$5,000	$5,000
Fan, wall exhaust, 2500 cfm	1	ea	$1,200	$1,200
Fan, recirc. fan coil, 10,000 cfm	1	ea	$10,000	$10,000
Smoke evac. fan & controls sys	1	ea	$7,500	$7,500
Fan, floor or base mt, 6 by 6, 120v	1	ea	$150	$150
Piping, black steel, screw, hung, 2"	1000	lf	$15	$15,000
Piping, black steel, welded, 12", insul	160	lf	$120	$19,200
Piping, black steel, welded, 8", insul	5000	lf	$60	$300,000
DI water piping, stainless, 4" w/ insul	2000	lf	$150	$300,000
Water piping, PVC, solvent weld, 4"	2000	lf	$20	$40,000
DI water piping, polypropylene, 2"	1000	lf	$15	$15,000
Copper piping, 2", soldered with valves	1250	lf	$20	$25,000
Borosilicate pipe, 2" acid waste	800	lf	$40	$32,000
PVC sch40, 6" w/ hangers	600	lf	$25	$15,000
PVC sch40, 2"	1500	lf	$14	$20,250
PVC sch40, 1/2"	600	lf	$9	$5,400
Grooved steel, 6" 150 psi victaulic	1000	lf	$40	$40,000
PVC, sch 10, 48" duct from hood	400	lf	$36	$14,400
FP water piping, 4", 150 psi, plastic	4000	lf	$10	$40,000
Galv. metal ductwork,	2000	lf	$9	$18,000
PVC lined metal duct, 4" by 14", slip jts	200	lf	$12	$2,400
Stainless steel duct, 4" by 14", welded	400	lf	$18	$7,200
Spiral duct, 12", flex, strap hangers	1000	lf	$8	$7,500
Plumbing				
Emerg. eye wash	2	ea	$75	$150
Eye wash with shower, plastic or metal	4	ea	$350	$1,400
Shower, ceramic tile surround, 3' by 4'	4	ea	$750	$3,000
Lavatories, sinks, etc.	14	ea	$175	$2,450
Water closets, uri's, white vc, flush v.	10	ea	$500	$5,000
Water drinking coolers, 5 gpm, inst.	3	ea	$450	$1,350
Drinking water fountains, installed	4	ea	$45	$180

Figure 4.20 **(d)** Continuation of multipage example of an estimate for a prototype plant.

Description	Quantity	Units	Cost per Unit	Price
ADA approved stall, shr, lav, wc, mirror	2	ea	$1,000	$2,000
Loop heat exch's, 10 gpm, 1200w/hr	2	ea	$3,500	$7,000
El. water heater, 200 Gal, 180 deg	2	ea	$15,000	$30,000
Point of use water heaters, 1500 w	6	ea	$150	$900
Accessories	25000	sf	$0.20	$5,000
Electrical, summary of takeoff				
Description	Quantity	Units	Cost per Unit	Price
Temporary service, 200 Amps, 2 loc's	2	ea	$750	$1,500
Primary svc, 13,200 volts, in conduit	1	lump sum	$450,000	$450,000
Power factor correction , 95%	1	lump sum	$140,000	$140,000
Main switchg'r, double ended, w/enclo.	1	lump sum	$250,000	$250,000
Motor control centers	1	lump sum	$3,000	$3,000
Distribution system, 480v, power	1	lump sum	$650,000	$650,000
Distribution at 208v, power	1	lump sum	$845,500	$845,500
Distribution at 110v, power	1	lump sum	$115,750	$115,750
El. generator, diesel w/tank , 150Kw	1	system	$40,000	$40,000
Lighting distribution panels	1	lump sum	$15,000	$15,000
Lighting circuits, conduit & wire, inst.	1	lump sum	$215,000	$215,000
Uninterruptible power system, 150A	1	lump sum	$165,000	$165,000
Controls, energy mgm't, security & HVAC	1	lump sum	$125,000	$125,000
Distributed controls system, dual, digital	1	lump sum	$1,600,500	$1,600,500
Systems programming, start up	1	lump sum	$45,000	$45,000
Fire alarm controls, smoke evac	1	lump sum	$25,000	$25,000
Line conditioning, point of use	10	ea	$1,500	$15,000
Surge protector	100	ea	$150	$15,000
Power towers, to 12', unistrut, cable&cond.	20	ea	$250	$5,000
Transformers, 480 to 208, 10 KVA	12	ea	$1,000	$12,000
Lightning protection system	100000	sf	$0.35	$35,000
Equipment ground system	100000	sf	$0.15	$15,000
480 V busway, w/ plugs, 2000A	2700	lf	$460.00	$1,242,000
Solid copper bus bars, 6000A,	400	lf	$600.00	$240,000
Cable in conduit, 500 MCM 15KV in tray	1000	lf	$54.00	$54,000
Cable in conduit conductors, 250 MCM	2000	lf	$25.00	$50,000
Cable in conduit, 5 KV w/PVC jacket, #2	5000	lf	$13.00	$65,000
Cable in tray conductor, #12, 3wire	15000	lf	$2.25	$33,750
Telephone wire, 4UTP, from closet to set	4000	lf	$1.20	$4,800
Data cable, plenum 16pr. UTP w/fiber	8000	lf	$3.00	$24,000
Data cable, coax only in cable tray	1000	lf	$2.00	$2,000
Security camera & cable, to closet	100000	sf	$0.50	$50,000
Multiplexer, rack mtd. w/monitor	1	ea	$2,000.00	$2,000
Magnetic door locks, interlocks	4	sets	$1,000.00	$4,000
Fixtures, 2 by 4 lay-in flou, high effic., 40w	1000	ea	$125.00	$125,000
Fixtures, incadescent task lights, 100w	40	ea	$75.00	$3,000
Fixtures, sodium vapor, 400w	60	ea	$585.00	$35,100
Fixtures, exterior, mercury vapor, 500 w	10	ea	$500.00	$5,000
Fixtures, exterior, parking on poles	12	ea	$1,250.00	$15,000
Fixtures, exterior ground, 35 w	10	ea	$300.00	$3,000
Fixtures, conference area, spots	14	ea	$250.00	$3,500
Fixtures, explosion proof	4	ea	$450.00	$1,800
Fixtures, emergency exit lights	25	ea	$125.00	$3,125
Fixtures, special	2	ea	$1,200.00	$2,400
				$0
Cable tray, ladder type, alum, heavy	5000	lf	$20.00	$100,000
Cable tray, ladder type, alum, light	2000	lf	$13.00	$26,000
TOTAL BUILDING HARD COST				**$17,028,213**

Figure 4.20 **(e)** Continuation of multipage example of an estimate for a prototype plant.

4.5 CONSTRUCTION

The two most common methods of building construction are to hire a general contractor (GC) or a construction management (CM) firm. General contracting is the traditional building system, where a firm, fixed-price contract is awarded for a fixed scope of work. Plans and specifications are usually completed prior to the beginning of the bid process. Contractors estimate their cost and markup before submitting fixed-price bids. The lowest acceptable bidder gets the award. Many of these firms have full-time employees, carpenters, masons, painters, or electricians. They have an office structure to support continuing business. Most will also have a pool of familiar subcontractors with whom they have established a timetested relationship. Some of the larger firms have in-house design capabilities. If the project is phased over years, or if there is planned expansion, it favors choosing a local GC firm or CM firm over a large national firm with no local ties.

Construction management firms are made up of managers and support staff and few actual workers. They often specialize in one building type or project size. Manufacturing facilities have become more sophisticated, with more piping systems and controls devices than hospitals. Due to time constraints, many recent projects have utilized CM processes. CMs are usually more experienced in multiple-bid packages and project expediting. Large CM firms have purchasing agents and contract managers who focus on their specialty. Their estimating divisions also have their fingers on the pulse of component costs. CM fees vary from 5 to 20% depending on the project, with an average of about 8% above cost. General conditions, utilities, and overhead for the job-site trailers are in addition to their fees.

The actual selection is made more difficult by the fact that the fees are usually fixed, meaning there is no direct cost competition. Of course, there are hybrids of these pure examples, GCs are always interested in doing cost-plus jobs, and CM firms often have a contracting branch for smaller projects. The choice is seldom clear without careful evaluation.

4.5.1 Selection Process

The first part in construction manager selection is to determine whether to advertise for competitive bids or to offer it to a limited number of preapproved firms at a fixed fee rate. Understanding the complexity and time constraints will help in making the decision. If the performance criteria are very high, such as for a clean-room-environment batch plant, it may be better to have a fixed fee and have the firms compete based on their ability to perform. If the budget is tight and time is not as critical, it may be better to develop a short list of preapproved CMs and let them submit sealed bids in a competitive situation, with fully completed plans and specifications, then award to low bidder. This logic is very general, however, and for each project careful consideration must be given to the selection method.

Either way, the next step is to write a notice of intent (call for bidders). You must first define the scope of the project. It should include a brief description of the facility to be built, along with the criteria to be used in selection. It must spell out the type of contractual agreement being sought, fixed price or guaranteed maximum price (GMP). On a GMP contract, the owner recovers all savings derived though the process of value engineering during construction. A fixed-price contract implies that total project costs are subject to change only through changes in the drawings or specs. A formal notice to construction managers would contain the following information:

1. Name and address of the owner, complete with the branches and main offices involved
2. Name, number, and location for the project
3. Gross area description of major facility components, complete with anticipated timetable and special requirements
4. Brief description of systems to be included
5. Phasing or bid alternatives desired
6. Specific expectations as to the role of the CM
   a. Cost feedback during completion of design documents
   b. Value engineering expectations
   c. Integral member of the design team
   d. Guaranteed maximum price (GMP)
7. Selection criteria (Defining the selection criteria is most important. Usually the fee is fixed, and it can be difficult to determine how much of a presentation is smoke and how much is real.)
8. General Information

It can be helpful to provide an explanation as to the method of notification for interviews and selection. If there are special requirements or circumstances which would be unique to this project, or are crucial to its ultimate success, they should be mentioned. The wording of the advertisement sets the tone for negotiation and selection of the CM. It must be specific. A project fact sheet should accompany all requests for bids, containing information as shown in the following example.

PROJECT FACT SHEET
PROJECT NAME AND NUMBER
OWNERSHIP, COMPANY NAME & ADDRESS

PROJECT DESCRIPTION:

This project consists of a complex of facilities for the manufacturing of composite electronics devices for use in converting photovoltaic DC energy into 120 volts AC. Phase One is currently planned for 50,000 gross square feet of factory space and 25,000 gross square feet of support spaces. It is projected to start in August 1998 and be complete within ten months. The main area will have a height of 40 feet with the support spaces at 25 feet. Primary structural system is anticipated to be post-and-beam steel framing with a ribbed metal panel exterior wall system.

The Second Phase is the remodeling of 250,000 GSF of existing manufacturing and office spaces. Phase 2 will start in July 1999, and will modify a brick and block facility built in 1950. The design of these facilities will require knowledge and experience with the following:

Special glass and plastic process piping installations
Clean environment construction
High-pressure stainless steel cooling loops
Large cooling towers, chillers & pumps
The effects of high-voltage harmonics, power correction
Conveyance systems design and installation
Distributed controls, integrated systems, fiberoptics
Fast-track projects

The first phase of the project will require the firm to work closely with the design team in order to provide input during the completion of plans and specifications. Long lead time items such as chillers, electrical switchgear, and water treatment equipment will be expedited through the use of early bid packages. After testing and startup of Phase 1, the existing factory will be upgraded to include connection to the Phase 1 Utility Plant system. The selected construction manager will provide value engineering during the design and construction processes. Upon negotiation of a Guaranteed Maximum Price (GMP) the CM will become a single point of responsibility for the performance of the construction contract for the project.

Firms will be evaluated and selected for interviews on the basis of their experience and ability. Experience in a proven track record for the firm and also for the individuals on the team that will be responsible for this particular project. The chosen firm will have illustrated record keeping ability, estimating, cost control, scheduling, quality control, and the ability to close out jobs.

The most important factors for comparison will be:

1. Knowledge and experience in the construction of large-capacity process piping systems
2. Knowledge and experience in the construction of facilities for the manufacture of composites
3. Experience in accelerated design & construction
4. Experience in GMP projects, not-to-exceed features
5. Experience or knowledge of local environmental, growth management, legislative or permitting processes

GENERAL INFORMATION:

All applicants will be notified of the results of the short list selections in writing. Finalists will be informed of the interview times and dates along with any additional project information which may be pertinent. The selection committee will make a final recommendation to the chairman of the board. The board will inform all finalists after successfully negotiating a contract with their highest rated firm.

Responses

Determination should be based on both the tangible and intangible aspects of those expressing interest. In order to obtain enough information to make such a selection, one must require that each firm complete a standard response form. The notice to contractors and fact sheets can be submitted through electronic mail to the distribution list rather than mailed. Since signatures are required, returns must be made through the mail. Soon there will be certifications and signatures done through electronic media which will be binding. This will save time and trees. After reviewing the completed packages, one may be able to compile a "short list" of six to eight firms for interviews. This is where one learns of the intangibles offered. The forms must be typed, and must include a copy of a license for contracting in the project state. A letter of intent from their surety company will acknowledge bonding for the job. The insurer should be licensed in the state and have an "A" rating or better.

The Sample Response Form includes the following:

1. Project name and number
2. Firm name and address

      a. Address of office in charge

      b. Telephone, fax, internet numbers

      c. Federal ID number

      d. Corporate charter in the state

3. Number of years the firm has been providing CM services

4. Applicant's personnel to be assigned

      a. List total number by skill group, department

      b. Name all key personnel, describe in detail experience and expertise of each member (Note: key personnel are to be assigned for the duration of the project unless excused)

5. Consultants, if any affiliated for this project & why

6. Experience, complete with:

      a. Owner name and phone number, if joint venture list all

      b. Project name

      c. Status, complete or in process, phase

      d. Completion date

      e. Project architect

      f. Services provided

      g. Approximate value of GMP or contract

      h. Role of the staff as submitted in 4 above

7. Services offered; does the firm also do developing, general contracting, or real estate, for example?

8. Description of cost control methodologies, what type and how often you update them. Examples of how they were used and what degree of accuracy was achieved in past projects.

9. The types of records, reports, monitoring systems, and information management systems used. Provide examples.

10. Methods to be used to avoid and prevent conflict

11. Methods to achieve quality control, with examples

12. Special considerations, why applicants firm is the best qualified for the job

The response form also contains these guarantees:

13. "It is hereby acknowledged that all information included herein is of a factual nature, and is certified to be true and accurate. All statements of future action will be honored."

14. "It is understood that if the information is found to be, in the opinion of the selection committee, substantially unreliable, this application may be rejected. The selection committee may reject any or all applicants and may stop the selection process."

15. "This completed package must be received by the owner at their published address on or before [for example] 3:00 PM MST on Wednesday, December First, 1997. The undersigned certifies that they are a principal or officer for the firm applying for consideration, and that they are authorized to make the above acknowledgements and certifications for the firm."

4.5.2 Construction Process

After selection of the best proposals, interviews should be scheduled for the limited number of firms that stood out. It is best not to exceed six; except for unusual circumstances, four is enough. After interviews, final selection should be made within a few days. The selected firm should then be notified and the work can begin.

Permitting and utilities connection can be the first of the challenges to face the team. Then shop drawings will be submitted and reviewed for compliance with the specifications, substitutions will be presented, and decisions made. Those with the longest lead time may be processed first. Detail review of shop drawings can prevent subtle changes which could significantly lower the quality of the finished product. Architects and engineers play a key role during this phase, as well as the owner's PM and team. Finally, it is time to order equipment and materials. Opportunities for value engineering alternatives can lead to significant cost savings. By this time the subcontractors will know actual costs for the anticipated products. Final negotiations often relieve schedule and cost problems. Both the owner and the contractor can benefit from honest, open exchange of ideas. The architect will issue change orders (COs) for any agreed-to deviations in scope or plans and specs. Each CO will have cost and schedule implications clearly spelled out.

Project coordination is the ultimate responsibility of the builder. The designers show line diagrams for three-dimensional systems. Experienced firms will often prepare complete coordination drawings prior to beginning installation of the first component. It costs a lot to go back and remove installed components; it costs very little to plan ahead. Usually the A&E did not have enough time between design and construction to evaluate sloping piping systems passing through areas with a lot of ducts or conduits. Conflicts can be avoided.

During construction the contractor will present payment requests for partial payment for completed portions of the work. They can be submitted monthly or as the project dictates. Fast-track projects could actually cost less if the payments were known in advance to be available every two weeks. The bidders would be able to reduce their exposure to material and labor costs.

After most construction is complete, a punch list will be generated which lists all known deficiencies. Upon completion of that list, the contractor can be considered to be substantially complete with the project. There may be some start-up or balancing of minor components remaining, but the major parts should all be up and running. At this time the contractor can expect to have around 5 to 10% of the total contract amount held in retainage, but the rest of the contract amount is payable. After final completion and project paperwork close-out, the retainage can be released. A complete test and balance report should be prepared to check the operation of the mechanical system. It is best to verify that there are no outstanding liens against the project, and that materials suppliers have been receiving payment throughout construction. If a GMP-type contract is used, all allowance figures will have to be accounted for and adjustments made to the contract sum. From this time forward, the contractor will still be obligated to correct any deficiencies or deviations encountered in the first year of operation without argument. Often the manufacturers of the system components offer longer warranties.

4.6 BIBLIOGRAPHY

Baumeister, Theodore, Eugene Avallone, Theodore Baumeister III, *Marks Standard Handbook for Mechanical Engineers,* McGraw-Hill, New York, 1978.

BCR for Reinforced Concrete, ACI 138.

CRSI Working Stress Design, ACI Code.

Greene, *Mind and Image,* University Press of Kentucky, Louisville, Kentucky, 1962.

Higgins and Morrow, *Maintenance Engineering Handbook,* McGraw-Hill, New York, 1994.

Littleton, *Industrial Piping,* McGraw-Hill, New York, 1962.

Lynch and Hack, *Site Planning,* MIT Press, Boston, Massachusetts, 1984.

McCormick, Sanders, *Human Factors in Engineering and Design,* McGraw-Hill, New York, 1987.

McGuiness, Stein, Reynolds, *Mechanical and Electrical Equipment for Buildings,* 6th Edition, John Wiley, New York, 1980.

Moffat, *Plant Engineer's Handbook for Charts and Tables,* Prentice-Hall, New York, 1974.

Placing Reinforcing Bars, CRSI.

Sanoff, *Methods of Architectural Programming*, DH & R.

Schodek, *Structures,* Prentice-Hall, New York, 1980.

Sheth, Vijay, *Facilities Planning and Materials Handling,* Marcel Dekker, New York, 1995.

Sleeper, Ramsey, *Architectural Graphics Standards,* John Wiley, New York, 1978.

Steel Construction Manual, American Institute of Steel Construction (AISC), 1973.

Stephens, Comer, *Internetworking with TCP/IP,* Prentice-Hall, Englewood Cliffs, New Jersey, 1993.

Swinburne, *Design Cost Analysis,* McGraw-Hill, New York, 1980.

Thumann, Albert, *Plant Engineer's and Manager's Guide to Energy Conservation,* Van Nostrand Reinhold, New York, 1977.

Urquhart, *Civil Engineering Handbook,* McGraw-Hill, New York, 1959.

White, Edward T., *Introduction to Architectural Programming,* Architectural Media, Tucson, Arizona, 1972.

5

Human Elements in the Factory

Jack M. Walker   *Consultant, Manufacturing Engineering, Merritt Island, Florida*

Timothy L. Murphy   *McDonnell Douglas Corporation, Titusville, Florida*

Jeffrey W. Vincoli   *J. W. Vincoli and Associates, Titusville, Florida*

Paul R. Riedel   *Dynacorp, Westminster, Colorado*

5.0   INTRODUCTION TO HUMAN ELEMENTS IN THE FACTORY

Jack M. Walker

To be effective, manufacturing strategy must recognize and address the interests of all three types of stakeholders, customers, investors, and employees. There is an increasing interest in producing goods and services that provide greater *customer* satisfaction. This shows up in the "new" considerations demanded in product design, product quality, competitive pricing, and continuing customer support. We are also quite aware of the importance of making a profit for our *investors* and the difficulty of achieving this in our global economy. This is seen in some of the drastic short-term cost-cutting solutions such as reducing the workforce, consolidating plants, as well as new improved machines, tools, and factory operations.

Written material, including this handbook, is available which talks about how best to achieve better product design, a better manufacturing process, better factory design, and better manufacturing operations for the long term. Computer systems that give us the ability to assess and control all of these functions are become more integrated, permitting us to solve problems by taking the positive actions needed to achieve total *customer* and *investor* satisfaction.

This chapter focuses on the interests of the third element, which in broad terms is the key to achieving satisfied customers and investors, the *employee* as a "stakeholder." To begin, we will examine a typical factory population, to get an idea of who the people are, and to talk about their individual and collective needs, and the skills they must have:

1. Of all "value-added" and "value-related" employees in the United States, 85% are employed by small companies, as opposed to the giants such as the automobile companies.
2. Most plants, even the very large ones, buy most of their parts and subassemblies

from outside suppliers. Ford, Chrysler, etc., usually perform only about 12 to 16% of the direct labor in-house. Other companies, such as McDonnell Douglas, Boeing, etc., typically perform 5 to 10% in-house. All subcontract the remainder to specialty shops that can do a better job for less money.

3. In a typical factory of 1000 employees, fewer than 40% are on the shop floor performing "touch labor"—which adds value directly to the product—by fabricating parts or performing assembly operations. Figure 5.1 shows the labor distribution. As we move toward more automation in the factory, the percentage of "touch" employees will decline even further.

Historically, a product was developed based primarily on function—something that would perform its intended purpose. The same pretty much described the production process, and the tools and machines involved. People were hired and expected to perform the required production tasks. Sometime later, skills training, mostly on the job, was needed in order to allow employees to perform the intended task. Still later, minor changes were made in the process to permit employees to perform assigned tasks somewhat better. This was usually to improve the cost or quality of the product.

Let's look at todays' shop worker. Most companies expect an entry-level worker to have a high school education, but this is simply not always the case. Many people now employed in small, low-technology shops are not even able to read and write. Some of them may need to learn English. Their opportunity for growth—and even continuing employment—is almost nonexistent. Learning to read and write, followed by a high school equivalency, is their only option to unemployment and welfare. These classes may be sponsored in some fashion by the employer, but only after the normal work shift, on the employees' own time. Classes in improving mathematical skills, blueprint reading, etc., may be given after hours by a company paid instructor. An entry-level machinist must have high school equivalency, reasonable math skills, and experience (at least in school) in basic machine operations such as drilling, turning, and milling—although it may be on older, manual machines. Classes on the newer computer-controlled machines may have to be furnished by the employer, and probably during working hours (with pay). We see this worker required to continually improve his or her skills, not only in the entry skill, but in

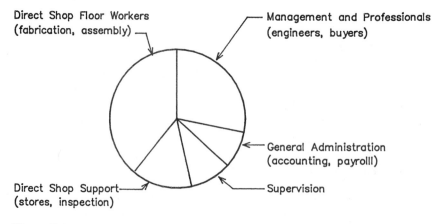

Figure 5.1 Labor distribution in a typical factory of 1000 employees.

developing new and broader skills. These skills will include the use of computers, knowledge of statistics, as well as the thought processes required to contribute to "continuing improvement" teams striving to improve quality and drive down costs. It boils down to fewer direct "value-added" workers in the manufacturing industry, but higher wages for the skilled ones that remain.

The other key people in manufacturing are those workers who support the production shop. They include engineers who design the products, establish the manufacturing process, design tooling, design and equip the factory, etc. There are also employees who purchase materials, maintain stock levels, and move parts around the factory. Add to these the workers in accounting, contract administration, quality control, training, safety, and others. In general, these workers have different training and education needs, but they must improve their individual job skills while broadening their knowledge of all the other operations in the factory. Advanced materials and processes, computer skills, communication skills, government requirements, and others are included.

The promise of lifetime employment security with one company may no longer be possible. Firms realize that they can no longer commit to providing jobs for the rest of an employee's life. They would perhaps like to, but in situations where they have to close down factories, combine divisions, or move them to another location, this is no longer a viable business option. On the other hand, workers are being given more power to expand their jobs and skills to permit them to be competitive, and therefore more useful to a company for a longer period of time. Some companies are sponsoring training programs for employees to expand their skills so that they can stay competitive in the job market—both within the company and outside the company. In return, corporations expect "bigger contributions" from a higher-skilled staff. Employees feel that they can "jump ship" for a better job, and still be able to return to the previous employer for advancement should conditions warrant it. Training is concerned primarily with *adapting people to machines or working situations*. We are not *just* teaching entry-level shop workers the elementary job skills, but rather trying to staff an entire plant with a population of employees requiring widely diverse skills.

Human factors engineering designs machines and working situations *to meet the capacities of people*. While training is essential, often the major gain is in job process improvement. Ergonomics, the "science of work," is a field of technology that considers human capabilities and limitations in the design of machines and objects that people use, the work processes that they must follow, and the environments in which they operate. It is a multidisciplinary system of how people interact with work tasks, workstations, and work area environments.

The productivity of an employee depends on the state of his or her work environment. If work area conditions are hazardous, or even perceived to be hazardous, morale will suffer and people will not perform effectively. Moreover, if conditions are so bad as to cause worker illness or injury, the consequences could affect the company severely. If the work environment provides a healthful and safe means to allow people to produce, then the employer will achieve profitable results.

Just about all of us have heard the statement that "safety is everyone's business," but few realize that this is more than just a cliché or a clever approach to safety management. The requirement for a safe work environment is actually a mandate on industry and its employees from the federal government. Safety and health regulations for general industry can be found in the U.S. Code of Federal Regulations. These codes, which address literally thousands of situations, working conditions, hazard control requirements, and worker safety and health protection standards, are all intended to make the work environment safer.

Although numerous areas are addressed and a minimum level of worker protection is ensured by implementation of these many standards, the major driving factor for worker safety is contained in Section 5 of the Occupational Safety and Health Act of 1970. This factor, better known as the "General Duty Clause," simply states that:

Each employer shall furnish to each of his employees employment and a place of employment which are free from recognized hazards that are causing or are likely to cause death or serious physical harm to his employees, and [each employer] shall comply with the occupational safety and health standards promulgated under this Act.

Worker's compensation is the name commonly applied to statutes that give protection and security to workers and their dependents against injury, disease, or death occurring during employment. The statutes establish the liability for benefits, which are usually financed through insurance bought by the employer.

Today, all 50 states have worker's compensation acts. The fundamental concepts of the compensation acts are to provide disabled employees with medical, economic, and rehabilitation support. The manner in which this support is handled has caused great distress between workers and their companies. Some employees feel that companies should do more for them, and many companies feel that they are providing more benefits than they can afford.

The escalating cost of worker's compensation is staggering. Payments for worker's compensation programs rose to an all-time record of $35.3 billion in 1990. The *average* cost of worker's compensation insurance per wage hour worked ranges from $1.00 in some states to more than $6.00 in others. The actual cost runs from less than 1% of wages to about 50% of wages, depending on the job classification and the history of claims payments. This can become a very significant element of payroll cost to a manufacturer, since the insurance company raises the worker's compensation insurance premium to cover the actual costs per year, plus the administrative cost of handling the claims.

The ME, along with his co-workers, needs to be aware of some of the problems and responsibilities regarding people in the factory today and the factory of the future. This chapter looks at ergonomics, safety, industrial hygiene, and worker's compensation. In the highly competitive environment of today's business world, success is achieved through the collective talents, skills, and contributions of all the individual employees of the organization. The business enterprise with the right combination of skills and clear, focused objectives will often outperform its competitors in every category. It cannot be disputed that people are a company's greatest resource. Therefore, it should be equally clear that the assurance of a safe and healthy workplace, within which this resource must perform, is as important to a company as any issue related directly to production, sales, or profit.

5.1 ERGONOMICS*

Jack M. Walker

5.1.0 Introduction to Ergonomics

The term *ergonomics* can be defined simply as the study of work. Ergonomics helps adapt the job to fit the person, rather than forcing the person to fit the job. Adapting the job to fit the worker can help reduce ergonomic stress and eliminate many potential ergonomic disorders. Ergonomics focuses on the work environment and items such as the design and function of workstations, controls, displays, safety devices, tools, and lighting to fit the employees' physical requirements and to ensure their health and well-being. It may include restructuring or changing workplace conditions to reduce stressors that cause repetitive motion injuries.

Major causes of many current ergonomic problems are technological advances such as more specialized tasks, higher assembly-line speeds, and increased repetition, plus a lack of ergonomically designed technologies. Consequently, workers' hands, wrists, arms, shoulders, backs, and legs may be subjected to thousands of repetitive twisting, forceful, or flexing motions during a typical workday. Some jobs still expose workers to excessive vibration and noise, eye strain, repetitive motion, and heavy lifting. In many instances, machines, tools, and the work environment are poorly designed, placing undue stress on workers' tendons, muscles, and nerves. In addition, workplace temperature extremes may aggravate or increase ergonomic stress. Recognizing ergonomic hazards in the workplace is an essential first step in correcting the hazards and improving worker protection.

Training is concerned primarily with adapting people to machines or working situations. Ergonomics helps design machines and working situations to meet the capacities of people. While training is essential, often the major gain is in the job process improvement.

Historically, a product was developed based primarily on function—something that would perform its intended purpose. The same pretty much described the production process, and the tools and machines involved. People were hired and expected to perform the required production tasks. Sometime later, skills training, mostly on the job, was needed in order to allow employees to perform the intended tasks. Still later, minor changes were made in the process to permit workers to perform assigned tasks somewhat better. This was usually to improve the cost or quality of the product.

Common sense, safety, the workers' general welfare—unfortunately, these all needed someone carrying a big stick to remind industry that this is good business. Considering only the costs of OSHA citations, willful acts of negligence on the part of employers can cost up to $70,000 per incident, accompanied by jail terms of up to 2 years, according to Raymond E. Chalmers, senior editor of *Forming and Fabricating* magazine. Industrial safety classes need to be a part of any management agenda. Good ergonomics depends on mobility, flexibility, and adaptability—blending together tool design and managerial attitudes, assembly-line specifications and company policies.

"It's not uncommon to think that by the end of the decade 50 cents of every worker's

---

*Note: The author would like to thank Timothy L. Murphy, Group Manager, Human Relations, McDonnell Douglas Corporation, and Alvah O. Conley, Mechanical and Safety Engineer in OSHA's Chicago office, for their assistance with this subchapter.

The examples in this subchapter are taken largely from MIL-STD-1472 and from MIL-HDBK-759. They are considered fairly accurate, and are used in the design of military hardware. The author recommends that the data be used with caution, and supplemented with more recent studies and in-plant measurements wherever possible. While the data presented do give a starting point for good ergonomic workplace design, they should not be considered the final authority should discussions between management and workers become elevated during contract negotiations regarding working conditions—rather, local studies and conditions should take preference.

compensation insurance dollar will go for soft-tissue disorders, unless we come up with some reasonable intervention strategies," according to OSHA's chief ergonomist, Roger Stephens. The good news, however, is that many soft-tissue disorders can be prevented through ergonomic intervention. By simply redesigning the job to fit the employee, not vice versa, workers can be spared a lot of pain and discomfort, and the company can save a lot of money without necessarily making a large investment.

The preferred method for control and prevention of work-related musculoskeletal disorders is to design the job to match the physiological, anatomical, and psychological characteristics and capabilities of the worker. In other words, safe work is achieved as a natural result of the design of the job, the workstation, and the tools; it is independent of specific worker capabilities or work techniques.

Worker selection or hiring based solely on physical capacities is generally illegal, as a result of the U.S. Federal Rehabilitation Act of 1973 and the Americans with Disabilities Act of 1990. However, once a worker is offered a job, he or she can be tested to determine his or her capabilities as a prelude to specific job placement within the firm. Good records can defend a company against complaints by OSHA or workers. Bad records mean bad faith. Companies that falsify records constitute the worst violators in the mind of OSHA and the workers.

5.1.1 Ergonomics

"Ergonomics is basically common sense," says Dr. Hal Blatman, who specializes in muscle pain and is an adjunct professor at Miami University's Center for Ergonomic Research. "Efficiency is the by-product of the work, and the company that manufactures no sore backs, shoulders, wrists, legs and behinds has a competitive edge over the ones that do," according to Blatman. As the science of work, ergonomics is a field of technology that considers human capabilities and limitations in the design of machines and objects that people use, the work processes that they must follow, and the environments in which they operate. It is a multidisciplinary system of how humans interact with work tasks, workstations, and work area environments.

The time and motion studies of Frederick W. Taylor, and of Frank and Lillian Gilbreth, laid the groundwork for the field. Ergonomics—then called human factors engineering— grew in importance during World War II as engineers sought help from psychologists and physiologists in designing military equipment adaptable to a wide variety of users. Since that time, ergonomics has also played a role not only in the design of better workplaces and tools, but also in the design of consumer goods—automobiles and stoves, for instance.

Industrial and organizational psychology is the field of applied psychology that studies the behavior of persons in organizational work settings. Its areas of concern are personnel selection, the mapping of organizational processes, training, improvement of employee morale and working conditions, and improvement of both individual and group productivity.

Ergonomics is concerned with making purposeful human activities more effective. Most of the activities studied can be called *work,* although there are topics such as "ergonomics of sport," "ergonomics in the home," and "passenger ergonomics."

The focus of study is the person interacting with the engineered environment. This person has some limitations which the designer should take account of. Complexity arises from the nature of people and the variety of the designed situations to be considered. These latter can vary from relatively simple ones such as chairs, the handles of tools, and the lighting of a bench, to highly elaborate ones such as aircraft flightdecks, control rooms in the process industries, and the artificial lifesupport systems in space or under the sea.

Ergonomic Essentials

Consultants practicing ergonomics hail from a variety of backgrounds—health sciences, industrial hygiene, safety, industrial engineering, psychology, and others. Regardless of an ergonomic consultant's primary area of expertise, two things are necessary to solve ergonomics problems in the workplace: an understanding of the *demands of the task* and an understanding of the *capabilities of the human operator,* said Jim Forgrascher, director of ergonomics for the Ohio Bureau of Workers' Compensation, Division of Safety and Hygiene.

Traditionally, those who practiced ergonomics specialized in studies of the human as a physical engine or the human as an information processor. The "physical engine" experts, often industrial engineers, studied human biomechanics and the physical elements of a task and workplace. The "humans as information processors," often experts in psychology, examined the reception, processing and sending of information from human to human, and from human to machine.

These two areas of expertise within ergonomics have blended and multiplied, creating many specialty areas within the field. Ergonomic specialties include biomechanics, environmental design, human-computer interaction, human performance, training and simulation.

The people studied can vary from fit young men in military systems to middle-aged housewives in the kitchen, from the disabled of many kinds, to the very highly selected such as racing drivers. Sometimes the relevant characteristic is essentially *biological* and unchanging except with age, as in the limits of dark adaptation, for example. Sometimes it *varies with sex and race*, for example body dimension. Sometimes it varies with degree of *economic and social development*, for example, acceptable working hours.

The limits or boundaries of ergonomics itself are not entirely agreed. For example, most ergonomists would agree that energy expenditure studies are within ergonomics, but the design of diets to provide the energy is not; that physical hazards such as heat stress are part of ergonomics, but that chemical hazards such as carcinogenic substances are not.

Ergonomic Tips

The following tips may resolve many ergonomic concerns in the workplace, without having to resort to the more formal anthropometric studies that are presented in the next subchapter.

1. *Avoid extreme reach.* Integrate all workplace components and surfaces so that frequent task motions occur in front of the body and between the hip (waist) and the shoulders. The optimum height is within 1 in. (2.5 cm) from the elbows. Having adjustable work surfaces and seats (if they are used) will make the integration possible and effective.

2. *Avoid excessive force.* Reduce muscular exertions (forces) by minimizing the weight handled, improving motion patterns, eliminating extreme joint motions (range of motion), avoiding awkward and maximum reaches, counterbalancing and suspending tools, eliminating differential heights among product/component transfer points, using mechanical aids (powered conveyors, gravity feed devices)

for moving products to and from the workplace, presorting and aligning work pieces (components), relaxing the fit tolerances for the assembly of components, and reducing pace or speed of motion (activity).

3. *Avoid static posture.* Job method should not impair circulation and should offer the muscles sufficient time to recover after exertion. Static loading can be eliminated through supporting body segments that are not involved in task motions. Seats and arm rests are examples of body supports. Use mechanical aids (jigs, special fixtures) for positioning and holding workpieces or components. The workplace should allow for flexible positioning of workpieces (adjustable height, angling, tilting, and rotation). Protective clothing should allow for freedom and ease of body movement and offer no impediment to circulation. Rest periods should be planned and scheduled to overcome any residual effects of static loading.

4. *Avoid stress concentration.* Job activities should be shared by more than one muscle group. In so doing, static loading and excessive force may be eliminated as a by-product. Use large contact surface for hand-held tools; use simultaneous arm motions. Change pinching tasks to gripping actions, wrist motions to arm motions, and arm motions to shoulder motions. Eliminate pressure at points of contact between the body and the work surface, tools, machines, or any hard surface. For standing jobs, use cushioned floor support or modify the job method to promote the use of "muscle pumps," especially in the lower extremities. Tight-fitting shoes and high heels (good examples of stress concentration) should be avoided.

5. *Avoid faulty human-machine interface.* Machine controls should be designed based on the limiting values of human strength, body joint range of motion, and speed and accuracy of the required response. Identification, grouping, and the relationship among controls as well as displays should conform to population norms and stereotypes. Eliminate the potential for ambiguous information feedback from the machine. Avoid having the machine operator process complex pieces of data before making an appropriate response.

Display (visual or auditory) designs should take into consideration the potential user population and their skills and degrees of familiarity with the equipment. Time, frequency of use, and the environment (noise, vibration, light under which equipment will be used) should also be considered.

6. *Avoid extremes of the environment.* Extreme environmental conditions are harmful because of the physiological and stress responses they elicit from the body, as well as for their potential to cause damage to body tissues. Regardless of the type of offending environment, from noise to vibration, from hot to cold, from brightly lit surfaces to high-intensity lights, the control strategies are the same. Basically, the following control options can be used separately or collectively.

(a) Apply engineering changes and modifications to bring the level of the offending source to an acceptable level. Examples include the substitution of different material or energy sources, reducing the intensity (level) of the current source, or bypassing the source all together through elimination of total enclosures.

(b) Separate in space and time the offending source from the exposed workers (receivers). Increasing distance from the source and shielding the receiver are examples of achieving the desired separation.

(c) Eliminate any potential harm through limiting or minimizing the exposure of those working close to the offending source. The use of personal protective equipment is an example of this control strategy.

Some of the common ergonomic problems, and common-sense solutions are listed in Figure 5.2.

---

1. Repetitiveness
  a. Use mechanical aids
  b. Enlarge work content by adding more diverse activities
  c. Automate certain tasks
  d. Rotate workers
  e. Increase rest allowances
  f. Spread work uniformly across workshift
  g. Restructure jobs

2. Force/Mechanical Stress
  a. Decrease the weight of tools/containers and parts
  b. Increase the friction between handles and the hand
  c. Optimize size and shape of handles
  d. Improve mechanical advantage
  e. Select gloves to minimize effects on performance
  f. Balance hand-held tools and containers
  g. Use torque control devices
  h. Optimize pace
  i. Enlarge corners and edges
  j. Use pads and cushions

3. Posture
  a. Locate work to reduce awkward postures
  b. Alter position of tool
  c. Move the part closer to the worker
  d. Move the worker to reduce awkward postures
  e. Select tool design for work station

4. Vibration
  a. Select tools with minimum vibration
  b. Select process to minimize surface and edge finishing
  c. Use mechanical assists
  d. Use isolation for tools that operate above resonance point
  e. Provide damping for tools that operate at resonance point
  f. Adjust tool speed to avoid resonance

5. Psychosocial Stresses
  a. Enlarge workers' task duties
  b. Allow more worker control over pattern of work
  c. Provide micro work pauses
  d. Minimize paced work
  e. Eliminate blind electronic monitoring

Figure 5.2 Examples of ergonomic interventions.

5.1.2 Anthropometry

It is important to have information about the people who are working in today's manufacturing industries. Their capabilities and limitations are among the most basic of our considerations in designing products, processes, and equipment. *Anthropometry* is the study of human body measurements. Important to the field of ergonomics, anthropometric measurements aid designers and manufacturers in the production of safe, efficient consumer goods that are comfortable to use.

The design of the workplace must ensure the accommodation of, and compatibility with, at least 90% of the workforce population. Generally, design limits should be based on a range of the 5th percentile to the 95th percentile values for critical body dimensions. For any body dimension, the 5th percentile value indicates that 5% of the population will be equal to or smaller than that value, and 95% will be larger. The 95th percentile value indicates that 95% of the population will be equal to or smaller than that value, and 5% will be larger. Therefore, use of a design range from the 5th to the 95th percentile values will theoretically accommodate 90% of the user population for that dimension. This is considerably different than the "average" worker.

When using the anthropometric data that is presented in this section, the following must be considered:

1. The nature, frequency, and difficulty of the specific worker task
2. The position of the workers body during performance of this task
3. Mobility or flexibility requirements imposed by this task
4. Increments in the design-critical dimensions imposed by the need to compensate for obstacles, projections, etc.
5. Increments in the design-critical dimensions imposed by protective clothing, packages, lines, padding, etc.

Anthropometric data is available from several sources. One source (MIL-STD-1472C) is a study of 14,428 military men and 1331 women from all branches of service, by the U.S. Army and Air Force. These data provide one representation of worker population—with a few notes of caution. According to Alvah O. Conley, a mechanical and safety engineer in OSHA's Chicago office, we must remember that the figures are for military personnel, and do not include the "real" worker population which contains more older workers, obese workers, and an increasing Asian population. This consideration would change some of the sizes of workers, etc., but would probably not affect measurements of range of motion, field of vision, and the like. The best source might be your personnel department, which could at least furnish the age, weight, height, and sex of your employees. This could be used to factor the anthropometric data that is available. Conley indicated that the standard data might be accurate for some regions of the country, while other regions would require more caution and consideration of the local workforce make-up. There may be an influence due to management style in a particular application, which can make a significant difference in motivation and discipline.

Under the U.S. Department of Health and Human Services, the National Institute for Occupational Safety and Health (NIOSH) develops and recommends criteria for preventing disease and hazardous conditions in the workplace. NIOSH also recommends preventive measures including engineering controls, safe work practices, personal protective equipment, and environmental and medical monitoring. NIOSH then transmits these recommendations to the Occupational Safety and Health Administration (OSHA), under the U.S. Department of Labor, for use in promulgating legal standards. Some of their studies and reports are available from NIOSH at no cost, while others can be ordered from the U.S. Department of Commerce, National Technical Information Service (NTIS), or the U.S. Government Printing Office, Superintendent of Documents.

Since some of the data are presented in the English system of measurement and others in the Metric system, a conversion table is included as Figure 5.3.

Body Dimensions

Studies of body dimensions are made by several authorities in the field, but some controversy exists as to the validity of any of them. The following data are presented as an example of military personnel body dimensions. Used as a baseline, these data can be modified by examining the population of your individual manufacturing facility and making proper adjustments. An approach would be to refer to the following data, and use the conservative limits that would apply to the most of the factory population. Figure 5.4 defines the standing body dimensions, and gives the values for men and women. Figure 5.5 is a similar chart for sitting body dimensions. Anthropometric data shown in Figure 5.6 shows the 5th and 95th percentile reaches for men and women. This describes the body in several positions.

TO CONVERT FROM:	TO:	MULTIPLY BY:
DEGREE (ANGLE) (deg)	RADIAN (rad)	1.745 329 E−02
FOOT (ft)	METER (m)	3.048 000 E−01
FOOT^2 (ft^2)	METER^2 (m^2)	9.290 304 E−02
FOOT^3 (ft^3)	METER^3 (m^3)	2.831 685 E−02
FOOTCANDLE (ft−C)	LUX (lx)	1.076 391 E+01
FOOTLAMBERT (ft−L)	CANDELA PER METER^2 (cd/m^2)	3.426 259 E+00
INCH (in. OR ")	METER	2.540 000 E−02
INCH^2 (in.^{2})	METER^2 (m^2)	6.451 600 E−04
INCH^3 (in.^{3})	METER^3 (m^3)	1.638 706 E−05
MINUTE (ANGLE) (min)	RADIAN (rad)	2.908 882 E−04
OUNCE−FORCE (ozf)	NEWTON (N)	2.780 139 E−01
OUNCE−INCH (ozf · in.)	NEWTON METER (N · m)	7.061 552 E−03
POUND (lb) AVOIRDUPOIS	KILOGRAM (kg)	4.535 924 E−01
POUND−FORCE (lbf)	NEWTON (N)	4.448 222 E+00
POUND−INCH (lbf · in.)	NEWTON METER (N · m)	1.129 848 E−01
SECOND (ANGLE) (sec)	RADIAN (rad)	4.848 137 E−06

PREFIXES						TEMPERATURE CONVERSION
NANO	n	10^{-9}	CENTI	c	10^{-2}	$^{\circ}C = \dfrac{5}{9} (^{\circ}F - 32)$
MICRO	μ	10^{-6}	KILO	k	10^{3}	$^{\circ}F = \dfrac{9}{5} \,^{\circ}C + 32$
MILLI	m	10^{-3}	MEGA	M	10^{6}	

NOTE: EACH CONVERSION FACTOR IS PRESENTED AS A NUMBER, BETWEEN ONE AND TEN, TO SIX DECIMAL PLACES. THE LETTER E (FOR EXPONENT), A PLUS OR MINUS SIGN AND TWO DIGITS FOLLOWING THE NUMBER, REPRESENT THE POWER OF 10 BY WHICH THE NUMBER IS TO BE MULTIPLIED.

FOR EXAMPLE: $3.048\ 000\ E{-}01 = 3.048\ 000 \times 10^{-1} = 0.3048000$

OR: $1.076\ 391\ E{+}01 = 1.076\ 391 \times 10^{1} = 10.76391$

EXAMPLES OF USE OF TABLE:
TO CONVERT 2 ft^3 TO m^3, MULTIPLY 2 BY 2.831 685 E−02
$$2 \times 0.028\ 316\ 85 = 0.056\ 634 \text{ m}^3$$
(TO CONVERT 2 m^3 TO ft^3, DIVIDE 2 BY 2.831 685 E−02)
$$(2/0.028\ 316\ 85 = 70.629\ 325 \text{ ft}^3)$$

A MORE COMPLETE LISTING AND DISCUSSION MAY BE FOUND IN ASTM E 380−76

Figure 5.3 Metric equivalents, abbreviations, and prefixes.

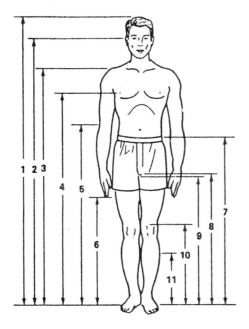

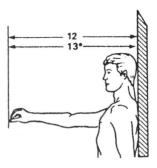

*SAME AS 12, HOWEVER,
RIGHT SHOULDER IS EXTENDED
AS FAR FORWARD AS POSSIBLE
WHILE KEEPING THE BACK OF
THE LEFT SHOULDER FIRMLY
AGAINST THE BACK WALL.

	PERCENTILE VALUES IN CENTIMETERS					
	5th PERCENTILE			95th PERCENTILE		
	GROUND TROOPS	AVIATORS	WOMEN	GROUND TROOPS	AVIATORS	WOMEN
WEIGHT (kg)	55.5	60.4	46.4	91.6	96.0	74.5
STANDING BODY DIMENSIONS						
1 STATURE	162.8	164.2	152.4	185.6	187.7	174.1
2 EYE HEIGHT (STANDING)	151.1	152.1	140.9	173.3	175.2	162.2
3 SHOULDER (ACROMIALE) HEIGHT	133.6	133.3	123.0	154.2	154.8	143.7
4 CHEST (NIPPLE) HEIGHT *	117.9	120.8	109.3	136.5	138.5	127.8
5 ELBOW (RADIALE) HEIGHT	101.0	104.8	94.9	117.8	120.0	110.7
6 FINGERTIP (DACTYLION) HEIGHT		61.5			73.2	
7 WAIST HEIGHT	96.6	97.6	93.1	115.2	115.1	110.3
8 CROTCH HEIGHT	76.3	74.7	68.1	91.8	92.0	83.9
9 GLUTEAL FURROW HEIGHT	73.3	74.6	66.4	87.7	88.1	81.0
10 KNEECAP HEIGHT	47.5	46.8	43.8	58.6	57.8	52.5
11 CALF HEIGHT	31.1	30.9	29.0	40.6	39.3	36.6
12 FUNCTIONAL REACH	72.6	73.1	64.0	90.9	87.0	80.4
13 FUNCTIONAL REACH, EXTENDED	84.2	82.3	73.5	101.2	97.3	92.7

Figure 5.4 Standing body dimensions.

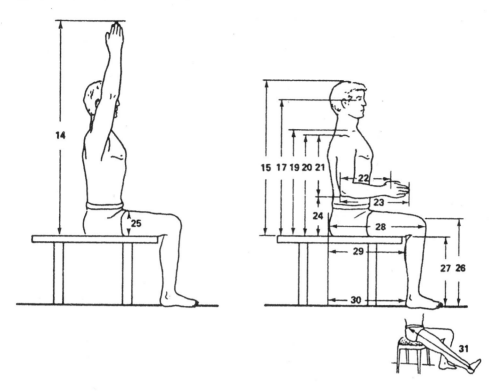

		PERCENTILE VALUES IN CENTIMETERS				
		5th PERCENTILE			95th PERCENTILE	
		GROUND TROOPS	AVIATORS	WOMEN	GROUND TROOPS	AVIATORS	WOMEN
SEATED BODY DIMENSIONS							
14	VERTICAL ARM REACH, SITTING	128.6	134.0	117.4	147.8	153.2	139.4
15	SITTING HEIGHT, ERECT	83.5	85.7	79.0	96.9	98.6	90.9
16	SITTING HEIGHT, RELAXED	81.5	83.6	77.5	94.8	96.5	89.7
17	EYE HEIGHT, SITTING ERECT	72.0	73.6	67.7	84.6	86.1	79.1
18	EYE HEIGHT, SITTING RELAXED	70.0	71.6	66.2	82.5	84.0	77.9
19	MID–SHOULDER HEIGHT	56.6	58.3	53.7	67.7	69.2	62.5
20	SHOULDER HEIGHT, SITTING	54.2	54.6	49.9	65.4	65.9	60.3
21	SHOULDER–ELBOW LENGTH	33.3	33.2	30.8	40.2	39.7	36.6
22	ELBOW–GRIP LENGTH	31.7	32.6	29.6	38.3	37.9	35.4
23	ELBOW–FINGERTIP LENGTH	43.8	44.7	40.0	52.0	51.7	47.5
24	ELBOW REST HEIGHT	17.5	18.7	16.1	28.0	29.5	26.9
25	THIGH CLEARANCE HEIGHT		12.4	10.4		18.8	17.5
26	KNEE HEIGHT, SITTING	49.7	48.9	46.9	60.2	59.9	55.5
27	POPLITEAL HEIGHT	39.7	38.4	38.0	50.0	47.7	45.7
28	BUTTOCK–KNEE LENGTH	54.9	55.9	53.1	65.8	65.5	63.2
29	BUTTOCK–POPLITEAL LENGTH	45.8	44.9	43.4	54.5	54.6	52.6
30	BUTTOCK–HEEL LENGTH		46.7			56.4	
31	FUNCTIONAL LEG LENGTH	110.6	103.9	99.6	127.7	120.4	118.6

Figure 5.5 Seated body dimensions.

	PERCENTILE VALUES IN CENTIMETERS			
	5th PERCENTILE		95th PERCENTILE	
	MEN	WOMEN	MEN	WOMEN
1. WEIGHT — CLOTHED (KILOGRAMS)	58.6	48.8	90.2	74.6
2. STATURE — CLOTHED	168.5	156.8	189.0	178.7
3. FUNCTIONAL REACH	72.6	64.0	86.4	79.0
4. FUNCTIONAL REACH, EXTENDED	84.2	73.5	101.2	92.7
5. OVERHEAD REACH HEIGHT	200.4	185.3	230.5	215.1
6. OVERHEAD REACH BREADTH	35.2	31.5	41.9	37.9
7. BENT TORSO HEIGHT	125.6	112.7	149.8	138.6
8. BENT TORSO BREADTH	40.9	36.8	48.3	43.5
9. OVERHEAD REACH, SITTING	127.9	117.4	146.9	139.4
10. FUNCTIONAL LEG LENGTH	110.6	99.6	127.7	118.6
11. KNEELING HEIGHT	121.9	114.5	136.9	130.3
12. KNEELING LEG LENGTH	63.9	59.2	75.5	70.5
13. BENT KNEE HEIGHT, SUPINE	44.7	41.3	53.5	49.6
14. HORIZONTAL LENGTH, KNEES BENT	150.8	140.3	173.0	163.8

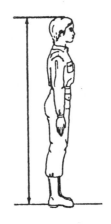

② STATURE (CLOTHED)

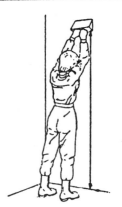

⑤ OVERHEAD REACH HEIGHT

④ FUNCTIONAL REACH, EXTENDED

⑨ OVERHEAD REACH, SITTING

Figure 5.6   Anthropometric data for common working positions.

Body Strengths

Static muscle strength data is shown in Figure 5.7. This includes both standing and seated. Arm, hand, and thumb-finger strengths are shown in Figure 5.8. Figure 5.9 shows leg strengths at various knee and thigh angles of the 5th percentile male. Finger- or hand-operated pushbuttons are shown in Figure 5.10. Where data are shown for male workers and comparable data are not available for female workers, the Army Research Institute

	PERCENTILE VALUES IN POUNDS			
	5th PERCENTILE		95th PERCENTILE	
STRENGTH MEASUREMENTS	MEN	WOMEN	MEN	WOMEN
A STANDING TWO–HANDED PULL:				
15 in. LEVEL				
MEAN FORCE	166	74	304	184
PEAK FORCE	190	89	323	200
B STANDING TWO–HANDED PULL:				
20 in. LEVEL				
MEAN FORCE	170	73	302	189
PEAK FORCE	187	84	324	203
C STANDING TWO–HANDED PULL:				
39 in. LEVEL				
MEAN FORCE	100	42	209	100
PEAK FORCE	113	49	222	111
D STANDING TWO–HANDED PUSH:				
59 in. LEVEL				
MEAN FORCE	92	34	229	85
PEAK FORCE	106	42	246	97
E STANDING ONE–HANDED PULL:				
39 in. LEVEL				
MEAN FORCE	48	23	141	64
PEAK FORCE	58	30	163	72
F SEATED ONE–HANDED PULL:				
CENTERLINE, 18 in. LEVEL				
MEAN FORCE	51	24	152	88
PEAK FORCE	61	29	170	101
G SEATED ONE–HANDED PULL:				
SIDE, 18 in. LEVEL				
MEAN FORCE	54	25	136	76
PEAK FORCE	61	30	148	89
H SEATED TWO–HANDED PULL:				
CENTERLINE, 15 in. LEVEL				
MEAN FORCE	134	54	274	173
PEAK FORCE	157	64	298	189
I SEATED TWO–HANDED PULL:				
CENTERLINE, 20 in. LEVEL				
MEAN FORCE	118	46	237	142
PEAK FORCE	134	53	267	157

Figure 5.7 Static muscle strength data.

256

Walker, Murphy, Vincoli and Riedel

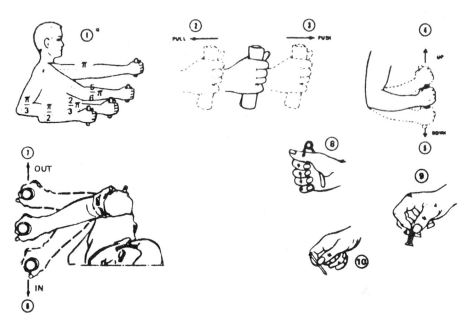

ARM STRENGTH (N)													
(1)	(2)		(3)		(4)		(5)		(6)		(7)		
DEGREE OF ELBOW FLEXION (rad)	PULL		PUSH		UP		DOWN		IN		OUT		
	L**	R**	L	R	L	R	L	R	L	R	L	R	
π	222	231	187	222	40	62	58	76	58	89	36	62	
$\frac{5}{6}\pi$	187	249	133	187	67	80	80	89	67	89	36	67	
$\frac{2}{3}\pi$	151	187	116	160	76	107	93	116	89	98	45	67	
$\frac{1}{2}\pi$	142	165	98	160	76	89	93	116	71	80	45	71	
$\frac{1}{3}\pi$	116	107	98	151	67	89	80	89	76	89	53	76	
HAND, AND THUMB-FINGER STRENGTH (N)													
	(8)			(9)			(10)						
	HAND GRIP			THUMB-FINGER GRIP (PALMER)			THUMB-FINGER GRIP (TIPS)						
	L	R											
MOMENTARY HOLD	250	260		60			60						
SUSTAINED HOLD	145	155		35			35						

*Elbow angle shown in radians.
**L = Left; R = Right.

Figure 5.8 Arm, hand, and thumb-finger strength (5th percentile male data).

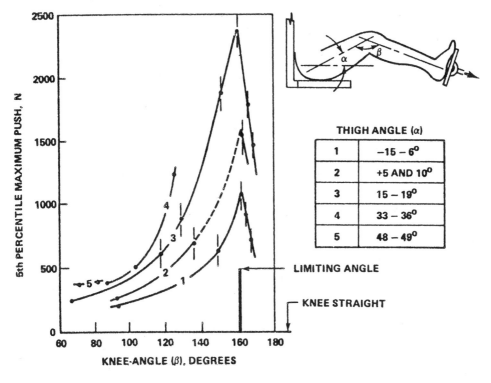

Figure 5.9 Leg strength at various knee and thigh angles (5th percentile male data).

of Environmental Medicine has performed a study of 1500 women, with the following results:

For upper extremities, women's strength is 56.5% of men's.
For lower extremities, women's strength is 64.2% of mens's.
For trunk extremities, women's strength is 66.0% of men's.

These numbers may serve as a preliminary design guideline until more up-to-date information becomes available.

NIOSH Lifting Equation (Excerpts from NIOSH Publications)

NIOSH convened an *ad hoc* committee of experts who reviewed the current literature on lifting, recommend criteria for defining lifting capacity, and in 1991 developed a revised lifting equation. Although this equation has not been fully validated, the recommended weight limits derived from the revised equation are consistent with or lower than those generally reported in the literature. The 1991 equation reflects new findings, provides methods for evaluating asymmetrical lifting tasks and objects with less than optimal hand-container couplings, and offers new procedures for evaluating a larger range of work durations and lifting frequencies than earlier equations. The objective is to prevent or reduce the occurrence of lifting-related low back pain (LBP) among workers. An additional benefit is the potential to reduce other musculoskeletal injuries associated with some lifting tasks, such as shoulder or arm pain.

Three criteria (biomechanical, physiological, and psychophysical) were used to define

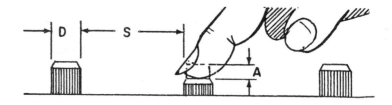

	DIMENSIONS		RESISTANCE		
	DIAMETER D				
	Fingertip	Thumb or Palm	Single Finger	Different Fingers	Thumb or Palm
Minimum	9.5 mm (3/8 in.)	19 mm (3/4 in.)	2.8 N (10 oz.)	1.4 N (5 oz.)	2.8 N (10 oz.)
Maximum	25 mm (1 in.)		11 N (40 oz.)	5.6 N (20 oz.)	23 N (80 oz.)
	DISPLACEMENT				
	A				
	Fingertip		Thumb or Palm		
Minimum	2 mm (5/64 in.)		3 mm (1/8 in.)		
Maximum	6 mm (1/4 in.)		38 mm (1-1/2 in.)		
	SEPARATION				
	S				
	Single Finger	Single Finger Sequential	Different Fingers	Thumb or Palm	
Minimum	13 mm (1/2 in.)	6 mm (1/4 in.)	6 mm (1/4 in.)	25 mm (1 in.)	
Preferred	50 mm (2 in.)	13 mm (1/2 in.)	13 mm (1/2 in.)	150 mm (6 in.)	

Note: Above data for barehand application. For gloved hand operation, minima should be suitably adjusted.

Figure 5.10 Pushbuttons, finger, or hand operated.

the components of the lifting equation. Figure 5.11 shows the criteria used to develop the new equation. Although the lifetime prevalence of LBP in the general population is as high as 70%, work-related LBP comprises only a subset of all cases. Some studies indicate that specific lifting or bending episodes were related to only about one-third of the work-related cases of LBP.

Since NIOSH and OSHA have released very few actual recommendations for industry to consider (or comply with), the author feels that an introduction to this specific study is worthwhile, not only to set a standard for lifting tasks, but as an indication of what will eventually be a massive set of recommendations and government regulations.

The NIOSH formula for calculation of recommended weight limit is as follows:

$$RWL = LC \times HN \times VM \times DM \times AM \times FM \times C$$

Component	Metric
LC = load constant	23 kg (normally the maximum)
HM = horizontal multiplier	(25/H)
VM = vertical multiplier	1–[0.003 (V–75)]
DM = distance multiplier	0.82 + (.5/D)
AM = asymmetric multiplier	1 – (0.0032A)
FM = frequency multiplier (from Figure 5.13)	
CM = coupling multiplier (from Figure 5.12)	

where

H =	horizontal distance of hands from midpoint between the ankles. Measure at the origin and the destination of the lift (cm)
V =	vertical distance of the hands from the floor. Measure at the origina and destination of the lift (cm)
D =	vertical travel distance between the origin and the destination of the lift (cm)
A =	angle of asymmetry—angular displacement of the load (left or right) from the sagittal plane. Measure at the origin and destination of the lift (degrees)
F =	average frequency rate of lifting measured in lifts/min. Duration is defined to be: ≤ 1 hr; ≤ 2 hr.; or, ≤ 8 hr. assuming appropriate recovery allowances

Coupling modifier: Loads equipped with appropriate couplings or handles facilitate lifting and reduce the possibility of dropping the load. Psychophysical studies that investigated the effects of handles on maximum acceptable weight of lift suggested that lifting capacity was decreased in lifting tasks involving containers without good handles. The coupling modifiers are displayed in Figure 5.12.

Frequency modifier: For the 1991 lifting equation, the appropriate frequency multiplier is obtained from Figure 5.13 rather from a mathematical expression. The committee concluded that the frequency multipliers provide a close approximation to observed and predicted effects of lifting frequency on acceptable workloads.

From the NIOSH perspective, it is possible that obese workers may exceed the energy expenditure criteria for lifts from below the waist. In addition, there are some circumstances in which local muscle fatigue may occur even though whole body fatigue has not occurred. This is most likely in situations involving lifting at high rates for longer than 15 min., or prolonged use of awkward postures, such as constant bending.

Discipline	Design criterion	Cut-off value
Biomechanical	Maximum disc compression force	3·4 kN (770 lbs)
Physiological	Maximum energy expenditure	2.2–4.7 kcal/min
Psychophysical	Maximum acceptable weight	Acceptable to 75% of female workers and about 99% of male workers

Figure 5.11 Criteria used to develop the NIOSH lifting equation.

	$V < 75\,cm\ (30\,in)$	$V \geq 75\,cm\ (30\,in)$
Couplings	Coupling multipliers	
Good	1·00	1·00
Fair	0·95	1·00
Poor	0·90	0·90

Figure 5.12 Coupling modifier (CM) for NIOSH lifting equation.

Visual Field

The worker's visual field is controlled by eye rotation, head rotation, or a combination of eye and head rotation. Nominal, optimum, and maximum values are given in Figure 5.14.

Range of Human Motion

Figure 5.15 gives the ranges, in angular degrees, for all voluntary movements the joints of the body can make, as illustrated in Figure 5.16. The designer should remember that these are maximum values; since they were measured with nude personnel, they do not reflect the restrictions that clothing would impose. The following general instructions apply to the dimensions in Figure 5.16.

1. The lower limit should be used when personnel must operate or maintain a component.

Frequency lifts/min	Work duration					
	$\leq 1\,h$		$\leq 2\,h$		$\leq 8\,h$	
	$V < 75$	$V \geq 75$	$V < 75$	$V \geq 75$	$V < 75$	$V \geq 75$
0·2	1·00	1·00	0·95	0·95	0·85	0·85
0·5	0·97	0·97	0·92	0·92	0·81	0·81
1	0·94	0·94	0·88	0·88	0·75	0·75
2	0·91	0·91	0·84	0·84	0·65	0·65
3	0·88	0·88	0·79	0·79	0·55	0·55
4	0·84	0·84	0·72	0·72	0·45	0·45
5	0·80	0·80	0·60	0·60	0·35	0·35
6	0·75	0·75	0·50	0·50	0·27	0·27
7	0·70	0·70	0·42	0·42	0·22	0·22
8	0·60	0·60	0·35	0·35	0·18	0·18
9	0·52	0·52	0·30	0·30	0·00	0·15
10	0·45	0·45	0·26	0·26	0·00	0·13
11	0·41	0·41	0·00	0·23	0·00	0·00
12	0·37	0·37	0·00	0·21	0·00	0·00
13	0·00	0·34	0·00	0·00	0·00	0·00
14	0·00	0·31	0·00	0·00	0·00	0·00
15	0·00	0·28	0·00	0·00	0·00	0·00
>15	0·00	0·00	0·00	0·00	0·00	0·00

Note:

‡ values of V are in cm; $75\,cm = 30\,in$.

Figure 5.13 Frequency modifier (FM) for NIOSH lifting equation.

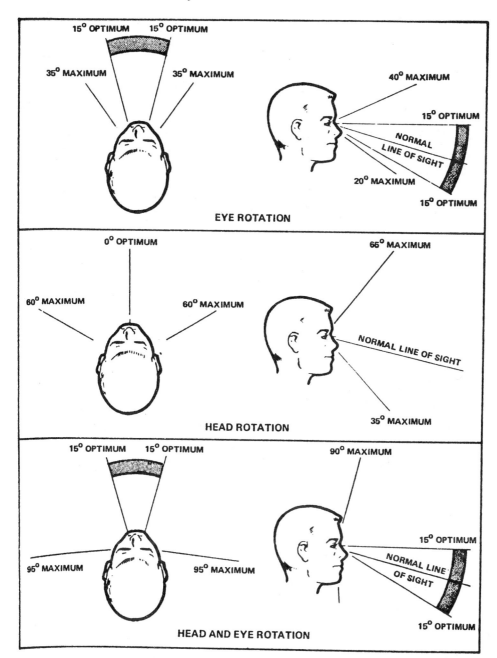

Figure 5.14 Vertical and horizontal visual field.

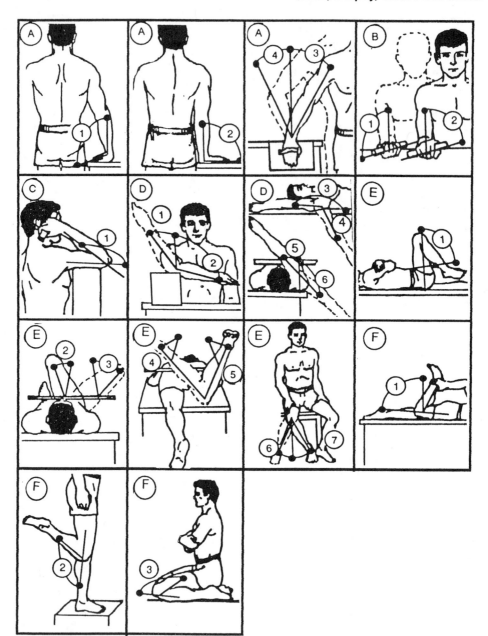

Figure 5.15   Range of human motion—definition.

2. The upper limit should be used in designing for freedom of movement.

All operating positions should allow enough space to move the trunk of the body. When large forces (more than 13.6 kg) or large control displacements (more than 380 mm in a fore-aft direction) are required, the operator should have enough space to move his or her entire body. Figure 5.17 is a continuation of the range of human motions. The following definitions may be useful in discussions of body positions and motions.

Flexion: Bending, or decreasing the angle between parts of the body
Extension: Straightening or increasing the angle between parts of the body
Adduction: Moving toward the midline of the body
Abduction: Moving away from the midline of the body
Medial rotation: Turning toward the midplane of the body
Lateral rotation: Turning away from the midplane of the body
Pronation: Rotation of the palm of the hand downward
Supination: Rotation of the palm of the hand upward

Body Member	Movement	Lower Limit (degrees)	Average (degrees)	Upper Limit (degrees)
A. Wrist	1. Flexion	78	90	102
	2. Extension	88	99	112
	3. Adduction	18	36	27
	4. Abdustion	40	54	47
B. Forearm	2. Supination	92	113	135
	2. Pronation	53	77	101
C. Elbow	1. Flexion	132	142	152
D. Shoulder	1. Lateral Rotation	21	34	47
	2. Medial Rotation	75	97	119
	3. Extension	47	61	75
	4. Flexion	176	188	190
	5. Adduction	39	48	57
	6. Abduction	117	134	151
E. Hip	!. Flexion	100	113	126
	2. Adduction	19	31	43
	3. Abduction	41	53	65
	4. Medial Rotation (prone)	29	39	49
	5. Lateral Rotation (prone)	24	34	44
	6. Lateral Rotation (sitting)	21	30	39
	7. Medial Rotation (sitting)	22	31	40
F. Knee Flexion	1. Prone	115	125	135
	2. Standing	100	113	126
	3. Kneeling	150	159	168

These values are based on the nude body. The ranges are larger than they would be for clothed figures.

Figure 5.16   Range of human motion—values.

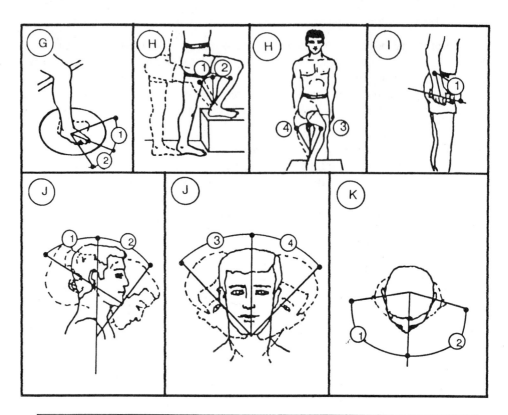

Body Member	Movement	Lower Limit (degrees)	Average (degrees)	Upper Limit (degrees)
G. Foot Rotation	1. Medial	23	35	47
	2. Lateral	31	43	55
H. Ankle	1. Extension	26	38	50
	2. Flexion	28	35	42
	3. Adduction	15	33	24
	4. Abduction	18	30	23
I. Grip Angle		95	102	109
J. Neck Flexion	1. Dorsal (back)	44	61	88
	2. Ventral (forwared)	48	60	72
	3. Right	34	41	48
	4. left	34	41	48
K. Neck Rotation	1. Right	65	79	93
	2. Left	65	79	93

Figure 5.17   Range of human motion—continued.

OSHA believes that ergo-hazards are prevented primarily by effective design of the workstation, tools, and job. Slump factor must be considered when selecting the range of movement for adjustable seats, as well as in locating displays, optics, and vision ports. Seated eye-height measurements may be reduced as much as 65 mm when personnel sit in a relaxed or slumped position. This slump factor should be considered when selecting the range of movement for adjustable seats, as well as in locating displays, optics, and vision ports.

5.1.3 Factory Applications

The work area environment consists of its total atmosphere. This includes climate, illumination, ventilation, noise, odor, vibration, congestion, isolation, and may include high elevations or subterranean locations. Each element of the work area environment will consciously or unconsciously effect the worker and his or her performance. Levels of illumination in the workplace are shown in Figure 5.18.

physiology

Ventilation requirements for persons working in enclosed spaces is shown in Figure 5.19. Vibration exposure criteria for both longitudinal and transverse directions with respect to body axis are shown in Figure 5.20.

The worker should not be expected to adjust to all these elements. Without some ergonomic controls, productivity suffers and individual employees may also sustain various work-related injuries and illnesses.

Workers are various sizes and ages, and may have physical or mental limitations. The work tasks that people perform require a variety of skill levels. The job assignment may be repetitive, or may require strength and physical endurance. Good and bad work practices on the part of the worker are sometimes more important than equipment selection by the ME Figure 5.21 shows a screwdriver used properly and improperly. Wrist posture is determined by the elevation and orientation of the work surface with respect to the worker and the shape of the tool. Stress concentrations should be distributed evenly over muscle eminences.

physical

Workstations usually consist of the tools and equipment required to perform specific tasks. The workstation could be a factory with assembly lines and large manufacturing equipment, or an office environment with office equipment. The workstation environment fits the worker properly if the worker feels comfortable and natural while performing the work task. If the work task requires excessive reaching, bending, lifting, twisting, stooping, or working with the arms over shoulder height, this unhealthy work environment will result in poor productivity. Figure 5.22 displays the preferred angle of inclination for ramps, stairs, etc.

Until changes in the design of the workstation or tooling are made, rotation of workers in an unfriendly environmental work area may be an option. Some managers believe that rotation of workers will help prevent an individual from being burdened with an unpleasant task. The real truth of rotation is that everyone shares in the misery. The solution to the problem is to correct the environment—which increases productivity for the entire shop. An electromyographic study of five jobs where job rotation had been introduced concluded that job rotation may be useful for reducing *stress* associated with heavy dynamic tasks rather than for reducing static muscular load in "light" work situations (Jonasson, 1988).

	* LUX (Ft-C)	
WORK AREA OR TYPE OF TASK	RECOMMENDED	MINIMUM
Assembly, general		
medium	810 (75)	540 (50)
precise	3220 (300)	2155 (200)
Bench work		
medium	810 (75)	540 (50)
extra fine	3230 (300)	2155 (200)
Electrical equipment testing	540 (50)	325 (30)
Screw fastening	540 (50)	325 (30)
Inspection tasks, general		
medium	1075 (100)	540 (50)
extra fine	3230 (300)	2155 (200)
Machine operation, automatic	540 (50)	325 (30)
Storage		
live, medium	325 (30)	215 (20)
live, fine	540 (50)	215 (20)
Office work, general	755 (70)	540 (50)
Reading		
large print	325 (30)	110 (10)
handwritten reports, prolonged reading	755 (70)	540 (50)
Business machine operation	1075 (100)	540 (50)

NOTE: As a guide in determining illumination requirements, the use of a steel scale with 1/64 in. divisions required 1950 lux (180 ft-c) of light for optimum visibility.
* As measured at the task object or 760 mm (30 in.) above the floor.

Figure 5.18   Illumination levels for work areas.

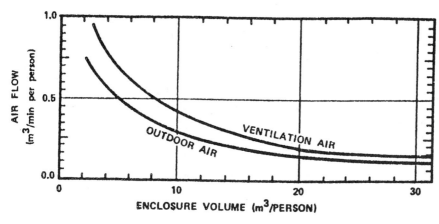

Figure 5.19 Ventilation requirements.

Factory Machines and Workplace Safety

In the past, progress in assembly conditions was necessitated by complaints, errors, accidents, absent workers, or other problems in operation of some operator-machine combinations. Education and training are two obvious means of improving workplace safety. Assembly-line operations need to be regarded as human line operations that create monotony, fatigue, excessive noise, and vibration. Noise levels for various types of work are given in Figure 5.23, plotted as ambient noise level versus distance from the speaker to the listener.

According to Dennis R. Ebens, President, Rockford (IL) Systems, Inc., updating power presses to comply with OSHA requirements means that employers must consider all five of the following requirements: safeguarding, lockout\tagout devices, starters, and covers.

One of the most common elements of work is activation of machinery or equipment by hitting control buttons. It is not uncommon for an operator to activate equipment every 4 or 5 sec. Allowing for rest intervals and other tasks, this may mean that the hands must perform the motion pattern more than 5000 times per day. It becomes imperative, therefore, that consideration be given to both the hand position and the force required to engage the control. Different controls may vary considerably in the amount of force required, and this should be given due attention in the selection of different devices. Touch-sensitive controls now available virtually eliminate forces. Rexcon Controls (Mt. Clemens, MI) has developed a button activated by sensors, requiring no "touch" at all.

Proper design of knobs and handles is becoming more important in the factory as we become aware of the ergonomic (and productivity) benefits of all elements of the work process. Figure 5.24 shows recommended knob dimensions. Palm grasp is preferred over thumb-and-finger encircles, and both are preferable to fingertip actuation. Figure 5.25 shows the recommended dimensions for handles of various designs. Gripping efficiency is best if the fingers can curl around the handle to any angle of 120° or more. The diameter of the handle increases as the weight of the item increases. The strength requirements of a task are affected by the size and shape of the handles, as also shown in Figure 5.25.

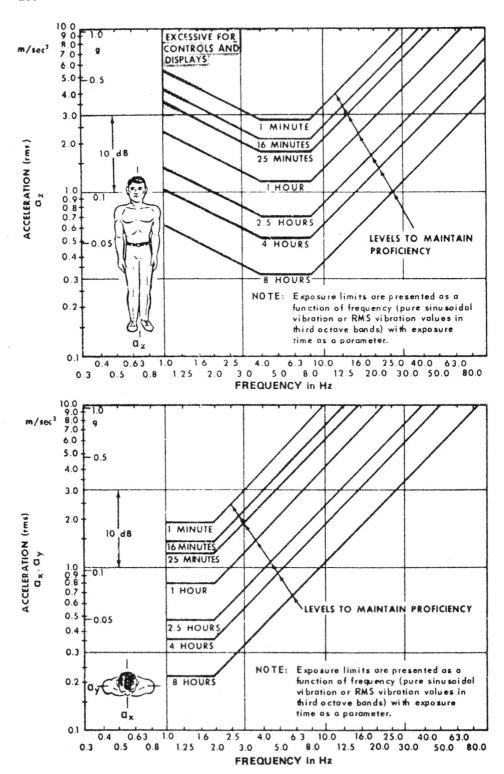

Figure 5.20 Vibration exposure criteria for longitudinal (upper curve) and transverse (lower curve) direction with respect to body axis.

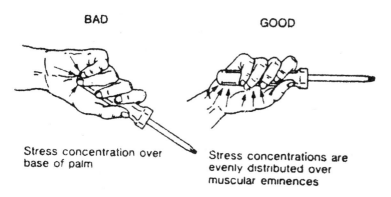

BAD   GOOD

Stress concentration over base of palm

Stress concentrations are evenly distributed over muscular eminences

Select tools which spread stress areas evenly over muscular eminences.

Figure 5.21 Wrist posture is determined by the elevation and orientation of the work surface with respect to the worker, and the shape of the tool.

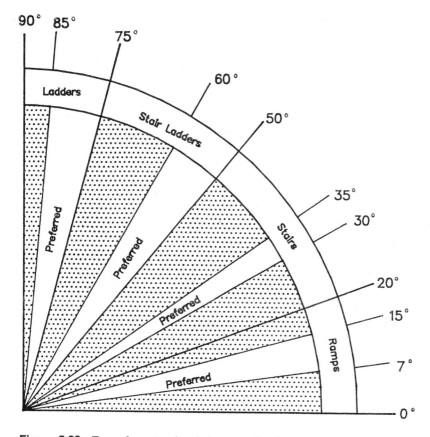

Figure 5.22 Type of structure in relation to angle of ascent.

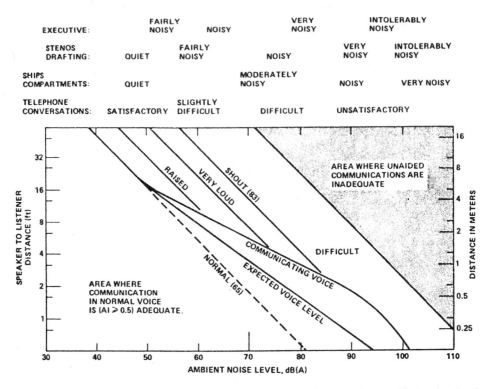

Figure 5.23 Permissible distance between a speaker and listeners for specified voice levels measured 1 m from the mouth.

Makers of hand tools are trying to make their tools easier on users. Redesign efforts aim particularly at avoiding stress and strain to the hand and wrist. One wrist disorder, carpal tunnel syndrome, strikes 23,000 workers a year and costs employers and insurers an average of $30,000 per injured worker. U.S. Labor Department figures show that the vast majority of hand and wrist disorders occur among tool-using workers.

Engineering Controls

Engineering controls are the preferred method of control, since the primary focus of ergonomic hazard abatement is to make the job fit the person, not force the person to fit the job. This can be accomplished by ergonomically designing workstations and tools or equipment.

Workstations should be ergonomically designed to accommodate the full range of required movements. Moreover, they should be designed so that they accommodate the workers who are actually using them to perform the job—not just the "average" or "typical" worker. Examples of good and bad designs for containers are shown in Figure 5.26. All edges that come in contact with the worker should be well rounded. Changes of this type are very low cost to implement, and are usually cost effective as well as ergonomically correct. Figure 5.27 shows some of the elements of workbench and jig design. The jigs should be oriented so that parts can be assembled without flexing the wrist.

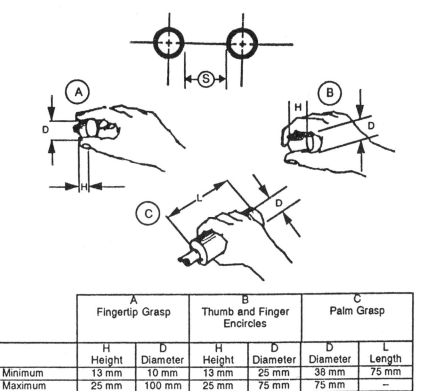

	A Fingertip Grasp		B Thumb and Finger Encircles		C Palm Grasp	
	H Height	D Diameter	H Height	D Diameter	D Diameter	L Length
Minimum	13 mm	10 mm	13 mm	25 mm	38 mm	75 mm
Maximum	25 mm	100 mm	25 mm	75 mm	75 mm	--

	S Separation		Torque	
	One Hand Individually	Two Hands Simultaneously	*	**
Minimum	25 mm	75 mm	--	--
Optimum	50 mm	125 mm	--	--
Maximum	--	--	32 mN•m	42 mN•m

*Up to/including 25 mm diameter.
**Greater than 25 mm diameter.

Figure 5.24 Recommended knob and handle design criteria.

The workstation should be designed to permit the worker to adopt several different but equally healthful and safe postures that still permit performance of the job; sufficient space should be provided for the knees and feet. Work tables and chairs should be height adjustable to provide proper back and leg support. Seat cushions can be used to compensate for height variation when chairs or stools are not adjustable. There should be definite and fixed space for all tools and materials. Machine controls should be reachable and equally accessible by both right- and left-handed operators.

Most hand tools are designed for only occasional use, not for repetitive use over prolonged periods. When acquiring tools for regular use in an industrial setting, an employer should consider the following ergonomic features:

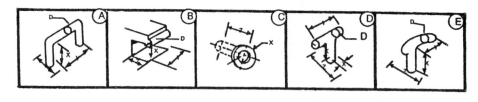

Type of handle:		(Bare Hand)			Dimensions (mm) (Gloved Hand)		
		X	Y	Z	X	Y	Z
A.	Two-Finger Bar	32	65	75	38	75	75
	One-Hand Bar	48	110	75	50	125	100
	Two-Hand Bar	48	215	75	50	270	100
B.	Two-Finger Recess	32	65	50	38	75	50
	One-Hand Recess	50	110	90	90	135	100
C.	Finger-Tip Recess	19		13	25		19
	One-Finger Recess	32		50	38		50
D.	T-Bar	38	100	75	50	115	100
E.	J-Bar	50	100	75	50	115	100

Curvature of Handle or Edge:

Weight of Item	Diameter (minimum)
up to 6.8 kg	D -- 6 mm
6.8 to 9.0 kg	D -- 13 mm
9.0 to 18.0 kg	D -- 19 mm
Over 18.0 kg	D -- 25 mm
T-bar Post	T. -- 13 mm

Figure 5.25 Recommended handle dimensions using bare and gloved hands.

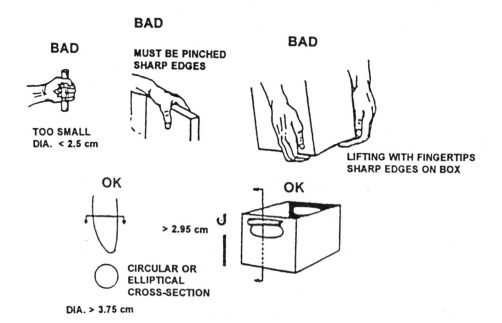

Figure 5.26 Strength requirements of a task as affected by the size and shape of the handles.

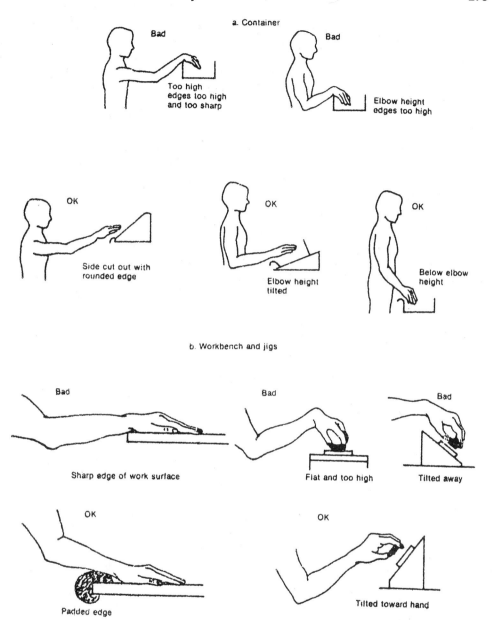

Figure 5.27 Good and bad designs for containers and work benches. Jigs should be located and oriented so that parts can be assembled without flexing the wrist.

Tools should be light in weight, and handles should be designed to allow a relaxed grip so the wrists can remain straight.

Tools should be designed for use with either hand and be of various sizes so they can be used by both women and men.

Tool handles should be shaped so that they contact the largest possible surface of the inner hand and fingers, and should be fitted to the functional anatomy of the hand. Tool handles with sharp edges and corners should be avoided.

Power tools should be used to reduce the amount of human force and repetition required. Whenever possible, the weight of the tool should be counterbalanced to make it easier to handle and to reduce vibration.

To reduce tool vibration, special absorbent rubber sleeves can be fitted over the tool handle.

Wrist posture is determined by the elevation and orientation of the work surface with respect to the worker and the shape of the tool. Figure 5.28 shows some good and bad designs for powered drivers and drills. An example of the old versus the new shape of hand drill motors is shown in Figure 5.29. Cooper Power Tools is one of the companies that is very conscious of the importance of ergonomic design of their equipment. The new grips are modeled on vibration, torque, noise, and human geometry data. The same type of equipment now comes in different size grips, for the different hand sizes of workers. Cooper's new drill line was designed by Biomechanics Corp. of America, a company that measures how users interact with products, tools, and workplaces. John Lawson, aerospace marketing manager at Cooper's power tool division, says; "There's more and more concern in the power tool industry for worker safety. Ergonomics is the future." Figure 5.30 shows one of Cooper's new drill motors.

Personal Protective Equipment

Personal protective equipment may also be necessary to help prevent or reduce ergonomic hazards. For example, when vibrating tools are used, vibration may be dampened by using

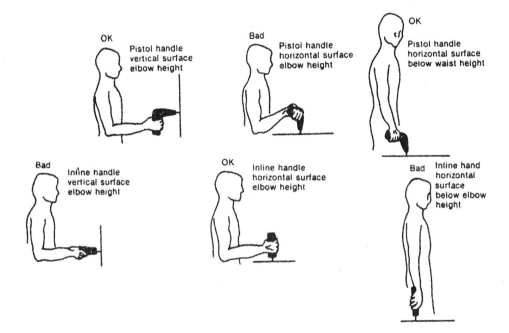

Figure 5.28   Good and bad designs for powered drivers.

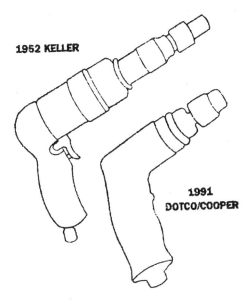

Figure 5.29 Comparison of 1952 drill motor modeled after the German Luger, and 1991 grips modeled on vibration, torque, noise, and human geometry data. (Courtesy of Dotco/Cooper, Inc.)

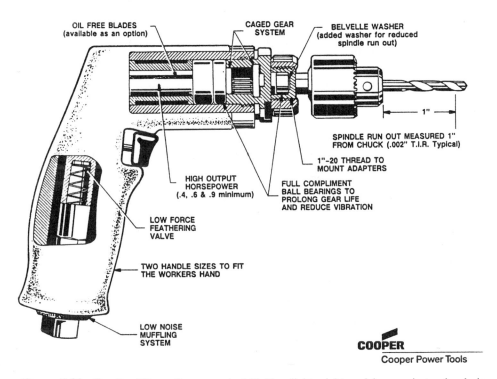

Figure 5.30 Details of Dotco ® ergonomic drill. New lightweight model comes in two handgrip sizes. (Courtesy of Dotco/Cooper, Inc.)

rubber-backed, low-pile carpet sections on the work surface. Gloves may also be worn to reduce the effects of vibration and force.

A long-term study in Norway found that employee turnover dropped to 10% from 40% after a plant installed new equipment. The company realized an 852% return on its initial capital investment—money saved primarily on training, recruitment, and employee sickness and injuries.

5.1.4 Office Applications

With the increased usage of computers, there has been a rise in work-related injuries and illnesses in this field. The same attention to a correct workstation should apply to office areas as in the manufacturing shops.

In order for equipment to be utilized most efficiently, it must be designed for the specific user population. We must design for people who, while actually working, will be under conditions of stress and fatigue from many causes. A performance decrement may arise while working not so much because the workers are basically unable to perform, but because the individual is overloaded—both physically and mentally.

Health and safety features are foremost in ergonomic design. Other considerations, such as the information processing and perceptual limitations of humans, must be taken into account in the design of communication devices, computer displays, and numerous other applications. We must design equipment and processes that are as simple as possible to use. The job should not require intellectual tasks, such as transforming data, which might prove distracting to the worker. While training and practice can improve a person's proficiency, this should not be considered a substitute for good design. Figure 5.31 shows a good layout for a computer workstation, with factors to consider.

NIOSH has done considerable research on the use of video display terminals (VDTs). They reviewed the possible effects of radiation, fear of possible cataracts, psychological

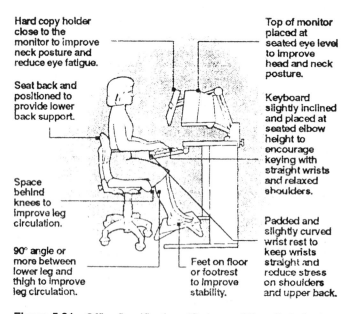

Figure 5.31 Office Specifications (Courtesy of Ergo Tech, Inc.)

stress and strain, reproductive concerns, and musculoskeletal strains. As a result of these studies, most problems were found to be nonexistent or minor—with the exeption of the ergonomic considerations. Their recommendations include the following.

1. *Workstation design:* Maximum flexibility should be designed into VDT units, supporting tables, and operator chairs. VDTs should have detachable keyboards, work tables should be height-adjustable, and chairs should be height adjustable and provide proper back support.
2. *Illumination:* Sources of glare should be controlled through VDT placement (i.e., parallel to windows as well as parallel and between lights), proper lighting, and the use of glare-control devices on the VDT screen surface. Illumination levels should be lower for VDT tasks requiring screen-intensive work and increased as the need to use hard copy increases. In some cases, hard-copy material may require local lighting in addition to the normal office lighting.
3. *Work regimens:* Continuous work with VDTs should be interrupted periodically by rest breaks or other work activities that do not produce visual fatigue or muscular tension. As a minimum, a break should be taken after 2 hr of continuous VDT work, and breaks should be more frequent as visual, mental, and muscular burdens increase.
4. *Vision testing:* VDT workers should have visual testing before beginning VDT work and periodically thereafter to ensure that they have adequately corrected vision to handle such work.

In a review of the literature, the World Health Organization found that "musculoskeletal discomfort was commonplace during work with VDTs," and that "injury from repeated stress to the musculoskeletal system is possible." It suggests that the primary emphasis for reducing musculoskeletal strain has been on improving the workstation environment by applying well-established ergonomic principles. However, some authorities suggest that ergonomically designed workstations are an incomplete prescription for preventing musculoskeletal discomfort in VDT work because they do not correct for a major contributory factor, namely, constrained postures. Some authorities feel that exercise programs designed to reduce this discomfort are valuable.

Some interesting work on "fitting the task to the man" was done by Dr. E. Grandjean, director of the Institute for Industrial Hygiene and Work Physiology, at the Swiss Federal Institute of Technology, Zurich. The frequency of bodily aches among sedentary workers in an office is shown in Figure 5.32, based on 246 persons questioned. The results are as follows.

1. Seat heights between 38 and 54 cm were preferred for a comfortable posture of the upper part of the body.
2. Aches in the thighs were caused mostly by putting the weight on the thighs while working, and only to a lesser extent by the height of the seat.
3. A desk top 74 to 78 cm high gave the employees the most scope to adapt it to suit themselves, provided that a fully adjustable seat and footrests were available.
4. Of those questioned, 57% complained of back troubles while they were sitting.
5. Of the respondents, 24% reported aches in neck and shoulders, and 15% in arms and hands, which most of them, especially the typists, blamed on too high a desk top (cramped hunching of the shoulders).

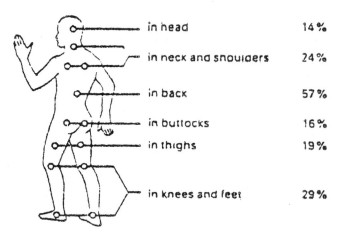

in head	14%
in neck and shoulders	24%
in back	57%
in buttocks	16%
in thighs	19%
in knees and feet	29%

Figure 5.32   Frequency of body aches among sedentary workers. (From Grandjean and Burandt.)

6. Of the respondents, 29% reported aches in knees and feet, mostly among small people who had to sit forward on their chairs because they had no footrests.
7. Regardless of their body length, the great majority of the workers liked the seat to be 27 to 30 cm below the desk top. This permitted a natural posture of the trunk, obviously a point of first priority with these workers.
8. The frequency of back complaints (57%) and the common use of a back rest (42% of the time) showed the need to relax the back muscles from time to time, and may be quoted as evidence of the importance of a well-constructed backrest.

Muscle groups commonly requiring relaxation of activation after continuous VDT work are shown in Figure 5.33. They are:

1. Chronically tensed scapular elevators require stretching and relaxation.
2. Spinal extensors of the lumbar, thoracic, and cervical regions are overstretched and require activation.
3. Muscles of the anterior thoracic region are shortened and require stretching.
4. Forearm flexors are chronically tensed and shortened, and require stretching and relaxation.

A 1986 study by the Army Corps of Engineers documented a 20.6% improvement in office employee productivity 1 year after ergonomic furniture was installed.

5.1.5   Ergonomic Checklists

Try checking off this list from Atlas Copco Tools (Farmington Hills, MI). "Yes" answers to most items means that an ergonomics effort is in order. "Don't Know" means that it is time for managers to take a look at their records and their assembly areas.

Do production reports show decreased or already low efficiency?
Do inspection and rejection reports indicate deteriorating product quality?
Is absenteeism increasing in certain areas—or everywhere?
Are accident rates up sharply?
Do medical reports show many back or repetitive stress injuries?

Figure 5.33 Muscle groups commonly requiring relaxation of activation after continuous VDT work. (From Grandjean and Burandt.)

Do workers on certain lines log frequent medical visits?
Is plant turnover high or increasing?
Is turnover in specific jobs high?
Does training new workers take too long?
Do workers make a lot of mistakes?
Is there too much material waste?
Are operators damaging or not maintaining equipment?
Are operators often away from their workstations?
Do workers make subtle changes in their workplaces?
Is the plant moving to two- or three-shift operation?
Are plant engineers ignorant about, or suspicious of, ergonomic principles?
Do you use incentive pay to increase productivity?
Are workers exercising hands, fingers, and arms to relieve stress?

A Sample Ergonomic Checklist to Identify Potential Hazards in the Workplace (Things to Look for as You Tour a Factory)

General Work Environment

Workforce characteristics
Age, sex, anthropometries (body size and proportions)
Strength, endurance, fitness
Disabilities
Diminished senses

Communication/language problems (e.g., non-English-speaking or illiterate workers)

Lighting
Climate
Noise level
Health and safety safeguards
Job and workstation design
    Location of controls, displays, equipment, stock
        Accessibility
        Visibility
        Legibility
        Efficiency of sequence of movements when operating or using
        Use of pedals
        Posture of workers
        Sitting, standing, combination
        Possibility for variation
        Stooping, twisting or bending of the spine
        Chair availability, adjustments
        Room to move about
        Work surface height
Predominately dynamic or static work
    Alternation possible
    Use of devices such as clamps or jigs to avoid static work
    Availability of supports for arms, elbows, hands, back, feet

Muscular Work Load/Task Demands

Repetitiveness
    Frequency
    Force
    Availability of rest pauses
    Possibility for alternative work
    Skill, vigilance, perception demands
    Efficiency of organization (supplies, equipment)
Use of hand tools
    Hand and wrist posture during use
    Work surface height
    Size and weight of tool
    Necessity, availability of supports
    Shape, dimensions, and surface of hand grip
    Vibratory or non-vibratory
Physical strength requirements
    Strength capabilities of employee
    Working pulses/respiratory rate
    Loads lifted, carried, pushed, or pulled
        Manner in which handled
        Weight and dimensions of objects handled

5.2 WORKER'S COMPENSATION

Timothy L. Murphy

5.2.0 Introduction to Worker's Compensation

Worker's compensation was traditionally called workman's compensation. It is the name commonly applied to statutes that give protection and security to workers and their dependents against injury, disease, or death occurring during employment. The statutes establish the liability of benefits—usually financed by insurance bought by the employer— including hospital, other medical payments, and compensation for loss of income.

Factory safety regulations should enable industry to produce and market products without causing the illness, injury, or death of employees. Workers have the right to return home to their loved ones in a healthy condition after a day's work. However, the proper working conditions were not always provided by all employers.

5.2.1 Compensation Systems

It was not until the latter part of the American Industrial Revolution in 1908 that the U.S. Congress passed the first Workman's Compensation Act to cover federal employees. Today, all of the 50 states, territories, commonwealths, and the District of Columbia have worker's compensation acts. The Federal Employees' Compensation Act covers federal civilian employees. The Longshoreman and Harbor Workers' Compensation act covers maritime employees. The fundamental concepts of the compensation acts are to provide disabled employees with medical, economic, and rehabilitation support. The manner in which this support is handled has caused great distress between workers and their companies. Some employees feel that companies should do more for them, and many companies feel that they are providing more benefits than they can afford.

Companies are responsible for providing medical care and financial support to their employees who have work-related injuries or occupational illnesses. Each state identifies the payment scale that should be followed while the injured or ill employee is recuperating. The company is also responsible for the rehabilitation of disabled employees. These employees must be rehabilitated so they can return to their previous job or be retrained for another position. Benefits covered by the worker's compensation act cannot be disallowed because a worker was either partially or fully responsible for the injury. The only exception to this rule is if the worker deliberately injured himself, or if he was under the influence of alcohol or drugs.

Claims

When an employee files a worker's compensation claim, he is filling out an application for benefits for an occupational injury or occupational disease/illness. A work-related injury is usually considered to be caused by a single incident. An occupational disease/illness is considered to have occurred from the result of exposure to an industrial hazard. A worker's compensation claim is valid when the injury or illness has occurred "in the course of" and "arising out of" employment. The phrase "in the course of" is defined as an injury or illness incurred during work time and at work. The phrase "arising out of" employment is defined as an injury or illness that was caused by working. Most states do not provide compensation for illnesses that are "ordinary illnesses of life," or illnesses that are not "peculiar to or characteristic of" the employee's occupation.

Four primary benefits are covered under worker's compensation: cash benefits, medical benefits, rehabilitation benefits, and death benefits. The most cost-effective monetary method of handling worker's compensation is to prevent an accident or illness from occurring in the first place. Consequently, companies that have inadequate safety programs usually pay a very high worker's compensation insurance premium, due to large payments to disabled employees.

5.2.2 Compensation

There are four basic disability compensation classifications:

1. *Temporary partial disability* (TPD): A disability caused by a work-related injury or occupational disease for a definite period that does not prevent the employee from gainful employment. A TPD employee is expected to regain most of his former capability.
2. *Temporary total disability* (TTD): A disability caused by a work-related injury or occupational disease for a definite period, as a result of which an employee cannot return to gainful or regular employment. A TTD employee will not be able to regain his former capability.
3. *Permanent partial disability* (PPD): A disability caused by a work-related injury or occupational disease where the employee is able to return to work but has not attained maximum medical improvement.
4. *Permanent total disability* (PTD): A disability caused by a work-related injury or occupational disease that completely removes the employee from employment, due to loss of use of designated part or parts of the body, or from amputation.

Cash Benefits

Cash (indemnity) benefits include both disability benefits and impairment benefits. *Disability* benefits are paid when there is a work-related injury or an occupational illness and a wage loss. *Impairment* benefits are paid only for certain specific physical impairments. Most states provide indemnity payments of 67% of the employees weekly wage with a maximum weekly benefit allowance. A few states will pay 100% of the employee's weekly wage. Most states require a waiting period of 3 to 7 days before payment of indemnity benefits. The waiting period is intended to eliminate minor medical and first aid cases.

Medical Benefits

Medical benefits include first aid treatment; services of physicians, chiropractors, dentists, outpatient nursing, and physical therapists; surgical and hospital services, medication, medical supplies, and prosthetic devices. Every state law requires the employer to provide medical care to work-related injured or occupationally ill employees. There is usually no dollar limitation or time limitation on medical benefits. The states are evenly divided over the practice of who is allowed to choose the attending physician. Some states supply an approved list of physicians.

Rehabilitation Benefits

Rehabilitation benefits are divided into two classes, medical rehabilitation and occupational rehabilitation.

Medical rehabilitation has two phases:

1. The acute phase, requiring hospitalization and/or surgical management
2. The postoperative phase, when multiple sessions per week are held with doctors, nurses, and/or therapists to restore the employee to his fullest physical capacity

Occupational rehabilitation is the transition phase in which evaluation of work capacity, training, counseling, and job placement are performed. Some workers may require retraining to enter completely new career fields. Rehabilitation for a seriously injured worker can be lengthy and extremely expensive. Many states have enacted a second injury fund to help promote the hiring of disabled workers. If a company knowingly hires a disabled worker and that employee reinjures himself, the state will help to reimburse the company for the extra worker's compensation costs. Every state receives funding from the Federal Vocational Rehabilitation Act to sponsor vocational rehabilitation for disabled industrial workers. Employers who choose modified work (light duty) for injured employees show a reduced total recovery period. The employees who work at modified workstations usually return to their former positions more quickly than those who remain out of work for the entire period. This method of reinstating employees in a timely manner rewards both the employee and the employer. If an employee is not working because of a work-related injury, the cost of that injury will be about 15 to 25 times greater than for an employee who goes for medical treatment and immediately returns to work.

Death Benefits

Death benefits are generally paid to a spouse until he or she remarries and to the children until a specified age, usually 18. Many states continue benefits for dependent children while they are attending college as full-time students. The benefits include a burial allowance as well as a portion of the worker's former weekly wages. In most states, the duration of benefits is unlimited. Death benefits are more than 14% of all total indemnity benefits. Although death is the ultimate personal loss, the economic loss associated with death cases is often less than for a permanent total disability.

5.2.3 Abuses in the System

Showing concern for an injured employee usually reduces the time that the employee is on disability leave. Most of the injured employees who seek legal counseling do so because no one from the company has talked with them and shown concern for them. In some states, a new industry has been created from their liberal worker's compensation systems. Many daytime television commercials are advertisements for worker's compensation attorneys.

Recruiters actually seek out potential worker's compensation candidates from the unemployment lines. The unemployed candidates are advised of their potential benefits from the worker's compensation system. If the candidates feel that they may qualify for worker's compensation benefits, the recruiter escorts them to a medical clinic. These clinics then process the people through a series of medical evaluations. When necessary paperwork is completed, a new candidate has joined the ranks of the state worker's compensation system. In the past, back injuries were most common. Recent experience shows a higher incident of other skeletal problems, most notably carpal tunnel syndrome. (see subchapter 5.1). Some states now recognize stress resulting not only from the workplace environment, but from traffic driving to work, as an injury or disease "arising out of employment."

5.2.4 Costs

The escalating cost of worker's compensation is staggering. Payments for worker's compensation programs rose to an all-time record of $35.3 billion in 1990. California leads the nation with $6.05 billion in medical and indemnity payments. Medical benefits account for approximately 30% of all worker's compensation costs.

The cost to a company for worker's compensation insurance premiums is based on payroll cost (wages). Recently, all U.S. wages and salaries are as follows (billions):

Salaries	$2,917
Supplemental	608
Composite	$3,525 per year

This says that 17% of payroll cost is due to the supplemental benefits, including worker's compensation. The gross average wages in the U.S. manufacturing industry are $470 per week. The median income for machine operators and assemblers is $350 to $400 per week.

There is a wide range of cost based on worker classification. Some examples from the state of Florida are as follows:

Clerical: $ 0.75/$100 of payroll
Electrical apparatus assembler: $ 4.40/$100
Iron and steel fabrication shop: $20.00/$100
Roofer: $48.00/$100

The average cost of worker's compensation insurance is $3.50 per $100 of payroll in Florida, and ranges from slightly over $1.00 in some other states to more than $6.00 in California. The variation depends on the average wages, types of work classification (industrial versus agricultural, etc.), and the history of claims payments.

Typical U.S. Payroll Costs

A new "start-up" company would pay rates in the above standard ranges. This is known as a *modification factor* of 1.0. As the insurance carrier has experience with the number and cost of claims, the "standard" rate for each employer is adjusted by changing the modification factor. The rates can be adjusted downward (or upward) based on the average of the previous 3 years' history. There is no cap on the upper limit that can be charged. An example of this can be made for a manufacturing plant with 1000 employees. Approximately half of the workers are employed in the shop. An average for a well-managed factory would be 160 to 170 claims per year. Of these, 120 or so claims are for very minor injuries, perhaps requiring a couple of stitches—and back to work. This cost might total $40,000 per year. The other 40 claims, when the worker goes to a clinic and stays off work for some time, may cost the company more than $600,000 per year in lost wages, medical bills, rehabilitation, etc. This can become a very significant element of payroll cost to a manufacturer, since the insurance company raises the worker's compensation insurance premium to cover the $650,000 per year, plus the administrative cost of handling the claims. This total of $750,000 per year can be a significant factor in the profitability of a small manufacturing firm, and warrants positive measures to keep it as low as possible.

5.2.5 Cost Reduction Methods

The most important positive actions that a company can take are to (1) reduce the number of claims and (2) lower the cost of medical expenses on each claim. Employee training and education are important elements contributing to reduced claims. Each job or work element in the manufacturing plant should be analyzed for proper ergonomic considerations (See Subchapter 5.1). An analysis of several medium-sized factories, where the work is relatively light and clean, shows that more than 40% of the "lost-time" cases are due to strain or overexertion; 25% are due to cumulative, repetitive job injuries; 15% are due to falls; and the remaining are due to all other causes. Looking at the cost of these categories, 40% are due to strain; 25% are due to falling; 17% are due to cumulative injuries, and 12% result from being struck by something (or striking something). Proper employee and supervisor training plus careful review of the process design can reduce both the number and severity of these lost-time cases.

Most states offer a 10% premium reduction if the firm has an organized safety program in place (see Subchapter 5.3). Also, a 5% reduction is available if the employer has a "drug-free workplace" program implemented (see Subchapter 5.4).

5.3 INDUSTRIAL SAFETY

Jeffrey W. Vincoli

5.3.0 Introduction to Industrial Safety

In the highly competitive environment of today's business world, success is achieved through the collective talents, skills, and contributions of all the individual employees of the organization. In fact, the business enterprise with the right combination of skills and clear, focused objectives will often outperform its competitors in every category. It cannot be disputed that people are a company's greatest resource. It is clear that people are literally and figuratively the lifeblood of any organization. Therefore, it should be equally clear that the assurance of a safe and healthy workplace, within which this resource must perform, is as important to a company as any other issue related to production, sales, and profit.

Just about all of us have heard the statement that *"safety is everyone's business,"* but few realize that this is more than just a cliché or a clever approach to safety management. The requirement for a safe work environment is actually a mandate on industry and its employees from the federal government. Safety and health regulations for *general industry* can be found in the U.S. Code of Federal Regulations (CFR) at Title 29 (Labor) Part 1910, and for the *construction industry* at Title 29, Part 1926. These codes, which address literally thousands of situations, working conditions, hazard control requirements, and worker safety and health protection standards, are all intended to make the work environment safer.

Although numerous areas are addressed and a *minimum level* of worker protection is ensured by implementation of these many standards, it is important to note that the major driving factor for worker safety is contained, not in the Code of Federal Regulations but in Section 5 of the Occupational Safety and Health Act of 1970. This factor, better known as the *General Duty Clause,* Section 5(a)(1), simply states that:

Each employer shall furnish to each of his employees employment and a place of employment which are free from recognized hazards that are causing or are likely to cause death or serious physical harm to his employees, and [each employer] shall comply with the occupational safety and health standards promulgated under this Act.

This section goes on, in Part (b), to state that *each employee* shall comply with safety and health requirements. It is from this legislation where the statement, "safety is every one's business," becomes a legal as well a moral obligation.

Having established this fundamental principal of industrial safety, this subchapter will briefly explain the required actions an organization must take to ensure compliance with the intentions of the law regarding employee safety and health. The reader should also note that many individual states have obtained approval from the Federal Occupational Safety and Health Administration (OSHA) to regulate their own state-wide safety and health programs, as long as these programs are *at least as stringent* as those mandated on the federal level. However, because so many individual state safety and health laws now exist, it would not be possible to discuss the particulars of each in the space provided here. This chapter will therefore focus on the Federal level, with the understanding that these requirements are the minimum standards which, in many cases, must be exceeded in order to ensure the proper protection and preservation of employee safety and health. As Figure 5.34 shows, those *proactive* companies that anticipate and exceed these minimum regulatory requirements will be the leader's of tomorrow's highly competitive market. Conversely, those who are simply *reactive* to the requirements will more than likely fail.

By establishing workplace objectives for accident and illness prevention, management can demonstrate both commitment to worker safety and action towards that commitment. Without exception, the realization of these safety objectives should be considered as important as those established for sales, productivity, engineering, quality, and so on. In fact, how could any organization justify placing the safety and health of its workers second

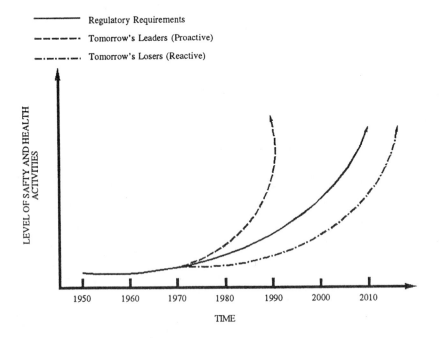

Figure 5.34   Staying ahead of the regulatory curve pays off.

to production and sales? The truth is, few employers who have tried this approach have been very successful in avoiding investigation by federal OSHA or state safety inspectors.

5.3.1 The Business of Safety

While the General Duty Clause does require *both* employers and employees to ensure safety at work, the language of the law makes it quite clear that the *employer* has a special responsibility to its workers to provide a safe and healthy workplace. Since it is no secret that we in the United States live in one of the most litigious societies on earth, employers who wish to remain in operation today must also be in the business of safety.

Under the law, the management of an organization becomes the responsible element for ensuring that the safety process within the organization is in compliance with applicable requirements. However, it must be clearly understood that compliance with these various codes and regulations basically equates to compliance with only the minimum acceptable requirements. Therefore, in the practice of modern manufacturing engineering, "design by code" is no substitute for intelligent engineering. Since codes only establish minimum standards, actual design must often exceed these codes to ensure an adequate level of safetyperformance. The successful management of the safety process equates to a *commitment* of company resources. Time, money, and talent are required not only to properly identify and control hazards in new and existing operations and processes, such resources must also be expended to eliminate, limit, or control these hazards to their absolute minimum. But the real *cost of safety* goes far beyond that which is readily identifiable as *direct* cost. There are also numerous *indirect* expenses which affect the profitability of any given enterprise as a result of poor or ineffective safety policies and programs. This concept is best illustrated using the familiar iceberg graphic as shown in Figure 5.35.

While the visible (or *direct*) costs are easily seen on the surface, there are often many more costs (*indirect*) resulting from inadequate safety performance which are not easily detectable at first glance but are there nevertheless. Only when an organization accepts the cost of safety as an important cost of doing business will the axiom "safety first" have any real meaning for the workforce. In a manufacturing setting, a combination of effort must often be employed to ensure that these established safety objectives are achieved. For example, engineering controls may have to be installed, personal protective equipment (PPE) might be required, and/or specialized training may have to be provided by the company.

In short, as with every other aspect of a successful business endeavor, safety in the workplace will never be realized without a firm commitment beginning with top management. Evidence of this commitment will be visible in every decision the company makes and in every action it takes. The management personnel responsible for the proper execution of this commitment throughout the organization must be clearly identified to each employee at every level through established policies. One way in which company safety policy is communicated to employees is through an effective *written* injury and illness prevention program, more commonly referred to as a *Safety manual*. Historically, when employees come to understand that their company has made a sincere commitment to the safety of the workforce, overall performance is typically improved and the organization works as a cohesive team toward company goals.

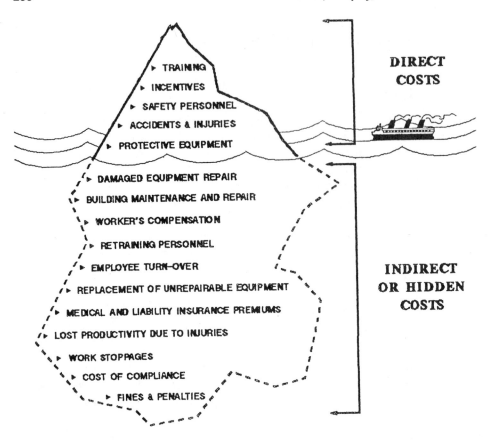

DIRECT
COSTS

- TRAINING
- INCENTIVES
- SAFETY PERSONNEL
- ACCIDENTS & INJURIES
- PROTECTIVE EQUIPMENT

- DAMAGED EQUIPMENT REPAIR
- BUILDING MAINTENANCE AND REPAIR
- WORKER'S COMPENSATION
- RETRAINING PERSONNEL
- EMPLOYEE TURN-OVER
- REPLACEMENT OF UNREPAIRABLE EQUIPMENT
- MEDICAL AND LIABILITY INSURANCE PREMIUMS
- LOST PRODUCTIVITY DUE TO INJURIES
- WORK STOPPAGES
- COST OF COMPLIANCE
- FINES & PENALTIES

INDIRECT
OR HIDDEN
COSTS

Figure 5.35   The true costs of safety.

5.3.2   Responsibility and Accountability

As stated in the previous paragraph, once management has *established objectives* designed to ensure the safety of their personnel, products, and services, and after they have *committed* the necessary and required resources to achieve these goals, they must also clearly *define those responsible* within their organization who will be held *accountable* for the success of the safety program.

The ME is often uniquely situated within the organizational structure to act as the primary implementing force behind effective safety practices. Whether the engineer is occupied with the complexities of designing factory layout, establishing operating specifications for specific machinery, retooling existing equipment, or writing process sheets, the responsibility to consider the impact on safety resulting from any of these actions is a key to the success of an overall safety program. For the ME, the ability to recognize obvious as well as insidious conditions that could potentially threaten the safety of personnel and/or property is of paramount importance. However, this is often a difficult thing to do, since most engineers are not specifically trained in the process of hazard recognition, evaluation, and control. This is where the concept of a *working team* becomes critical. Specifically, the manufacturing engineer should form a working partnership with the company safety professional. Together, regardless of the task, the hazards associated with production can be successfully identified, understood, and hopefully eliminated (or at least controlled). In most industrial settings, effective interdepartmental cooperation can mean the difference between the success and failure of the entire business enterprise. Therefore, in order to be truly effective, such working relationships should be a matter of company policy and not simply one of convenience. Practically speaking, this *teaming approach* to ensuring a safe and healthy work environment has proven to be the most effective course of action in the manufacturing setting.

The process of assuring safety at work is really a critical element of the manufacturing engineer's overall job. In fact, because engineering personnel are normally located within the typical organization as a line function, effective safety management becomes a primary task. For example, in a typical line-and-staff organization (as shown in Figure 5.36), the task of implementing safety requirements is always a line function. This means that, while company safety personnel may provide recommendations, advice, and assistance to the line managers/personnel regarding compliance issues,it is still the line managers and supervisors who have the authority and responsibility to implement those recommendations. The task of safety should therefore be approached with the basic understanding that workplace safety programs are implemented as a *line* responsibility.

In comparison, the company's professional safety representative performs a *staff* function only. The safety department is really a service organization that must be an effective member of any working team in order to ensure the appropriate infusion of safety concepts and considerations into work process. Among other things, the safety department representative will research applicable requirements, develop implementation strategies, work with line personnel to ensure proper implementation, monitor that implementation to ensure compliance, and conduct periodic assessments of the program to measure degree of success. But it is still the line managers, supervisors, and associated personnel, such as

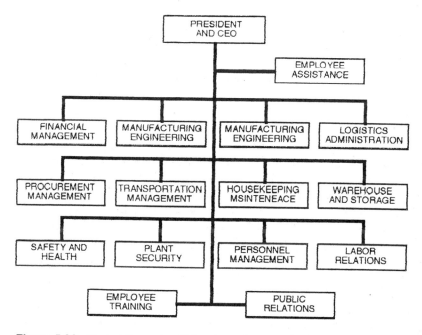

Figure 5.36 Typical line and staff functions of an organization.

the manufacturing engineer, who must actually implement and enforce the safety program in their designated work area(s). Therefore, safety as a task must clearly be the function of the line, or operational safety will not succeed. This approach to safety management is based on a fundamental concept which stipulates that line management (especially first-line supervisors, but including all levels from the top down) is absolutely responsible for all operations that occur within their assigned area(s). There are very few line managers or supervisors who would dispute this position or who would have it any other way. It is therefore logical that this responsibility include the safety of those operations and the employees required to perform them. This is an extremely important concept that must be clearly understood and accepted through all levels of the organization. As an example, Figure 5.36 shows how the safety department is located in a typical organizational chart. Notice how staff functions, such as safety, training, security, personnel, and labor relations, are distinct from the various line management responsibilities. This structure is generally found throughout industry.

5.3.3   Safety in the Manufacturing Process

Typical manufacturing operations occur in a work environment that contains many inherent hazards to both personnel and equipment. The very process of manufacturing requires the performance of hazardous tasks and procedures on a routine basis. These may include welding, painting, electrical work, heavy lifting, crane operations, radiation, robotics, machining, and the use of hazardous chemicals or even explosives. While safety professionals will examine the many work procedures that contain hazards to personnel and equipment and recommend controls to limit exposure to these hazards, it is the responsibility of the manufacturing engineer to implement these controls to ensure safe and successful operations. Simply stated, if the manufacturing engineer does not implement these safety recommendations on a daily basis, a safe workplace will not be achieved. To assist in with this objective, safety representatives can perform audits and inspections, which serve to verify the level of compliance in each work area. When deficiencies are identified in the manufacturing process, the task of abating these discrepancies falls to the engineering supervisor responsible for that work area. The supervisor should look to the company safety professional for advice and assistance in compliance issues. However, they should never assume that safety program requirements will be met by the safety department alone. Safety must be a *team effort*. In the manufacturing industry, or any other for that matter, it just does not work any other way.

To ensure safe operations and the safety of personnel in any manufacturing process, adequate *planning* for workplace safety and health is an important aspect of everyday business operations. To effectively reduce the costs and risks associated with occupational accidents and injuries, workplace *safety policies* must be well considered, clearly written, comprehensive, and, above all, properly disseminated through every level of the organization. To assist line management with their safety responsibilities and to ensure a consistent approach to safety program implementation, a company safety manual should be developed and distributed to all levels of the workforce. The safety manual should contain *general* as well as *specific* safety rules, regulations, policies, and stated objectives for all employees to understand and follow. Also, to ensure its adequacy, the safety manual should not be developed without thorough planning and effective evaluation of company tasks and operations.

Safety and health planning is considered effective when at least the following factors have been considered by management during the development of its safety policy:

1. General safety rules should be written and constructed so as to apply to every member of the organization. Rules to address requirements such as the use of personal protective equipment (PPE), appropriate clothing for the workplace, expected behavior and conduct while on company property, and special emergency procedures all fall within the general nature of written company safety policy. Employers and employees should perform regular reviews of these rules to ensure an up-to-date and fair approach to safety and health assurance. Also, whenever new processes or products are introduced into the work environment, new safety rules and procedures must be written to clearly define the hazards and controls required to perform any new tasks.

2. The safe and healthful work practices required for specific tasks and procedures must be developed and differentiated from any general rules, if applicable. In most cases, specialized training may be required on these particular safety considerations before employees are permitted to perform the job. These specific tasks or procedures may include examples such as laser operations, robotics, chemical processing, use of mechanical lifting equipment (cranes, hoists, forklifts), assembly and/or use of explosives and explosive devices, and other special equipment procedures and operations.

3. Procedures which explain the company's enforcement and award recognition policies will help ensure that all safety and health rules and work requirements are practiced adequately at all levels in the organization. Rewards or other positive reinforcement tools such as bonuses, incentives, or employee recognition programs provide for positive motivation for compliance with company safety rules and procedures. Likewise, when clear violations of established safety rules have occurred, the company must be equally prepared with written methods of disciplinary action. Unbiased and consistent enforcement of both negative (disciplinary) and positive (recognition) policies provide a clear message of the company's firm commitment to the safety process.

4. There should be a written plan to address all types of emergency situations which could possibly occur at the manufacturing enterprise. The plan should include a set of procedures which respond to each possible situation. In fact, some emergency procedures, such as those covering medical emergencies and fire evacuation, are required by OSHA laws.

5. There should also be adequate provisions in the safety plan to ensure employee participation and involvement. This includes a system for effectively communicating safety program plans and directives to employees at all levels in the organization. Safety plans, newsletters, incentive programs, poster contests, and safety committees are a few examples of methods used to obtain employee participation.

6. Management must include a system to assure that the proper record-keeping and safety program documentation requirements of OSHA are fulfilled. Examples of required record-keeping activities include employee medical records, injury and illness statistics, inspections and corrective actions, equipment maintenance logs, and employee safety training.

7. Considerations for accident prevention and investigation must be an integral

factor of the overall safety process. Since there is no way an organization can hope to prevent accidents, which only downgrade the efficiency of the business enterprise, without investigating those that do occur, there must be a comprehensive accident investigation and loss control program established as an integral part of company safety policy.

5.3.4 Developing a Specific Safety Program

In most cases, general safety rules focus on concerns which are germane to almost every type of work environment. These rules usually concentrate on *housekeeping* safety requirements, which form the basis of most generic safety and health programs. Hence, general rules will provide an excellent foundation on which to build a comprehensive occupational safety and health program. While some rules may seem to be nothing more than common sense, employers usually discover the hard way that their employees often have very different perceptions, understandings, and interpretations of common-sense issues. It is therefore always best to provide the workers with these requirements in a clear, written, and well-documented fashion.

Once the general rules for safe conduct and safe operations have been developed, the organization must then determine those specific rules and regulations that are required to ensure the safety and health of all workers. Before any selection of these specific safe work standards can be made, as much information as possible must be obtained regarding the current conditions of the workplace and the adequacy of existing safe work practices. This information will not only help identify existing problem areas, it will also help determine the requirements necessary to resolve them.

This safety assessment should be a team effort conducted by the engineering manager or supervisor responsible for ensuring work place safety and health, together with the company safety representative or professional occupational safety and health consultant. As a minimum, the workplace safety assessment should entail a detailed survey of the entire facility with a comprehensive evaluation of all required tasks and procedures. This survey should evaluate workplace conditions regarding:

Applicable safety and health regulations
Generally recognized safe work practices
Physical hazards of any equipment or materials
Employee work habits with regard to safety and health
A discussion with employees regarding safety and health

As shown in Figure 5.37 the safety survey should, as a minimum, also address the following five specific areas of concern (if applicable).

1. *Equipment safety:* The surveyor should develop a list of all the equipment and tools that are in the facility, to include their principal area(s) of use. Special attention should be given to inspection schedules, maintenance activities, and the physical layout of the facility. This information will greatly facilitate the development of specific safety program objectives.
2. *Chemical safety:* If it is not already available, a list of all chemicals (hazardous and nonhazardous) must be developed. The Material Safety Data Sheet (MSDS) for each chemical must be obtained and reviewed to ensure adequate precautions are being taken to preclude hazardous exposures. The surveyor should also attempt to identify and list the specific location(s) within the workplace where

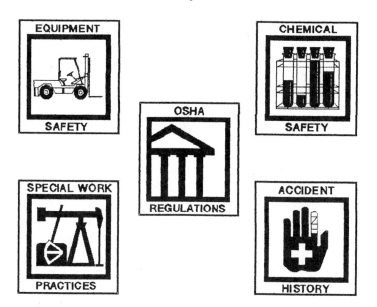

Figure 5.37   Specific areas of concern to investigate during a safety survey.

each chemical is being used, the processes involved with that use, and the possibility of employee exposure. This information will assist in the establishment of specific safety rules to prevent injury and illness resulting from the improper handling of hazardous chemical substances.

3. *Work practices:* All specific work practices involving the use of any equipment, tools, chemicals, and/or materials not previously identified must be examined in detail to ensure that appropriate and adequate precautions are in place. Special concentration on concerns such as the use of personal protective equipment (PPE), machine and equipment guarding, ventilation, noise, illumination, emergency procedures, and the use of appropriate tools for the job will help identify additional problem areas in need of attention. Also, any industry-specific safety hazards, such as potential exposure to biological hazardous wastes (medical industry), asbestos (automobile repair), or radiation (medical, nuclear, and aerospace industries) must be evaluated and properly addressed.

4. *OSHA requirements:* The safety professional should review all rules (state and federal) that are applicable to the specific operations, equipment, and processes in work at the subject facility. Since these rules normally establish only the minimum requirements for workplace safety and health, simple compliance may not always equate to absolute safe operations. Therefore, a detailed evaluation of the these minimum safety requirements, as they apply to the subject facility, will help determine the overall applicability of the rule as well as identify those areas where more attention is required to ensure a safe work place. Also, all OSHA-required recordkeeping activities should be evaluated for compliance with the minimum standards established in the CFR.

5. *Accident history:* The assessment would not be considered complete without an evaluation of the company's existing injury and illness prevention program. This will not only identify those aspects of the program that may be effective, it will

also help ascertain those areas where specific improvement may be needed. In this regard, the survey should examine the company's:

        Accident injury and illness rates
        Accident injury and illness dates
        Accident injury and illness types, by location
        Worker's Compensation costs
        Rates of employee turnover and absenteeism
        Company policy statements
        Documentation on previous safety/health activities
        Employee training records
        Existing company rules and regulations
        Guidelines for proper work practices and procedures
        Status of compliance with federal requirements
        Other safety and health related issues and programs

Once these areas have been properly researched, a site-specific safety and health program can be drafted to address all pertinent issues which must be considered to ensure a safe and healthy workplace.

5.3.5 The Safety Plan

After the rules and regulations have been developed and written into the company safety manual, management must still develop a plan which dictates how the safety program will be implemented. This Safety Plan describes specific levels of responsibility with regard to work place safety and health. It establishes accountability through the chain of command for all aspects of the safety program. Beginning with the chief executive and continuing through all levels to first-line supervisors and on to the employees, specific responsibilities for safety program implementation are outlined as company policy. Every single employee in the company should have access to the safety plan to ensure proper dissemination of policy and eliminate any potential for misunderstanding regarding workplace safety and health issues.

    The safety plan simply provides clear direction for implementing the safety program rules, regulations, and objectives established in the safety manual. Each document is complementary to the other. Together, they are the foundation of a comprehensive safety program.

5.3.6 The Benefits of Safety

It is no secret that a company is in business to make a profit. It is the primary objective of any organization to provide quality products and/or services to its identified market at a cost and schedule that are better than any of its competitors. This is the essence of a free enterprise, market economy system. For many years, safety compliance was viewed by industry as a sometimes avoidable obstacle in this process. However, with the right approach, safety can actually be a benefit to companies as they pursue their goal toward perpetual profits. As stated previously, safety should be approached as a team effort, with all elements of the organization working together toward the common goal of providing a safe and healthy workplace. Properly implemented, a comprehensive safety program will pay off in optimized and more efficient production. Typically, there will be fewer work stoppages due to unsafe conditions, injuries, accidents, damaged equipment/tools, and so on. Also, although the initial cost of complying with safety regulations may be considered relatively high, the actual expenditure for the safety effort will reduce over time and therefore effect a similar reduction in long-term implementation costs. Liability insurance, as well as medical and worker's compensation premiums, will also likely be reduced in the long term due to excellent safety performance. Employee turnover may also decline if the workers are convinced that their employer actually cares about safety in the workplace. All of these are examples of the positive impact that safety can have on the bottom line of any business enterprise. Clearly, employers cannot afford to ignore safety as a required aspect of successful business operations.

5.3.7 Summary

People are an organization's most precious resource. Without their collective contributions, success in today's highly competitive market is a hopeless objective. It should follow that the protection and preservation of employee safety and health should be the primary concern of any business endeavor. The federal government certainly maintains this view. In fact, an employer's responsibility to the safety and health of its workers forms the basis of the Occupational Safety and Health Act of 1970. Literally thousands of safety rules and standards are contained in the Code of Federal Regulations (CFR) for Labor (Title 29) which require specific employer actions to ensure employee safety and health. Compliance with these many mandated requirements is rarely achieved without management commitment, which often includes a costly expenditure of resources. But the cost of ensuring safety is, with few exceptions, a great deal less than the cost of ignoring these requirements. To facilitate the implementation of a consistent, comprehensive, and objective safety program, safety rules should be developed and written in a uniform safety manual. The manual must consider general safety requirements, which will typically apply across the organization, as well as very specific safety concerns and issues.

No safety program will succeed unless those elements of the organization responsible for its success are clearly identified. While most companies maintain a safety department within their organizational structure, the responsibility for safety program implementation and enforcement must be a team effort. Safety professionals provide a staff service to line management. They review and research requirements, develop rules and regulations based on those requirements, and recommend implementation strategies to management. They will also conduct assessments to determine program adequacy, and monitor program success through inspections and audits. While the responsibility for actual implementation of the safety program rests with those responsible for daily operations (i.e., the line managers), the safety professional must work together with each functioning element of the organization for the process to be successful. To clarify, these responsibilities should be described in a company safety plan, which establishes clear organizational accountability for safety program implementation.

Strategies for selecting specific rules were also presented and discussed. Without a doubt, every safety program should be different. But, for any program to work, there must first be a clear understanding of the reasons for establishing such requirements, those responsible for the success of the program, and an appreciation for those general and specific safety issues and concerns that each company must identify, understand, and address. Only then will an organization be on its way to achieving a safe and healthy workplace for all its employees.

5.3.8 Bibliography

Ray Boylston, *Managing Safety and Health Programs,* Van Nostrand Rienhold, New York, 1990

Mark Early, *The National Electrical Code,* 5th Ed., National Fire Protection Association, Quincy, MA, 1989.

Ted S. Ferry, *Modern Accident Investigation and Analysis, An Executive Guide,* John Wiley, New York, 1981.

Mark S. Larson and Stu Hann, *Safety and Reliability in System Design,* Ginn Press/Simon & Schuster, Needham Heights, MA, 1989.

National Safety Council, *Accident Prevention Manual For Industrial Operations,* 9th Ed., (Two Volume Set), National Safety Council, Chicago, 1988.

Jeffrey W. Vincoli, *Basic Guide to System Safety,* Van Nostrand Reinhold, New York, 1993.

Jeffrey W. Vincoli, *Basic Guide to Accident Investigation and Loss Control,* Van Nostrand Reinhold, New York, 1994.

5.4 INDUSTRIAL HYGIENE

Paul R. Riedel

5.4.0 Introduction to Industrial Hygiene

Of all the resources that a manufacturer has, the most valuable is its people. Without the people, the production stops. Therefore, the primary objective of the manufacturer should be to keep personnel creating, building, assembling, and producing every day, week after week, year after year. The productivity of an employee depends on the state of his or her work environment. If work area conditions are hazardous, or even perceived to be hazardous, morale will suffer and people will not perform effectively. Moreover, if conditions are so bad as to cause worker illness or injury, the consequences may severely impact the company. If the work environment provides a healthful and safe means to allow people to produce, then the employer will achieve profitable results sooner . The successful businesses of the future will be the ones that actively participate in sound industrial hygiene programs that protect their most valuable resource. Another good reason for the company to protect the workforce is the legal requirement. In 1970, the Occupational Safety and Health Administration (OSHA) was created under the U.S. Department of Labor. Many states and municipalities have also promulgated regulations. Failure to comply to legislation can result in fines and criminal penalties and may be very costly to the company.

Hazards are present in many manufacturing processes, some of which are unavoid-able. How do you eliminate nuclear hazards in a nuclear power plant, for instance? The answer often is that you cannot eliminate the hazard. The hazard can be controlled, however, so that the potential for injury or exposure is minimized.

This subchapter will discuss the following subjects: What kinds of hazards are present in the work environment? How do such hazards affect personnel? What are the proper ways to protect people from hazards? What programs should the manufacturer implement to ensure employee protection? What regulatory programs are required to be implemented? What is the regulatory agency that enforces laws for occupational health?

5.4.1 History of Occupational Environment

The recording of occupational hazards goes as far back as the fourth century B.C., when Hippocrates described the effects of lead poisoning on miners. The history of occupational hazards describes accidents and deadly diseases due to people's ignorance or indifference to workplace hazards. During the Industrial Revolution, many industrial dangers were created. While public awareness of occupational hazards has increased, work environments have also become more hazardous due to new industrial processes and increased chemical usage. In 1984, more than 2000 people were killed as a result of the release of a deadly chemical at a factory in Bhopal, India. Most catastrophies could have been avoided.

The employee assumes a certain amount of responsibility to ensure that he or she is protected from potential perils at work. However, it is the employer's responsibility to ensure that all employees are protected from occupational hazards that may cause impaired health or injury. These hazards may range from a minor, irritating odor in the office to a potentially life-threatening procedure on the factory floor. Under federal law, severe penalties have been imposed on companies where workers were subjected or exposed to uncontrolled hazards.

5.4.2 Occupational Hazards

Industrial hygiene is the science involved with identifying hazards in the work environment, evaluating the extent of the hazard, and providing recommendations to control such hazards. Simply put, the purpose of a sound industrial hygiene program is to protect the worker. The manufacturing industry involves hazardous work areas and dangerous work processes that can cause serious health effects. Occupational hazards can be divided into four groups: ergonomic hazards, biological hazards, physical hazards, and chemical hazards.

Ergonomics is the science that deals with the interaction of the worker and his or her work environment. Examples of such interaction may be the repetitive use of simple hand tools, tedious assembly-line work, or various activities in a meat-packing plant. An ergonomic hazard can be defined as a process which can cause physiological stress to the human body. For more details on ergonomics see Subchapter 5.1.

Any living organism that can cause disease is a *biological hazard*. These organisms, which include viruses, bacteria, and fungi, can be found anywhere if the right conditions exist. Occupational environments such as in hospitals, slaughterhouses, or farming, in which the source of all kinds of biological agents exist, will have a much higher degree of biological hazards than most manufacturing settings. However, microbial growth may be enhanced in any high-humidity environment. Indoor air-quality problems are often associated with high humidity or improperly designed or maintained air-conditioning systems. Generally, biohazards in the manufacturing plant will be minimized if humidity is controlled and good housekeeping is maintained.

Physical hazards present in the work environment include noise, temperature extremes, ionizing radiation, and nonionizing radiation. These agents at certain levels will severely damage the human body. Noise, which is defined as unwanted sound, will impair hearing to the point of deafness if the noise levels are high enough. One of the dangers of this hazard is that many people do not realize the permanent consequences of working constantly with high noise levels. Many workers think their hearing will return to normal when they leave the noisy work area. In fact, chronic exposure to high noise levels will cause irreversible hearing loss.

Minimizing the hazards of high noise levels can be achieved by engineering controls such as constructing enclosures around the process, isolating the operation from personnel, or by the utilization of ear muffs or ear plugs. A less serious problem than hazardous noise levels is a lower noise range known as speech interference noise. A good example of speech interference noise is a room full of computers. Although it is not damaging to the hearing system, this constant noise may be irritating to personnel and create communication difficulties which may adversely affect productivity. Eliminating speech interference noise levels can be accomplished either by reducing the noise level of the source or by installing acoustical walls in the general area.

Clearly, the danger of working in extreme temperatures can be serious. In the manufacturing world, heat stress is common. Heat stress on the human body is the combined effect of the hot environment and heat generated by the body's metabolism. The heat stress experienced by a worker performing exhaustive work in a hot environment will be significantly more than that of a person doing light work in that same environment because of increased metabolic heat load. Heat stress can cause heat cramps, heat exhaustion, and, most seriously, heat stroke. Ambient temperatures can be controlled by isolating existing hot processes or by providing good mechanical or natural ventilation. To keep the body cool, the metabolic heat load can be lessened by decreasing the workload or implementing scheduled work breaks. Personnel should wear light clothing in hot environments, to allow the body to cool itself. Another very important factor involved with heat stress is humidity. High humidity impairs the human mechanism to cool. Environments with lower humidity will decrease the physical hazards associated with heat stress. Similarly, cold environments can present serious consequences such as frostbite. Reducing the hazard is achieved by increasing the surrounding temperatures or by providing protective clothing for the worker. Certainly, for the sake of productivity, the best solution is to control the temperature, if possible.

Working in slight temperature variations, although not life threatening, may cause problems. Workers who are simply uncomfortable from being too hot or too cold may be less productive. The American Society of Heating, Refrigerating and Air Conditioning Engineers (ASHRAE) recommends temperature and humidity ranges for worker comfort in various types of working environments. See Chapter 4 for more information on building design.

The physical hazards of radiation, which are divided into ionizing and nonionizing radiation, may not be as familiar to the reader as other physical hazards. Nonionizing radiation is composed of ultraviolet light, visible light, infrared light, microwaves, and radio frequencies. The sources and uses of nonionizing radiation are becoming more prevalent in industry. The associated hazards for nonionizing radiation are a function of radiation frequency, intensity, duration of exposure, and proximity of the worker to the radiation sources. Overexposure to ultraviolet light can cause severe damage to the skin (i.e., sunburn) or eyes. Besides the sun, sources of ultraviolet radiation in the work environment include electric arc welding, mercury arc lamps, and xenon lights. The infrared energy hazard is heat. Occupational sources include high-temperature processes such as furnace operations or curing ovens. Microwaves and radio-frequency bands are notable in that they penetrate matter, causing a potential hazard to surface organs. The primary effects are thermal in nature.

Ionizing radiation differs from nonionizing radiation in that ionizing radiation has sufficient energy to ionize or remove electrons from their atomic orbits. The destructiveness of this type of radiation has been seen as a result of nuclear blasts and nuclear accidents such as in the Chernobyl catastrophe. The danger of ionizing radiation is due not only to the extensive biological damage it can cause, but also to the inability of the senses to detect its presence. Avoiding exposure to radiation can be achieved by reducing the dose, reducing the time of exposure, or by increasing the distance from the source.

Typically, the most significant hazards in the work area are *chemical hazards*. Depending on the specific use of the chemical, the hazard is caused by liquid, vapor, gases, dust, fumes, or mists.

What determines the dangers of chemicals? In other words, what is the likelihood that a particular substance will cause damage to the human body? The danger of being exposed to a chemical depends on its properties and the hazard potential associated with its uses. Significant properties of chemicals affecting safety and health include flammability, reactivity, corrosivity, and toxicity. The first three characteristics are serious safety concerns. The consequential results from flammable materials (e.g., acetone or methyl ethyl ketone) or highly reactive materials (e.g., nitroglycerine or perchloroethylene) are fires and explosion. Corrosive materials such as sulfuric acid, chromic acid, and phosphorus will destroy living tissues on contact.

Toxicity refers to the capability of a substance to cause damage to some biological mechanism. The human body has an amazing system to defend itself from chemicals that have entered the body. Toxic chemicals are either excreted or metabolized in such a way as to render them harmless. However, if the body is overburdened with a chemical, then damage will occur. Brief exposure to a highly toxic chemical or longer exposure to a less toxic chemical may cause severe and irreparable damage.

Some chemical hazards are obvious to the senses. For instance, you can immediately detect the suffocating odor of ammonia in a room, even at low concentrations. However, many hazards exist that you cannot see, smell, or hear. If you enter a room that has depleted oxygen levels, for example, you may not notice anything unusual, althogh in fact you could be in a deadly atmosphere. Some hazards can cause short-term health effects, while exposure to other hazards can occur for years before the onset of health impairment or disease. Exposure to ammonia or chlorine gas, for instance, can quickly cause burning of the eyes and throat. However, after a short period of time, the symptoms will subside. On the other hand, chronic exposure to low levels of lead fumes or some chlorinated solvents such as trichloroethylene may go unnoticed until irreversible health problems occur long after the initial exposure.

When discussing the dangers of chemicals, it is important to understand not only the physical properties of the materials, but also the hazard potential involved. *Toxicity* refers to how much damage a substance causes once it comes into contact with the human body. The *hazard potential* refers to how easily the material can get to the human body. This potential is determined, in part, by the particular process being used. For example, one process may call for heating the chemical in an open vat with frequent involvement by the worker. On the other hand, a second process may call for the same material to be heated in a closed system, without direct worker involvement. Under the latter scenario the hazard potential is much less than the former scheme. Even if a more toxic substance was used in the second situation than in the first, the hazard potential could be less. The open-vat heating process vapors could easily come into contact with the worker.

Another contributing factor to the hazard potential is the chemical's evaporation rate. Chemicals with high vapor pressures will volatize or evaporate faster, creating vapors in the air and thus increasing the potential for personal exposures. Other important hazard factors include the duration of the process, the general conditions of the room in which the process is located, and the health of the worker involved in the process. Obviously, longer processes increase the chance of worker exposure. An area that is confined or that has limited ventilation would contribute to a more hazardous situation than an area with good mechanical ventilation or good natural ventilation. Finally, the health of the worker is a very significant hazard factor. An individual who is already in poor health will be more susceptible to the external influence of chemicals. Control of these many factors contributing to hazard potential can minimize worker exposure.

Route of Entry

To understand these hazards, one must understand the various routes of entry that the hazardous material can take to inflict bodily damage. If the particular hazard never enters the body, no injury or illness will occur. The four major routes of entry are inhalation, absorption, ingestion, and injection.

The most significant route of entry is by *inhalation*. One reason is due to the rapidity that chemical vapors, mists, dust (including radioactive particles), or fumes can enter the body via the lungs and be absorbed into the bloodstream. Another reason for the importance of inhalation is that the worker can easily become exposed even though the source of the hazard is at a distance. As contaminants infiltrate the air, the hazard can expand throughout an entire building.

Absorption is an important route of entry that is responsible for causing skin diseases, which are the most frequent type of occupational illness. Some chemicals, such as phenol or toluene, are readily absorbed through the skin.

Ingestion occurs if the worker places contaminated hands to his mouth (i.e., eating or smoking). *Injection* of a chemical, usually from the result of an accidental spill, is rare in a manufacturing plant.

Airborne Contaminants

Airborne contaminants are gaseous or particulate. Gaseous contaminants include vapors from the evaporation of a liquid, and gases such as nitrogen, which are in a gaseous state under normal temperature and pressure. Particulate contaminants include dusts, mists, and fumes. Dusts are created from mechanical processes such as grinding or sawing. Mists are suspended liquid droplets which are generated from the condensation of vapors. Fumes are airborne solid particles that have condensed from volatized solids.

Control of Occupational Hazards

Recognizing hazards in the manufacturing industry is the first step in protecting workers, followed by evaluation or measurement of the extent of the hazard. The next step is to control such hazards by following an orderly approach and sound industrial hygiene principles. This approach should evaluate (1) substitution, (2) engineering controls, (3) administrative controls, and (4) personal protective equipment (PPE).

The best approach, from a health standpoint, is to use a material that is less harmful to the worker. Historically, manufacturers have used chemicals in specific processes that produce the most economical product. That process, however, may have exposed workers unnecessarily to highly toxic chemicals. The implementation of less toxic chemicals with minor process changes can often accomplish the desired results.

When substitution is not possible, engineering controls should be implemented to control the hazard locally. A pneumatic clipper or a punch press which produces hazardous noise, for example, cannot be substituted. The use of engineering controls to abate the high noise may include isolation of the noise source away from personnel, or surrounding the noise source with acoustical materials.

Local exhaust ventilation is another example of engineering controls. The principle is to contain and exhaust contaminates before they reach the worker. Engineering controls also include changing the original process to minimize the hazard. Brush application of chemicals, for example, is much less hazardous than spray application.

Administrative controls are used to protect not only workers involved specifically in hazardous operations, but also personnel in the general area who are not involved directly with the process. These control measures include personnel training in how to avoid exposure by using proper operating procedures. The rotation of employees to minimize exposure is another method.

The last mechanism of protecting the worker is the utilization of personal protective equipment such as respirators and protective clothing. PPE should be used when potential exposures exist and engineering controls are not feasible.

Process Design

As in every manufacturing process, the design phase must be complete and accurate before the construction phase begins. With regard to the sound industrial hygiene principles noted above, the manufacturing engineer should design the project or process to ensure that the welfare of the worker is protected and worker exposure is minimized. General ventilation and local exhaust ventilation should be designed to adequately control toxic vapors or particulates. Dangerous processes should be designed to be isolated from nonessential personnel. A less toxic material is the preferred material to be used. Finding an innocuous material which will adequately satisfy the job specification may require technological changes in the process, which could be expensive. However, using the safer product may be prudent. The use of asbestos has cost manufacturers millions of dollars in liability claims and reconstruction fees, in addition to loss of lives.

5.4.3 Implementing an Industrial Hygiene Program

Every manufacturing business with potential hazards, regardless of size, should implement an industrial hygiene program that is proactive. A properly implemented proactive program will prevent exposures and ensure that workers are adequately protected. For everyone, the goal should be to *control the hazard*. This program should, as a minimum, provide the following.

Routine Facility Inspections

Routine inspections, documented by the supervisor, should be performed on a scheduled basis to identify and evaluate hazardous or potentially hazardous operations. Although many work processes change very little, poor habits may develop that potentially spell trouble. Periodic evaluations of the operation from start to finish will uncover undesirable practices. The goal of the inspection should be "How can we do this safer and more efficiently?" not "Who is not working safely?" These inspections should also have full participation of the workers involved in the process, who understand what the real hazards are. Encouraging employees to provide suggestions and solve problems will increase the success of the industrial hygiene program.

Personnel Training

Training sessions should be given to every worker involved in hazardous or potentially hazardous operations. The goal of the training should be to prevent exposures by thoroughly identifying all potential and real hazards involved in the employees' work processes. This training should be performed by an individual who is knowledgeable in operations as well as in industrial hygiene. Sessions can be given in a classroom atmosphere or at the site of the hazardous process. Many companies now are utilizing videotapes to achieve personnel training. An advantage of videotapes is the relatively low cost. Videotapes, however, can be limited due to the one-way communication. Personal interaction between the instructor and the student enhances comprehension. Videotapes can be used in conjunction with interactive training sessions to ensure that the proper level of training is achieved. The employer should never conceal anything regarding occupational hazards from the employee. It is important that the employee has complete trust in the company when it pertains to something that may affect his or her health.

Exposure Monitoring

During hazardous operations, engineering controls and/or personal protective equipment may be utilized. Exposure monitoring can determine what airborne contaminants may be present in the worker's "breathing zone" during hazardous operations. For hazardous operations it is important to document that the proper level of protection is actually being utilized by employees.

Contingency Plans

The best way to handle exposure incidents or accidents is to prevent them from occurring. This, unfortunately, is not a perfect world. Episodes will occur during hazardous processes. A contingency plan should be implemented that precisely establishes the proper steps and procedures to take immediately after mishaps, spills, or accidents. Periodic simulated drills should be run to ensure that things go smoothly during actual emergencies. Formalized contingency plans and associated training will greatly reduce the likelihood of catastrophic results.

How to Implement These Programs

What is the best way to ensure that the proper industrial hygiene programs are in place? Who is going to implement and maintain these programs? One way is to keep an industrial hygienist on staff. Another way is to hire an industrial hygiene consultant. Finally, the company may rely strictly on assistance or information obtained directly from OSHA.

There are advantages to retaining an industrial hygienist (IH) on staff. The IH is readily available to provide technical knowledge required in setting up good process in the first place, and as a point of contact in case of contingencies or problems. The staff IH will be familiar with the company's work processes, and knowledgeable about the associated hazards. The IH on staff can build a working relationship with the employees and is able to reenforce the importance of following sound industrial hygiene principles.

The services of a consultant can be on a one-time basis or periodic, depending on the complexities of the company operations. An advantage of using a consultant over having an industrial hygienist on staff is lower cost. The consultant provides the necessary services only when requested. If the company is fairly small and simple, maintaining an IH on staff may not be necessary.

General industrial hygiene information can be obtained by contacting OSHA regional offices. The larger manufacturing companies typically have a number of potential hazards, and find that it is imperative to have an industrial hygienist on staff.

5.4.4 OSHA Programs

The subject of industrial hygiene would be incomplete without a discussion of OSHA. The purpose of OSHA, as stated in the act, is "to provide for the general welfare, to assure so far as possible every working man and woman in the Nation safe and healthful working conditions and to preserve our human resources." This act, which applies to all private employers, has broad powers of enforcement to reduce occupational hazards.

To carry out the mandates under the Occupational Safety and Health Act, OSHA has (1) established standards for hazards or procedures and (2) prescribed the necessary means to protect employees by referencing such standards. These health standards are called permissible exposure limits (PELs), and are used to identify personal exposures. These PELs are based on research and experimental data from animal studies, human epidemiological studies, and studies of the workplace. If personnel are exposed to airborne concentrations of contaminants exceeding the PEL, a personal exposure is considered to have occurred.

OSHA has established design criteria for industrial operations in addition to health standards. For example, OSHA defines the required ventilation rates for the design of various local exhaust ventilation systems. They have established procedures to enforce these standards to reduce hazards in the workplace. Such procedures include their on-site inspections, and company record-keeping requirements.

Not all aspects of the manufacturing industry are covered by specific OSHA standards. For example, no OSHA requirement exists for indoor air quality in offices. Guidelines and parameters that define indoor air quality have been written by scientific and engineering organizations. These guidelines, although not mandated by law, should be adhered to in order to ensure a healthful and safe environment. Where OSHA may not address some specific hazards, the act includes the General Duty Clause, which states: "Each employer shall furnish to each of his employees employment and a place of employment which are free from recognized hazards that are causing or are likely to cause death or serious physical harm to his employees." The employer has the onus of ensuring worker protection from hazards that OSHA has not specifically addressed. OSHA continues to create new standards and revise existing ones. Discussed below are important standards that have been implemented by OSHA. The manufacturing engineer should be familiar with existing applicable OSHA standards and be aware of newly created or revised standards as they appear.

Hazard Communication Standard (29 CFR 1910.1200)

The purpose of the Hazard Communication Standard is to ensure that all employees, in both the manufacturing and nonmanufacturing sectors of industry, are properly informed of the hazards used at work and how to protect themselves from such hazards. The standard establishes that chemicals produced or brought into this country must be evaluated as to their hazards. Manufacturers must provide the pertinent hazard information on the chemicals to their customers, who in turn must ensure that their employees are knowledgeable about the hazards. This means that the employee has the right to know of the hazards involved with the chemical he or she is using. Failure on the part of the employer to implement a Hazard Communication Program can result in serious penalties as well as avoidable exposure incidents.

The basic parts of a Hazard Communication Program are (1) a written hazard communication program, (2) a list of chemicals used in the work area, (3) a Material

Safety Data Sheet (MSDS) for each chemical used, (4) proper labeling of all chemicals used, and (5) training of all personnel who have a potential exposure to hazardous chemicals.

Respiratory Protection (29 CFR 1910.134)

The Respiratory Protection Standard requires employers to provide respiratory protection to employees when such protection is necessary to protect the employee from chemicals or hazardous atmospheres. The standard mandates the following: (1) written operating procedures on the proper selection and use of respirators based on the hazard involved; (2) training of individuals on the use, limitations, and care of respirators; and (3) routine evaluation of the hazardous operation including processes, workers' duties, and location of the hazardous area.

Noise (29 CFR 1910.95)

The Noise Standard is designed to protect workers from hazardous noise levels. OSHA has established PELs based on a combination of noise levels and duration of exposure. The OSHA PEL decreases as the length of time of exposure increases. For example, an 8-hr exposure of 90 decibels (dBA) would be equivalent to 2 hr of exposure at 100 dBA. The standard requires that engineering and administrative controls be implemented to keep exposure limits below these established PELs. The standard also requires that workers who are exposed to the action level of 85 dBA for 8 hr be enrolled in a hearing conservation program (HCP). The HCP includes exposure monitoring, audiometric testing, hearing protection, employee training, and record keeping.

Confined Space (29 CFR 1910.146)

The Confined Space Standard is designed to protect workers entering confined spaces. A confined space is a tank, vessel, silo, vault, or pit that (1) is large enough for any part of the body to enter, (2) has limited means of egress, and (3) is not designed for continuous occupancy. The standard has identified two types of confined spaces: non-permit and permit-required. Non-permit confined spaces do not contain any real or potential health hazards. Permit-required confined spaces, however, have one or more of the following characteristics:

1. A potential for a hazardous or toxic atmosphere
2. A potential for an explosion or flammable atmosphere
3. A potential to have an abnormal oxygen concentration (<19% or >23.5%)
4. A potential hazard of engulfment
5. Poor ventilation

The standard is lengthy and comprehensive, but the basic program elements include:

1. Identifying any confined spaces
2. Identifying permit-required and non-permit confined spaces
3. Establishing entry control measures
4. Establishing a written confined space program which includes permitting procedures, monitoring procedures to determine if the area is safe to enter, and emergency procedures for rescue
5. Employee training of attendants, entrants, and authorizing persons

Record Keeping

According to OSHA (29 CFR 1910.20), employee medical and exposure records must be kept by the employer. These records include environmental monitoring, Material Safety Data Sheets (MSDS), and biological monitoring. All employers are required to provide each employee access to his or her medical records. Medical records must be kept for the duration of the employment plus 30 years. According to OSHA (29 CFR 1904), employers with more than 10 employees must maintain records (OSHA No. 200 Log) of occupational illnesses and injuries. An incident that occurs on work premises is considered work related. The log, which must be retained for 5 years, tracks the date of the incident, the individual who received the illness or injury and his or her job, the kind of injury or illness, and lost work time. All employers are required to notify OSHA in the event of a fatality or the hospitalization of more than four employees.

OSHA Inspections

OSHA has the authority to inspect companies without advance notice. The employer has the right to refuse the OSHA compliance officer initial entry—but denying access is not recommended. OSHA will take appropriate legal procedures to gain access, and often send a team of inspectors, rather than a single inspector, to proceed with the inspection. It may be best to be cooperative, and have a contented compliance officer inspect the premises, than a team of OSHA inspectors finely examining the premises, thinking you are trying to hide something.

The standards established by OSHA are minimum requirements. Complying with these regulations will satisfy legal obligations. However, in today's competitive manufacturing industry, meeting legal requirements will not be enough. The successful manufacturing engineer will be the one who takes extra steps to stay ahead of the competition. These extra measures include an interactive and comprehensive industrial hygiene program designed to ensure that the employees are working in a safe and healthful environment that enables workers to achieve maximum productivity.

5.4.5 Bibliography

All About OSHA, U.S. Department of Labor, 1985.

ASHRAE Handbook, Fundamentals, I-P ed., American Society of Heating, Refrigerating and Air Conditioning Engineers, 1989.

Fundamentals of Industrial Hygiene, 3rd ed., National Safety Council, 1988.

The Industrial Environment—Its Evaluation and Control, U.S. Department of Health and Human Services, 1973.

Threshold Limit Values for Chemical Substances and Physical Agents, American Conference of Governmental Industrial Hygienists, 1993.

Industrial Ventilation, A Manual of Recommended Practice, 21st ed., American Conference of Governmental Industrial Hygienists, 1992.

6

Computers and Controllers

Allen E. Plogstedt [†] *McDonnell Douglas Corporation, Titusville, Florida*

with

Marc Plogstedt *ITEC Productions, Orlando, Florida*

6.0 INTRODUCTION TO COMPUTERS AND CONTROLLERS

Businesses today are going through a fundamental shift in the way they rely on information systems to improve their competitive advantage. Reengineering of the total business process makes these new processes more dependent on their information systems. The design of the new information systems could mean the difference between being first in class or being a so-so competitor. Information at the right place and the right time will allow decisions to be made at the lowest possible level in the organization, with the best results.

The mainframe computer system was the only choice of system design available in the 1950s and 1960s. Minicomputers became available in the 1970s, and the personal computer (PC) arrived in 1981. The explosion of PCs and network capabilities in the late 1980s has now provided the tool for businesses to expand the use of computers in an economical way.

6.0.1 Information Systems and the Manufacturing Engineer

The reliance of business today on their information systems will change the role of the manufacturing engineer (ME) dramatically. The ME will be asked to become part of this new order. As part of the new team approach to product development, MEs will need to have skills in information flow just as much as skills in managing the work flow in a manufacturing process. In progressive organizations, the ME may be part of a team to restructure the total manufacturing information flow and the integration of data into a computer integrated enterprise (CIE). This could affect how the ME generates, controls, and displays process planning to the workforce. Computer-aided process planning (CAPP) is being used by many organizations to speed the generation and distribution of planning to the floor. This can be a paperless process but is more likely to be a paper-"sparse" system. CAPP has shown dramatic process improvements (20 to 50%) in productivity. MEs must be knowledgeable about the interrelationships between ease of training and use

---

[†]Deceased.

of CAPP systems, and the requirements that CAPP puts on the CIE. The technology exists for presenting multimedia CAPP work instructions to a workforce. However, there is a cost for each of the features of that presentation.

Rework and scrap are non-value-added results of poor process control and misinformation. Correct, current information will eliminate most, if not all, of the errors that can creep into a process, either in business or manufacturing areas. To be first in class, a CIE must provide data to and from customers, vendors, and across the total factory. To accomplish your goals of meeting and beating the competition, the ME must look at all the factors that will affect price (and profit). Inventory and work in process must be reduced to an absolute minimum. This can be accomplished by the judicious design of computer interfaces (electronic data interchange, EDI) with vendors. With an MRP system, a just-in-time delivery system can and should be used to reduce inventory and work in process (see also Chapter 10).

6.0.2 Definitions of Terms for Information Systems

Application: A software program(s) that carries out useful tasks.

Architecture: The manner in which applications, software, and hardware are structured.

Batch: Sequential off-line processes.

Computer integrated enterprise: The complete assemblage of computers, network, and applications that support the entire manufacturing enterprise.

Central processing unit (CPU): The part of the computer in which the fundamental operations of the system are performed.

Client: The networked machine requesting services.

Data: Recorded information, regardless of form or method of recording.

Data dictionary: A repository of information describing the characteristics of the data elements.

Document image file: A digital file representation of a human interpretable document. Examples are raster image files and page description language files.

Document image standard: A technical standard describing the digital exchange format of a print/display file.

Electronic data interchange (EDI): A series of standards (X12) for the format and transmission of business forms data, such as invoices, purchase orders, or electronic funds transfer. EDI is intended for computer to computer interaction.

Government Open System Interconnection Protocol (GOSIP): A government standard that explicitly identifies the standards which apply to each of the 7 layer OSI model telecommunications protocol stacks.

Graphical file: Files containing illustrations, design data, schematics, in processable vector graphics or non-processable raster graphics.

Integrated data file: A digital data file which integrates text, graphics, alphanumeric, and other types of data in a single (compound) file.

Local area network (LAN): The LAN is geographically the smallest of the network topologies. LANs have three topologies in general use: the STAR, the BUS, and the RING.

Mainframe computer: The processor can be configured to receive inputs from all the various devices and networks in your computer integrated enterprise. All of the application manipulations are accomplished by the central processing unit. The major features of the mainframe is the very large I/O capability built into the processor. The random access memory in the CPU can be partitioned into various configurations to best operate with your applications. In most cases the computer installation requires special facilities, a controlled environment and an uninterrupted power supply.

Mini computer: Smaller, cheaper (than the mainframe) standalone computers that do not need all of the support systems to operate. In almost all cases these machines are used for a specific standalone application or process, with little or no networking.

MOTIF: A de facto standard graphical user interface that defines the look and feel for systems that utilize it.

Networks: Networks consist of the computers to be connected, the communication media for the physical connection between the computers, and the software for the network communications and management.

Personal computer: The personal computers evolved along two lines. Apple introduced the first practical personal computer in the mid 1970s. The IBM personal computer was introduced in 1981.

Protocol: Provides the logical "envelope" into which data is placed for transmission across a network.

Reduced instruction set computing (RISC): A computer chip architecture that requires all instruction be executable in one clock cycle.

Wide area network (WAN): The wide area network (WAN) has the capability to allow you to connect your LAN to other LANSs within your company or with LANs of vendors and customers. The speed of the WAN will be determined by the communication media used for the long distance transmission.

X-Windows: A public domain, standard software tool kit, developed and distributed by MIT, on which windowing applications can be built.

X.400: A standard electronic mail addressing scheme that is also the e-mail standard identified in GOSIP.

6.1 MAINFRAME SYSTEMS

The first electronic (electromechanical) computer was designed and built during World War II. In 1945, the Electronic Numerical Integrator, Analyzer, and Computer (ENIAC) went into use, and was used until 1955. Most of the analysis time for the ENIAC was spent on calculation of ballistics tables for artillery. Because of the extensive labor needed to operate and maintain the system, it had limited use in the business world.

The first commercial, programmable, electronic computer, the IBM 701, was introduced in 1953. Most large corporations began to rely on the mainframe computers to process their mission-critical business applications. As the need for more and better business data expanded during the last three decades, the capabilities, and the dependence of business on mainframe computers expanded at the same or a higher rate. These business applications are sometimes referred to as "legacy software systems." As the needs for other functions of the organization to use more and more of the capabilities of computing increased, ways and means to integrate these new systems into the legacy systems had to be developed. The concept of this new architecture evolved into the computer integrated enterprise (CIE), as discussed in Subchapter 6.5.

6.1.1   System Layout

The general layout of a typical mainframe system is shown in Figure 6.1. This is one of the simplest architectures to implement. The system consists of various modules connected together to form the complete mainframe computer system. The connection to local area networks, wide area networks, and any remote sites is made through the front end communications processor. This processor can be configured with various cards to receive inputs from all the various devices and networks in the CIE. All of the application manipulations are accomplished by the central processing unit (CPU). The major feature of the mainframe is the very large input/output (I/O) capability built into the processor. A very high-speed bus connects the 64-bit processor, memory, and the I/O bus together for the most efficient operation. The random-access memory in the CPU can be partitioned into various configurations to best operate with your applications. All of the peripherals are interfaced through controllers. Some of the controllers also contain memory for faster operation. An example of optimizing techniques which will improve the overall performance of the mainframe computer is setting up disk storage units for specific elements of data. The index is be stored on one disk, the library of subroutines, and the data is stored on individual disk units. This reduces the retrieval time for the data and speeds up the whole process.

The mainframe is connected to each user terminal via coaxial cable. In most cases the

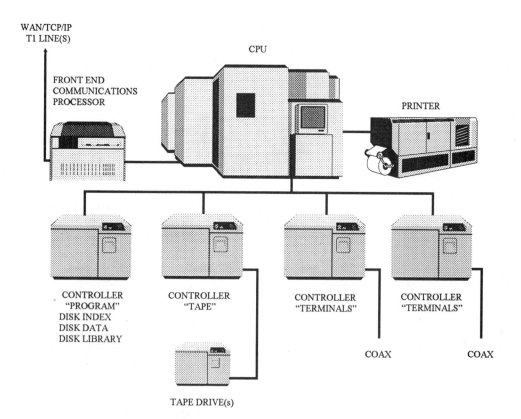

Figure 6.1 Mainframe system layout.

computer installation requires special facilities, a controlled environment, and an uninterrupted power source. As the need for additional terminals arises, more coax cables, and
controller boxes at the computer, must be added. The information is transmitted from the
mainframe to the terminals under IBM's System Network Architecture (SNA), and since
most installed mainframes are IBM, this has become the de facto standard. All processing
is accomplished at the CPU. The terminals are considered "dumb" terminals. All of the
screens are generated by the CPU, and little if any capability exists at the terminal to do
any processing. In some advanced terminals for this type of architecture, you can get some
graphical capabilities (the regular terminals are character only).

6.2 MINICOMPUTERS

During the first decade of the computer revolution, the mainframe had the computing field
to itself. In most cases the programs were built for the business applications of the
corporation. The need for productivity improvement on the factory floor caused the
manufacturing side of the business to keep the pressure on, and productivity improved at
a steady rate (2 to 4% per year). The productivity in the remaining portions of the business
(planning, support, and engineering) did not show steady improvement in productivity. The
pressure on the other functions of the organization set the stage for the second phase of
the computer revolution, the minicomputer. This arrived in the 1960s when DEC introduced computers such as the DEC PDP-8 and later the VAX series. The cost of these
computers was so much less than the mainframe that decisions on what and when to buy
minicomputers could be handled at a group or department level, not at the corporate level.
Minicomputers began to move into many areas of the organization; manufacturing,
engineering, and planning represent only a few. In almost all cases these machines were
used for a specific, standalone application or process, with little or no networking. The
productivity of the support functions, even after the introduction of the minicomputer, still
did not show the steady improvement that was accomplished in manufacturing. In
retrospect, one of the main reasons that the improvements did not materialize was "islands
of automation." Each of the organization entities did their own computing, but they could
not transfer data to other groups. This failing could not be remedied in a cost-effective
way until the introduction of two new technologies, the personal computer (PC) and
the network.

6.3 PERSONAL COMPUTERS

The first personal computer to have any real impact on computing was the ALTAIR,
available in kit form from MITS in 1974. The system used an Intel 8080 processor and
256 bytes of memory. The true beginning of the personal computer was the marriage of
Microsoft and IBM to develop the first PC machine in 1981. The Intel 8088 processor was
the heart of the machine. The IBM XT was a 16-bit machine with a minimum of 64K of
RAM and one disk drive. The IBM AT was introduced in 1982 using the 80286 processor.
The PC industry practically exploded due to the fact that IBM set up a series of PC
standards and allowed other manufactures to build PCs to the standards. Standards were
so important because they allowed software manufactures to produce software for IBM-
type machines that would run on any "clone." Over 100 million PCs have been built since
1981, and the production rate is increasing.

The opportunity of organizations to increase personal productivity was such a powerful lure that workers would do almost anything to get a PC for their own use. This allowed all departments to improve their capability to look at their data and manage their part of the business. The problem was that what was good for the financial department might be bad for the manufacturing department. They had no convenient way of transferring and using each others' data. Thus the original PC revolution did not meet the expectations of management for the total productivity improvements. In some cases the isolated use of the PC made communications between groups even worse than it had been before. At the time of the development of the first IBM PC, the cost of quality peripherals was exorbitant. No networking methods were built into the MS-DOS operating system available at the time. To help in this situation, "sneakernet" was used because nothing else existed. In this case data was copied from one PC onto a disk and transported by hand to another machine. A major problem was ensuring that documents were using up-to-date data. Just as important, was how documents could be kept from being stolen. Printers could also be shared by more than one PC by the use of a data switch. This was not very fast however, and did take a large amount of hardwires for the few PCs that could be connected to one printer. The real move forward was the development of MS-DOS 3.1 and the development of other networking systems (see Section 6.4.)

6.3.1 Systems and Processors

The personal computers evolved along two lines. Apple introduced the first practical personal computer in the mid-1970s. The IBM personal computer was introduced in 1981. The general layout of the PC is shown in Figure 6.2. The performance of the PC depends on the processor type, processor speed, bus type, cache (how much), and graphical interface. Over the years, many benchmark programs have been developed to test the performance of the various subsystems of the PC.

Apple

The Apple personal computer was designed from the start as a graphical interface. The machine was based on the 6502 processor. The "drag and drop" characteristics of the interface were patterned after the research work by Xerox. The users of this interface did not need much training to become proficient in the use of applications. Users did not have to use any command-line instructions to operate the applications.

Intel

The first Intel microprocessor was designed in the early 1970s. The first unit (4004) had about 1500 transistors on the chip. The first practical microprocessor from Intel was the 8080 in 1974. This chip contained about 8000 transistors. The Pentium processor released in 1993, contains over 3 million transistors on the chip. The evolution of the Intel microprocessor has for many years doubled the processing power every 18 months. A curve showing the growth in millions of instructions per second (MIPS) is shown in Figure 6.3. The 8080 could process only about 1 MIPS. The Pentium can process over 100 MIPS.

When IBM made the decision to release the design specifications of the PC, a myriad of other vendors began building IBM-compatible PC's. This drove the expansion of the IBM PC into high gear. In 1993, about 85% of the millions of PCs used Intel processors. Other microprocessors are appearing in the marketplace and will become a significant source in the late 1990s. In most cases these processors use a reduced instruction set

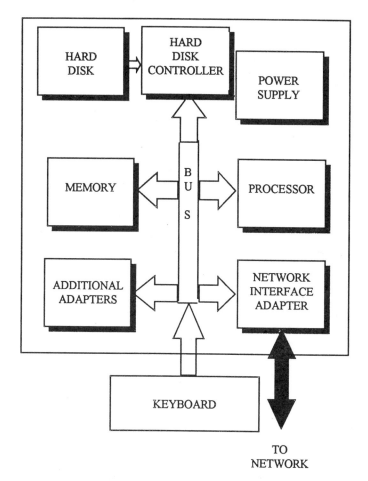

Figure 6.2 Personal computer layout.

computers (RISC) structure. The advantage of this type of machine is that it can operate on more that one set of instructions during the processor clock cycle.

6.3.2 Operating Systems

Computer software can be divided into two main categories: system programs, which manage the operation of the computer; and application programs, which solve problems for the user. The operating system is one of the fundamental system programs. The operating system has two main functions: providing the programmer with a "virtual machine" for ease of programming; and acting as a "resource manager" for the efficient operation of the total system. Some of the packaged operating systems also include a graphical user interface (GUI). The GUI provides the user with a simple-to-learn screen system which allows the program to be used with little or no instruction or programming knowledge. Examples are Windows 3.X from Microsoft for DOS, Presentation Manager for OS/2, and X-Windows or MOTIF for UNIX. Figure 6.4 shows how the various layers of the computer system are interconnected with each other. Without the operating system

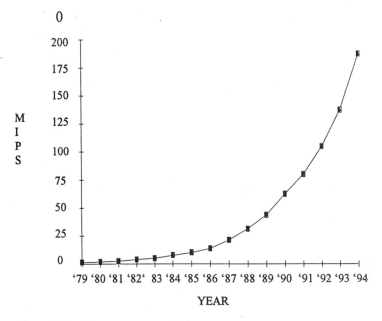

Figure 6.3 Microprocessor evolution.

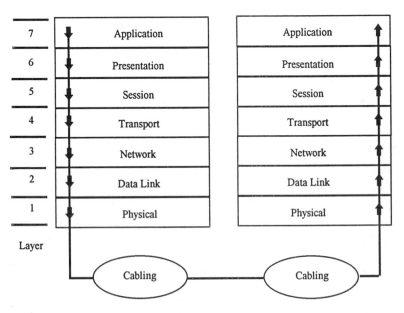

Figure 6.4 International Standards Organization (ISO) Open System Interconnect reference model.

shielding the programmers from the mundane tasks accomplished by the operating system, the development of new applications would be very difficult and time consuming.

The development of operating systems has progressed through four phases over the last 50 years. The first phase was from 1945 to 1955. As discussed in Subchapter 6.1, the first computer was developed during World War II and came into use in 1945. The operating system (it did not have a name at the time) for this massive computer was accomplished by plug boards. Each user had a plug board with which the operation of the very primitive computer could be controlled. If one of the vacuum tubes failed (which they did regularly), the program would not run correctly, and the user would have to start over after the machine was repaired. Most programs at that time were straightforward numerical calculations. Punched-card input devices were added in the early 1950s. Both programs and data were inputted via punched cards. Imagine what would happen if one of the program cards got out of order.

The invention of the transistor in 1947 and its introduction into computer systems in the mid-1950s changed the picture radically. With the size reduction and the reliability improvements of the transistor, the computer became a viable commercial product. The operating systems of the second phase (1955 to 1965) supported large mainframe university and corporate computer complexes. In this era, programs would be sent to the computer complex for processing. Since most of the time involved in using the computer was not in the processor, but in the input/output portion of the program, the use of batch processing came into vogue. The programs would be placed into a queue, and the processor would work on them in sequence. The input (to tape) was handled by a separate computer designed for that process, not the main processing computer. When the main processor had some free time, the next program would be read off the tape and the processor would carry out the program. After the program was complete, the output would be transferred to an output processor to print or store the output. In this way the main processor would be utilized to it fullest extent and its high cost justified.

The next phase in the development of operating systems occurred from 1965 to 1980. This phase was also driven by the advent of new hardware—the integrated circuit (IC). The IC allowed the mainframe computer to be enhanced to the point where it could handle multiple jobs in the processor at the same time via multiprogramming. When the processor paused for the input or output device to respond, the CPU would run part of another job for greater utilization. The other use of the IC was in the development of the minicomputer, such as the DEC PDP series of computers. These relatively low-cost machines found their way into engineering departments for calculations so the engineers no longer had to wait for the mainframe to be available. This is the time frame in which the UNIX operating system had it beginnings.

The introduction of the personal computer by IBM in 1981 started a revolution in the use of computers and the beginnings of distributed processing. Apple computer, on the market from 1977, had made inroads into the office environment. Wang had also developed a very good line of machines for the office environment. The problem with both of these computer systems in the long run was the proprietary operating systems and application software. The decision by IBM to make their computer design available to anyone was the key to the incredible expansion in the computer industry. Three major operating systems have been developed for IBM-compatible PCs: MS-DOS, OS/2, and UNIX.

The interface between the operating system and the user application is defined by a set of "extended instructions." The application invokes these extended instructions by the use of system calls. The system calls create, delete, and use various software elements managed by the operating system. In this way the application software, through these system calls, can provide the various functionalities of the computer to the application. While not a integral part of the operating system, important functionality is provided to the system by the editors, compilers, assemblers, and command interpreters.

MS-DOS

When IBM set out to develop the first true PC, the operating system in vogue at the time was CP/M from Digital Research. A new version of CP/M was in the works at the time IBM was developing the PC, but development was behind schedule. IBM had set the date for introduction of the PC, so another operating system had to be found. Tim Paterson of Seattle Computer Products had a CP/M-like operating system that was used for testing memory boards. The product 86-DOS was modified by Microsoft and Paterson and became MS-DOS (Microsoft Disk Operating System). The first version of MS-DOS occupied 12K of the IBM's 64K memory. The design of the IBM PC set up a separate use of hardware ROM for the BIOS (Basic Input/Output System). The BIOS contained the standard device drivers, thereby allowing MS-DOS to call them to do I/O. MS-DOS 6.0 now requires 2.4 Mbytes of hard disk memory instead of the 12 Kbytes in Version 1.

To operate MS-DOS, the user must envoke commands to the system via the command line. Through the command line one accesses the file system, system calls, utility programs, and other features. These procedures are similar to those in UNIX but more primitive. MS-DOS is not font case sensitive, as is UNIX. The user of MS-DOS must be familiar with the various command line calls and the format required for the operation of the system.

MS-DOS is not a multitasking system like OS/2 or UNIX. Normally the operating system will support one application at a time. After MS-DOS was in the hands of millions of users in the 1980s, it became obvious that if IBM and Microsoft were going to expand the PC into other uses, it needed an easier, more user-friendly interface. In the late 1980s, Microsoft and IBM were jointly developing the next operating system for the IBM PC's, OS/2. The first version of OS/2 was used on critical process computers, but did not enter the nonindustrial market. In parallel, Microsoft developed Windows, which is a graphic user interface (GUI) running on MS-DOS. This window environment gave the user some of the functionality of the Apple interface. When Windows 3.1 was introduced in 1991, it became an instant success. Windows supports cooperative multitasking and isolates the user from most of the command-line interface of MS-DOS.

OS/2 2.x

OS/2 2.x is a second generation replacement of IBM's PC DOS and Microsoft's MS-DOS operating systems. One of the main advantages of OS/2 2.x is that it frees the user/developer from the constraints of the DOS environment. The amount of RAM available to run applications is one of the greatest improvements. OS/2 2.x is not an extension of DOS; rather it is a completely new operating system which was designed to operate "mission-critical" business applications. The major advantages of OS/2 2.x are:

System integrity for applications
Preemptive multitasking and task scheduling
More available memory to run DOS programs (~620K)
Multiple concurrent DOS sessions

Ability to run concurrently OS/2, DOS, and Windows
Virtual memory
Fast, 32-bit architecture
Fast disk file access
High-performance file system (HPFS)
Presentation manager
National language support Object-oriented workplace shell (WPS)
Ability to run on IBM and IBM-compatible hardware

Presentation Manager is included in the OS/2 operating system. This GUI provides the user with icons and with drag-and-drop capability.

UNIX

UNIX is an interactive, time-sharing operating system. It was designed by programmers for programmers. It is not an operating system for the unsophisticated user. When the application shields the user from the operating system via a windowing environment, the system can be used by anyone. The UNIX operating system is generally used in the software development community, and in file servers. The major advantage of UNIX over MS-DOS and OS/2 is its capability for multiple concurrent users.

Another feature of UNIX, not present in MS-DOS, is its security system. Logging on to the system requires both your name and a password. This allows your files to be accessed only by you. As in MS-DOS, the interface with UNIX (without a GUI) is through the command line.

The UNIX operating system has been in use since 1974. Thompson and Ritchie of Bell Labs were the originators of UNIX. Thompson also wrote the C programming language for the development of UNIX. The development of UNIX over the years has followed a very crooked road. The original UNIX, licensed by AT&T, has been the basis for all the developments in UNIX. One of the major problems with UNIX is the fact that the machine-dependent kernel must be rewritten for each new machine. The machine-independent kernel is the same for all machines. Many universities used and modified the UNIX operating system. One of the major developers was the University of California at Berkeley, which caused problems for UNIX users and applications developers because their applications would not run on all versions of UNIX. This is a continuing problem. The first real effort to standardize UNIX was by the IEEE Standards Board. The name of this project was The Portable Operating System for UNIX (POSIX). This helped, but it did not solve the problem. Two groups of developers have been set up and are still competing: the open system and the basis protocol for the network that became TCP/IP (see Subchapter 6.9).

6.4 NETWORKS

6.4.0 Introduction to Networks

Networks consist of the computers to be connected, the communication media for the physical connection between the computers, and the software for the network communications and management. Until the development of MS-DOS 3.1, personal computers did not have the capability to utilize the full capability of a network. The capabilities added to MS-DOS 3.1 which allowed full use of a network were as follows.

File handles: An identifying number (handle) assigned by the PC operating system so that files can be located anywhere in the network. This leads to a new logical access for the files when they are stored on a file server.

File sharing: Allows programs to specify what kind of access mode they want to a file. This control allows the file to be set up as read only and or read/write.

Byte-level locking: Program can declare exclusive access to a range of bytes in a network file server and the network operating system to enforce that exclusive access.

Unique filenames: The network operating system can be requested to set up unique filenames so that the users do not ever use the same name.

Redirector support: Acts as the destination for all requests for remote logical drives. The program does not have to know if it is accessinga file on the local drive or on the network.

With these new features, MS-DOS 3.1 gave programmers the capability to write a standard set of programs for network services (interface). This has become the standard interface to the network industry.

6.4.1 Local Area Networks

As the name implies, a local area network (LAN) is geographically the smallest of the network topologies. In general, a LAN is limited to 1 to 2 miles in length and typically 10 to 16 Mbps. On some of the more advanced LANs, 100 Mbps is beginning to be used. Even with the advent of MS-DOS 3.1 and the network interface, personal computers were designed to be computers, not communication devices, which presents some limitations in response time and difficulties in expansion. Three LAN topologies are in general use: the STAR, the BUS, and the RING.

LAN STAR Topology

The layout for the STAR topology is shown in Figure 6.5. The terminals are connected by lines branching from a central node. The central node is a single point of failure for the entire network. Expansion is relatively easy, but it may require a large amount of cable to service even a small number of terminals.

LAN BUS Topology

The layout for the BUS topology is shown in Figure 6.6. Each of the terminals, or other devices, is connected to the bus (coaxial cable) via connectors. Expansion requires no rewiring, and no single node can bring down the network.

LAN RING Topology

The RING topology is a point-to-point connection topology that forms a circle; see Figure 6.7. In RING topology, each of the nodes retransmits the signal. Control is distributed, but no single failure in the ring will bring the entire RING down. The RING must be broken to install new nodes.

6.4.2 Wide Area Networks

The wide area network (WAN) has the capability to allow you to connect your LAN to other LANSs within your company or with LANs of vendors and customers. The speed of

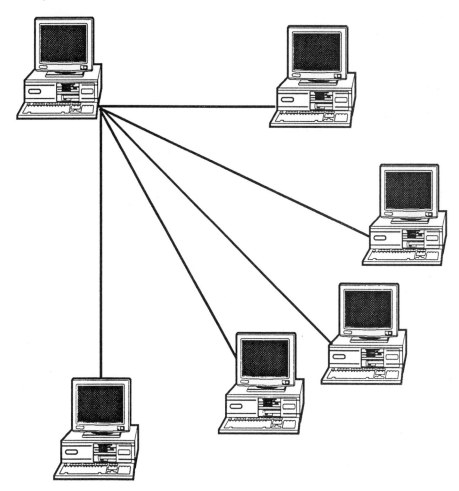

Figure 6.5 The LAN STAR topology.

the WAN is determined by the communication media used for the long-distance transmission. At the low end of the speed regime is the modem using regular phone lines. For some applications this analog signal method will suffice. In most cases, however, if you have more than one terminal and wish to transmit large amounts of data in a reasonable time, you need more than just a modem. Most of the data transmitted over long distances is sent over digital carrier signal lines, such as the T1 system, originated by AT&T. There are many new digital data transmission systems available today, and the selection of the proper one can have a great influence on the performance and cost of your commuinication networks.

6.4.3 Network Throughput

One of the most important factors to consider in the design of manufacturing support systems is the amount of data to be transmitted over the network and the response time of the data transfer. The speed of the network will depend on the type of

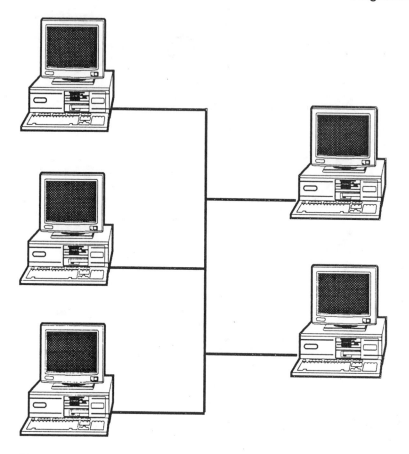

Figure 6.6 The LAN BUS topology.

connection medium (copper wire, coax, or fiber optic), the distance to be covered, and the network software [Ethernet, Token Ring, Fiber Distributed Data Interface (FDDI), and Apple Talk]. Text data does not take as much space as visual aids or photographs. For example, a book of a few hundred pages would be about 300,000 characters (bytes), while a full-color photograph (1024 × 768 pixels) would have 2,359,296 bytes. The transmission time of these two over a network would depend on the raw transmission data rate and the data throughput. The throughput is the limiting factor and is determined by the transport technology and the amount of overhead that must be added to transmit over the network (other data that describes and verifies the data that are sent. Using the Ethernet with a 10-bit data rate, the data rate is 1.25 Mbytes and with the overhead the effective throughput is 375,000 bytes/sec. Looking at the examples above, the 300,000-byte text file would be transmitted in less than 1 sec, and the 2.4-Mbyte photograph would require about 6 sec to transmit. Network loading and the advantages of where in the network various data should be stored are discussed in Subchapter 6.7. The following chart shows the data transfer capability of the various network transport products.

Type	In use	Proposed
Ethernet	10 Mbits/sec	100 Mbits/sec
Token Ring	4 and 16 Mbits/sec	
FDDI	100 Mbits/sec	
Apple Talk	115 Kbits/sec	

The actual cabling that connects the various components on the network come in many varieties. The three major types are:

Coaxial cable, where one or more concentric conductors are used
Twin twisted-pair cable, with two pairs of wires, each pair twisted together
Fiber optic cable, used for FDDI and can be used with Ethernet

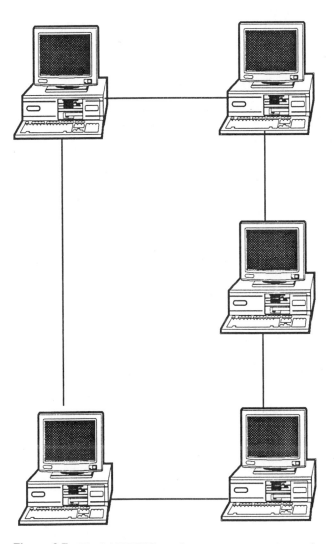

Figure 6.7 The LAN RING topology.

Coaxial and twisted-pair cables have been used for some time. The fiber optic cable is fairly new, but this technology is expanding at a very high rate as all the phone lines are being changed over, and the price of this transmission medium is dropping to the point where the material is very competitive with wire. The cost of installation of fiber optic is still higher than for the other types. The expanded capabilities and the immunity of the fiber optic to electrical interference make it a good choice for many applications.

6.4.4 Network Operating Systems

A network requires an operating system to control and provide the various services to the network users. The operating system should also provide management and maintain functionality for the network operations group. The capability of the network operating systems is expanding at a rate where a new version of the software, with expanded capabilities, is released about once a year. The descriptions provided here will indicate some of the important characteristics of each of the software packages at the time of publication. As each vendor sees the need to incorporate new features, the software packages will become more alike than different.

Banyan VINES

VINES is a UNIX-based operating system from Banyan Systems designed to operate on Intel processors 386 or later. The major feature of this network operating system is the capability to operate multiservers. VINES characteristics are as follows.

Banyan's global naming scheme (GNS), Street Talk, is the premier GNS.
Management of multiple file servers is easy.
It has a built-in mail system.
Large numbers of users (over 100) can be handled on one server.
An unlimited number of files may be open at one time.
Security is provided.
Up to 10 printers are supported per file server.
It is compatible with many multiuser database management software products.
Auto alerts for the administrator on system performance are provided.
It supports Apple Filing Protocol.

Microsoft's LAN Manager

LAN Manager is an OS/2-based application. Basic features of LAN Manager are that it is a mouse-driven point-and-click interface and it is relatively inexpensive. LAN Managers characteristics include:

Domain server style of global naming scheme
Direct queries to mainframe
OS/2 client-peer service
Autoreconnect when file server comes back on line
Administration features
System security

6.4.4.3 Novell's NetWare

NetWare was designed to be a network operating system from the ground up. It is not designed to run under UNIX or OS/2. This approach gives Novell the advantage of control of the total software package and therefore it can design in compatibility with a large array of third-party software packages and make the system compatible with all terminal operating systems. Features include:

Hardware independence
Extensive fault tolerance features
Supports OS/2, Named Pipes, and SQL products
Security features
Supports duplicate concurrent file servers for on-line operational backup
Applied programming interfaces (APIs)
TCP/IP implementation

6.4.5 Network Applications

The network for your computer system will provide various services and functions which were previously not available to your organization. The capability of moving data around in the organization is becoming indispensable. The use of network services for both software and hardware can be a large area for cost reduction. A network version of a software package such as Lotus 1-2-3 for 10 concurrent users is more cost-effective than buying 10 individual copies of the software. As many users as you have on the network can access the software, as long as no more than 10 are using it at the same time. As your usage increases, the probability that a user will have to wait to use the software increases. The network administration can, at relatively small expense, add more users to the network pack when waiting reduces productivity. General office and manufacturing areas may also be served by a network printer. The network operating system will provide the service so that your printing will be stored, directed, queued, and printed on the printer for your area. This is also a good cost-reduction measure. Most services such as printers are not in use 100% of the time for a single user, so group usage makes a lot of sense.

Sharing Files

Through the network you have the capability to share files with anyone on the network who has access to the file under the security system. Access to various files can be controlled by the user profile set up on the network operating system. Once this capability, and the users' profiles, are set up, you can share work across the network. This is especially valuable when various users across the enterprise need to input or review the same document. This capability can also be used to merge files or documents. If only one organization has access to write to a file, and keeps it up to date, the rest of the users can access to read that data so that all of the organization is using the same data. One of the features the network operating system must have is a file naming module, so that no two files ever have the same name. If the file is a database, then all of the users will be inputting and using the most up-to-date, on-line data in "real time." This is required for a good material requirements planning (MRP) system to operate effectively.

Sharing Data

The sharing of data requires that the applications you are using network compatible and control locking of data records in a file. Data is any information stored by an application (i.e., everything but the executable code). Sharing data gives your organization the opportunity to save various types of information after it has been generated, and then the whole organization can use it. This saves the regeneration of data and prevents the distribution and use of inconsistent data through the organization. In Subchapter 6.5 we discuss the advantages of inputting data into the system only once, with the system providing that data to all the elements of the system that require it.

6.5 COMPUTER INTEGRATED ENTERPRISE ARCHITECTURE

Once all the elements (i.e., mainframe, mini, PC) and all the necessary hardware and software parts for the network were in place, the computer integrated enterprise (CIE) became possible. In the mid-1980s, portions of the total system were available and some manufactures tried to build computer integrated manufacturing (CIM) systems. They were not very successful due to the fact that a tie between the planning system and the execution system was not available, especially on a real-time basis. Another problem, which has been the downfall of many good software systems is the fact that users were not involved in the design, and when the time for implementation came they fought it so the implementation of the system failed. Management of the enterprise must also embrace new management and organizational concepts to aid in removing walls between functional groups within the factory.

The CIE consists of a combination of various functional units to perform the desired tasks to operate the business. These include finance, purchasing, material control, manufacturing process control, and reporting. To perform these tasks the CIE must have certain capabilities. Processing (computer processing unit—CPU), input and output terminals, printers, and an interconnecting network among the units make up the minimum capabilities. As technology progressed, (see Subchapters 6.2 to 6.4), and the general use of computers became common in the 1950s, options for different architectures appeared. A typical architecture for a CIE system is shown in Figure 6.8. This example assumes that your organization has used computer systems for some time, and you have a mixture of types of hardware and legacy software programs in a multisite system. Medium to small, or new start-up, manufacturers have the luxury of designing a system from the ground up. See Figure 6.9 for an example of a new start-up system. With respect to multisite systems, Figure 6.10, the mainframe is at one location while other elements are at remote locations. The data is transmitted from location to location via hard-line (T1 lines, or other digital system) or in some cases by satellite. For the system to be a true CIE, and to be as productive as possible, the user should input data into the system only once. It is then up to the program logic to route and install the data into the correct tables and databases for processing and record keeping. Some of the processing will be handled at the local terminals and servers, while other data will be transmitted to the mainframe for processing. Examples of local processing might be CAPP for manufacturing, while the data from the manufacturing process (i.e., order status, labor charges, time keeping) would be collected locally and transmitted to the mainframe. One thing to remember is that, in a well-designed system, all applications should be available to the users at their local terminals. Timely information at the lowest level of the organization will allow your workers to make decisions and improve productivity. Correct, timely information is one of the most useful productivity improvements you can make.

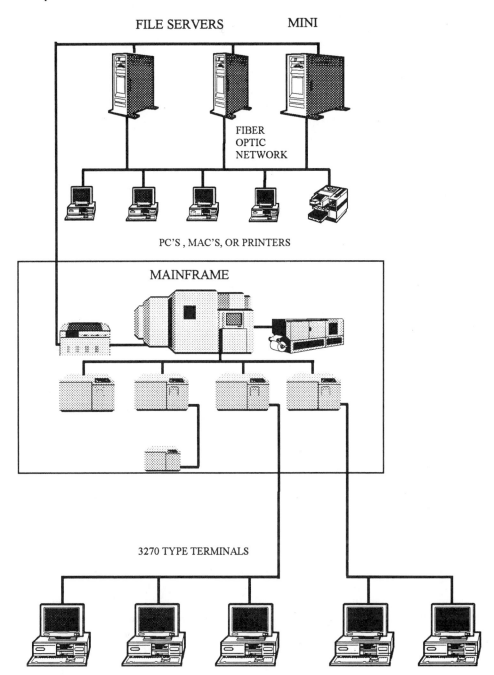

FILE SERVERS   MINI

FIBER
OPTIC
NETWORK

PC'S , MAC'S, OR PRINTERS

MAINFRAME

3270 TYPE TERMINALS

Figure 6.8   Computer integrated enterprise (CIE) system architecture.

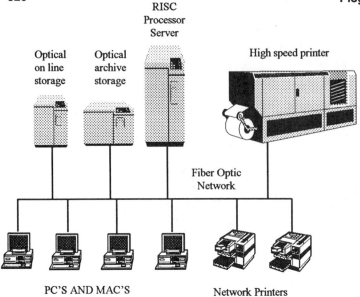

Figure 6.9   Start-up CIE system layout.

6.6 PROCESS CONTROLLERS

Marc Plogstedt

6.6.0 Introduction to Process Controllers

Electromechanical and electronic process controls have been used since the 1940s. Early systems used relays for their logical elements. The installation and maintenance of the systems was done by electricians who were not computer experts. The program logic was known as "relay logic." Process controllers today still use "relay logic" programming. The process controllers became electronic devices in the mid-1970s. Many companies provide a complete line of equipment to meet almost any need.

Controllers have evolved over the years from the simple devices of the early 1970s to the complex multifunctional devices of the 1990s. The first devices were operator controlled; the operator would activate a simple on-off switch, and the signal from the switch would activate a power relay to operate a motor or other controlled device. The power relay was used to provide the high power required by the controlled device, while the relay power supplied by the on-off switch could be low power and in most cases low voltage for simplicity in wiring. As more and more of the operator's functions in the process were taken over by the process controller, the power controller (relay) was replaced by solid-state devices. This provided the system with devices which were more reliable and required less power from the process controller.

With the advent of process controllers that could control many devices anywhere on the factory floor, the positioning of the power controllers becomes more of an issue. If it is a long distance from the process controller to the controlled device, is it better to control

MAINFRAME SITE

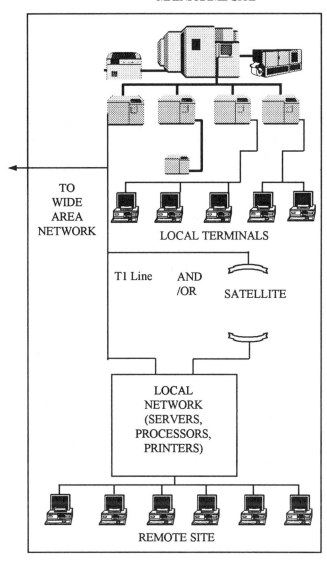

TO
WIDE
AREA
NETWORK

LOCAL TERMINALS

T1 Line   AND
             /OR   SATELLITE

LOCAL
NETWORK
(SERVERS,
PROCESSORS,
PRINTERS)

REMOTE SITE

Figure 6.10   Multisite CIE system.

all of the devices directly from the process controller, or should the power unit be controlled by a power module at the device—and the signals from the process controller be all that is sent over the long distance from the process controller to the power module? This configuration could simplify and lower the cost of the factory wiring required for the total system.

As more and more of the operator's functions are replaced by the process controller system, a need for the system to be set up with a master process controller processor and then feeding information to slave or satellite processors positioned at various points on the factory floor could occur. This configuration could be required if the response of the controlled device is fast enough that the master processor controller cannot service the device at the desired rate.

Once the process to be controlled has been defined, selection of hardware and the system design layout must be made to fit your organization and requirements. If you already have process controllers, your company will probably want to use the same brand for ease of inventory, maintenence, and training. If you do not have any installed base of equipment, then the decision on brand will depend on the availability of your required control elements and the evaluation of cost and performance of the various systems.

In selecting a system, one of the major criteria is not necessarily the installed cost, but rather the reliability of the system and ease of maintenance to keep the process operating. A one-day shutdown of the factory due to a problem with the control system could be more expensive than the total initial installation cost.

Safety is also very high priority. When a system is designed for a manufacturing process, the safety of the workforce is paramount. When a worker becomes involved with the controlled system, the system must sense any problem and shut down the process so that the worker will not be injured. An unusual example of the use of controllers is found in theme parks. Entertainment parks utilize process controllers to operate many of the rides. Since many people are involved in the controlled process, their safety is of paramount importance. Control of this kind of a process demands accuracy of many control functions at a very high rate. Since these types of systems usually are spread over a large area, multiple controllers with a connecting network are required for ride performance.

Another area for consideration is cost performance. Does your situation allow the setup of the control system in zones (possibly cheaper), or must you distribute the processors over the factory floor (higher performance)?

The following list includes some pertinent functions of process controllers:

An advanced instruction set including relay-type, timer and counter, math, data conversion, diagnostic, shift register, comparison, data transfer, sequence, immediate I/O, program control, and PID control

Use of sequential function chart to implement structured programming

Storage of memory files on a disk drive so that you can archive them or download them to other processors

EEPROM backup to protect critical control programs

Execution of a program file at repeated intervals with a selectable timed interrupt

Program fault routines to respond to system faults

Accessibility of date and time with the real-time clock/calendar

Storage of ladder program files on disk for printing on a serial or parallel printing device

A display of a data file starting with the requested address

Insert, append, change, or delete instructions or rungs without stopping the machine or process with on-line programming

Use of search functions to locate items within your program when the processor is in any mode of operation, or when the programming terminal is in off-line mode

Automatic switching from the primary processor to a backup processor when a fault is detected in an acceptable time frame (typically less than 50 msec) with backup communications

In addition, selection of the type and performance of the specific controller application is dependent on how many I/O elements you need to scan and the maximum time between scans. Table 6.1 shows the specifications of one of the Allen-Bradley controllers series.

6.6.1 System Design and Implementation

Implementation and maintenance of a process control system includes the following phases:

Definition of the system requirements
Development of the total system concept
Selection of the controller hardware
Design of the controller system
Simulation of the controller system
Implementation of the controller system
Debugging and tuning of the controller system
Final acceptance and stress testing of the controller system
Technical and maintenance manuals

For the controller system to meet all of the above objectives, the ME must be the leader of the team that follows the process from beginning to final product. One of the major problems encountered in the development of the controller system is that the design and development tasks are accomplished by engineers and designers, without the experience and knowledge of the personnel who have used the machines or the elements that will be controlled by the controller system. On small projects, and if the ME has many years of experience in the development of similar systems, he or she may be the only member of the team. More likely, a small group of experienced personnel from several disciplines will be required to develop the system.

As the ME, at the beginning of the project you must determine how much time will be spent on the design and simulation of the final system, and how much time should be spent testing and refining the system with real hardware and machines. If you are designing a large system that will be used at multiple locations, more time should be spent in the design and simulation stages of the process. At some point, however, the real hardware, code, and machines must come into the process for final refinement. The tools provided

Table 6.1 Specifications of a Typical Controller Series (Courtesy of Allen-Bradley)

Type	Base memory	I/O	Operating mode	Program scantime	I/Oscan
PLC-5/12	6K words	16 pt.-256 32 pt.-512	Adapter only	8 msec/K words typ.	1 msec local, 7 msec remote
PLC-5/15	14K words	512	Scanner/ adapter	Same as above	Same as above
PLC-5/25	21K words	1024	Scanner/ adapter	Same as above	Same as above
PLC-5/40	48K words	2048	Scanner/ adapter	2 msec/K words	0.25 msec local, 7 msec remote
PLC-5/60	64K words	3072	Scanner/ adapter	2 msec/K words	0.25 msec local, 7 msec remote

by the supplier of the controller hardware should be one of the major elements in hardware selection. The job of the design team can be made much easier if the tools are robust and user friendly. Development tools that allow the designer to test new code on the real system, "on line," is one of the most important elements, and will reduce the total development time for the project. The following suggestions may prove helpful:

Place the high power controllers as close as practical to the controlled device.
Follow rules on cable and raceway installation.
Always plan spares in the cable, and I/O at each rack.

Cable Installation

Conductor Category 1 consists of high-power AC I/O lines and high-power DC I/O lines connected to DC I/O modules rated for high power or high noise rejection. These lines can be routed with machine power up to 600 VAC (feeding up to 100-kp devices). Check local codes. Article 300-3 of the National Electrical Code requires all conductors (AC and/or DC) in the same raceway to be insulated for the highest voltage applied to any one of the conductors.

Conductor Category 2 consists of serial communication cables, low-power DC I/O lines connected to proximity switches, photoelectric sensors, encoders, motion-control devices, and analog devices. Low-power AC-DC I/O lines connected to I/O modules rated for low power such as low-power contact-output modules are included. All conductors must be properly shielded, where applicable, and routed in a separate raceway. If conductors must cross power feed lines, they should do so at right angles. Route 1 ft from 120 VAC, 2 ft from 240 VAC, and 3 ft away from 480 VAC. Route at least 3 ft from any electric motors, transformers, rectifiers, generators, arc welders, induction furnaces, or sources of microwave radiation.

Conductor Category 3 interconnect the PLC components within an enclosure. For PLC power cables, provide backplane power to DPLC components. This category includes processor periphal cables, connecting processors to their communication interface modules. All conductors should be routed external to all raceways, or in a raceway separate from any category 1 or 2 conductors.

The previous information illustrates some of the more important design criteria for process controllers. The specific requirements will be unique to your process. The information will help guide you through the design and evaluation process to select the right hardware and lay out the best system for your application. Keeping in mind that one of the most important requirements for a processor control system is ease of maintenance, you can provide, if desired, a terminal for use in the operations/maintenance center. This terminal can link into any of the processors on the network for system operation or maintenance. The design should also provide connections for portable terminals and annunciators at all racks for maintenance.

System Issues

1. *Environmental.* Hazardous, wet, or corrosive atmospheres will require the controller to be mounted in a sealed NEMA enclosure (see the Appendix for a listing of enclosure manufacturers). Depending on the ambient temperature range of the enclosure location, you may need local cooling for the long-term reliability of the controller. One way to reduce the internal temperature of the sealed enclosure is by the use of heat exchangers or a solid-state cooler. The only moving parts in

these devices are the external and internal circulation fans. By controlling the polarity of the current through the solid-state device, the unit can be used to remove or add heat to the enclosure.

2. *Lightning protection.* Lightning protection of data lines and network lines may be needed depending on the geographic location of the plant and the exposure of the lines to the elements. If the lines are near the exterior walls of the building or go outside, it is necessary to protect both ends of the data lines with lightning protectors. Small telephone-type protectors will work in most cases.

3. *Electrical noise.* Most factories have a number of rotating machines (motors) and controllers. These devices produce a large amount of electrical noise. If your process is of a critical nature, then noise in the system could be very costly. One of the best ways to eliminate the effects of electrical noise in the process controller system is by the use of fiber optics for transmission of data from rack to rack. This is more expensive in the short term but could be very cost effective for the smooth, uninterrupted operation of your process. The use of fiber optic cable can also be useful for a building-to-building network in a lightning-prone environment.

6.7 HUMAN INTERFACE TO THE COMPUTER INTEGRATED ENTERPRISE

6.7.1 Equipment Selection

The performance of the user, utilizing the terminal interface to the CIE, is very important so as to provide the user with timely response and accurate, up-to-date information from the CIE (see also Subchapter 5.1). User requirements will differ between the office environment and the factory environment. Most office users should have access to the integrated databases of the CIE, and will require a terminal (PC) with reasonable performance in the CPU and the graphics card. If the office user is using an analysis package or a CAD package, a coprocessor will probably be required. The performance of the factory (work instructions) units' performance will be determined by how much multimedia processing the units will be required to accomplish, and where that data is stored. A good policy to follow when selecting terminal hardware (PCs) is to test competing PCs with benchmark programs. We found a large variance in the performance of PCs which had the same processors (over 3 to 1). Also, the price of the hardware is not a good indication of the performance. In our case, the lowest-cost unit gave the best performance. When selecting the benchmark software, use ones that are designed for your operating environment (DOS, Windows, OS/2, or UNIX).

6.7.2 Computer-Aided Process Planning

The generation and the electronic display of computer-aided process planning (CAPP) work instructions are of great importance to the manufacturing engineer (see Chapter 9). The effectiveness of the interface will determine in great part the productivity improvements for both the author and the user of the planning. The acceptance level by the users of any new system will depend on their involvement during the design. To provide an effective interface, in a timely, cost-effective manner, you, along with the application developer, should use a rapid development process which includes the users. The planning author and the end user can provide insight into the use of the development software and the interface if they are involved in the development of a new system from the beginning. In most cases, the user interface will be some form of a graphical user interface (GUI). You may, for your application, use a multimedia GUI. Many development tools are available to design a working model (prototype) with limited capabilities in a very short time (weeks). This prototype can be used to check the "look and feel" of the interface. As the development progresses and the interface becomes more refined, the user should be brought in from time to time to check the progress. During the rapid development process, you and the application developer must be very careful not to include system errors in the design and then move them into production. These errors are most likely to be in the area of data relationship integrity. Another pitfall to avoid is trying to use too many application developers. Almost all firstclass software is developed by a small, dedicated group. All of the members of the group must understand that it is their responsibility to design, code, debug, and document their part of the program.

Providing process planning electronically to the shop floor worker gives you the opportunity to incorporate other features into the process that are not available in a paper system. When your product or process requires that certain employee skills, tools, and or materials must be certified to build the product, the interface, planning, and a relational database can be used for this control. This will prevent the use of uncertified employees, tools, and materials before they are used in the product, eliminating the need for rework. The control of the process will also provide a complete "as built" configuration of the product.

When you are involved in the design and development of a shop floor terminal (SFT) for the presentation of work instructions for a manufacturing process, the following are major issues that you and the development team should consider.

Is the software package a standard (generic) application, or a package designed for your application?

Are you going to use a keyboard, mouse, or touch screen for the interface (look at the merits and cost of each)?

Will the GUI allow the user only to navigate through the application under the control of the application?

How long would you expect the SFT to provide services with the network down? This will help determine how the work instructions are delivered to the SFT, and how much is stored on the SFT.

What level of education (training) does your workforce have?

How much time are you willing to utilize to train the workforce?

Should all of your displays meet the Swedish MRP I and II specification on lowfrequency radiation? If not, why not? It could be a liability issue in the future.

As an example, Figure 6.11 shows the navigation flow diagram of a GUI that will display CAPP (multi media). The process incorporated into this SFT provides for all of the functions required to operate and manage a work center. The worker logs onto the SFT at the beginning of the shift; the system compares the clock number and PIN for identification. If any problem occurs, the worker must get the supervisor to resolve it prior to start of a job. When he or she logs on for the first job, a list of the jobs available for that work center are shown in priority order. After selecting a job, the system checks to make sure the skill/certifications required for the job are met by that employee. The plan for the job

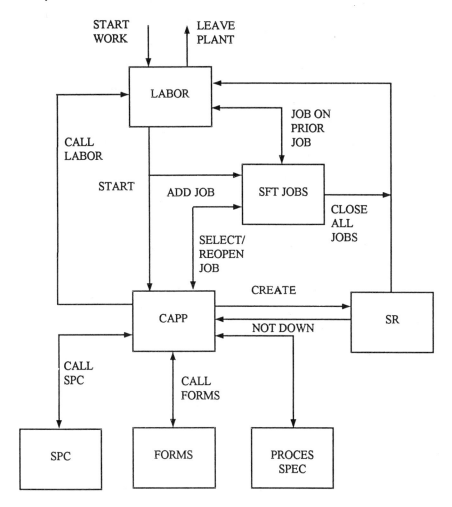

Figure 6.11 Navigation flow of a GUI that will display CAPP.

is then presented to the worker on an integrated screen. It contains the text, the visual aids, and the navigation buttons. This SFT utilizes touch screens. Proceeding through the work instructions, as material or product information is required, it is wanded into the SFT. The system checks to make sure that the material is the correct material for this configuration, and/or that the measured value on the required test information is within limits. All this information is stored in the relational database, and will be the "as built" information for this item.

You must have at least a rudimentary flow diagram prior to setting up the prototype GUI. As you work with the prototype and the user community, expand and refine the flow diagram until you have the final requirements defined. Spending a few weeks working with a group of users and the logic developers is a good investment. If this phase of the process is carried out correctly, the final production version of the GUI will require fewer revisions. After following this procedure, we needed to add only two revisions to the GUI during the first year of use. The total development time for the initial version of the GUI was only 3 months. Your application may look totally different from this example. It is highly dependent on the type of products you build, and the education and turnover of your workers. The workers using this GUI were able to fully utilize the system after less than 1 hr of training, and were up to full productivity in less than 1 week.

A major consideration in the design of the SFT with a GUI is where and how much data will be stored. With any network system, you must expect the system to be "down" from time to time. If the production workers are to be provided all the services of the SFT, decisions on where the information is stored must be made. The critical factors are how long the network could be down, and how soon new jobs can be expected at the SFT workstation. Another issue to consider, if you are using visual aids to support the work instructions, is the size of the files and when you will transmit them over the network. If your production cycle will allow, the SFTs could be updated at night when the network is not so busy. This can be accomplished automatically by setting up a data transfer log; at the appointed time, the data will be transmitted to the SFTs for local storage. If the various SFTs do not all support the same work content, each of them could be provided with only the jobs accomplished at that work center, to minimize local storage requirements. This functionality was incorporated into the example system described above. The SFTs should have the capability to command a data transfer if a process plan is changed and is required immediately.

6.8 CAD/CAM

6.8.0 Introduction to CAD/CAM

The computer has been used for some time in computer drafting. The capability to produce high-quality drawings by the computer has been utilized in the architectural and engineering fields for many years. This capability has also been used for electrical drawings for wiring. The use of the computer to support computer-aided design/computer-aided manufacturing (CAD/CAM) of products began in the late 1950s. P. J. Hanratty was one of the pioneers in the field of CAD/CAM. He did a lot of his earliest work for General Motors and McDonnell Douglas. Over the last 30 years, CAD/CAM software products have matured into a marketplace of over $3 billion per year. The software now has the capability to do not only the analysis on the structure using tools such as Nastran, but has at least the beginnings of solid modeling. One of the most exciting newer developments in the world of CAM is the use of simulation to test a process flow prior to building the factory. The optimization of the process and the improvements in process flow can be tuned during the simulation testing along with a graphical demonstration of the process to management. The use of simulation can save a large amount of time and money during development of the process or factory (see Subchapter 10.1).

From 1984 through 1994, a consortium of industrial, university, and research centers across Europe set up the European Strategic Program for Research and Development in Information Technology (ESPRIT). The output of this research is a set of interface requirements to facilitate the interconnection of the various modules of CAD/CAM systems. Most of the manufacturing software systems in Europe have been designed and built in this environment. Indications are that the European design environment will be more directed to the CAD/CAM systems than the U.S. environment, and many of the new programs are coming from Europe.

The capability and the complexity of the CAD/CAM software packages from the various vendors covers the spectrum from simple to very complex. The price per seat also varies from a few hundred dollars to tens of thousands of dollars per seat. The software package best suited for your application will depend on how you are intending to use it and for what purpose. The major benefits you can realize from an integrated CAD/CAM software package will occur only when your development cycle process changes to utilize the capabilities of the software. If you try to use these software packages without this shift in the processes, you will not see the major benefits. Some of the major suppliers of CAD/CAM software are:

Supplier	Product
Applicon	Bravo
Autodesk	Autocad Designer
Computervision	CADDS 5
Dassualt Systems	Catia Solutions
EDS Unigraphics	Unigraphics
Hewlett Packard	HP PE Solid Designer
Intergraph	I/EMS
Manufacturing and Consulting Services	Anvil-5000
Mechanical Advantage	Cognition
Parametric Technology	Pro/Engineer
Structural Dynamics Research Corporation	I-Deries

Computer modeling is one of the more dynamic areas of development in the CAD/CAM software world. To utilize this capability, the process in development of a product as mentioned above must change. The process of development of a new product, in general terms, follows.

6.8.1 General Problem-Solving Principles

To resolve all of the complex issues in a new product design, the following process steps of refinement activities must occur over and over again:

Finding a mental solution
Representing the solution
Analyzing the solution

When this process has processed to the point of decision, the final solution is compared to the requirements, the solution evaluated against those requirements, and a final decision made on the product. If the design does not meet the requirements, then the process must continue until the requirements are met. The above process will be used during all phases of the design process:

Definition of functions
Definition of physical principles
Preliminary design
Detail design

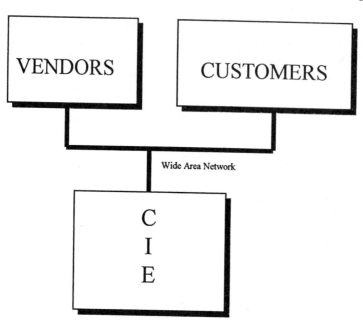

Figure 6.12   Extended enterprise information flow.

Prototype (if desired)
Production planning

The results obtained at the end of each phase of the design are the basis for the following phase. The result of the function definition phase is an abstract description of the product. This abstract product description is a function structure which describes how an overall function of the product is realized in terms of a combination of sub-functions. To realize the design, a solution principle must be found for each of the elements in the functional structure.

The definition of the initial shape characteristics during the preliminary design uses the physical principles. A complete design scheme is the result of the preliminary design. The documents generated during the detailed design are the basis for the NC programming, process planning, parts lists, and the other documentation for building the product.

As the total design matures, the design team might return to a previous step in the process due to further definition or other information which has become available during the design phase. Some of the decisions made in previous steps in the process may have to be modified with the changes in the detailed design caused by those changes. (See Chapters 1 and 2 for more on product development and product design.)

6.8.2   MODELING

One of the most powerful tools for the designer in the newer CAD/CAM software packages is the capability to do 2D and 3D modeling. The various packages from the vendors use some form of parametric or variational analysis, or a combination of both for design. The generation of the models depends on the modeling modes utilized. All modeling packages have some overlapping characteristics that allow the designer to show in the model how changing a variable will reconfigure the graphical model. The major types of models are:

Explicit Modeling: The geometry of the design is built without any relationships between entities. Some software packages combine wireframe, surfaces, and solid modeling in a single integrated package. Changes in an explicit model require the designer to redesign for that change, and may entail large changes due to effect on the total requirements.

Variable-Driven modeling: The design intent of the product is captured during the product's definition by capturing the relationships that exist within the model and the processes that define the model. By utilizing the intent of the design rather than the explicit geometry of the object, the designer will be able to edit and or modify the design for design alternatives or revisions of the design.

6.9 EXTENDED ENTERPRISE

For a manufacturing organization to fully utilize the capabilities of automation, it must understand the goals and capabilities of the extended enterprise. First, what do we mean by the extended enterprise? Referring to Figure 6.12, you can see the general interflow of information among customers, vendors, and your internal operations. Normally, the computer integrated enterprise is your corporation's computer systems. The extended enterprise is the shell surrounding the CIE to interconnect your systems to vendors and customers. The actual use of electronic data interchange (EDI) will depend on the type of business you are in. If you have a standard line of products, your customer's order will trigger the shipment of the product to the customer. In a large number of cases when you have standard products, the customer will expect quick delivery of the order. To be able to accomplish quick delivery, you must stock a reasonable amount of each product or have a very responsive manufacturing flow that will produce the product in a very short time.

6.9.1 Transmission Protocol

Transmission protocols are a set of standards which allow various computers to communicate with each other and control the transmission of data. The correct protocol must be selected for the type of transmission that will be made when you have set up an EDI procedure for your requirements. Some of the major issues that you and your implementor must resolve are: electronic signoff of POs, transfer of funds, and an audit trail of all business activities. The network protocols are shown in Figure 6.13, and system profiles (GOSSIP Version 1) are shown in Figure 6.14.

INTERNET	BANYAN VINES	OSI	TCP / IP						
5 Layer Model		7 Layer Model							
APPLICATION		APPLICATION	KERBOS	XWIN	SMTP	+-FTP- TELNET	TFTP NCS SNMP		
	DOS/WIN/OS2	PRESENTATION							
		SESSION							
TRANSPORT	VINES IPC,SPP	TRANSPORT	TCP				UDP		
INTERNET	VINES IP	NETWORK	IP	ICMP		ARP	RARP		
NETWORK INTERFACE	ETHERNET TOKENRING ARCNET NDIS X.25 BSC ASYNC SDLC HDLC	DATALINK	LOGICAL LINK CONTROL MEDIA ACCESS CONNECTION					DRIVER ADAPTER 802.3 ETHERNET 802.4 TOKEN BUSS 802.5 TOKEN RING 802.6 DIX THICKNET DIX+802.3 APPLETALK ARCNET FDDI SDLC BSC HDLC ISDN	
HARDWARE	SAME AS LIST AT FAR RIGHT →	PHYSICAL						UNSHIELDED TWISTED PAIR SHIELDED TWISTED PAIR COAX THIN-NET RG58 IBM TYPE 1 IBM TYPE 3 FIBER 62.5 MICRON FIBER 100/50 MICRON RS232C SATELLITE HYPERCHANNEL RADIO V.35	

Figure 6.13   Network protocols.

TRANSPORT PLATFORM (ALIAS OSI BRIDGES)

THREE LAYER (ALIAS OSI ROUTER)

FTAM / ACSE	FTAM / ACSE	FTAM / ACSE	FTAM / ACSE	FTAM / ACSE							
PRESENT*	PRESENT*	PRESENT*	PRESENT*	PRESENT*	X.400	X.400	X.400	X.400	X.400	X.400	X.400
SESSION	SESSION	SESSION	SESSION	SESSION	SESSION	SESSION	SESSION	SESSION	SESSION	SESSION	SESSION
TP4	TP4	TP4	TP4	TP4	TP4	TP4	TP4	TP4	TP4	TPO ++	TPO ++
CLNP	CLNP	CLNP	CLNP	CLNP	CLNP	CLNP	CLNP	CLNP	CLNP		
LLC	LLC	LLC	X.25	X.25	LLC	LLC	LLC	X.25	X.25	X.25	X.25
802.3	802.4	802.5	HDLC	HDLC	802.3	802.4	802.5	HDLC	HDLC	HDLC	HDLC
			V.35	RS232C				V.35	RS232C	RS232C	V.35

PRESENTATION * LAYER ——— ASN.1
** ISO CONNECTION ORIENTED NETWORK SERVER (CONS) USES X.25 PROTOCOL AT THIS LAYER

ISO 8473 CONNECTIONLESS NETWORK PROTOCOL
ISO 9542 END SYSTEM TO INTERMEDIATE SYSTEM ROUTING PROTOCOL
ISO 8208 X.25 PACKET LEVEL PROTOCOL FOR DATA TERMINAL EQUIPMENT
ISO 8802.2 LOCAL AREA NETWORK LOGICAL LINK CONTROL (LLC)
ISO 8202.3 LOCAL AREA NETWORK CSMA/CD - ACCESS METHOD AND PHYSICAL LAYER SPEC
ISO 8802.5 LOCAL AREA NETWORK TOKEN RING ACCESS METHOD AND PHYSICAL LAYER SPEC

Figure 6.14 End system profiles (GOSIP Version 1).

7

Practical Cost Estimating for Manufacturing

John P. Tanner   *Tanner and Associates, Orlando, Florida*

7.0 INTRODUCTION TO COST ESTIMATING

The manufacturing cost estimate is the key in new or follow-on business acquisition, and the continued growth of the company. It must be built around a sound, well-thought-out manufacturing plan. It must address the cost of facilities, equipment, tooling, materials, and support labor, as well as direct labor to actually fabricate and assemble the product. The estimate must be responsive to customer delivery requirements and production rates, and reflect manufacturing a product of the desired quality and reliability. The cost estimate is a prediction of what the product will cost to manufacture at some future point in time. Estimating is not an exact science, yet a good cost estimate will come very close to actual costs incurred. The accuracy of any cost estimate depends on:

1. The time allocated for the preparation of the estimate
2. The knowledge and skill of the estimator
3. The depth and completeness of preproduction planning
4. The amount of product description information available
5. The accuracy of the material estimate

The higher the percentage of labor cost in the estimate, the greater is the need for precise labor standards. If engineered standards or estimated standards are not available, the estimator must use historical data and his or her own judgment, experience, and knowledge to develop the labor estimate (see also Chapter 10—Work Measurement).

In many products, material and subcontract costs can be as much as 70% of product cost. This means that these costs must be examined very carefully at the time the

---

This chapter was condensed from a book manuscript prepared by John P. Tanner. The editor wishes to thank Mr. Tanner for generously allowing use of this material.

manufacturing cost estimate is prepared. For best results, direct quotes should be obtained from suppliers and subcontractors. Time constraints in completing the cost estimate may not allow sufficient time to solicit these quotes, forcing the use of historical cost data for the same or similar parts and materials, factored for inflation and anticipated cost growth.

Once the basic estimates of labor hours and material dollars has been put together, the judgment exercised in determining the initial costs and the rate of improvement to achieve the eventual cost will have a major affect on the final accuracy of the estimate. An important part of the estimator's job is to prevent ill-advised management decisions by making certain that the methodology, assumptions, and ground rules are understood. Management must look beyond the estimate and consider a bigger picture than the cost estimates at hand, but should understand the risks associated with arbitrary changes to a well-prepared manufacturing cost estimate.

Figure 7.1 shows the basic structure of a manufacturing cost estimate. The sum of direct labor and direct material is known as prime cost. When prime cost is added to manufacturing overhead, we obtain factory cost, or cost of goods manufactured. Total product cost, then, is the sum of selling expense, general and administrative expense, development cost, and factory cost contingencies. The addition of profit yields the price to the customer.

The problems encountered in developing a sound manufacturing cost estimate can be many, however, they can usually be categorized into the following seven categories:

1. Inadequate data on which to develop the cost estimate
2. Inadequate staff and time to prepare the estimate
3. Poor estimator selection
4. Careless estimating
5. Optimistic estimating
6. Inadequate preproduction planning
7. Management inertia

The seven problem categories are not listed in order of importance. Any one or several can be critical to the development of a sound manufacturing cost estimate, depending on the circumstances and the situation that prevails at the time.

The cost estimator may be a manufacturing engineer, an industrial engineer, or a manufacturing technical specialist with heavy experience in the manufacturing technology in which he or she is preparing cost estimates. Cost estimating is highly demanding work, often requiring extended overtime and short deadlines. It requires the ability to quickly formulate a preproduction plan, and to visualize work flows, equipment, and tooling. A labor estimate that comes close to actual incurred costs must then be accepted by management. Not only must this be done under considerable pressure, but it must handle last-minute changes to the requirements the estimate was built on, as well as management-directed changes. In many large companies, an independent cost estimate may be developed by the fiscal or marketing groups, and is used as a check against the more detailed analysis described in this chapter.

The cost of preparing the estimate must be borne by the company whether it results in new business or not. Most new contracts for manufactured products are awarded based on lowest cost, best delivery, and quality of the product, not necessarily in that order. Management must decide the "win probability" on any estimate for new business, and from that decide the effort to be expended in preparing the cost estimate. It may even decide not to submit a bid. Management should prepare a bid strategy and plan which

1.0   PROGRAM REQUIREMENTS

Program strategy and objectives. What are the issues? What needs to happen? What is the critical path? What are the assumptions and ground rules? Who is the customer? Is co-production involved? If so, what is the split?

1.1   PRODUCT/HARDWARE

Product definition. How will it change from concept, through development to production? How will configuration be controlled? Will a technical documentation package be provided? When? To what level? What quantities of deliverable hardware will be provided in development? Pilot Production? Production? What about spare parts? Will there be any GFE or CFE?

1.2   PROGRAM TIME PHASING

When will the program start? Development? Pilot production? Production? What are the key program milestones? What are anticipated peak production rates? When will they occur?

1.3   PROGRAM CONSTRAINTS

Potential problem areas and risks. New or advanced manufacturing processes and technologies. Unusual inspection, testing or acceptance requirements.

2.0   DEVELOPMENT PLAN

What will be accomplished in the development program? What deliverable hardware will result? Engineering built? Manufacturing built? How will producibility an DTUPC be addressed? How will the hardware change as it evolves through the development cycle?

3.0   MANUFACTURING PLAN

Fabrication and assembly sequences and flows. Estimated times. Equipment requirements..Tooling requirements. Overall block layouts. Space required. Manufacturing flows. Processes.

3.1   MAKE OR BUY PLAN

What is the rationale for the make or buy plan? Identify major items to be subcontracted.

3.2   TOOLING PLAN

Tooling philosophy. Tooling requirements. Is interchangeability a requirement? Will tooling masters and gages be required? How will tooling differ from development to pilot production to production? Will tooling be design tooling, shop aid tooling? Will NC be used? How will tools and tapes be controlled?

Figure 7.1   Structure of a cost estimate for a manufactured product.

3.3   MANPOWER PLAN

Skills requirements. Availability of manpower. Projected manpower needs of program..Anticipated training requirements. Support personnel needs.

4.0   FACILITIES PLAN

Identify new or additional equipment needed and estimated cost. Additional or existing building floor space requirements and estimated cost. Would include engineering, lab, manufacturing, test, storage, bunker and special process area. Identify any requirements for special or unusual facilities such as clean rooms, dark rooms, specially reinforced floors, ESD protection, etc. Provide estimated cost and time when facilities would be needed.

5.0   MATERIAL ACQUISITION PLAN

Identify long lead items, source selection plans. Who are major subcontractors and how will they be controlled and managed? What are plans for stores and kitting of material? What are the plans for dual sourcing? What is the material handling plan for receiving, stores, staging for production? How will vendor and subcontractor follow-=up and expediting be accomplished to ensure on-schedule delivery of material? How will engineering changes and shop overloads be handled?

6.0   SCHEDULES AND TIME-PHASING

6.1   PROGRAM MILESTONE SCHEDULE

Overview schedule of entire program showing all phases with key milestones and events, including follow-on work. Should show development, procurement, facilities, production, etc.

6.2   DEVELOPMENT SCHEDULE

Key events in engineering and development, in sufficient detail to clearly portray development time-phasing.

6.3   MANUFACTURING SCHEDULES

Initial low rate production and full production. Manufacturing buildlines, block release plan.

6.4   PROCUREMENT SCHEDULE

Long lead and subcontract deliveries. Should show entire procurement cycle from requisition through order placement to material kitting and issue from production.

6.5   FACILITIES AND EQUIPMENT SCHEDULE

Must show all key milestones for new equipment and facilities acquisition, including order placement, ship date if equipment, ground breaking if new construction, through to available for production use date.

Figure 7.1   Continued.

7.0 QUALITY ASSURANCE PLAN

What is the inspection and test plan for the product? What specifications and standards will apply? How will vendor and material quality levels be maintained? What is the plan for the rework and/or disposition of discrepant hardware?

8.0 NEW MANUFACTURING TECHNOLOGIES

Describe new manufacturing technologies associated with the program and plans for training and qualification in production, including equipment and process shakedown and de-bug. How and when this will be accomplished ahead of the production phase.

Figure 7.1 Continued.

includes reviews at critical stages in the preparation of the cost estimate. There are several good reasons for this:

1. To surface and correct any basic errors or omissions in time to make corrections before the estimate goes beyond the preliminary stage, when changes would be costly and time consuming.
2. The preliminary estimate may be sufficient to satisfy the requirements, and early management review ensures that no further cost estimating effort will be authorized beyond the preliminary stage.
3. Management review brings the best minds and talent available in the company to bear on the manufacturing cost estimate, serving as a check on the estimate and its assumptions.

Constraints of time and cost often leave no opportunity to explore and verify many of the premises and the assumptions used in preparing the estimate. In spite of this, cost estimates for manufactured products are prepared every day that accurately reflect manufacturing costs for the product and are truly competitive in bringing in new business for the company. In the following subchapters, the steps required to prepare a manufacturing cost estimate that is both accurate and competitive are explained. Examples and cost estimating data in practical form are also provided to help in preparing estimates.

7.0.1 Bibliography

Tanner, John P., *Manufacturing Engineering: An Introduction to the Basic Functions,* 2nd Ed., Rev., Marcel Dekker, New York, 1991.

7.1 UNDERSTANDING THE ESTIMATE REQUIREMENTS

7.1.0 Determination of Cost Estimate Requirements

In any engineering work, the solution is usually readily apparent once the problem is fully defined. The same holds true in cost estimating for manufacturing. If the cost estimate requirements are known and fully understood, preparation of the estimate is usually routine. To do this requires answers to the following questions:

1. Who is the prospective customer?
2. What is the bid strategy?
3. What are the requirements?

4. What are the assumptions and ground rules?
5. Is the product to be manufactured fully defined?
6. What are the potential problem areas and risks?
7. What are the key milestones, delivery requirements, and production rates?

Type of Solicitation

The type of solicitation is important in the formulation of a bid strategy, and in deciding the resources that will be committed to preparing the manufacturing cost estimate. A quotation for an additional quantity of parts to a current customer is one example. A quotation for a newly designed item, which has never been built in production before, is quite a different matter if your company is one of several submitting bids. Firms in the defense industry, whose primary customer is the overnment, can survive only by winning contracts awarded to the lowest bidder among several submitting bids. This winning bid must be supported by a detailed and valid cost estimate. Figure 7.2 shows the outline for a production program plan used by an aerospace company as the basis for its cost estimate and proposal to the government. Such an outline forces consideration of all the requirements.

The majority of manufacturing cost estimating involves job order production in the metal working and allied industries in the United States and the rest of the Western world. Castings, forgings, formed and machined parts, and assemblies are produced by thousands of small and medium-sized firms to exacting specifications within the limits of the estimated cost. Solicitations are usually *firm fixed price*, which means that if the cost estimate is in error on the low side, the difference is absorbed by the company.

A high estimate may mean extra profit for the company, or it may mean that the contract is awarded to another firm whose price was more in line with what the item should cost. On bids for an additional quantity of the same or a similar item ordered previously, customers will often expect a price reduction because of the learning curve effect, which implies continuous cost reduction the longer the product remains in production (see Subchapter 7.4).

Product or Hardware to Be Delivered

Adequate product definition is critical to developing a meaningful manufacturing cost estimate. This may include a set of engineering drawings that fully describes the product

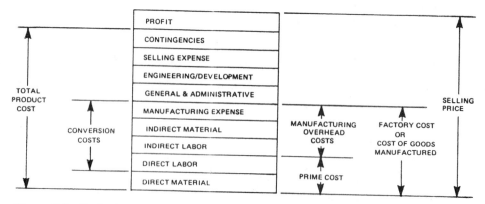

Figure 7.2 Production program plan checklist.

or parts to be produced. In many instances, shop process or routing sheets and sample parts are available. The product may currently be in production. Should this be the case, it is a relatively easy matter to determine the sequences of manufacture.

If the drawing package describes a product that has never been built in your plant before, a different approach is required to develop a sound manufacturing cost estimate. The drawing package must be broken down into piece parts, subassemblies and assemblies, and a parts list constructed. This will show (in the case of an assembled product) how the assembly goes together and will form the basis for the bill of material and the fabrication and assembly processes.

There are differing levels of detail and description provided by product technical data packages. Drawing packages may show the lowest level of technical detail down to the smallest piece part and assembly, each on a separate drawing. Other engineering drawings provide a minimum of detail, showing assembly and detail parts on the same drawing. Engineering drawings and other technical documentation cost money to prepare, and to update with the latest changes. Technical documentation packages provided to bidders for cost estimates and quotations do not always fully and correctly represent the product. There may be errors in dimensioning and tolerancing that were noted the last time the product was manufactured, but these changes were never picked up and documented by formal engineering change order to the drawings.

Many times in bidding the product as depicted in the engineering drawing package, Defense Department contractors, with no other product data or knowledge to go by, have seriously underbid production contracts on hardware and systems designed by other contractors simply because all the information needed to fabricate and assembly the product was not shown on the drawings. In preparing a manufacturing cost estimate, every effort should be made to make certain that the drawings are current and that all engineering change information is included.

Engineering drawings must often be supplemented by manufacturing engineering documentation such as process routings, methods sheets, visual aids, tool drawings, test procedures, and process specifications to determine what is really required to manufacture the product. If possible, this documentation should be provided by the prospective customer as part of the bid package. If your company is submitting a cost estimate for the manufacture of a build-to-print product that you have never produced before, it is imperative that whatever shop process documentation that exists be obtained to aid in developing the cost estimate. If not available, the bidder may have to prepare preliminary documents in order to form the basis of a bid.

Definition or Concept

The estimator must work with many kinds of drawings, specifications, and shop documents. Included will be specification control drawings similar to Figure 7.3, which clearly spell out all critical parameters of the jacketing material for a cable assembly, and may list approved or qualified sources for this material. Figure 7.4 shows a typical detail or fabricated part drawing for a 0.020-in. thick gasket which would be a stamped with a die. A typical assembly drawing is shown in Figure 7.5, for a voltage regulator assembly. Notes on such a drawing might include:

1. Prime and seal threads using Loctite.
2. Torque fasteners to 3 to 5 in.-lb.
3. Ink stamp assembly number as shown.

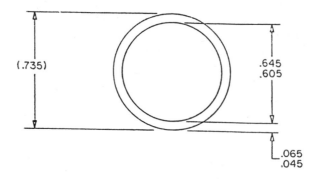

(.735)   .645
                          .605

                          .065
                          .045

NOTES- UNLESS OTHERWISE SPECIFIED
1   MATERIAL: SILICONE RUBBER CONFORMING TO
    ZZ-R-765, CLASS 2A OR 2B, GRADE 50, WHITE.
2. IDENTIFY PER MIL-STD-130 WITH MANUFACTURER'S
    PART NUMBER AND FSCM NUMBER; CONTROL NUMBER
    ENCLOSED IN PARENTHESIS; BAG AND TAG.
3. SHAPE AND CONCENTRICITY: ID AND OD SHALL
    BE NOMINALLY CONCENTRIC WITH A UNIFORM WALL
    THICKNESS. JACKET MAY BE ELLIPTICAL IN
    FREE FORM EXCEPT OPPOSING SURFACES SHALL
    NOT ADHERE TO EACH OTHER.
4. LENGTH SHALL BE FURNISHED CONTINUOUS AS
    SPECIFIED (10 FEET MINIMUM).

Figure 7.3 Specification control drawing (SCD) for a high-reflectance silicone cable jacket.

Column A in the parts list shows the number of items per assembly, column B is the part number, column C is the item name, and column D is the find number shown in the leader arrows on the drawing.

Often it becomes necessary to estimate the cost to produce an item which is not yet fully designed. Such estimates are made from sketches, concept drawings, or design layouts. Preparation of these estimates requires the estimator to fully understand the design concept envisioned by the designer. It is not uncommon for such preliminary designs and preliminary bills of material to grow by as much as 40% in complexity and parts count by the time the final manufacturing drawings are released. Figure 7.6 shows a design layout for a mortar round which, once fully designed, developed, and qualified, would be produced in high volume for the Army.

Delivery Requirements

The estimator must know if the customer's delivery schedule can be met, considering the lead time required for planning, tooling, obtaining the necessary parts and material, plus the number of days to actually manufacture the product. Analysis of the delivery requirements determines peak production rates, and whether one or two shift operations are needed. Perhaps the required delivery rate can be met only by extended overtime. All of these elements affect cost.

Analysis of shop flow times, material lead times, capacity of shop machines, and shop

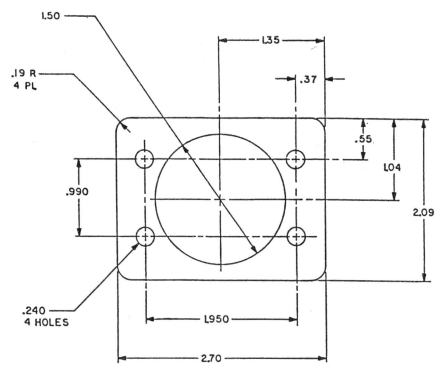

Figure 7.4 Gasket made from 0.020-in. material.

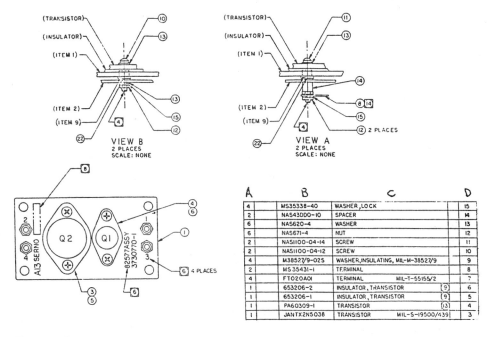

A	B	C		D
4	MS35338-40	WASHER,LOCK		15
2	NAS43DD0-10	SPACER		14
6	NAS620-4	WASHER		13
6	NAS671-4	NUT		12
2	NAS1100-04-14	SCREW		11
2	NAS1100-04-12	SCREW		10
4	M38527/9-025	WASHER,INSULATING, MIL-M-38527/9		9
2	MS35431-1	TERMINAL		8
4	FT020A01	TERMINAL MIL-T-55155/2		7
1	653206-2	INSULATOR, TRANSISTOR	[9]	6
1	653206-1	INSULATOR, TRANSISTOR	[9]	5
1	PA60309-1	TRANSISTOR	[13]	4
1	JANTX2N5038	TRANSISTOR MIL-S-19500/439		3

Figure 7.5 Voltage regulator assembly.

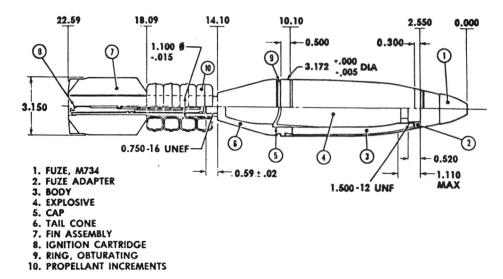

1. FUZE, M734
2. FUZE ADAPTER
3. BODY
4. EXPLOSIVE
5. CAP
6. TAIL CONE
7. FIN ASSEMBLY
8. IGNITION CARTRIDGE
9. RING, OBTURATING
10. PROPELLANT INCREMENTS

Figure 7.6 Design layout for a new mortar round.

and machine loads in the time period the proposed work would be performed can be crucial in developing any manufacturing cost estimate. A firm may manufacture to stock or inventory, based on a sales forecast. This offers the advantage of smoothing the shop workload and being able to plan well in advance for manufacturing operations. Delivery requirements requiring higher rates of production may require more units in work in a given time. This allows better labor utilization and production efficiency. This has to traded off against the cost of carrying inventory, as opposed to delivering for payment upon product completion. Lower production rates result in smaller lot sizes, increasing the number of setups for the same number of units. This lowering of production rates can have a profound adverse effect on the attainment of projected improvement curves. Setup and tear-down caused by small production lot sizes should be avoided as much as possible.

Special Provisions

A contingency factor should be applied to any manufacturing cost estimate in which the product is not fully defined or is still undergoing change. The amount of contingency is a judgment call which would:

1. Vary with the stage of product development
2. Depend on the newness of the program, the market application area, the technology, the industrial processes, and the organization
3. Depend on the time allowed for development
4. Consider the degree of development required
5. Vary with the general economic conditions

Figure 7.7 shows a contingency policy used by a large manufacturer of computers and point-of-sale terminals.

In addition to the importance of product definition, a complete bill of material or engineering parts list is vital to estimating manufacturing costs. This must be carefully reviewed, and the make-or-buy decisions for pricing purposes made. The make-or-buy plan

PHASES OF COST ESTIMATE	CONTINGENCY GUIDELINES AS % OF MLB
I. ENGINEERING DESIGN GOAL	30 - 50%
PRELIMINARY DESIGN "A" MODEL AVAILABLE IF REQUIRED	
II. ENGINEERING	20 - 30%
"B" MODEL IN FABRICATION PRELIMINARY B/M AVAILABLE PARTIAL DRAWINGS AVAILABLE	
III. PRELIMINARY PRODUCTION	10 - 20%
"B" MODEL TEST COMPLETED AND ACCEPTED MANUFACTURING B/M AVAILABLE MAJORITY OF DRAWINGS RELEASED	
IV. PRODUCTION	0 - 10%
B/M COMPLETE COMPLETE DRAWING PACKAGE AVAILABLE SPECIFICATIONS RELEASED	

Note: MLB=Material, Labor and Burden

Figure 7.7 Contingency policy.

determines the form the estimate will take, establishes the basis for material pricing, labor content estimate, and the long-range impact on facilities and capital equipment.

In government contracts there may be special provisions for placing work with small or minority businesses in economically depressed areas, or clauses that parts or materials can be obtained only from government-qualified sources. Careful attention should be given to such provisions specified in the request for quote when formulating a bid strategy.

Understanding of estimate requirements implies that management and the estimator understand all new processes and manufacturing technologies that may be introduced if the job is won. The costs that are associated with training, technical support, safety, toxic waste handling and disposal, etc., can be substantial and should be recognized. Such costs may force companies with limited resources to go outside for these products and services.

7.1.1   Estimate or Bid Proposal Strategy

Formulation of a bid or pricing strategy follows a thorough understanding of the requirements. Such a strategy would consider the probability of winning the job when competition is involved, or if no competition is involved. Marketing intelligence concerning the prospective customer and the competition will be a major factor in developing the bid or estimate strategy. *Win probability* determines the amount of company resources that will be devoted to preparing the cost estimate. A low win probability could only justify a minimal investment of resources, or a no-bid decision.

The bid strategy defines the rules and guidelines to follow in formulating a tooling philosophy, spelling out ground rules and assumptions, assigning personnel, and how close to shave the final price quotation. The bid strategy should then be made known to the estimators and all key people involved in preparing, reviewing, and approving the cost estimate. It will be the baseline for preparing and issuing a proposal authorization or directive, specifically spelling out who is going to do what and when. Such a proposal authorization or directive should contain the following information as a minimum.

1. What is to be bid, and who is to do what in preparing the cost estimate
2. A time-phased plan or schedule for preparing the estimate
3. A list of ground rules and assumptions
4. A make-or-buy plan
5. Specifcic description of any special provisions contained in the request for quote from the customer
6. The specifications that apply to the estimate, and in case of conflict, which takes precedence, the drawings or the specifications
7. Peak production rates that processes, equipment, tools, and facilities must be able to support

The cost estimators now have a well-thought-out and well-researched plan to follow, and management is providing the necessary leadership and direction to reach the goals determined to be advantageous for the manufacturing cost estimate at hand.

7.1.2 Estimate or Proposal Plan, and Time Phasing

It seems that there is rarely sufficient time to prepare the manufacturing cost estimate. The customer wants a response within days or even hours after requesting price and delivery, and management must have time to review and approve the cost estimate. This often creates a situation requiring long hours of overtime and much pressure on the estimating team. Temptation is strong to provide expedient answers and estimates that cannot be supported when analyzed in depth by management and supervision.

There should be a proposal or cost estimate schedule plan which highlights the critical milestones in the estimate preparation and review, and maximizes the available (limited) time that is allocated for this purpose. Figure 7.8 shows such a plan for an aerospace product. This represents a minimum plan. Where time is very limited, such as a few days to a week, the plan must be much more detailed, covering the actual steps in estimate preparation and review.

Resources to Be Allocated

A key decision of how much time and money to invest in preparation of the manufacturing cost estimate is required, since this investment may not result in winning the new business. A cost estimate for a manufactured product can be used for a number of reasons other than to establish the bid price of a product for quotation. These include:

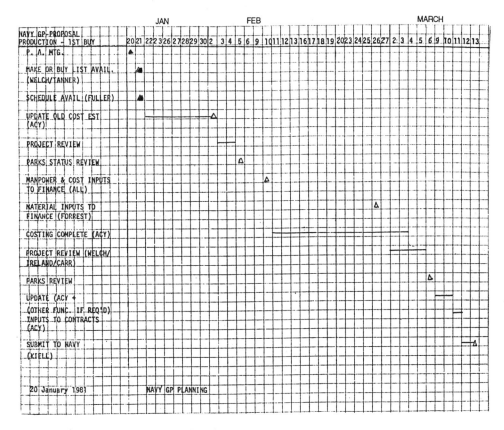

Figure 7.8 Cost estimate preparation plan.

1. To verify quotations submitted by vendors
2. To determine whether or not a proposed product or item can be manufactured and marketed profitably
3. To provide the basis for a make-or-buy decision
4. To determine the most economical methods, processes, tools, or materials for a manufactured product
5. As an aid in evaluating design alternatives
6. To determine whether or not to bid

Preparation of the cost estimate will also aid in the determination of resources to be finally expended in its preparation. One criterion used to determine the resources to be expended in cost estimate preparation is the value of the new business acquisition. Management may review a preliminary estimate and decide that no further time or effort should be expended. The preliminary estimate is deemed sufficient to satisfy the requirements.

Proposal or Estimate Milestones and Schedule Plan

Key events in preparation of the estimate should be listed, and a time span assigned for accomplishment. Key milestones for most cost estimates of a manufactured product include:

Availability of engineering drawings and specifications
Completion of the engineering parts list and the manufacturing bill of material
Preparation of the make-or-buy plan
Completion of the listing of needed shop equipment and facilities
Completion of the preproduction planning
Preparation of the listing of needed shop equipment and facilities
Preparation of the tool list
Identification of long-lead procurement items
Completion of material and subcontract pricing
Completion of manufacturing labor estimate, including support labor
Completion of the total manufacturing cost estimate
Scheduled estimate checks, reviews and approvals
Final pricing and costing
Submission to the prospective customer

Figure 7.8 shows other milestones, some of which may not be needed at all. In a small shop, the owner may have a relatively simple product line, allowing him or her to estimate the cost of new business with some degree of accuracy. The owner must make certain that he or she fully understands the bid requirements, and must cover the same proposal milestones as the larger organization, even though he or she does so in a less formal manner.

Management Control and Estimate Transmittal

As I. R. Vernon, in *Realistic Cost Estimating for Manufacturing*, a 1968 publication of the Society of Manufacturing Engineers, points out:

Fast, economical and accurate estimates require proper management control of the estimating function. Management establishes the type of estimating department that will best serve company needs, and then formulates the procedures and administrative controls necessary for efficient departmental operation.

Such controls include the screening of incoming requests for bids or estimates, usually by a committee of top managers, to make the bid-no-bid decision. If the decision is to bid, they set up the proposal schedule plan, assign personnel and resources to prepare the cost estimate, and establish the administrative routings and controls for review and approval.

An essential management control of the cost estimating function is the identification and analysis of previous estimate cost deviations versus actual costs. Records of these deviations should be maintained and plotted to determine trends and to pinpoint areas of weakness in the cost estimating function. There may be many reasons for cost estimates that are too high, too low, or simply unrealistic. One of the first areas to examine is arbitrary cuts or reductions made by management, or unrealistic contingency factors applied to an otherwise sound cost estimate which results in others winning the award. Other reasons why estimates are too high include:

1. Being deliberately high to discourage business the company does not want
2. The tendency to be overly cautious with new products, processes, technologies, etc., which the estimator is not familiar with or does not understand
3. Preproduction planning which calls for more processing steps or more sophisticated tooling than is actually required

4. Poor material pricing practices, such as failing to consider price breaks on quantity purchases of material
5. Overestimating labor costs by assuming higher-than-normal start-up costs, a shallower learning-curve slope than would actually be incurred, or by using loose labor standards in building the estimate (see Chapter 10)

High estimates sometimes result in greater profits than anticipated, but may also lose business that could have been profitable. Other problems caused by high estimates are loss of customer goodwill, greater investment of resources in cost estimating because more estimates must be processed to book a smaller volume of work, and eventually being priced out of the market.

Some of the more significant reasons for low cost estimates include:

1. Incomplete or incorrect product design information, or a new design that "grows" after the estimate, in parts count, in complexity, and in design changes that occur after the production cycle has started
2. Higher labor costs than anticipated, possibly due to delays that resulted in production stoppages, higher-than-anticipated rework due to design or tooling problems, or poor planning in the preproduction phase
3. Higher material costs than anticipated, due to such factors as unplanned material price increases, design changes that cause material requirements to change, and higher-than-planned scrap and line losses

How the Estimated Should Be Transmitted

When the cost estimate for a manufactured product is transmitted to the potential customer, it should always be by written quote or other formal means of transmittal. The price and delivery may be given verbally, but should always be followed with the confirming written quote. The formal quotation should list and explain all assumptions and contingencies, and for how long a time the price given in the quotation is valid.

Qualifications and Caveats

The transmitted price and delivery quotation should always be given in a manner that is responsive to the customer's request for price inquiry. For example, if the price quotation is a bid on government work, specific cost breakdowns, usually by work breakdown structure, are requested in addition to end-item price and delivery. Failure to provide this breakdown results in a nonresponsive proposal.

Cost estimates that are developed by companies that do business with the government often require a government audit before a final contract award. Such audits can be very upsetting for the company and the estimating department that does not have good records, cannot show step by step how costs were developed in the estimate, or that cannot show actual price quotations for major items to be purchased outside.

Final Price Negotiation

In most instances, the price and delivery quoted are either accepted by the customer, or the work is done by someone else. There is no final price negotiation. On the other hand, the larger the job in total dollar price, and the longer the production run, the more likely there will be a negotiation price. In such negotiations, profit margins and contingencies may be reduced in order to obtain the job. The estimated cost to actually do the work should never be part of the negotiating process.

The negotiating process is beyond the scope of this chapter, but it requires the best talent the company has in management and in cost estimating. They must have sufficiently detailed knowledge of the cost estimate and how it was prepared, be able to answer any questions, and face up to any challenge that is presented. A successful negotiation should result in a contract award very close to the price and delivery presented before the negotiations began. Additional negotiations may occur after contract award should there be a change in work scope, such as accelerated delivery, or an increase in number of units to be produced.

7.2 DEVELOPING THE MANUFACTURING PLAN

7.2.0 Review of Product Requirements

The manufacturing cost estimate must begin with a manufacturing plan. The thoroughness and accuracy of this plan determines to a large measure how good the estimate will be. The cost estimator must be able to concept a workable manufacturing plan for any product the firm manufactures. Such a plan must begin with an understanding of the product requirements.

As the cost estimator goes through the drawing package to do the preproduction planning, or what we know as the manufacturing plan, producibility problems may be apparent. In such cases, the problem should be noted, and if time permits, a cost trade study should be initiated. This is to determine if the desired change would generate sufficient savings to offset the cost of making the change. Producibility changes which take place after the drawings are released may not be as attractive as they appear initially, after the impact on schedule, tooling, retrofit, and the cost to change the engineering are all considered.

A checklist for reviewing producibility of an electronics product or assembly represents the kind of questions the cost estimator should be asking as he or she goes through the drawing package:

1. Does the dimensioning facilitate manufacturing, and are the tolerances realistic?
2. Is all marking and stenciling defined and visible?
3. Are assembly notes complete and definitive?
4. Is internal wiring critical? If so, is the location of the wiring specified?
5. Are test points and adjustments accessible?
6. Is harness development required? If so, can the harness be fabricated outside the unit, and installed later as a subassembly? Does the wire run list contain wire length information?
7. Does the design lend itself to mechanized or automated assembly techniques?
8. Does the design avoid the need for select-at-test component matching?
9. Are component parts accessible for assembly? Can all assembly and wiring be done without restriction?
10. Can required testing be performed without disassembling the unit?
11. If wire wrap is used, does the design facilitate automatic assembly?
12. Are standard connectors and assembly hardware used?
13. If circuit card assemblies are installed as part of the assembly, are they designed to plug in, or must theybe wired in?
14. Are there mechanical loads such that printed circuit epoxy glass boards are in compression?

15. Has consideration been given to using printed circuit flex cable, or molded ribbon wire cabling instead of hard wiring of the assembly?

As a minimum, the cost estimator should review the drawings of a fabricated product to ensure that they:

1. Are dimensioned and have datum surfaces that are compatible with accepted machining and fabrication practices
2. Have sufficient stock allowances on castings, forgings, and stampings to provide for any mismatch or distortion which may result from heat treating
3. Have maximum allowable tolerances on nonfunctional features and characteristics
4. Have realistic tolerances on functional characteristics
5. Have adequate provision for clamping and locating
6. Provide sufficient clearance and access for the assembly of all component parts
7. Call out standard parts and materials, which can be processed and assembled using general purpose machines, equipment and tools

The estimator must next prepare a manufacturing bill of material from the parts list, and the make-or-buy plan that has been established. Next the estimator proceeds to concept the steps and sequences of manufacture for each part, subassembly, and assembly to be made in house. Then, for each step in the process, he or she determines the machines, equipment, and tools that will be needed. Finally, he or she determines setup and run times for each step or sequence in the manufacturing process.

7.2.1 The Make-or-Buy Plan

Purchased parts and materials, as well as subcontracted items, can account for as much as 70% of manufactured product cost. It is therefore important that the initial make-or-buy analysis be made by a qualified cost estimating professional based on knowledge of the plant capability to manufacture or not to manufacture the items on the bill of material. For example, if the company is primarily an assembly house, then all fabricated and machined metal parts would be classified as "must buy" items, in addition to hardware, and bulk material items such as paints and solvents. Only those items that can be either made or bought will require a detailed analysis from cost and shop load standpoints. In larger companies, a formal make-or-buy committee with key people from all of the functions is chartered, chaired by the senior operations or manufacturing executive. This committee then makes final make-or-buy determinations based on recommendations from all disciplines concerned.

Must-Buy Items

Obvious "buy items" based on shop capability to perform the process of manufacture are the easiest decisions to make. However, management may decide to create the process or capability inhouse to have a degree of control not possible when the work is placed outside. If the requirement exists for only a short time, we may be unnecessarily committing company resources to a process capability that would stand idle much of the time.

Must-Make Items

There are usually a number of fabrication, assembly, and testing operations on any manufactured product that are critical to its manufacture and performance in the field. These are the operations and processes that must be done inhouse to ensure product integrity and control. Such operations usually include product final assembly and test, and fabrication and assembly of close-tolerance or mating parts, among others.

Items That Can Be Either Make or Buy

These items are the ones requiring investigation and analysis prior to the decision to make or buy. Usually decisions on these parts and assemblies are based on cost, promised deliveries, and shop capacity or load. Cost trade studies are made, quotes are obtained from various potential outside sources, and the final make-or-buy decision is made either by senior management or the make-or-buy committee.

7.2.2 Outlining the Manufacturing Sequences and Processes

The operation process chart is the best way to clearly visualize all of the steps in the manufacturing process. This chart depicts graphically the points in the process where materials are brought into the process, and the sequences of all manufacturing, inspection, and test operations. This chart may also contain such detail and information as the standard time for each operation, production equipment and tooling required for each operation, and applicable process specifications.

With the completed operation process chart, the entire process of manufacture can be visualized. The process can then be reviewed and analyzed for optimum sequencing, alternative methods of fabrication, assembly inspection and test, and most efficient methods of production. Equipment and tooling requirements at each step in the flow can be readily envisioned and recorded.

The Operation Process Chart

The principles of operation process chart construction are shown in Figure 7.9. A preprinted chart format is not recommended, because of the wide range of use and application. Symbols used in constructing the operation process chart include:

- ○ Operation: Occurs when an object is intentionally changed in any of its physical or chemical features, is assembled from another object, or is arranged for another operation, inspection, or test.
- □ Inspection test: Occurs when an object is examined or verified for quantity or quality in any of its characteristics.

All steps should be listed in proper sequence for each part or component, working vertically from top to bottom. The major assembly or component is always shown at the far right, and all other components are allotted space to the left of this main component or assembly. The presentation is similar to that of a mechanized assembly line, with parts and material, and subassemblies, fed into the top assembly in proper sequence and at the correct point in the process. The operation and inspection descriptions should be brief, utilizing shop terminology such as drill, tap, ream, weld, assembly, solder, and test.

Figure 7.10 shows an operation process chart for a printed circuit card assembly. Time values, when assigned to each operation, should be broken down into setup and run times. Other useful or amplifying information, such as production machines and equipment used, tools required, and special process provisions for each operation, completes the operation process chart.

The completed operation process chart highlights material, operations, inspections,

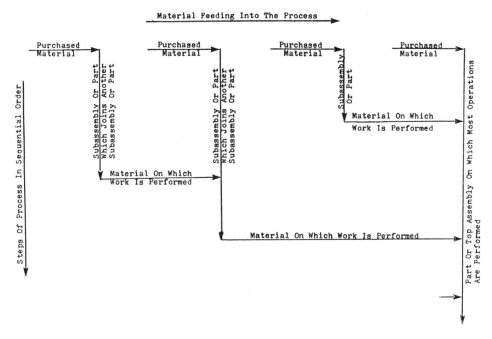

Figure 7.9 Principles of construction of an operation process chart.

tests, and time. The visualization process needed to construct the operation process chart is fairly straightforward. The primary requirement, besides knowledge of the principles of chart construction, is a working proficiency or knowledge of the various processes involved and the ability to read and correctly interpret the requirements of the engineering drawings. The process of visualizing the steps of the product manufacture, and construction of the operation process chart, defines the manufacturing plan for cost estimating.

New Manufacturing Technologies

Advanced manufacturing technologies which are new to the company should be identified in the manufacturing plan for the cost estimate. These technologies can mean new opportunities for the company to reduce costs and gain a competitive edge. They also present the risk of cost overruns, and significant schedule delays, if they are not properly researched and understood by management at the time the cost estimate is prepared.

Some new technologies in today's world include laser welding and soldering, fiber optics, composite structures, ceramics, electron beam welding, and more. As a minimum, the cost estimate should identify the cost of new equipment and facilities for each new technology; the personnel skills required; training that would be required for the existing workforce; the length of time that would be required to become qualified and proficient in the technology; any special safety or waste disposal requirements; and what long-term potential exists for work beyond the job that is being quoted. In addition, the possibility of subcontracting this work should be explored, especially if the need for the new process or technology is a short-term need. It makes little sense, for example, to set up a facility for coil winding, to invest in the winding machines, capacitance bridges, and Q-meters, to hire engineers proficient in magnetics, hire setup people who are qualified, train operators

POWER CIRCUIT BOARD (S/A-100) AMPLIFIER

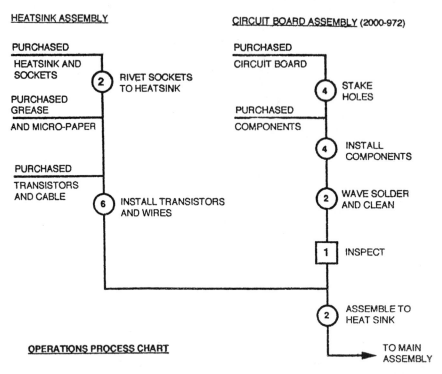

Figure 7.10 Operation process chart for a printed circuit board.

in coil winding, plus provide the facilities to varnish and encapsulate, when the coils can be purchased from a vendor who specializes in coil winding and has years of experience. This is true even if cost trade studies show that the coils would be produced cheaper in house.

7.2.3 Equipment and Tooling Requirements

Capital and other production equipment costs to support the production of a new manufactured product often can be quite high. This is especially true when no comparable equipment exists within the company. A good example of this would be a metal finishing or plating line, which would also require waste treatment and disposal capability. Many times the cost of new equipment to manufacture a new product, or to continue manufacturing an existing product line and remain competitive, is such that it cannot be written off over the life of a single or even several product lines. In such cases, management may choose to invest in the new equipment with the knowledge that some risk may be involved, or it may decide to subcontract the work to outside sources.

Selection of Manufacturing Equipment and Machines

Once the process has been defined, the cost estimator must determine what equipment and machines will be required at each step of the manufacturing operations. In most cases this can be done quickly and easily, as the manufacturing process is centered around existing production machines and equipment, or standard equipment which can be readily obtained (see Chapter 3). Should it be decided to purchase new equipment or machines for the job being estimated, a decision must be made as to how to pay for this equipment. It may be chargeable to the product being estimated, and written off against products delivered. It may be charged to overall plant equipment depreciation, which would be a lower cost to the product being estimated but would increase the overhead charge against all other products in the plant. In many cases, a review of the make-or-buy decision is in order, and the new equipment cost and charges must be compared to buying some of the parts from an outside supplier. A complete understanding of the accounting practices in the manufacturing plant is an important requirement for good decision making.

Identifying Tooling Required

Tooling requirements are identified concurrently with the machine and other production equipment requirements (see also Chapter 8). Working from the established tooling philosophy, the cost estimator can determine the tooling approach to take. He or she must know whether to plan the job around minimal shop-aid type tooling, or a full hard-tooling-type approach, or even a no-tooling approach, by fabricating on setup. It is essential that this tooling philosophy or approach be fully defined at the outset. Once the tooling philosophy has been determined, tools required at each step in the process are defined and listed. A drilling operation, for example, may require a drill jig and a holding jig or fixture, whereas an assembly operation may require weld fixtures, holding fixtures, and/or brazing fixtures. A careful part print and assembly print analysis is required, with special attention given to the manufacturing step or operation, part or assembly shape and geometry, and peak planned production rate. In planning tooling requirements, the cost estimator should keep in mind that tools fall into one of several categories:

Special-purpose tooling: Tooling peculiar to the product it makes. This usually required tool design, followed by special fabrication. It is good for the life of the life of the production run only. It can be considered a "one time, non-recurring charge" to the product, or pro-rated over the products delivered.

General-purpose tools: Tools which can be used in the production of many different products. No special design is required. This kind of tooling is commercially available and is much less expensive than special-purpose tooling. Some examples would include small hand tools, tool balancers, and standard assembly and motion economy devices. They may sometimes be charged to the product being estimated, but more often are considered part of the manufacturing overhead and need not be added to the product cost.

Expendable tools: Such items as cutters, end mills, drills, taps, broaches, and reamers. These tooling items are actually consumed over the life of the job or production run. They are usually considered part of manufacturing overhead, and are not charged to the particular job.

It should also be mentioned that the tapes, software, and the programming req____ for Numerically Controlled (NC) equipment is also a tooling cost. Tooling c____ hence product costs can be kept as low as possible by following the rules outl____

1. Utilize standard tools and tooling components whenever possibl____
2. When special tooling is required, utilize the design conc____ well-engineered tools.

3. When special tools and fixtures are required, use low-cost tooling materials in place of tool steel wherever possible.

7.2.4 Determination of Facility Requirements

The operation process chart, which describes the manufacturing process used to build the product, also defines the layout of the factory to support product manufacturing. The amount of floor space required is a function of the size of machines and of the equipment in the manufacturing process, the area required to work and service the machines and equipment, and the number of units of product in flow and in process at the planned peak production rate.

Once floor space requirements have been determined, the next step is to identify special facility requirements. These may include clean rooms, requirements for hoods, vents, and other similar openings in the roof or along outside walls, special power requirements, and requirements for sound and noise protection.

The operation process chart determines product flow and how dedicated areas would be laid out. If time permits, an overall block layout plan and most of the detailed area layouts should be developed. The primary purpose of this is to identify to management the impact on production facilities should the new job be won. No manufacturing plan should be considered complete until this has been done (see Chapter 3).

7.2.5 Putting the Total Plan Together

In summary, the manufacturing plan is the baseline for the manufacturing cost estimate. Identification of the process, the sequences of manufacture, the machines, the tools, and the facilities needed lead logically to the next steps in the cost estimate: the estimate of the setup and run times for each manufacturing operation, cost of material, etc.

7.2.6 Bibliography

Knox, C .S., *Cad/Cam Systems, Planning and Implementation,* Marcel Dekker, New York, 1983.
Maynard, H. B., *Industrial Engineering Handbook,* 3rd ed., McGraw-Hill, New York, 1971.
1984 N/C CAM Guidebook, Modern Machine Shop, January 1984.
Practical Machining Principles For Shop Application, The Machinability Data Center, Metcut Research Associates, Cincinnati, OH, 1981.

a government request for proposal early in the system evolution process. Figure 7.6 shows a design layout concept for an improved mortar round for the U.S. Army. There is sufficient information in this layout drawing to estimate the cost of this mortar round in production. This estimate must be as accurate as possible, since the decision to proceed with engineering development, and subsequent production of thousands of rounds, hinges on how well the manufacturing cost estimator does his or her job.

We next need to define the minimum product design information needed, and the form it should be in.

Product Definition

Before a significant investment is made in design and development of a new product, the product must show a real potential for sales, as evidenced by low estimated manufacturing costs, or because the new product wins by better performance in a competition with designs by competing companies. Companies that manufacture computer systems, or electronic devices or systems, or even commercial appliances and consumer-type products, may wish to select among several feasible new product design concepts to determine which would be the least expensive to manufacture. In such situations, sufficient information must be available to the manufacturing cost estimators to develop a valid cost estimate, yet the time and expense of developing and documenting a fully mature design is avoided.

In the high-technology companies, many new designs are concepted, yet few of these ever reach volume production. Anywhere from one to a few hundred units may be manufactured, and these units will change significantly in configuration between the first and the last unit built. In such companies, concept estimating is the norm for the manufacturing cost estimator. It is important that the estimator know and clearly understand what minimum engineering information is needed to prepare a valid concept estimate, and how to go about developing such an estimate. A concept estimate, more than any other kind, must of necessity draw heavily on the experience, ability to visualize, and ingenuity of the estimator.

Information Needed

Baseline design concept information can take many forms. It can range from reasonably good design layouts to freehand sketches on note paper. Often isometric or exploded view drawings are used in preference to design layouts. Figure 7.11 shows an exploded view of a fairly complex nutator assembly. This is an excellent way to show the parts and how they go together in an electromechanical assembly. A design layout would not provide sufficient information for the estimator to understand part configuration and how the assembly goes together. With this exploded view and a preliminary parts list, a reasonable manufacturing cost estimate can be developed.

Photographs of engineering models are also helpful, even though the final production design of the product may be somewhat different. A technique often used is to estimate the cost to manufacture the new product by similarity to existing products. A circuit card assembly, for example, may be 70% more complex and have greater parts density than a known circuit card assembly. A machined casting may be similar in configuration to one currently in production, but be only 80% as difficult to machine. In the manufacture of aircraft, the cost per pound of the airframe is one method used in determining the manufacturing cost of a new aircraft, long before the first drawing is released, yet the cost so estimated will come remarkably close to the actual cost of manufacture.

In addition to drawings, isometrics, sketches, and more, it is a mandatory requirement

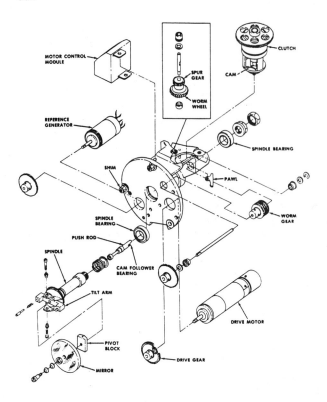

Figure 7.11   Exploded view of nutator assembly, showing sequence of assembly.

that there be a parts list or bill of material, reasonably close to the parts list that is released safter final product design. This preliminary bill of material is used by the estimator to determine the make-or-buy plan, and as the basis for developing product material and subcontract cost. If major parts or assemblies are to be subcontracted, preliminary procurement specifications and subcontractor statements of work must be developed.

   If preproduction units or models are to be built in a manufacturing shop rather than in an engineering model shop, the quantity to be built and the schedule dates when these units would be needed must also be included.

   Product test requirements (if any), all specifications and other requirements the product must meet, plus any new or unusual process requirements, must be provided by the designer to the cost estimator. New manufacturing technologies must be clearly identified, with details of the process requirements as well as any new specifications and standards that would apply.

   Finally, there must be a program schedule showing the major design and development milestones, including final release of manufacturing drawings. The schedule should also reflect the manufacturing start-up activities, including the preproduction and production planning milestones, tooling milestones, long-lead and other material procurement, the acquisition and bringing on line of any facilities or production equipment, low-rate initial production build, and the build-line up to rate (peak) production.

Producibility

The estimator's job is to learn as much as possible about the newly designed product, or product concept, and to assist the design team by providing cost estimates of the various design alternatives. The estimator must certainly point out design approaches which have caused manufacturing problems in the past. In short, the cost estimator, with his or her knowledge of manufacturing, should be an active and recognized participant in the design and development process. Cost trade studies should be made between design approaches such as castings versus forgings, sheet metal structure versus machined hogouts, chemical process finish versus paint, or hard-wired assembly versus flexible cable or ribbon cable.

7.3.1   Cost Estimate Development

Once the product has been defined sufficiently, and other necessary information such as parts lists, specifications, the plan for design and development, hardware requirements, and other vital information is available, the manufacturing cost estimator can begin to develop a preproduction plan, a make-or-buy plan, and a manufacturing bill of material. The purchasing department buyer or material estimator must immediately begin to price the *buy* and *subcontract* items on the bill of material, establish material procurement lead times, and identify any long-lead items. Any needed equipment and facilities over and above those available must be specified and priced.

The preproduction plan will identify equipment, facilities, and the tooling required. Nonrecurring start-up costs are estimated first by the cost estimator, and include initial manufacturing engineering, process planning, and tooling costs. Fabrication, assembly, inspection, and test labor to actually build the product are estimated next. The cost of recurring support labor such as manufacturing engineering, liaison engineering, and tooling maintenance must then be added to the recurring touch labor estimate.

Customer or internal delivery schedules will dictate peak rates for production, number of shifts, and multiples of equipment and tooling required. Time required for production start-up, and to progress from low-rate initial production to peak rate, will be determined by considering the material procurement, receiving, stocking, and kitting lead times, by tool design and fabrication lead times, setup and shake-down of new equipment and facilities, and time required to hire, train, and build up the necessary workforce.

The manufacturing labor estimate should be completely documented, with all notes, backup data, and calculations carefully retained. All documents and data should be clearly labeled and dated and should include even very rough, handwritten notes and sketches. This file should enable the cost estimator and others to go to the file long after the estimate was prepared, with the backup data available to support it, and answer questions about the estimate that may arise. Such data are very useful in preparing estimates for similar products, or to update an earlier estimate.

In developing a concept estimate for a new manufactured product, it is absolutely essential that *all* of the ground rules and the assumptions used in preparing the estimate be fully documented. This information tells management and others the premises and constraints used in preparing the estimate, and goes far in explaining why things are as they are. Some examples might include:

1.   The product baseline for the production estimate is the design concept as defined in the layout SK drawing dated 12/11/85.

2. Tooling and facilities are estimated to support a peak production rate of 500 units per month.
3. All work is to be performed on a single shift, ten hour work day, five days per week.
4. The estimate assumes the availability of the Harrison engine lathe and the DeVlieg mill.
5. All labor estimates were prepared using similarity to other products of similar size and configuration.

Similarity with Other Products

The similarity method is perhaps the most widely used way to estimate manufacturing labor costs when working with new product concepts. This method has the cost estimator select parts and assemblies that have been manufactured in the past that have features similar to the new product. Working from the actual or standard cost for the similar parts and assemblies, the estimator can develop the estimated cost for manufacturing the new product.

A circuit card assembly, for example, might be of the same overall size and shape as one currently in production, but the component density may be 50% higher. In this case, the assembly labor to load the new card may be 50% higher than the time it currently takes to load the existing production circuit card assembly. A machined casting may have some 25 drilled and tapped holes, while a new design casting may be similar in size, weight, and shape, but have only 5 drilled and tapped holes. The estimated reduction in time to machine the new casting might then be estimated at 20%.

The percentage that the new product differs from the similar known product is a judgment call on the part of the cost estimator. If there is any doubt as to the validity of the estimated percentage similarity, it should be verified with the shop supervisor who must actually do the work. Estimating by the similarity method is reasonably accurate, and enables the cost estimate to be completed rapidly. It is easy to support such estimates to management and in final pricing negotiations.

Weight, Volume, Parts Count

In the estimating of aircraft manufacturing costs, weight is a commonly used variable. Weight alone is seldom enough, and speed is almost always included as a second variable for aircraft airframes. According to a Rand Corporation report on *An Introduction to Equipment Cost Estimating* (December 1969), one cost estimating procedure for aircraft uses all of the following in their parametric equations:

Maximum speed at optimal altitude
Maximum speed at sea level
Ye͙ ͏f first delivery
      ͏rframe weight
       ͏in airframe weight from unit 1 to unit n
        f installed equipment
         ͏ight
          complexity factor

            following characteristics were considered for inclusion as part of the
            ͏re, although they were not used:

Maximum rate of climb
Maximum wing loading
Empty weight
Maximum altitude
Design load factor
Maximum Range

An airframe typically changes in weight during both development and production as a result of engineering changes. For example, the weight of one fighter aircraft structure assembly varied as follows:

Cumulative plane number	Airframe Unit weight (lbs)
1–11	8456
12–186	8941
187–241	8541
242–419	9193

Since labor hours are commonly associated with weight to obtain hours-per-pound factors, it is important to obtain weights that apply to each production lot when airframe weights by unit are not available.

Depending upon the industry, similar factors such as weight, parts count and other significant variables may be used to develop cost estimates.

Concept Estimate Preparation

The cost estimate for a new manufactured product may be developed in much the same manner as any other estimate, the difference being the degree of product definition and the lack of past actual costs to compare the current estimate with. Labor estimates for the fabrication, assembly, and test operations can be developed from standard times, from similarity to other products from weight or volume factors, or simply by professional judgment and experience on the part of the cost estimator. Estimates for support labor, tooling, facilities, etc. are developed as in any regular estimate. The primary difference in a concept estimate for a manufactured product and an estimate based on complete engineering documentation is the detailed listing of the ground rules, assumptions, and contingencies that were used in preparing the concept estimate.

7.3.2 Contingencies and Assumptions

In concept estimating, the contingencies and assumptions are the critical factors in preparing the cost estimate. This means that the estimate is only as sound and valid as the assumptions and premises used in its preparation. Examples of ground rules, assumptions, and contingencies are:

1. Assume the product baseline for estimating purposes is the current engineering concept.
2. Assume a peak production rate of 15 units per day, achieved 6 months after start- up.
3. All work to be performed on a 5-day, 8-hr/day standard work week.
4. No new machines or production equipment will be required.
5. Assume a minimal, shop-aid type tooling approach.

In addition to the above, there should be a projected engineering change curve indicating number of engineering changes to expect on a month-by-month basis after the engineering drawings are released.

Parts and Complexity Growth

As a new design for a manufactured product evolves and develops, it tends to grow more complex, and the total number of parts that make up the unit or product tends to grow, sometimes exponentially. This is due to a number of things, including the transition from concept to fully designed product. This brings out details and features of the product not envisioned at the concept state. Another may be a tendency to add functions not thought of at the concept state, or it may be found that the original design approach simply will not work. Consideration of customer requests or preferences are often factors.

Other factors that increase costs to manufacture are closer tolerances, a more expensive finishing process for the piece parts, or more elaborate testing of the finished product than was envisioned in the design concept stage. Rarely does the product get less expensive to manufacture as the design evolves. A concerted product producibility engineering effort during design and development will go far in eliminating much of this growth, but will never entirely eliminate it.

As a result of this tendency for parts count and complexity increase, the manufacturing cost estimate must either contain a cost factor increase to compensate for this growth, or state clearly that the estimate reflects only the cost to manufacture the concepted product, and any growth in the design is a change which must be costed out later.

Design Oversights

In the design and development of any new manufactured product, there will be errors or oversights which do not become apparent until the product is in production, or in the initial stages of production start-up. It may be found that the product does not perform as intended, or that it fails to pass qualification testing. Perhaps the decision was made to proceed with production on management risk ahead of production unit qualification tests in order to meet tight production schedules. Such problems can have major cost impacts, and may even cause the new product to be taken out of production for an indefinite period. Other design deficiencies may be of a less serious nature, but nevertheless will impact production and production costs. The lesser design oversights can be predicted with some degree of accuracy and usually consist of the following:

1. Design changes that are requested by manufacturing to ease fabrication and assembly. These might include loosening of tight tolerances, relocating a tapped hole to make it more accessible, adding a missing dimension, adding notes to the drawings, etc.
2. Design changes required to make the product perform as intended, sometimes called "make play" changes. These might include modifying the configuration of a moving part, adding a ground wire, adding a fuse to a circuit, or adding a production test requirement where none existed previously.
3. Record changes, which incorporate many of the changes described above into the formal drawing package.

In some of the high-technology industries, and the aerospace industry in particular, design changes well into production are the norm, not the exception. Samuel L. Young, in his paper on "Misapplications of the Learning Curve Concept," published by the Institute of Industrial Engineers in *Learning Curves, Theory and Application,* edited by Nanda and Adler (1982), states:

Production is characterized by poor starting documentation. The first planes or missiles are modified many times before they are acceptable. Engineering specifications are modified after the fact to reflect changes made first on the hardware. Many programs are put into production concurrently with development effort on identical hardware. It is not unusual to find hardware being produced simultaneously on both development and production contracts with the former far more costly than the latter effort. The whole approach results in extensive rework of previous efforts to make the end-items perform to specification.

As indicated earlier, such design changes are predictable, and must be considered by the manufacturing cost estimator at the time of the original concept estimate. Such predictions should take the form of an engineering change projection curve.

Engineering Change Projections

Projections of engineering changes with time at the start of a newly designed product into production can be made with a high degree of accuracy based on experience with previous programs. These change projections are normally prepared by the product design staff, and take the form of a graphical curve as shown in Figure 7.12. As would be expected, the number of changes would be high in the production start-up cycle and would gradually diminish with time until about the ninth month after initial drawing release, when engineering drawing changes would be minimal or nonexistent. The shape of this curve, and the time required to run its course, are functions of the product, the industry, and many other variables. That is why only the design department can draw this curve, based on knowledge of the firmness of the design at the time of drawing release, the condition of the drawing package, and the probability of major changes due to performance failures.

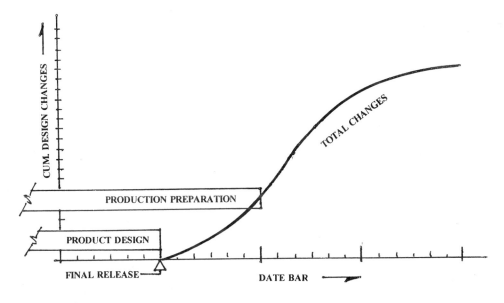

Figure 7.12 Design change curve.

Engineering drawing change curve projections are a basic requirement if the manu-facturing cost estimator is to develop a good estimate of manufacturing start-up costs. They are especially needed if the estimate is based on concepts for a new manufactured product. If such curve projections have not been done in the past, there will almost certainly be resistance to providing them on the current estimate. In spite of this, both management and the manufacturing cost estimator should insist that these projections be provided from this point forward.

7.3.3 Examples of Concept Estimates

HAIL Mortar Round

The high-angle improved-lethality (HAIL) mortar round is being developed for use with the improved 81-mm mortar system for the U.S. Army. This estimate involves costing of the 1984–1985 advanced development activities for the HAIL mortar round. Manufacture of rounds for the initial *test phases*, 70 rounds, will be done by a subcontractor. Then, during the advanced *development phase*, 765 rounds will be molded and assembled inhouse for delivery to the Army. A subcontractor will load the mortar bodies. Peak production rate will be 50 rounds per week.

The subcontractor who manufactured the initial 70 rounds will provide body molds for in-house manufacture of the 765 rounds. All other support equipment tooling will be provided in-house. The schedule allows 9 months to prepare in-house production facilities, and 6 months to manufacture the 765 rounds. Figure 7.2 is the design concept layout for the HAIL mortar round. The following ground rules and assumptions apply to this estimate:

1. Manufacturing labor standards were set at unit 1000 on an 85% learning curve, due to the developmental nature of the program.
2. Molding estimates are based on the subcontractor's current unit cost after approx-imately 50 units.
3. Supervision is estimated at 10% of assembly labor, and production control at 25% of assembly labor.
4. All estimated labor hours are based on an 8-hr work day, 5 days per week.
5. All changes to the concepted product base line will be costed and funded separately.

Nonrecurring labor hours were estimated as follows:

Manufacturing engineer: 3295 hr
Production control: 315 hr
Tooling: 1587 hr

Recurring labor hours were estimated to be:

Assembly: 1895 hr
Manufacturing engineer: 2927 hr
Production control: 433 hr
Tooling: 919 hr

Facilities requirements were determined to be as follows:

Inert manufacturing assembly: 5000 ft^2
Pyrotechnic assembly: 900 ft^2
Pyrotechnic bunker storage: 500 ft^2

Capital equipment, including material handling and special high-cost tooling not covered by contract, is as follows:

Capital Equipment

Order		Cost	Est. weeks lead
1.	Mold press	41200	12
2.	QC test, mold press	25750	12
3.	Compounding Equipment	20600	21
4.	Electrostatic paint booth	9000	32
5.	Lathe	1000	32
6.	Cooling tower	15500	21
7.	Curing oven	3000	32
8.	Digital scale	5000	24
9.	Air compressor	2500	24
10.	Arbor press (avail.)	—	—
	Subtotal	$132,550	

The following high-cost tooling items are priced for subcontract build, but are covered by procurement.

1.	Mold cavity inserts, 3 sets	$15,500	16
2.	Mold	40,000	16
3.	Tooling for lathe (spin crimp oper.)	8,000	16
	Subtotal	$63,500	

Estimated material handling requirements are as follows:

1.	Wire carts (6)	$ 3,000	16
2.	ESD tote boxes (40)	2,800	12
	Subtotal	$ 5,800	
	Total	$201,850	

All *material* requirements for manufacturing to be provided by the Engineering Department, and are estimated separately.

Commercial Computers and Data Processing

The method of estimating a new product is shown in the outline below. The outline is a general guide and will vary depending on the complexity of the product or system, schedule, quantity to be produced, and technology.

Request design data.
Prepare cost estimating outline.
    Milestones
    Purchasing
    OEM
    Production
    Test
    Field engineering

Analyze material and labor inputs.

Use independent judgment where voids exist.

Prepare MLB (Material, labor, and burden).

Determine learning-curve base for each category:

    Example:

        Material—Cum. Avg. or first-year block

        Labor—Production Cum. Avg. or T_{100}

        Test—Usually T_1

Extend MLBs, applying labor rate, published burden rates, and contingencies.

Calculate Cum. Avg. and Adj T_1 using published slopes for material and labor.

Cumulative average and adjusted T_1's are determined by applying a learning-curve slope to each cost item for material and labor.

    (1)   Determine slope for each cost item from tables.

    (2)   Calculate Cum. Avg. and true T_1 for each cost item.

    (3)   Combine cost items by slope, separating material and labor.

    (4)   Total Cum. Avg. and true T_1 for material and labor.

    (5)   Find Adj. T_1 by dividing Cum. Avg. by true T_1 and "roundingoff" to the nearest percent slope, thereby eliminating fractions.

The adjusted T_1 is the means of

    (1)   Calculating block averages

    (2)   Calculating 75%, 100%, and 125% quantities

    (3)   Simplifying MLB revisions

and results in the composite slope for material and labor used for monitoring costs after start of production.

Calculate block averages using Adj. T_1's for labor and material.

Document ground rules:

    Preliminary review—for supervisor and contributors

    First revision

    Management review includes:

        Pricing strategy

        Maintenance

        MLB review to the detail level

            Ex. How many PIBs?

        AWW?

        Unique costs

Note management recommendations and adjust final report accordingly.

Submit MLB to Office of Pricing.

7.3.4 Bibliography

Tanner, J. P., *Manufacturing Engineering, An Introduction to the Basic Functions,* Marcel Dekker, New York, 1985.

7.4 LEARNING CURVES

7.4.0 Introduction to Learning Curves

With the end of World War II in 1945, there was for the first time in history a mass of production cost data covering a wide variety of industries involved in quantity production.

This included many different types of products, with some of the same products being produced simultaneously by multiple suppliers. There was an extreme focus of both government and private industry to build good products for a low cost at high production rates for an extended period of time. Stanford University was tasked to summarize this cost data, and to look for significant factors and trends. The university hired some of the better people from industry who had been actively involved in this production. One of the most significant things that came out of this study was the "learning curve" phenomenon. They discovered that, with the proper effort and motivation, the labor cost per unit continued to decrease as production quantities increased. Different products and processes varied somewhat as to their slope, but they all seemed to match a standard logarithmic formula, of the form $Y = e^{-x}$.

For a manufactured product, the cost of value added (that is, labor hours per unit) declines predictably each time accumulated experience doubles. As an example, if the fifth unit built in a new process required 100 labor hours, and the tenth unit in the series required 75 hours, this 25% reduction represents what is commonly called a 75% learning curve. Similarly, if each doubling of experience brought about 20% reduction, the process would have an 80% learning curve. A learning curve is, in effect, a rate-of-improvement curve.

Labor reductions continue indefinitely as long as production continues. Such declines are not automatic; they require management and often capital investment. Learning curve effects can be observed and measured in any business, any industry, any cost element, almost anywhere. Key points about the learning curve concept are:

1. The learning curve itself is beyond question.
2. The curve results mostly from management action and management learning.
3. Costs will more surely decline if it is generally expected that they should and will.
4. Full understanding of all the underlying causes is not yet available.
5. The learning curve is so widespread that its absence is almost a warning of mismanagement or misunderstanding.

Not all products have the same learning curve. Two main factors affect curve slope:

1. The inherent susceptibility of a process to improvement.
2. The degree to which that susceptibility is exploited.

The learning curve results from, and is a measure of, combined effects of both worker and management learning, specialization, investment, and scale. The effect of each is, at best, an approximation. The history of increased productivity and industrialization is based on specialization of effort and investment in plant, equipment, and tools, and so is the learning curve.

7.4.1 Types of Learning Curves

The Stanford and Crawford curves, as shown in Figures 7.13 and 7.14, are two examples in general industrial use today. Though both curves embody the same basic principle, there are differences which can affect cost projections considerably. In addition, it is almost universal practice to assume that the learning curve is linear on log-log coordinates. A closer look, however, shows that long-cycle cost trends are not always straight. A typical S curve, as shown in Figure 7.15, is often found.

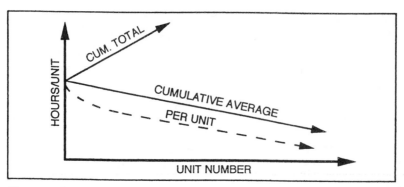

Figure 7.13 The Stanford learning curve. Cumulative average line is straight on log–log coordinates. Per-unit curve starts at higher slope, soon parallels cumulative average as units are produced. Often used in industrial applications.

Stanford Curve

The Stanford curve shows that the cumulative average (the ratio of total accumulated time to total units) starting with the first unit is a straight line on log-log coordinates. The per-unit line is derived from this average and runs approximately parallel from about the tenth unit.

Crawford Curve

The Crawford curve assumes that the unit line is effectively a straight line on log-log coordinates. The cumulative average line is derived from the unit line and runs approximately parallel from about the tenth unit. The Crawford curve results in higher projected values than the Stanford curve. The Stanford cumulative average line gives a better averaging effect than the Crawford concept. The Stanford curve is consistent with historical data and minimizes inflated cost projections.

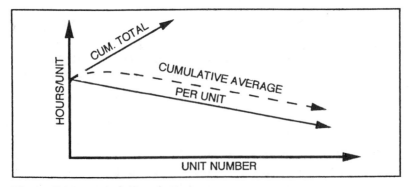

Figure 7.14 The Crawford learning curve. Per-unit line is straight on log–log coordinates. Cumulative average line starts at lower slope, soon parallels per-unit line. Gives higher projected values than Stanford concept.

The S Curve

The S-curve upward bulge, as seen in Figure 7.15, signifies production cost penalties from time compression that occurs during new product introduction. The amount of penalty can vary among products and companies, because of differences in the degree of time compression of each. Such differences can be caused by any of several elements influencing new product introduction, such as product design, tools and facilities, worker capability, supervision, and support services.

7.4.2 Learning Curve Mathematics

The mathematical development of the learning curve is not complex. Correlation and other statistical methods show that a graph of real performance data (hours per unit versus cumulative unit number) can be described with relatively high significance by the following equation:

$$Y = AX^k$$

where

Y = labor hours to produce first unit
A = cumulative average time at any unit, X
X = unit number
k = learning curve slope

The equation describes the theoretical learning curve. It can be used to describe learning curve unit time line, cumulative average time line, and total time line. Though the equation will describe actual data, the mathematical exactness of the theoretical learning curve will not permit the unit time, cumulative average time, and total time lines all to be straight lines on log-log graph paper of the equation type described.

Learning curves, therefore, may be considered in these three classes:

1. The cumulative average time line and total line are of type $Y = AX^\kappa$.
2. Only the unit time line is of that type.

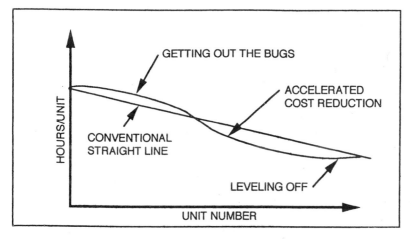

Figure 7.15 S-curve. True learning curve shape may not be straight on log–log coordinates, though straight line may approximate it well enough.

3.   All three lines are modifications of 1 and 2 above.

In practice, the learning curve is produced either graphically or in tabular form.

7.4.3 Practical Applications of the Learning Curve

The learning curve trend is a function of many variables. Factors which may affect curve slope include characteristics of the type of work to be done, program variations, and uncontrolled external factors. One of the more significant factors to consider in developing a cost estimate for manufacturing is the type of work. Different manufacturing processes can have different slopes. High-speed machining has a slope trend unlike that of conventional machining. Various assembly methods, wire harness fabrication, electronic assembly and wiring, final assembly, and testing all have different learning curve slopes. Consider the makeup of the end product with respect to different manufacturing processes in selecting learning curve slopes.

Also, to minimize distortion created by adding two or more learning curves with different slopes, segregate data as much as possible into appropriate subdivisions. If manufacturing processes such as machining versus sheet metal forming and fabrication cannot be separated, some other breakdown such as tail cone, center body, and/or forward body section might separate data of different process mixes well enough to minimize distortion.

Tooling and facility commitments affect both first unit cost and attainable learning curve slope. Influences on these commitments include the stage of product development, planned production quantities, lead-time constraints, limitations in available investment funds, and risk of not obtaining follow-on production.

For a minimum tooling approach, the expected first unit cost can be higher than if the tooling were designed for efficient rate production. Cost improvement for minimum tooling depends on improvement in shop personnel efficiency. If tooling is designed for efficient rate production, initial unit cost may still be high because of problems in shakedown. The learning curve should reflect rapid resolution of tooling problems. Speedy shakedown often results in a steep curve in early production which is impossible with minimal tooling.

Automatic test equipment in many applications can bring large cost reductions in production. If the equipment has been fully debugged before coming on line, its effect will show on the learning curve as a step reduction at the unit number where it starts to work.

Cause-and-Effect Relationships

Other factors affecting the learning curve include the following:

1.   Changes from development to pilot production and to full-rate production.
2.   Production rate constraints and dictated schedule changes. These are often caused by funding limitations which bring program stretch-outs. Higher rates allow more units in work at a given time, allowing better labor utilization. Lower rates mean fewer units in work, thus less efficient labor utilization. Lower rates also result in smaller lot sizes, increasing the number of setups required for the same number of units.
3.   Number of setups. Detail parts fabrication incurs both set-up and run costs. Setup costs occur each time a lot is fabricated, but are independent of the number of parts in the lot. Run costs are incurred each time a part is made. Average cost of

parts in a lot (total cost divided by the number of parts made) depends directly on the number of parts made with each setup.

4. Production interruptions for any reason cause a loss of learning.
5. Personnel changes and turnover.
6. Management learning and management action.
7. Timely resolution of manufacturing problems.

Pitfalls to Be Avoided

When applied correctly, the learning curve is one of many valuable tools available to the manufacturing cost estimator. However, some caution should be exercised in the selection and use of labor cost data in plotting the learning curve.

1. If a company receives a larger portion of its raw material in a finished state, its labor input per unit of product should decline. When this decline is merely the result of shifting labor input from the company to the plant of a supplier, it would be erroneous to consider the decline as a real reduction in labor hours. The same work is being performed in total, and no net saving has been realized.
2. It may be possible to generate a direct labor saving of 1000 hr by spending 2000 hr on additional tooling and manufacturing engineering. It is obvious that there would be no real savings as a result. It is clear from this that direct labor costs cannot be considered as separate from the changes in the other elements of cost.
3. There is always the danger of false labor savings resulting from the reshuffling of accounting records. The learning curve plots direct labor only. What happens when greater use is made of supervisory labor? This labor is not the same; it is classified as indirect. If the emphasis is placed on direct labor savings alone, without considering indirect labor, a distorted picture of direct labor cost savings will result.

Finally, it should be stated that there is a practical limit to the amount by which costs can ultimately be reduced. A common practice is to assume that after unit 1000, the standard time for producing the product will be reached, and beyond that point the learning curve effect is negligible. While this is not entirely true, it is the point where learning and improvement is at a minimum. The simple selection of unit 1000 is recommended especially when estimating costs for job order manufacturing, or where there is a very high labor content per unit.

The selection of the point where the work will be performed to standard time may vary as a function of the number of days from start-up, rather than the number of units produced from startup. An example of this would occur in high-rate production shops, where as many as 1 million parts may be produced in the process of proving out special production tooling. This could be done at the rate of 100,000 to 200,000 parts per day—and the production parts could be produced *at standard* following the debugging process from day 1 (or, in this case, from production unit 1). Again, depending of the process, the tooling, *and the factory*, there may need be a productivity factor applied to the standard. For example, routine operations of 90% productive to standard might be acceptable. In this case, the cost achieved would never drop below 111% of standard (standard ÷ .90 = 1.11). The reductions beyond this point would be due to changes in the process of doing the job, which would would reduce the standard hours (see Chapter 10).

Graphical Techniques

Most learning curve applications in cost estimating involving the direct labor cost for a manufactured product are performed graphically using log-log graph paper. A learning curve slope protractor is shown in Figure 7.16, with the values of the slope coefficient for slopes from 60% to 97%. To use the protractor, the following procedure should be followed:

1. Plot the cumulative average time per unit, or time per unit on log-log graph paper.
2. Fit a straight line through the data points.
3. Overlay the slope protractor on the line, and find the learning slope in percent.
4. Obtain values of the learning slope from the table.
5. By extending the line, the time for unit 1 can be read directly from the graph.

7.4.4 Application Examples

The learning curve is a primary tool of the Department of Defense and defense contractors in negotiating final prices for military hardware. In many applications, its usefulness in pricing decisions can extend even beyond direct labor costs. The curve is useful in make-or-buy analysis, especially when buying means ordering on a negotiated price basis.

Problem

Given an estimated first unit cost of 1200 hr,, and an anticipated 88% improvement curve slope based on experience with similar products. Calculate projected costs for the first 50 units and for 100 follow-on units.

Solution

This problem can be solved by using the cumulative average equation $Y = kx^{-n}$ after determining $-n$ by using the equation

$$-n = \log(p-100) \div \log 2$$

1. Determine $-n$, given that $p = 88\%$.

$$-n = \log(.88) \div \log 2$$
$$= -0.0555 \div 0.3010$$
$$= -0.1844$$

2. Substitute 1200 for k and -0.1844 for $-n$ in the cumulative average cost equation:

$$y = 1200 \times X^{-0.1844}$$

3. Solve for y at $x = 50$, and at $x = 150$.

Cumulative average for $50 = 1200 \times 50^{-0.1844} = 583.2$

Cumulative average for $150 = 1200 \times 150^{-0.1844} = 496.3$

4. Determine cost for 50 units.

Total for $50 = 50 \times 583.2 = 29{,}160$

5. Determine cost for first 150 units.

Total for $150 = 150 \times 476.3 = 71{,}445$

① Plot cumulative-
average -time-per-unit
or time-per-unit on
log-log graph paper.

② Fit a straight line to
data points.

③ Read time-for-one, (A)

④ Find learning percent
from protractor.

⑤ Obtain calues of learning
slope. (B) from table.

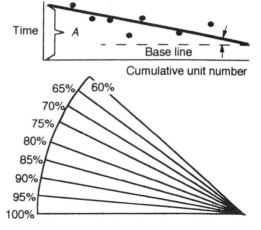

Time A
Base line
Cumulative unit number

65% 60%
70%
75%
80%
85%
90%
95%
100%

%	B	B+1	%	B	B+1	%	B	B+1	%	B	B+1
60	-0.737	0.263	70	-0.514	0.485	80	-0.322	0.678	90	-0.152	0.847
61	-0.713	0.287	71	-0.494	0.506	81	-0.304	0.696	91	-0.136	0.864
62	-0.690	0.310	72	-0.474	0.526	82	-0.286	0.714	92	-0.120	0.880
63	-0.667	0.333	73	-0.454	0.546	83	-0.269	0.731	93	-0.105	0.895
64	-0.644	0.356	74	-0.434	0.566	84	-0.255	0.748	94	-0.089	0.911
65	-0.621	0.379	75	-0.415	.585	85	-0.234	0.764	95	-0.074	0.926
66	-0.599	0.401	76	-0.369	0.604	86	-0.218	0.782	96	-0.059	0.941
67	-0.578	0.422	77	-0.377	0.623	87	-0.200	0.800	97	-0.044	0.956
68	-0.556	0.444	78	-0.358	0.642	88	-0.184	0.816			
69	-0.535	0.465	79	-0340	0.660	89	-0.168	0.832			

⑥ Substitute values of A, B, and $B + 1$ to find learning curve *

Learning curve	If unit times were plotted	If cumulative average times were plotted
⑦ Time for R - th Unit ($R = 1,2,3 \cdots$)	AR^B	$(B+1)AR^B$
⑧ Cumulative average time for R -th unit	$\dfrac{AR^B}{B+1}$	AR^B
⑨ Total time for R -th units	$\dfrac{AR^{B+1}}{B+1}$	AR^{B+1}

* For best accuracy R should be greater than 10

Figure 7.16 Learning curve slope protractor.

6. Determine cost of 100 units following 50 units by subtracting equation (5) from
equation (6).

$$71,445 - 29,160 = 42,285$$

Another common application of the learning curve is in determining the slope of the
curve, and the indicated first unit (T_1) value, when the hours for the first two blocks or
lots of production units are known.

$T_1 = 3\text{unit}$

Problem

Given that the first 3 units cost a total of 2539 hr, and the next block of 4 units cost 2669 hr, determine the theoretical cost of the first unit and the improvement curve slope indicated. Assume that the difference in block size did not affect the trend.

Solution

The cumulative average costs through 3 and through 7 units can be readily calculated and the cumulative average cost equation $Y = kX^{-n}$ can be solved to determine the first unit (k) cost, and the curve exponent ($-n$). The slope percentage equation $p \div 100 = 2^{-n}$ can then be used to determine the curve slope percentage.

1. Determine cumulative average costs at 3 and at 7 units.:

$$\text{At } X = 3, \ y = 2539 \div 3 = 846.3$$
$$\text{At } X = 7, \ y = (2539 + 2669) \div 7 = 5208 \div 7 = 744.0$$

2. Substitute these values into the equation $y - kX^{-n}$.

$$846.3 = k \times 3^{-n} \tag{1}$$

and

$$744.0 = k \times y^{-n} \tag{2}$$

3. To solve these two simultaneous equations, first divide equation (1) by equation (2) to eliminate.

$$(846.3 \div 744.0) = k \times 3^{-n} \div k \times 7^{-n} \tag{3}$$

or

$$1.1375 = (3 \div 7)^{-n} = 0.4286^{-n}$$

4. To solve for $-n$, take the logarithm of both sides of the equation:

$$\log(1.1375) = -n \log(0.4286)$$

Using a scientific calculator, this becomes:

$$0.05595 = -n(-0.36795)$$

5. Solve for $-n$

$$-n = 0.05595 \div (-0.36795) = 0.15206$$

6. Determine p

$$p \div 100 = 2^{-n} = 2^{-0.15206} = 90\%$$

7. Substitute $-n$ into Equation (1)

$$846.3 = k \times 3^{-0.15206} = k \times 0.8462$$

8. Solve for T_1 cost (k)

$$k = 846.3 \div 0.8462 = 1000 \text{ hr}$$

Problem

Assuming that the improvement trend in the previous problem will continue, calculate the cost of an additional 12 units.

Solution

The T_1 cost and n factor were determined in the preceding problem. The cumulative average method can be used to determine the cost of additional units.

1. Given that $T_1 = 1000$ and $-n = -0.15206$, calculate the cumulative average through $3 + 4 + 12$ or 19 units.

$$y = 1000 \times 19^{-0.15206} = 639.1$$

2. Compute the total cost through 19 units.

$$\text{Total cost } yx = 639.1 \times 19 = 12{,}142.9$$

3. Calculate the cost of the last 12 of these units, given that the first 7 cost 5208 hr, as determined in the preceding problem.

$$12{,}143 - 5{,}208 = 6{,}935 \text{ hr}$$

Note that the block average cost for thee additional units is

$$6{,}935 \div 12 = 577.9$$

As in the previous problem, both of these problems can be worked without calculations by using the learning curve on log-log paper. It is often recommended that both techniques be used as a method of checking or verifying an answer.

Problem

A common problem often faced by the manufacturing cost estimator is that of constructing a learning curve from historical labor cost data, and then converting the series of learning curves for each cost center or department into a composite learning curve for the total manufactured product. Figure 7.17 shows how this can be done by using a weighted average to determine the composite curve.

This particular application involved taking historical labor cost data from development and initial low-rate production to determine composite learning curve slopes for projecting rate production labor costs to manufacture and test receivers and antennas. The weighted average is determined by calculating the percentage the individual department labor cost is of the total cost, then applying that cost percentage against the learning curve for the department, and then obtaining the composite curve slope by totaling all of the weighted slope percentages of the different departments. Figure 7.17 also shows this calculation.

7.4.5 Bibliography

Gibson, D. C., *Manufacturing Improvement Curves; Concept and Application*, McDonnell Douglas Corporation, St. Louis, MO, 1981.

Nanda, R. and Adler, G. L., *Learning Curves, Theory and Application,* Industrial Engineering & Management Press, Institute of Industrial Engineers, Atlanta, GA.

Tanner, J. P., "The Learning Curve, A Line on Labor Cost," *Production Engineering*, May 1985.

Tanner, J. P., *Manufacturing Engineering: An Introduction to the Basic Functions*, Marcel Dekker, New York, 1985.

COMPOSITE LEARNING CURVE ANALYSIS

	LEARNING CURVE	LABOR HOURS		PERCENTAGE (%)	
DEPARTMENT	SLOPE (%)	ANT.	RCVR.	ANT.	RCVR.
Production	81	1,944	28,888	4.8	50.9
Quality Control	81	458	5,121	1.1	9.0
Magnetics	97	216	15,657		27.6
Mechanical Fabrication	81		1,683		2.9
Machine Shop	90	12,690	5,354	31.4	9.6
Antenna Lab	87	420		1.0	
Support Facilities	95	1,966		4.8	
Quality Control	95	592		1.5	
Product Test	95	1,486		3.7	
Integration Electronics	95	1,752		4.3	
SS Assembly	95	8,292		20.5	
Microelectronics Assembly	95	2,274		5.6	
Hybrid Assembly	97	4,261		10.7	
Top Assembly	87	4.088	12	10.6	
	TOTAL	40,439	56,715	----	----

Receiver Composite Slope = (.509) (.81) + (.09) (.81) + (.276) (.97)
                          + (.029) (.81) + (.096) (.90)

                        = $\boxed{87\%}$

Antenna Composite Slope = (.048) (.81) + (.011) (.81)
                          + (.314) (.90) + (.01) (.87)
                          + (.048) (.95) + (.015) (.95)
                          + (.037) (.95) + (.043) (.95)
                          + (.205) (.95) + (.056) (.95)
                          + (.107) (.97) + (.106) (.87)

                        = $\boxed{84\%}$

Figure 7.17 Converting a series of element learning curve histories into one representative composite curve for the entire product.

7.5 MATERIAL AND SUBCONTRACT ESTIMATING

7.5.0 Pricing the Bill of Material

In the majority of cost estimates for manufactured products, the cost of purchased parts, raw materials, commodity items, and items that are subcontracted constitute the biggest part of the cost estimate, often constituting 75% of the total cost of the product. The importance of accurately pricing the "buy" items on the bill of material cannot be emphasized strongly enough. The material estimator or buyer must thoroughly understand the potential pitfalls that commonly occur in estimating material costs:

1. Inadequate product specifications may result in prices for material at an incorrect quality level.
2. Incorrect delivery requirements may force the use of more expensive substitutes, or production changes that require more setups, or that cause production delays.
3. Incomplete product specifications may result in material estimates that do not cover the actual material costs involved in the final engineering design or product quality or reliability requirements.
4. Material price levels may change upward and exceed the estimated material costs. This is a particularly acute problem when estimating material costs for products with long manufacturing lead times, or for delivery at a future time.

5. Material price breaks may be anticipated and planned but not realized, because of delivery schedule changes or revisions to inventory policy.

The purchasing function is the primary support to the manufacturing cost estimator in determining material costs. For best results, direct quotes from vendors and subcontractors should be obtained. Where time to prepare the estimate is limited, purchasing must use cost history for commonly used purchased parts and material. If historical cost data is used, it must be factored for inflation and anticipated cost growth.

7.5.1 Estimating Standard Purchased Parts

Included in the category of standard purchased parts are items such as common hardware and fasteners, electronic components, certain types of bearings, gears, pulleys, belts, chain drives, electric motors, electrical connectors, wire, cable, clutches, batteries, power supplies, switches, relays, and similar items. These items can be purchased by ordering from catalogs or commercial specifications published by the manufacturer, and are usually available from stock inventory. They are sometimes priced from standard price lists provided by the manufacturer, with discounts or price breaks for quantity buys. In many instances such standard parts are available from local sources of supply such as distributors or manufacturer's agents or representatives. Figure 7.18 shows a typical line of standard fasteners available from manufacturers.

The call-out on the bill of material for standard parts may be by the manufacturer's part number, by a military standard (MIL) number call-out, or by a number assigned by the company which is a drawing or specification of the item. MIL and company specification numbers allow the buyer to order the part from any manufacturer whose product meets the specification or MIL requirements.

Estimating the cost of standard purchased parts using cost history or manufacturers' price lists presents a fairly low risk in the pricing of these items. Only high-dollar standard purchased parts should be supported with a recent quote from the manufacturer or distributor. High-dollar items are those with a unit cost of $5.00 or more. Figure 7.19 shows a representative bill of material call-out for a latch on a computer-generated bill of material indented parts list explosion. This list summarizes the total requirement for a standard or component part wherever it is used at all assembly levels of the manufactured product. Multiplied by the number of units to be produced and adding necessary overage factors gives the estimator a total quantity of the item to estimate.

There is much that the buyer can do to bring the cost of standard parts down below the list price level. He can, on his own initiative, improve costs by a number of tactics, including quantity buying, vendor price negotiation, material substitution, scrap reduction, etc. This should be taken into consideration in estimating the costs for standard purchased parts.

7.5.2 Estimating Raw Material Requirements

Much of the purchased material for any manufactured product must be ordered in the form of sheet, bar stock, blocks of material, in spools of wire and solder, or as drums of chemicals and solvents, in pint, quart, or 5-gal containers, etc. The material estimator must calculate the correct quantities of this raw material and bulk supplies or commodities to buy in order to manufacture the required number of units with a minimum of scrap or waste. To do this, the material estimator must list every fabricated part to be manufactured

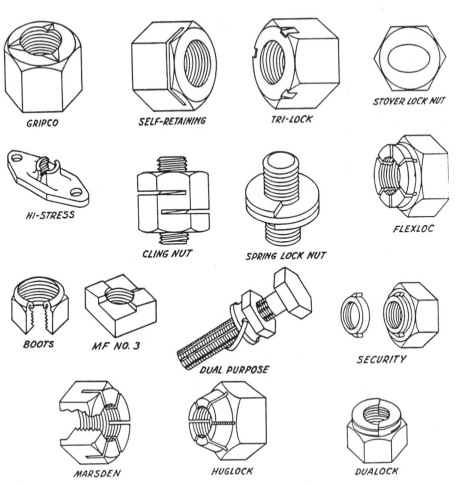

Figure 7.18 Standard hardware available from manufacturers.

inhouse by quantity required, size, weight, overall configuration, and type of raw material needed. If the product has been manufactured previously, there is a good chance that this information is already available on the process sheet, methods sheet, or work instructions for the part in question. If not, the cost estimator or material estimator must determine the requirement.

Determining Raw Material and Bulk Item Quantities

The estimator analyzes each item to determine the amount of raw material that must be purchased to manufacture the required number of units. The amount of raw material for a given piece part usually comes from determining the weight of raw stock used per piece. A part that is machined would include the finished dimensions of the part plus the amount of stock removed by machining. The overall dimensions of such a piece of stock would be determined and the volume calculated. The volume is multiplied by the density of the material (weight per unit volume) to obtain the weight. A piece that is irregular in shape

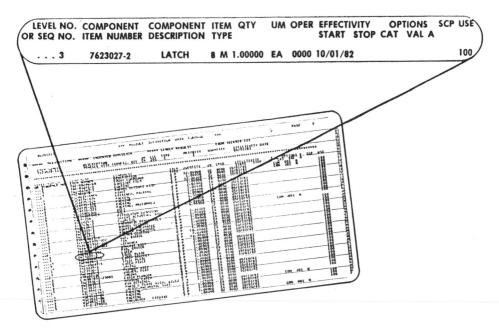

LEVEL NO. OR SEQ NO.	COMPONENT ITEM NUMBER	COMPONENT DESCRIPTION	ITEM TYPE	QTY	UM	OPER	EFFECTIVITY START STOP	OPTIONS CAT	SCP VAL	USE A
. . . 3	7623027-2	LATCH	8 M	1.00000	EA	0000	10/01/82			100

Figure 7.19 Representative computer generated bill of material calling out a standard latch to be purchased.

is divided into simple components, and the volumes of the components are calculated and added together to give total volume.

The total volume is multiplied by the density of the material to obtain the weight. The material cost is then obtained by multiplying the weight by the price per pound of the material. In the case of a sheet metal stamping fabricated from aluminum sheet stock, the procedure would be as follows:

1. Weight of sheet = gauge × width × length × density
2. Number of pieces per sheet = length of sheet ÷ length of multiple
3. Weight per piece = weight of sheet ÷ pieces per sheet
4. Piece part material cost = Piece part weight × aluminum price/lb.

When estimating bar stock, the length of the piece, plus facing and cutoff stock, should be multiplied by the weight or price per inch of the stock diameter. Scrap, butt ends, chips, etc., that are lost in processing must be considered in the estimate. These losses vary from 3% to 10% depending on the job, current shop practices, and the material itself.

Bulk items such as spooled core solder can also be easily estimated by determining the weight of solder used for each solder joint, then multiplying by the number of solder joints to arrive at the total weight of solder used per assembly. When this number is multiplied by the total number of assemblies, the total weight of solder needed is determined. Using the table shown in Figure 7.20, the number of feet of the required type of solder needed can be readily determined. An additional 10% should be added to the calculated requirement to cover losses, solder required to tin the soldering iron, etc.

When determining the amounts of raw material, bulk and commodity items, and the like, required for the job, the material estimator must consider that material for stampings

Diameter Inch	Area Sq. Inch	Lead Wire Ft. per Lb.	30/70 Solder Ft. per Lb.	40/60 Solder Ft. per Lb.	50/50 Solder Ft. per Lb.	60/40 Solder Ft. per Lb.	Tin Wire Ft. per Lb.
.032	.00080	254	297	310	325	340	396
.036	.00113	180	211	220	230	241	281
.040	.00126	161	188	196	206	216	251
.045	.00159	128	150	156	164	172	200
.050	.00196	104	122	127	133	129	162
.056	.00246	82.6	96.6	101	106	111	129
.063	.00312	65.1	76.2	79.4	83.3	87.2	102
.071	.00396	51.3	60.0	62.6	65.7	68.7	80.0
.080	.00503	40.4	47.3	49.3	51.7	54.1	63.0
.090	.00636	32.0	37.4	39.0	41.0	42.9	49.9
.100	.00785	25.9	30.3	31.6	33.2	34.7	40.4
.112	.00985	20.6	24.1	25.1	26.4	27.6	32.1
.125	.01227	16.6	19.4	20.2	21.2	22.2	25.9
.140	.01539	13.2	15.4	16.1	16.9	17.7	20.6
.160	.02011	10.1	11.8	12.3	12.9	13.5	15.8
.180	.02545	8.0	9.4	9.8	10.2	10.7	12.5
.200	.03142	6.5	7.6	7.9	8.3	8.7	10.1

Note: The number of linear feet per lb. of flux-core wire solder will be somewhat greater than the above figures for solid wire. Where the amount of flux is 1.1% by weight, linear footage is increased by 9% over solid wire, with 2.2% the increase is 15%, and with 3.3% the increase is 27%.

(Courtesy National Lead Co.)

Figure 7.20 Number of feet of solder required.

is bought by the sheet or reel, solder by the spool, paint by the gallon, and cleaning solvents by the 55-gal drum. Certain adhesives, potting compounds, and bonding agents are bought in 5-gal containers. In many instances, far more than the job requirement must be purchased because of this, often inflating the material cost for the job. This is especially so when there is no other use for the item except the current job. The offsetting consideration to this is that substantial price breaks are often available on these items when purchased in these amounts, and in these container sizes.

In foundry work, the manufacturing cost estimator calculates raw material costs as follows:

1. Determine the amount of metal required to charge the furnace by determining the ratio of finished casting weight to the weight of the metal charged into the furnace, based on previous experience to determine the shop yield factor.
2. Calculate the furnace charge per casting by dividing the casting weight by the yield factor.

Shop yield will vary with different casting materials, and consideration must also be given to metal losses due to spills, oxidation, gate cutoffs, and overruns. A good rule of thumb is to add 10% of the casting weight for these losses. Any metal not consumed in the finished casting or lost is returned to the furnace for remelting. This remelt metal is determined by subtracting the sum of shop yield and metal lost, from the amount of metal charged. By converting these values to percentages, with the amount of metal charged being 100%, we can express the amount of metal returned for remelting as a percentage.

In forgings and forged parts, the cost of material averages about 50% of the total cost. To determine material cost for a forged part, the estimator must first calculate shape weight using sketches or engineering drawings. The part can be divided into suitable geometric sections, the volume for each section obtained, and by adding the section volumes together obtain total volume. Multiplying this by the density of the forging material gives the shape weight of the forging, and by adding 4% to the shape weight, gives the net weight. The gross weight, which is the weight of forging stock required to actually make the forging, is determined next. This weight is found by adding material lost through flash, scale, sprue, tonghold, and cut waste to the net weight:

> *Flash* is excess metal extruded as a thin section that surrounds the forging at the die parting line to ensure that all parts of the die are properly filled. Flash width and thickness varies with the weight of the forging. Flash also includes punchout slugs from holes in the part. Punchout slugs vary with the dimensions of the punched holes and the thickness of the section through which the holes were punched.
>
> *Scale* is material lost due to surface oxidation in heating and forging. The amount of this loss varies with surface area, heating time, and material.
>
> *Sprue* is the connection between the forging and the tonghold.
>
> *Tonghold* is a projection used to hold the forging.
>
> *Cut waste* is stock lost as sawdust when bar stock is cut to length by a saw, plus bar end loss from length variations and short ends from cutting the stock to exact length.

Gross weight is then calculated by totaling the percentages estimated for each of these factors and adding to the net weight. These percentages should be determined from historical data and experience in your shop, but should be approximately as follows:

Flash = 1.5%
Scale = 5%
Sprue = 7%
Tonghold (included as part of sprue)
Cut waste = 5%

The direct cost of forging material then calculated by multiplying the gross weight by the cost per pound of the forging stock material.

Estimating Overages, Scrap, and Line Losses

Most firms have an established policy for estimating material scrap, overages, and line losses. Until such policies are established, or are determined from historical data, the following factors are given as good industry averages:

1. Purchased parts
Less than $1.00 unit price	15%
$1.01 to $5.00 unit price	10%
$5.01 to $10.00 unit price	2%
Over $10.00 unit price (determined by estimator)	

2. Forgings and castings
   Less than $100.00 unit price:
Sand Castings	10%
Pressure test castings	25%
Other forgings and castings	10%
Unit price $100 or over	5%

3. Raw material:

Titanium	10%
Aluminum, Steel, Magnesium, Rubber (sheet, plate, bar, rod, or tubing)	25%
Extrusions	10%
Wire	35%
Miscellaneous (plastics, fabrics, tapes, etc.)	20%

It is important to point out that these factors may vary considerably among companies, even in the same industry. In the metal forming and stamping industry, sheet stock utilization can well determine whether or not the job is profitable, and what the percentages for sheet stock overages will be. These same considerations apply in the utilization of sheet material for printed circuit card fabrication. Careful planning of the position of the blanks on the sheet stock means maximum stock usage and minimum scrap.

Line losses of expensive parts and components can drive scrap factors very high if not controlled. A scrap tag system that requires the shop to complete a scrap tag for each part lost or damaged in assembly can do much to ensure that line losses remain low and under control. Such systems make the estimator's job easier and leads to lower material costs through lower scrap factors.

Estimating Material for Tooling

The cost of material for tooling is a significant part of the total nonrecurring cost for any new program. Tooling material requirements are taken from the bill of material or parts listing on the tool drawing, or are estimated from the tooling concepts envisioned by the manufacturing cost estimator. Included are such items as tool steel, drill bushings, quick-release clamps, hold-down buttons, and similar items for the fabrication of special-design tooling. Tooling material also includes standard perishable tooling, such as drill bits, cutters, end mills, punches, reamers, broaches, and other similar items which are consumed over the life of the program.

If possible, these materials should be calculated exactly from tool drawings and perishable tool usage experience. If such documentation or tool history is not available, then tooling material costs on previously built similar tools should be used, and perishable tool usage estimated from experience on earlier jobs or programs that had similar requirements. It should be mentioned that the cost of special-design tooling can be held to a minimum by using inexpensive tooling materials such as wood, sheet metal, and aluminum instead of expensive tool steels to fabricate many tools such as holding fixtures, assembly jigs, and motion economy devices. Also, many times a standard, off-the-shelf tool or fixture with minor modification will do the same job as a special-design tool, at a fraction of the cost.

Tooling material may often be purchased directly by the tool design or manufacturing engineering organization without going through the purchasing, receiving, receiving inspection groups, and as a result would not carry the normal material overhead or burden.

7.5.3 Long-Lead and High-Dollar Items

One of the most important tasks in the preparation of any cost estimate for a manufactured product is the identification of the long-lead and high-dollar material items. This information is basic to the development of the production start-up schedule and will be a key factor in determining if the requested delivery schedule can be met. The material items which are the high-dollar items in the cost of product material are the ones that need to be worked by the buyer and material estimator to obtain firm quotes, and if possible, competitive quotes. How well this is done will have a major bearing on whether or not a competitive and winning cost proposal is finally submitted.

If time permits, qualified backup sources of supply should be located for each and every long-lead item on the bill of material. This ensures that the failure of any long-lead supplier to meet delivery of an item of acceptable quality will not shut down the production line. There is a certain amount of risk in any cost proposal or estimate. Risk that the work will cost more than the estimated cost, risk that promised deliveries will not be met, and risk of not getting the work at all because it was awarded to a lower bidder, or one who promised a better delivery schedule, are the major risks in any cost estimate for a manufactured product. Anything that can be done to develop a winning, low-risk cost estimate should be a prime consideration.

Any item with a delivery promise of 6 months or longer can be considered long lead. It is more difficult to define high-dollar items, except to say that any single item on the bill of material that costs significantly more than the others is high dollar.

7.5.4 Material Learning Curves and Inflation

There are numerous ways to apply learning curve theory to the purchase of material. This normally means that the vendor's experience or learning should reflect in the cost of the material, much as the in-house factory labor declines in cost with experience in manufacturing the product.

When the NCR Corporation buys material from a vendor, the cumulative volume is measured in NCR's *units of experience* (measurement units of the cumulative volume), which differ from the vendor's units of experience. They claim that the following can be proven:

1. NCR's slope for a purchased material may be steeper or flatter than the vendor's slope. This is due to differing units of experience and other factors.
2. NCR's slope can be estimated from historical data, by a projection by comparison of experience with like products, or from the vendor's learning curve, when known.

Figures 7.21 and 7.22 show how a buyer's learning curve is related to a vendor's learning curve. Note that the buyer's slope changes from year to year. Given historical data for a year or more, the costs for the next 5 years can be estimated (see Figure 7.22). The amount of the year-to-year change decreases as the technology ages and depends on many factors: differing units of experience, the time point at which the buyer defined unit 1, the year-to-year vendor volume, and more.

Learning curves such as those shown in Figures 7.21 and 7.22, reflect learning by the vendor only. The buyer can, on his own initiative, significantly improve his material costs by a number of tactics including quantity buying, vendor price negotiation, material substitution, and scrap reduction. Therefore, these curves represent a minimum slope achievable without any additional improvement by the material user. Further, even these minimum slopes are not automatically achieved unless the buyer goes after them and spurs the vendor to ride his learning curve.

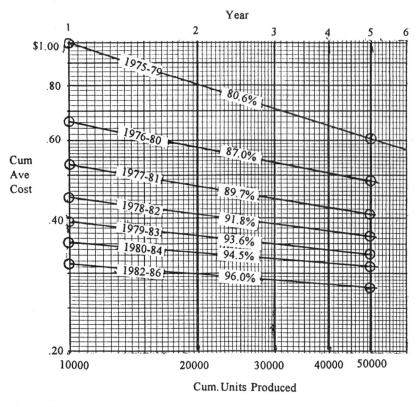

Figure 7.21 Buyers' cumulative average cost experience for a particular part number.

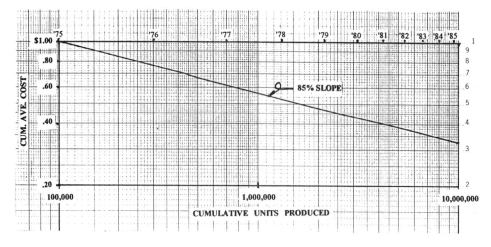

Figure 7.22 Suppliers' cumulative average selling price for a particular part number.

While a firm's material slopes may be higher or lower than the vendor's slopes, in reality, a firm's slope depends mainly on its own actions, not on the vendor's actions alone.

Material learning curve slopes are usually estimated by examining the history of past buys. Such costs include the average inflation experienced over the time period being analyzed, as well as the improvement factors not attributable to vendor learning, such as quantity buys, better negotiation of prices, switches to lower cost vendors, etc. It is often claimed that material learning should be based on costs before inflation. This is undesirable and impractical because:

1. It is well known that there is a wide divergence of inflation rates among commodities. How does the material estimator determine the appropriate year-by-year rate of inflation for each commodity? Who can say what the real inflation rate has been for MOS or TTL devices, for example? A detailed analysis of vendor operations and input materials would be required for such a determination.
2. Even if inflation is removed, the learning rate is not the same as the vendor learning rate because of the differing units of experience, and because buyer as well as vendor learning is included in the applicable learning curve slope.
3. It has been shown that learning curves can be fitted to data that includes inflation, and fitted to the buyer's units of experience.

The factors which are important in determining material learning curve slopes include:

1. Validating the slope with historical data or by comparison with similar materials at the same technological age
2. Using known or forecast prices (reflecting current adjustments for inflationary surges, price wars, etc.) for the basic cost estimate that is used for entering the learning
3. Correcting the slope for likely enduring changes in the environment (a dying technology will have little slope, a technology with surging usage will have an increasing slope, a continuous relative increase or decrease in average inflation, etc.)
4. Correcting the historical slope for technology aging

Using the techniques described above can have a significant impact on how competitive the material estimate for a manufactured product really is. Material may be the largest part of the estimated cost of the product, and as such, should be thoroughly analyzed.

7.5.5 Subcontract Estimating and Pricing

There can be many reasons for subcontracting part of the manufacture of a product, ranging from the use of the special expertise of the subcontractor to significant cost savings, usually in labor, by using a subcontractor. Whatever the reason, every consideration must be taken to ensure that the contractor will perform as promised at a reasonable cost.

With a subcontract, the outside source is doing more than providing material. He is providing material and a portion of the work to be performed on the manufactured product. He may provide a major subassembly, an operating part of the total system, or a specialized manufacturing process not available inhouse. The subcontractor must provide a detailed cost estimate for his product and services, which become part of the total material estimate for the manufactured product. This cost estimate will detail the subcontractor's start-up costs, including such things as special tooling and test equipment, lead times required, manufacturing labor, material, burden, and profit. His delivery schedule should tie-in directly with your production schedule, with his deliveries should feedi into your production at the point needed in the process.

Management and control of subcontractors is maintained by formal reporting requirements, by visits to the subcontractor's facility, by telephone reports, and by status review meetings. Such techniques provide real-time communications and data flow to provide effective means of assessing subcontract status, early detection of problems, and initiation of any needed corrective action. The cost of providing this information should be included in the subcontractor's cost estimate.

7.5.6 Bibliography

Metals Handbook Desk Edition, American Society for Metals, Metals Park, OH, 1985.

Tanner, J. P., *Manufacturing Engineering: An Introduction to the Basic Functions,* Marcel Dekker, New York, 1985.

Vernon, I. R., *Realistic Cost Estimating For Manufacturing,* The Society of Manufacturing Engineers, Dearborn, MI, 1968.

7.6 MANUFACTURING DIRECT LABOR ESTIMATE

7.6.0 Baseline Labor Estimating Techniques

Cost estimating for manufacturing can be done using any one or a combination of three general methods. The first method uses *cost history and statistical methods*. This technique, when properly applied, and where the required data exists, can be a reliable method of preparing the labor estimate. The problem with this method is that such data includes delays, shop inefficiencies, and time lost through a variety of reasons. Its use is recommended where standards do not exist, for budgetary quotes, and as a check against estimates developed using other methods.

The second method, using *similarity with other products*, assemblies, and parts, and/or using *estimator experience*, can be quite satisfactory when no previous production experience exists, when the product has never been in production before, or when the design is new and may still be in the concept stage. The accuracy of such estimates varies directly as the knowledge and experience of the estimator, and the time allocated for preparation of the estimate. The completeness of the product definition also plays a large part in determining the accuracy of the estimate.

The third method is by using *standard time data*. In this method, all of the possible elements of work are measured, assigned a standard time, and classified in a catalog. When a specific operation is to be estimated, the necessary standard time values are added together to determine the total time. There can be little doubt that the use of standard time data is the most accurate and reliable method of estimating manufacturing labor. The use of standard time data promotes consistency among cost estimators, and requires little in the way of estimator experience with the work being estimated (see also Chapter 10).

Historical and Statistical Data

Present and past costs may serve as starting points in preparing estimates for the future by recognizing the limitations outlined above, and modifying them by forecasting conditions at the time the job will be in production. Cost estimates that are derived primarily by projecting past and current labor cost history are generally made only for guidance or planning purposes, and ideally should be followed by a detailed labor estimate based on standard time data. The manufacturing cost estimator must have a clear understanding of the principles of the cost accounting system providing him the data used to prepare the labor estimate.

Standard Data Use and Application

Standard time data is a compilation of all the elements that are used for performing a given class of work with normal elemental time values for each element. Without making actual time studies, the data are used as a basis for determining time standards on work similar to that from which the data were determined. Its use in manufacturing cost estimating offers certain advantages over other estimating methods. These advantages include:

1. It is far more accurate than any other estimating method. It is based on work content, rather than how much work is to be done, and how long it will take to do.
2. It is easier to justify, because a series of individual elements and operations adding up to a given number of hours is easier to justify than one overall judgment of a given number of hours.
3. Standard data promotes consistency between estimates and cost estimators. Standard data, however, will show where there is a legitimate difference between similar equipment.
4. Estimator experience with the operation is not a requirement when using standard data. Such experience and knowledge, however, is extremely helpful.
5. Standard data coupled with learning curves can be used to estimate manufacturing labor for any production quantity. The cost estimator using experience to build an estimate in the job order 5 to 50 units range, but finds it difficult to estimate the same product at 1000 units. Standard data plus learning curves will cover the entire quantity spectrum.

The standards in this chapter are all based on standard data built up over the years by stopwatch time study, and using synthetic standards derived from MTM, MOST, Work Factor, and other predetermined time systems. Their application would be primarily in job order manufacturing quantities in the 25 to 2500 unit range. Figure 7.23 shows how the standard time data for the complete fabrication of any configuration of wire harness is organized and laid out for rapid determination of a standard time. It is simply a matter of determining the number of wires and wire ends, how each wire is terminated and routed, as well as marked, entering the data in the appropriate space on the form, performing the extensions, and making the additions to arrive at the standard time for fabricating the wire harness.

Estimating by Similarity and Estimator Experience

The majority of cost estimates for manufactured products in the United States today are prepared by experienced cost estimators using their professional judgment. Their experience, based on detailed knowledge of their product, shop processes, and methods, is perhaps the most important single requirement for a correct estimate. New products are estimated much the same way, by a comparison of similar products with the new product being estimated. In many small shops the estimating may be done by the shop foreman or even the owner of the business. There is no attempt to use data from time standards, or any of the other, more sophisticated methods, to prepare estimates for new business.

Such techniques, if successful, may be all that are required. The shop may be small, the product may be produced in job order quantities, or the process may be highly

PART NAME _____ OPERATION _____

PROCESS PLAN DATE _____ MATERIAL _____

% ALLOWANCE _____ STD. S/U _____ RUN TIME _____

ESTIMATOR _____ DATE _____

HARNESS ASSEMBLY EH

SET-UP DESCRIPTION	MINUTES	FR	TOTAL MINUTES
DIT-MCO HARNESS TEST TE			
Prepare for Test/Lot	6.0		
RUN TIME			
PREPARE WIRE ENDS MANUALLY			
Clock In on DPI/Step	.60		
Fill out C.A.T. Card & Stamp/Job	1.50		
C.A.T. Card, Stamp Only/Stamp	.04		
Find Wire Number on Plan/Wire	.11		
Check off Wire Number on Plan/Wire	.10		
Cut with Diagonal Pliers/Cut	.07		
Cut with Nail Clippers/Cut	.05		
Prepare Pigtail Normal Wire/Pigtail	.85		
Prepare Pigtail "H" Film Wire/Pigtail	.63		
Comb out Shield (Incl.Cover Removal)/Shld.	1.27		
Untwist Conductors (2 & 3 Cond.)/Wire	.07		
Measure Conductor & Cut/Conductor	.21		
Measure Strip Lgth.& Mark/Conductor	.15		
Hand Strip-Auto Strippers/Conductor	.07		
Hand Strip-Knife/Conductor	.43		
Twist Strands/Conductor	.06		
Hand Tin with Soldering Iron/Cond.	.15		
*Hand Crimp Lug, Taper Pin, etc./Crimp	.20		
*Hand Crimp (2) Wires, 1 Lug/Crimp	.25		
*Hand Crimp (3) Wires, 1 Lug/Crimp	.29		
Install Ferrule/Job, Set-Up	.27		
Install Ferrule/Ferrule, Run Time	1.15		
Install Solder Sleeve/Sleeve	.60		
Install Vartex Over Pigtail/Pigtail	.59		
Install Vartex over Floating Shld./Shld.	.10		
Install Vartex over Ferrule/Ferrule	.61		

RUN TIME DESCRIPTION	MINUTES	FR	TOTAL MINUTES
Spot Tie/Each Open	.26		
Congested	.76		
Lacing, Starting Knot/Each	.26		
Lacing, Continuous Lock Stitch/Inch	.33		
Tywrap/Each Open	.22		
Congested	.31		
Install Tywrap thru Anchor/Each	.34		

SEARCH NUMBERED WIRES

NO. OF WIRES	MINUTES	NO. OF WIRES	MINUTES	NO. OF WIRES	MINUTES	NO. OF WIRES	MINUTES
1	.03	17	2.58	33	8.67	49	18.37
2	.08	18	2.85	34	9.17	50	19.09
3	.15	19	3.14	35	9.68	51	19.82
4	.24	20	3.45	36	10.21	52	20.57
5	.33	21	3.77	37	10.75	53	21.33
6	.44	22	4.10	38	11.30	54	22.10
7	.57	23	4.45	39	11.87	55	22.89
8	.71	24	4.81	40	12.45	56	23.70
9	.86	25	5.18	41	13.05	57	24.51
10	1.02	26	5.57	42	13.66	58	25.35
11	1.20	27	5.97	43	14.28	59	26.19
12	1.40	28	6.38	44	14.92	60	27.02
13	1.61	29	6.81	45	15.62	61	27.89
14	1.83	30	7.26	46	16.29	62	28.78
15	2.06	31	7.71	47	16.97	63	29.68
16	2.31	32	8.19	48	17.66	64	30.59

For a trunk containing more than 120 wires:

$$\text{TIME} = \left[\frac{(N+2)(N-1)}{4} \right] + N \times .0277 \text{ Min.}$$

Where N = Number of wires in breakout

WIRE PREP - ENDS 12 GA & UP

MEASURE, CUT & STRIP, PLUS

	TIN	FREQUENCY	CRIMP (1) WIRE PER LUG OR PIN	FREQUENCY	CRIMP (2) WIRES PER LUG OR PIN	FREQUENCY	CRIMP (3) WIRES PER LUG OR PIN	FREQUENCY
SINGLE CONDUCTOR WIRE:								
Common Wire End - Open	.64		.70		.74		.79	
- Cong.	.82		.81					
Pigtail Shld. Norm.	2.08		2.13		2.18		2.22	
Float Shld. Norm.	1.58		1.64		1.68		1.73	
Pigtail Shld."H" Film	1.86		1.92		1.97		2.01	
Float Shld."H" Film	1.37		1.43		1.47		1.52	
TWO CONDUCTOR WIRE:								
Pigtail Shld. Norm.	2.78		2.90		2.99		3.08	
Float Shld. Norm.	2.29		2.41		2.50		2.58	
Pigtail Shld."H"Film	2.57		2.69		2.78		2.87	
Float Shld."H"Film	2.08		2.19		2.28		2.37	
MARK, CUT STRIP & TIN ONLY - SINGLE CONDUCTOR WIRE:								
Common Wire End - Open	.37							
- Congested	.55							
Pigtail Shield Normal	1.61							

Values are per wire, vartex over shields included.

Install Wire Marker (Non-Shrink.)/Marker	.20
Mark Conductor, Cut, Strip & Twist	.38
Wire Identification (Masking Tape)	.17
*For Wires smaller than 10 gage	
For 10 Gage and Larger, Use Values Below:	
Crimp Lug-Hydraulic Lugger/Lug	.69
Crimp Lug-Bnch. Lugger 6-14 Gage/Lug	.77
Crimp Lug-Bnch. Lugger 00-4 Gage/Lug	.80

HANDLING

Get Parts, Folder, etc./Job	1.8
Get, Locate & Secure Plug or P/A to Harness Board/Plug or Plug Assembly	.70
Get, Locate & Secure Sub-Assy. to Harness Board/Sub Assembly	1.02
Get, Locate & Fit Item to Harness Assy./Item	.12
Get, Locate & Secure Lacing Bar with 2 Nuts	.58
Get & Locate Loose Wire to Harn. Bd./Wire	.14
Remove Wired Plugs from Harn.Bd./Plug	.44
Remove Wired Lug from Stud/Lug	.05
Remove Wired Sub-Assy. from Harn.Bd./S-Assy.	.74
Remove Lacing Bar from Harness Board	.55
Remove Harn.Assy. from Bd./Junct.or 10" Lgth	.09
Place & Remove Complete Harness from Board for Fit Check/Junct. or 10" Length.	.18
Bag and Box Harness/Job	2.03
Search Out Wire in Bag of - 30 Wires	9.91
60 Wires	33.56
90 Wires	70.95
120 Wires	122.08

ROUTING

Route Wires Straight/Each 10" Length	.02
Route Thru Bends or Breakout/Bend or Brk.	.06
Route Thru Vartex/Inch	.04
Route Thru Transition/Wire	.18
Route Thru Grommet or Hole, Open/Wire	.05
Route Thru Grommet or Hole, Congested/Wire	.09
Route to Clip Spring or Equiv./Wire	.03
Dress Wires/Wire	.03

Figure 7.23 Wire harness fabrication standard data.

RUN TIME

HOOK UP AND SOLDER WIRES

DESTINATION	SOLDER TIME ONLY		1 WIRE/ TERMINAL		2 WIRES/ TERMINAL		3 WIRES/ TERMINAL		MINUTES	FR	TOTAL MINUTES
	MIN.	FR	MIN.	FR	MIN.	FR	MIN.	FR			
P.C. Pad	.05		.31								
Congested			.73								
Plug Pin 10-26G W/Weep Hole	.27		.64								
Plug Pin 10-26G	.27		.56		.85						
Congested			.92		1.54						
Plug Pin 4-8 G	.38		.84								
Congested			1.25								
Plug Pin 4/0-2G	1.00		1.45								
Congested			2.74								
Turret, Island	.09		.48		.71		.94				
Congested			.79		1.33		1.89				
Turret, Buss	.89		1.28		1.41		1.54				
Congested			.64								
Bifurcated.Isl.	.26		1.08		1.08		1.49				
Congested			1.28		2.00		2.87				
Bifurcated, Bus	.89		.50		1.41		1.54				
Congested			.76								
Eyelet Terminal	.11		.45		.65		.81				
Congested			.64		1.08		1.41				
Hook & Other	.07		.06		.59		.73				
Congested			.31		.90		1.16				
Use Assist Tool			.20		.06		.06				
Tin Plug Pin	.18		.19								
Hand Solder	.09										
Circuitry/Sq."											
Touch-up	.09										
Circuitry/Sq."											

RUN TIME

DESCRIPTION	MINUTES	FR	TOTAL MINUTES
COAXIAL CABLES			
Prepare One End	1.69		
Prepare One End & Install Connector	3.69		
Prepare Two Ends & Install Connectors/ Cable	7.37		
MISCELLANEOUS			
Insert Pinned Wire In Connector, Per Wire	.22		
Install Plug Seals, Per Seal	.09		
Install Prepared Lug Over Stud, Per Lug	.03		
Install Dust Cap, Each	.21		
Install Strain Relief & Tighten, Each	1.56		
Remove & Replace Potting Boot, Each	.36		
Install Plug in Vise & Remove, Per Lug	.16		
Install Common Points, Each	3.23		
Install Rubber Bands on Harness Bd., Each	.08		
Install Environmental Seal on Plug, Each	.54		
TORQUE MECHANICAL FASTENERS			
Change Socket on Torque Wrench/Socket	.16		
Set Required Torque/Gage	.05		
Torque Fastener/Fastener	.15		
Stripe Fastener/Stripe	.23		
POSITION HARNESS IN CHASSIS			
Harness Installation:			
Normal/Harness	.13		
Confined/Harness	.15		
Locate Straight Run:			
to 4" Per Run	.04		
Each additional 4" to 6" Per Run	.02		
Locate Bend of Breakout, Per Breakout	.06		

| | Reform & Locate Bend, Confined/Breakout | .12 |
| | Dress Branched Conductors/Conductor | .06 |

DIT-MCO HARNESS TEST- TE

Constant, Per Harness		3.51
Connect and Disconnect, Per Connector		.51
Test, Per Conductor		.10
Self Check Equipment, Per Harness		.031
	x Total R/T	

SUB TOTAL RUN TIME

PAGE #1 RUN TIME

ALLOWANCE

TOTAL STD. RUN TIME

DESCRIPTION	OPEN MINUTES	FR	CONGESTED MINUTES	FR
ROUTE, PREP, TIN, HOOK-UP, SOLDER, CLEAN & DRESS WIRES				
Turret Terminal (Island) 1W	.90		1.39	
Turret Terminal (Island) 2W	1.56		2.53	
Turret Terminal (Island) 3W	2.21		3.68	
Bifurcated Term. (Island) 1W	1.07		1.68	
Bifurcated Term. (Island) 2W	1.93		3.20	
Bifurcated Term. (Island) 3W	2.77		4.67	
Eyelet Terminal (Island) 1W	.92		1.36	
Eyelet Terminal (Island) 2W	1.50		2.28	
Eyelet Terminal (Island) 3W	2.08		3.21	
Hook Term. or Other 1W	.88		1.24	
Hook Term. or Other 2W	1.57		2.10	
Hook Term. or Other 3W	2.00		2.96	
VARTEX & MARKER INSTALLATION				
Cut, Install & Tape Wire Marker (Shrinkable)			.23	
Cut & Install Wire Marker (Non-Shrinkable)			.20	
Install Bundle Marker (Shrinkable)			.08	
Install Bundle Marker (Non-Shrinkable)			.57	
Install Marker Plat/Marker			.62	
Slide Short Vartex Over 1 Wire/Piece			.04	
Slide Short Vartex Over Solder Connection			.05	
Install Short Vartex, Under 36", Over			.08	
Wires (Non-Shrinkable)/Inch of Wire				
Install Long Vartex, Non-Shrinkable/Piece			8.07	
Install Long Vartex, Non-Shrink./In.of Wire			.13	
Install Long Vartex,Shrinkable/Piece			.14	
Shrink Short Vartex, Per Piece			.15	
Shrink Long Vartex, Gun Handling/Piece			.14	
Shrink Long Vartex, Per Inch			.23	
Wrap Vartex for Filler on Strain Relief,			11.36	
Per Piece				
Cover Exposed Shield with Vartex & Tie/			1.67	
Piece				

Figure 7.23 Continued

specialized and very predictable. In such cases there is no need for a better estimating approach. As the firm grows larger, and the cost estimating is done by more than one person, consistency among estimators may become a problem. As the product line becomes more diversified, or the volume of production work grows, estimating judgment and experience exercised by one or two knowledgeable individuals may not do the job.

The ideal cost estimating situation is the experienced cost estimator using standard time data to develop and validate the cost estimate, and professional judgment and shop knowledge to determine whether the cost estimate is reasonable and attainable. The final questions are whether or not everything is included, and whether the estimate can be sold to management and supported in negotiations with the customer.

7.6.1 Preparing the Labor Estimate

This section provides examples and procedures of manufacturing labor estimating for a variety of different processes and basic operations covering the broad spectrum of manufacturing. Included are casting and foundry operations, metal forming and fabrication, machining operations, mechanical assembly, electrical/electronic assembly, painting, metal finishing, inspection, and testing.

Casting and Foundry Operations

The cost of manufacturing sand castings consists of material, foundry tooling, molding costs, core making costs, grinding costs, cleaning and finish cleaning costs, heat treating or aging costs, inspection, and foundry overhead or burden costs. The most important factors in the cost of producing a sand casting include:

1. The cost of casting metal, and weight of metal poured per mold
2. Melting cost
3. The method of molding (a function of the number of castings produced)
4. The type of pattern used (also dependent on number of castings produced)
5. The weight of the casting
6. The number of castings per mold
7. The number of cores per casting and per core box
8. The core material and weight
9. The coremaking method
10. The number of risers used
11. Type and amount of finishing required

Estimating the cost of sand casting material and how to estimate foundry tooling, including molds, patterns, mold boxes, cores, risers, etc., is covered in Subchapter 7.5. The cost of the foundry labor to pour the molds, and to perform the grinding and the cleaning, is estimated as follows:

If we assume that the automotive cylinder block in Figure 7.24 weighs 100 lb, and we realize a shop yield of 55%, a remelt factor of 40%, and a metal lossage factor of 10% of finished casting weight, then the following weights are calculated:

Shop Yield = 100 lb

Pouring weight = finished casting weight ÷ 0.55

$$= 100 \div 0.55 = 182 \text{ lb}$$

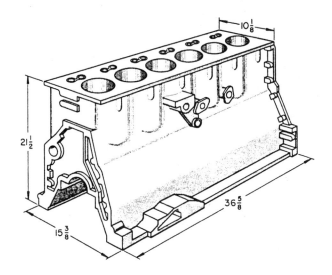

SAND CASTING OF AUTOMOTIVE CYLINDER BLOCK
Class 30, Gray Iron

COST ELEMENT	PERCENT OF TOTAL COST
Metal	13.6%
Core Material	6.8
Conversion	6.4
Molding	23.7
Coremaking	23.1
Grinding	10.5
Cleaning and Gaging	1.6
Finish Cleaning	5.1
Scrap Loss	9.2
Total	100.0%

Figure 7.24   Breakdown of sand casting costs for automotive cylinder block.

Remelted metal weight = pouring weight × remelt factor
= 182 lb × .40 = 72.8 lb

Lost metal = finished casting weight × metal loss factor
= 100 lb × 0.10 = 10 lb

The cost of the metal used in the finished casting is calculated from formulas as follows:

Poured metal cost per casting = pouring weight per casting × (Labor and Overhead + charged material cost)

Cost of metal in finished casting = poured metal cost per castin − (amount of melted metal × values of remelted metal)

If we assume that labor and overhead equal $0.07/lb, charged material equals $0.05/lb, and remelted metal equals $0.03/lb:

Poured metal cost per casting = 182 lb ($0.07 + $0.05) = $21.84

Cost of metal in finished casting = $21.84 − (72.8 lb × $0.03/lb) = $19.66

The estimated cost for cleaning and inspection of some 200 units of automotive cylinder blocks is $530.00 for labor and $25.00 for material. This means that the manufacturing cost for the 200 castings is $4923.00. To this of course, we would add foundry overhead or burden, general and administrative expense, and profit to arrive at the final selling price.

Metal Forming and Fabrication

Fabrication of sheet metal, whether steel or aluminum, accounts for most of the manufacturing today in the production of electronic chassis and cabinets. Many such structural units must withstand the unusual stresses of field military handling, and shipboard shock from naval gunfire, so strength in design and construction is of paramount importance. In the sheet metal manufacturing shop today, there is usually a central principal machine, such as an N/C punch, which is programmed to punch out the various parts that make up the chassis or cabinet assembly in the flat, using sheared blanks of material. These punched-out part shapes are then separated from the blank by notching out the material still holding the part in the blank.

The flat piece parts are then formed to the required shape on a press brake, and assembled by spot welding or riveting, to complete the fully assembled chassis or cabinet. The first series of operations involves shearing the sheet stock of material into correct blank sizes for the N/C punch. The estimated standard times for power shearing of blanks are as follows.

> Power shear setup:
>> Set stops for front or rear gauging, test cut and measure, adjust hold-downs as needed.
>
> Standard Time: 0.200 hr
>
> Run time analysis:
>> The maximum number of cuts required for a rectangular blank is four cuts for four sides. In actual practice the number of cuts is closer to one cut per, blank as each cut that is made frees at least one other blank. On the initial cuts, one cut is actually cutting the side of several blanks.
>
> Standard time: 0.250 hr average time/cut/piece

Unless otherwise noted, parts handling time is included in the run time. Standard times indicated include either large, medium, or small part size.

The N/C punch takes the sheared blanks and performs the blanking, notching, nibbling, and hole punching necessary to shape the individual piece parts to their required flat shape. This is a machine-controlled series of operations. Standard time for machine, turret, and tape reader setup is the only operator controlled time that is involved. Estimated standard times for N/C turret punch operations are as follows.

Table 7.1   Deburring Standard Times: Handling

```
                    GENERAL BENCH BURRING STANDARDS

Machine Parts - Forgings, Castings, Extrusions, Shims, Plates
               Regular and irregular forms (Aluminum Mag. and Steel)

   Tools      - Hand and Motor operated burring devices (Abrasive Wheels,
               Drums, Belts, Rotary and Hand Files)

(Types of burring Methods and Tools are listed below with their respective
Time Standards).


I.   Chart "A" -

     ┌─────────────────────────────────────────────────┐
     │  Set Up                           .15            │
     └─────────────────────────────────────────────────┘

Sign in, study print, obtain tools, clean bench and first piece
Inspection.



II.  Handling Time per part - P.U. and aside after each Burr opertion.
     Chart "B" -
```

Complexity	S	A	C
Weight of Part Length of Part	½# 0 to 6"	½ to 3# 6"to 24"	3 to 10# 24 and over

Per Job:
 	
A. In-Hand
    Toss Aside
    Stack
    Pack in Box
	
	.0008
	.0010
	.0019
		
		.0010
		.0014
		.0024
			
			.0018
			.0022
			.0030

B. In-vise P.U. and Aside	.0014	.0022	.0024
C. Clamped to Table Remove and Aside	.0026	.0042	.0060

and seam resistance welding to electron beam welding. Methods of applying the processes also vary widely. Workpiece thickness and composition usually determine the type of welding process that can be used. For example, the inert gas-tungsten arc (TIG) process is economical for welding light-gauge material, while the semiautomatic inert gas-metal arc (MIG) process is more economical for heavier-gauge materials and nonferrous materials such as aluminum.

Aluminum is not as readily welded as steel, but by using the proper alloy (1100, 5052, 6061) and cleaning prior to welding, the same results and efficiency can be obtained as with steel. Set-up time, including time for part handling, part positioning, and part alignment:

Standard time: 0.300 hr

Table 7.1 Deburring Standard Times (*Continued*): Types of Operations

Burring Time:
    Includes Burr edge and turn to next edge.
    (Code: S=Simple, A=Average, C=Complex)

III.
    Chart "C" –

TIME FOR EACH 10 LINEAR INCHES (EDGE)						
Complexity Code	S	A	C	S	A	C
Weight of Part Length of Part	½# 0 TO 6"	½# TO 3# 6" TO 24"	3# to 10# OVER 24"	½# 0 TO 6"	½# TO 3# 6" TO 24"	3# TO 10# OVER 24"
Material	Aluminum and Magnesium			Steel		
Burr Tool –						
Sanding Drum	.0009	.0012	.0016	.0012	.0015	.0021
Burr Knife	.0009	.0014	.0018	.0014	.0018	.0021
File–Hand	.0018	.0023	.0035	.0023	.0029	.0046
Drill Motor–						
Split Rod & Emery	.0012	.0014	.0019	.0014	.0017	.0023
Wire Brush Threads-10"	.0028	.0035	.0046	.0028	.0035	.0046
Lam. Cloth Burr	.0006	.0007	.0008	.0009	.0012	.0014
Bch.Press – "No Burr"						
Tool – Pac Web	.0007	.0009	.0012	.0009	.0012	.0014
Web (each)	.0005	.0007	.0009	.0007	.0009	.0012
Form 1/32 Radius–						
Sand Drum	.0030	.0036	.0051	.0036	.0048	.0073
Form 1/32 Radius–						
Hand File	.0060	.0072	.0096	.0144	.0168	.0180

TIME PER SQ. INCH			TIME PER SQ. INCH		
Length of Part	0 to 6"	6" to 24"	Over 26"	0 to 6"	6" to 24"	Over 24"
Material	Aluminum and Magnesium			Steel		
Surface Clean-up, Drill Motor & Split Rod- 250 / to 125 / Sand Casting Surface-	.0016	.0016	.0016	.0020	.0020	.0020
Sanding Drum	.0014	.0014	.0014	.0016	.0016	.0016
Hand Scrape Extrusion / Surface	.0012	.0012	.0012	.0016	.0016	.0016

Run times, including normal preheating for aluminum, steel, or magnesium:
    Weld 0.062 stock: 0.250 min/in.
    Weld 0.125 stock: 0.400 min/in.
    Weld 0.250 stock: 0.750 min/in.

Stress relieving using a heat-treat furnace is often required on some parts due to close tolerances or to highly stressed working parts:
    Set-up time: 0.100 hr/occurrence
    Run time: 0.300 min/assembly plus oven or furnace time

Table 7.1 Deburring Standard Times (*Continued*): Ferrous and Nonferrous Materials

	TIME PER INCH			TIME PER INCH		
Material	Aluminum and Magnesium			Steel		
Length of Part	0 to 6"	6" to 24"	Over 24"	0 to 6"	6" to 24"	Over 24"
Remove Forging Flash-Sand Drum 14 & 24ST.	.0072	.0072	.0096	.0080	.0100	.0120
Remove Forging Flash-Rotary & H-File 75ST.	.0140	.0140	.0180	.0150	.0160	.0200
Drill Press-Burr #40 Hole to 3" Diameter.	.0005	.0006	.0008	.0009	.0012	.0012
Machine Burr-Large Holes-per Circular Inch.	.0007	—	—	.0009	—	—
File (Rotary File or Wheel) to straight line: After Sawing	.0015	—	—	.0018	—	—
After Shearing.	.0006	—	—	.0009	—	—
Burr Ends (Rotary File or Wheel) after Sawing	.0005	—	—	.0009	—	—
Flange or Leg (Hand File or Wheel) after Milling or Profiling.	.0005	—	—	.0009	—	—

CALCULATION OF TIME PER PART

Burring: 1. Determine length of part and type of Burring operation.
2. Multiply the Burr operation from III (Chart "C" by Standard per Edge, Inch, Square Inch or Hole.

Handling: 1. Add Handling Time per Length of Part from II (Chart "B") to Burr Time to obtain Unit Time.

Example: Sand Drum Burr, Part 8" long, one edge up to 10", ½ to 3lb.
Burr Time -   .0012
Handling Time per Job -
Clamped to Table   .0042
Time per Part   .0054
Set-up from Chart I -   .15

Spot welding is defined as the welding of lapped parts in which fusion is confined to a relatively small circular area. It is generally resistance welding, but may also be gas-shielded tungsten-arc, gas-shielded metal-arc, or submerged arc welding. Spot welding is one of the most economical sheet metal assembly methods. It has the disadvantage of lack of structural strength, when compared to riveting, for example. A surge of electric current melts the two pieces of metal to fuse them together, and in so doing reduces the metal to the as-cast state, resulting in the weld having less strength than the surrounding heat treated alloy. Setup of a spot welder requires installing and adjusting the contact points, current, and timing. Positioning and adjusting the welding fixture which holds the parts is also included:

Set-up time: 0.400 hr

Spot-weld run time includes time for parts handling and the moving of the assembly from spot to spot, and depressing the foot pedal to activate the machine weld cycle at each spot:

Parts handling time: 0.200 min/operation
Welding time: 0.050 min/cycle

The assembly of minor machine parts using bolts, screws, threaded inserts, and rivets, also by adhesive bonding, eyelets, and nameplate assembly standard times are shown in Figure 7.26. Also shown are standard times for the positioning and alignment of parts for mechanical assembly. Setup time for these mechanical assembly operations is as follows:

Set-up time: 0.400 hr

Electrical/Electronics Assembly

In electrical/electronics assembly the work is very labor intensive, and standard times for cost estimating and other uses should be based on sound engineering work measurement if at all possible. The standard times which follow meet this criterion, although, as before, standards should reflect operations in your plant or facility. Industry standard times such as these can often be tightened up through more efficient methods, tooling, and/or automation.

The standard data in Figure 7.23 provides standard times for most common hand fabrication of wire harnesses and cables, including the hand tinning and soldering of wire

Description	Per occurrence (hr)
Install spline nut	0.008
Install thread inserts (Helicoil-Keenserts etc.)	0.030
Install bolt/screw, washer, and nut	0.014
Install bolt/screw, washer, nut, cotter key	0.025
Install bolt/screw, washer, stop nut	0.025
Install bolt to nut plate	0.010
Install rivet (includes drilling)	0.008
Install eyelet (insert and swage)	0.017
Install name plate per screw (includes drilling)	0.015
Adhesive (application) per square inch	0.001
Hand stamp (per letter and/or number)	0.003
Rubber stamp (ink) per application	0.001

Position and align parts for assembly:

Number of parts per assembly (with fixture or handheld)	Hours
Two (2)	0.030
Three (3)	0.070
Four (4)	0.100
Each additional part over 4	0.030

Figure 7.26   Standard times for minor parts assembly.

breakouts from a harness. The standard times which follow cover additional elements and tasks involved in the make ready, work, and put aside of job order electronic assembly operations. Get parts and prepped wires from a workbench stackbin or wire rack:

Standard time: 0.025 min/occurrence

Get tools from workbench, such as pliers, cutters, wire strippers, soldering iron and solder, and aside tools to workbench:

Standard time: 0.050 min/occurrence

Standard times for the hand assembly of printed wiring boards are given in Table 7.2.

Table 7.2 Standard Times for Hand Assembly of Printed Wiring Boards

Operation	Component	Task	Standard time (min)
Component preparation	Capacitor	Clean leads	0.13
		Tin leads	0.07
		Hand-form leads	0.17
	Resistor	Hand-form leads	0.10
	Diode, coil	Machine-form leads	0.05
	DIP/IC	Preform leads with arbor press	0.57
	Transistor	Pre-cut leads	0.30
	Sleeving	Sleeve component leads	0.20
Bench assemble	PC board	Get and aside, install and remove from fixture	0.30
	Buss wire	Prep, sleeve, and connect ends	0.95
	Capacitor	Install	0.13
	Coil	Install	0.15
	Connector	Install and solder	4.95
	Diode	Install	0.13
	Hardware	Install screws, nut, lockwasher	0.57
	Integrated circuit	Install dual-in-line package	0.35
	Resistor	Install	0.10
	Potentiometer	Install	
	Transistor	Install with pad	
		Cut and clinch (	
Final assembly tool handling		Get and aside ir solder	
Final assembly tool handling		Hand solder lead	
		Wave solder boa assembly	
		Cleaning board i	
		Bag and tag boa assembly	

Finishing and Plating

There are numerous finishing operations, ranging from mechanical buffing and polishing to sophisticated metal chemical processes for passivation and surface protection. *Buffing* is a form of surface finishing in which very little material is removed. The sole purpose is to produce a surface of high luster and an attractive finish. Abrasives such as aluminum oxide, emery, ferrous oxide, rouge, pumice, and lime are applied to the rotating face of a buffing wheel. A composition containing the abrasive is pressed with the work against the face of the buffing wheel. The abrasive is replenished periodically. Table 7.3 provides standard data for calculating standard times for buffing operations.

Polishing is a term commonly used to designate that branch of grinding which employs various types of yielding wheels, cushion wheels, and flexible belts, the surfaces of which are covered or impregnated with some sort of abrasive. Polishing is sometimes referred to as flexible grinding. For finer finishing, emery cloth is employed in preference to an abrasive wheel in many instances. Table 7.4 gives standard data for polishing operations.

Spray painting, utilizing a waterfall paint booth, as compared to a continuous conveyor that allows the operator to spray parts as they pass by, does require parts handling:

1. Handle parts to and from the turntable in the spray booth.
2. Spray part or parts on the turntable.
3. Handle parts to and from a drying rack, drying oven, or a heat bank conveyor.

The major time variables include part size, part configuration, the number of spray painting passes or coats, and the viscosity of the paint or primer. Painting time values assume a part with at least four sides to be painted (no inside surfaces); part size is assumed to be a 30-in. cube.

Part *masking and demasking* covers applying masking tape by rough measure, tearing from a roll, and applying to the part. It also includes removal of the tape after the part is painted. All time values are averages for the purpose of quick, easy application to cost estimates. To set up for a new paint type or color, the following elements of work must be accomplished:

1. Preparate equipment, including paint booth, turntable, and spray gun.
2. Secure paint and fluid tank.
3. Thin paint if required, and transfer to tank.
4. Secure air and paint lines, clean air lines, and attach the spray gun.
5. Attach nozzle, adjust spray gun, and try out.
6. After job is complete, clean above items with solvent and put away.

Standard Set-up time: 0.35 hours

Detail *spray painting* standard time values for estimating include the following:

> Parts handling time: 0.35 min/part
> Wash surface with solvent: 0.05 min/ft^2
> Spray paint (double this for 2 passes):   0.05 min/ft^2
> Brush paint areas not reached by spray painting:
> > Standard time: 0.25 min/ft^2
> > masking tape: 0.12 min/10-in.
> > masking tape: 0.26 min/10-in.

Table 7.3 Standard Times: Buffing Operations

Set-up

Includes - Get tools, prepare work station, check first part and record time.

Set-up	.15 Hrs.

Run Time: (Handling plus Buffing)

A - Handling Time Chart "A" - Includes pick up, positioning & asiding of part.

Part Size In Cubic Inches	Handling Time - Hrs. Per Part			
	Aluminum		Steel	
	Hand Held	Holding Device	Hand Held	Holding Device
To 12"	.002	.008	.003	.010
To 12" to 24"	.003	.008	.004	.010
Over 24" to 48"	.005	.012	.006	.015
Over 48"	.007	.012	.008	.020

B - Buffing Time Chart "B" - Includes periodic application of abrasive.

Type of Surface Finish Required	Hrs. Per Square Inch - Buffing	
	Aluminum	Steel
125	.0008	.0040
32 - 62	.0012	.0060

EXAMPLE:

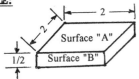

Material: Aluminum - 32 Finish
Surface "A" and "B" to be buffed-Hand Held

Size of Part - 2 Cubic Inches
Surface Area to be Buffed = 5 sq. inches

STEP I	- Set-Up			.15 Hrs.
STEP II	- Handling Time/Unit (from chart "A")	=	.002 Hrs.	
STEP III	- Buffing Time/Unit (from chart "B")			
	5 sq. inches x .0012	=	.006 Hrs.	
TOTAL RUN TIME (STEP II PLUS STEP III)		=		.008 Hrs.

Table 7.4   Standard Times: Polishing Operations

Set-up

Includes - Get tools, prepare work station, check first part and record time.

Set-up	.15 Hrs.

Run Time: (Handling plus Polishing)

A - Handling Time Chart "A" - Includes pick up, position & aside part.

Part Size In Cubic Inches	Handling Time - Hrs. Per Part			
	Aluminum		Steel	
	Hand Held	Holding Device	Hand Held	Holding Device
To 12"	.002	.008	.003	.010
Over 12" to 24"	.003	.008	.004	.010
Over 24" to 48"	.005	.012	.006	.015
Over 48"	.007	.012	.008	.020

B - Polishing Time Chart "B" - Remove scratches and marks.

Type of Surface Finish Required	Flex Wheel and Hand Motor and Emery Cloth Unit Time per Sq. Inch - High Polish	
	Aluminum	Steel
125	.0010	.0017
62	.0014	.0025
32	.0025	.0037

EXAMPLE:

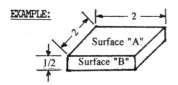

Material: Aluminum - 32 Finish
Surface "A" and "B" to be polished-Hand Held

Size of Part - 2 Cubic Inches
Surface Area to be Polished = 5 sq. inches

STEP I   - Set-Up   .15 Hrs.
STEP II   - Handling Time/Unit (from chart "A") =   .002 Hrs.
STEP III   - Polishing Time/Unit (from chart "B")
                     5 sq. inches x .0025   =   .0125 Hrs.
TOTAL RUN TIME (STEP II PLUS STEP III)   =   .0145 Hrs.

Assemble and remove masking plugs and stencils:
Standard time: 0.10 min/plug

Inspection and Testing

Inspection labor is normally estimated as a percentage of total manufacturing labor. These percentages can range from 5% to a high of 14%, depending on the requirements, the criticality of the part, product, or assembly, and much more. Often inspection is performed by the production workforce, and time is allowed for this in the standard and the cost estimate. A safe rule of thumb, where no previous actual data may exist, is to use 7% of fabrication labor, and 5% of assembly labor for inspection.

Test time is a function of the testing performed, the degree of test automation, whether or not troubleshooting is required, whether or not test data is required and must be recorded, whether the testing is destructive or nondestructive, whether it is mechanical or electrical, and a host of other considerations.

 Test labor can be measured by standardizing the operation, applying elemental time standards, adding time study allowances, and adding a time factor for troubleshooting and subsequent retesting. Some of the electronics testing industry average time data is provided as follows for cost estimating:

Test setup and make-ready/teardown times:
Pick up and lay aside unit: 0.35 min
Remove and reassemble covers: 0.95 min
Hook up and unhook alligator clip: 0.20 min
Hook up and disconnect plug or jack: 0.17 min
Assemble test adapter/fixture: 0.30 min
Test equipment adjustments:
Off and on toggle or rotary switch: 0.07 min
Adjust Variac: 0.11 min
Knob frequency adjustment: 0.13 min
Scope adjustment for phasing, etc.: 0.35 min
Observe scope and analyze pattern: 0.45 min
Unit under test adjustments:
Circuit check with probe/voltmeter: 0.07 min/point
Adjust coil tuning slug: 0.30 min
Adjust trimpots: 0.25 min

Troubleshooting and retest will vary greatly depending on the maturity of the product design, the quality of the assembled unit and its component parts, and more. This can range from 15% to 50% of normal test time. The percentage used must be left to the professional judgment and knowledge of the cost estimator.

7.6.2 Bibliography

Hartmeyer, F. C., *Electronics Industry Cost Estimating Data*, Ronald Press, New York, 1964.
Matisoff, B. S., *Handbook of Electronics Manufacturing Engineering*, Van Nostrand Reinhold, New York, 1978.
Tanner, J. P., *Manufacturing Engineering: An Introduction to the Basic Functions*, Marcel Dekker, New York, 1985.
Vernon, I. R., *Realistic Cost Estimating for Manufacturing*, The Society of Manufacturing Engineers, Dearborn, MI, 1968.

7.7 MANUFACTURING SUPPORT LABOR COSTS

7.7.0 Manufacturing Support Labor Requirements

Manufacturing support labor is a major cost driver in any cost estimate for a manufactured product. It includes the cost of the key functions needed to plan and tool for production of the item to be manufactured. Support labor includes manufacturing engineering, tool design and build, industrial engineering, supervision, and production planning and control. In some cases, the cost of support labor may be equal to or greater than the cost of the factory direct labor. In other cases, manufacturing support labor may be treated as an indirect cost and included in factory overhead. This is the case in many firms that do 100% of their business in the commercial or nondefense areas.

The amount of support labor required on any job is a function of the maturity of the product design, the technology or technologies involved, and the size of the company or program involved. In the aerospace and defense industries, support labor requirements are driven by government requirements for documentation, reports, and a host of other requirements that would not otherwise be needed. Manufacturing support labor may be estimated by first determining the tasks that must be accomplished, the time available to accomplish the tasks, and the level of support required.

Support labor costs are separated into recurring and nonrecurring, depending on whether the support provided is a one-time cost needed to plan and start-up production, or a continuing service to production over the life of the job. The design and fabrication of tools are nonrecurring support costs, but the maintenance and repair of the same tools are recurring costs.

7.7.1 Manufacturing Engineering

Manufacturing engineering includes the following activities and services:

1. Selection and design of manufacturing processes
2. Determination of sequences and methods for product fabrication, assembly, and testing
3. Selection and design of production equipment
4. Selection and design of tools and test equipment
5. Layout of factory buildings, machines, equipment, materials, and storage facilities
6. Determination of standard times for manufacturing operations
7. Selection and design of manufacturing systems and computer-aided manufacturing techniques
8. Manufacturing cost estimating, cost analysis, and cost trade studies
9. Manufacturing research and development
10. Review of product designs and specifications to ensure manufacturing producibility (or participation in concurrent engineering design team)
11. Management, coordination, and control of manufacturing operations

The primary recurring and nonrecurring tasks to be performed by manufacturing engineering must be identified, staffing levels estimated for each task, and then man-loaded and time-phased to determine the estimated manufacturing engineering labor hours.

Development

Manufacturing engineering support during development is for the review of product design as it evolves through the various phases from concept to design layout, and to ensure producibility in its final form when released to manufacturing. Manufacturing engineers also plan the building of any engineering models or preproduction units that are to be fabricated by the manufacturing shops. At the same time, the manufacturing engineers work on and develop the preproduction planning which is implemented after the product is released to production.

This manufacturing engineering effort during the product design and development phase has been often overlooked in the past. It should be estimated and funded as part of the engineering development program for any new product. As indicated earlier, the best way to estimate this effort is to determine the task requirement and the staffing that is required to perform the task. The total manufacturing engineering hours are the staffing level times the number of weeks or months needed to complete the project.

Preproduction Planning

Preproduction planning starts with the production plan prepared as the basis for the estimate, and modifying it and adding details to update it as the actual requirements of the job begin to develop. This includes selecting and designing the process of manufacture, planning the sequences of fabrication, assembly and test, selecting equipment and facilities needed, determining tooling requirements, methods and factory layout, and anything else needed to decide how the product will be manufactured. The preproduction planning effort should begin during the product development phase, and the method of estimating this effort is the same as for development, by man-loading.

Startup and Production Shakedown

Startup and shakedown involves the implementation of the preproduction plan and the initial low-rate production. This is followed by and concurrent with any modification or changes that might be required to debug the process and the tools. In this phase, tools and equipment are ordered, designed, built, and tried out. Work instructions, methods sheets, and operator visual aids are prepared. The factory layout and work flow are implemented, the operations are manned, and initial product manufacture takes place.

Figure 7.27 shows an example of detailed tooling planned for an orbital riveter to be used for a roll swaging operation on a circuit card. An example of detailed operator work instructions is shown in Figure 7.28. This shows part of the instructions for assembling a gear train in an electromechanical subassembly. The man-loading technique for estimating manufacturing engineering man-hours described in the previous sections can also be used in estimating start-up and shakedown support in this phase. It is also possible to estimate the number of tools to be ordered, methods sheets to be written, visual aids to be prepared, etc., to prepare a more definitive budget. Manufacturing engineering hours for these tasks may be estimated by multiplying the estimated time per tool, methods sheet, or visual aid by the total number of each.

Manufacturing engineering support for start-up and production shakedown is part of nonrecurring manufacturing support cost. As such, it should be possible to define the tasks in sufficient detail to prepare an accurate cost estimate.

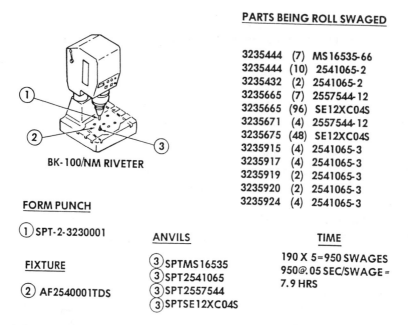

PARTS BEING ROLL SWAGED

3235444	(7)	MS 16535-66
3235444	(10)	2541065-2
3235432	(2)	2541065-2
3235665	(7)	2557544-12
3235665	(96)	SE12XC04S
3235671	(4)	2557544-12
3235675	(48)	SE12XC04S
3235915	(4)	2541065-3
3235917	(4)	2541065-3
3235919	(2)	2541065-3
3235920	(2)	2541065-3
3235924	(4)	2541065-3

BK-100/NM RIVETER

FORM PUNCH

① SPT-2-3230001

ANVILS

③ SPTMS16535
③ SPT2541065
③ SPT2557544
③ SPTSE12XC04S

FIXTURE

② AF2540001TDS

TIME

190 X 5 = 950 SWAGES
950@.05 SEC/SWAGE =
7.9 HRS

Figure 7.27 Orbital riveter tools, parts, and standard times.

Sustaining Support

Sustaining support is the recurring manufacturing support effort provided by the manufacturing engineer after the job is in production. It includes resolution of the production problems that always occur from time to time, incorporation of any engineering design changes which may occur after production has started, and improvements in the methods and tooling which may be realized after the job is on the floor and running. This sustaining manufacturing engineering effort is almost always estimated as a percentage of manufacturing labor. For a newly designed product that has never been in production, 10% of manufacturing labor is recommended. Again, this figure may be considerably higher or lower depending on the industry, the circumstances, and the degree of confidence in the product design.

For a product that has been in production previously and that has a mature design, the sustaining manufacturing engineering support can be as low as 3% of manufacturing labor. The manufacturing cost estimator should review the history of similar jobs to determine the correct percentage for the job being estimated.

7.7.2 Tooling

Tooling can be a significant element of cost in the production of a manufactured product. Tooling is primarily a front-end, nonrecurring cost. The tooling philosophy established by management during the early phases of the proposal or cost estimate will largely determine what the tooling should cost. A heavy up-front tooling investment may ensure a low-cost product in production. A minimal tooling effort may still be the most cost-effective approach, depending on the circumstances and the length of the production run. Certain

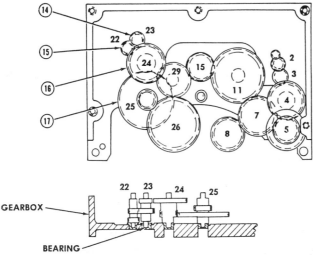

SEQ	PART NUMBER	DESCRIPTION	QTY	REQUIRED
14	39-00286-001	GEAR 23	1	POSITION TO HOUSING PER SKETCH
15	39-00287-001	GEAR 22	1	POSITION TO HOUSING PER SKETCH
16	39-00276-001	GEAR 24	1	POSITION TO HOUSING PER SKETCH
17	39-00275-001	GEAR 25	1	POSITION TO HOUSING PER SKETCH
18	—	—	—	SET HOUSING ASIDE AND POSITION COVER TO WORK AREA

Figure 7.28 Gear train assembly—operator work instructions.

types of manufacturing, such as investment casting and ordinary sand casting, are by their very nature tooling dependent, and as such will always request a large up-front investment in tooling. Other processes may not *require* a heavy tooling effort, but without an effective, up-front tooling investment, cannot be produced effectively in production. The length of the production run can be a big determinant in how much tooling is economically justified. A $2000 assembly fixture amortized over a production run of 200,000 units adds only 1 cent to the cost of each unit of product, but a simple holding fixture, costing $200 and needed for the assembly of a run of 200 units, adds $1.00 to the cost of each and every unit.

Unfortunately, most companies spend far more on tooling than is really necessary. In the metal fabrication and forming industry, for example, it is possible to utilize standard shop tooling on many occasions to accomplish the same job that is often tooled with special-design tooling. It cannot be emphasized too strongly how important it is for the manufacturing cost estimator to review each drawing of a part or sub-assembly. Although the tooling required may appear obvious or apparent, another method may exist, or perhaps shop-aid tooling can be improvised to avoid this apparent expense. Tooling costs, and hence product costs, can be kept as low as possible by utilizing standard tools and tooling components whenever possible, and by using inexpensive tooling materials.

Tool Design

The design of special-purpose tools is estimated by determining how many tools are to be designed, then subdividing this number into tools that are highly complex, tools that are of intermediate level of complexity and detail, and those which are the most simple and straightforward. Average time values for each of the three categories are:

> Highly complex: 24.0 to 40.0 hr/tool
> Intermediate: 16.0 to 24.0 hr/tool
> Simple design: 8.0 to 16.0 hr/tool

If a low-cost or minimal tooling approach is to be used, or one that utilizes little or no formal tool design documentation, then the cost estimator should figure 2.0 to 4.0 hr/tool to prepare a concept sketch or diagram to aid the toolmaker in building the tool. The time includes shop follow-up and liaison while the tool is under construction to clarify any details not clearly depicted on the tool drawing, and for the correction of any errors or design oversights.

Tool Fabrication

Tool fabrication includes the construction and tryout of the tool. The tool build hours almost always exceed design hours by an order of 3 to 1. In foundry tooling, for example, it includes building patterns, pattern plates, blow plates, and flasks, as well as various types of core-making tools.

Core-making tools make up most of the core estimate in foundry work. A pattern is set in the molding sand, and sand is packed around it to produce the impression into which the hot metal is poured to produce castings of a desired shape. Pattern plates separate the two halves of a pattern during molding, and flasks are the containers for the molding sand. Flask size, method of construction, and construction material are all key cost drivers in foundry tooling.

A core is a shaped projection of sand or other material inserted into the mold to create a cavity in the casting. Dry sand cores are formed separately and inserted after the pattern is removed, but before the mold is closed. A plan or layout of the cores in a casting is used to estimate the cost of core boxes, driers to support the cores during baking, blow plates, racks, special containers, fixturing including core-pasting fixtures, and ovens for core baking. The method recommended for estimating core costs uses historical data. The number of acceptable cores made over a given period of time is divided by the cost of making the cores, which is direct labor cost only, thus giving the cost per core.

Tools to be fabricated for most manufacturing operations other than foundry, investment casting, die casting, etc., are estimated on an individual tool basis. That is, the cost of every tool is estimated individually. As indicated earlier, a good rule of thumb for the estimate of fabrication cost is to use three times the design estimate. Tool fabrication cost also includes tool tryout and any subsequent tool modifications that may be required as a result.

Tool design and tool fabrication costs are nonrecurring support costs in the total manufacturing cost estimate. They are concerned only with special-purpose tooling peculiar to a given product, but do not include standard tooling, or perishable tools such as drills.

Tool Maintenance

Another element of manufacturing cost that is often overlooked by the estimator is the cost of tooling maintenance. This is a recurring element of cost and includes everything from the initial tool setup and cleanup if the job has been in production before and tooling exists, to repair, lubrication, painting, and other routine maintenance on operating tools used in production. This can be a major cost element if the production tooling includes automatic, special-design machines and their tools. It also includes the cost of tool preservation, and preparation for storage after the job is complete. Tool maintenance, if a direct charge, is estimated as a percentage of the total factory direct labor. It usually ranges from 3% to 5% of factory labor, depending on the degree of mechanization or automation of the process.

7.7.3 Industrial Engineering

Industrial engineering for manufacturing includes work measurement, standards setting and maintenance, along with methods improvement, operations analysis, and cost analysis. In many companies this is an indirect function that is included in plant overhead. If this function is treated as a direct charge, the cost estimator should handle it by man-loading the industrial engineering personnel working on the job being estimated, and calculate man-hours by multiplying the staffing by the time period during which the services will be performed, and converting this to hours. Industrial engineering support is a recurring cost. In many instances industrial engineering services are provided as part of the manufacturing engineering effort.

7.7.4 Supervision

First-line manufacturing supervision is often a direct charge in companies that produce manufactured products, especially in defense/aerospace industries. As such, it is a recurring cost to the job or program, and should be estimated as a percentage of the total factory labor hours on the job. Unless better information is available in your company, 10% of factory labor is the recommended norm for estimating supervision costs.

7.7.5 Production Planning and Control

Production planning and control includes manufacturing scheduling, material ordering, material control, material kitting, and issue to manufacturing, shop floor control, manufacturing status reporting and expediting, tool control, material handling and movement, work order release, packaging, and shipping. Even with today's sophisticated MRP software programs, the input and output is by people. The effectiveness of this function is critical to the success of manufacturing operations. Without an effective production planning and control system, manufacturing becomes a disjointed, uncoordinated group of activities. In a job order machine shop for example, the scheduling and loading of the machines is critical to meeting promised delivery on all jobs.

Production planning and control labor is often estimated as a flat percentage of total factory labor. This percentage can range from 7% to as much as 15%, depending on the functions and degree of control exercised over manufacturing and production operations. It is also possible to estimate production planning and control labor directly by listing the functions, man-loading the staffing, and then time-phasing to arrive at total hours.

7.7.6 Bibliography

Malstrom, E. M., *Manufacturing Cost Engineering Handbook*, Marcel Dekker, New York, 1984.

Tanner, J. P., *Manufacturing Engineering: An Introduction to the Basic Functions,* Marcel Dekker, New York, 1985.

Vernon, I. R., *Realistic Cost Estimating for Manufacturing,* The Society of Manufacturing Engineers, Dearborn, MI, 1968.

7.8 SUMMARIZING THE ELEMENTS OF COST

7.8.0 Cost Structure of the Final Manufacturing Estimate

In order to arrive at the final selling price of the manufactured product, costs beyond factory costs and profit margins must be determined. This cost and price structure for a manufactured product is shown in Figure 7.1. Prime cost is the sum of direct labor and direct material. Prime cost plus manufacturing overhead yields factory cost, or cost of goods manufactured. Total product cost is obtained by adding selling expense, general and administrative expense, engineering or development costs, and contingencies to the factory cost.

Several methods of *pricing* a product that is manufactured have been derived from the field of managerial economics. These pricing methods range from comparison of similar product prices to selling price based on total cost, with each product yielding the same percentage gross profit. Although many variations exist, there is usually some relationship between selling price and total cost. However, unit selling price is greater than total unit cost in all successful manufacturing firms. In cost estimates prepared for government audit after the contract has been awarded, the maximum profit is regulated by federal law.

In some companies the estimating department's responsibility for total manufacturing cost ends with establishing the direct labor, tooling, equipment, facilities, and material costs. The accounting group or marketing department adds wage and overhead rates from the various applicable cost centers, and performs the extension of the numbers to develop the final manufacturing cost estimate and price.

In structuring the final manufacturing cost estimate, care must be exercised in identifying those elements of cost which are recurring and those which are nonrecurring. In many instances the nonrecurring elements of cost, such as tooling, must be clearly identified to the prospective customer. Should a production contract result from the cost estimate, the tools so identified may become the property of the customer, to be delivered to the customer at contract completion. This mode of operation is fairly standard for contracts involving government work, and some of the major high-rate OEM companies.

If all of the *nonrecurring costs* are pro-rated into the final cost of the manufactured product, and no ownership of tools passes to the customer, it is still vitally important that management know and understand which costs are recurring and which are nonrecurring. Nonrecurring, or one-time costs include the following:

Tooling that is specially designed and peculiar to the product but is not consumed in manufacturing the product, such as milling fixtures, drill jigs, and cable harness form boards

Manufacturing engineering services to do the initial and final production planning for manufacturing the product. This would include process plans, methods sheets, factory layouts, work station layouts, manufacturing bills of material, assembly parts lists, routings, and more

Inspection plans and instructions, test procedures, test fixtures and tooling peculiar to the product

Engineering and development costs incurred in the design of the product, including manufacturing drawings, product performance specifications, parts lists and prototypes and models

Recurring costs include all costs that are contingent on the number of production units built, and include:

Labor and material to manufacture the product, including factory labor, inspection labor, and supervision and test labor, plus all materials and parts needed to build the product, even bulk and consumable items such as solvents, solder, paint, and plating chemicals

Perishable tools consumed in building the product, including cutting tools, drill bits, milling cutters, reamers, broaches, and similar items

Support labor, including manufacturing engineering, production control, engineering liaison, and other labor required to support the manufacture and assembly of the product on the production floor

Recurring and nonrecurring costs can be either direct or indirect. Production control labor, for example, would be both a direct charge and a recurring cost. Perishable tools are an indirect charge but also a recurring cost.

7.8.1 Computer-Assisted Cost Estimating

Without computer assistance, the manufacturing cost estimator must spend a great deal of time performing what are essentially clerical functions. These include researching data files, making many long and arduous arithmetical calculations and extensions, filling out spreadsheets and forms, running copies of originals, and much more. Not only is this very poor utilization of a skilled professional's time, it increases the possibility of error dramatically. With computer assistance, the cost estimating professional can spend the majority of time on basic estimating tasks that fully utilize experience and skill. These tasks include:

1. Obtaining a clear understanding of the requirements of the proposal or estimate
2. Concepting the plan and process of manufacture
3. Estimating the time required for each step or sequence of manufacture, including setups and run times
4. Determining equipment and tooling costs and alternatives
5. Calculating special costs, such as packaging and handling

Computer applications to manufacturing cost estimating also enable improved overall response time to customer and manufacturing needs, an overall improvement in cost estimate accuracy, and greater control and awareness on the part of management.

Computer systems are used in manufacturing cost estimating for data storage, to perform computations, and often to develop complete cost estimates. In data storage, actual cost data from previous jobs is much more accessible than when it is filed by conventional methods. The estimator can obtain process data, material costs, labor hours, and in some cases, burden rates to apply to the job being estimated. The computer can quickly and accurately perform any needed calculations, and prepare cost spreads and breakdowns needed to present the estimating data in the form and format required. Often, with proper programming and under the right conditions, the computer can do the entire estimate. Figure 7.29 shows an inexpensive computer cost estimating system for a small company built around a personal computer and a word processor.

7.8.2 Cost Estimating Examples

Many times the manufacturing cost estimating group is called upon to provide cost analyses, cost evaluations, and cost trade studies for the purpose of deciding which of several alternatives offers the greatest return on investment. The first example below determines the cost-effective breakeven point to change over from manual to automatic assembly of circuit cards used in the guidance system of a cruise missile. The study was critical in deciding if the proposed investment in automatic assembly equipment could be

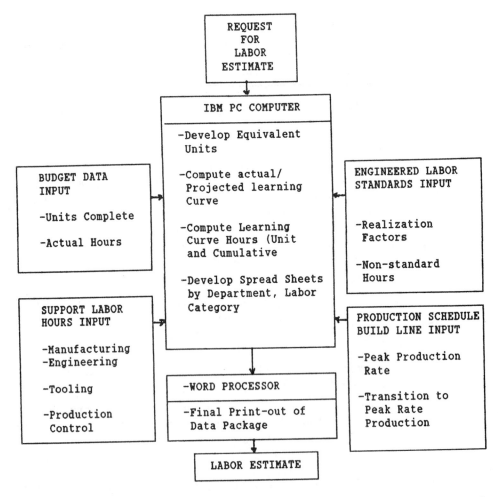

Figure 7.29 Computer-aided cost estimating system for small company utilizing a personal computer.

justified, considering the number of circuit card assemblies that remained to be built on the contract. Such studies performed by the manufacturing cost estimator perform a vital function in the decision- making process.

The second example is a commercial amplifier assembly manufactured as part of a background music system by a small company. The amplifier assembly, consisting of two printed circuit card assemblies wired into a small metal chassis, is a standard product of this company. These units have been produced many times in the past, and all necessary processes, tooling, methods sheets, and test equipment are in existence, ready to be put to work. The company has been requested to quote on building some 200 additional amplifier assemblies. The last build of these units occurred 4 months ago and was for 150 units.

It should be noted that the examples given in this chapter tend to emphasize the development and analysis of labor costs. This is because the skills of the manufacturing cost estimator are built around estimating primarily factory labor cost. This should not make the estimating of material costs any less important. As indicated earlier, material can account for up to 70% of product cost.

Circuit Card Assembly Auto Insertion Cost Trade

Basic Assumptions:

1. Eighteen boards per shipset, one hand insert board not included.
2. Fifty shipsets used as estimating quantity.
3. Rework estimated at 3%.
4. DIP inserter cost estimated at $79,000, with price increase by 5% in 6 months.
5. Axial leaded components requiring strain relief to be formed on the automatic lead former, and inserted using Man-U-Sert. Some 34% of axial leaded components require strain relief in leads.
6. Loaded labor rates used (includes manufacturing overhead):
   Assembly @ $11.66/hr
   Tooling @ $15.69/hr

Nonrecurring Tooling and Programming Costs:

1. Automatic sequencer: 18 programs × 7.0 hr/program × $15.69/hr = $1,977.00
2. VCD auto inserter: 18 programs × 7.0 hr/program × $15.69/hr = $1,977.00
3. Tooling: 4 tool orders × 1.5 hr/design × $15.69/hr = $94.00
   175 hr of tool fabrication × 4 tooling plates × $15.67/hr = $8,630.00
4. Total tooling and programming = $12,678.00
5. Man-U-Sert inserter: 10 programs × 2.5 hr/program × $15.69/hr = $392.00
6. Automatic DIP inserter: 17 programs × 7.0 hr/program × $15.69/Hour = $1867.00
7. Tooling: 4 tool orders × 1.5 hr/tool order × $15.69/hr = $94.00
   50 hr tool design + 125 hr tool fabrication × 4 tooling plates × $15.69/Hour = $8,630.00
8. Total tooling and programming = $10,591.00

VCD Automatic Inserter Method Standard Times:

1. Set up automatic sequencer: 0.050 hr
2. Load automatic sequencer: 0.033 hr/reel
3. Automatic sequencer run time: 0.00065 hr/component

4. Set up VCD inserter: 0.083 hr
5. Load and unload inserter: 0.0079 hr/Board
6. VCD inserter run time: 0.00065 hr/component

Man-U-Sert Inserter Standard Method Times:

1. Setup automatic lead former (including length and part number changes): 0.165 hr
2. Automatic lead former run time: 0.00035 hr/component
3. Setup Man-U-Sert inserter: 0.067 hr
4. Load parts trays, place on machine: 0.0065 hrs/tray
5. Load and unload circuit board: 0.0079 hr/Board
6. Run time to insert component, cut and clinch leads:

   Axial leaded component: 0.0038 hr/component
   Dual-in-line package (DIP):   0.0085 hr/component
   Radial leaded component (3 leads plus pad):   0.0106 hr/component

Automatic DIP Inserter:

1. Set up automatic DIP inserter: 0.083 hr
2. Load magazines: 0.0100 hr/tube/part number
3. Load/unload board assembly: 0.0079 hr/board assembly
4. Run time: 0.0006 hr/component

Equipment Setup Times:

Sequencer: Load program from tape, enter count required into the computer, load reels of components into sequencer, setup the reels.

VCD Automatic inserter: Load program from tape, load taped components onto reel, feed components into insertion head, verify sequence, assemble board holder to rotary table.

Man-U-Sert: Load program from disk, load trays into rotary bins, adjust board holder arms.

DIP inserter: Load program from tape, load magazines in proper sequence on machine, feed component into insertion head, verify insertion sequence, assemble tooling plate to rotary table.

Analysis

All components are currently programmed for the Man-U-Sert manual inserter. Some (592) components could be sequenced and run on the VCD automatic axial lead component inserter. This would leave 137 axial leaded components to be run on the manual inserter because of the requirement for strain relief when the leads are formed.

Some (299) components that are now inserted manually could be run on an automatic DIP inserter. Of the 19 different board configurations in the program, 18 are loaded on the Man-U-Sert, and 1 is totally hand loaded.

Proposed Method

1. Use automatic sequencer, and VCD automatic inserter:
   18 programs and 4 tooling plates required
   All components (axial leaded) would have to be purchased on reels

2. One (1) Man-U-Sert program would be required.
3. Automatic DIP inserter to be used for integrated circuits:
   17 programs and 4 tooling plates required.

(The programming time could be reduced by one-half with the use of a program generator and a host computer.)

Seven different board configurations require work not covered in the analyses detailed above. These include adding jumper leads, piggyback components, removing circuit paths from the printed circuit board, and hand cleaning and soldering of the solder joints required to make these changes. The estimator's detailed analysis of this work is not shown, but a review of the bid-file actual working papers shows 6 additional pages of calculations for setup and run times for these three circuit board assembly methods. The costs in hours for each of the three assembly methods are:

Assembly Method	Setup time	Run time	Total time
Man-U-Sert	2.4870	294.68	297.167
Man-U-Sert + VCD	6.669	239.97	246.634
Man-U-Sert+ VCD+ DIP Inserter	7.1150	134.67	141.78

These figures include lead strain relief, rework at 3%, and additional work described earlier. The labor hours derived from standard times are then applied to an 85% learning curve to adjust for learning that would be realized for quantity increases. This curve, Figure 7.30, shows the actual learning curve currently being realized for the Man-U-Sert method of hand assembly.

Finally, the breakeven curves are drawn for both of the proposed new methods of assembly. These lines indicate the quantity of circuit card assemblies that would have to be produced using the proposed methods to pay back the investment required in machines, tooling, and software programs, plus methods and operations sheets required for new method implementation. These plots are shown in Figure 7.31 for the Man-U-Sert + VCD inserter versus Man-U-Sert only, and in Figure 7.32 which shows the DIP inserter + VCD inserter + Man-U-Sert vs. Man-U-Sert only. The breakeven point for Figure 7.31 is 19,500 circuit board assemblies, and for Figure 7.32 is 95,000 circuit card assemblies. The second breakeven point in Figure 7.32 results if a DIP inserter is found to be surplus in another division of the company.

Job Order Assembly Cost Estimate

Basic Assumptions

1. Assume an 85% learning curve based on previous production experience.
2. Standard time per unit attained at unit 1000.
3. Some learning would transfer from previous production run, allowing point-of-entry on the learning curve at unit 12.
4. All necessary processes, methods, tools, test equipment, and work instructions currently exist.
5. Amplifier units will be manufactured in two lots of 100 units each.
6. Material required is either already on hand or on order, thus allowing almost immediate start-up of production operations.

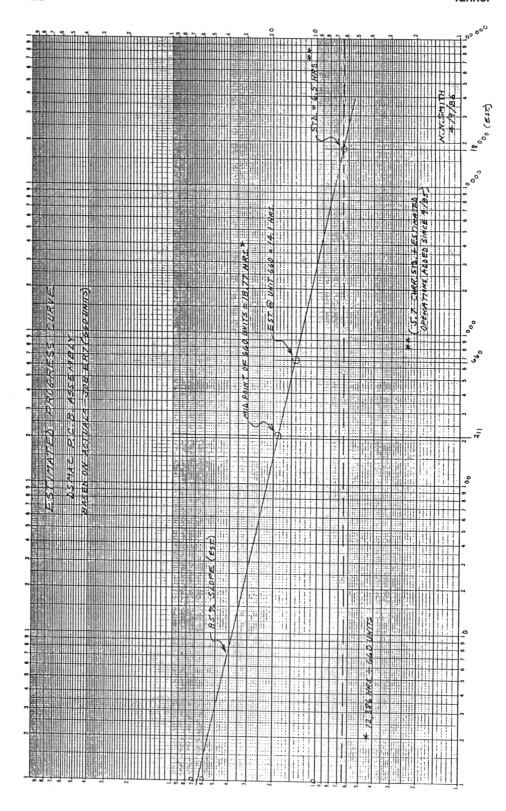

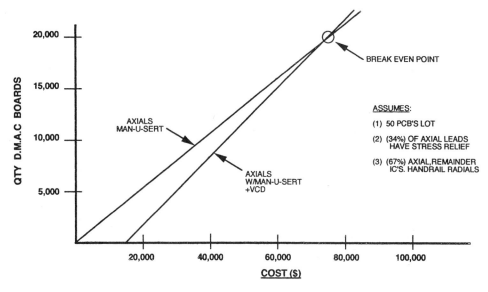

Figure 7.31 Breakeven curve when variable center distance (VCD) inserter is added to Man-U-Sert operation.

7. Make-or-buy plan provides for buying all circuit cards, component parts, wire, connectors, metal chassis, cover, hardware, terminal boards, heatsinks, sockets, lugs, and consumables such as solder, grease and paint.
8. Off-line operations to be performed in house include chassis paint, mark and drill, terminal lug installation on terminal boards, all wire cutting, stripping, and marking operations.
9. Standard times for all production operations currently exist, and are derived from original time studies of operations.

Preproduction Plan

The operation process charts shown in Figures 7.33 through 7.35 fully describe the plan of manufacture for the amplifier units. These three operation process charts provide the information that is required to prepare the base labor estimate. By totaling the standard minutes shown on the charts for the various steps and then dividing by 60, we have the total standard time in decimal hours for the two subassemblies and the main amplifier assembly:

(2000-972) Circuit card assembly	0.065 hr
(200-973) Circuit card assembly	0.47 hr
(S/A-100) Amplifier assembly	1.27 hr
Total	2.39 hr

If we assume that the units will be produced in two lots of 100 each, then there will be two setups in addition to the run time shown above. Setup time is estimated to be 4.50 hr.

Figure 7.30 Learning curve for original printed circuit board assembly using Man-u-sert.

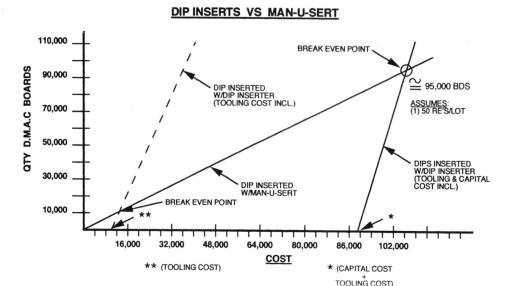

Figure 7.32 Breakeven curve when DIP inserter and VCD inserter are added to Man-U-Sert operation.

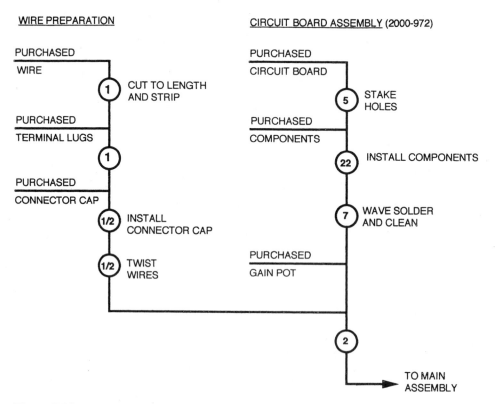

Figure 7.33 Operation process chart for main circuit board assembly.

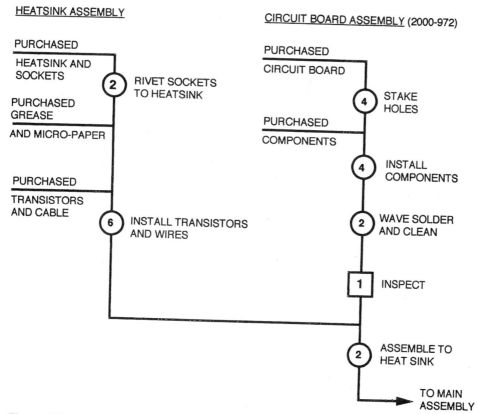

Figure 7.34 Operation process chart for power circuit board.

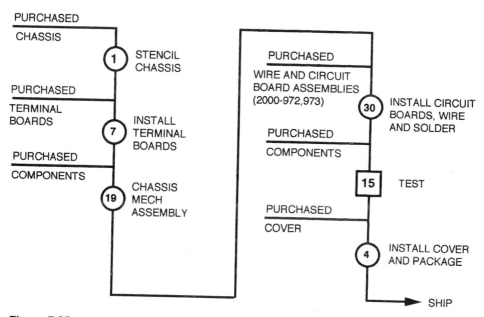

Figure 7.35 Operation process chart for main amplifier assembly.

With two setups we would have a total of 9.00 hr for setup and 478.00 hr of run time for the amplifiers, giving a standard time of 2.44 hr per unit for the 200 units.

Learning Curve Application

Previous production of the amplifier units followed an 85% learning curve, with standard being attained at unit 1000. Since the amplifiers have been manufactured previously, some learning would transfer to these units, allowing a point of entry on the learning curve at unit 12, giving an average time per unit of 5.15 hr. This represents what the run of 200 units will actually cost to produce. It should be kept in mind that the 85% learning curve slope is a composite slope which includes factory mechanical and electrical assembly labor, inspection labor, and test labor. If this customer has purchased units built earlier, he will expect a lower price this time.

Summary of Amplifier Costs

We next multiply the total labor hours by the hourly rate to obtain factory labor in dollars. The factory support labor is determined and multiplied by its hourly rate to determine support labor cost. Both of these labor dollar numbers are then multiplied by the factory burden or overhead rate to obtain indirect dollar costs, and are added to the factory and support labor dollars to obtain total dollar labor costs. Total material dollar costs are determined from the priced bill of material, and from the cost of available residual inventory already on hand. We then add material overhead or burden to give burdened material dollars. The sum total of burdened labor dollars and burdened material dollars gives the prime manufacturing cost for the units.

Selling price of the units to the customer is determined by adding the prime manufacturing cost to the general and administrative expense plus profit. This example is somewhat simplified in that no consideration is given to contingency factors, delivery schedule, or peak production rate requirements, all of which could add to the final selling price of the amplifier units.

After a careful check of all calculations, and review by management, the labor hours to manufacture the 200 units may be spread or time-phased over the planned build rate or schedule to determine monthly and even daily and weekly planned hourly expenditures. At this point, the selling price of the 200 amplifier units should be ready for transmission to the prospective customer.

7.8.3 Bibliography

Malstrom, E. M., *Manufacturing Cost Engineering Handbook,* Marcel Dekker, New York, 1984.
Matisoff, B. S., *Handbook of Electronics Manufacturing Engineering,* Van Nostrand Reinhold, New York, 1978.
Tanner, J. P., *Manufacturing Engineering: An Introduction to the Basic Functions,* Marcel Dekker, New York, 1985.
Vernon, I. R., *Realistic Cost Estimating for Manufacturing,* The Society of Manufacturing Engineers, Dearborn, MI, 1968.

8

Product Tooling and Equipment

Clyde S. Mutter   *CMS Defense Systems, Titusville, Florida*

8.0   INTRODUCTION TO TOOLING AND EQUIPMENT

Webster's dictionary defines the word *tool* as "any hand implement, instrument, etc., used for some work, or any similar instrument that is part of a machine." You could ask a hundred people what they think of when they hear the word "tool," and get answers anywhere from a screwdriver to a highly exotic instrument. The number and types of tools appear to be endless. The manufacturing engineering task in the *tooling world* of manufacturing is to understand all these tooling choices and select the right type of tool for each job. Think of the ME as the conductor of an orchestra, who must pick the correct instruments to produce the results that a critical audience is expecting

Planning and tooling are essential in bringing a product design into production. Tooling ensures proper fit and helps guarantee a repeatable process. Planning establishes the processes and establishes the timing at which the parts come together to make the finished product. Tooling and planning are two areas which can significantly affect the cost of a product.

Product tooling can range from inexpensive hand tools to highly sophisticated automatic assembly machines, from motion economy devices for simple parts assembly to highly complex milling fixture for precision machining. Product tooling can mean molds for die casting, or dies for extrusion. Tooling requirements for any manufacturing operation depend on the process requirements, the size and type of product, and the quantity of products to be manufactured. The idea and principles discussed in this chapter apply to virtually any type of process or product.

When production levels are high over a period of several years, extensive and sophisticated tooling is easily justified. When production levels are low, or limited to a single run, the tooling must still perform its intended functions, but it is difficult to economically justify anything but the lowest costs. A $2000 assembly fixture amortized over a production run of 200,000 units adds only 1 cent to the cost of each unit of product, but a simple holding fixture, costing $200 and needed for the assembly of a production run of 200 units, adds $1.00 to the cost of each unit. The factory costs of low-volume or short-run products therefore may depend not only on the direct labor and material costs of production, but also to a very large degree on the costs of tooling. Tooling costs, and hence product costs, can be kept as low as possible by following the rules outlined below.

1. Utilize standard tools and tooling components whenever possible.
2. When special tooling is required, utilize the design concepts of previously well-engineered tools and fixtures.
3. When special tools and fixtures are required, use low-cost tooling materials in place of tool steel wherever possible.
4. Maintain close communication among the manufacturing engineer, the tool designer, and the toolmaker through all phases of the tooling program.

The planning and definition subchapter discusses the task of the manufacturing engineer, with the help of others, that decides the type of tooling or equipment to be used. This is a very important step and must be given much thought. If the product is new to the company, establishing a tooling philosophy is a good first step. The tooling philosophy is the guideline of the type of tooling to be used to build a product. Elements to be considered for setting the tooling philosophy include:

1. Quantity/rate product is to be built
2. Length of time product will be marketed
3. Estimated cost/value of product
4. Product design
5. Cost trade-off studies for tooling
6. Skill level of production workers
7. Profit margin required
8. Method of moving parts from station to station

The tooling philosophy and the production plan go hand in hand, with many people helping make the decisions.

Design, documentation, and control is an area that is often overlooked or minimized budget-wise by upper management. One of the reasons to invest in tooling is to have a repeatable process. Without control of the tool design and the physical tools, the process is in jeopardy. As the tools become more complex, the documentation becomes even more critical for building and maintaining them.

Tool construction and maintenance is as important as the design. There is nothing worse than having an expensive tool in the tool room for repair. It should be on the production line making money. Building the tool with proper materials and workmanship, plus a good preventive maintenance program, saves both time and money.

Selecting the proper hand tool or other standard tool to do a job is a process all manufacturing engineers need be aware of. The cost of automating some operations may be very expensive, whereas the hand-eye coordination of a skilled worker, with the proper hand tool selection, can perform the same function at a much lower cost, with better reliability and the added advantage of flexibility. A tool should never be designed if it can be purchased off the shelf.

Proper coordination during the planning phase should define the type of assembly tooling required. This is followed by refining the many small details that must be considered.

The assembly task may require only a minor holding device, or a large, complex piece of equipment to maintain alignment of key features of the product. Consideration must be given to the load/unload portion of the assembly job, as well as the movement of parts to the workstation, and the subsequent transport of the assembled product to the next position.

Special machinines and high cost go hand in hand. Special machines should never be ordered or designed without a great deal of thought and deliberation. However, a good special machine can make a very significant contribution to the success of a company. The ability to envision the machine function and process, being aware of the latest in off-the-shelf machine components such as part positioners, sensors, computer controls, etc., will be a great help in knowing what can be easily achieved. Computerized simulations of the workplace, and the overall factory flow of the product, may be valuable assets in this field (see Subchapter 11.1).

Inspection and test tooling is another area where off the shelf equipment has made great strides. In the last few years, the amount of instrumented equipment modules which are easily incorporated into a tool has greatly increased. This reduces the cost to make tooling for inspection or testing requirements into state-of-the-art, easy-to-maintain, and quick-to-build equipment.

8.1 PLANNING AND DEFINITION

8.1.0 Introduction to Planning and Definition

During the production planning phase for the manufacture of any new product, or the continued production of existing lines of products, it must be decided what approach will be used for tooling. In many instances, the process itself dictates the tooling requirements, leaving little room for deviation. This is the case for foundry and investment casting operations, as well as die casting, injection molding, and similar processes. In the high-rate production of automobiles and appliances, tooling is the key to successful production of a quality product, and little is spared to ensure that the tooling is appropriate and will do the job.

However, many companies spend far more on tooling than is necessary. In the metal fabrication and forming industry, for example, an experienced manufacturing engineer can specify standard shop equipment on many occasions to accomplish the same job that is tooled with special-purpose tooling.

This is also true in metal removal operations. In the past, precision machining often required elaborate fixtures to ensure accuracy and productivity. Modern numerical controlled (NC) equipment may require only a holding fixture in addition to any special cutters, drills, or form tools. It is important to review each product drawing and ask yourself, even though the tooling approach *seems* apparent, what other methods exist, and whether standard tooling can be modified or improvised to avoid this expense.

8.1.1 Tooling Preplanning

The tooling philosophy should reflect a low-cost yet fully adequate tooling approach. In an increasingly more competitive world, the company that is successful is the one that goes a step beyond the standard tooling approach, and finds ways to do things better and cheaper. This requires the knowledge, experience, and resourcefulness of the manufacturing engineer. The tooling preplanning should determine how the low-cost tooling philosophy will be carried out.

Reviewing the Product

The first and most important phase of any new tool or fixture design is the pre-planning accomplishment by the manufacturing engineer. The ME must review the entire fabrication or assembly operation, define the tool or fixture requirements, and consider such factors as:

How the part or assembly will be held or clamped in the tool or fixture

What orientation and accessibility the tool must provide the machine or operator so that the work may be performed most efficiently

What relationship the tool will have to the workplace (if an assembly tool), parts bins, hand and power tools, and other equipment required for the operation

The spatial relationship between the product, part, or assembly, and the operator to provide an efficient and comfortable working position for that operator (assembly tool)

To understand the importance of tooling preplanning, it is necessary to explore in more detail the tooling requirements, how the tool is conceptualized, critical tool features, tool design standards, and the importance of maintaining a standard tool inventory.

Defining Tooling Requirements

The requirements for the tool must be thoroughly understood by the manufacturing engineer, since this is basic to all else that now follows. The ME must thoroughly understand *why* a tool is needed, what *function* it will perform, and what *process* parameters it must support. The first step is to prepare some type of process flow chart on the assembly steps, followed by preparation of the parts fabrication process sheets (see Chapters 2 and 7).

In the injection molding field as an example, an understanding of all the conditions that prevail during the molding operation should ensure that necessary precautionary measures are incorporated into the mold design to safeguard the expected features in the product. The function of an injection-mold type of tool is to receive molten plastic material ranging in temperature from 350 to 750° F at pressures between 5000 and 20,000 psi. In the injection process, the plastic comes from a heated nozzle and passes through a sprue bushing into feed lines, or runners, and then through a gate into a cavity. The cavities are maintained at lower temperatures, at which solidification takes place. The range of temperatures and pressures depends on the type of plastic material. The plastic is held in the cavity of the mold for a prescribed time, until full solidification takes place. At this point, the mold opens, exposing the part to the ejection or removal operation. From the general concept of the molding operation, it is recognized that it is important to design a mold that will safely absorb the forces of clamping, injection, and ejection. Furthermore, the flow conditions of the plastic path must be adequately proportioned in order to obtain, in cycle after cycle, uniformity of product quality. Finally, effective heat absorption from the plastic by the mold has to be incorporated for a controlled rate of solidification prior to removal from the mold. The tool designer must incorporate all details that are conducive to good molded parts (see Chapter 17).

Investment castings are made in both small lots and very large lots, and range from the simplest design to extremely intricate configurations. Many different methods of tooling are possible. If the requirements are small or moderate, a single-cavity tool may be sufficient to make the patterns (see Figure 8.1). The pattern rate from a single-cavity,

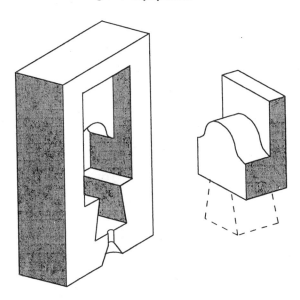

Figure 8.1   Single-cavity mold for investment casting patterns.

simple die can vary from 20 to 50 parts per hour. A highly automated single-cavity die with hydraulic openers and compressed-air pattern removal can produce as many as 250 patterns per hour. Automated dies, machined from steel or aluminum, are used when thousands of patterns a month are necessary. Once this tooling is ready for production, it is expensive to make changes. For this reason the product designer must be sure of the product design before the manufacturing engineer proceeds with development of high-rate production tooling. If an area of possible design change is known, the manufacturing engineer should be involved so that provisions for such changes can be incorporated in design of the die at the original concept stage. An alternative to this traditional approach is to include rapid prototyping in the process (see paragaph 8.1.2, and Subchapter 1.4).

A well-thought-out tooling concept for a special tool to open the insulation track in aluminum window sash bars is shown in Figure 8.2. This was an especially troublesome problem in the manufacture of aluminum window frames, as the extruded shape of the sash bar was such that the insulation track was necked down at the sash bar cutoff operation, requiring an extra manual operation with a screwdriver to open the track. The new tool cut the time for this operation in half. The concept sketch tells the tool designer exactly what kind of tool to design and what the tool is expected to do.

Figure 8.3 shows a concept for a "lazy susan" fixture to hold and turn large window frames while caulking and insulation strips are applied to hold the glass panels in place. Previously this required two operators to physically pick up the frame and turn it 180° on the workbench. Now one operator can turn the window frame by depressing a foot pedal, without leaving the workplace.

An excellent way to assist the manufacturing engineer in developing a tooling concept during the preplanning phase is to provide a manual that catalogs, illustrates, and describes tooling that was successful in the past. Such a manual also shows applications of standard tools and tooling components that the company has in its tooling inventory. These would include such items as small vises, stops, quick-acting clamps, ball joints or swivel

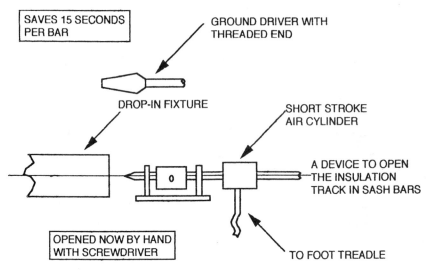

Figure 8.2   Tool for spreading window sash bar.

fixtures, rotary fixtures, air cylinders, small electric motors, punch-and-die sets, and so on. Reference to such a manual provides the manufacturing engineer with ideas that might help with current tooling requirements, and could uncover a design or application of a previously used tool that might be used with minor modification for the new tooling application.

Communication

It is one thing for the manufacturing engineer to know the tool required for the fabrication or assembly operation, but often another thing to adequately communicate these concepts to the tool designer. Concept sketches such as those shown in Figures 8.2 and 8.3 go a long way toward conveying this information, as does the written tool order used in many companies. Nothing takes the place of direct verbal communication between the manufacturing engineer and the tool designer. The tool designer must understand how the operation

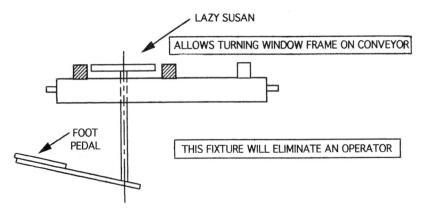

Figure 8.3   "Lazy Susan" tool for turning large window with foot pedal.

is to be performed and must discuss with the manufacturing engineer precisely what is required of the tool. The tool designer must learn critical dimensions and features, as well as desired tool locating points and surfaces, and stops and clamps required. The expected life of the tool in terms of anticipated total production of the product should be discussed, as this is the basis for the degree of durability required of the tool. For example, tooling made of wood or plastic components in place of tool steel can, in some cases, provide the required durability for low-volume, short-run production and reduce the total cost of tooling by 40% or more. In many shops, the tool designer prepares the concept sketches as part of the tool definition process. This aids in the mutual understanding of the tooling requirements, the advantages and disadvantages, a preliminary cost estimate, and schedule impact of the various choices available for an operation.

8.1.2 Rapid Prototyping

A few years ago, product design for plastic parts, and for cast or forged metal parts, was a long, drawn-out exercise. The design engineer would have a master machinist hog-out a part from a drawing or sketch. The part would be evaluated and changed, a new part made, evaluated, etc. In the meantime, the manufacturing engineer and the rest of manufacturing waited for the product design. Today's rapid prototyping techniques have made changed this process. Now, using CAD/CAM, models can be made quickly. The flexibility is available to adjust the design and see the results in record time without affecting schedules or cost. Manufacturing can now proceed earlier with expensive hard tooling, without concern about last-minute engineering changes. As an added bonus, the models can be used to prove-out other tools.

Armstrong Mold Corporation (East Syracuse, NY) is one firm that specializes in rapid prototyping. Armstrong has the ability to quickly produce prototype plaster mold metal castings and cast thermoset plastic parts—often within 1 to 2 weeks for simple parts and 1 to 2 months for complex parts.

The latest rapid prototyping techniques include stereolithography, laminated object manufacturing, and CAM hog-outs from solid stock, as well as traditional pattern making and craft skills. With all of these skills and methods available, premium-quality parts can be produced in a short time at relatively low cost. Some of the advantages of rapid prototyping are:

Paperless manufacturing—work directly from CAD files
Functional parts as quick as 1 week
Produces models which can be used as masters for metal castings and plastic parts in small or large quantities
Fit and function models available to detect design flows early in development

CAD file formats available include:

IGES, DFX, CADL, STL
Floppy diskettes, 3.5 or 5.25 in.
Tape: mini data cartridge, DC 2000 style

Stereolithography

Steolithography can be used to create three-dimensional objects of any complexity from CAD data. Solid or surfaced CAD data is sliced into cross sections. A laser that generates an ultraviolet beam is moved across the top of a vat of photosensitive liquid polymer by a computer-controlled scanning system. The laser draws each cross section, changing the liquid polymer to a solid. An elevator lowers the newly formed layer to recoat it and establish the next layer's thickness. Successive cross sections are built layer by layer, one on top of another, to form the three-dimensional plastic model.

Laminated Object Manufacturing

Laminated object manufacturing (LOM) is another method of creating three-dimensional objects of any complexity from CAD data. Solid or surfaced CAD data is sliced into cross sections. Thin plastic sheet materials are laminated one on top of another. A single laser beam cuts the outline of each specific layer. The process continues until all layers are cut and laminated, creating a three-dimensional, woodlike model.

CAM Hogouts

Three-dimensional objects can be cut from a variety of solid stocks (metal, resin, plastic, wood) directly from two- or three-dimensional CAD data using multiaxis CNC machining centers.

8.1.3   Tooling for Special Machines or Robots

Within recent years, in addition to special machines, the robot has become a very useful tool for manufacturing. It has found its place in material handling, spot welding, seam welding, spray painting, machine tool tending, and numerous other applications. In all these cases, sensors are employed to exclude uncertainties—that is, to detect them and react appropriately. In the context of assembly, these uncertainties can be traced to two factors: uncertainties from tools (feeders, fixtures, and machines) and uncertainties in the parts themselves (manufacturing tolerances). There are two basic ways to handle these uncertainties in a manufacturing environment:

1. Avoid all uncertainties in the assembly planning phase.
2. Detect uncertainties with the aid of sensors.

The first possibility is the traditional approach, which gives rise to inflexible, expensive, and time-consuming systems, especially in the construction of specialized feeding devices. Such a solution makes quick product changes uneconomical because of long changeover times.

A compromise between these highly sophisticated and therefore expensive systems and conventional grippers can sometimes be achieved by mounting some sort of compliant device between a conventional gripper and the robot. These compliant devices consist of two metal plates connected by elastomeric shear pads, and are known as remote center compliance devices (RCCs). RCC devices can compensate positioning faults resulting from reaction forces in the assembly phase (see Subchapter 8.6).

8.2   DESIGN, DOCUMENTATION, AND CONTROL

8.2.0   Introduction to Design, Documentation, and Control

Many factors enter into the proper design of jigs, fixtures, and other tools that are not apparent to the casual observer. A finished tool may seem to be nothing more than rough castings or built-up pieces of steel with a few seats and straps to hold a part in place. The study of the manufacturing engineer to provide a proper sequence for fabrication or assembly, the care with which the tool designer lays out those straps to hold the work without springing it, and the accuracy with which the toolmaker sets the seats and bores the holes do not show on the surface. The real value can be easily determined, however, when the tool is put into use.

It will be found that tool design is not a haphazard, hit-or-miss proposition. Each manufacturing concern must decide upon the quality of work that it wishes to produce, and it must adapt certain standards of design and tolerance which will produce that quality. This subchapter will explore the duties and constraints of the tooling functions that are responsible for designing, building, and controlling the tooling in modern industry.

8.2.1 Control of Tooling

After the decision has been made to design and build a tool, the next step is to actually perform the work. Some type of system is needed which controls the tool through every aspect of the tool's life. A tool order is the normal way that companies authorize a tool to be built. A common system or procudure for processing the tool order is shown in Figure 8.4. In larger companies, much of this may be done by computer, but the steps are essentially the same. The tool order form will vary from company to company, but the data required is quite consistent. The tool order will contain the following information:

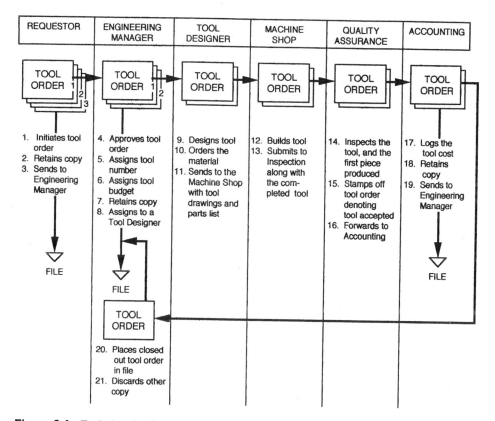

Figure 8.4 Typical tool order control system.

1. Description of the tool function
2. Product part number to be used in the tool
3. Production rate the tool will support
4. Quantity of tools required
5. Life expectancy of tool
6. Special conditions the tool will see (hot, cold, wet, etc.)
7. Safety considerations
8. Expected skill level of user

Tool Orders

The tool order is the primary document used by the manufacturing engineer in most companies to formally authorize the design and fabrication of tools. Figure 8.5 shows a tool order format for a small to medium-sized company which contains all of the necessary information blocks to convey information to the tool designer for the design of the tool. In addition to conveying information, it is a formal work-authorizing document for design and toolroom work. It is used by accounting to identify and collect costs by individual tool for a given job. All tool order forms should be similar to this form.

When the tool drawing is completed and checked, the manufacturing engineer should review the design and accept the design concept. At that time, he or she should estimate the cost of building the tool and review the economics of the tool and the manufacturing

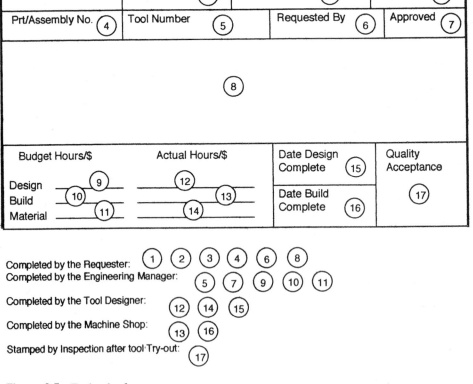

Figure 8.5 Tool order format.

operation to determine whether a more or less elaborate tool would be more economically feasible. On certain projects, the budget for tooling may have been established as the driving factor. In other cases, the product fabrication or assembly cost may permit spending more money on tooling.

Although the tool drawing should be so complete that no other communication with the toolmaker is needed, a few minutes of explaining the function of the tool is usually time well spent. If the toolmaker knows what is expected of the tool, he or she can normally do a better job of building that tool, and can often offer suggestions for improvements or cost reductions during the design phase.

Upon completion of the tool, both the tool designer and the manufacturing engineer should observe the tool try-out or, if shop conditions permit, try out the tool themselves by performing one or more complete operations with production parts. Any required debugging or modification can be accomplished before starting manufacturing operations. All changes and modifications made during tool build and tool try-out should be recorded so that the tool master drawing can be updated to agree with the final configuration of the tool.

Once a tool is complete through tool fabrication, tool tryout, and final acceptance, control of the tool is still required. All tools require some form of maintenance and may have to be recalibrated periodically. Do not expect the production floor to remember to do these tasks. When the tool control system works properly, the *system*, with recall tickets or orders, will do the job.

8.2.2 Design Considerations

Once the necessary concept and descriptive information have been transmitted to the tool designer via the tool order, concept sketch, and verbal communication from the manufacturing engineer, it becomes the tool designer's responsibility to communicate adequately with the toolmaker. The primary medium of that communication is the formal tool drawing. Beyond following standard drafting practices, tool designers should consider the following to assist in communications and reduce tooling costs:

1. Tool drawings should describe the tool completely and also specify optional construction in noncritical areas wherever possible.
2. Tool tolerances should be specified as loosely as possible, with every attempt made to avoid extremely tight tolerances.
3. Tool drawings should be dimensioned completely, so that scaling is not required by toolmakers.
4. Noncritical screw and dowel locations should not be dimensioned; instead, a note such as "locate approximately as shown" should be indicated.
5. A complete material list should be included on the first sheet of the tool drawing.
6. A list of standard sizes and shapes of materials stocked in the toolroom should be available to each tool designer, and materials should be specified from this list whenever possible.
7. Clamps, stops, locating pins, rest buttons, and hold-down screws should be located so that the part or assembly will not be distorted or have its surface damaged.
8. The tool drawings should specify marking the tool number, part number, and any other numbers to allow positive identification of that tool in the future.
9. The current revision letter or number of the part or assembly should be noted on the tool drawing and also on the tool.

The details of tooling design depend on several factors, such as cost, quantity of work, and the ingenuity of the designer. There are some suggestions to bear in mind, with differences of opinion as to whether or not they are listed in the order of their importance:

Simplicity
Rigidity of clamping devices
Sequence of operations
Interferences in tool itself
Interferences with the machine on which used
Clearances for work and hands
Avoidance of chip pockets
Locating points should correspond with dimensions on part drawing and with locating
    points on other tools for the same part
Convenience and speed in operation
Accuracy of work produced
Durability
Economy of construction
Stock sizes of material used

In making the drawing, it is necessary to consider the accuracy as to scale, the correctness of the projections, proper representation of the fixture and the work, and its reproduction qualities. The dimensions on the drawings must be accurate and sufficient without unnecessary repetition, legible, and contain the tolerances in understandable form. Dimensions which are out of scale should be underlined to call attention to that fact. Drawings must also contain information or specifications as to materials, finish, the kind of heat treatment if any, finish of surfaces by grinding or otherwise, necessary notes, instructions for marking part numbers, and the title of the part to be used in the tool.

After all the general rules of design are covered, more decisions are to be made. Some companies require every detail of a tool be drawn, which adds to the design time, but reduces the fabrication time by allowing many people, including a machinist, to work on the tool at the same time. When duplicate tools or replacement details are to be made, we can be assured they are all alike. Other companies prefer the design to be as basic as possible with only a few key dimensions provided. This saves design time, but requires the tool to be built completely by one toolmaker.

The most important consideration in design of tooling for production is the *product*. The dimensions, finish, quantity, and ultimate use of the part in the completed product will usually dictate the *processes* available to achieve the desired results. This process definition will then aid in determining the machine or other equipment to be used with the tool. The following are some examples of the choices, and the relatively complex factors, involved in defining the tooling requirements—and therefore the details of the *tool design*.

8.2.1 Points to be Considered in Design

Patterns

Where patterns are to be made there are four major considerations:

Economy of construction
Ease of molding

Equality of sections to avoid unequal shrinkage
General appearance

Wherever possible, stock patterns should be used.

Drilling

A part requiring 14 operations of drilling, reaming, countersinking, and counterboring to be done in one jig, can be handled in several ways. The operations can be performed on two eight-spindle gang drills, on three six-spindle machines, or even on four four-spindle machines by passing the job from machine to machine and returning it to the original station on a conveyor. This, of course, requires several duplicate jigs and is one method for large production. The same jig can be used on any one of the machines by carrying out eight, six, or four operations, according to the number of spindles, and by finishing up with a change of tools in quick-acting chucks for the remaining operations. Any of these methods may be said to have good points in that all operations are performed in a jig at one setting—all holes being necessarily in proper relation with each other. One method ties up machine tools, and the others expend labor hours. If, for the same part, two or three jigs are made for the eight or six spindle machine, a more flexible arrangement results. It is possible to group the larger holes in one jig and the smaller in another, gaining time by using a faster speed for the small holes. With modern NC machines the drill jig may sometimes be replaced with a simple holding fixture, where the accuracy and repeatability is supplied by the machine.

Press Blanking

In the metal fabrication industry, there are many choices available for stamping out the flat pattern of a sheet metal part. As shown in Figure 8.6, a simple rectangular blanking tool can be made from four pieces of low-cost hardened steel measuring 6 in. by 6 in. Two of the four pieces will act as a base for spot-welding the punch and die portions, and for mounting the two guide pins. One other, which will become the punch portion, is cut to the exact size of the hole it will punch. It is then spot-welded to the 6 in. by 6 in. base. The remaining piece is now sawed or punched, leaving a cutout in the piece of the desired size plus 0.007 in. This becomes the die, and the 0.007 in. is for die clearance (clearance will vary according to the material to be blanked). This piece is now mounted to the other base, again by spot-welding. Two guide pins are inserted in opposite corners, or in the center of the cutout in the female portion of the base. Two holes of the same diameter as the pins, plus 0.003 in. for clearance, are added opposite the pins in the other base, which

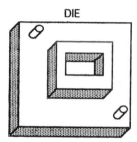
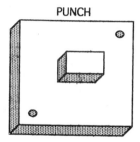

Figure 8.6 Simple blanking tool.

is the punch. Slide one onto the other, using the pins for alignment, and the blanking tool is complete. It can then be mounted in a standard die set, which maintains alignment and provides for attachment to the punch press. The guide pins can now be removed.

This tool is capable of piercing 800 large holes per sharpening. It can be sharpened twice before being discarded. It would take 6 hr to make this economy tool. For comparison, a tool with the same function made in a tool shop by a more standard method using more durable materials and a dedicated die set would cost many times more, plus a number of weeks for delivery. This tool would have great value, in that similar tools could be made in any good sheet metal shop by a journeyman mechanic. The part tolerances would be about the same as for a permanent tool. Under normal conditions, this would be plus or minus 0.005 to 0.010 in. The real differences of tool life, automatic feeding, and the like may not be as important for the part as lead time or initial cost.

Brake Forming

On many press-brake bending jobs, solid blocks of urethane and other elastomers can be used to replace conventional machined dies. The savings, of course, are great. The value of the urethanes lies in their ability to transmit kinetic energy. Urethane, unlike rubber, is virtually incompressible. If a blank is forced into a block of the material by a punch, the energy of the punch is transmitted almost evenly in all directions. If the material is confined by a retainer, the energy is reflected back by the retainer walls, since it has no place else to go. It will force the blank against the punch with uniform pressure. The advantages of urethane tooling can be summarized as follows.

1. Low tool cost. Urethane is normally used in block form with minor relief to direct metal flow. It can be machined with ordinary shop equipment.
2. Minimum setup time.
3. Extreme flexibility.
4. Nonmarring of surface finish.
5. Excellent part definition when properly used.
6. Production of sharper bends or smaller radii than with metal dies.
7. Sharp reduction of springback and wrinkling.
8. No compensation needed for variation in stock thickness.

Urethane is now accepted as a standard die material, and its use is standard practice in most progressive sheet metal shops. Techniques of this type may eliminate designing and building dedicated punch-and-die sets to do the same job.

8.2.2 Documentation

It was only a few years ago that tool documentation meant a decision between a drawing on Mylar or vellum, and using pencil or ink. While there are still occasions for manual drafting, most tool design is accomplished using Unigraphics, CADD, AUTOCADD, or one of several other well-known computer-aided design programs. Drawings are now received electronically, and tool design data can be sent directly to N/C machines to make the tool (see Chapter 6). Newer and better programs are available every day and most are well within the price range of even the smallest company. It is important to remember that while the computer is an aid to design preparation, the same human thought processes are required for good design. Computers can be very good tools when used correctly, or expensive toys when not. A computer cannot improve the design of a tool; it just makes it

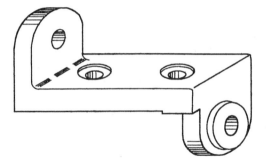

Figure 8.7 Basic isometric sketch of part.

easier for the designer to work. One important consideration for using CAD in the design of tooling is the ability to make changes or modifications. The basics of a good design are still in the hands of the designer. Standards have been established on how the basic designs are arranged, to allow the toolmaker to be able to read a drawing without misinterpreting the design intent.

Arrangement of Views

For drawings in orthographic projection, the third-angle system, known in Europe as American projection, has been in practically universal use in the United States for many years and is continued as the American standard. A brief discussion of this practice will be based on sketches of the object shown as Figure 8.7. In third-angle projection the top view is placed directly above the front view, the right-side view to the right of and facing the front view (Figure 8.8).

Figure 8.9 shows the relative positions of the six possible principal views of an object sometimes needed to describe a part or tool; front, top, right side, left side, and bottom. A bottom view, or "view looking up," can be used to advantage instead of a top view when the shapes or operations to be shown are on the underside of the part. For example, in a normal punch-and-die drawing, the arrangement of views would be as in Figure 8.10, with the view of the bottom of the punch placed in the position of the bottom view and the top

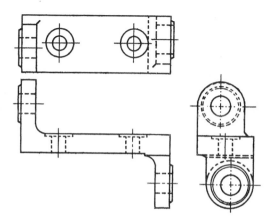

Figure 8.8 Normal third-angle projection.

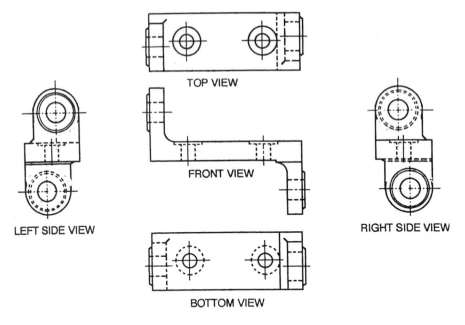

Figure 8.9 Additional views of a part are sometimes needed.

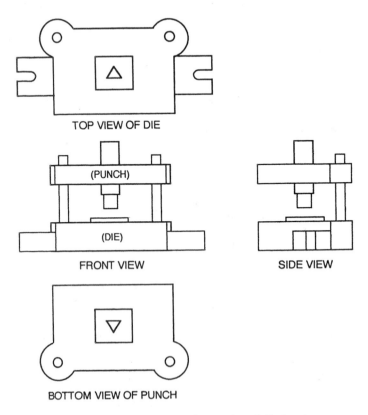

Figure 8.10 Normal arrangement for punch-and-die drawings.

of the die in the position of the top view, each facing the front view. In case of lack of space, this arrangement can be modified by placing the drawing of the bottom of the punch to the right of and in line with the top view of the die, as if it were turned over from the top view (Figure 8.11). In drawings where any such arrangements of views are employed, the views should always be carefully titled to aid in reading.

For objects for which two side views can be used to better advantage than one, these need not be complete views of the entire object, if together they describe the shape of the object (Figure 8.12). Only those views should be drawn that are absolutely necessary to portray the shape of the part clearly. Often two views will suffice, and many cylindrical parts may be portrayed adequately by one view if the necessary dimensions are indicated as diameters.

After the design is documented, a decision is made on whether tool usage and maintenance instructions are required. Complicated tools and machines always require these two items as part of the documentation package.

8.2.4 Tool Construction

As discussed in Section 8.2.1, tooling is controlled and tracked per schedule by the tool order system. It is important to have this system to ensure that the tooling arrives on the production line at the correct time. Another item needed is the actual cost of the tool. If the tool is fabricated by a vendor, this information is available from the purchase order. When the tool is fabricated in house, it is much more difficult without a

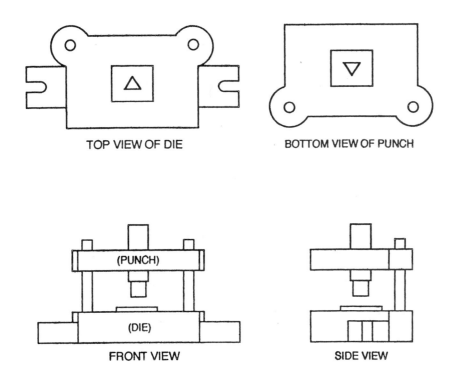

Figure 8.11 Alternative punch-and-die drawing arrangement.

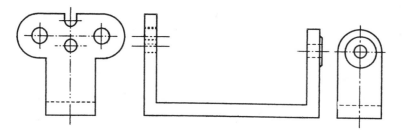

Figure 8.12 One or two views may be sufficient to describe an object.

good system and people willing to use it. Listed below are a few reasons that the cost information is valuable.

1. As costs are collected against each tool, they should be compared to the estimates. We can see if the estimates were correct, or if the estimating technique needs to be corrected up or down. This exercise should be never-ending, and over the years the estimates will become quite good.
2. Knowing the real cost of tooling is important when duplicates of the same tool are required. Estimates of the cost can then be made with a great deal of confidence.
3. Having a large file of actual cost data to support your estimates will prove invaluable, since future cost estimates can be made with a great deal of confidence.
4. The data may also be used to help determine how competitive your in-house tool fabrication shop is versus having the work done by a vendor. This can be done by comparing costs on similar type tools, and the results are often valuable.
5. Knowing the true cost of the tools on a production line is necessary to determine product cost.
6. Tools are tangible company assets and may play a role in the tax structure of a company. The actual dollar value of company tooling can be depreciated each year, or sometimes expensed in the year the tool cost is incurred. This difference may be important in future bidding for the same product.

8.2.5 Acquisition of Special-Purpose Machines

Acquisition often becomes a major task, whether the machine (or robot) is to be used in a fabrication shop or in assembly of the product. Most special machines are designed and built by companies that specialize in this type of work. In the writer's experience, the simpler, more straightforward machines may be built in-house, but usually not the larger, more complex ones.

The task of bringing a new machine on line may logically be divided into two phases. The first phase is prior to placing a contract with the outside firm, and the second phase is after placement of the purchase order.

Most of these machines are quite costly. Their timely installation and subsequent performance is very important to the success of the product cost, delivery schedule, and perhaps the success of the company. Therefore, all levels of company management will require accurate and timely reports on the progress of the project.

Activities Prior to Purchase Order Placement

Activities prior to order placement are probably the most important activities, requiring the most analysis and in-house user coordination. Some things to consider are:

Assign a project manager or task leader.
Make a detailed schedule of activities.
Prepare good flow charts of the operations.
Coordinate with product design, tool design, quality, safety, maintenance, and training experts.
Involve the production shop and the purchasing agent.
Prepare a detailed machine specification.
Send the specification to several suppliers.
Review the supplier proposals for technical approach, price, and delivery time.
Select a Supplier.
Coordinate with the supplier to agree on final changes to the technical statement of work, price, delivery, etc.
Place purchase order.

Note that most of these same concerns would apply if the user plans to design and build the machine in house.

Activities After Order Placement

Close coordination with the machine builder is usually valuable during final design, construction, debugging, and prove-out. Travel cost should be added to the price of the machine during evaluation. Items to consider during this period include:

Update master schedule.
During design, consider the maintenance problem.
Watch for safety issues.
Make sure that any computer software is documented.
Check final drawings for "as built" configuration.
Review the training manuals and maintenance manuals.
Insist on accurate status reports during build.
Plan on debugging run at the supplier.
Provide plenty of prove-out parts. These must be *exactly* representative of the production parts.
Maintenance personnel should observe the final operation and prove-out at the supplier.
Final acceptance should be in your plant after final prove-out runs.
Never pay out more money in progress payments than justified by the machine progress.

8.3 CONSTRUCTION AND MAINTENANCE

8.3.1 Tool Construction

Proper construction materials and tool fabrication are established by the tool design, but it always helps to perform a liaison with the toolroom. Listed below are a few topics which should be reviewed as tool construction progresses:

Is the tool too heavy for its use? (Too light?)

Has the environment the tool will see been addressed (expansion/contraction, rust/corrosion)?

Have all the wear points been addressed?

Will the tool hold up on the production floor?

Will there be any high-maintenance details?

Should the tool be painted?

After fabrication of the tool, the following items should be checked:

Have all sharp edges and corners which could injure the operator been removed?

Do all the moving parts function as designed?

Are there any high-maintenance details for which spare parts should be ordered?

8.3.2 Tool Maintenance

After spending large amounts of money and time on the design and construction of a tool, it may be ignored until it breaks. Most companies have systems, procedures, or policies on tool maintenance, but often the system fails. The tool is turned over to production, where the driving goal is to produce an item, and there seems to be little time for anything else. The missing principle all must believe is that keeping a well-functioning tool on line is more important than continuing to use a tool until it stops, and then wait for repairs.

Before a tool is released to the production floor, the manufacturing engineer, tool designer, and quality engineer should decide the type of maintenance and frequency required. Many tools will require very little maintenance—perhaps a quick trip through the toolroom for a cleanup. Tools that require some sort of regular maintenance should be the focal point. A computer can be one of the best tools when setting up a maintenance system. Establish the maintenance that needs to be done, the time frame (weekly, monthly, etc.), and program the computer to extract work tickets at the proper time for maintenance. Tools requiring lubrication are the ones most neglected. A lubrication chart, which shows the correct type of lubrication needed, how often, and when the task was performed, should be attached to each tool. Any number of systems can be established, but without discipline and management's complete backing, none will give the desired results.

8.4 HAND TOOLS

8.4.0 Introduction to Hand Tools

Hand tools are tools that most of us are very familiar with. We all have used a screwdriver, wrench, hammer, etc. Visiting a hardware store or thumbing through a hand tool catalog will remind you that there are hundred of different types of tools. A well-stocked tool crib or use of a catalog library will save time and money. Use all the resources at your disposal to find the right tool for the job before having a special one designed and built. Hand tool salespeople in your area can be helpful if given the chance. The wise engineer uses this cheap resource to full advantage.

8.4.1 Hand Tools

Many hand tools are now air, electric, or pneumatic driven. Proper setup on the production floor is essential for proper function and tool life. For air tools, you must ensure that the tool obtains the correct air pressure and oil mixture; electrical tools need proper grounding and correct voltage; pneumatic types need proper line pressure. Working with experts from the fields of industrial hygiene and industrial safety is recommended to assist in the selection of new hand tools. Ergonomics should be included in this selection. Plant layout and facilities engineering are also involved in providing the correct infrastructure to support the use of various hand tools. A maintenance schedule with a recall system should be set up to clean, functionally check, calibrate, and replace worn parts.

A computer is the perfect device to use when setting up a maintenance system, and in scheduling maintenance requirements. At the time of purchase, most power hand tools come with an instruction booklet which provides care and maintenance information. This information may be entered into the computer along with a date-recall program. When the maintenance schedule time is reached, the computer will print out a recall ticket with all the data required. This system will help plan the spare parts required, and prevent any unnecessary duplication. It can also help in the planning for future overhead cost requirements by allowing you to see the condition and age of the equipment presently available.

8.4.2 Tool Cribs

An inventory of commonly used standard tools and tooling components should be built up specifically to meet the requirements peculiar to your shop and your plant operations. This inventory would include standard cutters, end mills, broaches, reamers, drills, taps, and similar metal-cutting tools if your manufacturing operation is machining. Also included in an inventory for a machine shop would be a variety of different chucks, collets, and hand tools such as files, scribes, deburring tools, vernier calipers, and micrometers.

Tooling components such as drill bushings, quick-release clamps, tool balancers, and small vises would also be included. The manufacturing engineer should determine what the standard tooling inventory should contain, and should specify the items to be stocked, from whom they should be ordered, and how many of each to order. Expendable tools and cutters should be placed on "max-min," which means that when the actual inventory of an item reaches a certain minimum quantity, an order is immediately placed for a pre-determined number of the items to bring the inventory back up to the desired level.

Inventory

The standard tool inventory is usually stocked and issued from a central tool crib. This tool crib can be a part of the manufacturing engineering department, but more than likely is under the production control department jurisdiction. The tool crib will also stock items such as shop rags, work gloves, lubricating and cutting oils, C-clamps, machinist's scales, safety glasses, and shop aprons.

If the plant operations include structural and mechanical assembly, the standard tool inventory will include welding rod, standard clamps and holding fixtures, ball-peen hammers, power nut runners, combination wrench sets, socket wrench sets, screwdrivers, adjustable wrenches, pliers, clamping pliers, rubber mallets, power screwdrivers, rivet squeezers, and welding and brazing equipment. If operations include electronic assembly and wiring, the standard tooling inventory will include soldering irons and soldering iron tips, solder pots, temperature-controlled soldering irons, soldering aids, heat sinks, spools of cored colder, side-cutting pliers, needle-nose pliers, long-nose pliers, chain-nose pliers, assorted small brushes, and plastic squeeze bottles. Also included would be such items as holding fixtures and power arms.

8.5 TOOLING FOR PARTS FABRICATION

8.5.0 Introduction to Tooling for Parts Fabrication

Producing parts for the assembly line may seem rather straightforward. Looking a little deeper, you will find it very complex. A production part can have as many different and varied process steps as a complicated assembly operation. The cost of producing these parts will also have a major impact on the final cost of the assembly being produced. The manufacturing engineer must always be on the watch to reduce or eliminate any machining or process steps. A good understanding of all machine shop equipment, methods and practices is a must (see Section III). With this knowledge as a base, sound decisions can be made. One of the worst problems on the assembly line is late or poor-quality production parts.

8.5.1 Study the Part to Be Made

When you are reviewing a part for fabrication, look at it as the most important part in the assembly. Learn everything there is to know about the part. Listed below are some good questions which should be answered before deciding on tooling:

1. Why was the part designed the way it was?
2. What does it interface with?
3. If there are attach holes or mating fits, has a tolerance analysis been performed?
4. What are the total number of parts that will ever be needed?
5. Is the part slated for redesign in the future? If so, when?
6. Will the part be sold as a spare or replacement part?

8.5.2 Tool Planning Process

With an understanding of the history and future of the part, you can plan the fabrication process as economically as possible. The plan should start when the raw stock is picked and continue to the delivery of the part to the production line. Break down each process and list all the equipment available to you to perform the process. Cost trade-off studies may be performed to decide whether a machined part requires a conventional or an N/C machine. Each piece of equipment requires different types of tooling, setup, and machine time. For a sheet metal part, the choice may be between a die or forming on a press brake and other assorted hand operations. Once the cost trade-off studies are completed and the equipment decision is made, the fabrication plan for the part can be finalized. Using the plan, discuss with the quality engineer where the best inspection points will be for the part. Critical holes or surfaces may need to be inspected long before the part is completed, or some type of in-process quality control plan established.

8.5.3 Ordering the Tool

The tooling required for the part can now be ordered. Each tool order will describe in detail which operations have already been performed, what operation is to be done, and which machine the tool is to be used on. Most tooling for parts fabrication is straightforward when the tool designer uses good design practices (see Subchapter 8.2). Examples are back-up of the part to eliminate chatter during milling, and locating the part in the tool from targets or key starting positions the way the design engineer dimensioned it on the product drawing (see Chapter 1).

Punch Press Dies

Dies are another story, since most dies are sent out to be built by a die shop. Tool drawings are usually not provided. Extra care must be given to this type of tooling. Even after finding a die shop with a good reputation, you will still need to ensure that all necessary data is given to them to get your desired results. Writing a total *scope of work* is a good way of doing this. This will not only help the vendor understand the task but will also ensure that the purchasing department knows what your wishes are. The following is an example of a scope of work for a combination die (Courtesy of McDonnell Douglas Florida Missile Plant).

SCOPE OF WORK FOR COMBINATION DIE
  1.0 SCOPE
       This specification covers the requirements to design, build, ship, install, and debug, a combination die to fabricate Flat Pattern Blank of AWM8020 (Det. 3) Rev. "F."
  2.0 DESIGN REQUIREMENTS
       2.1 Format drawings are not required. Vendor to furnish prints of all shop drawings with strip layout, die stations and regrind instructions.
  3.0 APPROVALS
       3.1 MDMSC-FMP required approval points are:
            3.1.1 Vendor to supply (25) sample parts for MDMSC-FMP and approval prior to die shipment.
       3.2 Final approval will occur only after an in house tryout.
            3.2.1 Vendor to assist in initial setup of die at MDMSC-FMP with in house tryout.
            3.2.2 Final approval will be given from 3 parts, which meet flat pattern specifications, that are taken from a random sampling of 10 parts.
  4.0 TOOL REQUIREMENTS-GENERAL
       4.1 Die must be constructed to provide maximum safety for the operator and meet or exceed:
            4.1.1 MDC TFIM 70.101.
       4.2 All components are to be of first line quality and conservatively rated for the given application.
       4.3 Die to be identified as "PROPERTY OF MDMSC-FMP" and "CDAWM8020-3TDS."
       4.4 Die to be capable of producing a minimum of 78,000 parts.
       4.5 Die to have quick change punches with supplier part number.
       4.6 Die set to be 4 post ball bearing precision.
       4.7 Die to have miss-feed detector.
       4.8 Die to have stock pusher.
       4.9 Die to have scrap chopper and chute.
       4.10 Die to have spring loaded strippers.

        4.11   Feed arm 6″ high to bottom of material.

        4.12   Strip/coil stock alum. alloy 6061T-6, .032x7.5 wide.

        4.13   Top of die shoe to conform to sketch on Page 3.

5.0   TOOL DESCRIPTION

        5.1   The tool is to be a progressive die to produce flat pattern to print revision as stated above.

        5.2   The part is to be run length wise thru the die.

        5.3   Production rate 300 min. per week.

6.0   PRESS DESCRIPTION

        6.1   Type: Komatsu 110 ton press

        6.2   Model: OBS 110-3

        6.3   Serial No.: 11813

        6.4   Stroke: 4.0

        6.5   Shut Height: 14.96

        6.6   Slide Adjustment: 3.94

        6.7   Slide Stroke: 5.97

        6.8   No. of Slide Stroke: 32-65 SPM

7.0   SYSTEM SPECIFICATIONS

        7.1   MDMSC-FMP will supply part material required for tool tryout. The Vendor should notify MDMSC-FMP when material will be required.

8.6 TOOLING FOR ASSEMBLY

8.6.0 Introduction to Tooling for Assembly

The techniques described in this subchapter are concerned with minimizing the cost of assembly within the constraints imposed by the design features of the product. The specific focus of this section is to discuss the assembly tooling required to support the assembly process selected. During the normal course of developing a new product design, the manufacturing process must be developed progressively and concurrently. Chapter 7 describes a preliminary manufacturing plan that is required in order to estimate manufacturing cost and pricing. This plan should be finalized as the ultimate product design (see Chapter 1) is in work. Information that was used to develop the manufacturing costs, including quantity, production rate, delivery schedule, and the like, should be utilized in establishing the tooling philosophy. The factory layout, as described in Chapters 3 and 4, also influences tooling requirements.

Assembly cost is determined at the product design stage. The product designer should be aware of the nature of asembly processes and should always have sound reasons for requiring separate parts, and hence longer assembly time, rather that combining several parts into one manufactured item. Each combination of two parts into one will eliminate an operation in a manual assembly workstation, and usually an entire workstation on an automatic assembly machine (see Subchapter 1.3).

The assembly process should be developed early in a program, since the lead time for assembly tooling may be greater than the time needed to make the detail part fabrication tools. Also, in a normal program, the final piece part assembled configuration may be the last design drawing completed. The assembly process will often dictate the critical parameters needed of some of the key parts, and will therefore influence the tooling required for parts fabrication.

8.6.1 Assembly Tooling Systems

Assembly tooling can be divided into automated or manual systems. Production process flow charts should help define the system. Assembly tooling can also be grouped by requirements for performing work operations, making checks or tests, or by transportation and movement. Special requirements may consider painting or other finish processes. As the assembly process matures, the need for duplicates of some of the assembly tooling will become apparent. This may be the result of "choke points" in the assembly line. Subchapter 11.1 discusses the technique of computer sumulation of production assembly flow. Manual assembly differs widely from automatic assembly due to the differences in ability between human operators and any mechanical method of assembly. An operation that is easy for an operator to perform might be impossible for a special-purpose workhead or robot.

8.6.2 Manual Assembly

In manual assembly, the tools required are generally simpler and less expensive than those employed on automatic assembly machines. The downtime due to defective parts is usually negligible. The direct operator labor portion of manual assembly costs remains relatively constant and independent of the production volume. Manual assembly systems also have considerable flexibility and adaptability. Sometimes it will be economical to provide the assembly operator with mechanical assistance in order to reduce the assembly time. Assembly tooling for manual operations will require careful consideration of ergonomics (see Subchapter 5.1). The problem of repetitive operations causing poor worker performance and injuries is well understood today, and is becoming an increasingly important factor in both the processes and the tooling provided. The selection of pneumatic squeezes and other power tools should be considered as the assembly tool develops, since the solution may lie partly in the assembly tool design and partly in the ancilliary equipment selected.

Concept sketches of the planned assembly process and assembly tooling are an important part of understanding the details involved in setting up the production line. Computer simulation, as described Subchapter 11.1, is a valuable technique for analyzing the assembly line. Choke points, requirement for duplicate stations, and the like will become apparent. Options can be simulated, and final decisions made as to the final production process. Manual assembly can be performed by an operator at a fixed work-bench, or by multiple operators utilizing some sort of transfer device to pass work progressively down a line. Parts may be provided in tote pans or supplied by single-purpose parts feeders (see Figure 8.13).

All resident production experience should be involved in the assembly process. The product design, quality, shop supervision, safety, industrial hygiene, and other elements should provide suggestions and recommendations from their point of view. The manufacturing engineer often acts as a coordinator, or team leader, in developing the optimum process required for a world-class manufacturer. With this information and process definition, the assembly tool designer now has a good basis for actual design of the assembly tooling.

8.6.3 Automated Assembly

Automated assembly can be broadly broken down into special-purpose machines which have been built to asemble a specific product, and programmable assembly machines, where the workheads are more general-purpose and programmable (see Figure 8.14).

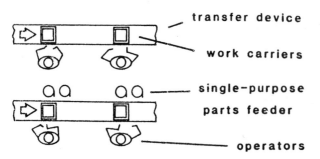

Figure 8.13 Manual assembly systems.

Special-Purpose Machines

The assembly machines consist of a transfer device with single-purpose workheads and parts feeders at the various work stations. The transfer device can operate on an indexing (synchronous) principle or on a free-transfer (nonsynchronous) principle. These special purpose machines are expensive and involve considerable engineering development before they can be put in service. The downtime due to defective parts can be a serious problem unless the parts are of relatively high quality. Also, it must be appreciated that these machines are designed to work on a fixed cycle time and are therefore inflexible in their

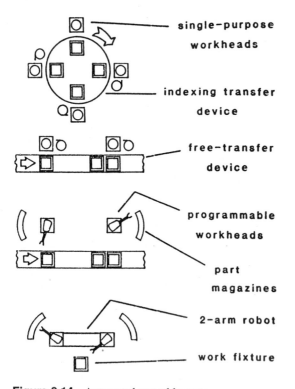

Figure 8.14 Automated assembly systems.

rate of production. If they are underutilized and cannot be used for any other purpose, this will result in an increase in asembly cost.

Programmable Assembly Machines

Programmable assembly machines can allow for more than one assembly operation to be performed at each workstation. This provides for considerable flexibility in production volume, greater adaptability to product design changes, and different product styles. For lower production volumes, a single robot assembly workstation may be preferable. Parts are normally available at the workstations in manually loaded magazines, since the use of parts-feeding devices of the type employed for special-purpose machines is usually not economical. In the case of an automated assembly process, end effectors must be designed and built.

Most automated equipment today will require a software program. This may be a document of programming steps for a programmable controller (see Chapter 6), or digitized programs on a floppy disk or other electronic media for a computer. In any case, these will become an important part of the "tool documentation" and is often controlled as such.

Within recent years, the robot has become a more useful tool for manufacturing. Most of the current assignments of robots are rather simple and repetitive. The robots are employed by mass producers of consumer goods, such as the automotive and appliance companies. An area where the robot has had very little impact so far is assembly. Most assembly opertations are being conceived for human assemblers who have two dextrous hands, vast assembly experience, and an intricate sensory system. Without such features, robots can perform only very simple, stack-type assemblies.

In order to reach a point in a three-dimensional space, the robot must have three degrees of freedom. If, however, the robot wants to do useful work on an object located at this point, the end effector must have six or more degrees of freedom. The kinematic, dynamic, and control principles for the present industrial robots are amply covered in other literature and are not discussed in this chapter.

With simulation, it is possible to display a pictorial image of the plant on a graphic display, and the manufacturing engineer can observe the creation of a workpiece through its different production stages. Once the layout of the manufacturing floor and equipment has been determined, programming of the machine tools, robots, and other facilities can commence.

Programming

There are explicit and graphical programming tools available for robots. A user-oriented approach for programming is to direct the motions of the robot with task-oriented instructions. This method is of interest for planning and programming of assembly work. For planning and controlling the work of a robot-based manufacturing cell, a centralized data management system must be provided. It is the data repository for modeling and programming the robot and supervising its actions. See Chapter 4 for the infrastructure required.

Sensors and Grippers

Sensors and grippers constitute the interface between a robot and its environment. The modern intelligent robot systems need to be supported by various advanced gripper and sensor systems. Sensors are, in principle, transducers that transform physical properties (the input signal) into electrical (output) signals. These properties can be electrical, magnetic, optic (surface reflection or transmission), or mechanical (distance, position, velocity, acceleration).

Sensors have a key function to reduce uncertainties concerning the robot itself as well as its environment. Complex tasks for robots (e.g., assembly operations) need a combination of sensors in a manner similar to the use of multiple sensors by human beings for the same tasks. Sensors that simulate the human ability of vision or touch are now coming on the market. For this reason, task-specific sensor configurations are needed for many applications. In manufacturing environments, sensors can be used to measure information about materials, amounts, geometry, place/location, and time.

In all these cases, sensors are employed to exclude uncertainties—that is, to detect them and react appropriately. In the context of assembly by robots, these uncertainties can be traced to two factors: Uncertainties from tools (robots, feeders, and fixtures) and uncertainties in the parts themselves (manufacturing tolerances). Two ways to handle these uncertainties in a manufacturing environment are:

1. Avoid all uncertainties in the assembly planning phase.
2. Detect uncertainties with the aid of sensors.

The first possibility is the traditional approach, which gives rise to inflexible, expensive, and time-consuming systems, especially in the construction of specialized feeding devices. Such a solution makes quick product changes uneconomical because of long changeover times.

The gripper of a robot is the only part that has mechanical contact with the object; its main functions are to grip objects, hold objects, and release objects. Gripping establishes a defined position and orientation relative to the robot during the material transfer route and assembly operation. When releasing the object, the relationship between gripper and object is given up at a specific point.

Generally, a conventional gripping device used for industrial robots is a specialized device that handles only one or a few objects of similar properties (shape, size, weight, etc.). When a single gripper alone cannot cope with the variety of parts to be handled, a multiple gripper, a gripper change system, or a jaw change system can be used.

Grippers can be equipped with sensors for monitoring the gripping functions. An integrated sensor system can monitor the internal state of a gripper (e.g., jaw distance), and the structure of the environment (e.g., object distances). Various gripper systems have been developed that can adapt to the shape of objects. Such grippers are built with flexible fingers with several passive joints, which close around the object to grasp it.

The design of so-called dextrous hands, which imitate the versatility of the human hand, feature multiple fingers with three or more programmable joints each. This kind of hand allows gripping objects of different geometries (cube, cylinder, ball, etc.). Fine manipulations of the object can also be realized by the gripper system itself. This fine manipulation ability results in a task separation similar to that of the human hand-arm system.

A compromise between these highly sophisticated, and therefore expensive, systems and conventional grippers can be achieved by mounting some sort of compliant device between a conventional gripper and the robot. These compliant devices consist of two metal plates connected by elastomeric shear pads and are known as remote center compliance devices (RCCs). RCC devices can compensate positioning faults resulting from reaction forces in the assembly phase.

The design, construction, and prove-out of many of these devices can be very costly and require a great deal of experimentation. As the industry continues to advance, it should become less of a risk to cost and schedule, and therefore more valuable to an "agile manufacturing" plant. Smaller job lots, or continuous manufacturing with variable production rates, should be able to take advantage of a greater degree of automation when the automation is more "flexible." However, for today, a careful study is required in order to justify full automation.

9

Computer-Aided Process Planning

Alexander Houtzeel *Houtzeel Manufacturing Systems Software, Inc., Waltham, Massachusetts*

9.0 AN OBSERVATION

The majority of the world's most successful competitors in the manufacturing industry have achieved their success by continually improving their existing design and manufacturing processes rather than by using limited funds to develop new ones. These improvement efforts are possible only with systems that manage design and manufacturing data efficiently. *Nobody* needs to reinvent the wheel.

Computer-aided process planning, or CAPP, is a term that has been associated with achieving this manufacturing data management. CAPP has been around for close to 20 years and has recently experienced a revival in the manufacturing industry with the advent of sophisticated workstations, relational databases, and other advanced software technologies. The following is an observation of current manufacturing trends, the evolution of the manufacturing engineering (ME) department, and how process planning, in particular computer-aided process planning, is essential for today's advanced manufacturing engineering. Also included is an overview of the history of CAPP, how it started, how it has evolved, and a brief study of recent CAPP installations at four successful manufacturing companies.

9.1 THE CHALLENGE OF CURRENT MANUFACTURING TRENDS

In mass production, where few if any variations of the same product are allowed, the extra investment in design for manufacturability (including quality control) can be amortized over large lot sizes. An effective interface between design engineering, manufacturing engineering, tool design and production is required right from the beginning of the product concept if a company is to be competitive on a worldwide scale. However, since today's consumers require greater product variety at competitive prices and high quality, the costs of the design and the manufacturing process can only be amortized on much smaller lot sizes. And, as is typical in a highly competitive environment, this must be accomplished with much shorter lead times. Given this scenario, three key issues become critical.

Low cost and small lot sizes. The efficient management of the design and manufacturing process becomes a difficult challenge where it concerns smaller lot sizes. This is because the changeover from one product variety to the next typically takes a long time in the mass production environment. How does one speed up the changeover for smaller lot sizes? Modern manufacturing and assembly tools must be designed so that changeovers can be accomplished very quickly and efficiently to better use the capital invested in that machinery.

Short lead times. The ability to fill niches in the market with high-quality products requires a tightly interwoven design and manufacturing team and easy access to previously acquired design and manufacturing experience. Reinventing the wheel not only introduces longer lead times but also increases the costs of the design and manufacturing process and puts a strong burden on quality control.

High quality and long-term reliability. The quality of a product is no longer determined simply by its performance at the moment of delivery (the "defect rate"), but also by how well it will work during the promised product life cycle. A product which is delivered without deficiencies but becomes ineffective within 6 months will not entice buyers to purchase that same product again.

Manufacturing trends have placed new, more difficult demands on the manufacturing industry. Now, the only way for manufacturing companies to stay competitive is to create as flexible and advanced a manufacturing environment as possible. This means taking the time to reevaluate the roles of design and manufacturing engineering, and the tools these departments should use to become as flexible and responsive as possible.

9.2 RESPONDING EFFECTIVELY TO THE KEY ISSUES

An effective response to these issues is certainly not a mindless throwing of money onto the problem with hopes that it will go away. Instead, a careful analysis of the design and manufacturing organization is required, along with a clear definition of the company's objectives. Only after this has been accomplished can one begin to consider investment in support systems and, if necessary, in equipment to make the workforce more efficient in preparing for the future.

9.2.1 Sound Engineering

In today's market, no product can expect to be sold effectively and successfully if it is not rooted in sound design and manufacturing engineering. This means not only maintaining a well-educated design and manufacturing staff, but also a set of systems to help these technical personnel to be as effective as possible. This includes providing them with the ability to draw easily on previous experience and to exchange information with others.

9.2.2 Design and Manufacturing Interaction

In far too many companies, the engineering (product design) department seems to be physically and intellectually separate from the manufacturing engineering department. Generally, a company's design group creates products with little or no consideration for

Reasons for using CAPP

how to efficiently manufacture, assemble, and service them. They leave these problems to the manufacturing engineers. More often than not, the concept of designing for manufacturability is still alien to manufacturing companies.

Based on observations made by Dr. Ohno of the Toyota Corporation, it is 10 times more expensive to have a design mistake fixed by the manufacturing department once a part has been released for production than to have the error eliminated by the design team. It costs 100 times more when the mistake has to be fixed by the assembly department, and 1000 times more when the field engineer has to repair the problem at the customer site.*

9.2.3 Fewer But More Complex Parts

As a general rule, the more parts are involved in a product assembly, the more difficult is to maintain quality. This is due to the cumulative tolerances involved in making a large number of parts fit together. If two similar assemblies, one consisting of two parts and the other of 10 parts, should have the same final tolerance, then the parts in the 10-part assembly have to be machined with five times tighter tolerances.

Obviously, this has very substantial cost effects. Generally speaking, it pays to design a product consisting of few but more complex parts rather than one with many simple parts. Fewer yet more complex parts mean more sound engineering, more knowledgeable designers and manufacturing engineers, and better machine tools that can cope with these parts.

9.2.4 Flexible Manufacturing and Numerical Control

How does a company deal with quick changeovers when manufacturing in smaller lot sizes? The major advantage to flexible manufacturing is that a lot of the preparatory work can be done before a changeover, such as numerical control tapes, tools, etc. The vast majority of work (machining work in particular) can be prepared using numerical control (NC) technology. Not only does numerical control technology allow for the quick implementation of changes, one can also better control the entire manufacturing process, thus ensuring consistent quality. Once a numerical control program has been written by the NC programmer, the machine will continuously make the part in that fashion, provided that the machine tool is always running at its expected standard. By changing the numerical control tapes, other products can be made with a minimum of change time.

9.2.5 Efficient Manufacturing Engineering

No matter how good and complex a new design is, if it is not produced in an optimum way (due to inefficient production or inconsistent quality), then the company is not going to be successful. The driving force in any production environment is the tight control of the manufacturing process. To gain such control, the manufacturing engineering department must generate effective process plans to manufacture the product and also have easy access to existing plans and past experience.

Manufacturing engineering departments should be involved in the product design cycle. They should be included in the discussion of how to best design a part for manufacturing and assembly with the design engineers. Once the design is finalized, ME should serve as the control center for determining how the parts are going to be

---

*George D. Robson, *Continuous Process Improvement*, Free Press, New York, 1991.

made and assembled. The span of control should include such areas as establishment of the manufacturing process, numerical control methodology, assembly, tooling, etc.

9.3 THE EVOLUTION OF THE MANUFACTURING ENGINEERING DEPARTMENT

The activities of the ME department vary considerably from one facility to another. Some companies still work under the "make fit to print" mentality; others have instituted very complex organizations in charge of the manufacturing process control. It is worth considering how the role of manufacturing engineering has evolved over the last 25 years.

9.3.1 The Traditional ME Function

Traditionally, the ME department looked at the drawings received from the engineering department and essentially wrote on them, "make fit to print." In other words, there was no ME function. Furthermore, in many instances the foreman of the manufacturing or production department would figure out the best way to do things based on his workers' experience. All too often, manufacturing engineering instructions would be totally ignored and the foreman decided to manufacture the product in an entirely different way. Luckily, that time has passed for most companies, and a more organized approach has evolved. Now at many companies the ME department writes the manufacturing instructions, generates the numerical control tapes, and sends the complete package to the manufacturing department to make the product.

Slowly but surely, a new conception of the ME department is emerging, one that can control the cost of the manufacturing operation. It is ME decisions which determine what the manufacturing process is going to be and, consequently, what capital assets will be used to produce that part.

9.3.2 Management and Dissemination of Manufacturing Information

In an environment where design for manufacturing and design for assembly are increasingly important, it is obvious that the experience either directly available from manufacturing engineering or from a database becomes very important to the design engineering department. Only with knowledge of the past can a design department come up with the best product design, manufactured at the lowest cost, resulting in the highest quality. It is for this reason that the ME department's role should be expanded by including it in the design process. Furthermore, the ME department's traditional area of control, i.e., production activities, should also be expanded to encompass the following: generation of work instructions for manufacturing parts and numerical control tapes, setup of interfaces with the material requirements planning (MRP) and material resources planning (MRP II) systems, and production control, ensuring that the proper tools are available from the tool department at the right moment, ordering material, generating time standards, etc. In other words, manufacturing engineering is at the center of a spider web that extends to a great variety of manufacturing functions. This expanded control is the only way to maintain quality and costs in a manufacturing company.

9.3.3 The Tools of the Trade

ME departments, of both the traditional and the modern sort, convey information to the various players in the production process in the form of different "documents" such as the process plan, work instructions, tool instructions and the routing forms.

The most inclusive of these documents is the *process plan*. Ideally, a process plan represents a complete package of information for fabrication or assembly of a detailed part or product. This package may include work instructions for the shop floor, a manufacturing bill of material, a quality control plan, tool planning, effectivity of the process plan for particular part (assembly) numbers, and links to such systems as MRP, time standards, engineering and manufacturing change control, "as-built" recording systems, shop floor control and data collection systems, etc. The result of this creative process is the process plan. The package may also include multimedia information, such as text, graphics, photographs, video, or sound.

Work instructions comprise of a set of documents released to the shop floor to produce or assemble the detailed part or product; they are part of the total process planning package. The work instructions are mostly defined as a set of operations with operation numbers. Depending on the type of industry, the work instructions can encompass detailed work procedures for individual steps within an individual assembly (manufacturing) operation, or may contain only summary information for operations. It should be noted that many companies use the expression "process plan" when they really mean work instructions.

The *routing* is a summary of mostly one-liners corresponding to the individual work instructions. The routing normally includes setup and run times. Routing information is most often transferred to an MRP system for scheduling and work-order release purposes and to the shop floor as a traveler with the detailed work instructions.

With ME departments ever expanding their role in the manufacturing (and design) process, more powerful tools are required to provide easy access to in-house manufacturing information.

9.3.4 Computer-Aided Process Planning

The creation of a set of work instructions for the shop floor to produce a detailed part or assemble a product has always been a basic requirement. The "electronic pencil" approach is a great improvement on the old ways of paper-driven process planning, because one has the advantage of word processing on a computer. However, the electronic pencil lacks (1) the sophisticated retrieve/modify capabilities of a modern computer-aided process planning (CAPP) system, thus forgoing existing company manufacturing experience; (2) the potential to connect to other existing systems within a company; and (3) the multimedia capabilities of current state-of-the-art CAPP systems.

With the advent of computers, the generation of a *complete* process planning package has become easier, but it also requires a substantial integration effort in order to link different software systems together (sometimes on different computer platforms). An effective CAPP system should draw on a variety of different software and database functions to create the process planning package.

CAPP is an interactive software tool used primarily in the manufacturing engineering department to create a coherent set of work instructions that the shop floor (fabrication or assembly) then uses to create a part or product. This tool provides the capability to retrieve and modify existing process plans, or to create a process plan from scratch (using sets of standardized instructions). CAPP also enables the process planner to communicate with a variety of other data resources to complete the set of work instructions.

Although CAPP provides the same functions as the process planner used to perform manually, many new aspects become important to consider:

Management of engineering and manufacturing changes

Use of previous experience through retrieval of similar and standard process plans

Linkage to different systems such as time standards, tool control, MRP, etc.

Inclusion of computer graphics, images, photographs, or videos to illustrate a manufacturing process

Identification of procedures to best design a product with the lowest possible manufacturing and assembly cost

9.3.5 Manufacturing Data Management

As a result of product service requirements, product liability fears, government regulations, or plain old good sense, more and more companies are trying to keep better, more retrievable records. "There is gold in them files"—because these records contain the company's best manufacturing experience and ought to be instantly retrievable before process planners go about reinventing the wheel. What product data management (PDM) intends to do for the engineering process, manufacturing data management (MDM) proposes to do for manufacturing—i.e. maintenance and retrieval of the entire process planning package.

Depending on how far a company's ME department has extended its responsibilities over the design and manufacturing process, a level of computer-aided process planning or manufacturing data management should be applied accordingly. In other words, those companies that maintain an ME department that is limited in scope will most likely also maintain a very limited computer-aided process planning function. Alternatively, those companies which have worked to integrate the ME function into the design and manufacturing process have installed, or are seeking to install, an elaborate and all-encompassing CAPP/MDM system.

9.4 GETTING TO A PROCESS PLAN: WHAT MUST HAPPEN FIRST?

Several important actions are required before a process plan hits the shop floor. How effectively these actions are taken is crucial in generating the most useful and error-free process plan.

9.4.1 The Review Cycle

In most companies, designs are not automatically released to manufacturing engineering. If a product or part requires a new process or a very difficult process, a substantial review may be needed to examine the manufacturability of the product design both during the design cycle and after the release to manufacturing engineering. In many companies this procedure is formalized in specific committees for value engineering. In others, the chief tool engineer and the chief production engineer review and discuss exceptional parts with the design department and come up with a set of changes to the product design to improve manufacturability.

In more advanced environments, this review process (involving both design and manufacturing engineers) is initiated at the very beginning of the product definition (even before one part or product has been designed) and is continued throughout the design cycle. The advanced retrieval capabilities of a CAPP system are very useful in such reviews because such a system provides easy access to previous experience. (See Chapter 1.)

9.4.2 Part Analysis

Once a part is released from the review cycle by the chief production engineer or manufacturing engineering manager, the process planner analyses the new part to determine if a process plan for this type of part already exists. Depending on the data management or CAPP system available to the planner (or lack thereof), he or she will employ certain retrieval methods to check if information on similar parts is stored in a database or, if operating in a very advanced CAPP environment, if a set of rules exists to generate the process plan automatically. This analysis determines whether previous experience can be included in the process plan or whether one has to start from scratch. Clearly, this analysis can be greatly expedited by the presence of an efficient retrieval system that provides easy access to past manufacturing experience.

The following short list below of common analysis and retrieval methods sums up how efficient (or inefficient) this stage of production can be.

Eyeballing. In this straightforward and most often used method, the planner visually scans the drawing to determine if the part or assembly is retrievable either by name or number, or if he or she remembers working on a similar part before. Here, the burden is on the planner and his or her memory.

Black book. The process planner may look at the drawing and go through scribbled notes on past process planning projects in a "black book" (personal notebook) to see if he or she can find some similar experience that can be applied to the part in question.

Classification and coding. A more structured approach to retrieval is by part features that can possibly lead to generative process planning. The process planner may have a computerized classification and coding system available to identify the basic part features and relate those to families of parts already in the database. In this way, he or she can determine if the part is related to a certain "part family" for which there is already a process plan available. In some cases, a new part may be so similar that a set of rules already in the database will generate the process plan automatically. If the part is not immediately familiar, the system may still recognize enough of its features to retrieve a standardized process plan from the database.

9.4.3 Process Plan Generation

Process plans typically have three main sections, a header, an operation sheet, and call-outs to reference material.

The header contains all the "name plate" information on a particular part (name of the part, part number, lot size, etc.) to meet a process planner's identification and retrieval needs. Other items that should be considered essential to retrieval purposes, and thus should be included in the header, are group technology classification code, engineering or manufacturing revision, plant code, preferred or alternate routing, material, quantities, type of plan, status, name of the process planner, name of the customer, etc. If a company is using a relational database, users can retrieve process plans by using a combination of fields in the header. Fields in the header may also be subject to validation.

The operation sheet provides a detailed description of the machine tools and manufacturing processes required to make a part. Generally, the sheets are divided into columns to organize the information into appropriate categories. The first column typically contains an operation number that identifies the sequence of different operations. The numbers may just be inserted chronologically by the process planner, or the CAPP system being used may provide a resequencing function for different operations.

The second column usually contains a number representing the machine or work center. Essentially this number represents a front-end access code to profile tables that will assist the planner in retrieving more detailed information on machining or assembly operations. Information in these profile tables may include a list of different operations that are possible in a given shop (milling, drilling, etc.), available machine tools and their functionalities, the locations of machine tools, costs per hours, and standard setup and tear-down times.

The next column on the operation sheet is usually the operation description column, which provides a detailed outline of how the work is to be preformed. This column can accept text from the machine/work center profile tables, or word processing may be employed to enter text from scratch for entirely new operations. Some advanced systems also have the capability to make small CAD sketches in this column as well as provide photographs, animated computer images, or even videos.

The call-out to reference material enables CAPP system users (both the process planner and the person on the shop floor) to call up standard documents while working with a process plan. Reference material may comprise customer specifications, military specs, welding and assembly descriptions, etc. Some or all of the reference material may have to be made available to the shop.

Most companies have their own process plan layouts that have been in existence for many years. If a company decides to introduce a computer-aided process planning system, it should spend time evaluating whether a new process plan layout is desirable. Often, the "dinosaur" process plan forms of old have a lot of unnecessary information in their layout. A company must decide whether or not that information will only waste valuable storage space in a database.

9.4.4 Security, Sign-Off, Change Tracking

Given the proprietary information that resides in a CAPP system, most CAPP users insist on a security system that will prevent errors and tampering with data. In many companies, an elaborate system that includes multiple levels of security codes has been incorporated into the computer-aided process planning environment.

Once a process plan has been created, it is customary for several persons in the organization to review and sign off on the plan. These persons may be the chief of the manufacturing engineering department and possibly the foremen in the shop. The process planning system must contain a good sign-off capability including parallel as well as serial sign-off functions.

Finally, CAPP systems should provide a tracking function, such as effectivity and change control, to account for engineering changes and manufacturing changes. This is especially true for those companies that are concerned with product liability and those who typically make frequent changes to a part during the design and manufacturing processes.

9.4.5 Dissemination of Information

Once a process plan is finished and the required sign-offs have taken place, information from the process plan should be available to any of the following organizations within a manufacturing company—although different departments may receive different "info packages":

> MRP or production control, because ultimately this group will release the process plan to the shop floor for production or assembly
>
> The tool room, to assemble the tool kit and make special tools, jigs, and fixtures
>
> The shop floor (at the behest of MRP) when the part is ready for manufacturing or assembly
>
> The materials department, again triggered by the action of MRP, for release of materials
>
> Quality control
>
> The NC department, although this is usually part of the ME department
>
> The work measurement and time standards group, if this happens to be a separate group
>
> Support functions such as maintenance and purchasing services
>
> Ad-hoc status reports to management

9.5 VARIANT VERSUS GENERATIVE PLANNING

CAPP technology has evolved into two methodologies: variant planning and generative planning. A variant CAPP system generally has the ability to retrieve previous process plans by similarity, either directly from a relational database or through a group technology feature retrieval system. After retrieval of process plans for similar parts, the planner can review these and then make changes, if necessary, for the next application. In most situations, companies may have preferred process plans that are stored according to families of parts. The advantages of such a variant CAPP system are substantial over the older, "electronic pencil" variety of CAPP. The retrieval capabilities inherent in a variant CAPP system prevent a company from reinventing a process plan that already exists in a variant CAPP system's database.

With the advent of "artificial intelligence" in the early 1980s, it seemed logical to explore the concept that process plans could be generated automatically based on sets of "rules" that could reside in an advanced database system. These rules would be set up according to a company's manufacturing experience and then installed into a database. With such a system, a process planner would simply identify the features on a part and then automatically call up a corresponding set of rules for the manufacture of each of those features. The objective was to generate a process plan from scratch through this feature identification process. However, like the miraculous folding bicycle, this idyllic product seemed to work only during product demonstrations.

Generative CAPP systems do exist. However, they have been successfully implemented only in those companies where an enormous in-house effort has been made. A vast number of company personnel and company funds must be allocated to the project of analyzing and defining the individual rules for the manufacture of each part feature resident in a company's collection of parts. Rules are determined by shop practices which have evolved over years of experience and are quite different for each manufacturing company and for each of the available machine tools on the shop floor. Although generative systems are fascinating from an academic perspective, their practical implementation is only rarely truly cost effective.

9.6 THE FIRST 15 YEARS OF CAPP

A brief look at the history of CAPP and how CAPP has advanced during its years of use in the manufacturing world will help to identify what today's industry has come to expect from its process planning function.

The Organization for Industrial Research, Inc. (OIR) appears to have been the first company to develop a commercial computer-aided process planning system. The General Electric Light Equipment division in Cleveland became its first customer in 1979. OIR went on to install over 150 systems before it was merged into Computervision in 1984. Although OIR's Multicapp system contained elaborate group technology classification and retrieval capabilities, most customers used it (or are still using it) as an electronic pencil. However, the aerospace and defense industries, both then and now, needed more than just an electronic pencil. Regulations required them to keep elaborate manufacturing records. Consequently, these industries and several other advanced manufacturers developed mainframe-based complex manufacturing data management systemsall in-house. In the late 1980s, several other companies developed commercial CAPP systems. None of these, however, came close to fulfilling the requirements of the aerospace/defense industry, nor did they offer multimedia capabilities. The evolution of CAPP had come to a crossroads. It seemed as though those companies that wanted highly sophisticated CAPP functionality were going to have to rely on their own in-house sources to develop and meet their CAPP needs. Those interested in developing commercial CAPP systems were forced to reassess how they might come up with a commercial CAPP offering that was functional and sophisticated enough to challenge the in-house development option.

9.6.1 Time to Regroup: What Does Industry Really Want?

In 1991, one developer of commercial CAPP systems organized a set of meetings, in both the United States and Europe, with several leaders in the manufacturing industry for the sole purpose of determining what companies envisioned for CAPP in the future. The representatives from these companies all had at least 6 to 10 years of experience in computer-aided process planning. These meetings generated an extensive report on the status of CAPP and a detailed list of functional specifications for an advanced CAPP system that would truly meet the needs of the manufacturing industry. The following are highlights from this report on CAPP.

1. Most in-house-developed CAPP systems were much better integrated with other preexisting manufacturing software than their commercially available counterparts. However, the process planners, i.e., the daily users, found the commercial systems much more user friendly than the in-house-developed systems.
2. All participants initially maintained that their company's requirements were different from anybody else's. After 3 days of discussion, the general consensus was that at least 85% of everyone's requirements were the same. User interfaces and links to other systems, however, were different for most participants. It was

determined that an acceptable solution to this problem would be a generic, but tailorable, CAPP system, where user interfaces and links to other systems could easily be set up within a customer-defined macro and then linked to the basic system.

3. Since more and more people are able to read less and less, a multi-media process planning system with the capability to incorporate text, graphics, photographs, and video would be beneficial.

4. To manage the manufacturing data and perform sophisticated retrievals to utilize the best available company experience, a relational database management system was absolutely required.

5. Several participants had invested substantial efforts in the development of artificial intelligence-based generative process planning systems. All had come to the conclusion that such systems are technically feasible but provide unacceptable low returns on investment, especially with regard to general detail parts or assembly operations. However, with groups of very similar parts or assembly operations, such as the manufacture of turbine blades for a steam turbine, an AI-based system could be profitable. All participants opted for variant-based process planning systems equipped with standard process plans and sophisticated retrieval capabilities, including retrieval by combinations of part features.

6. Since the maintenance of existing mainframe CAPP and other manufacturing data management systems has become increasingly expensive (and is now starting to be outdated), the UNIX-based client-server environment was thought to be the solution for the near future, with PC networks feasible at a later state—once proprietary manufacturing data were proven to be adequately protected on such PC networks.

7. It was solidly determined that "electronic pencil" as a method of process planning is not a promising method for the future, since it only allows a company to create work instructions without any referral to preexisting manufacturing experience. It enables the user to make the same mistakes as in earlier paper-driven systems, only much faster.

8. CAPP systems of the future would have to have interactive links to the design (CAD) database to make the drive towards concurrent engineering a reality. Currently, one or two commercial CAPP vendors are developing links to engineering data management systems (sometimes called PDM or EDM).

9. If CAPP is going to be a spider in the middle of the manufacturing data web, links have to be provided to MRP, time standards, engineering and manufacturing change control, shop floor control and data collection, etc.

9.7 SOME EXAMPLES OF WHAT IS REALLY HAPPENING

More than 3 years after the 1991 meetings, it is worthwhile to take a reality check and compare some recent CAPP implementations with the specifications of the "CAPP of the Future" project.

9.7.1 A Large Subcontractor to a Commercial Aircraft Company Leaves the Mainframe and Goes Server-Based

In 1992, a company that had once been almost exclusively involved in production of military aircraft decided to create a commercial aircraft division geared toward the production of major parts for the Boeing 747. The new aircraft division needed to place itself aggressively in the market. A review of all the costs associated with the 747 product revealed that the data processing costs were very high relative to other costs. This meant a multimillion dollar reduction in only 18 months, or a two-thirds reduction in data processing costs. It became apparent that this cost reduction could not be accomplished by operating the existing mainframe running their so-called legacy systems. These dated back to the late 1960s and racked up high overhead costs. The company decided to switch from the centralized mainframe processing approach to a server-based distributed processing system.

Use of Off-the-Shelf Products

The company decided to go with UNIX-based off-the-shelf software and was willing to change some of its business processes where necessary, in order to move to a less expensive platform and stay within the vendor's software capabilities. That meant that "tailorability" of the off-the-shelf software became a critical issue. Even more important, it meant that the various vendors had to enter into a partnershiplike relationship with the company with each other, in order to ensure a successful endeavor and a smooth transition. Above all, this required the integration of a commercial state-of-the-art CAPP system with off-the-shelf MRP, shop floor order control, and manufacturing change control and time standards systems, all using the same relational database system.

The company's new CAPP system was implemented in two phases:

During the first phase, when the mainframe was still in operation, the new server-based system was interfaced with the company's legacy systems so that the mainframe system could still provide detailed part and subassembly plans to the shop floor and support final assembly. This link also provided for a one-time downloading of existing process plans from the legacy system to the new CAPP system. In this way, the new CAPP system maintained the legacy system while enabling planners on the new system to exploit existing text for the creation of new plans.

In the second phase, the CAPP system was integrated with the other new UNIX applications. These included MRP and factory management to produce shop-floor orders for subassembly and fabrication.

The new CAPP system was in place and operating within 60 days after placement of the order, and since then it has been used to perform all process planning at the company's commercial aircraft division. The legacy systems were finally phased out in August 1994, and the mainframe system has been discontinued. The installation and integration of the systems took approximately 18 months, with total costs below budget and a return on investment in less than 2 years.

Ancillary Benefits

Although the main reason for replacing the mainframe system had been to reduce the cost of data processing, there have been considerable ancillary benefits. Some of the changes in processes have already started to pay off enormous dividends. With the legacy system, it took approximately 57 people 50 days to produce a bill of material (BOM) that could be used for ordering. The BOM resided in two different systems that did not match, and therefore validation was a problem. With the implementation of the new CAPP and MRP systems, the company has been able to cut that down to 10 days and do it with about 6 people. The validation and checking groups were taken out of the loop because the system validates itself as entries are made.

In effect, the new approach and particularly the installation of the CAPP system has resulted in the rethinking of the entire flow of manufacturing information. Currently, well over 400 users (50 authors and 350 viewers) employ the CAPP system daily, not only for detail and assembly process planning, but also for tool planning, manufacturing change management, and, in the future, for time standards. Currently the company uses 20 UNIX-based production servers and one server which contains the CAPP database.

It all has added up to a successful conversion from military to commercial operations that leads company employees to speculate about the future: "We want to be a world-class manufacturer of commercial aircraft products. This new systems architecture certainly provides the information management infrastructure to meet our goals."

9.7.2 U.S. Army Arsenal Selects Multiple Vendors for Joint Development of a Very Advanced CAPP System

The arsenal manufactures and assembles heavy arms (such as howitzers) in a vertically integrated manufacturing facility using sophisticated production technology. Among others, it uses a mainframe-based MRP II system, a Tandem-based tool management system, and extensive in-house information on machinability data. In an effort to improve and build upon their current in-house technology, the arsenal recently defined a set of tools to assist process planners, methods, and standards personnel in all phases of their planning activity.

According to their system definition, the new CAPP system was to be client-server based with links to existing systems and with the following capabilities:

Variant process planning that would include expert technology for individual manu-
   facturing "steps"
Process planning for operations and detailed manufacturing steps within an operation
On-line machinability database
On-line time standards database
CAPP graphics and interfaces to existing CAD systems

Furthermore, the arsenal defined a wide range of highly sophisticated CAPP requirements that demanded an extension of the basic CAPP system functionality. These specifications included:

User interfaces, including text, graphics, and images
Elaborate checking and sign-off capabilities
Detailed revision history, tracking, and mass update

Vendors Make a Joint Effort to Meet Arsenal's Advanced CAPP Needs

Considering the commercially available CAPP systems and their off-the-shelf capabilities, substantial and expensive customization efforts would be required of each vendor to meet the arsenal's requirements. To reduce these individual customization efforts and to take advantage of available off-the-shelf capabilities, two normally competing vendors in the computer-aided process planning market decided to make a joint proposal in response to the arsenal's solicitation.

The developers jointly set up the new arsenal CAPP system as follows:

One vendor would supply the general process planning structure including multimedia user interface, sign-off capabilities, revision history and tracking, relational database installation, links to the other arsenal systems and platforms, and supply of a B-size scanner.

The second vendor would supply the manufacturing technology and time standards software.

The two vendor systems would be seamlessly integrated using one graphical user interface and one relational database.

Flexibility Enables Users to Go from Manufacturing Operations to Manufacturing "Steps"

This CAPP system is currently being installed at the arsenal and will capitalize on the advanced features of both commercial packages. It is an extremely flexible system, able to call up manufacturing information from a wide variety of sources, to retrieve and generate not only manufacturing operations but manufacturing steps, thus providing feeds, speeds, and time standards for part features.

This joint effort marks the first between two competing vendors and will ultimately produce one of the most sophisticated process planning systems to date, with a projected return on investment for the arsenal in 2 to 3 years.

9.7.3 The Automation Division of a Large Machine Tool Manufacturer Purchases a CAPP System with Group Technology Retrieval Capabilities

Cutting Down on Time to Market

How can we cut down on time to market? This was a challenge that the Assembly Automation Division of the company had to meet. This manufacturer of customized automation assembly systems, known for installations such as the body assembly line at the GM Saturn plant and the chassis line for the new Ford Explorer, was faced with customer demands for completed installations in only 6 months. For a company that was used to completing an installation in a 9- to 18-month time frame, such time-to-market competitiveness was going to require significant changes in their manufacturing procedures.

Repetitive Tasks as a Stumbling Block

The Assembly Automation Division singled out excessive repetition in design and manufacturing tasks as a major stumbling block to improving the rate of product turn around at their division. Process planners at the plant were re-creating process plans from scratch with each new order. They had no effective retrieval system available that would enable manufacturing engineers to work with existing process plans, simply copy-and-edit, and thus drastically cut down on generation time. The company decided to install a flexible process planning system as well as an up-to-date group technology system.

The Installation of the CAPP System

The company's implementation of a commercially available process planning system was initiated as a two-step process. Step one was the installation of standard process planning software, including screen layouts according to the company's specifications, multimedia capabilities, and links to other company systems (on IBM AS 400). The second step encompassed the installation of the group technology (GT) classification system for metal forming and metal cutting parts and analysis of the process planning database. A random sampling of 10% of the company's most recent parts was taken, i.e., parts that had been produced in the last year. The company generates about 12,000 to 13,000 different parts per year; 1250 of these parts were used for analysis to establish part families and standard routings. With the classification of the sample completed, the process planning database (more specifically, the parts' features) was broken down into part families with similar part features. Parts were sorted according to simple parameters: round/nonround, type of material, similar features, size, and proportion similarities. All of the analysis was performed by the CAPP vendor and outside the company facility, enabling the plant to go on with its business with little interference. The company was called upon only at key points in the project to review and approve the classification work.

From Part Families to Standard Routings

With part families established, the final stage of analysis and classification was to set up standard routings for these parts. Again, the company served as an editor in the process, making only minor changes to the standard routings that were produced by the CAPP vendor. A total of only 21 standard routings were set up. These routings managed to cover over 70% of all parts, both lathe and machining center parts, at the division! What did this mean for the company in terms of cutting down time to market? These standard routine "templates" are now set up in the new CAPP database for planners to call up, edit, and generate with each new order. Reinvention of the wheel has been cut down dramatically. Furthermore, as the company's manager of manufacturing engineering points out, these templates ensure that planners generate routings in a consistent and standardized way. And they can do so with little training and great ease: The CAPP and GT systems all employ a user-friendly point-and-click approach which makes them easy to learn and use. One of the immediate benefits is that parts can be directed to machine tools best suited to the required manufacturing operation.

It is important to note that the company does not produce a "standard" product. Each installation from the Integrated Automation Division is custom-designed. But with each customized installation comes a myriad of similar parts. The division has cut down its similar parts manufacturing effort significantly (all within 6 months)—thus enabling the group to focus on the heart of the work, the customization of unique parts and the optimization of similar parts.

How CAPP Affected Design Engineering

The new CAPP system has helped the company isolate standard routings which enable planners to recycle existing process plans. This same system could also be used by designers, to recycle designs, though possibly at a later stage, because the system's flexibility enables it to be fine-tuned to specific design needs. Both existing designs and manufacturing methods can be retrieved from the same database, thereby increasing communication and efficiency between the two departments.

9.7.4 A Large Steam Turbine Manufacturer in England Is Getting Closer to a Nearly Generative Process Planning System for Turbine Blades

Since 1980, this company has used an advanced computer-aided process planning system for preparing the complete information package necessary to manufacture detailed parts;

it is minicomputer-based and runs on a relational database. The use of CAPP has been a very profitable venture for the company over the last 14 years. However, the time had come for a change to a new generation CAPP system in the turbine blade department, one that operated in a client-server environment but retained a direct link to the old minicomputer-based system. This link, much like the implementation at the commercial aircraft division, would provide the process planners with the ability to retrieve and modify old process plans without having to download the entire old database.

Turbine blades require a complex machining process, but the process for one blade is very similar to that required for another. Consequently, a turbine blade process planning system that is driven by parameters from CAD which then feed a group technology-type work cell appears to be an excellent solution. On the other hand, this solution does not seem to be even remotely feasible for the manufacturing of other parts of a steam turbine, such as housings, valve assemblies, etc.

9.8 WHERE TO GO FROM HERE?

In the late 1970s, defense and aerospace companies were looking for a solution to meet their advanced process planning needs. Commercial CAPP systems of that time offered an electronic pencil solution, and in some cases, advanced GT techniques, which served the needs of many in the manufacturing industry but did not come close to meeting the needs of aerospace and defense. Thus, the aerospace/defense industry turned inward, developing highly complex in-house systems on the mainframe—at great cost and with very expensive maintenance overhead. Both schools of process planning; the electronic pencil variety and the sophisticated in-house system, have come to a crossroads: Both approaches are in need of major change. Now, most industry leaders have decided to move away from both the mainframe systems and the outdated electronic pencil, toward a UNIX-based client-server environment. Furthermore, virtually all have decided to purchase, if possible, commercially available off-the-shelf software, even to the extent that they are willing to "slightly" modify some of their business practices to fit the off-the-shelf products.

Given their substantial in-house expertise, the companies which are purchasing software are quite selective, preferring to mix and match several vendors to obtain the best technical solution rather than go with a one-stop purchase which may be less optimal. Such a policy does require extensive cooperation between vendors and the buying company. The principal reasons for this move to a client-server environment are:

Cost reduction
Local empowerment of the information service activities
Rethinking of the entire design and manufacturing information structure

Electronic pencil process planning, although still used in many companies, is no longer a wise investment. In the paper age of process planning, the electronic pencil provided a substantial improvement. Its return on investment was very attractive—but that was more than 10 years ago.

As companies start to realize that the intelligent management of product and manufacturing data can be a determining factor in their productivity, the protection of that data also becomes imperative. This an important reason why UNIX-based systems with good data security are the selection of choice for most large companies; however, new PC operating systems which also provide this capability may make PCs equally attractive, especially for smaller companies. A note of caution: It is believed that the new PC operating systems working in a complex client-server network are going to be easier to run than UNIX-based systems, but this may very well be a fallacy.

In the near future, true generative process planning systems (not some sales pitch about it), may not be an attractive solution except where it concerns very narrowly defined groups of parts. On the other hand, feature-driven variant process planning systems with parameter-driven standard process plans should result in large paybacks. This approach also fits well with feature-driven CAD systems. In other words, the gradual linking of product data and manufacturing data management may soon become a realistic venture. At least one or two commercial CAPP system vendors are working along these lines with CAD system suppliers.

Real process planning systems do not come cheap, especially in terms of financial investment and personnel commitment within a company. But there is ample evidence to show that CAPP's return on investment is larger than the ROI on a large machine tool—and the price is about the same!

9.8.1 Getting Started in CAPP

Having reviewed how today's CAPP implementations have helped four companies better meet the challenges of the manufacturing industry, it is worth summing up some of the key steps involved in preparing to install the right CAPP system for a particular manufacturing site.

To make computer-aided process planning part of an integrated design and manufacturing solution, companies will have to put considerable thought into their long-term goals and how a system can be best integrated into existing facilities. Obviously, no one "cookbook" will supply every possible approach for every company's manufacturing environment. In fact, the purchase of a system should come only after a company has had many discussions, both with in-house personnel and outside experts, on what their needs really are. The following steps should be considered.

1. *Form a Discussion Group.* A philosophy should be established for the company's long-term integrated design and manufacturing solution. On the basis of these discussions, which should involve top management, objectives should be set for fulfilling the philosophy. During this process, the discussion group should consider discussing their ideas and decisions with other companies that are in the same situation. Participation in several seminars, both external and internal, is typically necessary in order to sell the intended objectives to company personnel.

2. *Define Information Flows.* Once a company has established its objectives for an integrated system, the information flows among different departments should be defined (i.e., between the machine shop and manufacturing engineering, between manufacturing engineering and design, etc.). It makes sense to create an elaborate data flow chart to give a bird's-eye view of the different needs of each department and how they will function in the new environment. One can save a lot of money if these data flows are identified ahead of time, highlighting such details as who needs it, what information is needed when, etc. This information flow plan includes the definition of a software linkage system that will link all the different databases together.

3. *Establish Functional Specifications.* Once the information flows are defined, the individual departments have to identify which functional capabilities they need to deal with the information flows. Sometimes one system may serve several departments, or it may be designed only for a single department. For example, the computer-aided design needs of

the manufacturing engineering department are usually much less sophisticated than those of the design department. Consequently, the company may have to define two separate systems. At this point the manufacturing engineering department should begin to lay out what it needs from a computer-aided process planning system.

4. *Take Inventory.* Since most companies have already installed several design and manufacturing systems (some simplistic, others more sophisticated), it is vital to review the functionality of existing systems and analyze how these systems could be used with the new, integrated solution. Several important decisions have to be made, some of which will be politically unpopular. Some systems simply cannot be integrated into the total picture and should therefore be abandoned. One should avoid a scenario where new and old systems exist concurrently without being able to communicate. In some cases, companies may attempt to link old and new systems with "bailing wire." A problem arises when the people who designed the oddball connection leave the company, the entire system breaks down, and the company's integration solution is in jeopardy. A thorough inventory should be taken to determine what systems should stay and what should go.

5. *Decide Whether to Build or Buy.* Once the functional specifications have been set for the required system, the company must decide if the system should be developed in house or purchased. Companies that have small software departments should not attempt to develop their own systems, but rather purchase the necessary software. For larger companies, economics must be considered. When a capable software department is available (which is often the case with larger companies), the company could easily develop its own systems and probably better gear these new systems toward the functional specifications and data flows required. However, in-house software development is very expensive in both the initial development stages and thereafter in maintenance. It also tends to depend on a few gifted individuals who, when they leave the company, may leave the organization stuck with software that is very difficult to maintain. The alternative is to find software developed by outside parties. These packages tend to be somewhat cheaper. However, third-party software rarely meets all of a company's needs. And, there is always the risk that a vendor will go out of business. A financial analysis may be of use when deciding what to do.

9.8.2 Software and Hardware

Given the aforementioned functional specifications for an advanced computer-aided process planning system, the software requirements are as follows:

Word-processing text and graphics capabilities, and intelligent look-up tables (profit tables)

Linkage to other systems and relational or object-oriented databases

Retrieval capabilities of relational databases using similarities based on group technology

Expert system capability to automatically generate process plans from scratch for certain parts

Rapid response times and user friendliness

Extensive help capabilities

A company that is new to CAPP may decide to start with just text and graphics and then expand to other capabilities later on. It should be recognized, however, that by following this path, one can easily tall into the trap of simply "automating the file cabinet" and believing it is CAPP. So if a company decides to install a system incrementally, it should be sure that it can be expanded without having to rewrite the entire system.

Given the rapid decline in the cost of computer hardware, one should opt for a system with software that is portable across different platforms. The CAPP application should look the same and react the same on different platforms, be it PCs, workstations, mainframes, or a combination of these. This will make training much easier and will also broaden the system's range of application.

With prices of high-end PCs and low-end UNIX workstations converging, prescribing one machine over the other as the platform of choice is difficult. Clearly, PCs, work-stations, or a combination of both in a network environment will be far more cost efficient than running computer-aided process planning solely on a mainframe. Not only is this more efficient from an asset utilization point of view, but also better system response times provide for better usage of personnel.

The option to use X-Windows may improve the process planner's efficiency; however, if the already installed hardware is unable to support X-Windows without incurring major update costs, that option may be less attractive.

Local area networks are of major importance, but hidden expenses should be considered. A simple local area network function means that if one person requests an existing process plan for modification or review from the file server, then the plan will not be available to another person. This is the standard practice in many manufacturing environments. Supporting more complex data transactions will increase cost substantially.

9.9 AN AFTERWORD ON GROUP TECHNOLOGY

An awareness of the similarities of parts and products has benefited design and manufacturing for many years, usually resulting in substantially higher production throughputs, faster design turnarounds, and better utilization of equipment on the shop floor. The manual search for similarities of parts was very tedious and time consuming until the 1960s when computers started to be used to search for these similarities. At that time, group technology (GT) became a practical approach for analyzing a part population for standardization and for machine shop layout.

In the mid-1970s, the grouping of similar parts was enhanced by computerized classification and coding systems. However, despite these pioneering efforts, the results were only marginally successful, in part because the use of similarities for design and manufacturing standardization is a long-term effort. Since then, group technology has become an increasingly attractive choice, especially since the availability of relational databases and object-oriented databases has made the retrieval of similar parts much easier.

Group technology can lead to several applications. In the short-term, GT provides information for daily operations, retrieval of product or part design, and retrieval of manufacturing and assembly experience of parts and products before the engineer starts the expensive process of reinventing the wheel. In the long term, GT can provide the analysis of retrieved information, not only for standardization in design, fabrication, or assembly, but also as a tool to successfully introduce design for manufacturing and assembly (DFMA), and the introduction of "concurrent engineering" (see Chapter 1). GT also can be used to better employ the available assets in the machine shop. In others words, standardization in the manufacturing and assembly departments can lead to a rational organization of the production department and provide a tool to analyze what machine tools should be purchased in the future and how machine shops should be laid out.

9.9.1 People and Group Technology

Group technology is a tool box that provides knowledgeable people with the means to come up with cost-efficient solutions. Consequently, its introduction in a company should be accompanied by a clear set of objectives, timetables, and an understanding and commitment by the people involved—including personnel ranging from top management down to the people on the shop floor. Group technology will *not* be successful if it evolves as an edict from the top, nor will it make any inroads if it is simply a back-room activity in some department at the bottom. Departments which *should* be affected by the implementation of group technology include:

The product design department, as a tool to retrieve previous design/manufacturing information to either modify or take as is. GT can also serve as a tool to standardize certain design approaches

The manufacturing engineering department, as a tool to retrieve, modify develop and use manufacturing process information

Both the design and manufacturing engineering departments, as an integration tool to implement the principle of "design for best manufacturing" and "design for best assembly" in order to improve the quality of the product and lower the cost

The materials purchasing department, as a tool to reduce the variety of materials which are used to produce a product

The production, production control, and manufacturing engineering departments, to lay out the machine shop, possibly in so-called group technology work cells, and as a method to purchase machine tools (see Chapter 3)

10

Work Measurement

Kjell Zandin   *H. P. Maynard & Company, Inc., Pittsburgh, Pennsylvania*

10.1 WHY WORK MEASUREMENT?

Work measurement in primitive forms has been around for hundreds of years. According to some sources, the Egyptians applied some form of work measurement when they were building the pyramids. It seems fairly obvious to us, since the work of stacking blocks of rock can be considered repetitive. In the Middle Ages, the armor and weapons makers of Catalonia based their prices on the basis of material, quality, and the time it would take to make the product. The time was based on experience. More recently, there have been specific demands for work measurement created by the need to increase productivity (during and after the two World Wars) or to establish a basis for incentive payments.

Our society is filled with measurements of all kinds. We measure the distance between two cities or places, and the weight of steaks and other foods in a grocery store. We measure the temperature of the air and of people. We measure the dimensions of parts and components to make sure they will go together. We measure the pressure in automobile tires. We measure how much it rains or snows. We measure the length and weight of a newborn baby and of ourselves throughout life. We even measure how fast someone can drink a beer and how far a golfer can hit a ball, or a football player can carry it. Thousands upon thousands of other more or less useful examples of our measuring spree could easily fill an entire chapter. Practically everything we are, we do, and we own is related to a yardstick.

Measurement is information to us. Without measurements, the world could not function. We base costs, plans, controls, and many decisions on measurements. We need that information to improve the conduct of a business. For instance, we tell a machine operator what to do through blueprints filled with dimensions, tolerances, and other significant facts about a part or product, but seldom how to make it, indicating the process, equipment, tools, and method to use, as well as the time (work standard) for the job. We expect the operator to use the most efficient tools and methods with little or no help.

We expect the operator to perform at a high rate. But how can we know what the performance is if we do not measure the work to be done?

Even though we readily accept the science of measurement, as well as being measured in different ways, there is one area where measurements have been either controversial or rejected: *the measurement of work.* However, many industries and organizations have gradually, through improved measurement techniques, education of its people, and commitment to sound management principles, turned the work measurement practice into a logical and natural part of its business conduct. The regards and benefits have often been both dramatic and lasting.

> *As with any measurement, the measurement of work brings knowledge. Through this knowledge, factual decisions and improvements can be made and control can be exercised.*

The purpose of work measurement is to provide management, as well as the person performing the job, with information about how much time it should take to perform a task according to well-defined work conditions and a specified method. How management uses that information is a different and undoubtedly more significant matter.

Basically, proper measurement of work facilitates and improves management. Using simple or nonscientific forms of work measurement, the time values may, however, be inaccurate or may not match the conditions for the performed work. On the other hand, *scientific work measurement* not only facilitates, but provides a sound basis for management. In general terms, work measurement is the basis for planning, evaluation of performance, and estimation of costs. In essence, the manager looks at work measurement as a basis for his or her ability to forecast with confidence. The key word here is *confidence,* because if the time for a job is not accurate enough, very often the wrong decisions will be made.

Work measurement is used to establish time standards for individual operations or jobs. Time standards apply primarily to manual operations, even though machine-controlled operations are measured as well, using a different method.

Work measurement is a tool that every manager should use to do a better job—because work measurement brings *knowledge* in the form of definitions of work conditions, method instructions, and reliable, consistent time values. As we shall see, these time values can be very useful in many areas of management.

10.2 METHODS OF MEASURING WORK

10.2.1 Non-Engineered Time Standards

The fastest and least expensive way of measuring work is to *guess.* Very often, one may have little or no information available on which to base a guess. Time standards created by guessing are usually very inaccurate and inconsistent. All they provide is a number, unsupported by the conditions under which the work is being done.

By using *historical data* in combination with experience for the particular area, time standards may become less inaccurate and less inconsistent compared to pure guesses. In this case, they are called *estimates* because they are *estimated* by someone who is familiar with the operations or activities to be measured. Estimates are quite often based on a subjective evaluation of the work content supported by available data as the basis for a time standard. Invariably, people over-estimate just to make sure that the job can be done in the time allotted. Therefore, one tends to get "loose" standards by using the *estimating* method. Accuracy and consistency usually do not meet established requirements for "good" standards.

Another method of establishing estimated time standards is *self-reporting*. Typically, if a person believes that a job will take 45 minutes to do, the self-reporter will allow 1 hour, just to make sure that the job can be completed within the allotted time and perhaps also to allow time to perform the task at a pace below normal.

All the methods above can be categorized as *nonengineered standards* because no backup documentation supports the time standards. That means that each time a product or part, a job method, or a work condition changes, a new time standard must be determined. Since no backup data are available, it is not possible just to change a time standard based on the change in work conditions.

Stopwatch Time Studies and Work Sampling

Although there are cases when acceptable backup data are produced, very often *stopwatch time studies* fall into the category of non-engineered standards as well. The same applies for another watch-based system: *work sampling*. These commonly used methods will likely produce more accurate and consistent standards than the estimating methods. However, the watch-based systems have many drawbacks.

Both the stopwatch time study and work sampling studies require an analyst to observe one or more operators performing the work. In order to obtain a reasonably accurate time standard, a large number of observations must be recorded. This means that the analyst has to spend extensive time in the workshop just observing what other people are doing. By applying these methods, one will only be able to get a "picture" of the current work content as performed by the operators who are being observed. If anything within this work content changes, a new study (picture) will have to be taken.

The most controversial drawback is, however, the subjective rating of the performance of the operators being observed. Many arguments, grievances, and arbitrations have resulted from disagreements about a "fair" judgment of the performance level of the operator. Why then, is the stopwatch time study method so popular? Basically, the method is easy to learn, although the performance rating takes substantial and continuous effort to master. In the manager's eyes, using a watch is the logical way to find out how long a job should take. We all use watches to set standards for ourselves and others.

Work sampling studies are useful to determine utilization and reasons for downtime and delays. However, such studies will only provide data that relate to the period during which the study is made.

Finally, there is a non-engineered method that has been used extensively for setting standards in areas such as maintenance. This method is called *benchmark comparison*. Because of the nature of the work, the accuracy requirements are not as stringent as in many other situations. Therefore, a set of scientifically determined time standards called *benchmarks* are developed for typical operations and used as a basis for the determination of time standards for operations that are similar but sometimes very different from the benchmarks. As with the other non-engineered methods, no backup data are available.

Non-engineered methods are frequently used mainly with the purpose of quickly establishing a *time value*. Some people consider it satisfactory just to have a time value for scheduling, cost estimating, budgeting, etc., rather than develop realistic and consistent time standards that are based on established work conditions and defined work methods and that one can have confidence in. Without backup data, non-engineered standards are *low-quality time standards*.

10.2.2 Engineered Work Measurement

Prior to starting the manufacturing of a part, component, or product, a detailed engineered drawing is almost always produced. This drawing contains the size and shape of the part, dimensions, tolerances, surface finishes, material, etc., so that the operator will have a very accurate specification of *what to manufacture*. Engineered work measurement will produce an equally detailed "drawing" of *how to manufacture* or assemble the part, component, or product.

Engineered work measurement uses a *predetermined motion time system* (PMTS) as a basis for the measurement. All necessary time values have been predetermined (based on very detailed time studies) and are therefore available to be used by anyone who is trained in the procedures on how to apply these time values. Since all the work measurement time values are predetermined, *preproduction* standards can be established. This means that the time standard for an operation can be established long before the operation will be performed.

Motion-Based Systems

There are a number of predetermined motion time systems that will result in documented engineered standards of high quality. These systems can be categorized into *motion-based systems* such as Methods Time Measurement (MTM) and Work Factor (WF). The time elements in these systems have been determined for basic human motions such as reach, grasp, move, position, etc. These systems were introduced in the late 1940s. MTM was published in a book by the late Harold B. Maynard in 1948 which has made MTM the most-used predetermined motion time system in the world. Because of their detail, these systems are time consuming and tedious to apply.

Element-Based Systems

Simplified versions of these systems have been developed as *element-based systems*. Examples of such systems are MTM-2, MTM-3, General Purpose Data (GPD), Universal Standard Data (USD), Universal Office Controls (UOC), Universal Maintenance Standards (UMS), etc. Usually these so-called "standard data systems" have been designed for use in specific application areas. These element-based systems are also considerably faster to use than the motion-based systems. However, in the majority of cases, they are being applied only in a manual mode. Further information on MTM is available in the book Engineered Work Measurement, 4th ed., by Delmar W. Karger and Franklin H. Bayha (Industrial Press, New York, 1987).

Activity-Based Systems

The third category of predetermined motion time systems includes an *activity-based system* called MOST (Maynard Operation Sequence Technique). The MOST system consists of logical sequence models that cover all the motions included in the activity of moving an object from one location to another. The concept of MOST was developed in 1967, and the complete BasicMOST System was introduced in 1972. The MOST system is considerably faster to use and easier to learn than the motion-based and element-based systems.

For these reasons, MOST is a user-friendly work measurement system. More detailed information on MOST systems and the computerized version, MOST Computer Systems, will be presented in Subchapter 10.4.

10.3 METHODS OF DEVELOPING, ADMINISTERING, AND MAINTAINING TIME STANDARDS

All non-engineered systems, and in some cases also engineered systems, are being used to establish a time standard directly for manufacturing and other operations without any intermediate steps. This method is called the *direct measurement method.* An *operation* is defined and the work measurement technique is applied to determine the time standard for it. With large numbers of time standards, e.g., 50,000, 100,000, or 1,000,000 or more, the direct method becomes very time consuming and costly to apply. Administering and maintaining that many "direct" time standards becomes an unmanageable task, even if a computer is being used. However, when only a few time standards are needed or when time standards for a specific situation or product are required, the direct method is acceptable and even preferable.

Engineered standards are in practically all cases being established by using an *indirect measurement* method. Instead of applying the work measurement technique to establish a time standard for an operation directly, the work measurement technique is used to develop *standard data,* or parts of operations, also called *suboperations.* Such data units or building blocks can be used in different combinations to determine time standards for operations. It is quite possible to establish 100,000 or more time standards from a base of 200 to 700 suboperations. The task of measuring these suboperations using a work measurement technique and subsequently maintaining a large database of time standards becomes much more manageable.

Therefore, a predetermined time system can be applied either for direct work measurement of defined operations or can be used as a basis for standard data (suboperations). In the case of short-cycle, "unique" operations such as subassemblies, the direct approach is preferred. On the other hand, if a great variety of the operations are being performed at a work center, the standard data approach is the most efficient and economical method. A *worksheet* composed of standard data units, each backed up by a work measurement analysis, will provide a fast and simple way to calculate standards. Initially, the desired accuracy level for the resulting time standards should be determined and the worksheet designed accordingly. This means that the tighter the accuracy requirements are, the more data units (suboperations) and the more decisions have to be made in order to set a time standard. A multipage, detailed worksheet will take more time and cost more to use than a single-page worksheet with few elements designed for a lower accuracy level. Consequently, the economics of setting standards is a direct function of the required accuracy of the output.

For instance, if the required accuracy is ±5% with 95% confidence over an 8-hour period, the worksheet may consist of 75 different elements, while a ±10% accuracy with 90% confidence over a 40-hour period may produce a worksheet with only 10 to 15 elements. The difference in application time will be substantial, and since standard setting normally is an ongoing activity, the cost-saving potential is considerable.

In order to further simplify and expedite the standard-setting process, decision models based on expert system technology can be used. In such a case, only simple questions regarding the parts or products to be manufactured need to be answered. The selection of suboperations is made automatically by the computer using the AutoMOST program.

10.4 MOST® WORK MEASUREMENT SYSTEMS

10.4.1 Introduction to MOST

Because industrial engineers are trained that with sufficient study any method can be improved, many efforts have been made to simplify the analyst's work measurement task. This has, for instance, led to a variety of work measurement systems now in use. These achievements also led us to examine the whole concept of work measurement to find a better way for analysts to accomplish their mission. This induced the information of a new approach later to be known as *MOST—Maynard Operation Sequence Technique*.

10.4.2 The MOST Concept

To most of us, work means exerting energy, but we should add, to accomplish some task or to perform some useful activity. In the study of physics, we learn that work is defined as the product of force times distance ($W = f \times d$) or, more simply, work is the displacement of a mass or object. This definition applies quite well to the largest portion of the work accomplished every day, such as pushing a pencil, lifting a heavy box, or moving the controls on a machine. Thought processes, or thinking time, are an exception to this concept, as no objects are being displaced. For the overwhelming majority of work, however, there is a common denominator from which work can be studied, the displacement of objects. All basic units of work are organized (or should be) for the purpose of accomplishing some useful result by moving objects. That is what work is. *MOST is a system to measure work*; therefore, *MOST concentrates on the movement of objects*.

Work, then, is the movement of objects—maybe we should add, following a tactical production outline. Efficient, smooth, productive work is performed when the basic motion patterns are tactically arranged and smoothly choreographed (methods engineering). The movement of objects follows certain consistently repeating patterns, such as reach, grasp, move, and positioning of the object. These patterns can be identified and arranged as a sequence of events (or subactivities) manifesting the movement of an object. A model of this sequence is made and acts as a standard guide in analyzing the movement of an object. It should also be noted that the actual motion contents of the subactivities in a sequence vary independently of one another.

This concept provides the basis for the MOST *sequence models*. The primary work units are no longer basic motions as in MTM (Methods Time Measurement), but fundamental activities (collections of basic motions) dealing with moving objects. These activities are described in terms of subactivities fixed in sequence. In other words, to move an object, a standard sequence of events occurs. Consequently, the basic pattern of an object's movement is described by a universal sequence model instead of an aggregate of detailed basic motions synthesized at random.

Objects can be moved in only one of two ways: Either they are picked up and moved freely through space, or they are moved in contact with another surface. For example, a box can be picked up and carried from one end of a workbench to the other, or it can be pushed across the top of the workbench. For each type of move, a different sequence of events occurs; therefore, a separate MOST activity *sequence model* applies. The use of tools is analyzed through a separate activity sequence model which allows the analyst the opportunity to follow the movement of a hand tool through a standard sequence of events, which is in fact, a combination of the two basic sequence models.

Consequently, only three activity sequences are needed for describing manual work. The *BasicMOST work measurement technique*, therefore, is comprised of the following sequence models:

The *General Move sequence*, for the spatial movement of an object freely through the air

The *Controlled Move sequence*, for the movement of an object when it remains in contact with a surface or is attached to another object during the movement

The *Tool Use sequence*, for the use of common hand tools

A fourth sequence model, the *Manual Crane sequence* for the measurement of moving heavy objects by using, for instance, a jib crane, is also part of the BasicMOST system, although it is used less frequently than the three first sequence models.

10.4.3 Sequence Models

General Move Sequence Model

General Move is defined as moving objects manually from one location to another freely through the air. To account for the various ways in which a General Move can occur, the activity sequence is made up of four subactivities:

A, action distance (mainly horizontal)
B, body motion (mainly vertical)
G, gain control
P, placement

These subactivities are arranged in a *sequence model* consisting of a series of parameters organized in a logical arrangement. The sequence model defines the events or actions that always take place in a preset order when an object is being moved from one location to another.

The *General Move sequence model*, which is the most commonly used of all available sequence models, is defined as follows:

A	B	G	A	B	P	A
Action distance	Body motion	Grasp	Action distance	Body motion	Placement	Action distance

These subactivities, or *sequence model parameters* as they are called, are then assigned time-related index numbers based on the motion content of the subactivity. This approach provides complete analysis flexibility within the overall control of the sequence model. For each object moved, any combination of motions could occur, and using MOST, any combination can be analyzed. For the General Move sequence, these index values are easily memorized from a brief data card (Figure 10.1). A fully-indexed General Move sequence, for example, might appear as follows:

$A_6 B_6 G_1 A_1 B_0 P_3 A_0$

ABGABPA

Index × 10	A — Action Distance		B — Body Motion		G — Gain Control		P — Placement		Index × 10
	Parameter Variant	Keyword	Parameter Variant	Keyword	Parameter Variant	Keyword	Parameter Variant	Keyword	
0	≤ 2 in. ≤ 5 cm	CLOSE					Hold Toss	THROW TOSS / CARRY PICKUP	0
1	Within reach				Light object Light objects simo	GRASP (optional)	Lay aside Loose fit	MOVE PUT	1
3	1–2 steps	1 STEP 2 STEPS	Bend and arise 50% occ.	PBEND	Non Simo Obstructed Heavy/Bulky Interlocked Blind Collect Disengage	GET DISENGAGE FREE COLLECT	Adjustments Light pressure Double placement	PLACE REPLACE	3
6	3–4 steps	3 STEPS 4 STEPS	Bend and arise	BEND			Care Precision Blind Obstructed Heavy pressure Intermediate moves	POSITION REPOSITION	6
10	5–7 steps	5 STEPS 6 STEPS 7 STEPS	Sit or stand	SIT STAND					10
16	8–10 steps	8 STEPS 9 STEPS 10 STEPS	Through Door Climb on or off Stand and bend Bend and sit	DOOR CLIMB/DESCEND STAND AND BEND BEND AND SIT					16

Figure 10.1 Index values for the parameters used in the General Move sequence model.

where A_6 = Walk three to four steps to object location
         B_6 = Bend and arise
         G_1 = Gain control of one light object
         A_1 = Move object a distance within reach
         B_0 = No body motion
         P_3 = Place and adjust object
         A_0 = No return

This example could, for instance, represent the following activity: "Walk three steps to pick up a bolt from floor level, arise, and place the bolt in a hole."

General Move is by far the most frequently used of the three sequence models. Roughly 50% of all manual work occurs as a General Move, with the percentage running higher for assembly and material handling work, and lower for machine shop operations.

Controlled Move Sequence Model

The second type of move is described by the *Controlled Move sequence*. This sequence is used to cover such activities as operating a lever or crank, activating a button or switch, or simply sliding an object over a surface. In addition to the A, B, and G parameters from the General Move sequence, the sequence model for Controlled Move contains the following subactivities:

M, move controlled
X, process time
I, align

As many as one-third of the activities occurring in machine shop operations may involve Controlled Move sequences. A typical activity covered by the Controlled Move sequence is the engaging of the feed lever on a milling machine. The sequence model for this activity might be indexed as follows:

$A_1 \ B_0 \ G_1 \ M_1 \ X_{10} \ I_0 \ A_0$

where A_1 = Reach to the lever a distance within reach
         B_0 = No body motion
         G_1 = Get hold of lever
         M_1 = Move lever up to 12 in. (30 cm) to engage feed
         X_{10} = Process time of approximately 3.5 seconds
         I_0 = No alignment
         A_0 = No return

Tool Use Sequence Model

The third sequence model comprising the BasicMOST technique is the *Tool Use* sequence model. This sequence model covers the use of hand tools for such activities as fastening or loosening, cutting, cleaning, gauging, and recording. Also, certain activities requiring the use of the brain for mental processes can be classified as Tool Use, such as reading and thinking. As indicated above, the Tool Use sequence model is a combination of General Move and Controlled Move activities. It was developed as part of the BasicMOST system, in order to simplify the analysis of activities related to the use of hand tools. It will later become obvious to the reader that any hand tool activity is made up of General and Controlled Moves. The use of a wrench, for example, might be described by the following sequence:

$$A_1 \ B_0 \ G_1 \ A_1 \ B_0 \ P_3 \ F_{10} \ A_1 \ B_0 \ P_1 \ A_0$$

where A_1 = Reach to wrench
        B_0 = No body motion
        G_1 = Get hold of wrench
        A_1 = Move wrench to fastener a distance within reach
        B_0 = No body motion
        P_3 = Place wrench on fastener
        F_{10} = Tighten fastener with wrench
        A_1 = Move wrench a distance within reach
        B_0 = No body motion
        P_1 = Lay wrench aside
        A_0 = No return

10.4.4   Elements and Characteristics of MOST

Time Units

The time units used in MOST are identical to those used in the basic MTM (Methods Time Measurement System) and are based on hours and parts of hours called TMUs (Time Measurement Units). One TMU is equivalent to 0.00001 hour.

The time value in TMU for each sequence model is calculated by adding the index numbers and multiplying the sum by 10. For our previous General Move sequence example, the time is

$$(6 + 6 + 1 + 0 + 0 + 3 + 0) \times 10 = 170 \text{ TMU}$$

corresponding to approximately 0.01 minutes. The time values for the other two examples are computed in the same way. The Controlled Move totals to

$$(1 + 0 + 1 + 1 + 10 + 0 + 0) \times 10 = 130 \text{ TMU}$$

and the Tool Use to

$$(1 + 0 + 1 + 1 + 0 + 3 + 10 + 1 + 0 + 1 + 0) \times 10 = 180 \text{ TMU}$$

All time values established by MOST reflect the pace of an average skilled operator working at an average performance rate. This is often referred to as the 100% performance level that in time study is achieved by using "leveling factors" to adjust time to defined levels of skill and effort. Therefore, when using MOST, it is not necessary to adjust the time values unless they must conform with particular high or low task plans used by some companies. This also means that if a time standard for an operation is properly established by using either MOST, MTM, or stopwatch time study, the TMU values should be identical or almost identical for the three techniques.

The analysis of an operation consists of a series of sequence models describing the movement of objects to perform the operation. See Figure 10.2 for an example. Total time for the complete MOST analysis is arrived at by adding the computed sequence times. The operation time may be left in TMU or converted to minutes or hours. Again, this time reflects pure work content (normal time without allowances) at the 100% performance level. The "final" *time standard* will include an allowance factor consisting of P (personal time), R, or F (rest or fatigue factor), and D for unavoidable delays (often determined by a work sampling study). Therefore, if the normal time = 1.0 hr and the allowance factor (PR&D) = 15%, the *final time standard* = 1.15 hr.

Application Speed

MOST was designed to be considerably faster than other work measurement techniques. Because of its simpler construction, under ideal conditions, Basic MOST requires only 10 applicator hours per measured hour. (MTM-1 requires 300 to 400 applicator hours per measured hour.)

Accuracy

The accuracy principles that apply to MOST are the same as those used in statistical tolerance control. That is, the accuracy to which a part is manufactured depends on its role in the final assembly. Likewise, with MOST, time values are based on calculations that guarantee the overall accuracy of the final time standard. Based on these principles, MOST provides the means for covering a high volume of manual work with an accuracy that can be determined and controlled.

Method Sensitivity

MOST is a method-sensitive technique; that is, it is sensitive to the variations in time required by different methods. This feature is very effective in evaluating alternative methods of performing operations with regard to time and cost. The MOST analysis will clearly indicate the more economical and less fatiguing method.

The fact that the MOST system is method sensitive greatly increases its worth as a work measurement tool. Not only does it indicate the time needed to perform various activities, it also provides the analyst with an instant clue that a method should be reviewed. The results are clear, concise, easily understood time calculations that indicate opportunities for saving time, money, and energy.

Documentation

One of the most burdensome problems in the standards development process is the volume of paperwork required by the most widely used predetermined work measurement systems. Whereas the more detailed systems require between 40 and 100 pages of documentation, MOST requires as few as 5. The substantially reduced amount of paperwork enables analysts to complete studies faster and to update standards more easily. It is interesting to note that the reduction of paper generated by MOST does not lead to a lack of definition of the method used to perform the task. On the contrary, the method description found with the MOST system is a clear, concise, plain-language description of the activity. These method descriptions can very well be used for operator training and instruction.

Applicability

In what situations can MOST be used? Because manual work normally includes some variation from one cycle to the next, MOST, with its statistically established time ranges and time values, can produce times comparable to more detailed systems for the majority of manual operations. Therefore, MOST is appropriate for any manual work that contains variation from one cycle to another, regardless of cycle length. BasicMOST should not be used in situations in which a short cycle (usually up to 10 sec or 280 TMU long) is repeated identically over an extended period of time. In these situations, which by the way do not occur very often, the more detailed MiniMOST version should be chosen as the proper work measurement tool. In fact, MiniMOST was developed to cover highly repetitive, short-cycle work measurement tasks. At the other end of the spectrum, MaxiMOST was developed to measure long-cycle (2 min or more), nonrepetitive operations such as heavy assembly, maintenance, and machine setups.

10.4.5 Application of the Sequence Models

The General Move Sequence Model

The *General Move* sequence deals with the spatial displacement of one or more object(s). Under manual control, the object follows an unrestricted path through the air. If the object is in contact with or restrained in any way by another object during the move, the General Move sequence is not applicable. Characteristically, General Move follows a fixed sequence of subactivities identified by the following steps:

1. Reach with one or two hands a distance to the object(s), either directly or in conjunction with body motions.
2. Gain manual control of the object(s).
3. Move the object(s) a distance to the point of placement, either directly or in conjunction with body motions.
4. Place the object(s) in a temporary or final position.
5. Return to workplace.

These five subactivities form the basis for the activity sequence describing the manual displacement of the object(s) freely through space. This sequence describes the manual events that can occur when moving an object freely through the air and is therefore known as a *sequence model*. The major function of the sequence model is to guide the attention of the analyst through an operation, thereby adding the dimension of having a preprinted and standardized analysis format. The existence of the sequence model provides for increased analyst consistency and reduced subactivity omission.

The sequence model takes the form of a series of letters representing each of the various subactivities (called parameters) of the General Move activity sequence. With the exception of an additional parameter for body motions, the General Move sequence is the same as the above five-step pattern:

A B G A B P A

where   A = action distance
           B = body motion
           G = gain control
           P = placement

Parameter Definitions:

Action distance, A. This parameter covers all spatial movement or actions of the fingers, hands, and/or feet, either loaded or unloaded. Any control of these actions by the surroundings requires the use of other parameters.

Body motion, B. This parameter refers to either vertical (up and down) motions of the body or the actions necessary to overcome an obstruction or impairment to body movement.

Gain control, G. This parameter covers all manual motions (mainly finger, hand, and foot) employed to obtain complete manual control of an object(s) and to subsequently relinquish that control. The G parameter can include one or several short-move motions whose objective is to gain full control of the object(s) before it is to be moved to another location.

Placement, P. This parameter refers to actions at the final stage of an object's displacement to align, orient, and/or engage the object with another object(s) before control of the object is relinquished.

Parameter Indexing

Index values for the above four parameters included in the General Move sequence model can be found in Figure 10.1. Definitions of all available index values for the four General Move parameters can be found in *MOST Work Measurement Systems* [1]. The definitions for A - Action Distance have been included below as an example.

Action distance, A. Action distance covers all spatial movement or actions of the fingers, hands, and/or feet, either loaded or unloaded. Any control of these actions by the surroundings requires the use of other parameters.

$A_0 < 2$ *in. (5 cm).* Any displacement of the fingers, hands, and/or feet a distance less than or equal to 2 in. (5 cm) will carry a 0 index value. The time for performing these short distances is included within the G and P parameters. Example: Reaching between the number keys on a pocket calculator or placing nuts or washers on bolts located less than 2 in. (5 cm) apart.

A_1 *within reach.* Actions are confined to an area described by the arc of the outstretched arm pivoted about the shoulder. With body assistance—a short bending or turning of the body from the waist—this "within reach" area is extended somewhat. However, taking a step for further extension of the area exceeds the limits of an A_1 and must be analyzed with a A_3 (one to two steps). Example: With the operator seated in front of a well laid out work bench, all parts and tools can be reached without displacing the body by taking a step.

The parameter value A_1 also applies to the actions of the leg or foot reaching to an object, lever, or pedal. If the trunk of the body is shifted, however, the action must be considered a step (A_3).

A_3 *one to two steps.* The trunk of the body is shifted or displaced by walking, stepping to the side, or turning the body around using one or two steps. Steps refers to the total number of times each foot hits the floor.

Index values for longer-action distances involving walking on flat surfaces as well as up or down ladders can be found in Figure 10.1 for up to 10 steps. This will satisfy the need for action distance values for most work areas in a manufacturing plant. Should longer walking distance occur, however, the table can be extended. All index values for walking are based on an average stop length of $2\frac{1}{2}$ ft (0.75 m).

General Move Examples

1. A man walks four steps to a small suitcase, picks it up from the floor, and without moving, further places it on a table located within reach.

$$A_6 \, B_6 \, G_1 \, A_1 \, B_0 \, P_1 \, A = 150 \text{ TMU}$$

2. An operator standing in front of a lathe walks six steps to a heavy part lying on the floor, picks up the part, walks six steps back to the machine, and places it in a 3-jaw chuck with several adjusting actions. The part must be inserted 4 in (10 cm) into the chuck jaws.

$$A_{10} \, B_6 \, G_3 \, A_{10} \, B_0 \, P_3 \, A_1 = 330 \text{ TMU}$$

3. From a stack located 10 ft (3 m) away, a heavy object must be picked up and moved 5 ft (2m) and then placed on top of a workbench with some adjustments. The height of this stack will vary from waist to floor level. Following the placement of the object on the workbench, the operator returns to the original location, which is 11 ft (3.5 m) away.

$$A_6 \, B_3 \, G_3 \, A_3 \, B_0 \, P_3 \, A_{10} = 280 \text{ TMU}$$

The Controlled Move Sequence Model

The *Controlled Move* sequence describes the manual displacement of an object over a controlled path. That is, movement of the object is restricted in at least one direction by contact with or an attachment to another object.

The Sequence Model

The sequence model takes the form of a series of letters representing each of the various subactivities (called *parameters*) of the Controlled Move activity sequence:

A B G M X I A

where   A = action distance
         B = body motion
         G = gain control
         M = move controlled
         X = process time
         I = align

Parameter Definitions

Only three new parameters are introduced, as the A, B, and G parameters were discussed with the General Move sequence and remain unchanged.

Move controlled, M. This parameter covers all manually guided movements or actions of an object over a controlled path.

Process time, X. This parameter occurs as that portion of work controlled by processes or machines and not by manual actions.

Align, I. This parameter refers to manual actions following the controlled move or at the conclusion of process time to achieve the alignment of objects. The index value definitions for the above parameters (M, X and I) can be found in the textbook, "MOST Work Measurement Systems" [1].

Controlled Move Examples

1. From a position in front of a lathe, the operator takes two steps to the side, turns the crank two revolutions, and sets the machining tool against a scale mark.

   $A_3 \, B_0 \, G_1 \, M_6 \, X_0 \, I_6 \, A_0 = 160$ TMU

2. A milling cutter operator walks four steps to the quick-feeding cross lever and engages the feed. The machine time following the 4-in. (10-cm) lever action is 2.5 sec.

   $A_6 \, B_0 \, G_1 \, M_1 \, X_6 \, I_0 \, A_0 = 140$ TMU

3. A material handler takes hold of a heavy carton with both hands and pushes it 18 in. (45 cm) across conveyor rollers.

   $A_1 \, B_0 \, G_3 \, M_3 \, X_0 \, I_0 \, A_0 = 70$ TMU

4. Using the foot pedal to activate the machine, a sewing machine operator makes a stitch requiring 3.5 sec process time. (The operator must reach the pedal with the foot.)

   $A_1 \, B_0 \, G_1 \, M_1 \, X_{10} \, I_0 \, A_0 = 130$ TMU

The Tool Use Sequence Model

The *Tool Use* sequence is composed of subactivities from the General Move sequence, along with specially designed parameters describing the actions performed with hand tools or, in some cases, the use of certain mental processes. Tool Use follows a fixed sequence of subactivities occurring in five main activity phases:

1. Get object or tool.
2. Place object or tool in working position.
3. Use tool.
4. Put aside object or tool.
5. Return to workplace.

The Sequence Model

The five activity phases form the basis for the activity sequence describing the handling and use of hand tools. The sequence model takes the form of a series of letters representing each of the various subactivities of the Tool Use activity sequence:

Get object or tool	Place object or tool	Use tool	Aside object or tool	Return
A B P	A B P		A B P	A

where   A = action distance
      B = body motion
      G = gain control
      P = place

The space in the sequence model—"Use Tool"—is provided for the insertion of one of the following Tool Use parameters. These parameters refer to the specifications of using the tool and are

F $=$ Fasten
L $=$ Loosen
C $=$ Cut
S $=$ Surface treat
M $=$ Measure
R $=$ Record
T $=$ Think

Tool Use Examples for "Fasten and Loosen"

1. Obtain a nut from a parts bin located within reach, place it on a bolt, and run it down with seven finger actions.

$$A_1 \ B_0 \ G_1 \ A_1 \ B_0 \ P_3 \ F_{10} \ A_0 \ B_0 \ P_0 \ A_0 = 160 \text{ TMU}$$

2. Obtain a power wrench from within reach, run down four 3/8-in. (10-mm) bolts located 6 in. (15 cm) apart, and set aside wrench.

$$A_1 \ B_0 \ G_1 \ A_0 \ B_0 \ (P_3 \ A_1 \ F_6) \ A_1 \ B_0 \ P_1 \ A_0 \ (4) = 440 \text{ TMU}$$

3. From a position in front of an engine lathe, obtain a large T-wrench located five steps away and loosen one bolt on a chuck on the engine lathe with both hands using five arm actions. Set aside the T-wrench from the machine (but within reach).

$$A_{10} \ B_0 \ G_1 \ A_{10} \ B_0 \ P_3 \ L_{24} \ A_1 \ B_0 \ P_1 \ A_0 = 500 \text{ TMU}$$

10.4.6 The MOST Systems Family

In addition to the BasicMOST system, several application-oriented versions of MOST are now members of the MOST systems family: MiniMOST, MaxiMOST, and ClericalMOST. A new version, MegaMOST, is under development for future applications.

The MiniMOST System

BasicMOST was not designed to measure short-cycle operations, although the original BasicMOST version can be applied to nonidentical operations of 10 sec or less and still meet the accuracy criteria. Therefore, the MiniMOST version of MOST was developed to satisfy higher accuracy requirements that apply to very short-cycled, highly repetitive, identical operations. Such operations may be from 2 to 10 seconds long and often are being performed over long periods of time. MiniMOST consists of two sequence models:

General Move: A B G A B P A
Controlled Move: A B G M X I A

These sequence models are identical to the two basic sequence models in the BasicMOST version. There is one major difference, however. The multiplier for the index value total is 1 for MiniMOST. Therefore, if the sum of the applied index values is 64, this is also the total TMU value for the sequence model. Another difference compared to BasicMOST is that distances in MiniMOST are measured in inches. The application speed of MiniMOST is about 25:1 under "ideal" conditions compared, to about 10:1 for BasicMOST.

The definitions and descriptions of the parameters and elements in MiniMOST have been excluded because of space considerations. The second edition of *MOST*

Work Measurement Systems [2] includes a complete review of MiniMOST Work Measurement Systems.

The MaxiMOST System

In order to satisfy the need for a fast, less detailed, but still accurate and consistent system for the measurement of long-cycle, nonrepetitive, nonidentical operations, MaxiMOST was developed. MaxiMOST consists of five sequence models with a multiplier of 100. The sequence models are:

Part handling
Tool/equipment use
Machine handling
Transport with powered crane
Transport with wheeled truck

MaxiMOST has a measurement factor of 3-5:1 (analyst hours to measured hours) and is therefore a very cost-effective technique to use in a large number of cases where minute details are unnecessary or even detrimental to proper work instructions. The recommendation is to use MaxiMOST for nonidentical cycles that are 2 min or longer.

The definitions and descriptions of the parameters and elements in the MaxiMOST system have been excluded because of space considerations. The second edition of *MOST Work Measurement Systems* [2] includes a complete explanation of the MaxiMOST Work Measurement System.

ClericalMOST

The MOSTClerical system is based on three sequence models identical to those in BasicMOST:

General move
Controlled move
Tool/equipment use (two data cards)

MegaMOST

The main purpose of adding a version of MOST on the 1000 multiplier level is to simplify and accelerate the standard setting for long (over 20 min), nonrepetitive operations in areas such as assembly and maintenance.

While MiniMOST is totally generic and BasicMOST is about 60-80% generic, MaxiMOST is primarily tool-oriented, and MegaMOST is part and operations-oriented. MegaMOST will be adopted for automated calculation of standards by the computer.

10.4.7 Procedures for Developing MOST Time Standards

A standard MOST Calculation Form should be used for all analysis work using BasicMOST. (Similar forms have been designed for use with MiniMOST and MaxiMOST.) As can be seen from the included example, Figure 10.2, this form consists of four sections:

1. A header identifying the activity to be measured and the work center (area) in which it is being performed
2. A step-by-step method description (left half)

	MOST® Calculation	CODE	
⋙M⋙		DATE 7/29/91	
	PROD AREA *ELECTRONIC ASSEMBLY*	SIGN *A. A.*	
	OPERATION	PAGE 1/1	

TITLE *INSTALL CONNECTOR ON PC-BOARD AT WORK STATION*

ACTIVITY • OBJECT • IN, ON, FOR • PRODUCT EQUIPMENT • TOOL • TO, AT • WORK AREA

CONDITIONS *EDGE CONNECTORS ONLY*   **PER** *BOARD*

NO.	METHOD DESCRIPTION	NO.	SEQUENCE MODEL							FR	TMU
1	POSITION EDGE CONNECTOR TO BOARD	1	A₁	B₀	G₁	A₁	B₀	P₆	A₀		90
		3	A₁	B₀	G₁	A₁	B₀	P₃	A₀	2	120
2	ALIGN CONNECTOR TO ACCURATE LOCATIONS	4	A₁	B₀	G₁	A₁	B₀	P₁	A₀	4	160
		7	A₁	B₀	G₁	A₁	B₀	P₃	A₀		60
3	PLACE SCREW TO HOLE IN CONNECTOR F2		A	B	G	A	B	P	A		
			A	B	G	A	B	P	A		
4	MOVE WASHER TO SCREEN ON BOARD F4		A	B	G	A	B	P	A		
			A	B	G	A	B	P	A		
5	FASTEN NUTS 2 SPINS USING FINGERS F2		A	B	G	A	B	P	A		
			A	B	G	A	B	P	A		
6	FASTEN 2 SCREWS 5 SPINS USING SCREWDRIVER		A	B	G	A	B	P	A		
			A	B	G	A	B	P	A		
7	PLACE BOARD TO RACK	2	A₀	B₀	G₀	M₃	X₀	I₁₆	A₀		190
			A	B	G	M	X	I	A		
			A	B	G	M	X	I	A		
			A	B	G	M	X	I	A		
			A	B	G	M	X	I	A		
			A	B	G	M	X	I	A		
			A	B	G	M	X	I	A		
		5	A₁	B₀	G₁	A₁	B₀	P₁	F₃	A₀B₀P₀A₀	2	140
		6	A₁	B₀	G₁	A₁	B₀	(P₃ A₁ F₁₀) A₁ B₀ P₁ A₀			(2)	330
			A	B	G	A	B	P	A B P A		
			A	B	G	A	B	P	A B P A		
			A	B	G	A	B	P	A B P A		
			A	B	G	A	B	P	A B P A		
			A	B	G	A	B	P	A B P A		
			A	B	G	A	B	P	A B P A		

TIME = .65   minutes (min.)   **1090**

3/93 ⁱH. B. Maynard & Company, Inc.

Figure 10.2 Standard calculation form used for analysis work using BasicMOST.

3. Preprinted sequence models in three groups: General Move, Controlled Move, and Tool Use
4. A field for the time value or time standard for the activity (bottom part)

Note: The activity time or standard does not include any allowances at this stage. Prior to applying this time standard, the time value on the form should be multiplied by the appropriate allowance factor (PR&D), thereby constituting the standard time for the operation.

A frequency factor (Fr) for each sequence model can be specified in the column next to the TMU value column for the sequence model. Normally, the space provided on one page of the MOST Calculation Form will allow for analyses up to approximately 1 minute.

In all situations where MOST is being used, the "top-down" approach should be followed. A two-step decision model can be put to use, with the first generation being "Is it appropriate and practical to do direct measurement?" If the answer is "yes," the work should be measured by using the MOST Calculation Form. If the answer is "no," the operations or activities for the plant, department, or work center should be broken down into logical suboperations. Each suboperation is then measured using MOST and placed on a worksheet for the calculation of time standards. In some instances (e.g., for long-cycle operations), suboperations may have to be broken down still one more level and later combined into sub-operations before assigning them onto the worksheet.

By following the top-down approach, the database with standard data (suboperation data) will remain compact and more manageable than if the conventional "bottom-up" procedure is applied. In most cases, only 200 to 500 suboperations will be required to calculate any number of time standards.

MOST is an application-oriented or user-friendly system that will require some unlearning and rethinking by its users that are experienced in conventional work measurement. It is a new concept, not only regarding work measurement, but also in the application areas.

10.4.8 MOST Computer Systems

The logical sequence model approach lends itself very well to a computerized application. Therefore, in 1976 the first lines of code were written in an effort to develop a software program that would advance the state of the art of work measurement. While other computerized systems use element symbols or numerical data as input, MOST Computer Systems use method descriptions expressed in plain English. In other words, MOST Computer Systems are a language-based system. Today the computerized MOST program reminds one of an "expert system," although this term was not commonly known when the development started.

Computerized MOST Analysis

The input for a computer MOST analysis consists of (1) *work area data* and (2) a *method description*. Based on this information, the computer produces a MOST analysis as output; i.e., the computer actually completes the work measurement task "automatically." A simple but representative work area layout sketch is also part of the output. A typical example of a work area description is shown in Figure 10.3, and the MOST analysis for an operation performed in that work area is shown in Figure 10.4.

In designing the program, the basic philosophy of establishing a time standard as a direct function of the work conditions was followed. The computer was therefore programmed to produce a time standard based on well-defined and complete user work conditions. The computer was also programmed not to allow the change of a time value without a change of the underlying work conditions. A change of distance, for instance, or "gain control" or "placement" of an object or a body motion, will result in a different standard. This discipline has proven to increase the uniformity and consistency of the method descriptions and analyses. Equally important is the fact that one does not have to read both the method description and the MOST index values to interpret an analysis. A

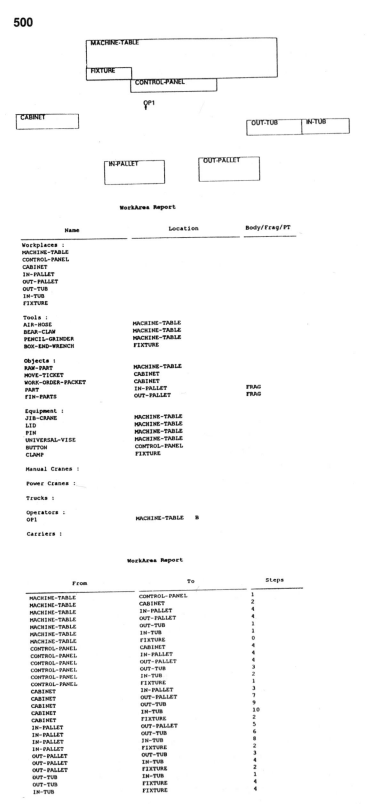

WorkArea Report

Name	Location	Body/Frag/PT
Workplaces :		
MACHINE-TABLE		
CONTROL-PANEL		
CABINET		
IN-PALLET		
OUT-PALLET		
OUT-TUB		
IN-TUB		
FIXTURE		
Tools :		
AIR-HOSE	MACHINE-TABLE	
BEAR-CLAW	MACHINE-TABLE	
PENCIL-GRINDER	MACHINE-TABLE	
BOX-END-WRENCH	FIXTURE	
Objects :		
RAW-PART	MACHINE-TABLE	
MOVE-TICKET	CABINET	
WORK-ORDER-PACKET	CABINET	
PART	IN-PALLET	FRAG
FIN-PARTS	OUT-PALLET	FRAG
Equipment :		
JIB-CRANE	MACHINE-TABLE	
LID	MACHINE-TABLE	
PIN	MACHINE-TABLE	
UNIVERSAL-VISE	MACHINE-TABLE	
BUTTON	CONTROL-PANEL	
CLAMP	FIXTURE	
Manual Cranes :		
Power Cranes :		
Trucks :		
Operators :		
OP1	MACHINE-TABLE	B
Carriers :		

WorkArea Report

From	To	Steps
MACHINE-TABLE	CONTROL-PANEL	1
MACHINE-TABLE	CABINET	2
MACHINE-TABLE	IN-PALLET	4
MACHINE-TABLE	OUT-PALLET	4
MACHINE-TABLE	OUT-TUB	1
MACHINE-TABLE	IN-TUB	1
MACHINE-TABLE	FIXTURE	0
CONTROL-PANEL	CABINET	4
CONTROL-PANEL	IN-PALLET	4
CONTROL-PANEL	OUT-PALLET	4
CONTROL-PANEL	OUT-TUB	3
CONTROL-PANEL	IN-TUB	2
CONTROL-PANEL	FIXTURE	1
CABINET	IN-PALLET	3
CABINET	OUT-PALLET	7
CABINET	OUT-TUB	9
CABINET	IN-TUB	10
CABINET	FIXTURE	2
IN-PALLET	OUT-PALLET	5
IN-PALLET	OUT-TUB	6
IN-PALLET	IN-TUB	8
IN-PALLET	FIXTURE	2
OUT-PALLET	OUT-TUB	3
OUT-PALLET	IN-TUB	4
OUT-PALLET	FIXTURE	2
OUT-TUB	IN-TUB	1
OUT-TUB	FIXTURE	4
IN-TUB	FIXTURE	4

Figure 10.3 Typical work area description as part of the work area data in ComputerMOST.

Sub-Operation - Method/TMU Report

```
Sub-Op ID :    547                        Status :    Private
Description :  LOAD PART IN FIXTURE AT MULTI SPINDLE VERTICAL
               DRILL

   Activity :  LOAD/UNLOAD                   Obj :    PART
 Prod/Equip :  FIXTURE                       Tool :   WRENCH
       Size :                                Orig :   MACHINE
      WA No :                                Other :

Unit of Measure :  PART                      OFG :    2
    Workarea ID :  9
Starting Operator: OP1
Starting Location: MACHINE-TABLE          Total Time :   770 TMU

   Applicator :  MCS                         Issue :    1
  Create Date :  02/16/96              Effect Date :    02/16/96


Step              Method Description              TMU
_____
1    PLACE PART FROM IN-PALLET TO FIXTURE
                   A6  B0  G1  A3  B0  P3  A0       1.00     130
2    PUSH CLAMPS  AT FIXTURE F 2
                   A1  B0  G1  M1  X0  I0  A0       2.00      60
3    FASTEN 2 NUTS AT FIXTURE 5 SPINS USING
     FINGERS
          A1  B0  G1  A0  B0  (P1  A1  F10 )A0 B0  P0  A0  (2) 1.00  260
4    FASTEN 2 NUTS AT FIXTURE 4 ARM-TURNS USING
     BOX-END-WRENCH AND ASIDE
          A1  B0  G1  A0  B0  (P3  A1  F10 )A1 B0  P1  A0  (2) 1.00  320
```

Figure 10.4 ComputerMOST analysis for an operation performed in the Figure 10.3 work area.

review of the method is adequate. The index values and the time standard are by-products and direct functions of the method.

How is it possible for the computer to generate a MOST analysis from the input of "only" work area data and a method description? How does the computer select the right sequence model and the correct index values? As explained above, and as can be seen from the example in Figure 10.3, all "action distances" and "body motions" are specified as part of the work area data. Therefore, the A and B parameters in the sequence models are assigned an index value from the work area information.

Three additional variables remain to be determined: (1) sequence model selection, (2) the index value for the G parameter, and (3) the index value for the P parameter. This required information has been compounded into one word: *a keyword*. This keyword, always found in the beginning of each method step, has been chosen from a list of commonly used English activity words such as MOVE, PLACE, and POSITION. For instance, the keyword PLACE means the "general move sequence model" and a combination of G_1 and P_3 to the computer. MOVE indicates the same sequence model with a G_1 and P_1 combination and POSITION, a G_1 P_6 combination. A GET preceding MOVE, PLACE, and POSITION will render G_3 P_1, G_3 P_3, and G_3 P_6 respectively. (The keywords for General Move are indicated in the table in Figure 10.1.)

Similar keywords are available for all sequence models in MOST Computer Systems. The knowledge of approximately 30 to 50 keywords for BasicMOST will provide the analyst with a sufficient "vocabulary" to be able to perform most of the analysis work.

Since both the work area data and the method description are entered within a well-structured format, it is possible to dictate this information using a hand-held tape recorder. A person can, in most situations, talk as fast as or faster than an operator can perform an assembly or a machining operation. Therefore, data collection becomes much more efficient. The conventional hand writing of methods is usually cumbersome and inefficient.

While the "dictation" of a method in principle requires the observation of just one cycle, the "writing" of the same method requires observation of several cycles. The information on the tape is then transcribed by the analyst or a typist on a computer terminal as input to the program. In the future, when a voice recognition system becomes available for practical applications, this "intermediate step" can be eliminated. In fact, a TalkMOST system is presently under development that will let the analyst communicate directly with the computer. The input of information will be done by voice and the output either by voice (computer), or screen, or printed.

Data Management

The major advantage of a computerized application of MOST lies in the databases (suboperations and standards). These are accumulated as a result of the MOST analysis work and calculation of standards: The filing, searching, retrieving, and updating of the data become extremely efficient and fast compared to a manual system. Some functions requiring manipulations of data, such as mass updating, simulations, and history of standards, are very impractical or impossible to execute manually, while the computer can perform them routinely and quickly.

A complete database system for filing and retrieving suboperations and time standards is the backbone of MOST Computer Systems. The database has, to date, been "pushed" to handle over 1 million standards.

The filing system for the database also uses the "word" concept. All suboperation data is filed and retrieved under welldefined words in five categories: activity, object/component, equipment/product, tool, and work area origin. The filing system for standards is in all cases customized to fit the user's requirements and includes such conventional header items as part number, operation number, work center number, etc.

The Complete MOST Computer System

MOST Computer Systems is a complete program for measuring work and calculating time standards as well as documenting and updating these standards. It consists of a Basic Program and a set of Supplementary Modules. The Basic Program includes the following features:

1. Work area generator
2. Work measurement (Basic, Mini, and MaxiMOST)
3. Suboperation database
4. Time standard generator
5. Standards database
6. Mass update
7. Documentation of work conditions (Work Management Manual)
8. Auxiliary data
9. Data transfer (electronically to other systems)

Supplementary modules and application programs include (often these programs will be customized to meet the specific requirements of a user):

1. Machining data (feeds, speeds, and process times)
2. Welding data
3. Line balancing
4. Station assignment

5. Process planning
6. Cost estimating
7. AutoMOST
8. ErgoMOST

AutoMOST

AutoMOST is an *expert system* that can be integrated into MOST Computer Systems to simplify and accelerate several program functions. For instance, the calculation of standards can be either fully automated by transferring part-related data from the bill of material (BOM) or other system; or semi-automated in which case a standard is generated after one answers some simple questions regarding the part characteristics. Specific decision models are developed for each user reflecting its product and part configuration as well as other work conditions. The manual selection of suboperations from a worksheet can thus be eliminated. AutoMOST by itself or in combination with voice recognition will raise the level of system automation substantially.

ErgoMOST

The risk of injury in industry is at an unacceptable level in many situations. In order to identify specific problem areas, ErgoMOST has been developed to analyze the workplace design and the work method with regard to force, posture, and repetition, three of the most critical factors causing cumulative trauma disorders (CTD) (see Subchapter 5.1). After "dangerous" motions have been identified and improvements have been made to the workplace, ErgoMOST can analyze the effect of the improvements.

The objective with MOST Computer Systems was to adapt the system to cover all possible aspects of establishing time standards in a wide range of situations. Another objective was to make the updating and maintenance of standards efficient and simple. Our intention was also to stimulate industrial engineers in industry, services, and in universities and colleges to adopt a positive attitude toward a fundamental and widely used discipline: the *measurement of work*.

10.5 APPLICATION OF TIME STANDARDS

There are numerous reasons for applying work measurement. Often, however, a company or a manager has a specific purpose for investing in an advanced computerized work measurement system, for instance, productivity improvement or cost control. What everyone seems to have in common is that they want to improve the *accessibility to* and *quality of information*, whether it is related to planning, cost control, budgeting or productivity. *Estimates or historical data are simply not good enough in today's competitive industry!* Scientific methods have to be applied to provide accurate and consistent information about shop floor operations prior to their performance. Let us briefly review some of the most important justifications for work measurement that managers have to consider.

10.5.1 Productivity Improvement

The first and foremost reason for implementing a work measurement program is (and should be) to improve productivity and/or reduce costs. Simply put, the job of any manufacturing engineer is to *improve productivity.*

The industrial engineering principles and methods of achieving productivity improvements or cost reductions are many; the industrial engineer has, however, no exclusivity on productivity improvements—everyone in the company should contribute.

A work measurement system becomes an important tool for the industrial engineer and others to accomplish the productivity improvement or cost reduction goals of the company. An overall productivity improvement (or cost reduction) in the workshops of 20–30% or more is often the result of a work measurement project in a U.S. company. The improvement potential is even greater, however, in areas such as industrial engineering, process planning, and cost estimating. It is not uncommon for output (or productivity) to double or triple as a result of the installation of an appropriate computerized work measurement system with the proper application functions attached.

10.5.2 Incentive Plans

The studies showed that about 40% of the companies using work measurement have adopted a wage incentive plan, which obviously requires realistic, accurate, and consistent time standards as a base. Another 40% have measured daywork systems, which need reliable backup standards as well. Without proper work measurement, such wage systems will be either out of control and/or inequitable and unfair to the workers who are financially dependent on them.

On the other hand, a well-designed incentive plan supported by high quality standards and work instructions is probably the most powerful means and motivator to maintain a high level of productivity. Therefore, if incentives or measured daywork are in place, proper work measurement is necessary.

10.5.3 Operations Management

It is almost inconceivable that some managers accept the responsibility for managing a factory operation without having access to accurate and consistent data to accomplish planning and scheduling, to determine and evaluate performance, and to establish and control costs. However, the majority of managers have learned to appreciate the benefits of work measurement and have committed themselves to the use of proper work standards as part of their management philosophy and system. This is primarily a question of understanding what modern work measurement is and what it can do for an organization.

10.5.4 Cost Estimating

In today's competitive marketplace, it is important not only to exercise control of the costs of producing a component or product, but also to be able to establish an *accurate* cost estimate *quickly*. Quotes and bids must reflect the *real* cost, so that the desired margin can be attained. If the bid is too high, the order may be lost; and if it is too low, money may be lost (see Chapter 7).

Not only will an engineered work measurement system generate a low cost level through the methods engineering effort, it will also provide a realistic cost estimate for the manufacturing of a product or component. If a computerized work measurement system is in use, a cost estimate can be issued promptly. In addition, simulations of costs based on product modifications and variations can be made to adapt a quote rapidly.

The scope of work measurement applications is even broader. Process planning, scheduling, equipment justifications, budgeting, line balancing, manpower or staffing requirements, and product design are other focus areas for work measurement. Each of these areas can become a principal area of application for an individual company, because in order to satisfactorily perform one of these functions, realistic and consistent data (work standards) are a requirement.

In all the application areas indicated above, the use of expert system technology has proven to be of great benefit to the user. Through the development of customized decision trees for the particular application area, the user can obtain the desired output just by answering a few simple questions pertaining to the product or component being processed. A knowledge engineer identifies and collects the "best" method from domain experts and subsequently develops the necessary tree or trees. This way, expert knowledge is captured, distributed, and preserved for everyone in the company to use. Employees can become "experts" with a minimum of training by applying the system. Also, the consistency and quality of data improves substantially.

Because of the availability of expert systems and other advanced computer technology, work measurement has become much more accessible both to the user and management. The conclusion is that companies have many good reasons for using work measurement. The fundamental and prevalent reason is, however, *To satisfy the desire and need to know and to predict with confidence.*

10.6 SUMMARY AND FUTURE TRENDS

During the 1950s and 1960s, the work measurement field became inflated with "conventional" derivatives of the original MTM system (MTM-1). That trend has continued to some extent, with one exception: MOST. In the mid-1960s, we believed that a new approach, a more practical and user-friendly method, and perhaps more important, a faster and simpler technique, was necessary to maintain a reasonably high level of interest in work measurement. MOST seems to have been the answer. Over 20,000 persons representing more than 4000 organizations have become certified MOST users. MOST has been translated into at least 15 languages and is in use in more than 35 countries around the world. MOST satisfies all the criteria of simplicity, speed, accuracy, consistency, applicability, integrity, and universality that can be put on a modern work measurement technique and system. MOST Computer Systems represents the state of the art in the areas of work measurement and time standards. And users enthusiastically endorse and support MOST.

Traditionally, the manufacturing industry has been and still is the most active user of work measurement. The main reason for this is the emphasis on productivity improvements and incentive systems. Billions of dollars have been saved as a result of the appropriate application of work measurement.

A renaissance in work measurement has been noticeable during the past few years in the defense industry because of a military standard (MIL-STD-1567A) issued by the Department of Defense in 1983. Since then, defense contractors have been obligated to comply with this standard on major contracts. Compliance includes fully documented time standards (80% coverage) based on a recognized predetermined motion time system. MOST Systems has very been used successfully by a large number of defense contractors to satisfy the requirements of MIL-STD-1567A.

Also, service industries have shown an increased interest in work standards for staffing, personnel planning, budgeting, and the like.

Despite the efforts by industry to increasingly mechanize manufacturing operations, the measurement of work done by people is here to stay for many more years. The advantages of knowing and being able to plan from realistic and consistent standards are just too great to dismiss. A good example is the well-known NUMMI project, a successful joint venture between General Motors and Toyota. A great deal of its success has been attributed to the "back-to-basics" standardization that was a cornerstone of the project and which was documented in the article by Paul S. Adler. [7]

However, the work measurement and standard setting disciplines have to become simpler, faster, and more integrated with other functions to attract the attention they deserve. MOST Systems and MOST Computer Systems have met these criteria, but more can be done and more will be done. Today's computer technology has reached a level that cannot be ignored by work measurement specialists. If they take advantage of this technology, time standards can and should become a logical and integral part of any business system, as is the case in many companies already.

The general trend in industry is automation. Therefore, it is obvious that we will see fully automated procedures (using AutoMOST) for calculating and updating time standards based on data developed and maintained by industrial engineers. A direct link to a CAD or MRP II system with the purpose of producing process plans and cost estimates based on these standards will likely become a reality within the next few years. Finally, it will be necessary to simplify presently used work measurement systems, including MOST, to such a degree that everyone in a company can learn how to use them and benefit from them in their ongoing effort to make continuous improvements.

> **Work measurement: the science that brings more and better knowledge to people about work and how to improve work.**

10.7 REFERENCES

1. Kjell B. Zandin, *MOST® Work Measurement Systems*, Marcel Dekker, New York, 1980.
2. Kjell B. Zandin, *MOST® Work Measurement Systems: BasicMOST, MiniMOST, MaxiMOST*, 2nd ed., Marcel Dekker, New York, 1990.
3. *Maynard's Industrial Engineering Handbook*, 4th Ed., McGraw-Hill, New York, 1992.
4. David I. Cleland and Bopaya Bidanda, *The Automated Factory Handbook*, TAB Professional and Reference Books, Blue Ridge Summit, PA, 1990.
5. Delmar W. Karger and Franklin H. Bayha, *Engineered Work Measurement*, 4th ed., Industrial Press, New York, 1987.
6. Benjamin W. Niebel, *Motion and Time Study*, 9th ed., Richard D. Irwin, Inc., Burr Ridge, IL, Boston, MA, and Sydney, Australia, 1993.
7. Paul S. Adler, *Time-and-Motion Regained*, Harvard Business Review, Boston, MA (January–February 1993).

11

Control of Production and Materials

Stephen C. Cimorelli*   *Learjet, Inc., Wichita, Kansas*

with

Greg Chandler   *Hubbel Premise Wiring, Inc., Marion, North Carolina*

11.0 INTRODUCTION TO CONTROL OF PRODUCTION AND MATERIALS

This chapter deals with the control of production and material. It is not intended to provide exhaustive coverage of the subject; rather, it seeks to impart a working knowledge of the major elements of the field to manufacturing engineers who must apply them in actual practice, and who must work side-by-side with production and material control and other professionals.

A sizable number of systems and techniques have been devised to aid those responsible for production and material control. This chapter introduces and discusses the more widely accepted of these. The authors believe that by focusing on the approaches which readers are likely to encounter, and by exploring their applications rather than specific technical details, readers will be better prepared to recognize and apply them to their own situations.

In order to understand and appreciate the importance of these controls, some understanding of the business context in which they are applied must first be gained. Beginning with Subsection 11.0.1, this introduction seeks to provide that context.

Two of the most prominent organizations dedicated to this chapter's subjects are the American Production and Inventory Control Society (APICS) and the Oliver Wight Companies. Many of the concepts and ideas presented in this chapter were developed by these organizations, both of which offer a wide range of literature and educational programs for those interested in developing a deeper understanding and appreciation of

---

*Current affiliation: Manufacturing Systems Analyst, Square D Company, Asheville, North Carolina.

this field. Specific references are found in the bibliography. The remainder of Sub-chapter 11.0 in particular is based largely on APICS Systems and Technologies literature.

11.0.1 Corporate Strategy

Manufacturing companies, like all companies, exist for the purpose of making money. They do this by designing and producing products which satisfy the wants and needs of their customers in a way which is superior in some form to that of the company's competitors. These *competitive advantages* are spelled out in the company's marketing strategy, which in turn supports the overall corporate strategy, mission, and objectives. Likewise, a *manufacturing strategy* is needed to enable the company to achieve these higher-level goals. This is not a one-way street, however. Manufacturing excellence in certain areas can be leveraged to create or gain competitive advantage. See Figure 11.1 for the relationships of the strategies and objectives.

A company's *mission statement* sets the overall focus for the firm. It defines what business the firm is in, and it provides a unifying thread tying all products together. For example, a mission statement which says, "We are in the light rail business" might limit production to small, lightweight trains, while a company "in the mass transportation business" might have a product line as diverse as trains, buses, subways, airplanes, and many other types of vehicles.

Within this focus, *objectives* for running the business must be set to ensure that the goal of the business—making money—is relentlessly pursued. Typical objectives are (1) financial, related to such things as profit, cash flow, and return on investment; and (2) marketing, related to market share, sales targets, or penetration of new markets. Such objectives may be felt in significant ways in the manufacturing organization as they trigger supporting goals, and drive, or limit, new investment in manufacturing systems and technologies.

Corporate strategy defines what resources will be deployed and what actions will be

Figure 11.1 Manufacturing strategy must be part of the overall company mission, objectives, and strategy.

taken to accomplish corporate objectives, ultimately allowing the company to accomplish its mission.

Marketing strategy deals with three primary subjects: market niches, market share, and competitive advantage. Of the three, the most important to the manufacturing organization is *competitive advantage*—the ability or characteristic of the firm which makes customers choose it over other firms offering similar products. This is not to say that market niches and market share are unimportant. It means simply that manufacturing strengths play a large role in providing a competitive advantage in a particular market niche, thereby increasing market share. Typical areas where manufacturing companies seek to gain competitive advantage include price, quality, delivery speed, delivery reliability, flexibility, product design, and after-market service. It is not necessary for a company to gain competitive advantage in all areas, but at least one area must be demonstrably better than the competition for the company to stand out. In order to identify which of these a company should compete upon, they should be evaluated in terms of *order winners* and *qualifiers*. Qualifiers are those characteristics needed just to play the game, while order winners are those characteristics which make customers choose one company over all others.

11.0.2 Manufacturing Concepts

Before moving on to manufacturing strategy, a few additional concepts and definitions bear reviewing: (1) general manufacturing environments, (2) product volume and variety, and (3) product life cycles.

Manufacturing Environments

Manufacturing environments may be grouped into three general categories—job shop, repetitive, and continuous flow. In a *job shop*, products are typically made one at a time or in very small quantities. Parts are routed between machines and work centers which have been organized by the type of work performed, such as sawing, milling, turning, etc. This arrangement offers a high degree of flexibility in that a great variety of products may be produced, but sacrifices speed and efficiency. Many manufacturers choose to set up certain departments, such as machine shops, around a job shop concept, and seek to improve efficiency by producing parts in *batches*. This batching allows the time required to set up a machine or work center to be spread across multiple parts, rather than being charged to a single part. A *repetitive manufacturer*, by contrast, organizes the plant around product lines, allowing parts to flow through predefined processes. While this arrangement is faster and more efficient in producing selected products, it is less flexible than a job shop. The final category is the *continuous flow* manufacturer. An example of continuous production is an oil refinery, where continuous production of multiple products (gasoline, motor oil, etc.) flows from a primary raw material—crude oil. Regardless of the environment, manufacturing engineers are instrumental in the design and layout of the manufacturing processes.

Product Volume and Variety

Product *volumes* reflect the quantities of products planned to be produced, usually expressed in terms of product families. The concept of product families and its usefulness to manufacturing professionals, especially those engaged in production and inventory management, will be discussed in greater detail in Subsection 11.2.2. Product *variety* deals

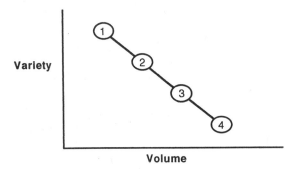

Figure 11.2   Product/Variety relationship chart. [3]

with the number of end products which the firm produces, or is capable of producing. A general relationship between volume and variety is represented in Figure 11.2, where product variety decreases as volume increases.

To better understand this relationship, consider some parallels between the volume/variety matrix and manufacturing environments. In position 1, exhibiting high variety and low volume, a job shop environment would seem preferable. Continuing down the volume/variety positions, one would likely encounter batch production at position 2, repetitive at position 3, and continuous at position 4, where variety is very low but volume is high.

Product Life Cycle

The third concept is product life cycle. Figure 11.3 depicts this cycle. In the introduction phase, when a new product is being test marketed, variety may have to change rapidly to adjust to real versus anticipated customer demands. As a result, low volumes of each product configuration are typical. As we enter the growth phase, design stabilizes, resulting in somewhat lower variety and higher volumes. In the maturity phase, assuming the product is successful and enjoys good customer acceptance, volumes are typically at their peak, with variety being very low. Finally, a product moves into a phase of decline in which customers begin to seek alternatives to the product. In this phase, a company may add a

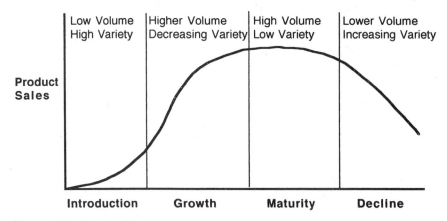

Figure 11.3   Product life cycle chart showing the influence on manufacturing requirements. [3]

myriad of features and options to satisfy these desires, thereby increasing variety again but reducing the volume of any given model.

Again, some parallels to manufacturing environments may be drawn. The introduction phase often requires the flexibility of a job shop, followed by a move toward batch in the growth phase, then repetitive production during product maturity. Finally, a return to batch production may be required in the decline phase to satisfy the increased variety demanded by customers. While these parallels aid in our understanding of the various concepts, this is not to imply that products move naturally through all three "environments." In fact, few product lines compete in more than two. Notice also that the continuous flow environment was never reached. Rarely will this occur, since continuous flow is a special case where production, once begun, proceeds at a constant preset rate and is either "on" or "off."

11.0.3 Manufacturing Strategy

Manufacturing strategy may be defined as a long-range plan to deploy manufacturing resources in such as way as to support the corporate strategy. In other words, it defines what *actions* will be taken, and what *resources* will be deployed to achieve manufacturing's objectives. For example, if the company has defined "delivery speed" as a competitive advantage, manufacturing decisions to lay out machines and work centers in a repetitive process line would be appropriate, while a decision to employ a job shop layout would not.

Such decisions may be categorized as structural or infrastructural. *Structural* or "hard" decisions relate to "brick and mortar," equipment, and technology decisions. Structural decisions tend to support the "resource deployment" element of strategy. Since these decisions typically require significant expense, they are made infrequently, usually by upper management. Manufacturing engineers are often deeply involved in providing technical expertise to these decision makers. *Infrastructural*, or "soft" decisions, on the other hand, relate to people, organizations, and systems. These decisions support the "actions" element of manufacturing strategy and are made by middle managers on a more frequent basis.

A second element of manufacturing strategy deals with capability and the *timing of capacity changes*. These capacity strategies are defined as lead, lag, and tracking. In a *lead strategy*, capacity is added before it is needed, enabling a firm to take advantage of a new market opportunity, or to build up inventory prior to a cyclic upturn in demand. This approach is somewhat risky, since anticipated demand may not materialize, leaving the firm with an underutilized asset and a negative effect on the bottom line. By contrast, the *lag strategy* adds capacity only after an increase in demand has been demonstrated. This approach is very conservative and enables a firm to use existing capability to its fullest, but may make entering a new market, or recovering from unanticipated capacity problems, very difficult if not impossible. Finally, the *tracking strategy* is a compromise between the other two. In this approach, the firm attempts to keep market demand and capability closely matched by adding or removing capacity on a much more frequent basis through actions such as overtime, out-sourcing, etc. Here again, manufacturing engineers are often key players in developing and executing these capacity strategies.

The final element of manufacturing strategy is *plant focus*. A plant may be focused on one of three things: product, process, or order winners. *Product focus* implies that a plant's production is limited to a given product or product line. For example, a plant may produce only one printer, or a small family of similar printers. *Process focus* limits the plant to a process, or a closely related set of processes. Typical examples might be a lumber mill or a foundry. Many different products may be produced, but they are all produced from common processes. Finally, there is a focus on *order winners*. In this case, manufacturing resources are organized in a manner which allows a given marketing strategy to be carried out. If delivery reliability is identified as the order winner, our printer manufacturer may choose to carry a high level of finished goods inventory, or may carry excess capacity to ensure that no bottlenecks in production are encountered.

11.0.4 Summary

A company's manufacturing strategy—decisions and actions affecting the deployment of resources—must support its corporate and marketing strategies. The manufacturing strategy may provide competitive advantage in a given market niche, and must consider issues of manufacturing environment (job shop, repetitive, continuous flow), product volume and variety, and product life cycles. Specific issues of manufacturing strategy include structural decisions (plants, equipment), infrastructural decisions (people, organizations, systems), capacity timing issues (lead, lag, tracking), and plant focus (product, process, orderwinners).

As discussed earlier, this introduction seeks to provide a context for the remainder of Chapter 11—control of production and materials. It is in no way a comprehensive discussion of the issues presented. Further study of the materials referenced in the bibliography, especially APICS Systems and Technologies [3], is recommended for the reader seeking to understand these subjects in greater detail.

11.1 CAPACITY PLANNING AND SIMULATION

Greg Chandler

11.1.0 Introduction to Capacity Planning and Simulation

The manufacturing sector has spent a great deal of time and money trying to define, calculate, measure and control "capacity." Many companies have tried to clarify its meaning by prefacing it with adjectives such as: rated, nominal, maximum, dedicated, demonstrated, and so on. All of these prefixes attempt to capsulize and connote to the user what that particular company means by the word "capacity." The Seventh Edition of the *American Production and Inventory Control Society Dictionary* [1] defines capacity as:

1. The capability of a system to perform its expected function.
2. The capability of a worker, machine, work center, plant, or organization to produce output per time period. Capacity required represents the system capability needed to make a given product mix (assuming technology, product specification, etc.).

Implicit within the definition is the fact that capacity is the *planned* amount that *could* be produced during a given time period. The prefixes presented earlier attempt to narrow the plan down to something less than that presented in the definition. Inherent within any definition of capacity is the need to capture the quantity produced over time.

When a company's products are in high demand, the importance of precision in the calculation of capacity is greatly reduced. The function of capacity calculation is merely ensuring that there is sufficient capacity available. The only way to go wrong is to provide for too little capacity. This situation is further eased if the cost of money is low. The accuracy required, and the difficulty involved in calculation, increases with the increase in competition for the same market.

The 1990s have seen a tremendous increase in the global marketplace, competition stiffening and more companies, including the "Blue Chip" names, trimming their staffs and moving toward lean, agile manufacturing. The new emphasis is on companies being more flexible, responsive, and proactive to the marketplace and customers' needs. In light of these changes, companies need to have people and systems that behave in a similar fashion.

Enter simulation. Even though it has been around, in one form or another, for over 50 years, it is having a resurgence in popularity due to its flexible nature (also recent improvements in computer software and hardware). Flexible businesses modify their operational parameters frequently and need ways to predict, understand, and control those changes quickly and with varying degrees of precision. The question of the plant capacity to produce a new product can be an important one. Simulation in its variety certainly fits the bill. This subchapter will attempt to show how simulation can be used to help predict and help control capacity.

11.1.1 Evolution of Capacity Calculation Methods

Since there are as many different definitions of capacity as there are companies defining it, there have been numerous techniques and tools for calculating and predicting it. Some of these methods have proven themselves more useful than others. Different industries seem to have a preference for a given technique or a type of calculation tool.

Initially, a company may choose to predict and/or control capacity by the use of the "intuitive" method. The company will rely on the experience-based, educated guess of a knowledgeable employee (usually in management). This technique is fraught with prejudice and bias, inaccuracies and guesses not founded in factual data or calculations. Sometimes the company may not realize that the method they are using is the intuitive method. The capacity "calculator" will get the answer set in his mind, discuss his ideas with others in management, incorporate their thoughts and insight into his position, and fabricate a plan to justify his position. This will usually result in a "buy-in" from management, since they will see some of their thoughts come back to them from the capacity expert.

Another, more sophisticated approach to capacity calculation is the use of some simplistic mathematical techniques. These mathematical techniques try to eliminate the problems associated with the intuitive guesses while incorporating some real system data. By adopting these methods, the company realizes the shortcomings of intuitive guessing and attempts to remove some of these biases. This Subchapter will not go into any of the particular techniques but instead list the types of techniques that could fall into this category. These types of techniques are characterized by the use of simple algebraic methods that incorporate time available, time required, scrap factors, equipment/personnel utilization, efficiency, downtime, and so on. As example of such a formula is shown below:

$$\text{Capacity} = \frac{\text{M1 avail} + \text{M2 avail} + \text{M3 avail} + \text{Mx avail}}{\text{Time required on each machine}}$$

where M1 avail = the time machine 1 is available for use, etc.

The calculations tend to be very simple in nature and limited in scope, since they can become lengthy quite quickly. Spreadsheet programs often helps in this class of calculations, since they handle large sets of calculations quickly. The formula above can be used to incorporate all items passing across each resource within a department to accumulate the aggregate capacity as well as for individual capacity calculations.

The next step in the evolution of capacity calculations is usually the use of complex mathematical techniques and/or computers. These two ideas are related because they tend to happen together, yet they are distinct since they are unrelated in their view of capacity calculations. They need not coexist within the same organization and often do not.

From the outgrowth of simple mathematical techniques comes the recognition that the capacity calculation need to include more of the overall operations. The company will realize that the interrelations *between* departments play a dramatic role in overall plant capacity. What good is it to have a single department that can produce at a rate of twice any other, since the items flowing to the customer are limited by the slowest flow path? Calculating each individual department's capacity, while useful in a micro analysis, adds nothing to the understanding of plant-level capacity.

This need to include more information within the capacity calculations does not in itself mean that the company must use complex mathematical techniques, but could demand a more effective method of calculation itself, hence the adoption of a computer. The techniques will still be the same as before, simple algebraic relationships, but their magnitude begs for the use of the computer to handle the significantly larger quantity of formulas and relationships. After a few times of calculating these by hand, the company may resort to the intuitive method if it does not migrate to the computer. The computer provides the ideal tool for the organization, calculation, recalculation, and presentation of the larger set of data.

The use of complex mathematical techniques is usually closely related to two things: the adoption of a computer for capacity calculations, or the recognition of the limitations of the simple mathematical techniques. The complex techniques include those types of methods using calculus, statistics, and queuing theory. There are several software programs available which use queuing theory formulas such as the following:

L = avg # of units in system = $\lambda / (\mu - \lambda)$

Wq = avg time a unit waits in queue = $\lambda / \mu \, (\mu - \lambda)$

Lq = avg time unit is in queue = $\lambda^2 / \mu \, (\mu - \lambda)$

where λ = mean number of arrivals per time period, and μ = mean number of items served per time period.

These techniques offer a vastly improved interdepartmental relationship representation, but themselves offer new and often restrictive limitations. For example, these formulas deal with the system at a "steady state." Have you ever seen your factory in a steady state? These techniques, because of their complex nature, require much more data gathering, emphasis on limiting assumptions, and difficulty in presentation of results. They tend to be less trusted by management, who may not be fully aware of their specific mathematics and calculation techniques. As a result, the company may go back to using the computer and simple mathematical techniques.

The epitome of today's capacity calculation techniques is *simulation*. Its ease of use and flexibility give it the inherent ability to be used for a wide range of capacity calculation situations, from simple to complex. It can be used to formulate a simple model, the creation of which clarifies interrelationships, or to model complex systems beyond the ability of all but the most complex of mathematical computations. The software itself can range from simplistic spreadsheetlike packages to more complex programming languages.

11.1.2 Simulation Basics

This subsection will define simulation as: A time based, statistical data driven program that tracks the behavior of the system while entities flow through the system and modify or get modified by it.

That definition implies that only in this type of capacity calculation technique is the activity of an object flowing through the system tracked over time. All the other techniques merely guess or estimate what happens over time by the use of assumptions or averages. It is this ability to track changes over time in conjunction with the use of statistical distributions instead of averages, that allows the simulation to be run through several iterations, each one selecting different values from the distributions. This approach allows the model to behave much in the same way as the real world; i.e., each unit flowing through takes a slightly different time to have an operation performed upon it.

Simulation Model Review

To illustrate how simulation software works, let us use a simple machine shop system of a single machine and a single waiting line (queue) and go through a cognitive example. The flow information is that work pieces (entities) arrive every so often. If the machine is idle, they get right on and their machining begins immediately. If an entity is already on the machine, the parts wait in line for their turn.

Someone has collected data and finds that the time between entity arrivals (called the interarrival time) at the machine follows a normal distribution with a mean of 6 min and standard deviation of 1.5 min. Also, the data collected show that the time to process the entity on the machine can be described as a normal distribution with a mean of 5.1 min and a standard deviation of 2.6 min. From these data one might think that since the machining time is less than the interarrival time, there would not be the need for a large waiting line. However, since the standard deviation, hence variability, for the machining time exceeds that of the interarrival time, there will be occasions where entities (parts flowing through the shop) will pile up in the waiting line.

We will assign a random number stream to both of the distributions. These random numbers allow the model to have the variable behavior similar to the "real world." Then, we set the initial conditions for the system to begin with the queue empty and the machine idle. When the simulation run begins, the computer will select a random number between 0 and 1. It will match this number against the arrival distribution which yields the simulation clock time when the first entity arrives into the system. It will do this for each random number in the arrival stream until the time the next entity due to arrive exceeds the run-time limit we have set. It will store each of these clock times and release an entity into the shop according to this file. The computer will then repeat this for the machine time. When it selects the processing time, it stores it as a duration of time an entity spends on that machine after it gets through with the entity in front of it. Remember, this is all done by the computer, so it happens very quickly!

Now the computer has stored two sets of clock times. One is a simulated time when the entities will arrive into the system, and the other is the duration of time the entity will spend being processed on the machine. I called the interarrival time a "simulated time," by which I mean that the time in the computer model is not linked to the real passage of time. This allows the computer to simulate the passage of time without having to wait for it to happen. It has the effect of compressing time and can simulate years of operation in a matter of seconds. In the following description I will refer to the clock time, which means the computer's simulated clock time.

Now the computer begin the model tracking. It looks at the first interarrival time and jumps the clock ahead to that time (after all, no sense waiting around—nothing happens until the first entity arrives). Since the first entity finds an idle machine, it hops on and begins to have its machining performed. Meanwhile, the computer model recorded that the time entity 1 spent in the queue was 0. The program then sets the machine to busy, looks at the processing time file, and adds that to the current clock time and stores it. It will then compare that "finish machining" time to the next arrival time to see what happens next, an arrival or a machining finish. It jumps ahead to the next event (we will say it is an arrival time) and updates the system statistics it is tracking, such as:

The average time an entity spends in queue
The maximum time an entity spends in the queue
The total time an entity spends in the system
The maximum queue length
Machine utilization
Maximum and minimum throughput for a given time period (8 hr shift)
Average throughput

The model repeats these steps until the end of the simulation time (or, alternatively, until a certain number of entities have been machined). It then has tracked one full simulation run. The time it takes to run this model depends on the hardware, software, and the model written. For most leading simulation software programs, this simple model can run a year's worth of time (2000 hr) in under 10 sec!

Even with this simple model, one can experiment to see how a change affects the outcome of the model. If the mean interarrival time increases/decreases or the processing time increases/decreases, what happens to the system measures? At what interarrival time would we have to add a second machine to keep machine utilization between 80% and 90%? We can quickly modify this system, run it for 10 sec or so, and analyze the output to determine the answers. Remember, though, that the computer model is selecting times based on random numbers. They will be "fast times" or "slow times," depending on how the dice roll! Because of this, it is advisable to run several iterations of the same conditions to approximate the real system.

Steps of a Simulation Study

The general outline shown in Figure 11.4 describes the steps involved in a simulation study. The process is a sound process to investigate the operation and interrelationships of the manufacturing system even if a simulation model is never built. Each of the steps will vary in the time needed to perform them based on experience, degree of detail needed to answer the issues, the model addresses, and the number of "what ifs" modeled. The largest amount of time is going to be spent during the formulation and data collection steps, but attention should also be focused on the validation, verification, production runs, and output analyzation steps.

The first foundation step in simulation studies is the formulation/planning step. This step is where the objectives of the study are documented. It includes the criteria for

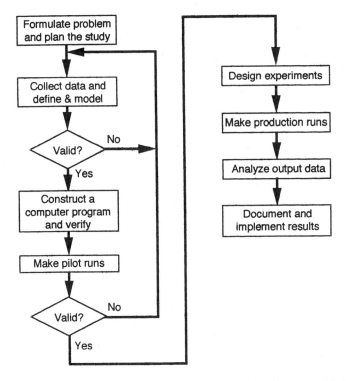

Figure 11.4 Iterative steps of a simulation study. (Courtesy of Industrial Engineering and Management Press.)

determining if one manufacturing system design is better than another. Some of these criteria may include throughput, time of an entity in the system, time of an entity in the queue, time a machine is blocked (unable to work on the next entity until the current entity is removed), time a machine or cell is starved (time spent waiting for an entity), or the size of in-process inventories. This step should also include a schedule (time allotment) and labor projection for the study. This step is not unique to simulation studies. All thorough studies should include some degree of this basic step to limit more detail given than needed, so that more time can be devoted to the unique simulation elements.

The next step is data collection and model definition. The data can be collected on the existing system or estimated for a new system. The data will need to be compiled into statistical distributions for use in the simulation packages. This use of distributions, along with time tracking elements, is the distinguishing feature of simulation and is a key element in the study. Many texts and software programs exist to help one turn raw data into statistical distributions. Several of the simulation software systems allow the use of discrete data as a distribution, as well as the standard distributions, such as the normal, exponential, gamma, and others. Most of the simulation packages will use the time between arrivals as a distribution instead of the number of arrivals per time period.

Next comes the validation step. This frequently overlooked step is an important element in the study, since it is the first chance everyone gets to review the operation of the model. This step allows the people who are working in the actual shop operations to see if the data and model definition will accurately reflect the behavior of the system. If it is accurate enough for the questions defined in the initial step, and credible, management will believe the recommendations that come out of the model building/running process. This step is important because it keeps the people in the system and decision makers involved and understanding the progress and principles of the model. At this stage of the game it is much easier to comprehend the model/data than "walking them through" the simulation code later on. If they have understanding and confidence now, the faith in the construction of the computer model will be easier to establish.

The next step is the creation of the computer simulation model. This step varies in the requirements of the software and time. The important item in this step is that the model mimic the conceptual model in the step above. During this phase, the analyst must decide which kind of simulation system to use (i.e., a general language such as FORTRAN, a simulation language such as SIMAN or SLAM, or a simulator such as WITNESS), and if animation of the model is desired. Subsection 11.1.5 provides more detail about model types and the use of graphical animation in a simulation model.

After the model is created, one will need to perform pilot runs. Notice that "runs" is plural. Since there are statistical distributions for time to perform tasks in the model, the computer will sample from them, sometimes with a fast processing time and sometimes with a slow time. It is this feature, coupled with the need to debug the model, that makes it necessary to run the model several times to see the long-run average operation of the system. These pilot runs need to be validated again, just as the conceptual model was. Tied closely to validation is verification. Verification is the process by which the model is checked to see if it is operating as designed. Both of these elements, validation and verification, help to instill credibility in the model, its output, and any subsequent recommendations.

Next, experiments must be designed for the model. These experiments and their production runs provide the output from which analysis leads to recommendations. The design of these experiments will include some of the following:

The length of time to run the model.
The number of runs to perform for each alternative.
The initial conditions of the system when the model begins.
The point in time when system statistics begin to be collected.
What system statistics are to be monitored?
Whether the animation will be watched, or the output reports simply reviewed?

As with anything, the next task is to analyze the output. This task can be eased by the use of secondary software programs (spreadsheets, statistical analysis/graphing, etc.). Do not short-change this step, because many times it will lead to more alternative system designs and experiments. Sometimes the analysis can point out subtle bugs in the model that must be corrected before the output is valid to be used. If the system exists in real life, the data collected for model input and the people working in the system can help validate the output.

One last note that should not be overlooked: No matter how accurately the model was created, it will not behave exactly the same way as the real system. For once changes are made to the real system, the random nature of the real world will cause different outcomes. The model approximates the behavior of the real world. Hopefully, this approximation is close to the long-run outcome of the real system in action.

11.1.3 Simulation Model Types

Now that the operation of a simulation program and the steps in a study have been explained, a discussion of the different types of software will be addressed. Since simulation software packages are used to mimic a wide variety of operations, the software packages are themselves diverse. They can be grouped into three categories: continuous, discrete event, and combined. There are other subcategories, such as: general purpose languages, specific application languages, and simulators. Each of these will be covered in the following paragraphs.

Simulation software for the modeling of continuous operations, such as food processing, chemical plating, etc., involves the use of differential equations as opposed to "static" kinds of calculations. Most of the software manufacturers make products that have the ability to model both continuous and discrete event systems.

Discrete event simulation software can be used to model the vast majority of the manufacturing operations in use today. These systems are characterized by the entities flowing through the system having definite time spans at different operations. The entity will change its state at fixed times, i.e., welding, painting, testing, etc., as opposed to electroplating, which happens over time. Within this type of software, the modeler defines the steps involved inside the system being modeled and uses the language to perform those operations.

General-purpose languages allow the modeler the freedom to construct specific program code to mimic the system being studied. This class of software is extremely flexible in its ability to model different type of operations and systems. The price paid for this flexibility is the need for skill in the modeler's ability. Since there are no preset routines as there are in the simulators, all the needs (whether general or specific) must be met by the modeler.

Specific application languages are losing ground in the simulation software field. This class is categorized by its ability to handle specific model items easily, i.e., automated guided vehicles (AGVs), conveyors, or automated storage and retrieval systems (AS/RS). These systems are being phased out as more of the general languages are adding commands that handle a majority of the common data collection/input requirements of the specific packages.

Simulators as a class represent a vast array of packages. They are characterized by their ease of use, usually through the use of pull-down menus and answering questions generated by the program. They help the modeler build a genetic model by assuming basic information is needed, such as interarrival time, operations times, and the use of common statistical distributions. They will also have preprogrammed routines for some of the specific items as mentioned under specific application languages. The shortcoming of this class of software is when your model has special needs, since rarely will one be able to get into the simulation code to program that function. Some allow the ability to use an outside language such as FORTRAN to write specific code, but this requires the modeler to add code for the tracking of the statistical data usually handled by the software package.

The line between simulators and general-purpose languages is becoming blurry since the general language programs are trying to incorporate the helpful features of the simulators. Also, the simulators are improving their ability to handle special situations within their simulation language and not make the modeler drop out to another language.

Each of the classes has its pros and cons. It is up to the modeler and the team to analyze each of the packages relative to their needs, resources, and skills. It may well be that more than one package can be used to address different operations, departments, or goals of the model. For example, one could use a rough-cut type of simulator to identify the bottleneck operation, machine, or department and analyze its sensitivity to changes in certain variables, then use a general-purpose language for detailed analysis of that specific area. No two modelers, models, or studies will be exactly the same. Each needs to be treated as an individual, but usually there are some common elements that should be shared.

11.1.4 Hardware Requirements

Simulation programs are offered in many different styles, languages, approaches, and cost brackets. Likewise, the hardware required to run a simulation program spans the spectrum from minimal personal computers through the powerful PCs, minicomputers (or work-stations), up to mainframes.

As discussed in the previous subsection, simulation software varies in its approach to modeling, complexity of models handled, flexibility, and ease of use. Similarly, the hardware used must be analyzed in accordance with the intended user of the model, accessibility for the user and modeler, and the availability of secondary analysis software.

The first step in selecting the hardware may have been performed when the software was chosen. If, for example, the software runs only on a particular hardware platform or requires an operating system that can run on only a few platforms, the final choice of hardware is somewhat limited. This subsection will not deal with particular brand names, but rather with classes of machines and operating systems. The hardware choice may be narrowed down to what currently exists within the organization. This could be due to the lack of funding or merely a misunderstanding of the special requirements of simulation and animation software.

If the person running the model is someone other than the modeler who created it, the user's abilities with a computer will influence the hardware choice. Today's high-end PCs often have operating systems and user interfaces that novice computer users find easier to use than some of the more cumbersome mainframe platforms. If, on the other hand, the end user has a particular preference, expertise, or familiarity with a given platform, there is a distinct advantage in staying with the familiar. No matter how good the model is, no matter how accurately it predicts the behavior of the system, if the user cannot use the hardware on which it resides, it will not be utilized. It stands to reason that if the user is comfortable with the hardware, he or she will be more likely to use, believe, and even defend the model's output.

Also taken into consideration must be the availability of the hardware itself. If the computer will be used frequently by others or is slowed down by running several programs simultaneously (multitasking), the modeler will have a difficult time producing a model and the users will be less likely to use the model simply because it is hard to find an opening on the machine. This situation can be avoided if those individuals involved in the purchase decision are aware of the time requirements of model creation and use. Computers are inexpensive enough today that the benefit generated by the simulation model vastly overshadows the cost of dedicated hardware.

The last, and often overlooked, aspect of selecting simulation hardware is that of the availability of secondary analysis software. Secondary analysis software includes the programs that assist the input, output, analysis, or presentation of the simulation model, i.e., anything used other than specific simulation software. Simulation programs can generate vast quantities of data as output. If the modeler so chooses, he or she can have each action that happens to an entity recorded in a data file. From this data one can analyze various aspects of the system's behavior. In a simulation that lasts a year or more (simulated time), that could generate a tremendous amount of data to sort through and perform statistical manipulations on. A good statistical software program that can receive input from a file created by the simulations program would be worth its weight in gold. Usually the modeler does not realize the usefulness of such a package until after hours have been spent in the tedious pursuit of statistical information. The secondary analysis software is not limited to statistics, but also could encompass database management software for the organization of experiments, word processors to help in model code creation or data file creation, graphics programs for creating presentations on results and recommendations, and others depending on specific circumstances and software. Indeed, the simulation computer hardware may need to host an entire suite of software to ease the input and analysis drudgery and free the modeler/user to create effective models that will solve problems and predict system behavior.

Personal Computers

The 1980s saw an explosive growth of the personal computer technology and market, so that the distinction between the high-end PCs and minicomputers has blurred. Some might say that PCs have penetrated well into the mini's territory. In terms of speed, graphics, and multitasking, there has indeed been a overlap in general software areas, but in simulation software there still seems to be a gap (however slight) between PCs and minis.

The first hardware choice may be a PC, since they are well entrenched in the business world. A vast majority of the simulation programs run on this platform. Some programs output ASCII files that can be imported into other software programs for analysis and presentation. Indeed, some simulation software accept two-way passage of information between popular programs that read and write ASCII files. The PC is a familiar object within many organizations and will not compound resistance which may already exist for a "new" technology. Training for operations of PCs is minimal, since many people within the company already have experience with other software or at least could assist in training new users. Secondary analysis programs exist for the PC in abundance and may currently exist on those platforms. PCs also enjoy a low price/performance ratio, which should please even the most frugal controller. PCs have the option of running different operating systems (DOS, Windows, OS2, UNIX, etc.), which could enhance the operation or increase the model size and complexity for greater modeling flexibility.

Minicomputers

Minicomputers have their niche within the simulation field. Some of the simulation language programs can utilize the power, speed, and size provided by these larger machines. Within this class of machines exists the characteristics sought after by the computer-aided design (CAD) engineering crew, so one can be assured that the mathematically demanding simulation software will find a good home here too. Computers are, after all, mathematical manipulators, and simulation programs are a good match for their capabilities. Workstations have the edge over PCs in multitasking, so the modeler can be analyzing one simulation run while the computer is performing the current simulation experiment. This is an underestimated feature that will, if used well, shorten the time required to prepare a recommended course of action.

Mainframe Computers

Mainframes, too, play a role in the complex world of simulation. Their prevalence, speed, multistation capacity, and link with other information within the business allow them to have the first thought in simulation. Indeed, in the 1960s, IBM wrote the software GPSS (General Purpose System Simulation) language and shipped its mainframes with a free copy. For years, it was many people's first look at simulation and what it can do. Since mainframes can multitask at far greater levels than PCs or minis, a business could benefit by several people modeling and using models simultaneously. Mainframes also can process large batches of simulation runs and record vast amounts of data for later, detailed analysis. Often, mainframes are used to run the accounting and MRP functions within the business. Therefore, the simulation modelers' access to usage of specific company data can be eased via this link. Many tedious hours of data reentry can be avoided by having the mainframe create files of information structured so that the simulation model can read and use it. This data reentry represents one of the most difficult, time-consuming areas to create and debug. Who, in their right mind, enjoys poring over thousands of records searching for the slipped decimal point or swapped digits?

The drawbacks to mainframe simulation usage are often common with the other hardware platforms: lack of secondary support software, slow response time (not to be confused with execution time), and difficulty in usage of the overall operating system. These aspects should not, in themselves, scare someone away from use of this platform, but rather be a basis of particular simulation environment.

11.1.5 Successful Implementations

This subsection will present some samples of the diverse and successful uses of simulation in capacity analysis. These few are by no means a complete listing; rather, this is a sampling, showing the variety and magnitude which should encourage readers to open up their imagination and investigate/apply simulation wherever it is possible.

The U.S. Air Force (USAF) contracted for a study to determine the number of air crews to receive training for their strategic bombers. The goal was to have 100% crew readiness based on five crew members per plane on standby for 12 hr each. The training and readiness pay was very costly. If, for example, it was decided that 10 crews per plane was the appropriate number and that was too low, the personnel would cycle through the standby duty too often and possibly suffer fatigue. If the number were too high, we (the taxpayers) would spend millions of tax dollars needlessly. The study showed that a lower number of crews, used on a certain schedule, would achieve the preparedness goal, yet save $20,000,000 per year in training costs.

McDonnell Douglas Missile Systems Company, makers of the Tomahawk Cruise Missile, used simulation to determine when within a 5-year window another ordnance storage bunker was needed. There were over 10 other programs sharing the storage bunkers as well. The Department of Defense had specific rules governing what class of items could be stored with other items. There were also federal, state, and local regulations that added to the storage complexity. Also, the quantity of missiles produced varied each year. McDonnell Douglas has an annual capital budget cycle, and it took about 10 months to build a storage bunker, so the company needed to know of the additional requirement 2 years before the required date. The manual calculation methods were difficult, were based on estimates, took several months to complete, and were low confidence builders. The simulation model showed that one of these $400,000 bunkers was not needed within the

5-year time frame. It improved the confidence level in the outcome, and showed how much more product growth could be tolerated before a bunker was needed. An animation of the operations was used to demonstrate the model to both McDonnell Douglas management and the Navy program management.

A major city hospital used simulation to determine the quantity, mix, and schedule of doctors, nurses, and other support staff for the entire facility. This included over 450 personnel in many different functions. The model showed the sensitivity to peak demand times and its impact on the manpower schedule. The model predicted when and what type of personnel were needed, which delayed significant hiring surges.

11.1.6 Additional Reading

Law, A. M., "Introduction to Simulation: A Powerful Tool for Analyzing Complex Manufacturing Systems," *Industrial Engineering*, May 1986.

11.2 SCHEDULING

Having established a clearly articulated manufacturing strategy, we now turn our attention to the subject of scheduling. Recall that the manufacturing strategy describes *actions* involving the *deployment* of manufacturing resources. These actions must be communicated and understood by the entire organization in a way which allows all functions to act in harmony with each other. Activities and events must be properly timed and synchronized in order to be performed most effectively. We call this level of scheduling *major activity planning*.

11.2.1 Major Activity Planning

The *American Heritage Dictionary* defines a schedule as "a production plan allotting work to be done and specifying deadlines," while the *APICS Dictionary* states a schedule is "a timetable for planned occurrences, e.g., shipping schedule, manufacturing schedule, supplier schedule, etc." [1]. For our purposes we will need to distinguish between *activities* and *events*. An *event* is something which occurs at a point in time, having a duration of zero. An *activity* occurs over a span of time; i.e., it has a duration of greater than zero. For example, the activity "develop engineering drawing for wing flap" must precede the event "release wing flap drawing to manufacturing." In this example, an engineering activity is seen in the larger context of an internal supplier/customer relationship, with design engineering as the supplier of a drawing to their customer, manufacturing. Proper documentation of this relationship on a schedule, extended to all major interactions between internal suppliers and customers, allows all business functions to act in proper synchronization with each other, ultimately delivering goods to the firm's (external) customers in time to satisfy their needs (delivery schedule). For the manufacturing engineer, some important activities/events might include the following: release engineering drawing, create manufacturing bill of material, release planning documents (routings and work instructions) to shop floor, begin first article production, begin rate production, etc. Middle management might be concerned with schedules reflecting routine business activities such as budget preparation, performance reviews, vendor negotiations, etc.

Because of the varying needs of people at different levels in the organization, more than one schedule is required. In fact, many firms employ a multilevel scheduling approach which presents appropriate information to each intended audience, still with the goal of synchronizing activities throughout the business. This "vertical integration" of activities allows top management to communicate the timing of overall business activities to the entire organization, and enables progressively lower levels to carry out those tasks in harmony with each other.

In the subsections which follow we will explore this vertical integration in more detail within the context of production and material planning. Specifically, we will introduce a top-down model beginning with a production plan and continuing down through a master production schedule (MPS), material requirements plan (MRP), and detailed schedule.

11.2.2 Production Planning

Continuing through Subchapter 11.2.6, we will focus our attention on scheduling activities related specifically to production and material control. An underlying assumption throughout is that a manufacturing resources planning (MRPII) philosophy is employed, and a corresponding MRPII computer systems is in place (see Subchapter 11.5). Then, in Subchapters 11.3 and 11.4, brief discussions of two other philosophies will be developed: just in time (JIT) and the theory of constraints (TOC). For now we will deal with scheduling in an MRPII environment.

The top-level production schedule in MRPII is called the *production plan*. A production plan defines the overall production rates for each product family (a group of products having similar characteristics). It takes into account all sources of demand, such as customer orders, sales forecasts, spare parts production, inter-plant production requests, etc. It is developed by senior management as the result of a "sales and operations planning" process, typically on a recurring monthly basis. As such, the production plan becomes managements "steering wheel" on the entire MRPII system. (See Chapter 7.)

The *APICS Dictionary* [1] defines sales and operations planning (previously called production planning) as:

the function of setting the overall level of manufacturing output (production plan) and other activities to best satisfy the current planned levels of sales (sales plan and/or forecasts), while meeting general business objectives of profitability, productivity, competitive customer lead times, etc., as expressed in the overall business plan. One of its primary purposes is to establish production rates that will achieve management's objective of maintaining, raising, or lowering inventories or backlogs, while usually attempting to keep the work force relatively stable. It must extend through a planning horizon sufficient to plan the labor, equipment, facilities, material, and finances required to accomplish the production plan. As this plan affects many company functions, it is normally prepared with information from marketing, manufacturing, engineering, finance, materials, etc.

This definition introduces another plan, referred to as the *business plan*. A business plan specifies, among other things, the amount of money that must be made through the sale of product. The production plan must support these sales objectives, but specifies sales in terms of product rather than dollars.

To summarize, the sales and operations planning function is the process by which senior management becomes regularly and personally involved in MRPII. Decisions made are reflected in the production plan, which in turn steers the master schedule and ultimately drives the entire formal MRPII system.

11.2.3 Master Production Scheduling

The master production schedule (MPS) defines what the company intends to produce, expressed in terms of specific configurations, dates, and quantities. These requirements in turn drive material requirements planning (MRP). According to the APICS, the master schedule "must take into account the forecast, the production plan, and other important considerations such as backlog, availability of material, availability of capacity, and management policy and goals" [1]. Therefore, any product intended to be produced, whether for sale or other purposes, must be scheduled in the MPS.

At its heart, the MPS is the tool used by master scheduler to balance supply and demand. *Demand* represents the need for a particular product or component, and may come from a number of sources: customer orders, spare parts, management risk production/procurement, forecasts, etc. (see Figure 11.5).

Once demand has been established, what is needed is *supply* to balance the equation. Each unit of demand must be matched with an MPS order which satisfies it in terms of specific configuration, quantity, and date. These orders fall into three categories:

1. Released orders—approved orders which have already been released to production
2. Firm planned orders—approved orders which have not yet been released to production
3. System planned orders—orders which are not yet approved, but drive lower-level planning activities in MRP

The resulting supply plan must exactly balance the demand plan—no more, no less. See Figure 11.6 for this principle.

Although this may at first seem unduly restrictive, it is in fact quite reasonable and practical. Remember that the formal MRPII system is the *repository* for all production activity information. It represents management's decisions for producing products, and the production and purchasing organizations' ability to carry out those decisions. In other words, demand placed on the system with no corresponding supply represents a situation where a management decision to accept demand is not being carried out by production and/or purchasing. Similarly, supply entered into the system with no corresponding demand represents production and/or purchasing's commitment to produce product with-

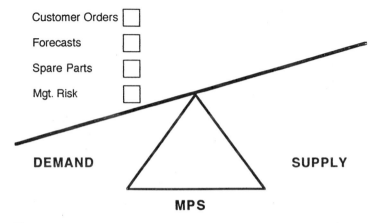

Figure 11.5 Chart showing the demand side of the MPS equation.

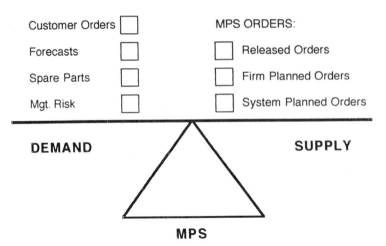

Figure 11.6   Chart showing a balanced supply and demand in the MPS equation.

out management concurrence. The point is to capture and communicate all production decisions in a way that systematically informs the entire organization of all production decisions and permits each department to respond accordingly.

Returning briefly to production planning, we see that senior management participates in a similar process. The sales and operations planning meeting provides the forum for reviewing current performance against the sales plan (high-level demand plan defined in the business plan) and the production plan (high-level supply plan). Resulting decisions, in the form of revised production plans, are entered in the MPS and ultimately communicated to the rest of the organization through the formal system. In this way, senior management "steers" the entire production and inventory control process. Managing these changes is the subject of the next subsection.

Master Scheduling Approaches

At the beginning of this chapter, three scheduling environments were presented: job shop, repetitive, and continuous flow. Three other categories are useful for describing manufacturers, especially in the context of master scheduling. These are make to order, assemble to order, and make to stock. A *make-to-order* manufacturer produces products only upon receipt of a customer order. Products are generally made from a combination of standard parts and materials, plus custom-designed parts to meet the specific requirements of each customer. An *assemble-to-order* manufacturer is similar, but maintains stocks of completed subassemblies and combines these into custom configurations, also upon receipt of an order. Finally, the *make-to-stock* manufacturer produces standard finished products for "off-the-shelf" consumption. Figure 11.7 shows typical product structures for each case. Notice from the figure that in all three, the narrowest section of the product structure is master scheduled. This is done so as to minimize the manual planning involved. Other planning and schedules systems, such as MRP, can perform the more tedious detailed scheduling from that point down. But what about the levels above these points?

In the assemble-to-order case, a method known as final assembly scheduling (FAS) is often employed, where a customized schedule is created for each custom-ordered product. The FAS, or finishing schedule as it is sometimes referred to, schedules the operations

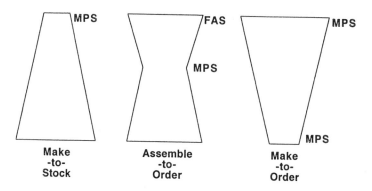

Figure 11.7 Different product structures influence the type of scheduling systems required.

needed to finish a product from the point where the MPS left off. The role of the master schedule in this environment is to ensure that sufficient supplies of the needed subassemblies are maintained to be "pulled" by the FAS.

In the make-to-order case, notice that two levels of MPS are needed. At the bottom, raw materials and base components are master scheduled, usually based upon anticipated consumption rates. This is much easier than attempting to anticipate quantities of a far greater number of possible custom combinations at the top. However, as customer orders are received, a specific customized product structure can be created, and the intermediate levels (subassemblies, options, etc.) may than be planned and scheduled. The scheduling of levels below the MPS falls to material requirements planning (MRP).

11.2.4 Materials Requirements Planning

Material requirements planning (MRP) is a process for planning all the parts and materials required to produce a master scheduled part (refer to Subsection 11.2.3). The resulting time-phased requirements take into account parts and materials already on hand or on order, subtracting these from the total requirement for a given time period, to arrive at net requirements. Each resulting MRP planned order defines, as a minimum, the part number, quantity required, and the date.

A detailed description of the MRP process is given in Subchapter 11.5. For now we will limit the discussion to a fairly conceptual level and focus on the role of MRP in an overall scheduling context. Let us begin with the process itself. Figure 11.8 depicts the overall process, beginning with a master schedule. Associated with each requirement in the MPS is a bill of material (BOM) which defines the parts and subassemblies required to produce the MPS part, then the parts required to produce each subassembly, and so on down to purchased parts and raw materials. MRP "explodes" the bill of material, converting the orders at a given level into a set of gross requirements for parts and materials at the next level. Next it subtracts available inventory from the gross to arrive at a net requirement for each item. MRP then creates planned orders to satisfy these requirements, and repeats the process for all remaining BOM levels.

Figure 11.9 shows a simplified BOM for a two-drawer file cabinet. Notice that two drawer assemblies are needed for each cabinet, and that two drawer sides are needed for each drawer assembly. If the MPS contains orders for five cabinets, MRP will calculate a gross requirement of 10 drawer assemblies (5 cabinets × 2 drawers/cabinet = 10 drawers).

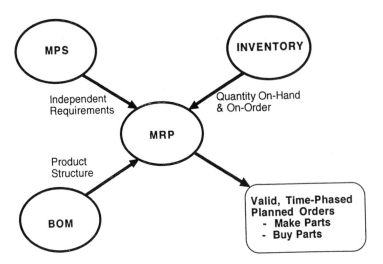

Figure 11.8 Material requirements planning (MRP).

Now let us assume that we already have three drawer assemblies in inventory. MRP will subtract these from the gross requirement to arrive at a net requirement of seven, and create planned orders to build seven drawer assemblies. The process continues down to the next level, exploding the seven drawer assemblies into their component parts, netting at that level, covering the net requirement with planned orders, then continuing down until the entire bill of material has been exploded. But what about the timing element?

Let us assume that final cabinet assembly takes 3 days, and each drawer assembly

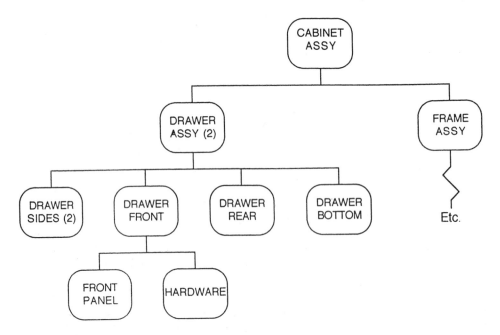

Figure 11.9 Simplified bill of material for a two-drawer file cabinet.

takes 2 days. These "lead times" are used to establish the completion dates that each part must meet in order to satisfy the next-higher-level requirements. In our example, if the MPS order (for five completed cabinets) is due on day 10, then the drawers must be completed 3 days earlier, on day 7. Likewise, the parts required to assemble the drawers must be completed 2 days prior to that, resulting in a due date of day 5.

Two types of demand are evident in this example: *dependent* and *independent*. Let us define these terms:

> *Independent demand*—Demand for an item which is unrelated to the demand for other items in the product structure. Demand from customer orders, for replenishing finished goods stock, and spare parts are all considered independent. Since independent demand cannot be calculated, it must often be forecast.
>
> *Dependent demand*—Demand which is derived directly from another item or end product defined in the bill of material. Because this demand is driven by some other requirement, it should be calculated rather than forecast.

We would consider the end-item cabinet assembly to be independent demand. Demand for the cabinet frame and drawer assemblies would be considered dependent, because their requirements depend on the demand for the cabinet assembly. Note, however, that a given item may have both dependent and independent demand. For example, if a spare part order is received for a drawer assembly, this order would represent independent demand, while the demand for drawer assemblies resulting from cabinet assembly orders is dependent.

Supply and Demand Balancing

In Subsection 11.2.3 we discussed the need to balance demand from various sources (customer orders, spare parts, etc.) with the company's production plan and master production schedule. MRP continues this supply/demand balancing process down to the lowest levels of the bill of material. Beginning at the top (e.g., a cabinet assembly), the MPS defines a *supply plan* for finished goods. MRP translates this data into demand for items at the next BOM level. After netting available inventory, it calculates planned orders to supply the remaining requirements. This supply plan in turn becomes the demand placed upon the next lower level, and so on down the structure. In the case of combined dependent and independent demand, MRP must satisfy both sets of demand simultaneously. As changes occur, such as new or cancelled orders, inventory adjustments, scrapped parts, etc., MRP dynamically replans to maintain a balance between supply and demand at every level.

There are, however, some self-imposed limitations to this process. In general, the nearer an MRP order is in the future, the more human intervention is required. The farther out in the future, the more decisions may be automated. MRP systems generally define three types of orders, distinguished by how far in the future they are planned.

> *System-planned order*—Orders far enough in the future that they are planned entirely by the MRP system. Also referred to as MRP planned orders.
>
> *Firm-planned order*—Orders which are near enough in the future to warrant human intervention, but not close enough to release into production.
>
> *Released orders*—Orders which have been released to production. Released orders fall outside MRP control.

In the case of near-term orders (firm planned and released), it is probably not wise to allow a computer to make decisions regarding timing and quantity of orders. In all likelihood, resources have already been deployed or are being deployed to execute these orders. The impact of a change at this point in time is severe. People (inventory planners) must therefore make these decisions. In the next subsection we will discuss how the MRP system assists planners in making these decisions.

Management by Exception

A basic tenet of good management is to handle routine things routinely, and manage the exceptions. Routine decisions are best handled by following tight, clear policies and procedures. A good procedure will define "decision rules" to be followed in specific situations. This approach provides consistency and control; attributes which are vital to the effective management of production and inventory. It also frees managers and workers to apply their skills and intelligence to the exceptions.

MRP employs just such an approach. By automating the routine decision rules for planning parts and materials, MRP frees people's time to deal with exceptions and near-term requirements. MRP communicates these occurrences in the form of "action messages." In each case, the system is recommending an action that it would take if it had control of the order. Let us examine some of the more common ones.

- *Firm message*—Recommends that the planner convert a system planned order to a firm planned order. By executing this action, the planner takes control of the order from MRP. If not converted, MRP will retain complete control over the order.
- *Expedite message*—Recommends that the planner accelerate an existing firm-planned order; i.e., reschedule the order to an earlier date. Also referred to as a "reschedule in" message.
- *De-expedite* message—Recommends that the planner slow down an existing firm-planned order, i.e., reschedule the order to a later date. Also referred to as a "reschedule out" or "defer" message.
- *Cancel message*—Recommends that the planner cancel a firm-planned order.
- *Release message*—Recommends that the planner release a firm-planned order to production. By executing this action, the planner removes the order from further scrutiny by MRP.

In all but the first and last messages, MRP is communicating an imbalance in supply and demand. In the case of reschedules (expedite, de-expedite), the total planned quantity is balanced but the timing is off, resulting in future periods of either too much or too little supply. In the case of a cancel message, the total planned quantity (supply) exceeds the total requirements (demand); i.e., an imbalance of supply over demand.

The final message is a recommendation to release an order to production, removing it from MRP's view. At this point, control of the order falls to the shop floor control system, which is the subject of the next subsection.

11.2.5 Detailed Scheduling (Shop Floor Control System)

Detailed scheduling as defined in an MRPII context is the assignment of start and/or finish dates to each operation on a routing. Within MRPII this level of scheduling falls in the domain of the shop floor control system. In this section we will review the detailed scheduling process itself, as well as introduce some basic elements of production activity control (PAC). Subsection 11.2.6 will discuss PAC in greater detail.

Every shop order (released MRP order) has an accompanying document known as a *routing*, which describes how a given part is manufactured. Each step of the manufacturing process is given an operation number. Accompanying each operation is brief description of the work to be performed, tooling to be used, and standard setup and run times. These are the basic elements, although other data such as queue times, reference documents, and required operator skills are often included as well. Routings also call out nonmanufacturing steps, such as picking of parts from stores, movement of parts between operations, inspection operations, and test requirements (see Chapter 9).

In addition to the routing, a set of detailed work instructions is provided. These documents are often referred to by such names as planning documents, process sheets, and assembly manuals, and are produced by manufacturing engineers (see Chapter 9).

Forward Scheduling

The detailed scheduling process differs significantly from that performed by MRP. MRP uses a "backward scheduling" algorithm, which defines due dates by which orders must be completed in order to satisfy the requirements at the next higher bill of material level. Shop floor control (SFC), on the other hand, often employs a forward scheduling algorithm. Starting at today's date and calculating forward, each operation is assigned a start and/or finish date. Depending on the precision required, some systems schedule operations by date and time (hour, minute, etc.). Figure 11.10 depicts a sample calculation.

A typical routing will begin with instructions to pick parts from stores. These parts

Sample Detailed Scheduling Calculation

Pick	Move	Queue	S/U	Run	Move	Queue	S/U	Run	Inspect	Stock

Pick = 2 days
Queue = 5 days
Move = 1 day
S/U = 4 hours
Run = 0.2 hours/part
Inspect = 4 Hours
Stock = 2 days

Oper. No.	Description	Hours	Ord. Qty.	Dur (Hrs)	No. Days
1	Pick	16.0		16.0	2.0
2	Move	8.0		8.0	1.0
3	Queue	40.0		40.0	5.0
4	Setup	4.0		4.0	0.5
5	Run	0.2	100	20.0	2.5
6	Move	8.0		8.0	1.0
7	Queue	40.0		40.0	5.0
8	Setup	4.0		4.0	0.5
9	Run	0.2	100	20.0	2.5
10	Inspect	4.0		4.0	0.5
11	Stock	16.0		16.0	2.0
			Totals	180.0	22.5

Oper No.	Desc.	No. Days	1	2	3	4	5	6	7	8	9	10	11	12	13	14	15	16	17	18	19	20	21	22	23
1	Pick	2.0	░	░																					
2	Move	1.0			░																				
3	Queue	5.0				░	░	░	░	░															
4	Setup	0.5									░														
5	Run	2.5										░	░	░											
6	Move	1.0													░										
7	Queue	5.0														░	░	░	░	░					
8	Setup	0.5																		░					
9	Run	2.5																			░	░	░		
10	Inspect	0.5																					░		
11	Stock	2.0																						░	░

Figure 11.10 Sample calculation of forward scheduling.

are then moved to a staging area where they sit in queue until assigned to a worker. Some preparation is generally required (setup) before actual work on the part begins. Once begun, work normally continues (runs) until it is completed for all parts in the order. Therefore the amount of time required to process the order through run operations is the run time multiplied by the order quantity. Parts are then moved to the next staging area and the queue–setup–run sequence is repeated. Finally, the order is inspected and returned to stock. An actual routing typically has many more operations, as well as intermediate inspection points, test operations, etc. However, this simplified routing will suffice for our example.

Assuming that today is considered day 1, operation number 1 (pick), which has a duration of 2 days, will start at the beginning of day 1 and complete the end of day 2. Operation number 2 (move) will start at the beginning of day 3 and complete at the end of the same day. The order will then sit in queue until the end of day 8, until it is ready to be worked. The work area (machine, work bench, etc.) is set up, and by the middle of day 9 is ready to begin processing the order. The total order requires 20 hr (2.5 days) to complete, finishing at the end of day 11. Continuing this calculation we see that the order is completed and parts returned to stock by the middle of day 23.

For the order in our example to be completed on time, the MRP lead time must be at least 23 days. In practice, lead times are often inflated slightly to allow for unanticipated delays in their processing. Our sample order might have a lead time of 25 days. If the MRP planner had released this order just in time to complete on day 25, then according to our calculations, the order would be completed 2 days early, resulting in two days of *slack time*. We will see in the following section how this slack time can be used as one means of prioritizing orders. Before continuing, some observations about the above example will help to illustrate two general principles.

First, notice the relationship between setup and run time. Since the setup time is very large compared to run time, it is necessary to produce a sizable quantity in order to hold down the cost per part. For example, if only one part was produced, the cost of labor (or machining) for that single part would be 4.2 hr (4 hr setup + 0.2 hr run). If 10 parts were produced, the total time required would be 6 hr $(4.0) + (10 \times 0.2)$, or 0.6 hr per part. Manufacturers must use caution however, not to produce too many parts, since parts must then be stored. Storage and related costs must be considered when determining order lot sizes. This subject is discussed in more detail in Subsection 11.6.3. (A better alternative is to reduce the setup time, thereby avoiding the need to produce large batches. This approach is advocated by *just in time*, and is discussed in Subchapter 11.3.).

Secondly, notice that a large portion of the time that this order spends in the shop is tied-up sitting in work queues. In our example, 44% $(80 \div 180)$ of the order's time is spent in queue. This is not unusual. In fact, in many companies, especially job shops, orders spend more time in queue than in all other operations combined, making the management of work queues one of the most important elements of production activity control. One method of controlling queues is called input/output control and is discussed in subsection 11.2.7. The relationship between queue time and lead time follows.

11.2.6 Production Activity Control

In this subsection we will deal with five related subjects: control of queue size, priority sequencing rules, dispatching, shop floor data collection, and performance measurement. A sixth major area of production activity control—capacity management—is covered in

Subsection 11.2.7, as well as in Subchapter 11.1. We begin the discussion with control of queue size.

In the previous subsection it was pointed out that queue time makes up a large percentage of an order's lead time, especially in a job shop environment. In order to manage queue size effectively, an essential ingredient in overall queue management, an understanding of this relationship is vital.

Let us begin by looking at the relationship between order lead time and work-in-process (WIP) levels (the number of parts/orders in various stages of completion in the shop). Figure 11.11 illustrates the point. Assume that one order per day is scheduled to complete, beginning on day 5. If the lead time is 5 days as depicted in the top chart, the first order must begin on day 1. By day 5, when the fifth order has been released, there will be a total of five orders in the shop. This level will remain as long as work is released and completes at the same rate. The bottom chart shows what would happen to WIP levels if the lead time could be reduced to 3 days. To satisfy the first completion requirement on day 5, the first order would need to begin on day 3. By day 5, and for all subsequent days, a total of three orders will be in work, a reduction in WIP of 40%.

So how is this reduction in lead time accomplished? Recall our earlier discussion on the length of time that orders spend in queue. The more orders that are in WIP, the longer an order entering a queue will have to wait its turn to be processed.

By arbitrarily reducing the total lead time from 5 to 3 days, we will succeed in

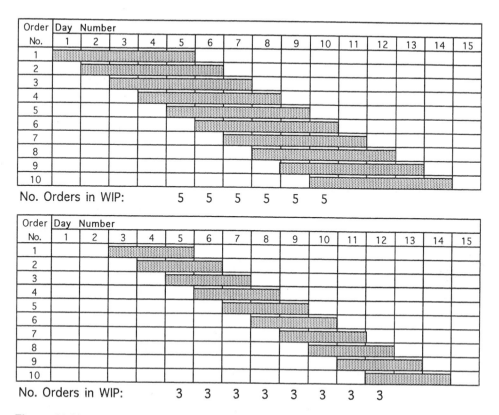

Figure 11.11 Effect of lead time on work-in-process levels.

reducing the number of orders waiting their turn, thus reducing queue time. Notice that no reduction in actual operation times was required. By simply recognizing the circular effect of queue size on lead times, of lead time on WIP levels, and of WIP levels on queue size (Figure 11.12), it is possible to reduce the number of orders open at any given time, thereby simplifying the management of the shop (and producing considerable savings to the company) while maintaining the required completion dates.

Production control managers who fail to recognize the effects described above tend to react to poor delivery performance in just the opposite way. If deliveries are not being met, the rationale is that they do not have sufficient time to complete the orders they are given. To compensate, they increase lead times. This of course has the opposite effect than they anticipated. By increasing lead times, WIP increases, queue times lengthen, and the burden of managing these increased queues only makes delivering on time more difficult. Those responsible for production scheduling, especially at the MRP level, must diligently guard against this practice. Since manufacturing engineers are often responsible for establishing leadtimes, they play a key role in this process.

Priority Sequencing Rules

In the previous subsection we saw how proper control of lead times influences queue size. But how are the queues themselves managed? What tools and techniques allow production control management to determine relative priority of all the orders in a given queue? This subsection briefly presents same of the more common approaches.

> *Operation due dates:* Each operation is assigned a due date, typically using the forward scheduling technique discussed in the previous section. Orders with operations having the earliest due date are selected first.
>
> *First-in, first-out (FIFO):* Orders are worked in the sequence in which they entered the queue; i.e., the oldest order (first in) is worked first (first out).
>
> *Critical ratio:* A calculated ratio using the formula (time remaining) ÷ (work remaining), where time remaining is the difference between today's date and the order's due date, and work remaining is the sum total of all remaining operation

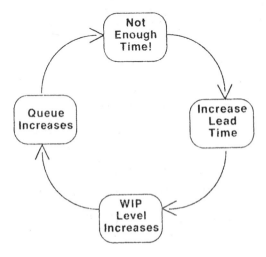

Figure 11.12   Circular effect of lead time, WIP, and queue.

times. A ratio of less than 1.0 indicates that the order is behind schedule (i.e., time remaining is less than work remaining); a ratio equal to 1.0 indicates that the order is on schedule; a ratio greater than 1.0 indicates that the order is ahead of schedule (i.e., time remaining is greater than work remaining). A negative critical ratio indicates that the order is already past due. Orders with the smallest critical ratio are selected first.

Slack time: The amount of time between the order completion date (calculated by forward scheduling) and the order due date. Positive slack indicates ahead of schedule, zero is on schedule, negative is behind schedule. Orders with the least slack (or most negative) are selected first.

Shortest operation next: Orders whose next operation are shortest are selected first. The rationale is to process the greatest possible number of orders, thereby minimizing the number of orders in queue. This rule must be used with caution, and is probably best used as a tie breaker in combination with some other rule(s).

Most operations next: Orders with the greatest number of remaining operations are selected first. The rationale is that orders with fewer operations are easier to schedule into available work centers, therefore the more difficult orders (i.e., those with more operations) should be scheduled first.

Informal: Numerous informal priority sequencing rules exist, including hot lists (maintained manually to override the formal system), colored tags (e.g., assign orange-tagged orders top priority), and "he-who-screams-the-loudest."

Dispatching

Dispatching is "the selecting and sequencing of available jobs to be run at individual workstations and the assignment of those jobs to workers." A *dispatch list* is "a listing of manufacturing orders in priority sequence. The dispatch list is usually communicated to the manufacturing floor via hard copy or CRT display, and contains detailed information on priority, location, quantity, and the capacity requirements of the manufacturing order by operation. Dispatch lists are normally generated daily and oriented by work center" [1].

In Subsection 11.2.5 we discussed the process of detailed scheduling. The resulting operation-level schedules are prioritized using a variety of sequencing rules, and these priorities are communicated to manufacturing via dispatch lists. In order for schedules and priorities to be maintained in a timely manner, some way of communicating the status of individual orders must be provided. This is the role of shop floor data collection.

Shop Floor Data Collection

The shop floor data collection system provides the means of collecting information on a wide variety of shop floor activities, including the following:

Employee time and attendance
Order status
Scrap and rework
Labor accumulated against orders
Work center capacity output
Machine downtime

Data collection can be done manually, but as bar coding and other data collection technologies advance, more and more manufacturers are turning to automated systems for this purpose. Computer terminals located throughout the factory may be used to query the status of a particular order on-line at any time. Other information such as work center queues, capacity data, etc., might also be available.

Within the context of this chapter, we will focus primarily on data related directly to shop orders (status and labor). We will deal with the subject of order status first. A typical order "life cycle" is depicted in Figure 11.13.

A typical scenario might employ a bar-coded shop order traveler, usually a heavy card stock paper document identifying the order number, part number, and operation numbers matching the information contained on the order's routing. These data elements are usually printed using standard text, accompanied by their bar-code equivalents. When an order is released, it may be assigned a status of "released." As the order progresses through its life cycle, its status will change repeatedly and might reflect any of the following:

In work
Held for . . . (various reason codes)
Complete
Closed
etc.

In addition, operation status information might also be reported, typically indicating the last operation completed. This information, however, does not appear by magic. It comes only through the disciplined efforts of many people—machinists, parts movers, inspectors, etc. Each worker has a responsibility to report his or her progress on orders as they are processed, thereby providing this information through the system. These data are then available to support production management decisions such as priority sequencing and dispatching.

As orders are completed, returned to stock, and closed, inventory balances are updated, and MRP takes these new balances into account in its next planning cycle. This closed-loop process is discussed further in Subchapter 11.5.

In addition to status information, most data collection systems record the amount of time charged to each operation. While labor is almost universally collected, some systems also permit both labor and machine (or other process) time to be collected. This information is used for various purposes such as capacity planning (refer to Subsection 11.2.7 and Subchapter 11.1), inventory costing, and performance reporting, which is the next subject.

Performance Reporting

Three primary performance indices are of interest to manufacturing engineers—efficiency, utilization, and productivity. Definitions and a brief example of each follows:

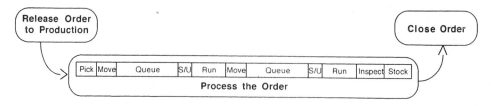

Figure 11.13 Life cycle of a typical order.

Efficiency: A measure of the actual output versus the standard output. If the standard time to produce a part is 4 hr and the part is actually produced in 5 hr, the efficiency is $4 \div 5 = 0.8$, or 80%.

Utilization: A measure of the percentage of a resource's available time which is actually used. If a resource, say a machine, is available 8 hr/day, 5 days/week, and is actually used for 35 hr in a given week, its utilization for that week is $35 \div 40 = 0.875$, or 87.5%.

Productivity: An overall measure of work center efficiency comparing the total standard hour output of a work center to the "clock time" expended. If a work center completes a total of 38 standard hours in a 40 hr week, its productivity for that week is $38 \div 40 = 0.95$, or 95%.

While these measures have much validity in measuring the capabilities of a given work center, they must be applied with caution. All three indices measure localized performance, with little or no regard for the overall effectiveness of the total manufacturing operation. In fact, it may be desirable for some work centers to operate well below 100% in any of these measures, especially if the work center is a "nonbottleneck." "Bottleneck" resources, on the other hand, must strive for 100% in all three measures to maximize the overall productivity of the operation. These principles are central to the theory of constraints discussed in Subchapter 11.4.

11.2.7 Levels of Capacity Planning

Throughout Subchapter 11.2 we have dealt with scheduling without regard to capacity. In other words, we have implicitly assumed that unlimited capacity was available to produce whatever was scheduled. In reality, however, capacity is a very real constraint in manufacturing companies, and must be considered when developing schedules.

In this subsection, four levels of capacity planning will be addressed:

1. Resource requirements planning
2. Rough-cut capacity planning
3. Capacity requirements planning
4. Input/output control; finite loading

Figure 11.14 identifies these levels, the purpose of each, and their relative position in the scheduling hierarchy. A few definitions must be introduced before continuing.

Infinite loading: Assigning work to a work center without regard to capacity. The resulting calculated load may show periods of overload and/or underload as compared to available capacity; i.e., infinite loading reflects how much capacity is needed to perform the schedule.

Theoretical capacity: The calculated maximum capacity, usually based on the standard hours a work center is able to produce in a given time period. Theoretical capacity makes no adjustments for nonproductive time such as routine maintenance, repair, shutdown, inefficiency, etc.

Demonstrated capacity: The proven capacity of a work center, demonstrated over a period of time. Usually based on actual standard hour output per day, week, etc.

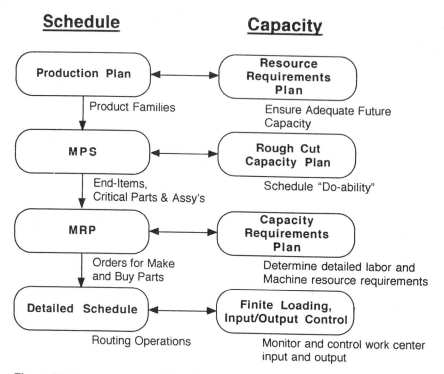

Schedule   Capacity

Production Plan ⟷ **Resource Requirements Plan**

Product Families   Ensure Adequate Future Capacity

MPS ⟷ **Rough Cut Capacity Plan**

End-Items, Critical Parts & Assy's   Schedule "Do-ability"

MRP ⟷ **Capacity Requirements Plan**

Orders for Make and Buy Parts   Determine detailed labor and Machine resource requirements

Detailed Schedule ⟷ **Finite Loading, Input/Output Control**

Routing Operations   Monitor and control work center input and output

Figure 11.14 Levels of schedule and capacity planning/control.

Figures 11.15a and 11.15b provide an overall context for these definitions. A typical model used to represent capacity and load is the "bathtub diagram" shown in Figure 11.16. Let us use this diagram to examine a "microview" of a work center first. As orders are released and moved through the shop, production control assigns them to the proper work center as depicted by the spigot. If the work center is a "gateway" (the first work center on a routing), production control may exercise some discretion over when to release the order, thereby controlling the *input rate*. Upon entering a work queue, orders become part of *work in process* (WIP) and increase the work center's load, as shown by the water level in the bathtub. The rate at which WIP is processed (i.e., orders are completed) is a function of the work center's capacity. This capacity is also somewhat adjustable through actions such as overtime, worker reassignments, etc., as depicted by the stopper in the tub's drain. The elapsed time between when an order enters a work queue and when it leaves is the manufacturing lead time for that operation.

Now let us take a "macroview" of Figure 11.16 as representative of an entire plant. If work is released to the plant at a higher rate than it is completed, work will begin to pile up on the shop floor. In addition, the overall load will increase, thereby increasing the actual lead time of each order. (Recall the discussion related to Figure 11.12 in Subsection 11.2.6 regarding the circular effect of WIP, queue time, and lead time.) Conversely, if the output rate is greater than the input rate over a period of time, work will dry up, leaving manufacturing resources idle. Maintaining a balance between the input and output rates is largely a production control responsibility. However, as we will see in the following

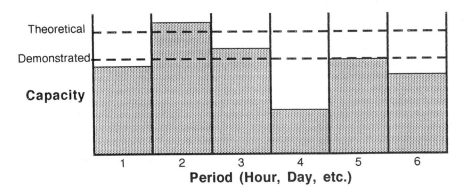

(a)

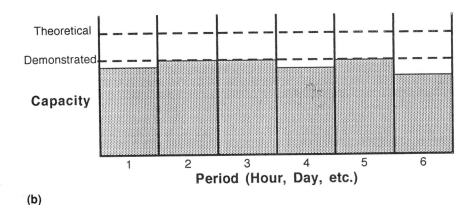

(b)

Figure 11.15 (a) Infinite shop capacity loading. (b) Finite shop capacity loading.

paragraphs, concern for this balance must be demonstrated at all scheduling and capacity planning levels.

Resource Requirements Planning

As discussed in Subsection 11.2.2, a production plan defines the overall production rates for each product family (a group of products having similar characteristics). The associated level of capacity planning is known as *resource requirements planning* (RRP) (see Figure 11.14). At this level, the primary concern is to ensure that adequate resources will exist to support the production plan, with consideration given to long-term, "brick and mortar" and overall manufacturing capacity. Units of capacity might include the following:

Total plant square feet
Warehouse volume
Total machining hours

The horizon for these decisions is considered longterm, typically 2 to 5 years with quarterly or longer time periods.

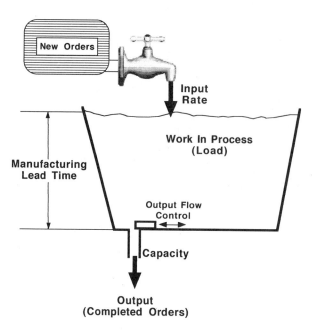

Figure 11.16 Capacity and shop load model.

Figure 11.17 depicts an overview of the RRP process. For each product family, a *bill of resource* must be defined which identifies the resources the company wishes to plan and the amount of each resource required to produce a typical product in that family. In our example, "sawing," "painting," and "assembly" have been identified. The units of capacity (U/C) for sawing and assembly are in hours, while painting is in square feet. This will allow the company to calculate future requirements for sawing and assembly hours, perhaps to identify the need for a new saw, additional assembly workers, added shifts, etc. Calculating painting requirements in square feet might indicate that the company is concerned that existing paint booths are not large enough to accommodate the projected production rates, and management wishes to know when an additional booth needs to be added. (See Chapter 3.)

The "Bill of Resource Summary" table summarizes all three bills of resource. This table is then multiplied by the "Production Plan" table to calculate the resource requirements. The "Resource Requirements Plan for Sawing" table shows the results of this calculation on the resource "Sawing." This process is repeated for the remaining two resources (not shown in the figure).

Notice that no lead times were defined in the bill of resource. While most RRP systems allow for lead-time offsetting of resources, it is fairly common to make a simplifying assumption that the resources will be used in the same time periods that endproducts are produced. In our example, which uses yearly time periods, this assumption is quite valid. We will see in the next subsection how lead time offsetting affects capacity planning.

Once resource requirements are known, the next step is to compare them to available resources. If the current sawing capacity is 2000 hr per year, the RRP plan indicates an overcapability situation beginning in 1997. The company might decide to add a saw in that

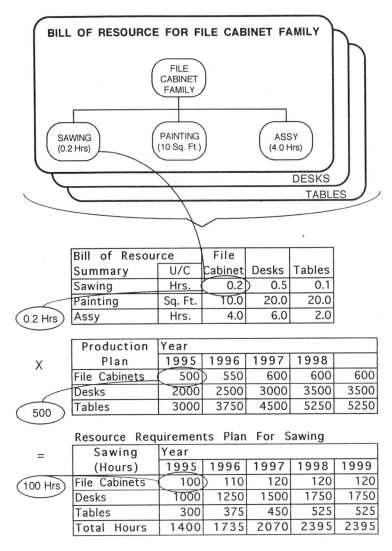

Figure 11.17 Resource requirements planning (RRP) process.

period, perhaps adding an additional 1000 hr of capacity. The resulting resource plan is depicted in Figure 11.18.

Rough-Cut Capacity Planning

The process of RCCP is identical to that described above, except that it almost always takes lead-time offsetting into account and uses smaller time periods. Recall from Subsection 11.2.3 that the master production schedule (MPS) defines what the company intends to produce, expressed in terms of specific configurations, dates, and quantities. The purpose of the roughcut capacity plan is to validate the MPS, i.e., to ensure that it is attainable, or "do-able." When RRP is used to identify the need for major resource

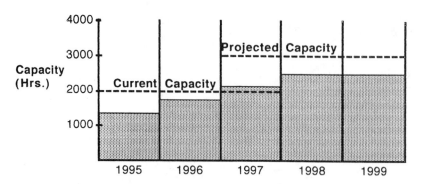

Figure 11.18   Resource graph for sawing, showing current and planned capacity versus projected workload.

adjustments, RCCP is used primarily to ensure that *existing* resources are used effectively. However, not all resources are planned at this level. Only those resources considered critical are planned by RCCP. A critical resource is generally thought of as a known or suspected bottleneck which is likely to be a limiting factor in attaining the MPS. The bills of resources used at this level generally come from an analysis of the BOM structure for each MPS item, and the routings associated with every part in the BOM. Extracting the load and time-phasing information for critical resources allows the bill of resources to be developed. Other resources might be added to this basic structure at management's discretion. Examples might include the following:

> Hours on a lathe which is used by multiple products
> Workers with a critical skill, such as precision welding
> Cash
> Hours required to pick parts from stores
> Packaging operation(s)

The horizon for these decisions is considered medium term, typically 6 months to 2 years. Figures 11.19 and 11.20 show the RCCP process, adding the effect of lead-time offsetting of resources. For the sake of simplicity, only cabinets are shown. An actual case would include all products in the MPS.

Beginning with Figure 11.19, let us walk through a calculation of sawing requirements for all cabinets. As indicated in the figure, 3 weeks prior to completing a cabinet A assembly, 0.2 hr of sawing is required. The MPS table in Figure 11.20 shows a requirement for eight such cabinets in period 4. Therefore, 3 weeks earlier (period 1), the requirement for sawing is 8×0.2 hr = 1.6 hr. Repeating this process for all products in the MPS and summing the totals results in the total numberof hours indicated at the bottom table of Figure 11.20.

As in the RRP example above, the next step is to compare these totals against the sawing capacity. The graph in Figure 11.20 shows the resulting capacity plan, assuming two capacity levels, demonstrated and theoretical. Any overload conditions should be corrected before running MRP against the MPS. This may require minor adjustments in capacity through overtime, off-loading work, etc., or may require the MPS to be modified. Loads falling between the two capacity lines are generally managed by adding capacity,

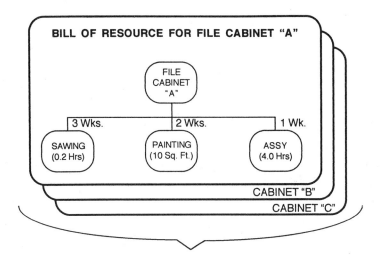

Resource Setback in Weeks				
4	3	2	1	0
	Sawing◄ 0.2 Hrs	Painting◄ 10 Sq. Ft.	Assy ◄ 4.0 Hrs	Cabinet "A"
Sawing◄ 0.5 Hrs		Painting◄ 20 Sq. Ft.	Assy ◄ 6.0 Hrs	Cabinet "B"
		Painting◄ 20 Sq. Ft. Sawing◄ 0.1 Hrs	Assy ◄ 2.0 Hrs	Cabinet "C"

Figure 11.19 Rough-cut capacity planning (RCCP) of critical or bottleneck resources resulting in a time-phased bill of resource.

while loads above the theoretical (maximum) capacity almost always require revision to the MPS. In our example, no overloads exist, indicating the MPS is attainable.

Once an attainable MPS is established, MRP may be run. Its associated level of capacity planning is discussed next.

Capacity Requirements Planning

Once MRP has run, and produced a new set of planned orders, capacity requirements planning (CRP) may begin its task. In addition to MRP planned orders, CRP takes into account all released and open orders, considering the status of each to determine present location and next operation (see Subsection 11.2.6, shop floor data collection). The first step in CRP is to determine the start and/or finish dates and times for each order's remaining operations. Recall the detailed scheduling approach discussed in Subsection 11.2.5, where detailed operation-level schedules were developed by the shop floor control system using a forward scheduling calculation. CRP backward scheduling is identical to this process, only in reverse.

The purpose of CRP is to provide a forecast of load at each work center, enabling

Master Production Schedule (MPS)										
Week Number										
	1	2	3	4	5	6	7	8	9	10
Cabinet "A"	8	8	8	8	8	8	10	10	10	10
Cabinet "B"	32	32	32	32	32	32	40	40	40	40
Cabinet "C"	30	30	30	30	30	30	30	30	30	30

8 x 0.2 Hrs. = 1.6 Hrs
3 Week Offset

40 x 0.5 Hrs. = 20 Hrs
4 Week Offset

Rough Cut Capacity Plan For Sawing										
Week Number										
	1	2	3	4	5	6	7	8	9	10
Cabinet "A"	1.6	1.6	1.6	2	2	2	2			
Cabinet "B"	16	16	20	20	20	20				
Cabinet "C"	3	3	3	3	3	3	3	3		
Total Hrs.	20.6	20.6	24.6	25	25	25	5	3	0	0

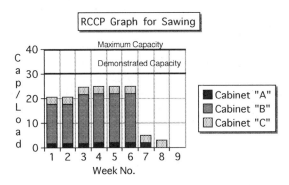

Figure 11.20   Example of the RCCP process.

production control to manage the current and anticipated workload at each. A sample work center load report is shown in Figure 11.21. Notice that, in addition to the load graph, detailed information on orders is provided to show the sources of each period's load. This information is critical when decisions much as moving work between periods must be made. Notice also that an "infinite loading" approach was used; i.e., no attempt was made to limit the load in each period to the available capacity. This is fairly typical in CRP. Finite loading is generally reserved for the shop floor control system, as discussed in the next subchapter.

Notice also that dates are shown as numeric values representing manufacturing days. While this is fairly common in older systems, most newer software converts these values to their corresponding calendar dates and displays them using normal date conventions, e.g., 940304, 3/4/94, 4-Mar-94, etc.

Finite Loading and Input/Output Control

In Figure 11.15 we saw a comparison of infinite versus finite loading. The finite load graph is repeated in Figure 11.22 with further detail to show the process involved. The loads moved from periods 2 and 3 represent operations containing sufficient load hours to bring the total down to the capacity level. Notice that work was moved forward into an

Drilling Load vs. Capacity

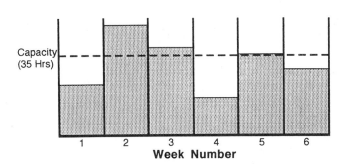

Work Center:		Drilling				
Capacity: 35 Hours/Week						
Week	Load Hours	Part Number	Start Date	Finish Date	Order Number	Order Status
1	10	ABC	230	231	123	In Work
	5	DEF	232	232	234	In Work
	8	GHI	233	234	345	Released
Total	23					

2	4	JKL	235	235	456	Released
	12	MNO	236	237	567	Planned
	.	.	.	.	.	.
	.	.	.	.	.	.
	.	.	.	.	.	.
Total	.					

Figure 11.21 Capacity requirements planning (CRP) example taken from the work center report, showing drilling requirements and capacity.

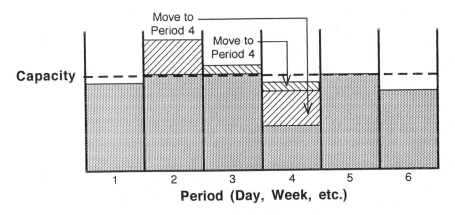

Figure 11.22 Example of the shop floor control system adjustment of the infinite loading of work orders, to balance capacity by moving orders earlier in the queue, and achieve a more balanced workload.

underloaded period. If the loads depicted by this graph were developed by a CRP backward scheduling calculation, this would likely mean that the operations moved will finish behind schedule, since the process of backward scheduling results in "just-in-time" due dates and operation dates. If, on the other hand, the loads were developed by SFC using a forward scheduling calculation, moving work forward would more likely consume some of the order's "slack time," without jeopardizing the order due date.

Assuming that we now have a viable capacity-balanced schedule, the next step is to ensure that each work center's load is properly managed. One method for accomplishing this is called *input/output control*.

The input/output report is used to monitor the actual versus planned load entering a work center and the actual versus planned load leaving it. Its purpose, therefore, is to monitor planned versus actual work flow through the work center. Its horizon is considered short-term, typically covering 4 to 8 weeks in the past (i.e., actual flow) and 6 to 8 weeks in the future (i.e., planned flow). A sample input/output report is shown in Figure 11.23.

A few observations will aid in our understanding of this report. First notice that, over the previous 5 weeks, total actual input exactly matches total planned input, resulting in a cumulative delta of zero hours. While week-to-week fluctuations occurred, overall the flow into the work center matched the plan. Likewise, the actual output is reasonably close to its plan, adding only 4 hr to the cumulative delta.

Second, notice that at the beginning of our tracking period a backlog of 25 hr existed, versus a goal of 40 hr. By planning more input than output over the next 5 weeks, we sought to increase the backlog to the desired level. However, because less output was produced than was planned (i.e., cum delta of 4 hr), the actual backlog is 4 hr greater than desired. If a cumulative input delta had also existed, the difference between planned and actual backlog would be offset by that amount as well.

	Week	Actual					Planned								
		-5	-4	-3	-2	-1	0	1	2	3	4	5	6	7	8
Planned Input		25	25	25	25	25	25	25	25	30	30	30	30	30	30
Actual Input		20	22	30	30	23									
Cum. Delta (Plnd-Actl)		5	8	3	-2	0									
Planned Output		20	20	20	25	25	25	25	25	30	30	30	30	30	30
Actual Output		18	20	15	25	28									
Cum. Delta (Plnd-Actl)		2	2	7	7	4									
Actual Backlog	25	27	29	44	49	44									
Planned Backlog		30	35	40	40	40	40	40	40	40	40	40	40	40	40

Desired Backlog: 40 Hours

Figure 11.23 Sample input/output report showing the planned and actual receipt of orders in a workcenter, versus the planned and actual output.

11.2.8 Summary

Throughout this subchapter we have examined the basic elements of scheduling and capacity planning employed in an MRPII environment. A production plan sets the overall production rates for families of products over a relatively long time frame and provides the input to the resource requirements planning (RRP) system. The resulting resource plan is used by upper management to support long-range decisions regarding "brick and mortar" and other major capacity adjustments. The master production schedule (MPS) translates the production plan into a set of requirements, expressed in specific configurations, quantities and dates, over a medium time frame. Its associated level of capacity planning, known as rough cut capacity planning (RRCP), tests the MPS against available capacity on critical resources. Once an attainable MPS is established, material requirements planning (MRP) explodes these requirements down a bill of materials (BOM) product structure, netting against available inventory, to create a set of planned orders for both manufactured and purchased parts. capacity requirements planning (CRP) calculates the resulting work center loads, taking into account planned, open, and released orders. Finite scheduling may be used to further adjust schedules at a routing operation level, allowing only as much work to be scheduled as each work center can produce. Input/output control can be used to monitor work flow through a work center and identify the need for capacity adjustments in the immediate time frame.

11.3 JUST-IN-TIME

The American Production and Inventory Control Society (APICS) defines *just-in-time* (JIT) as:

A philosophy of manufacturing based on planned elimination of all waste and continuous improvement of productivity. It encompasses the successful execution of all manufacturing activities required to produce a final product, from engineering to delivery and including all stages of conversion from raw material onward. The primary elements of zero inventories (synonym for JIT) are to have only the required inventory when needed; to improve quality to zero defects; to reduce lead times by reducing setup times, queue lengths, and lot sizes; to incrementally revise the operations themselves; and to accomplish these things at minimum cost. In the broad sense it applies to all forms of manufacturing job shop and process as well as repetitive [1].

Because a just-in-time philosophy encompasses all a manufacturing company's activities, it is wise to start small and build on success rather than attempt an all-out conversion. In *Just-in-Time—Surviving by Breaking Tradition* [6], Walt Goddard recommends approaching a JIT implementation in three steps: Begin with people issues, then move on to physical plant issues, and finally deal with changes which the earlier improvements will require in the way of computer systems (see Chapter 6).

It is important to recognize that the just-in-time journey is not linear, but circular. It is a philosophy of continuous improvement, not a destination. No matter how good a company becomes, how excellent its quality, how short its lead times, how good its prices and delivery performance, it can always improve. In fact, it must improve to stay ahead of the competition, which is always in hot pursuit.

In this subchapter we will briefly examine each of the key elements introduced in the definition given above. Readers should note that this is a thumbnail sketch of JIT only. Additional sources of information on JIT are found in the bibliography.

11.3.1 People Issues

When dealing with people issues, it is advisable to start with education. Put the users in charge, and organize them in teams. Provide professional facilitation training and detailed subject matter training to the team leaders to ensure they are prepared to handle team meetings, and can address both the people and technical issues that are sure to arise. Cross-train all workers so they can perform multiple tasks, not just in specific manufacturing jobs, but in teamwork skills, process simplification, and other improvement techniques. Such a "whole-person" approach utilizes all the knowledge and capabilities of employees, provides improved flexibility for the company, and benefits employees at the same time.

11.3.2 Elimination of Waste

At its heart, JIT is the relentless pursuit and elimination of waste. Waste is defined in this context as anything which does not add value to the product, such as poor quality, double handling and storage of materials, large inventories, long setup times, and paperwork. The goals of JIT are to produce quality products, on time, at the best price, and with the shortest possible lead time. Waste in any form interferes with a company's ability to meet these goals. Beginning with lead times, we will discuss each of these elements and introduce methods for reducing or eliminating them.

11.3.3 Reduced Lead Times

In Subsection 11.2.5 (Figure 11.12), we examined the circular effect of lead times, work in process (WIP), and queue times. If lead times are reduced, lower work in process levels are required to produce at a given rate, in turn reducing queue times. Because queue times (prior to JIT) account for the largest single element of a part's lead time, reducing queue time further reduces lead time. We noted, however, that no impact on actual processing time—made up of setup and run times—was involved in this cycle. Here we turn our attention to setup reduction as a key element in reducing lead times.

To better understand setup time and ways in which this time may be reduced, we must start with a clean definition. We will define setup time as follows, further breaking it down into two elements:

Setup time: The time from making the last good part *A* to the first good part *B*.
Internal setup: Activities which can only be performed when a machine is not running.
External setup: Activities which can be performed while a machine is running.

The traditional approach to dealing with setups has been to assume that they are fixed. To compensate, the setup time (and cost) must be amortized, or spread, among many parts by producing in batches. A time-honored technique for determining batch size is the *economic order quantity* (EOQ). Figure 11.24 shows this approach graphically. Two costs are considered. First is the cost per part (descending line), indicating the effect of amortizing setup on an increasingly large batch. Second is the carrying cost (ascending line), which recognizes that storage of parts, once produced, is not free. The EOQ is the point at which the combined total cost (sum of the two lines) is minimized.

Because the bottom of the total cost curve is relatively flat, the calculated EOQ is frequently adjusted by management decree to "make at least 3 month's worth," or similar

Setup Time (Hrs.)	10
Run Time (Hrs.)	0.1
$/Hour Rate	$12.00
Carrying Cost Rate	$2.50

Batch Size	Cost Per Part	Carrying Cost	Total Cost
1	$121.20	$2.50	$123.70
2	$61.20	$5.00	$66.20
3	$41.20	$7.50	$48.70
4	$31.20	$10.00	$41.20
5	$25.20	$12.50	$37.70
6	$21.20	$15.00	$36.20
7	$18.34	$17.50	$35.84
8	$16.20	$20.00	$36.20
9	$14.53	$22.50	$37.03
10	$13.20	$25.00	$38.20
11	$12.11	$27.50	$39.61
12	$11.20	$30.00	$41.20
13	$10.43	$32.50	$42.93
14	$9.77	$35.00	$44.77
15	$9.20	$37.50	$46.70
16	$8.70	$40.00	$48.70
17	$8.26	$42.50	$50.76
18	$7.87	$45.00	$52.87
19	$7.52	$47.50	$55.02
20	$7.20	$50.00	$57.20

Cost Per Part =
((S/U + BS x Run) x $/Hr) / BS

Carrying Cost =
BS x Carrying Cost Rate

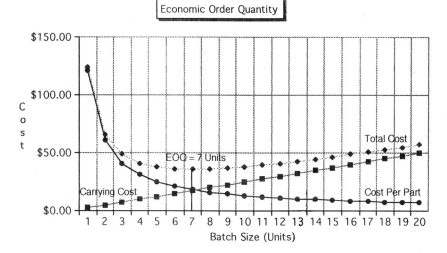

Figure 11.24 Example of the time-honored system of economic order quantities—prior to considering JIT or some other system.

policy decisions. Such approaches ignore other important considerations, however. First, large batches increase the amount of time orders sit in queue, increasing lead times and decreasing flexibility. Second, quality problems are typically not found until parts from the batch are used in a later operation. By then, so much time has elapsed since they were produced that determining the root cause is far more difficult than if it had been identified at the source. A third factor is the cost of loss, damage, or obsolescence. The longer a part is held in inventory, the greater is the chance that it will be lost or damaged, or that it will be superseded by a new design. Even if design changes can be delayed until on-hand inventory is used up, the company pays a price in customer satisfaction by not being able to bring design improvements to the marketplace quickly.

A far better approach is to limit batch sizes, ideally to one, but to at least begin down the path. The place to start is with setup reduction. In the above example, if we reduce the setup time to 30 min (0.5 hr), we see that the EOQ becomes 2, and the additional cost to build just one is only 50 cents (Figure 11.25). It is now feasible to build only the quantity

Setup Time (Hrs.)	0.5
Run Time (Hrs.)	0.1
$/Hour Rate	$12.00
Carrying Cost Rate	$2.50

Batch Size	Cost Per Part	Carrying Cost	Total Cost
1	$7.20	$2.50	$9.70
2	$4.20	$5.00	$9.20
3	$3.20	$7.50	$10.70
4	$2.70	$10.00	$12.70
5	$2.40	$12.50	$14.90
6	$2.20	$15.00	$17.20
7	$2.06	$17.50	$19.56
8	$1.95	$20.00	$21.95
9	$1.87	$22.50	$24.37
10	$1.80	$25.00	$26.80
11	$1.75	$27.50	$29.25
12	$1.70	$30.00	$31.70
13	$1.66	$32.50	$34.16
14	$1.63	$35.00	$36.63
15	$1.60	$37.50	$39.10
16	$1.57	$40.00	$41.58
17	$1.55	$42.50	$44.05
18	$1.53	$45.00	$46.53
19	$1.52	$47.50	$49.02
20	$1.50	$50.00	$51.50

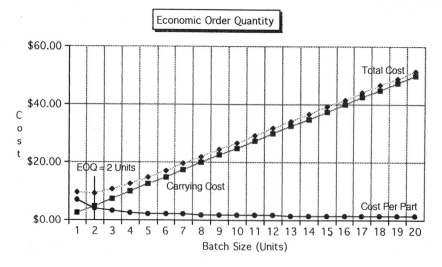

Figure 11.25 Reducing setup time for a batch reduces cost per part and total cost—showing that EOQ is now 2 units (rather than previous example of EOQ = 7).

needed, with less lead time than was previously required to build the entire batch, thereby avoiding all the "waste" previously attributed to batch production.

The question remains, though, how do we achieve the setup reduction? A key approach is to perform as much of the setup as possible in parallel with the current job's run time. In other words, convert as many of the setup steps as possible to "external setup" and perform them all ahead of time. Then, minimize the remaining "internal" setup steps.

Some specific techniques for reducing setup times are outlined by Walt Goddard in *Just-in-Time, Surviving By Breaking Tradition* [6]:

1. Standardize the external setup actions; replace adjustable gauges with permanent ones.
2. Put all probe and blow-off hoses on one side of die.
3. Put a bench at side of press at the same level as the press opening, to hold the next die.
4. Color code all hose connections: air, hydraulic, water, etc.
5. Use parallel operations—deliver all components to support die setup (Use a check-off sheet to ensure all are present prior to setup).
6. Design a quick locating system—positioning pins and holes, with quick fasteners.
7. Standardize all die receptacles.
8. Add tonnage monitor on press to detect two pieces in die before damaging dies.
9. Involve tool-and-die designers in setup reduction programs so all new designs incorporate quick change-over concepts.
10. Use two-way radio between setup man and lift-truck operator who removes and delivers dies.
11. Photograph completed operation as a guide for setup man: location of tables, wrenches, baskets, etc.
12. Review material flow charts with reduced movements in mind.
13. Make as many of the setup activities internal to the run time. That is, do as much of the setup as possible without shutting down the machine.
14. Standardize all bolt sizes.
15. Code parts on dispatch list for major or minor setups to aid scheduling.
16. Standardization and use of common parts in the product will reduce the number of different parts required. If design engineering does not design a new part, no setup is required for it.

Remember to start small. The idea is to strive for continuous improvement, not to reach the ultimate solution in one quantum leap. Setup reduction goals, however, should be aggressive. When dealing with machine setups, for example, the goal should be for "single minute exchange of die" (SMED). Any set time less than 10 min (i.e., measured as single minutes, versus tens of minutes, hours, etc.) qualifies as SMED. A good analogy for those who consider this an impossible goal is that of a racetrack pit crew. While it might take the average person 15 min to change a tire and gas up, the pit crew takes only 15 sec! This comes from teamwork, detailed study of the process, proper tools, practice, and attitude. It can be done. And the payback is well worth the effort.

11.3.4 Reduced Inventory

A direct result of reduced lead times is the ability to reduce inventory. At the finished-goods level, if a product can be produced within the delivery lead time required by customers, why carry finished goods inventory? The same holds true of internal customers. If parts can be built one-for-one as the next process calls for them, why build them ahead of time only to pay for storage, additional handling, and potential costs of obsolescence, loss, or damage?

Many companies use inventories to hide problems associated with poor scheduling, poor quality, and large setups. A good analogy is a river with a rocky bottom. The traditional approach is to raise the water level in the river, thus hiding the problems. This of course is expensive and wasteful, but far easier than dealing with the problems head-on. The JIT alternative is to slowly lower the water level (i.e., reduce inventory), exposing the rocks (problems). Deal with each problem as it occurs, and not just the immediate symptom. Address and correct the root cause. Then lower the water some more. Another approach is to dive below the surface to identify problems before they surface. In either case, the intent is to solve root-cause problems, so they are gone for ever.

11.3.5 Zero Defects

As the name implies, a just-in-time system requires that the right part be delivered to the right place just as it is needed. But if the part is of poor quality, on-time delivery is worthless. Quality has been defined by Phil Crosby as "conformance to requirements," and by J. M. Juran as "fitness for use" [6]. The former, "conformance to requirements," is of primary interest to internal customers who must install or use the parts previously produced in a subsequent operation. The latter, "fitness for use," is often the final customers' ultimate yardstick. Unless a product does what the customer wants it to do in a particular application, it fails this criterion, even though it may conform to a set of technical requirements.

The goal of a quality improvement program should be to strive for zero defects. Along the path, manufacturers must first seek to replace traditional definitions of "acceptable quality," measured in percentage points, to parts per million. Where 98% quality might have been acceptable yesterday, today's target is six-sigma, or just over 3 parts per million. One method used to great benefit by JIT practitioners to ensure high quality is *statistical process control* (SPC). What SPC does is to shift the focus from inspection of completed parts to monitoring the process used to produce parts. If a process is capable of producing quality parts, and the process is kept in control, then all the parts produced by the process will satisfy the "conformance to requirements." Figure 11.26 shows three sample SPC control charts. The basic elements include upper and lower control limits (UCL/LCL), within which the process must remain to be considered "in control." (Note: These limits identify the natural variability of the process, not the design tolerances of the parts being produced. The UCL and LCL must be within these design tolerances. What is being measured is "process control," so that even if a process has gone out of control, it may not have produced bad parts yet.) Also included is a center line and observations occurring over a span of time. Notice that two out-of-control conditions exist. The first is depicted by the middle chart, showing a single point out of the specified control limits. The second, depicted by the bottom chart, shows a situation where seven sequential points have occurred between the center line and the upper control limit. Such a situation is not "statistically random" and should be interpreted as the process being out of control (see Chapter 12).

A natural result of SPC is to allow workers to monitor their own processes and the quality of the parts they produce. This is referred to as "operator verification,"

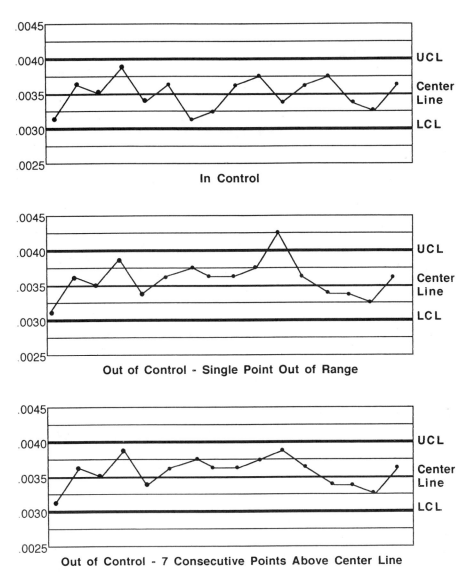

Figure 11.26 Sample SPC control charts.

"self-inspection," and "quality at the source," among other names. Such responsibility is often coupled with authority to stop the production line when a quality problem is found. While this may sound extreme, it services to focus everyone's efforts on solving the immediate problem, preferably at the root-cause level so the problem is unlikely to occur again.

If SPC is to ensure the production of quality parts, the natural variation in the processes must be held within the specified control limits. Perhaps the best way to accomplish this is with routine preventive maintenance (PM). Virtually all machine manufacturers specify PM programs for their machines. Follow them. Schedule the PMs routinely and do them religiously. This is key to ensuring that processes stay in control. An excellent practice is to make machine operators responsible for at least some PM activities. They know from experience how the machine should look, feel, and sound. Who is better, then, to keep them running at peak performance?

Yet another way to improve quality is to use foolproofing, or fail-safe techniques. The intent is to ensure that a process can be performed only one way—the right way! Some examples include [6]:

Use checklists or monitoring devices to detect parts missing on assembly. Do not allow them to pass to the next operation until corrected.

Use locator pins to ensure parts are aligned correctly, or dies are installed properly during machine setup.

Perform weighing operations to detect missing or extraneous parts or materials.

One final subject dealing with quality involves design engineering. No amount of effort, skill, or desire on the part of manufacturing can overcome the effects of poor product design. For this reason, design engineering must develop designs with manufacturing limitations in mind. This *concurrent engineering* or *design for manufacturability* as it is often termed, is critical to the success of a just-in-time effort (see Chapters 1 and 2). Some examples to clarify the possibilities follow.

Replace multiple sheet metal components with a single molded part.

Replace highly complex parts with simpler parts, taking advantage of "compensating tolerances."

Move options higher in the bill. Avoids adding high-percentage options as a "standard" and reconfiguring to customer requirements later.

Move difficult-to-install components lower in the bill, where access may be easier.

Move design engineering offices close to production floor.

While some of these may sound contradictory, the message is to find what works best in a given situation. There is no cure-all. Each case is different, and must be addressed by the responsible functions and teams.

11.3.6 Valid Schedules

If an objective of JIT is to have the right part at the right time, then it stands to reason that schedules, which specify time and place, must be valid. In Subchapters 11.2 and 11.5 the subject of scheduling is addressed from an MRPII perspective. It has been suggested, erroneously, that MRPII and JIT are incompatible. Nothing could be further from the truth. In fact, most successful practitioners of JIT use MRPII very successfully, and consider the resulting schedule validity that MRPII brings as critical to their success with JIT. Where the difference often lies is at the material planning level (i.e., "little MRP"). While many companies practicing JIT use MRP to schedule orders, an alternative scheduling technique that others have found beneficial is based on the "Toyota production system." The Toyota system is identical to MRPII from the business plan level down to MPS. It is not until the MRP level that the systems diverge.

Where MRP plans orders based on future demand, the Toyota system employs a pull-scheduling approach using *kanbans*, a Japanese term meaning "visual signal," *Kanbans* can take the form of cards, golf balls, squares painted on the floor, or any other visual signal indicating the need for a particular part. *Kanban* is similar to a two-bin, reorder point system (see Subsection 11.6.4). An empty bin (or square, card, etc.) signals the need to replenish what has been used. *Kanbans* are most effective in stable, repetitive environments, where lead times are fixed and demand is regular.

The primary limitation of *kanban* is the lack of future visibility. For example, if an engineering change causes a part to be superseded by another, continuing to replenish the old part could be disastrous. To compensate, Toyota explodes the BOM against the master schedule to provide a 90-day, summarized forecast of all components. When a particular part is no longer needed, a special *kanban* is inserted into the system to signal when its production is to cease.

The bottom line, regardless of the scheduling approach employed, is to make only what is needed by the immediate consumer. If nothing is needed, *do not produce*. Doing so only builds excess inventory, which is of course waste.

A useful technique employed by both systems (MRP and *kanban*) for stabilizing production rates, a critical element of JIT, is known as *mixed-model scheduling*. The goal of mixed-model scheduling is to make some of every product every day. For example, if the ratios of products A, B, and C are, respectively, 60%, 30% and 10%, a mixed-model schedule for a total of 10 items per day might look like this:

Day 1	Day 2	Day 3
A,A,B,A,A,B,C,A,A,B	A,A,B,A,A,B,C,A,A,B	A,A,B,A,A,B,C,A,A,B
(6 A's, 3 B's, 1 C)	(6 A's, 3 B's, 1 C)	(6 A's, 3 B's, 1 C)

Mixed-model scheduling is usually employed at the master schedule (MPS) level. Prerequisites to such scheduling are the ability to produce in small order quantities and to switch production quickly from one product to another. As discussed earlier, this is the result of having short or nonexistent setups.

11.3.7 Vendors

The key point to be made when bringing vendors on board in a JIT program is to lead by example. A company must begin by getting its own house in order, with active programs aimed at quality, setup reduction, dependable scheduling, and order quantity reduction. The best indication to a vendor that you are prepared to lead the way is to provide stable, valid schedules over a long enough period of time to establish your own credibility, *before you even approach them*. Next, provide them with education. Tell them what it's all about and prepare them for what they are about to see. Then bring them in and show them what you have done internally that has enabled you to provide the schedule stability they have already witnessed. However, don't wait too long. You don't have to be expert at JIT, just far enough along to show concrete improvement and the commitment to continue.

The goal in bringing vendors on board is to replace the traditional adversarial role with vendors, and between competing vendors, with partnerships between the company and a greatly reduced vendor base. The benefit for the company is in having vendors who are able to deliver quality parts, just in time, at a better price. The benefits to the vendors who survive are the same, providing a competitive edge over their competition which their own marketing people can use to good advantage. In addition, the remaining vendors will enjoy a larger share of your business, since they will no longer compete with others for the same parts and materials.

Not all vendors will be willing to come along, and not all should be considered. First weed out those with chronic quality or delivery problems. Stack the deck in your favor and take only those with a high likelihood of success through the process. This process in itself may take a year or more. Use this time for your own in-house efforts, then begin the process of bringing the survivors on board. Because the vendor base is reduced, buyers can take on a new role, becoming "vendor managers," who will assist the companies they work with in their own JIT efforts, ranging from SPC to setup reduction to valid scheduling. Rather than searching for additional "just-in-case" suppliers, and expediting when problems occur, they can spend their time developing the partnering relationships so critical to success with JIT.

Let us look ahead now to a few problems (opportunities!) which will have to be addressed once vendors are successfully delivering just in time. First is the issue of transportation. When vendors are delivering more frequently (i.e., daily or more frequently versus monthly), managing the traffic in receiving can become a problem. A good way to overcome this is to use public carriers with either a central delivery point and vendors delivering to the carrier, or a milk-run approach where the carrier makes frequent pickups from each of the vendors' sites and then delivers to the company at a prearranged time. Carriers must also be educated in JIT philosophy, so they understand the importance of on-time delivery. The consolidation and partnering arrangements made with vendors apply to carriers as well.

Second, is the potential impact on the accounting department. The goal should not be to overwhelm them with paper. While it may be acceptable with infrequent deliveries to process all shipments through an existing accounts payable process, daily or hourly deliveries might quickly bury accounting in paper. The best approach is to involve the accountants early on in the implementation, and devise ways to resolve such problems ahead of time. One approach is to do away with individual orders and invoices, replacing them with a once-a-month invoice. Some companies go so far as to eliminate this process entirely. Since they know how many of a vendor's parts are needed to complete each end product, they simply total the number of products sold and inwork once a month and pay the vendor via electronic funds transfer for that amount of product.

Similarly, if receiving inspection is to keep pace, simplifying or eliminating traditional processes is a must. Applying the same logic as worker self-inspection, once a vendor has demonstrated the ability to produce consistently high quality, they might be allowed to bypass receiving inspection altogether. Combined with JIT delivery of small lots, the logical extension of this is to allow them to deliver directly to the shop point of use, bypassing not only receiving inspection but the warehouse as well. (See Chapters 3 and 4; where multiple receiving docks may be required.)

Finally, consider electronic links between the company and its vendors. The goal here is the elimination of paper (waste, since it adds no value). In its simplest form this might involve giving vendors inquiry-only access to your scheduling system, enabling them to view on-line the current and future needs for their products. More advanced capabilities might include electronic PO placement, invoicing, and even electronic funds transfer to pay for parts received. (Subchapter 6.0 shows the need to tie in suppliers to your computer networks in order to achieve a computer-integrated-enterprise.)

11.3.8 Plant and Processes

An objective of JIT is to move a company's manufacturing as much toward a process-flow, or repetitive environment as possible. Another is to enhance visibility between operations so that communication between work centers is enhanced, and problems, once identified, can be more readily dealt with.

One method of accomplishing these objectives is the use of manufacturing cells. Cellular manufacturing brings machines and workstations together to make similar parts or products. [Cellular manufacturing is related to group technology (GT), in that GT classifies parts in such a way as to identify candidate parts for a given cell] (see Chapter 9). In a traditional job-shop layout, parts are routed between machines and workstations which have been arranged by function. Many companies' managers are astounded to learn just how far a part travels and how many non-value-added steps are involved in production. By bringing machines together in cells, these problems are avoided. Work enters a cell and is processed sequentially across the appropriate machines, leaving the cell as a completed part. Reductions in distance traveled from thousands of feet, or even miles, to tens of feet, and in lead times from weeks to hours are not uncommon (see Chapter 3).

A resulting benefit also comes in the form of simplified scheduling and tracking. Without cellular manufacturing, each step or groups of steps in the process may have required a new part number with its own level in the bill of material. Individual shop orders are certainly required in this situation. With cellular manufacturing, none of this is needed. A single part number moving through the cell is all that is required. And because the part moves through so quickly, the need to track intermediate steps in the process is eliminated. Cellular manufacturing, then, has the effect of "flattening" bills of material, simplifying scheduling, reducing paperwork, and reducing waste.

A second change in plant layout frequently employed is the use of U-shaped assembly lines. Where assembly lines exist, arranging them in a U-shaped configuration has the dual effect of improved visibility and reduced space, both of which improve communication between workers on the line, enhancing teamwork and product quality. See Chapter 3 for more detail.

11.3.9 Computer Systems

The previous discussions lead to some conclusions about a company's operating software, such as MRPII [6]. First, to accommodate reduced lead times, time periods must be daily or smaller. The ability to replan daily or more frequently is also a must, preferably in a manner similar to net-change MRP (see Subsection 11.5.4). To accommodate point-of-use inventory, the ability to identify multiple inventory locations, not just bin locations in a common stockroom, will be needed. If paperwork, such as shop orders, is to be reduced or eliminated, electronic or other means of maintaining inventory balances, labor collection, and other performance-related data must be provided. One such approach is known as "inventory back flush" or "post-deduct" logic. Rather than a traditional order-based approach, where inventory is pulled from stock, matched to an order, parts produced, and finally returned to stock to be received, back flushing automatically performs these transactions when an order is completed and moved to its "consuming" location.

11.3.10 Summary

JIT manufacturing is the relentless pursuit and elimination of waste. Waste comes in many forms, including poor quality, invalid schedules, excess inventory, unnecessary material handling, excess floor space, and anything else which adds cost but does not add value to a company's products. A number of tried-and-true approaches for reducing or eliminating each of these elements of waste have been demonstrated by JIT pioneers, and are available for companies wishing to begin their own journey.

The benefits are tremendous. Combined, these approaches, and the attitudes they foster, improve every competitive element of a manufacturing company. Quality is improved through statistical process control and shortened feedback; delivery performance is improved through reduced lead times, smaller order quantities and valid schedules; cost is lowered through the elimination of waste.

Workers contribute to a process of continuous improvement through a whole-person concept which treats them as the experts they are, not just as cogs in a division-of-labor wheel. They are given the education and tools required to do the job right, and the authority to stop the line when something goes wrong. Problems are addressed at the root-cause level, not hidden with excess inventory.

Successful JIT implementations begin by addressing the people issues, then move to physical changes in the plant's processes, and finally to ensuring that the company's computer systems support the requirements of the new processes. All functions participate, including the company's vendors, who are dealt with as partners, not as adversaries.

JIT is a journey, not a destination. No matter how good a company becomes it can always improve. JIT provides the banner under which to proceed.

One final word of caution: Since a successful JIT program hinges on people, don't sabotage your efforts by rewarding initial gains with a layoff. That is the surest way to guarantee the program's failure. Instead, take the opportunity to cross-train workers, provide new skills, or work on the next JIT improvement. When the benefits begin to take hold, customers will notice. Higher quality at less cost and superior delivery performance in less time than the competition will bring new business, avoiding the need for layoffs.

11.4 THEORY OF CONSTRAINTS

In his book, *The Goal* [4], Eliyahu M. Goldratt introduces, in novel form, the principles of his *theory of constraints* as applied to a manufacturing environment. Also known as optimized production technology (OPT), the application of the theory's principles allows the novel's main character to turn his failing plant into the corporation's most profitable. Following is a brief discussion of the major elements of *The Goal*. Readers are encouraged to read the book, which brings these principles to life in a way not otherwise possible.

11.4.1 The Goal

The goal of a manufacturing company is to make money. However, this goal is sometimes achieved in the short term, to the detriment of the company's long-term viability, by selling off assets, for example. Therefore, the goal should be "to make money now as well is in the future." Goldratt introduces three measurements to monitor progress toward this goal: throughput, inventory, and operational expenses. He defines these as follows [9]:

Throughput: The rate at which the system generates money through sales.
Inventory: All the money the system invests in purchasing things the system intends to sell.
Operating expense: All the money the system spends in turning inventory into throughput.

To achieve the goal, throughput must be increased while simultaneously reducing both inventory and operating expense.

Further study reveals some interesting differences between these definitions and their more standard business definitions. For example, throughput is traditionally measured at the point where finished goods are completed. By contrast, applying the above definition, unsold finished goods are considered inventory; increasing this inventory works counter to the goal and should be avoided. A second example might be inventory carrying costs. Carrying costs typically include such expenses as the cost of capital invested in inventory, taxes, insurance, obsolescence, warehouse space, etc. These costs are traditionally prorated and applied to each part in inventory. In the definition above, these costs are all considered operating expense.

11.4.2 Troop Analogy

An excellent example used in the book to illustrate these points is the analogy of a scout troop on a hike. In this example, the following definitions apply:

Throughput: The rate at which the entire troop (the system) progresses; i.e., the distance covered by the slowest scout.

Inventory: The distance between the first and last scout. As the first scout passes a given spot, this spot is added to inventory. The system retains this inventory until it is converted to throughput (passed by the last scout).

Operating expense: All the energy the system (scout troop) spends in turning inventory (distance between first and last scout) into throughput (distance covered by the last scout).

Let us add two more definitions before continuing the analogy:

Dependent event: An event which cannot occur until some prior event occurs.

Statistical fluctuation: Random variations in the time to perform an operation.

Returning to the example, imagine that the scouts are allowed to hike at their own pace. Before long they will have spread out, with the faster hikers in front and the slower in the rear. Using our definitions, the system will have created throughput only at the rate of the slowest hiker (named Herbie in the book), and will have consumed a great deal of energy and inventory in the process.

To correct this situation, we need to make each hiker dependent on another, and limit the distance between hikers by controlling the rate at which they walk (i.e., limit statistical fluctuations). This is best accomplished by moving Herbie to the front of the line, and instructing the remaining scouts to line up from slowest to fastest, and not to pass each other (i.e., make them "dependent" on each other).

Again, imagine the troop hiking under these rules. With Herbie at the front, any gaps which open between scouts can be readily closed, since all other hikers are able to increase their pace to catch up. Since we have done nothing to change Herbie's pace, the system will continue to generate throughput at the same rate, but with two significant differences. First, inventory (distance between the first and last hiker) will be considerable less, and second, less energy (operating expense) will be expended since the faster hikers will not be allowed to charge ahead.

We have now succeeded in accomplishing two of our objectives: decrease inventory and operating expense. But what about the third? In the story, Herbie carries a large backpack full of goodies. To increase his speed, and thereby the entire system's throughput, his load is distributed to other scouts. The result is easy to imagine. All three measurements (throughput, inventory, operating expense) move in the desired direction, in keeping with the goal.

Let us look at some additional definitions and apply them to our troop analogy.

Bottleneck: Any resource whose capacity is equal to or less than the demand placed on it.

Nonbottleneck: Any resource whose capacity is greater than the demand placed on it.

In our example, Herbie is the bottleneck. All the other scouts are by definition non-bottlenecks, because each contains some amount of excess capacity.

11.4.3 Balance Flow with Demand

A traditional approach to managing capacity is to balance capacity with demand. In theory, by not paying for "unnecessary" capacity, manufacturing costs can be reduced. This theory, however, ignores the problem of statistical fluctuation described above. Without this extra capacity, the gaps between dependent operations cannot be closed, leading to higher than necessary work-in-process inventory. Instead, what should be sought is to balance flow with demand. The goal should be to make the flow through the bottleneck equal to demand from the market. Two principles must be kept in mind to accomplish this: (1) Make sure the bottleneck's time is not wasted, and (2) make the bottleneck work only on what will contribute to throughput today, not 9 months from now.

How is the time of a bottleneck wasted? One way is for it to sit idle during breaks, shift changes, meals, etc. Another is for the bottleneck to produce parts which could be made at a nonbottleneck resource. Yet a third is for it to process parts which are already defective, or which will become defective in a subsequent operation. In other words, don't waste the bottleneck's time on parts unless you are sure they are good and will stay that way. All three time wasters must be relentlessly eliminated. The point to remember is this: *A minute wasted at a bottleneck is gone forever. It can never be replaced!*

To further emphasize this point, Goldratt suggests that the per-hour cost of a bottleneck is the total expense of the system divided by the number of hours the bottleneck produces. This approach provides a financial perspective on the reality that the *entire system is controlled by the bottleneck*. It should be managed as though it cost as much as the total system!

But what about nonbottlenecks? How should they be managed? First, we must challenge the notion that all machines must be kept busy all the time. When a non-bottleneck is producing a part that is not immediately needed, it is not increasing productivity, it is creating excess inventory, which is against the goal. Instead, we should *expect* nonbottlenecks to be idle part of the time, allowing their excess capacity to be used to close gaps created by statistical fluctuation. This applies to all resources, including people!

The APICS dictionary identifies a five-step approach to applying the theory of constraints [1]:

1. Identify the constraint (bottleneck) of the system.
2. Exploit the constraint (i.e., tie it to market demand).

3. Subordinate all nonconstraints (i.e., make sure they support the needs of the constraint and never hold it up).
4. Elevate the constraint (i.e., increase its capacity).
5. If the constraint is broken in step 4 (i.e., if by increasing its capacity it is no longer the constraint), go back to step 1 to identify the next constraint.

11.4.4 Summary

The goal of a manufacturing company is to make money now and in the future. To do so, manufacturers must seek to increase throughput while simultaneously reducing inventory and operating expense. To accomplish this, identify the bottleneck in the manufacturing system, then balance the flow through the bottleneck with demand from the market. Subordinate all other resources (nonbottlenecks) to the bottleneck by allowing them to produce only as much as is needed to keep the bottleneck supplied. Doing anything more only increases inventory and/or operating expense, in conflict with the goal.

11.5 MATERIALS REQUIREMENTS PLANNING (MRP) SYSTEMS

In Subchapter 11.2 we presented a top-to-bottom scheduling approach with the assumption of an MRPII system being in place. In this subchapter we will provide some historical context on the evolution of MRP systems and further describe the processing of the basic MRP "engine." Two basic replanning techniques—regenerative and net change MRP—will be presented, and we will end with a brief discussion of a technique for measuring the effectiveness of MRPII, the ABCD checklist.

11.5.1 Evolution of MRP

Prior to the 1960s, manufacturing companies depended primarily on reorder-point techniques to plan parts and material requirements. While some manufacturers sought an alternative approach, most directed their energies toward finding the most economical order quantities (EOQ) and on developing machine loading techniques to "optimize" the utilization and efficiency of factory resources. Those who did seek an alternative pioneered the way for today's MRP systems. When material requirements planning (MRP) was first developed, computers were just being introduced into manufacturing companies. The development of material plans was therefore done manually, requiring a month or more from top to bottom of the product structure. It was not until the 1970s, when computers became commonplace, that computerized MRP came into its own.

The basic MRP system was nothing more than a tool for breaking down a master production schedule (MPS) into its component parts by exploding a bill of material. After accounting for available material, planned orders were generated, creating a time-phased plan for ordering manufactured and purchased parts. As the order release dates arrived, orders were issued to the factory and to vendors, to be managed by the "execution" functions—production and purchasing. This "order launch and expedite" approach led to the creation of hot lists, shortage reports, and expediting as the major elements of a production control system. What was missing was some way to report actual status of the released orders, both in the shop and with vendors.

As orders are processed, material is issued and received, master schedules are revised, and other real-life changes occur, the material plan needs to be revised. By capturing this information and feeding it back to MRP, a new plan may be generated which takes these changing conditions into account. The MRP system then issues action messages to production and inventory control workers recommending appropriate intervention.

The advent of shop floor control systems, capacity requirements planning, and purchasing systems allowed for the basic MRP system to be upgraded. When all these elements were combined, the term "material requirements planning" was no longer sufficient to describe the new system, and the term "closed-loop MRP" was introduced. Further development brought capabilities such as "what-if" simulations, various levels of capacity planning and control, and the incorporation of financial data requiring yet another term. Oliver Wight coined the term MRPII, changing the meaning of the acronym from material requirements planning (MRP) to manufacturing resources planning (MRPII). Figure 11.27 depicts a fairly typical MRPII system overview.

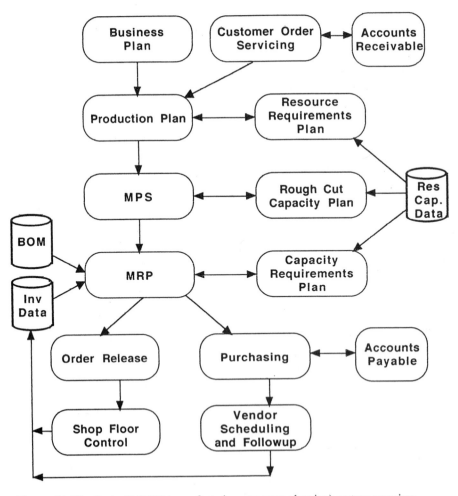

Figure 11.27 Typical MRPII (manufacturing resources planning) system overview.

11.5.2 MRP Processing

The basic planning "engine" in MRP systems is the material requirements planning subsystem itself. In this subsection we will examine the process by which MRP develops its time-phased material plans. Figure 11.28 provides an example.

Beginning with the top table, we will proceed through each row and review the processing steps involved. Recall from Subsection 11.2.4 that our goal is to translate a

Lot Size = 50
Lead Time = 2
Safety Stock = 0

Part Number "A" - Preliminary Record

Periods	1	2	3	4	5	6	7	8
Gross Requirements	20	25	25	10		30	20	10
Scheduled Receipts		50						
Projected On Hand	10	35	10	0	0	-30	-50	-60
Projected Available (30)								
Net Requirements								
Planned Order Receipts								
Planned Order Releases								

Bill of Material

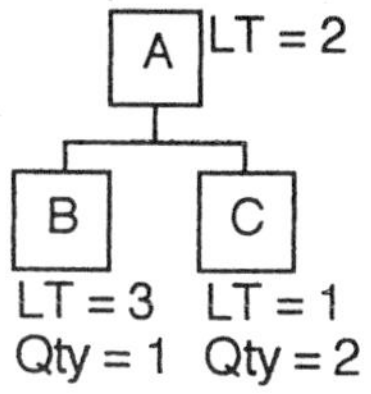

A LT = 2
B LT = 3 Qty = 1
C LT = 1 Qty = 2

Lot Size = 50
Lead Time = 2
Safety Stock = 0

Part Number "A" - Completed Record

Periods	1	2	3	4	5	6	7	8	Balance Check
Gross Requirements	20	25	25	10		30	20	10	
Scheduled Receipts		50							
Projected On Hand	10	35	10	0	0	-30	-50	-60	
Projected Available (30)	10	35	10	0	0	-30/20	0/0	-10/40	
Net Requirements	0	0	0	0	0	30	0	10	
Planned Order Receipts						50		50	100
Planned Order Releases			50		50				-60

40

Lot Size = 30
Lead Time = 3
Safety Stock = 25

Part Number "B" - Completed Record

Periods	1	2	3	4	5	6	7	8	Balance Check
Gross Requirements				50		50			
Scheduled Receipts			30						
Projected On Hand	35	35	65	15	15	-35	-35	-35	
Projected Available (35)	35	35	65	15/45	45/45	-5/25	25/25	25/25	
Net Requirements	0	0	0	10	0	30	0	0	
Planned Order Receipts				30		30			60
Planned Order Releases	30		30						-35

25

Figure 11.28 Material requirements planning (MRP) subsystem showing details of the part records.

master production schedule into a time-phased material plan (i.e., planned orders) for component parts and materials. We begin with an assumed master schedule for part A as shown in the "Gross Requirements" row of the top table. The second row, "Scheduled Receipts," indicates that an order is already in production for a quantity of 50 parts and is scheduled to complete in period 2. The next two rows calculate future inventory balances, beginning with a starting quantity of 30 parts. The "Projected On Hand" row ignores future additions to inventory resulting from planned orders, while "Projected Available" includes these quantities. The respective formulas are:

Projected On Hand in Period 1: $(POH)_1 = SI - (GR)_1 + (SR)_1$

Projected On Hand in Period n: $(POH)_n = (POH)_n - 1\,(GR)_n + (SR)_n$

Projected Available in Period 1: $(PA)_1 = SI - (GR)_1 + (SR)_1 + (POR)_1$

Projected Available in Period n: $(PA)_n = (PA)_{n-1} - (GR)_n + (SR)_n + (POR)_n$

where POH = Projected On Hand
       PA = Projected Available
       SL = Starting Inventory (Initial Projected Available)
       GR = Gross Requirements
       SR = Scheduled Receipts
       POR = Planned Order Receipts
       n = Period number 2, 3, 4, . . .

Working through a few periods for "Projected On Hand" results in the following:

$(POH)_1 = SI - (GR)_1 + (SR)_1 = 30 - 20 + 0 = 10$

$(POH)_2 = (POH)_1 - (GR)_2 + (SR)_2 = 10 - 25 + 50 = 35$

$(POH)_3 = (POH)_2 - (GR)_3 + (SR)_3 = 35 - 25 + 0 = 10$

etc.

"Projected Available" is similar, except that it is used as the "trigger" for planning a new order, and then takes the planned order receipt quantity into account when projecting future inventory quantities. If the resulting quantity falls below the specified "Safety Stock," MRP plans a new order, then recalculates the Projected Available balance. Working through a few rows, we see the following:

$(PA)_1 = SI - (GR)_1 + (SR)_1 + (POR)_1 = 30 - 20 + 0 + 0 = 10$

$(PA)_2 = (PA)_1 - (GR)_2 + (SR)_2 + (POR)_2 = 10 - 25 + 50 + 0 = 35$

$(PA)_3 = (PA)_2 - (GR)_3 + (SR)_3 + (POR)_3 = 35 - 25 + 0 + 0 = 10$

.
.
.

$(PA)_6 = (PA)_5 - (GR)_6 + (SR)_6 + (POR)_6 = 0 - 30 + 0 + 0 = -30$ (first pass)

Since this calculation is less than the safety stock of zero, MRP must plan an order. The quantity and timing of the order are determined by the lot size (50) and lead time (2). If the order is to be received in time to satisfy the requirement it must be released 2 periods earlier; i.e., in period 4. Recalculating with the planned order quantity results in the following:

$$(PA)_6 = (PA)_5 - (GR)_6 + (SR)_6 + (POR)_6 = 0 - 30 + 0 + 50 = 20$$

We now continue through the remaining periods:

$$(PA)_7 = (PA)_6 - (GR)_7 + (SR)_7 + (POR)_7 = 20 - 20 + 0 + 0 = 0$$

$$(PA)_8 = (PA)_7 - (GR)_8 + (SR)_8 + (POR)_8 = 0 - 10 + 0 + 0 = -10 \text{ (first pass)}$$

MRP plans a second order for 50 units, released in period 6 for receipt in period 8, then recalculates Projected Available:

$$(PA)_8 = (PA)_7 - (GR)_8 + (SR)_8 + (POR)_8 = 0 - 10 + 0 + 50 = 40 \text{ (2nd pass)}$$

The remaining row, "Net Requirements," is shown only when the Projected Available balance drops below the specified Safety Stock. Its value is the difference between the Safety Stock and the first-pass calculation of Projected Available. In other words, it is the quantity needed to bring the Projected Available balance back up to the Safety Stock level. In our example, only periods 6 and 8 show net requirements. Their values reflect the quantity required to bring negative Projected Available balances up to zero.

Once MRP completes planning a "parent" part number, its next step is to "explode" the resulting requirements down to the next level in the bill of material. The "Gross Requirements" of each "child" are equal to the "Planned Order Release" of the "parent" multiplied by the "child's" quantity. In our example, since only 1 B is required for each A, the Gross Requirements for B equal the Planned Order Releases of A. The Gross Requirements for C, on the other hand, will be double these quantities (i.e., 100 in periods 4 and 6 each), since 2 C's are required for each A.

MRP then performs its calculations on the next-level parts, resulting in Planned Order Releases, explodes to the next level, and continues down the bill of material until all levels have been calculated.

One final check is useful to further our understanding of this process, and to ensure that the calculations performed at each level are correct. The "Balance Check" shown at the right of the two completed records show the total Planned Order quantity minus the projected inventory shortfall (i.e., negative on hand). The difference represents how much inventory should remain after satisfying the shortfall, and should therefore match the final Projected Available balance. This is the case in both our tables.

11.5.3 Lot for Lot Versus Fixed Order Quantity

In the above example we illustrated how MRP operates using a fixed order quantity. (See Subchapter 11.6 for a variety of lot sizing rules which arrive at fixed order quantities.) This is only one of the ways in which material can be planned. MRP also has the ability to determine its own lot sizes based upon a lot-for-lot calculation. In this approach MRP processing is identical to that described above, except only the net requirement is planned, as opposed to some predetermined fixed quantity. Figure 11.29 shows the results of lot-for-lot planning as applied to part A.

Notice that no "remnant" material is planned. In other words, only the immediate need is satisfied, with no material remaining at the end of a period to cover demand in the next.

Lot Size = Lot-for-Lot	Part Number "A" - Completed Record									Balance
Lead Time = 2	Periods									Balance
Safety Stock = 0		1	2	3	4	5	6	7	8	Check
Gross Requirements		20	25	25	10		30	20	10	
Scheduled Receipts			50							
Projected On Hand		10	35	10	0	0	-30	-50	-60	
Projected Available	30	10	35	10	0	0	-30/0	-20/0	-10/0	
Net Requirements		0	0	0	0	0	30	20	10	
Planned Order Receipts							30	20	10	60
Planned Order Releases					30	20	10			-60
										0

Figure 11.29 Detail records for lot-to-lot planning in MRP.

Because of this, inventory is minimized, making lot-for-lot production the favored technique in a JIT environment (see Subchapter 11.3).

11.5.4 Measuring System Effectiveness: The ABCD Checklist

In 1977, Oliver Wight [8] created the first ABCD checklist to aid companies in determining how effectively they were using MRPII. The original list consisted of 20 items, grouped into three categories: technical, focusing on design; data accuracy, to determine the reliability of key data; and operational, designed to gauge employees' understanding and use of the system. The checklist has evolved over the years and is now accepted as the industry standard for measuring MRPII. In its present form, the checklist is organized into five basic business functions [8]:

1. Strategic planning processes
2. People/team processes
3. Total quality and continuous improvement processes
4. New product development processes
5. Planning and control processes

Each function is scored individually, resulting in a rating of A, B, C, or D. As an example of how ratings should be interpreted, the following is excerpted from the first and last items listed above:

Strategic Planning Process

Class A: Strategic planning is an ongoing process and carries an intense customer focus. The strategic plan drives decisions and actions. Employees at all levels can articulate the company's mission, its vision for the future, and its overall strategic direction.

Class B: A formal process, performed by line executives and managers at least once per year. Major decisions are tested first against the strategic plan. The mission and/or vision statements are widely shared.

Class C: Done infrequently, but providing some direction to how the business is run.

Class D: Nonexistent, or totally removed from the ongoing operation of the business.

Planning and Control Processes

Class A: Planning and control processes are effectively used company wide, from top to bottom. Their use generates significant improvements in customer service, productivity, inventory and costs.

Class B: These processes are supported by top management and used by middle management to achieve measurable company improvements.

Class C: Planning and control system is operated primarily as a better method for ordering materials; contributing to better inventory management.

Class D: Information provided by the planning and control system is inaccurate and poorly understood by users, providing little help in running the business.

11.6 INVENTORY MANAGEMENT

Inventory management is the business function which deals with the planning and control of inventory [2]. Inventory is commonly thought of as materials, parts, and finished goods, which allow a company to satisfy their customers while enabling manufacturing to produce these products in an efficient manner. Typical classifications of inventory include raw materials, work in process, and finished goods, each of which serves a particular purpose or function, such as:

Covering random or unexpected fluctuations in demand—normally handled with safety stock.

Covering seasonal fluctuations—off-season production to cover demand during high-sales periods.

Providing a "buffer" between factory and distribution centers, i.e., "transportation" inventory. Usually the amount of inventory equals the demand rate multiplied by the transportation time.

Providing a "buffer" between manufacturing operations—cushions variations in production between sequential work centers.

The primary goals of inventory management include:

Minimize inventory (dollars) in order to increase return on investment: ROI = net income ÷ total assets (includes inventory).

Provide high customer service levels—cover off-the-shelf demand for goods, or provide faster delivery than would be possible by producing "from scratch" upon receipt of an order.

Stabilize production rates—allow for stabilization of workforce and equipment utilization.

Generate profits through the sale of finished goods.

Some typical measures of how effectively the inventory management function is performed are:

Total value of inventory—goal is to minimize.

Inventory turns (the rate at which products are sold versus the average level of inventory maintained)—goal is to maximize.

Inventory turns = cost of goods sold ÷ avg. value of inventory on hand.

Example: If 2500 items with a standard cost of $100 were sold in a year, then the cost of goods sold = $250,000. If the average value of inventory on hand = $50,000, then inventory is "turned over" 5 times per year (250,000 ÷ 50,000).

Periods on hand (inventory "coverage" measured in periods)—the goal is to maintain a desired level to cover projected sales.

Example: If $50,000 of inventory is on hand, and 750 units with a standard cost of $25 are projected to be sold each month (cost of sales = 750 × $25/ month = $18,750), then POH = $50,000 ÷ $18,750/month = 2.67 months.

In the subsections which follow we will describe some general concepts and approaches which allow inventory to be managed in such a way as to contribute to the goals and functions presented above. We will also present some techniques to answer the three fundamental questions which must be answered by inventory management [2]:

1. What to order?
2. How much to order?
3. When to order?

11.6.1 General Concepts

ABC Analysis

ABC analysis of inventory, also know as inventory stratification, is the process of categorizing inventory into classes. ABC classifications are determined based on the Pareto principle of the 80/20 rule, which states that 80% of inventory value will be contained in 20% of the parts. The three classes are defined as follows:

A items: The "top" 20% of items, representing 80% of inventory cost.
B items: The "middle" 30% of items, representing 15% of inventory cost.
C items: The "bottom" 50% of items, representing 5% of inventory cost.

In Subsection 11.6.2 we will see how ABC classification is used in maintaining inventory accuracy through a process known as cycle counting. Other uses include determining appropriate types of inventory systems, prioritizing inventory, and controlling ordering policy and procedures.

Inventory Controls

Three general control mechanisms are generally acknowledged in reference to inventory: financial, operational, and physical. Financial and physical will be dealt with here; operational control deals with inventory accuracy and is the subject of the next subsection.

Financial controls deal with inventory accounting, which is the "function of recording and maintaining inventory status information" [2]. Two general inventory accounting approaches are applied: (1) perpetual, in which all inventory transactions are recorded as they occur, providing up to the minute information; and (2) periodic, in which inventory is counted only at specified times. The latter is less expensive, but provides less current information.

General methods of determining inventory value include:

Last-in, first-out (LIFO):   Inventory value is assigned to items as they are consumed, based on the cost of the most recently received unit.

First-in, first-out (FIFO):   Inventory value is assigned to items as they are consumed, based on the cost of the oldest unit on hand.

Moving-average unit cost: Inventory value is recalculated whenever parts are added to inventory. The cost per part is calculated by taking the current average cost multiplied by the quantity on hand prior to adding the new part(s). Then the value of the new part(s) is added to this total and a new average is calculated for the combined quantity.

Standard cost: The "normal" cost of an item, including labor, material, overhead, and processing costs. Standard costs are typically used to determine anticipated costs prior to production for management control purposes. They are later compared to actual costs and revised as necessary to ensure their usefulness for future applications.

Physical controls ensure that inventory records are maintained accurately with regard to such things as quantity, location, and status. Controls are achieved through techniques such as part numbering conventions, lot sizing and replenishment rules, stock room security, and appropriate inventory handling practices. Each of these subjects will be dealt with in further detail in subsequent subsections.

11.6.2 Inventory Accuracy

Many of the problems encountered in a manufacturing company may be attributed to (or at least compounded by) inaccurate inventory records. Parts shortages, poor schedule performance, low productivity, and late deliveries are but a few. In an MRP environment, inaccurate inventory adversely affects the systems' ability to perform netting, resulting in suspect material plans, schedules, and capacity plans. Having less inventory on hand than records show leads to costly expediting, or in the extreme, work stoppage. Having more inventory than is recorded leads to the purchase of unnecessary additional inventory, thus reducing profitability. Accurate recording of inventory allows these problems to be avoided. Likewise, having accurate records of obsolete inventory allows a company to dispose of it, freeing up space for active parts, and possibly generating additional cash. Two general methods of achieving inventory accuracy are cycle counting and periodic physical review.

Cycle counting is the continuous review and verification of inventory record accuracy through routine counting of all items on a recurring basis. Errors encountered are immediately reconciled. Cycle counting provides a high level of record accuracy at all times, eliminating the need for an annual physical inventory, with a minimal loss of production time. Since errors are detected in a timely manner, causes of the errors are much more likely to be found and corrected in addition to simply correcting the inventory records.

Periodic inventories, by contrast, are typically done on an annual basis, requiring the plant and warehouse to shut down for the duration of the review. No correction of the causes of errors is possible, and no permanent improvement in record accuracy results. While the total time spent conducting the inventory may be less than with cycle counting, none of the benefits result.

The frequency of cycle counting is typically controlled by a part's value, often by using its ABC classification. Because A items represent the greatest value (80% of total inventory dollars), they are counted most frequently. And since they represent relatively few part (20% of part count), the effort is generally not too great. In addition to frequency, tolerance in counting errors is also generally much tighter than with lower-valued items. Tolerances allow for some "acceptable" level of error, to avoid spending more time (i.e., cost) trying to reconcile an error than the parts in question are worth. A typical cycle count strategy, based on ABC class codes might be as follows:

ABC Classification	Count Frequency	Tolerance
A	Weekly, monthly	0%
B	Monthly, quarterly	2%
C	Quarterly, semiannual	5%

11.6.3 Lot Sizing Rules

Lot sizing rules answer the question, "how much to order." These rules are as valid for computerized replenishment systems such as MRP or automated reorder point systems as they are for manual approaches (see Subsection 11.6.4). This subsection presents some of the most common lot sizing rules.

Fixed quantity: These techniques are demand-rate based, attempting to determine an average quantity which will satisfy all future demand over a predetermined horizon, typically 1 year. Whenever an order is triggered, the specified fixed quantity is ordered. Fixed quantities almost always produce remnants (inventory carried over from one period to the next).

For example, if annual demand is anticipated to be 24,000 units, then monthly quantities of 2000 units might be desired. However, if the vendor produces in lots of 2500, then this quantity would likely be used instead; perhaps on a less frequent basis, or skipping several months each year.

EOQ: A specialized fixed order technique which attempts to minimize the combined cost of producing (or ordering) and carrying inventory. Subsection 11.3.2 (Figure 11.24) presents a graphical model of EOQ. The equivalent formulas are shown in Figure 11.30.

Lot for lot: This approach orders only the discrete quantity required to cover the period in question. Because the order quantity equals the period's demand, no remnant results. Unlike fixed quantity techniques, lot-for-lot order quantities very from order to order. (Refer to Subsection 11.5.3 for additional discussion.)

Period order quantity: This technique uses the EOQ to establish the number of periods to be covered by a single order. Remnants are only carried over within the periods covered, not between orders. Order quantity varies with each order. An example is shown in Figure 11.31.

Other, less frequently used techniques include the following.

Least total cost: Attempts to minimize the total ordering cost by selecting order quantities where setup costs and carrying costs are most nearly equal.

Least unit cost: Calculates the combined ordering (or setup) and carrying costs of trial lot sizes, then divides by the lot size. Selects the lot size which minimizes the unit cost.

Dynamic programming: mathematical optimization models which select an order size which minimizes the total ordering costs over the planning horizon.

1. To determine quantity in units

$$EOQ = \sqrt{\frac{2US}{IC}}$$

Where

U = Annual usage in units (units / yr)

S = Setup or ordering cost ($)

I = Inventory carrying cost (% / yr)

C = Unit cost ($ / unit)

Example: U = 10,000 units / yr, S = $100, I = 10%, C = $50

$$EOQ = \sqrt{\frac{2 \times 10,000 \times 100}{0.10 \times 50}} = \sqrt{\frac{2,000,000}{5}} = 632 \text{ Units}$$

2. To determine quantity in dollars

$$EOQ = \sqrt{\frac{2AS}{I}}$$

Where

A = Annual usage in dollars ($ / yr)

S = Setup or ordering cost ($)

I = Inventory carrying cost (% / yr)

Example: A = $500,000 / yr, S = $250, I = 18% / yr

$$EOQ = \sqrt{\frac{2 \times 500,000 \times 250}{0.18}} = \sqrt{\frac{250,000,000}{0.18}} = \$37,268$$

Figure 11.30 Equivalent formulas for determining EOQ in number of units and dollars.

11.6.4 Replenishment Systems

Replenishment systems seek to answer the "when to order" question introduced at the beginning of Subchapter 11.6. A useful model for aiding our understanding of replenishment is the *order point* system depicted in Figure 11.32. In this example the demand rate is variable, as indicated by the changing slope of the heavy "sawtooth" line. As inventory is consumed (downward-sloping line), it eventually reaches the order-point level and an order is placed. Assuming an order size based on economic order quantity (EOQ) and instantaneous replenishment, inventory will be replenished (light vertical line) and demand

Example: If Annual Demand = 250, and EOQ = 60, then

No. Orders / Yr = 250 ÷ 60 = 4.2 (Round to 4)

12 Months / Yr ÷ 4 Orders / Yr = 3 Months / Order

Period	1	2	3	4	5	6	7	8	9	10	11	12	Total
Net Req't	25	20	18	18	15	12	15	10	20	27	30	40	250
Planned Order	63			45			45			97			250

Figure 11.31 Example of period order quantity.

will continue to consume it (perhaps at a new rate). Since instantaneous replenishment cannot occur in reality, demand will continue to consume inventory during the order lead time, further reducing it. The goal is to maintain inventory at or above the safety stock level, dipping below only in the event of unanticipated demand (such as the unusually steep demand shown in the second cycle). Upon receipt of an order, inventory is increased by the amount received, returning to a level greater than the order point and the cycle repeats.

The order-point level is calculated as follows [2]:

$$OP = DLT + SS$$

where

OP = order point
DLT = demand during lead time
SS = safety stock

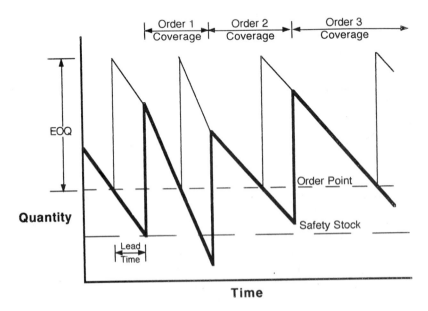

Figure 11.32 A model depicting the material replenishing system, assuming the demand rate is variable for this part.

For example; If demand averages 100 units/week, lead time is 3 weeks, and 2 weeks of safety stock are to be maintained, then

$$OP = 3 \text{ weeks} \times 100 \text{ units/week} = 300 \text{ units}$$
$$DLT = 2 \text{ weeks} \times 100 \text{ units/week} = 200 \text{ units}$$
$$SS = 300 + 200 = 500 \text{ units}$$

Other replenishment techniques include the following:

Periodic review: Inventory levels are reviewed on a periodic basis, and replenished when a predetermined order point level is reached. This review may be based on physical inspection or review of system records. Periodic review is used when recording individual withdrawals is difficult, or for items with limited shelf life.

Visual review: Inventory is reordered based on a visual review of inventory level. This may be as simple of painting a line on the inside of a barrel and reordering only if the line is visible at the time the review is conducted. Visual review is generally applicable only to low-value items; e.g., nuts and bolts.

Two-bin: Inventory is reordered when a bin of parts is emptied. The second bin covers demand during the lead time and remains in use until it empties, being replaced in turn by the first bin. Also used to control low-value items.

Material requirements planning (MRP): Time-phased material plan which determines planned orders by exploding a master production schedule against a bill of material and accounting for inventory already on hand or on order. Orders are planned in advance, providing a proactive approach to inventory replenishment as opposed to the reactive approaches described above. Refer to Subsections 11.2.4 and 11.5 for further information.

11.6.5 Warehousing

Physical Control

Effective physical control of inventory is the final subject of this subchapter. Since the majority of inventory in manufacturing companies is typically maintained in controlled stockrooms or warehouses, we will focus the discussion on those areas. As companies move to reduce inventory through JIT and other initiatives, a greater percentage of inventory will be kept in flow on the factory floor. However, many of the discipline and controls discussed here are equally valid in such a decentralized environment. A brief overview of some fundamental principles follows.

Secure storerooms ensure that only a limited number of responsible, accountable individuals have access to inventory. Stock keepers must be well educated in the importance of inventory accuracy and the practices for maintaining accurate inventory records.

Timely recording of all issues and receipts is critical to maintaining inventory record essential for determining net space requirements for items, to ensure that inventory is available to support customer orders and production requirements, and to avoid carrying excess or obsolete inventory in stock.

Location of all inventory is known. Inventory location is as critical as any other element of record accuracy. Various storage and location methods are presented below.

Stock keeping performance is measured to ensure that adequate control is maintained and an atmosphere of continuous improvement is fostered. As errors are found and recorded, they should be immediately corrected, along with the cause of the error, to avoid recurrence.

Storage Location Methods

Three general methods for assigning storage location for parts and materials are fixed, random, and zoned. The APICS dictionary describes these as follows [1]:

Fixed location storage: A method of storage in which a relatively permanent location is assigned for the storage of each item in a storeroom or warehouse. Although more space is needed to store parts than in a random location storage system, fixed locations become familiar and therefore a locator file may not be needed.

Random location storage: A storage technique in which parts are placed in any space that is empty when they arrive at the storeroom. Although this random method requires the use of a locator file to identify part locations, it often requires less storage space than a fixed location storage method.

Zoned location storage: A combination of fixed and random location storage in which items are assigned to random locations within a predesignated zone.

11.7 REFERENCES

1. *APICS Dictionary*, 7th ed., APICS Educational and Research Foundation, American Production and Inventory Control Society, Inc., Falls Church, VA, 1992.
2. APICS Inventory Management Certification Review Course, American Production and Inventory Control Society, Inc., Falls Church, VA, 1987.
3. APICS Systems And Technologies Certification Review Course, American Production and Inventory Control Society, Inc., Falls Church, VA, 1987.
4. Eliyahu M. Goldratt and Jeff Cox, *The Goal*, North River Press, Great Barrington, MA, 1984.
5. Thomas J. Peters and Robert H. Waterman, *In Search of Excellence*, Harper & Row, New York, 1982.
6. Walter E. Goddard, *Just-In-Time, Surviving by Breaking Tradition*, The Oliver Wight Companies, Essex Junction, VT, 1986.
7. Vollmann, Berry, and Whybark, *Manufacturing Planning and Control Systems*, 2nd ed., Business One Irwin, Homewood, IL, 1988.
8. *The Oliver Wight ABCD Checklist for Operational Excellence*, 4th ed., Oliver Wight Publications, Inc., Essex Junction, VT, 1993.
9. Eliyahu M. Goldratt and Robert E. Fox, *The Race*, North River Press, Croton-on-Hudson, New York, 1986.

12

Control of Quality

Robert L. Lints   *Quality Assurance Systems, St. Louis, Missouri*

12.0   INTRODUCTION TO CONTROL OF QUALITY

Ultimate customer satisfaction with today's complex products has its roots in the quality of the design process. Quality must become an inherent characteristic of the design at its inception. The utilization of a multidisciplined review team, working in a concurrent engineering environment, has been proven to be an effective means of accomplishing this goal, and the preliminary phase of design development is not too early for implementation. This melding of technical and administrative professionals, ranging from operators and inspectors on the production floor, through supporting departments, to the individual participation of executive management, is the key to achieving customer satisfaction in the marketplace. This practice is commonly referred to as *total quality management*, (TQM), and is being implemented in one form or another by most major manufacturing and service industries today, regardless of their product line.

Today, many businesses find that procurement costs for raw materials, detail parts, assemblies, and outside services run in excess of 50% of their sales dollar. In addition, factory down time, schedule slips, rework and replacement costs, and lost time working with suppliers to prevent problem recurrence elevate the need for improved supplier quality to a high-priority item. This points to the need to assure that all outside procurement activity is included under the TQM umbrella and is an integral part of the up-front concurrent engineering activity. The prevention of supplier quality problems which impact just-in-time (JIT) delivery of quality materials, problems which surface during later production runs, or problems which may result in customer warrantee claims and product recall actions must be identified up front to control costs and assure customer satisfaction when products reach the field. Recent reports of automotive recalls attributed to supplier problems bear mute testimony to the importance of establishing a clear understanding of specification requirements with suppliers, and the controls in place to assure they will be met.

Once the integrity and producibility of the design have been validated by the concurrent engineering team and the design has been subjected to development testing, another phase of team quality assurance comes into play. This phase of quality assurance involves the preparation and implementation of a *transition to production* plan. Development and timely completion of this plan will assure a smooth and trouble-free start of production, an area where many have faltered in the past when insufficient attention has been given to past lessons learned. The transition plan is truly a quality assurance plan and a defect prevention plan combined, as it will include a detailed review of past lessons learned, a detailed review of potential risks which may be encountered during the transition, and specific plans to progressively reduce and eliminate those risks. The Department of Defense manual titled *Transition from Development to Production* is available to the public and may be found helpful in establishing a transition plan tailored to a specific application.

With design integrity assured, procurement sources carefully selected and controlled, and an effective transition to production plan in place, we are ready to proceed with production and the steps necessary to assure that a quality product will be produced and delivered to the customer. Here, TQM comes into play again. The customer is defined as "the next person to receive the output of the process." In this context, the customer is not necessarily the *final* customer. In a sequence of operations where there are many individual steps, there will be *many* customers, with each one in a position to judge the quality of the "product" being delivered to him or her from the previous step. An assembler, as an example, expects to receive detail parts for assembly properly formed, free of burrs, and painted the proper color. Applying the principles of TQM to this example includes providing the proper training, work instructions, specifications, tools, work environment, clarification of responsibility, motivation, and recognition. When this applies to both the "supplier" *and* the assembler, we will have a supplier intent on producing a quality product and a "customer" who will provide feedback to the supplier if his or her expectations are not met. This activity will result in correction of the process and ultimate satisfaction of the customer. When carried on throughout the whole process, we have the process control necessary to assure delivery of a quality product satisfying the *ultimate* customer's expectations.

The aim of the quality control program should be prevention of defects by improvement and control of processes in all aspects of the business. Regular monitoring of the process will inevitably result in steady improvement through timely process adjustment and result in a corresponding reduction in cost. The following subchapters will provide greater insight into the techniques which have proven to be effective in reaching this goal.

12.0.1 Additional Reading

Feigenbaum, A. V., "Total Quality Control," *Harvard Business Review*, November–December, 1958.

Juran, J. M. and Gryna, F. M., *Quality Control Handbook*, 4th ed., McGraw-Hill, New York, 1988.

Shuster, "Profound Knowledge: Source, Character, and Application," Proceedings of the 1st International Symposium on Productivity and Quality Improvement, Norcross, GA, Industrial Engineering Management Press, 1992.

"Transition from Development to Production," Department of Defense Handbook on Manufacturing Management, 2nd ed., Defense Systems Managements College, Fort Belvoir, VA, 1984

12.1 QUALITY ENGINEERING AND PLANNING

12.1.0 Introduction to Quality Engineering and Planning

The quality engineering and planning function provides the technical and administrative guidance necessary to plan the overall quality initiatives for the company, to assist in their implementation, and to oversee achievement of the desired objectives and goals. Companywide interface and collaboration with all organizational elements is required for the successful accomplishment of this task. The planning function is a key part of quality engineering and planning responsibility. This includes the necessary indoctrination and training of participants at all levels to establish a basic understanding and acceptance of the principles of Total Quality Control (TQC).

Typical quality engineering and planning functions include the following:

Prepare overall quality plans	Establish test plans and procedures
Prepare inspection plans	Establish classification of characteristics
Establish quality metrics	Establish quality cost estimates
Prepare manuals and instructions	Establish supplier control program
Act on material review board	Establish supplier certification plans
Participate in concurrent engineering	Evaluate and assist in supplier selction
Provide statistical QC support	Conduct internal and external QC audits
Establish sampling plans	Establish metrology designs and controls
Perform process capability sutdies	Assure C/A on customer complaints
Prepare QC input to process specs	Assist in benchmarking quality
Plan preventative action program	Monitor handling/packaging processes
Monitor corrective action (C/A) progress	Report quality status to all levels

12.1.1 Participation in Concurrent Engineering

The successful completion of a project or production run, which will result in the ultimate customer being satisfied with the quality of the product received, may well depend on the initial concurrent engineering participation during the design process. The following are some of the more important areas specifically addressed by quality engineering and planning during this process:

Drawing Review

Quality engineering participation in the preliminary design process provides a timely check for such things as clarity of views and dimensioning, proper call-out of specifications for materials and processes, and inclusion of required specifications for nondestructive inspection processes. Drawing review, and correction at the preliminary stage, minimizes costly drawing revisions after final release.

Review of Lessons Learned

The preparation, maintenance, and use of a file called "lessons learned" is invaluable in the prevention of repetitive discrepancies in new or updated products. The recording of problems encountered, the root cause of the problems, and the corrective action taken to eliminate them can be reviewed by the concurrent engineering team and positive steps taken to avoid the same pitfall in the new design. A new design should not be released without this important review.

Assessing Inspectability and Testability

A major responsibility of quality engineering during the initial design reviews is the evaluation of the inspectability and testability of the product. Special attention must be given to the accessibility of design characteristics for inspection with standard measuring instruments, nondestructive inspection (NDI) probes, or other special fixtures or gages. The accessibility of test points and fixture locating pads must also be considered. If equipment currently available will not be adequate, then consideration must be given to design changes, if practical, or to the design and procurement of required special equipment. Requirements for nondestructive inspection assure that design configuration or structure will not preclude application of probes, or in the case of radiographic inspection, the alignment of the X-ray beam or the placement of film. If special inspection or test equipment will be required, adequate fabrication or procurement time should be provided for equipment to be available when needed. Where production volume or the nature of the product dictates that automated equipment is required, even greater time for design and procurement may need to be included in the manufacturing flow plans.

Classification of Defects

Quality engineering has the additional responsibility to coordinate with the designer and other specialists to establish the importance of each individual characteristic of the design to the quality and operational integrity of the product. This activity is referred to as *classification of defects* and provides the basis for determining the seriousness of any defects, and the attention which must be given in subsequent inspection or testing. The characteristics are normally classified as critical, major, or minor. A *critical* defect is broadly defined as one which is likely to result in unsafe operation for personnel using or observing its use, or a defect which will preclude the product, or those used in conjunction with it, from performing its primary function. A *major* defect is one which, although not as serious as a critical defect, is likely to result in failure of the product or materially reduce its usability. It follows, then, that a minor defect is one that will not materially reduce the usability of the product and will have little effect on its operation. Classifications of defects are used as a basis for establishing inspection plans for either 100% or sampling inspection, and are often placed on blueprints and work instructions as aids to inspection. When placed on blueprints or other documents, they are generally placed adjacent to the blueprint note or characteristic using the letter C or M. On some blueprints, these letters may be enclosed in a symbol. The minor classification is not generally shown on documents, but may be listed in specific inspection or test instructions.

Determining Process Capability

Quality engineering can provide valuable assistance in the selection of equipment which will consistently provide products with the least variation from the specification mean. Manufacturing facilities often have more than one piece of equipment which may be used to produce a product When this situation exists, a process capability study may be conducted using data collected on quality control charts from past operations, or by gathering data on current production runs. Data thus gathered can be statistically analyzed to provide guidance to the manufacturing engineer as to which equipment will consistently produce products within specification limits and minimize the risk of costly scrap and rework.

Process capability as referred to in this section is a measure of the reproducibility of product, where the *process* is the unique combination of materials, methods, tools, human, and machine, and *capability* is the statistical ability of the process based on historical records. In other words, *process capability* is the inherent ability of the process to consistently turn out similar parts independent of external forces. It is expressed as a percent defective or as a statistical distribution and is most often calculated from data gathered using quality control charts. The application of quality control charts is explained further in Subchapter 12.7.

Inherent process capability is the ratio of component tolerance to 6 σ and is expressed by the equation

C.P. = tolerance ÷ 6 σ

The process capability (CPK) is measured in relation to the specification mean and is expressed by the following equation:

CPK = the lesser of:

$$\frac{(USL-mean)}{3\,\sigma} \text{ or } \frac{(mean-LSL)}{3\,\sigma}$$

where   USL = upper specification limit
        LSL = lower specification limit

A CPK of between 0 and 1.0 indicates that some part of the 6 σ limits falls outside of specification limits. A CPK larger than 1.0 means that the 6 σ limits are falling completely within specification limits.

A review of the process capability report shown in Figure 12.1 indicates that the output of machines 2, 4, and 5, with CPKs of 0.68, 0.40, and 0.45, respectively, vary outside specification limits and the process requires adjustment. The CPKs of machine 1 (1.97) and machine 3 (3.07) are well within specification limits and machine 3 has the smallest variation from mean (N). Machine 3 should be first choice for the production run.

Procurement Source Selection

Source selection is an important consideration at the beginning of any project, whether it be for raw material, detail parts, finished assemblies, or services. The concurrent engineering approach is invaluable in that it brings technical as well as administrative expertise together to evaluate all aspects of the procurement i.e. supplier design capability, manufacturing capability, quality assurance record, cost and schedule performance, as well as management commitment to implementation of TQM principles.

Quality engineering, working closely with procurement personnel, maintains current records of past supplier performance and has available, through industry-wide evaluation programs, quality ratings on many suppliers other than those whom a particular company may have dealt with in the past. Most ratings available through these industry sources have been calculated using formulas based on the supplier's recent performance in the areas of quality, schedule, and cost.

In the case of large procurements, or of critical items, it is wise to supplement available ratings with source surveys and concurrent engineering conferences to assure supplier understanding of specifications and quality requirements. These conferences take on added significance when just-in-time (JIT) deliveries are planned and the receipt of first-time quality is necessary for smooth operation within your facility. They are also an

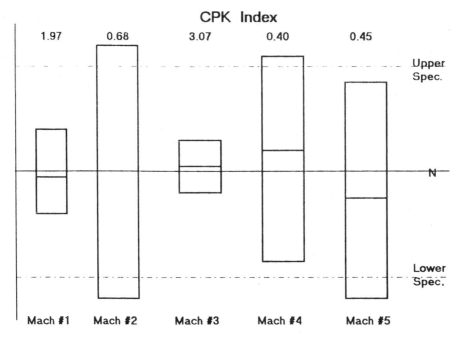

Figure 12.1 Process capability report.

excellent time to obtain supplier input relative to any cost savings that may be obtained though modification of specification requirements.

12.1.2 Process Flow Diagrams

Figure 12.2 illustrates a typical top-level manufacturing flow diagram for a simple fabrication and assembly process. Development of similar flow diagrams for the entire manufacturing process as a part of early concurrent engineering activity is an important planning tool. Lower-level, or *sub flow* diagrams should be prepared for each major element of the process in conjunction with the development of specific work instructions for the subprocesses. This planning effort provides insight for the location and type of equipment required to perform and control the process, highlights the need for special inspection and test equipment, including the need for automated equipment, and assures that proper attention has been directed to the location and type of inspection to be conducted.

Completion of the flow diagrams also aids in highlighting those operations where special training and supplementary work instructions may be required. Manufacturing operators and inspectors must be trained and understand the process they are expected to control, and their authority to keep it in control. They must understand how their piece of the process fits into the total picture and what actions they may take to adjust or correct the process. Posting the flow diagrams in the work area enhances operator and inspector understanding and the importance of their contribution. Inspection points, or stations, should always be highlighted on the flow diagrams as shown in Figure 12.2. A more in-depth discussion of inspection methods for control of processes will be found in Subchapter 12.2.

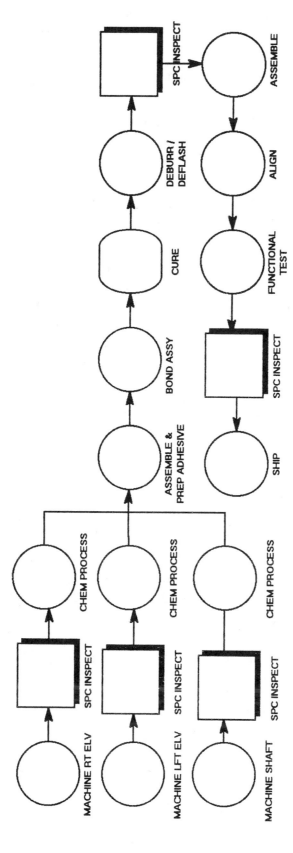

Figure 12.2 Typical manufacturing flow chart with inspection points highlighted.

12.1.3 Quality Planning

The quality planning function is responsible for the preparation and implementation of all quality operating plans, from the top-level company plan for assuring contractual compliance and customer satisfaction to the lowest-level instruction sheets used in data gathering and other record keeping. The following elements of quality planning are among the most important.

The Quality Plan

Preparation of the company quality plan should begin with a review of contractual requirements for quality and a review of company policies, procedures, and goals which have been established by management to assure customer satisfaction with the products they receive. The quality plan will describe overall policies, procedures, and interfaces required to meet company objectives, including company plans for implementing the principles of total quality management (TQM). Implementing direction is generally contained in a series of lower-level manuals, directives, and specific work instructions The quality plan is often required to be submitted as a deliverable item on some government contracts.

The Inspection Plan

The inspection plan is generally a subset, or lower-level plan prepared by quality planning in conjunction with the company quality plan. The inspection plan will include a detailed description of the overall approach to product inspection, including a copy of the manufacturing flow plan which shows the location of all inspection and test points established during the concurrent engineering review of the manufacturing process. Details of the type and level of inspection (100%, sampling, etc.) which will be conducted at each point, the requirement for statistical quality control charts or other data recording, and the preparation of inspection records of acceptance or rejections will be included. Since data gathering is an important part of the inspection process, specific instructions should be included as to *what* data are to be recorded, *when* are they to be recorded, *who* is to record them, *where* and in what form they are to be recorded, *why* they are being recorded, and *when* they may be disposed of. Data must not be collected for the sake of data! There must be a valid reason for all data collection. Configuration identification of product as manufactured, as inspected in process, as tested, as inspected at final inspection, and as shipped is a good example of data that must be carefully recorded and maintained in accordance with the inspection plan.

The Process Control Plan

Process control planning should begin with a review of the manufacturing flow plan to establish the specific processes planned for use and an evaluation of historical data on the acceptance rate of each process. If data available on existing processes indicate a need for improvement, if historical data are unavailable or inconclusive, or if a new process is to be used, consideration should be given to conducting a process capability study as a first step in establishing process control. Processes to be studied should be prioritized based on their sequence in the flow plan. Manual or automated quality control charts can be used to gather the necessary data on initial runs. Several excellent software programs are currently available on the market for use with machining stations, which will gather data and provide automated computation of the machine capability in relation to part tolerance limits. The procedure for establishing these charts, their use, and the evaluation of data should be an integral part of the process control plan. Additional information on process capability studies may be found in Subchapter 12.1, and further information on the use of process control charts may be found in Subchapter 12.7.

Handling, Packaging, and Shipping

The manner in which products are handled, packaged, and shipped is often the cause for many avoidable scrap and rework actions within the production facility, and results in many customer complaints and loss of goodwill among customers who receive the products—a direct reflection on the quality of the product as seen by the customer. Quality planning, working in conjunction with production and manufacturing engineers responsible for the design and development of handling, packing, and shipping containers and procedures, can provide data on lessons learned from prior problems with similar products. In so doing, they will aid in pinpointing the need for design improvements, better handling and packing instructions, and specific training needs of personnel directly involved in these operations. In addition to the attention directed to handling and packaging concerns, quality planning receives input on all quality customer complaints related to improper count of quantities received versus quantities ordered, and of discrepant product received. These complaints are entered into the corrective action system, investigated, and action initiated to preclude recurrence. This action often involves the preparation of improved instructions and check sheets for packaging, shipping, and inspection personnel, and the addition of improved equipment, for example, digital scales to aid in counting. Quality planning also has the responsibility of assuring that proper correspondence is forwarded to the customer in response to a complaint, as an aid to improved customer relations.

12.1.4 Supplier Quality Control

Today, many industries are committing over 50% of their sales dollars to the procurement of outside raw materials, detail parts, assemblies, and special services. This, combined with the advent of JIT procurement practices, makes the assurance of supplier quality of paramount importance.

In large organizations, the quality department often has a group charged with the specific responsibility for the preparation and implementation of a supplier quality plan. In other organizations, this responsibility may be assigned to quality engineering and planning. Regardless of where the plan originates, it is important that it be prepared and put in place as a directive which has been coordinated with all major departments which may be involved with, or affected by, procurement practices and the receipt of products obtained through them.

Supplier quality control must begin with the very first supplier contact made by the purchasing department buyer. A common understanding must be established of the quality requirements included in the statement of work, the specifications to be met, authorized materials and processing sources which may be used, supplier understanding and application of statistical process control in his facility, type of inspections to be conducted, i.e., supplier 100% or authorized sampling, inspection by the procuring agency at the supplier's plant, or on-site witnessing of final inspections and tests, and requirements for data to be supplied at the time of shipment. In certain instances, where proprietary processes may be involved, it may also be necessary to have contractual coverage for site visits. On larger or critical procurements, it is wise to hold concurrent engineering meetings with the supplier to assure common understanding of requirements and to determine the need for any special training or other assistance the supplier may need in fulfilling his requirements. Establishing a supplier team arrangement will go a long way toward smooth procurement and receipt of quality products.

Rating supplier performance not only provides incentive to the supplier to have a better that average rating, but it provides the buyer with an evaluation tool in monitoring the performance of suppliers and an aid in placing future procurements. Today, most supplier performance measuring systems compute a composite score based on quality, schedule, and contract administration data. Quality is emphasized in this rating system, and quality performance usually accounts for 50% of the total composite rating. Typically, the quality portion of the rating is based on factors such as quality costs, responsiveness to corrective action requested on quality problems, number of data delinquencies, and number and frequency of repetitive rejections. Additional factors utilized are schedule performance (late or early), number of rejected parts and/or shipments, number of shipments held up upon receipt due to missing documentation, and number of invoice reconciliation problems. These latter factors comprise the other 50% of the composite rating. The resultant ratings are quantified in dollars and reported to those with a "need to know" as the "cost of doing business" with that specific supplier. In many organizations, ratings are maintained in on-line computer systems which provide the buyers with real-time performance data on their suppliers. Ratings are usually published quarterly and provided to individual suppliers for review and action as necessary. Individual supplier conferences are called with suppliers who need to improve their ratings in an effort to improve the supplier base. Special recognition awards are often presented to those suppliers with outstanding ratings, and they are given special consideration in the placement of future orders.

12.1.5 Control of Nonconforming Material

Nonconforming material generated within the production facility, generated and identified at a supplier's facility, or detected during incoming inspection and test must be conspicuously identified and segregated from good production material until a disposition has been made by authorized personnel. The use of red identification tags securely fastened to the material or special banding identifying the material as nonconforming, along with appropriate identifying paperwork, are common ways of accomplishing identification. The next step is to remove the material from the normal production flow and place it in a controlled area with access restricted to authorized personnel. If the material requires special environmental storage, for example, refrigeration, special accommodations need to be made within the freezers to provide identifiable restricted storage. Similar accommodations need to be made for fluids and gasses normally stored in large external tanks or drums. Material must be held in these restricted areas until its usability has been determined by authorized personnel.

Authority for disposition, or determining acceptability for use, is established with certain authorized inspection personnel for minor nonconformances (see Subchapter 12.1) Procedural coverage, limits for approval, and specific instructions for this action must be included in the quality manual or other authorized operating procedures. Similar control procedures for the complete material review board must also become a part of company procedures. Authority for the disposition of major defects is restricted to an established material review board (MRB) consisting of a design engineer and a quality engineer.

Government contracts may require that a customer or government representative also be a member of this board. It is the responsibility of the MRB to review the noted defects, evaluate the effect they might have on the ability of the product to satisfactorily fulfill the purpose for which it was designed, and mutually agree upon its acceptability "as is," prescribe rework which will make it acceptable for use, or determine that it should be scrapped. Their disposition is recorded on the rejection documentation and becomes a historical record retained as permanent record of that product. Following disposition and the completion of the required paperwork, the rejection tags are removed from the material and it is returned to the normal production flow and treated as normal acceptable material.

12.1.6 Cost Account Planning and Budgets

The estimation of costs for the performance of required tasks is required for bid preparation on all proposed work. The responsibility for the preparation of these estimates and their coordination with the project business office or others charged with preparation of the total estimate is usually assigned to quality engineering and planning. When historical-cost performance data are available on similar projects that had a comparable set of tasks, it is possible to factor these estimates up or down to arrive at the required estimate for the new work. In estimating quality tasks, including inspection and test effort, such estimates are often based on a percentage of manufacturing effort for the inspection-related tasks, and a "level-of-effort" for other quality tasks. While this approach is sometimes necessary in the interest of time, it does not provide the basis for preparation of specific work packages and cost accounts to be used later in managing and evaluating cost/schedule/budget performance relative to planned tasks. For maximum control of costs, a bottom-up estimate should be prepared starting with (1) the definition of the work to be performed and (2) the identification of specific work packages. Individual work packages should be estimated at the level where the task is to be performed, that is, the level or classification of personnel who will be necessary to complete the task, the time period required for completion of the task, and the place in the production flow where the task will be performed, i.e. the planned start and completion dates. These work packages will provide a clear picture of the job at hand and may later be integrated into cost accounts for management evaluation and budget control.

12.1.7 Quality Audits

The primary purpose of quality audits is to determine individual or organizational compliance with an established plan, procedure, process, or specification. The audits should themselves follow an audit plan, and should be conducted by professionals who are trained auditors familiar with the plan or process being audited. With few exceptions, supervisory personnel in the area being audited should be given advance notice of the impending audit in order to gain their full cooperation. Quality audits generally fall into three classifications: procedural, hardware, and system audits. Each will be discussed in the following paragraphs.

Procedural Audits

Procedural audits are reviews to determine compliance with published manuals, bulletins, work instructions, process specifications, and other contractual documents which provide regulation and direction for the manner in which work is to be performed.

They may originate with the customer, the local or federal government, or internally as a matter of company policy or directives. Audits of this nature are most often conducted following a unique set of check sheets prepared in advance of the audit, which contain the elements of the procedure to be checked. These check sheets are later used to summarize audit findings.

The auditor, or audit team, should begin all audits by meeting with supervisory and management personnel responsible for the area to be audited to brief them on the purpose and scope of the audit. This meeting should be used to establish the audit ground rules and, in the case of proprietary processes, to obtain specific authorization for the audit. At the conclusion of the audit, an exit briefing should be held with the same personnel to review findings and establish definitive dates when any corrective actions required will be implemented. Depending on the nature or seriousness of certain findings, it may be advisable to plan a follow-up audit to verify implementation of required actions.

Hardware Audits

Hardware audits are designed to closely examine the product being produced and are most often conducted on the final product by an audit team. This audit team may be comprised of a design engineer, a manufacturing engineer, and a quality engineer. This audit is a more complete audit than a procedural audit in that it involves a detailed review of the production process, including handling of the product during the process flow, and may involve disassembly of a selected item for internal inspection if this can be accomplished by the producer without damaging the product. The same ground rules for entrance and exit briefings stated above for procedural audits apply to hardware audits.

Systems Audits

Systems audits are broad in scope and are conducted infrequently due to the time and cost associated with planning, coordinating and conducting an audit of this scope. The more narrowly focused procedural and hardware audits have been found to be a more cost effective way of monitoring performance. However, over projects that last 5 or 6 years, another type of audit, or evaluation, has found favor with many organizations as a means of self-evaluation of their *quality posture* and comparison with others in the industry. This is the activity associated with the Malcolm Baldrige National Quality Award. This award was established by Congress in 1987 in recognition of the need for improved competitiveness and improved quality across the broad range of industries in the United States. Since that time, there has been a growing interest in the competition for the National Quality Award, and a growing group of organizations utilizing the Baldrige criteria for self-evaluation and improvement refardless of their desire to enter the competition. The appeal of the criteria is their relation to "total quality" concepts. In-depth evaluations are required in seven major categories: leadership, information and analysis, strategic quality planning, human resource development and management, management of process quality, quality and operational results, and, last but not least, customer focus and satisfaction. Ground rules, information, and forms for use with the Baldrige criteria are available from the American Society for Quality Control (ASQC). Local and national Baldrige certified examiners are used in the evaluation of organizations that choose to enter the competition for the award, but any organization will find the criteria valuable for use by internal teams as a means of self improvement.

12.1.8 Continuous Improvement Programs

The aim of the *continuous improvement plan* should be the prevention of defects through improvement and control of processes. Once undertaken, regular monitoring of the process will inevitably result in steady improvement as a result of process adjustments, and with this improvement will come corresponding reductions in the cost of scrap and rework. It follows that since quality engineering and planning play key roles in defect prevention and data analysis, they will be instrumental in formulating and guiding the company-wide effort on continuous improvement plans. The principle elements of an effective continuous improvement plan are discussed below.

Data Analysis

Plans for continuous improvement must begin with a careful analysis of historical quality records to pinpoint those processes with the greatest variation. A study of area and individual process control charts and investigations already underway will aid in selecting and prioritizing those processes with the greatest potential for gain. Special "continuous improvement charts" should be made for the processes selected, and these charts should be placed in the area that will give high visibility to all achievements.

Plan Implementation

The key element in any continuous improvement plan is the training and attention to detail of the participants. Everyone involved must have a sense of serving the customer and be ready, willing, and able to "buy in." They must understand the goals to be reached, participate in their formulation, and take part in the setting of interim targets with which to measure progress. They must understand the "tools of the trade"—that is, the use of the statistical quality control charts which form an important part of the data gathering on the process, and they must understand how to interpret the charts relative to the need for process adjustments (see Subchapter 12.7). It has been found that reaching this point involves both formal training and on-the-job training to give the participants a thorough understanding of this phase of process improvement and their responsibility and authority in *making it happen*. It has also been found that indoctrination in the accomplishment of continuous process improvement and the actual achievement of desired results is often a slow process in itself, requiring regular coaching to preclude a slackening of effort when desired results are elusive.

Feedback Action and Recognition

Good communication is important in most endeavors, but it is of vital importance in the conduct of a continuous improvement program. Two-way communication is the mode of operation and must flow between all levels of management, customers, suppliers, and the operators on the production floor. Timely feedback to the process source, whether good or bad, is important in the evaluation of the effectiveness of past actions and the formulation of further process adjustments. Of equal importance is the coaching and recognition of accomplishments of both group and individual achievers. Attainment of targets on or ahead of schedule, a significant breakthrough, etc., should be worthy of special recognition via posters, pictures in company papers, special award ceremonies, and, as a minimum, a handshake from top management. Selection of those to be recognized has special meaning when the recommendations originate with peer groups or co-workers and awards are presented by management in the work area. Recognition of accomplishment keeps the wind in the sails of the continuous improvement program.

12.1.9 Additional Reading

Assistant Secretary of Defense, *Transition from Development to Production*, DoD 4245.7M, Department of Defense, 1985.

Crosby, P. B., *Quality Is Free*, McGraw-Hill, New York, 1979.

Crosby, P. B., *Quality Without Tears*, McGraw-Hill, New York, 1984.

Crosby, P. B., *Running Things*, McGraw-Hill, New York, 1986.

Deming, E. R., *Out of Crisis*, Management Center, Cambridge, MA, 1986.

Hall, R. W., *Attaining Manufacturing Excellence*, Dow-Jones-Irwin, 1987.

Juran, J. M., and Gryna, F. M., *Quality Planning and Analysis*, 3rd ed., McGraw-Hill, New York, 1993.

Malcolm Baldrige National Quality Award Criteria, U.S. Department of Commerce and National Institute of Standards and Technology, 1993.

Mohr-Jackson, "Quality Starter vs. Quality Advanced Organizations," *Quality Management Journal,* 1994.

Shuster, *Teaming for Quality Improvement*, Prentice-Hall, Englewood Cliffs, NJ, 1990.

12.2 INSPECTION

12.2.0 Introduction to Inspection

Through the 1930s, "inspection" was considered to be an operation that occurred at the end of the line to visually, and occasionally dimensionally, check work performed by the production operators. Since these operators worked for the production foreman, it was considered his responsibility to make sure their work was properly completed, or to have it *fixed* before it left his department. This task was often time consuming for the foreman to accomplish, so he added some "inspectors" to his staff to aid in the *sorting*. This idea of inspection continued in many companies until the late 1940s or early 1950s when the need for better control of in-process losses due to scrap and rework began to be recognized. Initially, some organizations began to add in-process inspection stations and assigned an ex-production "salvage foreman" to review findings at these stations to cause needed corrections. This action was shortly followed by the addition of trained quality engineers to replace the salvage foreman, and the establishment of an inspection department under the control of an inspection manager. This latter action was a natural outgrowth of the recognition that until that time, "The fox had been guarding the chicken house."

In today's environment, with many companies adopting some tailored version of the total quality control (TQC), or total quality management (TQM) concepts, we see a much-improved approach to assignment of individual responsibility for customer satisfaction and the modification of the production flow to include and identify specific inspection points to aid in total process control. These points are addressed in the following paragraphs.

12.2.1 Inspection at Source

The decision to inspect or test purchased parts or assemblies at the supplier's facility may be based on several concerns. Among the more prevalent are the following.

1. Internal structure, wiring, components, etc., may be covered or sealed within an assembly, making inspection at a later point difficult if not impossible.
2. Specialized inspection or test equipment used by the supplier may be costly to duplicate. This is particularly true where environmental acceptance testing is required.
3. Size or sensitivity of the item being purchased may cause added risk if the item has to be returned to the supplier for any reason.
4. Transportation time to and from the supplier's facility might affect schedules adversely.

Inspection manuals or instructions should clearly prescribe the manner in which source inspection is to be conducted, and source inspection personnel should be thoroughly trained. Since source inspection may be performed by personnel resident at the supplier's facility or by transient inspectors, it is most important that ground rules for supplier interface and the completion of source inspection reports be well established. In all cases, access to the supplier's facilities for the purpose of source inspections should be a part of the buyer's initial negotiations and become a part of the resulting purchase order agreement.

12.2.2 Receiving Inspection

Incoming inspection is the norm in many industries, and is conducted upon receipt of everything from raw materials to finished subassemblies and assemblies. The extent of inspection upon receipt is usually determined by the quality history of the supplier, the complexity of the item, and the quantity or lot size received in the individual shipment. Inspection and/or acceptance testing may be conducted on 100% of the parts received, or on a random sample, and may include visual, dimensional, mechanical, chemical, electrical and, in certain situations, environmental testing.

The receiving inspector should be provided with an inspection data package consisting of a copy of the purchase order, the blueprint or specification, copies of all related company specifications and procedures, and a history data card showing past receiving inspection records on the particular part and any "Alert" notices received. In some larger companies, the inspector is also provided with a kit of required gages or instruments needed for the inspection, to preclude the necessity of making one or more trips to the gage crib for the gages. As in other inspection operations, it is important that they be trained in the type of inspection to be conducted, and in the completion of all required documentation.

12.2.3 Setup Inspection

Setup inspection is a preventative action-type inspection designed to give assurance that machine surfaces are clean, holding fixtures are in good condition and properly placed, machine settings have been set or adjusted to predetermined positions, material has been properly loaded in the fixtures or will be fed into the machine at proper speeds and feeds, and any lubrication or coolant flow has been properly set. Depending on the particular assignment of responsibility in an organization, this inspection may be conducted by the line foreman, a lead man, or a floor inspector. Regardless of who performs this inspection, it is wise to have a setup check sheet prepared for each unique operation, which is used as a guide in completing the inspection and a record that it was completed. When conducted in conjunction with a "first article" or "first piece" inspection, which is usually completed by an inspector, setup inspections prove their value in precluding repetitive rework or scrap due to inadequacies or oversights in the setup procedure and establish a solid base for the process that follows.

12.2.4 Operator Self-Check

Today's production operators are being asked for their input on how to improve the quality of the products they produce and how to better satisfy their customer. And management is listening to their advice; they are rewarding the operators for their participation, giving them greater responsibility for the production of a quality part, and giving them the authority to make it happen. Along with the responsibility and authority goes additional training to give them the "tools of the trade" and a greater overall perspective of the big picture. In many instances, the operators complete training programs which result in their becoming "certified operators," who have authority to *inspect* their own work and stamp acceptance records with their "certified operator's stamp." Application of this approach is most common for in-process-type inspections and process control, with final inspection prior to shipment usually being conducted by members of the inspection department.

12.2.5 Roving Inspection

Roving Inspection, otherwise referred to as patrol inspection, audit inspection, or floor inspection, is most often used where production operations are spread out over a large floor area, or where the product is large and not readily moved to a separate inspection area. It is usually conducted by experienced inspectors who have been specially trained and are familiar with the shop layout, the product, and the product flow. These inspectors perform preplanned in-process inspections and maintain surveillance over gage, equipment, and instrument calibrations, environmental controls, and stores areas. Their records of acceptance and rejections in the area become a part of the total inspection database. Roving inspection is sometimes used to check the results of operator self-check operations on a random basis.

12.2.6 Process Inspection

Process inspection is a specialized surveillance of an identifiable element of a total production process to assure that it is being operated in accordance with specifications and variations from norm are being held to a minimum. The element might be a flow soldering operation, a chemical processing operation, heat treating, plating, forming, braiding, bonding, laser cutting of material patterns, and so forth. In cases where these processes are located in the same area, as in a chemical processing area, or an environmentally controlled area, a single certified operator or inspector may be assigned the task of assuring that the operation is being conducted in accordance with procedures and is maintaining control. The latter is generally monitored through the use of process quality control charts which are maintained by personnel in the area, and in some instances, also posted by the floor inspector to add an independent control point to the chart. Where individual process elements are located at various parts of the operation, the use of certified operators assigned to the particular area or a roving inspector best meets the need.

12.2.7 Bench Inspection

Certain inspections are best conducted at a specific inspection station, where the product to be inspected can be spread out over a large surface or passed over or through an inspection fixture or gage. This is also true where a product is passed down a conveyer belt and either manually or automatically inspected on a section of the conveyer or bench. At times, as with optical inspection of bearings for surface finish blemishes, it is necessary to pass the product over a special light table to readily detect defects while maintaining high productivity. Cases such as these are best suited to bench inspection.

12.2.8 Nondestructive Inspection

The inspection of materials, and parts fabricated by special processes such as forging, casting, molding, and welding, requires the use of a variety of nondestructive inspection techniques to detect and evaluate surface and subsurface defects. Most common among these techniques are radiography, ultrasonics, eddy current, magnetic particle, and several types of penetrant inspection. Description of the use and techniques utilized in these inspections is beyond the scope of this chapter, but it is important to note that the application of all of the techniques mentioned requires that personnel who have been specially trained and are certified in the use of each special technique must be available to plan, establish techniques, and conduct these inspections. In determining inspection personnel needs, it is important to keep this requirement in mind. It is also necessary to consider the availability of the special equipment required to conduct these inspections. If equipment and certified personnel are not available, there are commercially available companies that specialize in nondestructive inspections which may be called upon to meet this need.

12.2.9 Automated Inspection

Increasing use is being made of automated inspection to reduce hands-on labor costs, reduce cycle time, assure uniform inspection results (eliminate personal judgment in evaluation), and reduce monotony, stress, or eye strain. Automated inspection should be considered when you are faced with inspection of repetitive characteristics on a large volume of parts. Automated inspection and testing is also of particular value in the process control of electronic equipment, where designs accommodate ready access to test points. However, careful consideration must be given to the advantages to be realized before investing in special-purpose automated equipment. Some flexibility may be realized through the application of the latest developments in microcomputers, artificial intelligence (AI), computer-aided manufacturing and inspection setups, as well as through the use of robotics which may be fitted with universal probes.

12.2.10 Software Inspection

The identification, proofing, and control of computer software in manufacturing, inspection, and test is of paramount importance, since the operational performance will be no better than the software which controls it. Software code must be carefully checked and verified through proofing trials. Configuration must be maintained from the outset and all design changes must be reflected in appropriate configuration updates. Software libraries must be maintained in secure areas, and only authorized personnel should be permitted to remove software from the library for installation in production or inspection equipment.

12.2.11 Sampling Inspection

Experience has proven that 100% inspection, or even 200% or 300% inspection, will not yield risk-free assurance that defects present within a large lot of material will be detected. The principal reason for this is that monotony and physical fatigue set in and affect the effectiveness of personnel performing the inspection. The use of automated inspection and computer-assisted inspection equipment has helped materially to reduce this problem in instances where cost trade studies have justified the procurement. In cases where this is not true, and in cases where inspection or test may be destructive, statistical sampling provides a more economical approach, and may even prove to be more accurate than 100% inspection.

There are several approaches that may be used in sampling, and many variations to each. There are *single sampling* plans where the decision to accept or reject a lot of material is based on the evidence obtained from one sample, there are *double sampling* plans where it is possible to put off the decision until a second sample has been evaluated, and there are *multiple* or *sequential* sampling plans which defer the decision to accept or reject until several samples have been taken from the lot. There are plans for lot-by-lot inspection by attributes, and variables. When an attribute plan is used, the decision to accept or reject the lot is based on a "go" or "no go" basis—that is, whether the number of defective parts in the sample exceeds the acceptance number for the plan. With variable plans, acceptance is based on the acceptability of individual characteristics on the sample parts, or on selected specific characteristics. This type of inspection is more expensive than attribute inspection, but in certain instances it is desirable to gain more information on a particular characteristic for guidance in product improvement, and variable-type plans meet this need. There are also plans for continuous sampling for use with large production runs.

The majority of sampling plans in use today are based on desired achievement of a predetermined acceptable quality level (AQL). The AQL is the maximum percent defective that, for the purpose of sampling inspection, can be considered satisfactory as a process average. It is the designated value of percent defective that the consumer has indicated will be accepted most of the time by his incoming inspection plans. In other words, an AQL plan favors the producer as it gives assurance of probable acceptance by the consumer. The AQL value may be specified in contracts, or may be established internally by quality engineering and planning for guidance in the application of sampling plans.

Another quality value, referred to as the average outgoing quality limit (AOQL), covers plans which may be utilized in conjunction with sampling when rejected lots of product can be 100% inspected and have rejects replaced with good product. AOQL plans assume that the average quality level of multiple lots will not exceed the AOQL because all rejected product detected during sampling and the 100% screen have been replaced with good product.

Sampling plans prepared by Harold Dodge and Harry Romig utilize still another classification index, the lot tolerance percent defective (LTPD). The LTPD represents an allowable percentage defective which favors the consumer due to a decease in the risk of accepting a lot at or below the lower quality limits.

Sampling tables are based on operating characteristic curves. The tables specify the size of the sample to be randomly selected based on the lot size, the inspection level desired, and the selected AQL. When these parameters have been selected, reference to the table will establish the acceptance and rejections numbers governing the acceptability of the sample, and thus the lot, for various sample sizes, lot sizes, and given AQLs.

Due to the number of sampling plans available, it is not possible to adequately address selection of plans in this handbook. However, many texts have been published on this subject which include descriptions of the various plans available. A partial list of texts and military specifications for sampling is included at the end of this subchapter under Additional Reading.

12.2.12 Final Acceptance Inspection

This inspection is the last visual, dimensional, and operational evaluation of the product before it is delivered to the purchaser or ultimate customer. At final, the routing traveler or other documentation should be reviewed to assure that all previous operations have been completed and accepted by authorized personnel. The inspection is then completed in accordance with final inspection check sheets and instructions prepared by inspection supervision or quality planning. The inspection kit and data package used by the inspector should include the top-level drawing and any lower-level drawings necessary for the final inspection, as well as copies of all related specifications and process specifications. It should also include any inspection aids and necessary charts to be used in the inspection. If a log book, instruction book, manual, or other documentation is to be delivered with the product, these items should be available for acceptance at this time.

Final inspection also serves as a final process control point, since it provides a positive evaluation of the effectiveness of process controls built into the production process flow. Feedback to preceding control points provides a closed-loop corrective action system for continuous process improvement.

12.2.13 Shipping Inspection

There are many horror stories told by customers who waited patiently for the product needed to keep the line moving, only to open the carton upon receipt and find the product inside damaged beyond use. The receipt of a damaged product has never been known to result in customer satisfaction with the quality of the product purchased. Rather, the memory of the damaged product lingers long after it has been replaced and due apologies sent. The message is clear: Careful attention must be given to the design of packages or containers to be used for shipping, and these designs must be submitted to testing in all environments which the package may reasonably be expected to encounter in service. In addition, packing and handling instructions must be written, and personnel must be trained in proper packing procedures to assure the product is not damaged during the packing operation or in transit. Specific tests plans must be prepared and both the product and the empty container inspected for conformance to design requirements before the testing is conducted. Inspection of both items is again necessary following the tests to assure that the container adequately protects the product and is capable of reuse if it has been designed as a reusable container.

12.2.14 Data and Records

One of the most important functions of the inspection department is the generation of acceptance and rejection records, which become the basis for future action throughout the company. Standard forms, usually in multiple copies, are used for this purpose, and inspection personnel are trained to describe clearly any nonconformances found in order to facilitate further disposition of the material. Inspection is also responsible for the collection and review of all production work orders to assure that all planned work has been properly completed and accepted prior to final inspection. These documents are combined with all other certifications, log records, and check sheets and forwarded to a central data center. A more detailed discussion of data and records collection and management will be found in Subchapter 12.6.

12.2.15   Training and Certification

With the increased use of computerized inspection equipment capable of extreme precision in multiple planes and computer-assisted gaging, specific training of inspection personnel becomes a necessity. The more complicated coordinate measuring equipment, and special-purpose electronic harness and board inspection equipment, often requires training at the equipment manufacturer's facility, or special in-house training programs. Once trained, it is important that personnel regularly use the equipment on which they were trained, and have periodic refresher training to practice techniques that are not utilized frequently. Similar training needs exist for inspection personnel responsible for environmental control of fluids, atmospheres, furnace and refrigeration equipment, and processing equipment. In the case of certain nondestructive inspection solutions, regular inspections must be conducted for contamination, and tank additions must be properly controlled. Personnel who are required to establish operating techniques, perform setups, and operate radiographic or ultrasonic inspection equipment, and other similar equipment, require special training for certification and must have qualifying experience for many applications.

Personnel handling and inspecting electronic components must receive specific training to avoid damage to the hardware as a result of static electricity. Proper handling procedures, proper clothing, and special workstations which include provision for personal grounding are required, and personnel must be trained and understand the importance of their use.

Personnel working in areas which must be maintained at specific cleanliness levels, as in space systems and electronics clean rooms, must also be trained and provided special clothing for work in these controlled areas. Training, monitoring, and control of these special areas is critical to the production of reliable, quality hardware.

12.2.16   Additional Reading

Bowker, A. H., and Goode, H. P., *Sampling Inspection by Variables*, McGraw-Hill, New York, 1952.

Dodge, H. F., "Sampling Plans for Continuous Production," Industrial Quality Control Magazine, American Society for Quality Control, November, 1947.

Dodge, H. F., and Romig, H. G., *Sampling Inspection Tables—Single and Double Sampling*, John Wiley, New York, 1944.

Foster, R. A., (1975) *Introduction to Software Quality Assurance*, 3rd ed., R. A. Foster, Library of Congress Card 75-28630, 1975.

Juran, J. M., *Quality Control Handbook*, McGraw-Hill, New York, 1951.

Shewhart, W. A., *Economic Control of Quality of Manufactured Product*, D. Van Nostrand Co., Princeton, NJ, 1931.

12.3 TESTING

12.3.0 Introduction to Testing

Testing is that part of the inspection acceptance task which determines the functional acceptability of the product. It may be accomplished onsite at a supplier's facility if the product is large and difficult to move or if test equipment is unique, expensive, or otherwise unavailable. Product testing covers the broad spectrum of items from minute electronic components to major space-age systems, from the smallest mechanical fasteners to major structures, and from the simplest fluids and fabrics to raw materials rolled, cast, forged, extruded, and formed, both before and after special processing. Today, most testing is performed in house by the purchaser or by a special laboratory, contractor, or agency under contract to the purchaser. The latter is particularly true when specialized equipment and expertise are required, as in environmental and vibration testing and product safety testing. There are a number of companies today that specialize in this type of testing. Whatever the scope of testing, a specific test plan should be prepared and carefully followed to assure the quality of the product to be delivered.

12.3.1 Test Planning

Test planning starts with the development of a written test plan prepared by a test engineer. It is the responsibility of the test engineer to develop the concepts and procedures to be used for the test, based on the specific product specifications which must be tested for conformance to requirements. Plans may be prepared for development testing, for simple in-process testing prior to close-up of a component, or for complex testing of a major assembly. In preparing development test plans, the test engineer may consider the use of fractional factorial concepts in the design of the test in order to obtain the greatest amount of information in the shortest period of time. The fractional factorial approach provides for testing individual characteristics in combination with one or more other characteristics during the same test. Statistical analysis of test data thus obtained not only provides valuable information to the design engineer, but provides the test engineer with additional information which he or she may use in the cost-effective design of production tests.

Plans for testing production items should specify the following elements:

1. Test location
2. Environments to be used in test
3. Specific equipment to be used
4. Test sequence and procedures to be followed
5. Check sheets for recording test results
6. Pass/Fail criteria for each step of test
7. Summary record of complete test results
8. Signature or identification stamp of tester and date

It is important that the product to be tested undergo an inspection prior to the start of the test to establish and record both its configuration and conformance to specifications. This is especially true during development of the product, when configuration revisions may be the order of the day, but holds true for all production testing. Following completion of the tests, it is also important that the product be inspected to assure that defects have not been induced during the testing. This inspection may be a simple visual inspection, or may require a more extensive check, but it is important to assure the quality being passed on to the next "customer."

12.3.2 Test Equipment

Test equipment must be included in the metrology calibration and control system of the production facility and be subject to established calibration cycles. All gauges, instruments, meters, and recording devices must be calibrated against working standards traceable to the National Bureau of Standards' masters. They must be identified with current calibration labels, and conformance to the required calibration cycles must be confirmed prior to the start of the test. In addition, all control cabinets for test equipment should be sealed with tamper-proof seals following calibration, to preserve the integrity of the equipment.

Design of test equipment and test equipment fixtures must be carefully reviewed to assure that the product will not be unintentionally stressed by forces induced through the moment arms of the fixture, or by clamps used to secure the product to the test fixture. This is particularly important in vibration testing where improper fixture design may subject the product to destructive amplitudes (see Section 8.0).

12.3.3 Test Software Control

Test software must be inspected and controlled in the same manner as other production or deliverable operational software. All software media, such as punched cards, magnetic tape, punched paper or Mylar tape, cassettes, disks, and computer programs must be under the control of a software library accessible only to authorized personnel.

Test software must be designed and prepared by software test engineers and subjected to design review by software technical specialists to identify risk areas. This design review will assure that the software will test all significant parameters, and establish traceability to specification requirements. Following design review, all software must be subjected to tests witnessed by a quality assurance software specialist. Any anomalies or discrepancies noted must be documented and the documentation held open until the anomaly or discrepancy has been investigated and any needed corrections made. Following acceptance, the software may be released and placed under configuration control. It must be prominently identified with the current change letter and placed in the software library with other working and master media. The software now becomes subject to change control and must be submitted to a software change control board for technical review of any desired change. If approved by the change board, the software media may be changed and subjected to the same acceptance and validation process as the original software for approval.

12.3.4 Conduct of Tests

All testing should be accomplished in accordance with the written and approved test plan and in the sequence dictated by the work order. (Some contracts may require customer review and approval of the test plan.) Development testing may be conducted by authorized test engineers or development technicians with monitoring by roving inspection personnel. Personnel assigned to conduct production acceptance tests should be trained production operators or inspectors, rather than test engineers, and acceptance tests should be witnessed by inspection when not specifically conducted by inspection. In production facilities where training and certification of production operators has been incorporated, many tests are conducted by certified operators without benefit of inspection witness. However, final acceptance tests are normally performed or witnessed by inspection. On some government and commercial contracts, the customer may also request to be present to witness acceptance tests. Records of all acceptance tests become a part of the product record and are retained in the central record center.

12.3.5 Additional Reading:

Juran, J. M., and Gryna, F. M., *Quality Planning and Analysis*, 3rd ed., (p. 191), McGraw-Hill, New York, 1993.

12.4 RELIABILITY

12.4.0 Introduction to Reliability

The standard definition for reliability established years ago by those in the electronics industry states: "Reliability is the *probability* of a device performing its purpose *adequately* for the *period of time intended* under the *operating conditions encountered*." This definition has stood the test of time, especially when considering design reliability, but it is currently being reexamined in the eyes of "the customer," who often equates reliability with dependability, and in the eyes of those producers who are alert to consumer demands. If this definition were to be written by the customer, it would read: "Reliability is the *assurance* that a device will *satisfy my expectations* for *as long as I expect to keep the product* and *under the operating conditions which I may need to rely upon it*." In other words, "I can depend on it!"

It is generally recognized that inherent product reliability is dependent on the excellence and maturity of the design process, and that reliability cannot be *inspected* into the product. At the same time, the customer is showing an increased awareness of the importance which manufacturing workmanship plays in the quality and reliability of the end product. The competition in the automotive field between American and foreign-made automobiles and between manufacturers of electronic components here and offshore provides positive evidence of this trend.

12.4.1 Reliability in Manufacturing

The key to assuring that designed-in reliability is not degraded during manufacturing is indoctrination and training. Everyone who is associated with the manufacturing process must be indoctrinated with the concept of "first-time-quality" and *accept* this concept as a personal commitment. Achievement of this goal at each step must be the result of training, coaching, and personal attention to detail in all aspects of planning the work flow, the tooling, the procedures, and the work instructions which will be used in production, with special attention paid to past lessons learned. This must be followed by the establishment of a positive system of manufacturing process control. It is here that the greatest assurance against any degradation of the designed-in reliability will be gained. The use of statistical process control charts to monitor critical processes by either *items produced* or *time samples*, provides the most efficient means of accomplishing this task. Additional details on the training and use of statistical control charts can be found in Subchapter 12.7.

12.4.2 Reliability Testing

One of the lessons learned from the past is that bench testing and even 100% inspection leave an element of risk that defects undetected in the design, and latent defects introduced during production, will find their way to the field. Reliability testing of the final product was introduced to aid in minimizing this risk. The precise form of reliability testing chosen will depend on the product. Most plans include subjecting the product to the extremes of the environmental elements expected to be encountered during use for a period of time which statistical evaluation has proven to be effective as a screening measure. It is important that products be operationally tested both before and after exposure to these environments in order to assess the effectiveness of the testing as a reliability enhancement tool. The complexity of electronics in the manned space program, and the need for maximum assurance of their reliability for the mission, focused attention on the need for reliability testing. It was on this program that the value of temperature cycling and random vibration were proven to be cost-effective means of reducing the risk of latent failures occurring in the field. Reliability failure investigations established that foremost among the causes for these failures were all types of poor solder joints, shorts and opens in circuitry, damaged insulation, inadequate mounting of components, and damage induced in multilayer boards and harnesses due to installation procedures or mishandling.*

Temperature cycling of completed circuit boards and/or black boxes has been found to provide an effective method of exposing latent defects or weaknesses in electronic assemblies. The writer's experience has shown that 3 to 10 thermal cycles provide the best screen, with the number of cycles depending on the complexity of the unit under test. Over 70% of latent defects are normally detected during the first three cycles when hardware is subjected to cycling between -40° C and $+75^\circ$ C with a temperature rate of change of 10° C per minute. Soaking at temperature extremes is required only to permit temperature stabilization of internal parts within $\pm 5^\circ$ of the extreme. The use of "power on" during the thermal rise provides some additional screening power, but the added cost of test setup makes this addition of marginal value. Likewise, monitoring during test should be used only if additional test data are needed for evaluation. For some complex equipment, operational testing at temperature extremes (as a minimum, on–off cycles) may be considered, but use of these added tests needs to be evaluated against the complexity of the equipment and its intended use. Operational tests at ambient temperature should always follow thermal cycling, with results monitored and investigated by reliability engineering to assure that needed corrections are implemented as part of the continuous improvement program.

Subjecting completed units or assemblies to random vibration has also been found to be an effective way of reducing the risk of delivering hardware with latent defects. Vibration testing may be accomplished at ambient temperatures, or in combination with temperature cycling if equipment is available and ultimate use indicates the advisability of such testing. However, under most circumstances, testing at ambient temperatures provides

---

*A visit to the IBM Laptop computer plant in Austin, Texas, revealed that each completed computer must pass a continuous power-on, 24-hr functional test. This operation is fully automated, including the insertion of the test floppy disk and a printout at the end showing that the computer is accepted—or if there is an anomaly, which test sequence or component failed, and what specific rework steps are to be taken.

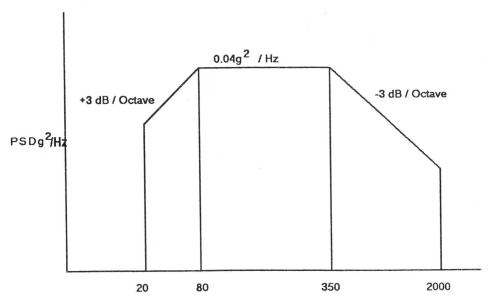

Figure 12.3 Typical power spectral density characteristics versus frequency used for vibration testing of electronics assemblies.

an adequate screen. Random vibration has been proven to be more effective than sinusoidal excitation (±2g at fixed, nonresonant frequencies between 20 and 60 Hz), which was used for years for vibration testing.

Vibration equipment and fixtures used for mounting hardware to the vibration table must be carefully designed to assure that moments will not be induced which will overstress the item under test. In addition, care must be exercised to assure that hardware is securely mounted to the shaker table and that the axis of vibration is perpendicular to the circuit boards in the unit. When this has been accomplished, the unit should be subjected to the power spectral density (PSD) characteristics shown in Figure 12.3 for 10 min. If the circuit boards or major components of the item are located in more than one plane, the duration of vibration should be 5 min in each plane. No significant difference in effectiveness has been detected whether temperature cycling is performed before or after the vibration, but as in temperature cycling, a complete operational test should be performed after completion of the environmental exposure to assure the effectiveness of the test and the delivery of reliable hardware.*

---

*A survey of a factory in England revealed normal (low-cost) precautions during fabrication and assembly of electrical and mechanical components. This was followed by a very sophisticated environmental test including high- and low-temperature cycling and vibration in a "white" spectrum for 24 hr, while the components were also undergoing an electronic functional test that reproduced the take-off, manuevering, and landing of a French fighter aircraft—the final customer. Although good products were delivered, a better manufacturing process control would have been more cost effective.

12.4.3 Additional Reading

Calabro, S. R., *Reliability Principles and Practices*, McGraw-Hill, New York, 1962.

12.5 COST OF QUALITY

12.5.0 Introduction to Cost of Quality

For years, the cost of quality was equated to the labor cost of inspectors on the payroll. Then, some enlightened pioneers began to consider that the cost of repairing or reworking product that was improperly made should also be considered when assessing the cost of quality. Today, we find a much broader focus on what *really* constitutes the cost of quality, and an across-the-board commitment to do something about it. This change of attitude was brought about by many things, but one of the more important was an appreciation of just how much, in dollars and cents, the cost of quality amounted to, and its relation to other business expenses. Industry surveys made in recent years have established that quality costs, as a percentage of sales, range from a low of 1% for small, uncomplicated products to a high of 25% for complex, man-rated space hardware. Typically, products produced using basic mechanical processes run in the range of 3–5%, with costs for precision industries running in the range of 10–12%. It is recognition of these facts that has resulted in recent attention to the collection and monitoring of costs and the establishment of specific programs and goals for the reduction of quality costs.

12.5.1 Elements of Quality Costs

The approach most commonly used to identify Quality Costs for collection, evaluation, and reduction allocates costs into four major cost categories; (1) Prevention, (2) Appraisal, (3) Internal Failure Costs, and (4) External Failure Costs. It is emphasized that these costs exist in all area of the company and must be collected, monitored and controlled on that basis. Cost collections should be the responsibility of the accounting department to assure proper allocation and reporting.

Cost categories, and typical activities which should be considered in them are listed here for guidance. They may be tailored to suit your unique organizational structure:

Prevention

The cost of ALL activities specifically designed to prevent poor quality in products AND services. Some examples are the cost of new product review, design review, quality planning, review of specifications and specific work instructions, supplier capability surveys, supplier performance measurement, process capability evaluations, equipment calibration, tool control, preventive maintenance, environmental controls, audits, education and training.

Appraisal

The cost associated with measuring or evaluating products or services to assure conformance to quality standards or performance requirements. Costs in these categories might include receiving and/or source inspection and test of purchased equipment and parts, environmental testing, supplier surveillance, process controls, roving inspection, bench inspection, final inspection, test inspection, and shipping inspection.

Failure costs are costs resulting from products or services not conforming to requirements or customer needs. They are divided into Internal and External Failure costs:

Internal Failure Costs

Those occurring prior to delivery or shipment of the product, or the furnishing of a service to the customer. Examples are the cost of material and labor expended on material which is scrapped; rework, reinspection, retesting, dispositioning of discrepant material for possible use; engineering changes or redesign; purchase order changes; and corrective action activities.

External Failure Costs

Those occurring after delivery or shipment of the product, and during or after furnishing a service to the customer. Examples of these costs are the cost of processing customer complaints, customer returns, warranty claims and product recalls.

12.5.2 Quality Cost Metrics

A key element in the process of measuring, documenting and improving processes is the selection of parameters to be measured, the baseline from which to measure progress, and the metrics of measurement. Of equal importance, is the establishment of these parameters, baselines, and metrics in conjunction with those responsible for the element being measured. In order for metrics to be an effective tool for control of costs and process improvement, those being evaluated must have input in establishing the measurement criteria and agree that the selection of parameters is representative and fair.

Metrics are the measurable characteristics of products or processes used to assess performance and track progress in attaining improvement. They facilitate measurement of *all* processes, both product *and* administrative. For proper control of the cost of quality, both should be utilized. When this is done, the relative importance of costs in various areas is highlighted, and provides the information necessary for sharpening focus on areas needing improvement.

The first step in establishing a metrics chart or graph for any area is the selection of the parameter to be measured and establishment of the baseline from which to measure progress. This is usually done based on past performance records, even though these records may not have been used for this purpose in the past. This is followed by selection of the measurement technique to be used and the frequency of measurement. Now comes the most difficult task: selection of an improvement goal that will be both realistic and attainable while still requiring the "process owners" to *stretch* to attain the goal. The tendency in human nature is to set the goal *low* in order to assure attainment and look good on the next performance review. A good deal of coaching is required here to assure that the improvement goal is, in fact, a goal which will require considerable *stretching*, and yet be a goal *that the process owners are willing to accept!* An understanding of *why* the goal is important to them personally, as well as why it is important to the company, is critical to goal attainment and the success of any improvement program,especially to a goal that requires a good deal of effort on the part of the process owners.

Once agreement has been reached on the metrics to be used, a decision on the manner and frequency of distribution of the information must be made. The frequency should be tailored to the format, content, and type of presentation of data: that is, detailed process charts and graphs, area charts, departmental summary charts, and top-level management reports. The important thing is that the charts, graphs, or reports must be very visible to the process owners, i.e., posted prominently in their work area, and equally visible to upper-level management. Management review and recognition of progress is vital to the continuation of process owner interest.

The selection of the metrics to be used in various areas must be selected based on the unique size and organizational structure of the company, but it is important that all areas be included if a true picture of the contribution of quality costs to the bottom line is to be obtained.

12.5.3 Additional Reading

Crosby, P. B., *Quality Is Free*, McGraw-Hill, New York, 1979.
Juran, J. M., and Gryna, F. M., *Quality Planning and Analysis*, McGraw-Hill, New York, 1970.
Juran, J. M., and Gryna, F. M., *Quality Control Handbook*, 4th ed., McGraw-Hill, New York, 1988.

12.6 DATA MANAGEMENT

12.6.0 Introduction to Data Management

One of the most important functions in any organization is the maintenance of historical performance records which become the basis for future action. Company procedures should provide specific direction as to who is responsible and who has the authority to generate these records, what methods and forms are to be used in the collection process, why the records are being generated, who will use them and for what purpose, where they are to be maintained, and for how long they are to be retained.

12.6.1 Types of Data in the System

There are many types of quality data generated during most development and production programs to aid in both day-to-day and trend evaluation of product quality and progress in quality improvement. The specific data maintained will vary by industry, but some of the more common types are summarized below.

Acceptance and rejection records. Records of the number of items in a production lot as well as the number accepted and the number rejected are maintained for every inspection performed, whether at the source, at receiving, in process, at final, at test, or at shipping. If the use of sampling procedures ise authorized and utilized in the inspection, the sample size selected is also recorded. These records provide the basis for performance evaluation and process control of all product, whether purchased or produced within the facility, and are key inputs to the computation of the cost of quality.

Research and development (R&D) test results. These records should include identification of the exact product configuration being tested, as well as inspection data taken both before and after test, in order to provide adequate information for test evaluation.

Data on supplier tests and inspections. Some purchase orders may require the supplier to maintain all data and provide it upon request, while others may require certain data to be provided as a data item at time of delivery to the purchaser. If the tests or inspections are witnessed by representatives of the purchaser's organization, all records should be so noted.

Environmental test data. Environmental tests may be conducted at the component, subassembly or assembly level. Data collected is essentially the same, regardless of level, and should include identification of the product being tested, the environment utilized, the specific test equipment used, calibration date of equipment, date of the test, and the identification of the person performing the test. Date and test personnel identification are common requirements on all inspections and tests.

Incoming (receiving) inspection data. In addition to the normal visual and dimensional inspection results, incoming data should include material certifications from the supplier, laboratory test results, review of supplier data requirements specified in the purchase order, and the results of any functional or nondestructive tests and inspections performed.

Process inspection data. Included in this category are records associated with processes such as heat treating, plating, other surface treatments, and all other special processes used in the production operation. Records of equipment uniformity surveys, temperature cycles, periodic solution checks, and safety checks should form a part of these data files.

Process control data. These files should include statistical quality control charts used for process control, all automated data collection charts used for temperature and humidity control, contamination control records on both air and fluids, and charts or automated data print-outs from computer controlled equipment.

Final inspection and test data. In addition to acceptance inspection results, this data will include log books or other deliverable data along with a complete accounting of the configuration of the product at time of shipment.

Data on scrap and rework. Special reports on investigations and corrective measures taken as part of continuous improvement efforts, both in-house and at supplier facilities, will be required.

Data on product returns, recalls and warranty replacements.

12.6.2 Data Collection

Procedures and forms to be used for the collection of inspection and test data should be specified in company manuals. For years, information has been recorded on multicopy forms which are forwarded to an information center for distribution to other organizations within the company that must take action. Typically, these forms are distributed to purchasing, production, production control, quality engineering and planning, design engineering, production engineering, reliability, accounting, and any other department affected by the acceptance or rejection of material. Following individual department action, the documents are updated to reflect status and are forwarded to a central control area, usually in quality control, for final action and/or filing. This, or similar collection methods, are still in usein many industries, but the availability of portable computers, workstation data entry terminals, bar-code readers, and the marked reduction in the cost of this equipment, is bringing real-time data collection and retrieval to the forefront (see Chapter 6.0).

Today, many companies are transmitting test data directly from automated test equipment to mainframe or desktop computers which are networked to provide real-time information on quality status to all personnel with a need to know. Portable hand-held computers and digital measuring devices enable certified operators and mobile inspectors to enter data as they are processing or inspecting material. This permits timely investigation of any problems and expedites incorporation of solutions. In addition, computerized data entry prompts individuals to enter needed information and will not let them unconsciously omit data entry. This has been a major problem with paper systems: Key information is omitted and often permanently lost because it was not entered at the time of rejection, even when space was provided on the form for the data. Computerized data entry has the added advantage that data is entered into the systems only once (there is no transcription of data from paper to computer) and may be analyzed and summarized in reports in a cost-effective and timely manner.

12.6.3 Operations Data Center

"Central files" is another name frequently associated with the area within a company where all data comes together, is cataloged and filed for either current or future reference. These files often become large, requiring a staff of trained personnel to properly maintain them and adequately service company reference needs. In some instances, the data being retained or used in specific processes may be company proprietary, as in the case of competition-sensitive processes or formulas; or the data may be classified, as it the case of certain work for the government. Special procedures for handling such records must be maintained, and limited-access storage facilities are required to protect this data adequately. Likewise, access to computerized data files must be protected by an authorizing password system to limit access. Computer disks and tapes require regular backup and limited-access storage in an area free from magnetic interference.

12.6.4 Records Retention

Many contracts contain clauses specifying the period of time that inspection records must be retained, either in active files or in "permanent" files. Since record storage is costly, procedures governing records retention must be carefully prepared. Records should not normally be maintained in active files any longer than necessary to permit adequate trend analysis, and to provide information in response to field problems and prompt servicing of customer complaints. This will vary by industry, but 2 years is probably a good norm. Permanent files will vary depending on the product and company desires to maintain historical files on certain items. However, records should not be maintained for the sake of records, and in most cases should be disposed of within 7 years.

12.7 STATISTICAL PROCESS CONTROL

12.7.0 Introduction to Statistical Process Control

Statistical process control is, as the name implies, the control of a specific process through the use of numbers. In this instance, *control* means to operate within prescribed boundaries with a minimum of variation from an established goal. The *process* selected may be as simple as a single element in a production flow—as applying a label to a product—or it may be a complex combination of operations consisting of many human/machine interfaces. In any case, it is necessary to gather numerical data on the process fluctuations, establish operating boundaries based on statistical evaluation of the data, and evaluate the results to determine the process adjustments necessary to minimize the fluctuations and maintain control within the boundaries.

Successful process control requires evaluation of data gathered in an effort to understand the causes of variation. Experience has shown that process variations are either caused by system management action, (or lack of it), or they are caused by factors within the control of the operator. Typical causes of process variation are machine wear, inadequate maintenance, fixtures which are not foolproof, power surges, variations in material, lack of adequate training, and inattention by the operator. Historically, the ratio of causal factors has been found to be about 80% within management's ability to correct, and 20% which the operator may correct. Process improvement, then, is dependent on identification and removal of those detrimental factors which may be corrected by management before the process starts, and continuation of the process monitoring/correction cycle while the process is running, through use of process control charts.

12.7.1 Selection of Charts for Process Control

Two types of charts are commonly used for process control in production operations: charts for measuring variable data, and charts for measuring attribute data. Each type serves a unique purpose in process control. Their applications are described below.

Variable Charts

Variable charts are commonly referred to as X-bar and R charts ($\bar{X}$ and R) charts and are particularly useful for several reasons:

1. Most processes have characteristics that are capable of being recorded numerically.
2. A measured value such as "2 pounds, 7 ounces," or "7mm," has more value than a simple evaluation of compliance.
3. Fewer pieces need to be checked when exact measurements are taken in order to determine variation.
4. Since decisions can be reached sooner when exact data are available, it is possible to get near-real-time feedback to the operator for needed correction—a key point in effective process control.
5. Variables charts can explain process data in terms of two important categories: shortterm variation, that is, piece to piece; and typical performance, at a point in time. These two characteristics provide for very sensitive process control tracking.
6. Because smaller samples need to be taken than with other systems, flow time can be improved and costs lowered even though additional time is needed to take exact measurements.

X-Bar Charts

X-bar charts plot the averages of each subgroup (collection of a given number of consecutive pieces) over a specific period of time. This produces an estimate of the process average over time. It also provides a picture of long-term variability which could include tool wear, machine drift, or adjustments.

The *R* Chart

The *R* chart is used in conjunction with the *X*-bar chart and plots the ranges (smallest value to largest value) of each subgroup of measurements taken over a period of time. This measures and quantifies an estimate of short-term variability. Short-term variability is defined as the inherent process variability within each subgroup which is independent of tool wear, machine drift or adjustments.

The best applications for *X*-bar and *R* charts are

First run jobs, where specific data will aid in design tolerance evaluation
Where specific data is needed to determine the cause of variation in a troublesome
    operation
When obtaining data is expensive or destructive
When operations are consistently running out of control and causes seem varied

To establish *X*-bar and *R* charts, calculate the average and range of each subgroup of 4 or 5 items. Plot *X*-bar and *R* where:

X = average measurement of each subgroup

R = range of measurements in each subgroup

n = number of measurements in each subgroup

Calculate the control limits using the following equations:

UCL for $X = \overline{X} + A_2\overline{R}$ LCL for $X = \overline{X} - A_2\overline{R}$

UCL for $R = D_4\overline{R}$ LCL for $R = D_3\overline{R}$

where: $\overline{X}$ is the process average and $\overline{R}$ is the range average for k subgroups
    The factors to use in the above equations are given below:

n	4	5	6	7	8
D_4	2.28	2.11	2.04	1.92	1.86
D_3	0	0	0	0.076	0.136
A_2	0.73	0.58	0.48	0.42	0.37

Attribute Charts

Charts based on attribute data are often referred to as "go no-go" charts because they accept or reject an entire item on the basis of measurement of a single characteristic. These charts are less complicated to use than variable charts and therefore find frequent use in process control during production. They are particularly useful when it is desired to obtain a picture of operating trends without the cost of variable data, or when variations are known to be operator controllable.

The three most common attribute charts are the *p chart* which plots *percent defective*; the *c chart*, which plots *number of defects*; and the *u chart*, which plots *defects per unit*. The *p* chart plots *p* as the number of defectives in each subgroup divided by the number inspected in each subgroup (*n*). Upper control limits (UCL) and lower control limits (LCL) are calculated with the following equations:

$$\text{UCL}p = \bar{p} + \frac{3\sqrt{\bar{p}(1-\bar{p})}}{\sqrt{n}} \qquad \text{LCL}p = \bar{p} - \frac{3\sqrt{\bar{p}(1-\bar{p})}}{\sqrt{n}}$$

where p – bar = the total number of defectives ÷ total number inspected.

Such p charts are often initially plotted from data which was taken for purposes other than control charts. Subgroups of data consisting of four or five units taken from production on a daily or weekly basis may be used. When this is done, n may vary from point to point on the control chart. If this is the case, one can compute new control limits for the individual n's. One can also estimate the average n and draw a single set of limits, or draw several sets of limits for different n's, but computing new limits for the individual n's is recommended.

The procedure for establishing a p chart is as follows:

1. Obtain data from 20–25 subgroups if possible.
2. Calculate p for each subgroup.
3. Calculate p-bar, which will became the centerline for the chart.
4. Calculate the upper and lower control limits using the above equations.
5. Draw the centerline, upper and lower control lines on the chart.
6. Plot additional p values on the chart taken from subsequent subgroups.
7. Interpret data and take corrective action as necessary.
8. A minimum of 25 consecutive points should be within limits for control.

Two conditions must be satisfied for the effective use of c charts and u charts: (1) The possibilities of nonconformities must be theoretically infinite; and (2) the probability that a particular type of nonconformance will occur must be small. When these conditions exist, the c chart may be effectively used for number of defects and the u chart for defects per unit. The procedure for establishing these charts is essentially the same as that illustrated above for the p chart. To establish a c chart, plot c, the defect count in the subgroup, and calculate the upper and lower control limits using the following equations:

$$\text{UCL}c = \bar{c} + 3\sqrt{\bar{c}} \qquad \text{LCL}c = \bar{c} - 3\sqrt{\bar{c}}$$

where c = total number of defects ÷ total number of subgroups.

To establish a u chart for defects per unit, plot u, total defects in each subgroup, divided by the number of units in each subgroup (n). Calculate the control limits as follows:

$$\text{UCL}u = \bar{u} + \frac{3\sqrt{\bar{u}}}{\sqrt{n}} \qquad \text{LCL}u = \bar{u} - \frac{3\sqrt{\bar{u}}}{\sqrt{n}}$$

12.7.2 Chart Analysis

There are seven typical patterns that may become evident in data plotted on control charts:

The natural pattern. This pattern is stable, it shows no trends, runs, shifts, or erratic swings, most points are distributed equally on either side of the centerline, and there are few, if any, points outside of the control limit, although a few may approach control limits. The natural pattern is the desirable pattern.

Cycles. The cycle pattern reveals short up and down trends that repeat themselves. Patterns such as this indicate that an assignable cause exists, because repetition is not a characteristic of a random pattern.

Freaks. A "freak" is an individual point which differs greatly from all other measurements. Freaks should not be ignored if they persist, as they may be an indication of the start of a larger problem.

Instability. Instability manifests itself by large swings or fluctuations on first one side of the centralline and then the other. If the unstable condition is on the high side, it may be an indication of operator inattention or poor maintenance of equipment. Other instability may be the result of variations in subgroup sizes, variations in lot quality (very good or very bad), or simply nonrandom selection of samples in the subgroups.

Stable mixtures. The characteristic of the stable mixture pattern is that data tend to fall on either side of the central line, with very few points near the middle. This is usually caused by mixed lots, as from different runs at earlier operations, or from different raw material sources.

Sudden shifts in level. Sometimes there are sudden shifts in level, as from a series of points all above the central line to a series all below the line. The shift is usually attributed to a definite change in the process, such as an operator change, a material change, a machine adjustment, or a similar assignable cause.

Trends. The trend pattern, uniform and continuous movement up or down, is perhaps the most obvious pattern. It is usually the result of drift induced by machine wear, operator fatigue, gradual loosening of holding fixtures, or drift in test equipment.

As noted above, the aim of process control is to attain the naturally stable pattern, and then to take action to narrow the variation in the pattern so as to bring the points as close to the centerline as is economically feasible. Continuous monitoring, detailed analysis, and effective corrective action are necessary to bring this about. Charts should be updated regularly, as a minimum once a month (depending on the production rate), and kept current at all times. Control charts, like other tools in the hands of the craftsperson, must be properly maintained and used with understanding to achieve optimum results.

13

Principles of Structural Mechanics

Jack M. Walker   *Consultant, Manufacturing Engineering, Merritt Island, Florida*

13.0 INTRODUCTION TO MECHANICS

This chapter is intended for those who may not have an extensive background in structural mechanics, or who have not used some of that knowledge recently and need a refresher. The working manufacturing engineer today is involved in product design, tool design, process selection and development (both parts fabrication and assembly), materials, and many other areas that require an analytical evaluation for proper understanding. In the field of metallurgy, for example, terms such as stiffness, stress, strength, elasticity, etc., are the language involved—and sometimes we tend to use some of the terms incorrectly. In the fabrication of sheet metal parts, the elongation and yield strength are key elements. In machining, shear may play an important role in understanding cutting principles, and selecting the proper cutting tool for a specific material.

To be able to understand the discussion of problems in materials and processes, it is essential to have a knowledge of the basic principles of forces in equilibrium. All of the elements of today's complex structural considerations are not introduced here. Rather, the intent is to establish a set of definitions and relationships that are needed to understand terms such as tension, shear, compression, and bending. In order to use terms such as stiffness, yield strength, deflection, stress, and strain correctly, the author feels that mathematical relationships offer the only logical approach to definition and understanding.

Care has been taken to avoid the use of advanced mathematics; a knowledge of arithmetic and high-school algebra is all that is needed. It is the desire of this author to present, to those having little or no knowledge of mechanics, some simple solutions to everyday structural problems.

13.1 DEFINITIONS OF FORCE AND STRESS

13.1.1 Force

A *force* may be defined as that which exerts pressure, motion, or tension. We are concerned here with forces at rest, or in equilibrium. If a force is at rest, it must be held so by some other force or forces. As shown in Figure 13.1, a steel column in a building structure supports a given load, which, due to gravity, is downward. The column transfers the load to the footing below. The resultant upward pressure on the footing equals the load in magnitude, and is called the reaction. The two forces are opposite in direction, have the same line of action, and are equal in magnitude. The system is in equilibrium; that is, there is no motion.

The unit of force is usually pounds or kilograms. In practice the word "kip," meaning a thousand pounds, is frequently used. Thus, 30 kips might also be written 30,000 lb.

13.1.2 Stress

Assume that a short column has a load of 100,000 lb applied to its end (see Figure 13.2a). The load, P, is distributed evenly over the cross section $X–X$. To calculate the area of the cross section in Figure 13.2b,

$$Area = 3. \times 3.33 = 10 \text{ in.}^2$$

Or, if it was a circular cross-section, as shown in Figure 13.2c,

$$Area = \pi r^2$$

where r (radius) $= 3.568/2 = 1.784$ in., or

$$A = 3.14 \times (1.784)^2 = 10 \text{ in.}^2$$

In our example, the load P is distributed evenly over the cross-sectional area A. We can say that 100,000 lb is distributed over 10 in.^{2}, or $100,000 \div 10 = 10,000$ lb acting on each square inch. In this instance, the unit *stress* in the column is 10,000 lb/in.^{2} (psi).

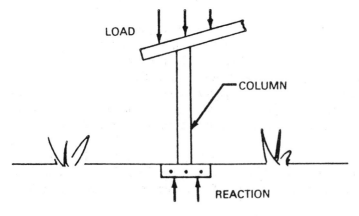

Figure 13.1 Example of forces in equilibrium. The load on the column from the roof is reacted by the footing.

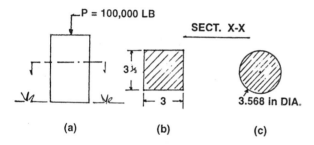

Figure 13.2 Example of the compressive stress in a short column dependent on the load applied and the column cross section.

A stress in a body is an internal resistance to an external force, or the intensity of the forces distributed over a given cross section.

13.2 TYPES OF STRESSES

The three primary stresses that we will discuss are *tension, compression,* and *shear.* Unless noted otherwise, we will assume that forces are axial, and that the stresses are uniformly distributed over the cross-sectional area of the body under stress.

We normally call the load, or external force, P; the area of the cross section A; and the unit stress f. As discussed above, the load divided by the section area will give the unit stress. This is stated as a fundamental principle:

$$f = P/A, \text{ or } P = Af, \text{ or } A = P/f$$

13.2.1 Deformation

Whenever a force acts on a body, there is an accompanying change in shape or size of the body. This is called *deformation.* Regardless of the magnitude of the force, the deformation is always present, although it may be so small that it is difficult to measure even with the most delicate instruments. It is often necessary to know what the deformation of certain members will be. For example, a floor joist in the second story of the house shown in Figure 13.3 may be large enough to support a given load safely, but it may *deflect* (or deform) to such an extent that the floor will vibrate, or bend, and cause the plaster in the ceiling below to crack.

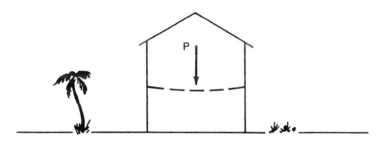

Figure 13.3 Example of the deformation of a floor joist (beam) under load (deflection).

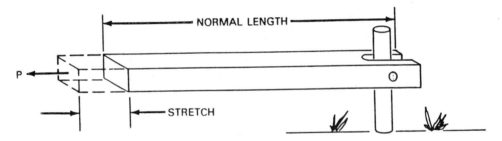

Figure 13.4 Example of the deformation in a bar under tension.

13.2.2 Tension

When a force acts on a body in such a manner that the body tends to lengthen or pull apart, the force is called *tensile*. The stresses in the bar due to the tensile force P are called tensile stresses (see Figure 13.4).

Example

A wrought-iron bar with a diameter of $1\frac{1}{2}$ in. is used in a roof truss. If the tensile force supported is 20,000 lb, what is the unit stress? (See Figure 13.5.)

Solution

To find the cross-sectional area of the bar, we square the radius and multiply by 3.1416, $A = \pi r^2$.

$$A = 3.1416 \times (0.75)^2 = 1.76 \text{ in.}^2$$

The load (P) is 20,000 lb. These are the data; what we are looking for is the unit stress. The formula giving the relation among these three quantities is $f = P/A$.

$$f = 20{,}000/1.76 = 11{,}363 \text{ psi, the unit stress in the bar}$$

13.2.3 Compression

When the force acting on a body has a tendency to shorten it, the force is called *compressive*, and the stresses within the member are compressive stresses. (See Figure 13.6.)

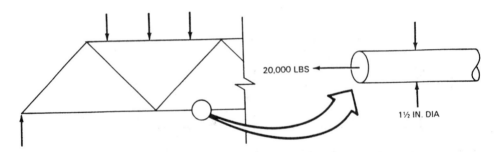

Figure 13.5 Wrought-iron bar in a roof truss under tension.

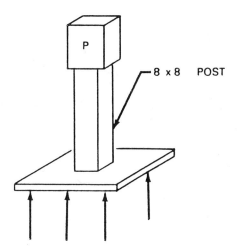

Figure 13.6   Short timber post under compression due to load P.

Example

Suppose we have a short timber post, $7\frac{1}{2}$ in. $\times 7\frac{1}{2}$ in. in cross section, which has an allowable stress of 1000 psi and we wish to know what load it will safely support.

Solution

The cross-sectional area, A, is $7\frac{1}{2} \times 7\frac{1}{2}$, or 56.25 in.^{2}. From the data, the allowable unit stress for this timber, f, is 1000 psi. Substituting in the formula, $P = Af$, we find:

$$P = 56.25 \times 1000, \text{ or } 56{,}250 \text{ lb}$$

which is the maximum safe load.

Note: This example is to illustrate the definition of pure compressive strength. When the member in compression is relatively short in relation to its crosssection, the above is correct. However, as the length increases or the member becomes more "slender" in relation to its cross section, it may tend to buckle before it actually fails in compression. This "long column" failure is discussed in Subchapter 13.6.

13.2.4   Shear

A *shearing stress* occurs when we have two forces acting on a body in opposite directions, but not on the same line. Forces acting like a pair of scissors, tending to cut a body, is an example. Figure 13.7 shows two plates held together by a rivet. The forces P acting on the

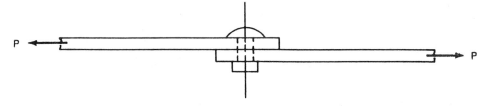

Figure 13.7   Example of shearing forces on a rivet caused by the forces in the plates.

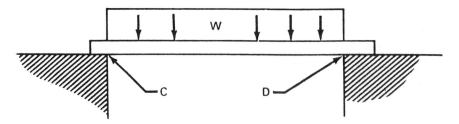

Figure 13.8 Example of shearing forces on a beam at points C and D due to distrubuted load W.

plates tend to shear the rivet on an area equal to the cross-sectional area of the rivet at the plane of contact between the two plates.

Example

The forces P in the plates shown in Figure 13.7 are each 5000 lb, and the rivet has a diameter of ¾ in. What is the shearing unit stress?

Solution

A ¾ in rivet has a cross-sectional area of 0.4418 in.^{2}. The basic formula for shearing stress is $\tau = P/a$. By substituting the known quantities,

$$\tau = 5000/0.4418 = 11,317 \text{ psi}$$

which is the average shearing stress in the rivet.

The load, W, rests on a beam which in turn is supported on walls at its ends. There is a tendency for the beam to fail by shearing at points C and D (see Figure 13.8).

13.2.5 Bending

Figure 13.9 illustrates a simple beam with a concentrated load P at the center of the span. This is an example of bending or flexure. The fibers in the upper part of the beam are in compression, and those in the lower part are in tension.

These stresses are not equally distributed over the cross section. (A more complete discussion will be presented later in this chapter.)

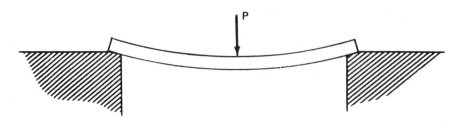

Figure 13.9 Bending of a beam due to concentrated load P.

13.3 HOOKE'S LAW

Robert Hooke was a mathematician and physicist living in England in the seventeenth century. As a result of experiments with clock springs, he developed the theory that deformations are directly proportional to stresses. In other words, if a force produces a certain deformation, twice the force will produce twice the amount of deformation. This law of physics is of the utmost importance, though unfortunately Mr. Hooke did not carry his experiments quite far enough, for it may be readily shown that Hooke's law holds true only up to a certain limit.

13.3.1 Elastic Limit

Suppose that we place a bar of structural steel with a section area of 1 in.^{2} in a machine for making tensile tests (see Figure 13.10). We measure its length accurately and then apply a tensile force of 5000 lb. We measure the length again and find that the bar has lengthened a definite amount, which we will call X in. On applying 5000 lb more, we note that the amount of lengthening is $2 \times X$, or twice the amount noted after the first 5000 lb. If the test is continued, we will find that for each 5000 lb, the length of the bar will increase the same amount noted when the first unit of 5000 lb was applied; that is, the deformations are directly proportional to the stresses. So far Hooke's law has held true, but (and this is the significant point) after we have applied about 36,000 lb, the length increases more than X in. for each additional 5000 lb. This unit stress, which varies with different materials, is called the *elastic limit*. The "proportional limit" is the largest value of the stress for which Hooke's law may be used. It may be defined as the unit stress beyond which the deformations increase in a faster rate than the applied loads.

Here is another phenomenon that is noted. If we make the test again, we will discover that if any applied load less than the elastic limit is removed, the bar will return to its original length. If the unit stress greater than the elastic limit is removed, we will find that the bar has permanently increased its length. This deformation is called the *permanent set* or plastic deformation. This fact permits another way of defining the elastic limit: It is that unit stress beyond which the material does not return to its original length when the load

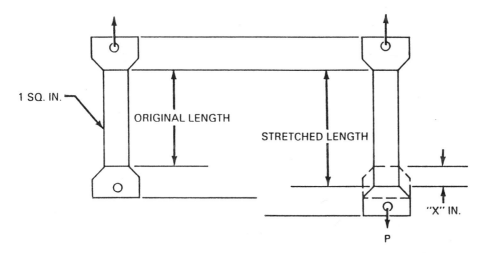

Figure 13.10   Increased length of a steel bar due to applied tensile force P.

is removed. The property which enables a material to return to its original shape and dimensions is called *elasticity*.

Another term used in connection with these tests is the yield point. It is found during tests that some materials, steel for instance, show increases in deformation without any increase in stress; the unit stress at which this deformation occurs is called the *yield point*. The yield point, although slightly higher than the elastic limit, is almost identical to the elastic limit. Nonductile materials, such as cast iron, have poorly defined elastic limits and no yield point.

13.3.2 Ultimate Strength

If a load of sufficient magnitude is applied to a test specimen, rupture occurs. The unit stress within the bar just before it breaks is called the *ultimate strength*. For the steel bar referred to in the previous pages, failure occurs at a unit stress of about 70,000 psi.

13.3.3 Stress and Strain

We pointed out earlier that stress always implies a force per unit area, and is a measure of the intensity of the force. *Strain* refers to the elongation per unit length of a member in stressed condition. Strain should never be used in place of the terms "elongation" and "deflection." Most of the above terms are best shown on a stress–strain diagram (see Figure 13.11).

13.3.4 Modulus of Elasticity

We have seen that if a bar is subjected to a force, a deformation results. Also, if the unit stress in the bar does not exceed the elastic limit of a material, the deformations are in direct proportion to the stresses. The key to computing the magnitude of the deformation

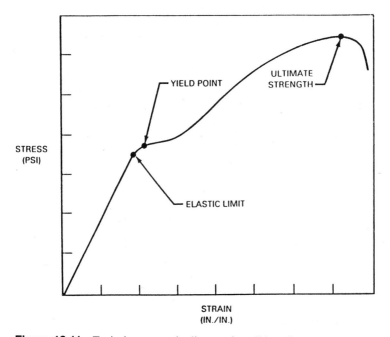

Figure 13.11 Typical stress-strain diagram for mild steel.

lies in the *stiffness* of a material. The number that represents the degree of stiffness of a material is known as the *modulus of elasticity*. We represent this quantity by the letter E and define it as the unit stress divided by the unit deformation (strain), or

$E = f/s$

where E = Mod of Elasticity

f = unit stress

s = unit deformation

Example

Suppose we place a steel bar 1 in.^{2} in cross section in a testing machine and apply a tensile force of 1000 lb. Its length becomes greater (although we cannot see it with the naked eye). If we apply the same force to a piece of wood having the same dimensions as the steel bar, we find that the deformation is greater—probably 20 times greater. We say that the steel has a greater degree of stiffness than the wood (see Figure 13.12).

From our discussion on unit stress earlier, we saw that

$f = P/A$

And, from Figure 13.12, where L represents the length of the member and e the total deformation, s the deformation per unit of length would equal the total deformation divided by the original length, or

$s = e/L$

Now, since the modulus of elasticity, $E = f/s$, this becomes

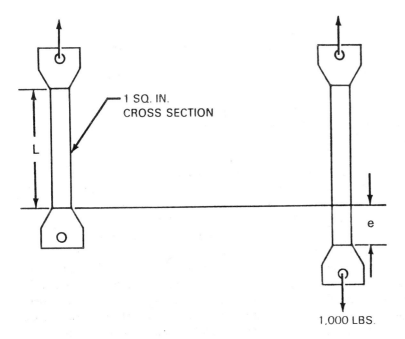

Figure 13.12 Example of stress and strain for calculation of modulus of elasticity E.

MATERIAL	ELASTIC LIMIT (PSI)		ULTIMATE STRENGTH (PSI)			MOD. OF ELASTICITY (PSI)	WEIGHT (LB/CF)
	TENSION	COMPRESSION	TENSION	COMPR	SHEAR		
Structural Steel	36,000	36,000	70,000	70,000	55,000	29,000,000	490
6061 Aluminum	35,000	35,000	38,000	38,000	30,000	10,000,000	170
Timber	3,000	3,000	10,000	8,000	500	1,200,000	40
* (Perp. to grain)					3,000		

Figure 13.13   Average physical properties of some common materials.

$$E = \frac{P/A}{e/L} \quad \text{or} \quad E = \frac{P \times L}{Ae}$$

which can also be written as:

$$e = PL/AE$$

where e = total deformation in inches
  P = force in pounds (axial load)
  L = length in inches
  A = cross-sectional area in inches
  E = modulus of elasticity in pounds per square inch.

Remember that all of the above is valid only when the unit stress does not exceed the elastic limit of the material. Figure 13.13 gives typical values for some common materials.

13.4 MOMENTS

You have probably heard the term *moment* used with problems in engineering. A force of 100 lb, an area of 16 in.^{2}, or a length of 3 ft can readily be visualized. A moment, however, is quite different; it is a force multiplied by a distance! A moment is the tendency of a force to cause rotation about a certain point or axis. The moment of a force with respect to a given point is the magnitude of the force multiplied by the distance to the point.

The following examples may help explain moment.

Example 1

Two forces are acting on the bar, which is supported at point A, as shown in Figure 13.14. The moment of force P_1 about point A is 8 ft $\times$ 100 lb, or 800 ft-lb. This force tends to produce clockwise rotation about point A (the direction in which the hands of a clock revolve), called a *positive* moment. The other force, P_2 has a lever arm of 4 ft with respect to point A, and its moment is 4 ft $\times$ 200 lb, or 800 ft-lb. P_2 tends to rotate the bar in the opposite direction (counterclockwise), and such a moment is called *negative*.

In Figure 13.14 the positive and negative moments are equal in magnitude, and equilibrium (or no motion) is the result. Sometimes it is stated: "If a system of forces is in static equilibrium, the algebraic sum of the moments is zero." In Example 1, if the system of forces is in equilibrium, the sum of the downward forces must equal the upward forces. The reaction at point A, therefore, will act upward and be equal to 200 lb plus 100 lb, or

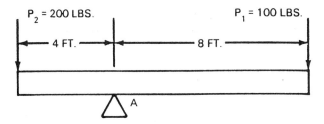

Figure 13.14 Simple beam used to illustrate the moments about the support point A in Example 1.

300 lb. We could say: "If a system of parallel forces is in equilibrium, the algebraic sum of the forces is zero."

Example 2

The beam shown in Figure 13.15 has two downward forces, 100 and 200 lb. The beam has a length of 8 ft between supports, and the supporting forces (called *reactions*) are 175 and 125 lb.

Check

1. The sum of downward forces must equal the upward forces:
       $100 + 200 = 175 + 125$
       or, $300 = 300$ (It's true!)
2. The sum of the moments of the forces tending to cause clockwise rotation (positive moments) must equal the sum of the moments of forces tending to produce counterclockwise rotation (negative moments), about *any* center of moments.

   Check the moments about point *A*. The force tending to cause clockwise rotation about this point is 175 lb, its moment is 175×8, or 1400 ft-lb. The forces tending to cause counterclockwise rotation *about the same point* are 100 and 200 lb, and their moments are $(100 \times 6) + (200 \times 4)$ ft-lb. Now we can write:

$$(175 \times 8) = (100 \times 6) + (200 \times 4)$$

$$1400 = 600 \times 800$$

$$1400 \text{ ft-lb} = 1400 \text{ ft-lb}   \text{(we lucked out again!)}$$

Note: If you wonder where the force of 125 lb went in writing this equation, the 125 lb

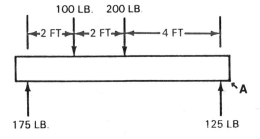

Figure 13.15 Simple beam used to demonstrate calculation of moments in Example 2.

force has a lever arm of 0 ft about point *A*, and the moment of the force becomes $125 \times 0 = 0$. In future problems, when we write equations of moments, we can therefore omit writing the moment of the force acting through the point we have selected, because we know it can cause no rotation about the point, and its moment is zero.

13.5 BEAMS

A *beam* is a structural member resting on supports, usually at its ends, which supports transverse loads. The loads acting on a beam tend to *bend* rather than lengthen or shorten it. In general, there are five types of beams, depending on the position and number of supports, as shown in Figure 13.16.

The two kinds of loads that commonly occur on beams are called *concentrated* (*P*) and *distributed* (*W*). A concentrated load acts at a definite point, while a distributed load acts over a considerable length. Both types are shown in Figure 13.16a. A distributed load produces the same reactions as a concentrated load of the same magnitude acting through the center of gravity of the distributed load.

13.5.1 Stresses in Beams

Figure 13.17a represents a simple beam. Examination of a loaded beam would probably show no effects of the load. However, there are three distinct major tendencies for the beam

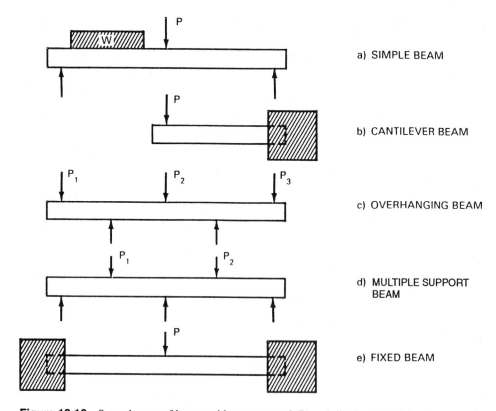

Figure 13.16 Several types of beams with concentrated (P) and distributed (W) loads.

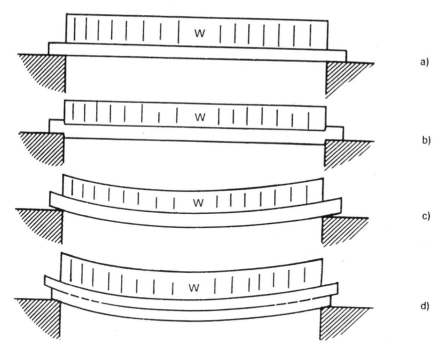

Figure 13.17 Example of a beam with a distributed load W showing vertical shear, bending, and transverse shear.

to fail. First, there is a tendency for the beam to fail by dropping down between the supports, as shown in Figure 13.17b. This is called *vertical shear*. Second, the beam may fail by *bending* as shown in Figure 13.17c. Third, there is a tendency for the fibers of the beam to slide past each other in a horizontal direction, as shown in Figure 13.17d. This is known as *horizontal shear*, or, in composites, *interlaminar* shear. Some of the most common beams and loads are shown in Figure 13.18. The values of V, the maximum shear, M, the maximum bending moments, and D, the maximum deflection, are given. If the loads are in units of pounds, the vertical shear (V) is also in pounds. When the loads are in pounds and the span in feet, the bending moment will be units of foot-pounds. Particular attention

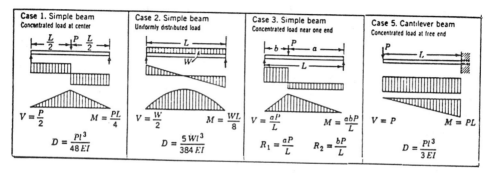

Figure 13.18 Examples of calculation of moment, shear, and deflection for different types of beams.

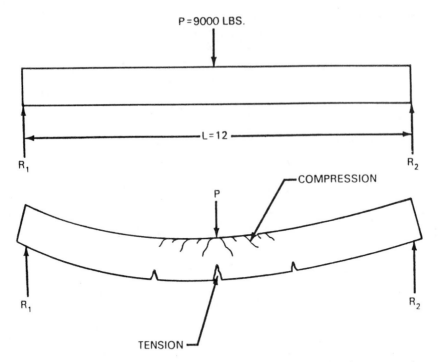

Figure 13.19 Example of bending of a simply supported beam showing compression and tension of the outer surfaces.

should be given to l in the formulas for maximum deflection. In this case, the span length l is given in inches—and the resulting deflection in inches.

13.5.2 Theory of Bending

Figure 13.19 shows a simple example of bending of a beam that is supported at each end with a load concentrated at the center. The load will cause the beam to deflect at the

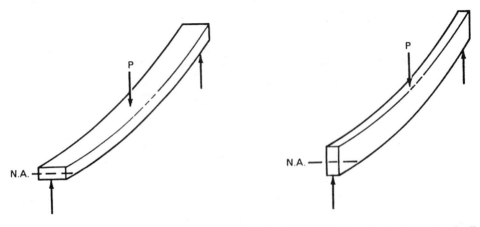

Figure 13.20 Example of the deflection in a wood 2 × 4 beam turned "flat-wise" and "edge-wise."

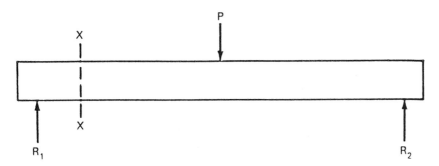

Figure 13.21 Beam with a section cut X–X for analysis.

center. The deflection stretches the fibers on the lower surface, and compresses those on the upper surface.

Somewhere between the compressive stresses in the upper fibers and the tensile stresses in the lower fibers is a place where there is neither. This is known as the *neutral axis*, (NA), the location of which depends on the cross section of the beam. If a 2-in. × 4-in. bar as seen in Figure 13.20 is used as the beam, we know that there will be a difference in the amount of deflection if the 2×4 is placed "flat" or if it is placed "on edge." The stress due to this bending is not a uniform stress, but varies in intensity from a maximum at the extreme fiber to zero at the neutral axis.

We discussed previously that the sum of the moments about any point on a beam must be zero—or, the positive moments (clockwise) must equal the negative moments (counterclockwise). If we cut the beam shown in Figure 13.21 at section $X - X$, and look at the left end in Figure 13.22, we see the following.

Call the sum of the compressive stresses C, and all the tensile stresses T. The *bending moment* in the section of beam above about point A is equal to $R_1 \times X$. For our example to be in equilibrium, the *resisting moments* must be equal. The resisting moment about point

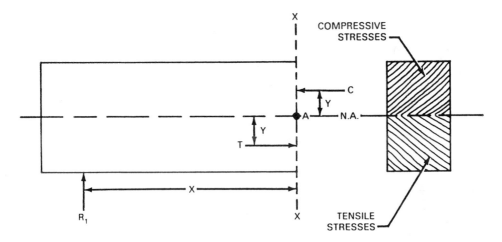

Figure 13.22 Example of the compressive and tensile stresses at section X–X.

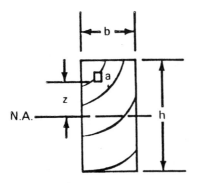

Figure 13.23 Cross section of a beam used in calculation of moment of inertia, I.

A is $(C \times Y) + (T \times Y)$. The bending moment tends to cause a clockwise moment (+), and the resisting moment tends to cause a clockwise moment (−). We see that:

$$+ [R_1 \times X] - [(C \times Y) + (T \times Y)] = 0$$

$$\text{or } R_1 \times X = (C \times Y) + (T \times Y)\}$$

This is the theory of bending in beams. For any type of beam, we can compute the bending moment, and if we wish to select (design) a beam to withstand this tendency to bend, we must have a member with a cross section of such shape, area, and material that the resisting moment will have a magnitude equal to the bending moment.

The maximum stress is given by the flexure formula;

$$f_b = My/I$$

where f_b = bending stress
M = applied bending moment
y = distance from the neutral axis to the extreme fiber
I = moment of inertia

This is also often expressed as

$$f_b = M/S$$

Where $S = I/y$, called the section modulus.

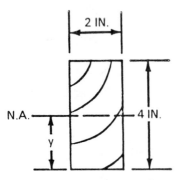

Figure 13.24 Cross section of a 2 × 4 used in calculations of edgewise moment of inertia.

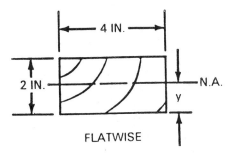

Figure 13.25 Cross section of a 2 × 4 used in calculation of flatwise moment of inertia.

13.5.3 Moment of Inertia

No attempt will be made to derive moment of inertia in this chapter, but we will discuss its use. We can say, however, that moment of inertia is defined as the sum of the products obtained by multiplying all the infinitely small areas (a) by the square of their distances to the neutral axis (see Figure 13.23).

For a rectangular beam, $I = bh^3/12$. From our 2 × 4 beam example earlier, we can see the cross section and moment of inertia calculations as Figure 13.24 and Figure 13.25. From our flexure formula, $f_b = M/S$, we can see that the allowable bending stresses in our beam (f_b) due to moment M is as follows:

Flatwise 2 × 4   *Edgewise 2 × 4*
$f_b = M/2.7$ $f_b = M/5.35$

What all this really says is that the properties of a beam vary to a great degree with the size and shape of the cross section. In our example of the 2 × 4, the edgewise 2 × 4 will withstand twice as much load as the same beam turned flat. Also, if we take the same cross-sectional area and convert it into a 1 × 8 beam on edge, we will see much more flexural strength than in a 1 × 8 turned flatwise.

The maximum stresses (both tension and compression) in a beam occur at the outer surfaces, which are the maximum distances from the neutral axis. These stresses diminish toward the NA, where they are zero. The NA is at the *centroid*, or center of area, of any cross section (see Figure 13.26). Values (or formulas) for I, y, and S can be obtained from a number of standard reference handbooks for standard member cross sections.

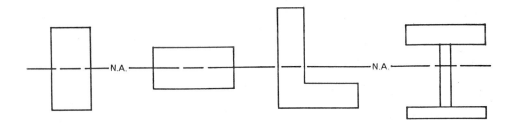

Figure 13.26 Location of the neutral axis of different cross sections used as beams.

13.6 LONG COLUMN FAILURE

When a column is subjected to compressive forces at its ends, one obvious but highly important result is that the distance between those ends is forcibly reduced. We discussed this in Subchapter 13.2. The action of a column under load can perhaps be most clearly understood by considering the implications of this shortening effect. One immediate result is that internal stresses are developed in each element of the column, their magnitudes depending on the amount and type of shortening involved, the elasticity of the material, and the original dimensions of the member, and their resultant being such in any given column that end loads of definite magnitude will be held in equilibrium by these internal stresses.

With the ideal column, i.e., one that is perfectly straight, perfectly homogeneous, and with the line of action of the end loads passing exactly through the centroid of each crosssection, the action under load will correspond with the calculations we performed in Subchapter 13.2, on compression. The practical column never conforms completely to the characteristics of the ideal, unless it is a very stocky member, such as a cube, which cannot bend or twist appreciably under load. When this short column is subjected to loads within the elastic limit, such columns act like springs and will return to their original length after the load is removed. However, if the load is increased above this point and plastic flow occurs, the column will not return to its original length, due to the permanent deformation caused by plastic flow. Continued increase in load will eventually cause failure in a short column. In some cases of plastic flow failure of short columns, the member will be mashed out of all resemblance to its original shape but still remain in stable equilibrium. Since there is no true ultimate load in a case of this kind, failure is assumed to take place as the permanent deformation exceeds some arbitrary amount. In practice, this "plastic flow failure" is assumed to take place when the average compressive stress equals the column yield stress in Table 13.1. Columns therefore, should be designed so that there will be no plastic flow (permanent deformation) under service conditions. It is convenient to call the end loads corresponding to any given shortening the "equilibrium load" for that shortening.

For a long, slender column, the "imperfect" column begins to exhibit different characteristics as the load is increased. Due to the column's imperfections, the degree to which its ends are held rigid or fixed, and the geometric shape of the column cross section, the long, slender column will tend to bend as well as shorten longitudinally. So long as the bending is not permitted to cause stresses beyond the proportional limit of the material at any point in the member, such columns continue to act as springs and will return to their original positions when the end loads are removed. However, if forcible movement of the ends toward each other continues beyond a certain point, the compressive strains of some of the fibers on the concave side of the column will exceed the proportional limit and there will be plastic flow of the material. At this point, as the load is increased, the long column will buckle rapidly, and lose most (or all!) of its structural integrity. We can visualize that the longer, or more slender, the column becomes, the less strength it will have in compression (due to buckling).

13.6.1 End Restraints

The load which a column can carry is influenced by the restraint imposed by the structure on the ends of the column. The tendency of the column to rotate or to move laterally would be different if the load were transmitted through frictionless pins or knife-edges, as shown

Table 1 Ultimate Allowable Stress for Various Metallic Elements

Material	Tension — Ultimate Strength	Tension — Yield Strength[1]	Tension — Modulus of Elasticity	Compression — Block Compression	Compression — Column Yield[2]	Torsion — Modulus of Failure	Torsion — Propor. Limit	Torsion — Modulus of Rigidity	Torsion — Fatigue Limit[3]	Shear — Ultimate Strength	Bending — Modulus of Failure[4]	Bending — Fatigue Limit[5]	Bearing — Ultimate Strength[6]	Weight Per Cu. Ft.	Weight Per Cu. In.
Mild Steel, S.A.E. 1025	55	36	28,000	55	36	50	20	10,000	…	35	55	25	90	490	0.2833
Alloy Steel, not Heat-treated	65	45	29,000	65	36	55	25	11,000	…	40	65	30	110	490	0.2833
Chrome-Molyb., Normalized	95	75	29,000	95	79.5	80	40	11,000	…	55	95	45	140	490	0.2833
Chrome-Molyb. (X-4130) Welded after Heat-Treatment	80	60	29,000	80	60	70	35	11,000	…	50	80	15	125	490	0.2833
Alloy Steels, Heat Treated	100	80	29,000	100	80	90	55	11,000	…	65	100	50	140	490	0.2833
Alloy Steels, Heat Treated	125	100	29,000	125	100	110	65	11,000	…	75	125	65	175	490	0.2833
Alloy Steels, Heat Treated	150	135	29,000	150	130	125	80	11,000	…	90	150	78	190	490	0.2833
Alloy Steels, Heat Treated	180	165	29,000	180	145	145	95	11,000	…	105	180	85	200	490	0.2833
Corrosion-Resistant Steel, Cold-Worked	185	140	26,000	185	155	155	105	10,000	30	115	200	94	220	490	0.2833
Annealed, or Near Welding	125	65	26,000	125	50—110	…	50	10,000	55	90—125	…	75	…	490	0.2833
17-ST Aluminum Alloy, Sheet	55	35	26,000	80	30	50	15	10,000	…	70	…	15	…	490	0.2833
Extruded Shapes	55	32	10,300	55	36	…	…	3,850	…	33	55	15	75	174	0.101
Unstretched Tube	50	32	10,500	50	32	50	15	3,850	…	30	55	15	75	174	0.101
Cold-Stretched Tube	55	30	10,500	55	34.5	50	15	3,850	…	33	62	15	75	174	0.101
24-ST Aluminum Alloy, Sheet	62	40	10,300	62	42.5	…	…	3,850	…	33	57	14	90	173	0.100
Extruded Shapes	57	42	10,300	57	40	…	…	3,850	…	37	62	14	83	173	0.100
Unstretched Tube	62	40	10,500	62	42	…	…	3,850	…	34	57	14	90	173	0.100
Cold-Stretched Tube	62	42	10,300	62	50	…	…	3,850	…	37	62	14	90	173	0.100
17-SRT Aluminum Alloy Sheet	55	42	10,500	55	50	…	…	…	…	37	62	14	75	174	0.101
24-SRT Aluminum Alloy Sheet	65	50	10,300	65	42	…	…	…	…	33	62	…	93	174	0.100
17-ST Alclad Sheet	50	28	10,300	50	50	…	…	…	…	39	65	…	68	174	0.101
17-SRT Alclad Sheet	50	37	10,300	50	28	…	…	…	…	30	…	…	68	174	0.101
24-ST Alclad Sheet	56	37	10,300	56	37	…	…	…	…	34	56	…	82	173	0.101
24-SRT Alclad Sheet	58	46	10,300	58	46	…	…	…	…	35	58	…	83	173	0.100

[1] Stress at which set is 0.002 in. per in.
[2] Nominal value for use in short column formulas.
[3] Maximum alternating torsional stress to withstand 20 × 10⁶ cycles.
[4] Nominal value for shapes not subject to local buckling. See Art. 6 : 1 for round tubes.
[5] Maximum alternating bending stress to withstand 300 × 10⁶ cycles for steel, 500 × 10⁶ cycles for aluminum alloys.
[6] For connections involving no relative movement between parts.

in Figure 13.27a, or if the ends were connected rigidly to a structure so stiff that the axis of the column was fixed in both direction and position, as shown in Figure 13.27b. In Figure 13.27a, the "restraint coefficient," c, would be unity or 1; while in Figure 13.27b, where we have a "theoretically fixed-ended column", the restraint coefficient of 4 might be applicable. In Figure 13.27c, the value of c would be 2.05; and in Figure 13.27d, the value of c would be 0.25. All these conditions are purely theoretical, and apply strictly only to "ideal" columns.

Experience over a number of years has indicated that $c = 2$ may be used for the design of tubular members where the joints are very rigid, as in welded structures, and $c = 1$ for theoretical pin-ended columns.

About two centuries ago, a Swiss mathematician, L. Euler, did some significant work in this area. While his work is certainly not sufficient background in buckling phenomena to use in sophisticated design problems, it does serve as a working model to better understand the fundamentals.

If the end loads on a column are so large that they produce internal stresses in excess of the elastic limit of the material, they will cause plastic as well as elastic strain, and the member will not completely regain its original length when the external load is removed. Another possibility is that during a part of the process of loading the external load may exceed the resisting force developed as a result of the elastic strains. Thus the entire load, P, may be imposed on the loaded end of the column at the start of the shortening process. At this instant the elastic strain, and therefore the resisting stress, is zero. The resisting force is built up as the column shortens and the elastic strain is produced, and does not become equal to P until the shortening s is equal to PL/AE. This makes the resulting work done greater than can be stored as strain energy, due to the corresponding strain. The peculiarities of the practical column are readily discernible in the testing laboratory, but

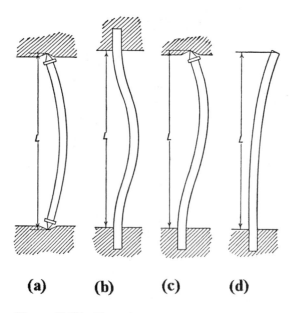

(a)   (b)   (c)   (d)

Figure 13.27 Theoretical end restraint conditions for a long column.

are impossible to treat mathematically (with exactness). Returning to Euler again, the following expression seems to approximate the load P that should not be exceeded in safe design (usually called P_e):

$$P = \pi^2 EI/L^2$$

By dividing both sides of the above equation by the area, A, and substituting the radius of gyration, ρ, for $\sqrt{I/A}$, we have the following:

$$f_c = \pi^2 E/(L/\rho)^2$$

The quantity L/ρ appears frequently in column formulas and is usually called the *slenderness ratio*. In general, if we do not exceed P_e in column design, the member probably will not fail—if we do exceed the Euler load P_e, it may fail. Table 13-2 shows column formulas, in terms of allowable stress, for both short and long columns, as well as the

Table 2 Column Formulas for Allowable Stress

Material	c	Short Columns	Transitional L/ρ	Long Columns
Spruce	1	$5{,}000 - 0.5 \, (L/\rho)$	72	$13 \times 10^6/(L/\rho)^2$
	2	$5{,}000 - 0.25 \, (L/\rho)$	102	$26 \times 10^6/(L/\rho)^2$
Aluminum Alloy				
17-ST Tubes	1	$34{,}500 - 245 \, (L/\rho)$	94	$101.6 \times 10^6/(L/\rho)^2$
Not Stretched	2	$34{,}500 - 173 \, (L/\rho)$	133	$203.2 \times 10^6/(L/\rho)^2$
17-ST Tubes	1	$42{,}500 - 335 \, (L/\rho)$	84.6	$101.6 \times 10^6/(L/\rho)^2$
Cold-Stretched	2	$42{,}500 - 237 \, (L/\rho)$	120	$203.2 \times 10^6/(L/\rho)^2$
24-ST Tubes	1	$50{,}000 - 427 \, (L/\rho)$	78	$101.6 \times 10^6/(L/\rho)^2$
Not Stretched	2	$50{,}000 - 302 \, (L/\rho)$	110	$203.2 \times 10^6/(L/\rho)^2$
24-ST Tubes	1	$50{,}000 - 427 \, (L/\rho)$	78	$101.6 \times 10^6/(L/\rho)^2$
Cold-Stretched	2	$50{,}000 - 302 \, (L/\rho)$	110	$203.2 \times 10^6/(L/\rho)^2$
Carbon Steels				
S.A.E. 1015	1	$27{,}000 - 0.660 \, (L/\rho)^2$	143	$276 \times 10^6/(L/\rho)^2$
Y.P. = 27,000 p.s.i.	2	$27{,}000 - 0.330 \, (L/\rho)^2$	202	$552 \times 10^6/(L/\rho)^2$
S.A.E. 1025	1	$36{,}000 - 1.172 \, (L/\rho)^2$	124	$276 \times 10^6/(L/\rho)^2$
Y.P. = 36,000 p.s.i.	2	$36{,}000 - 0.586 \, (L/\rho)^2$	175	$552 \times 10^6/(L/\rho)^2$
Alloy Steels [3]				
S.A.E. 2330, 3120, etc.	1	$60{,}000 - 3.144 \, (L/\rho)^2$	98	$286 \times 10^6/(L/\rho)^2$
Y.P. = 60,000 p.s.i.	2	$60{,}000 - 1.572 \, (L/\rho)^2$	138	$572 \times 10^6/(L/\rho)^2$
Y.P. = 75,000 p.s.i.	1	$75{,}000 - 4.913 \, (L/\rho)^2$	87	$286 \times 10^6/(L/\rho)^2$
	2	$75{,}000 - 2.457 \, (L/\rho)^2$	128	$572 \times 10^6/(L/\rho)^2$
Y.P. = 85,000 p.s.i.	1	$85{,}000 - 6.311 \, (L/\rho)^2$	83	$286 \times 10^6/(L/\rho)^2$
	2	$85{,}000 - 3.155 \, (L/\rho)^2$	118	$572 \times 10^6/(L/\rho)^2$
Y.P. = 100,000 p.s.i.	1	$100{,}000 - 8.735 \, (L/\rho)^2$	76	$286 \times 10^6/(L/\rho)^2$
	2	$100{,}000 - 4.367 \, (L/\rho)^2$	107	$572 \times 10^6/(L/\rho)^2$
Y.P. = 130,000 p.s.i.	1	$130{,}000 - 14.761 \, (L/\rho)^2$	66	$286 \times 10^6/(L/\rho)^2$
	2	$130{,}000 - 7.381 \, (L/\rho)^2$	94	$572 \times 10^6/(L/\rho)^2$
Y.P. = 155,000 p.s.i.	1	$155{,}000 - 20.98 \, (L/\rho)^2$	61	$286 \times 10^6/(L/\rho)^2$
	2	$155{,}000 - 10.49 \, (L/\rho)^2$	86	$572 \times 10^6/(L/\rho)^2$
Chrome-Molyb. X-4130	1	$79{,}500 - 51.78 \, (L/\rho)^{1.5}$	91	$286 \times 10^6/(L/\rho)^2$
Normalized Round Tube	2	$79{,}500 - 30.79 \, (L/\rho)^{1.5}$	129	$572 \times 10^6/(L/\rho)^2$

transitional L/ρ. The c used in this table is described in the previous paragraph on end restraints.

13.7 STRESSES IN PRESSURE VESSELS

We are often involved in the design and fabrication of cylindrical structures, used as launch tubes, rocket encasements, or rocket motor cases. The example in Figure 13.28 is for a spherical tank, or bottle. The tensile stress in the outer fibers of the sphere due to internal pressure is a function of the internal diameter and the wall thickness. The hoop stress, S,

$$S = pD/t$$

If the pressure is 1000 psi, the wall thickness is 0.050 in., and the radius is 1 in.,

$$S = (1000 \times 2)/(0.050) = 40,000 \, \text{psi}$$

However, if we increase the inside diameter to 4 in. and keep the pressure and wall thickness the same,

$$S = (1000 \times 4)/(0.050) = 80,000 \, \text{psi}$$

In other words, the larger the diameter, the higher the hoop tensile stress in the sphere due to the same pressure. If we double the wall thickness, the hoop stress is cut in half. (This makes sense, because we spread the load over twice the cross-sectional area in the wall.)

Note: The same formula applies to longitudinal stress in a closed tube or cylindrical tank (see Figure 13.29).

Longitudinal stress $= S^1 = pD/t$

The longitudinal stress in an open cylinder is only one-half the hoop stress:

Longitudinal stress $= S^2 = pD/2t$

In the design of metal cylinders, the wall thickness must be established to withstand the hoop stress. The cylinder is then twice as strong in the longitudinal direction as is really

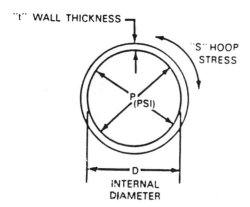

Figure 13.28   Cross section of a spherical pressure vessel.

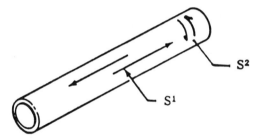

Figure 13.29   Cylindrical pressure vessel (rocket launch tube).

required to hold the pressure. Since metals are essentially isotropic (they have the same mechanical properties in all directions) in strength, the product of operating pressure and volume (capacity) (PV) per unit weight of material is less in the cylinder than in the sphere for a given peak stress.

In composite cylinders, the reinforcement can be oriented and proportioned so that the hoop strength of the cylinder wall is actually twice the longitudinal strength. Consequently, the PV per unit weight of material is the same in the cylinder as it is in a sphere for a given peak stress, and much higher than for the metal cylinder. This unique weight-per-unit-volume relationship, independent of size and shape, has obvious importance in rocket motor case design, or any cylindrical pressure vessel.

14

Materials Characteristics of Metals

Jack M. Walker *Consultant, Manufacturing Engineering, Merritt Island, Florida*

with

Robert S. Busk *International Magnesium Consultants, Inc., Hilton Head, South Carolina*

14.0 INTRODUCTION TO MATERIALS CHARACTERISTICS OF METALS

Modern industry is dependent on a knowledge of metallurgy. Nearly every kind of manufacturing today is affected by the behavior of metals and alloys. Therefore, anyone who plans a career in modern industry will find a working knowledge of metallurgical processing to be a valuable asset.

Today's manufacturing engineer may not need to be a materials engineer or metallurgist, in addition to all the other skills that he or she uses in the broader role that we have discussed in several chapters of this handbook. However, to understand the forming, chip cutting, and processing principles involved in fabricating parts of metal, and in order to participate in a product design team, introductory background information is essential.

Some of us are quite familiar with the terms and properties of many of the common metals, while others of us are specialists in different fields and need an overview of the subject. The approach the author has taken in this chapter is to introduce the materials most commonly used, and provide an explanation of the properties that make a particular material or alloy a desirable choice for a specific application. It will make a difference in the machines selected, the design of the tooling, and the cost of the part fabrication and finishing. Different materials require different heat treatments, and different surface finishes. Subchapter 14.1 discusses metallurgy, 14.2 introduces iron and steel (ferrous metals), 14.3 talks about aluminum and other nonferrous materials, and 14.4 describes the peculiarities of magnesium.

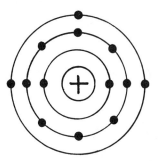

Figure 14.1 Diagram of the aluminum atom, showing 13 electrons around the nucleus.

14.1 FUNDAMENTALS OF METALLURGY

Metallurgy is the art and science which is concerned with metals and their alloys. It deals with the recovery of metals from their ores or other sources, their refining, alloying with other metals, forming, fabricating, and testing, and a study of the relation of alloy constituents and structure to mechanical properties.

The manufacturing engineer is more interested in physical metallurgy, which is concerned with the structure, properties, and associated behavior of metallic products. The properties and behavior of metals are based on their inherent crystalline structure. They do not react as amorphous (shapeless) aggregates of atoms, with a general equality of properties in all directions. They act as crystals with preferred directions of strength, flow, cleavage, or other physical characteristics, and have many limitations due to the oriented character of their particles.

14.1.1 Crystalline Structure

To illustrate crystalline formation, consider a metal in fluid state, in the process of slowly cooling and solidifying. To begin with, we have a solution of free atoms. These atoms consist of a dense nucleus surrounded by several electrons. Figure 14.1 shows a diagram

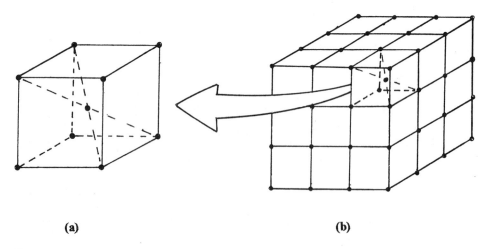

(a)   (b)

Figure 14.2 The body-centered unit cell (a); and a series of these unit cells connected in a small space lattice (b).

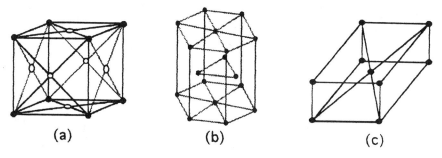

Figure 14.3 The face-centered (a); close-packed hexagonal (b); and body-centered tetragonal(c) unit cells.

of the aluminum atom. With continued cooling, the atoms bond together in groups to form unit cells. A group of unit cells tends to collect as cooling continues, and forms branches, called *dendrites,* which resemble an unfinished frost pattern.

Each type of metal has its own unit cell and space lattice formation. The most common are the following four basic types, depending on the metal. Figure 14.2a shows the *body-centered unit cell,* and Figure 14.2b shows a series of these unit cells connected in a small space lattice. The *face-centered, close-packed hexagonal, and body-centered tetragonal space lattices* are shown in Figure 14.3. With continued cooling, the space lattices combine in groups to form crystals, and the crystals group to form grains. This process is shown in Figure 14.4. The molten state is depicted in Figure 14.4a, the formation of a unit cell in Figure 14.4b, the progressive formation of dendrites in Figures 14.4c and 14.4d, and the final crystals within the grain in Figures 14.4e and 14.4f. A grain is any portion of

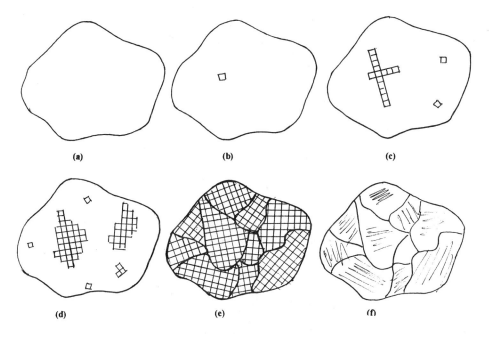

Figure 14.4 Formation of grains, starting with molten metal in (a), and the progressive cooling to achieve a solid metal grain in (f).

a solid which has external boundaries, and an internal atomic lattice structure that is regular.

Body-Centered Structures

The unit cell of a body-centered cubic space lattice consists of eight atoms in a square cube, plus one more in the center of the formation; see Figure 14.2a. The body-centered cubic metals are at least moderately ductile. Metals included in this group are chromium, molybdenum, tantalum, tungsten, vanadium, columbium, and iron below 900°C. They always slip in planes of densest atom packing—diagonally from edge to edge through the center atom.

Face-Centered Structures

As shown in Figure 14.3a, face-centered structures have an atom at each corner of the unit cube and an atom at the center of each face. Thus, the total number of atoms in the basic unit cell of a face-centered cubic space lattice is 14. With the application of pressure, they tend to "slip" in the most dense atomic plane—the face diagonals. This generates 12 slip systems, consisting of the four planes and three directions. Examples of metals in this configuration are as follows:

Aluminum
Copper
Gold
Iron (between 900°C and 1400°C)
Lead
Nickel
Silver
Platinum

All face-centered cubic metals are freely plastic.

Close-Packed Hexagonal Structures

Zinc and cadmium are examples of close-packed hexagonal structures. They are freely plastic and slip horizontally. Magnesium, titanium and zirconium have limited plasticity, while "unworkable" metals include beryllium and osmium. They are shown in Figure 14.3b.

Body-Centered Tetragonal Space Lattice

The body-centered tetragonal space lattice is almost identical to the body-centered cube, as seen if Figure 14.3c. However, the faces of this structure are rectangular instead of square. The major example of this lattice structure is martensitic iron, which is the hardest, strongest, and most brittle type of iron. Tin is unique in that it has a tetragonal-shaped structure and is very ductile. Only iron and tin are polymorphic—they change their type of crystal at *specific temperatures*. Many other metals undergo transformations with the application of heat, which has an important effect on plasticity.

14.1.2 Grain Size and Characteristics

The size of the grain has a profound effect on strength, hardness, brittleness, and ductility. If metal is cooled from the molten state very, very slowly, the colonies have much more time to add on members. Therefore, if metal is cooled slowly, these colonies will have time to grow larger and larger, and very large grain size will result.

On the other hand, if metal is cooled very rapidly, many more colonies will immediately start to spring up. Then, the size of each colony is limited because so many colonies are formed. Therefore, while slow cooling produces a large grain size, rapid cooling produces a small grain size.

The larger grains are easier to tear or break or fracture. Those with small grain size have high resistance to fracture. A small crack has more difficulty moving across a series of small grains than across one large, open field. In summary, the smaller the grain size, the greater is the strength; the larger the grain size, the less is the strength. Since strength, hardness, and brittleness are three inseparable partners, small grain size not only will yield better strength characteristics, but will also result in a harder and more brittle material. On the other hand, if ductility is more important than strength, a larger grain size is desirable.

The ease with which metals yield to applied loads by slip processes enables them to be formed into sheets, wire, tubes, and other shapes. When deformation occurs below a certain critical temperature range, the hardness and strength properties increase, with a corresponding decrease in plasticity. This is *strain hardening*. With further deformation, plasticity may be reduced to the extent that fractures occur.

Deformation processes are used not only for shaping metals, but also for improving strength properties. After cold-working, plasticity (and softness) may be restored by heating to above the limiting temperature.

The term *plasticity*, as applied to metals, refers to the ability of metals to retain a change in shape brought about through deformation using pressure. This is an important metallurgical concept and is the basis for understanding formability and machinability. All metals are crystalline and owe their plasticity to the simplicity and high degree of symmetry of their crystalline structure.

14.2 FERROUS METALS

Historically, the principal raw materials used in steel making are ferrous scrap and pig iron, a product of the blast furnace. In the production of pig iron, iron ore containing from 50% to 60% iron, coke for fuel, and limestone for fluxing off ore impurities and coke ash are charged into the top of the blast furnace. A preheated air blast is introduced near the bottom of the furnace, burning the coke and forming carbon monoxide gas, which, in turn, reduces the iron oxides in the ore, leaving the iron in the metallic state. Molten iron dissolves considerable amounts of carbon, silicon, phosphorus, and sulfur; as a result, the pig iron contains these "impurities" in such quantities that it is extremely hard and brittle, making it unsuitable for applications where ductility is important. Figure 14.5 shows a typical blast furnace.

14.2.1 Steel Making

Steel making may be described as the process of removing impurities from pig iron and scrap and then adding certain elements in predetermined amounts so as to impart the desired properties to the finished metal. The elements added in the process may be, and in many cases are, the same ones that have been removed, but the amounts or proportions are different. Figure 14.6 shows the steel-making process. Starting with the raw materials

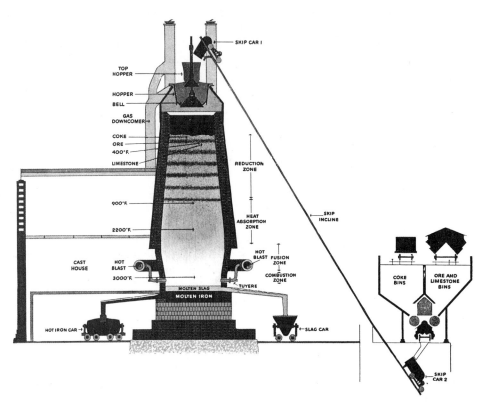

Figure 14.5 Cross section of a blast furnace in which pig iron is made. (Courtesy of Bethlehem Steel.)

on the left side, the farther the steel progresses to the right of Figure 14.6, the more expensive it becomes. However, the finished part or product may be of lower cost by using the more finished metal—as is the case with formed sheet metal stampings as opposed to castings.

Historically, virtually all steel was made in either the open-hearth furnace or the electric furnace, with small percentages made by the basic oxygen process and in the Bessemer converter. All processes use pig iron in varying proportions, from 60% to less than 10%. Figure 14.7 shows a cutaway diagram of a typical open-hearth furnace. The steel produced from one furnace charge, known as a "heat," will usually weigh from 100 to 400 tons, although both larger and smaller furnaces are currently in operation. A furnace is charged with scrap, limestone, and ore. The pig iron or "hot metal" is usually added in the molten state, after the scrap is partially melted. During subsequent refining, nearly all of the manganese, phosphorus, and silicon are oxidized and retained in the slag. The carbon is generally removed by oxidation to a percentage approximating that desired in the finished steel. At this point the heat is tapped into a ladle. To obtain the desired analysis, ferromanganese and other alloying materials are added as needed. In some cases they are added to the molten bath just prior to tapping; in others they are added to the ladle as the heat is being tapped. Aluminum or ferrosilicon also is generally added to the ladle to "deoxidize" the steel, as discussed later. The heat is then poured into ingot molds and solidifies into steel ingots. Recent advances in steel making allow a continuous billet to be

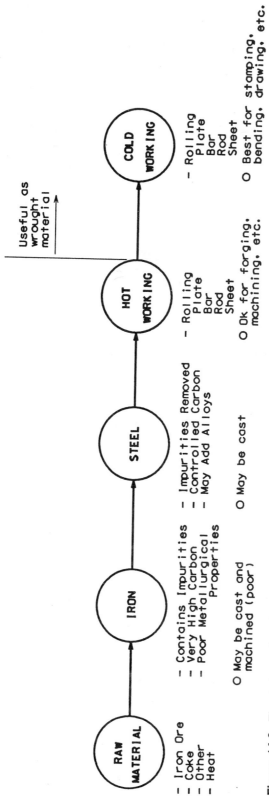

Figure 14.6 The steel manufacturing process.

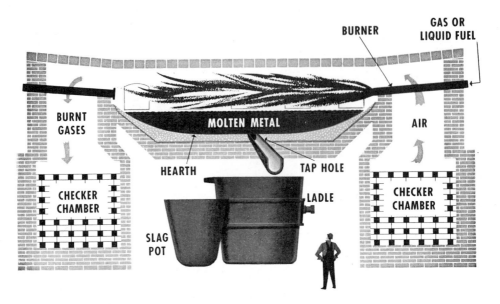

Figure 14.7 Simplified cutaway diagram of a typical open-hearth furnace, viewed from the tapping side. (Courtesy of Bethlehem Steel.)

produced by cooling the metal as it is being poured—not in an ingot but as a continuous slab of metal.

Special steels, particularly high-alloy steels, stainless steels, and tool steels, all of which have an expanding use in industry, are frequently made in electric furnaces. These vary in capacity from a few hundred pounds up to 100 tons or more. Electric arcs between the electrodes and the metal bath furnish the heat. The advantage of this type of furnace is that it is kept closed, and operates with a neutral atmosphere. Oxidizing and reducing agents can therefore be applied as required, allowing close control of the chemical elements in the steel. See Subsection 14.2.2 for a discussion of the increasing utilization of "mini-mills" using a high percentage of scrap metal and direct-reduction iron.

Definitions of Carbon and Alloy Steels

It is sometimes difficult to draw a clear dividing line between carbon and alloy steels. They have been arbitrarily defined by the American Iron and Steel Institute as follows.

Carbon Steel

Steel is classed as *carbon steel* when no minimum content is specified for aluminum, boron, chromium, cobalt, columbium, nickel, titanium, tungsten, vanadium, or zirconium, or any other element added to obtain a desired alloying effect; when the specified minimum for copper does not exceed 0.40%; or when the maximum content specified for any of the following elements does not exceed the percentages noted: manganese 1.65%, silicon 0.60%, copper 0.60%.

By far the most important element in steel is carbon. The following are general rules to classify steel based on its carbon content:

Wrought iron	Trace to 0.08%
Low-carbon	0.10 to 0.30%
Medium-carbon steel	0.30 to 0.70%
High-carbon steel	0.70 to 2.2%
Cast iron	2.2 to 4.5%

Note: All carbon above 2.2% is uncombined with iron, and is present in the form of graphite. This presents planes of easy cleavage, which accounts for the easy breakage of cast iron.

Alloy Steel

Steel is classified as *alloy steel* when the maximum of the range specified for the content of alloying elements exceeds one or more of the following limits: manganese 1.65%, silicon 0.60%; copper 0.60%; or in which a definite range for a definite minimum quantity of any of the following elements is specified or required within the limits of the recognized commercial field of alloy steels: aluminum, boron, chromium up to 3.99%; cobalt, columbium, molybdenum, nickel, titanium, tungsten, vanadium, zirconium, or any other alloying element added to obtain a desired alloying effect.

In addition to differences in the steel-making processes, factors such as segregation, the type and amount of deoxidizers used, and variations in chemical analysis all profoundly affect the properties of the steel.

The Steel Ingot

Many hundreds of shapes and sizes of steel ingots have been developed over the last century. The cross section of most ingots is roughly square or rectangular, with rounded corners and corrugated sides. All ingots are tapered to facilitate removal from the molds. Depending on the type of steel, ingots may be poured big end up or big end down, as will be discussed under "types of steel."

All steel is subject to variation in internal characteristics as a result of natural phenomena which occur as the metal solidifies in the mold. Liquid metal, just above the freezing point, is less dense than solid metal just below it; that is, there is a shrinkage in volume during solidification. Hence a casting or an ingot is given a "sink head," large enough to supply the extra metal needed in the desired shape when frozen. If this extra metal can be fed in while freezing is going on, the frozen metal will be solid; otherwise it will have voids at locations where feeding has been cut off by early freezing at thin sections between those locations and the sink head. In ingots, unless the metal freezes wholly from the bottom up and not simultaneously from the sides, there will be a "pipe" (see Figure 14.8). The flow of metal in the sink head to fill the void must not be cut off; else "secondary pipe" or "spongy centers" will result. The extent of the piping is dependent upon the type of steel involved, as well as the size and design of the ingot mold itself. Pipe is eliminated by sufficient cropping during rolling.

Another condition present in all ingots to some degree is nonuniformity of chemical composition, or segregation. Certain elements tend to concentrate slightly in the remaining molten metal as ingot solidification progresses. As a result, the top center portion of the ingot, which solidifies last, will contain appreciably greater percentages of these elements than the average composition of the ingot. Of the elements normally found in steels, carbon, phosphorus, and sulfur are most prone to segregate The degree of segregation is influenced by the type of steel, the pouring temperature, and ingot size. It will vary with position in the ingot and according to the tendency of the individual element to segregate.

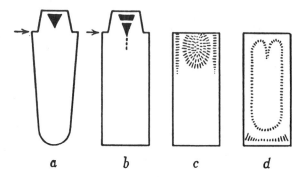

Figure 14.8   Pipe or blowholes in cast steel ingots.

Types of Steel

The primary reaction involved in most steel-making processes is the combination of carbon and oxygen to form a gas. If the oxygen available for this reaction is not removed prior to or during casting (by the addition of ferrosilicon or some other deoxidizer), the gaseous products continue to evolve during solidification. Proper control of the amount of gas evolved during solidification determines the type of steel. If no gas is evolved, the steel is termed "killed," because it lies quietly in the molds. Increasing degrees of gas evolution result in semikilled, capped, or rimmed steel.

Rimmed steels are only slightly deoxidized, so a brisk effervescence or evolution of gas occurs as the metal begins to solidify. The gas is a product of a reaction between the carbon and oxygen in the molten steel which occurs at the boundary between the solidified metal and the remaining molten metal. As a result, the outer "rim" of the ingot is practically free of carbon. The rimming action may be stopped mechanically after a desired period, or it may be allowed to continue until the action subsides and the ingot top freezes over, thereby ending all gas evolution. The center portion of the ingot, which solidifies after rimming ceases, has a composition somewhat above that of the original molten metal, as a result of the segregation tendencies discussed above.

The low-carbon surface layer of rimmed steel is very ductile. Proper control of rimming action will result in a very sound surface during subsequent rolling. Consequently, rimmed grades are particularly adaptable to applications involving cold-forming and where the surface is of prime importance.

The presence of appreciable percentages of carbon or manganese will serve to decrease the oxygen available for the rimming action. If the carbon content is above 0.25% and the manganese over 0.60%, the action will be very sluggish or nonexistent. If a rim is formed, it will be quite thin and porous. As a result, the cold-forming properties and surface quality will be seriously impaired. It is therefore standard practice to specify rimmed steel only for grades with lower percentages of these elements.

Killed steels are strongly deoxidized and are characterized by a relatively high degree of uniformity in composition and properties. The metal shrinks during solidification, thereby forming a cavity or pipe in the extreme upper portion of the ingot. Generally, these grades are poured in big-end-up molds. A "hot-top" brick is placed on top of the mold before pouring and filled with metal after the ingot is poured. The pipe formed is confined to the hot-top section of the ingot, which is removed by cropping during subsequent rolling. The most severe segregation of the ingot is also eliminated by this cropping.

While killed steels are more uniform in composition and properties than any other type, they are nevertheless susceptible to some degree of segregation. As in the other grades, the top center portion of the ingot will exhibit greater segregation than the balance of the ingot. The uniformity of killed steel renders it most suitable for applications involving such operations as forging, piercing, carburizing, and heat treatment.

Semikilled steels are intermediate in deoxidation between rimmed and killed grades. Sufficient oxygen is retained so that its evolution counteracts the shrinkage upon solidification, but there are is no rimming action. Consequently, the composition is more uniform than that of rimmed steel, but there is a greater possibility of segregation than in killed steels. Semikilled steels are used where neither the surface and cold-forming characteristics of rimmed steel nor the greater uniformity of killed steels are essential requirements.

Capped steels are much the same as rimmed steels except that the duration of the rimming action is curtailed. A deoxidizer is usually added during the pouring of the ingot, with the result that a sufficient amount of gas is entrapped in the solidifying steel to cause the metal to rise in the mold. With the bottle-top mold and heavy metal cap generally used, the rising metal contacts the cap, thereby stopping the action. A similar effect can be obtained by adding ferrosilicon or aluminum to the ingot top after the ingot has rimmed for the desired time. Action is stopped and rapid freezing of the ingot top follows. Rimming times of 1 to 3 min prior to capping are most common.

Capped steels have a thin, low-carbon rim which imparts the surface and cold-forming characteristics of rimmed steel. The remainder of the cross section approaches the degree of uniformity typical of semikilled steels. This combination of properties has resulted in a great increase in the use of capped steels.

14.2.2 Steel Rolling

Vladimir B. Ginzburg, author of *High-Quality Steel Rolling—Theory and Practice* (Marcel Dekker, New York, 1993), has the following comments on steel rolling in his book:

In each stage of the development of steel rolling technology, there have been specific challenges to be met by both steel producers and designers of rolling mill equipment. In the past three decades, the most important challenges have included increasing production rates, conserving energy, increasing coil weights, and reducing the finishing gauge.

These goals have been gradually achieved by the majority of steel producers, however the two main challenges that remain for steel producers today are improving quality and reducing production costs. Although these goals have always been considered in the past, they are now looked at in a completely new perspective because of the following three factors:

Excess capacity for production of flat rolled products
Entry of developing nations into the marketplace
Entry of mini-mills into flat rolled production

This excess capacity for the production of flat rolled steel products has created an extremely competitive environment among the world's leading steel producers. The producers who meet the customer's high quality standards and at the same time maintain low production costs have a distinct advantage over their competitors. However, current competition is not just between the steel producers of the major industrial countries.

The competition is continuously intensifying as a growing number of developing countries enter into the steel producing market. The impact of the developing nations on the world steel marketplace is not just in capacity alone, but in improved product quality and lower production costs. These developments can be directly attributed to the huge investments that the developing nations have made in modern rolling mill technology for their steel producing plants. The most recent and possibly most influential development is the transfer of the flat rolled steel production process from integrated steel mills that use iron ore and coal as the prime sources of their steelmaking process to mini-mills that utilize steel scrap and direct reduced iron. The rate of this transfer will depend on the capability of the integrated steel producers to defend their market position by further improving product quality and reducing production costs.

Effect of the Rolling Process on the Customer Goods Manufacturer

The steel may be hot rolled to final size (when this is not too small), or the reduction in size may be completed by the addition of cold rolling or cold drawing. The internal structure and the properties of the steel may be adjusted by the temperature at which hot rolling is done, by the rate of cooling from the hot-rolling temperature, by reheating and cooling at a controlled rate (called *annealing* when the cooling is slow, *normalizing* when the cooling is done in air and is more rapid), by the amount of reduction in cold working and by low-temperature annealing of cold worked steel.

By such heat treatments and by the use of a trace of vanadium in unkilled steels, or by the use of vanadium, aluminum, titanium, etc., in killed steels, the grain size can be varied from coarse to fine, with marked influence on behavior. The directional effects of rolling in one direction only can be minimized by cross-rolling, although the through-plate direction retains its differences. Cast metals have closely similar properties in different directions, but wrought metals seldom have. Rolled metals frequently show somewhat different tensile and yield strengths when the test specimen is taken longitudinally (in the direction of rolling) or transversely (at right angles to the rolling direction). The ductility is likely to be markedly lower in the transverse direction, and the properties in the "through-plate" direction tend to be spectacularly poorer.

Pure tensile stress in a part is rare. When stresses come in two or more directions, behavior under these biaxial or triaxial stresses cannot always be predicted from a knowledge of behavior under simple tension. Hence, combined stresses with components other than pure tension are working on material whose properties are not evaluated by the conventional tensile test. Application of the stresses to be met in service, both in magnitude and in pattern, to the particular material to be used, in full size and in its exact geometry, is often required for certainty of behavior. This testing is vital in product development today.

Handbook values are usually given only for longitudinal specimens of wrought metal; for small, fully quenched specimens of heat-treated steel, and for fully fed specimens of cast metal. It needs to be noted whether or not a handbook or reported test value is on a true sample, one really representing the material as it is to be used. A value determined for an as-rolled or a normalized steel in one thickness will not necessarily hold for the same steel in another thickness, nor is the strength in one direction necessarily the same as in another.

Composition of Steel

Steel is an iron-base alloy whose strength is due primarily to its carbon content. Small amounts of manganese, and frequently of silicon and a trace of aluminum, are also present in steel. The carbon in steel ready for most commercial uses is present as iron carbide (Fe_3C), called *cementite*. The iron matrix is called *ferrite*. The carbon may be present in plates of ferrite, in a structure called *pearlite*. Steels with a predominant ferrite matrix are called *ferritic*. If pearlite predominates, they are *pearlitic*.

At the high temperatures used for hot rolling or forging, or to prepare the steel for being hardened by quenching, the steel has a different crystal structure, called *austenite*, in which the carbon is in solid solution, interspersed in the iron, but not combined as in cementite. Upon cooling, the austenite of ordinary steels transforms to ferrite and carbide. With the presence of sufficient amounts of certain alloying elements, austenite can be retained upon cooling without transformation. Stainless steel with 18% chromium and 8% nickel, and manganese steel with about 1.25% carbon and 13% manganese, are familiar examples of austenitic steel. Figure 14.9 illustrates the heat-treating process that produce ferrite, austenite, and martensite in steel. In Figure 14.9a, ferrite is transformed to austenite and back again in stages between the lower transformation temperature and the upper transformation temperature during slow cooling. In Figure 14.9b, below the the lower transformation temperature, ferrite exists without austenite. Above the upper transformation temperature, austenite exists without ferrite. In Figure 14.9c, martensite is transformed to austenite and back again in stages between the lower transformation temperature and the upper transformation temperature during fast cooling.

Space Lattice Structures in Iron and Steel

Iron is unusual in that it can take three different space lattice structures. As iron goes through a temperature change, its atoms realign themselves into new geometric patterns. This has a great effect on the strength, hardness, and ductility of the iron.

Ferritic iron or ferrite takes the body centered cubic lattice structure formation. Ferrite is basic iron at room temperature which has not previously been heat-treated.

Austenitic iron or austenite takes the face-centered cubic lattice structure. Austenite is the structure that iron takes at elevated temperatures. In other words, if ferrite is heated, it gradually becomes austenitic when high temperature is reached. As it is becoming austenite, the atoms are reshuffling within the crystal, realigning themselves into a new space lattice formation.

Martensitic iron or martensite has the body-centered tetragonal crystal lattice structure. Martensite is iron at room temperature that has previously been heated and suddenly quenched. The heating and quenching operation serves to produce this third geometric pattern. Heating and sudden quenching tend to harden metal. Therefore, martensite is the strongest, hardest, but most brittle of the three iron structures.

The *lower transformation temperature* is the temperature at which the body-centered cubic structure starts to change to the face-centered cubic structure. It is the temperature at which ferrite starts to change to austenite. The *upper transformation temperature* is the temperature at which the body-centered cubic lattice structure has changed completely to face-centered cubic. It is the temperature at which no ferrite exists. All of the iron structure above the upper transformation temperature is austenite.

14.2.3 Steel Sheet Properties

Figure 14.10 is a load-versus-elongation curve for a typical steel sheet. Tensile testing is a common method of determining the mechanical properties of metal. A sample, taken from a roll of steel, is placed under tension and pulled until it fails. Data obtained from the test are used to plot a load-elongation (stress-strain) curve that shows the yield point, yield-point elongation, total elongation, ultimate tensile strength, and other properties. All

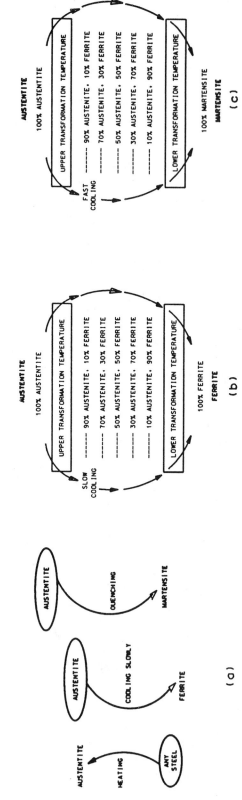

Figure 14.9 The heating and cooling processes that produce ferrite, austenite, and martensite in steel.

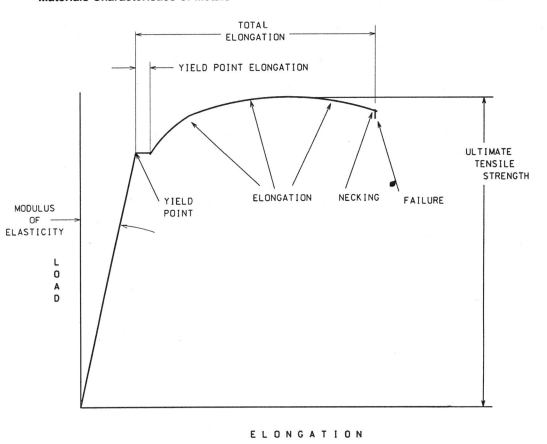

Figure 14.10 Stress-strain curve for typical wrought steel sheet.

metal forming takes place after reaching the yield point but before the ultimate tensile strength is reached. This means that a metal with a relatively low yield point and high ultimate strength would be the easiest to form. On the other hand, if the yield point is quite high in relation to the ultimate strength, careful measures must be taken to prevent overstressing the metal, or it will fail in the forming process. The yield point, which must be exceeded to produce a permanent shape change in the metal, begins when elastic deformation ends and plastic deformation starts. Beyond this point, the steel yields discontinuously (repeated deformation followed by work hardening) up to the point where the load on the steel rises continuously. This yield-point elongation can produce strain lines if it exceeds 1.5% of total elongation.

After yield-point elongation ends, the steel thins in two dimensions—through the thickness and across the width. A good way to measure ductility is by the percent elongation in the gage length of a broken test sample. In low carbon sheet steels, for example, elongation is usually between 36 and 46% in a 2-in. gage length. The amount of springback in forming aluminum alloys is generally less than in forming low-carbon steel, and this must be considered in tool design. The amount of springback is roughly proportional to the yield strength of the metal. The slower rate of work hardening of aluminum alloys permits a greater number of successive draws than is possible with steel.

Achieving Overall Economy

Figure 14.11 is a block diagram of the principal parties which are involved in producing consumer goods based on steel products. At the top of the hierarchy is the consumer, who, through the market mechanism, dictates both the quality and the price of goods. Thus, for example, if the consumer demands a better fit of car panels, the car manufacturer may consider the following approaches:

Demand tighter geometric tolerances of the coils being supplied by steel producers
Modernize production equipment so that the quality of the assembly process can be improved without tightening the geometric tolerances of the supplied coils
Distribute improvements in quality fairly between the car manufacturers and steel producers

If the car manufacturer decides to tighten the tolerances of the purchased coils to 50% of the standard value, then the steel producer, in turn, would have to consider the following similar approaches:

Demand that the machinery supplier install mill equipment that will produce coils with geometric tolerances within 25% of the standard value
Improve maintenance, operating practices, and quality control so that the desired tolerances can be obtained without modernization of the mill equipment
Distribute improvements in quality fairly between the steel producers and machinery suppliers

Tremendous efforts have been made in the last decade by all parties to improve the quality of products that are produced by their own facilities. There are, however, questions which one may ask. The first question is if the burden of improving quality has been fairly

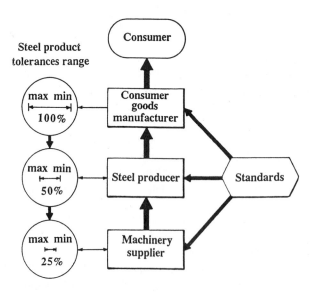

Figure 14.11 Hierarchy of subordination and distribution of tolerances between consumer goods manufacturer, steel producer, and machinery supplier. (From V. Ginzburg, *High-Quality Steel Rolling*, Marcel Dekker, New York, 1993.)

distributed between all parties so that the overall economy is maximized. If the answer is no, then who is going to make this fair distribution of responsibilities for improvement of quality? Obviously, in the long run, it will be enforced by the market system. In the short term, however, it is very difficult to prevent the passing of a disproportionate burden by the parties located at the higher hierarchical levels of the production process to the ones at the lower levels.

From this point of view, standards may play an important role by stipulating technically and economically reasonable tolerances that would be required to be achieved for each particular application. Ginzburg shows comparative analyses of the Japanese JIS standards, the ASTM standards, the German DIN standards, and the new ISO standards for some of the parameters of flat-rolled products. An example of the thickness tolerances of cold-rolled, high-strength sheets for automotive applications is shown in Figure 14.12, comparing Japanese and U.S. standards. The manufacturing engineer must carefully compare tolerances as well as price in the selection of a material for production.

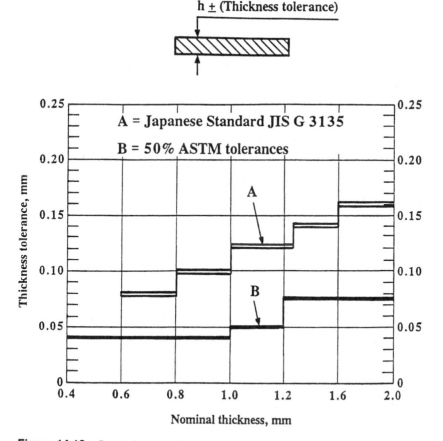

Figure 14.12 Comparison of thickness tolerances of cold rolled high strength steel sheets for automobile applications. (From V. Ginzburg, *High-Quality Steel Rolling*, Marcel Dekker, New York, 1993.)

Figure 14.13 plots the manufacturing cost being reduced as the quality (and cost) of the finished steel is increased—showing the optimum quality level somewhere in between.

14.2.4 Steel Designation Systems

Steel is composed primarily of iron. Most steel contains more than 90% iron. Many carbon steels contain more than 99% iron. All steel contains a second element, which is carbon. Many other elements, or alloys, are contained in most steels, but iron and carbon are the only elements that are in all steels. The percent carbon in steel ranges from just above 0% to approximately 2%. Most steels have between 0.15 and 1.0% carbon.

Steels with the least carbon are more flexible and ductile (tend to deform appreciably before fracture), but they are not as strong. However, as the carbon content increases, so do strength, hardness, and brittleness.

Each type of steel has a name, usually consisting of four numbers. The first two digits refer to the alloy content. The last two digits refer to the percent carbon in the steel. In 5147 steel, for example, the "51" tells you that the steel has a lot of chromium in it. In 2517 steel, the "25" indicates that there is an unusual amount of nickel in this steel. Similarly, the "10" in 1040 steel, tells you that the steel has very little alloy content except carbon. The last two digits (or three digits) indicate the percent carbon that the steel contains. In 1040 steel, for example, the "40" tells you that there is 0.40% carbon in the steel. In 1018 steel, the "18" indicates that there is only 0.18% carbon in it; thus, it is a very low carbon steel. An 8086 steel contains approximately 0.60% carbon, which makes it a medium-carbon steel. Table 14.1 relates the alloy content in steel to the first two digits of its name. Table 14.2 shows some examples of common steels with their carbon percentages, major alloying ingredients, and tensile strengths.

14.3 NONFERROUS METALS: ALUMINUM

Aluminum is made by the electrolysis of aluminum oxide dissolved in a bath of molten cryolite. The oxide, called alumina, is produced by separating aluminum hydrate from the

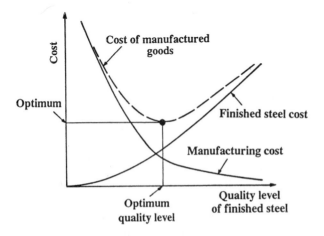

Figure 14.13 Quality level/cost relationship in manufacturing plants utilizing finished steel products. (From V. Ginzburg, *High-Quality Steel Rolling*, Marcel Dekker, New York, 1993.)

Table 14.1   Relation of the Alloy Content in Steel and
the First Two Digits of Its Name

Steel Numerical Name	Key Alloys
10XX 11XX 13XX	Carbon Only Carbon Only (free Cutting) Manganese
23XX 25XX 31XX	Nickel Nickel Nickel-Chromium
33XX 303XX 40XX	Nickel-Chromium Nickel-Chromium Molybdenum
41XX 43XX 44XX	Chromium-Molybdenum Nickel-Chromium-Molybdenum Nickel-Molybdenum
46XX 47XX 48XX	Chromium Chromium Chromium
50XX 51XX 501XX	Chromium Chromium Chromium
511XX 521XX 514XX	Chromium Chromium Chromium
515XX 61XX 81XX	Chromium Chromium-Vanadium Nickel-Chromium-Molybdenum
86XX 87XX 88XX	Nickel-Chromium-Molybdenum Nickel-Chromium-Molybdenum Nickel-Chromium-Molybdenum
92XX 93XX 94XX	Silicone-Manganese Nickel-Chromium-Molybdenum Nickel-Chromium-Molybdenum Manganese
98XX XXBXX XXLXX	Nickel-Chromium-Molybdenum Boron Lead

impurities associated with it in naturally occurring deposits of bauxite, and calcining to drive off the combined water.

The electrolytic process was discovered only a hundred years ago. In the short period since, the production has risen until aluminum now stands third in tonnage among the nonferrous metals, and the volume of aluminum produced is second only to that of steel.

In order to understand better the internal structural modifications that occur as a result of the various alloying ingredients and heat-treating operations, some knowledge of the

Table 14.2   Alloy Content of Several Typical Steels

STEEL	TYPE OF STEEL	TENSILE STRENGTH x 1000psi	C	Mn	P	S	Si	Ni	Cr	Mo	V
1025	Plain Carbon	60- 103	.22- .28	.30- .60	.04max	.05 max					
1045	Plain Carbon	80- 182	.43- .50	.60- .90	.04max	.05 max					
1095	Plain Carbon	90- 213	.90- 1.0	.30- .50	.04max	.05 max					
1112	Free Cutting Carbon	60- 100	.13 max	.70- 1.00	.07- .12	.16- .23					
1330	Manganese	90- 162	.28- .33	1.60- 1.9	0.035	0.04	.20- .35				
2517	Nickel	88- 190	.15- .20	.45- .60	0.025	0.025	.20- .35	4.75- 5.25			
3310	Nickel Chromiun	104- 172	.08- .13	.45- .60	0.025	0.025	.20- .35	3.25- 3.75	1.40- 1.75		
4023	Molybdenum	105- 170	.20- .25	.70- .90	0.035	0.04	.20- .35			.20- .30	
52100	Chromium	100- 240	.98- 1.1	.25- .45	0.035	0.04	.20- .35		1.30- 1.60		
6150	Chromium Vanadium	96- 230	.48- .53	.70- .90	0.035	0.04	.20- .35		.80- 1.10		.15 min
8840	Nickel Chromium Molybdenum	120- 280	.38- .43	.70- .90	0.04	0.04	.20- .35	.85- 1.15	.70- .90	.20- .30	
4140	Chromium Molybdenum	95- 125	.38- .43	.75- 1.00	0.035	0.04	.20- .35		.80- 1.10	.15- .25	

physical changes taking place during solidification is necessary. When molten aluminum is cooled, its temperature drops until it reaches 1220.4°F, at which point the material gives up its latent heat of fusion and begins to solidify. Aluminum, as well as most of the other easily worked metals, crystallizes as a face-centered cubic structure, which possesses more effective slip planes than any other structure. As cooling continues, additional crystals form, building on the first ones and producing the larger units, called "grains". The temperature will remain nearly constant at 1220.4°F until the entire mass has solidified. Then the temperature again drops as cooling continues. The solidified metal is thus composed of grains, which are in turn composed of crystals.

Figure 14.14 indicates the relation between time and temperature as a pure metal is allowed to cool from the molten state, represented by point A. As its temperature falls, it reaches a point B, where the metal begins to solidify or freeze. Note that the curve indicates that the temperature remains at this value for a period of time. This is because the change from a liquid to a solid is accompanied by the release of heat, the mechanism of the operation being such that just enough heat is released to balance that being lost, thus retaining the temperature of the metal constant during the period this solidification is taking place. Therefore the curve is level from B to C. As soon as the metal has solidified completely, its temperature again falls gradually as it is allowed to cool, represented by the sloping line D.

It should be noted that only a pure metal follows this type of curve, and each different metal has a different solidification or freezing point; that is, the level portion of the curve or "plateau" occurs at a different temperature.

Now let us see what happens when we melt two pure metals together—say, aluminum and copper—and allow them to cool. We find that we have a curve of an entirely different shape because the combination of the two metals has a freezing "range" instead of a freezing "point"; that is, the material begins to freeze at one temperature and continues to freeze while the temperature falls to a lower value before all of it has solidified. This is shown by the dashed portion of the curve at F, where the curve slopes from E to G. The

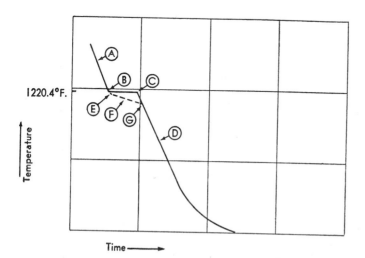

Figure 14.14 Time versus temperature for a pure aluminum metal. (Courtesy Reynolds Metal Company.)

combination of aluminum and copper does not freeze completely at a single temperature because the mixture formed by the two metals behaves in an entirely different manner than a pure metal such as copper or aluminum.

At point E, the crystals forming out as the molten metal is just beginning to solidify consist of an alloy of almost pure aluminum. As the temperature falls, crystals with appreciable amounts of copper begin forming. With continued dropping temperature, the crystals forming contain more and more copper. Thus, at E, the alloy particles freezing out may contain 99.9% aluminum and 0.1% copper. Just below E, the particles freezing out of solution may contain 98% aluminum and 2% copper. Similarly, particles containing 97% aluminum and 3% copper will freeze out at a lower temperature, and so on. At G, the entire mass is solidified and the temperature drops along the same type of curve as before.

Thus, as the temperature falls, the material freezing out of solution at any particular moment corresponds to the alloy of aluminum and copper that freezes at that temperature.

14.3.1 Alloys of Aluminum

High-purity aluminum, while it has many desirable characteristics, has a tensile strength of only about 9000 psi. Even though this strength can be doubled by cold working, the resulting strength is still not high and the alloy is not heat treatable. The small amounts of iron and silicon present in "commercially pure" aluminum increase the annealed condition strength by about 45%. Addition of 1.25% manganese, in addition to the impurities iron and silicon, produces a strength some 75% greater than the annealed pure metal. Addition of 2.5% magnesium increases the strength to about three times that of pure metal, and four times the pure metal strength after cold working.

Pure aluminum has inferior casting qualities. The improvement in casting qualities of aluminum alloys as compared to the pure metal is perhaps even greater than the improvement in their mechanical properties. The alloy that first gained general use contained 8% copper in addition to the impurities normally present in commercial aluminum. By reducing the copper below 5% and adding a rather large percentage of silicon, a new series of alloys came into use, followed by a group of alloys in which silicon was the only added element—or at least the major addition.

The alloys that respond to heat treatment with improvement in their physical properties all contain at least one constituent whose solid solubility is substantially greater at elevated temperatures than at room temperature. For best results this element is added only in amounts that are completely soluble at temperatures below the melting point of the alloy, although small amounts of other alloying elements may be added. Wrought alloys are available in heat-treated tempers of 80,000 psi, and casting alloys are available which can be heat-treated to strengths of nearly 50,000 psi.

Intermetallic Compounds

The addition of soluble elements to aluminum to produce aluminum alloys exerts a pronounced influence on the behavior of the material during the cooling period. In the molten state, certain of the alloying elements combine with each other and with the aluminum to form complex compounds called *intermetallic compounds*. These have characteristics entirely different from those of the elements of which they are composed.

Some of the intermetallic compounds may be dissolved in the molten aluminum alloy. Their presence lowers the solidification (or freezing) temperature of the molten metal just as the addition of alcohol to water produces a solution that freezes at a lower temperature than pure water. The freezing point of an alcohol-water solution is determined by the relative proportions. Likewise, the freezing point of molten aluminum alloy is dependent upon the amount and type of dissolved constituents present.

There are many different compounds in aluminum alloy. Consequently, the material does not have one solidification point. Instead, it solidifies throughout a temperature *range*. Aluminum alloys start to freeze at a temperature just below 1220.4°F and are not completely solidified until a still lower temperature is reached. The points at which solidification starts and ends, referred to as the *liquidus* and *solidus temperatures*, are dependent on the constituent elements and the amount of each that is present in the alloy. For commercially pure aluminum, the liquidus is 1215°F and the solidus is 1190°F. Addition of alloying elements changes these figures considerably. For example, an alloy containing about 4% copper, 0.5% magnesium, and 0.5% manganese posses a liquidus temperature of 1185°F and a solidus of 955°F.

In freezing, the first crystals to form, at approximately 1220°F, are of pure aluminum. Just below this temperature, a solution containing a major percentage of aluminum with only a minute amount of dissolved compounds will freeze. This substance solidifies separately from the primary crystals and deposits around the original pure crystals. As freezing progresses, the proportion of aluminum in the remaining molten matter becomes smaller, and the dissolved compounds form a larger and larger portion. These substances also solidify separately from the primary crystals according to their respective solidus temperatures and deposit against the core and branches of the dendrites already formed around the original pure aluminum crystals.

It is easy to see how successive layers containing increasing amounts of soluble elements solidify on the previous portion as freezing progresses. Finally, a temperature is reached at which the last portion of the melt freezes. The last metal to solidify contains a large portion of the elements added to the aluminum. Generally, this material is hard and brittle. Freezing last, these brittle components are thus concentrated at the grain boundaries (and between dendrites). The mechanical properties of an aluminum alloy are determined by the shape and size of the grains, the layer of brittle material between the dendrites, the type and amount of dissolved compounds, the shape and distribution of the insoluble constituents, and the amounts of these materials present. In the as-cast alloy, these factors combine to produce points of weakness. This inherent weakness can be overcome in two ways—by proper heat treatment (homogenizing) or by mechanically working the material.

There is another phenomena that occurs in casting a molten alloy into a mold. Freezing initially takes place at the surfaces of the mold, thus forming an outer solid shell with the center portions remaining liquid. Like other metals, aluminum alloys contract considerable and lose volume when passing from the liquid to the solid state. This contraction of the outer shell tends to exert hydrostatic pressure on the liquid metal inside the shell. This pressure acts to actually squeeze out the alloy-rich liquid mixtures from between the grains and dendrites. The result is that examination of a cast ingot shows few particles of the alloy-rich compounds on the outer surface. Also, a step-by-step analysis of the ingot from the center outward reveals that certain alloy percentages increase greatly as the outer skin is approached.

Homogenizing

The resulting ingot can be made homogeneous by a special preheating technique appropriately called *homogenizing*. For this purpose, the metal is heated within the range of 900 to 1000°F and held at that temperature for a period of time—sufficient for "solid diffusion" to take place. Solid diffusion is a term used to denote the diffusion or spreading out or dissolving of one intermetallic compound into another when both are in the solid state. While it is well known that certain liquids can dissolve certain solids, as water dissolves salt, it is also true that some solids can dissolve other solids—and that is what occurs here.

While holding at temperature, not only do the alloying elements and compounds diffuse evenly throughout the ingot, but also the so-called cored structure of the grains is diffused evenly. You will recall that during solidification, the first crystals to form are almost pure aluminum and succeeding layers contained more and more of the alloying materials. Thus the inside crystals near the core of a grain are very different from the outer crystals of that grain—producing a cored structure. Homogenizing, however, allows these crystals of alloying materials to diffuse evenly throughout the structure and thus corrects the undesirable cored arrangement. Since the phenomenon of solid diffusion proceeds almost imperceptibly at room temperature and increases speed with temperature, homogenizing is done at as high a temperature as possible—just below the melting point of the compounds present in the aluminum alloy.

Plastic Deformation

When a metal is subjected to sufficient stress, such as that produced by plastic deformation, slippage occurs along definite crystallographic planes. The number of planes on which this slippage can take place is entirely dependent on the crystal structure of the metal. Aluminum, as well as most of the other easily worked metals, crystallizes in the form of a face-centered cubic structure, which possesses more effective slip planes than any other structure. Metals that possess this crystal structure can therefore be plastically deformed severely before rupturing occurs. (See Subchapter 14.2 for further explanation of crystal structure.)

When aluminum is subjected to plastic deformation, slippage takes place along the slip planes that are most favorably oriented with regard to the direction of the applied stress. As slippage continues, the planes which are slipping change their positions in such a manner that they become less favorably oriented to the applied stress than other planes. Slippage then begins along these other planes. As the degree of plastic deformation progresses, the planes continue to change their orientation and the metal becomes increasingly difficult to work.

The changing of the positions of the slip planes, often referred to as *rotation* of the slip planes, produces a condition wherein a substantial number of planes have the same orientation. This condition is known as *preferred orientation*, and is one of the reasons why some material forms "ears" when deeply drawn. The direction of preferred orientation in cold-worked aluminum depends on the thermal treatments, the degree of plastic deformation, and the direction of the applied stresses which produced the orientation.

Fragmentation

The slipping of the planes naturally causes fragmentation of the grains. Increasing the degree of cold working increases the amount of fracturing that takes place, with the grains becoming elongated. The amount and directions of the elongation are closely associated with the reduction of cross-sectional area and the direction of working. Slippage along the slip planes is restricted in several ways. The distortion of the space lattice by atoms of other elements in a solid solution or by mechanical strain restricts slippage. The presence of insoluble or precipitated constituents can exert a keying effect which also restricts slippage. Small grain size is still another factor, due to the interference of the grain boundaries. There are many other factors, such as interatomic cohesion forces, but the above are the major ones that are closely associated with plastic deformation.

Cold-working aluminum increases the tensile strength, the yield strength, and the hardness, but decreases ductility properties such as percent elongation, the impact strength, and the formability. Excessive cold working will result in reaching a point where excessive pressures are required for further reduction, or where fracturing of the metal structure occurs. This comment applies to both the heat-treatable as well as the non-heat-treatable aluminum alloys. For this reason, annealing cycles are inserted at points in the fabricating cycle where cold-working stresses have been built up by large reductions during rolling or other mechanical work.

Recrystallization

When cold-worked material is heated to a sufficiently high temperature, the fragmented particles produced by the cold-working process form new, unstrained grains, provided sufficient cold work has been performed on the material. This is recrystallization. The high-energy points created during the cold-working process serve as points of nucleation for the formation of the new grains. The formation of the new grains removes a substantial amount of the effects of the cold work, tending to produce properties similar to those originally possessed by the material. The degree of cold work is important. If an insufficient amount is present, recrystallization will not take place. When just enough cold work is present to cause recrystallization at the temperature used, the resulting material will possess a very coarse grain size. The presence of a substantial amount of cold work promotes the formation of fine-grained material. The fundamentals of recrystallization are as follows.

1. Increasing the degree of cold work decreases the temperature necessary for recrystallization.
2. Increasing the length of time at temperature decreases the recrystallization temperature.
3. The rate of heating to and through the recrystallization temperature affects the size of the grains formed.
4. The degree of cold work and the temperature employed affect the size of the grains formed.

Grain Size

It is usually desirable to have a material possessing a medium-to-fine grain size for severe drawing operations. While large-grained material actually has a greater capacity for plastic deformation than fine-grained material, such material also has a greater tendency to deform locally or "neck down," and may produce an undesirable appearance known as *orange peel*. The final grain size of a recrystallized material is dependent on the size of the grains after recrystallization and upon grain growth. These, however, are influenced by many factors, such as:

Original grain size
Degree of cold work
Heating rate
Final temperature
Length of time at temperature
Composition

14.3.2 Heat Treatments

Heat treatment involves heating the aluminum alloy to a point below the melting point, where the alloying ingredients (zinc, copper, etc.) are in solution with the aluminum. The grain size is small, and the alloying ingredients are evenly distributed within the grains, along the aluminum lattices. When the metal has "soaked" for a time at this temperature to permit even distribution of the alloying ingredients, it can be *quenched* rapidly in water (or other cooling medium). The purpose of suddenly dropping the temperature in this manner is to prevent certain constituents from precipitating out, which they would do if the part were cooled slowly. Quenching from any particular temperature range tends to retain in the metal the structure present just before quenching. The result will be a slightly hard, fine-grained material. If the material is held at the elevated temperature for too long a period of time, the grains would continue to grow, and the resulting structure will be rather coarse grained, and will not develop the optimum properties we are looking for.

The fast cooling to near room temperature upon quenching produces a "supersaturated" condition, where the material has already dissolved in it more of the constituents than it normally can carry at that temperature. The metal can be held in this condition (called the *W condition*) for some period of time by placing it in a freezer, to permit subsequent forming before the metal reaches full hardness. Such a condition is obviously unstable. The result is that certain constituents begin to separate out, or *precipitate*, from the main mass of the aluminum alloy. This precipitation occurs at room temperature with many of the alloys (for example the copper alloys), and is known as *natural aging*. Other alloys must be heated slightly to bring this precipitation to completion within a reasonable length of time. This is called *artificial aging* and is common with the zinc alloys. In either case, this controlled reprecipitation is aimed at providing the correct size, character, and distribution of precipitated particles in the aluminum to produce maximum strength and other desired mechanical properties. The precipitation of the major alloying ingredients from the solid, homogeneous solution provides maximum "keying" to resist slipping or deformation.

An example of some heat treatments of a zinc-based aluminum alloy (7075) is as follows:

7075-0   Soft, annealed condition, with a minimum ultimate tensile strength of 33,000 psi and a minimum tensile yield strength of 15,000 psi.

7075-W   Solution heat treated at 870°F, and quenched in water. Metal is quite soft, and can be held in freezer for subsequent forming operations.

7075-T6   Precipitation heat treatment at 250°F (artificially aged), and allowed to cool at room temperature. Has minimum ultimate tensile strength of 78,000 psi and tensile yield strength of 69,000 psi.

7075-T73   Additional precipitation heat treatment at 350°F (overaged) to improve fatigue life and reduce stress cracking. Minimum ultimate tensile strength of 67,000 psi and yield strength of 56,000 psi.

Aluminum alloys hardened in this manner can be made soft and easily workable again by an *annealing* treatment. Annealing takes place when the metal is heated to about the same temperature as for solution heat treating, and then slowly cooled. This produces a precipitate in the form of large particles outside the grains along the grain boundaries and not inside the crystals. In this manner, minimum keying effect results, and the material is "soft" because the crystals easily slip along their slip planes.

14.3.3 Wrought Versus Cast Products

Commercial aluminum alloys are divided into two general types—wrought alloys and casting alloys. The wrought alloys are designed for mill products whose final physical form is obtained by mechanically working the material. This mechanical work is done by rolling, forging, drawing, or extruding the material. Wrought aluminum mill products include forgings, sheet, plate, wire, rod, bar, tube, pipe, structurals, angles, channels, as well as rolled and extruded shapes.

Casting alloys are used in the production of sand, permanent mold, or die castings—processes in which the molten alloy is allowed to solidify in a mold having the desired final size and shape.

Aluminum Association Designation System

Wrought aluminum and wrought aluminum alloys are designated by a four-digit index system. The first digit of the designation serves to indicate the alloy group. The last two digits identify the aluminum alloy or indicate the aluminum purity. The second digit indicates modifications of the original alloy or impurity limits. Table 14.3 shows the aluminum alloy groups. Mechanical properties may be obtained from other reference sources.

14.3.4 Bibliography and Additional Reading (Subchapters 14.0–14.3)

The ABC's of Aluminum, Reynolds Metals Company, Richmond, VA, 1962.
Alcoa Aluminum and its Alloys, Aluminum Company of America, Pittsburg, PA, 1960.
Aluminum Heat Treating, Reynolds Metals Company, Richmond, VA, 1960.
Aluminum Standards and Data, The Aluminum Association, New York, 1976.
Brandt D. A., *Metallurgy Fundamentals*, Goodheart-Wilcox, South Holland, IL, 1972.
Ginzburg, V. B., *High-Quality Steel Rolling*, Marcel Dekker, New York, 1993.
Gillett, H.W., *The Behavior of Engineering Metals*, John Wiley, New York, 1951.
Jensen, J. E. Ed., *Forging Industry Handbook*, Forging Industry Association, Cleveland, OH, 1970.
Modern Steels and Their Properties, Bethlehem Steel Co., Bethlehem, PA, 1964.
Tanner, John P., *Manufacturing Engineering*, Marcel Dekker, New York, 1991.

Table 14.3 Aluminum Alloy Groups

Aluminum—99% minimum and greater	1xxx
Aluminum alloys, grouped by major alloying element:	
Copper	2xxx
Manganese	3xxx
Silicon	4xxx
Magnesium	5xxx
Magnesium and silicon	6xxx
Zinc	7xxx
Other elements	8xxx

14.4 NONFERROUS METALS: MAGNESIUM

Robert S. Busk

14.4.0 Introduction to Magnesium

Magnesium, with inexhaustible raw material sources, utilizes three outstanding properties in many of its applications. It has the lowest density of any structural metal and therefore is used to produce products of light weight, such as aircraft engine parts, automotive parts, wheels, computer parts, portable tools, and materials handling equipment. Its large electronegative potential finds use in the cathodic protection of steel structures such as domestic hot water heaters. Its high damping capacity is used to produce vibration tables, and stable platforms for instruments subject to vibration damage.

Subsection 14.4.3 lists works referenced throughout this subchapter. However, two of the references are also valuable for general reading. Reference [4] contains excellent information on the chemistry, metallurgy, and practical handling of molten magnesium and its alloys. Reference [5] contains complete physical and mechanical property data, design criteria for magnesium, and detailed information on the machining, joining, forming, and finishing of magnesium and its alloys. It is well worth having on hand before starting any project involving the fabrication of magnesium.

Magnesium is produced from the magnesium ion found in ores such as sea water, the Great Salt Lake, dolomite, carnalite, or magnesite. Reduction is by electrolysis of magnesium chloride, or by reaction of silicon with magnesium oxide. There are two types of electrolytic cells in use, those based on the Dow Chemical Co. design [1], and those based on the I.G. Farben design [2,3]. General descriptions of electrolytic reduction are given in Ref. [4], p. 25, and Ref. [5] p. 3. Two methods of reduction of magnesium oxide by silicon are also in general use. The first was developed by L. M. Pidgeon [6], the second by Pechiney [7]. General descriptions of silicon reduction are also given in Ref. [4], page 25, and Ref. [5], p. 3. The pure metal is sold in various grades with total magnesium content varying from 99.80 to 99.98% [8]. Impurities that are controlled for specific effects are aluminum, copper, iron, lead, manganese, nickel, and silicon. All other impurities are only found in very small quantities.

14.4.1 General Characteristics of Magnesium

Physical Properties

For a complete listing of the physical properties of pure magnesium, see Ref. 5, p. 150. Those properties that are of most common significance for manufacturing purposes are the coefficient of friction (0.36), density (1.74 g/cm^3 at 20°C), electrical resistivity (4.45 ohm—m $\times 10^{-8}$ at 20°C), melting point (650°C), specific heat (1025 J/kg-K at 20°C), thermal conductivity (154 W/m-K at 20°C), and thermal expansion (25.2 unit strain $\times 10^2$).

Safety

The following is taken largely from Ref. 5, p. 37.

Magnesium can be ignited, upon which it will burn with a brilliant white light at a temperature of about 2800°C. However, it is only magnesium vapor that will burn. Thus, the metal must be heated to a temperature high enough to produce a sufficient quantity of vapor to support combustion. As a practical matter, this means that the metal must be melted. Because the heat conductivity of magnesium is high, all of a massive piece must be raised to the melting point for the piece to burn freely. Thus it is very difficult to start a fire with a massive piece such as a casting. Even if a torch is used to raise a part of the casting to the ignition temperature and burning starts, it will cease when the torch is removed, because of heat conduction to other parts of the casting and consequent lowering of the temperature below the ignition temperature. On the other hand, finely divided magnesium, such as some kinds of machining scrap, and very thin ribbon, cannot conduct the heat away from a source and burning can be initiated even with a match, which has a flame temperature of about 900°C.

As is true of all oxidizable materials, an air suspension of fine magnesium powder can explode, and this is the most serious hazard associated with the flammability of magnesium. Explosion will occur only if the powder is fine enough to remain suspended in the air for an appreciable period of time. As a practical matter, this means that the powder must be at least as fine as about 200 mesh, which has an nominal particle size of 74 μm.

Since magnesium will react very slowly with water, even at room temperature, to produce hydrogen, the large surface area associated with finely divided machining chips can, when wet, produce sufficient hydrogen to result in a hydrogen explosion hazard. The worst case is a large amount of damp powder, since the temperature will then rise, leading to still greater hydrogen production and even to ignition of the magnesium itself. Powder should be stored dry. If it must be wet, the amount of water should be copious enough to prevent a temperature rise, and means should be provided for hydrogen dispersal.

Those areas of manufacturing that should take precautions regarding magnesium ignition are melting (e.g., for casting), machining, welding, and heat treating. Specific precautions will be given in the sections treating each of these operations.

Once started, a magnesium fire can be extinguished by using the normal strategies of cooling, removing oxygen, or letting the magnesium be totally consumed. Machining chips that have started to burn can be cooled effectively with cast-iron chips, and this technique is used quite successfully. Since magnesium is an active chemical, it will react with oxygen preferentially to many other materials. Hence, any extinguisher that contains oxygen as part of its chemical makeup will probably support the combustion of magnesium rather than stop it. This is true of water, so one should not use water to cool magnesium without being aware of the fact that, during cooling with water, reaction will occur, producing heat and hydrogen. The hydrogen, in turn, may react explosively with the air. Water can be used, but only in large quantities in order to cool the mass of magnesium below the melting point, with recognition of the hydrogen explosion hazards that are being introduced. Water should never be used for extinguishing a fire of finely divided material, since the vigorous reaction with water will scatter the chips and spread the fire. Oxygen can be excluded by covering the magnesium with a nonreactive material such as melting fluxes, G-1 powder, Metal-X, or other proprietary materials. Any oxygen-containing material, such as sand, should be avoided. See also Ref. [9].

Alloys of Magnesium

The most common alloying elements for magnesium are aluminum, zinc, and manganese. In addition, silicon, rare earth metals, and yttrium are used for higher strength, especially at elevated temperatures. If aluminum is not present as an alloying element, zirconium is added for grain refinement.

14.4.2 Manufacturing Processes

See reference [5] for a detailed discussion of typical uses, effects of alloying elements, properties, and design criteria for the use of magnesium alloys. The purpose of this subsection is to point out specific effects the properties of magnesium have on manufacturing practices.

Casting

Magnesium alloy castings are produced by all the standard techniques of sand, permanent-mold, low-pressure, high-pressure (die), and investment casting. In general, the practices used for other metals apply also to the handling of magnesium. See Refs. 4 and 19–25 for specific details. However, there are three properties of magnesium that profoundly affect all casting processes: (1) Liquid magnesium does not attack steel. (2) Liquid magnesium will ignite if exposed to air. (3) Magnesium has a lower density than almost any other metal or impurity.

Because the liquid metal does not attack steel, the common material for handling liquid magnesium alloys is steel. Plain carbon steel is the most common, although stainless steel is sometimes used where excessive oxidation of carbon steel is a problem. If stainless steel is chosen, a type such as 430, which is low in nickel, should be used. Steel pots are used for melting; steel pumps and steel lines for conveying liquid magnesium from one point to another; steel equipment for metering exact quantities for casting; and steel hand tools for miscellaneous handling of the molten metal. Pumps of various kinds are suitable, including electromagnetic pumps. However, the most common is a simple, rugged, centrifugal pump generally built by the casting shop itself. The secret for building a good centrifugal pump for molten magnesium is to use generous, even sloppy, clearances. An air motor is best for this kind of pump since, in contrast to an electric motor, it is little affected by the heat of the molten metal.

Because molten magnesium will ignite if exposed to air, it must be protected from such exposure. The classic method is to cover the molten metal with a liquid flux, consisting of a mixture of chlorides, fluorides, and magnesium oxide [4,15]. A more efficient, cleaner, and less expensive protection is by the use of small quantities of SF_6, or of SF_6 and CO_2, in air [16–18].

The reactivity of magnesium with other oxides, such as silicon dioxide, requires that sand used for molds be protected from the molten magnesium by mixing inhibitors with the sand, such as sulfur, boric oxide, and ammonium or potassium borofluoride [4,19].

Because of the low density of magnesium, impurities in molten metal tend to sink rather than float. Thus, sludge builds up on the bottom of the melting container and must periodically be removed by dredging. An advantage of flux over SF_6 is the ability of the former to trap impurities, thus making dredging easier. If scrap, for example, is melted, the high content of impurities can easily be removed with the use of flux, but can be removed when SF_6 is used for protection only by filtering [27].

A new casting technique has been developed for magnesium, called *thixomolding* [28–30]. Magnesium pellets are fed to a screw conveyer that advances them through a heated chamber to raise the temperature to a point between the liquidus and solidus while simultaneously shearing the metal by the action of the screw. This produces metal in a "thixotropic" condition [31], which is then injected, by advancing the screw, into a die to produce a part, the process closely resembling the injection molding of plastics. A considerable advantage for the caster is that there is no need to handle molten magnesium.

As with all molten metal, there is a severe explosion hazard if tools that are wet with water are immersed below the surface. The rapid expansion of the water as it first expands to the gaseous state and then further expands as its temperature rises will empty molten metal from the container with explosive force. Any tools must be perfectly dry before immersion.

Wrought Semiproducts

Although magnesium alloys can be rolled, extruded, and forged in much the same way as other metals [4, 5, 32–34], there are certain characteristics peculiar to magnesium that affect the technology.

The crystal structure of magnesium is hexagonal close-packed. (See Subchapter 14.1 for a discussion of unit cells and lattice formations.) The major plane of deformation by slip is the basal, and all of the crystallographic directions of deformation, regardless of the plane, lie in the basal plane [33]. This crystallographic mechanism results in a preferred orientation such that the basal plane of sheet lies in the plane of the sheet; of extrusions, in the plane of extrusion; and of forgings, in the plane of major deformation. Since slip deformation does not allow any deformation out of the basal plane, deformation of polycrystalline material without cracking would be impossible if slip were the only mechanism available. Other mechanisms, such as twinning, grain-boundary sliding, and kinking, also exist. Twinning takes place when the direction of compressive stress is parallel to the basal plane, the twinning resulting in a reorientation of the basal plane in the twin to an angle of about 86° to the original basal plane. Slip on the twin then allows deformation out of the original basal plane at an angle of 86°.

Since grain boundary sliding and kinking are both more easily activated as the temperature increases, and as the grain size becomes smaller, both temperature of deformation and grain size control become important. Magnesium is usually rolled at temperatures exceeding 400°F, and both extruded and forged at temperatures exceeding 600°F. Rolling is sometimes done at room temperature, but total deformation before annealing at about 700°F is limited to about 30%, and reduction/pass to about 4%.

Because twinning occurs when compression is applied to the surface of the sheet, coiling results in twinning on the compressive side of the coil, but not on the tension side. This leads to many problems in forming and handling, so sufficient tension must be applied to the up-coiler to prevent twinning during coiling. Roller leveling, while practical from a manufacturing standpoint, results in extensive twinning in the sheet, which results in a lowered tensile yield strength. For this reason, anneal flattening is generally preferred.

Forming

All of the forming methods used for metals, such as bending, drawing, spinning, dimpling, and joggling, are used with magnesium. Good details for these processes are given in Ref. 5 and in the works cited therein. The differences between forming of magnesium and other metals are all related to the desirability of forming at elevated temperatures and to the need for controlling grain size and the effects of twinning.

Joining

Joining of magnesium is common, using welding, brazing, adhesive bonding, and mechanical attachment such as riveting. Soldering is not recommended. Good details for all of these are given in Ref. 5, and in the works cited therein. Protection of the molten metal from oxidation is required for both welding and brazing. Arc welding using shielded helium or argon gas is suitable. Fluxes are used for brazing.

Machining

Magnesium is the easiest of all metals to machine. For example, 1 horsepower is required to remove 1.1 in.^{3} of steel, but is sufficient to remove 6.7 in.^{3} of magnesium. Reference 5 should be consulted for tool design, machining practice, and safety precautions.

Heat Treatment

Recommended practices for heat treating magnesium products are given in Ref. 35. Included in that reference is a recommended procedure for combatting a fire in a heat-treat furnace.

Finishing

There are three characteristics of magnesium that influence finishing procedures.

1. Magnesium is inert in strong caustic solutions. Therefore cleaning of surface contamination such as grease and oil is best accomplished using strong caustic solutions. If embedded material must be removed by removing some of the magnesium, acidic solutions must be used.
2. Magnesium develops hydroxide or hydroxycarbonate surface films when exposed to air. Paints such as epoxies, which are resistant to high-pH surfaces, should therefore be used.
3. Magnesium is more electronegative than all other common metals when exposed to salt solutions. It is thus the anode in galvanic couples and will corrode while protecting the cathodic material if exposed to salt water while connected electrically to the cathode (see Subchapter 19.1). It is necessary to protect joints so that this does not happen in service [36]. Good details on finishing methods are given in Ref. 5 and the works cited therein.

14.4.3 Bibliography

1. Hunter, R. M., *Trans. Electrochem. Soc.* 86:21 (1944).
2. Høy-Petersen, Nils, Magnesium production at the Porsgrunn Plant of Norsk Hydro, *Proc. Int. Magnesium Assoc.*, (1979).
3. Strelets, Kh. L., Electrolytic production of magnesium, translated from the Russian, obtainable from the *International Magnesium Association*.
4. Emley, E. F., *Principles of Magnesium Technology*, Pergamon Press, Elmsford, NY, 1966.
5. Busk, R. S., *Magnesium Products Design*, Marcel Dekker, New York, 1987.
6. Pidgeon, L. M., and W. A. Alexander, *Trans. AIME*, 159:315 (1944).
7. Trocmé, F., The development of the magnetherm process, *Trans. AIME*, (1974).
8. Standard specification for magnesium ingot and stick for remelting, ASTM Standard B92, *Annual Book of ASTM Standards, V. 02.02,* ASTM, Philadelphia, PA.
9. *Storage, Handling, and Processing of Magnesium*, National Fire Protection Association Bulletin 48.
10. ASTM Standard B80, Standard specification for magnesium alloy sand castings, *Annual Book of ASTM Standards, V. 02.02,* ASTM, Philadelphia PA.
11. Standard specification for magnesium-alloy sheet and plate, ASTM Standard B90, *Annual Book of ASTM Standards, V. 02.02,* ASTM, Philadelphia, PA.
12. Standard specification for magnesium-alloy forgings, ASTM Standard B91, *Annual Book of ASTM Standards, V. 02.02,* ASTM, Philadelphia, PA.

13. Standard specification for magnesium-alloy die castings, ASTM Standard B94, *Annual Book of ASTM Standards, V. 02.02*, ASTM, Philadelphia, PA.

14. Standard specification for magnesium-alloy extruded bars, rods, shapes, tubes, and wire, ASTM Standard B107, *Annual Book of ASTM Standards, V. 02.02*, ASTM, Philadelphia, PA.

15. *Melrasal Fluxes*, Magnesium Elektron Limited Bulletin 498.

16. Couling, S. L., F. C. Bennett, and T. E. Leontis, Melting magnesium under air/SF$_6$ protective atmosphere, *Proc. Int. Magnesium Assoc., 1993*.

17. Couling, S. L., Use of Air/CO$_2$/SF$_6$ mixtures for improved protection of molten magnesium, *Proc. Int. Magnesium Assoc., 1979*.

18. Busk, R. S., and R. B. Jackson, Use of SF$_6$ in the magnesium industry, *Proc. Int. Magnesium Assoc. 1980*.

19. *Molding and Core Practice for Magnesium Foundries*, Dow Chemical Company Bulletin 141-29, 1957.

20. Berkmortel, John, and Robert Hegel, Process improvement & machine development in magnesium cold chamber die cast technology, *Proc. Int. Magnesium Assoc. 1991*.

21. Fink, Roland, Magnesium hot chamber improvements, *Proc. Int. Magnesium Assoc. 1993*.

22. Holta, O., O. M. Hustoft, S. I. Strømhaug, and D. Albright, Two-furnace melting system for magnesium, NADCA, Cleveland, OH, October 18–21, 1993.

23. Øymo, D., O. Holta, Om M. Hustoft, and J. Henriksson, Magnesium Recycling in the Die Casting Shop, American Society for Materials, "The Recycling of Metals," Düsseldorf, May 13–15, 1992.

24. *Recommended Practice for Melting High-Purity Magnesium Alloys*, International Magnesium Association.

25. *Magnesium Die Casting Manual*, Dow Chemical Company.

26. *Permanent Mold Practice for Magnesium*, Dow Chemical Company Bulletin 141-101.

27. Petrovich, V. W., and John Waltrip, Fluxless refining of magnesium scrap, *Proc. Int. Magnesium Assoc. 1988*.

28. Erickson, Stephen C., A process for the injection molding of thixotropic magnesium alloy parts, *Proc. Int. Magnesium Assoc. 1987*.

29. Frederick, Paul, and Norbert Bradley, Injection molding of thixotropic magnesium: update, *Proc. Int. Magnesium Assoc 1989*.

30. Carnahan, R.D., F. Decker, D. Ghosh, C. VanSchilt, P. Frederick, and N. Bradley, The thixomolding of magnesium alloys, in Mordike, B.L. and F. Hehmann, Eds., *Magnesium Alloys and Their Applications*, DGM Informationgesellschaft, 1992.

31. Flemings, M. C., R. G. Riek, and K. P. Young, Rheocasting, *Material Sci. Eng.*, 25, 1976.

32. Ansel, G., and J. O. Betterton, The hot and cold rolling of Mg-base alloys, *Trans. AIME* 171, 1947.

33. Roberts, C. Sheldon, *Magnesium and Its Alloys*, John Wiley & Sons, New York, 1960.

34. *Magnesium Forging Practice,* Dow Chemical Company, 1955.

35. Standard practice for heat treatment of magnesium alloys, ASTM Standard B661, *Annual Book of ASTM Standards, V. 02.02*, ASTM, Philadelphia, PA, 1992.

36. *Preventive Practice for Controlling the Galvanic Corrosion of Magnesium Alloys*, International Magnesium Association.

15

Conventional Fabrication Processes

Jack M. Walker *Consultant, Manufacturing Engineering, Merritt Island, Florida*

15.0 INTRODUCTION TO CONVENTIONAL FABRICATION PROCESSES

More steel is produced than any other metal, with aluminum second in volume and in use in fabricating products. The weight of aluminum is approximately one-third the weight of steel. Steel is approximately 0.3 lb/in.^{3}, aluminum is 0.1 lb/in.^{3}, and magnesium is 0.06 lb/in.^{3}. Figure 15.1 is a nomogram for calculating weights of steel and aluminum stock. In this chapter we concentrate our discussions on parts fabricated from these two materials. Steel is made by heating iron ore, and casting into pigs or ingots, or continuous casting into slabs. Aluminum is made from bauxite, and converted into molten aluminum pigs or ingots using a complex chemical and electrolytic process. At this point, both metals are in their least expensive form, but suitable only for making rather poor castings. Gray cast iron is superior to "raw" aluminum for casting at this time. From this point on, the processes for steel and aluminum are quite similar. Gray cast iron is the lowest-cost ferrous material, since little work has been done to refine either its form or its material content and properties. As you continue through the process of removing impurities, adding alloying ingredients, hot rolling and cold rolling, it appears that carefully controlled sheet metal has the greatest labor content, and therefore may have the highest cost per pound. The material form is one factor that enters into the cost of a finished metal part, but it can seldom offset the fabrication expense of the total part cost.

Metal manufacturers have traditionally relied on several basic forming techniques, such as casting, forging, machining, and sheet metal stamping, to impart the desired geometric shape to their products. In recent years, because of stiffer industrial competition, the development of new alloys, shortages of certain metals, and the increase in energy costs, these traditional processing methods have been critically analyzed and

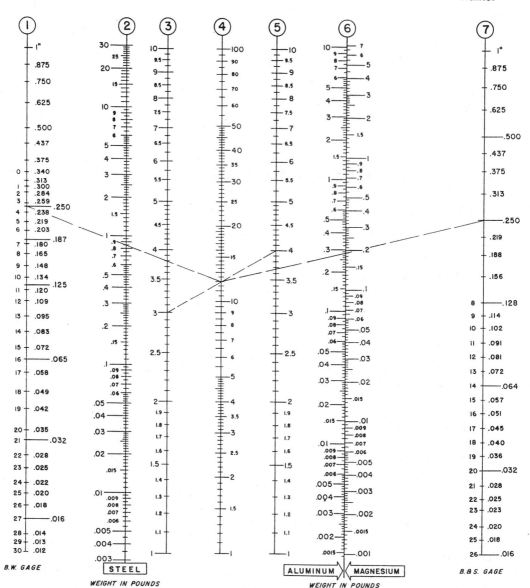

THE ABOVE NOMOGRAM PROVIDES A SIMPLE MEANS FOR CALCULATING AND COMPARING WEIGHTS OF MAGNESIUM, ALUMINUM, AND STEEL PARTS AND, INCIDENTALLY, FOR GENERAL MULTIPLICATION AND DIVISION.

1. LOCATE LENGTH AND WIDTH OF PART ON LINES ③ & ⑤ RESPECTIVELY. A STRAIGHT LINE CONNECTING THESE POINTS INTERSECTS LINE ④ INDICATING AREA OF PART IN in^2.

2. FOR MAGNESIUM AND ALUMINUM LOCATE THICKNESS OF PART ON LINE ⑦. CONNECT THIS POINT BY STRAIGHT LINE WITH THE VALUE FOUND ON LINE ④ AND READ THE WEIGHT IN POUNDS ON LINE ⑥.

FOR STEEL USE LINE ① FOR THICKNESS AND READ THE WEIGHT VALUE ON LINE ②.

3. FOR PARTS WHOSE WIDTH OR LENGTH EXCEEDS 10" (OR THICKNESS EXCEEDS 1") USE THE NOMOGRAM WITHOUT REGARD TO DECIMAL POINT UNTIL THE WEIGHT IS OB-TAINED. THEN MOVE THE DECIMAL POINT OF YOUR RESULT AS REQUIRED.

IF ONE, TWO, OR ALL THREE FACTORS ARE MULTIPLIED BY 10, MULTIPLY THE ANSWER BY 10, 100, OR 1000 RESPECTIVELY BY MOV-ING THE DECIMAL POINT ONE, TWO, OR THREE PLACES TO THE RIGHT.   FOR EXAMPLE :

A MAG. PART 3"x 4" x .25" WEIGHS .19 POUNDS, A PART 3"x 40"x.25" WEIGHS 1.9 POUNDS, A PART 30"x 40" x .25" WEIGHS 19 POUNDS , A PART 30"x 40"x 2.5" WEIGHS 190 POUNDS .

4. FOR GENERAL MULTIPLICATION AND DI-VISION USE LINES ③ ④ ⑤.

Figure 15.1   Nomogram for calculating weights of steel, magnesium, and aluminum stock.

reevaluated. It is becoming very desirable to produce the final product in fewer processing steps and with as little waste as possible. Several techniques for the manufacture of components to "net shape" or to "near net shape," based on the firm foundation of the traditional processes, are being developed to meet these challenges of today and the future.

This chapter introduces the conventional processes and equipment for sheet metal fabrication in Subchapter 15.1, machining in Subchapter 15.2, extrusion and forging in Subchapter 15.3, and casting and molding in Subchapter 15.4.

15.1 SHEET METAL FABRICATION PROCESSES

15.1.0 Introduction to Sheet Metal Fabrication Processes

Sheet metal stampings are generally the lowest-cost parts to produce. Both the machinery and labor are relatively low in cost, and production rates can be quite high. Most of the equipment used in the forming of steel and other metals is suitable for use with aluminum alloys. Because of the generally lower yield strength of aluminum alloys, however, press tonnage requirements are usually lower than for comparable operations on steel, and higher press speeds can be used. Similarly, equipment for roll forming, spinning, stretch forming, and other fabrication operations on aluminum need not be so massive or rated for such heavy loading as for similar operations on steel.

15.1.1 Blanking

Shears

One of the most common and versatile machine tools used in sheet metal work is the vertical shear, or square shear. For small-quantity jobs, this is the most common blank-

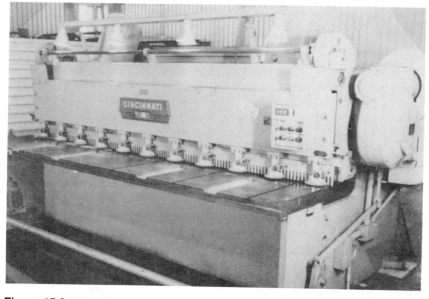

Figure 15.2 Example of vertical shear, or square shear, used for making straight cuts in sheet stock.

ing process. Figure 15.2 show a common shear. The back stops are adjustable, and can maintain cutoff lengths of ±0.010 with care, and ±0.030 even with old, worn equipment. A clearance of 6–8% of the thickness of the sheet is normally used between shear blades, although for thin sheet and foil, smaller clearances give a better edge with no burr or curvature.

Punch Press

For larger parts runs, or for blanks requiring other than straight sheared sides, the punch press and blanking dies are commonly used. Figure 15.3 shows a production punch press for relatively small parts that can run at 300 strokes per minute with material fed automatically from coil stock in precut widths. Die-cut blanks can hold tolerances of ±0.001 with good equipment. The correct clearance between punch and die is essential to obtain a good edge with a low burr. Clearance is dependent on alloy, temper, and gauge. Recommended clearances for the more common aluminum alloys are shown in Figure 15.4. The table shows that the required clearance increases with higher mechanical properties of the metal.

Turret Press

The turret punch has come into its own with the development of computer numerical control (CNC) fabrication equipment. The most common application is to cut the blank edges on a square shear, and transfer this blank to the fabricator for all the internal cutouts and special shapes. The turrets can hold as many as 42 different punch–die sets, which can be programmed X–Y to tolerances of ±0.004 at rates up to 330 hits per minute. Station-to-station indexing time is as low as 0.5 sec. Figure 15.5 shows one of Strippit's 20-ton models, programmed on a PC. With some loss in accuracy, the latest machines allow one

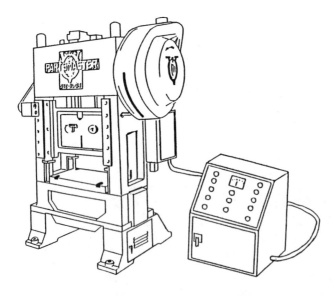

Figure 15.3 High-speed production punch press.

		Clearance ① per side			Clearance ① per side
Alloy	Temper		Alloy	Temper	
1100	O H12, H14 H16, H19	0.050t 0.060t 0.070t	5083	O H112, H323, H343	0.070t 0.075t
2014	O T4, T6	0.065t 0.080t	5086	O, H112 H32, H34, H36	0.070t 0.075t
2024	O T3, T361, T4	0.065t 0.080t	5154	O, H112 H32, H34, H36, H38	0.070t 0.075t
3003	O H12, H14 H16, H18	0.050t 0.060t 0.070t	5454	O, H112 H32, H34	0.070t 0.075t
3004	O H32, H34 H36, H38	0.065t 0.070t 0.075t	5456	O, H321 H323, H343	0.070t 0.075t
5005	O H12, H14, H32, H34 H16, H18, H36, H38	0.050t 0.060t 0.070t	6061	O T4 T6	0.055t 0.060t 0.070t
5050	O H32, H34 H36, H38	0.050t 0.060t 0.070t	7075	O W T6③	0.065t 0.080t 0.080t
5052	O H32, H34 H36, H38	0.065t 0.070t 0.075t	7178	O W T6③	0.065t 0.080t 0.080t

t = Thickness of sheet in inches

Blanks over 0.080 in. thick should be sheared 1½t oversize on each side and be machined to size.

Figure 15.4 Die clearances for blanking common aluminum alloys. (Courtesy Aluminum Corporation of America.)

to rotate oversized sheets and fabricate parts up to double the machine throat capacity. Large sheets can also be repositioned, permitting fabrication of parts greater in length than the table size.

Dimensioning Practices

If there is a single area where the manufacturing engineer can accomplish the greatest benefit in producibility and economy of manufacture, it is in ensuring the appropriate detailing practices on drawings. Following are a few basic guidelines.

First, select a meaningful datum in the body of the part—passing through a hole center, if possible—rather that using an edge or corner of the part. This avoids problems of possible misalignment of the part, distortion from clamping, etc. It allows for more precise measurement by avoiding measurements from edges that may be tapered and therefore dimensionally uncertain. It facilitates accurate inspection, and it avoids unnecessary accumulation of tolerances.

Second, on related hole patterns, dimensioning and tolerances should be within this pattern, with only one dimension linking to the general datum. Better quality control and function of the product can be expected.

Figure 15.5 Example of a turret punch. (Courtesy Strippit, Inc. Akron, NY.)

Third, highlight the truly significant dimensions. Critical dimensional relationships can be protected, if they are known.

Blanking Pressures

The blanking operation is usually performed on a single-action press employing a punch and die with sharp cutting edges. The dimensions of the blank correspond to the dimensions of the die.

The blanking or shearing load is calculated using the following equation:

$$P = Lts$$

where P = load
         L = peripheral length of the blank
         t = thickness of the material
         s = shear strength of the material

As an example, to blank a circle 4.75 in. in diameter from material 0.050 in. thick with a 40,000-psi shear strength requires 15 tons of load.

$$P = \pi 4.75 - 0.050 - 40{,}000$$

$$= 30{,}000 \text{ lb} = 15 \text{ tons (minimum press capacity)}$$

The shear strength of the commonly used aluminum alloys ranges from 9000 to 49,000 psi, whereas that of low-carbon steel is from 35,000 to 67,000 psi. Because of the generally

BLANKING PRESSURES

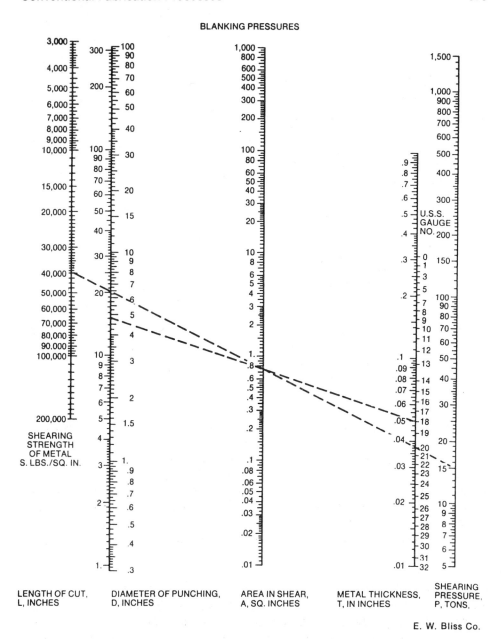

Figure 15.6 Blanking pressure nomogram.

lower shear strength of aluminum alloys, lower-tonnage presses are required than for comparable operations with steel. For easy and quick determination of the load, refer to the nomogram in Figure 15.6.

Dayton Rogers Manufacturing Company uses the following formula for calculating blanking tonnage requirements:

$$T = P \times Th \times C$$

where: T = pressure required in tons
             P = perimeter of blank in inches
        Th = thickness of material
         C = constant (see common ones below)

Example
0.050 CR steel, half-hard; cutting edge of 12 linear inches

$$T = 12 \times 0.050 \times 32$$

$$= 20 \text{ tons required}$$

Constants

$$
\begin{aligned}
\text{Aluminum—Soft} &= 11 \\
\text{T4/T6} &= 15 \\
\text{Steel—HR/Cold rolled} &= 27 \\
\text{Half hard} &= 32 \\
\text{Stainless Steel—Annealed} &= 37 \\
\text{Half-hard} &= 50 \\
\text{4230 AQ} &= 40 \\
\text{Brass—half-hard} &= 22 \\
\text{hard} &= 25
\end{aligned}
$$

Figure 15.7 gives the approximate pressures required for punching round holes in mild sheet metal.

To reduce the load required for blanking, the face of the die can be ground at an angle so that the cutting edge around the die opening is on a slanted plane, and a peak is formed across one of the diameters of the die face (see Figure 15.8). The maximum difference in height between the peak and the lowest point on the face should not be more than the thickness of the metal to be blanked; the minimum difference should not be less than one-half the thickness of the material (referred to as incorporating "shear" into the die).

Fine Blanking

There are several techniques for improving the edge finish of a blank. One method, referred to as *finish blanking*, incorporates a small radius on the cutting edge of the die. Another method, applicable to thicknesses from approximately 1/64 to ½ in., is called *fine blanking*. The process involves clamping the material securely throughout the entire blanking operation. A small radius is employed on the die, with almost zero die clearance, and a counterpressure is used against the blank, as shown in Figure 15.9. The clamping action, which is produced by a V knife edge on the pressure pad around the punch at a distance of about 1/16 in. from the punch, stops any lateral movement of the stock during blanking. This action, combined with the very small clearance, small-radius die, and counterpressure, produces a blank with a smooth sheared edge, very little burr, good flatness, and close tolerances. Because fine blanking is akin to extruding, some aluminum alloys and tempers are difficult to blank with a smooth edge. Alloys 2024-T6 and 7075-T6, for example, produce a relatively rough edge because of their hardness and tendency to

PRESSURE IN TONS

Gauge	28	26	24	22	20	18	16	14	12	10	5/32	3/16	1/4
Material thickness	.0149	.0179	.0239	.0299	.0359	.0478	.0598	.0747	.1046	.1345	.1562	.1875	.250
Hole Diameter													
.125	.2	.2	.2	.3	.4	.5	.6	.7	1.0	1.3	1.5	1.8	2.5
.1875	.2	.3	.4	.4	.5	.7	.9	1.1	1.5	2.0	2.3	2.8	3.7
.250	.3	.4	.5	.6	.7	.9	1.2	1.5	2.1	2.6	3.1	3.7	4.9
.3125	.4	.4	.6	.7	.9	1.2	1.5	1.8	2.6	3.3	3.8	4.6	6.1
.375	.4	.5	.7	.9	1.1	1.4	1.8	2.2	3.1	4.0	4.6	5.5	7.4
.4375	.5	.6	.8	1.0	1.2	1.6	2.1	2.6	3.6	4.6	5.4	6.4	8.6
.500	.6	.7	.9	1.2	1.4	1.9	2.3	2.9	4.1	5.3	6.1	7.4	9.8
.5625	.7	.8	1.1	1.3	1.6	2.1	2.6	3.3	4.6	5.9	6.9	8.3	11.1
.625	.7	.9	1.2	1.5	1.8	2.4	2.9	3.7	5.1	6.6	7.7	9.2	12.3
.6875	.8	1.0	1.3	1.6	1.9	2.6	3.2	4.0	5.7	7.3	8.4	10.1	13.5
.750	.9	1.1	1.4	1.8	2.1	2.8	3.5	4.4	6.2	7.9	9.2	11.0	14.7
.8125	1.0	1.1	1.5	1.9	2.3	3.1	3.8	4.8	6.7	8.6	10.0	12.0	16.0
.875	1.1	1.2	1.6	2.1	2.5	3.3	4.1	5.1	7.2	9.2	10.7	12.9	17.2
.9375	1.2	1.3	1.8	2.2	2.6	3.5	4.4	5.5	7.7	9.9	11.5	13.8	18.4
1.0000	1.3	1.4	1.9	2.4	2.8	3.8	4.7	5.9	8.2	10.6	12.3	14.7	19.6

General Information:

Formula determining punching pressure for round holes in mild sheet steel: D × Thickness × 25 = Pressure in Tons.

Formula for determining blanking pressure in mild sheet steel: Shear Length × Thickness × 25 = Pressure in Tons.

Formula for stripping pressures: Shear Length × Thickness × 3500 = Pressure in Pounds.

Figure 15.7 Blanking Pressures for round holes. (From John Tanner, *Manufacturing Engineering*, Marcel Dekker, New York, 1982.)

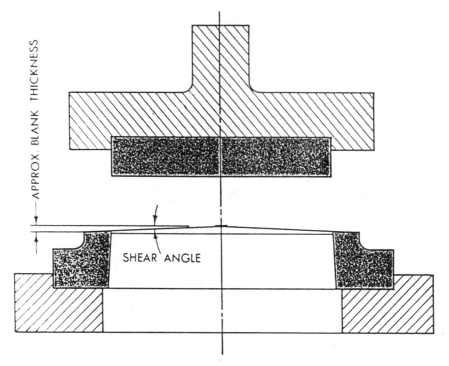

Figure 15.8 Example of "shear" ground into a punch to reduce blanking pressure.

fracture rather than shear. Punch speeds of one-half to one-third that of conventional blanking are used. Therefore, special presses have been developed to give a fast punch advance followed by a decreased punch blanking speed. The production rate may vary from 10 to 100 parts/min; the actual rate depends on the thickness of the stock and the geometric shape of the blanked part. Punch and die clearance for ordinary stamping is usually about 5% of the dimension across the die. For fine blanking, this clearance is about 1%—sometimes less than 0.5%. With larger presses, parts can be fine-blanked from plate as thick as 0.75 in. (19 mm).

Laser Cutting

Current trends toward just-in-time (JIT) manufacturing, shorter part runs, and limited product life cycles have increased the use of laser cutting machines in production and prototype fabrication. Laser cutters are constantly evolving, as manufacturers find new and innovative ways to apply this technology.

Often the capabilities of lasers and turret punches can be combined. Turret presses are very fast and generate acceptable accuracy when punching many holes of the same or different diameters. Lasers are particularly accurate and economical for profiling irregular exterior contours. These capabilities can be combined to produce accurate, complex parts at acceptable production rates by using each machine to perform that part of the cutting operation for which it is best suited.

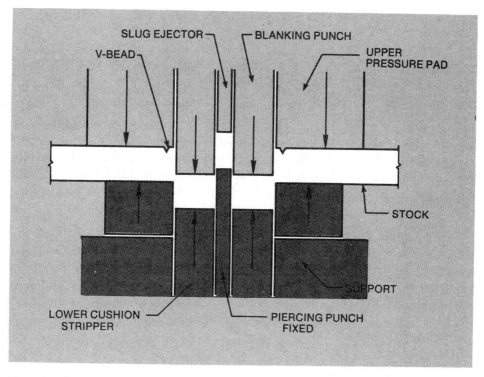

Figure 15.9 Cross section of a fine blanking die set.

Laser Operation

Lasers can be operated in either the continuous wave (CW) or pulsed mode. CW operation is faster and generates a smoother edge. It is inherently less accurate because of thermal workpiece expansion due to the higher power levels reaching the work.

Where there is a need for intricate or very close-tolerance cutting, the pulsed mode generates less heat but produces a very finely serrated edge. The finest quality of the workpiece is a carefully balanced compromise between speed, workpiece cooling, and edge condition.

Lasers are most productive when applied to mild steel and stainless steel and are more difficult to employ on aluminum. Aluminum and certain other metals, such as zinc and lead, continue to reflect light when molten. This quality scatters the beam, requiring more power. In addition, aluminum and copper alloys conduct heat away from the cutting area, which again means that more power is required.

Laser Considerations

In addition to production economics, precision, and edge condition, the knowledgeable manufacturing engineer considers these characteristics of laser-produced parts:

Localized hardening. Lasers cut by melting or vaporizing metal. This action can create problems when cutting heat-treatable materials, as the area around the part will become case hardened. Laser-cut holes in stainless steel or heat-treatable steel alloys which require machining (tapping, countersinking, or reaming) can be particularly troublesome. By the same token, this characteristic can benefit a product that must be case hardened for wear resistance.

Edge taper. The laser is most accurate where the coherent light beam enters the workpiece. As the beam penetrates the part, the light scatters, creating an edge taper condition *similar but opposite* to "breakout" in a sheared or pierced part. (The hole on the side of the workpiece from which the laser beam exits is generally smaller in diameter than the entrance hole). Thus one must carefully consider the final use of the part and, in some cases, may need to specify from which side the part should be cut.

Minimum through-feature size. The cutting beam is focused down to approximately 0.010 in. (0.2 mm) and is therefore capable of cutting holes and features with radii approximating 0.030 in. (0.76 mm). The limits applicable to piercing or blanking with a punch and die, such as the relationship between hole size and material thickness, or the minimum distance between features to avoid distortion, do not apply when laser cutting.

However, some limitations do exist, and are also related to the material thickness. See Figure 15.10 for an illustration of the minimum through features which are possible using lasers. Laser cutting allows for through features to be one-sixth to one-eighth the size needed for die piercing. Also, since no mechanical force is applied, the width of material remaining between cutout features may be very narrow without distortion occurring during metal removal. A typical application is tightly spaced venting holes on a visually important surface.

15.1.2 Piercing

When the diameter of the punch becomes too small to hold an ejector pin in its center, the clearance between the punch and die is decreased to about 5% or less of the metal thickness to ensure that the slugs are not lifted with the punch on the return stroke. The hole in the die below the cutting edge should be tapered to permit the slug to fall freely. When working gauges of aluminum alloy up to about 0.081 in., a taper of 3/8° is ample. For greater thicknesses, the taper angle should be increased up to twice this amount. A gang punch is often used for simultaneous punching of a large number of holes. In this case, it is advisable to "step" the individual punches slightly to stagger their entry into the metal. If the punches are close together, stepping will also prevent "crowding" of the metal and the deflection of thin punches. The relative lengths of the stepped punches will depend on the

minimum through-features			
material thickness range		minimum hole diameter and slot width achievable	
in.	mm	in.	mm
0–0.075	0–1.9	0.010	0.25
0.075–0.090	1.9–2.3	0.015	0.38
0.090–0.125	2.3–3.2	0.020	0.50
0.125–0.156	3.2–4.0	0.025	0.64
0.156–0.187	4.0–4.8	0.030	0.76

Figure 15.10 Through features possible by laser cutting. (Courtesy Dayton Rogers Manufacturing Co.)

gauge of the metal to be perforated. The difference in length between one punch and the next shorter one should be slightly less than the thickness of the metal. If the difference is too large, jerky operation will result. The longer punches should normally be on the outside, surrounding the smaller punch, such as drawing or blanking. Punches of large diameter, however, should always be longer than those of small diameter, regardless of position, to prevent distortion of the perforation and chipping of the smaller punches. Recommended minimum ratios of punched hole diameters to stock thickness are shown in Figure 15.11.

15.1.3 Forming

True bending is done in a straight plane, such as in the use of a press brake and brake dies. Forming is a better term when the bend line is not straight, requiring stretching or shrinking of a flange. Total wear of tools used in forming aluminum is somewhat less than with steel. This results in part from the lower force levels involved, and in part from the smoother surface condition that is characteristic of aluminum alloys. Accordingly, tools can sometimes be made from less expensive materials, even for relatively long runs. However, a higher-quality surface finish is generally required on tools used with aluminum alloys, to avoid marking. The oxide film on the surface of aluminum alloys is highly abrasive, and for this reason many forming tools are made of hardened tool steels. As a rule, these tools, even if otherwise suitable, should not be used interchangeably to form steel parts, because this use could destroy the high finish on the tools.

Suggestions that will assist in successful bending are:

1. Clean bending tools thoroughly, removing particles of foreign material.
2. Remove burrs, nicks or gouges at ends of bend lines which can initiate fractures. (Heavy plate should be chamfered or radiused on the edges at the bend lines to reduce the possibility of cracks.)
3. Avoid nicks, scribe lines, or handling marks in the vicinity of the bend.
4. Employ rubber pads, flannel, or other intermediate materials between tools and aluminum where high finish standards must be maintained. (Recent developments in coating tools with low-friction materials may be useful.)

P = Punched Hole Diameter
    (0.062 min. dia.)
T = Stock Thickness

Material Ultimate Tensile Strength (PSI)	Ratio P to T
32,000	P = 1.0T
50,000	P = 1.5T
95,000	P = 2.0T

Figure 15.11 Minimum ratios of punch hole diameters to stock thickness. (Courtesy Dayton Rogers Manufacturing Co.)

5. Apply a light oil coating on tools and bend lines to minimize scoring and pickup.
6. Form metal across the direction of rolling.

Figure 15.12 shows forming characteristics for forming carbon steel strip. Minimum permissible bend radii for aluminum are shown in Figure 15.13. The minimum permissible radius varies with the nature of the forming operation, the type of forming equipment, and the design and condition of the forming tools. Minimum working radius for a given material or hardest alloy and temper for a given radius can be ascertained only by actual trial under contemplated conditions of fabrication. Figure 15.14 shows the calculation of flat pattern bend allowance development for various sheet thicknesses and bend radii. The approximate load per lineal foot required to make a 90° bend in sheet metal is shown in Figure 15.15.

Dimensioning Practices for the Press Brake

Practical experience has shown that dimensioning and measuring practices must be understood and agreed on by all parties to a achieve a mutual, workable standard. Formed

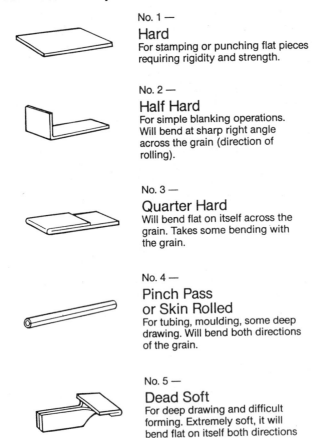

Cold Rolled Tempers

No. 1 —
Hard
For stamping or punching flat pieces requiring rigidity and strength.

No. 2 —
Half Hard
For simple blanking operations. Will bend at sharp right angle across the grain (direction of rolling).

No. 3 —
Quarter Hard
Will bend flat on itself across the grain. Takes some bending with the grain.

No. 4 —
Pinch Pass or Skin Rolled
For tubing, moulding, some deep drawing. Will bend both directions of the grain.

No. 5 —
Dead Soft
For deep drawing and difficult forming. Extremely soft, it will bend flat on itself both directions of the grain.

Figure 15.12 Forming characteristics of carbon steel strip.

Radii required for 90° bend in terms of thickness (T) Approximate Thickness						
TYPE	.016	.032	.064	.125	.187	.250
2S0, 3S0 52S0	0	0	0	0	0	0
2S¼H-½H, 3S¼H, 24S0*, 61S0	0	0	0	0	0-1T	0-1T
3S½H, 52S¼H	0	0	0	0-1T	0-1T	½T-1½T
2S¾H, 52S½H	0	0	0-1T	½T-1½T	1T-2T	1½T-3T
3S¾H, 61SW	0-1T	0-1T	½T-1½T	1T-2T	1½T-3T	2T-4T
2SH, 52S¾H, 61ST	0-1T	½T-1½T	1T-2T	1½T-3T	2T-4T	2T-4T
3SH. 52SH	½T-1½T	1T-2T	1½T-3T	2T-4T	3T-5T	4T-6T
24ST*	1½T-3T	2T-4T	3T-5T	4T-6T	4T-6T	5T-7T

*Alclad 24S can be bent over slightly smaller radii than the corresponding tempers of the uncoated alloy.

Figure 15.13 Minimum possible bend radii for aluminum. (Courtesy Dayton Rogers Manufacturing Co.)

sheet metal parts present a unique problem in that angular tolerances as well as flatness conditions interact with single plane dimensions because of the flexibility of sheet metal, especially the thinner gauges. To achieve consistent results when measuring formed parts, a standard must be established on where and how dimensions are to be taken.

Form dimensions should be measured immediately adjacent to the bend radius in order not to include any angular and flatness discrepancy. See Figure 15.16a for a simple illustration. Figure 15.16b is a typical stress-strain curve, showing the *elastic springback* in the straight portion of the curve (within the elastic limit), and the *plastic strain*, or permanent deformation, beyond this point—shown as a curved line. The deformation of a formed flange of a sheet metal part varies with the type of forming, the material, the height of the flange, etc.

Feature-to-feature dimensions on formed legs of any length on flexible parts will be assumed to be measured in constrained condition, holding the part fixtured to the print's angularity specification. This standard is appropriate for the majority of thin sheet metal parts and results in a functional product. This is always true when the assembled part is, by design, held in the constrained condition. For the most economical production, dimension the part in a single direction whenever possible. Because of the sequential nature of the forming process, and the fact that dimensional variation is introduced at each bend, dimensioning in a single direction parallels the process and helps to control tolerance accumulation. It is generally recommended that dimensioning be done from a feature to an edge. Feature-to-feature dimensions may require special fixtures or gauging.

Elements of Forming

Figure 15.17a shows some of the elements of formed stampings. A stretched flange is easier to form than a flange that needs to be shrunk. It may tend to thin out, but it will not buckle as a shrink flange will. The continuous corner is especially difficult, and usually is

WORKING FROM THE INSIDE, THE FIGURES SHOWN ARE THE CORRECT BEND ALLOWANCES.
SEE EXAMPLE FOR CLARITY

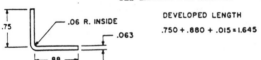

DEVELOPED LENGTH

.750 + .880 + .015 = 1.645

MATERIAL THICKNESS	INSIDE RADIUS OF BEND													
	.016	.030	.048	.063	.094	.125	.156	.188	.219	.250	.312	.375	.437	.500
.016	.004	.003	.010	.016	.029	.043	.056	.070	.083	.097	.124	.150	.177	.204
.020	.007	.000	.007	.014	.027	.041	.054	.067	.080	.094	.121	.148	.175	.201
.025	.010	.004	.004	.010	.023	.037	.051	.064	.077	.091	.118	.144	.171	.198
.030	.013	.007	.001	.007	.020	.034	.047	.061	.074	.088	.115	.141	.168	.195
.032	.014	.008	.000	.006	.019	.033	.046	.059	.073	.086	.113	.140	.167	.194
.036	.017	.010	.003	.003	.016	.030	.044	.057	.070	.084	.111	.137	.164	.191
.040	.020	.013	.006	.001	.014	.027	.040	.054	.067	.081	.108	.135	.161	.188
.045	.022	.015	.009	.002	.011	.025	.038	.052	.065	.078	.105	.132	.159	.186
.048	.025	.018	.011	.004	.008	.022	.035	.049	.062	.076	.103	.130	.156	.183
.050	.027	.020	.013	.007	.007	.020	.033	.047	.060	.074	.101	.127	.154	.181
.060	.032	.026	.019	.013	.001	.014	.028	.041	.054	.068	.095	.122	.148	.175
.063	.036	.029	.022	.015	.003	.011	.025	.038	.052	.065	.092	.119	.145	.172
.075	-	.036	.029	.022	.009	.005	.018	.031	.045	.058	.085	.112	.139	.165
.080	-	.039	.033	.026	.012	.000	.014	.028	.041	.055	.081	.108	.135	.162
.090	-	-	.040	.033	.020	.006	.007	.021	.034	.047	.074	.101	.128	.155
.125	-	-	-	.056	.042	.029	.015	.002	.011	.025	.051	.078	.105	.132
.160	-	-	-	.076	.063	.049	.036	.022	.009	.004	.031	.058	.085	.112
.190	-	-	-	-	.083	.070	.056	.043	.030	.016	.011	.038	.064	.091

NOTE: FIGURES TO THE LEFT OF THE HEAVY LINE ARE TO BE ADDED
       FIGURES TO THE RIGHT OF THE HEAVY LINE ARE TO BE SUBTRACTED

Figure 15.14 Calculation of flat pattern bend allowance development. (From John Tanner, *Manufacturing Engineering*, Marcel Dekker, New York, 1991.)

limited in flange height to prevent buckling. Figure 15.17b shows the drawing of the part as it would be seen by the manufacturing engineer, the tool designer, the toolmaker, and the inspector. The preferred dimensioning system is shown.

15.1.4 Deep Drawing

The force exerted by the punch varies with the percent of reduction, the rate of strain hardening, and the depth of the draw. Figure 15.18 shows the drawing force to take about 43% reduction in 5052-0 aluminum alloy. It can be seen that maximum load occurs at about 50% depth of draw. For most alloys, peak loads occur at one-half to two-thirds depth of draw. Aluminum, like other metals, strain-hardens during draw operations and is changed to a harder temper with a corresponding increase in tensile and yield strength. Figure 15.19 shows the effect of drawing on the mechanical properties of 3003-0 and 5052-0 alloys.

If a part is to be drawn successfully, the force exerted by the face of the punch must always be greater than the total loads imposed on that portion of the blank between the blankholder and the die. Also, the metal between the edges of the punch and die must be strong enough to transmit the maximum load without fracturing. This set of conditions establishes a relationship between blank size and punch size for each alloy-temper

APPROXIMATE LOAD PER LINEAL FOOT REQUIRED TO MAKE RIGHT ANGLE BEND IN 3003-H14, TONS

Thickness of metal, t	Width of female die opening, w, in.																										
	1/4	3/8	1/2	5/8	3/4	7/8	1	1-1/8	1-1/4	1-3/8	1-1/2	1-3/4	2	2-1/4	2-1/2	2-3/4	3	3-1/2	4	4-1/2	5	5-1/2	6	7	8	10	12
1/16		1.7	1.2	0.91	0.73	0.58																					
3/32				2.3	1.8	1.5	1.3	1.1																			
1/8					3.4	2.8	2.4	2.1	1.8	1.6	1.4																
5/32							3.9	3.3	2.9	2.6	2.4	2.0	1.7														
3/16								5.0	4.5	3.9	3.5	3.0	2.5	2.0													
1/4									6.8	5.5	4.7	4.1	3.6	3.2	2.9												
5/16											7.7	6.6	5.9	5.3	4.7	3.9	3.3										
3/8													10	9.0	7.8	7.1	5.9	5.0	4.4								
7/16															11	9.6	8.3	6.8	5.9	5.2	4.8						
1/2																	14	11	9.4	8.2	7.3	6.4	5.7				
5/8																		15	13	12	11	9.5	7.8	6.6			
3/4																			20	18	16	14	12	10	8.2		
7/8																				22	19	17	14	12	10	7.5	
1																							27	22	19	17	11
1-1/4																							44	36	31	23	19
1-1/2																									46	36	28

Conversion factor for various alloys

Alloy	3003-H14	3004-H32	3004-H34	5052-H32	5052-H34	5083-H113	5154-H32	6061-T4	6061-T6	2024-O	2024-T3	5456-H321
factor	1.00	1.4	1.6	1.5	1.7	2.1	1.8	1.6	2.0	1.2	3.2	2.3

NOTES: 1. Given values are for a male die having a radius, r, equal to the thickness of the metal, t. For the alloys for which a larger radius is used because of minimum bend requirements (see Table 3-1), the loads will be greater. These loads can be approximated by using an effective width of opening, determined by subtracting the value of (r-t) from the actual die opening in the above table. For example, if it were desired to bend a 5-ft long, 1/8-in. thick 6061-T6 sheet to a radius of 2t in a die with a 1-3/8-in. opening, the effective opening would be 1-3/8—1/4—1/8=1-1/4. Therefore, the required load would be 5 x 1.8 x 2.0 =18 tons.

2. The factor for any other alloy can be determined by dividing the typical tensile strength by 22 ksi.

Figure 15.15 Approximate force per lineal foot required to make a 90° bend. (Courtesy of Aluminum Corporation of America.)

684

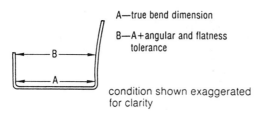

A—true bend dimension

B—A+angular and flatness
tolerance

condition shown exaggerated
for clarity

(a)

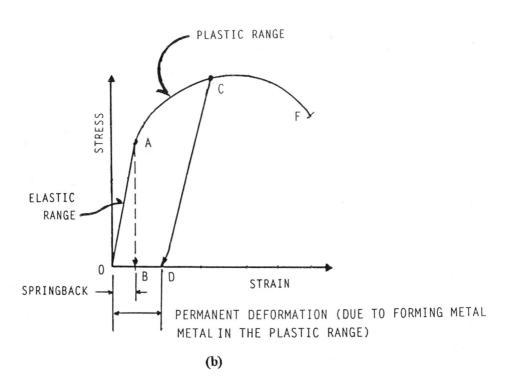

(b)

Figure 15.16 (a) Measurement of form dimension adjacent to bend radius, (b) Typical stress-strain curve showing elastic springback.

combination, and dictates the minimum punch size that can be employed in both draw and redraw operations. For circular parts, this relationship, stated in terms of percent reduction of the blank, allows the punch diameter to be 40% smaller than the blank diameter and still produce good parts consistently in the first draw operation, and 20% and 15% smaller for second and third operations, respectively, when drawing annealed tempers of most alloys (Figure 15.20). The full hard tempers of low-strength alloys often can withstand up to 50% reduction, while high-strength alloys, which work-harden rapidly, are limited to 30–35% reduction for first draw operations. In the latter case, redrawing may involve intermediate annealing operations. The edge condition of drawn parts is shown in Figure 15.21. The top view shows "earring," which is a function of material properties as described in Chapter 14.

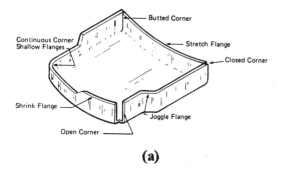

(a)

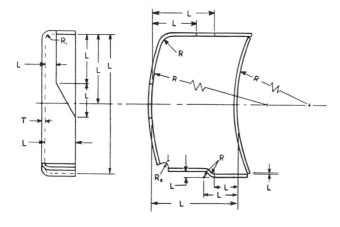

Preferred dimensioning and points to measure:

L = Linear dimensions; corner radius

R = Radii;

R^1 = Typical inside bend or

R^2 = Radius in flat blank

T = Material thickness

(b)

Figure 15.17 (a) Elements of formed stamping, (b) Drawing of the part as designed. (Courtesy Dayton Rogers Manufacturing Co.)

15.2 MACHINING

15.2.0 Introduction to Machining

The basic machine tools were introduced in Chapter 2, to help explain the meaning of some of the "do's and don'ts" of design for machining. This subchapter goes a little deeper into the machine functions and operations. Chip formation, cutters, feeds and speeds, power requirements for machining, etc., are defined and discussed. We discuss the software systems that are so critical in converting a product design (although it may be in digital

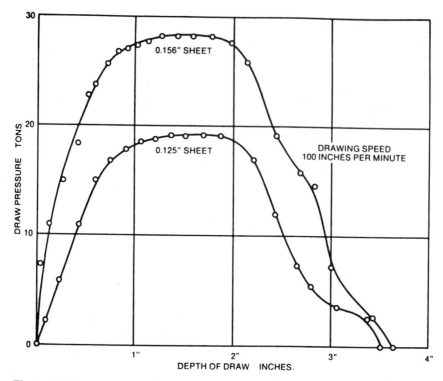

Figure 15.18 Effect of depth of draw on drawing force (43% reduction on 10.0-in.-diameter blank, 5052-O alloy). (Courtesy Aluminum Corporation of America.)

Alloy	Number of draws	Tensile strength, psi	Yield strength, psi	Elongation, in 2 in. percent
3003	0	16,000	6,000	30
	1	19,000	17,000	11
	2	22,000	21,000	9
	3	23,500	22,000	8
	4	24,500	22,500	8
5052	0	28,000	13,000	25
	1	34,500	32,000	6
	2	39,500	36,000	6
	3	43,000	37,000	6
	4	44,000	38,000	6

Specimens taken from sidewall at top of shell.

Figure 15.19 Effect of deep drawing on mechanical properties of aluminum. (Courtesy Aluminum Corporation of America.)

DIE DIMENSIONS FOR DRAWING CYLINDRICAL SHAPES

First draw.........................Punch diameter plus 2.2 times thickness of blank
Second draw.......................Punch diameter plus 2.3 times thickness of blank
Third and succeeding draws..........Punch diameter plus 2.4 times thickness of blank
Final draw of tapered shells..........Punch diameter plus 2.0 times thickness of blank

REDUCTIONS IN DIAMETER FOR DEEP SHELLS[1]

Operation	1100, 3003, 3004, 3005, 5005, 5050, 5052, 5457, 6061	2014, 2024, 5083, 5086, 5154, 5456
Blank (D)	—	—
First draw (D_1)	0.40D	0.30D
Second draw (D_2)	$0.20D_1$	$0.15D_1$
Third draw (D_3)	$0.15D_2$	$0.10D_2$
Fourth draw (D_4)	$0.15D_3$	—

[1]Based on annealed blanks.

Figure 15.20 Relationship of blank size to punch size, in terms of percent reduction. (Courtesy Aluminum Corporation of America.)

form on CAD) into machine codes, and actually cutting a part in the shop. Due to page limitations, there is not sufficient space to introduce all of the new machines and machining centers on the market today. It is to the point now where one cannot always distinguish between a milling machine and a lathe, since many machines are able to do both functions, as well as drilling, reaming, broaching, etc. References to some of the many books available are supplied. The author's recommendation is first to understand the product requirements; then to define the machining requirements; and finally to contact machine suppliers to obtain the latest machine tool information for a particular project.

In this subchapter, the basic principles of the machining processes are described and fundamental definitions given, after which chip formation and the process conditions are discussed. Material removal can be based on four fundamental removal methods, which illustrate the relationship between the imprinting of the information and the energy supply. Figure 15.22 shows the classification of mass-reducing processes in terms of the process and methods of material removal. The mechanical processes of turning, milling, and drilling are the main subjects of this subchapter. Blanking, punching, and shearing were covered in Subchapter 15.1, and some of the more unconventional processes will be introduced in Chapter 16.

Mass-reducing processes are used extensively in manufacturing. They are characterized by the fact that the size of the original workpiece is sufficiently large that the final geometry can be circumscribed by it, and the unwanted material is removed as chips, particles, and so on (i.e., as scrap). The chips or scrap are a necessary means to obtain the desired geometry, tolerances, and surfaces. The amount of scrap may vary from a few percent to 70–80% of the volume of the original work material. Most metal components have, at one stage or another, been subjected to a material-removal process.

Owing to the rather poor material utilization of the mass-reducing processes, the anticipated scarcity of materials and energy, and increasing costs, development in the last

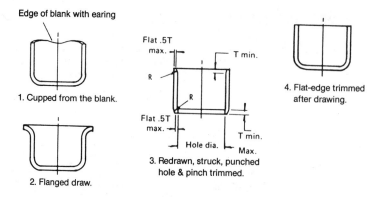

Edge of blank with earing

1. Cupped from the blank.

2. Flanged draw.

Flat .5T max.

R

R

T min.

Flat .5T max.

Hole dia.

Max.

T min.

3. Redrawn, struck, punched hole & pinch trimmed.

4. Flat-edge trimmed after drawing.

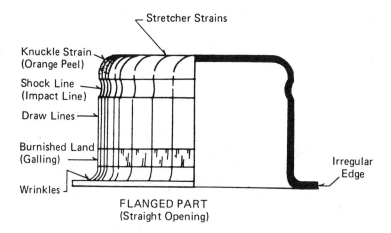

Stretcher Strains

Knuckle Strain (Orange Peel)

Shock Line (Impact Line)

Draw Lines

Burnished Land (Galling)

Wrinkles

Irregular Edge

FLANGED PART
(Straight Opening)

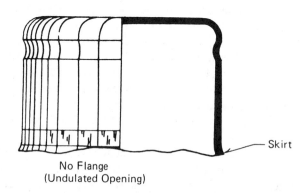

Skirt

No Flange
(Undulated Opening)

Figure 15.21 Edge condition of deep-drawn parts. (Courtesy Dayton Rogers Manufacturing Co.)

Category of basic process	Fundamental removal method	Examples of processes
Mechanical	I	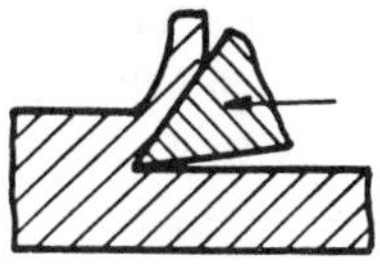 Cutting: Turning Milling Drilling Grinding, etc.
	II	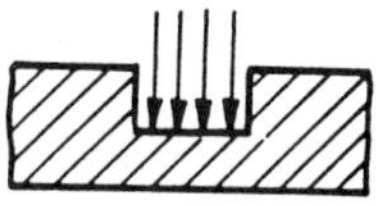 Water jet cutting Abrasive jet machining Sand blasting, etc.
	III	 Ultrasonic machining
	IV	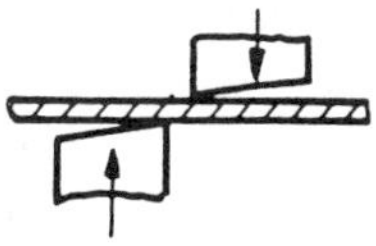 Blanking Punching Shearing
Thermal	II	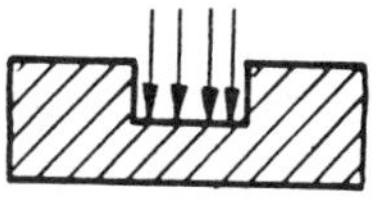 Thermal cutting (melting) Electron beam machining Laser machining
	III	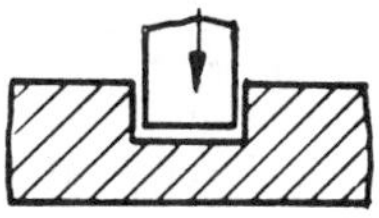 Electrodischarge machining
Chemical	II	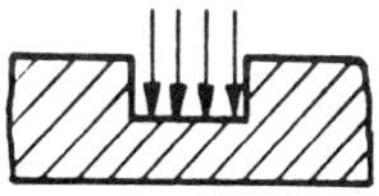 Etching Thermal cutting (combustion)
	III	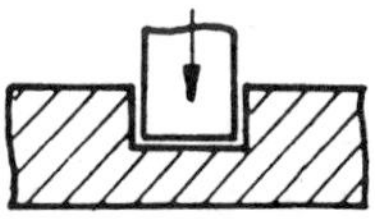 Electrochemical machining

Figure 15.22 Classification of mass-reducing processes. (From Leo Alting, *Manufacturing Engineering Processes,* Marcel Dekker, New York, 1982.)

decade has been directed toward an increasing application of *mass-conserving* processes. These include casting, forging, powder metal, and deforming processes resulting in a "near net shape" product, without extensive metal removal. However, die costs and the capital cost of machines remain rather high; consequently the mass-reducing processes are in many cases the most economical, in spite of the high material waste. Therefore, it must be expected that the material-removal processes will maintain their important position in manufacturing for the next several years. Furthermore, the development of automated production systems has progressed more rapidly for mass-reducing processes than for mass-conserving processes.

15.2.1 Machining Fundamentals

The unwanted material in mass-reducing processes, based on mechanical removal method I in Figure 15.22, is removed by a rigid cutting tool, so that the desired geometry, tolerances, and surface finish are obtained. Most of the cutting or machining processes are based on a two-dimensional surface creation, which means that two relative motions are necessary between the cutting tool and the work material. These motions are defined as the *primary motion*, which mainly determines the cutting speed, and the *feed motion*, which provides the cutting zone with new material. As examples (see Figure 15.23), in turning the primary motion is provided by the rotation of the workpiece, and the feed

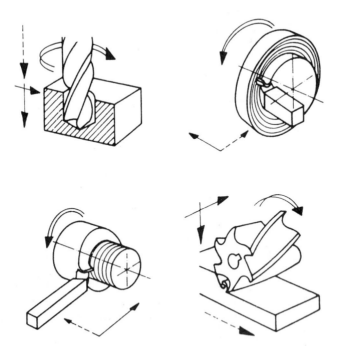

Examples of primary motions (⟹), feed motions (--->), and
positioning or adjustment motions (⟶)

Figure 15.23 Primary motions and feed motions in machining. (From Leo Alting, *Manufacturing Engineering Processes,* Marcel Dekker, New York, 1982.)

motion is a continuous translation of the cutting tool. In milling the primary motion is provided by the rotating cutter, and the feed motion by moving the workpiece.

Cutting Speed

The cutting speed v is the instantaneous velocity of the primary motion of the tool relative to the workpiece (at a selected point of the cutting edge). The cutting speed for these processes can be expressed as

$$v = \pi dn \text{ (in m/min)}$$

where v is the cutting speed in m/min, d is the diameter of the workpiece to be cut in meters, and n is the workpiece or spindle rotation in rev/min. Thus v, d, and n may relate to the work material or the tool, depending on the specific kinematic pattern of the machine.

Feed

The feed motion f is provided to the tool or the workpiece, and when added to the primary motion leads to a repeated or continuous chip removal and the creation of the desired machined surface. The motion may proceed by steps or continuously. The feed speed v_f is defined as the instantaneous velocity of the feed motion relative to the workpiece (at a selected point on the cutting edge).

Depth of Cut (Engagement)

In turning, the depth of cut a (sometimes called back engagement) is the distance that the cutting edge engages or projects below the original surface of the workpiece. The depth of cut determines the final dimensions of the workpiece. In turning with an axial feed, the depth of cut is a direct measure of the decrease in radius of the workpiece; and with a radial feed the depth of cut is equal to the decrease in the length of the workpiece. For milling, the depth of cut is defined as the working engagement a_e and is the radial engagement of the cutter. The axial engagement (back engagement) of the cutter is called a_p. In drilling, the depth of cut is equal to the diameter of the drill.

Chip Formation

The cutting process is a controlled interaction among the workpiece, the tool, and the machine. This interaction is influenced by the selected cutting conditions (cutting speed, feed, and depth of cut), cutting fluids, the clamping of the tool and workpiece, and the rigidity of the machine. Figure 15.24 illustrates this interaction. The clamping of the tool and the workpiece are not discussed here, and it is assumed that the machine possesses the necessary rigidity and power to carry out the process.

The chip formation mechanism is shown in Figure 15.25a. It can be seen that the shear deformation in the model is confined to the shear plane AB, extending from the tool cutting edge to the intersection of the free surfaces of the workpiece and chip. In practice, shearing is not confined to the plane AB, but to a narrow shear zone. At low cutting speeds the thickness of the zone is large, but at practical speeds the thickness is comparable to that shown in Figure 15.25b and can be approximated to a plane. The angle ϕ that the shear plane forms with the machined surface is called the *shear angle*.

The chip can be considered as built up of thin layers, which slide relative to each other as in Figure 15.25c. These layers can be compared to a stack of cards pushed toward the tool face. High normal pressures exist between the chip and the tool, causing high frictional

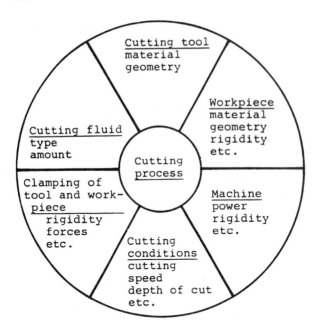

Figure 15.24   Interaction among the workpiece, the tool, and the machine in chip cutting. (From Leo Alting, *Manufacturing Engineering Processes,* Marcel Dekker, New York, 1982.)

forces and resulting in a chip with a smooth rear surface. The influence of friction is not shown in Figure 15.25c.

In the cutting process, the properties of the tool, the work material, and the cutting conditions (h_1, γ, and v) can be controlled, but the chip thickness h_2 ($< h_1$) is not directly controllable. This means that the cutting geometry is not completely described by the chosen parameters. The cutting ratio or chip thickness ratio, which is defined by

$$r = h_1/h_2 \quad (<1)$$

can be measured and used as an indicator of the quality of the cutting process.

The shear angle ϕ can be expressed by the rake angle γ and the inverse cutting ratio λ_b as seen in Figure 15.25c:

$$\tan \phi = \cos \gamma/(\lambda_h - \sin \gamma)$$

The inverse cutting ratio (also called the chip compression) and the rake angle determine the shear angle ϕ. The smaller ϕ is, the larger h_2 is, which means that the shear zone increases in length (i.e., the force and power requirements increase). Consequently, a large shear angle gives the best utilization of the supplied power. The chip compression must thus be kept as small as practically possible, since this increases the shear angle and, consequently, decreases the power consumption.

Hard work materials give lower chip compression values than do soft materials but require higher cutting forces. Friction increases the chip compression and can be reduced by introducing suitable cutting fluids. Chip compression can be reduced further by increasing the cutting speed or the feed. These increases in cutting speed and feed have an upper limit, however, because the tool life decreases, which might have a greater economic

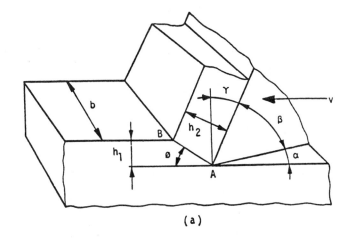

(a)

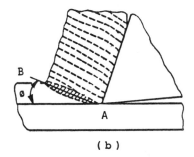

(b)

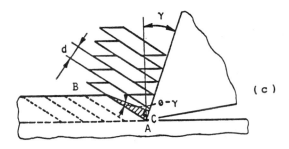

(c)

Figure 15.25 Chip formation mechanism. (From Leo Alting, *Manufacturing Engineering Processes*, Marcel Dekker, New York, 1982.)

effect than the resulting increases in material-removal rate. The actual shear angle ϕ can be determined experimentally by measuring h_2.

Types of Chips

Much valuable information about the actual cutting process can be gained from the appearance of the chip, as some types of chips indicate more efficient cutting than others. The type of chip is determined mainly by the properties of the work material, the geometry of the cutting tool, and the cutting conditions. It is generally possible to differentiate three types of chip: (1) discontinuous (segmental) chips, (2) continuous chips, and (3) continuous chips with built-up edges.

Discontinuous chips, shown in Figure 15.26a, represent the cutting of most brittle materials such as cast iron and cast brass, with the stresses ahead of the cutting edge causing the fracture. Fairly good surface finish is, in general, produced in these brittle materials, as the cutting edge tends to smooth the irregularities. Discontinuous chips can also be produced with more ductile materials such as steel, causing a rough surface. These conditions may be low cutting speeds or low rake angles in the range of 0–10° for feeds greater than 0.2 mm. Increasing the rake angle or the cutting speed normally eliminates the production of discontinuous chips.

Continuous Chips, shown in Figures 15.26b and 15.26c, represent the cutting of most ductile materials that permit shearing to take place without fracture. These are produced by relatively high cutting speeds, large rake angles ($\gamma = 10$–$30°$), and low friction between the chip and the tool face. Continuous and long chips may be difficult to handle; consequently the tool must be provided with a chip breaker, which curls and breaks the chip into short lengths. The chip breaker can be formed by grinding a stop or a recess in the tool, or brazing a chip breaker onto the tool face.

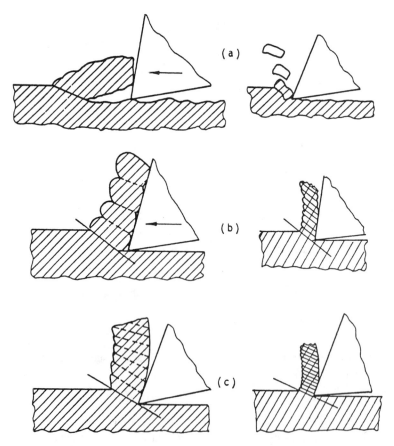

Figure 15.26 Types of chips. (From Leo Alting, *Manufacturing Engineering Processes,* Marcel Dekker, New York, 1982.)

Continuous chips with built-up edges represent the cutting of ductile materials at low speeds where high friction exists on the tool face. This high friction causes a thin layer of the underside of the chip to shear off and adhere to the tool face. This chip is similar to the continuous chip, but it is produced by a tool having a nose of built-up metal welded to the tool face. Periodically, portions of the built-up edge separate and escape on the chip undersurface and on the material surface, resulting in a rough machined surface as shown in Figure 15.27a. The built-up edge effectively increases the rake angle and decreases the clearance angle Figure 15.27b. At sufficiently high cutting speeds, the built-up edge normally disappears, and this upper limit is called the *free machining cutting speed*. A hard

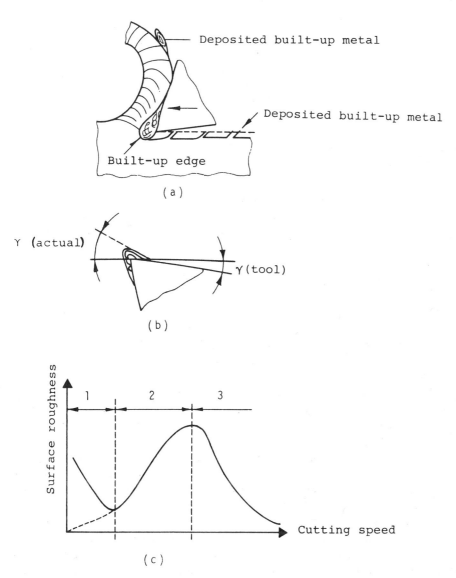

Figure 15.27 Chip cutting creating a built-up edge on the tool face. (From Leo Alting, *Manufacturing Engineering Processes*, Marcel Dekker, New York, 1982.)

material will generally have a lower free machining speed than a softer material. At increasing feed, the curve in Figure 15.27c will shift to the left. In most processes, cutting speeds above the free machining speed are chosen, but for broaching, for example, it is sometimes necessary to approach the minimum.

Tool Material

Chip formation involves high local stresses, friction, wear, and high temperatures; consequently the tool material must combine the properties of high strength, high ductility, and high hardness or wear resistance at high temperatures. The most important tool materials are carbon tool steels (CTS), high-speed steels (HSS), cemented or sintered carbides (CC), ceramics (C), and diamond (D).

Carbon Tool Steel

Plain carbon steels of about 0.5–2.0% C, when hardened and tempered, have a high hardness and strength, and can be used as hand tools for cutting softer materials at low speeds. The wear resistance is relatively low, and cutting-edge temperatures must not exceed about 300°C. This material is used now only for special purposes and has generally been replaced by the materials below.

High-Speed Steel

High-speed steels are alloyed steels that permit cutting-edge temperatures in the range of 500–600° C. The typical alloying elements are tungsten, chromium, vanadium, and cobalt. The higher cutting-edge temperatures make it possible to increase the cutting speed by about 100 over carbon tool steels—hence the name high-speed steels. This steel is used quite extensively in twist drills, milling cutters, and special-purpose tools and is, in fact, the most common tool material.

Sintered Carbide

Sintered (or cemented) carbides are produced by powder metallurgical processes. Sintered carbides of tungsten carbide with cobalt as a binder are hard and brittle and are used in cutting cast iron and bronze. If titanium carbide is added or used as the main constituent, the strength and toughness can be increased, and these materials can be used in cutting hard materials. A large variety of sintered carbides exist, and each is generally developed to fulfil the requirements of effective cutting of different material groups.

Carbide cutters are very hard, and they permit an increase in cutting speeds of about 200–500% compared to high-speed steel tools. But it must be remembered that they have a relatively low ductility and, consequently, care must be taken to avoid high-speed impacts such as those that occur during interrupted cutting operations. Sintered carbides are, in general, used as "throwaway" inserts supported in special holders or shanks. The inserts may have from three to eight cutting edges, and when one edge becomes dull, the insert is indexed to a new cutting edge. This procedure continues until all edges are used, when a new insert is substituted.

During recent years, coated sintered carbide tools have been developed, which allow both higher cutting speeds and higher temperatures. Production rate increases of about 200% are obtainable compared to conventional sintered carbides. Titanium carbide, titanium nitride, aluminum oxide, and so on, can be used as coating materials to prolong the life of the tool.

Ceramics

Ceramic tool materials have been developed within the last couple of decades. The material most frequently used is aluminum oxide, which is pressed and sintered. For light finishing cuts the cutting speeds obtainable are two or three times greater than the cutting speeds for sintered carbides. Ceramics are used mainly where close tolerances and high surface finish are required. The ceramics are produced as throwaway inserts or tips.

Diamond

Diamond is the hardest of all tool materials and is used mainly where a very high surface finish is required as well as close tolerances.

The Work Material

When an economical machining operation is to be established, the interaction among the geometry, the material, and the process must be appreciated. It is not sufficient to choose a material for a product that merely fulfills the required functional properties; the suitability of the material for a particular process must also be considered. It must have properties that permit machining to take place in a reasonable way; these properties are collectively called its *machinability*.

The term machinability describes how the material performs when cutting is taking place. This performance can be measured by the wear on the tool, the surface quality of the product, the cutting forces, and the type of chip produced. In most cases, tool wear is considered the most important factor, which means that a machinability index can be defined as the cutting speed giving a specified tool life. Machinability tests are carried out under standardized conditions (i.e., specified quality of tool material, tool geometry, feed, and depth of cut).

The machinability of a material greatly influences the production costs for a given component. In Table 15.1, the machinability for the different material groups is expressed as the removal rate per millimeter depth of cut when turning with carbides. The table can be used only as a general comparative guide; in actual situations accurate values must be obtained for the particular material.

The machinability of a particular material is affected primarily by its hardness, composition, and heat treatment. For most steel materials, the hardness has a major influence on machinability. A hardness range of HB from 170 to 200 is generally optimal. Low hardness tends to lead to built-up edge formation at low speeds. High hardness, above

Table 15.1 Removal Rate per Millimeter Depth of Cut for Different Groups of Materials When Turning with Carbides

Material	Removal rate/mm depth of cut (mm^2/min)
Construction steel	47,000–63,000
Tool steel (annealed)	15,000–37,000
Stainless steel	17,000–43,000
Cast steel	20,000–27,000
Cast iron	13,000–23,000
Copper alloys	50,000–63,000
Brasses	60,000–70,000

$HB = 200$, leads to increased tool wear as seen in Figure 15.28a, which gives the machinability as the cutting speed for a tool life of 30 min (T_{30}) for hardened and tempered alloy steel. Sometimes it is preferable to accept a lower tool life when machining hard materials (HB from 250 to 330) instead of annealing and rehardening the material.

The heat treatment of the work material can have a significant influence on its machinability. A coarse-grained structure generally has better machinability than a fine-grained structure. The distribution of pearlite and cementite has a definite influence too. It should be mentioned, however, that hardened plain carbon steels (0.35% C) with a martensitic structure are very difficult to machine. Inclusions, hard constituents, scale, oxides, and so on, have a deteriorating effect on the machinability, as the abrasive wear on the tool is increased.

Figure 15.28b shows machinability as a function of hardness for different material groups. The machinability is again defined as the cutting speed giving a tool life of 30 min. From the figure it can be seen that hardened and tempered materials—in spite of their higher hardness—have machinabilities approximately as high as the softer materials in turning and milling. In drilling, increased hardness results in poorer machinability.

Surface Quality (Roughness)

In a machining process, a specific geometry is produced, which also implies that a surface of satisfactory quality must be achieved. A machined surface always deviates from the theoretical surface. The real surface looks like a mountain landscape. Figure 15.29a shows

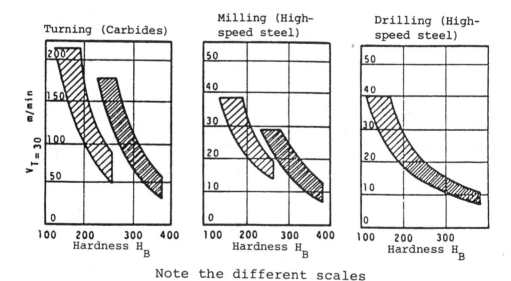

Figure 15.28 Machinability as a function of tool life for different materials. (From Leo Alting, *Manufacturing Engineering Processes,* Marcel Dekker, New York, 1982.)

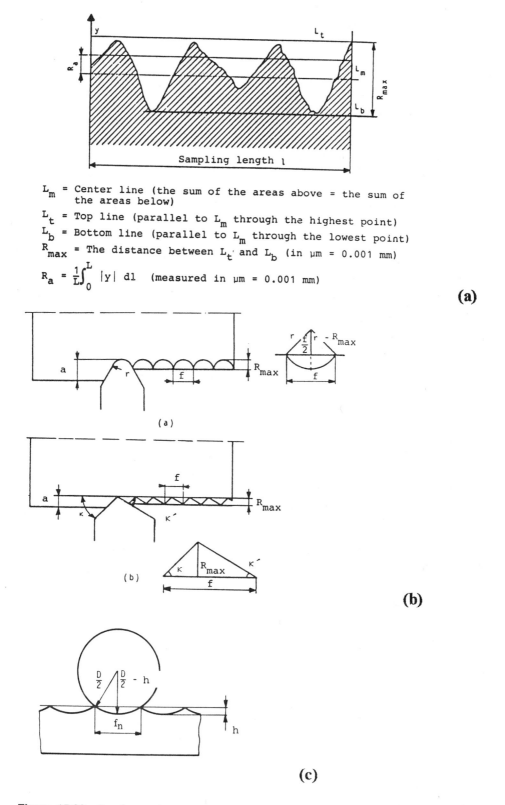

L_m = Center line (the sum of the areas above = the sum of the areas below)

L_t = Top line (parallel to L_m through the highest point)

L_b = Bottom line (parallel to L_m through the lowest point)

R_{max} = The distance between L_t and L_b (in μm = 0.001 mm)

$R_a = \frac{1}{L}\int_0^L |y|\ dl$ (measured in μm = 0.001 mm)

(a)

(a)

(b)

(b)

(c)

Figure 15.29 Roughness of a machined surface (surface finish). (From Leo Alting, *Manufacturing Engineering Processes,* Marcel Dekker, New York, 1982.)

definitions of roughness height and arithmetic average; Figure 15.29b shows that the corner radius of the cutting tool and the feed determine the surface roughness in turning. Roughness height in face milling is shown in Figure 15.29c. It can be concluded that the roughness decreases (i.e., the surface quality improved) when the feed is decreased, the nose radius is increased, and both the major cutting-edge angle and the minor cutting-edge angle are reduced. Furthermore, increasing cutting speeds and effective cutting lubricants can improve the surface quality. Typical roughness values are shown in Table 15.2 for the various machining processes.

15.2.2 The Lathe

The turning process is characterized by solid work material, two-dimensional forming, and a shear state of stress. The work-piece (W) is supported [e.g., clamped in a chuck (C) and supported by a center] and rotated (the primary motion, R). Through the primary motion (R) and the translatory feed (T_a = axial feed for turning and T_r = radial feed for facing) of the tool (V) the workpiece is shaped. See Figure 15.30.

Turning is used primarily in the production of various cylindrical components, with a nearly unlimited number of external and internal axial cross-sectional shapes (including tapers, threads, etc.). Facing is used both for regular and irregular shapes. Turning is the most extensively used industrial process. The material should not be too hard (HB 300) and should possess a minimum of ductility to confine deformation mainly to the shear zone. Turning provides close tolerances, often less than ±0.01 mm. Tighter tolerances may be obtained. The surface quality is good, normally in the range of $3 < R_a < 12$ μm. A wide variety of lathes are on the market: for example, the engine lathe, the turret lathe, single and multispindle screw machines, automatic lathes, and CNC lathes.

The engine lathe shown in Figure 15.31, forms the basis of our introduction to lathe work. The tool post shown probably contains a single cutting tool, fixed to the carriage, and therefore capable of movement fore and aft, and left or right. The old machine would be fitted with cranks to permit the operator to change the travel of the cutting tool. He would also be capable of changing the rotational "speed" of the spindle, and the "feed" of the carriage by changing the lead screw speed. This was sometimes done by changing the drive belt to different steps of the drive pulleys. Later, this was done by changing gears—first manually, then by "shifting" a lever to a transmission set of gears. This required an experienced machinist to make accurate parts in a "reasonable" time. It is used today for toolroom work or making one-of-a-kind parts.

Table 15.2 Typical
Roughness Values
(Arithmetic Mean Value R_a)
for Different Processes

Process	Roughness, R_a (μm)
Turning	3–12
Planing	3–12
Drilling	3–25
Milling	1–10
Grinding	0.25–3

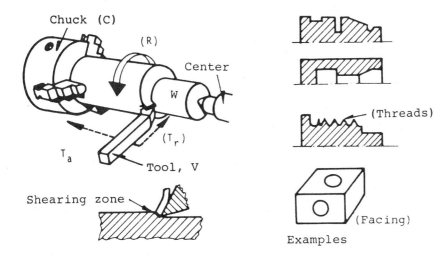

Figure 15.30 The turning process. (From Leo Alting, *Manufacturing Engineering Processes*, Marcel Dekker, New York, 1982.)

The lathes in use in production shops today are progressively improved versions of this machine. First, by mounting a turret containing several types of cutting tools on the carriage, the operator could rotate the turret to a new position and perform different operations more quickly. A second turret could be added opposite the one on the cross-slide, or a turret could be mounted on the tail stock. See Figure 15.32 for examples of various lathe cutting tools. The addition of cams and gears allows automatic change in feed or speed, and indexing the turrets to different positions. After proper set-up for a particular part to be machined, a less skilled operator can now make repeatable parts.

It can be seen that by providing power to the turret tool holders, we can use drills, reamers, milling cutters, etc., and perform operations with the spindle not rotating. (For example, we can drill and tap a hole through the diameter of the cylindrical part that had been turned.) Now, however, the problem of "control" is much more complex. Electrical limit switches replace cams, gears, and hydraulic valves. The next logical step is to replace limit switches with controllers which can simply be programmed to sequence the desired functions. With today's microprocessors, PCs are being used on the more complex programs and machines, for both basic housekeeping functions as well as tool path programs.

A "lathe" today looks like any other machine in the shop. It probably has a nicely painted housing around it, may have two turrets that can perform operations simultaneously, both of them with power drives, powerful electrical motors with controllable speeds, and numerical controls (NC) with computers or advanced controllers to act as the "director" of this multipurpose machine. Rather than just cutting cylinders or tapers, the electronic controls permit cutting an infinite number of contours and shapes. Bar stock can be fed automatically through the headstock, or a programmable robot can load and unload piece parts to the machine and to a moving conveyor. Milling and drilling can be performed with either the piece part rotating or stopped. Measurements of power consumption of the various operations can alert the operator that a cutting tool is getting dull or has broken. Measurements of various features of the part can be measured automatically, and the machine will adjust that parameter to correct the "out of tolerance" trend before the

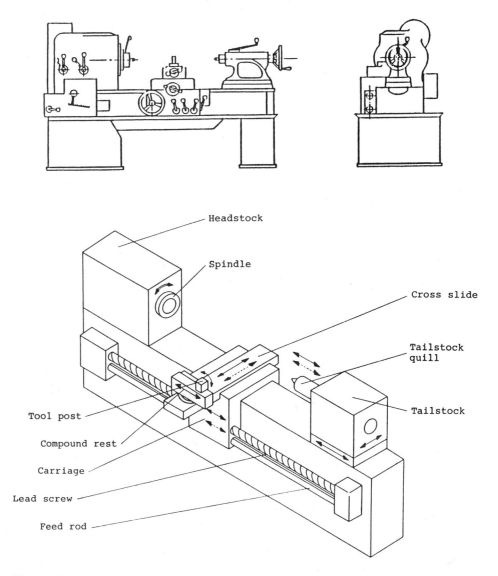

Figure 15.31 The engine lathe. (From Leo Alting, *Manufacturing Engineering Processes,* Marcel Dekker, New York, 1982. Courtesy Cincinnati Machine Tool Co.)

parts are actually made wrong. This can feed a statistical process control (SPC) program in parallel for verification of the piece-part accuracy. One of the latest lathes that the author purchased is even capable of performing broaching from the powered tool turret, while the spindle is stopped. Figure 15.33 is an example of such a "lathe" today. It can automatically feed bar stock through the spindle, or act as a "chucker" by changing the collet to a face plate of sorts to mount castings or forgings. This particular machine is outfitted with a five-axis Fanuc robot to load and unload parts.

With all the advancements made in the "lathe," there are several other very important actions that have been taken. Material selection is now a very precise process, and cutting

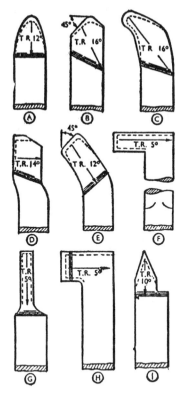

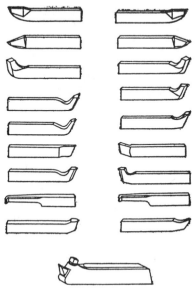

A selection of lathe tools. *A*, round-nose bore; *B*, a better shape; *C*, a roughing tool; *D*, knife tool; *E*, tool for sliding and surfacing motions; *F*, square-nose boring tool; *G*, parting tool; *H*, recessing tool; *J*, screw-cutting tool.

A further selection of lathe tools, showing the great variety they may have.

Figure 15.32 Examples of lathe cutting tools. (From Leo Alting, *Manufacturing Engineering Processes,* Marcel Dekker, New York, 1982.).

tools have been characterized and improved. Without these advances, the higher feeds and speeds in use today would be impossible. The machines themselves have been designed to be much more rigid, and can hold tighter tolerances consistently. In general, the higher the cutting speed, the lower the forces required, and the better the finish.

The computer programs that control today's NC machines can be generated on the floor by an operator using CNC; or developed in an office environment and sent to the floor controls through a coax or fiber optics cable by distributed numerical control (DNC). When the product design is in digitized format, using one of the many computer-aided design (CAD) programs available, many of the piece part machining programs can *almost* be prepared automatically using computer-aided manufacturing (CAM) software. Although CAD/CAM has a long way to go to become more universally used, it is *technically* available today. The machine manufacturers and software companies have not yet agreed to a standard that would make all machines capable of running all program instruction sets. The difficulty of actually converting the different product design CAD programs to programs that direct the machine to make a part should not be minimized. The wide

**THREE AXES CNC TURNING/MILLING CENTER LJ-103M
WASINO**

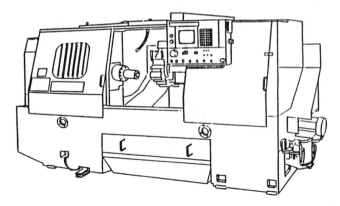

Figure 15.33 Wasino Turn-Mill Center (lathe). (Courtesy McDonnell Douglas Corp.)

variation of hardware and software available today is changing rapidly—along with the computer industry as a whole. Conversion can be a difficult problem.

Figure 15.34 shows the definitions of turning, and mathematical relationships of the different elements of lathe work.

15.2.3 The Milling Machine

The milling process is characterized by solid work material, two-dimensional forming (one-dimensional forming may be used in a few cases), and a shear state of stress. The workpiece (W) is clamped on the table (B), which is given a translatory feed (T), that together with the primary motion (R) of the cutter (V) provides the many geometric possibilities. The milling process—through the various types of cutters and the wide variety of machines—is a versatile, high-production process. Through various accessories (dividing head, attachments, etc.), many different special shapes can be produced. The milling process comes close to turning in extensive industrial use, since the geometric possibilities are enormous and the removal rate high. See Figure 15.35 for examples of a horizontal mill on the left, and a vertical mill on the right.

The hardness of the material should not be too high (HB 250–300), and a minimum of ductility is advisable. The obtained tolerances are normally good (±0.05 mm), and the surface quality high ($1 < R < 10$ μm). A wide variety of milling machines is available: for example, the plain column-and-knee type (general purpose), the universal column-and-knee type, the bed type, and the planar type.

Much of the previous discussion on lathes applies to milling machine progress. In a sense, the milling machine is the opposite of the lathe, in that it provides cutting action by rotating the tool while the sequence of cuts is achieved by reciprocating the workpiece. The sequence of consecutive cuts is produced by moving the workpiece in a straight line, and the surface produced by a milling machine will normally be straight—in one direction at least. A milling machine, however, uses a multiple-edged tool, and the surface produced by such a tool conforms to the contour of the cutting edges. If the milling cutter has a straight cutting edge, a flat surface can be produced in both directions. The workpiece is usually held securely on the table of the machine, or in a fixture clamped to the table. It

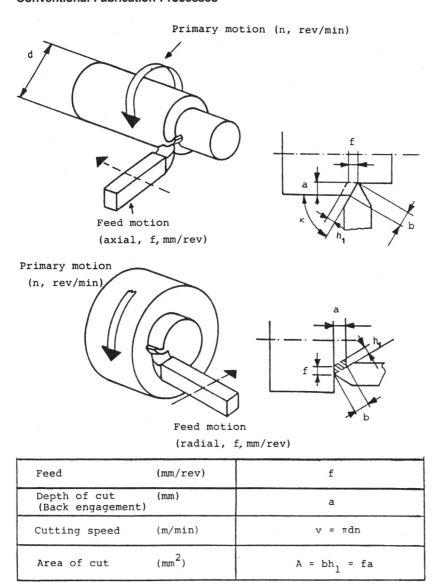

Feed	(mm/rev)	f
Depth of cut (Back engagement)	(mm)	a
Cutting speed	(m/min)	$v = \pi dn$
Area of cut	(mm^2)	$A = bh_1 = fa$

Figure 15.34 Mathematical relationship of the elements of lathe work. (From Leo Alting, *Manufacturing Engineering Processes,* Marcel Dekker, New York, 1982.)

is fed to the cutter or cutters by the motion of the table. Multiple cutters can be arranged on a spindle, separated by precision spacers, permitting several parallel cuts to be made simultaneously. Figure 15.36 shows some of the typical cutters in use.

The material is moved in an *x-y* direction for each pass, and can be moved toward or away from the cutting tool to change the depth of cut. The plain column-and-knee-type horizontal milling machine shown in Figure 15.37 has been a "standard" for many years. The universal milling machine resembles the plain-type mill shown in Figure 15.37. The chief difference lies in the fact that the table is supported on and carried by the housing,

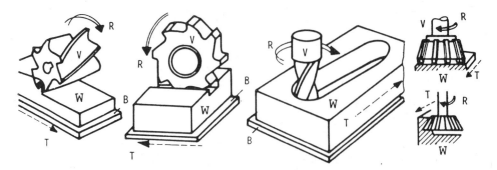

Figure 15.35 Examples of horizontal and vertical milling machines. (From Leo Alting, *Manufacturing Engineering Processes,* Marcel Dekker, New York, 1982.)

which swivels on top of the saddle. Thus, the table can also be rotated in the horizontal plane. This arrangement permits cutting helices, for milling flutes in twist drills or milling cutters. With an indexing arrangement, it is useful for cutting gear teeth.

The vertical milling machine derives its name from the position of the spindle, which is located vertically and at right angles to the top surface of the table. The vertical milling machine is especially adapted to operations with end mills and face mills; for profiling interior and exterior surfaces; for milling dies and metal molds; and for locating and boring holes. Figure 15.38 shows one of the "old standard" vertical mills.

There are special-purpose production milling machines of various types. A duplex mill, for example, has two horizontal spindles mounted on independently adjustable spindle carriers, which slide on two headstocks placed on opposite sides of the bed. Two identical or two different milling operations can be performed simultaneously on one or more workpieces.

The improvement history of milling machines is much like that of the lathe, and perhaps even more features and capabilities are now available. Figure 15.39 shows a modern vertical milling machine. It is truly difficult to identify today's machining centers as either "a lathe" or "a milling machine." The question becomes more one of the capability of a particular machine to make a particular part, versus the cost of producing the part on another type of machine. Of course, the lathe still primarily rotates the material to be cut, and the mill primarily rotates the cutters.

For years the milling machine has been considered a three-axis machine, with the cutter fixed (either horizontal or vertical), the table moving past the cutter in the x axis, toward the cutter (90° to the table motion) as the y axis, and up and down movement of the table called the z axis. With today's machines, it is possible to tilt or rotate the table, or tilt the cutter spindle—or both—creating five- and six-axis machines. Some grinders have six axes of motion. The use of multiple axes simplifies the number of cutters required and permits smooth transitions, but greatly increases the complexity of the programs and the training of the programmers. Figure 15.40 shows the definitions of milling, and mathematical relationships of the different elements of mill work.

15.2.4 Drilling

The drilling process is also characterized by solid work material, two-dimensional forming, and a shear state of stress. The workpiece (W) is clamped on a table (B) and the tool (V)

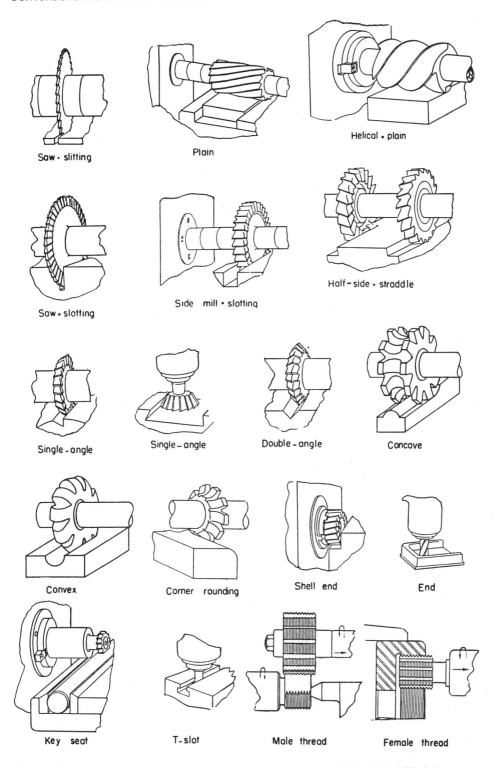

Saw · slitting   Plain   Helical · plain

Saw · slotting   Side mill · slotting   Half – side · straddle

Single - angle   Single – angle   Double – angle   Concave

Convex   Corner rounding   Shell end   End

Key seat   T-slot   Male thread   Female thread

Figure 15.36 Typical milling cutters in use. (Drawings courtesy of Illinois Tool Works.)

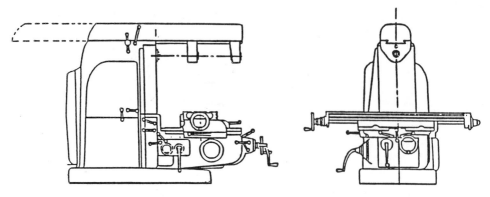

Figure 15.37 Column-and-knee-type horizontal milling machine. (Courtesy Cincinnati Machine Tool Co.)

is given a rotation (the primary motion, R) and a translatory feed (T). In drilling on lathes, the workpiece is rotated and the feed is applied to the tool. See Figure 15.41. The drilling process is used primarily to produce interior circular, cylindrical holes. Through various tools (twist drills, combination drills, spade drills, gun drills, etc.), different hole shapes can be produced (cylindrical holes, drilled and counterbored, drilled and countersunk, multiple-diameter holes, etc.). Drilling is an important industrial process.

The hardness of the material should normally not exceed HB = 250. For diameters less than 15 mm, the normal tolerance is around ±0.3 mm. Finer tolerances may be obtained, but finishing is often carried out by a special reaming process. The surface roughness is typically $3 < R_a < 25$ μm.

Many types of drilling machines are available: for example, bench, upright, radial, deep-hole, and multispindle drilling machines. Figure 15.42 shows the definitions of drilling, and mathematical relationships of the different elements of drill work.

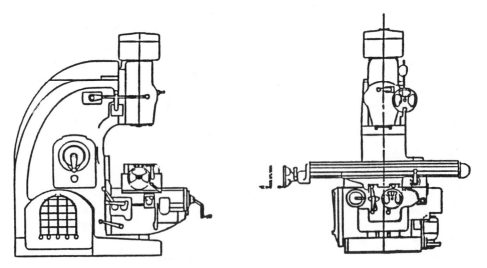

Figure 15.38 Example of an "old standard" vertical milling machine. (Courtesy of Kerney & Trecker Machine Tool Co.)

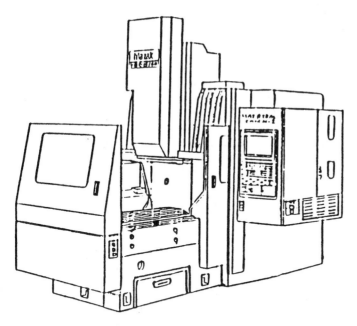

Figure 15.39   Example of a modern vertical milling machine.

15.3   EXTRUSION AND FORGING PROCESSES

15.3.0   Introduction to Extrusion and Forging Processes

In the previous two subchapters, we discussed sheet metal fabrication (15.1) and conventional machining processes (15.2). While these processes are important to the manufacturing industry, there are other methods of making parts—and with the correct consideration of some of these alternates, distinct advantages are possible.

In Chapter 2 we introduced many of the conventional processes of forging and casting, in order for the manufacturing engineer and product designer to understand the basic characteristics of each process. The advantages and disadvantages of the processes include material, fabrication cost, mechanical properties, and tolerances.

In this subchapter we look in more detail at some of the *mass-conserving* processes as defined by Leo Alting in *Manufacturing Engineering Processes*, Marcel Dekker (New York, 1982). There are advantages to fabricating a product in a minimum number of process steps. If a part can be made in one operation, by the closing of a mold, for example, and obtain the desired net shape—or even its "near net shape"—and still possess the desired mechanical properties, tolerances, etc., we can reduce or eliminate costly machining finishing operations. In other words, there is more than one way to skin a rabbit—once you have decided you want rabbit fur!

The author was once faced with the problem of buying small precision aluminum investment castings, made to close tolerances at a cost of $2.40 each. This purchase was for an assembly that had a very aggressive low-price target, and a quantity of 100,000 pieces. One option was to buy a low-cost plaster molding, of the same material, for $0.70, and finish machine it to obtain the required final shape and tolerances. Our final decision was to buy the low-cost shell molding, and "straighten" it by pressing it in a warm

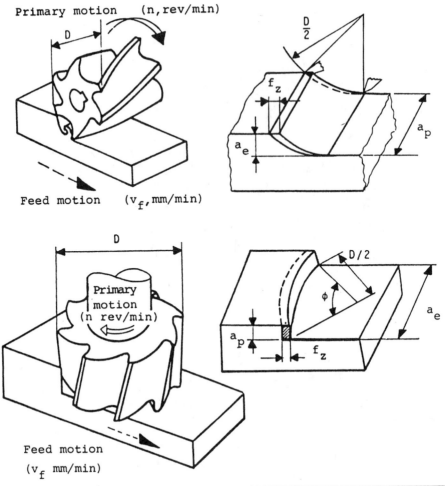

Feed speed (for table)	(mm/min)	v_f
Feed (per revolution of cutter)	(mm/rev)	$f = v_f/n$
Feed (per tooth)	(mm/tooth)	$f_z = v_f/nz$
Cutting speed	(m/min)	$v = \pi Dn$
Removal rate V	(cm^3/min or mm^3/min)	$V = a_e a_p v_f$

Figure 15.40 Mathematical relationships of milling work. (From Leo Alting, *Manufacturing Engineering Processes,* Marcel Dekker, New York, 1982.)

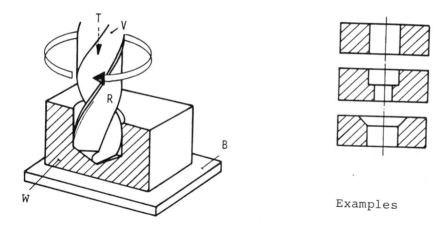

Examples

Figure 15.41 The drilling process. (From Leo Alting, *Manufacturing Engineering Processes*, Marcel Dekker, New York, 1982.)

mold—exceeding the yield strength in compression. This process corrected any dimensional errors of thickness, warpage, surface finish, and dimensions between planes. Total cost was less than $1.00 for material and labor—including the steel forming die amortization. Side benefits were the increased strength and stiffness of the part. Later production of the same part by a different supplier was by machining from aluminum plate on an NC milling machine, which was done at a cost of over $5.00.

The message of this subchapter is to look at some of the "near net shape" processes in the general field of pressing and casting. While they may not be applicable in all cases, they are certainly worth looking at. An example is the progress made in cold forging, warm forging, and pressing in general. On the other side, squeeze casting, oversize investment castings, and the like offer some of the mechanical properties originally thought to be obtainable only from wrought alloys. Replacing the variations in processes due to an operator with the precision process control of pressures, time, temperature, etc., by some type of electronics permits us to accomplish operations on a production basis that were previously not cost effective. The electronics may range from a low-cost programmable logic controller to CNC or DNC computer controls.

Newer, more accurate cost accounting systems now let us see the "real cost" of production. Things such as energy cost, work-in-process inventory, cycle time, worker's compensation insurance cost, parts cleaning, etc., are now visible in the modern factory as well as direct labor and material costs.

While Subchapter 15.2 covered the *mass-reducing* process of machining, this subchapter is more concerned with the *mass-conserving* processes, which come closer to a "near net shape" achieved with the primary operation.

15.3.1 Form and Structure of Fabrication Processes

Figure 15.43 shows the morphological structure of material processes. This figure shows the material flow, the energy flow, and the information flow. The *material flow* deals with the state of the material for which the geometry and/or the properties are changed, the basic processes that can be used to create the desired change in geometry and/or properties, and the type of flow system characterizing the process. The material may be solid, liquid,

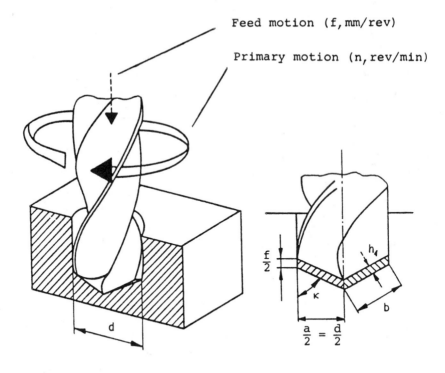

Feed	(mm/rev)	$f (= f/2$ per cutting edge)
Depth of cut (Back engagement)	(mm)	$a = d$
Cutting speed	(m/min)	$v = \pi d n$
Area of cut	(mm^2)	$A = b \cdot h_1 = \frac{f}{2} \cdot a$ per edge

Figure 15.42 Mathematical relationships of the elements of drilling. (From Leo Alting, *Manufacturing Engineering Processes*, Marcel Dekker, New York, 1982.)

granular, or gaseous. To carry out the basic processes described under material flow, energy must be provided to the work material through a transmission medium. The second part of Figure 15.43 shows this *energy flow* required. This consists of the tool and die system, and the equipment system. The tool and die system describes how the energy is supplied to the material and the transfer media used. The equipment system describes the characteristics of the energy supplied from the equipment and the type of energy used to generate this. The right side of Figure 15.43 shows the *information flow*, which is the impressing of shape information on the work material. The principles on which information impressing is based can be analyzed in relation to the type of process (material flow), the material, and the basic process.

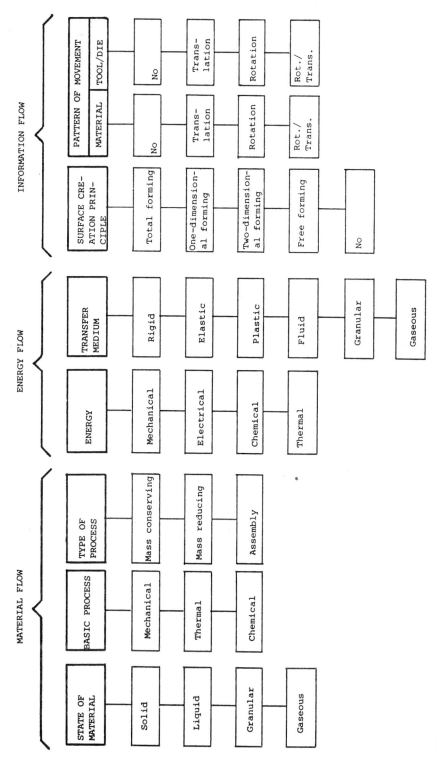

Figure 15.43 Morphological structure of material processes. (From Leo Alting, *Manufacturing Engineering Processes*, Marcel Dekker, New York, 1982.)

Figure 15.44 shows examples of information impressing by mass-conserving processes with solid materials. The basic principles of surface creation include four possibilities:

Free forming. Here the medium of transfer does not contain the desired geometry (i.e., the surface/geometry is created by stress fields).

Two-dimensional forming. Here the medium of transfer contains a point or a surface element of the desired geometry, which means that two relative motions are required to produce the surface.

One-dimensional forming. Here the medium of transfer contains a producer (a line or a surface area along the line) of the desired surface, which means that one relative motion is required to produce the surface.

Total forming. Here the medium of transfer contains (in one or more parts), the whole surface of the desired geometry, which means that no relative motion is necessary.

Figure 15.45 shows examples of shape impressing on liquid materials. Figure 15.45a shows examples of processes where shaping and stabilizing are separate, and Figure 15.45b shows examples of processes where shaping and stabilizing are integrated.

15.3.2 Engineering Materials Properties

In Chapter 14 we discussed some of the basics of metallurgy, and properties of metals that we would probably be working with in the manufacturing industry. As we consider the selection of materials *and processes* for a particular product, we need to look at the basic information a little differently. Figure 15.46 shows typical tensile test diagrams obtained at room temperature and slow test speed. The vertical axis is the applied stress, and the horizontal axis is the strain, or deformation. The straight portion of each curve for the different materials defines the proportional limit, with the elastic limit being the point where increases in stress create a strain (deformation) that does not recover completely when the load is released. Within this proportional limit (where strain is truly proportional to the amount of applied stress), the ratio of stress divided by strain is called the modulus of elasticity (E), and defines the stiffness of the material. We could say that alloyed steel is "stiffer" than aluminum. (E for steel is approximately 30,000,000 psi, and E for aluminum is about 10,000,000 psi.)

We also know that plastic deformation is required for any permanent change in shape of a metal, and we must exceed the proportional limit in our forming processes in order to form a part. Figure 15.47 is an important characteristic of metals in our use of mass-conserving processes. In Figure 15.47a we can see that the rate of applying strain increases the strength of the material in tension (or compression) and reduces the amount of deformation possible. In Figure 15.47b, we can see that increasing the temperature (at the same rate of strain as in Figure 15.47a), reduces the strength of the material and increases the ductility, or amount of deformation allowed before failure. Figure 15.48 shows the properties of some aluminum alloys at elevated temperatures. Typical deformation velocities for various processes are shown in Figure 15.49.

Figure 15.50 shows the melting points of various metals. This is the point where all the alloying ingredients in the metal (or alloy) are in solution. This temperature is a good reference point for casting. Figure 15.51 shows the lowest possible recrystallization point for four different metals. We must exceed this point in order to obtain some of the more desirable characteristics of the metal during the annealing and hardening processes. Also

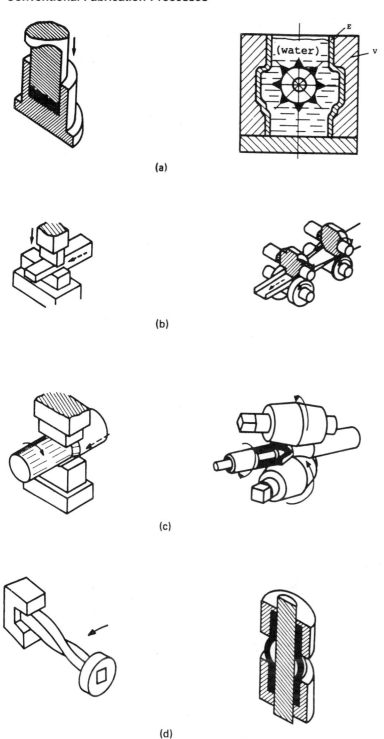

(a)

(b)

(c)

(d)

Figure 15.44 Information impressing by mass-conserving processes with solid materials. (From Leo Alting, *Manufacturing Engineering Processes*, Marcel Dekker, New York, 1982.)

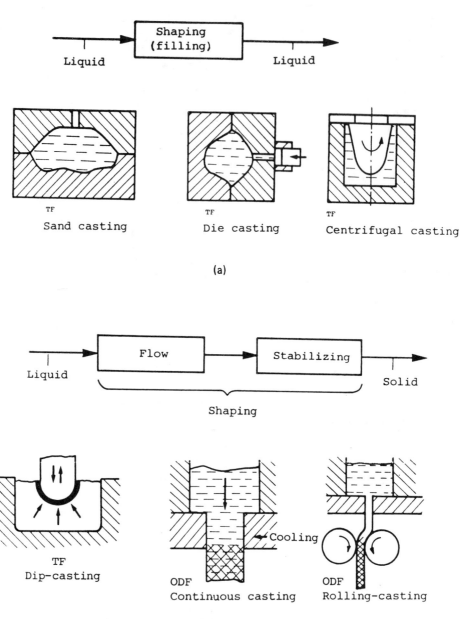

Figure 15.45 Shape impressing on liquid materials.

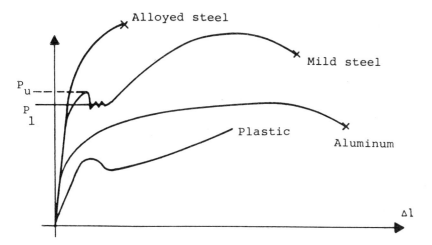

Figure 15.46 Typical tensile test diagrams obtained at room temperature and slow speed. (From Leo Alting, *Manufacturing Engineering Processes,* Marcel Dekker, New York, 1982.)

shown is the upper limit for hot working (e.g., forging). Figure 15.52 shows physical and mechanical properties of some structural materials.

15.3.3 Aluminum Extrusions

Extruded aluminum alloy shapes are produced by slowly forcing cast or wrought cylindrical billets, heated to plastic condition that is approximately 600–800°F, under hydraulic pressure through a steel die opening of desired cross section. Billet diameters range from 4 to 16 in. and larger. Pressures up to 5500 tons are required to push the hot metal through the die opening. A schematic illustration of extrusion equipment and tools is shown in Figures 15.53. In subsequent operations the aluminum shapes, which emerge from the extrusion press in lengths up to 80 ft, are heat treated, straightened, and cut to desired length.

Since extrusion dies are relatively inexpensive (often costing less than $1000), it will generally pay to design special extruded sections to meet the specific requirements of the

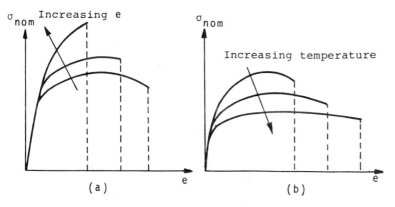

Figure 15.47 Results of the rate of increasing (a) strain, and (b) temperature on strength and ductility. (From Leo Alting, *Manufacturing Engineering Processes,* Marcel Dekker, New York, 1982.)

Alloy and Temper	Property	75°F	300°F	400°F	500°F	600°F	700°F
3S-H14	Ultimate Strength (psi)	21,500	17,500	14,000	10,000	5,000	3,000
	Yield Strength (psi)	19,000	12,500	8,000	4,000	2,500	2,000
	Elongation % in 2 in.	16	17	22	25	40	60
14S-T6	Ultimate Strength (psi)	70,000	47,000	18,000	11,000	6,500	4,500
	Yield Strength (psi)	60,000	40,000	12,000	8,500	5,000	3,500
	Elongation % in 2 in.	13	15	35	45	65	70
17S-T4	Ultimate Strength (psi)	62,000	40,000	22,000	12,000	6,500	4,500
	Yield Strength (psi)	40,000	30,000	17,000	9,500	5,000	3,500
	Elongation % in 2 in.	22	16	28	45	95	100
24S-T4	Ultimate Strength (psi)	68,000	43,000	26,000	14,000	7,000	5,000
	Yield Strength (psi)	48,000	37,000	22,000	10,000	5,000	3,500
	Elongation % in 2 in.	19	17	22	45	75	100
61S-T6	Ultimate Strength (psi)	45,000	32,000	19,000	7,000	4,000	3,000
	Yield Strength (psi)	40,000	30,000	16,000	5,000	2,500	2,000
	Elongation % in 2 in.	17	18	25	65	90	105

Figure 15.48 Elevated temperature effect on mechanical properties of aluminum. (Courtesy Aluminum Corporation of America.)

Process	Tool/die velocity (deformation velocity) (m/s)
Tension test	$10^{-6}-10^{-2}$
Hydraulic press	$2 \times 10^{-2}-3 \times 10^{-1}$
Tube drawing	$5 \times 10^{-2}-5 \times 10^{-1}$
Sheet rolling	$5 \times 10^{-1}-25$
Forging	2–10
Wire drawing	5–40
High-velocity forging	20–50
Explosive forming	30–200

Figure 15.49 Deformation velocities of various processes.

structure to be built, rather than to sacrifice efficiency by using available standard sections nearest the desired shapes. Extruded sections can be "tailored" economically to meet the needs of the design from the standpoint of strength and stiffness. Metal can be placed where it will do the most good, and wall thicknesses of sections in aluminum extrusions may vary within broad limits. Typical mechanical properties of aluminum extrusion alloys are given in Figure 15.54.

Manufacturing Possibilities

Extruded aluminum shapes offer many interesting possibilities as to sizes, weights, strengths, and types of shapes that can be produced. The maximum cross-sectional dimensions generally considered commercial are governed by a circumscribing circle approximately 12 in. in diameter; a limited number of 5500 ton presses can handle sections circumscribed by a 17-in. circle. The circumscribing circle is the smallest circle that

a. Pure metals (°C)

Iron	1535	Lead	327
Copper	1083	Tin	232
Aluminum	660	Magnesium	650
Nickel	1455	Chromium	1850
Zinc	419		

b. Alloys (°C)

Stainless steel (18% Cr, 9% Ni)	1400–1420
Brass (35% Zn, 65% Cu)	905– 930
Bronze (90% Cu, 10% Sn)	1020–1040
Aluminum–bronze	1050–1060
Aluminum (1% Si, 0.2% Cu)	643– 657

Figure 15.50 Melting points of various materials.

Metal	Lowest recrystallization temperature (°C)	Melting point (°C)	Upper limit for hot working (°C)
Mild steel	600	1520	1350
Copper	150	1083	1000
Brass (60/40)	300	900	850
Aluminum	100	660	600

Figure 15.51 Lowest possible recrystallation temperatures for various metals.

completely encloses a shape. Section thicknesses may vary from about 0.050 in. to several inches, depending on requirements.

The permissible ratio of thickness to width of a section depends on the alloy. The softer alloys can be extruded to thinner sections than can medium- and high-strength alloys. For example, the minimum thickness for a shape about 8 in. wide is 3/32 in. for alloys 3S and 63S, but 1/8 in. for alloy 14S. The wider the shape, the more the minimum thickness may be varied.

Minimum and maximum weights per foot of aluminum extrusions are governed by the limits of the extrusion ratio, which normally should not be smaller than 16:1 nor greater that 45:1. The extrusion ratio is the ratio of cross-sectional area of the cast extrusion ingot to the cross-sectional area of the extruded shape. Within these limitations the weight per foot of aluminum extrusions can normally range from an ounce or less to about 20 lb, which corresponds to a cross-sectional area of over 15 in.^{2}.

Small castings or forgings, such as brackets, clamps, or hinges, can often be changed to aluminum extrusions with considerable cost advantage if their dimensions are symmetrical about one plane. For instance, 2 in.-wide coupling clamps for aluminum irrigation pipe are cut from 63S-T6 alloy extrusion as shown in Figure 15.55. A cut between the two heavy bosses changes the solid ring into a tightening clamp. Forging stock for aluminum die forgings is often extruded in cross sections designed to produce the desired metal flow during the forging operation and to reduce forging and trimming costs. Figure 15.56 shows a typical example.

In the author's experience with the high-rate production of pyrotechnics, rocket motors, etc., the use of extrusions for the machining of rocket motors is a requirement for safety and reliability. Normal rolled aluminum stock, which is lower in cost than an extrusion, occasionally has inclusions that might permit a weak side wall in the motor—causing an occasional failure upon rocket motor ignition. A large gas generator used 7075 aluminum extruded stock, cut into 6-in. lengths, as the billet for forging. Again, the consistency of the extrusion aided in the reliability of the generator case.

One of the problems that an extruder has in maintaining tolerances in an extruded cross section is in the straightening process that follows extrusion. The extrusion die, of course, can be machined very accurately, which permits the metal coming through the orfice to be quite accurate. However, the length of the extrusion (40 to 80 ft) may exhibit some waviness down its length. This is clamped at each end, and pulled to slightly exceed the tensile yield strength. Upon release, the extrusions are quite straight, and when cut into lengths for shipment, the waviness is within the advertised limits. A more serious problem is in the cross-section tolerances. For instance, a narrow U-shaped extrusion with a relatively thick base and thin, long legs will not look the same after straightening. Normally, the legs tend to close up at the top (or unsupported) end of the U. The straightening operation may occur immediately after extrusion, as in the case of non-heat-treatable aluminum. However, it is always required after heat treatment and quenching, before the aging process. The best policy here is to talk with the extruder, to make certain that you both understand the risks, and that you are able to use the product. Many times the extruder will refuse to make a quotation for an extruded shape because of this problem. If your finished part is quite short in length, neither waviness nor angle of the legs may cause a problem in your particular product.

Dimensional Tolerances

Although extruded aluminum shapes minimize and often eliminate the need for machining, they do not possess the dimensional accurace of machined parts, and the dimensional tolerances to which extrusions are commercially furnished must be taken into account. These tolerances, shown in Figures 15.57 and 15.58, generally cover straightness, flatness, twist, and cross-sectional dimensions, such as section thickness, angles, contours, and corner and fillet radii.

The tolerances on any given dimension vary somewhat depending on the size and type of the shape, relative location of the dimension involved, and other factors. Figure 15.59 illustrates many tolerances as applied to an arbitrary section used as an example.

15.3.4 Precision Aluminum Forging

All forgings fall into two general classes: hand forgings and die forgings. *Hand forgings* are sometimes called *open die forgings*. As the name suggests, the metal is not confined laterally when being forged to the desired shape. The forger manipulates the stock between repeated squeezes of the hydraulic press or ring roller, or blows of a hammer, in progressively shaping the forging to the desired form. These forgings have some of the desired grain-flow characteristics, and require less machining than when making a finished part out of a billet or large bar stock. However, they do require significant machining to achieve a finished part.

The next step in producing a more complete part is *blocker-type forgings*. These are generously designed, with large fillet and corner radii and with thick webs and ribs, so that they can be produced in a single set of finishing dies only. Producing such forgings may typically require a unit pressure of 10 to 15 tons per square inch of projected plan area, depending on the alloy and the complexity of the design. This is less pressure than is necessary to make more intricate forgings. The projected plan area of the forging is used to arrive at the estimated total tonnage required.

A blocker-type forging generally requires machining on all surfaces. Economics may dictate such a design if quantity requirements are limited or if the finished part tolerances necessitate complete finishing. A *blocker-type* forging is an end product and should not be confused with a *blocker forging*, which is a preliminary shape requiring a subsequent finishing die operation to attain its final shape. See Figure 15.60a for an example of a blocker-type forging, and Figure 15.60b for an example of a conventional forging.

Conventional forgings are the most common of all die forging types. A conventional forging is more intricate in configuration than a blocker-type forging, having proportionately lighter sections, sharper details, and closer tolerances, and thus is more difficult to forge. The design differences between the two types are illustrated graphically in Figure

Material	Tensile Strength, psi	Yield Strength, psi	Elonga-tion, Percent In 2 Inches	Shear Strength, psi	Modulus of Elasticity, psi	Specific Gravity
BRASS (35% Zinc)						
Hard	76,000	45,000	7	43,000	15,000,000	8.46
Annealed	45,000	12,500	50	33,000	15,000,000	8.46
BRONZE (5% Tin)						
Hard	81,000	75,000	10	......	16,000,000	8.86
Annealed	47,000	19,000	64	......	16,000,000	8.86
COPPER						
Hard	50,000	45,000	6	28,000	17,000,000	8.90
Hot Rolled	34,000	10,000	45	23,000	17,000,000	8.90
IRON						
Gray Cast	30,000	25,000	0.5	44,000	14,000,000	7.10
Wrought Plate	51,000	31,000	21	42,000	28,000,000	7.65
MAGNESIUM						
Wrought, J1 Alloy	44,000	32,000	14	20,500	6,500,000	1.80
Sand Cast, C Alloy	39,000	14,000	10	20,000	6,500,000	1.82
MONEL METAL (67% Ni, 30% Cu)						
Hard	110,000	100,000	8	87,000	26,000,000	8.80
Annealed Sheet	80,000	35,000	40	46,000	26,000,000	8.80
PHENOLIC SHEET						
Laminated Fabric-Base	9,500	......	2	10,000	1,000,000[3]	1.33
STEEL						
Carbon Cast	75,000	42,000	24	60,000	30,000,000	7.86
Structural, Hot Rolled	60,000	38,000	30	45,000	28,000,000	7.85
Stainless 18-8, Annealed	90,000	40,000	55	67,000	29,000,000	7.90
Stainless 18-8, Cold Rolled	150,000	125,000	15	112,000	29,000,000	7.90
WOOD						
Hard Maple	10,000[5]	......	1.5	1,500	1,600,000	0.67
ZINC						
Die Cast	40,000	26,000	5	31,000	....	6.64
ALUMINUM						
3S-O	16,000	6,000	30	11,000	10,000,000	2.73
3S-H18	29,000	27,000	4	16,000	10,000,000	2.73
24S-O	26,000	11,000	19	18,000	10,600,000	2.77
24S-T4	64,000	42,000	19	40,000	10,600,000	2.77
52S-O	28,000	13,000	25	18,000	10,200,000	2.68
52S-H38	41,000	36,000	7	24,000	10,200,000	2.68
61S-O	18,000	8,000	22	12,500	10,000,000	2.70
61S-T6	45,000	40,000	12	30,000	10,000,000	2.70

NOTES:

[1] Values shown are typical approximate values.

[2] Per Cent of International Annealed Copper Standard.

[3] Approximate average value. Initial modulus values of reinforced plastics vary over wide range.

Figure 15.52 Physical and mechanical properties of some structural materials.

Weight, Lb./Cu.In.	Melting Range (F.°)	Electrical Conductivity, Percent of Copper (2)	Thermal Conductivity, (at 212°F.) C.G.S. Units (6)	Coefficient of Thermal Expansion (68°—212°F.), °F. x 10–6	Specific Heat (68°—212°F.), Cal/g/°C.	Specific Heat (68°—212°F.), Cal/Cm3/°C.
0.306	1660-1715	26	0.29	10.2	0.091	0.769
0.306	1660-1715	26	0.29	10.2	0.091	0.769
0.320	1750-1920	18	0.19	9.9	0.09	0.797
0.320	1750-1920	18	0.19	9.9	0.09	0.797
0.322	1949-1981	100	0.93	9.3	0.092	0.819
0.322	1949-1981	100	0.93	9.3	0.092	0.819
0.257	2000-2400	2	0.12	5.6	0.13	0.923
0.277	approx. 2800	16	0.17	6.5	0.114	0.872
0.065	950-1150	13	0.19	14.4	0.249	0.448
0.066	760-1110	12	0.17	14.8	0.249	0.453
0.318	2370-2460	3.6	0.06	7.8	0.128	1.126
0.318	2370-2460	3.6	0.06	7.8	0.128	1.126
0.042	320(4)	. . .	0.0007	14.0	0.35	0.465
0.284	2670-2750	11	0.13	6.5	0.177	1.392
0.283	2765	12	0.14	6.5	0.114	0.895
0.288	2600-2680	2.4	0.04	9.6	0.12	0.948
0.288	2600-2680	2.1	0.04	9.6	0.12	0.948
0.024	. . . .	. . .	0.0004	3.5	0.55	0.368
0.240	717	27	0.27	15.2	0.010	0.066
0.099	1190-1210	50	0.46	12.9	0.23	0.628
0.099	1190-1210	40	0.37	12.9	0.23	0.628
0.100	935-1180	50	0.45	12.9	0.23	0.637
0.100	935-1180	30	0.29	12.9	0.23	0.637
0.097	1100-1200	35	0.33	13.2	0.23	0.616
0.097	1100-1200	35	0.33	13.2	0.23	0.616
0.098	1080-1205	45	0.41	13.1	0.23	0.621
0.098	1080-1205	40	0.37	13.1	0.23	0.621

(4) Beginning distortion.

(5) 3-ply Plywood, parallel to grain faces.

(6) For aluminum alloys, data given at 25° C.

(a)

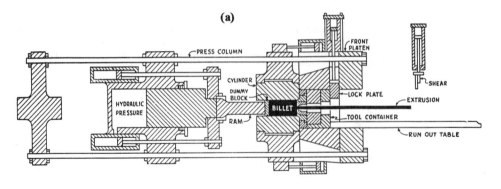

(b)

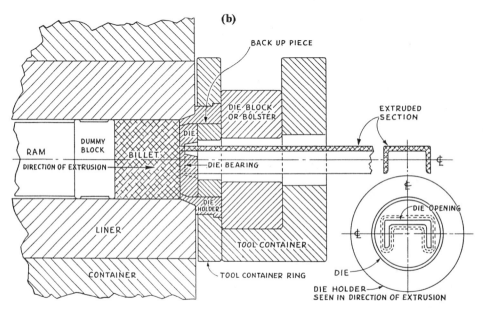

Figure 15.53 Schematic of extrusion press used for producing aluminum shapes: (a) A hydrauli-cally operated ram pushes hot aluminum through the die opening. (b) Steel tools for producing aluminum shapes consist of die, backup block, die holder, die block, and tool container. (Courtesy Reynolds Metals Company.)

15.61a and 15.61b. A conventional forging requires only partial final machining. A typical unit pressure of 15 to 25 tons per square inch of plan area is required, and usually a blocking operation is required prior to the finishing operation.

The manufacturing engineer and designer must evaluate the cost difference: A blocker-type forging has a lower die cost but will be heavier, requiring more extensive machining; a conventional forging has a higher die cost but will be lighter, requiring much less machining. Only a cost comparison by the customer can determine which type of forging will give lowest total cost.

A *precision forging* denotes closer than normal tolerances. It may also involve a more intricate forging design than a conventional type, and may include smaller fillet radii, corner radii, and draft angles, and thinner webs and ribs. The higher cost of a precision forging, including increased cost of dies, must be justified in the reduced machining required for its end use. This type of forging may typically require forming pressures of 25 to 50 tons per square inch of plan area.

The Forging Industry Association (FIA) defines a precision forging as "any forging which, by reason of tolerance, draft angle, web-to-rib ratio, or other specific requirement, falls outside the design suggestions applicable to conventional forgings" (from *Forging Industry Handbook*, Forging Industry Association, Cleveland, Ohio, 1970). While this definition is all-inclusive, it suits the wide capabilities of the precision forging process. Only recently adopted, this definition eliminates some persistent confusions. As long ago as World War II, parts were produced that today would be classified as precision forgings. As the use of such parts increased, names such as "net forgings," "no-draft forgings," "pressing," and others came into use. The FIA definition standardizes terminology and includes all these inexact, and sometimes inaccurate, labels. The "standard" tolerances were used to introduce forging design in Chapter 2 of this handbook.

Other types of forgings are "can and tube" forgings and "no-draft forgings." These are special forms, and really a continuation of the "chain" of increasingly precision forgings introduced above. They will not be covered in any detail in this subchapter, since the peculiarities must be discussed with the forger as the product design evolves. However, the general field of *impacts* as discussed in the next subchapter of this handbook, encompasses most of the more widely used special forgings, pressings, etc., that are generally grouped as impacts because of the higher punch velocities used in the forming process.

Precision Forgings or Conventional Forgings?

Precision forgings are used in assemblies where the basic characteristics of forgings are desirable or required, but where a conventionally forged part would require extensive machining. Although most precision forgings are more expensive than conventional forgings, they are less expensive than the same part machined from a conventional forging, or from a hand forging, bar or plate. The appropriate comparison is between the full costs of the parts ready for use. Most precision forgings are designed to eliminate the need for machining, aside from drilling attachment holes for installation. But for extremely complex designs, it may be more economical to precision forge only those sections of the part that are expensive to machine, and allow for machining the remaining sections. The intent is always to achieve the lowest cost yet highest quality for the finished part. The precision forged part shown for cost comparison in Figure 15.62a and 15.62b does not show an unusually favorable cost in comparison to other methods of production. It represents an average finished part. For actual cost comparisons, Alcoa and other reputable forging houses will quote parts as precision forgings and also as other types of forgings, if requested.

Mechanical Properties

Aluminum precision forgings are ordered to the same specifications, quality assurance provisions, and mechanical property levels that apply to conventional forgings. Nevertheless, many users feel that precision forgings used without machining have better mechanical properties, fatigue characteristics, and resistance to stress corrosion cracking. This superiority is attributed to the high degree of work during forging, the grain orientation, parting plane location, and metallurgical advantages retained when the "as forged" surfaces are not removed. See Figure 15.63 for minimum mechanical properties of popular aluminum precision forging alloys per specification QQ-A-367g.

Alloy and Temper	Specific Gravity	Weight, lb. per cu. in.	Approximate Melting Range (°F).	Electrical Conductivity[1]	Thermal Conductivity at 25°C. C.G.S. Units[2]	Coefficient of Thermal Expansion per °F. $\times 10^{-6}$	
						68°–212°F.	68°–572°F.
3S-F	2.73	.099	1190–1210	41	.38	12.9	13.9
14S-O	2.80	.101	950–1180	50	.46	12.9	13.6
14S-T4	2.80	.101	950–1180	..	..	12.8	13.6
14S-T6	2.80	.101	950–1180	50	.37	12.8	13.6
24S-O	2.77	.100	935–1180	30	.45	12.9	13.7
24S-T4	2.77	.100	935–1180	30	.29	12.9	13.7
61S-O	2.70	.098	1080–1205	45	.41	13.1	14.1
61S-T4	2.70	.098	1080–1205	40	.37	13.1	14.1
61S-T6	2.70	.098	1080–1205	40	.37	13.1	14.1
63S-T42	2.70	.098	1140–1205	50	.46	13.0	14.0
63S-T5	2.70	.098	1140–1205	55	.50	13.0	14.0
63S-T6	2.70	.098	1140–1205	55	.50	13.0	14.0
75S-O	2.80	.101	890–1180	..	..	13.1	14.4
75S-T6	2.80	.101	890–1180	30	.29	13.1	14.4

Notes: (1) Percent of International Annealed Copper Standard.

(2) C.G.S. units = calories per second, per square centimeter, per centimeter of thickness, per degree Centigrade.

Specific heat of commercially pure aluminum is 0.226 cal/g/°C for the temperature range of 68° to 212° F. Values for the commercial alloys differ slightly from the value of the pure metal.

(a)

Alloy and Temper	Tensile Strength (1), psi	Yield Strength in Tension, psi	Elongation, Percent In 2 inches	Shear Strength, psi	Endurance Limit (2), psi	Brinell Hardness (500 kg. Load, 10mm. Ball)
3S-F	16,000	6,000	40	11,000	7,000	28
14S-O	27,000	14,000	18	18,000	13,000	45
14S-T4	62,000	42,000	22	38,000	18,000	105
14S-T6	70,000	60,000	13	42,000	18,000	135
24S-O	27,000	11,000	22	18,000	13,000	47
24S-T4	68,000	48,000	19	41,000	18,000	120
61S-O	18,000	8,000	30	12,000	9,000	30
61S-T4	35,000	21,000	25	24,000	13,500	65
61S-T6	45,000	41,000	17	30,000	13,500	95
63S-T42	22,000	13,000	20	14,000	: : : : : :	42
63S-T5	30,000	25,000	12	18,000	: : : : : :	65
63S-T6	35,000	30,000	12	22,000	: : : : : :	73
75S-O	33,000	15,000	17	22,000	: : : : : :	60
75S-T6	82,000	72,000	11	49,000	21,000	150

Notes: (1) The modulus of elasticity of all aluminum alloys is approximately 10,300,000 psi. Poisson's Ratio is about 0.33.

(2) The endurance limit values are based on 500,000,000 cycles of reversed stress using the R. R. Moore type of machine and specimen.

(b)

Figure 15.54 (a) Physical properties of aluminum extrusion alloys. (Courtesy Reynolds Metals Company.)
(b) Mechanical properties of aluminum extrusion alloys. (Courtesy Reynolds Metals Company.)

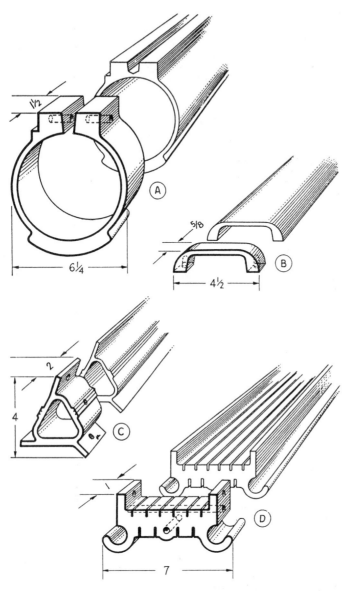

Figure 15.55 Short parts cut from extruded shapes are often cheaper than castings, forgings, or parts machined from bar stock. (a) Clamp for coupling aluminum irrigation tubing, (b) Drawer pull, (c) Tripod clamp, (d) Loom part. (Courtesy Reynolds Metals Company.)

Size

The usual method of determining overall precision forging size limitation is to compare its plan view area and pressure required against the rated capacity of the producer's largest forging equipment. For further discussion, plan view area (PVA) can be used as the unit of measure. The PVA is the length multiplied by the width ($L \times W = PVA$) when viewing the forging in the same direction as forging pressure will be applied. As a "rule of thumb," 30 tons of pressure per square inch of PVA are required in precision forging. Very few

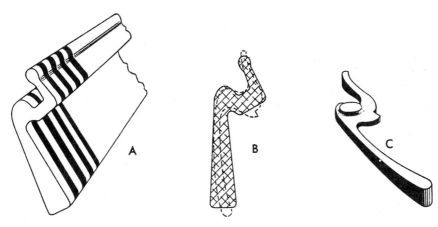

Figure 15.56 Extruded stock for forging reduces cost of forging and trimming operation. (a), (b) Extruded blanks, (c) Finished forging. (Courtesy Reynolds Metals Company.)

precision forgings are in the over-200-in.^{2} PVA range, due to press tonnage limitations. Very few parts are limited by bed size, press stroke, and "daylight" in the press. However, since new, larger presses are placed in service periodically, it is best to contact your forging supplier.

Tolerances

Terminology and definitions shown in Figure 15.64 are a good reference for some of the terms that follow.

Rib: Thin members, normal to web, confined within other forging members

Web: Thin panel members essentially parallel to plan view of forging or dimensioned by length and/or width

Wall: Outside members essentially normal to webs

Length: Usually considered the longest dimension of the forging

Width: Usually considered the next longest dimension of the forging and approximately at right angles to the length dimension

Recommended design proportions and tolerances for precision aluminum forgings are as follows:

Draft: $0° + \frac{1}{2}° - 0°$.

Edge radii: 0.06 in. + 0.03 in. − 0.06 in.

Fillet radii: 0.13 in. + 0.03 in. − 0.06 in. (for parts heights from web up to 1.500 in.) 0.25 in. + 0.03 in. − 0.13 in. (for parts heights from web exceeding 1.500 in.).

Web thickness: See Figure 15.65a.

Wall or rib thickness: See Figure 15.65b.

Mismatch tolerance: In general, mismatch tolerance is the misalignment of one die with another. In the production of precision forgings, the tooling may consist of several pieces or sections fitted together. Thus, mismatch may occur within a precision forging. Wall thickness tolerance is normally specified to include mismatch tolerance. However, other members of the forging such as the boss in section A–A of Figure 15.64 also must be recognized as requiring a separate mismatch tolerance of 0.00 to 0.015 in.

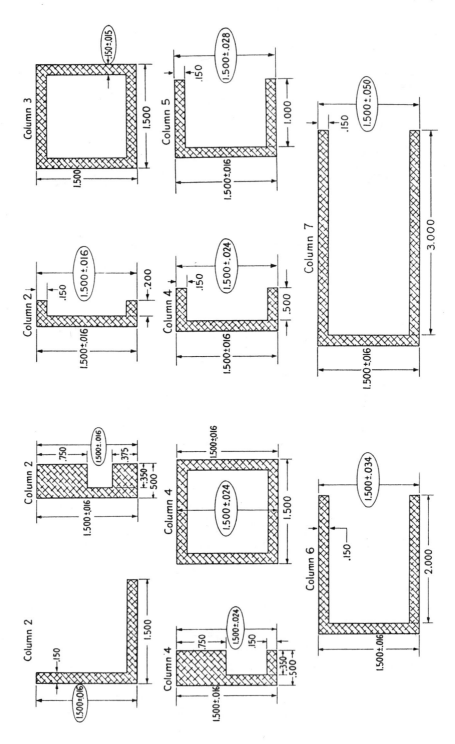

TOLERANCE (¹)—Inches Plus or Minus

Specified Dimension, Inches	For dimensions taken at a point where 75 percent or more of the dimension is metal		Allowable deviation from specified dimension, where more than 25 percent of the dimension is space (², ³)			
	All except those covered by column 3	Wall thickness(⁴) completely enclosing space 0.11 sq. in. and over	.250-.624 inch from base of leg	.625-1.249 inches from base of leg	1.250-2.499 inches from base of leg	2.500 inches or more from base of leg
Column 1	Column 2	Column 3	Column 4	Column 5	Column 6	Column 7
Up thru .124	.006		.010	.012	.014	.016
.125– .249	.007		.012	.014	.016	.020
.250– .499	.008		.014	.016	.018	.022
.500– .749	.009	Plus or minus 10%	.016	.018	.020	.026
.750– .999	.010	max. ±.060 min. ±.010	.018	.020	.022	.030
1.000– 1.499	.012		.020	.022	.026	.034
1.500– 1.999	.016		.024	.028	.034	.050
2.000– 3.999	.024		.032	.036	.048	.064
4.000– 5.999	.034		.042	.050	.064	.088
6.000– 7.999	.044		.054	.062	.082	.112
8.000– 9.999	.054		.064	.074	.100	.136
10.000–11.999	.064		.074	.088	.116	.160
12.000–13.999	.074		.084	.100	.134	.184
14.000–14.999	.080		.090	.106	.142	.196

NOTE: See examples on opposite page (Fig. 122)

NOTES:
(¹) The tolerance applicable to a dimension composed of two or more component dimensions is the sum of the tolerances of the component dimensions if all of the component dimensions are indicated.
(²) At points less than ¼ inch from base of leg, the tolerances shown in Column 2 are applicable.
(³) Where the space is completely enclosed (hollow shapes), the tolerances in Column 4 are applicable.
(⁴) Where the dimensions specified are outside and inside, rather than the wall thickness itself, tolerance on wall thickness shall be plus or minus 10 percent of mean wall thickness, max. ±0.060, min. ±0.010.

Figure 15.57 Standard tolerances for aluminum extrusions (cross-sectional dimensions). (Courtesy Reynolds Metals Company.)

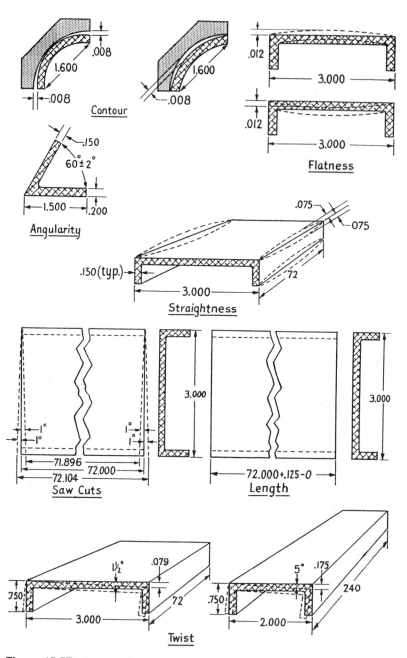

Figure 15.57 Continued.

Type of Tolerance	Dimension to Which Tolerance Applies (1)	Tolerance
STRAIGHTNESS (2)	Circumscribing Circle Diameter (3): Up through 1.499" Up through 1.499" 1.500" and up	$\pm$ .0500" per foot (4) for minimum thickness up through .094" $\pm$ .0125" per foot for minimum thickness .095" and up $\pm$ .0125" per foot
TWIST (2)	Circumscribing Circle Diameter (3): Up through 1.499" 1.500"—2.999" 3.000" and up	$\pm$ 1° per foot $\pm$ ½° per foot; 5° total $\pm$ ¼° per foot; 3° total
CONTOUR	Deviation from specified contour	$\pm$ .005" per inch of chord width ($\pm$ .005" minimum) (5)
CORNER AND FILLET RADII	Sharp Corners Specified Radius up through .197" Specified Radius .188 and up	$+$ 1⁄64" $\pm$ 1⁄64" $\pm$ 10 percent
ANGLES	Minimum Specified Leg Thickness Under .188" .188" to .750" .750 to solid	$\pm$ 2° $\pm$ 1½° $\pm$ 1°
FLATNESS		$\pm$ .004" per inch of width ($\pm$ .004" minimum)
SURFACE ROUGH-NESS (6)	Section Thickness: Up through .063" .064"—.125" .126"—.188" .189"—.250" .251" and up	.0015" maximum depth of defect .002" maximum depth of defect .0025" maximum depth of defect .003" maximum depth of defect .004" maximum depth of defect
SQUARENESS OF SAWCUTS		$\pm$ 1°
LENGTH	Specified Length up to 10' Specified Length 10' to 30' Specified Length 30' and up	$+$ 1⁄8" $+$ ¼" $+$ ½"

NOTES:

(1) These tolerances are applicable to the average shape. Wider tolerances may be required for some shapes, and closer tolerances may be possible for others.

(2) Not applicable to annealed (0 temper) material.

(3) The smallest circle that will completely enclose the shape.

(4) When weight of shape on flat surface minimizes deviation.

(5) Applicable to not more than 90° of any arc.

(6) Includes die marks, handling marks, polishing marks.

(7) Tangent values used in calculation of twist and angularity tolerances:

Tan	30'	.0087	2° 30'	.0437
	1° 0'	.0175	3° 0'	.0524
	1° 30'	.0262	4° 0'	.0699
	2° 0'	.0349	5° 0'	.0875

Figure 15.58 Standard tolerances for aluminum extrusions (straightness, flatness, twist, contours, radii, angles, roughness). (Courtesy Reynolds Metals Company.)

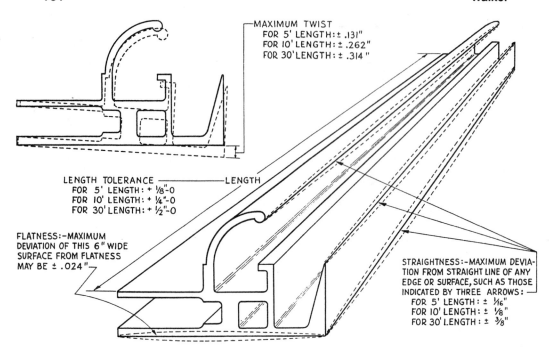

MAXIMUM TWIST
FOR 5' LENGTH: ± .131"
FOR 10'LENGTH: ± .262"
FOR 30'LENGTH: ± .314"

LENGTH TOLERANCE ———————— LENGTH
FOR 5' LENGTH: + 1/8"-0
FOR 10' LENGTH: + 1/4"-0
FOR 30'LENGTH: + 1/2"-0

FLATNESS:—MAXIMUM
DEVIATION OF THIS 6" WIDE
SURFACE FROM FLATNESS
MAY BE ± .024"

STRAIGHTNESS:—MAXIMUM DEVIA-
TION FROM STRAIGHT LINE OF ANY
EDGE OR SURFACE, SUCH AS THOSE
INDICATED BY THREE ARROWS:
FOR 5' LENGTH: ± 1/16"
FOR 10' LENGTH: ± 1/8"
FOR 30' LENGTH: ± 3/8"

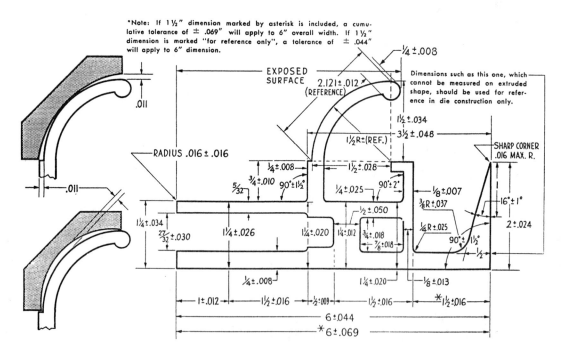

*Note: If 1 1/2" dimension marked by asterisk is included, a cumu-
lative tolerance of ± .069" will apply to 6" overall width. If 1 1/2"
dimension is marked "for reference only", a tolerance of ± .044"
will apply to 6" dimension.

Figure 15.59 Example of the extruded shape tolerances described in Figures 15.57 and 15.58. (Courtesy Reynolds Metals Company.)

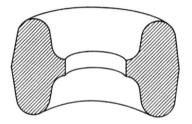

BLOCKER-TYPE FORGING – REQUIRES ONLY ONE DIE OPERATION AND A TRIMMING AND PIERCING OPERATION TO REMOVE SURPLUS METAL.

(a)

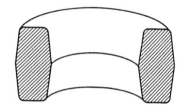

CONVENTIONAL FORGING – REQUIRES TWO DIE OPERATIONS AND A TRIMMING AND PIERCING OPERATION TO REMOVE SURPLUS METAL.

(b)

Figure 15.60 Examples of (a) a blocker-type forging, and (b) a conventional forging. (From *Aluminum Forging Design Manual*, The Aluminum Association, New York, 1975.)

Thickness tolerance (die closure): See Figure 15.65a.

Length and width tolerance (does not apply to wall thickness): See Figure 15.65b.

Normal straightness tolerance in any plane of the forging: 0.016 in. per 10 in.

Step tolerance (joggle) of ±0.010 in. does not include straightness.

Contour tolerance: ±0.015 of basic contour over the entire length.

Flash entension: 0 to 0.03 in. in the plane of the metal where interference is not normally encountered (see Figure 15.64).

Angular tolerance: ±0° 30′.

Surface finish: 125 rms (except flash extension).

Grain direction: parallel to length dimension, unless specified otherwise.

Parting plane: location optional with forger, unless specified otherwise.

15.3.5 Impacts

An impact is "a part formed in a confining die from a metal slab, usually cold, by rapid single-stroke application of force through a punch causing the metal to flow around the punch and/or through an opening in the punch or die" (The Aluminum Association definition).

Basic Process

Impacting combines extrusion and forging and has therefore been called variously impact extruding, cold pressing, extrusion forging, cold forging, extrusion pressing, impact extruding, and the like. Today the process is called simply impacting and the parts so formed, impacts.

Although the design and fabrication of the punches and dies used is a highly skilled area, the process itself is relatively simple. The aluminum slug to be impacted, its volume carefully predetermined with an accuracy of anywhere from 1% to 10%, is placed in a die. A single-stroke punch—sometimes called a ram or mandrel—comes down at high speed. The developed pressure extrudes the aluminum through designed openings. These orfices may be in the bottom of the die, between the die and the punch, or sometimes in the sides of the die. The aluminum that is not extruded is held between the bottom of the punch and the inside of the die. This portion of the metal is forged.

Plastic flow begins when the yield point of the aluminum is exceeded, and extrusion

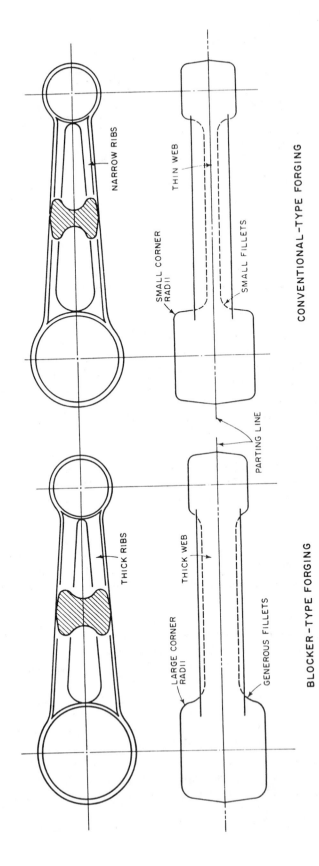

NARROW RIBS

SMALL CORNER RADII

THIN WEB

SMALL FILLETS

CONVENTIONAL-TYPE FORGING

THICK RIBS

THICK WEB

PARTING LINE

LARGE CORNER RADII

GENEROUS FILLETS

BLOCKER-TYPE FORGING

Figure 15.61 Design differences between conventional and blocker-type forgings. (From *Aluminum Forging Design Manual*, The Aluminum Association, New York, 1975.)

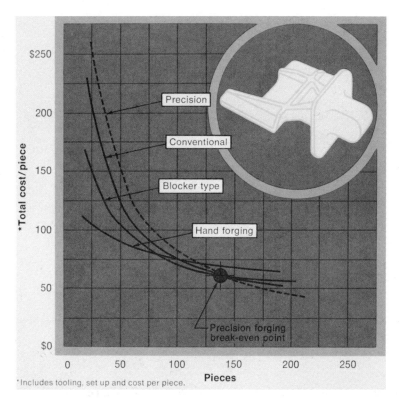

Figure 15.62 Cost comparison example of precision versus conventional forgings. (From *Alcoa Aluminum Precision Forgings,* Aluminum Company of America, Pittsburgh, PA, 1950.)

| Alloy and temper | Longitudinal[a] | | | Transverse | | | |
	Tensile strength (KSI)	Yield strength (KSI)	Elonga-tion (%)	Tensile strength (KSI)	Yield strength (KSI)	Elonga-tion (%)	Maximum cross section when heat treated (in.)
2014-T6	65	55	6	64	54	3	4
2219-T6	58	38	8	56	36	4	4
6061-T6	38	35	7	38	35	5	4
6151-T6	44	37	10	44	37	6	4
7075-T6	75	65	7	71	62	3	3
7075-T73	66	56	7	62	53	3	3
7175-T66	86	76	7	77	66	4	3
7175-T736	76	66	7	71	62	4	3
7079-T6	72	62	7	70	60	3	6

Figure 15.63 Minimum mechanical properties of aluminum precision forgings. (From *Alcoa Aluminum Precision Forgings,* Aluminum Company of America, Pittsburgh, PA, 1950.)

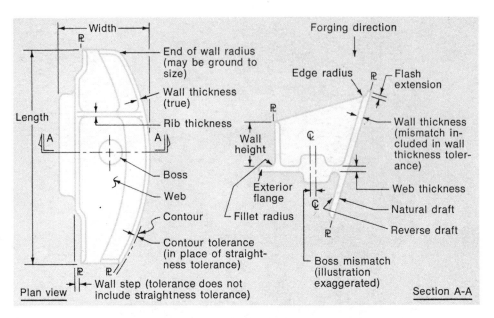

Figure 15.64 Example of a part showing terminology and definitions of tolerances. (From *Alcoa Aluminum Precision Forgings,* Aluminum Company of America, Pittsburgh, PA 1950.)

ensues when the pressure has increased to approximately 7 to 15 times initial yield pressure. The pressure necessary for impacting depends on the alloy as well as the complexity of the shape to be formed: the harder the alloy, the greater the pressure. Alloy 7075, for example, requires two to three times as much pressure as alloy 1100 when cold impacted. Pressures required for aluminum alloys vary between 25 and 110 tons/in.^{2} of punch area, and punch deforming velocities in the 20- to 50-m/sec range.

Impact Types

There are three types of impacts: reverse, forward, and combination, named after the principal direction in which the aluminum flows under pressure developed by the punch.

Reverse impacting is used to make shells with a forged base and extruded sidewalls. A blank of material (slug) to be extruded is placed in a die cavity and struck by a punch, forcing the metal to flow upward around the punch, through the opening between the punch and the die, to form a simple shell (see Figure 15.66). Outside diameters can be stepped, but the inside diameter should be straight. Short steps, however, if necessary, can be incorporated on the inside, near the bottom of the part.

The clearance between the punch and the die determines the wall thickness of the impact. The base thickness is determined by adjustment of the bottom position of the press ram and should be a minimum of 15% greater than side wall thickness as impacted. In general, the side walls should be perpendicular to the base. Multiwall shells, internal or external ribs, and circular, oval, rectangular, square, or other cross sections can be produced. Advantages of reverse impacts include:

1. Single operation, resulting in low cost
2. Simplified tooling

WALL/RIB HEIGHT THICKNESS GUIDELINE

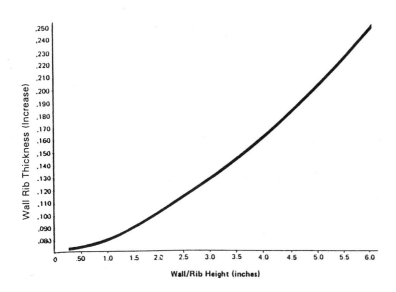

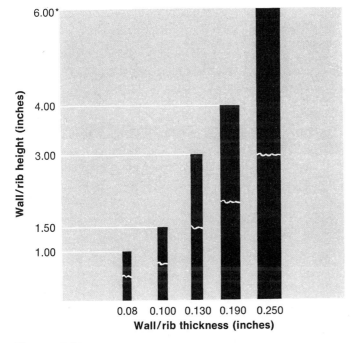

Figure 15.65 Proportions and tolerances of web, wall, or rib thicknesses for precision aluminum forgings. (From *Alcoa Aluminum Precision Forgings,* Aluminum Company of America, Pittsburgh, PA, 1950.)

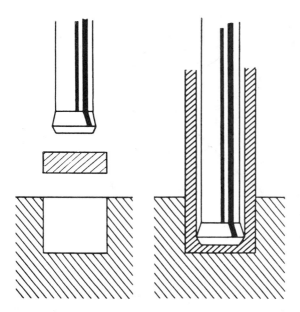

Figure 15.66 The reverse impact process. (From *Aluminum Impacts Design Manual,* The Aluminum Association, New York, 1982.)

3. Ease of removal
4. Inside and outside bottom contours easily achieved
5. Irregular symmetrical shapes and broader range of shapes possible

In the reversing process there is a tendency for the punch to "wander" in producing the longer pieces, making consistency in wall thickness more difficult to maintain. See Figure 15.67 for specific design recommendations.

Forward impacting, sometimes called the Hooker process, somewhat resembles conventional extruding in that the metal is forced through the orfice of a die by the action of a punch, causing the metal to flow in the direction of punch travel (see Figure 15.68). The punch fits the walls of the die so closely that no metal escapes backwards. The method is used for forming round, nonround, straight, and ribbed rods and thin-walled tubing with one or both ends open, and with parallel or tapered side walls. Some large parts, such as transmission shafts, may be made by forward impacting.

Hollow or semihollow parts with a heavy flange and multiple diameters formed on the inside and outside are often made by forward impacting. Some of the advantages of forward impacting are:

1. Improved wall tolerances
2. Greater length-diameter ratios
3. Improved concentricities
4. Ease of producing thinner sections

However, shapes are more limited than in reverse impacting, and bases must be plain, as no forging action takes place against the bottom of the die.

A *combination impact* is, as the name implies, a combination of forward and reverse metal flow. This method is used to produce complex-shaped parts (see Figure 15.69). The

Outside Diameter (inches)	Alloy and Wall Thicknesses (inch)			
	1100	6061	2014	7075
1	0.010	0.015	0.035	0.040
2	0.020	0.030	0.070	0.080
3	0.030	0.045	0.105	0.120
4	0.040	0.060	0.140	0.160
5	0.050	0.075	0.175	0.200
6	0.060	0.090	0.210	0.240
7	0.075	0.110	0.245	0.280
8	0.100	0.130	0.280	0.320
9	0.110	0.145	0.315	0.360
10	0.125	0.165	0.350	0.400

Based on length-to-diameter ratio of 8:1 or less. Wall thicknesses less than those shown may be obtained by secondary operation.

Figure 15.67 Design recommendations for reverse impacts. (From *Aluminum Impacts Design Manual,* The Aluminum Association, New York, 1982.)

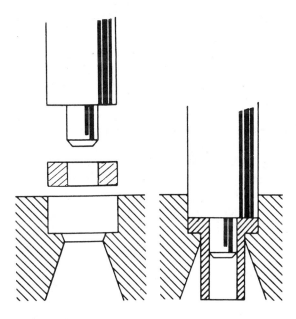

Figure 15.68 The forward impact process. (From *Aluminum Impacts Design Manual,* The Aluminum Association, New York, 1982.)

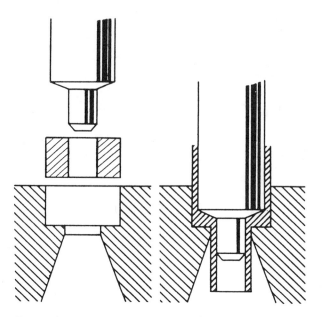

Figure 15.69 Combination impact process. (From *Aluminum Impacts Design Manual,* The Aluminum Association, New York, 1982.)

metal is confined inside the cavity between the upper and lower punches, forcing it to flow both up and down. If a solid slug is used, a web is left; if a hollow slug is used, a tubular part is formed.

By incorporating a cavity in the punch, the metal is allowed to flow upward into the punch, until the cavity is filled. Further punch movement causes forward extrusion of the remaining metal.

One of the useful examples of combining processes was observed by the author at the Royal Schelde Nuclear Equipment Division in the Netherlands. There, an existing small extrusion press had been modified, to make a part for a nuclear power plant boiler water system. The tooling consisted of a formed male punch on the press ram and a formed female die in the die holder. The die had an orifice of approximately 10 mm, with provisions to extrude a tube with approximately 1-mm wall thickness in the center of the small bathtub fitting formed by the punch and die. The billet was heated, and the press speed increased to its maximum, which was less that the normal velocity for forward or combination impacting. The die was opened, the part removed, and the 5.0-m length of tube was coiled carefully to form the cooling coil. This tubing replaced a part that was previously made in two pieces and then electron-beam welded, and it eliminated any question of leaks, additional tests, welding, etc. The temperatures and pressures were controlled to obtain improved physical properties in the small housing formed on the end of the tubing. The process was not normal "extruding," nor was it really "impacting," but it was an excellent example of understanding processes and metal flow characteristics, in order to make a part that performed its required function!

Impact Design Considerations

Good impact design practice includes full consideration of alloy selection, tool design and construction, impact production, lubrication, and possibly heat treatment of the impact following manufacture, in addition to good engineering design principles. It is essential that all contemplated impact designs falling outside recommended practices be discussed with an impact engineer. Basic design guidelines affecting cost, tool life, dimensional accuracy, and repeatability are:

1. Keep the design simple.
2. Avoid designs that are not symmetrical around the punch.
3. Make circular sections in planes perpendicular to the punch axis.
4. Avoid dimensional tolerance closer than necessary.
5. Use thin sections.
6. Use lowest-strength alloys applicable.

Reverse Impacting

Press capacity and type of impact determine the maximum length that can be produced on a given press. In reverse impacting, length is determined by the inside diameter of the shell, the mechanical properties of the alloy, the reduction in area, and the stroke of the press. To avoid column failure of the punch, maximum shell length should not exceed eight times the inside diameter of the part (see Figure 15.70). While parts exceeding this 8:1 ratio have been reverse-impacted, it is usually more economical to add an ironing operation to obtain greater length.

A small outside corner radius or chamfer should be provided on the part bottom to promote an even flow of metal and to prevent "dead metal" in the corner of the die, which could cause slivers, poor surface finish, and possible separation in shear at the outside corners. See Figure 15.71 for examples. In Figure 15.71a, the inside radii, where the side

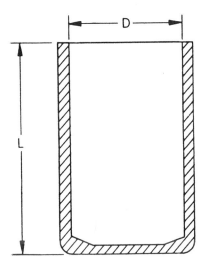

The length to diameter ratio runs from 6 to 10.1 depending on the alloy area reduction and column strength of the punch. Higher ratios may be obtained by secondary operations such as ironing.

Figure 15.70 Example of reverse impacting maximum length-to-diameter ratio. (From *Aluminum Impacts Design Manual,* The Aluminum Association, New York, 1982.)

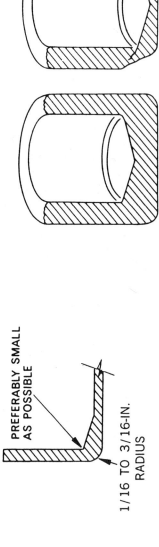

PREFERABLY SMALL
AS POSSIBLE

1/16 TO 3/16-IN.
RADIUS

The inside radius between sidewall and base should be kept to a minimum. If additional strength is needed at this point it may be secured by tapering the bottom as shown. An outside radius should be provided to promote the even flow of metal.

Even distribution of metal flow is of great importance. The reverse impact at left is correctly proportioned while the drawing at the right depicts a base that is thinner than the wall, causing defects.

Figure 15.71 Examples of inside radii and bottom thickness of reverse impacts. (From *Aluminum Impacts Design Manual*, The Aluminum Association, New York, 1982.)

wall and base join, should be kept as small as possible. A sharp inside corner reduces friction during metal flow and improves surface finish and concentricity. If stress concentration at the junction of the base and side wall is a major consideration, additional metal can be provided there without resorting to a large inside radius. In Figure 15.71b, the bottom's thickness at the base of the wall must be at least 15% greater than the thickness of the wall to prevent shear failure. Even distribution of metal flow is of great importance. The reverse impact on the left side of Figure 15.71b is correctly proportioned, while the drawing on the right side depicts a base that is thinner than the wall, causing defects. Practical minimum wall thicknesses for reverse impacts are shown in Figure 15.72.

Forward Impacting

OD-to-ID ratio is not as great a limiting factor for forward impacting as it is for reverse impacting. The tube length is the main limitation. Irrigation tubing with 6-in. OD and an 0.058-in. wall has been produced in lengths up to 40 ft by hydraulic direct impacting. Tubing in 6061 alloy with 3/8-in. OD and an 0.035-in. wall has been made in 14-ft lengths.

Secondary Operations

In some instances it is desirable to form or machine a part that has already been impacted. This process can be done without penalty. Metal so shaped has all the desirable characteristics—good grain flow, dimensional accuracy, smooth surface, and strength. Figure 15.73 illustrates two fairly common punch-and-die designs used for secondary operations. In

Outside Diameter (inches)	Alloy and Wall Thicknesses (inch)			
	1100	6061	2014	7075
1	0.010	0.015	0.035	0.040
2	0.020	0.030	0.070	0.080
3	0.030	0.045	0.105	0.120
4	0.040	0.060	0.140	0.160
5	0.050	0.075	0.175	0.200
6	0.060	0.090	0.210	0.240
7	0.075	0.110	0.245	0.280
8	0.100	0.130	0.280	0.320
9	0.110	0.145	0.315	0.360
10	0.125	0.165	0.350	0.400

Based on length-to-diameter ratio of 8:1 or less. Wall thicknesses less than those shown may be obtained by secondary operation.

Figure 15.72 Practical wall thicknesses for reverse impacts. (From *Aluminum Impacts Design Manual,* The Aluminum Association, New York, 1982.)

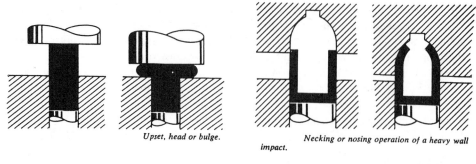

Upset, head or bulge.   *Necking or nosing operation of a heavy wall*
                                                            impact.

Figure 15.73 Examples of secondary operations of upsetting and bulging of impacts. (From *Aluminum Impacts Design Manual,* The Aluminum Association, New York, 1982.)

Figure 15.73a an upset, head, or bulge is shown. In Figure 15.73b, a necking or nosing operation of a heavy wall impact is shown. Figure 15.74 is a drawing of the ironing operation used to reduce thickness of the impact's walls, increase its length, and ensure a smooth, uniform surface throughout. This result is accomplished by making the clearance space between the punch and the die slightly less than the thickness of the shell wall. The bottom of the shell retains its original thickness. This technique is often used to control tolerances after heat-treating the impact.

Tolerances

Practically all factors involved in any fabricating process affect tolerances, hence each is design-related either directly or indirectly. For example, eccentricity (a design consideration) helps reduce tool wear (a production factor). Some of the principal factors affecting tolerances are:

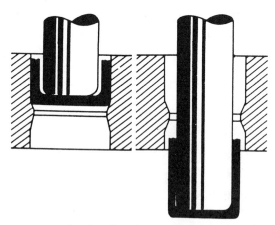

Ironing: Drawing or draw wipe. Used for sizing and obtaining parts with a larger ratio of length to diameter.

Figure 15.74 Secondary process of ironing an impact. (From *Aluminum Impacts Design Manual,* The Aluminum Association, New York, 1982.)

1. Tool wear
2. Irregular metal flow
3. Size of part
4. Shape of part
5. Alloy and temper
6. Wall thickness

Figure 15.75a shows commercial tolerances in general use for base, web, or flange thickness of aluminum impacts. Typical surface finishes are shown in Figure 15.75b. The finish depends on the alloy, lubricant, configuration, and tool design. Wall tolerances, as a general rule for 1100 alloy, can be held to ±10% of wall thickness for walls 0.200 in. and ±7% for walls between 0.250 and 1.00 in.

Mechanical Properties

Figure 15.76 gives *typical* mechanical properties of some aluminum impacts, while Figure 15.77 shows the *minimum* mechanical properties of aluminum impacts that can be expected in production at 75°F.

While all of the discussion has been with reference to aluminum impacts, other nonferrous metals can also be impact extruded. These metals include lead, tin, zinc (heated to 300°F), magnesium (heated to 400°F), etc. For special applications, impacts in fine silver have been successfully made.

15.4 CASTING AND MOLDING PROCESSES

15.4.0 Introduction to Casting and Molding Processes

Up to this point we have discussed the fabrication of parts or products by forming sheet metal, machining, forging, extrusion, and the like. In some cases these processes seem to have become blended, or modified, or combined, making it somewhat difficult to classify the production process clearly. An example is the lathe that can perform milling—and vice versa. Another is the forging, pressing, impacting, and extrusion area, where innovative combinations appear in all kinds of industries, using most types of materials. We have shown examples and limitations of several processes, all using a form of solid material. In some cases, the material was heated just below the liquid temperature; however, the processes started and finished with a solid metal.

In Chapter 2 we talked about the primary means of casting liquid metals into molds, in order to be able to influence the product design in the direction of a producible configuration. As we now introduce other important processes using liquid or powder

(a)   BASE, WEB OR FLANGE

Diameter (inches)	Tolerance (± inch)
Up to 3	0.012
3 to 5	0.015
5 to 7½	0.020
Over 7½	0.030

(b)   TYPICAL, AS-IMPACTED PRODUCTION FINISHES

Alloy	Interior (micro inch)	Exterior (micro inch)
1100	16-32	32-63
6061	63-125	125-250
7075	125-250	125-250

Figure 15.75 Commercial tolerances and finishes of impacts. (From *Aluminum Impacts Design Manual,* The Aluminum Association, New York, 1982.)

Alloy and Temper	Tensile Strength [1]-KSI		Elongation in 2 in. Percent	Brinell [2] Hardness
	Ultimate	Yield		
1100-F	24.0	22.0	5	44
-0	13.0	5.0	35	23
-H112	24.0	22.0	5	44
2014-T4	62.0	42.0	20	105
-T6	70.0	60.0	11	135
2618-T61	64.0	54.0	10	125
3003-F	29.0	27.0	4	55
-0	16.0	6.0	30	28
-H112	29.0	27.0	4	55
6061-F	28.0	25.0	5	52
-0	18.0	8.0	25	30
-H112	28.0	25.0	5	52
-T4	35.0	21.0	22	65
-T6	45.0	40.0	12	95
-T84	43.0	39.0	6	85
6066-T6	57.0	52.0	10	115
6070-T6	55.0	51.0	10	115
7001-T6	94.0	85.0	7	147
7075-T6	83.0	73.0	10	145
-T73	73.0	62.0	12	135

[1] Specimen axis parallel to direction of grain flow.

[2] Brinell hardness is usually measured on the surface of an impact using a 500 kg load and a 10-mm penetrator ball.

NOTE: *These are not guaranteed values and therefore should not be used for design.

Figure 15.76 Typical mechanical properties of some aluminum impacts at 75°F. (From *Aluminum Impacts Design Manual,* The Aluminum Association, New York, 1982.)

materials or alloys as the input material, it becomes even more difficult to classify some of the processes. Any metal that can be melted and poured can be cast, and the size and range of parts regularly produced by presently available methods are greater than that of any other process affording similar results. Generally, casting methods may be divided into those using nonmetallic molds and those using metallic molds. However, to permit a logical breakdown regarding design, these processes can best be subdivided according to the specific type of molding method used in casting, such as sand, centrifugal, permanent mold, etc.

Alloy and Temper	Thickness	Tensile Strength [1]-KSI		Elongation % min. in 2 in. or 4D [2]	Brinell Hardness 500 kg Load 10-mm ball
		Ult	Yield		
		Min.	Min.		Minimum
1100-F*	All	____			
-0	All	11.0	3.0	15	____
-H112	All	15.0	7.0	2	____
2014-T4	All	55.0	32.0	10	100
-T6	All	65.0	55.0	6	125
2618-T61	All	58.0	45.0	4	115
3003-F*	All	____			
-0	All	14.0	5.0	14	____
-H112	All	19.0	9.0	2	____
6061-F*	All	____			
-0	All			10	____
-H112	All	22.0	16.0	2	____
-T4	All	26.0	16.0	12	50
-T6	All	38.0	35.0	6	80
-T84	.040-.093	35.0	30.0	4	75
	.094 & over	38.0	35.0	4	75
6066-T6	All	50.0	45.0	4	100
6070-T6	All	48.0	43.0	4	95
7001-T6	All	88.0	78.0	5	135
7075-T6	All	75.0	65.0	5	135
-T73	All	66.0	56.0	5	125

[1] Specimen axis parallel to direction of grain flow.

[2] "D" equals specimen diameter. Elongation requirements do not apply to material thinner than 0.062 inch (nominal).

* For wrought products in F temper, there are no mechanical property limits.

Figure 15.77 Minimum mechanical properties of some aluminum impacts at 75°F. (From *Aluminum Impacts Design Manual,* The Aluminum Association, New York, 1982.)

We can discuss patterns, mold making, shrink rates, pouring, cooling or chilling, and good and bad product design features. Powder metal and sintering is a separate field (or is it?), which involves *heating after solidifying* rather that solidifying after heating, as in conventional "casting" processes. It has a relationship to "forging," but is really not part of the normal forging industry. Making parts of rubber and synthetic rubber falls in somewhere, but again, it is really not a close family member of the processes discussed up to this point. Are rubber parts "castings," "forgings," "sintered," or none of the above? Ceramics are playing an increasingly important part of the world of manufacturing, and are related to the processing of metal parts. A combination of pure oxide ceramic and a metal constituent, both of which have good high-temperature properties (as opposed to the older clay-based ceramics), have brought about the development of highly specialized technical ceramics suited to unusual industrial applications.

All of the above processes have a place in today's complex industrial manufacturing arena, and deserve at least a proper introduction in a handbook of manufacturing engineering. Space limits prevent an in-depth coverage of some of the processes, but Subchapter 15.5 lists sources for further study of the detailed technical knowledge that may be needed by many readers.

15.4.1 Sand Casting

Sand casting is the oldest and most familiar method of casting. In this process the castings are made by pouring molten metal into either green sand molds or molds made of baked sand. Refer to Chapter 2 for a further description of the sand casting process.

15.4.2 Centrifugal Casting

Centrifugal casting entails pouring a measured quantity of molten metal into a mold that is then rotated rapidly. The rotation of the mold forces the molten metal outward to give intimate contact between the metal and the mold. Spinning is continued until all the metal poured into the mold has solidified. Pouring must be done quickly to prevent chilling and laps. There are three standard types of centrifugal casting methods in general use:

True centrifugal casting, where the mold is rotated about its own axis without using a central core. If the mold is partially filled, a hole appears along the center of rotation of the casting, the diameter of the hole being determined by the amount of metal used. The weight of casting produced to that of metal poured approaches 1:1.

Semi-centrifugal casting, where central cores are used to give irregular shapes to the central hole. A measured amount of metal is poured so the mold space between the core and the outer wall is filled completely. Proper design of the mold is essential so that directional solidification of the metal is retained. Sand or plaster cores are usually employed. Differentially heated or cooled outer molds may be needed to control the direction of solidification.

Centrifuged casting, where irregular shapes can be obtained that would not be possible if the parts were rotated on their own axes. In this process a number of molds are arranged about a central sprue like the spokes of a wheel. Molten metal is fed into the castings through radial gates. The process is similar to semicentrifugal casting, except that several castings are produced at once, and the molds are not spun on their own axes. Directional solidification is a problem in this process, but it can be solved by proper location of the castings, proper gating, and mold temperature control.

Process Data

Centrifugal casting molds may be made of a variety of materials, including steel, cast iron, sand, and graphite. Various wall coatings can be used, such as a mixture of graphite and sodium silicate in water.

Most aluminum casting alloys suitable for other processes can be used for centrifugal casting. The alloys should be poured at about 100°F less than with static casting. Alloys with short solidification ranges are preferable to those with wide freezing ranges.

15.4.3 Permanent Mold Casting

In permanent mold casting, the molten metal is poured by gravity into heated metal or graphite (permanent) molds. Sand casting is essentially a batch process, while permanent mold casting is suitable for quantity production of a continuous nature. This production line approach requires a different arrangement of foundry equipment, metal handling methods, and production procedures compared to those used with sand casting. A simple permanent mold usually consists of two halves that, when closed, form the mold cavity. Either metal or sand cores can be used—the process is designated as semipermanent mold casting when sand cores are used. The mold is heated before pouring and held at a constant temperature during pouring. Some castings require either heating or cooling of the mold between pouring operations; in others, the molten metal keeps the mold at the desired temperature. Permanent molds must be prepared for use by coating with refractory material. The refractory coating serves two purposes:

1. The solidification rate in different sections of the mold can be controlled by varying the coating thickness.
2. The mold metal is protected from contact with the molten metal.

Solidification will start at those sections where the coat is thinnest, due to the faster heat dissipation at those points. The refractory coating is applied heavily along thin sections and through the gates and risers. Thin coats are used on areas of wide cross section to promote faster solidification there. By properly applying the coating (by brushing or spraying), the rate of solidification throughout the casting will be uniform. The coating can be formulated from almost any refractory material finely ground in a water suspension with a suitable binding agent.

Permanent mold castings present different problems of gating and risering. The longest dimension of a permanent mold casting is usually vertical, as contrasted to horizontal positioning in sand casting. Sprues, risers, gates, and runners must all be designed as part of the mold structure. The difficulty and expense of making radical changes in metal molds makes it essential that the entire feeding system of the mold be finalized before the mold is made. It is wise to build a mold with initially undersized gates and risers. Then, by making experimental castings, gradually increase the sizes of the feed channels until the best possible casting is obtained. This method obviates the possibility of having oversized channeling to begin with, and keeps to a minimum the amount of metal that will be poured for each casting.

With permanent mold casting, a carefully established and rigidly maintained sequence of operations is essential. Every step in the foundry, from charging the furnace to removal of the cast piece from the mold, must be systematized. If any of the factors (pouring temperature, mold temperature, pouring rate, solidification rate) are thrown out of balance, the resultant castings may end up as scrap.

15.4.4 Die Casting

In die casting the molten metal is forced into a metal mold under considerable pressure, which is applied on the metal either pneumatically or hydraulically. Die casting gives low-cost production of large numbers of thin-sectioned parts. Close tolerances and extremely smooth surfaces can be produced without subsequent machining and finishing; also, small complex coring is possible, saving many drilling operations.

Such coring is not possible with sand or permanent mold casting. Intricate parts, not practical with other casting methods, can be produced easily by die casting. Satisfactory die casting depends on:

1. A suitable die casting machine

2.   A properly designed die

The die casting machine consists of a substantial, rugged frame designed to support and open and close the die halves in perfect alignment. Usually the fixed, or ejector, half of the die is mounted on a stationary platen. The other half, or cover, is mounted on a movable platen. The two halves of the die must move together accurately and must be locked together with sufficient force to overcome the separating force developed as the metal is injected. Either a toggle linkage arrangement or hydraulic rams are used to lock the dies.

Besides the basic die halves, most dies have moving cores and other features that allow the production of complex castings. In operation, the cores, slides, and other moving die parts are operated by hydraulic action synchronized with the opening and closing of the main die halves. These added features complicate die design but enable the production of variegated surfaces.

Metal injection into the die cavity is by means of either the goose neck or the cold chamber method. In *goose neck injection*, the molten alloy is forced into the mold by means of pneumatic pressure. In *cold chamber injection*, a hydraulically actuated plunger forces the molten metal from a cylindrical "shot" sleeve into the mold. The process is designated as "cold chamber" because the molten alloy is ladled into the shot sleeve just before it is forced into the mold. Pressures in the two methods vary; for the goose neck method it is usually about 750 psi, while the cold chamber method runs at 3000 to 20,000 psi.

Metal enters the die in the cold chamber process in a semimolten condition and forces the air out ahead of the metal. In the goose neck system, the metal enters the mold in a completely molten state and tends to mix with air in the die cavity. This tendency to produce porous castings has resulted in goose neck equipment becoming obsolete; it is now largely replaced by cold chamber machines.

Die Considerations

Heat-treated alloy steel dies are needed for die casting aluminum. These dies start to check after long usage due to the thermal shock from the molten metal. Initial tooling cost is high due to skilled labor needed to make a die. However, the productivity of a die casting machine is high and will bring the unit cost down when the production run is long, and where the part has sufficient complexity and need for precision to warrant die casting. To best utilize this process, consult die casting experts such as those available through the American Die Casting Institute, 366 Madison Avenue, New York, NY 10017.

Frequently, die cast parts improve products through production simplification and more sales appeal. Inserts of other materials can easily be placed in the die cavity and permanently molded in the finished piece. Die cast parts can be bulk finished by barrel finishing, automatic polishing and buffing, continuous chemical treatment, or painting. Die castings offer freedom of shape and an unlimited range of surface ornamentation with but a slight increase in tooling or finishing cost.

15.4.5 Plaster Molding

The most common specialty casting process employs plaster molds. This process is a refinement of sand casting, in that the sand is replaced by plaster, giving the finished casting a smoother surface and allowing greater accuracy in the dimensions of the molded part. A plaster mold is used for just one casting, since it is necessary to destroy the mold to remove the casting. The process is usually confined to castings under 2 lb. Gypsum plasters are the type most often used to make plaster molds. It is essential that all water be removed from the molds before casting.

The equipment required for plaster molds is more expensive than that for sand molds. Metal match plates and metal core boxes of extreme accuracy are used. Most plaster castings are poured by gravity, but moderate pressure gives improved casting detail.

The aluminum alloys used with plaster molds must be carefully selected. The refractory nature of the plaster results in a slow solidification time, with resultant lowering of mechanical properties. This refractory quality, however, enables thin and intricate sections to be cast. Due to their excellent fluidity, aluminum-silicon alloys are best for plaster casting.

15.4.6 Investment Casting

Investment casting, based on the "lost wax" process, allows for the intricacy of design of sand casting and the precision of die casting. To make an investment casting mold, a refractory-type plaster is poured around an expendable wax (or low-melting-temperature plastic) pattern. As the plaster sets, it is dried in a oven and the wax pattern is melted out. As in plaster casting, the molds are used only once, since they must be broken to remove the casting. More often, in today's shops, the refractory material is sprayed over the wax patterns and dried in an oven, melting out the wax pattern as above. This method allows the use of a robot to spray the refractory material the same way each time, improving repeatability and lowering costs.

A master mold is required to make the wax patterns. The number of castings required determines the permanency of the master mold. When large numbers of castings are to be made, the master molds should be made of metal, and an injection molding process should be used for the production of the wax patterns. This process is expensive due to the number of steps required, the need for skilled operators at each step, and the slow production rate compared to other casting processes. The use of plaster or other refractory material limits the choice of materials to those suited to plaster casting.

The molten metal may be poured under pressure, or poured with the mold in a vacuum chamber. Extremely sharp details can be obtained in the cast piece. The accuracy of this process is very high—even greater than that achieved with die casting because there are no moving parts in the mold. Naturally, the ultimate accuracy of the method depends on the accuracy of the master mold used to make the wax patterns. Because of the slow cooling of this type casting, it is imperative that the metal be thoroughly fluxed with chlorine gas in order to eliminate pinhole porosity.

15.4.7 Powder Metal

For lack of a better, more concise term for the pressing and sintering of metal powders into machine parts of all varieties, the generally recognized "powder metallurgy" or just "PM" is normally employed. The first porous metal bearings were marketed following World War I, and with the outbreak of World War II, the vast potential of powder metallurgy began to be realized for its value not only as a method of fabricating parts whose physical characteristics are impossible to produce otherwise, but as a large-volume mass-production process having excellent speed and material economy. Beyond the well-known oilless or self-lubricating bearings and similar parts, there is a tremendous field of machine parts in production. Some of these include clutch friction facings, internal and external splines, rollers, external and internal gears, ratchets, piston rings, bushings, magnets, and the like.

Metallurgy

Several general classes of metal-powder structure can be set up. In the first, consolidation during sintering is primarily a particle-to-particle cohesion or contact fusion of particles containing a melting constituent. In the second, one of the powders acts as a melting medium, bonding or cementing together a high-melting-point constituent. In the third class, consolidation of a fairly high-melting-point metal is achieved as in the first category, but lacking high density, the compact is impregnated with low-melting-point metal.

Production Steps

The procedures employed in producing PM parts are generally as follows:

1. Selection of the powder or powders best suited for the part being designed as well as for the most rapid production
2. Wet or dry mixing of powders where more than one powder is to be used
3. Pressing in suitable dies
4. Low-temperature, short-time sintering, usually referred to as presintering, for increasing strength of fragile parts, removing lubricants or binders, etc.
5. Machining or otherwise forming of presintered parts
6. Sintering green compacts or presintered parts to obtain the desired mechanical properties such as proper density, hardness, strength, conductivity, etc.
7. Impregnating low-density sintered compacts, usually by dipping in molten metal so as to fill all pores or by allowing the impregnant to melt and fill the pores during sintering
8. Coining or sizing operations, cold or hot, when necessary, to attain more exacting dimensional tolerances and also improve properties
9. A hot-pressing operation to replace the usual pressing and sintering

Compacting

Pressing or compacting of PM parts generally requires anywhere from 5 to around 100 tons of pressure per square inch. Common garden-variety parts are produced with 20 to 50 tons/in.² pressure. It is interesting to observe, however, that the final density of a PM product is not determined only by the pressure under which it is cold-pressed or briquetted. Rate of pressure application, particle size, type of material, sintering time and temperature, occluded gas, etc., also have effects on final density and size.

Press stroke also presents certain broad limitations as to part size. The compression ratio between the volume of powder in a die before and after pressing is dependent on loading weight, particle size, form and composition, metal hardness, and pressure used. With most common metals and alloys, this ratio is usually 3 to 1, but it may vary from 5 to 1 up to 10 to 1 with fine powders, and 2 to 1 up to 4 to 1 with medium-size powders. Combined with the compression ratio is the general limitation that diameter-to-length ratio of parts be restricted to a maximum of 3 to 1.

Equipment

Presses may be mechanical, hydraulic, or a combination of both. Small parts that can be made at high speed with relatively low pressures are best produced in mechanical automatic presses with single– or multiple-cavity dies. Such presses for average parts usually are built in pressure capacities ranging from 100 to 150 tons, although sizes to 1500 tons are available. Presses are generally of the single-punch type or of the rotary multiple-punch type. Single-punch presses are either of the single-action type that compresses with the top punch only, or of the double-action type that employs movement of the lower punch simultaneously with the upper to obtain more uniform compacting and automatic ejection. As many as two upper telescoping punch movements and three or four lower punch telescoping punch movements are used along with side core movements for complex designs. However, it is generally difficult to press powder into reentrant angles, sharp corners, or undercuts. An average die produces 50,000 pieces before wear necessitates refitting or replacement.

Where the ordinary process for handling PM parts does not result in a satisfactory density or the pressures required are extremely high, *hot pressing* can often be used. In this method, the powders are heated in the dies and pressure is applied to form the part. The hot-pressing method yields high and nearly ideal density and greater strength at relatively low pressures. Carbide parts up to 100 in.^{2} in cross section, the greatest dimension of which can be 18 in. with a length of 8 in., have been produced by hot pressing, especially parts too large for regular cold pressing and sintering, and thin-wall parts that tend to go out of round.

Ordinarily, to obtain desirable density and precision, parts of materials other than carbides are *coined* or *sized* after sintering, but naturally this step adds to the cost. Compacts up to 10 in.^{2} in cross section are readily coined in hydraulic presses at a rate of about 4 to 6 pieces/min with hand feeding, and up to 10 pieces/min with automatic feeding. Dies may produce 100,000 to 200,000 pieces before replacement.

High-Density Powder Metallurgy

High-density powdered metals can be produced by a number of techniques, including powder forging, powder rolling, liquid metal infiltration, and liquid-phase sintering. Although powder forging has a relatively long history, it is only recently that the process has become important commercially. A typical flow diagram for the hot forging process is shown in Figure 15.78. This is a diagram of the Federal Mogul process called Sinta-Forge. Material improvements have included development of clean special grades of low-alloy powder. Iron-carbon and iron-copper-carbon alloys are another addition to the supply. Figure 15.79 is a compilation of representative mechanical properties and combinations of alloys available.

The properties of powder-forged steel are intermediate between the properties found for forgings between the horizontal and transverse direction of the forging. This is illustrated in Figure 15.80. The advantage of the powder-forged alloy is the consistent properties in longitudinal and transverse directions. This has been found particularly useful in powder-forged connecting rods and gear wheels. It appears that from a material cost saving and manufacturing process simplification, powder forging will continue to grow and replace machined parts and drop forgings.

Hot Isostatic Pressing of Metal Powders

Use of hot isostatic pressing (HIP) in conjunction with PM processing has produced parts with 100% density and properties equivalent to those of wrought alloys. Although the principal use of the HIP process has been for producing large billets for subsequent mechanical working and improving the properties of castings, it is expected that it will be used for much smaller parts on a high-volume production basis. The principal barrier to

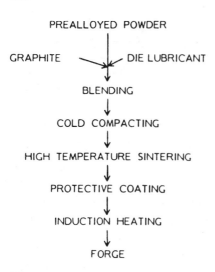

Figure 15.78 Typical flow diagram of the hot-forging process for PM (Sinta-Forge process of Federal Mogul Corp.).

the widespread use of HIP is the necessity for obtaining a closed surface rather than the porous interconnected condition that exists in conventional as-pressed parts.

Injection Molding of Metal Powders

An alternative to the use of pressure techniques to form a green compact is the use of injection molding. This technique uses the technology of plastic injection molding by combining metal powders with a polymer binder to produce a slurry-type mixture that can be injected into a complex die. Unlike plastics, however, the size of the parts produced by this method is somewhat restricted. Complex geometries including undercuts, holes, and reentrant angles may be produced using appropriate slide tooling on the die. Following injection, a binder removal operation at relatively low temperatures (400–600°F) is done.

Material group	Density (g/cm^3)	Tensile strength (N/mm^2)	Elon-gation (%)	Examples of applications
Iron and low-alloy compacts	5.2–6.8	5–20	2–8	Bearings and light-duty structural components
	6.1–7.4	14–50	8–30	Medium-duty structural parts, magnetic components
Alloyed steel compacts	6.8–7.4	20–80	2–15	Heavy-duty structural parts components
Stainless steel compacts	6.3–7.6	30–75	5–30	Components with good corrosion resistance
Bronzes	5.5–7.5	10–30	2–11	Filters, bearings, and machine components
Brass	7.0–7.9	11–24	5–35	Machine components

Figure 15.79 Representative Mechanical properties and combination of PM alloys available. (From Leo Alting, *Manufacturing Engineering Processes,* Marcel Dekker, New York, 1982.)

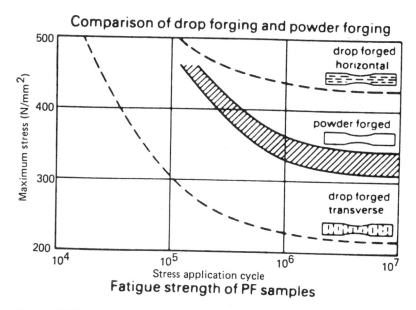

Figure 15.80 Properties of powder-forged steel in comparison with horizontal and longitudinal properties of conventional forgings. (From R. Bolz, Production Processes, Industrial Press, New York, 1963.)

Next, a sintering operation is conducted that is much the same as that for conventional PM processing. In injection molding of PM, shrinkage between the original injection molded dimensions and final sintered dimensions may be as much as 20%; therefore, careful tool design is needed. The polymer-PM blend must produce a predictable shrinkage preform.

The range of materials applicable to this process is limited only by the use of any fine powder material that can be sintered. These include steels, both stainless and carbon steel, and alloy steels as well as superalloys, tungsten carbides, and so on. The technique has also been used with ceramic materials such as aluminum oxide.

15.4.8 Ceramics/Cermets

Among the earliest objects fashioned by man, ceramic products have withstood the test of centuries and still find applications in numerous fields. Exceptional properties of ceramics have increased their use and their availability and have brought about the development of highly specialized technical ceramics suited to unusual industrial applications. Properly chosen and suitably designed, technical ceramics can fill a real need in the design of industrial equipment.

Failure to consider the use of ceramics in mechanical equipment generally stems from rather meager knowledge of these materials and lack of data with which to design such parts to ensure not only better performance but lower production costs.

Types of Ceramics

Traditionally, ceramic materials have been composed largely of naturally occurring clays, alone or in admixture with various amounts of quartz, feldspar, and other nonmetallic minerals.

High-clay ceramics account for the largest tonnage of manufactured ceramics. For convenience, these may be said to have a clay content in excess of 50%. The ceramics are characterized by high shrinkage—approximately 25% by volume—during drying and firing and therefore exhibit the widest size variation. To a large extent, the production tolerances are a direct function of the clay content. Parts produced include almost all structural clay products, clay-based refractories, most chemical porcelains and stoneware, and electrical and mechanical porcelains.

Low-clay ceramics include steatite and other low-loss dielectrics, special and super-refractories, and special porcelains. Steatite normally carries over 80% talc (hydrated magnesium silicate), which is bonded with ceramic fluxes. Shrinkage is low, and close tolerances can be held.

Various methods are used for forming clay-type ceramic bodies prior to firing. Practical production design must take cognizance of not only the tolerance effect of the method required but also the general range and limitations imposed. The various forming methods are either the wet or dry process types, shown in Figure 15.81. The *plastic wet process* is probably the oldest and most diversified general process, consisting of extrusion, casting, pressing, throwing, or jiggering. Figure 15.82a shows four of the processes, and Figure 15.82b shows an extruded blank that was machined after extruding and drying, but before firing. The *dry process* is different only in that a semimoist granular powder is used, and results in parts with much less dimensional variation. Pressing in metal dies is the only method of forming used; this method is adaptable to high production rates with automatic presses. The process is generally limited to parts of 2 to 4 in., but parts up to 14 in. have been made.

Clay-free ceramics are entirely devoid of clay. Included in this group are the pure oxide types and the so-called cermets, which have properties between metals and ceramics. A combination of pure oxide ceramic and a metal constituent, both of which have good high-temperature properties, the metal-ceramics have been developed to meet the need for

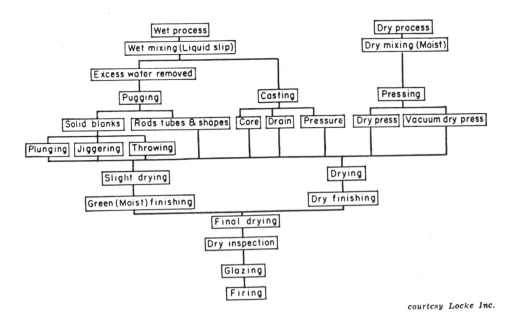

courtesy Locke Inc.

Figure 15.81 Forming methods for clay-type ceramic bodies prior to firing.

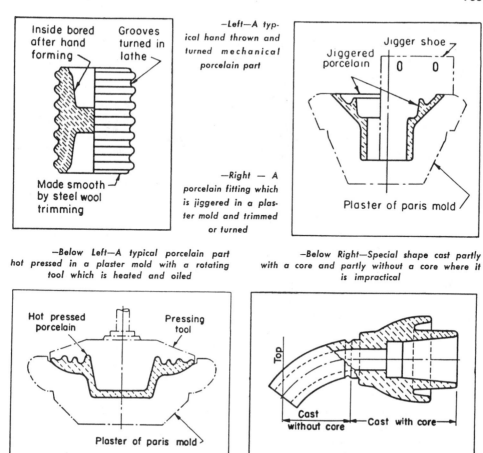

Figure 15.82 Four of the common forming methods using the plastic wet process for clay-type ceramics. (Courtesy of Westinghouse Electric Corp.)

a material capable of withstanding operating temperatures to 2400°F or more. The metal constituent is employed to provide thermal conductivity and shock resistance, while the ceramic provides resistance to deformation under high stress at high temperatures.

Some of the most recent innovations in cutting tool technology have been taking place in ceramics and cermets. Ceramic cutters, which first became available in the late 1940s, are characterized by high wear resistance, good hot hardness, and a low coefficient of friction. Additionally, ceramics are inherently inexpensive materials providing a cost advantage compared to carbide materials. As with all cutting tool innovations, increased productivity resulting from higher material removal rates and from less frequent tool changes is the driving force behind implementation. The use of ceramic cutting tools has been fastest in Japan, where 5–7% of the cutting tools are ceramic, compared to about half this usage in the United States in 1985.

Ceramic cutters are not without limitations. The poor mechanical and thermal shock resistance of ceramics limits their usefulness to applications with low feed rates and noninterrupted cuts, and to machine tools capable of great rigidity and high speeds.

Three types of ceramic cutting tools predominate today: aluminum oxide (Al_2O_3), sialon (a combination of silicon, aluminum, oxygen, and nitrogen), and silicon nitride (Si_3N_4). All three types are manufactured by cold pressing the ceramic powder and then sintering at high temperature in an oven.

Aluminum oxide, the most common ceramic cutting tool material, has been commercially available since the 1950s. It is most often applied when machining cast iron softer than 235 BHN and steel softer than Rc34. It also performs best for finishing, semifinishing, and noninterrupted cuts. When applied to cast iron, single-point cutting tools are able to achieve production removal rates of 1442 mm^3/min at a cutting speed of 610 m/min, a depth of cut of 5.84 mm, and feed rate of 0.41 mm per revolution.

Sialon, which has been available since 1981, offers a high degree of resistance to mechanical shock and exhibits very good hot hardness. These properties, in turn, provide uniform wear and lower chances of chipping failure compared to aluminum oxide. Unlike aluminum oxide, sialon can withstand severe thermal shock resulting from the use of coolants. It cannot be used on carbon steels due to chemical incompatibility.

Silicon nitride (Si_3N_4) cutting tools exhibit some of the same attributes of sialon, but with the addition of increased toughness, the ability to withstand higher chip loads, and higher cutting speeds. Used most often for the high-speed milling and roughing of cast iron and nickel-based alloys, silicon nitride is able to machine gray cast iron at speeds in excess of 1500 m/min.

Cermet tools are made of approximately 70% Al_2O_3 and 30% titanium carbide (TiC). Unlike ceramic tools, cermets are manufactured by hot pressing, that is, simultaneously pressing and sintering powder materials. The resulting cermets are much tougher and more fracture resistant than ceramics, allowing them to be used for machining cast iron materials harder than BHN 235 or steels harder than Rc34. The high toughness and hardness inherent with cermets also provide additional advantages of cutting speeds 600% faster than carbides, together with the ability to perform interrupted cuts.

A new class of cutting tool materials, known as ceramic/ceramic composites, are composed of an Al_2O_3 matrix reinforced with thousands of tiny silicon carbide whiskers measuring only 0.02 μm in diameter by 1 μm in length. The addition of these reinforcing fibers acts to distribute cutting forces throughout the matrix and to assist in carrying heat away from the cutting edge. This action results in a 200% increase in fracture toughness compared to cermets, and this process is suitable for applications requiring either roughing or finishing cuts.

15.4.9 Rubber Products

Like the plastics, rubber and rubberlike synthetics provide almost unlimited possibilities as components of machine assemblies. Again, it is a case where specific properties, largely unattainable with any other materials, are a design necessity and some knowledge of rubbers, their advantages and limitations, and their manufacture into parts is invaluable. As a rule, it is possible to design or redesign a rubber part to closely approach the ideal conditions for economical production without affecting or destroying the intended functional characteristics.

Materials

The rubber materials that can be used for molded parts now range from natural stocks from various parts of the world to a full array of rubberlike synthetics. The crude or synthetic-base material is compounded with curing agents, antioxidants, accelerators, lubricants, etc., and is thoroughly kneaded on mixing mills. It emerges as large slabs for storage pending subsequent manufacturing operations. This "green" or uncured compounded rubber can vary from a soft, gummy state to a hard, leathery condition. In no case is it liquid. (We are omitting discussion of some of the silicone compounds, which can be cast in a liquid or semiliquid state—making them more in the adhesive or plastic material field, rather that grouped with rubbers.) The nature of this uncured stock largely determines just how it needs to be shaped or prepared to fit a mold cavity for proper flow. A soft compound can be roughly shaped and laid in or adjacent to the mold cavity, whereas a hard stock requires careful tailoring and direct placement in the cavity. Characteristics of rubber and synthetic materials are shown in Figure 15.83.

Preliminary Processing

To obtain uncured stock of controlled size and shape for molding, the compounded material is processed by one of several preliminary operations: (1) slabbing, (2) calendaring, or (3) extruding (termed "tubing" in the rubber field). In slabbing, the compounded rubber is loaded into a mill, consisting of two large steel rolls that operate at slightly different speeds, which kneads and heats the compound and reduces its plasticity. After sufficient milling, the rubber is cut off in slabs, the gauge of the stock being determined by the roll spacing. Cutting templates for "slabbing off" pieces to proper contour for molding are often employed.

The calendar prepares the rubber compound in thin sheets that can be held to close tolerances. In calendaring, a ribbon of rubber is fed from a warmup mill to the calendar rolls, from which it emerges as a sheet or strip. Rubberizing or "frictionizing" of fabric is also done in a calendar. Extruding or tubing is carried out in a screw-type extruder, the shape and size of the stock being determined by the contour of the die. After cooling in a water tank, the extruded raw stock is cut to necessary lengths for molding.

Molding Methods

There are four practical molding processes: (1) compression molding (most widely used), (2) transfer injection molding, (3) full injection molding, and (4) extrusion molding.

Compression molding consists of placing a piece or pieces of prepared stock in the heated mold cavity, bringing the halves of the mold together under pressure of 500 to 1000 psi, and curing (see Figure 15.84). Heat for curing is usually supplied by the heated platens of the press used. Depending on size, it is possible to mold from one to as many as 360 pieces per mold.

Material	Desig- nation (Gov't)	Tensile Strength Pure Gum	Black Rein- forced	Hardness (Durom- eter)	Mixing Eff. (%)	Extrud- ing (%)	Calen- dering (%)	Cohesion	Molding
Natural	.....	3000	4500	any	100	100	100	Excellent	Excellent
Buna S	GRS	400	3000	any	85	90	90	Fair	Fair
Buna N	GRA	600	3500	any	50	50	50	Poor	Excellent
Butyl	GRI	3000	3000	28 to 85	90	90	100	Good	Good
Thiokol	GRP	300	1500	35 to 90	75	75	50	Poor	Good
Neoprene	GRM	3500	3500	25 to 90	100	75	90	Good	Good

Figure 15.83 Characteristics of rubber and synthetic rubber materials. (From R. Bolz, *Production Processes,* Industrial Press, New York, 1963.)

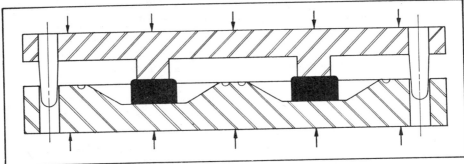

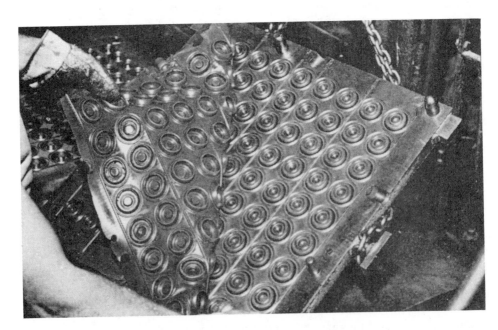

Figure 15.84 Cross section of compression mold and views showing mold loading and removal of cured parts. (Courtesy Goodyear Tire & Rubber Co.)

Transfer-injection molding permits the use of a single piece of prepared compound. Intricately shaped parts can be molded with improved efficiency over the compression method. The prepared piece is placed in a charging cavity in the mold and forced at high pressure through runners or channels into the mold cavity (see Figure 15.85). Usually the mold is opened and closed by hydraulic pressure, a separate plunger being used for injection.

Mull injection molding uses an extrusion head as an integral part of the molding machine. Lengths of extruded compounded rubber stock are fed directly into the extruder, which in turn injects or forces the material into the mold cavity or cavities. These units (see Figure 15.86), are entirely automatic in operation with the exception of stripping the finished parts from the mold. Cavities are laid out so that the stock is injected into a central canal that branches into two or more feeders and thence to the cavities. High pressure and turbulence developed during injection result in high temperature, reducing the curing period materially. Resultant savings in molding justify the use of this equipment in many instances.

Extrusion molding can be used to extrude a wide variety of uniform-cross-section parts to the desired shape. Very intricate sections are practicable. Screw-type feed is employed for forcing the stock through the die, as shown in Figure 15.87. Unlike the other molding methods, extrusion does not permit curing during the cycle. After extruding and removal from the water cooling tank, extruded stock is cured under temperature and pressure by means of steam, and after cooling, it is cut to the necessary lengths. Extruded shapes require fair volume production for economy, since at least 100 lb of stock are required to put an extruder into operation. As examples, a tube of 6 in. diameter and 1/8 in. wall can be produced in lengths to about 50 ft; sections under 3/16 in. in diameter can be extruded to lengths of about 500 ft.

Parting Lines

As in all molding methods, placement of the mold parting or partings is extremely important, not only for ensuring simplest possible mold design and operation, but also for simplifying flash removal and finishing. Any particular location on a part that for design purposes should be free from flash, should be so indicated on the drawing. Because nonfills result in rejects, molded rubber parts invariably have overflow flash, and molds are designed to accommodate this condition. Circular flash, as shown in Figure 15.88a, is readily removable, automatically and cheaply. The part in Figure 15.88b is difficult to trim, whereas the one molded vertically in Figure 15.88c is much more economical.

Metal Inserts

Where inserted pieces are to be completely imbedded in rubber, it is difficult to ensure positive location. As a rule, rubber flow in the mold varies and exact position of full floating inserts is impossible to predict. Thus, where inserts are employed, the parts should be designed so that the mold ensures positive positioning (see Figure 15.89).

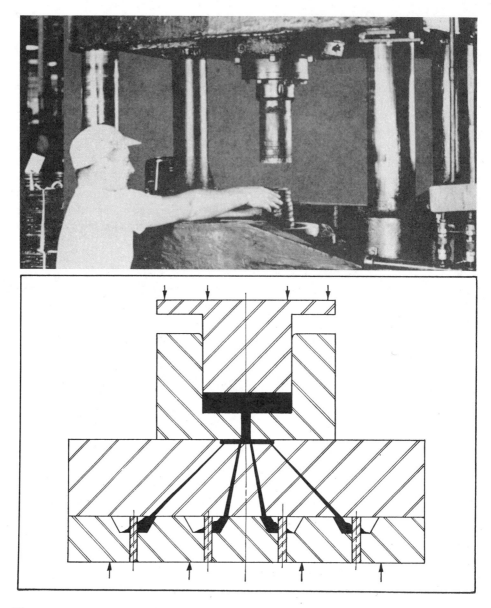

Figure 15.85 View through a transfer injection mold for molding of rubber compounds. (Courtesy Goodyear Tire & Rubber Co.)

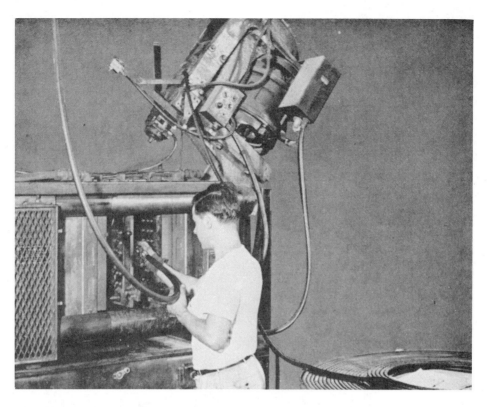

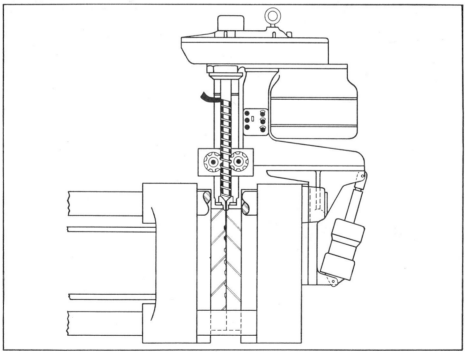

Figure 15.86 Screw-type injection molding of lengths of extruded compounded rubber stock, and cross section through extrusion head and die. (Photo and drawing courtesy Goodyear Tire & Rubber Co.)

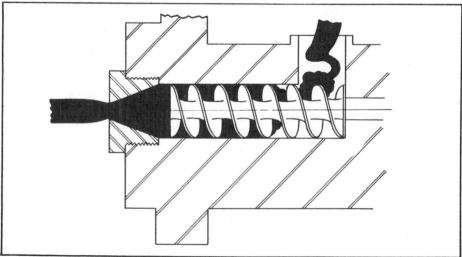

Figure 15.87 Extrusion molding of uniform-cross-section parts. (Photo and drawing courtesy Goodyear Tire & Rubber Co.)

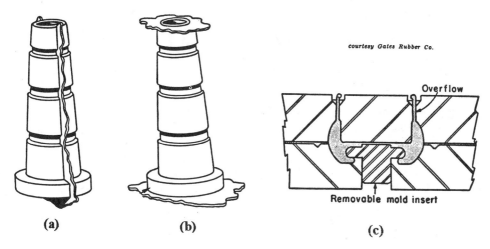

Figure 15.88 Parts molded in (a) are difficult to trim; circular flash is readily removed in (b); while (c) is much more economical.

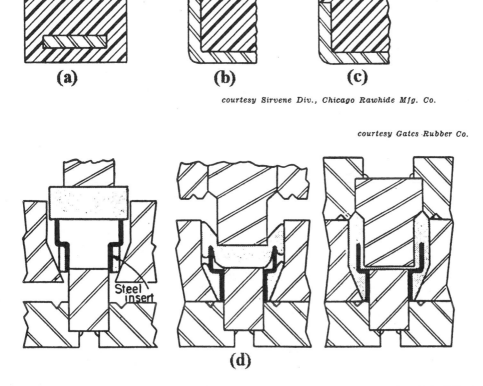

Figure 15.89 Totally imbedded metal inserts (a) are difficult to position. Design (b) is difficult to trim, and (c) is preferred for economy. Imbedded inserts should be designed to be positioned by the dies positively as shown in (d), where the insert is located by the center pin.

15.5 BIBLIOGRAPHY and REFERENCES

ABC's of Aluminum, Reynolds Metals Company, Richmond, VA, 1962.

Alting, Leo, *Manufacturing Engineering Processes*, Marcel Dekker, New York, 1982.

Aluminum Forging Design Manual, The Aluminum Association, New York, 1975.

Aluminum Impacts Design Manual, The Aluminum Association, New York, 1980.

Bolz, Roger, *Production Processes*, Industrial Press, New York, 1963.

Casting Aluminum, Reynolds Metals Company, Richmond, VA, 1965.

Design Manual for Aluminum Precision Forgings, Aluminum Company of America, Pittsburgh, PA, 1950.

Designing for Alcoa Die Castings, Aluminum Company of America, Pittsburgh, PA, 1955.

Designing with Aluminum Extrusions, Reynolds Metals Company, Louisville, KY, 1952.

Facts and Guideline Tolerances for Precision Aluminum Forgings, Forging Industr Association, Cleveland, OH, 1975.

Forging Industry Handbook, Forging Industry Association, Cleveland, OH, 1970.

Forging Product Information, Kaiser Aluminum, Oakland, CA, 1959.

Forming Alcoa Aluminum, Aluminum Company of America, Pittsburgh, PA, 1973.

Gillett, H. W., *The Behavior of Engineering Metals*, John Wiley, New York, 1951.

Ginzburg, Vladimir B., *High-Quality Steel Rolling*, Marcel Dekker, New York, 1993.

How to Design Metal Stampings, Dayton Rogers Manufacturing Company, Minneapolis, MN, 1993.

Investment Casting Handbook, Investment Casting Institute, Chicago, IL, 1968.

Modern Steels and Their Properties, Bethlehem Steel Company, Bethlehem, PA, 1964.

Tanner, John P., *Manufacturing Engineering*, Marcel Dekker, New York, 1991.

Tijunelis and McKee, *Manufacturing High Technology Handbook*, Marcel Dekker, New York, 1987.

16

Nontraditional Machining Methods

Lawrence J. Rhoades   *Extrude Hone Corporation, Irwin, Pennsylvania*

with

V. M. Torbilo   *Ben-Gurion University of the Negev, Beer-Sheva, Israel*

16.0 INTRODUCTION TO NONTRADITIONAL MACHINING METHODS

The term *nontraditional machining* refers to a variety of thermal, chemical, electrical and mechanical material removal processes. The impetus for the development of nontraditional machining methods has come from the revolution in materials, the demand for new standards of product performance and durability, the complex shapes of products engineered for specific purposes, and considerations of tool wear and economic return. Nontraditional machining methods have also been developed to satisfy the trend toward increased precision and to create improved surface conditions. Because nontraditional machining processes can provide new ways of satisfying the demands of nascent technological advances in many areas, design engineers need not limit ideas to traditional machining methods. A new horizon of choices has opened up for the design of products.

There are several reasons for choosing nontraditional machining methods. One of the most important of these is the trend toward using engineered materials that are difficult to machine with conventional processes. High-temperature and high-performance metal alloys are often difficult to machine using traditional methods, but nontraditional machining processes generally work well with these materials. Because of the varying degrees of hardness they present, composites can be extremely difficult to machine using conventional methods. However, due to the selective machining and low applied forces of some nontraditional machining methods, composites can be machined with accuracy. Nontraditional machining methods are also ideal for machining ceramics and a host of other difficult materials.

Another reason for choosing nontraditional machining methods is that the features to be machined are often difficult or impossible to do with traditional methods. Complex geometric shapes, or shapes derived by the needs of flow, can present special machining problems. Nontraditional machining methods deal with these problems, offering full-form machining as opposed to single-point machining and providing the ability to reach normally inaccessible places. Nontraditional machining methods can create and machine special holes. These methods can machine shaped cross sections, tapered holes, and holes where there is a high length-to-diameter ratio. Nontraditional machining methods can also provide special profile cutting for thin slots and small internal corner radii.

Burr-free machining, which may be essential in some applications, can be achieved by nontraditional machining methods. Low applied forces can prevent damage to the workpiece that might occur during traditional machining. Surface conditions that are left by some of the processes that fall into the category of nontraditional machining can ensure that the workpiece meets specific demands. The selective machining provided by nontraditional machining methods can be particularly interesting in the milling of composite materials in which there are combinations of components and some of the materials may be required to be left emerging from the field of a different material. Tool wear advantages that are offered by some of the nontraditional machining processes allow for continuous machining with, at least in theory, zero tool wear. All of these unique advantages of nontraditional machining methods open almost limitless possibilities of design and application. Figures 16.1 through 16.5 show the range and applications of nontraditional

CUTTING MECHANISM	PROCESS	3-D FORM "SINKING"	HOLE "DRILLING"	CUTTING	NON-TRADITIONAL EQUIVALENTS OF TRADITIONAL MACHINING (PROFILE MILLING, TURNING, GRINDING)
THERMAL REMOVAL	EDM	RAM EDM	SMALL HOLE EDM	WIRE CUT EDM	EDM PROFILING / ELECTRICAL DISCHARGE GRINDING
	LASER		LASER DRILLING	LASER CUTTING	LASER TURNING / LASER MILLING
	PLASMA ARC			PLASMA ARC CUTTING	PLASMA ARC TURNING
DISSOLUTION	CHEMICAL	CHEM MILLING	CHEM MILLING / PHOTO ETCHING	CHEM MILLING / PHOTO ETCHING	
	ELECTROLYTIC	SALT WATER ELECTROLYTES	"STEM" "ELECTROSTREAM" "CAPILLARY" DRILLING		
HYBRID BOTH MECAHNICAL ABRASION AND ELECTROLYTIC REMOVAL	ELECTROLYTIC GRINDING			ECG "CUT OFF"	ECG FORM GRINDING
MECHANICAL ABRASION	ULTRASONIC	ULTRASONIC "IMPACT GRINDING"	ULTRASONIC DRILLING	ULTRASONIC "KNIFE" CUTTING	USM PROFILE MILLING / ROTARY ULTRASONIC MACHINING
	ABRASIVE WATER JET		AWJ DRILLING	AWJ CUTTING / ABRASIVE SUSPENSION JET CUTTING	AWJ "TURNING" / AWJ "MILLING"

Figure 16.1 Nontraditional equivalents of traditional machining.

	EDM	RAM EDM
THERMAL REMOVAL	LASER	
	PLASMA ARC	
DISSOLUTION	CHEMICAL	CHEM MILLING
	ELECTROLYTIC	SALT WATER ELECTROLYTES
HYBRID	ELECTROLYTIC GRINDING	
MECHANICAL ABRASION	ULTRASONIC	ULTRASONIC "IMPACT GRINDING"
	ABRASIVE WATER JET	

Figure 16.2 Summary of nontraditional 3-D form sinking processes.

	EDM	SMALL HOLE EDM
THERMAL REMOVAL	LASER	LASER DRILLING
	PLASMA ARC	
DISSOLUTION	CHEMICAL	CHEM MILLING PHOTO ETCHING
	ELECTROLYTIC	"STEM" / "ELECTROSTREAM" / "CAPILLARY" DRILLING
HYBRID	ELECTROLYTIC GRINDING	
MECHANICAL ABRASION	ULTRASONIC	ULTRASONIC DRILLING
	ABRASIVE WATER JET	AWJ DRILLING

Figure 16.3 Summary of nontraditional hole "drilling."

THERMAL REMOVAL	EDM	WIRE CUT EDM
	LASER	LASER CUTTING
	PLASMA ARC	PLASMA ARC CUTTING
DISSOLUTION	CHEMICAL	CHEM MILLING PHOTO ETCHING
	ELECTROLYTIC	
HYBRID	ELECTROLYTIC GRINDING	ECG "CUT OFF"
MECHANICAL ABRASION	ULTRASONIC	ULTRASONIC "KNIFE" CUTTING
	ABRASIVE WATER JET	AWJ CUTTING ABRASIVE SUSPENSION JET CUTTING

Figure 16.4   Summary of nontraditional cutting processes.

THERMAL REMOVAL	EDM	EDM PROFILING ELECTRICAL DISCHARGE GRINDING
	LASER	LASER TURNING LASER MILLING
	PLASMA ARC	PLASMA ARC TURNING
DISSOLUTION	CHEMICAL	
	ELECTROLYTIC	
HYBRID	ELECTROLYTIC GRINDING	ECG FORM GRINDING
MECHANICAL ABRASION	ULTRASONIC	USM MILLING ROTARY ULTRASONIC MACHINING
	ABRASIVE WATER JET	AWJ "TURNING" AWJ "MILLING"

Figure 16.5   Summary of nontraditional "profile milling," "turning," and "grinding" processes.

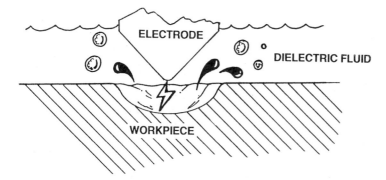

Figure 16.6 Schematic of the thermal removal process of electrical discharge machining (EDM). Sparks between the electrode and the workpiece perform the material removal.

machining processes. A closer look at some of these processes will show their strengths in general and for specific tasks.

16.1 THERMAL REMOVAL PROCESSES

In thermal removal processes, high-intensity heat is focused on a small area of the workpiece, causing it to be melted and vaporized. One kind of thermal removal process is electrical discharge machining (EDM). In EDM, sparks between the electrode and the workpiece perform the material removal (see Figure 16.6). Two other types of thermal removal processes are laser machining and plasma arc machining. In both of these thermal removal processes, a directed energy beam performs the cutting (see Figure 16.7). In all of the thermal removal processes, not all of the material removed is vaporized. Much of the material removed is melted. Most of the melted material is expelled from the cut by the turbulence of the adjacent vaporization or by the flow of an "assist gas" used in the

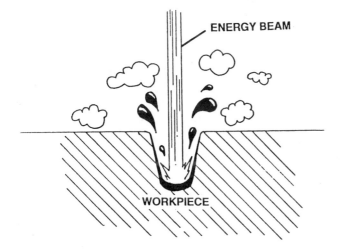

Figure 16.7 Schematic of laser and plasma arc thermal removal processes. The directed energy beam performs the cutting.

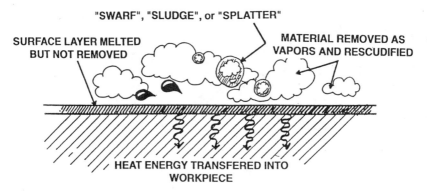

Figure 16.8 In the thermal removal processes, some material remains and resolidifies on the surface.

process. Some material remains and resolidifies on the surface, cooling rapidly as heat is transferred to the subsurface material. The remaining "recast layer" is likely to have microcracks and residual tensile surface stresses, encouraging those cracks to widen when the material is fatigued. Beneath the recast layer there is typically a "heat-affected zone," where the material's grain structure may have been altered (see Figures 16.8 and 16.9). A closer look at thermal removal processes will show in detail how they work.

16.1.1 Electrical Discharge Machining

In electrical discharge machining (EDM), carefully controlled sparks (electrical discharges) are generated between an electrode and the workpiece. The electrode material chosen and the characteristics of the generated sparks are designed so that much more material is removed from the workpiece than from the electrode. The EDM machine

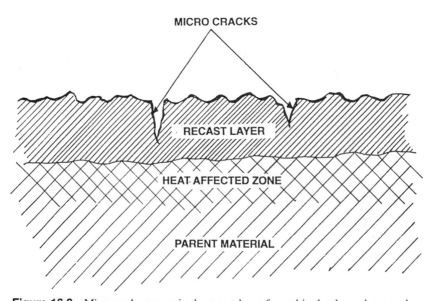

Figure 16.9 Microcracks appear in the recast layer formed in the thermal removal processes.

typically has a servo control system built in, and it is this control system that manipulates the electrode and the workpiece relative to one another to maintain a controlled spark gap. A flow of dielectric fluid is maintained in the spark gap. This fluid is used to provide a nonconductive barrier. It is also used to cool and resolidify the removed material into "swarf" particles and to flush this swarf from the machining gap to prevent uncontrolled arcing.

In RAM EDM, the form of a preshaped electrode is "eroded" into the workpiece material. Sparks occur between the electrode and the workpiece at the smallest gap, removing microscopic amounts of material with each spark. In the 3-D form sinking of RAM EDM, often the shaped electrode is manipulated under servo control. This manipulation, along with the high frequencies at which the voltage pulses are discharged, enables sparking to occur along the entire face of the electrode. The electrode then progressively advances into the workpiece, generating a uniform spark gap around itself, and the shape of the electrode is gradually reproduced on the workpiece (see Figure 16.10).

Nevertheless, in EDM the electrode wears as it removes material from the workpiece, and the electrode may need to be replaced or reshaped to machine multiple workpieces. Often more than one electrode is required to "rough" and "finish" machine a single workpiece with deep features. The servo-controlled manipulation of the electrode within the cavity being machined improves flushing and provides more uniform electrode wear. Flushing is important in EDM, especially when deep, complex shapes have to be reproduced, for the spark gap must be kept free from machining debris that can cause uncontrolled arcing.

Another kind of electrical discharge machining is small-hole EDM. When EDM is used to produce holes, typically a small-diameter wire, rod, or tube is held in an electrode holder and advanced into the workpiece (see Figure 16.11). Electrode wear is compensated by advancing or refeeding the electrode material through the holder to reestablish proper electrode length. If necessary, the electrode ends are trimmed to restore the desired tip shape. To enhance productivity, small-hole EDM is employed using a number of electrodes held simultaneously to drill multiple holes precisely. If a single hole is to be drilled, the electrode can be rotated during machining to improve flushing and distribute wear uniformly. Using tube electrodes permits flushing through the tube. Small-hole EDM allows precise drilling operations in a variety of materials.

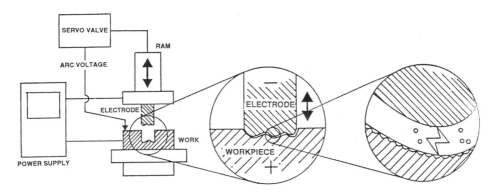

Figure 16.10 In RAM EDM, the form of a pre-shaped electrode is eroded into the workpiece material.

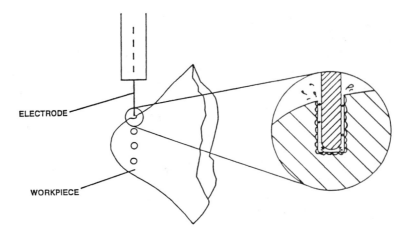

Figure 16.11   Example of using the EDM process to produce small-diameter holes.

Wire EDM has proven to be a major development in electrical discharge machining. In wire EDM, the electrode used for cutting is a small-diameter wire, usually brass or copper with a diameter of 0.05 to 0.25 mm. Servo control systems are employed for guiding the movement of the tool and the workpiece relative to one another and controlling the direction of the machining. The small-diameter wire is held taut between two spools. Fresh wire is continuously fed into the cut to make up for electrode wear and encourage flushing. The dielectric fluid used in traveling wire EDM is typically deionized water because of low viscosity and high dielectric constant. The dielectric fluid is usually injected into the machining zone coaxially with the wire. Cutting speeds have increased by a factor of more than 10 since wire EDM was introduced, and those speeds are now in the range of 140cm^2/hr. Accuracies can routinely be held to ±0.01 mm. Wire EDM is the most widely used unattended machining operation, with overnight "lights out" operation a routine practice even in small manufacturing companies in this country and throughout the world (see Figure 16.12).

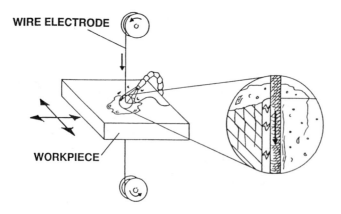

Figure 16.12   Process of cutting shapes using wire EDM.

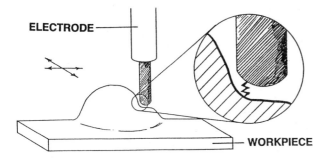

ELECTRODE

WORKPIECE

Figure 16.13 EDM profile milling where the electrode is used like a milling cutter, with the machine manipulating the tool and workpiece relative to one another under servo control.

Electrical discharge machining can also be applied to milling and grinding operations (see Figures 16.13 and 16.14). EDM can be used as a profile milling operation similar to conventional milling. The electrode is used like a milling cutter. The servo control of the milling machine manipulates the tool and the workpiece relative to one another for accurate milling. The process is not very fast, but it is quite flexible. Another electrical discharge machining process is EDM grinding. In EDM grinding, the electrode is a wheel, typically graphite, with power connected to it. The electrode is rotated while the workpiece and wheel are moved relative to one another under servo control. The electrode wear is distributed across the entire wheel diameter, and redressing can easily be performed as in conventional grinding. Generally the work is done submerged within the dielectric fluid.

Electrical discharge machining is a thermal removal process that is versatile and precise. It has a wide range of applications in machining a variety of materials. It can be controlled with an accuracy that many traditional machining methods lack, and as the trend toward increased precision and hard materials grows, EDM will become even more integral to manufacturing.

16.1.2 Laser Beam Machining

Another thermal removal process is laser beam machining. In laser beam machining, an intense beam of collimated, single wave-length, in-phase light is focused by an optical lens

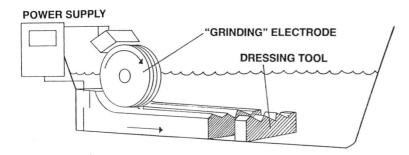

POWER SUPPLY

"GRINDING" ELECTRODE

DRESSING TOOL

Figure 16.14 EDM grinding where the electrode is used like a grinding wheel, with the machine manipulating the tool and workpiece relative to one another. Dressing tool shown can be used similar to redressing a conventional grinding wheel.

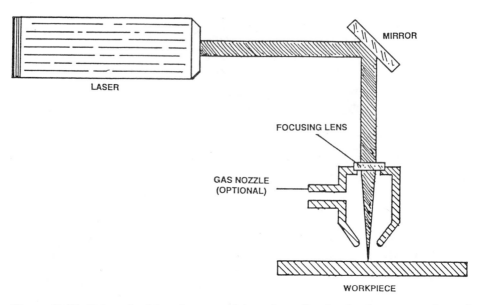

Figure 16.15 Schematic of laser beam machining using a directional, coherent, monochromatic beam of light to achieve precision in cutting and drilling.

onto the workpiece point to be machined. The light absorbed by the workpiece is converted to heat, which melts and vaporizes the workpiece material. Molten material is evacuated from the cut by the adjacent vaporization turbulence which typically occurs in drilling operations, or by the use of an "assist gas" in cutting operations (see Figure 16.15). Laser beam machining uses a directional, coherent, monochromatic beam of light to achieve precision in cutting and drilling. The intensity of this light produces a tremendous amount of heat at the point of application to the workpiece, and laser beam machining can take place at relatively high speeds.

A number of different types of lasers are used in laser beam machining, each with certain advantages for different operations or applications. The most commonly used lasers for machining are Nd:YAG lasers, which have certain advantages for hole drilling due to their higher pulse energy, and CO_2 gas lasers, which have certain advantages in cutting since they are capable of delivering much higher average power. Lasers may be operated in either pulsed or continuous-wave (CW) modes. The most powerful CO_2 lasers, however, are operable only in CW mode. CO_2 lasers can have output power generally ranging from 100 to 2000 W when pulsed and from 250 to 5000 W in CW mode. Some lasers are capable of an output power of 25,000 W.

A 1250-W CO_2 laser can cut mild steel at speeds ranging between 40 and 140 cm^2/min, depending on material thickness. At a thickness of 12 mm, the cut can be 40 cm^2/min. At a thickness of 2 mm, a 1250-W CO_2 laser can cut at a rate of 140 cm^2/min. Aluminum is generally cut at about one-half the speed of carbon steel with a CO_2 laser because of the high thermal conductivity of the aluminum. In cutting applications, the laser beam may be "transmitted" and "switched" by using mirrors to manipulate the beam, or with Nd:YAG lasers by the use of a combination of fiber optic cables and switches. Below is the typical average power range and maximum pulse energy for the two types of lasers most commonly used in laser beam machining.

	Nd: YAG	CO_2
Typical average power range	100–400 W	250–5000 W CW
		100–2000 W pulse
Maximum pulse energy	80J	2 J

In the application of laser beam machining, it is possible to perform the operations of drilling and cutting with speed and precision. For instance, percussion (i.e., repeated pulse) drilling of Inconel 718 with a 250-W Nd:YAG laser can produce accurate holes 12 mm deep in under 10 sec and holes 25 mm deep in 40 sec. Length-to-diameter ratios are limited to about 30 or 40 to 1 with conventional Nd:YAG lasers. However, other laser technologies allow the length-to-diameter ratio to be higher. Frequently, gas is used to assist laser beam machining. A coaxial columnar flow of gas (oxygen, air, or inert gas) at pressures ranging from 1 to 6 bar expels molten metal from the cut. Oxygen assists in cutting steel and other materials at an increased rate because of the oxidation reaction with the metals.

Laser beam machining uses thermal energy to drill and cut with speed and precision that traditional machining methods often cannot duplicate. Its applications are becoming more varied as new techniques are developed.

16.1.3 Plasma Arc Machining

The final thermal removal process to be discussed here is plasma arc machining. Like electrical discharge machining and laser beam machining, it employs thermal energy to accomplish its work. In plasma arc machining, gas heated to very high temperatures by a high-voltage electric arc partially ionizes and consequently becomes electrically conductive, sustaining the arc. When gas is heated to the degree that electrons become ionized (electrically charged), the gas is called a plasma. Primary gases used for plasma arc machining may be nitrogen, argon-hydrogen, or air. The gas is forced at a high rate of speed through a nozzle and through the arc. As the gas travels, it becomes superheated and ionized. The superheated gas reaches temperatures of 3000 to 10,000 K. A hot tungsten cathode and a water-cooled copper anode provide the electric arc, and the gas is introduced around the cathode. It then flows out through the anode. The size of the orifice at the cathode determines the temperature, with small orifices providing higher temperatures. The ionized particle stream is consequently a high-velocity, well-columnated, extremely hot plasma jet, supporting a highly focused, high-voltage, "lightninglike" electric arc between the electrode and the workpiece. With such high temperatures, when the plasma touches the workpiece, the metal is rapidly melted and vaporized. The high-velocity gas stream then expels molten material from the cut.

In plasma arc machining, frequently a swirling, annular stream of either water or a secondary gas is injected to flow coaxially with the plasma arc. The use of water can serve several purposes. It increases the stability of the arc and increases cutting speeds. Water injected coaxially can also cool the workpiece and reduce smoke and fumes. It can also increase nozzle life. Sometimes a secondary gas is introduced surrounding the plasma stream. The choice of the secondary gas depends on the metal being cut. Hydrogen is often used as a secondary gas for the machining of stainless steel or aluminum and other nonferrous metals. Carbon dioxide gas can be used successfully with both ferrous and nonferrous metals. Oxygen is often introduced as a secondary gas surrounding the plasma stream, adding the heat from the exothermic oxidation reaction with steel and other materials to assist in the cut (see Figure 16.16).

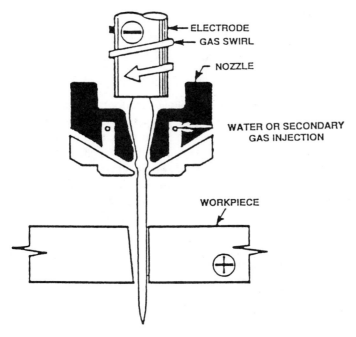

Figure 16.16 In plasma arc machining, gas heated to very high temperatures by a high-voltage electric arc partially ionizes and consequently becomes electrically conductive, sustaining the arc.

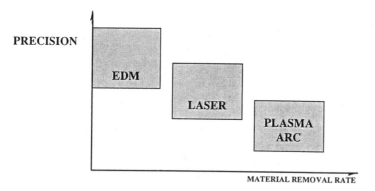

Figure 16.17 Relation of precision and material removal rate of the major thermal removal processes.

Plasma arc machining is widely used for sheet and plate cutting. It is incorporated in many CNC sheet metal punching machines. With plasma arc machining, cutting speeds of 700 cm^2/min and higher can be achieved. Accuracies are limited, however, to ±0.1 mm at best, and a taper of 2° or more is normally generated. A recast layer and heat affected zone of roughly 0.5 mm depth is typical. Nevertheless, plasma arc machining is a fast and effective method of machining in many applications.

Electrical discharge machining (EDM), laser beam machining, and plasma arc machining with differentials of an order of magnitude offer precision and machining speeds that many conventional processes do not (see Figure 16.17). In cutting, for instance, wire EDM provides cutting speeds of about 40 cm^2/hr. Laser beam machining provides cutting speeds on average of 70 cm^2/min. Plasma arc machining provides cutting speeds of about 700 cm^2/min. The precision, however, ranges from about ±0.1 mm at best for plasma arc machining to ±0.01 mm as typical for wire EDM. The trade-off of speed for precision might be a determining factor in choosing the right kind of thermal removal process for the specific application.

16.2 CHEMICAL MACHINING

Chemical milling is accomplished simply by dipping the workpiece into a tank with strong etchants. Nitric acid, hydrochloric acid, and hydrofluoric acid are the most commonly used etchants in chemical machining. The areas where no machining is desired are masked, typically with an elastomerlike silicone rubber or with an epoxy. Periodic interim steps are used to monitor remaining material or, if necessary, to remask or to cover undercut areas. Progressive masking and etching can produce complex shapes in difficult-to-form materials, even very large ones, with potentially no applied forces (see Figure 16.18).

Photo etching is another variant of chemical machining. Photo etching uses a photo-resistant maskant to generate intricate 2-D patterns in thin, flat metal sheets. The metal sheets used in photo etching have a thickness range from 0.01 to 1.5 mm. The process of photo etching uses a relatively mild etchant such as ferric chloride, which is typically applied by spraying as conveyorized parts pass through a spray chamber. Photo etching is widely used to produce circuit boards and other sheet materials. It can produce thousands of holes at once with a high degree of accuracy.

16.2.1 Electrochemical Machining

Electrochemical machining (ECM) removes metal by anodic dissolution. ECM electrolytes are normally safe-to-handle common salt water solutions. However, the sludges produced in the electrochemical machining of certain materials, notably chromium, can be poisonous and must be processed or disposed of carefully and responsibly. Electrochemical machining uses high amperage (500–20,000 A, typically) and low voltage (10–30 V) DC power with relatively high electrolyte flow rates through the machining gap (e.g., 3000–6000 cm/sec). Metals removed by ECM quickly become a metal hydroxide, insoluble sludge, which is washed from the machining gap by the electrolyte and is removed by some filtration process—settling, centrifuge, or some other type of filtration. The form machined is a near mirror image of the cathode (tool), with variations resulting from electrolyte flow, from hydroxide concentration, and from temperature variations.

There are compelling advantages to electrochemical machining processes. One is that there is virtually no tool wear. Another advantage to ECM is that work can be done at

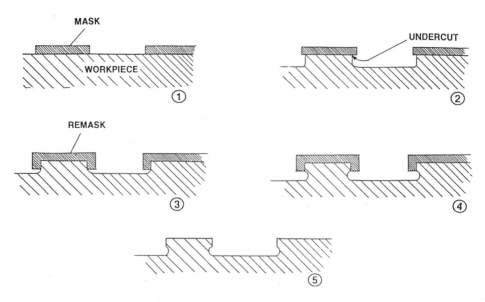

Figure 16.18 Chemical milling is accomplished simply by dipping the workpiece into a tank with strong etchants. The areas where no machining is desired are masked, typically with an elastomerlike silicone rubber or with an epoxy. Progressive steps of masking and etching are shown as 1 through 5 for a complex part.

relatively high speeds. For example, a 15,000-A machine can cut 25 cm^3/min, regardless of material hardness. Electrochemical machining normally provides excellent surface conditions with no thermal damage. Nevertheless, there are some drawbacks to using electrochemical machining. The cost of equipment is relatively high because corrosive-resistant materials are required for the corrosive salt environment in which ECM operations take place. Machining tight accuracies can often require extensive tool shape development. This results from the changing conductivity of the electrolyte as it passes across the machining gap due to the heat and metal hydroxide added from the machining process, causing the electrolytic machining gap to vary as much as 0.02 to 0.2 mm. Pulsed power ECM reduces this variance. Finally, sludge handling and removal, particularly when chromium alloys are used, can be a concern.

Electrochemical machining can provide both speed and precision in many applications. ECM is well suited to machining the complex shapes and difficult materials used in turbine engines, and it is commonly used to machine compressor blades. It is also widely used, with stationary electrodes, as a deburring method. There is great potential in the ECM process to influence design and production.

16.2.2 Stem, Capillary, and Electrostream Drilling

Some commonly applied variations of electrochemical machining include stem, capillary, and electrostream drilling. These processes use strong acids as electrolytes. Stem drilling uses titanium tubes with conventional low voltage, generally 8–14 V. Electrolyte is pumped down the central bore of the titanium tube and out through the gap formed between the wall of the tube and the hole being dissolved in the workpiece. Capillary drilling uses straight glass tubes with a platinum wire inside each to conduct electricity to the electrolyte stream. In capillary drilling, the voltage is higher, generally in the range of 100–300 V. Again it is the electrolyte that dissolves the hole in the workpiece as it is pumped through the glass tubes and out through the gap. Electrostream drilling uses glass nozzles, shaped glass tubes without wires. A much higher voltage, generally in the range of 600–900 V, allows the electrolyte stream to cut without a conductive tool.

With stem, capillary, and electrostream drilling, speed and precision are key benefits. Cutting speeds range from 0.75 to 3 mm/min with precision, and multiple holes can be machined simultaneously. All of these processes can produce long, small holes:

	Diameter (mm)	Depth (mm)
Stem	0.5	75
	6.0	1000
Capillary	0.15	6
Electrostream	0.5	25

16.2.3 Electrochemical Grinding

Electrochemical grinding (ECG) is a hybrid process combining electrolytic dissolution and mechanical grinding. In ECG, a grinding wheel with conductive bonding material and protruding nonconductive abrasive particles does the cutting. Electrolyte is carried across the machining gap with the surface of the wheel (see Figure 16.19). The grinding wheel is the cathode (tool), and the workpiece is the anode. The nonconducting particles protruding from the wheel act as a spacer between the wheel and the workpiece to allow a constant gap for the flow of the electrolyte. The electrolyte is fed through a tube or nozzle to flow into the machining gap.

In electrochemical grinding, the feed rate and voltage settings determine the relative

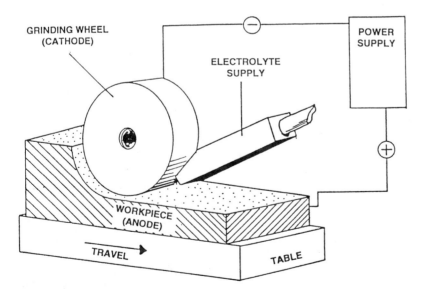

Figure 16.19 Electrochemical (ECG) grinding is a hybrid process combining electrolytic dissolution and mechanical grinding.

roles of the electrolytic and mechanical action. With high electrolytic action, in other words, higher voltage and lower feed rates, wheel wear is minimized and more of the cutting is accomplished by electrochemical action. With lower voltage and higher feed rates, there will be more mechanical cutting. Although higher rates of mechanical cutting produce more wheel wear, they also improve machining accuracy. In electrochemical grinding operations, generally 90% of the stock is removed electrolytically. Then low-voltage finish passes can be made to produce high accuracy and sharp edges that electrolysis alone cannot achieve.

There are real advantages to electrochemical grinding. ECG can machine dissimilar materials and composites well, since the process itself is a composite process. Electrochemical grinding uses electrolytes and machining conditions that are prone to surface passivation, thus minimizing unwanted "stray" etching, since grinding action will mechanically remove the nonconductive "passivation layer," thereby exposing only the desired areas to continued electrolytic action. Electrochemical grinding can machine without burrs, can have very low grinding wheel wear since most of the removal is by electrolytic action, is capable of extremely high accuracy, and leaves stress-free surfaces.

16.3 MECHANICAL PROCESSES

Nontraditional machining methods employ mechanical processes just as traditional methods do. These processes include ultrasonic machining, abrasive water jet machining, and abrasive flow machining. Each of these processes has certain advantages in the machining of different materials and shapes.

16.3.1 Ultrasonic Machining

In ultrasonic machining, a tool is vibrated along its axis at its resonant frequency. The material, the size, and the shape of the tool are typically chosen to have a resonant frequency in the range of 20 kHz, which is above the human hearing range and may therefore be deemed ultrasonic. The vibration amplitude of the tool is typically 0.01 to 0.05 mm. The ultrasonic machining process erodes holes or cavities in hard or brittle material by means of the cutting action of an abrasive medium. Ultrasonic impact grinding uses an abrasive/water slurry. This slurry is drawn into the machining gap between the vibrating tool and the workpiece. The abrasive particles are propelled or hammered against the workpiece by the transmitted vibrations of the tool. The particles then microscopically erode or "chip away" at the workpiece.

Rotary ultrasonic machining uses an abrasive surfaced tool that is rotated and vibrated simultaneously. The combination of rotating and vibrating action of the tool makes rotary ultrasonic machining ideal for drilling holes and performing ultrasonic profile milling in ceramics and brittle engineered materials that are difficult to machine with traditional processes. Ultrasonic assisted machining adds ultrasonic vibrations to conventional drilling, turning, and milling operations.

Ultrasonic machining is ideal for certain kinds of materials and applications. Brittle materials, particularly ceramics and glass, are typical candidates for ultrasonic machining. Ultrasonic machining is capable of machining complex, highly detailed shapes and can be machined to very close tolerances (± 0.01 mm routinely) with properly designed machines and generators. Complex geometric shapes and 3-D contours can be machined with relative ease in brittle materials. Multiple holes, sometimes hundreds, can be drilled simultaneously into very hard materials with great accuracy.

Ultrasonic machining can be used to form and redress graphite electrodes for electrical discharge machining. It is especially suited to the forming and redressing of intricately shaped and detailed configurations requiring sharp internal corners and excellent surface finishes. Low machining forces permit the manufacture of fragile electrodes too specialized to be machined conventionally. Redressing can be accomplished quickly, typically in 2 to 10 min, often eliminating the need for multiple electrodes. One electrode can be used for roughing, redressed for semifinishing, and redressed again for final finishing. Because of this advantage to ultrasonic machining of electrodes, EDM parameters can be selected for speed and finish, without regard to electrode wear. Minutes spent in ultrasonic electrode redressing can save hours of EDM time while also improving final finish and accuracy.

A variation of ultrasonic machining is ultrasonic polishing. Ultrasonic polishing can uniformly polish and remove a precise surface layer from machined or EDMed workpieces by using an abradable tool tip. This process uses a sonotrode (see Figures 16.20 for machining and 16.21 for polishing) that has a special tip that is highly abradable, such as graphite. By vibrating the abradably tipped tool into the workpiece, the tool tip takes the exact mirror image of the workpiece surface and uniformly removes the surface layer from the workpiece material, improving the surface finish and removing undesirable surface layers. The polishing action occurs as fine abrasive particles in the slurry abrade the workpiece surface, typically removing only slightly more material than the surface roughness depth (e.g., 0.01 mm from a 1-μm Ra EDMed surface). The extent of polishing required is determined by the initial surface roughness and the finish required after polishing. Typical surface improvements range from 3:1 to 5:1. A variety of materials including tool steels, carbides, and ceramics, can be successfully processed with ultrasonic polishing.

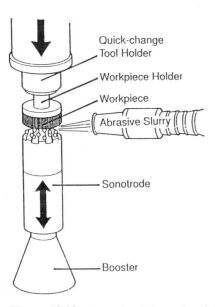

Figure 16.20 Example of ultrasonic machining with a sonotrode and abrasive slurry.

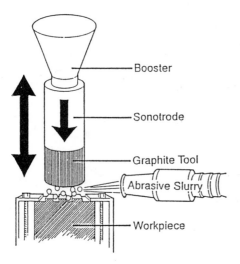

Figure 16.21 Example of ultrasonic polishing with a graphite tool on the sonotrode with an abrasive slurry.

16.3.2 Abrasive Water Jet Machining

Another of the mechanical processes of nontraditional machining methods is abrasive water jet machining (AWJ). Abrasive water jet machining cuts by propelling high-velocity abrasive particles at the workpiece. The propulsion is done by "entraining" abrasive particles into a small-diameter, high-pressure water stream. The process is widely used for cutting, with limited use for drilling, turning, and milling.

In abrasive water jet machining, high-pressure water is fed into a tube and through a small-diameter water orifice, producing a high-velocity water jet. Abrasive particles are fed into the mixing chamber of the nozzle through another tube. The abrasive is "entrained" into the high-velocity water stream as they both pass through a mixing tube to the workpiece (see Figure 16.22). Water pumping pressures of 2000–3000 bar are typically used, with orifice exit velocities of about Mach 3 in a 0.2- to 0.3-mm diameter stream. Introducing about 8% (by weight) of garnet abrasive into this stream accelerates the abrasive particles to perform the cutting. However, it also disperses the stream and wears both the mixing tube and the abrasive particles. The following table shows the speeds of abrasive water jet machining.

	MAXIMUM CUTTING SPEED (mm/min.)			
Thickness (mm)	Aluminum	Carbon Steel	Stainless Steel	Alloy718
3	1250	750	600	500
6	750	500	400	300
12	450	300	250	150
25	200	150	100	40
50	150	75	55	5
100	100	25	25	—

Abrasive water jet machining is a relatively fast and precise mechanical process of nontraditional machining. Its primary use is for sheet cutting operations. It is a method of

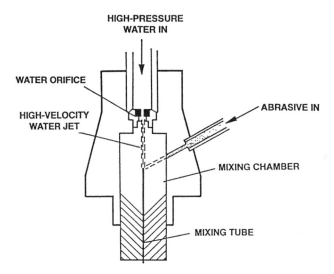

HIGH-PRESSURE
WATER IN

WATER ORIFICE

HIGH-VELOCITY
WATER JET

ABRASIVE IN

MIXING CHAMBER

MIXING TUBE

Figure 16.22 Principle of abrasive water jet machining.

choice for creating prototypes. Abrasive water jet machining has many applications in the aerospace industry. It also has several applications in the automotive and appliance industries.

Another variation of abrasive water jet machining that is currently under development is abrasive suspension jet machining (ASJ). In ASJ, the abrasive and the suspension carrier are propelled together through a diamond nozzle. ASJ's more uniform stream velocity and stream coherency promise improved precision over abrasive water jet machining, and cutting efficiency is high enough to use pressures 75% lower than with AWJ. Abrasive materials can be continuously reused with abrasive suspension jet machining, dramatically reducing waste generation. When it is perfected, abrasive suspension jet machining will broaden the applications for abrasive jet machining.

16.3.3 Abrasive Flow Machining

Abrasive flow machining (AFM) hones surfaces and edges by extruding a pliable abrasive-filled medium through or across the workpiece. In this process, two vertically opposed cylinders extrude abrasive media back and forth through passages formed by the workpiece and the tooling (see Figure 16.23). Abrasive action occurs wherever the media enters and passes through the most restrictive passages. The major elements of the process include the tooling, which confines and directs the media flow to the appropriate areas; the machine, which controls the media extrusion pressure, flow volume and, if desired, the flow rate; and the media, which determines the pattern and aggressiveness of the abrasive action that occurs. By selectively permitting and blocking flow into or out of workpiece passages, tooling can be designed to provide media flow paths through the workpiece. These flow paths restrict flow at the areas where deburring, radiusing, and surface improvements are desired. Frequently, multiple passages or parts are processed simultaneously.

The machine controls the extrusion pressure. The range of useful pressures extends from low pressures, down to 7 bar, to high pressures, in some cases, over 200 bar. Increasing extrusion pressure generally increases process productivity. However, there

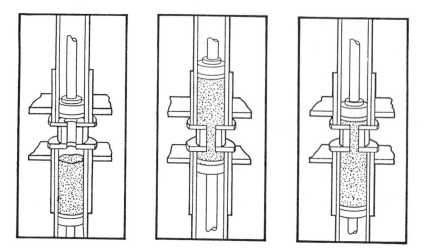

Figure 16.23 Principle of abrasive flow machining.

may be reasons to choose a lower extrusion pressure. If the workpiece is fragile, lower extrusion pressures may be necessary. Lower extrusion pressures might also be beneficial for lower tooling costs, or when there is the desire to machine multiple parts or large-area parts.

The machine also controls the volume of the media flow. The volume of flow per stroke (in cubic inches or centiliters) can be preset, as can the number of cycles. Each cycle is typically two strokes, one up and one down. Following the dimensional change caused by smoothing the rough surface peaks, stock removal is directly proportional to media extrusion volume. This permits precise control of the minute enlargement of flow passages while maintaining shape within a fraction of the surface layer removed.

Abrasive flow machining can controllably remove undesirable surface layers remaining from thermal machining processes such as electrical discharge machining and laser beam machining. Abrasive flow machining can also improve finishes by an order of magnitude. For example, 1-μm Ra EMDed surface is improved to 0.1-μm, or a 2.5-μm Ra finish is improved to 0.25-μm (see Figure 16.24). Required dimensional change to achieve this improvement is slightly greater than the original total roughness per surface. For example, an EDM surface with a 1-μm Ra surface typically has a total roughness (Rt) of about 8-μm. Apparent stock removal (dimensional change) will be about 10-μm (0.01 mm).

Abrasive flow machining is used in a variety of applications. In aircraft turbine

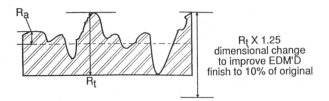

Figure 16.24 Abrasive flow machining can improve surface finishes by an order of magnitude (e.g., micron Ra EDMd surface is improved to 0.1 micron Ra).

engines, it is used to machine compressor blades, blisks, impellers, turbine blades and vanes, disks, casings, and other components, with more applications being discovered frequently. Abrasive flow machining is ideal for dies and molds. It can be used for extrusion dies, cold heading dies, tableting and compacting dies, forging dies, and die casting dies. It can also be used for plastic injection molds and glass molds. Other applications include electronic components, medical components, and high-precision pumps, valves, and tubes. The use of AFM in finishing diesel and automotive components is growing rapidly.

16.4 BURNISHING

V. M. Torbilo

16.4.0 Introduction to Burnishing

Burnishing is one of the methods of finish machining, yielding significant improvements in the service properties of machined parts. It provides efficient machining of machined parts made of most of the engineering metallic materials, including the high-strength alloys of practically any hardness. The method is used in machine production, especially for finishing of precision and critical parts.

Burnishing is a method of finishing and hardening machined parts by plastic deformation of the surface. The plastic deformation of the processed surface is accomplished by the pressure of a sliding tool (burnisher) with a rounded working surface (Figure 16.25). During burnishing, the surface roughness caused by the previous machining is flattened and leveled, and the surface acquires a mirror like finish. The surface layer strength increases and compressive residual stresses are generated. After burnishing, the surface becomes smooth and clear of metallic splinters or abrasive grains that usually occur during abrasive machining. Combination of properties of the burnished surface determines its high working specifications including wear resistance, fatigue strength, etc.

16.4.1 The Burnishing Process

The burnishing method is rather simple in its basic operation. The burnishing tool consists of a working element and a metallic holder. The working element is made of superhard

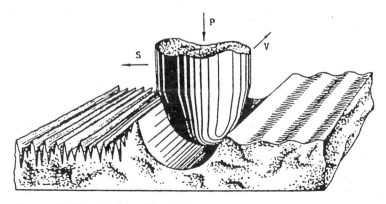

Figure 16.25 Principle of burnishing.

materials and has a rounded form, usually spherical. The tool is held in a tool holder on a regular metal-working machine. Burnishing is most often performed on turning machines (Figure 16.26). Burnishing can also be performed on machines of other types such as milling, boring, or drilling machines. CNC machine tools and machining centers can also be used in the burnishing process.

Burnishing has the following main features: use of superhard materials as a deforming element (most often, diamonds), a small radius on the deforming element (0.5–4.0 mm), and sliding friction between the deforming element and the workpiece being processed. The high hardness of diamonds and other superhard materials provides an opportunity to burnish almost all metals susceptible to plastic deformation in the cold state, whether they are relatively soft or hardened as high as Rockwell C 60–65.

Due to the small rounding radius of burnishers, the contact area between them and a workpiece is small (less than 0.1– 0.2 mm^2). This allows the creation of high pressures exceeding the yield limit of the processed material in the contact area at comparatively small burnishing tool forces (50–260 N). It also reduces the requirements for the ridigity of manufacturing equipment. However, the small contact area in burnishing combined with slow feed rates limits process efficiency. Sliding friction between a burnisher and a workpiece increases the quality of the burnished surface.

Since there is no metal cutting in burnishing, and the burnisher's hardness is much higher than that of the processed material, the burnishing process can be regarded as a process of motion of a rigid rounded deforming element (indentor), forced into the processed surface and deforming it. In the area of tool–workpiece contact, some physical processes occur. The main ones are surface-layer deformation, friction between the tool and the workpiece, heating of the tool and the workpiece, and wear of the former. These physical processes establish the quality of the surface layer, and determine the efficiency and economy of this manufacturing process, considering the machinability of different metals and the associated tool life of the burnisher.

16.4.2 Materials Suitable for Burnishing

Almost all industrial metals and alloys, that are subject to plastic strain in the cold condition can be burnished. The results of burnishing depend to a great extent on a material's type, properties, and machinability. The following criteria characterize machinability at burnishing: (1) surface smoothability; (2) hardenability; and (3) wearability, denoting the material's ability to wear the tool's working surface.

Almost all types of steels (of different chemical composition, structure, and hardness

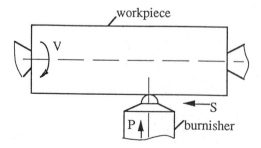

Figure 16.26 Principle of burnishing on a lathe.

up to HRC 62–64), including nickel-based alloys, burnish well. Aluminum and copper alloys have good machinability from the standpoint of burnishing. Gray cast iron is generally known to be a low-plastic material, therefore its burnishing is inefficient. But some types of cast iron, such as high-strength cast iron and alloy cast iron, can be burnished.

Electrodeposited coatings (chrome, nickel, etc.) can be treated by burnishing. There are some specific features in burnishing of coatings. Burnishing of titanium alloys is usually not practical because of the great adhesive interaction between the processed material and the burnisher's working surface.

16.4.3 Tools, Toolholders, and Machines

A burnisher is a metallic holder in which a working element made of superhard material (diamond, synthetic corundum, carbide) is fixed. Several shapes of the burnisher's working part are utilized: the spherical surface, the side surface of a cylinder, the surface of a circular torus, and the cone surface burnisher. The spherical shape is the most versatile. It allows one to burnish outer and inner round surfaces and flat surfaces.

Burnishing can be performed on standard universal and special machine tools—turning, boring, planing, milling, etc.—with normal and high precision. Higher precision is required for burnishing with rigid fixing of a burnisher to the machine. Burnishing on lathes is most common. At this point, particular attention should be paid to the value of the spindle's radial concentricity (not more than 0.01–0.02 mm runout), the rigidity of the support, and vibration resistance. A number of feeds, beginning with about 0.02 mm/rev, should be provided on a machine tool. The machine should also be equipped with a lubricating-cooling device.

Burnishing tools are mounted on metal-cutting machines with the help of holders. Two main types of holders are distinguished by their method of a tool attachment, either rigid or elastic. Holders with a rigid or fixed tool have a very simple design and differ little from mounting a cutting tool on a lathe cross slide, as shown on Fig.16.27. Several designs of holders using an elastic element to maintain pressure have been developed, including hydraulic and electromagnetic. This tool has more flexibility of use, since they tend to hold the required pressure between the surface of the part and the burnishing tool face, even though the part may have some eccentricity, or other minor surface deviations. See Figure 16.28 for a common example using a spring as the force element to hold pressure.

16.4.4 Basic Methods of Burnishing

Burnishing with rigidly fixed tools creates a solid kinematic link between the tool and the workpiece, as in turning, for example. The burnisher is fixed on a machine tool in the same way as a cutter, and its position in relation to the workpiece is determined only by the machine's kinematics and the elasticity of the manufacturing system. With rigid burnishing, the burnisher is indented into the surface for a predetermined depth, which varies from several microns to several hundredths of a millimeter. The depth depends on the plasticity of the material, its surface roughness, and the burnishing tool radius. The advantage of burnishing with a rigidly fixed tool is that it offers an opportunity to increase the precision of dimensions and shape of the workpiece by redistribution of volumes of plastically strained metal. By rigidly fixing the burnisher, however, the value of the burnisher's indentation, and hence the burnishing force, can vary considerably because of the workpiece's beat. Therefore the machine tool (for example, the lathe) must have precision

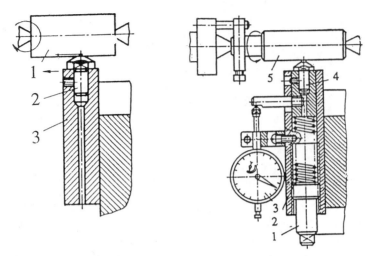

Figure 16.27   Example of rigid burnishing: (1) workpiece, (2) burnisher, (3) rigid holder. [left]
Figure 16.28   Example of elastic burnishing: (1) screw, (2) spring, (3) indicator, (4) burnisher, (5) workpiece. [right]

spindle runout, and very good rigidity in the bearings, frame, ways, crossslide, and work holder. The part must also be rigidly mounted, and the burnishing tool precisely and rigidly mounted. Burnishing with a rigidly fixed tool can be recommended for processing of especially precise machined parts for high-precision machines.

Burnishing with an elastic tool mount is a simpler and more convenient method. With this method of burnishing, the tool is pressed elastically to a workpiece with the help of a spring, or in some other way. The force pressing the burnisher depends on the plasticity of the material, its surface roughness and the burnisher's radius. It is controlled easily and should be kept constant during burnishing. In this process there is no rigid kinematic link between the workpiece and the tool, and the position of the latter in relation to the workpiece is determined only by the surface itself. With an elastically fixed tool, the errors of the shape are copied and are not corrected. Only smoothing of the surface roughness and surface hardening take place. Simplicity of setting and comparatively moderate requirements for precision and rigidity of machine tools and workpieces are the advantages of this method of burnishing.

Vibroburnishing

In vibroburnishing the burnisher is imparted an oscillating movement in the direction of feed in addition to the normal direction and movement of feed. Oscillations can be imparted mechanically, by imparting oscillating movement to the burnisher, or by generating ultrasonic oscillations. With vibroburnishing, a net of sinusoidal grooves forms on the piece surface. The microrelief of the burnished surface can be regulated within wide limits by changing the conditions of burnishing (the speed of the workpiece rotation, feed, amplitude and frequency of the tool's oscillations, burnishing force, and the burnisher's radius). Vibroburnishing allows one to create a regulated surface microrelief and to raise the wear resistance of machine parts by improving the lubrication of contacting surfaces and preventing their seizure.

16.4.5 Burnishing Tool Life and Wear

The wear of burnishers is influenced by a combination of factors. Brittle damage (chipping) is the main type of damage to the working surface of the tools in burnishing of ground, hardened steels. Hard particles of carbides in steel, and also the abrasive particles left in the surface after grinding, push the burnisher's surface and chip the particles. Besides this main process of damaging the burnisher surface, friction-fatigue damage also takes place. The fatigue damage is worsened by the presence of defects in the structure. Microcracks, pores, cavities, etc., play the role of stress concentrators and generate fatigue damage. The accumulation of fatigue damage leads to the cracking and chipping of the surface, which amplifies the abrasive wear of the latter. Friction-fatigue wear, not saturation by the abrasive particles, seems to be the main type of wear in the burnishing of soft materials.

Thermal damage due to graphitization of the diamond surface is important only at high temperatures (>500–600°C). Under the usual burnishing conditions, when the contact temperature is 200–400°C, diamond thermal damage seems to be scarcely noticeable. Table 16.1 shows the burnishing tool life in processing materials of different hardnesses.

16.4.6 Burnishing of Various Surface Forms

Outer cylindrical surfaces are most frequently burnished. Burnishing is used for finishing of cylindrical surfaces of shafts, bars, pistons, piston and crank pins, bearing rings, and many other pieces. Burnishing of outer cylindrical surfaces is usually performed on turning machines.

High burnishing speed and high efficiency in burnishing cylindrical continuous pieces such as shafts, piston pins, etc., can be attained by centerless burnishing. The rigidity of the manufacturing system in centerless burnishing is much higher than with ordinary burnishing on-centers, and allows one to increase the burnishing speed significantly. Centerless burnishing is performed on special automatic machines.

Holes

A technology for burnishing holes 20–200 mm in diameter and up to 500 mm deep has been developed. The burnishing of the holes in pieces made of high-strength and hardened steels, which are difficult to process by other finishing methods, is especially effective.

Table 16.1 Tool Life of a Diamond Burnisher

Processed materials	Burnishing length to a blunting L, in km
Aluminium	up to 1000
Bronzes	150–200
Non-hardened steels	150–250
Hardened steels	50–120
Cast iron	50–80
Hard metal coatings	10–20
Cemented carbides	0.2–2.0

Burnishing of the holes can be performed on turning machines, boring and drilling machines, and in machining centers.

Flat Surfaces

One can perform burnishing of flat face surfaces of round workpieces and flat linear surfaces. In the first case, the workpiece rotates (processing on turning machines). In the second case a tool (or a workpiece) has forward motion (processing on a planer), or the tool has a rotational motion (on milling machines).

Contoured Surfaces

Burnishing of contoured and conic round surfaces is most frequently used. The diamond burnisher can also roll on the workpiece surface grooves, the radius of which is determined by the diamond's working part. Roller paths of bearing rings can also be processed by burnishing.

Gears

The high quality of teeth of critical gears can be provided by gear burnishing. It can be used to improve the quality and service life of gears.

Threads

Burnishing is sometimes applied for finishing of trapezoidal threads. The lateral sides of the profile of a trapezoidal thread are burnished by tools with a cylindrical working surface. Burnishing is applied as well for finishing the threads of semicircular profiles that are applied in lead-screw nuts, which are widely used in machine tools and actuators for the aircraft industry.

16.4.7 Surface Finishes (Roughness)

A burnished surface differs from surfaces processed by other finishing methods by its structure. After burnishing, an even and solid surface is formed that is distinguished by a mirror luster. The roughness of a burnished surface comprises a combination of irregularities that were formed by the process of burnishing, whose spacing equals the feed, with crushed initial irregularities. Surfaces of a similar roughness height (Figure 16.29) that were received by different finishing methods differ in the shape of the irregularities and in their service properties.

A ground surface has irregularities in the shape of sharpened projections and peaks, the rounding radius of them is equal to $R = 0.07 - 0.10$ mm. Polished and superfinished surfaces have a more blunted shape of irregularities with $R = 0.2 - 0.4$ mm. For a burnished surface a smoothened, rounded shape of irregularities is typical with $R = 1.0 - 3.5$ mm.

The bearing area curve gives an indication of the bearing capability of the surface. It characterizes the filling of the profile irregularities along the height and the bearing capability of the surface at different levels of profile height. In Figure 16.30 the bearing area curves for surfaces processed by different finishing methods are presented. When the surface finish height is similar, surfaces processed by different methods have different bearing capabilities. The burnished surface has the highest, and the ground one the lowest.

A quantitative description of the bearing capability is the bearing length ratio t_p where p is the level of the profile section. The values of the parameters t_{10}, t_{20}, t_{30}, that describe

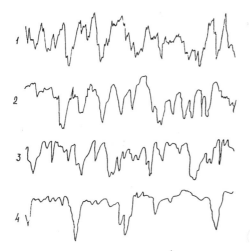

Figure 16.29 Profiles of Ra=0.1μm, processed by: (1) grinding, (2) polishing, (3) superfinishing, and (4) burnishing.

the properties of the upper part of the surface roughness layer are the most important ones from the point of view of the service properties of the surface. A comparison of the values of the parameter t_p for some finished surfaces (see Table 16.2) shows the considerable advantage of the burnished one.

The main factors that affect the surface roughness of a burnished surface are the properties of the burnisher (the material, the shape and condition of the work surface, and its radius), the properties of the workpiece (hardness, surface roughness, and stability), the processing regimes (force, feed, velocity), and the kind of applied lubricant. These factors predetermine the nature and intensity of the contact processes (deformation, friction, heating) that form the surface.

The hardened steels are burnished effectively when the initial surface roughness does not exceed Ra = 1.5–2.0 μm. When the initial surface roughness is Ra = 1.5–2.0 μm, a

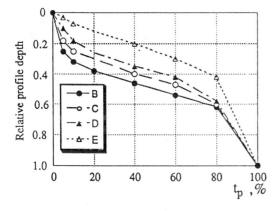

Figure 16.30 The bearing curves of surfaces Ra = 0.1 μm: (B) grinding, (C) polishing, (D) superfinishing, (E) burnishing.

Table 16.2 Values of Bearing Length Ratio t_p for Some Finishing Methods

Finishing method	t 10, %	t 30, %
Burnishing	5–10	30–50
Superfinish	5–8	25–45
Polishing	4–8	20–42
Lapping	5–6	22–27
Grinding	2–3	12–23

stable reduction of the surface roughness occurs and a surface roughness of Ra = 0.03–0.30 μm is achieved, respectively. When soft materials such as copper and aluminum alloys are burnished, surfaces with an initial surface roughness of Ra = 0.5–5.0 μm can be burnished effectively. In this process a surface roughness of Ra = 0.04–0.30 μm can be achieved.

Accuracy

The accuracy of the burnishing depends on the processing method. Elastic fixation allows the burnisher to copy the errors of the blank shape. Because the variation in the burnishing force caused by the beating of the processed blank is small, the reduction and the change of its size occurs uniformly under elastic fixation of the tool. The size of the blank changes due to crushing of initial irregularities and redistribution of the metal in the surface layer.

After the burnishing, the outer diameter of the processed workpieces decreases and the diameter of the holes increases. The change in the dimensions depends mostly on the initial roughness of the processed surface and is usually found by the equation

$$\Delta d = (1.2 - 1.4) \text{ Rz in.}$$

where Δd is the change in the workpiece diameter, and Rz in. is the height of the irregularities before the burnishing.

The necessity of assigning an allowance for the burnishing operation depends on the tolerance value for the processed surface. As a rule, the tolerance for the final dimension is significantly higher than the value of dimension change during burnishing. Therefore there is usually no need to assign a special allowance. When processing with micrometer tolerances, the allowance for dimension change that can be calculated by above-mentioned equation has to be predetermined.

When burnishing with a rigid tool, forced correction of shape errors occurs in the transverse as well as the longitudinal sections of the workpiece. All the error indexes are reduced two to four times. The value of the possible shape correction is within the limits of the initial irregularities crushing (micrometers and parts of micrometers), but rigid burnishing imposes increased demands on the accuracy and rigidity of the machine tool and the accuracy of the tool and workpiece setting.

16.4.8 Changes in the Surface Layer

During burnishing, the surface layer of the metal undergos momentary deformation and heating; in other words, it is subjected to a kind of thermomechanical processing. As a result, structural and phase changes may occur, wnich in the final analysis affect the strength and the service properties of the surface layer. The temperature that develops during burnishing is usually lower than the critical ones that cause structural and phase transformations, and therefore deformation plays the leading role in changing the properties of the surface layer.

After burnishing, the shattering of the grains and formation of a disperse structure takes place in the thin surface layer, as well as stretching of grains located near the surface in the direction of the deformation and creation of an oriented structure or texture. Thus, in the process of burnishing hardened low-tempered steels of a martensite structure, significant property changes of the metal's thin surface layer take place. Under the effect of plastic deformation, the dislocation density increases significantly. Creation of disperse carbides, crushing of grains and blocks (coherent scattering areas), their deorientation under the effect of grid distortions, and microstresses that occur during dissociation of the residual austenite create additional barriers for dislocation movement, increase the resistance of surface layer to the plastic deformation, and strain-harden it. The thickness of the deformed layer is usually 0.02–0.04 mm.

When steels are martempered, normalized, or annealed, phase transformations do not occur in the surface layer after burnishing. This is explained by the high stability of the ferrite-pearlite and sorbite structure. Brushing and stretching of surface layer grains is evident. The deformation rate of the grains is maximal at the surface and decreases with the depth. The thickness of the deformed layer is usually 0.2–0.4 mm.

Strain Hardening of the Surface Layer

The structure-phase transformations that occur in the surface layer during burnishing cause it to strengthen (strain harden); the hardness, strength, and yield strength increase, but the plasticity decreases. Usually the strain hardening of the surface is estimated by the increment of its hardness. The main indexes of strain hardening are; first, the strain hardening rate,

$$\delta = (H_b - H_{in})/HD_{in}$$

where H_b, H_{in} are the hardness of the burnished and initial surfaces, respectively, and second, the thickness of the strengthened (strain-hardened) layer, h_s.

The main factors that affect the strengthening of a burnished surface are the initial properties of the processed workpiece (the material, its structural condition, the hardness), the tool properties (the radius of the burnisher, its wear rate), burnishing conditions (force, feed, speed), and the type of lubricant.

The structural condition of the material affects its strain hardening strongly. For example, Figure 16.31 shows the variation in microhardness with the thickness of the surface layer after burnishing for the carbon steel AISI 1045 in different structural conditions. Maximal strain hardening rate occurs during the burnishing of steels with ferrite, austenite, and martensite structures, and minimal strain hardening rate occurs when the steels have a sorbite and troostite structure. Table 16.3 shows the strain hardening rate of steels and alloys in different structural conditions as achieved during burnishing.

Pressure is the second main factor that affects strengthening. The value of the pressure during burnishing determines the intensity of plastic deformation and affects the surface layer strain hardening properties to the greatest extent. As can be seen from the graphs in Figure 16.31, the surface hardness increases significantly in relation to the initial value, and it increases when the force of burnishing increases.

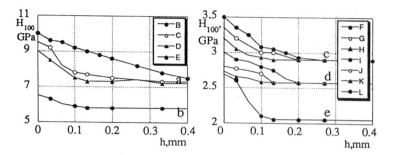

Figure 16.31 Surface layer microhardness H_{100} versus steel structural condition and burnishing force. Structural condition: (a) martensite, (b) troostite, (c) sorbite, (d) ferrite + pearlite, (e) ferrite. The burnishing force P=100 N(D, H, K); P=200 N(C, G, J); P=300 N(B, E, F, I, L). Workpiece material is carbon steel AISI 1045.

When the pressure rises, the strain-hardened layer thickness increases almost linearly. As this takes place, strain hardening of deeper and deeper metal layers occurs. In the general case, the thickness of the strain-hardened layer is 0.1–0.3 mm for hardened steels and 0.2– 0.5 mm for nonhardened steels and nonferrous alloys. The lower limit of the listed figures is for low pressures and the upper one is for higher ones.

Residual Stresses

During burnishing, the surface layer of the processed workpiece is subjected to intensive plastic deformation, but its heating is not high. When hardened steels of a martensite structure are burnished, the phase transformations that take place are reduced to almost complete residual austenite dissociation with its conversion to martensite. These effects cause the development of compressive residual stresses in the surface layer after burnishing. Figure 16.32 shows typical diagrams of residual stresses that are formed after burnishing.

Significant compressive stresses, which are close to the elasticity strength of the material, are formed in the surface layer. The depth of their occurrence is 0.15–0.35 mm, depending on the material and the burnishing conditions. Maximal tangential stresses occur not on the surface, but at a certain depth. The axial stresses are maximal at the

Table 16.3 Strain Hardening Rates of Different Materials

Material	Structure	Relative pressure P_o	Strain hardening rate δ
Steel	Austenite	0.95–1.00	0.50–0.60
Steel	Ferrite + pearlite	0.95–1.00	0.35–0.45
Steel	Pearlite	0.85–0.95	0.25–0.35
Steel	Sorbite	0.80–0.90	0.20–0.30
Steel	Troostite	0.80–0.90	0.20–0.30
Steel	Martensite	0.80–0.90	0.35–0.45
Aluminum alloys		0.90–1.00	0.15–0.20
Copper alloys		0.90–1.00	0.15–0.20

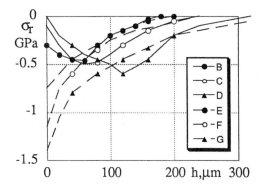

Figure 16.32 Tangential (B, C, D) and axial (E, F, G) residual stresses in surface layer. $P=50$ N(B, E); 100 N(C, F); 200 N(D, G). Alloy steel AISI 5140, HRC 52.

surface and gradually decrease as one moves away from it. At a depth greater than 0.1 mm, the tangential and axial residual stresses become practically identical.

16.4.9 Surface Improvements Due to Burnishing

Wear Resistance

The burnished surface is characterized by a combination of the following properties: low roughness (RA = 0.04–0.32 μm), large bearing capacity (t_p up to 60%), hardening with great value (20–50%) and depth (0.2–0.4 μm), residual compressive stresses, and absence of abrasive particles charged into the surface. Such a surface is likely to have good working properties, particularly in friction conditions.

Figure 16.33 gives some results of comparative wear tests on ground, polished, and burnished sample rollers (Ra = 0.10–0.14 μm). The burnished rollers were the least worn. Full-scale field tests of certain wares with burnished machine parts (in particular, truck compressors with burnished piston pins and crankshafts, ball bearings of a turbodrill) demonstrate that burnishing reduces wear significantly (an average of 35–45%). Especially effective is the burnishing of sealing surfaces.

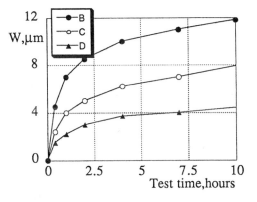

Figure 16.33 Wear of ground (B), polished (C), and burnished (D) surfaces. Workpiece material carbon steel AISI 1045, HRC 60.

Seizure Resistance

Seizure of friction metallic surfaces is an unfavorable phenomenon, deteriorating the work of a friction pair and resulting in machine parts damage. The seizure resistance of friction surfaces depends greatly on the method of finishing of a harder piece. The finishing methods create various microreliefs of the friction surface, differing in the asperities' shape, bearing capacity, etc. Investigation results show that burnishing provides a smooth enough surface and significantly improves the seizure resistance of a friction surface.

Fatigue Strength

Fatigue strength at cyclic loadings depends mostly on the condition of the surface layer, as the surface is usually loaded to a greater extent and fatigue failure most often begins from the surface. Fatigue strength is known to depend on the surface roughness, since the cavities between the projections, especially those with a sharp bottom, are the stress concentrators. Burnishing reduces the surface roughness, makes it smoother, and flattens some surface defects—marks, scratches, etc. Fatigue strength also depends on the strength and structural metal strength, and eliminates or reduces the action of structural stress concentrators—burns, microcracks, etc. Finally, fatigue strength depends very strongly on residual stresses. It is well known that the compressive stresses are favorable from the point of view of the material fatigue strength. As was said above, significant residual compressive stresses are created by burnishing in the surface layer. Comparative tests of a number of carbon and alloyed steels showed that the fatigue limit increased 20–40% as a result of burnishing.

Corrosion Resistance

Many machined parts in their working environment are subjected to the corrosive action of air or liquid media. Numerous tests have shown that burnishing, by providing good surface smoothness and compressive residual stresses, improves machined parts' corrosion resistance.

Contact Ridigity

Contact rigidity is the ability of the surface layers of machined parts, in contact with others, to resist strain under the action of loading forces. Contact displacements under the action of work loading greatly influence the precision working of machines and instruments, the precision of the machine's processing and assembly, etc. Contact rigidity is determined by the properties of the material and the surface quality of the conjugate parts. The surface hardness and roughness are the most important parameters for the contact rigidity. The higher the surface hardness and bearing capacity of roughness, the smaller will be the contact strain and the higher the contact rigidity. Comparative tests of contact rigidity showed that burnishing reduces the contact strain by half. Thus burnishing, by improving the quality of the surface layer, greatly improves contact rigidity.

16.5 CONCLUSIONS

Nontraditional machining methods meet the needs of emerging technologies by providing many new choices for design engineers and manufacturers. They can be applied in a variety of ways, just as traditional machining processes can. These methods offer advantages in dealing with certain kinds of materials and in the performance of intricate milling, machining, and drilling operations. Nontraditional machining methods offer an arsenal of new tools that can, in turn, offer new manufacturing strategies to harness the benefits promised by advanced engineered materials. These nontraditional machining methods can also be important in generating and machining the complex shapes and features of tomorrow's products. Design, material selection, and manufacturing options that are constrained by traditional machining limitations can be overcome by the special capabilities of nontraditional machining methods.

16.6 BIBLIOGRAPHY

Farrar, F., How diamond burnishing can help the engineer looking for improved surface finishes, *Ind. Diamond Rev.* 28:552 (1968).

Hull, E. H., Diamond burnishing, *Machinery* 5:92 (1962).

Khvorostukhin, L. A., *Increase of Machine Parts Bearing Capacity by Surface Hardening*, Mashino-stroenie, Moscow, 1988.

Torbilo, V. M., *Diamond Burnishing*, Mashinostroenie, Moscow, 1972.

Schneider, Y. G., *Service Properties of Machine Parts with Regular Microrelief*, Mashinostroenie, Leningrad, 1982.

17

Nonmetals: Plastics

Bruce Wendle *Advanced Input Devices, Coeur d'Alene, Idaho*

17.0 INTRODUCTION TO PLASTICS

This chapter on nonmetalics provides some guidelines on the right way to develop plastic applications. It will not make you a plastics engineer, but it will give you a place to start. Additional references are listed at the end of the chapter.

One of the biggest mistakes is to try to replace metal applications with plastics on a one-for-one basis. Plastic cannot be substituted on the drawings for a metal part in the hope of reducing cost, weight, or obtaining the same structural properties. The metal part may be overdesigned and loaded with labor-intensive functions which are not needed in a plastic part. A good metal design is normally made up of several parts fastened together to accomplish a specific function. A good plastic design, on the other hand, can usually be combined into fewer parts and often is able to accomplish other functions at the same time.

A well-designed plastics application is usually lighter in weight, more functional, sometimes transparent, and often lower in cost than a metallic application designed for the same function. In addition, plastics can be modified in any number of ways to provide a higher degree of functionality. Glass fibers can be added to provide additional stiffness and reduce shrinkage. Pigments can be added to the part to achieve almost any color desired. Molybdenum sulfide can be added to provide natural lubricity, and any number of fillers can be added to reduce costs.

The variety of ways that plastics can be processed or formed also increases their usefulness. Small to medium-sized parts can be made by injection molding, large sheets can be formed by vacuum forming, or medium to large parts can be produced by hand layup or spray techniques to produce parts such as boat hulls. Other processes such as blow molding and rotational molding provide ways to make hollow parts. Profiles can be produced by extrusion. Nearly any shape or size of part can be manufactured using one of these techniques (see Figures 17.1 and 17.2 for some examples).

In this chapter we will discuss the various properties of plastics that make them unique, touch on the technology of tooling, and cover some of the design rules needed for

Figure 17.1 The first carbon matrix composite vertical stabilizer for a Boeing 777 jet liner is hoisted into position. (Courtesy of Boeing Commercial Airplane Company.)

a successful plastic project. In addition, an effort will be made to provide additional sources of information to assist in the design process.

Some of the problems with plastics will be discussed as well. Such areas as volume sensitivity, high tooling costs, and property deficiencies will be covered. The important thing to realize is that plastics, when used correctly, can provide you a functional, inexpensive part that will get the job done.

17.1 DEFINITIONS

The world of plastics has its own language. As with any technical field, it has a unique jargon. We will define the important terms, and describe why each is important. For a more complete list we recommend *Whittington's Dictionary of Plastics* (Technomic Publishing Co.).

To understand any list of plastic words and definitions, you must first know that plastic materials come in two distinct forms, thermoplastic and thermosetting materials (see Figure 17.3). A *thermoplastic* is any resin or plastic compound which in the solid state is capable of repeatedly being softened and reformed by an increase in temperature. A *thermosetting* material, on the other hand, is a resin or compound which in its final state is substantially infusible and insoluble. Thermosetting resins are often liquids at some state in their manufacture or processing, and are cured by heat, catalysis, or other chemical

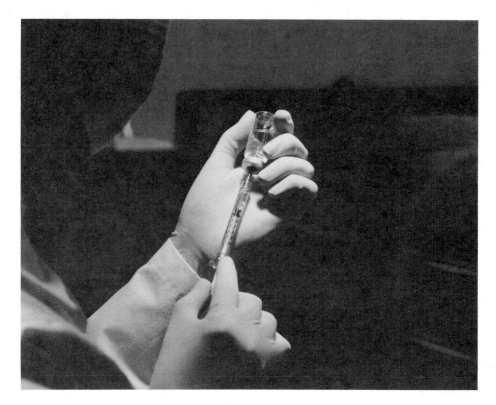

Figure 17.2 High-pressure syringe made from G.E. Lexan GR resin. (Courtesy of General Electric Company.)

means. After being fully cured, thermosets cannot be resoftened by heat. Some plastics which are normally thermoplastic can be made thermosetting by means of cross-linking. Some common terminology defining the processing of plastic follows.

Injection molding: A method of molding objects from granular or powdered plastics, usually thermoplastic, in which the material is fed from a hopper to a heated chamber where it is softened. A ram or screw then forces the material into a mold. Pressure is maintained until the mass has hardened sufficiently for removal from the cooled mold. (See Figure 17.4.)

Extrusion: The process of forming continuous shapes by forcing a molten plastic material through a die.

Thermoforming: A method of forming plastic sheet or film into a three-dimensional shape, in which the plastic sheet is clamped in a frame suspended above a mold, heated until it becomes softened, and then drawn down into contact with the mold by means of a differential pressure to make the sheet conform to the shape of a mold or die positioned below the frame.

Vacuum forming: A form of thermoforming in which the differential pressure used is a vacuum applied through holes in the mold or die.

Blow molding: The process of forming hollow articles by expanding a hot plastic element against the internal surface of a mold. In its most common form, the

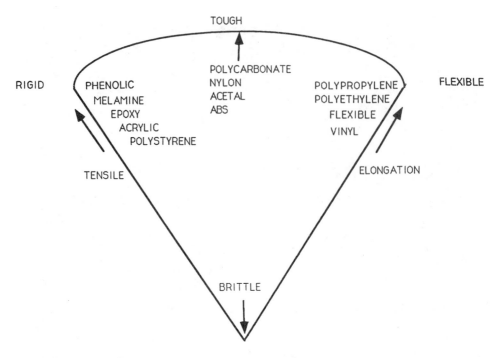

Figure 17.3 Physical characteristics of plastic materials. (From R. D. Beck, *Plastic Product Design*, Van Nostrand Reinhold, New York, 1980.)

Figure 17.4 Large, 1500-ton clamp injection molding machine capable of shooting 50 lb of plastic. (Courtesy of Cincinnati Milacron.)

plastic element used is in the form of an extruded tube (called a parison). Many variations of the process exist, including using two sheets of cellulose nitrate bonded together instead of the parison. Sometimes the parison is a preform made by injection molding.

Rotational molding: The process whereby a finely divided, sinterable, powdered plastic is sintered, then fused against the walls of a mold. The process forms hollow articles by rotating the mold containing the powdered material about one or more axes at a relatively slow speed until the charge is distributed on the inner mold walls by gravitational forces and hardened by heating and then cooling the mold.

Layup molding: A method of forming reinforced plastic articles by placing a web of the reinforcement fibers, which may or may not be preimpregnated with a resin, in a mold or over a form. When a dry fiber is used, fluid resin is applied to impregnate and/or coat the reinforcement. This is followed by heating and curing the resin. When little or no pressure is used in the curing process, the process is sometimes called contact pressure molding. When pressure is applied during curing, the process is often named after the means of applying pressure, such as bag molding or autoclave molding. A related process is called spray-up, in which a "chop gun" is used to apply the reinforcement and resin at the same time. The techniques above are used with thermoset resins.

Tooling: In the case of plastics, the hollow form into which a plastic material is placed and which imparts the final shape to the finished article.

Mold: A form of tooling, usually made from metal or composite. It normally consists of a concave, or female, section shaped to the outside of the part (known as the cavity), and a convex, or male, section shaped to the inside of the part (known as the core). For some molding processes, only the cavity or the core is utilized. (See Figure 17.5.)

Sprue: In injection or transfer molding, the main channel that connects the mold's filling orifice with the runners leading to each cavity gate. The term is also used for the piece of plastic material formed in this channel.

Runner: In an injection or transfer mold, the feed channel that connects the sprue with the cavity gate. This term is also used for the plastic material formed in this channel.

Cavity: The female portion of a mold. This is often the side into which the plastic material is injected. In injection molding, this is usually the movable side of the mold, which opens after the part has solidified. (See Figure 17.6.)

Core: Usually the male side of a mold. This is often the side of the tool that is fixed to the injection molding machine, and from which parts are ejected. However, cores can also be used to create undercuts, and are often moved mechanically or hydraulically in planes different from the normal open and close directions of the mold.

Platen: A steel plate used to transmit pressure to a mold assembly in a press. In some cases heat is often transferred through this plate as well.

Nozzle: In injection or transfer molding, the orifice-containing plug at the end of the injection cylinder or transfer chamber which contacts the mold sprue bushing and conducts molten resin into the mold. The nozzle is shaped to form a seal under pressure against the sprue bushing. Its orifice is tapered to maintain the desired flow of resin, and sometimes contains a check valve to prevent backflow, or an on–off valve to interrupt the flow at any desired point in the molding cycle.

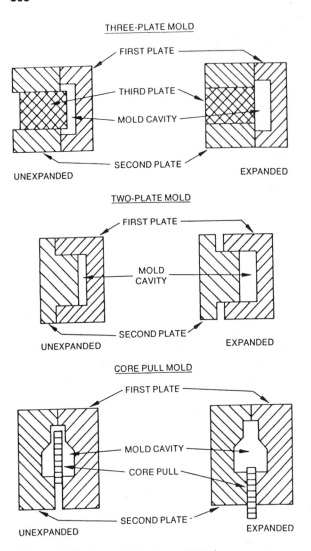

Figure 17.5 Typical injection mold design concepts.

Die: A steel block containing an orifice through which plastic is extruded, shaping the extrudate to the desired profile; or the recessed block into which plastic material is injected or pressed, shaping the material to the desired form. The term *cavity* is more often used.

Polymer: The product of a chemical reaction (polymerization) in which the molecules of a simple substance (monomer) are linked together to form large molecules. The terms polymer, resin, and plastic are often used synonymously, although the latter also refers to compounds containing additives.

These terms and many others make up the vocabulary of the plastics industry. As you gain experience in the field, you will find others and will no doubt make up some of your own.

Figure 17.6 A tool maker works on the cavity of a large mold. (Photo courtesy of SPL.)

17.2 DESIGNING WITH PLASTICS

Regardless of whether you are working with thermosets or thermoplastics, many of the design rules are the same. Thermoplastics are more forgiving materials, and do not cross-link when exposed to heat or pressure. Cross-linking generally is built into thermosetting polymers by the addition of branching or added cross-structure to the long molecular chains that make up the polymer's backbone, much like adding rungs to a ladder.

Thermoplastic materials may not be as stiff or have as high a modulus as thermosets. Thermoplastics are generally more pliable and have higher impact strengths. Thermosets often vary depending on which fiber (glass, carbon, etc.) is used with them. (See Figure 17.7.)

17.2.1 Design Ground Rules

1. The cardinal rule for all plastics design is, wherever possible, to keep the wall section constant. Unless you are designing a foamed product, this is a "must" for good plastics design. Use only the thickness necessary to get the job done. Too much material generates higher material costs and increases cycle time. It also increases the chance of warping problems. In plastics, if a cross section does not cool uniformly, chances are that the thicker, slower cooling area will warp and throw the part out of tolerance. Sometimes coring ribs or pins are called for just to core out these thick sections. (See Figure 17.8 for minimum thicknesses.)

2. Another good rule is to avoid undercuts where possible. This is not always possible. The plastics tooling industry has come up with many clever ways to

PLASTIC	SPECIFIC GRAVITY	MOLD SHRINKAGE IN/IN	TENSILE STRENGTH 10 PSI	FLEXURAL MODULUS 10 PSI	DEFLECTION TEMPERATURE 164 PSI(DEG F)	THERMAL EXPANSION 10 IN/IN (F)
ASTM	D792	D995	D638	D790	D648	D696
ABS	1.05	0.006	6	0.32	195	5.3
W /30% GLASS	1.28	0.001	14.5	1.1	220	1.6
ACETAL	1.42	0.02	8.8	0.4	230	4.5
W /30 % GLASS	1.63	0.003	19.5	1.4	325	2.2
NYLON 6/6	1.14	0.018	11.6	420	170	4.5
W/ 30 % GLASS	1.37	0.004	26	170	490	1.8
PPO	1.06	0.005	9.5	490	265	3.3
W/30 % GLASS	1.27	0.002	21	1.3	310	1.4
POLYCARBONATE	1.2	0.006	9	0.33	265	3.7
W/30 % GLASS	1.43	0.001	18.5	1.2	300	1.3
POLYESTER TP	1.31	0.2	8.5	0.34	130	5.3
W/ 30 % GLASS	1.52	0.003	19.5	1.4	430	1.2
POLYETHYLENE	0.95	0.02	2.6	0.2	120	6
W/ 30 % GLASS	1.17	0.003	10	0.9	260	2.7
POLYPROPYLENE	0.91	0.018	4.9	0.18	135	4
W /30 % GLASS	1.13	0.004	9.8	0.8	295	2
POLYSTYRENE	1.07	0.004	7	0.45	180	3.6
W /30 % GLASS	1.28	0.001	13.5	1.3	215	1.9
POLYSULFONE	1.24	0.007	10	0.4	340	3.1
W /30 % GLASS	1.45	0.003	18	1.2	365	1.4

Figure 17.7 Table illustrating the change in properties of a thermoplastic material when glass fiber is added. (From R. D. Beck, *Plastic Product Design*, Van Nostrand Reinhold, New York, 1980.)

create undercuts, but keep in mind that they all cost money. The best way to get around the problem is just to design the undercuts out, unless you must have them for function.

3. Add generous fillets (radii) to all inside corners. All plastics materials are notch sensitive in one way or another, so in order to avoid areas of high stress, keep the sharp corners to minimum. Generous fillets also help the material to flow more evenly around sharp corners in the tool. Figures 17.9 and 17.10 show design standards for high-density polyethylene and polycarbonate.

4. Allow for sufficient draft on all vertical walls. Sometimes, because of fit or function, this is difficult to do. However, a part that locks in the mold will not help either you or your molder. Most molders would like to have as much draft as possible. The designer, on the other hand, would like to design all parts with zero draft. Often a compromise is in order. Anywhere from ½° to 2° common. Be sure that if the part is going to be textured, there is enough draft on the vertical walls to remove the part. A degree of draft is required for every 0.001 in. of texture depth.

5. Metal inserts can be molded into thermoplastic parts as shown in Figure 17.11.

6. Another good rule is to let the tool maker and molder know where the critical surface areas on the parts are located. If some areas are going to be seen by the consumer, indicate these areas on the drawings. If a gate vestage is not acceptable, indicate this to the molder. Textures and special treatment areas should be so designated. If an area is going to require close tolerances, be specific. There are many ways that a molder can give specific treatment to an area, but these areas must be known before the tool is built.

SUGGESTED WALL THICKNESSES FOR PLASTIC MOLDING MATERIAL		
THERMOPLASTIC MATRIALS	MINIMUM/IN	MAXIMUM /IN
ACETAL	0.015	0.125
ABS	0.03	0.125
ACRYLIC	0.025	0.25
CELLULOSICS	0.025	0.187
FEP Fluoroplastic	0.01	0.5
NYLON	0.015	0.125
POLYCARBONATE	0.04	0.375
POLYESTER TP	0.025	0.5
POLYETHYLENE(LD)	0.02	0.25
POLYETHYLENE(HD)	0.035	0.25
EVA	0.02	0.125
POLYPROPYLENE	0.025	0.3
POLYSULFONE	0.04	0.375
MOD PPO	0.03	0.375
POLYSTYRENE	0.03	0.25
SAN	0.03	0.25
PFV(RIGID)	0.04	0.375
POLYURETHANE	0.025	1.5
THERMOSETTING MATERIALS		
ALKYD-GLASS FILLED	0.04	0.5
ALKYD-MINERAL FILLED	0.04	0.375
DIALLYL PHTHALATE	0.04	0.375
EPOXY GLASS	0.03	1
MELAMINE-CELLULOSE	0.035	0.187
UREA-CELLULOSE	0.035	0.187
PHENOLIC-GENERAL PUR.	0.05	1
PHENOLIC-FLOCK FILLED	0.05	1
PHENOLIC-GLASS	0.03	0.75
PHENOLIC FABRIC	0.062	0.375
PHENOLIC-MINERAL	0.125	1
SILICONE GLASS	0.05	0.25
POLYESTER PREMIX	0.04	1

Figure 17.8 Suggested wall thicknesses for plastic moldings. (From R. D. Beck, *Plastic Product Design*, Van Nostrand Reinhold, New York, 1980.)

17.2.2 Design Checklist

A list of design rules follows:

1. Wall sections should be constant and of a minimum thickness.
2. Avoid undercuts wherever possible.
3. Add generous fillets (radii) to all inside corners.
4. Avoid sharp transitions in wall design.
5. Allow for draft wherever possible.
6. Indicate desired surface, i.e., textured, polished, and ejection pin location.
7. Consider the interface of all joining walls. Thickness at joining rib wall should be 75% of the wall it joins.
8. Core out all thick sections to minimize cycle time and material.
9. Design for the particular process selected.

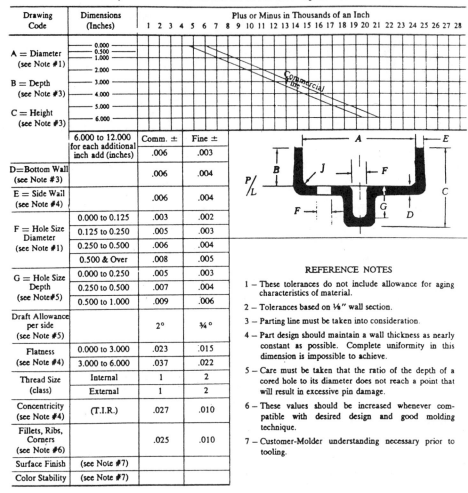

STANDARDS AND PRACTICES OF PLASTICS MOLDERS

Engineering and Technical Standards
HIGH DENSITY POLYETHYLENE

NOTE: The Commercial values shown below represent common production tolerances at the most economical level. The Fine values represent closer tolerances that can be held but at a greater cost.

Drawing Code	Dimensions (Inches)		Plus or Minus in Thousands of an Inch
			1 2 3 4 5 6 7 8 9 10 11 12 13 14 15 16 17 18 19 20 21 22 23 24 25 26 27 28
A = Diameter (see Note #1) B = Depth (see Note #3) C = Height (see Note #3)	0.000 0.500 1.000 2.000 3.000 4.000 5.000 6.000		Commercial line

Drawing Code	Dimensions (Inches)	Comm. ±	Fine ±
	6.000 to 12.000 for each additional inch add (inches)	.006	.003
D=Bottom Wall (see Note #3)		.006	.004
E = Side Wall (see Note #4)		.006	.004
F = Hole Size Diameter (see Note #1)	0.000 to 0.125	.003	.002
	0.125 to 0.250	.005	.003
	0.250 to 0.500	.006	.004
	0.500 & Over	.008	.005
G = Hole Size Depth (see Note#5)	0.000 to 0.250	.005	.003
	0.250 to 0.500	.007	.004
	0.500 to 1.000	.009	.006
Draft Allowance per side (see Note #5)		2°	¾°
Flatness (see Note #4)	0.000 to 3.000	.023	.015
	3.000 to 6.000	.037	.022
Thread Size (class)	Internal	1	2
	External	1	2
Concentricity (see Note #4)	(T.I.R.)	.027	.010
Fillets, Ribs, Corners (see Note #6)		.025	.010
Surface Finish	(see Note #7)		
Color Stability	(see Note #7)		

REFERENCE NOTES

1 – These tolerances do not include allowance for aging characteristics of material.

2 – Tolerances based on ⅛" wall section.

3 – Parting line must be taken into consideration.

4 – Part design should maintain a wall thickness as nearly constant as possible. Complete uniformity in this dimension is impossible to achieve.

5 – Care must be taken that the ratio of the depth of a cored hole to its diameter does not reach a point that will result in excessive pin damage.

6 – These values should be increased whenever compatible with desired design and good molding technique.

7 – Customer-Molder understanding necessary prior to tooling.

Figure 17.9 Engineering and technical standards for high-density polyethylene. (From *Standards and Practices of Plastics Molders*.)

STANDARDS AND PRACTICES OF PLASTICS MOLDERS	Engineering and Technical Standards POLYCARBONATE

NOTE: The Commercial values shown below represent common production tolerances at the most economical level. The Fine values represent closer tolerances that can be held but at a greater cost.

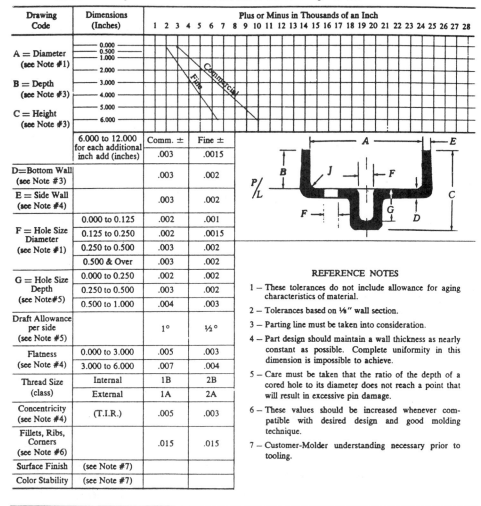

Drawing Code	Dimensions (Inches)		Plus or Minus in Thousands of an Inch 1 2 3 4 5 6 7 8 9 10 11 12 13 14 15 16 17 18 19 20 21 22 23 24 25 26 27 28
A = Diameter (see Note #1) B = Depth (see Note #3) C = Height (see Note #3)	0.000 0.500 1.000 2.000 3.000 4.000 5.000 6.000		

Drawing Code	Dimensions (Inches)	Comm. ±	Fine ±
	6.000 to 12.000 for each additional inch add (inches)	.003	.0015
D=Bottom Wall (see Note #3)		.003	.002
E = Side Wall (see Note #4)		.003	.002
F = Hole Size Diameter (see Note #1)	0.000 to 0.125	.002	.001
	0.125 to 0.250	.002	.0015
	0.250 to 0.500	.003	.002
	0.500 & Over	.003	.002
G = Hole Size Depth (see Note#5)	0.000 to 0.250	.002	.002
	0.250 to 0.500	.003	.002
	0.500 to 1.000	.004	.003
Draft Allowance per side (see Note #5)		1°	½°
Flatness (see Note #4)	0.000 to 3.000	.005	.003
	3.000 to 6.000	.007	.004
Thread Size (class)	Internal	1B	2B
	External	1A	2A
Concentricity (see Note #4)	(T.I.R.)	.005	.003
Fillets, Ribs, Corners (see Note #6)		.015	.015
Surface Finish	(see Note #7)		
Color Stability	(see Note #7)		

REFERENCE NOTES

1 — These tolerances do not include allowance for aging characteristics of material.

2 — Tolerances based on ⅛" wall section.

3 — Parting line must be taken into consideration.

4 — Part design should maintain a wall thickness as nearly constant as possible. Complete uniformity in this dimension is impossible to achieve.

5 — Care must be taken that the ratio of the depth of a cored hole to its diameter does not reach a point that will result in excessive pin damage.

6 — These values should be increased whenever compatible with desired design and good molding technique.

7 — Customer-Molder understanding necessary prior to tooling.

Figure 17.10 Engineering and technical standards for polycarbonate. (From *Standards and Practices of Plastics Molders*.)

METAL INSERTS IN THERMOPLASTIC MATERIALS				TEST DATA	
			ABS		POLYCARBONATE
THREAD SIZE	LENGTH	ROTATION	TENSILE	ROTATION	TENSILE
	INCH	INCH-LBS.	LBS.	INCH-LBS	LBS.
NO. 0	0.115	50 IN-OZ *	79	70 IN-OZ *	135
	0.188	65 IN-OZ *	162	52 IN-OZ *	258
NO. 2	0.115	67.5 IN-OZ	79	110 IN-OZ	135
	0.118	81 IN-OZ	162	168 IN-OZ *	258
NO. 4	0.135	13	147	20 *	230
	0.219	17	258	23 *	417
NO. 6	0.15	23	220	33*	341
	0.25	25	370	45	661
NO. 8	0.185	37	304	52	538
	0.312	45	469	88 *	910
NO. 10	0.225	51	448	86	773
	0.375	68	726	125	1388
NO. 12	0.265	60	508	96	937
	0.438	78	809	140	1520
1/4"	0.3	102	700	157	1283
	0.5	116	1157	231	2073
5/16"	0.335	155	739	259	1555
	0.562	214	1435	345	3128
3/8"	0.375	220	940	383	2065
	0.625	229	1743	520	3638
	* INDICATES SCREW FAILURE				

Figure 17.11 Metal inserts in thermoplastic materials. (Courtesy of Helicoil, Inc.)

Above all, listen to the advice of the molder and tool builder. They may not know much about your widget, but they do know how to save money when building a plastic part. There are very few configurations that cannot be built in plastics, but some may cost you more money and time than an alternative design.

17.3 MATERIAL AND PROCESS SELECTION

It is difficult to determine the material and subsequently the process needed for a given application. Plastic products are often a compromise between design goals and technical feasibility from the manufacturing point of view. Before approaching a plastics molder, a consultant, or a plastics engineer, the following questions should be answered.

Product Requirements Checklist

1. End-use temperature—What will the application see in general use, and what is the highest temperature to which it will be exposed? Any low-temperature exposure should be indicated as well.
2. Environment—Where will the application be used? Will it be exposed to any chemicals and for how long? Will it be exposed to sunlight? How long?
3. Flammability—What are the flammability requirements of the application? Fire retardant packages are available, but their addition can cause loss of critical properties.

4. Part strength—What loads will the part need to withstand? Are they static or dynamic? Does the part get impacted?
5. Life cycle—How long is the application expected to last?
6. Light transmission—Is the application transparent, translucent, or opaque?
7. Part combination—Is the part under consideration next to another part which could be combined with it?
8. Part usage—What volume of parts do you expect to use? This is important, because the cost of the tooling may be prohibitive if the volume is too low.
9. Government regulations—Does the part have to meet any government or agency regulations?

With the answers to these questions, a determination of the optimum material and process to use can be made.

17.3.1 Polymer Selection

The choice of which material to use in a product application is always difficult to make. The first question is whether to use plastics at all, or is some other material better suited? Economics plays a major part in this decision.

Tooling for any plastic material is a large part of the cost. With metal or wood, the part can be manufactured in small numbers with little or no tooling, but this can be very labor intensive. Parts can be machined from plastics as well, and this option should be considered, especially for prototypes or models and for low-volume production. (See Subchapter 17.4.)

Part size will often dictate both which process and which material to choose. Very large parts usually are made from a fiber/matrix composite using thermoset material, and produced using a layup technique. This choice will provide a large, rigid part with excellent physical properties.

If the part has a complex shape or contains ribs and holes, then the choice is probably going to be one of the molding techniques. With a hollow part such as a container, blow molding or rotational molding will probably be the choice.

When you get into thermoplastics, a variety of forming and molding techniques are available. The large variety of polymers, each with its own set of properties, makes the decision difficult.

If the part is going to be transparent, a material such as polycarbonate, styrene, or acrylic in highly polished tooling is often the only choice. Forming the part from transparent sheet is also a possibility, depending on the configuration.

The environment in which the part is going to perform will also affect your decision. High-heat applications, where the continuous-use temperature may get above 300° F, will dictate the use of materials such as polyetherimide or polyarylsufone. High-impact applications will force you to consider a material such as polycarbonate and will probably force you to reconsider any glass fibers or other fillers and additives.

17.3.2 Modification of Plastic Polymers

There are 50 or so basic polymers available to use in plastics applications. In recent years, the introduction of numerous combinations of these polymers, and the addition of fillers, additives, and modifications to basic polymers, has been nothing short of miraculous.

Copolymerization, the combining of basic polymers into families of copolymers, has been proceeding at a rapid rate. The synergistic effect often seen with this approach has given us materials with outstanding and unusual properties.

The filling of materials with additives such as glass and carbon fibers, calcium carbonate, mica, lubricants, pigments, and numerous others has changed the physical properties and appearance of many of our basic resins.

The addition of glass or carbon fibers to a thermoplastic usually increases the stiffness of the polymer but often reduces the impact strength. The length of fiber is also a factor, with the longer length (up to 0.250 in.) being about the longest that will go through an injection molding machine without breaking up.

The art of coloring plastics has progressed until now it is very much a science in its own right. Color analysis computers are available now which can read colors accurately, and color labs can reproduce colors to near perfect standards. Environmental problems with some colors have caused the industry to examine and reevaluate some to the pigments which have been commonly used. As an example, color pigments containing lead and cadmium are being reformulated because of their toxic nature.

It should also be noted that putting pigments into polymers often reduces physical properties. Sometimes, the natural darker color of the basic polymer needs to be covered up with pigment to match a lighter color, causing heavy loading and a loss in physical properties.

Flammability and smoke generation are also properties demanding careful evaluation. These properties are getting careful attention from government agencies responsible for health and safety in many different fields. As an example, commercial aircraft builders are under a strict mandate to eliminate polymers that burn rapidly or give off toxic fumes when combusted. This problem is often eliminated with flame-retardant additives that are blended into the plastic raw materials.

17.3.3 Secondary Operations

For all types of plastic materials, the secondary operations performed on them can be expensive and time consuming. Most of the processes available produce a part requiring residual secondary work to be done. Some, such as injection molding, leave relatively little flash to clean or surface treatment to apply to the part. Others, such as vacuum forming, normally produce products that need a great deal of secondary work before they are ready for use in production.

When bonding or joining is required, a choice of whether to use mechanical fastening or adhesives must be made. These actions require specialized treatment, and such things as compatibility and stress formation must be taken into consideration. Metal inserts can be added to provide threaded holes or threaded studs. (See Figure 17.11.)

Various adhesive systems work with specific plastics; others create problems due to solvent contamination and adhesion failures. It is always best to check with your material suppliers before using a specific adhesive system with a given material.

Machining of various plastic materials can usually be accomplished with the use of the recommended feed, speed, and cutters. Removing plastic materials by such techniques as sanding can be done, but may create problems due to the abrasive material surface rapidly loading up with melted plastic, or the plastic surface becoming imbedded with abrasive particles.

Painting of plastic materials has developed into a science. Adhesion problems are quite common, especially with materials such as polyethylene and polypropylene. Contamination is usually caused by solvents in the paints, and may cause stress cracking. Some materials require a barrier coat before applying the color or final coat. In many cases the pigments are added to the molding resin and the parts are integrally colored. In some cases the parts are molded in the same color as the paint applied to them, to prevent show-through in the final finish.

There are many coatings available for coating transparent products such as sunglass lenses which give the clear plastic better abrasion properties and some protection against chemical attack. These systems are a direct result of space technology and have increased the markets for plastic products considerably.

17.4 TOOLING

The production of any plastic application requires many difficult decisions. The choice of tooling is one of those areas. It is normally the most expensive part of any application and is filled with pitfalls. It is often said that tooling is the most important part of any plastic project. If the tooling is not right, the part which comes from it cannot be right.

The tooling for almost any part design can be designed and built, but the cost may be prohibitive. Undercuts and hard-to-mold areas are easier to eliminate in the part design stage than they are after metal has been cut. Redesigning a part after the tool has been made is expensive and time consuming. The best approach is to contact a good tool maker and listen to him. He can save you untold dollars, time, and gray hairs.

There are different classes of tooling. Because of this, you need to know the volume of parts you plan to build with the tool in question. Aluminum tooling is usually good enough to produce a limited quantity of parts, but damaged tooling will not produce any parts. Hardened tool steel is more expensive but will be good for larger volumes.

The important rule here is to get good advice when it comes to tooling and then listen to it. Also, it is important to give your mold maker enough time to give you a first class tool. Often the demand for a short lead time on a product dooms it to failure because the mold maker was not given enough time to do a good job.

Another good suggestion is to get your tool maker to give you weekly updates on progress. Nothing is so discouraging as to get right up to the date of mold delivery and find out that the tool maker is not nearly finished. Also, leave time in the schedule for "tweaking" a tool after the first mold trial. A tool is like an expensive piece of machinery and will often need some modification before it produces perfect parts. Recent introduction of CAD/CAM systems has improved the ability of tool makers to produce a good part the first time, but you can still expect problems to crop up before you have a part ready for production.

You may want to take advantage of computer-aided design systems to simulate the filling of a plastic part. The information derived from one of these systems, such as Moldflow, can be very useful in assisting the mold designer in laying out the tooling before metal is cut. Other programs to help designers improve the cooling characteristics of a tool and help eliminate warping are also available. Moldflow is available from Moldflow Pty. Ltd, Kalamazoo, Michigan.

Above all, treat the tooling arrangement as you would any business deal. Put everything in writing and have good documentation. Be sure the molder and the tool maker both know what you expect. Often added features or changes are given via verbal orders, and this can only lead to problems. Update your drawings or data sets regularly, and don't be afraid to write everything down. It could save you both money and time. (See Chapter 8 for information on rapid prototyping tooling.)

17.5 BIBLIOGRAPHY

Many books are available to further your knowledge of the plastics industry. A short list follows.

Tooling

Donaldson, Le Cain, and Goold, *Tool Design*, McGraw Hill, New York, 1957.
Menges, G., and Mohren,P., *How to Make Injection Molds,* Hansen/Gardner, Cincinnati, OH, 1986.
Michaeli, W., *Extrusion Dies for Plastics and Rubbers*, 2nd ed., 1992.

Design

Beck, R., *Plastic Product Design,* Van Nostrand Reinhold, New York, 1980.
Designing with Plastic, The Fundamentatls, Hoeschst Celanese Design Manual TDM-1, Hoeschst Celanese, 1989.
Dym, J.B., *Product Design with Plastics,* Industrial Press, New York, 1983.
Plastic Snap-Fit Joints, Miles, Inc., Pittsburgh, PA, 1992.
Rosato, D.V., *Designing with Plastics and Composites, a Handbook,* Chapman & Hall, New York, 1991.

Structural Foam

Wendle, B., *Structural Foam*, Marcel Dekker, New York, 1985.

Extrusion

Richardson, P., *Introduction to Extrusion*, 28th Society of Plastics Engineers, Brookfield, CT, 1974.

Blow Molding

Rosato, D.V., *Blow Molding Handbook,* Hansen/Gardner, Cincinnati, OH, 1989.

Thermosets

Whelan, T. and Goff, J., *Molding of Thermosetting Plastics,* Chapman and Hall, New York, 1990.

General

Wendle, B., *What Every Engineer Should Know About Developing Plastic Products,* Marcel Dekker, New York, 1994.

Further information is available through the following:
Society of Plastic Industries   (202) 371-5200
Society of Plastic Engineers   (203) 775-0471
American Society for Testing and Materials   (215) 299-5400
American Mold Builders Association   (708) 980-7667
Directory of Moldmaker Services   (202) 371-0742
Members Directory Canadian Association of Moldmakers   (519) 255-9520

18

Composite Manufacturing

John F. Maguire   *Southwest Research Institute, San Antonio, Texas*

with

Don Weed   *Southwest Research Institute, San Antonio, Texas,*

Thomas J. Rose   *Advanced Processing Technology/Applied Polymer Technology, Inc., Norman, Oklahoma*

18.0   INTRODUCTION AND BACKGROUND

It is difficult to find a truly satisfactory definition of composite material. The *American Heritage Dictionary* (Houghton Mifflin, Boston, 1981), comes close, with "a complex material, such as wood or fiber glass, in which two or more complementary substances, expecially metals, ceramics, glasses, and polymers, combine to produce some structural or functional properties not present in any individual component." The problem with even a good definition, of course, is that it is all-encompassing, so that every material in the universe could in some sense be defined as a composite. This loss of exclusivity diminishes the usefulness of the definition. For our present purposes, we shall restrict attention to that subset of materials known as fiber-reinforced advanced polymeric composites. In these materials a reinforcing fiber is embedded in an organic polymeric resin. The fiber acts as a structural reinforcement and the resin binds the fibers together. This transfers loads, and provides structural and dimensional integrity.

Advanced composite materials, developed in the latter half of the twentieth century, may well provide a key enabling technology for the twenty-first century. These materials are strong, light, and corrosion resistant, offering considerable technical advantages in aerospace, automotive, offshore petrochemical, infrastructure, and other general engineering applications. Composite components may be made by laminating or laying-up layers of composite material, each ply consisting of one or more patterns, which may be as large as 4½ ft wide by 9 ft long. These patterns are cut from a continuous roll of cloth, or from sheets, with standard widths measuring up to 4½ ft. Composite material wider than 12 in.

is referred to as *broadgoods*. If a material is narrower, it is usually a *unidirectional tape*. In such a tape the fibers run in the longitudinal direction along the length of the tape. There are no *fill* fibers, and the material is stabilized with backing paper to permit handling prior to use.

There are many additional types of processes, which utilize the fiber as strands, flakes, etc., that will be discussed in this chapter. Today there are projects in automobiles, bridge construction, and a host of sporting goods applications. Figure 18.1 shows a representative cross section of products which are currently manufactured using composite materials [1].

The strength and stiffness of these materials far exceeds those of metals [2], as shown in Figure 18.2, and it is this combination of strength and stiffness coupled with light weight which lies at the heart of the performance advantage. While the need for lightweight materials of exceedingly high strength and stiffness is apparent in aerospace applications (see Figure 18.3), it is not quite so obvious that a strong, lightweight plastic might offer advantages, say, in the construction of a bridge or a submarine. The average density of a submarine must be equal to that of seawater, regardless of the material from which the pressure hull is constructed. However, the drag on a submarine increases dramatically when the radius of the pressure hull is increased, so that one arrives quickly at a situation where the payload advantage of a larger radius is more than offset by the need for a much bigger power plant. One very effective way to increase the payload or endurance is to decrease the weight of the pressure hull while keeping the radius constant, and it is in this

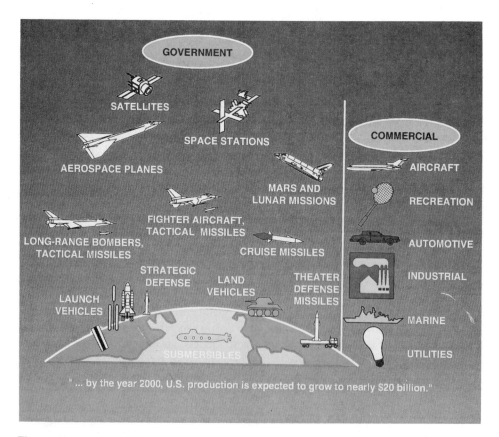

Figure 18.1   Representative cross section of products containing composites.

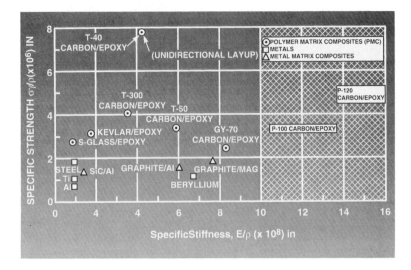

Figure 18.2 Strength and stiffness of composite materials and metals.

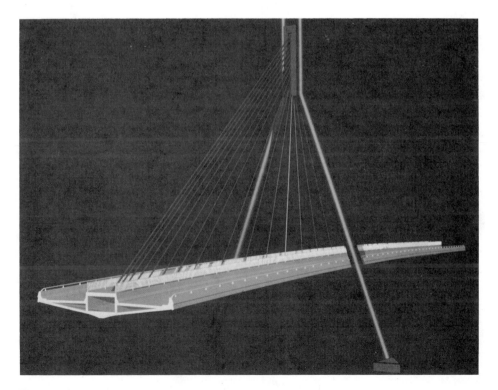

Figure 18.3 Design concept for an all-composite bridge.

area that composite materials offer significant promise. Similarly, since most of the material in a bridge is needed to hold up the bridge itself rather than carry the loads which cross it, composite materials offer much more efficient structures with far greater spans. Figure 18.3 shows the design concept for an all-composite bridge. For automobile applications, reduced weight translates into increased fuel efficiency, reduced emissions, and greater payload—demonstrating the significant benefits to be accrued from greater utilization of composite materials [3].

Before contemplating the fabrication of detail components or major subassemblies using composites in a true production manufacturing environment, it is wise to quantify the advantages in terms of cost. If we define the savings in dollars per pound of fabricated structure across various industries, we arrive at a plot as shown in Figure 18.4. In this figure the savings per pound are plotted as a function of the potential market. In satellite applications the savings are huge ($10,000 per pound) but the actual total amount of material used in the manufacturing is minuscule. On the other hand, in automobile or marine applications the savings are far more modest but the tonnage is enormous due to the production rates.

The level of sophistication required to make a given component or subassembly is not a constant, but depends to a large extent on the standards and operating practices of the particular industry. For example, in the aerospace industry the structural integrity is of the utmost importance and cosmetic considerations may be somewhat secondary. On the other hand, in the automobile industry, while the less stringent mechanical performance criteria must be met, there is the additional requirement for a very high degree of surface finish for cosmetic reasons. This requires consideration of aesthetic qualities which have little to

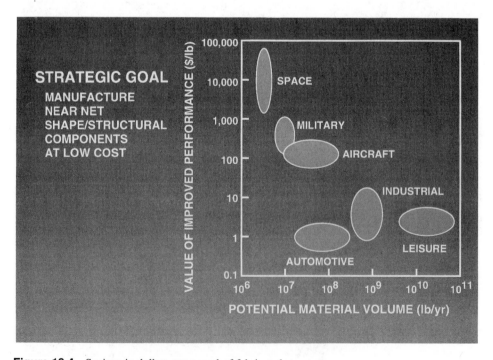

Figure 18.4 Savings in dollars per pound of fabricated structure.

do with mechanical or functional design requirements, but which may have a high impact on cost. The challenge, then, is not to be able to fabricate a given structure per se, but rather to design and develop a product and implement cost-effective manufacturing technologies which will allow the technical benefits afforded by composite structures to translate into competitive advantages in the commercial marketplace. This requires a two-pronged attack on current practices. First, it will be necessary to produce structural plastics which will have properties close to those of metals [4], and which can be processed using exceedingly low-cost manufacturing technologies such as injection molding [5]. This approach will require fundamental work in the structure of matter and will be of a long-term nature. Second, more effective manufacturing technologies and processes must be utilized in order to reduce the cost associated with current technologies.

18.0.1 Design/Manufacturing Interface

In today's competitive cost environment, the producibility of a new design is not merely important—it is the key to survival as a leader in the field of manufacturing. Many authorities would agree that perhaps as much as 90% of the ultimate cost of a product may be predicated by the design, and only 10% can be influenced by the manufacturing process. However, by considering the known production processes, the quality and reliability inherent in the process, and the material selections available within the constraints of product function, we can optimize a design early in the development of a new or revised product. This is certainly true in the field of composites.

In common with other manufacturing processes, the fabrication of composites may be conveniently discussed in terms of the labor, materials, tooling, and equipment requirements. Quite straightforward design changes can sometimes lead to great cost savings in manufacturing. Structures which would be perfectly acceptable in metals may be difficult or impossible to fabricate in composites. For example, the flanges on the outer duct of an aircraft engine may be fabricated in metal using standard shop practices, but are a real challenge in carbon fiber-reinforced plastic. Here it is good to bear in mind that there may be little to be gained but a good deal to be lost by dogmatic adherence to a particular "all-composite" philosophy. If a particular structure or part of a structure would be easier to fabricate in metals, then the smart compromise is to combine the materials to meet design requirements at minimum cost. Failure to recognize this can lead to program delays and very substantial cost overruns. In the development of the world's first all-composite small transport airplane, the Lear Fan 2100, the rigid adherence to an "all-composite dogma" resulted in the very cabin door hinges being designed and fabricated from carbon fiber-reinforced epoxy! These hinges failed twice during pressure testing of the fuselage, resulting in significant delays and cost to the program, all for the rather minimal advantages to be gained by the weight reduction due to the door hinges! Therefore, before locking in on a particular fabrication technology, one should ask whether a hybrid approach, possibly combining metals and composites, would lead to acceptable performance at lowest cost. (See Chapter 1 on product development.)

18.0.2 Materials

The composite materials most commonly used in the field can be made up of at least two constituents:

1. The *structural constituent*, usually the reinforcement, used to determine the internal structure of the composite, such as fibers, particles, lamina, flakes, and fillers

2. The *body constituent*, or matrix, used to enclose the composite structural constituents and give it its bulk form, such as epoxy, polyester, polyimide, vinyl ester, and bismalemides

Fibers

Of all composite materials, the fiber type (specifically the inclusion of fibers in a matrix) has generated the most interest among engineers concerned with structural applications. The fibers, coated and uncoated, typically control the strength and stiffness characteristics, formability, and machining characteristics of the laminate. The more commonly used fibers today include glass, carbon, and Kevlar. See Figure 18.5 for stress-strain diagrams for some common fibers.

Glass

Glass is the most widely used reinforcing material, accounting for more than 70% of the reinforcement for thermosetting resins. Forms of glass fiber materials include roving (continuous strand), chopped strand, woven fabrics, continuous-strand mat, chopped-strand mat, and milled fibers (0.032–0.125 in. long). The longer fibers provide the greatest strength; continuous fibers are the strongest. Glass does not burn, and it retains good mechanical properties: up to approximately 50% of its strength up to 700°F, and 25% of its strength up to 1000°F. Moisture resistance is excellent, and glass fibers do not swell, stretch, disintegrate, or undergo other chemical changes when wet. The sizing or chemical

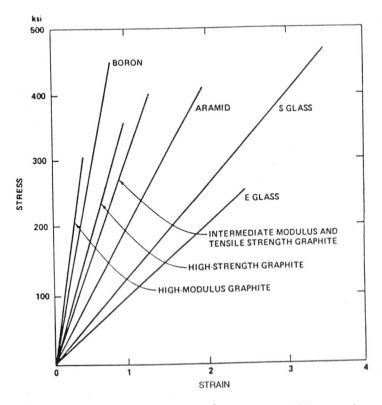

Figure 18.5 Stress-strain diagram for fibers used in hybrid construction.

treatment applied to a glass fiber surface is designed to provide compatibility with the type of resin matrix used. It also improves handling characteristics of the glass fiber, such as the ability to control tension, choppability, and wet-out. See Figure 18.6 for glass fiber nomenclature.

E glass was the first glass developed specifically for production of continuous fibers. It is a lime-alumina-borosilicate glass designed primarily for electrical applications. It was found to be adaptable and highly effective in a great variety of processes and products, ranging from decorative to structural applications. It has become known as the standard textile glass. Most continuous-filament glass produced today is E glass.

S glass is a high-tensile-strength glass. Its tensile strength is 33% greater and its modulus almost 20% greater than that of E glass. Significant properties of S glass for aerospace applications are its high strength-to-weight ratio, its superior strength retention at elevated temperatures, and its high fatigue limit. S glass costs up to 20 times more than E glass, depending of the special form of the fibers.

Carbon

The terms *carbon* and *graphite fiber* are frequently used interchangeably. The correct term in most cases is carbon, because commercially produced fibers do not exhibit the characteristic X-ray parameters of graphite. Due to a unique combination of properties, carbon fibers are the ideal reinforcement for lightweight, high-strength, and high-stiffness structures. High-performance carbon fibers are available in a range of properties, product forms, and prices. Continuous yarns or tows contain from several hundred to several thousand filaments per strand, and generally fall into two categories:

1. High strength (350–500×10^3 psi), intermediate-modulus (30–50×10^6 psi) fibers
2. High-modulus (50–75×10^6 psi), intermediate strength (250–350×10^3 psi) fibers

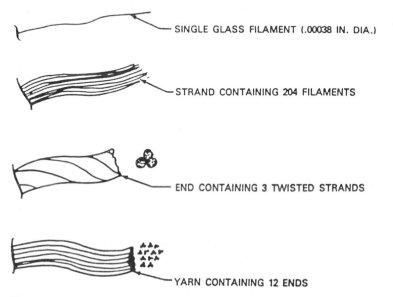

SINGLE GLASS FILAMENT (.00038 IN. DIA.)

STRAND CONTAINING 204 FILAMENTS

END CONTAINING 3 TWISTED STRANDS

YARN CONTAINING 12 ENDS

Figure 18.6 Definition of forms of glass fibers in the process of making a yarn.

The high-modulus fibers are generally more costly, and the higher-filament-count tows are lower in price than yarns containing fewer fibers.

Aramid (Kevlar)

Introduced commercially in the 1970s, Kevlar aramid is an aromatic long-chain polyamide polymer, the fibers of which may be produced by spinning using standard textile techniques. The low-density, high-tensile-strength, low-cost fiber produces tough, impact-resistant structures with about half the stiffness of graphite structures. The fiber was originally developed to replace steel in radial tires and has found increasing use in the belts of radial car tires, and carcasses of radial truck tires, where it saves weight and increases strength and durability.

Kevlar 29 is the low-density, high-strength aramid fiber designed for ballistic protection, slash and cut resistance, ropes, cables, and coated fabrics for inflatables and architectural fabrics.

Kevlar 49 aramid fiber is characterized by low density, high strength, and high modulus. These properties are the key to its successful use as a reinforcement for plastic composites in aerospace, marine, automotive, sports equipment, and other industrial applications.

Matrices

The matrix, usually resin as a binder, determines the transverse mechanical properties, interlaminar shear characteristics, and service temperature of the laminate. The matrix (or body constituent) serves two very important functions: (1) It holds the fibers in place, and (2) under an applied force, it deforms and distributes the stresses to the high-modulus fiber constituent. Both influence the selection of shop processes and tool design. The matrix may be a thermoplastic or thermosetting type of resin.

Thermosets include the epoxies, bismaliemides, and polyimides as well as some of the lower-cost resins such as the phenolics, polyesters, and vinylesters. Generally, they require the addition of a catalyst in order to cure. The type of catalyst influences the pot life of the mix, and whether heat is required to achieve full cure strength. Once cured, they cannot be resoftened by heat for re-forming. Although thermosetting plastics do not have much greater tensile strength that the thermoplastics, they are much more effective as the temperature environment increases.

Thermoplastics are softened by heating, permitting the forming of sheet material which retains its formed shape after cooling. They can be reheated and re-formed if necessary. Most injection molding is accomplished with thermoplastics, and they are seeing a resurgence in various applications in the structural plastic field. Many applications in the automotive field utilize thermoplastics, and the marine industries are seeing increasing utilization.

The materials selection depends on the design requirements of the component (i.e., the mechanical, thermal, and environmental conditions). Table 18.1 gives a selection of common fibers and Table 18.2 shows a selection of resins. As for the matrix materials, the choice for aerospace applications (and a few others) is dictated largely by the operating temperature. Figure 18.7 shows a selection of properties of resins which might be considered for various temperatures of operation. Clearly, there is some overlap here, and the temperature regimes are not sharply defined.

For operating temperatures of thermosetting resins up to 200°F, the toughened epoxy systems are acceptable and are most easily processable. Up to about 250°F, the cyanate esters have acceptable properties and are generally low-viscosity materials which are

Table 18.1 Selection of Commonly Used Fibers

Fiber	Fiber Properties		Unidirectional Composite Properties		
	Diameter (μm)	Tensile strength	Tensile modulus (Pa)	Tensile strength Mpa (V%)	Tensile Modulus Gpa (V%)
Carbon					
T-300 (Union Carbide)	7.0	3447	230	1482(60)	141(60)
Hitex 46 (Hitco)	6.1	5688	317	2696(64)	184(64)
P-100 (Union Carbide)	9.6	2413	760	—	—
Organic					
Kevlar 49	11.9	3169	124	1380(60)	76(60)
Polybenzoxazole	10 - 12	3447 - 4482	310 - 358	—	—
Ceramic					
S-2 Glass (Owens-Corning)	19	4585	87	1903	52
Silicone carbide - Nicalon (Dow) (ceramic grade)	10 - 20	-2758	193	—	—
Newtel 480(3M)	10 - 12	2241	193	—	—

amenable to resin transfer operations. In the region 200–350°F, the bismaleimides are preferable. These classes of resins are addition thermosets and do not generate volatiles during cure. In the region 350–550°F it is necessary to use polyimide resins. These are condensation-type thermosetting resins and lose about 10% of their weight and many times their volume as volatiles (usually alcohol and water) during the cure. The management of such large amounts of volatiles increases the complexity of making components with low (less than 2%) void content. At the time of writing there are no commercially available polymers with reliable long-term integrity when operated for long times above 600°F.

18.0.3 Composite Manufacturing Technology Overview

The primary subject of this chapter is to introduce the various materials and processes commonly used to manufacture products utilizing composites. After determining the

Table 18.2 Selection of Commonly Used Resins

Resin	Tensile Strength (MPa)	Tensile Modulus (GPa)	$T_g(K)$
Thermosets			
Epoxy(TSMDA)	103.4	4.1	463
Bismaleimide	82.7	4.1	547
Polyimide	137.9	4.8	630
Thermoplastics			
Polyphenylene sulfide	65.5	4.3	366(555mp)
Polyetheretherketone	70.3	1.1	400

Property	Epoxy	Cyanate	Bismaleimide
Tensile Strength, MN/m²	48-90	69-90	35-90
Tensile Modulus, GN/m²	3.1-3.8	3.1-3.4	3.4-4.1
Tensile Strain at Break, %	1.5-8	2-5	1.5-3
G_{IC}, J/m²	70-210	105-210	70-105
Specific Gravity	1.2-1.25	1.1-1.35	1.2-1.3
Water Absorption, %, Saturated at 100 °C	2-6	1.3-2.5	4.0-4.5
Dry	150-240	230-260	250 - -
Water Saturated at 100°C	100-150	150-200	200-250
Coefficient of Thermal Expansion, ppm/°C	60-70	60-70	60-65
TGA Onset, °C	260-340	400-420	360-400
Dielectric Constant at 1 MHz	3.8-4.5	2.7-3.2	3.4-3.7
Dissipation Factor at 1 MHz	0.02-0.05	0.001-0.005	0.003-0.009
Cure Temperature, °C	150-220	177-250	220-300
Mold Shrinkage, mm/mm	0.0006	0.004	0.007

Figure 18.7 Properties of resins which might be selected based on operating temperatures.

physical properties demanded by the product, the selection of a fiber and matrix to be used probably occurs next. However, this is closely tied with manufacturing process selection. There are many methods available to combine a fiber and the matrix. The selection of the best process is a major contribution of the manufacturing engineer as a part of his or her activities on the design/development team. The task of combining a fiber with a matrix can range from the preparation of a dry fiber preform, and then adding the resin, to preparation of the resin, and then adding the fibers. The range of molds and other tooling available today is quite extensive, as is the increasing technology of process controls, and the need for greater automation in the entire process. The following subchapters cover the major processes.

18.1 FABRICATION WITH PREPREG BROADGOODS

The drawbacks of the manual layup approach with broadgoods fabrication have been well recognized for a long time. There has been much work and many millions of dollars devoted to attempting to automate this part of the process [6]. Notwithstanding these efforts, the problem has remained stubbornly intractable and real progress is not likely in the near future; though predicting the time scale of innovation is notoriously dangerous.

The basic problem is that sticky cloth is an inherently disorderly medium. Robots and computer vision systems do an exceedingly poor job of recognizing the various shapes which such a material may assume. The tackiness which is required for successful layup is a further impediment to the efficient implementation of pick-and-place operations for a robotic end effector. Also, the current generation of these devices is nowhere near "smart" enough to provide the level of pattern recognition, dexterity, or tactile feedback which will be required to genuinely solve the layup problem within the existing technology base. On the other hand, the combination of the human eye and finger is exceedingly well adapted to the following operations:

1. Picking up a piece of floppy, sticky cloth
2. Recognizing that it is slightly wrinkled
3. Smoothing out the wrinkle
4. Placing the pattern shape on a mold surface and providing the tactile feedback to position the pattern on an active geometry around a tight radius, i.e., packing it into a delicate radius with the human finger

These types of tasks are difficult or impossible for the current generation of pattern-recognition algorithms, vision systems, and robotic end effectors. For example, without belaboring the point, try writing a foolproof algorithm for "identify the wrinkle." It is something of an exercise in humility to recognize that while we have sent men to the moon, we are nowhere close to solving the latter problem. This is unfortunate, because it is one of those problems that looks easy (there are no mathematical formulas) but is very difficult and is therefore of the class on which a great amount of money can be spent with little or no real return.

The ME, therefore, needs to exercise particular caution when contemplating automated layup manufacturing techniques with prepreg broadgoods as a means of cost reduction. In this subchapter we will attempt to provide an overview of the steps required to actually manufacture a composite detail component using technology which exists today. No attempt has been made to cover every possible combination and permutation. The topic of "composite manufacturing" covers a complete section in the New York City Library, and any attempt to cover this much material in a chapter of any reasonable length would be impossible. However, the material covered in the field is characterized more by the breadth and scope of the subject than by its depth, so it is hoped that a discussion of the major topics will be valuable in providing a useful source and some insights for the practicing engineer. Also, the intent has been to provide enough information for the ME to make informed choices of the various manufacturing options and proceed to fabricate an actual component. The major technologies are covered in a self-contained fashion, and the reader is referred to the literature for more detailed discussion [7]. An overview of the basic manufacturing process is shown in Figure 18.8. We will limit our present discussion to the fabrication of detail components from preimpregnated broadgoods (prepreg cloth), using hand layup and autoclave cure. This is a baseline technology and accounts for a large share of the today's production volume. The extension to unidirectional tape is straightforward. The term *prepreg* describes a reinforcement or carrier material which has been impregnated with a liquid thermosetting (or thermoplastic) resin and cured to the B-stage. At this stage, the prepreg is dry or slightly tacky and can be re-formed into a mold. In the prepregging process, the reinforcing material in web form is drawn through a bath of liquid

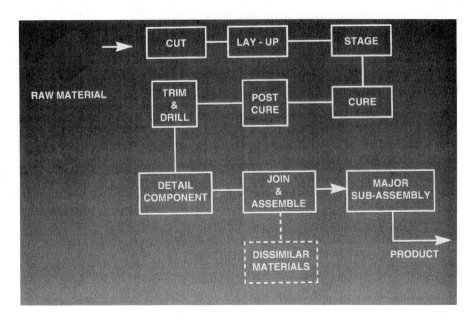

Figure 18.8 Overview of prepreg broadgoods manufacturing process.

resin. Excess resin is removed as the web leaves the bath to provide a controlled resin content in the reinforcement. The saturated material passes through an oven, where the resin is partially cured, usually to the point where the resin becomes firm but not fully cured. The web is cut into sheet lengths, or wound onto rolls to facilitate processing by the fabricator. It must be kept frozen in order to prolong shelf life.

18.1.1 Planning

The materials are purchased to conform to an incoming materials specification. It is important to pay attention to developing a good functional specification in order to discriminate against low-quality materials. An example of the criteria addressed in functional specifications is shown in Figure 18.9.

In order to conform to good shop practice, it is essential to develop sufficiently detailed manufacturing planning documentation. This is a list of operations which detail the various steps in the fabrication process (see reference Chapters 7, 9, and 11). An example of the manufacturing process for an airfoil component is shown in Figure 18.10. In developing the fabrication process, the first step is to define the pattern shapes which will be used in the layup sequence. It is desirable to minimize waste material during the cutting operation, and a rough rule of thumb is that offal should not exceed about 10%. Note that if we were to cut circular patterns in a close-packed fashion, only about 90% [exactly $100\pi/(2\sqrt{3})$] of the material could be utilized.

While it is not possible to determine mathematically what the optimum nesting pattern should be, it is possible to use commercially available software to calculate the material utilization for a given pattern and thereby heuristically iterate to an acceptable nest. Since the same pattern may be cut many thousands of times, even minor gains in materials

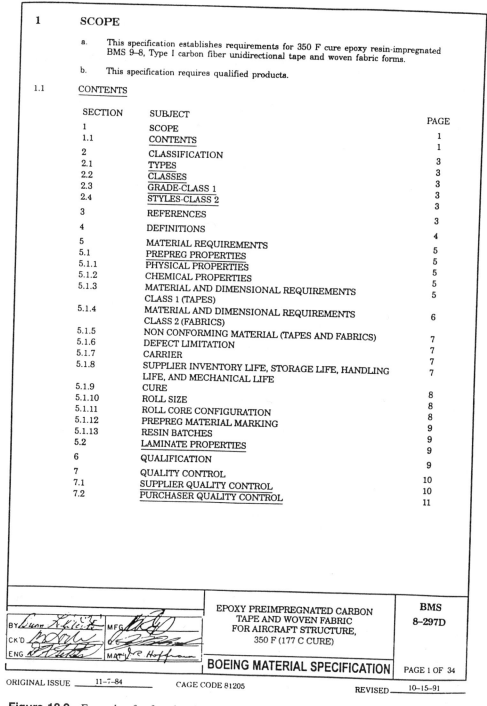

BY	MFG	EPOXY PREIMPREGNATED CARBON TAPE AND WOVEN FABRIC FOR AIRCRAFT STRUCTURE, 350 F (177 C CURE)	BMS 8-297D
CK'D			
ENG	MAT'L		
		BOEING MATERIAL SPECIFICATION	PAGE 1 OF 34

ORIGINAL ISSUE ___11–7–84___   CAGE CODE 81205   REVISED ___10–15–91___

Figure 18.9 Example of a functional specification for an epoxy-impregnated carbon fiber tape and woven fabric.

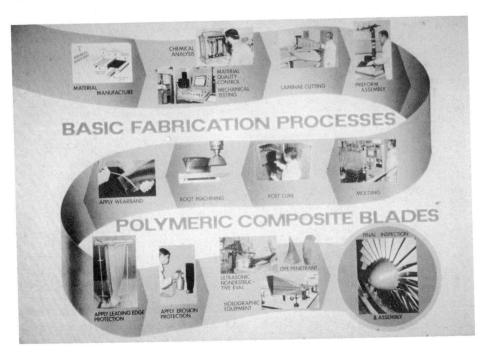

Figure 18.10 Fabrication process for an airfoil component.

utilization at this stage can translate into significant gains over the production lifetime of a major component. Figure 18.11 shows an acceptable nest for an airfoil component.

Pattern nesting of either broadgoods or tapes may be satisfactorily accomplished using a computer software system which takes into account warp orientation. Commercial systems are available from GGT, CAMSCO Division, and Precision Nesting Company, among others. Some of the more convenient systems allow direct porting of existing computer-aided design data to the nesting and cutting programs. Digitization of the pattern shape may be accomplished by tracing lines on a drawing with a hand-held electronic probe. Alternatively, automatic scanning of drawings by an electro-optical system is another method available for transferring information into the computer.

18.1.2 Cutting

The basic need in the cutting operation is to obtain a well-defined sharp edge in the minimum amount of time [8,9]. A number of technologies have been developed to do this, including lasers, water jets, reciprocating knives; and die boards using either a clicker press or an incremental (progressive) feed press. Each of these techniques has its strengths and weaknesses, and each can claim specialty applications in which it performs particularly well. In general, the approach which will provide the best trade-off in a particular application depends in large measure on the complexity and size of the component. For small components where the pattern shapes will conveniently nest on a die board, the clicker or roller-type die press will provide a cost-effective technology which provides a cut of excellent quality.

The manual baseline technology for cutting is simple: a roll of material and a pair of

Figure 18.11 Pattern nest for an airfoil component made from impregnated broadgoods.

scissors and/or a Stanley knife. When the manufacturing production rate increases just a little, it is essential to consider more efficient cutting technologies.

Many of the cutting technologies described here provide computer-controlled manipulation of the cutting medium to produce the individual ply shape. The coordinate system sketched in Figure 18.12 shows the convention used for the cutting systems. The ideal prepreg cutting system would produce full-sized patterns, having clean-cut edges, give 100% fiber-cut along the ply periphery, cause no fraying, not alter the chemical composition of the matrix in any way, and leave no scrap material.

It is important to balance the cutting capacity against overall production requirements. The very advanced nesting and cutting technologies which have been developed for the garment industry have capacities far in excess of that which might be required in a small or medium-sized shop. There is no point in having a sophisticated, automated, computerized cutting system which can cut a month's pattern shapes in an afternoon and then sits

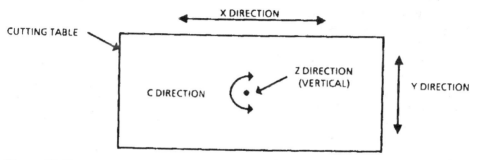

Figure 18.12 Coordinate convention used in systems for cutting prepreg broadgoods.

idle the rest of the time. The textile industry has been very successful in automating the cutting operation—so successful, in fact, that just a few of their cutters in full-time production could supply the total current need of cutting fabric for the entire U.S. advanced composite industry.

While a wide range of methods have been used successfully in industry for the cutting of a variety of prepreg broadgoods and unidirectional tape materials, the following methods cover the most important practical technologies.

Lasers

High-power lasers are currently used in a wide range of industries for cutting and drilling holes in such diverse materials as metal, glass, ceramics, wood, and human flesh. One laser which is used commonly for cutting applications is the carbon dioxide laser operating at about 10,000 cm^{-1} in the mid-infrared region of the electromagnetic spectrum. Only a few companies (including SAAB) have seriously evaluated the laser as a cutting technology for prepreg graphite material in a production environment. The method has not gained widespread acceptance due to the inability to cut individual plies without curing the edges in the immediate vicinity of the cut. Also, the power densities have to be exceedingly high if an attempt is made to cut multiple stacks of materials.

Advantages	Disadvantages
Very good cut edge, especially on thin-gauge metals	Can be used only on a single ply of graphite material
Capable of cutting extremely intricate pattern shapes	Noxious fumes when cutting composites
May be automated using mirror-driven technology	Causes curing of the resin immediately adjacent to the cut edge
Excellent hole-drilling capabilities in metals	Requires very high power densities and therefore careful isolation of the beam

Water Jets

Water jet cutting is a method that uses a thin, high-velocity stream of water as the "cutting tool." In order to produce this high-velocity water jet, pressures of 30–60×10^3 psi are used to force the water through an orifice (0.1–0.5 mm diameter) in a sapphire or other hard material. The resultant water jet is supersonic and therefore extremely noisy. Automated prepreg cutting can be achieved with only a two-axis control system since the water stream is fairly well collimated, making it relatively insensitive to z-axis positioning. A number of commercially available systems come equipped and provide a z axis for added flexibility, if required. The sapphire orifice produces a fully symmetrical water stream because of its circular shape, eliminating any need for a rotational c axis. Since the cutting action is performed by the high-velocity water, there is no "wear" of the "cutting tool." However, the sapphire orifice does require periodic replacement because of gradual water erosion.

Epoxy resins absorb moisture readily, so care must be taken with water jet cutting to minimize the amount of water contacting the resin. Water jet cutting has been used in Europe successfully; however, concerns over the moisture-absorption problem have prevented widespread acceptance in the United States. Cutting requires some means of supporting the prepreg material while allowing the water stream to pass through. One simple method uses a support for the prepreg which is made from a thin honeycomb structure, with the cell walls aligned with the direction of the water stream. The walls of the cells support the material adequately while at the same time allow the jet to penetrate the material. As time goes on, the honeycomb material needs to be replaced.

An alternative method of prepreg support [10] using a conveyor belt system with a movable slit in the surface material is shown in Figure 18.13. The four-roller system forms a unit in which the rollers are kept in the same relative position to each other, as shown. Each roller spans the full width of the support surface. The prepreg material lies on the top of the support surface. When the roller unit is moved back and forth in the x direction, the slit moves with it following the jet of water as it traces the ply pattern on the prepreg. The effect is to leave the prepreg material stationary on the support surface while the slit is moved beneath it. In this manner the slit is always maintained directly underneath the jet stream. The water jet head translates in the y direction.

Typical water velocity at the nozzle is about 2000 ft/sec. In most advanced applications the cutting fluid contains an abrasive which allows the cutting of extremely hard materials. In such applications the difficulty is to maintain convergence of the beam. Novel tip geometries can be used to do this in, for example, the system developed at Southwest Research Institute. The advantages and disadvantages of utilizing water jet cutting technology are as follows.

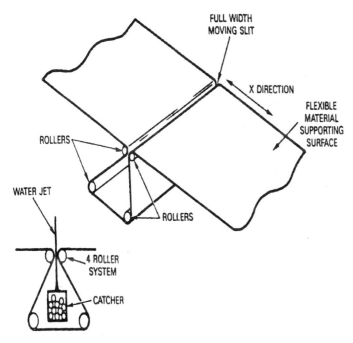

Figure 18.13 Example of conveyor belt with full-width moving slot used to support pre-pregbroadgoods during abrasive water jet cutting.

Advantages	Disadvantages
Small, clean, and accurate cut	Both a protective top film and backing paper must be left on during cutting
Minimal amount of waste	Too slow for high-volume production work
Excellent on soft, delicate, or unidirectional material	Unit takes up lots of space
Toughness of material virtually irrelevant	Cuts composite material well to a maximum of 10 plies
No deterioration of material being cut	Some question as to leaking of the resin in the immediate vicinity of the cut

Reciprocating Knife Cutting

Reciprocating knife cutting, an automated computerized cutter and nesting system, is now a fairly popular cutting option. The cutting rate is high (20 in./sec), and the system can handle a broad range of woven broadgoods. A batch of broadgoods materials, typically 54 in. wide, can be thawed and stacked up to about 10 layers deep on the power-head table of the cutter. While the cutting operation is in progress on one such table, the cut pattern shapes may be removed and kitted on a second table and the board loaded for the next cutting operation. In this way one automated cutter can adequately meet the needs of even a large production facility, often with capacity to spare.

The reciprocating knife cutter utilizes a cutting blade mounted on a reciprocating motor. The motor is mounted on a gantry above an X,Y table. The material is then placed over a length of the table and a thin polymeric cover ply is placed on top. Sometimes a vacuum may be applied to hold the plies in place. The gantry-mounted head then cuts specific pattern shapes under the control of a computer.

There are a number of variations on the basic reciprocating knife technology which can be useful in particular applications. The underlying difficulty with the reciprocating knife is that in some cases it does not yield a clean edge. An acute angle is required relative to the warp or fill direction. In such cases an ultrasonic head may provide an acceptable solution to the problem.

Advantages	Disadvantages
Shapes excellent for woven fabric broadgoods, handles many pattern shapes	Ends of material tend to fray when cutter incidence angle is greater than about 100° or less than 80°.
High cutting rate may be maintained	Knife replacement fairly frequently

The simple die board and press is a cheap, reliable, and effective cutting tool. Figure 18.14 shows a commercially available, production-type press. It will be noted that die boards can be shifted around on automated roller systems, permitting this function to be semiautomated in high-volume operations requiring efficient materials transport. Also, from the technical point of view, the die board produces excellent-quality cuts even on unidirectional tape.

Figure 18.14 Commercially available production-type press used for die-cutting prepreg broadgoods.

Roller Presses

The roller press utilizes a hardened steel roller which is positioned in the head of the press. The material to be cut is pressed between the head of the press and the cutting edge of the steel rule die. The cut material is then unloaded from the die board. New prepreg sheets or tape are then placed on top of the steel rule die to repeat the operation.

Advantages	Disadvantages
Fast—cuts many different-shaped patterns in only a few seconds	Tends to bend steel rule dies
Point-to-point cutting action	Pulls material through—accuracy of cuts can be unreliable

Incremental Feed Presses

The incremental or progressive feed press incorporates a vertically mounted head on a steel rule die which is fed a section at a time through the press. An electronic base is mounted under the head, and this base transmits pressure to the material between the base and the cutting edge of the die. After the cut is made, the head retracts a few inches, which allows the die board to be fed incrementally a specified distance under the cutting head. The operation is repeated until all the material is cut. Upon completion of the operation the operator lifts out the patterns, removes the scrap, and the die board shifts automatically to the opposite end of the press, at which point it is reloaded with material. Incremental press cutting with a steel rule die gives clean, accurate cuts without the aid of a top sheet and cuts well without backing paper. The incremental press is a self-contained, compact unit which can be easily interfaced with a microprocessor and is one of the fastest methods of cutting multiple patterns. The steel rule die board, when used in conjunction with the press, lends itself readily to the concept of a kit. No extra kitting operations would be necessary. There are several models of excellent-quality presses on the market today. Many are equipped with feed tables and die-board storage retrieval systems and 50–200 tons working pressure.

Advantages	Disadvantages
Clean cuts, fast	Available only in specific tonnages
Cuts material well without backing paper	
Die board contains a complete kit	
No cover sheet needed on material	

18.1.3 Layup Stage

For the reasons mentioned in the introduction to this subchapter, there is no fully automated production layup facility in the world. This would be a facility where arbitrary patterns are placed on arbitrary tool surfaces in an active production facility. There are a number of facilities where experimental systems are at various stages of development, and portions of the handling, cutting, and layup tasks are partially automated. One problem is that it is not good enough to have 90% reliability, which is very good given the nature of the considerable challenge, and therefore one still needs a person to watch the robot or to free up the fouled system when a few plies get mangled! (See Chapter 21 on assembly automation.)

However, in those cases where the design requires unidirectional tape, the geometry is not very active, and the component large (e.g., wing skins), automated tape laying may be a viable technology. Figure 18.15 shows a tape-laying machine laying up the wing skin for a large aircraft. This is one of a number of such machines that are commercially available.

Finally, it is worth noting that there have been a number of attempts to improve the efficiency of the manual layup operation. Here the most promising approach is that of an optically assisted ply-locating system. This idea was first anticipated by Joe Noyes while he was technical director of the Lear Fan project. Studies conducted by one of the authors about 15 years ago resulted in a functional specification and a request for a quotation for such a system. Such systems are now commercially available and have shown considerable promise as a means of increasing the efficiency of the layup process. The basic mode of operation is as follows.

1. Using an existing kit, the system is taught the location and shape of pattern #1 using a joystick to guide a laser spot around the outer edge of the ply.
2. When the learning step for pattern #1 is complete, the pattern shape is then projected onto the tool surface to ensure that the projected pattern and tool alignment are correct.
3. When step 2 is complete, a new pattern is placed on the tool surface and the process is repeated until the complete kit has been learned.

This device reduces layup times by providing a direct reference datum to ply placement. Also, there is a considerable advantage in that the planning documentation is paperless and no written instructions need to be followed. Recent trials with optically assisted

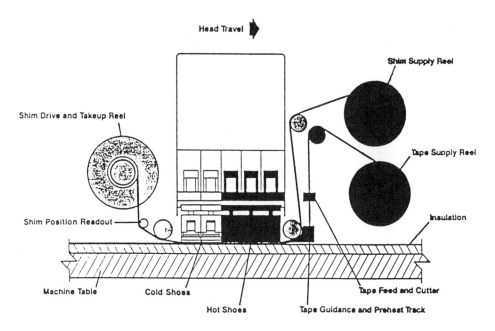

Figure 18.15   Tape-laying machine laying up large aircraft wing skin.

ply-locating systems indicate that layup efficiency may be increased by approximately 30% using this approach. There are a couple of extensions of the technology which, no doubt, will soon become available, given that the initial technology has now been adopted.

The first extension is the coupling of the ply-locating system with a ply-dispensing system. The basic idea here is that when the pattern shape is projected onto the tool surface, an interlock operates which makes the projected pattern physically accessible to the operator. The operator lays up the ply in the usual fashion using the laser image as a guide. When the pattern layup is complete, the operator prompts the system for release of the next ply. Before an additional ply is made available, a grid is projected onto the tool surface and a vision system (both visible and infrared) confirms that the previous ply has been positioned correctly. Also, common-occurrence inclusions under the ply, or backing paper left in place, can be detected in this way while there is still a possibility for rework (i.e., before the irreversible cure operation has been carried out).

18.1.4   Cure

The objective of the cure/postcure operation is to assure that the laid-up, properly staged, and debulked parts are cured. The parts are converted from a stack of resin-impregnated cloth in which there is no dimensional stability, to a hard, dimensionally stable near-net-shape component. In this operation it is important to recognize that the process engineer is engaging in an activity which bears on many disciplines. For example, the need to fabricate a component of acceptable dimensional tolerance is certainly an activity familiar to the mechanical engineer. Also, the resin system is in most cases a thermosetting resin, which, by definition, is undergoing a set of chemical reactions which take it from a tacky fluidlike state to a hard, impenetrable, solidlike or glassy state. The control of such reactions is the realm of the chemist or chemical engineer. On the other hand, such control is actually implemented as the control of a large pressurized oven, and this is often handled by an electrical engineer with a background in control theory. This mix of skills is indeed highly interdisciplinary, and it is interesting to note the differing perspectives which are brought to this part of the processing operation.

An important feature of the curing operation is that it is generally irreversible (i.e., there is no way to reverse the chemical reaction sequence), so after this stage in the process, no rework is possible. If for some reason the cure operation leads to a "bad" part, then all labor and material costs up to this point are lost. It is important, therefore, that a good physical and chemical understanding of the cure process be achieved so that cure cycles are designed using an adequate knowledge base. The criticality of the cure cycle depends to a significant degree on the nature of the resin system. For some very mature systems, such as the toughened epoxy resins, the materials, through years of chemical development, have been designed to be exceedingly tolerant. Other systems, such as the polyimides, are notoriously variable and require careful design and control of the cure process.

Epoxy Resins

An epoxy resin cures by means of an addition reaction. Basically, the catalyst, which is usually a Lewis acid, complexes with the oxygen atom in the epoxy ring and thereby activates the carbons in the ring. This activated complex is susceptible to rapid reaction with the nitrogen atom of the hardener. The resulting adduct may undergo a facile equilibration to form an intermediate with both the secondary amine and alcohol groups. This process then continues with the multifunctional epoxy and amine molecules until a densely cross-linked structure is formed. Notice that there is no elimination of a small molecule such as water, methanol, or HCL, so the cure process itself does not evolve volatiles.

This means that in an epoxy resin the vacuum schedule can be very straightforward; namely, the vacuum hose is attached to the bag and a full vacuum is applied at the start of the cure schedule and kept on for the complete cycle. In the usual epoxy resin system there are only two minor sources of volatiles. The first is the small amount of air which is dissolved in the resin. As the material heats, this air is released and, if it is not evacuated, it will give rise to voids which will be more or less uniformly distributed throughout the laminate. The second and more serious cause of voids in an addition-type resin system is simply the air which, of necessity, becomes entrapped in the small-volume elements which arise due to the shape of the weave of the material. In a typical system, such as an eightharness satin weave material, small voids are formed when the fiber tow in the fill (or weft) direction goes over the tow running in the warp direction. This air should be removed as far as possible—thus vacuum must be applied.

Insofar as there is a need to determine the cure heatup rate, hold time, and other processing variables, this is best done by an initial investigation using differential scanning calorimetry (DSC). Differential scanning measures the enthalpy change which occurs during the chemical reaction of the resin. The measurements can be made isothermally or, more usually, on a temperature ramp. When an epoxy resin cures, heat is given out (i.e., the reaction is exothermic). The ordinate in a DSC trace is the differential power required to maintain the same temperature in a pan containing the sample relative to a reference pan which contains no sample. In isothermal operation the resin is heated rapidly to the required temperature and held isothermally. The time evolution of the heat change is measured as the area under the exothermal peak. If the measurement is made at high temperature, the reaction can be brought to completion and this is usually denoted $H\infty$.

Now, at some lower temperature, or during some temperature ramp, the same procedure can be used to determine the enthalpy evolved up until time t, denoted as $H(t)$. The usual definition of degree of cure of an epoxy is

$$\alpha = H(t)/H_\infty$$

(i.e., the degree of cure α is defined as the amount of heat which has been given out relative to the maximum possible amount of heat which could be given out). It is noted, without further comment, that this widespread definition ignores the fact that the enthalpy is a function of state, so that the quantity $H(t)$ depends on the path taken (i.e., on the heating rate). In any case, one should confirm that the heat rate, ramp, and temperature holds are consistent with the attainment of some high (normally 90%) degree of cure.

From a practical point of view it will usually be the case that the actual cure cycle will be more dependent on the thermal response characteristics of the autoclave than on the chemical kinetics of the resin system. For this reason cure cycles are usually considerably longer than would appear warranted on purely kinetic arguments. This whole question is one of considerable complexity and bears on issues of chemical stability, nonlinear kinetics, and structure/property relationships, which are far outside the scope of the present chapter. Suffice it to say that caution should be exercised and incremental improvement in shortening cycles should be the norm. There has been a good deal of recent progress in designing intelligent cure cycles [11]. Figure 18.16 gives a cure cycle for a number of common epoxy material systems.

Bismaleimides

The bismaleimides are also addition-type resins and the cure considerations, apart from the kinetic and maximum hold temperature, are very similar to those for epoxies. A common cure cycle for bismaleimides is shown in Figure 18.17.

Polyimides

Polyimides differ fundamentally from either epoxy and bismaleimide systems in that they are condensation resins which emit considerable amounts of volatiles during the cure process. For example, the most widely used commercial polyimide, PMR-15, loses some 10% by weight (and many times the part volume) during cure. This poses considerable

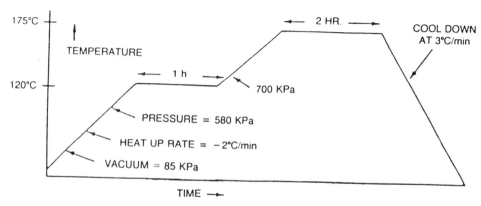

Figure 18.16 Cure cycle or a number of common epoxy material systems.

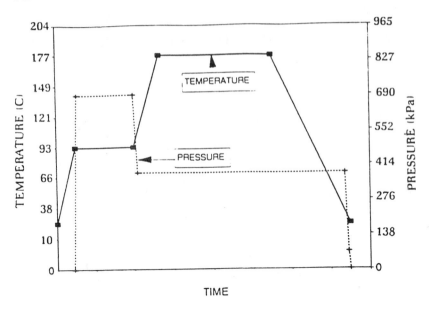

Figure 18.17 Common cure cycle for bismaleimides.

challenges in designing cure schedules, bagging, and breather systems to accommodate the large amounts of volatiles which are released.

In this type of condensation system, of which PMR-15 is prototypical, the lone pair of electrons of nitrogen attacks the carbonyl function of either the ester and/or anhydride with elimination of a methanol molecule to form a polyamic acid. This polyamic acid will undergo a rapid ring closure to the imide. In this reaction mechanism it is clear that there are a number of competing equilibria among the anhydride, the esters, and the hydrolyzed ester. These equilibria will shift with temperature and pH, so the complete mechanism is one of significant complexity. In any case, the important fact is that in the "first" part of the cure cycle a polyimide is formed. In the case of PMR-15 this polyimide is end-capped with a norbornene group to form a more or less stable, thermoplasticlike material in the first part of the cycle. Upon raising the temperature further, the norbornene group causes a cross-linking reaction to take place which gives the final polymer.

In this resin system there are a number of isomers, so a total of some 20 or more distinct chemical compounds are present in the monomer state. There are a number of possible reactions. These include attack by the nucleophile on the back side (Sn2 type) of the methyl group to yield a carboxylic anion and methyl-substituted secondary amine. Also, there is the possibility of nucleophilic attack on the bridging carbonyl of benzophenone which will yield a Schiff base adduct. In this regard it is interesting to note that the fully cured resin, if prepared in a void-free state, has a deep majenta red color which is reminiscent of an azo dye. The important factor to bear in mind from the point of view of part fabrication is that these side reactions can give rise to oil-like materials which can vaporize during cure or postcure or can give rise to chemical structures with inherently low thermal stability. While much research has been done, there is still a good deal of work needed in these areas. Again, at this time it behooves the engineer to err on the side of caution and process materials according to accepted standards.

Figure 18.18 shows a typical cure cycle for a polyimide. In this cure cycle there is an

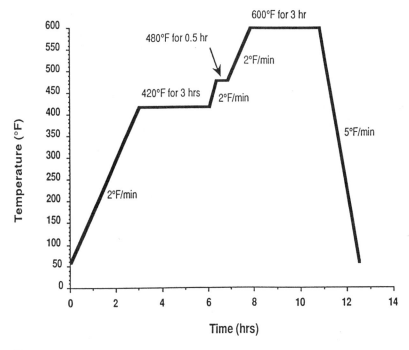

Figure 18.18 Typical cure cycle for a polyimide.

initial hold at 430°F which allows imidization to complete, after which a ramp is initiated to the high-temperature hold. The vacuum schedule is critical to obtaining void-free laminates with polyimides. On the one hand, it is necessary to engineer a "slow boil" so that the volatiles will escape through an efficient breather schedule, while on the other hand it is necessary that the rate of volatile evolution not be so vigorous as to displace plies with the laminate. A later subchapter will deal with sensors which can help with control of this situation. For most components, which have fairly thin cross sections, the standard bagging schedule and cure cycle should prove adequate. However, for thick-section polyimides the processing challenges will be greater and special procedures, including reverse bag pressurization, should be considered.

Reverse Bag Pressurization

In order to understand the basis for the reverse bag pressurization technique, it is necessary to consider the physics of pressure distribution of a bagged part in an autoclave. Figure 18.19 shows a schematic of the situation. The autoclave pressure acts on the bag and causes physical compaction of the laminate. At equilibrium the force exerted by the bag is exactly compensated by the resistance to further compaction offered by the laminate, breather plies, release plies, and so on. Note that the pressure *under* the bag (i.e., the pressure felt by the resin) is the saturated vapor pressure of the resin at that temperature. If full vacuum is applied under the bag, volatiles will escape until only those voids which are trapped in interstitial spaces which are in highly tortuous escape routes remain. It will probably not be possible to remove these voids by application of further vacuum. However, if the resin is still in a fluid state and the vacuum is not removing further volatiles, the void size may be shrunk further by applying pressure under the bag. This increases the hydrostatic

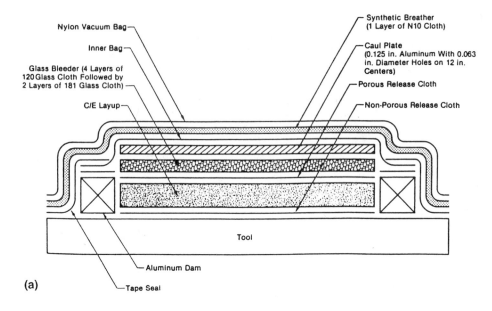

(a)

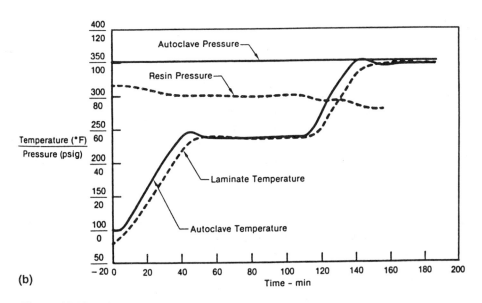

(b)

Figure 18.19 (a) Bagged prepreg broadgoods composite layup in an autoclave, and (b) schematic of pressure distribution in an autoclave during typical heat cure cycle.

pressure on the resin, and any bubble in it will tend to shrink or possibly dissolve in the resin. Of course, it is essential that the hydrostatic pressure applied under the bag be substantially less than the autoclave pressure; otherwise, the bag will blow off! This technique can be used to routinely process thick (1 in.) laminates. Figure 18.20 shows the reverse bag pressurization system developed and employed at Southwest Research Institute.

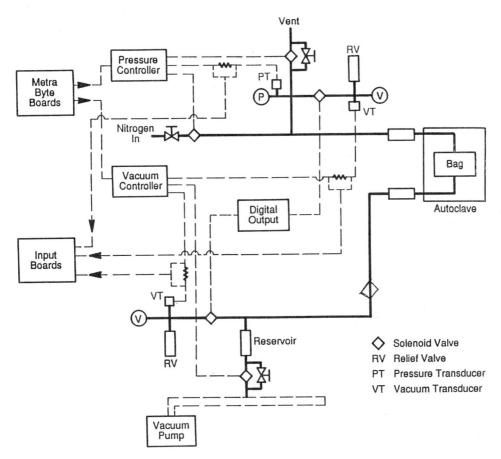

Figure 18.20 Computer-controlled vacuum and pressurization system. (Courtesy of Southwest Research Institute.)

18.1.5 Trim and Drill

A wide variety of methods is available for machining composite components. In general, the presence of a hard fibrous reinforcement leads to excessive wear, and diamond-coated cutting tools are considered the industry norm. Even here the tool life is not notably long.

Conventional machinery such as drilling, grinding, milling, and turning equipment requires contact between the hard surface of the cutting tool and the softer material to be cut. Two major problems arise with cutting composite materials. Firstly, these materials are not ductile, so the picture of a metal as an idealized plastic body which yields in shear either on the shear plane or in the so-called shear zone around the cutting edge will not, even approximately, be met [12,13]. The cut edge invariably shows fiber damage, and possibly delamination to a depth which is a function of the pit geometry, material, speed, and material feed rate.

In addition to the traditional mechanical contact methods a number of nontraditional methods are in routine use in the composite industry (see also Chapter 16 on nontraditional machining). Water jet cutting is a particularly useful technology for cured laminates, and more particularly cured laminates with honeycomb cores. These are particularly difficult to cut with a good edge. These methods include the following.

Water Jets and Abrasive Water Jets

When applied to cured laminates, the use of water jets and abrasive water jets is essentially identical to that described previously for prepreg. For laminate cutting, a jet of water, often containing an abrasive medium, is forced through a sapphire orifice as small as 0.1 mm in diameter with water pressure in excess of 50×10^3 psi. If abrasive water jet cutting is used, including the patented, highly focused jet technology developed at Southwest Research Institute, very hard tough materials can be cut, such as SiC/Al, boron/aluminum, and fiber-reinforced ceramics. For cutting cured graphite/epoxy laminates, an orifice of about 0.754 mm with water pressure of about 50×10^3 psi has been found to be most satisfactory. An overview of the water jet cutting process is shown in Figure 18.21.

A number of more or less exotic technologies such as ultrasonic, electron beam, and

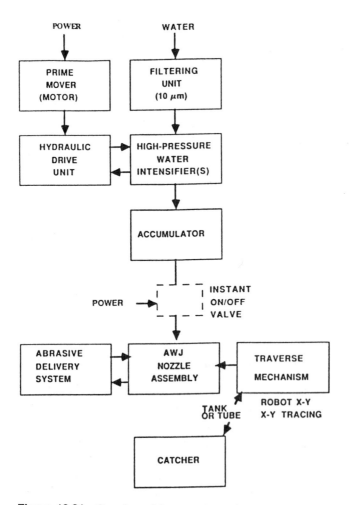

Figure 18.21 Overview of the water jet cutting process.

electric discharge machining have all been tried, but none of these has found the wide-spread acceptance and general applicability of the methods described (see Chapter 16).

18.2 FABRICATION WITH CONTINUOUS FIBERS

The goal of the composite designer and manufacturer is to place the primary load-bearing members (filaments) and the matrix together in the reinforced composite in such a way as to achieve maximum efficiency in the structure. The fact that the individual filaments are primarily loaded unidirectional in tension makes it possible to consider using continuous filaments in the following methods.

18.2.1 Filament Winding

For special geometries such as spheres or centrosymmetric parts, filament winding may be a convenient route to fabrication. As the name suggests, this approach consists essentially of winding a single or a small number of fibers onto a rotating mandrel. By adjusting variables such as the fiber direction and angle, it is possible to fabricate a wide variety of parts using this technology.

Filament winding is a process whereby a rotating mandrel is wound with roving, yarn, or tape, in a given angular orientation. The resin-impregnated, filament-wound parts are then cured in an oven or by other means. These parts are often fabricated for high-pressure applications such as gas storage cylinders or rocket motors. Filament winding is easily automated and is relatively inexpensive compared with hand layup or tape-laying technology (see Figure 18.22).

Winding Methods

From the raw materials standpoint, three methods of winding can be used. Parts can be produced by: (1) wet winding, in which the roving is fed from the spool, through an impregnating resin bath, and onto the mandrel; (2) prepreg dry winding, in which preimpregnated B-staged roving is fed either through a softening oven and onto the mandrel, or directly onto a heated mandrel; or (3) postimpregnation, in which dry roving is wound on the mandrel and the resin is applied to the wound structure by brushing, or by impregnating under vacuum or pressure. This last technique is usually limited to

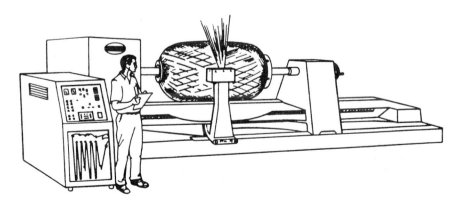

Figure 18.22 Typical filament winding machine.

relatively small parts, such as shotgun barrels, since thorough impregnation without the use of pressure or vacuum is difficult.

At the present time, wet winding is by far the most common method used. It is lowest in terms of materials cost; and for those producers equipped with plastics-formulating facilities, it offers the benefits of flexibility of resin formulation to meet specific requirements for different parts. Tension of the roving must be altered as the diameter of the part increases if accurate control of the resin-glass content is required. If the winding tension is not altered, resin content varies directly with the diameter.

Winding Patterns

Two basic patterns can be used in winding, each having a number of variations. Each pattern can be used by itself, or combined with the other to provide the desired type of stress distribution in the part.

Circumferential winding involves "level" winding of circumferential filaments. By this method, as shown in Figure 18.23, the impregnated roving, either in single- or multiple-strand band, ribbons, or tape, is layed down on the rotating mandrel at approximately 90° to the axis of rotation. The movement of the carriage that is directing the roving onto the mandrel advances the band of roving a predetermined amount, each revolution depending on the total thickness of wrap desired. These "circs" provide strength only in the hoop direction; longitudinal strength is supplied by applying longitudinal rovings, bands of rovings, or woven fabrics, by hand or machine. When such longitudinal reinforcements are applied by hand, usually a unidirectional tape is preferred to provide sufficient pretensioning of the filaments. In the Strickland B process, used by Brunswick, longitudinal rovings are machine applied and can produce open-end structures. Circumferential winding provides maximum strength in the hoop direction, but does not permit winding of slopes over about 20° when wet winding, or 30° when dry winding. Nor does circuferential winding permit effective integral winding-in of end closures.

Helical winding is the second widely used winding pattern (see Figure 18.24). In this technique, the mandrel rotates while an advancing feed places the roving or band of roving on the mandrel. In helical winding, however, the feed advances much more rapidly than in circumferential winding; the result is that the rovings are applied at a angle of anywhere from 25° to 85° to the axis of rotation. In helical winding, no longitudinal filaments need be applied, since the low winding angle provides the desired longitudinal strength as well as the hoop strength. By varying the angle of winding, many different ratios of hoop to longitudinal strengths can be obtained. Generally, in helical winding, single-circuit winding is used: i.e., the roving or band of roving makes only one complete helical revolution around the mandrel, from end to end. Young Development has a system for multicircuit

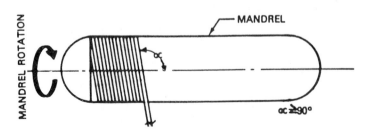

Figure 18.23 Circumferential filament winding.

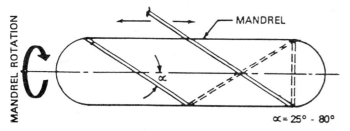

Figure 18.24 Helical filament winding.

winding, which permits greater flexibility in angle of wrapping and in the length of the cylinder. According to Young, the optimum helix angle of wrapping to provide a balanced, closed cylindrical structure is 54.75°. The netting analysis used to derive this angle is summarized in Figure 18.25. (See Chapter 14 for calculation of stresses in a cylinder.)

All pressure vessels, as well as most other devices that are filament wound, require end closures of one type or another. The most common method has been to design a relatively large collar or flange on the end of a metal closure and wind over it, making the closure integral with the vessel. For winding an integral end closure, a modified elliptical or "ovaloid" configuration is used. Such a design not only loads the fibers properly in service, it also can be formed using only one angle of winding; a hemispherical end shape would require several winding angles. The ellipse is shown in Figure 18.26. Since winding of this type of end requires a low helix angle, circumferential windings are added to

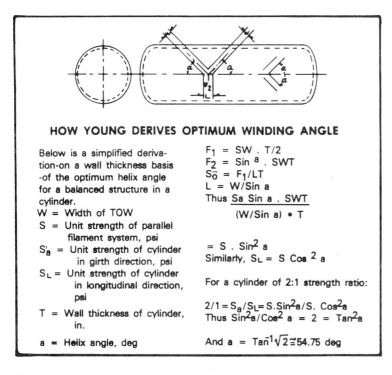

HOW YOUNG DERIVES OPTIMUM WINDING ANGLE

Below is a simplified derivation-on a wall thickness basis -of the optimum helix angle for a balanced structure in a cylinder.

W = Width of TOW

S = Unit strength of parallel filament system, psi

S_a' = Unit strength of cylinder in girth direction, psi

S_L = Unit strength of cylinder in longitudinal direction, psi

T = Wall thickness of cylinder, in.

a = Helix angle, deg

$F_1 = SW \cdot T/2$

$F_2 = \sin a \cdot SWT$

$S_a = F_1/LT$

$L = W/\sin a$

Thus $\dfrac{Sa \sin a \cdot SWT}{(W/\sin a) \cdot T}$

$= S \cdot \sin^2 a$

Similarly, $S_L = S \cos^2 a$

For a cylinder of 2:1 strength ratio:

$2/1 = S_a/S_L = S \cdot \sin^2 a / S \cdot \cos^2 a$

Thus $\sin^2 a / \cos^2 a = 2 = \tan^2 a$

And $a = \tan^{-1}\sqrt{2} \cong 54.75$ deg

Figure 18.25 Young's multicircuit filament winding angle netting analysis.

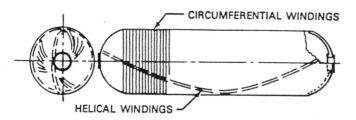

Figure 18.26 Modified elipse, or ovaloid filament winding-in of end closures for a cylindrical vessel.

provide the required hoop strength in the cylinder. Most products can be made on a two-axis machine, or perhaps three at most. However, state-of-the-art machines today may have six-axis mobility, and seventh and eighth axes are possible with existing knowhow.

Speeds depend on the drive motor and the gearing. Generally, the maximum rate for moving the carriage across the mandrel is in the range of 3 ft/sec, with a top spindle rotation of 100 rpm. Larger machines are naturally slower, and a typical production unit might operate 2–3 ft/sec at 30 rpm.

18.2.2 Braiding

In the braiding operation, the mandrel is not rotated, but the fiber carriers, which are mounted on a wheel normal to the mandrel axis, rotate around the mandrel axis. The unique feature of the braiding technique is the over-and-under process of adjacent strands of yarn as shown by the "dancing of the maypole" in Figure 18.27a. The process shown in Figure 18.27b illustrates that by adjusting the braid angle, θ_B, the hoop and longitudinal strength can be optimized. The greater the braid angle, the greater will be the hoop strength.

With larger-diameter structures, longitudinal yarn is interwoven at intervals around the circumference. This increases axial and bending strength. Longitudinals may be used on any diameter where additional stiffness is required (see Figure 18.28). Most braiding is accomplished using a *balanced braid*; i.e., the widths of yarn lie flat and adjacent so that no gaps or bunching occurs between widths. The braid angle is controlled by adjusting the number of carriers, the speed the carriers travel, and the feed rate of the mandrel through the braider.

Examples of a small braiding machine, usually in the range up to 36 to 48 carriers, are shown in Figure 18.29. They are often mounted with the wheel parallel to the floor and the mandrel traveling up and down. Figure 18.29a shows the use of a mandrel, while 18.29b shows making a braided sock without using a mandrel. The material in Figure 18.29b will be used as a preform, and will require curing in a mold of some type to achieve the desired shape.

For larger diameters and lengths, parts may be braided on the larger machine shown in Figure 18.30. The machine shown has 144 yarn carriers and provisions for 72 longitudinal tows. The wheel diameter is approximately 10 ft, and is shown with a traverse mechanism that moves the mandrel back and forth under the direction of a programmable controller and limit switches on the traverse mechanism. The feed and speed can be changed by the operator. The second braiding wheel is used as a yarn reloading station, and is easily pushed into position in front of the mandrel by the operator, since the weight is supported by air bearings. Figure 18.31 is a view of the details of the braider mechanism, showing the horn gears which drive the carriers around the circumference of the braid-

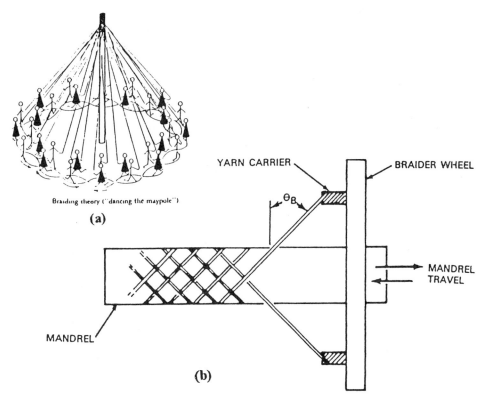

Figure 18.27 Introduction to the braiding process as described by (a) "dancing the maypole"; and (b) in basic machine operation. (Courtesy McDonnell Douglas.)

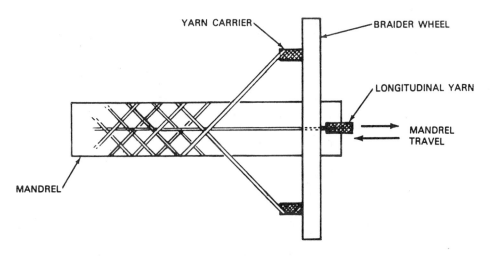

Figure 18.28 Braiding with the addition of longitudinal tows. (Courtesy McDonnell Douglas.)

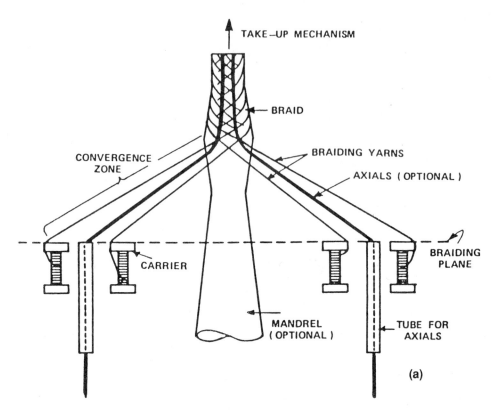

Figure 18.29   Examples of a horizontally mounted small braiding operation: (a) using a mandrel; (b) without using a mandrel. (Courtesy McDonnell Douglas.)

ing wheel. The number of carriers and the track in which they travel dictate the diameter of the braider.

Example of an Automated System

The braided, seamless encasement (which may be round or otherwise) is then impregnated with epoxy resin, cured at elevated temperatures, and the mandrel removed for further use. The proprietary impregnation and cure system used by McDonnell Douglas Corporation at their Titusville, Florida missile production facility is as follows.

1. Braid two layers of 6-end glass yarn over a hollow tubular mandrel, with a thin random fiber mat between the layers (automated system with two braider wheels in tandem, and automatic mandrel transport).
2. Place a premeasured piece of frozen epoxy resin and catalyst on the dry braided tube.
3. Insert the braided-over mandrel with epoxy into a special tubular chamber with a calrod heater in the center and a silicone bladder around the inside wall of the chamber.
4. Start the programmable controller (the system is fully automatic).

   Pull 28 in. vacuum on inside of mandrel and ends of braid.
   Start heating inside of mandrel with calrod (melts the resin).

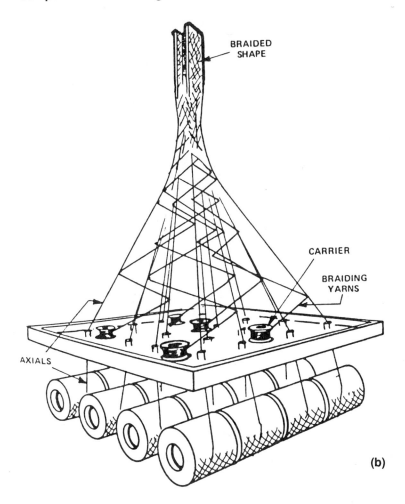

BRAIDED
SHAPE

CARRIER

BRAIDING
YARNS

AXIALS

(b)

Apply 100 psi air pressure on outside of bladder (forces resin into braid).
Increase temperature up a programmed ramp to 250°F (minimum viscosity).
Hold temperature for 30 min.
Turn off heater and flow water through center core to cool mandrel.
Turn off vacuum and pressure. Eject mandrel and B-staged braid.

5. Remove mandrel and recycle.
6. Cure tube at 300°F in oven to full cure.

Nearly 300,00 rocket tubes (3 in. diameter, 26 in. long) were made with this system. Four people produced 100 tubes per 8-hour shift through this operation (not including the additional operations that were required to complete the launch tube assembly).

Glass, carbon, aramid, ceramic, polyester, and polyethylene fibers are being braided— having come a long way from the fabrication of shoe strings, the original use for the process. Figure 18.32 shows tennis rackets braided with glass and graphite prepreg. The original rackets were made from braided graphite only, but were unsuccessful due to the extreme stiffness of the racket. This transmitted the forces to the player's wrist, causing

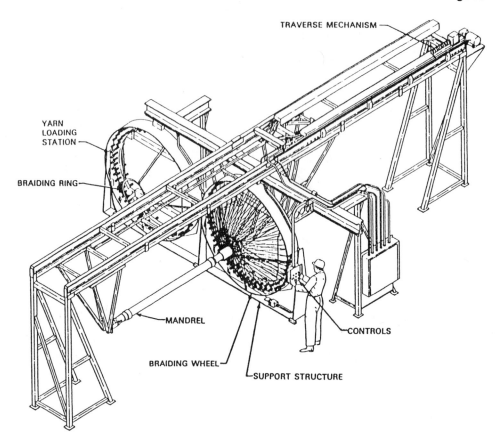

Figure 18.30 Example of a large-diameter production braider setup. (Courtesy McDonnell Douglas.)

damage to his or her arm from hitting the ball. The addition of glass acted as a damper, and made the product very successful.

18.2.3 Pultrusion

Pultrusion is a technique whereby the fibers are pulled through a heated die as shown in Figure 18.33. This figure has an H cross section to emphasize the fact that this technology is very useful in the production of beams (I beams or H beams) as well as square or cylindrical beams. In Figure 18.34, the addition of broadgoods to the rovings and an injection system for the resin is shown. In those applications where there is a need to produce "bar stock" at exceedingly low cost, this is by far the preferred route. Such methods can be used to advantage in the production of ribs and stiffeners within the aerospace industry, and in the production of stock materials for the civil engineering, infrastructure, and sporting goods fields.

Pulforming

Pulforming is a process that was developed to produce profiles that do not have constant cross-sectional shape, but that do have a constant cross-sectional area at any point along

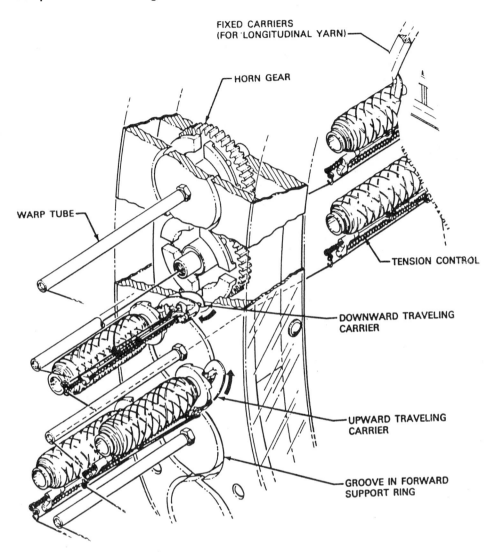

FIXED CARRIERS
(FOR LONGITUDINAL YARN)

HORN GEAR

WARP TUBE

TENSION CONTROL

DOWNWARD TRAVELING
CARRIER

UPWARD TRAVELING
CARRIER

GROOVE IN FORWARD
SUPPORT RING

Figure 18.31 Details of the braider wheel mechanism, showing the upward and downward carriers in their continuous figure 8 track; the warp tubes for feeding in longitudinal tows; and the adjustable yarn tension control device. (Courtesy McDonnell Douglas.)

the length of the profile. The materials are pulled from reinforcement creels and impregnated with resin, and in some cases combined with a bulk molding compound charge. The material can be preheated using RF energy as in the pultrusion process.

At this point, the process technology departs from the conventional pultrusion process. Beyond the fiber impregnation area is a horizontal table on which is mounted a continuous ring of open female molds. As it operates, the table rotates like a carousel, pulling the wetted fibers through the process. The second mold half, or in some cases a flexible steel belt, is closed or held against the bottom mold. Since the mold and the belt are heated, the material within the closed mold is pulled and cured accepting the contoured mold profile.

When a two-part mold is used, it opens at the completion of the cycle, moves to the

Figure 18.32   Braided tennis racket.

side, and is redirected to the front of the machine to repeat the process. The finished, cured product continues its path and moves into a cutoff saw which is synchronized to cut at the end of each part. To produce curved pulformed parts, a heated steel belt is used to close the mold. The radius of the part determines the number of molds utilized in this process. A curved part made in this manner could be an automotive leaf spring. Other common closedmold products include hammer handles and axe handles.

18.3   FABRICATION WITH CHOPPED FIBERS

In addition to the ratio of fiber to resin in the composite, the orientation, length, shape, and composition of any of the fibers selected determines the final strength of the composite—and the direction in which the strength will be the greatest. There are basically three types of fiber orientation. In one-dimensional reinforcement, the fibers are parallel and have maximum strength and modulus in the direction of fiber orientation. In two-dimensional (planar) reinforcements, different strengths are exhibited in each direction of fiber orientation, as in broadgoods. The three-dimensional type is more isotropic, but has greatly decreased reinforcing values (all three dimensions are reinforced, but only to about one third of the one-dimensional reinforced value). The mechanical properties in any one direction are proportional to the amount of fiber by volume oriented in that direction. As fiber orientation becomes more random, the mechanical properties in any one direction become lower.

The virgin tensile strength of glass is around 500,000 psi, and the tensile strength of the various matrices is in the range of 10,000–20,000 psi—therefore the fiber and its orientation is much more important than the matrix selection as far as strength of the composite is concerned. Factors such as cost, elongation, strength at temperature, and general "workability" in the manufacturing process may become more important in matrix selection.

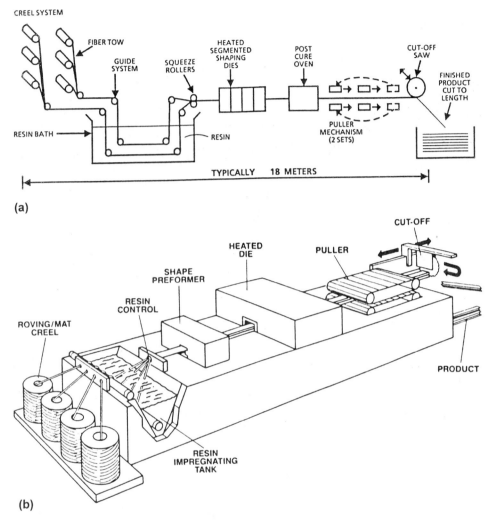

Figure 18.33 Example of pultrusion of an H beam, using dry rovings, and a liquid bath: (a) process schematic; (b) machine set-up.

18.3.1 Sprayup

In sprayup, the roving is fed into a specially designed "chopper gun" which chops the roving into approximately 1-in. lengths and simultaneously sprays a predetermined amount of resin on the fibers and into the mold. Sprayup is best suited to low to medium production, but has a greater production rate (faster mold turnover), produces more uniform parts with skilled operators, and can utilize more complex molds than hand layup of broadgoods. Factory installations can be highly mechanized; while portable equipment allows in-field repair, onsite fabrication, and product maintenance.

In this process, reinforcement fibers (usually glass) are deposited simultaneously in a mold by specialized spraying equipment. Hand or automatic spray guns/dispensing devices, either airless (hydraulic) or air-atomization types, are in common use. A marriage

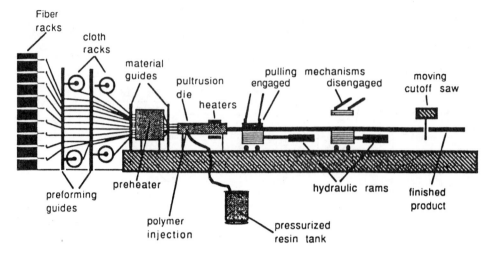

Figure 18.34 Example of pultrusion with fiber rovings and broadgoods. Resin is injected under pressure after preforming and preheating the material, and before final cure in the pultrusion-forming die.

of both, called air-assisted, augments airless with external air to shape and improve the pattern. This utilizes the best features of both types, and is rapidly becoming the standard. Glass roving passes through a chopper, is chopped to predetermined length, and is projected to merge with an atomized resin-catalyst stream. The stream precoats the chop, and both are deposited simultaneously on the mold surface. Special equipment controls fluid volumes and pressures, catalyst-to-resin ratio, fiber volume, and chop length. The deposited composite is rolled with a hand roller to remove air and to lay down fiber ends. Composite deposition is dependent upon operator skill. Multiple passes can build up nearly any required thickness. Typical glass fiber-reinforced uses are boats, automobile components, electrical consoles, aircraft parts, helmets, and tanks.

Special automatic units (vertical, horizontal, rotary—or universal, articulating robots) greatly enhance quality and production rates. Panels in various widths made in endless lengths, or seamless necked tanks, can be made completely by machine. Waist and shoulder pivot, elbow extension, and wrist pitch-yaw-rotate movements can be combined with a transverse axis to provide a total of seven axes. Products produced by robots include

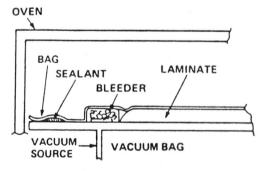

Figure 18.35 Example of vacuum cure system for layup or sprayup. Polyvinyl acetate (or other film) is placed over the wet part, and joints are sealed with plastic.

snowmobile hoods and car fenders. Custom automatic mechanical (nonrobotic) package installations can include conveyors, exhaust systems, resin supply systems, pumping stations, and a traversing carriage with spray gun and chopper. Controls allow automatic gel-coat and sprayup with automatic rollout.

A vacuum bag, or a film of polyvinyl acetate, polyethylene, or nylon film, may be used to cover the sprayup (see Figure 18.35). The film is sealed at the edges, vacuum applied, and the part is cured—either at room temperature or in an oven—depending on the matrix used. Atmospheric pressure forces out entrapped air, improves resin distribution, lays down fiber ends, and glazes the surface. Physical properties are enhanced, and surfaces away from the mold are improved. The most commonly used matrix is polyester, although others can be applied if required.

18.3.2 Compression Molding

Compression molding in general is the process of inserting a measured amount of molding material into one half of a mold (or die), and closing the mold under pressure in a press. The die is heated, the part formed, and the mold opened for part ejection. We will discuss one of the more common methods of making composite parts with this method.

Sheet Molding Compounds

The basic process for making sheet molding compounds (SMCs) was developed and refined at the Dow Corning Research Labs in Ohio. A thin plastic film is continuously unrolled onto a wide (36–48 in.) belt conveyor. The film passes under a series of chop guns, where the glass roving is cut to $1/4$–2 in. lengths, and deposited dry onto the moving film. As this dry, random fiber mat progresses down the moving belt, liquid polyester resin is applied to the fiber mat. A second plastic film is automatically placed over the wetted mat, and at the next workstation the material enters a series of corrugated rollers which knead the material, mixing the resin into the random fibers. It is then rolled up into a large roll and partially cured. The material is frozen and shipped to the fabricator as SMC. It is then unfrozen as needed for production, and cut into strips or other shapes needed to make the particular part. The pot life of the SMC at this point is quite reasonable, and in most cases it does not have to be refrozen to halt the cure. At ambient temperature the material at this stage feels like thick leather, or rubber sheets, and can be handled with some amount of automation. In practice, the sheets are then stacked onto half of the heated mold, to achieve the desired thickness (after removing the film backing which prevents the SMC from sticking to itself), and the mold is closed in a press. Although the shape of the precut SMC need not be exact prior to compression molding, the total weight must be correct, and the various pieces of SMC placed in the mold where tests have shown they are needed to achieve the correct thickness and fiber location.

18.3.3 Transfer Molding

There are at least two types of transfer molding. One system utilizes a fiber preform in a closed mold, and a resin is pumped in under pressure. The other method heats and forces a fiber matrix into the mold. Both are cured in the mold and then ejected.

Resin Transfer Molding

In classical resin transfer molding (RTM), the fiber reinforcement is constructed as a somewhat rigid preform assembly. One popular technique is to make a form that approximates one side of the finished part from a porous material, such as common screenwire. A chop gun, as described earlier under sprayup techniques, is used to build up a thickness of fibers on the wire premold, making a preform. The binder may be a version of the final matrix to be used in the production part, or more often simply an inert binder that holds the preform in the approximate shape desired. Oven cure of the preform will speed up the drying process, permit handling, and allow reuse of the wire form. The fiber preform is then inserted into the lower half of the mold (die), and the mold is closed and locked. Application of vacuum to the cavity is followed by injection of the heated resin matrix into the mold, where it is held under pressure until the resin is set up by the application of additional heat, or the appropriate time has passed.

There are many applications of this technique, using a dry filament-wound or braided preform, or by laying broadgoods into the cavity prior to injecting the resin. The resin content tends to be somewhat high in many applications, and is totally dependent on the fiber density of the preform. However, it does permit placing the fibers in the product at the point where they are most needed.

Other Types of Transfer Molding

Preforms of sheet molding compound (SMC) or bulk molding compound (BMC) are placed under a hydraulic ram in a cylindrical pot above the mold cavity. The material is heated to the point of softness, and the warm material is forced through a small gate into the mold cavity, which has been preheated to 430–600°F. Increased shear at the gate during transfer molding induces random fiber orientation, and the resulting components have more isotropic properties.

This process has been used for molding thermosets such as phenolics for some time. More recently, the same process has been applied to polyimides. Both SMC and BMC forms of PMR-15 polyimide can be used in transfer molding because prior imidization removes all volatiles. Materials must be volatile-free for transfer molding of this type, since the mold is a closed system with limited venting capacity. With sufficiently large runners and gates, transfer molding is done successfully with relative long-fiber SMC (to ½ in.). Strips are cut and rolled up, imidized in an oven, preheated, and inserted into the transfer pot. Successful transfer molding may offer economical shorter molding cycles than compression molding, ranging from 8 to 20 min.

18.4 REFERENCES

1. Suppliers of Advanced Composites Materials Association (SACMA) Personnel Needs in the Advanced Composites Materials Industry: An Assessment, 1988; for a market assessment, see Peter D. Hilton and Peter W. Kohf, *Spectrum: Chemical Industry Overview Portfolio,* Arthur D. Little Decision Resources, September 1987.
2. Watts, A. A. (ed.), *Commercial Opportunitues for Advanced Composites,* ASTM Special Technical Publication 704, American Society for Testing Materials, Philadelphia, 1980.
3. Economy, J., High-strength composites, in *Biotechnology and Materials Science,* Good, Mary L., (ed.), American Chemical Society, Washington, DC, 1988.
4. Maguire, J. F., P. Paul, and M. Sablic, Computer-aided molecular design, synthesis, and magnetic processing of polymer composites, Final Report 06-9730, Southwest Research Institute, San Antonio, TX, 1995.
5. Lee, Stewart M. (ed.), *International Encyclopedia of Composites,* VCH, New York, 1990.
6. Klein, Allen J., Automated tape laying, *Advanced Composites,* pp. 44–52, (January 1989).

7. Strong, A. B., *Fundamentals of Composites Manufacturing,* Society of Manufacturing Engineers, Dearborn, MI, 1989.
8. Leonard, LaVerne, Composites cutting comes of age, in *Adv.Composites,* pp. 43–46 (September 1986).
9. Cook, R .J., Waterjets on the cutting edge of machining, SAMPE Int. Symp., 31: 1835 (1936).
10. Hall, Terence F. W., Manufacturing automation/polymer composites, in *International Encyclopedia of Composites*, Vol. 3, pp. 133–142, VCH, New York, 1910.
11. Maguire, J. F., Peggy L. Talley, Sanjeev Venkatisan, Mark Wyatt, and Tom Rose, Cure sensing systems for the efficient and cost-effective manufacture of parts fabricated in composite materials, Final Report 06-3658, Southwest Research Institute, San Antonio, TX, 1995.
12. Lee, E. H., and B. W. Shaffer, *J. Appl. Mech.,* 18: 4 (1951).
13. Everstine, G. C., and T. G. Rogers, *J. Composite Materials*, 5: 94 (1971).

19

Finishing

Frank Altmayer   *Scientific Control Laboratories, Chicago, Illinois*

with

Robert E. Persson   *EG&G, Cape Canaveral, Florida*

Jack M. Walker   *Consultant, Manufacturing Engineering, Merritt Island, Florida*

19.0 INTRODUCTION TO FINISHING

Metals are usually found in nature as ores in the oxide form, their lowest energy level state. When man refines a metal, nature begins at once to return it to its lowest energy level. Today we understand the mechanism of corrosion, and the treatments and protective finishes required to make it feasible to use steel and other metallic structures in environments where they would not otherwise have an economic life.

This chapter starts with an explanation of the corrosion mechanism, and the various types of corrosion. This is followed by a discussion on the cleaning and surface preparation processes, that are necessary to prepare the metal for its ultimate finish. Electroplating, the most widely used protective and decorative finish, is introduced in considerable detail. The various processes are explained, and the process steps outlined for them. The final section covers the other decorative and protective finishes including paint, powder coating, metal spraying, and porcelainizing.

19.1 CORROSION

Robert E. Persson and Jack M. Walker

19.1.0 Introduction to Corrosion

Metals are usually found in nature as ores in the oxide form, their lowest energy-level state. When humans refine a metal, nature begins at once to return it to its lowest energy level. *Corrosion* is defined as the adverse reaction of a refined metal with its environment. The rate of corrosion on any specific metal varies because of temperature, humidity, and chemicals in the surrounding environment. Variations of intensity of these factors have great impact on the rate of corrosion. Rules of thumb indicate that a temperature increase of 10°C doubles the chemical activity if a suitable electrolyte is present. An increase in humidity or conductivity increases the rate of corrosion accordingly. When the environment contains several factors that aid corrosion, the corrosion rate is usually greater than the resultant of the forces would indicate. For example, Figure 19.1 illustrates corrosion rate differences resulting from the two environmental variables of oxygen concentration and temperature.

The contamination of expensive chemicals and the corrosion of valuable structures and equipment is going on about us every day. It has been estimated that the direct cost to industry and home approaches $10 billion annually in the United States. When the indirect costs are added, this becomes $15 billion! The corrosive effects on all metal products is a very large concern to everyone in the manufacturing industries. Several types of corrosion, however—those associated with specific areas such as high temperature, nuclear activity, and the like, are not discussed in this chapter. Most of the problems that arise in the manufacturing industries are of an electrochemical nature. We need to have a clear understanding of the processes by which a useful item is reduced to a collection of rusty or corroded scrap if we are to be effective in controlling corrosion. Intelligent selections

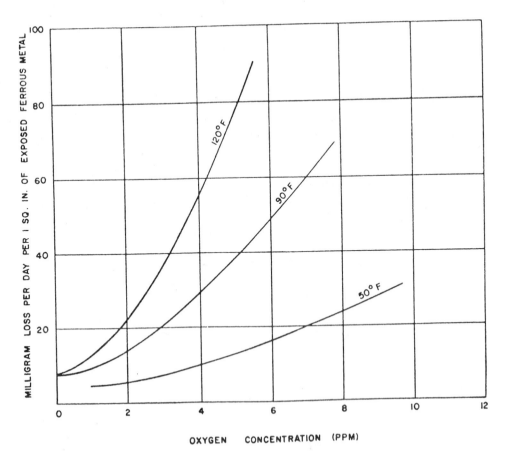

Figure 19.1 Corrosion rate differences with changes in temperature and oxygen content.

of metals and protective coating systems during the design and manufacturing phases are an important part of the manufacturing engineer's task. Continuing maintenance of facilities, equipment, and products is greatly reduced by correct design and initial manufacture.

19.1.1 Principles of Corrosion

Corrosion is almost always a detrimental reaction, but there are some exceptions. One that serves humanity is the wet-cell automobile storage battery. To understand the principles of corrosion, we will examine the operation of this familiar object in some detail.

The Automobile Storage Battery

A wet-cell storage battery is a box holding a number of lead plates. Half of the plates are made of metallic lead, and the others of lead oxide. If we were to hook a wire to two of these plates and insert a lightbulb in a circuit, as shown in Figure 19.2, nothing would happen without the addition of an electrolyte. This is a mixture of battery acid and water. Battery acid is a 30% solution of sulfuric acid. The electrolyte conducts current inside the battery in addition to taking part in a chemical change.

When sulfuric acid is mixed with water, it undergoes a change that makes it possible to function as an electrolyte. A particle of acid is split into ions. These are charged chemical particles capable of conducting an electrical current. Each particle of acid splits into hydrogen ions which carry positive electrical charges, and sulfate ions which carry negative electrical charges. In dilute solutions of sulfuric acid, all of the acid present undergoes this change. Equal amounts of positive and negative charges are developed. The ability of an electrolyte to conduct current depends directly on how many ions are available. Pure water has relatively few ions and is a poor conductor of electricity. It has a pH of 7.0, which means that 1 part in 10 million parts is present as hydrogen ions and hydroxyl ions. The absence of available ions to carry electricity in pure water accounts for the addition of sulfuric acid to make the automobile storage battery electrolyte.

Everything in nature, including metals, has some tendency to dissolve in water and hence in dilute acid solutions. In the case of our battery, we have lead in two forms. Of these two, metallic lead has the greater tendency to dissolve in battery acid solution and goes into solution in the form of lead ions. These are *positively* charged electrical particles. Since opposite charges attract, the positive lead ions are attracted to the *negative* sulfate

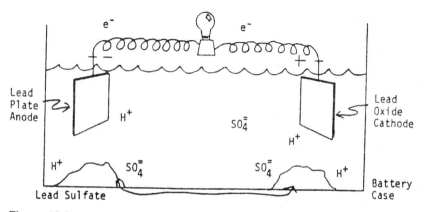

Figure 19.2 Wet-cell storage battery with lightbulb in circuit.

ions and produce the chemical compound lead sulfate. A deposit of sulfate builds up on the lead plates of a storage battery as it discharges. The lead in the battery plate was *neutral* electrically, and small particles go into solution as positive lead ions. There is then a surplus of negative electricity (electrons) left on the plate. These electrons flow through the wire circuit to light the electric bulb in the external circuit. Electrons are negative particles of electricity and subsequently flow to the lead oxide plate, where further chemical changes take place. Upon receipt of electrons, the lead oxide is converted to lead ions and oxide ions. The lead ions combine with passing sulfate ions to form lead sulfate. In Figure 19.2, we show two piles of lead sulfate in the bottom of the battery. In actual practice, the lead sulfate is deposited on the plates. This action can be reversed by driving an outside source of electrical current in the opposite direction and restoring lead, lead oxide, and sulfuric acid as ions. The lead plate is the positively charged anode, and the lead oxide plate is the negatively charged cathode.

Sulfuric acid is not unique in its ability to undergo a change upon solution in water to form ions. Indeed, all soluble salts, alkaline or caustic compounds, and mineral acids share this ability to split into active ions. Other mineral acids, such as hydrochloric, nitric, phosphoric, chromic, etc., liberate the same hydrogen ion in solution with water as sulfuric acid. The hydrogen ion is the essence of acidity. Sulfuric acid is used in storage batteries rather than the other popular mineral acids simply because of its low cost and low volatility.

This same phenomenon explains the reduction of steel and other metals into useless rust and corrosion products. Corrosion is an electrochemical change identical in fundamental principle to the transformation of lead and lead oxide to lead sulfate in a storage battery.

19.1.2 Surface Corrosion

We can now set up an electrolytic cell or *corrosion battery* to describe the type of rusting suffered by steel in normal atmospheric or marine environments. This corrosion battery exists whenever metal is exposed to a combination of oxygen, water, and ions. This combination, the "big three" of corrosion, is not hard to find. All the *oxygen* we will ever need is available in the atmosphere, and all exposed metal is constantly in contact with it. The atmosphere, except for desert areas, is usually *moisture* laden. Either moisture vapor or dew droplets are in contact with the iron surface. Smog, salt spray, marine locales, industrial soot and fumes, splash and spillage, soil contamination, and other sources can provide minerals, salts, acids or alkaline materials which will dissolve in available moisture to produce *ions*.

Corrosion on a Steel Plate

Figure 19.3 shows the tiny "corrosion battery" which will produce rust when current flows. We have shown a greatly magnified portion of the iron surface. Adjacent areas, not visible to the eye, act as anode and cathode. On and just above the surface is moisture loaded with ions, to serve as an electrolyte. Gaseous oxygen is dissolved in the electrolyte, and is freely available to the surface of the metal. Note that, in this case, the electrolyte provides the external circuit. The internal circuit is provided by the metal part itself, since it is highly conductive from point to point. In Figure 19.4, our corrosion battery is operating:

1. At the anode, an atom of metallic iron is converted to *iron ions*, and immediately reacts with water to form rust.
2. *Current flow* is set up between the anode and cathode.

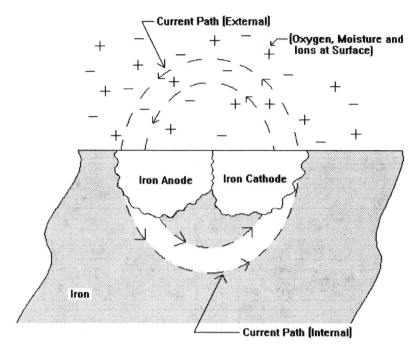

Figure 19.3 Typical corrosion battery.

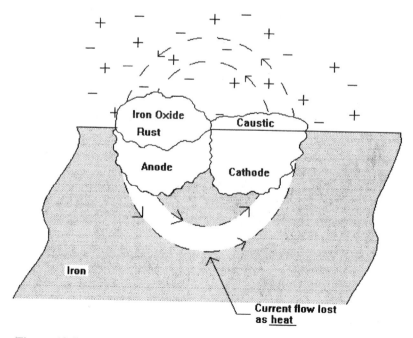

Figure 19.4 Corrosion battery in operation.

3. The small currents generated here are dissipated as *heat* through the adjacent body of metal.
4. At the cathode, receipt of the current results in transformation of oxygen to *caustic ions*. This results in a highly alkaline solution on the surface, as the caustic ion dissolves in the moisture on the surface.
5. At and above the surface, current is carried by the dissolved ions.
6. As the iron anode is eaten away, a growing deposit of rust forms.
7. This tiny battery is duplicated millions of times over a large iron surface, so that eventually the eye sees what appears to be a uniform layer of rust building up.
8. Note that the iron cathode is not attacked. However, it performed its function in providing the locale at which oxygen could undergo its necessary change to a caustic.
9. This process will continue until (a) all the iron is converted to rust, or (b) something happens to weaken or break the circuit.

Edges and other areas subjected to mechanical work (shearing, drilling, sawing) are prone to be anodes, and will corrode preferentially. Particular care must be paid to protection of edge areas or other sharply prominent parts of a metallic structure.

Crevice Corrosion

The presence of crevices or pockets in the design of a steel structure presents a special corrosion problem and one that is extremely difficult to combat successfully. Crevices have a tendency to corrode at a far greater rate than the adjacent flat metal areas.

We may use a simplified version of the "corrosion battery" to examine crevice corrosion, as shown in Figure 19.5. The essentials of the corrosion processes—oxygen, water, and ions—must be present. Crevices certainly hold soluble salts and liquid solutions of these salts, so more than enough water and ions are present to form an electrolyte. The remaining ingredient involved is oxygen. As we have observed in the iron "corrosion battery" and in the galvanic "corrosion battery," the change of oxygen to caustic is

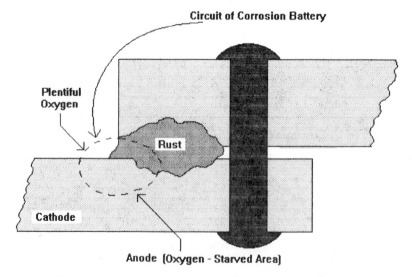

Figure 19.5 Crevice corrosion with two areas of metal.

associated with the cathode area (the area protected). If the two areas of metal are involved in a "corrosion battery," such as in Figure 19.5, the area exposed to the most oxygen will become the cathode, thereby forcing the other area (oxygen starved area) to act as an anode and be corroded. This is the situation that exists in crevice corrosion. The metal in the crevice is an oxygen-starved area and is therefore subject to corrosion and acts as an anode.

It is known from field experience that it is very difficult to stop this type of corrosion. Application of a coating over the crevice is usually not satisfactory for two reasons:

1. Trapped salts and rust laden with corrosion products and water will promote early blistering beneath the coating, leading to premature film failure.
2. The sharp edges of the crevice are likely to be inadequately covered with the coating, therefore edge rusting can start promptly. Film rupture along the edges, followed by progressive adhesion loss, will result in unsatisfactory coating performance.

The most practical way to eliminate crevice corrosion is to avoid structural design details that create crevice conditions (lap welds, lap sections riveted together, etc.) In existing structures, crevices should be filled with appropriate mastic or fillers, and proper welding techniques should be observed. Weld spatter, pockets, cracks, and other crevice-type spots in the area to be coated should be eliminated.

Pitting Corrosion

Pitting corrosion takes a heavy toll on machinery and equipment due to the concentration of the electrical cycle on a very small anodic area. When protective films break down or pinholing occurs, local corrosion or pitting may follow. An anode will be formed at the point where the film break occurs and, being in contact with the external electrolyte, will thus bear the full onslaught of any attack by a surrounding cathodic area. See Figure 19.6. Generally, the smaller the area that is in contact with the electrolyte, the more rapid and severe the pitting will be.

Strong Acid Corrosion

A corrosion battery of great industrial importance and frequent occurrence is that formed by the action of *strongly acidic* materials on iron or steel. See Figure 19.7. Briefly, a vigorous reaction occurs to dissolve the metal into solution at the anode, and transforms hydrogen ion (H^+) into hydrogen gas bubbles (H_2) at the cathode. The reaction involved

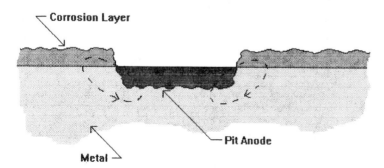

Figure 19.6 Pitting corrosion caused by a breakdown in the protective coating, or "pinholing" in the coating.

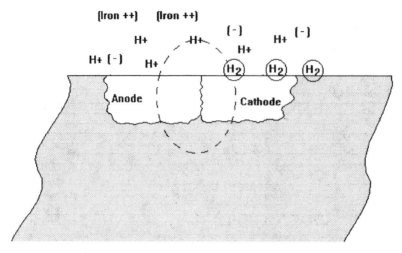

Figure 19.7 Corrosion battery caused by exposure to a very strong acid solution.

here is a vigorous one, and will continue until all of the available metal is consumed, or until all available acid is used up.

The dissolved metal in this battery will not form rust (at least not initially), but will remain dissolved in the acid electrolyte as an iron salt as long as sufficient moisture is present and as long as an excess of acid is present. When the available acid is consumed (for example, after an acid splash on steel), and the moisture evaporates somewhat, copious rust will then form and become deposited over the corroded areas of the metal surface.

19.1.3 Internal Corrosion

The corrosion phenomena discussed in Subsections 19.1.1 and 19.1.2 apply generally to all metals. However, some metal alloys have additional corrosion problems. When the alloy consists of two or more metal elements that are widely separated on the galvanic scale, as discussed in Subsection 19.1.4, the corrosion battery may develop at the boundary layer between grains, or within the grain.

As an example, the initial strength of the heat-treatable aluminum alloys is improved by the addition of alloying elements such as copper, magnesium, zinc, and silicon. In the heat-treatment process, the initial heating is designed to put the soluble element or elements in solid solution with the aluminum. This is followed by rapid quenching, which momentarily "freezes" the grain structure. In some cases, this is followed by a second controlled heating at a slightly elevated temperature. The precipitation of the alloying constituents into the boundary layer between grains essentially "locks" the grains in position to each other. This increases the strength, and reduces the formability.

Intergranular Corrosion

The aluminum alloys that are given high strength by heat treating contain considerable amounts of copper or zinc. If those containing copper are not cooled rapidly enough in the heat-treating process to keep the copper uniformly distributed, there is a concentration of copper at the grain boundaries, setting up a potential difference there. Corrosion occurs in damp air, eating into the boundaries, causing "intergranular corrosion" and resultant brittleness, with lowered fatigue resistance from the notches thus formed. Precise control of the heating and quenching temperatures and times is a minimum requirement. See Figure 19.8.

Stress Corrosion

The aluminum alloys that are rather high in zinc or magnesium are subject to *stress corrosion*. This behavior is not confined to aluminum alloys but is quite prevalent in many alloys that are constantly loaded and simultaneously subjected to a corrosive environment. Even slow general corrosion may be accelerated by stress, but accelerated local attack may occur either at grain boundaries or in the grain. Some magnesium alloys display stress corrosion, and in some corrodents, even steel shows such attack. In an oversimplification, one may look at this behavior as due to the tensile stress tending to pull the grains apart more and more as corrosion occurs between them. See Figure 19.9.

Corrosion Fatigue

All corrodible alloys are also subject to *corrosion fatigue*. Corrosion pits or intergranular penetrations naturally act as stress raisers. Corrosion, simultaneous with repeated stress, produces far more and more rapid damage than these stress raisers would inflict were corrosion not going on after they were formed. The intermittent release of stress, or its reversal, permits the pits and fissures to close a bit, and if corrosion products are trapped therein, they may act like chisels to help pry the metal apart. Steel, which in the absence of corrosion has an *endurance limit*—a stress whose repetition it will withstand indefinitely—has no such definite limit under corrosion fatigue. Sometimes the endurable repeated stress for even a reasonable life before replacement is necessary is only some 10% of the repeated stress it would withstand indefinitely in a noncorrosive environment.

To defeat intercrystalline corrosion, stress corrosion, and corrosion fatigue, the attack-

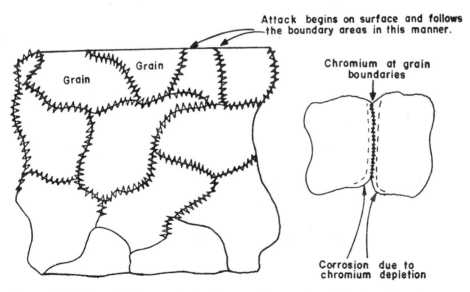

Figure 19.8 Corrosion battery showing intergranular corrosion of aluminum alloy due to copper migration to the grain boundary layers as a result of improper heat treatment.

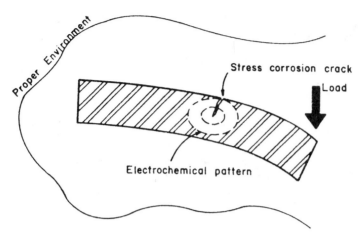

Figure 19.9 Stress corrosion crack in metal due to material being constantly loaded and simultaneously subjected to a corrosive environment.

able metal must be kept out of contact with the corrodent, by a complete and impervious coating. Subchapters 19.3 and 19.4 describe some of these coatings and their application processes.

19.1.4 Galvanic Corrosion

Situations often arise where two or more *different* metals are electrically connected under conditions permitting the formation of a corrosion battery. A situation then exists where one metal will be corroded preferentially in relation to the other metal to which it is physically connected. This is termed *galvanic corrosion*. Three ingredients are required— an electrolyte, a material to act as an anode, and another to act as a cathode—in addition to the metals. The electrolyte is the medium in which ionization occurs. The electrons flow from the anode to the cathode through a metal path. The loss of metal is always at the anode. Figure 19.10 illustrates the electrical and chemical interplay of electrochemical corrosion of two different metals. While this is similar to the corrosion battery discussed previously, it is usually faster acting and more severe.

Anodic reactions are always *oxidation* reactions which tend to destroy the anode metal by causing it to go into solution as ions or to revert to a combined state as an oxide. Cathodic reactions are always *reduction* reactions and usually do not affect the cathode material. The electrons that are produced by the anodic reaction flow through the metal and are used in the cathodic reaction. The disposition of the reaction products is often decisive in controlling the rate of corrosion. Sometimes they form insoluble compounds that may cover the metal surface and effectively reduce the rate of further corrosion. At other times the reaction products may go into solution or be evolved as a gas, and do not inhibit further reaction. Galvanic corrosion is an extremely important corrosion process and one that is frequently encountered. An understanding of it will be helpful in rounding out our knowledge of corrosion processes. The principles of galvanic corrosion may actually be utilized to advantage in the cathodic protection of surfaces by using sacrificial metal anodes or inorganic protective coatings.

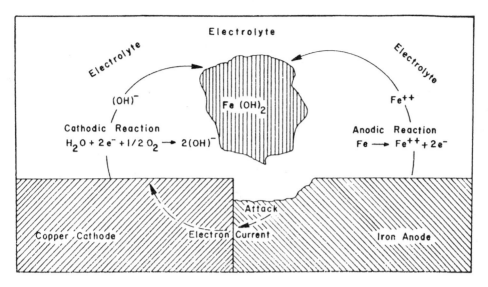

Figure 19.10 Example of galvanic corrosion between iron and copper.

The Galvanic Scale

Corrosion occurs at the rate which an electrical current, the *corrosion driving current*, can get through the corroding system. The driving current level is determined by the existing electrode potentials. Electrode potential is the tendency of a metal to give up electrons. This can be determined for any metal by measuring the potential between the specimen (metal) half-cell and the standard (hydrogen) half-cell. Tabulating the potential differences between the standard (hydrogen) half-cell and other elements opens an extremely important window to view one part of the corrosion spectrum. Such a tabulation, known as the *electromotive series,* is illustrated in Figure 19.11.

Utilizing the electromotive series, an engineer can determine the electrical potential between any two elements. This electrical potential is the algebraic difference between the single electrode potentials of the two metals. For example:

Zinc and copper:
$+0.76 - 0.34 = +1.10$ V potential
Iron and copper:
$+0.44 - 0.34 = +0.78$ V potential
Silver and copper:
$-0.8 - 0.34 = -0.46$ V potential

The electromotive series forms the basis of several possibilities for controlling and decreasing corrosion rates. It provides the data required to calculate the magnitude of the electrical driving force in a galvanic couple. The electrical driving force of an iron and copper couple may be thought of as promoting the following activities:

Oxidation of the iron (the anodic reaction)
Flow of electrons through the solid iron and copper couple
Cathodic reaction on the copper (reduction), where the electrons are used
Current (ionic) of $Fe^{2}+$ and $(OH)^-$ in the electrolyte

LITHIUM	- Li, Li+	+3.02 volts *
POTASSIUM	- K, K+	+2.92
SODIUM	- Na, Na+	+2.71
MAGNESIUM	- Mg, Mg++	+2.34
ALUMINUM	- Al, Al+++	+1.67
ZINC	- Zn, Zn++	+0.76
CHROMIUM	- Cr, Cr++	+0.71
IRON	- Fe, Fe++	+0.44
CADMIUM	- Cd, Cd++	+0.40
COBALT	- Co, Co++	+0.28
NICKEL	- Ni, Ni++	+0.25
TIN	- Sn, Sn++	+0.14
LEAD	- Pb, Pb++	+0.13
HYDROGEN	- H_2, H+	0.00
BISMUTH	- Bi, Bi++	-0.23
COPPER	- Cu, Cu++	-0.34
MERCURY	- Hg, Hg++	-0.80
SILVER	- Ag, Ag++	-0.80
PLATINUM	- Pt, Pt++	-1.2
GOLD	- Au, Au+	-1.7

+indicates valences

*Oxidation reaction voltages

Figure 19.11   Electromotive series.

It is possible to arrange many metals and alloys in a series, known as the *galvanic scale*, which describes their relative tendency to corrode. Figure 19.12 is a listing of some of the industrially important metals and alloys, including the ones which are most frequently encountered. Bearing in mind that a metal located higher on the scale will corrode preferentially and thereby protect a metal lower on the scale from corrosion attack, the example shown in Figure 19.12 may be set up.

Tendency to Corrode

When copper and zinc are connected together, the zinc will dissolve or be corroded preferentially, thus protecting the copper. Since the metal attacked is defined as the anode, the zinc will then serve as an anodic area. The copper will be the cathodic area or the metal protected. The function of each will be identical to that found in the corrosion battery set up for the rusting of iron. The intensity with which two metals will react in this preferential manner may be measured by the distance between the two metals in the galvanic scale. Thus, between brass and copper there would be only a weak tendency for the brass to corrode; whereas zinc or magnesium would dissolve very readily to protect silver. It may be noted in Figure 19.12 that magnesium is at the top of the scale and will have a tendency to corrode in preference to any other metal shown on the galvanic scale. Conversely, platinum, which is extremely inert, never corrodes preferentially.

Rate of Corrosion

While the tendency to corrode depends on the *kinds* of metal coupled together, the *rate* at which the corroding anode is attacked depends on the relative area of anodes and cathodes

Magnesium	Copper
Zinc	Aluminum Bronze
Alclad 3S	Composition G Bronze
Aluminum 3S	90/10 Copper-Nickel
Aluminum 61S	70/30 Copper Nickel- Low Iron
Aluminum 63S	70/30 Copper-Nickel- High Iron
Aluminum 52	Nickel
Low Steel	Iconel (nickel-chromium alloy)
Alloy Steel	Silver
Cast Iron	Type 410 (Passive)
Type 410 (Active)	Type 430 (Passive)
Type 430 (Active)	Type 304 (Passive)
Type 304 (Active)	Type 316 (Passive)
Ni-Resist (corrosion-resisting,	Monel (nickel-copper alloy)
nickel cast iron)	Hastelloy (Alloy C)
Muntz Metal	Titanium
Yellow Brass	Gold
Red Brass	Platinum

Figure 19.12   Galvanic series.

hooked together. Thus, in Figure 19.13, if we couple a small magnesium anode to a large area of steel (as in protection of a ship's hull), the anode area (being small compared to the cathode area) will corrode very readily. This is due to the entire impact of the galvanic current being concentrated on a small area of active metal.

Conversely, if the cathode area is small compared to the anode area, the corrosion of the anode will be relatively slow, since the demand on the anode is spread thin and any local spot loses little metal. The areas of each metal involved are those in electrical contact and not just the areas of metal in physical contact. The area of metals in electrical contact will be determined by those areas which are in contact with an external conductive circuit (electrolyte).

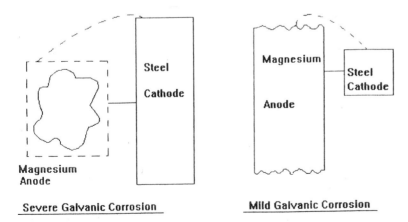

Figure 19.13   The rate of galvanic corrosion is dependent on the relative area of the anode and cathode.

Rivets as Fasteners

The use of rivets of one metal to fasten together plates of a different metal is an excellent example of the possible effects of galvanic corrosion. See Figure 19.14. When steel plates (anodes) are fastened with copper rivets (cathodes), only a very slow corrosion of the steel occurs, since the galvanic corrosive effect is spread out over a large area of steel. On the other hand, if copper plates (cathodes) are fastened with steel rivets (anodes), rapid rusting of the rivets will occur. The small area of rivets will be attacked by all the galvanic current generated by the large copper plates and could corrode rapidly.

19.1.5 Methods of Corrosion Control

The tendency of metals to corrode is a natural occurrence which must be recognized as inevitable. The corrosion engineer's job is to control these destructive effects at minimum cost. There are five principal methods in use:

1. Altering the environment
2. Using corrosion-resistant materials of construction
3. Cathodic protection
4. Overdesign of structures
5. Placing a barrier between the material and its environment

Each of these methods has certain characteristic advantages and disadvantages and certain areas of use where it is most economical. Since an industrial plant is a composite of many and various types of corrosion areas, no single method should be classified as a universal cure-all. Each situation must be studied individually and a decision reached based on such factors as available downtime, the possibility of equipment obsolescence, operating temperatures and cycles, appearance, environment, etc. For each separate corrosion

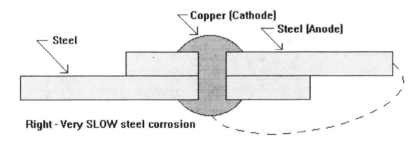

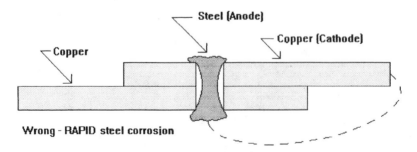

Figure 19.14 Effects of galvanic corrosion shown in rivets.

problem, it is necessary to weigh these individual factors and to pick the corrosion tool which provides the most economical means of protection.

Often this choice must be made in the design stage of a plant, so it is important that the corrosion or maintenance engineer be closely consulted during this period. If this practice is not followed, the architects and engineers may find that they have eliminated various economical weapons of corrosion control from the arsenal.

Altering the Environment

Altering the environment usually involves controlling accidental discharge of corrosion vapors, or the addition of inhibitors to liquids in a closed system. The use of chemical inhibitors is normally restricted to water supply or water circulating systems, steam and condensate lines, and brine systems. Since they are strictly in immersion solution, their usefulness in the maintenance field is definitely limited. In addition, proper caution must be exercised in the selection of quantities and types of chemicals. Improper selection or maintenance of inhibitor systems can often accelerate rather than retard corrosion. However, if used properly within their limited scope, they provide a simple and relatively low-cost solution to corrosion control.

Corrosion-Resistant Materials

Largely because of their price and structural qualities, steel and iron are the most widely used metals in industrial construction. Unfortunately, as has been discussed previously, these materials tend to corrode or revert to their oxides more rapidly than other types of metal. Therefore, the corrosion or design engineer may turn to the more inert metals or alloys to retard the corrosion process. In extremely severe exposures, this may be the only feasible answer. High-temperature operations combined with highly corrosive chemicals may produce a situation too severe for any other type of structural material or protection. In such cases, the relatively high initial cost of these metals is easily justified by their long life span.

Among the most common metals alloyed with steel or iron are chromium, copper, nickel, and molybdenum. Of the metals used in their natural state, aluminum is the most reasonably priced and widely used, while such rare metals as titanium and tantalum are employed only under the most severe conditions. The decision as to the use of this type of protection versus other means of control will depend to a great extent on the severity of exposure and the ultimate cost of the alternative methods. In the majority of plant maintenance, exposures are only mildly corrosive and the use of alloys and rare metals as construction materials is not economical.

In addition to alloys, many corrosion-resistant plastic materials are now available. Thermoplastic materials such as polyvinyl chloride and polyethylene are used in pipe and fume ducts. Glass-reinforced epoxy and polyester compounds are employed in process piping with a higher temperature limit, and can also be used in tanks or reaction vessels.

Cathodic Protection

A third tool available to the corrosion engineer is cathode protection. We have discussed how two metals can be coupled together to produce galvanic corrosion. While we are often concerned with reducing the corrosion of the active metal, it is possible to take advantage of the fact that the cathode metal is protected while the anode corrodes. By deliberately coupling two dissimilar metals together, we can prevent corrosion of the less active (cathode) at the expense of the other metal.

Therefore, to protect a steel surface, we would choose a more active metal, that is, one higher on the galvanic scale. Magnesium is commonly used for this purpose. When it is coupled electrically to steel, a magnified corrosion battery is set up in which the magnesium, because of its greater activity, becomes the anode and the steel the cathode. In so doing, the magnesium anode corrodes preferentially, leaving the steel cathode intact.

The same results can be achieved by providing the current from an external source applied to the metal we wish to protect. On such systems, generators, rectifiers, or batteries may be used as the DC source and, to prevent rapid disintegration of the anode, usually an inert metal will be selected. This electrical current and voltage must be of the correct amount—too little or too much will worsen the corrosion problem.

In the protection of marine equipment, hot water tanks, and underground and underwater pipe lines, cathodic protection has found its greatest use. Simplicity is certainly one of its prime values, and its effectiveness in the presence of a good electrolyte is unquestionable. In dry or damp areas its usefulness is limited, and it is replaced or complemented by protective coatings, or other methods.

Overdesign

Overdesign of structures refers to the common use of heavier structural members or thicker plates in anticipation of corrosion losses. This method is often used unwittingly, and excessive plate thickness is frequently specified as a matter of habit or custom, where lighter weights could be used if corrosion were prevented. There is now a trend toward the use of lighter structural members when suitable protective methods are employed.

The principal disadvantage of overdesign, or built-in corrosion allowance, is that neither the exact length of life nor the replacement cost of a corroding material can be predicted. The cost and the effectiveness of other methods can be determined much more accurately.

Barriers

Barriers are corrosion-resistant materials which can be used to isolate a material of construction, such as steel or concrete, from a corrosive environment. Examples are acid-proof brick or tile in conjunction with suitable cements, grouts, and bedding compounds, plastic sheeting, troweled-on resinous membranes, and spray-applied protective coatings. These barriers are the principal weapons in the arsenal of the corrosion engineer. Although each has its own area of use, careful analysis of an individual corrosion problem is necessary in choosing the most effective and most economical system.

Not only is it important to select the proper barrier for a given situation, of equal importance is the proposed surface preparation and the application technique. A variety of equipment and procedures is available for these purposes, and the ultimate performance of any coating or lining hinges on the correct decision and the proper follow-up during the application. In the following sections, these points will thoroughly investigated. Faying surfaces of two metals presents one of the most difficult areas for protective coatings, and is even worse when friction is involved. Use of inert barrier sheets may be most effective—even when there may be welding performed later. Another difficult task is in the proper coatings of steel springs, where friction is inherent in their function. A zinc-rich primer with an ethyl silicate carrier may be effective.

Corrosion Maintenance Plan

The importance of good design and proper protective coating systems cannot be overemphasized! However, an element sometimes overlooked is the maintenance plan. This can start with inspections to identify the type and location of corrosion, and to establish the beginning of a database. Problems can then be classified as to their severity, as very urgent, or to be corrected next month, or next year, as an example. This permits the preparation of a comprehensive maintenance plan, where all items subject to corrosion can be scheduled for maintenance at a specific interval—and eventually on a specific date.

19.1.6 Bibliography

Baumeister, Theodore, *Marks Handbook for Mechanical Engineers*, McGraw-Hill, New York, 1979.

Bendix Field Engineering Corporation, *Corrosion Control*, NASA Report MG-305, under NASA Contract NAS5-9870, 1968.

Gillett, H. W., *The Behavior of Engineering Metals*, John Wiley, New York, 1951.

The National Association of Corrosion Engineers (NACE), P.O. Box 218340, Houston, TX 77218-8340, publishes several books and articles devoted to corrosion, including:

   Munger, C. G., *Corrosion Prevention by Protective Coatings*

   Van Delinder, L. S., *Corrosion Basics—An Introduction*

The Steel Structures Painting Council (SSPC), 4516 Henry Street, Pittsburgh, PA 15213, also publishes books and other information, including the following:

   Steel Structures Painting Manual, Vol. 1, Good Painting Practice

   Steel Structures Painting Manual, Vol. 2, Systems and Specifications

19.2 CLEANING AND PREPARING METALS FOR FINISHING

19.2.0 Preparing Metals for Plating

Before any part can be electroplated, it must be properly prepared. In fact, the "secret" to being a successful electroplater or finisher is in the knowledge of how to clean and prepare a substrate for coating. An improperly prepared substrate typically results in poor adhesion of the plated metal, resulting in peeling and/or blistering. When prepared properly, the plated metallic layer will have about the same adhesion to the substrate as the individual metal atoms within the substrate have to themselves. There are some exceptions to this, notably plated metals on plastics, and metals that have a perpetual layer of tenacious oxide, such as aluminum, magnesium, titanium, and stainless steel.

Preparing a part for plating is typically performed in a series of tanks and rinses referred to as the *preplate cycle*. The preplate cycle typically is customized to the type of substrate that is to be plated. It may involve degreasing the part, then cleaning and acid pickling it, or it may be far more complicated. The process steps that are common to most parts are discussed below.

19.2.1 Precleaning for Plating

Most metal parts are covered with grease or oil during their manufacture, by operations such as stamping, drilling, forging, buffing, and polishing. Heat-treated steel parts may have a "scale" of iron oxides (rust) that would interfere with the appearance and adhesion of the plated coating. Other ferrous-based parts may have some rust due to unprotected storage. This "soil" must be removed by the metal finisher to avoid contamination of the processing solutions and to obtain adhesion of the coating to the plated part. *Precleaning* refers to the processing a metal finisher may perform on parts before they are routed through the plating line. Precleaning is typically accomplished in one or a combination of the following procedures.

Vapor Degreasing

A vapor degreaser typically consists of a stainless steel tank with a compartment at the bottom for boiling one of several solvents, such as trichloroethylene, perchloroethylene, methylene chloride, Freon, or 111—Trichloroethane. The boiling solvent creates a vapor zone within the walls of the tank. The vapors are condensed using cooling coils near the top of the tank. There are several variations of how to vapor degrease, but the basic principle involves hanging the greasy parts in the vapor zone on stainless steel hooks or wire. Larger degreasers for small parts are automated and the parts enter in steel trays and are routed through the equipment automatically. Since the parts entering the degreaser are cooler than the vapor, the solvent condenses on the parts and flushes off the oil/grease. The parts emerge from the degreaser in a relatively dry state, although some solvent may be trapped in pockets or drilled holes. The solvent/oil mixture returns to the boiling chamber, where the solvent is reboiled, while the oil/grease remains in the mix. Eventually, the oil/solvent mix must be removed, replaced, and disposed of. See Figure 19.15.

Since vapor degreasing equipment utilizes organic solvents, and since many of these solvents are either currently banned, to be banned, or possibly banned in the future, many companies are eliminating vapor degreasers and substituting parts washing systems that use aqueous cleaners. These parts washing systems are designed to remove oil and grease from a specific type of substrate. Parts washers designed to degrease multiple types of substrates invariably cause problems with one of those substrates.

Vapor degreasing using presently exempted solvents and modern systems designed to emit extremely low amounts of solvent are available today and may allow vapor degreasing to continue in the future. For removing waxes and polishing/buffing compounds, vapor degreasers currently have little or no competition.

Pickling/Descaling/Blasting/Shot Peening

Heavily rusted or scaled parts need to be processed through an operation that removes the heavy oxide from the surface of the part. There are a number of ways to accomplish this,

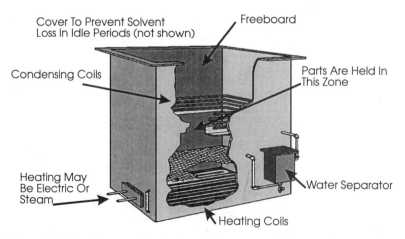

Figure 19.15 Cut-away of typical vapor degreaser used for removing organic soils.

including blasting with dry ice, nutshells, sand, or other "grit," pickling in strong solutions of acids, or descaling in an alkaline descaling solution consisting of concentrated sodium hydroxide and potassium permanganate at high temperature. Pickling and descaling is normally performed "off-line," because they are time consuming operations.

Steel and stainless steel parts that have been subjected to heat-treating processes or are designed to carry high tensile loads are typically shot-peened prior to finishing to relieve surface stresses that can enhance hydrogen pickup and result in hydrogen embrittlement. The peening process also imparts a slight compressive stress into the surface of the part, enhancing fatigue resistance. Shot-peening is conducted using equipment designed to impact the surface of the parts with media that may consist of steel, stainless steel, or other composition of shot. Impact force, medium geometry, and composition are all important parameters to control.

Soak Cleaning

A soak cleaning tank typically consists of a steel tank containing a cleaning solution consisting of strong alkalies, various other ingredients and detergents mixed with water at temperatures from 160 to 200°F. The parts to be cleaned are racked or put in plating barrels and are immersed in the cleaning solution; the oil and grease are either emulsified or converted to soaps through saponification. The parts emerge from the soak cleaner coated with hot cleaner, which then is either rinsed off before further processing or is dragged into the next process. Soak cleaners are normally incorporated into the plating process line and are usually the first tank the parts go into, so they may also be considered part of the preplate operations described below. See Figure 19.16.

19.2.2 Preplate Operations

Precleaning removes the bulk of the oils, greases, rust, scale, or other soils present on the parts. In "the old days," when most plating solutions contained large concentrations of cyanide, precleaning often was all that was required prior to plating. Modern metal finishing operations have replaced many cyanide solutions with noncyanide chemistries and have reduced cyanide concentrations in those processes where cyanide remains, so precleaning is not enough to prepare the parts for plating. The preplate operations must be tailored to the type of metal processed and the conditions of the surface of

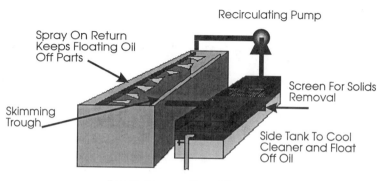

Figure 19.16 Typical soak cleaning tank with continuous oil removal.

the parts. The object of the preplate process is to remove the last traces of surface soils and to remove all oxides from the surface. The following preplate processing steps are typical.

Cleaning

Degreased parts still require additional cleaning to remove traces of soils left behind by the precleaning step. This is normally accomplished by an electrocleaning operation. Electrocleaning systems consist of a heated steel tank that contains a solution similar to the soak cleaner. The tank is equipped with a rectifier and steel or stainless steel electrodes hanging from bus bars on either side of the tank. Cleaning is normally the next process step after degreasing or soak cleaning. If the parts were vapor degreased, they are either racked on plating racks or scooped/shoveled into plating barrels for cleaning. The racks or barrels are immersed into the electrocleaner and a DC current is passed through the parts. The current decomposes the water into two gases, oxygen and hydrogen. It is these gases, which are discharged in finely divided bubbles, that do the cleaning. If the parts to be cleaned are negatively charged during this process, then hydrogen bubbles (commonly referred to as "direct" cleaning) are generated on the parts. If the parts are positively charged (commonly referred to as "reverse" cleaning), then oxygen bubbles perform the cleaning task. In each case the opposing electrode generates the other gas. The choice of direct versus reverse cleaning depends on the type of metal to be cleaned and its tendency to react with oxygen to form an oxide or the tendency for direct cleaning to deposit smuts. Both direct and reverse cleaning are sometimes performed either in sequence, with a periodic reverse rectifier, or through use of a rectifier and a reversing switch. Following electrocleaning, the parts are rinsed in water. See Figure 19.17.

Acid Dip/Pickeling

Parts that have been electrocleaned still have a thin alkaline film remaining, even after prolonged rinsing. This film must be removed for adequate adhesion to take place in the plating tank. Additionally, the parts may have a thin oxide film, either formed during electrocleaning or formed by exposure of the clean metal to air. This oxide also must be removed in order to obtain adequate adhesion of the plating. Lastly, some metals contain alloying elements that interfere with good adhesion. An example is lead which is added to brass to enhance the machining properties of the brass. The lead in the brass forms oxides

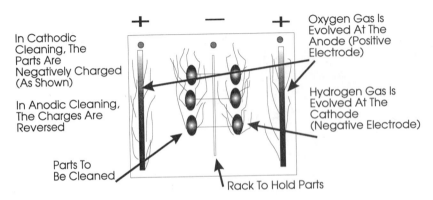

Figure 19.17   Illustration of the electrocleaning process.

that are not removed by the acids that are normally used before plating, such as sulfuric and hydrochloric. A special acid must be used to remove lead oxide from the surface of such brasses. The acid dip must therefore be of a chemistry that will neutralize alkali, and remove all surface oxides present on the part to be plated. After the acid dip, the parts are rinsed in water.

Special Dips

Some parts are made of metals that re-form oxides as soon as the metal is exposed to air. An example is aluminum. Aluminum parts can be cleaned and acid dipped before plating and the plating will still not adhere, because the aluminum forms an oxide by reacting with the air as soon as the part is removed from the acid, rinsed, and exposed to air. A special dip is therefore needed to prevent this from happening. The cleaned and acid-dipped aluminum is dipped into a solution of sodium hydroxide and zinc oxide (often other ingredients are added) and water. In this solution, a controlled galvanic reaction occurs, where some of the aluminum dissolves and, at the same time, some of the zinc coats the aluminum with a very thin film of zinc. The part that leaves this dip (called a zincate) is now coated with zinc, so there is no aluminum surface to react with the air.

Strikes

Some parts are made out of metals that react galvanically with certain plating solutions. An example is a zinc die casting or an aluminum part that has been dipped into a zincate. If we want to plate the zinc (or zinc-coated) part, most plating solutions will chemically or galvanically attack the zinc. We therefore must use a specially designed solution called a *strike* to apply a thin protective coating of metal that will not react with the plating solutions. For zinc, or zincated aluminum, such a strike typically is a cyanide copper strike solution (described later). Some metals cannot be adherently plated without first applying a thin strike deposit from a specialized strike plating solution. An example is stainless steel, which has a rapidly forming oxide that must be simultaneously removed in a special nickel strike solution, while a thin film of nickel is deposited over the stainless steel to prevent the re-formation of the oxide. The nickel strike solution is purposely formulated to yield a thin deposit while generating a large amount of hydrogen that reacts with the oxide on the stainless steel.

Postplating Processes

Some plated parts are further processed to yield additional corrosion protection or to change the color of the deposit. Examples of such further treatment include the application of waxes or lacquers to enhance tarnish resistance and chromate conversion coatings following zinc, cadmium, or other plated deposits to yield chromate films that range in color from transparent to olive drab green. Brass plating is often treated with various chemical solutions to turn the brass to different colors ranging from green to black (even red is possible). All such subsequent treatments typically involve dipping the rack or plating barrel in one or more chemical solutions in various tanks and then rinsing off those solutions. Such solutions often contain ingredients such as nitric acid, sodium dichromate, selenium, arsenic, antimony, or other hazardous ingredients. The processing tanks and associated rinses may be incorporated into the plating line, or the operation may be carried out "off-line."

19.2.3 Typical Cleaning Process Sequences for Plating

The following are typical processing sequences for commonly plated metals (vapor degreasing is not included, but may be required).

Leaded Brass

Process step	Temperature	Time (sec)
Cathodic clean	140	10–30
Rinse		
Reverse clean	140	10
Rinse		
Fluoboric acid	70	5–20
Rinse		
Copper strike	140	180–300
Rinse		
Copper plate	140	As needed
Rinse/dry		

Note: For nonleaded brass, and copper or copper alloys, substitute acid salts or sulfuric acid for fluoboric acid.

Zinc Alloy Die Castings

Process step	Temperature	Time (sec)
Soak clean	140	120
Anodic clean	140	5–10
Rinse		
Dilute acid	70	To first gassing
Rinse		
Copper strike	140	180–300
Rinse		
Plate		

Note: A high pH nickel plate may be an alternate to the cyanide copper strike, but may pose some operational problems.

Case-Hardened/High-Carbon Steel

Process step	Temperature	Time (sec)
Soak clean	180	120
Anodic clean	180	120
Rinse		
Hydrochloric acid	70	30
Rinse		
Cathodic clean	180	120
Rinse		

Hydrochloric acid	70	30
Rinse		
Anodic etch		
(25% sulfuric)	70	60
Rinse		
Plate		

Note: A woods nickel strike or sulfamate nickel strike may be substituted for anodic etch.

Aluminum or Magnesium Alloys

Process step	Temperature	Time (sec)
Soak clean	140	30
Rinse		
Nitric acid	70	20
Rinse		
Zincate	70	10
Rinse		
Copper strike	70	180–300
Rinse		
Plate		

Notes: Add bifluoride salts to nitric if the aluminum alloy contains silicon. If the parts are magnesium, substitute a solution of 10% ammonium bifluoride in 20% phosphoric acid for the nitric acid, and substitute a pyrophosphate zincate for the normal zincate. Pyrophosphate zincate contains 1–1.6 oz/gal zinc sulfate plus 10–12 oz/gal sodium pyrophosphate. The bath operates at 170–180°F, and the immersion coating forms in 3–5 min.

Zincate may also be applied in "double step." The first zinc coating produced is dissolved off in nitric acid. The part is then rinsed and zincate coated a second time. For unexplained reasons, this two-step process often enhances adhesion.

400 Series Stainless Steel, Inconel, Hastelloys

Process step	Temperature	Time (sec)
Soak clean	180	60–120
Rinse		
Cathodic clean	180	60–120
Rinse		
Hydrochloric	70	30
Rinse		
Anodic in woods-nickel strike	70	10
Cathodic in woods-nickel strike	70	30
Rinse		
Plate		

300 Series Stainless Steel, Monel, Tungsten Steel

Process step	Temperature	Time (sec)
Soak clean	180	60–120
Rinse		
Cathodic clean	180	60–120
Rinse		
Hydrochloric acid	70	30
Rinse		
Anodic in sulfuric acid 50%	70	10
Rinse		
Cathodic in woods-nickel strike	70	30
Rinse		
Plate		

Beryllium and Tellurium Copper Alloys

Process step	Temperature	Time (sec)
Brite dip	70	10–20
Rinse		
Hydrochloric 50%	70	10–20
Rinse		
Ammonium persulfate 8–32 oz/gal	70	15–45
Rinse		
Cyanide copper strike	140	180–300
Rinse		
Plate		

Note: Brite dip contains 2 gal sulfuric acid plus 1 gal nitric acid plus 1/2 fluid oz hydrochloric acid, no water. Optional substitute: 20–30% sulfuric acid at 160–180°F.

Bronze Alloys Containing Silicon or Aluminum

Process step	Temperature	Time (sec)
Cathodic clean	180	60–120
Rinse		
Hydrochloric 15%	70	30
Rinse		
Nitric acid 75% containing 10 oz/gal ammonium bifluoride	70	2–3
Rinse		
Copper strike	140	180–300
Rinse		
Plate		

Titanium Alloys

Process step	Temperature	Time (sec)
Blast clean		
Hydrochloric 20%	70	30
Rinse		
Electroless nickel		
Plate	700	300–600
Rinse		
Diffuse	1000	1800
Sulfuric 20%	70	30
Rinse		
Plate		

19.2.4 Cleaning and Preparing Metals for Painting

While the same cleaning processes detailed for electroplating could theoretically be employed to prepare metals for painting, anodizing, and other metal finishing processes, the type of finishing and the base metal to be finished will usually dictate some variation in the cleaning procedure or method.

To yield maximum performance of painted films on metallic substrates, we must overcome surface problems that are detrimental to good adhesion and corrosion resistance. If a metal part is simply cleaned and painted, the smooth surface of the cleaned metal does not allow any "anchoring points" for the paint to hold onto, should it become chipped or scratched. If such a painted metal part is chipped or scratched in service, the resulting corrosion between the paint film and the metal substrate can lift the paint off in large sheets. This is called *creepage* and leads to rapid failure of a painted part. We must therefore utilize a preparation cycle that not only cleans the surface of the base metal, but also creates a surface that is conducive to enhanced adhesion, even after the paint becomes damaged (scratched or chipped). Such enhancement is obtained when a preparation cycle first cleans the surface and then deposits a porous film that will allow the paint to seep into the pores and "anchor" itself onto the surface. The mechanism use to produce the porous surface will depend on the metal substrate.

Steel Substrates

The most common method of cleaning and preparing steel substrates for painting is by employing a "five-stage washer." This equipment first cleans off surface oils and greases using alkaline cleaning. Following a rinse, the steel surface is either sprayed or immersed in an acidic chemical solution that converts the surface of the steel into a crystalline material. Depending on the chemicals employed, the crystals can be thin and composed mostly of iron phosphate, or moderately heavy and composed mostly of zinc phosphate. Following a rinse, the crystalline surface is "sealed" using a dilute chromate solution. The sealing process neutralizes residual acidity and imparts a small amount of hexavalent chromium into the crystals, which enhances the adhesion of the paint to the crystals.

Aluminum and Magnesium Substrates

There are numerous methods for preparing aluminum and magnesium substrates for painting (after a mildly alkaline cleaning to remove surface oils), including use of phosphoric-based primers, application of phosphates, and anodizing. The most common method of preparing aluminum and magnesium surfaces for painting is to clean the surface with mildly alkaline cleaners, followed by a preparation cycle that results in the conversion of the surface to a chromate film.

For aluminum substrates, the preparation cycle following alkaline cleaning may employ an alkaline etching step, and will usually employ chemicals for desmutting the aluminum prior to application of the chromate. The chromate film may be either of the chromate-phosphate or the chromate-oxide type. The chromate-phosphate film can be applied by spray or immersion, yielding an iridescent green to gray film. The chromate-oxide film can also be created by immersion or spray and is considered to be superior to the chromate-phosphate film in corrosion resistance performance. See Figure 19.18.

Each of the chromate films may be "sealed" by immersion in a dilute chromate solution.

Magnesium substrates are typically vapor degreased, cleaned in a mildly alkaline cleaner, acid pickled, and either anodized or immersed in chromate conversion solutions (most of which have been developed by Dow Chemical specifically for use on magnesium).

Copper and Copper Alloys

Copper alloys can be painted after alkaline cleaning, followed by acid pickling or bright dip, followed by application of a chromate conversion coating.

Lead and Lead Alloys

Lead and lead alloys require alkaline cleaning (or vapor degreasing), followed by pickling in dilute fluoboric acid. Following rinsing, the parts can be painted directly.

Stainless Steels and Nickel Alloys

Stainless steels and nickel alloys can be painted after alkaline cleaning (or vapor degreasing). Roughening the surface by grit blasting or sanding will promote adhesion. On

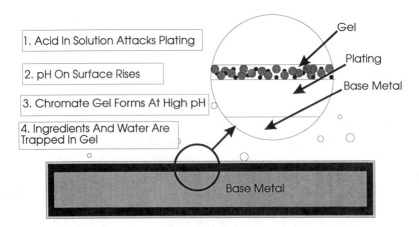

Figure 19.18 Formation of chromate film.

stainless steels, passivating treatments or application of a commercial "black oxide" conversion coating will also enhance paint adhesion (corrosion is not an issue, since these alloys are corrosion resistant without the paint).

Zinc Alloys

Zinc alloy parts are typically vapor degreased, followed by alkaline cleaning, rinsing, immersion in a mild acid, rinsing, and then immersion in a chromate conversion solution. Alternately, the zinc part can be converted to a crystalline phosphate film, using a five-stage parts washer (see steel; the alkaline cleaner would need to be a milder version).

Galvanized parts can be alkaline cleaned and then primed with paints formulated with phosphoric acid. Alternativly, the galvanized parts can be processed for phosphating.

19.2.5 Cleaning for Other Purposes

While it is difficult to provide specific cleaning guidance for purposes other than those already covered, in general, use of any of the above methods prior to resistance welding, soldering, brazing, or other metal-working operations will result in satisfactory work. Metals that are highly reactive, such as aluminum and magnesium, should be cleaned in mildly alkaline solutions, followed by application of chromate films, unless the operation is carried out immediately after cleaning.

19.3 ELECTROPLATING

19.3.0 Introduction to Electroplating

This subchapter on electroplating contains information on the design of products for cost-effective, high-quality plating; the advantages and disadvantages of using various substrates; selection of the proper plating materials and processes; and the use of electroless plating. The various processes and chemistries are definitized, with sufficient detail to permit understanding and application of the many choices. However, consulting with the plater during design of the product and process is highly recommended.

19.3.1 Factors in Product Design That Affect Electroplating

Choosing the Right Substrate

By choosing the right substrate to produce a part, the design engineer can lower final product cost and help protect the environment by reducing the generation of pollutants and wastes. In general, the fewer processing steps that a substrate requires to achieve the final appearance, the less expensive it will be to process. Also, by avoiding certain metals that require strong or unusual chemicals/acids, the design engineer can help generate less waste (also lowering costs).

The designer must be aware of the difficulties imposed on the electroplater by intricate designs and use of metal alloys in the manufacturing of parts that are difficult to plate. The electroplating process does not coat a part uniformly, due to concentrations of electricity that occur at corners and sharp edges. The plater can alleviate this to some extent through use of conforming anodes and shielding. However, this normally drives up costs and increases pollution loading. Exotic alloys create problems for electroplaters in properly cleaning the parts prior to plating (see Subchapter 19.2). The designer should minimize/avoid use of alloys such as stainless steels, inconel, hastalloy, titanium, and combinations of metals, as these are extremely difficult to process for electroplating. If such metals or combinations of metals must be used, the designer should locate and consult with the plater *before* completing his/her design in order to minimize costs and difficulties. (See also the subsection Design Parameters that Affect Plating Uniformity, p. 892.)

The following are typically plated metals and alloys, problems associated with processing them, and the possible reasons for specifying them.

Zinc/Zinc Die Castings

Zinc die casting is an inexpensive method of manufacturing a part from a low cost basis metal. The die castings are deflashed, polished, and buffed prior to cleaning and plating, adding significant labor costs to the product. The die casting process must be carefully designed (especially the "gating") and controlled (temperature and pressure) to minimize air entrapment, which results in excessive casting porosity. Castings with excessive surface porosity cannot be successfully plated (they will evidence blistering and peeling). Zinc die castings are typically plated first with a cyanide copper "strike." This initial plated layer promotes adhesion of subsequent plated metals. Noncyanide solutions for initially plating zinc die castings are available, but are only in the "trial" phase at this time. If a presently die-casted zinc part can be manufactured using steel, the overall cost of the finished product will normally be reduced due to elimination of manual labor involved in polishing/buffing and increased ease of cleaning and plating.

Aluminum/Aluminum Die Castings

Numerous aluminum alloys can be used in manufacturing. All aluminum alloys are highly "reactive" when cleaned, forming surface oxides immediately upon contact with air. These surface oxides inhibit adhesion of electroplated metals. Therefore, aluminum alloys are typically cleaned and processed in a manner that will form an initial immersion coating that does not form oxides upon contact with the air. Two such processes are the zincate process and a proprietary process called the Alstan (trademark of Atotech USA) process. These processes form thin layers of zinc (zincate) or tin (Alstan) on the surface of the cleaned aluminum. The electroplated coating is applied over the zinc or tin immersion coating. There are other methods of obtaining adhesion of electroplate on aluminum, including anodizing the aluminum prior to plating, but the immersion coatings are most commonly used and are the least expensive to apply. Problems with aluminum die castings are similar to those discussed above for zinc die castings. If a part can be manufactured from steel instead of aluminum die cast, significant savings can be realized in finishing the part.

Stainless Steels

All stainless steels have a tendency to form surface oxides upon contact with the air, and therefore pose similar problems to those discussed above for aluminum. The method of processing to obtain adhesion, however, differs in that stainless steel parts are cleaned and first plated in either a sulfamate nickel strike solution or a Woods nickel strike solution. These solutions are specially formulated to created high volumes of hydrogen gas, which removes surface oxides while simultaneously depositing a thin layer of nickel, which can then be plated with other metals. Because stainless steel has much lower conductivity that most other plated metals, the designed part should allow for a larger number of electrical contact points (see Subsection 19.6.2).

Common Steel

Common, mild, low-carbon steel is the most easily cleaned and electroplated substrate. However, these steels can be subjected to machining, stamping, drilling, and heat treating, which can significantly alter the ease of plating. Manufacturing operations that use lubricants and corrosion-inhibiting fluids should use only those types of fluids that are compatible with the electroplating process. Especially to be avoided are any lubricants/fluids that contain silicones. Corrosion-inhibiting products can also create problems for the plater. Products such as calcium stearate or calcium sulfonate are not easily removed and lead to adhesion problems. Heat-treating steel causes the formation of heat-treat scale, which must be removed through mechanical means to prevent adhesion problems after plating. Heat treating can also alter the structure of the steel, making it much more difficult to prepare for plating, especially case hardening. Case-hardened alloys must be cleaned in a double cycle or plated with a Woods nickel or sulfamate nickel strike in order to obtain adhesion. Case-hardened steels and steels hardened above Rockwell C 40 should be stress relieved at 400°F for 1 hr prior to processing for plating. Avoid use of leaded steels, as such alloys cause considerable adhesion problems for a plater. Leaded steel alloys must be pickled in acid containing fluoride salts to remove surface oxides.

Cast Iron

Cast iron can contain enough graphite (carbon) to be impossible to plate in a *cyanide* zinc plating solution (due to low hydrogen overvoltage potential on carbon). Such parts will require a copper deposit first, or an *acid* zinc plating process can be used. Cast iron also contains silicon as an alloying element. These alloys will require acid pickling in acid containing fluoride salts or hydrofluoric acid.

Copper/Copper Alloys

Copper and copper alloys develop an adherent tarnish over prolonged storage that is extremely difficult to remove. Tarnished parts are typically "bright dipped" to remove the oxide/tarnish present on these alloys. The bright dipping process produces a significant amount of toxic nitric oxide fumes, and the solution itself must be periodically disposed of, increasing the cost of plating such parts. Parts manufactured from copper/copper alloys should be coated with a tarnish-preventative material such as a chromate or organic antitarnish product. Heat-treated copper alloys should be carefully treated to create as little oxide scale as possible. Tellurium (0.5%) is added to copper to increase machinability. Beryllium (0.5 – 2.5%) is added to yield high hardness upon heat treatment. Avoid manufacturing parts from tellurium or beryllium containing copper alloys if at all possible, as these alloys are extremely difficult to clean and prepare for plating. Brass often has lead added to enhance machining properties. Leaded brass must be processed through fluoboric acid in the electroplating line, to obtain adhesion. Bronze alloys often contain silicon or aluminum. These alloys require the plater to use a nitric acid dip containing ammonium bifluoride, to promote adhesion of plated deposits.

Titanium

Parts manufactured from titanium alloys require an electroless nickel deposit that is diffused at 800°F for 1 hr, in order to obtain adhesion. Few platers are equipped to prepare and plate onto titanium parts, so a premium can be expected on the price for plating them.

Nickel Silver

Nickel silver actually contains no silver (copper 55–66%, nickel 15–30%, balance zinc). Some alloys of nickel silver contain high concentrations of lead. Such alloys must be pickled in acid containing fluorides to remove lead oxide from the surface. Special treatments, such as cathodic charging in sulfuric acid or a Woods nickel strike are also commonly required.

Powder Metallurgy Products

Avoid the plating of parts fabricated by the powder metallurgy process, as such parts yield high numbers of rejects due to adhesion problems originating from high levels of porosity, which traps processing solutions. If powder metallurgy must be used, the parts should be vacuum impregnated prior to processing for plating.

Design Parameters That Affect Plating Uniformity

A part that is to be electroplated should be designed to allow for uniform plating thickness over the part geometry, low liquid retention upon withdrawal from the processing tanks, ease of racking, good electrical contact between the part and the rack, and ease of handling. Following are some guidelines in this regard.

Part Geometry

Electricity concentrates along sharp edges, ribs, corners, and other points. Conversely, recesses, deep troughs, slots, and other depressed areas are deficient in electric current. The amount (thickness) of plating obtained is directly proportional to the amount of electricity that a specific area of the part obtains during the plating process and thickness is normally directly related to corrosion resistance. One can therefore expect excess thickness at high current densities, and thickness deficiency (poor corrosion resistance) at low current densities. Sharp edges and points should be reduced as much as practical. Gently curving surfaces, grooves, and serrations yield more uniform plating. Edges should be beveled/rounded to a radius of at least 1/64 in. (0.5 mm), 1/32 in. (1 mm) being preferred. The inside/outside edges of flat-bottomed grooves should be rounded off and their depth should be limited to 33% of their width. Avoid V-shaped grooves. Other indentations should also be limited to a depth of 33% of their width. The depth of "blind holes" (holes that do not go all the way through the part) should also be limited to 33% of their width. Avoid blind holes with very small diameters [less than 1/4 in. (6 mm)]. Apply countersinks to drilled and threaded holes. The height of fins and other projections should be reduced as much as possible, with rounding at the base and tips by at minimum 1/16 in. (1.5 mm) radius.

Avoid manufacturing a part from different types of metals or metals with distinctly different treatments. For example, if a steel stamping is made from mild steel with a case-hardened steel rod attached, during processing, the mild steel will need to be treated with harsh chemicals to be able to plate the rod, and severe etching of the mild steel or poor plating on the rod will result.

Influence of Manufacturing Processes on Electroplating

Certain methods of manufacturing a product can cause trouble for a plater, which translates to higher costs, rejects, and waste generation.

Welding

Welding should be performed using welding material that matches the basis metal as closely as possible. The weld must be pore free. Avoid lap welding, unless the lap can be completely pore free. Pores in welds and porous laps will trap processing liquids, contaminating the process solutions and yielding adhesion and appearance problems. Parts that are welded at high temperatures can develop a scale that will require blasting or pickling to remove. Weld spatter must be avoided, as these spots will have a reduced amount of corrosion resistance. Spatter should be removed by grinding/sanding.

Brazing

Brazing yields the same basic problems as welding, except that the creation of dissimilar metals cannot be avoided. In such cases, the plater must be informed as to what was used to braze the components together, so that he or she can adjust the preparatory cycle accordingly or make other modifications to the cleaning of the part.

Soldering

Soldering should be performed *after* plating as much as possible. If soldering must be performed before plating, the operation should be carefully controlled in terms of temperature and fluxing, to yield as pore free a joint as possible. Avoid use of silver solder, if possible, as this requires extra preparatory steps by the plater. Remove excessive amounts of flux from the parts before sending them to the plater.

Drawing

The drawing operation utilizes lubricants that can be either easy or difficult for the plater to remove. Consult with the plater you intend to use to determine which cause problems. Drawing at the wrong speed, with poorly maintained equipment, or without adequate lubrication can create surface fissures that trap plating chemicals or can force lubricant deep into surface defects, yielding blistering.

Annealing

Annealing at the wrong temperature or in the wrong atmosphere can leave oxides on the surface that are very difficult to remove.

Case Hardening

Case hardening yields a high-carbon surface that yields large amounts of smut upon cleaning and pickling. Careful control of the carbon content of the case to the minimum that will still yield the desired case will reduce plating problems.

Shot Peening

Use of shot media that leave as little residue on the surface of the part is very important to the plater. Glass beads leave residual glass on the surface that can be very difficult to remove. Cast iron can leave graphite residues on the surface, which can be difficult to remove. Steel, ceramic, or stainless steel shot usually results in a surface that is easier for the plater to prepare for plating.

Chosing the Correct Plated Coating

The choice of what metal(s) are to be plated onto a part is usually made by the design engineering team at the manufacturing site. Plated and chemically applied coatings are typically applied to enhance one of the following properties:

Corrosion resistance
Appearance
Abrasion resistance
Intrinsic value
Solderability
Rubber bonding
Wire bondability
Electrical contact resistance
Reflectivity (UV, visible, infrared)
Diffusion barrier
Lubricity
High temperature resistance
Susceptibility to hydrogen stress cracking

The designer will usually find that there is not much choice as to which combination of coatings is the best for his or her application, but usually there are *some* choices. The plating that will provide the best compromise of cost versus the above benefits will usually be the one specified for the part. The designer must utilize knowledge of the galvanic properties of the metal combinations that he or she will be creating, the corrosion characteristics, and any of the other properties mentioned above. Often, a combination of metallic coatings will be applied to achieve a combination of the benefits available.

The designer can utilize any of the numerous military, ASTM, or corporate specifications that are presently available for most any type of part contemplated (contact the U. S. Government Printing Office for copies of specifications and/or an index). If necessary, the designer will create a unique specification for a part, which will detail the type of plating, the thickness range that each plated layer is to have, special properties (hardness, solderability, etc.) the plating is to have, and the subsequent coatings, if any, that are to be applied after plating.

If the designer is not careful in properly specifying the plating to be performed, there is a great likelihood of the parts failing to meet expectations. For example, the designer cannot simply specify "zinc plating" if he or she is looking for highly conductive zinc plating. There are three process chemistries that can yield zinc plating, and only the cyanide system yields highly conductive zinc. If the plating comes from the acid or alkaline system, it may not have sufficient conductivity for the application. It is always best to consult a knowledgeable person in the field to review a specification before proceeding to produce the part.

The engineer must develop a knowledge of the types of process chemistries that a plated metal/alloy can be produced in, and the properties of that deposit versus those obtained from other chemistries.

Table 19.1 is a summary of plated metal/alloys, the process chemistries most commonly available, features of those chemistries, and the properties most commonly sought from those deposits. There are numerous other metals and alloys that can be plated to obtain specific benefits. These include precious metals other than gold and silver (rhodium,

Table 19.1 Summary of Plated/Metal Alloys, the Process Chemistry, Features of Those Chemistries, and Beneficial Properties of the Deposits

Deposit	Available chemistries	Beneficial property of chemistry	Beneficial property of deposit	Common use
Zinc	Cyanide	High throw	Pure high conductivity	Hardware Fasteners Automotive
	Acid	Leveling	Bright appearance	Same
	Alkaline noncyanide	High throw	Fine grain	Same
Zinc-cobalt	Acid	Leveling	High corrosion resistance	Same
Zinc-cobalt	Alkaline	Uniform alloy composition on complex shapes	Same	Same
Zinc-nickel	Acid	Leveling	Same	Same
Zinc-nickel	Alkaline	Less corrosion on complex shapes	Same	Same
Zinc-tin	Alkaline	—	Solderability High corrosion resistance	Same
Cadmium	Cyanide	Pure deposit	Solderability Corrosion resistance No mold growth Lubricity	Same
Cadmium	Acid	Avoids cyanide	Appearance	Hardware Fasteners Automotive
Copper	Cyanide	Fine grained Can plate onto zinc	Pure	Cables Pennies Carburizing-stop-off Automotive
Copper	Alkaline non-cyanide	Avoids cyanide	— Die castings	Hardware Some zinc
Copper	Acid-sulfate	Leveling or high throw	Appearance	Printed circuit boards
Copper	Pyro-phosphate	High throw	—	Printed circuit boards
Brass	Cyanide	High throw	Appearance Hardware Vulcanizing	Furniture
Bronze	Cyanide	High throw Lubricity	Appearance Bearings	Hardware
Nickel	Acid	Leveling Bright	Decorative Hardness	Consumer Items Electroforming Electronics Jewelry Tools

Table 19.1   Continued

Deposit	Available chemistries	Beneficial property of chemistry	Beneficial property of deposit	Common use
Chromium	Acid-hexavalent	—	Appearance Hardness	Automotive Plated plastics Printed circuit boards Consumer items Wear surfaces
Gold	Cyanide	High throw	Pure	Jewelry Electronics
Gold	Acid	No cyanide	Hardness	Jewelry Electronics
Gold	Neutral	No cyanide Low corrosivity	Alloys	Jewelry Electronics
Gold	Sulfite	No cyanide	Pure	Jewelry Electronics
Silver	Cyanide	High throw	Pure	Jewelry Consumer items Electronics
Silver	Non-cyanide	No cyanide	Pure	Same
Tin	Alkaline	High throw	High solderability	Electronics Consumer items
Tin	Acid	Bright	Appearance	Consumer items
Tin-lead	Acid (fluoboric)	High throw	Solderability	Electronics Printed circuit boards
Tin-lead	Acid (sulfonic)	Uniform alloy composition over complex shapes	Solderability	Electronics Printed circuit boards

palladium, ruthenium, platinum), and some uncommon "common" metals and alloys such as bismuth, iron, and Alballoy (copper-tin-zinc). There are also "composite coatings" that can be plated. For example, one can plate a nickel-cobalt alloy containing finely dispersed particles of silicon carbide to enhance abrasion resistance. This, or some other composite, may one day be a substitute for chromium plating.

Hydrogen Stress Cracking

More than eleven major theories on the mechanism involved in hydrogen stress cracking—commonly referred to as hydrogen embrittlement—have evolved, yet no single mechanism explains the complete characteristics of this phenomenon.

The following background information should be helpful to the designer.

1. Steels are susceptible to hydrogen stress cracking in varying degrees depending on the composition, microstructure, and type/amount of defects.
2. Embrittlement occurs at all strength levels, but is most prevalent in steels with strength levels above 200,000 psi.
3. Under tensile load, hydrogen stress cracking is manifested mainly as a loss of

ductility. There is no influence on the yield point or the plastic properties up to the point where local necking starts. Hydrogen apparently prevents the local necking from continuing to the normal value of hydrogen free steel.

4. In static loading with notched specimens, hydrogen causes delayed failures at loads as low as 20% of the nominal tensile strength.
5. No macroscopic plastic flow occurs during static loading.
6. Unnotched tensile specimens exhibit hydrogen stress cracking at higher yield strengths than notched specimens.
7. Hydrogen stress cracking effects are diminished or disappear at low temperatures (around $-100°$ C) and high strain rates.
8. Embrittlement increases with increased hydrogen content.
9. For a given hydrogen content, embrittlement increases with increased stress concentration.
10. The surface hardness of affected metals is not altered.
11. While it is most critical to the proper functioning of high-strength steels, hydrogen stress cracking also manifests itself in other metals/alloys:

Copper: Pure copper alloys are not subject to the problem, while alloys that contain oxygen have been reported to be hydrogen embrittled.

High-temperature alloys: Nickel, cobalt-based, and austenitic iron-based exotic alloys show no apparent susceptibility to the problem, while body-centered cubic alloys containing cobalt or titanium are susceptible.

Stainless steels: Austenitic alloys (200 – 300 series) are not susceptible, while martensitic (400 series) alloys are. Ferritic (400 series) alloys exhibit varying degrees of susceptibility depending on the amount of work hardening (work hardening appears to reduce the effects).

Steels: Alloys with Rockwell hardnesses above C 35 and tensile strengths above 140,000 psi are susceptible to the problem. Additionally, reports of low-strength, plain carbon steel also being embrittled by hydrogen are in the literature. Under the right conditions, almost any steel alloy can exhibit susceptibility.

Mechanism

Hydrogen embrittlement involves most or all of the following conditions:

1. Stress
2. Adsorption
3. Dissociation
4. Dislocations
5. Steel microstructure/composition
6. Other variables such as strain rate, temperature

The preliminary step is the adsorption of the hydrogen onto the iron surface. The adsorption of hydrogen does not occur unless the iron surface is chemically clean. This is why plain carbon steel cylinders can be used to store high-pressure hydrogen without catastrophic failure.

Once the steel contains hydrogen gas, it must dissociate to nascent hydrogen for embrittlement to proceed. If the hydrogen is absorbed during acid pickling or plating, it is already dissociated to the nascent state. Experimenters have shown that large increases of crack growth occurred when a steel specimen was exposed to partially ionized hydrogen gas versus nonionized.

Under stress, the hydrogen within the steel diffuses to regions of maximum triaxial stress. If microcracks are present, maximum triaxial stress occurs just beneath the crack tip.

Microcracks present may be filled with hydrogen gas which can dissociate, enter the iron lattice structure, and also diffuse to regions of triaxial stress. If no microcracks are present, hydrogen gas can be transported by dislocations which can pile up at inclusions and second phases forming microcracks. Competing actions therefore deliver hydrogen to areas where the damage can occur: diffusion and mobile dislocations.

By adsorption, hydrogen ions are supplied to areas of high triaxial stress.

In the region of high triaxial stress, the formation of dislocations is accelerated. Numerous pileups occur just below the crack tip, forming voids filled with hydrogen gas which can instantly dissociate and diffuse to dislocations outside the void, adding to the growth of the voids.

As the stress intensity decreases due to void growth, decohesion (cleavage) takes over, resulting in catastrophic failure.

Inclusions are not necessary to the mechanism because dislocation pileups assisted by hydrogen can substitute for microvoids. Steels of low strength are not as susceptible because these steels are too ductile for the cleavage mechanism to take over from microvoid coalescence. The reversibility of the effect can be explained by the fact that if nascent hydrogen is given enough energy to form gas and diffuse out of the steel before dislocation pileups and diffusion of nascent hydrogen to areas of stress can occur, then embrittlement will not result. Since diffusion plays a major role, low temperatures reduce the effect.

Finishing Processes Influencing Hydrogen Embrittlement

Any aqueous process that contains hydrogen ions, or yields hydrogen on the surface of a susceptible alloy, can result in embrittlement. These include:

    Acid pickling
    Cathodic cleaning
    Plating

Plating processes that are not 100% efficient can result in hydrogen embrittlement. Zinc electroplating from cyanide baths results in the most embrittlement, with cyanide cadmium next. This is probably because cyanide plating baths do an excellent job of cleaning the surface of the steel, and the mechanism requires a relatively clean surface.

Chromating usually does not increase embrittlement unless the plated deposit is too thin to accept a chromate film adequately.

Minimizing Hydrogen Stress Cracking

Although it is not totally preventable, hydrogen embrittlement can be minimized by use of the following techniques.

1. Plating at the highest current density allowable, to improve current efficiency.
2. Plated deposits approaching 0.0005 in. or more should be applied in two steps, with a bake after 0.0002 in. followed by a second plate to the final thickness and a second bake.
3. Stress-relieve all highly stressed steels and any steels with a hardness greater than Rockwell C 40 prior to processing.

4. Bake hydrogen out of the steel after prolonged acid pickling.
5. Avoid cathodic electrocleaning.
6. Use inhibited acids.
7. Substitute mechanical cleaning for chemical cleaning.
8. Keep brightener content in plating baths low.
9. Bake parts as soon as possible after processing, usually within an hour and within 15 min for alloys above Rockwell C 40.
10. Bake times should be:

    0.0002 in. or less: 4 hr
    0.0002–0.0003 in.: 8 hours
    0.0003 in. or more: 24 hours

    The bake temperature should be $375 + 25 - 0°F$. Baking times must be at temperature: Do not start from cold.
11. Minimize the amount of descaling required by controlling the heat-treatment process.
12. Remove surface defects (tumble, deburr, shot peen) prior to processing to reduce the number of areas of stress.

19.3.2 The Electroplating Process

Electroplating is a process for coating a metallic or nonmetallic substrate with a metallic coating through the use of a combination of electricity and a chemical solution that includes ions of the metal in the coating.

To conduct the process, we first need to purchase some *hardware*. Simple electroplating hardware consists of:

A rack or barrel to hold parts
A tank
Electric cables or copper bus
A rectifier
Filtration equipment (may be optional)
Agitation equipment (may be optional)
Ventilation equipment (may be optional)
Plating solution
Other processing tanks for cleaning, rinsing, acid pickling

Plating Methods

Plating can be performed using any of three main methods of part handling, each requiring different hardware.

Rack Plating

Rack plating is sometimes referred to as still plating and is used whenever the parts are too large, delicate, or complicated to be barrel plated. Rack plating is much more expensive than barrel plating because of the labor involved in putting the parts on the rack and taking them off after they are processed. Rack plating is performed by hanging the parts to be plated on racks, which are typically plastic-coated copper or aluminum rods, with stiff wires that hold the parts in place protruding at various intervals. Racks come in numerous designs and are most often constructed by outside vendors and sold to the plater. Some small parts are racked simply by twisting a thin copper wire around them. The wire, with perhaps 20 – 50 pieces hanging on it, is then handled as a "rack." During plating, the part of the rack that makes electrical contact with the part being plated is also plated, so after several cycles these contacts have a lot of metal buildup. The racks are then sent through a stripping solution that removes the excess metal, or the plater physically removes the excess metal using pliers or a hammer. Chemical rack strippers are usually strong solutions of cyanide or acid and can be difficult to waste-treat. Noncyanide and regenerative strippers are available for some processes, but are expensive to use or strip very slowly. See Figure 19.19.

Barrel Plating

Barrel plating is the most efficient and least costly method. Plating barrels of varied designs are purchased from manufacturers of such products. The basic barrel consists of a hexagonal cylinder, closed at both ends, with perforated walls made of polypropylene. A "door," which is held in place with plastic-coated clips, is installed in one wall of the barrel to allow entry and exit of the load. Electrical contact between the saddle on the tank and the parts inside the barrel is made by a copper or bronze rod attached to the barrel, which sits in the electrified saddle. The rod has a cable attached, and this cable is routed inside the plating barrel through the end of the barrel. Sometimes two cables are used, one entering each end of the barrel. The barrel has a hole in the center that allows the cable to enter. The end of the cable, inside the barrel, has a stainless steel ball attached, called a *dangler*. This dangler makes contact with the parts inside the barrel by gravity. There are other methods of making electrical contact inside the barrel, including rods and button contacts, but the dangler is the most commonly used. Parts to be plated are scooped or shoveled into the barrel. The load is often weighed to make certain that the parts are plated uniformly. As a general rule, the barrel is never filled beyond one-half of its total volume. As plating proceeds, a motor mounted either on the plating tank or on the barrel turns the barrel at 3 – 5 rpm, through either a drive belt or a set of gears mounted on the barrel. If

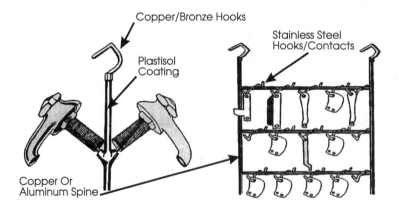

Illustrations Courtesy of Belke Mfg. Chicago IL

Figure 19.19 Parts being prepared for rack plating.

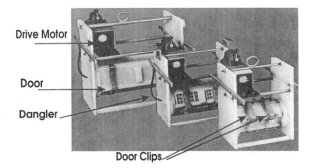

Drive Motor

Door

Dangler

Door Clips

Figure 19.20 Small plating barrel.

the barrel is not rotated during plating, the top of the load will be plated and the bottom part will remain bare. See Figure 19.20 for typical portable horizontal barrels. Other designs that allow automatic loading and unloading, such as oblique and horizontal oscillating barrels, are also commonly employed (see Figure 19.21). Barrel plating can be performed by manually transferring barrels from one station to the next, or through use of automated processing lines.

Continuous/Reel-to-Reel Method

The continuous or reel-to-reel method of plating is highly efficient and competes effectively against all other methods when the parts are small, uniform, of simple geometry, and amenable to being stamped from a thin strip of metal. This method is used to electrogalvanize (zinc plate) steel strip that is used to stamp automobile bodies and to plate brass or copper strip for stamping electrical connectors for telecommunications. In this method of plating, the parts to be plated consist of long strips of metal that are rolled up on a wheel. The wheel is mounted on the equipment and the strip goes through a sequence of rollers directing it through various processing tanks, including the plating tank. The strip may partially dip into the plating tank, or it may be completely immersed. Electrical contact is made through metal brushes, rollers, or by a principle called bipolarity, which does not actually contact the strip. The strip may travel at speeds ranging from 50 to 1000 or more

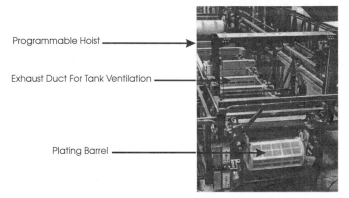

Programmable Hoist

Exhaust Duct For Tank Ventilation

Plating Barrel

Figure 19.21 Automated horizontal barrel plating line.

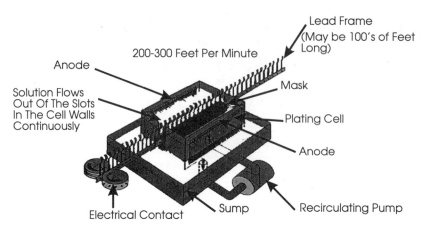

Figure 19.22   Continuous reel-to-reel plating.

feet per minute. At the other end of the continuous strip plating line, a second wheel takes up the processed strip. See Figure 19.22.

Rack and barrel plating are typically performed by manual and automated methods. In manual operations, platers transfer the racks or barrels from one processing tank to another by hand or with electric hoists. In automated operations, racks or barrels are transferred using programmable robots. Continuous strip plating operations are always automated, as shown in Figure 19.23.

Brush Plating and Other Methods

Numerous other techniques are also used to electroplate substrates. One of these is called brush plating. To perform brush plating, the plating solution is formulated with ingredients that create a paste or thick liquid. A "brush" consisting of an anode covered with either a sponge or layers of absorbent material is used to hold the plating paste onto the substrate to be plated. The part to be plated is connected to the negative electrode of the rectifier, while the brush is connected to the positive electrode. The area to be plated is stroked with a back-and-forth motion (the area must be cleaned prior to plating). See Figure 19.24.

Brush plating is an ideal method for covering selected areas of large parts that must

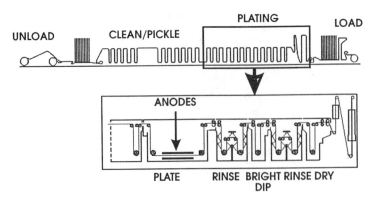

Figure 19.23   Continuous strip plating schematic.

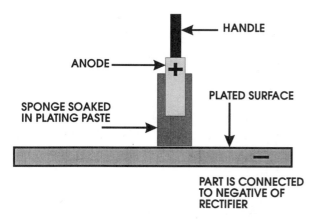

Figure 19.24   Schematic of brush plating method.

be repaired, or plating large parts without the need for a large tank and large volumes of plating solution. For example, an entire sports car has been gold plated using this method. To immerse even a small fender into a conventional gold plating solution would require a plating tank containing hundreds of gallons of plating solution at over $500.00 per gallon of solution, while with brush plating, the car was plated using only a few gallons of solution.

Another method of plating is called out-of-tank plating. This method is commonly used to plate deep cavities that are difficult to plate in conventional tanks, due to solution flow problems. In out-of-tank plating the plating solution is pumped into the cavity and flows continuously out of the cavity, returning to a holding tank, which may hold hundreds of gallons of solution (which keeps the chemical constituents in balance). Plating is performed with an internal anode.

Hardware

The hardware required to perform plating is typically purchased from a company that specializes in producing this equipment, although some platers produce their own hardware. Let us take a closer look at the hardware used for plating.

Tank

The plating tank must resist chemical attack from the plating solution. Tanks containing cyanide plating solutions are often made of bare steel. Tanks for other plating solutions are typically made of steel with PVC lining, polypropylene, polyvinyl chloride, or polyethylene. Plating tanks should not have any wall perforations below the liquid line, to prevent accidental discharge of contents. A rack plating tank typically has three copper bus bars mounted on top of the tank. One bus is in the center and is used to hang the parts in the plating solution. The other two bus bars are located near the walls of the tank, and are used to hang anodes or "baskets" for anodes.

A barrel plating tank has the same anode bus, but there is no center bus. Instead, the tank typically has four "saddles" made of copper or bronze, mounted to the lips of the tank, so that the barrel contact rods can sit firmly in the saddles. At least one of the four saddles has a cable or copper bus attached to it for contact with the rectifier.

Rectifier

The rectifier for plating converts AC current to DC. The rectifier is typically installed near the plating tank, but it may be located in another room. In either case, cable or a bus bar is used to connect from the positive terminal of the rectifier to the anode bus bar on the tank. The negative terminal of the rectifier is connected by cables or copper bus to the saddle of barrel plating tanks or to the cathode bus of rack tanks. Rectifiers generate heat as a by-product. This heat must be removed by using either a fan (air cooled) or by circulating cooling water through the rectifier (water cooled). The water used to cool the rectifiers can be routed to other plating operations such as rinsing. See Figure 19.25.

Filter

Some plating solutions require continuous filtration; others do not. A general rule is that alkaline solutions can usually operate satisfactorily without a filter, whereas acidic solutions need filtration to avoid particulates suspended in the solution from being incorporated into the coating, yielding "roughness."

Agitation

Most plating processes require some form of solution agitation to deliver the brightest, densest, most uniform deposit, and to plate at higher current densities without "burning." A common method of agitation is to move the cathode rod back and forth in the solution using a motor attached to the cathode rod. This is called cathode rod agitation. A second common method is to install an air sparger in the bottom of the plating tank and use low-pressure air bubbles to perform the agitation. A less common technique is to use a prop mixer. Plating tanks for barrel plating are not agitated, because the rotation of the barrel provides sufficient solution movement.

Ventilation Equipment

Various metal finishing process solutions can emit vapors, fumes, and mists. To maintain a safe working environment, such emissions are captured using ventilation equipment. There are numerous methods of ventilating a process tank, including side-draft, push-pull,

Figure 19.25 Bank of air-cooled rectifiers located above process tanks.

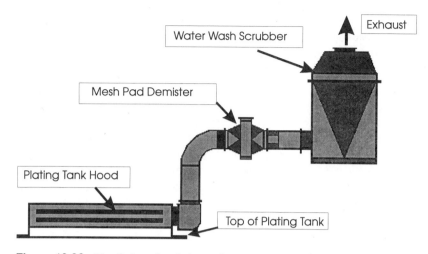

Figure 19.26 Ventilation of a plating tank.

pull-pull and four-sided ventilation. Highly sophisticated systems may even totally enclose a process tank during use. Depending on the nature of the emissions, the ventilation system may be connected to a system for removing the emissions prior to discharge of the air to the outside of the building. These systems may employ water wash scrubbing, mesh pad impingement systems, or combinations of both, as shown in Figure 19.26.

Plating Lines/Departments

Except for vapor degreasing, which normally is performed off-line, plating operations are normally incorporated into a sequence of tanks, called a line. A plating line may be designed to produce a single coating or a number of coatings. The process line usually contains tanks lined up in sequential order. Automated lines may or may not contain tanks in sequential order.

A zinc plating line, as shown in Figure 19.27, may therefore consist of 13 tanks, each containing a chemical processing solution or rinse water; soak clean, electroclean, rinse, acid, rinse, zinc plate, rinse, bright dip, rinse chromate, rinse hot water rinse, dry. If the line is for barrel plating, each tank may have one or more stations, that is, places to put a barrel. A six-station zinc plating tank can plate six barrel loads of parts at one time. To economize, some shops may have one cleaning line that services several plating lines.

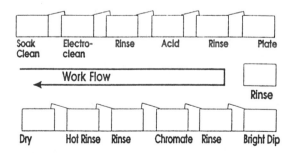

Figure 19.27 Zinc plating line showing typical process sequences.

There also are tanks for rack stripping, stripping rejects, purifying contaminated solutions, or holding solutions that are used only sporadically. The entire lineup of tanks and lines creates the shop layout, with parts entering the plating department from one direction, traveling through the process lines, and then leaving the plating department.

Electroplating Process Summary

The plating step may be a single plate or a series of different deposits. If a series of deposits is to be applied, there usually is a rinse and an acid dip between the different plating steps. For example, a zinc die casting may be plated with a cyanide copper strike, followed by a cyanide copper plate, followed by a semi-bright nickel plate, followed by a bright nickel plate, followed by a thin deposit of chromium. No rinsing or acid dipping may be required between the cyanide copper strike and the cyanide copper plate, or between the semi-bright nickel and the bright nickel, because the chemistries of the sequential baths are similar. However, there will be a rinse and an acid dip between the cyanide copper plate and the semi-bright nickel plate, because one solution is alkaline and the other is acidic. There will be a rinse between the bright nickel plate and the chromium plate to prevent contamination of the chromium plating solution with nickel. There may or may not be an acid dip, since both solutions are acidic. After all plating has been performed, the parts are rinsed and dried before packaging them for shipment.

The plating solution normally contains water and a number of ingredients that determine if the coating produced is dense, bright, hard, a certain color, or has a number of other desirable properties that can be obtained through chemistry. It also contains an ingredient that forms ions of the metal to be plated when added to water. For example, one ingredient in a Watts nickel plating solution is nickel sulfate. When nickel sulfate is added to water, it dissolves and forms nickel ions and sulfate ions, just as when one adds salt to water, it dissolves to form sodium and chloride ions. The dissolved nickel ions can be converted to nickel metal by passing a direct current through the plating solution using a rectifier (which converts AC current to DC), the anode as the electrode with positive polarity, and the part to be coated as the electrode with negative polarity. The conversion of the nickel ions to nickel metal will occur on the surface of the negative electrode, where excess electrons (which make the electrode negatively charged) "reduce" the nickel from ions to metal. While nickel ions are converted to metal at the cathode (negative electrode), at the anode, nickel metal is converted from the metal back to the ions. Ideally, for each nickel atom plated out at the cathode, a new ion is formed at the anode to replace the one plated out. This is the case in most plating solutions, but with some solutions the anode is not converted to metal ions.

19.3.3 Typical Plating Solution Chemistries

We will now briefly discuss the ingredients of the most commonly used plating solutions.

Zinc

Zinc is the most commonly plated metal, normally applied over ferrous substrates for the purpose of enhancing the corrosion resistance. Zinc can be plated from a number of different chemistries, but the three most common are cyanide, alkaline noncyanide, and acid chloride.

Cyanide Zinc

Cyanide baths are favored when high thicknesses are required or when parts are to be plated and then deformed. Still widely used, cyanide zinc plating solutions contain

Zinc cyanide: 2 – 8 oz/gal
Sodium cyanide: 1 – 6 oz/gal
Sodium hydroxide: 10 – 14 oz/gal
Organic brightener: as required
pH: 14+

Operating conditions are

Temperature: 70 – 90°F
Anodes: zinc/steel baskets
Plating current density: 10 – 50 ASF
Agitation: cathode rod/none
Filtration: not required
Color of solution: straw colored, sometimes has floating oily layer
Odor of solution: aldehyde

Alkaline Noncyanide

Alkaline noncyanyde baths can be substituted for cyanide baths without the need for major equipment modifications. The deposits tend to become brittle as the thickness increases, and some parts that have been heat-treated yield poor adhesion. Ingredients are

Zinc oxide: 1 – 2 oz/gal
Sodium hydroxide: 10 – 15 oz/gal
Sodium carbonate: 0 – 3 oz/gal
Additives: as required
pH: 14+

Operating conditions:

Temperature: 70 – 90° F
Anodes: zinc/steel baskets
Plating current density: 5 – 45 ASF
Agitation: cathode rod/none
Filtration: not required
Color of solution: pale yellow
Odor of solution: sharp odor

Acid Chloride

Acid chloride baths yield the brightest deposit. The process requires excellent cleaning and corrosion-resistant equipment, however, and thick deposits tend to be brittle. Ingredients are

Zinc chloride: 4 – 12 oz/gal
Potassium chloride: 14 – 20 oz/gal
Ammonium chloride: 3 – 5 oz/gal
Boric acid: 4 – 5 oz/gal
pH: 4.5 – 6.0

Operating conditions are

Temperature: 70 – 115° F
Anodes: zinc/titanium baskets
Plating current density: 10 – 150 ASF
Agitation: air or cathode rod
Filtration: required
Color of solution: pale yellow
Odor of solution: none; sharp with air agitation

Cadmium

The vast majority of cadmium plating is performed from the cyanide-based chemistry. The sulfate chemistry has made small inroads, but often does not adequately cover heat-treated steel parts that have high surface hardness. The sulfate process also requires a much higher degree of cleaning, or adhesion becomes marginal.

Cyanide Cadmium Solutions

The cyanide-based cadmium plating solution contains

Cadmium oxide: 3 – 5 oz/gal
Sodium cyanide: 10 – 20 oz/gal
Sodium carbonate: 3 – 14 oz/gal
Sodium hydroxide: 2 – 8 oz/gal
Brighteners/additives: none
pH: 14+

Operating conditions are

Temperature: 70 – 90° F
Anodes: cadmium/steel baskets
Plating current density: 5 – 90 ASF
Agitation: cathode rod/none
Filtration: not required
Color of solution: pale yellow
Odor of solution: aldehyde

Acid Sulfate Solutions

Acid sulfate-based cadmium plating solution contains

Cadmium chloride: 1 – 1.5 oz/gal
Ammonium sulfate: 10 – 15 oz/gal
Ammonium chloride: 1.5 – 3 oz/gal
Brighteners/additives: none
pH: 5 – 6

Operating conditions are

Temperature: 70 – 100° F
Anodes: cadmium, titanium baskets
Plating current density: 2 – 15 ASF
Agitation: cathode rod

Filtration: required
Color of solution: pale yellow
Odor of solution: sharp

Copper

Copper is plated from three popular chemistries, cyanide, acid sulfate, and pyrophosphate. Recently, patented alkaline noncyanide copper plating processes have been developed and marketed by at least three companies, but they are troublesome and expensive to operate.

Cyanide Copper Process

There are several cyanide copper plating processes, but they can be divided into two basic chemistries; a strike bath and a "plate" or high-speed bath. The ingredients are

	Strike	Plate
Copper cyanide	1.5–2.5 oz/gal	4–6 oz/gal
Sodium cyanide	3–4 oz/gal	8–12 oz/gal
Sodium carbonate	3–14 oz/gal	3–14 oz/gal
Sodium hydroxide	0–1.5 oz/gal	2–4 oz/gal
Brightener/additives	(none)	As required
pH	10-10.5	12–14

Note: Potassium salts are most often used in high-speed cyanide copper plating solutions, at approximately the same concentrations.

Operating conditions:

Temperature: 140–160° F
Anodes: oxygen-free, high-conductivity (OFHC) copper
Plating current density: 10–100 ASF
Agitation: cathode rod
Filtration: required
Color of solution: pale yellow
Odor of solution: pungent

Acid Sulfate Processes

Two main acid sulfate chemistries are used in electroplating copper. One is termed "conventional" and is often used as an underlayer for plated plastic, or in applications where a high degree of leveling (smoothing of scratches) is desired. The second process is called a "high-throw bath," and is used mostly by printed wiring board manufacturers because of its ability to produce uniform thicknesses on the outside of a circuit board and on the inside of tiny holes drilled into the board. The ingredients are

	Conventional	High throw
Copper sulfate	26–33	10–14 oz/gal
Sulfuric acid	6–12	20–30 oz/gal

Chloride	40–80	40–80 ppm
Brighteners/additives	Yes	Yes
pH	<1	<1

Operating conditions are

> Temperature: room
> Anodes: copper containing 0.02–0.06% phosphorus, bagged
> Plating current density: 20–200 ASF
> Agitation: air
> Filtration: continuous
> Color of solution: deep cobalt blue
> Odor of solution: no specific smell, inhaled mist may yield sharp odor/burning of nose

Pyrophosphate Copper Solutions

Pryophosphate copper plating solutions are used almost exclusively by printed circuit board manufacturers. Their major benefits include low copper concentration and the ability to deposit an even thickness over complex geometries, such as the top of a circuit board versus the inside of a drilled hole. The majority of these solutions have been replaced with bright throw acid sulfate systems because the pyro baths have much more difficult chemistries to analyze and control. The baths may be found in some job shops as substitutes for cyanide copper strike baths on zinc die castings or for copper-striking zincated aluminum. Ingredients are

> Copper pyrophosphate: 10–12 oz/gal
> Potassium pyrophosphate: 40–45 oz/gal
> Potassium nitrate: 1–1.5 oz/gal
> Ammonia: 0.15–0.5 oz/gal
> Additives: variable
> pH: 8–9

Operating conditions are

> Temperature: 125–135° F
> Anodes: oxygen free high conductivity/titanium baskets
> Plating current density: 10–90 ASF
> Agitation: air
> Filtration: yes
> Color of solution: iridescent blue/purple
> Odor of solution: no specific

Brass

Currently, there are no commercially viable processes for plating brass, and alloys of copper and zinc, other than from a cyanide-based chemistry. Brass is applied mostly for decorative purposes, wherein it is subsequently stained to yield an "antique" or colored finish. Brass is also applied to enhance adhesion of rubber to steel. Brass tarnishes readily, so most often it is finished off with a coat of lacquer. Ingredients are

> Copper cyanide 4–8 oz/gal
> Zinc cyanide: 1–2 oz/gal

Sodium cyanide: 2–4 oz/gal
Sodium carbonate: 3–14 oz/gal
pH: 10–11.5

Operating Conditions:

Temperature: 125–135° F
Anodes: brass of same alloy composition as plated
Plating current density: 5–15 ASF
Agitation: cathode rod
Filtration: yes
Color of solution: pale yellow
Odor of solution: no specific odor

Bronze

Bronze (80% copper, 20% tin) can be plated only from a cyanide-based chemistry. The plating equipment is identical to that for copper or brass plating. Ingredients are

Copper cyanide: 4.6 oz/gal
Potassium cyanide: 10.3 oz/gal
Potassium stannate: 5.6 oz/gal
Potassium hydroxide: 1.7 oz/gal
Sodium potassium tartrate: 6 oz/gal
pH: 12+

Operating conditions:

Temperature: 150–160° F
Anodes: copper/carburized steel, graphite, or stainless steel
Plating current density: 20–100 ASF
Agitation: cathode rod
Filtration: yes
Color of solution: pale yellow
Odor of solution: pungent

Zinc Alloys

Alloys of zinc have been the major focus for a good substitute for cadmium plating. If an alloy of zinc contains a small amount of a more noble metal, such as nickel, tin, cobalt, or iron, the zinc retains its cathodic relationship with steel, but the alloying metal reduces the activity of the coating so that it corrodes sacrificially at a slower rate, thereby enhancing corrosion protection over plain zinc. Numerous zinc alloy processes are being touted as the best cadmium alternative, including zinc-nickel, zinc-cobalt, zinc-tin, and zinc-iron. Of these, zinc-nickel appears to be a favorite at this time, while some zinc-cobalt installations have been made. The others are either too expensive or do not produce a pleasing enough appearance to be applicable for anything other than as a paint undercoat for automobile body panels. The equipment for plating zinc alloys is identical to that used for nickel plating (see next page).

Zinc-Nickel

Zinc-nickel alloys can be plated from both alkaline and acidic chemistries, with the alkaline process the most often favored.

Alkaline Zinc-Nickel

Ingredients are

> Zinc: 1–2 oz/gal
> Sodium hydroxide: 12–17 oz/gal
> Nickel: 0.1–0.2 oz/gal
> Additives: as required
> pH: 14+

Operating conditions:

> Temperature: 70–90° F
> Anodes: zinc and steel
> Plating current density: 10–45 ASF
> Agitation: cathode rod
> Filtration: yes

Zinc-Cobalt

Ingredients are

> Zinc chloride: 10–12 oz/gal
> Potassium chloride: 26–33 oz/gal
> Cobalt: 0.25–0.50 oz/gal
> Boric acid: 3–4 oz/gal
> pH: 5–6

Operating conditions are

> Temperature: 65–95° F
> Anodes: zinc, bags
> Plating current density: 1–40 ASF
> Agitation: air
> Filtration: yes

Nickel

Nickel is most often plated from the Watts chemistry, although there are numerous other formulations, including a specialized Woods nickel strike that is used to obtain adhesion on stainless steels. The Watts bath is used to obtain bright or semi-bright deposits for decorative applications. In decorative applications where deposit appearance and corrosion resistance are highly important, as on the exterior of an automobile, two or more layers of nickel from Watts baths are applied. The most common such layered nickel plating is referred to as "duplex nickel" and consists of two layers of nickel. The first layer, called semi-bright nickel, contains no sulfur-bearing brighteners, and the second layer is a fully bright nickel deposit containing a controlled amount of sulfur-bearing brightener.

The duplex nickel is normally topped off with a thin coating of chromium plating. The bright nickel layer corrodes in favor of the semi-bright layer, protecting it galvanically and delaying the onset of corrosion of the base metal.

Another often-used nickel plating formulation is the sulfamate based chemistry. It is used in electroforming or other applications where a nickel deposit containing no or low internal stress is desired.

Nickel is also used to plate "composite" deposits, where the plated nickel contains finely dispersed diamond dust or other abrasives such as silicon carbide. Such composite coatings are used to create long-lasting cutting tools.

Watts Nickel

Watts nickel plating chemistry can contain a variety of additives to control pitting, yield leveling, and produce brightness from a medium (semibright) to a full mirror-bright deposit. Additives are normally patented products sold by suppliers along with the plating chemicals. Ingredients are

Nickel sulfate: 30–45 oz/gal
Nickel chloride: 4–12 oz/gal
Boric acid: 4–6 oz/gal
Additives: as required
pH: 3–5

Operating conditions are

Temperature: 125–135° F
Anodes: nickel or nickel containing 0.02% sulfur or others, bagged
Plating current density: 25–100 ASF
Agitation: cathode rod or air
Filtration: yes
Color of solution: deep green
Odor of solution: no specific odor

Woods Nickel Strike

Woods nickel strike is purposely designed to generate high volumes of hydrogen gas while depositing only a thin layer of nickel, even at the highest current densities. The major use is to obtain adherent thin nickel deposits that can then be plated with other metals. Ingredients are

Nickel chloride: 30 oz/gal
Hydrochloric acid: 16 fl oz/gal
pH: <1

Operating conditions:

Temperature: 70–90° F
Anodes: nickel
Plating current density: 100–300 ASF
Agitation: none
Filtration: none
Color of solution: dark green
Odor of solution: sharp hydrochloric fumes

Sulfamate Nickel

The sulfamate nickel chemistry is used mainly for electroforming purposes, although some electronic applications requiring a low-stress nickel underplate for gold overplates also use this bath. The equipment for sulfamate plating is identical to that used for Watts baths, with the exception that the sulfamate process will typically have a purification compartment attached to the tank, incorporated into the tank, or alongside the tank. The purification compartment is about one-fifth the size of the plating tank, and the solution is recirculated through the compartment, using the filtration system. In the purification compartment, electrolytic nickel anodes and dummy electrodes plate out metallic contaminants and the polarization occurring at the anodes decomposes some of the sulfamate ions into stress-reducing compounds. Ingredients are

  Nickel sulfamate: 60–70 oz/gal
  Magnesium chloride: 1–2 oz/gal
  Boric acid: 6–7 oz/gal
  Additives: as recommended
  pH: 3–5

Operating conditions are

  Temperature: 125–135°F
  Anodes: sulfur depolarized nickel
  Plating current density: 20–140 ASF
  Agitation: air
  Filtration: yes
  Color of solution: deep green
  Odor of solution: no specific odor

Sulfamate Nickel Strike

An alternative strike solution that can be used to activate stainless steel is composed of

  Nickel sulfamate: 320 g/L
  Sulfamic acid: 150 g/L

Operating conditions:

  Temperature: 50°C
  Anodes: electrolytic nickel
  Plating current density: 50 ASF
  Agitation: none
  Filtration: yes
  Color of solution: deep green
  Odor of solution: no specific odor

Chromium

Chromium plating generally falls into two categories; "decorative" and "hard." Both categories can and are plated from the same chemistries based on hexavalent chromium, while decorative chromium can also be plated from one of several trivalent chemistries. The terms "decorative" and "hard" are confusing and really mean "thin" and "thick." All chromium plates have a hardness in the same range (900–1100 Vickers). Decorative chromium is a very thin layer of chromium applied over a substrate that has been bright-nickel-plated. The appearance of decorative chromium is to a large extent, due to the appearance of the nickel. The chromium is so thin (3–20 millionths of an inch) that is essentially transparent. Decorative chromium plating equipment is identical to that for hard chromium. An exception is the trivalent decorative chromium equipment, which typically has no exhaust system and requires continuous filtration.

"Hard" chromium should be called engineering chromium, because it is usually applied when a hard, wear-resistant metallic coating is required on a part that is subject to abrasive forces during service. A typical example is the chromium applied to hydraulic shafts for heavy equipment, on the piston rings of internal combustion engines, and on the shafts of landing gear for aircraft. A typical chromium plating tank is constructed of steel with a PVC lining. It is equipped with heating elements and an exhaust system to remove the chromic acid fumes from the workers' breathing zone.

Hexavalent Chemistries

The hexavalent chromium plating chemistries fall into two categories, conventional and mixed catalyst. The conventional is a simple chemistry that anyone can mix up and use. The mixed catalyst chemistries are patented processes that have a few advantages, including faster plating, fewer problems caused by current interruption, and fewer problems plating onto passive nickel deposits. They also tend to be more difficult to control and expensive to operate. In hard chromium applications, mixed catalyst baths also tend to etch steel in areas where plating is not intended, making masking more critical.

Conventional chemistry ingredients are

Chromium trioxide: 30–35 oz/gal
Sulfate: 0.3–0.35 oz/gal

Operating conditions:

Temperature: 125–135°F (for hard-plating applications, temperatures may be 140°F)
Anodes: lead
Plating current density: decorative 100–150 ASF, hard 150–250 ASF
Agitation: none (hard may use some air agitation)
Filtration: no
Color of solution: deep dark red-brown
Odor of solution: no specific odor

Mixed catalyst ingredients are

Chromium trioxide: 30–45 oz/gal
Sulfate: 0.15–0.18 oz/gal
Fluoride: 0.13 oz/gal
Note: Fluoride may be present as one or more of a variety of fluoride-containing compounds.

Operating conditions are the same as for conventional chemistry.

Trivalent Chemistries

Trivalent chemistries were developed in response to concerns about the detrimental effects of hexavalent chromium on the environment and on workers' health. Hexavalent chromium is a powerful oxidizer that readily attacks human tissues and has been linked in some studies to lung cancer. Trivalent chromium has a much lower toxicity level, is not an oxidizer, and to date has not been linked with cancer. Platers have been slow to accept trivalent chromium chemistries as substitutes for hexavalent chemistries because the former tend to plate deposits that are noticeably darker or not consistently of the same color. Since trivalent baths are used only for decorative applications, this is a major drawback, but solution manufacturers have made great progress toward solving these problems. A major benefit from trivalent processes is that these baths contain very low concentrations of chromium (about 1/5th as much), and the chromium can be waste-treated without a reduction step, so waste treatment and sludge disposal costs are reduced. Equipment generally consists of a rubber- or plastic-lined steel or plastic tank, an air agitation system, a heating and cooling system, and a filtration system. Ingredients (Atotech process) are

> TC additive: 52 oz/gal
> Chromium: 2.7 oz/gal
> TC stabilizer: 8% vol
> TC-SA: 1.2% vol
> TC regulator: 1 ml/liter
> pH: 3.2

Operating conditions:

> Temperature: 70–90°F
> Anodes: graphite (Atotech) (lead in membrane cell Enthone-OMI)
> Plating current density: 90–200 ASF
> Agitation: air
> Filtration: yes
> Color of solution: deep blue-green
> Odor of solution: no specific odor

Gold

Gold can be plated from three major chemistries; alkaline cyanide, neutral, and acid. All three chemistries utilize gold from potassium gold cyanide salts. A proprietary gold plating process that does not utilize potassium gold cyanide is on the market, but is expensive to operate and limited in alloying capability. Most gold plated is an alloy of gold and some other metal or combination of metals such as nickel, cobalt, copper, and silver. Gold can be plated in any commercial karat desired. The neutral and acid gold plating chemistries utilize chelating agents to perform the tasks normally performed by cyanide, control of metallic impurities and alloying elements. While these chelates could cause waste treatment problems, they rarely enter the wastewater treatment system in high concentrations, since most gold plating operations have meticulous recovery systems to use as little rinsewater as possible and recover the plating chemicals.

Alkaline Cyanide Chemistry

Alkaline Cyanide chemistry is most often used to apply a thin film of gold over bright nickel in decorative applications such as jewelry. The plating tank is a typical layout with a plastic or lined steel tank, filtration, cathode rod agitation (optional), and heating elements. Ingredients are

> Potassium gold cyanide: 0.1–0.5 oz/gal
> Potassium cyanide: 1–1.5 oz/gal

Potassium carbonate: 3–30 oz/gal
Additives: none
pH: 10

Operating conditions:

Temperature: 125–150°F
Anodes: gold or stainless steel or platinized titanium
Plating current density: 1–35 ASF
Agitation: cathode rod
Filtration: yes
Color of solution: dark yellow
Odor of solution: no specific odor

Neutral Gold Plating Chemistry

Neutral gold plating chemistry is favored for barrel plating applications of high-purity gold. Ingredients are

Potassium gold cyanide: 1–1.5 oz/gal
Monopotassium phosphate: 10–12 oz/gal
Potassium Citrate: 8–10 oz/gal
pH: 6–6.5

Operating conditions:

Temperature: 125–135°F
Anodes: platinized titanium
Plating current density: 1–3 ASF
Agitation: cathode rod/recirculation pump
Filtration: yes
Color of solution: pale yellow/clear
Odor of solution: no specific odor

Acid Gold Plating

Acid gold baths can produce a variety of gold deposits, including the hardest, most wear-resistant ones. They are favored for plating of printed circuit board connectors and in the semiconductor industry. Ingredients are

Potassium gold cyanide: 0.5–1.0 oz/gal
Citric acid: 4–6 oz/gal
Ammonium citrate: 4–6 oz/gal
pH: 3–5

Operating conditions:

Temperature: 90–140°F
Anodes: platinized titanium or platinized niobium
Plating current density: 1–5 ASF
Agitation: cathode rod
Filtration: yes
Color of solution: range from clear to purple
Odor of solution: no specific odor

Silver

While noncyanide silver plating chemistries based on sulfites or succinimides have been available for some time, almost all silver plating is presently being performed in the cyanide chemistry. The noncyanide baths are far more expensive to install and operate and cannot tolerate contamination to the same degree as can the cyanide process. The plating equipment for cyanide silver plating is typically a lined steel or plastic tank equipped with a filter and cathode rod agitation. Silver can also be barrel plated. Ingredients (for rack or barrel plating) are

Silver cyanide: 1–4 oz/gal
Potassium cyanide: 2–4 oz/gal
Sodium carbonate: 3–14 oz/gal
Potassium hydroxide: 1–2 oz/gal
Potassium nitrate: 0–2 oz/gal
pH: 12–14

Operating conditions:

Temperature: 70–90°F
Anodes: silver
Plating current density: 1–40 ASF
Agitation: cathode rod
Filtration: yes
Color of solution: dark brown/black
Odor of solution: organic

Tin

Tin can be plated from more than four major chemistries; the alkaline stannate process, the fluoborate, the proprietary halogen and sulfonate processes, and the sulfuric acid-based process. The alkaline stannate, flouborate, and sulfate chemistries are most often encountered in job shops. The alkaline stannate process typically consists of a heated steel tank, while the sulfuric acid-based process uses a plastic- or PVC-lined steel tank and has filtration. The alkaline process produces a matte, pure tin deposit that has excellent solderability, while the sulfate process produces matte or bright deposits with lesser or marginal solderability but superior appearance. The alkaline process is a bit more difficult to operate than the sulfate process. The alkaline bath is favored for barrel plating applications, although the flouborate bath can also be used.

Alkaline Stannate Process

Ingredients are

	Rack	Barrel
Potassium stannate	13.3	26.6
Potassium hydroxide	2	3
Potassium carbonate	3–14	3–14
Additives	None	None
pH	>14	>14

Note: ¼ to ½ oz/gal of cyanide is sometimes added to reduce the effects of metallic contaminants.

Operating conditions:

Temperature: 150–180° F
Anodes: Tin
Plating current density: 1–100 ASF
Agitation: cathode rod
Filtration: no
Color of solution: pale yellow/clear
Odor of solution: no specific odor

Fluoborate-Based Chemistry (for Rack and Barrel Plating)

Ingredients are

Tin (from concentrate): 4–6 oz/gal
Fluoboric acid: 25–35 oz/gal
Boric acid: 3–5 oz/gal
Additives: as recommended, but required
pH: <.1

Operating conditions:

Temperature: 90–120° F
Anodes: Tin
Plating current density: 1–80 ASF
Agitation: cathode rod
Filtration: yes
Color of solution: pale yellow
Odor of solution: no specific

Sulfate-Based Chemistry

Ingredients are

Stannous sulfate: 2–6 oz/gal
Sulfuric acid: 1–3 oz/gal
Additives: as recommended, but required
pH: <1

Operating conditions:

Temperature: 55–85° F
Anodes: tin
Plating current density: 1–25 ASF
Agitation: cathode rod
Filtration: yes
Color of solution: pale yellow
Odor of solution: sweet

Tin-Lead

Tin-lead is applied to electronic components that require high solderability. The plating hardware is a typical plating setup with continuous filtration optional. Two basic chemistries are used in plating the alloy: the fluoboric and the (proprietary) sulfonic acid-based chemistries. These baths are most commonly found in printed circuit board manufacturing shops and job shops specializing in plating for electronics. The electronics industry uses a "high-throw" formulation to allow for plating inside drilled holes. Others use a conventional bath that yields the best solderability. The fluoborate bath is made by mixing liquid fluoborate concentrates with water. Fluoboric bath ingredients are

	Conventional	High throw
Tin (from concentrate)	7–8 oz/gal	1.6–2.7 oz/gal
Lead (from concentrate)	3–4 oz/gal	1.1–1.9 oz/gal
Fluoboric acid	13–20 oz/gal	47–67 oz/gal
Boric acid	3–5 oz/gal	3–5 oz/gal
pH	<.1	<.1

Operating conditions:

> Temperature: 70–90° F
> Anodes: lead-tin alloy
> Plating current density: 15–25 ASF
> Agitation: cathode rod
> Filtration: yes
> Color of solution: pale yellow
> Odor of solution: sweet

19.3.4 Electroless Plating Processes

One major drawback to electroplating is the nonuniform coating produced, because electric current tends to concentrate on sharp edges, corners, and points. Electroless plating processes are used when it is necessary to obtain a very uniform coating on complex geometries, because these processes do not depend on electricity delivered from a rectifier. As the name implies, the coating is produced without an outside source of current. The reducing electrons are provided chemically.

Some electroless deposits are also more corrosion resistant than their electroplated counterparts. The following are two of the most often applied electroless processes.

Electroless Nickel

Electroless nickel is applied to numerous complex electronic and industrial components for the high degree of wear resistance and corrosion protection. An example is the sliding plates that "mold" the hamburger patties served in fast-food restaurants. Two plates slide against each other to form the cavity that is used to "injection mold" the patty. The plates slide against each other at lighting-fast speeds. Electroless nickel is also used in the plating of plastics, to provide the first metallic layer on the plastic to yield conductivity for subsequent deposits.

The electroless plating process normally consists of two plating tanks and a nitric acid storage tank. Each plating tank contains heating elements, an air sparger, and a recirculating filter. The solution will eventually deposit nickel on everything it contacts, so periodically the tank walls and associated equipment must be stripped with nitric acid (thus the nitric storage tank). The plating solution has a finite life (8–14 "turnovers"), after which, it must be waste treated or disposed of through a commercial disposal firm. The electroless nickel plating solution contains strong chelating agents that interfere with a conventional wastewater treatment system, so they must be treated separately using electrowinning, proprietary treatment methods, or special chemical treatments. The rinsewater from electroless nickel operations is usually segregated and treated separate from other rinsewater.

Electroless nickel is typically plated from one of two basic chemistries, yielding either a nickel-phosphorus alloy (most common) or a nickel-boron alloy. Each alloy can be plated from a number of different solutions to yield varying alloy compositions. Following are two such chemistries.

Nickel-Phosphorus Alloys

Nickel sulfate	21 g/L	11.8g/L
Acetic acid	9.3 g/L	
Lactic acid	27 g/L	
Molybdic acid		0.009 g/L
Propionic acid	2.2 g/L	
Lead acetate	0.001 g/L	
1,3 Diisopropyl thiourea		0.004 g/L
Sodium hypophosphite	24 g/L	22.3 g/L
pH	4.6	5.5
Temperature	95°C	95°C

Nickel-Boron Alloys

Nickel chloride	30 g/L
Sodium hydroxide	40 g/L
Ethylene diamine	86 ml/L
Sodium borohydride	0.6 g/L
Thallium nitrate	0.007 g/L
Sodium gluconate	15 g/L
Diethyl amine borane	1.0 g/L
Lead Acetate	0.02 g/L
pH	13–14
Temperature	90°C

Electroless Copper

The major use for electroless copper is in the manufacture of printed wiring boards. The electroless copper is used to apply a thin coating of copper over the top side and into the drilled holes of the boards. The drilled holes are initially nonmetallic, since the boards are made of epoxy-fiberglass. With the electroless copper, the holes become conductive for further plating.

Equipment for electroless copper plating usually consists of a polypropylene or PVC tank and filter. Some baths operate at room temperature, so heating is not required; others require heating. The rinsewater and spent electroless copper often contain chelating or complexing agents, so waste treatment is difficult.

Two typical compositions for an electroless copper solution are

Copper sulfate	13.8 g/L	5 g/L
Rochelle salts	69.2 g/L	25 g/L
Sodium hydroxide	20 g/L	7 g/L
MBT	0.012 g/L	
Formaldehyde	38 ml/L	10 ml/L
Temperature	50°C	25°C

Other Electroless Processes

There are numerous other electroless plating solutions in the literature, although they are rarely used. The following solutions may be encountered:

Electroless Cobalt

Cobalt sulfate: 30 g/L
Ammonium chloride: 84 g/L
Sodium hypophosphite: 20 g/L
pH: 10
Temperature: 95° C

Electroless Silver

Silver cyanide: 1.34 g/L
Sodium cyanide: 1.49 g/L
Sodium hydroxide: 0.75 g/L
Dimethyl amine borane: 2 g/L
Temperature: 55° C

Electroless Gold

Potassium gold cyanide: 5.8 g/L
Potassium cyanide: 13 g/L
Potassium hydroxide: 11g/L
Potassium borohydride: 21.6 g/L
Temperature: 75° C

19.3.5 Typical Postplating Operations

After plating, parts are often further treated in various chemical solutions to enhance the appearance or corrosion/tarnish resistance of the plated coating. Examples of such operations include chromate films on zinc, cadmium, and copper plates, and stains produced on copper or copper alloy deposits. Application of lacquers, waxes, and other organic topcoats are also popular postplating methods for improving shelf and service life of parts. Chromate "conversion" coatings are so popular that almost every zinc and cadmium plating line has chromating tanks and rinses built into it, and most all zinc and cadmium plated parts have some form of chromate film on top of the metal deposit.

Chromating

A chromate is a very thin complex film created by "converting" a small amount of the top surface of the plated metal into the film, thus the term conversion coating. The chromate film is formed by immersing the plated deposit into an acidic solution containing a variety of chemicals depending on the color and corrosion resistance to be obtained. If, for example, we want the zinc to turn bright, reflective, and with a hint of blue (typically called a blue-bright dip), the solution will contain nitric acid and potassium ferricyanide, along with some trivalent chromium compounds. If we want a yellow iridescence, hexavalent chromium in the form of sodium dichromate may be added. If we want olive drab green, even more dichromate along with sulfates may be added. If we want a black coating, a small amount of silver nitrate is added (finely divided silver particles create the black color).

In each case, the mechanism for forming the chromate film is the same: The plated metal is attacked by the acid in the chromate dip, releasing hydrogen as a by-product of the attack. As the hydrogen is released from the metal surface, the pH of the solution near the metal surface rises high enough to "deposit" a film of metal hydroxides and other trapped ingredients from the solution. The film at first is a delicate gel, but quickly hardens into a thin coating only a few millionths of an inch thick (or less).

The chromate film protects the plated metal from corrosion by acting as a barrier layer against corrosive atmospheres. In general, the more hexavalent chromium is trapped in the film, the higher will be the coloration, and the better the corrosion resistance.

Since the chromate functions by attacking and dissolving some of the plated metal, eventually the solution becomes so contaminated with plated metal that it stops producing acceptable coatings. At this point the chromate becomes "spent" and must be waste treated.

Rinses following chromating operations are not recoverable (since recovery would only hasten the demise of the chromating solution) and therefore must be routed to a wastewater treatment system before discharge to sewer.

There are several variations on the chromating process, including some films that are applied with reverse current and others that are applied and then "leached" to remove the coloration. The most commonly applied chromates are the blue brights, followed by the yellow and then the black. The military favors the olive drab for its color and high corrosion resistance.

A significant amount of research is being made into substitutes for chromates that do not contain hexavalent chromium, although not much is on the market now.

Bright Dipping

Some operations utilize a "bright dip" after zinc or cadmium plating. This dip can be used by itself or before chromating. It is simply a very dilute solution of nitric acid (0.25–0.50% vol). The bright dip removes a thin organic film from the surface of freshly plated zinc or cadmium, rendering the deposit far brighter than before. This is not a chromate, nor is it a true conversion coating.

Other Postplating Operations

Numerous other postplating operations can be performed, including application of lacquer, wax, dyes, or stains. Invariably, such operations involve additional tanks, equipment, and chemical solutions.

19.3.6 Bibliography

The following materials were sources of information included in this subchapter and are recommended for additional reading:

Intensive Training Course in Electroplating, Illustrated Lecture Series, American Electroplaters and
    Surface Finishers Society, Orlando, FL.

19.4 COATINGS

Jack M. Walker

19.4.0 Introduction to Coatings

In keeping with the premise that manufacturing engineers working today cannot possibly be familiar with the total factory operation and all manufacturing processes, this subchapter on coatings is an introduction to the principles of coatings. The modern protective coatings are an outgrowth of the paint which started out as an artist's material. From the early history of humanity we find that it was the artist who gathered the materials and developed the methods for making paint. Since early types of paint were made by artists, paint and varnish making was in itself an art for many centuries.

Decoration seems to have been the original purpose of paint, but in time its power to protect the vulnerable surfaces of maufactured objects became of equal importance. With the industrial development which took place in the eighteenth and nineteenth centuries, paint began to emerge as a commercial material. However, the basic ingredients were still the natural resins and oils traditionally used by artists. These materials had a certain value as protective coatings, but were limited in effectiveness.

At the turn of the twentieth century, industrial development began to mean heavier demands on coatings, and scientists began to investigate the traditional art of paint making with an eye toward improvement. During the last 30 years, steadily increasing effort has virtually revolutionized the manufacture of coatings. Today coatings are available which resist attack by nearly all chemicals and corrosive conditions. These protective coatings, which are usually distinguished from paints that are primarily decorative, are valuable engineering materials. They often make it feasible to use steel structures, or other materials, in environments where they would not otherwise have an economic life.

19.4.1 Wet Coatings

Paint can be defined as any fluid material that will spread over a solid surface and dry or harden to an adherent and coherent colored obscuring film. It usually consists of a powdered solid (the *pigment*) suspended in a liquid (the *vehicle*). The pigment provides the coloring and obscuring properties. The vehicle is the film-forming component which holds the pigment particles together and attaches them to the surface over which they are spread.

Coverage of Coatings

If every drop in a gallon (U.S.) of liquid protective coating can be applied to a surface without any loss due to the application equipment, the material will cover 1604 ft^2 at a thickness of 1 mil (0.001 in.). In most instances, the liquid contains a volatile solvent which, upon evaporating, reduces the thickness of our film. For instance, if the wet coating contains 50% by volume of solvent, the film after drying will be 0.5 mil thick rather than 1 mil thick. The former is called the wet thickness, while the latter is the dry thickness.

Knowledge of the volume percent solvent, or more usually, the volume of nonvolatiles which we call the *percent volume solids*, enables us to calculate the dry theoretical coverage for a gallon of paint as follows:

Theoretical coverage per U.S. gallon = (1604 × % volume solids) ÷ (100 × dry film thickness in mils)

Obviously, no applicator can get every drop of material out of the container, nor can one avoid leaving some of it in or on the application equipment. More important, there will be considerable losses, particularly in spray painting, due to air movement as well as missing the target. The magnitude of these losses will vary depending on what is being coated, the application equipment, and air movement. With the same coating, Painter A, spraying a flat wall indoors with an apparatus in good condition, will get considerably more mileage than Painter B, spraying 2-in. channel steel in a 20-mile wind with a badly maintained spray gun which must be unplugged every 10 min. Some contractors figure on a 20% loss during spraying, while others use 30% in estimating material requirements.

Components of Wet Coatings

Most protective coatings contain a volatile component, the solvent. Its purpose is to keep the coating in a liquid condition suitable for uniform application and bonding to the target or substrate. The other major components of a coating are shown in Figure 19.28. They include the resin or binder, and pigments.

Pigment Functions

Color—aesthetic effect, hides substrate.
Protection of resin binder—absorb and reflect solar radiation which can cause breakdown of binder (chalking).
Corrosion inhibition—chromate salts and red lead in primers as passivators. Metallic zinc, when in high enough concentration, gives cathodic (sacrificial) protection.
Film reinforcement—finely divided fibrous and platey particles which increase hardness and/or tensile strength of film.

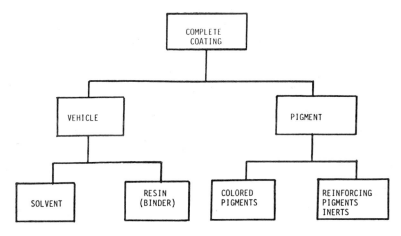

Figure 19.28 The major components of a wet coating.

Nonskid properties—particles of silica or pumice which roughen film surface and increase abrasion resistance.

Sag control—so-called thixotropic agents which prevent sagging of the wet film and also reduce the tendency of other pigments to settle in the container during storage.

Hide and gloss control—increasing color pigment concentration improves hide, while an increase in either color or other pigmentation decreases gloss.

Increased coverage—properly selected "filler pigments," sometimes called "inerts," can increase the volume solids (or coverage) of a coating without reducing its chemical resistance. There is a limit to how much filler pigment can be used with a given resin composition. This constraint is termed the *critical pigment volume concentration* and indicates the volume of pigment which can be bound by the resin without leaving voids in the film.

The pigments, both those which contribute color and those which have other functions, must be uniformly dispersed and completely "wetted" in the resinous binder to function properly.

Resinous Binder Functions

The binder, in addition to an ability to "glue" the pigments together in a homogeneous film, must be capable of wetting and adhering to the substrate, preventing penetration of aggressive chemicals, and maintaining its integrety in the corrosive environment. Since there is a wide variety of environments, it should not be surprising that there is a wide variety of binders, each with its special niche in the coatings industry.

The Vehicle

To be capable of being applied in an even, essentially void-free film which will wet and adhere to the substrate, the binder-pigment mixture must be in a highly fluid condition. The use of low-molecular-weight liquids (solvents) to dissolve or to increase the fluidity of the binder is usually necessary. The combination of such solvents with the binder is termed the *vehicle*.

Each type of binder has specific solvent combinations which are most efficient in producing the desired application and film-forming properties. There is no universal solvent for protective coatings. The best solvent for one type of coating is water, while for another a combination of expensive, toxic organic compounds may be required. Use of the wrong solvent, or as it is usually called, "thinner," will cause precipitation of the binder. The result may range from instant gelation of the coating to films with substandard properties in application or performance. Some of the common solvents used in the formulation and in the thinning of protective coatings are listed below.

Mineral spirits—often called the painter's naphtha; high-boiling-point petroleum product used for oil and alkyd vehicles.

Aromatics—compounds of the benzene family, including toluene, xylene, and higher boiling homologs; major solvent for chlorinated rubber, coal tar, and certain alkyds; used in combination with other solvents in vinyl, epoxy, and polyurethane vehicles.

Ketones—compounds of the acetone family including methyl ethyl ketone (MEK), methylisobutyl ketone (MIBK), and cyclohexanone; most effective solvent for vinyls but sometimes used in epoxy and other formulations.

Esters—have distinctive, usually pleasant odors; commonly used in epoxy and polyurethane vehicles.

Esthers—high-boiling solvents sometimes used to achieve vehicles with high flash points.

Alcohols—good solvents for highly polar resins such as phenolics and acetals. Also used in solvent-based self-curing silicates.

Water—only thinner for latex and emulsion coatings, as well as for certain inorganic zinc silicate primers.

Binders

In order to produce a film which will perform satisfactorily in a given environment, the coating, after application, must convert to a dense, solid membrane. Some materials can accomplish this simply by releasing their solvents, while others must go through a series of complicated chemical reactions, sometimes requiring the application of heat after evaporation of the solvent.

The ability of a resin to form a dense, tight film is related directly to its molecular size and complexity. The polymers which are capable of forming such films simply by the evaporation of solvent are initially of very high molecular weight and are not capable of further chemical reaction. Because of their large size, these polymers must be kept in a dilute solution, and coatings based on them have a low volume of solids. Resins of low molecular weight, although requiring chemical conversion to attain polymer structures of suitable size, have the advantage of being able to produce higher solids combinations.

Five general types of binders are used to formulate protective coatings for the chemical processing and marine industries. These include lacquers, drying oil types, co-reacting compounds, condensation coatings, and inorganic vehicles.

1. *Lacquers* are based on high molecular weight, chemically resistant polymers which form films by evaporation of solvents; low volume solids which sometimes are increased by the addition of lower molecular weight resins or plasticizers. Examples include:

    Polyvinyl chloride copolymers—ranging from low-film-build materials (1–2 mils per coat) used as tank linings to high-build (5–10 mils per coat) maintenance coatings.

    Chlorinated rubbers—formulated in a variety of combinations with modifying resins to attain higher solids; chemical and water resistance varies with type and amount of modifier used; usually limited to applications at 2–3 mils per coat, although some newer materials are capable of building at 4–6 dry mils.

    Polycrylics—have excellent color and gloss stability in outdoor weathering; water and chemical resistances are not as high as for vinyl copolymers or chlorinated rubbers; often used in mixture with polyvinyl chloride copolymer or as final coat.

2. *Drying oil types*—low-molecular-weight resins capable of converting to tough films through intermolecular reactions with oxygen catalyzed by the presence of metal soaps, such as cobalt octoate. Examples include

    Natural oils, such as linseed, tung, and soya oils—which contain reactive sites which are activated by oxygen in their molecules; very slow in converting to cured films.

Alkyds—natural oils which are chemically modified by reactions with synthetic aids or alcohols to improve rate of cure, chemical resistance, or ultimate hardness; degree of modification is designated by terms *long oil* (relatively minor modification), *medium oil* (moderate modification), and *short oil* (considerable modification).

Varnishes—natural oils or alkyds containing dissolved or reacted resins such as chlorinated rubber or phenolic to increase hardness and chemical resistance.

Epoxy esters—a type of alkyd in which a high-molecular-weight epoxy resin is used in the chemical modification; usually have much higher degree of chemical resistance than normal alkyds but have poor exterior weathering properties.

Uralkyds (Urethane oils)—a type of alkyd modified by reaction with tolylene diisocyanate which imparts excellent abrasion resistance, but also detracts from exterior weathering properties.

Synthetic polyesters—contain no drying oil, but curing mechanism is similar to that of the latter; organic peroxide added to the resin at the time of application; major use is in fabricating glass-reinforced structures such as piping but are occasionally employed with chopped glass or glass mat as tank linings; have advantage of being 100% solids since the solvents, styrene and vinyl toluene, co-react with the polyester resin upon the addition of the peroxide.

Note: The drying oil types, except for the synthetic polyester, convert to their cured state by taking up atmospheric oxygen; consequently, the rate of cure is most rapid at the air surface and slowest at the substrate interface. Because of the difference in cure rate through the thickness of the film, care must be taken to avoid *wrinkling* resulting from excessive film thickness or too rapid recoating.

3. *Coreacting types*—unlike the lacquers and drying oils, these materials have two or more separately packaged components which are combined just before application. Reaction rates between the components are modified somewhat by the presence of solvents. However, the coreacting coatings, after mixing of the components, have limited pot lives which are reduced by increased application temperatures and usually extended by lower ones. Examples include epoxy and polyurethane coatings.

Epoxy coatings—which cure at ambient temperature are based on low molecular weight resins containing the reactive *epoxy group* at each end of the molecule. This group has a particular affinity for ammonia derivatives called *amines*. The amine group may be strung along a short molecule, the *polyamine*, or may be more widely scattered on a long chain, the *polyamide*. In both cases, the amine-containing chain ties the epoxy resin molecules together in a dense cross-linked structure. The film obtained with the polyamine is usually more dense, having better solvent and chemical resistance but decreased flexibility, and it is of a more brittle character than that of the polyamide. Modification of these *curing agents* by prereacting them with a portion of the epoxy or other resins or chemicals is sometimes undertaken by coatings manufacturers to improve cure properties. These modified amines are called *amine adducts*.

The water resistance and, to a degree, the acid resistance of the various room-temperature epoxy coatings can be improved by the incorporation of high concentrations of selected coal tar resins into the formulas. These *epoxy coal tar* coatings are often used in areas where a high order of water resistance is required without the need of the solvent resistance or appearance of the base epoxy system.

Epoxy coatings are characterized by hard films of limited flexibility with a high order of chemical and, except for the coal tar modifications, solvent resistance. They are, however, much more susceptible to attack by oxidizing agents such as chlorine gas, peroxides, and nitric acid then are coatings based on vinyl copolymers or chlorinated rubber. A major weakness is their high rate of chalking in exterior exposures.

Polyurethane coatings are based on the reaction of a group of chemicals, the *diisocyanates* with resins or chemicals containing alcohol or amine substituents in their structures. The diisocyanates have a high order of toxicity and must be chemically modified to permit their use in protective coatings. The most common practice is to make a *urethane prepolymer* by reacting two molecules of the diisocyanate with one molecule of a resin or chemical compound containing two alcohol groups in its structure. The urethane prepolymer is capable of further reaction with resins or chemical compounds containing amine groups (a very rapid reaction) or alcohol groups (much slower). A third possible reaction which is sometimes useful but often a nuisance is with water. If a free isocyanate group comes in contact with water, it will decompose to form an amine and simultaneously release carbon dioxide gas. Neighboring isocyanate groups will react immediately with the amine to initiate the promotion of a complex polymer structure. This reaction with moisture is the basis for the one-package moisture-curing polyurethane.

Another approach to polyurethane coatings is to mix the prepolymer with a resin containing alcohol groups at the time of application to form a two-package polyurethane. The choice of the second resin will determine to a large degree physical properties such as hardness and flexibility in the cured film.

When properly formulated, applied, and cured, urethane coatings have outstanding toughness and abrasion resistance with chemical properties similar to the epoxies. Unlike the epoxies, they do not have good adhesive properties to steel and concrete. Epoxy primers are commonly used as part of a polyurethane system.

4. *Condensation (heat condensing) coatings* are based on resins which interact to form cross-linked polymers when subjected to temperatures of 350–400° F. The use of such materials is limited to tank linings and to objects or structures which can be handled in an oven. Many of the so-called powder coatings as well as high-temperature silicones fall into this category. (See Subchapter 19.2).

The oldest of the condensation coatings is the *pure phenolic*. Applied in several coats to a thickness of 6 mils and baked to a temperature of 375° F, the phenolic film becomes extremely hard and chemically resistant. It is, however, extremely brittle and is rapidly attacked by caustic solutions.

5. *Inorganic silicates* are binders which do not contain organic (carbon) structures in their composition. For this reason they are noncombustible and are unaffected

by sunlight dust in *zinc inorganic primers*. There are several approaches to silicate binders, but each appears to depend on the development of an extremely adhesive form of silica during its cure. Examples follow.

Postcured inorganic silicates are based on water solutions of *alkali silicates* pigmented with zinc dust or zinc dust and a metal oxide. The dust or powder is mixed into the liquid at the time of application. A 2½- to 3-mil film spray dries very rapidly, usually within 1 hr. At this point it is very hard but remains water soluble. Application of an acid curing solution is necessary to achieve conversion of the silicate to the insoluble silica.

Self-curing, water-based silicates are mixtures of alkali silicates and colloidal silica pigmented with zinc dust or zinc and a metal oxide. As with the postcured materials, the separately packaged powder is dispersed in the vehicle at the time of application. The development of water insolubilization for this type of film is dependent on the absorption of carbon diozide from the atmosphere during the curing process.

Self-curing ammonium silicates are water-based dispersions of colloidal silica containing additives which are capable of generating ammonia during the cure of the film. The pigmentation, packaging, and mixing procedures are identical to those of postcured inorganic silicates and self-curing, water-based silicates. The initial hardness of the film is considerable lower than that of the latter coatings. However, as moisture is absorbed from the atmosphere, the curing mechanism proceeds to produce a film which is ultimately harder than the other types of inorganic films. Since ammonia is generated during the curing process, ammonium silicate films should not be topcoated until they have weathered to their ultimate hardness.

Self-curing, solvent-based silicates are organic esters which gradually hydrolyze upon exposure to moisture to produce a binder which appears to be essentially identical to those of the water-based types. A major advantage of this type of film is its almost instant resistance to rain or flowing water. Although most of the commercially available inorganic coatings in this category are cold as two separately packaged components, one-package products are proving to be practical.

The inorganic zinc primers are recognized as having many of the benefits of galvanizing while providing a much more suitable surface for topcoating than the latter. They should not be confused with the so-called zinc-rich organic primers.

The use of metallic zinc as the pigmentation in a variety of vehicles, most notable in epoxy-polymide combinations, does result in primers having outstanding anticorrosive properties. However, field experience has demonstrated that coating systems employing these zinc-rich organic primers do not have the longevity of those based on the inorganic zinc silicates.

Primer Coats

A primer should meet most, if not all, of the following requirements:

1. Good adhesion to the surface to be protected when the latter has been cleaned or prepared according to specification
2. A satisfactory bonding surface for the next coat
3. The ability to stifle or retard the spread of corrosion from discontinuities such as pinholes, holidays, or breaks in the coating film

4. Enough chemical and weather resistance by itself to protect the surface for a time period in excess of that anticipated before application of the next coating in the system
5. Under certain conditions, notable tank linings, chemical resistance equivalent to the remainder of the system

Four types of primers which are commonly used over steel may be compared in their abilities to meet these requirements. See Table 19.2.

Intermediate coats may be required in a system to provide one or more of the following properties:

1. Adequate film thickness of the system (body coat)
2. A uniform bond between the primer and the topcoat (tie coat)
3. A superior barrier with respect to aggressive chemicals in the environment (may be too deficient with respect to appearance or physical properties to be a satisfactory finish coat)

Finish coats are the initial barriers to the environment, but are also the surfaces which are seen by your management, the public, and the ultimate consumer. There are situations, however, where the barrier to the environment will be primarily a function of the body or primer coat, while the finish coat serves to provide a pleasing appearance, a nonskid surface, a matrix for antifouling agents, or other specialized purposes. Obviously, the chemical resistance of the finish coat in one of these situations must be sufficient to ensure its remaining intact in the environment.

Surface Preparation for Wet Coatings

Industrial, commercial, and weekend painters have one thing in common: They resent the time or expense required to prepare surfaces properly for the application of protective or appearance coatings. However, if they ignore this most important step, the whole operation is likely to be a waste of time and money. A protective barrier, unless it is thick enough to be a self-supporting structure, must be uniformly bonded to the substrate for the following reasons:

1. Sufficient attachment is necessary to prevent dislodgement under the gravitational and mechanical forces to which it is exposed.
2. Under immersion or condensation conditions, water vapor will penetrate a barrier and condense on any unbonded surfaces. Progressive disbonding is almost sure to occur.

Cleanliness

Primers spread on the surfaces which they contact. If these are dirt, dust, scale, rust, oil, or moisture, the bond of the coating system to the structure can be only as good as the bond of the contaminant to the real surface. Furthermore, primers are formulated to stick to metals, concrete, wood, and masonry rather than to surface contaminants. The need for scrupulous surface cleaning prior to any coating has become more important since spraying has replaced brushing as the most common method of wet material application. Understanding of the corrosion factor is discussed in greater detail in Subchapter 19.1, and cleaning in Subchapter 19.2.

Table 19.2 Major Types of Primer Coatings

Primer type/ requirement	Alkyd or oil	Mixed resin	Resin identical to topcoats	Inorganic zinc
Bonding to surface	Usually have the ability to wet and bond to most surfaces and are somewhat tolerant of substandard surface preparation.	Adhesive properties are major consideration of formulation. Not quite as tolerant of substandard surface preparation as oil types.	Adequate for proposed use when surface properly prepared.	Outstanding adhesion to properly cleaned and roughened surfaces.
Adhesion of topcoats	Satisfactory for oil types. Usually unsatisfactory for vinyls, epoxies, and other synthetic polymers. Are softened and lose integrity, by attack from solvent systems of these topcoats.	Formulated for a specific range of topcoats.	Usually part of specific generic system. Maximum permitted dry time before application of second coat must be observed.	Fits into wide range of systems. Tie Coat may be required. Specific recommendation should be obtained for immersion systems.
Corrosion suppression	Limited Alkali produced at cathode and corrosion battery attacks film (saponification) and causes disbonding. Results in spread of underfilm corrosion.	Usually formulated with good resistance to alkali undercut and contain chromate pigment for a degree of corrosion inhibition.	Often contain chromate pigments for degree of corrosion inhibition. Resistance to alkali undercut is variable.	Outstanding ability to resist disbonding and underfilm corrosion. Anodic property of metallic zinc protects minor film discontinuities.
Protection as single coat	Limited by severity of exposure.	Limited by severity of exposure.	Limited by severity of exposure.	With very few exceptions, will protect without topcoat.
Chemical	Typical of alkyds.	Usually of lower order of resistance than that of topcoat.	Typical of system.	Not resistant to strong acids and alkalis. Has outstanding solvent resistance.

Improving the Bonding Surface (Profile)

Mechanical roughening of a surface by the use of abrasives or acids can provide improvements in the degree of bond which will be developed by the initial coating material—whether this is a primer or some other coating. These processes expose fresh, chemically clean surfaces which are easy to wet. Furthermore, roughening increases the actual contact area. Imagine two surfaces, one smooth steel and the other sandblasted, with exactly the same dimensional area. Although both have exactly the same measured area, the sandblasted specimen has a much greater surface area for bonding with the coating. If the primer bond to the smooth surface has a strength of 2 lb, its bond to the sandblasted surface may be twice the actual number of square inches of surface, with a strength of 4 lb. If the sandblasting quadrupled the surface area, the bond strength should be 8 lb. Acid etching of steel, although less effective than sandblasting, also increases the bondable surface area. The latter can be further improved by the application of a phosphate conversion coating. In such a process, iron, zinc, or manganese dihydrogen phosphate is applied by dipping or spraying to clean steel. When this is done properly, a film of tightly adherent iron phosphate crystals forms at the surface to promote a more receptive substrate for the coating. The use of phosphate conversion coatings is very common to the product finishing industry and other assembly-line types of operations such as coil coating.

Application of Wet Coatings

The most important elements in producing good results and avoiding problems in application of wet coatings are described in the following paragraphs.

Environmental Conditions

The ideal time for painting is when the weather is warm and dry with little wind. Obviously, many coating projects cannot be delayed until these ideal conditions prevail. Also, many shops either do not have an environmentally controlled spray booth, or the product cannot be brought inside where conditions can be better controlled.

Under conditions of *high humidity*, condensation of moisture is likely to occur on surfaces. Condensation on the substrate interferes with bonding of the coating. Condensation on the surface of a freshly applied coating may alter its curing process.

Very low humidities can be a problem with water-based products. Rapid "flash-off" of the water may result in film cracking. It can also cause poor curing rates with certain types of inorganic coatings.

At *low temperatures*, the film thickness of high-build or *thixotropic* coatings becomes more difficult to achieve. Curing reactions slow down or stop for many materials. Water-based products may freeze. Solvents evaporate more slowly. Furthermore, when the relative humidity is high, condensation is likely to develop.

Although heat has many beneficial effects in the application of coatings, *high temperatures* often increase overspray (dry fallout), trapped air or solvent bubles, and in the case of zinc inorganics, the incidence of film cracking. High temperatures also reduce the pot life of catalyzed materials.

Wind is a nuisance, particularly in spray painting. The material can be deflected from the target as it leaves the spray gun. Solvent tends to flash off, creating excessive dry spray at edges of the spray pattern. Lap marks become more evident. Dirt and other debris may become embedded in the wet film. The velocity of air inside a spray booth can create the same problems, as will be discussed later in this subchapter.

Condensation becomes a problem when humidities are high and surface temperatures are low. Unfortunately, on large-scale painting projects, which must be done outside, primers are often applied late in the workday, and sometimes at night. Abrasive blasting is a slow process, while applying a primer by spray goes very rapidly. Because of this wide difference in work rates, the contractor may take 6 hours of an 8 hour day to prepare the surface for 1 to 1½ hours of primer application. Table 19.3 illustrates the relationship of air temperature, metal temperature, and percent relative humidity to condensation. The best procedure is to paint only when the surface temperature is at least 5°F above the temperature where condensation will form (the dew point).

Many of the application, drying, and curing problems created by weather conditions, either outside or inside the factory, can be reduced by lowering the viscosity of the material by the addition of the proper thinner. However, the limits shown in the product's application instructions should not be exceeded without checking with the paint manufacturer. A *thinner* is simply a mixture of solvents that are compatible with the resins in the coating. Thinning can provide these benefits:

1. Improved flow and uniformity in application of the material
2. Reduced overspray, lap marks, bubble entrapment, and film "mud cracking" caused by rapid solvent flash-off

With some materials, heating or warming has an effect similar to thinning. Several types of heating devices are available for use with spray equipment. The use of very thick lacquer, sprayed hot, is common in the furniture industries. This preheating permits a high-solids material to be sprayed easily in a single coat.

Since thinning reduces the volume solids of a coating, film build may be difficult to obtain. In that situation, reducing the thickness per coat and increasing the number of coats will result in a better job. This is often true in both cold and extremely hot weather. Thinner films permit easier escape of solvent under both conditions. Bubbles and pinholes in hot weather and extremely slow hardening rates of thick film in cold weather are the result of the solvent's difficulty in escaping at its ideal rate.

Table 19.3 Percent Relative Humidity Above Which Moisture Will Condense on an Uninsulated Metal Surfaces

METAL SURFACE TEMP.	— SURROUNDING AIR TEMPERATURE °F																
	40	45	50	55	60	65	70	75	80	85	90	95	100	105	110	115	120
35°F	60	33	11														
40		69	39	20	8												
45			69	45	27	14											
50				71	49	32	20	11									
55					73	53	38	26	17	9							
60						75	56	41	30	21	14	9					
65							78	59	45	34	25	18	13				
70								79	61	48	37	29	22	16	13		
75									80	64	50	40	32	25	20	15	
80										81	66	53	43	35	29	22	16
85											81	68	55	46	37	30	25
90												82	69	58	49	40	32
95													83	70	58	50	40
100														84	70	61	50
105															85	71	61
110																85	72
115																	86

Methods of Application—Spray Painting

The object in spray painting is to create a mist of atomized (finely dispersed) coating particles which will cling to the target in a uniform pattern and then flow into a continuous, even film. The three most common types of spray painting are

1. Conventional (air atomization)
2. Airless (hydraulic pressure)
3. Electrostatic

Air Atomization

The spray gun, which is the primary component in a spray system, brings the air and paint together. This is accomplished in such a way that the fluid is broken up into a spray which can then be directed at the surface to be coated. There are two adjustments in most spray guns: one which regulates the amount of fluid which passes through the gun when the trigger is pulled back; and a second which controls the amount of air passing through the gun, thus determining the width of the fan. In the external-mix spray gun, which is the most widely used type, the air breaks up the fluid stream outside the gun after being directed through a specially designed air cap. The number, position, and size of the holes in the air cap determine the manner in which the air stream is broken up. This in turn governs the breakup or atomization of the fluid stream. See Figures 19.29, 19.30, and 19.31 for effects of fan and operator on the spray pattern.

The fluid leaves the gun through a small hole in the fluid tip. A needle which is operated by the gun's trigger controls the flow of material through this tip. As with air caps, fluid tips are manufactured in different sizes to accommodate various materials, the diameter of the orifice in the tip being the differentiating factor. Various coatings require different types of air caps and fluid tips in order to be properly atomized. Therefore, it is usually wise to follow the coating manufacturer's recommendations as to the proper spray fan, air cap, and fluid tip. As an example, vinyl, epoxy, and chlorinated rubber-based coatings can be successfully sprayed with a DeVilbiss MBC or JGA-type spray gun equipped with a 78 or a 765 air cap and an E fluid tip and needle. (The letter E denotes the diameter of the orifice.) Binks and others make similar equipment.

There are two methods for bringing fluid to the gun: suction or pressure feed. For *suction feed*, the gun is usually fitted with a 1-qt cup holding the fluid to be sprayed. When the trigger is pulled, suction is developed at the tip of the gun, drawing fluid out of the cup and up to the nozzle where it is sprayed. This type of setup has severe limitations: the gun must be operated only pointing horizontally, it will not spray viscous material, spraying is slow, and the cup must be filled frequently. See Figure 19.32.

With *pressure feed*, the fluid to be sprayed is forced to the gun under pressure. The pressure type has several advantages: Material of higher viscosity can be sprayed, a heavier coat can be applied, and spraying is much faster. Although a 1-qt pressure cup is sometimes attached to the gun for small applications, the majority of pressure-fed spray operations are conducted with a separate pressure pot. This system ensures better pressure control and permits faster spraying. See Figure 19.33.

The pressure pot is a closed chamber of usually 2–10 gal capacity which contains the fluid to be sprayed. A large air hose, preferably 1-in. ID connects the pressure pot to the air source. A fluid hose and an air hose connect the pot to the spray gun. Air is directed into the pot through a regulator which maintains the proper pressure upon

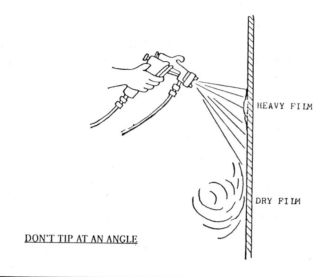

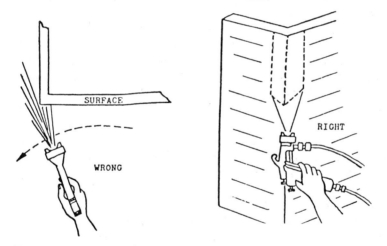

Figure 19.29   Paint spray gun techniques.

the fluid. The fluid is forced under pressure out through the fluid hose to the gun. The adjustment of the pressure regulator on the pot determines the amount of fluid available to the spray gun. Air is also bypassed from the main source into the air hose leading to the gun.

It is extremely important that correct size fluid and air hoses are used. For best results, the gun should be equipped with a 1/2-in.-ID fluid hose and a 5/16-in. air hose. Smaller air hoses should be avoided, since they cause excessive pressure losses. For example, if an MBC gun with a 765 air cap is connected by 50 ft of 1/4-in.-ID hose to an air supply of 90 psi air pressure, the actual pressure at the gun will drop to 50 psi when the trigger is

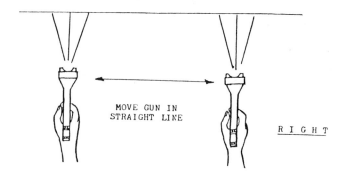

MOVE GUN IN
STRAIGHT LINE

R I G H T

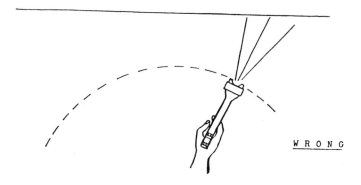

W R O N G

DO NOT SWING IN ARC

Figure 19.30 Paint spray gun motions.

opened. This is not enough pressure for most coatings. However, when 50 feet of the recommended 5/16-in. air hose is used, under the same conditions, the pressure will drop only to 75 psi, which in most cases is satisfactory.

Hydraulic (Airless) Spray

With hydraulic (airless) equipment, no air is used for atomization. The spray pattern is formed simply by forcing the material under high pressure through a very small orifice in the spray gun. As the material leaves the orifice, it expands and is broken up into fine droplets.

The outstanding advantage of this type of equipment is the absence of overspray. Therefore, smoother applications can be made, especially to corners and crevices, and material loss to wind or other air flow is negligible.

Adaptability to various types of material is obtained by providing a series of interchangeable fluid tips for the gun, each of which has a fixed orifice and fan. Since the size and shape of the orifice determines the breakup of the material, the width of the fan, and the delivery rate, it is important that the proper tip be selected for spraying a particular coating. The only other adjustment is the pressure which is applied on the fluid. An air-

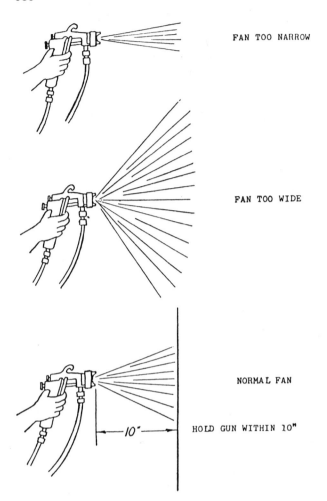

FAN TOO NARROW

FAN TOO WIDE

NORMAL FAN

HOLD GUN WITHIN 10"

Figure 19.31   Spray gun fan widths.

driven pump with a ratio of approximately 28:1 or higher is commonly used to supply this pressure. Thus, if 100-psi air pressure is supplied to the pump, the resulting pressure on the material at the gun will be 2800 psi. As with the fluid tip, the pump pressure must be regulated to meet the application characteristics of various materials.

Ventilation of Work Areas

A few simple rules must be observed in the handling of coatings in enclosed areas where flammable solvents are a factor. Adequate ventilation is the most important safety rule. This is true whether materials are aliphatic hydrocarbons with low toxicity or aromatic hydrocarbons with more dangerous physiological properties. Ordinarily, no fire hazard will exist if the solvent vapor concentration is kept below 1% by volume.

Recommended ventilation rates for enclosed spaces are shown in Table 19.4, and general ventilation procedures for using suction blowers at the lower part of the enclosed area is shown in Figure 19.34.

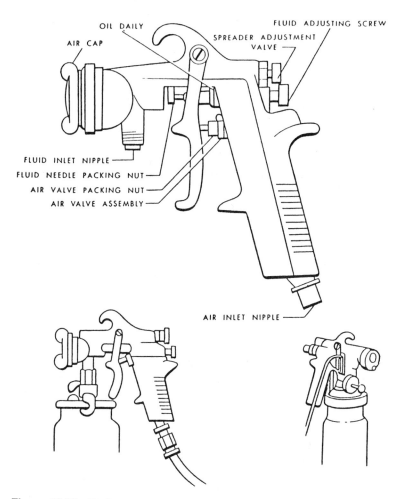

Figure 19.32 Typical spray gun showing suction cup and pressure cup.

19.4.2 Powder Coating

For industrial products, powder coating is rapidly becoming the "finish of choice" for many applications. While the principle is straightforward and easily understandable, success or failure lies in the detailed process control. The principle involved is to utilize the electrostatic phenomenon of attraction of oppositely charged particles. While the part to be coated could be charged and the powder particles sprayed onto it, the more popular system is to ground the product and charge the powder particles as they are sprayed (dry) on it. The product is then heated, which fuzes the dry "paint" particles to make an adherent and coherent film.

The processes to be controlled in order to make this system viable—from a performance and cost point of view—start with product design. While the particles flow and are attracted to the sides and back of the product as well as the front facing the gun, there is a difference in the corrona or force field. Sharp points and deep pockets are of concern in the powder coating process. The material, or substrate, also has its influence. It must be

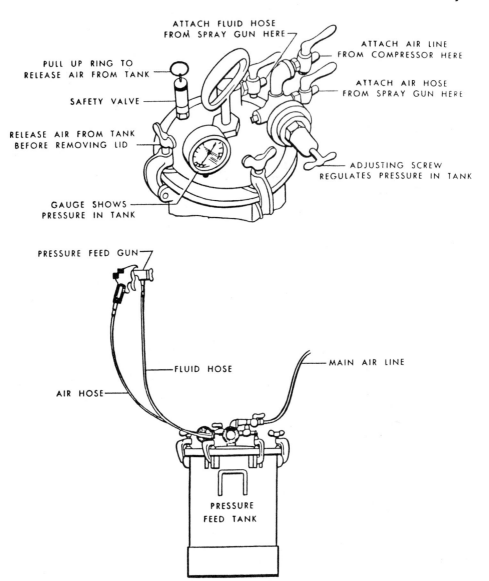

Figure 19.33 Typical pressure pot setup.

clean and dry, as in any other coating system. It usually needs a chemical coating after cleaning to improve adhesion and add corrosion protection. The size of the particles of powder and the type of powder affect the performance of the process and the finish coating. Perhaps the most important factors are the time and temperature control required for each step in the process. Although parts can be hand cleaned and hand sprayed in small shops or short production runs, the powder coating process lends itself to a conveyor of sorts, and precise controls on speed, cleanliness, and temperature at each step. The process we will

Table 19.4 Recommended Ventilation Rates for Safe Application of Coatings in Enclosed Spaces

(1) Size of enclosed space, cu ft	600	1200	2000	5000	10,000	25,000	50,000	100,000
(2) 1 percent of enclosed space, cu ft	6	12	20	50	100	250	500	1000
(3) Typical coating application rate, gal per hr	5	5	10	10	10	20	20	50
(4) Volume of solvent vapor produced, cu ft per hr, @ 25 cu ft vapor per gal of coating material. (Line 3 × 25)	125	125	250	250	250	500	500	1250
(5) Air changes per hour to keep vapor concentration below 1 percent by volume* (Line 4 ÷ Line 2)	20	10	12	5	2.5	2.0	1.0	1.25
(6) Rate of air change, cfm, to keep vapor concentration below 1 percent by volume (Line 5 × Line 1 ÷ 60 min/hr)	200	200	400	41	41	83	83	28
(7) Recommended air movement through suction fan, cfm, to keep concentration far below 1 percent vapor by volume	1000	2000	2000	3000	5000	10,000	15,000	20,000
(8) Air changes per hr at recommended air movement (Line 7 × 60 min/hr ÷ Line 1)	100	100	60	36	30	24	18	12

*1 percent concentration of vapor in air is below the lower explosive limit of most solvent mixtures used in paint and coatings (except turpentine which has LEL of 0.8 percent)

discuss has the following steps, although all of them may not be necessary for some applications:

1. Cleaning
2. Rinsing
3. Phosphatizing
4. Rinsing
5. Sealing
6. Drying
7. Coating
8. Curing

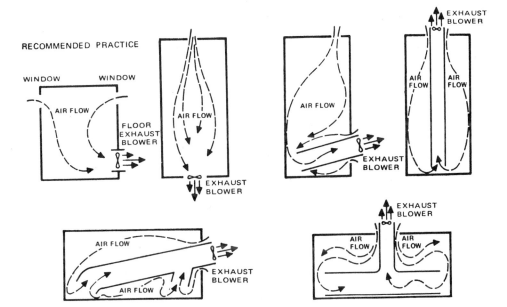

Figure 19.34 Recommended ventilation procedure draws clean air from opening at top by exhausting air from lower part of enclosure.

Cleaning

What Is Soil?

Soil is simply matter that is out of place. A *soil audit* should be performed to determine the cleaning required prior to powder coating. It should include the following:

What soils are incoming?
In-house applied?
Substrate type(s).
Substrate flow.
Process control.

Today you can demand a Material Safety Data Sheet from your metal supplier. What oil or preservative do they apply? Is it heat sensitive? Does it contain waxes? When you buy steel from many vendors, do you have control over the quality and consistency? When you determine that you can clean the incoming soils, make sure your vendors do not change soils without notifying you.

With in-house metal fabricating, what soils do you apply? Do you use rust inhibitors, forming oils, coolants, or lubricants? If so, pretest all these soils in your pretreatment system. Once you determine you can successfully clean these soils in a fresh or aged state, do not make changes unless you pretest again.

How many different types of metal substrates make up your product or products? Combinations of ferrous and nonferrous metals require different chemistries to clean effectively without metal attack. The use of zinc-bearing metals such as die-cast or galvanized metals may require posteffluent treatment.

Effective pretreatment and process control cannot be accomplished unless there is control over incoming soils, in-house applied soils, and substrates in use.

Is the Part Clean?

If the part is clean, the powder will adhere. In most cases, if the chemical vendor can produce a clean part, a phosphate treatment will be sufficient prior to powder coating. Definitions and tests of cleanliness follow.

Clean surface: one which is free of oil and other unwanted contaminants.
Organic soils: oily, waxy, films such as mill oils, rust inhibitors, coolants, lubricants, and drawing compounds. Alkaline cleaning solutions are most effective on organic soils: *Alkalines clean organics*.
Water break-free surface: all organic soils have been removed. The parts exiting the last pretreatment stage prior to drying will show a uniform sheeting of the rinse water, indicating an organically clean surface. If the part exhibits a surface which resembles a freshly waxed car surface, there will be beads of water, indicating that the part is not organically clean.
Inorganic soils: rust, smut, heat, scale, and other inorganic particulates that reduce adhesion and gloss. They most commonly can be found after allowing the part to pass through the dry-off oven. Check for smut and other loosely adherent inorganic solids by using a clean white towel and wiping it over the dry surface. Smut is the black gritty substance found on weldments and hot-rolled pickled and oiled steel. Sometimes it is impossible to remove all soot, carbon, and smut without some form of mechanical or abrasive cleaning. Poor cleaning is most

often found on or near weldments, or in areas which receive poor spray impingement to the part. Acidic cleaning solutions are most effective on inorganic soils: *Acids clean inorganics.*

Mechanical Cleaning

Mechanical or abrasive cleaning is very suitable where steel surfaces have been subject to abuse such as severe corrosion and oxidation, or where steel surfaces exhibit large amounts of heat scale or controlled oxidation. This is especially true for the loosely adherent accumulations found in improperly stored steel, or hot-rolled steel of poor quality. These contaminants make it very difficult to achieve any form of quality adhesion. Three types of mechanical surface preparation have somewhat wide acceptance.

1. *Air/media blast.* Sandblasting, the most common type of air/media blasting, is a combination of compressed air and sand or other media. Sandblasting can be accomplished automatically or manually. Problems associated with employee safety, particularly silicosis, have decreased widespread use in open-air environments. A specially designed cabinet or enclosed area along with air-induced breathing apparatus has ensured continuing growth of this method of mechanical surface preparation.

2. *Water/media blast.* This method is gaining popularity because of the reduction of silicosis-associated problems. Wet spot blasting of weldments has become accepted because of these reasons:

    During the welding process, oily soils are carbonized, creating an impossible cleaning condition using three- or five-stage washers.

    The surfaces of weldments are basically inert to the development of a conversion coating because of the scale and glassing developed. Wet spot blasting units are relatively inexpensive, low in labor requirements, and prove to provide the best possible substrate conditions prior to phosphatizing.

3. *Centrifugal wheel (airless) process.* This airless process is quite popular for larger, heavier-bodied parts where rust and scale must be removed. Centrifugal wheel cleaning is most often done in enclosed cabinetry. The medium is normally steel shot of varying sizes, depending on the substrate profile required. For optimum results, centrifugal wheel cleaning should employ additional chemical pretreatment to ensure quality long-term finish life.

Chemical Cleaning and Phosphatizing

Fremont Industries, Inc., of Shakopee, Minnesota, is one of the respected firms specializing in pretreatment systems to clean and phosphatize parts prior to powder coating. Their 5-stage cleaning and phosphatizing system is shown in Figure 19.35. A six-stage system most often includes a final deionized water mist rinse.

Stage 1: Cleaning. Typically, alkaline cleaning produces a metal surface that is free of organic and inorganic reactive soils. These cleaning products incorporate detergents and surfactants to wet the soil; alkaline builders to dergrade, emulsify, and saponify organics; and water conditioners to soften and control contaminants.

Stage 2: Fresh water rinsing. The purpose of stage 2 is to flush all remaining organic soil from the part, neutralize alkalinity, and prevent pH contamination to stage 3.

Stage 3: Phosphatizing. Iron phosphatizing is the most common form of conversion coating in general industry for powder coating. The clean and rinsed part enters

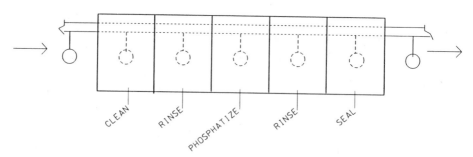

Figure 19.35 Typical five-stage cleaning and phosphatizing precoating system for powder coating. (Courtesy of Fremont Industries, Inc.)

the phosphate stage and receives a uniform acidic attack. Chemical reactions occur at the substrate solution interface. Most five-stage iron phosphates deliver 40–70 mg/ft^2 of coating.

Stage 4: Fresh water rinsing. The purpose of stage 4 is to flush any remaining phosphate solution, and prevent the subsequent stage from being clorinated.

Stage 5: Seal rinsing. The purpose of final seal rinsing is to remove any unreacted phosphate and other contaminants, to cover bare spots in the coating, to prevent the surface from flash corrosion, and to extend the salt spray performance. The selection of the type of seal rinse, whether a deionized water rinse, a chromic acid rinse, or a reactive rinse, depends on the type of substrate as well as the type of powder coating system planned.

Nonferrous Metals

Nonferrous materials require slightly different chemical treatment, although the principles are the same. One approach includes cleaning of organic soils without attack to the nonferrous metal. This is followed by a rinse and the application of an acidic cleaner to slightly etch or remove the oxide layer of the metal. Careful consideration must be given to the choice of both the alkaline cleaner and the acidic material. The limiting factor is the alloy and the amount of etch that can be done without overattack and the subsequent development of smut.

The second approach for pretreatment of nonferrous metals includes a third step of conversion coating. This approach is common when the metal finisher runs a combination of ferrous and nonferrous metals through the same system. The etch portion and conversion coating portion are usually accomplished in the same stage by incorporating fluoride accelerators into the iron phosphate bath. The ultimate deposited coating is a combination of surface etch and a combination conversion coating of the dissolved alloys bound in the particular nonferrous substrate.

The third approach to nonferrous metal pretreatment for powder coating is the chromate process, associated with the highest-quality underfilm corrosion protection. Chromate conversion coatings serve as effective pretreatments for powder coatings. These pretreatments are used extensively on aluminum and also find application with zinc and magnesium Two types of chromate coatings are in use: chrome oxide (amorphous chrome) and chrome phosphate. Coatings formed with chrome oxide are based primarily on hexavalent chrome, which is extremely corrosion resistant. Chrome phosphate coatings contain primarily trivalent chrome, which is less corrosion resistant than hexavalent, but generally more mechanically sound and stress durable than chrome oxide types.

The parts should be dried at this point, usually in an oven. One aproach is to duct the hot air exiting the curing oven, and controlling the dry-off oven temperature by varying the fresh air intake. The moisture laden air is vented outside. Another approach is to provide the dry-off oven with its own gas burner, separate from the curing oven. This allows an option to preheat the parts prior to application of powder. In some cases this can solve some of the problems related to Faraday cage and outgassing. Use of air knives to remove moisture can reduce the time and temperature for drying the parts, unless they are too small to withstand the air velocity.

Additional preparation prior to powder coating can include masking of areas not needing the coating. Also, finished bolt holes, threads, etc. may require plugs or caps to keep the powder out. There is a wide assortment of products to aid in this process step, and usually they are rated as to the temperature they will withstand. Tapes may be made of paper, polyester, or glass cloth. In general, the cost of a 400°F tape is twice the cost of one that will withstand 300°F, and half the cost of one that will withstand 500°F. The selection must be based on the correct temperature for the product powder cure temperature, ease of application and removal, and cost of the masking material. Some masking materials will withstand chemical or abrasive cleaning operations, and others will withstand temperatures from 150°F to 600°F.

One of the more important remaining steps is the racking arrangement, including hooks. This is necessary to present the part to the powder coating operation correctly, and to provide a continuous ground to the part. The measured resistance between the part to be coated and a ground wire connected to the booth ground should not exceed 1 M2. Since the hooks will become coated as well as the part, the cured coatings must be removed from the hooks and racks in order to maintain this ground path. This is often accomplished in a "burn-off" oven, followed by brushing or some other cleaning step, prior to reuse. In many cases it is more economical to replace the hooks rather than clean them.

Powder Coating Process

The powder coating spray booth may be a simple one, using a manual powder spray gun. Binks makes such a gun, using low-voltage cable and an integral-cascade generator that maximizes tip voltage. The gun's two-position trigger enables the operator to easily control voltages to combat Faraday cages. Pattern-shaping air can also be controlled at the gun. Other suppliers have somewhat similar products available.

Most finishers utilize an automatic system with an overhead conveyor, which carry the racked parts through the pretreatment phase, drying (and preheating if required), and on into the powder application booth. This booth contains several coronacharging spray guns, which apply the powder the same way on each part.

Each spray booth has reclaim modules to collect the overspray powder, using filters similar to an industrial vacuum cleaner. The booth, and particularly the reclaim filters, must be cleaned thoroughly prior to changing color in the system. The reclaimed powder is then processed for reuse, usually with about 50% new powder added each time.

Curing Ovens

There are as many solutions to curing ovens as there are in the other sections of the powder finishing system. Gas convection ovens are probably the most common today. They generally operate at 350° to 425°F, and can cure parts in 15 to 30 min, depending on the coating and the mass of the part. In production operations, the conveyor carries the parts through the oven, where the length of the oven, the conveyor speed, the temperature required to fuze the coating, and the mass of the parts are influencing factors.

With new product and process technology under continuous development, production equipment is required to give precise control over product quality while operating efficiently with minimum downtime. Infrared radiation, with wavelengths expressed in micrometers (microns) can be accurately measured, controlled, and applied to the product. Figure 19.36a illustrates the electromagnetic energy spectrum in which infrared is centered. Figure 19.36b shows the infrared theory of quickly bringing the powder coating up to curing temperature, and Figure 19.36c shows the proper infrared wavelength for the most efficient process.

Every organic material has a unique energy spectral absorption curve. There will be peak wavelengths where the material absorbs very well, and valley wavelengths, where it is almost transparent to the applied energy. Most materials have absorption peaks falling between the wavelengths of 2.0 and 5.0 µm. Peak effeciencies are achieved on most products by matching the emission wavelength of the heater to the absorption wavelength of the product.

Casso-Solar Infrared (Pomona, New York) uses the proprietary selective wavelength method (SWM). For example, with a water-base coating on a polyethylene film, the goal is to transfer maximum energy to the water in the coating, with minimal energy pickup by the film substrate. Water, with peak absorption wavelengths between 2.5 and 2.6 µm, will receive energy from an infrared heater operating at an emitter temperature of 1600°F with efficiencies of greater than 80%. The substrate film, with a peak absorption wavelength at 3.4 µm, will be almost transparent to the emission energy, preventing heating of the film and subsequent distortion.

On thick materials, the SWM theory can be applied in reverse. Penetration within a material can be realized by selecting an off-peak emission wavelength, allowing the radiant energy to pass through the outer layers of the material. With the trend toward new high-solids coatings, 100% solids coatings, and powder coatings, the infrared energy can be absorbed in the coatings directly, without substantial heat absorption by the substrate, saving power and cooling time. For sensitive solvent coatings, short-wavelength infrafed is often utilized to penetrate the coating, heating from the inside out, eliminating blistering and other surface defects.

Troubleshooting Basics for Powder Coatings*

As mentioned in the introduction to this subsection on powder coatings, the basic principles are simple of powder coating are simple, and quite understandable. However, the success or failure of a system may lie in close attention to the details actually involved in the system.

---

*Adapted with permission from a February 1995 article in Powder Coating Magazine. Matt Matheny, Technical Services Manager of O'Brien Powder Products (Houston, TX), is the author, and was most helpful in the preparation of this subsection.

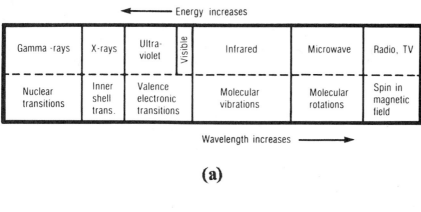

(a)

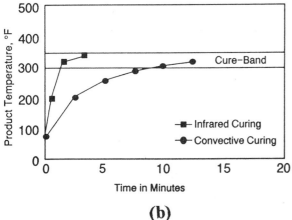

(b)

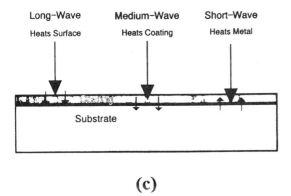

(c)

Figure 19.36   (a) Electromagnetic scale showing the position of infrared radiation relative to other types. (b) Product temperature versus time for curing a powder in a convection oven compared with an infrared oven. (c) Selection of proper infrared wavelength. (Courtesy Casso-Solar-Corp.)

The application of any chemical coating, especially in a relatively young and evolving technology, tends to be fairly problematic. Coaters simply have so many variables to control that chances are anyone who sprays powder will sooner or later have some problems. Arming yourself with some basic tools and guidelines to overcome these problems will help you run a smooth powder coating day to day.

Basic Troubleshooting Tools

The test kit outlined in Table 19.5 will help trace and prevent many of the problems you will encounter. Although it is true that you could spend the entire family inheritance on sophisticated, state-of-the-art troubleshooting equipment, you will find that you can accomplish a lot with just the basic tools in this kit.

Film Thickness Problems

Both thin and thick films can cause problems, as shown in Table 19.6. Aside from these problems, excess film build simply costs money—the larger the job, the more impact there is on cost. For example, if you spent $3000 on powder to coat a job at 3 mils thick when 2 mils would have met your requirements, you just gave away $1000!

Even veteran powder coaters and other experts can be fooled by a coating's visual appearance and perceived thickness. To remove doubt, use a film thickness gauge. Most gauges cost from $175 to $2900, depending on style and options, but it is more difficult to prevent and diagnose coating problems without one.

Film Thickness Troubleshooting

As a first step in troubleshooting, you should get into the habit of checking film thickness first. In many cases, this will seem unnecessary (for instance, when film color is incorrect). Many times, however, film thickness will turn out to be the unexpected cause of a problem or at least a contributing factor. The problem with incorrect film color, for example, could be caused by the substrate showing through a thin coating, which affects the perceived color.

Considering that film thickness measurement is probably the easiest and fastest test you can do with powder coatings, it makes little sense not to check it. Even if a coating problem requires the assistance of your suppliers, you should have all film thickness data

---

Table 19.5 Basic Troubleshooting Test Kit for Powder Coatings. (From Powder Coating Magazine.)

Film thickness gauge
Small bottle of MEK
Cotton swabs with wooden stems
Razor knife
Tape with strong adhesive backing
Jeweler's magnifying lens (10X loupe)
Clean white cloth, tissues, or gloves
Commercial test panels
Sharpie marker
Razor blades
Ohmmeter

Table 19.6 Thin and Thick Films Can Cause These Problems. (From Powder Coating Magazine.)

Thin films	Thick films
Reduced corrosion resistance	Reduced impact resistance
Reduced chemical resistance	Reduced flexibility
Pinhole rusting	Reduced chip resistance
Reduced electrical insulation	Inconsistent or incorrect
Inconsistent or incorrect	appearance, texture, and gloss
appearance, texture, and gloss	Orange peel
Orange peel	
Coating seeds	
Reduced edge coverage	
Inconsistent or incorrect color	

on hand before calling them, especially when calling your powder coatings supplier. Some of the most common factors in film thickness control are listed in Table 19.7.

Curing Problems

During the powder coating baking prosess, a chemical reaction called *cross-linking* occurs. When complete, this chemical reaction provides a fully cured thermoset powder coating film with the physical attributes that were designed into it. Incomplete cross-linking, or cure, provides a final product with reduced physical properties, depending on the actual degree of undercure. In short, a fully cured powder coating prevents a host of postapplication problems.

As with film thickness measurement, cure is a cornerstone of powder performance, yet easy enough to check at the start of most problem investigations. A strong solvent, such as methyl ethyl ketone (MEK), provides a fast and easy way to measure the chemical resistance of a powder coating. Because chemical resistance generally develops in relation to the degree of cross-linking that has occurred, it provides a workable assessment of cure. See Table 19.8 for instructions on how to use MEK to test cure. Because of exceptions, you must compare your test results against a fully cured powder coating to determine if the results are normal or not. A dulling of gloss at the test point is normal. Rapid softening of the coating usually indicates undercure. The quicker and more pronounced the softening, the less is the cure. Heavy discoloration of the cotton swab will

Table 19.7 Common Factors in Film Thickness Control. (From Powder Coating Magazine.)

Line speed
Electrical grounding of parts
Part presentation when coated
Gun charging problems
Gun kilovolt settings
Operator training
Powder particle size
Powder charging characteristics (formulation)

Table 19.8 Solvent Cure Test for Powder Coatings.
(From Powder Coating Magazine.)

Equipment required

1. A few milliliters (or ounces) of clean MEK, available at most paint and hardware stores
   Note: A few ounces of MEK will be enough to do hundreds of tests
2. Cotton-tipped swabs with wooden stems
   Note: Swabs with plastic stems dissolve in strong solvents

Instructions

1. Wet the swab generously in the MEK bottle.
2. With the thoroughly soaked swab, rub a small, 1-inch-long area of the coated surface with 50 double rubs. A double rub is once up the coated surface and once down the coated surface, as if erasing a pencil mark. Try to use a force similar to erasing with a pencil.
3. After 50 double rubs, most cured coatings will permit very little or no removal of the film on the cotton swab.
   Note: Urethanes, textures, and other select powder coatings may not resist strong solvents well, even when fully cured. Therefore, it's important to establish how a fully cured coating will react to MEK before conducting this test. To establish such a baseline, get a fully cured standard panel from your powder coatings supplier or bake a coated part two or even three times to ensure that you have a cured coating to use for comparison.

〈Test kits are available from O'Brien Powder Products〉

---

be evident. In cases of severely undercured films, the coating can be completely removed from the metal substrate.

When the coating is close to fully cured, it can be difficult or even impossible to determine the exact degree of cure. Fortunately, powder coatings that are close to fully cured seldom create significant problems. Remember, failure to compare all test results with a fully cured coating of the same powder can lead to false conclusions.

Oven Temperature

The most frequent problem regarding oven temperature is usually not a lack of knowledge regarding the correct temperature range to use for a powder, but rather a blind faith in the oven temperature readout. Powders usually have a cure curve developed by the coating supplier. Figure 19.37 shows typical temperature/time curves for an epoxy, and Figure 19.38 for a polyester. The lower the temperature, the longer cure time is required. More often than not, oven readouts tend to be at least a little off from the actual average oven temperature. This is due to normal drifting of the calibration and to the fact that most oven temperature probes are located in only one spot in the oven, which may or may not reflect the overall average temperature (usually it doesn't).

Most ovens have hot and cold spots that are not registered on a single probe located at the other end of the oven. Although it is convenient if the oven readout and the actual oven temperature are the same, it is not mandatory in curing powder-coated products successfully. Just follow two simple rules for curing success: (1) If the powder is turning yellow or brown, reduce the oven temperature setting (ignore the readout) or decrease the

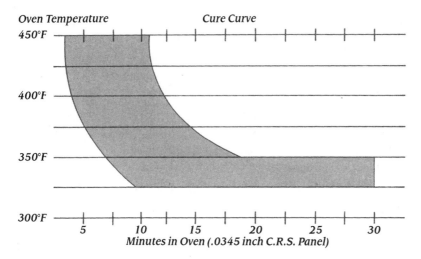

Figure 19.37 Typical epoxy powder temperature versus time cure curve. (Courtesy Morton Powder Coatings.)

oven bake time; and (2) if the powder is not cured (fails the MEK testing), increase the oven temperature (ignore the readout) or increase the bake time.

Metal Mass

Most powder manufacturers report cure schedules for their products at metal temperature. For example, if a powder is supposed to be cured for 10 min at 400°F and you are coating a 16- gauge steel part, you will probably need to add another 5–10 min to the bake time to allow the metal to come up to temperature and to achieve full cure of the powder coating.

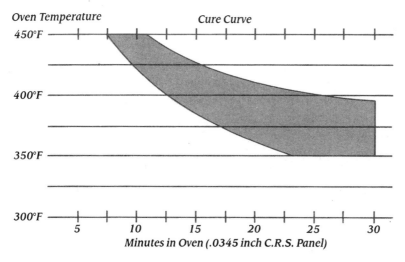

Figure 19.38 Typical polyester powder temperature versus time cure curve. (Courtesy Morton Powder Coatings.)

Metal Thickness

As you might expect, the energy required to heat two drastically different metal cross sections is also different. If your production setup permits it, the best solution for this problem is to preheat the part first in a range between 250 and 350°F, apply the powder, and then bake at the time and temperature required for the thinnest metal section. The next best option is to reduce your oven temperature as low as permissable for the powder you are using and then extend the bake time as long as required to cure the thickest metal section. You may still be unsuccessful with this approach, however. Some yellowing of the thin section could still occur. In that case, you may need to contact your powder supplier for other options, such as changing to a powder with increased heat stability.

Adhesion Problems

If a coating is to meet performance expectations, it must adhere to the substrate. Because the substrate composition and its preparation have a profound effent on adhesion, a method to evaluate adhesion is useful for troubleshooting and in-process quality control. Some measure of adhesion can be done by cutting through the coating in a latticed pattern and then trying to pull the film away with tape. Table 19.9 describes this test procedure.

If failure occurs during the test, look at the back of the removed coating for evidence that might provide clues. Red-brown rust from steel parts or white oxidation from aluminum and galvanized parts is frequently found on the back of the coating. If you are testing freshly coated parts, these oxides were present before the parts were coated. Continue your investigation, examining parts about to be coated and working backwards through the process as needed. If you are testing parts that were coated some time ago, you will have to conduct a more complex investigation. Of course, it is still possible that the parts had oxidation present when coated but just were not caught at that time. If the coating has blisters, is still intact with visible pinhole rust, or both, then this is a strong possibility. If pinhole rust is visible, the powder coating may just be too thin or have had contaminants present when the coating was applied.

For more demanding application requirements or for more test details, consult the procedures in American Society for Testing and Materials Test D-3359. This procedure also provides a useful pictorial table for classification of test results. For a copy, call ASTM at (215) 299-5400.

Table 19.9 Powder Coating Adhesion Test Procedure. (From Powder Coating Magazine.)

1. Use a razor knife to cut through the coating to the substrate in a latticed pattern. Make six or more cuts in each direction. Each parallel scribed line should be 2 millimeters apart for coatings that are at least 2 mils thick. For coatings thinner than 2 mils, space the lines 1 millimeter apart. A steel straightedge is useful for making these cuts.
2. Brush off loose bits of coating.
3. Apply adhesive tape firmly over the lattice. Use a pencil eraser to rub the back of the tape. This ensures good contact with the coating. Then, quickly pull back the tape at about a 180 degree angle.

Contamination Problems

Many clues to coating problems are not easily visible to the naked eye. A jewler's lens (10× loupe) is especially useful in identifying foreign matter on an intact film or on the back of a film that has peeled off a substrate. The root of a crater (fish-eye) problem may also become apparent by closely examining the bottom of a film depression for a contaminant.

Powder-coated panels that have not been cured can be hung in strategic places around the coating area, collected later, and then cured and checked for contaminants. Panels should be marked for later identification of the test sites. Airstreams from outdoors and other plant operations are especially suspect. The most common contaminant types and sources in a powder coating operation are discussed next.

Cross-Contamination (from Within the Powder Coating System)

Inadequate cleanup between different powders is a major culprit. A quick test for determining the source of colored specks from another powder is to thoroughly clean a manual spray gun and then spray a sample of virgin powder right from the box. Next, spray a powder sample from the hopper. If the virgin powder looks good and the hopper powder looks contaminated, reclean the powder system and recharge it with virgin powder.

Dirt and Powder Falling from the Conveyor

When dirt and powder fall from your conveyor, you should notice most of the contaminants at the top of the parts. If this is the case, clean the conveyor and install shields to protect the parts.

Surface Preparation Problems

Regardless of the mechanical and chemical processes you have selected, it is important to monitor them closely to get the best and most consistent results. If you need help determining the best approach for your application, contact your powder supplier, your pretreatment supplier, or both.

The adhesion test described previously can be used to alert you to serious surface preparation problems. A part with more subtle surface preparation problems may pass initial adhesion tests, such as the cross-hatched method, but will eventually cause premature failure of the coating while in use. Detection of these subtle problems requires special testing, such as salt spray (ASTM B-117) or hot-water immersion (ASTM D-870). If performed at temperatures of 180 to 200°F, the hot water test has the advantage of being fairly quick to perform—usually 2–6 hr—and can be done with inexpensive equipment right in your plant.

The most common problems associated with chemical pretreatment systems are poor adhesion and premature corrosion failure. Frequently these problems are caused by the following.

> *Residual soils* may be caused by (1) conveyor line speed that exceeds the design limits of the cleaning system, causing low dwell time, (2) inappropriate cleaner for the soils present (bear in mind that the soils on your metal may have changed), and (3) incorrect temperatures—120–130°F is best for good cleaning unless you are using a low-temperature cleaner. In that case, high temperatures can be detrimental.

To determine the temperature that emulsifies the soils present on parts, immerse an uncleaned part into a container of water and begin heating it. Use a thermometer to watch the temperature rise. Keep an eye on the point where the water line touches the part. At some point, the water will become hot enough to visibly loosen the soils, causing globules to float to the surface.

Flash Rust can be caused by (1) excessive line speeds that prevent adequate exposure to the sealer in the final rinse, (2) line stops that overexpose parts to chemicals or allow them to dry off between stages, and (3) lack of sealer in the final rinse. When using a solvent-type cleaning system or an iron phosphate conversion process, wiping with a clean, white cloth is an ideal way to check parts cleanliness before powder coating.

Aluminum oxide: When coating aluminum, remember that a natural oxide is present on the surface that will interfere with adhesion if it is not removed. If using a combination iron phosphate and cleaner, be sure it is designed for steel and aluminum.

Inadequate rinsing is one of the great sins of metal cleaning. It is caused by increased line speeds that reduce rinse-stage dwell time and inadequate rinse-water overflow (excessive dissolved solids in the final rinse). Simple tests can include slowing down the production line or hand rinsing parts in deionized water.

If you suspect that your surface preparation system is causing a problem, clean test parts with clean rags dipped in a solvent, such as MEK, instead of running the parts through your normal cleaning process. If this fixes the problem, you should focus your investigation on the surface preparation system.

Powder Fluidization Problems

Consistent application of powder starts with consistent powder flow through the spray guns. Consistent powder flow requires good powder fluidization If the powder bed has dead spots or blow holes (geysers) or just does not fluidize well, a smooth flow through the hoses and guns is jeopardized. The most common causes of poor fluidization are as follows.

Moisture from the compressed-air supply. If your powder is initially free of soft lumps but starts forming lumps while fluidizing, it probably means that your air supply is contaminated with moisture, oil, or both. Compressed-air dryers and filters are recommended for all powder coating operations. A white cloth can be used to detect oil or water in the coating operation's compressed air source. After blowing the compressed air into a cloth for a couple of minutes, you will see that oil- or water-laden airstreams leave wet or discolored deposits.

Lumps in the powder. Aside from moisture in the air supply, lumps can be caused by exposure of the powder to high temperatures (above 75°F) during storage or transportation. Usually, sieving the powder before use breaks up or removes the lumps. A common window screen will often suffice. The best way to check for lumps is to fluidize the powder well and run your hand through it. While you have your hand in there, check the bottom and the corners of the bed for dead spots.

Plugged porous plate in the fluidizer. Anything less than clean, dry air will eventually plug the porous plate in the bottom of the fluidizer. You will notice the air pressure rise to maintain good fluidization. If possible, examine the bottom of the porous plate for plugging, oil residue, and so on. If the powder does not fluidize well, try using another fluid bed to see if the problem goes away.

Grounding Problems

Because electrostatic equipment is used to apply most powder coatings, adequate electrical ground is required. The measured resistance between the part to be coated and a ground wire connected to the booth ground should not exceed 1 mΩ. Take measurements inside the booth where parts are actually coated. Following are the most common problems associated with poor ground.

> *Poor powder transfer efficiency.* This can be characterized by slow or low film build.
>
> *Electrical arcing from the gun to the part or hanger.* This indicates a grounding or gun problem and is a safety hazard. Under specific conditions, electrical arcing can start fires or cause explosions. This situation should be corrected immediately.
>
> *Premature electrostatic rejection of the powder from the part.* In premature electrostatic rejection, powder is repelled from the part, usually leaving pits in the powder surface. With some powder formulations, this problem can occur with good grounding. In those cases, reduce the gun's kilovolt setting. In general, electrostatic rejection occurs with all powders if spraying time is excessive. Poor grounding aggravates the condition, accelerating it.
>
> *Bare areas on the part that are difficult to coat.* This occurs expecially at edges, around holes, and near hooks. The most common cause for these grounding problems is dirty hangers and hooks. Cleaning or replacing these fixtures should be part of your regular maintenance program.

19.4.3 Metallizing by Flame Spraying

Corrosion is constantly taking a heavy toll of costly equipment and structures. Metallizing, a cold process of coating with flame-sprayed metal, presents a flexible, practical, and very effective means of protection against corrosive attack. The great advantage of the metallizing process is that it employs portable equipment. See Figure 19.39 for equipment layout for flame-spraying using metal wire. This makes it adaptable for applying protective metallic coatings to large, complex assemblies and structures. It is not economically feasible to attempt hot-dip galvanizing or electroplating of assembled structures, expecially in the field, but pure metal protective coats of aluminum or zinc can be applied to such structures by metalizing.

The Process

Metallized coatings have unique structures and desirable properties. They are formed by layers of thin flakes as the atomized globules of the sprayed metal strike the target area with considerable speed and impact. Initially, each particle is a tiny casting which has solidified very rapidly and acquired an oxide film during its short journey from the gun nozzle to the metal surface. On impact, the particles are instantly deformed into thin platelets or flakes. Their impact energy permits them to conform to the surface contour and bond to the prepared surface by mechanical interlocking with surface irregularities. Where the oxide film is disrupted on impact, it is reestablished by cold welding and metal-to-metal contact is regained.

The bond is almost entirely mechanical in nature. The base metal presents a degree

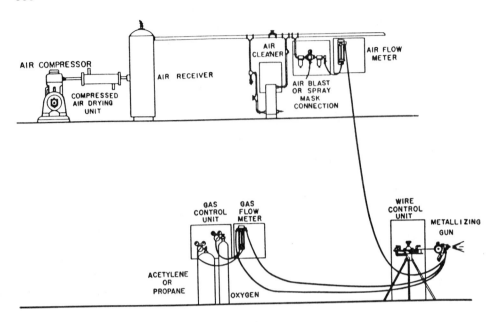

Figure 19.39   Equipment layout for flame spraying.

of roughness depending on the method used for surface preparation. Most of the sprayed metal particles which strike the surface are sufficiently plastic to conform to and interlock with the surface irregularities. Although only a small amount of fusion may take place between the particles of sprayed metal and the base metal, the overall effect of such fusion is very significant. This unique structure results in a change of the physical characteristics of the metal. The ductility, elongation, and tensile strength of the sprayed metal are greatly reduced when compared with the same metal in cast form.

Materials

All metals that are available in wire form can be sprayed, and several of them are quite effective in controlling corrosion. These include stainless steel, aluminum, zinc, cadmium, lead, tin, nichrome, nickel, nonel, tantalum, molybdenum, silver, and gold. These metals are applied for decoration, wear resistance, or corrosion control. The corrosion control that can be obtained from properly applied sprayed metals can be rated as good to excellent even for a harsh environment. (See Table 19.10.) Tests have demonstrated that metallized aluminum sealed with vinyl is an excellent corrosion-control method for salt-water exposure. These coatings provide protection against corrosion not only by covering the surface with a corrosion-resistant metal, but also by cathodic action since both aluminum and zinc are anodic to iron and steel. Twelve-year tests by the American Welding Society of sprayed aluminum and zinc coatings on steels showed good results on all panels. However, the sprayed metal that was sealed with a wash primer and vinyl top coat was in the best condition. In salt-water immersion tests, the staying power of zinc sprayed metal ran a very poor second to aluminum. The zinc oxide coating does not appreciably retard the oxidation rate; the zinc continues to sacrifice or oxidize. The aluminum-sprayed coating oxidizes to form an inert film (Al_2O_3), which greatly retards further attack. The film is stable unless broken or ruptured.

Table 19.10 Examples of Environments Where Flame Sprayed Metals Are Used. (I is Goldstone, CA; IV is Bermuda.)

Environment		Physical Limits	
Class	Type		
I	Dry (Minimum corrosion rate)	Rel. Humidity - 15% or less Temperature Range below 0° F to 130° F	95% of time
II	Normal	Rel. Humidity - 15% to 50% Temperature Range below 0° F to 120° F	95% of time
III	Humid	Rel. Humidity - 41% to 100% Temperature Range 32° to 120° F	95% of time
IV	Harsh (Maximum corrosion rate)	Rel. Humidity 41% to 100% Temperature Range 35° F to 120° F One or more saline aerations/24 hrs.	95% of time

Metallizing has the advantage that much thicker coatings of the protective metals can be applied more rapidly than other conventional metal coatings applied by hot dipping or electroplating. This is significant because corrosion protection is often in direct relation to the protective coating thickness. The life of a flame-sprayed zinc coating is almost directly proportional to the thickness of the coat. Aluminum and cadmium, the other two metals commonly sprayed for corrosion control, also show a fairly direct relationship between the life of the coating and its thickness. Because of the porosity of metallized coatings, it is customary to apply a greater thickness of protective metal than for dipped or plated coatings.

The best corrosion-control system to protect steel may be the aluminum sprayed coating, sealed with a wash primer, and vinyl paint. Wash primer coatings are useful because they reduce the danger of unfavorable reaction from a doubtful paint combination. They have the added advantage of displacing moisture to some extent and should always be used if there is any question of moisture in the sprayed metal. Unfortunately, some wash primers are too reactive for use on sprayed metal. Formulations suitable for galvanized sheet may be too acidic for use with sprayed coatings of zinc. No wash primer should be used until it has been positively determined that it is suitable for use on sprayed metal. In no instance should a wash primer containing more than 4% H_3PO_4 be used. In the case of aluminum, since it is slightly porous, the sealing materials fill up the pores of the coating to establish a permanent barrier.

Surface Preparation

In metallizing, as in painting, the preparation of the substrate is critical. In preparing a surface for flame-spraying aluminum, there are no options regarding the grade of surface preparation; it must be white-blasted with washed, salt-free angular silica sand or crushed garnet that will pass through mesh sizes 20 to 40. In addition, the white-blasted surface must be perfectly dry before the flame-sprayed metal is applied. The drying process can be accomplished by the metallizing gun merely by releasing the metal-feeding trigger and drying the desired section with the gas flame. Whenever there is any doubt whether a surface is dry, it should be warmed before flame spraying. The ambient temperature may be well above the dew-point temperature but the metal itself, because of shadows, the mass involved, etc., may be below the dew point, creating the danger of an invisible moisture film forming on the metal. Heating a surface to 100°F will prevent atmospheric moisture. Some highly experienced flame-spray craftspersons prefer to heat the surface from 175 to 200°F before starting to metallize; this results in an improved bond. In addition to atmospheric moisture condensation, the higher preheat prevents water vapor originating in the combustion of the gas-oxygen flame from condensing on the base metal or the preceding pass of sprayed metal.

19.4.4 Porcelain Enameling

Enamel is a vitreous glaze, of inorganic composition (chiefly oxides), fused on a metallic surface. Glass is particularly resistant to corrosion by atmospheric influences and chemicals, and has a smooth and very strong surface. But glass is fragile. When the good properties of glass are combined with the strength of steel or cast iron, the objects made from these materials (kitchen utensils, bathtubs, pipes, basins, laundry equipment, etc.) have excellent service properties. The name "vitreous enameling" or "porcelain enameling" is applied to such materials.

In particular cases the two materials supplement each other so well that entirely new material properties are obtained. Certain parts of jet engines and marine propulsion engines are enameled in order to make the surfaces resistant to high temperatures. Some of these applications are now considered to be in the field of ceramics, however enameling still has wide usage as a finish for various products.

Enamel has been known since ancient times, when it was used (as it still is) for ornamental purposes on precious and nonferrous metals. In the last few hundred years, however, it has been used chiefly for improving the surface properties of steel and cast-iron objects and protecting them against corrosion.

The Material

An enamel consists of glass-forming oxides and oxides that produce adhesion or give the enamel its color. A normal enamel may consist, for example, of 34 (23) parts borax, 28 (52) parts felspar, 5 (5) parts flourspar, 20 (5) parts quartz, 6 (5) parts soda, 5 (2.5) parts sodium nitrate, 0.5–1.5 parts cobalt, manganese, and nickel oxide, and (6.5) parts of cryolite. The figures not in parentheses relate to a ground-coat enamel, while those in parentheses relate to a cover enamel, to which 6–10% of an opacifier (a substance which makes the enamel coating opaque, e.g., tin oxide, titanium silicate, antimony trioxide) and a color oxide is added. This mixture is ground to a very fine powder and melted. The hot melt is quenched by pouring it into water, and the glasslike "frit" that is thus produced is ground fine again. During grinding, water (35–40%), clay, and quartz powder are added. Opacifiers and pigments may also be added. The enamel "slip" (thick slurry) obtained in this way must be left to stand for a few days before use.

The Process

The metal objects to be enameled are heated thoroughly, pickled in acid, neutralized in an alkaline bath, and rinsed. Next, the ground-coat enamel slip is applied to them by dipping or spraying and the material is fired at 850–900°C, so that it fuses to form a glass coating. The ground-coated objects are then provided with one or more coats of cover enamel, each coat being fired at 800–850°C in a muffle furnace. See Figure 19.40.

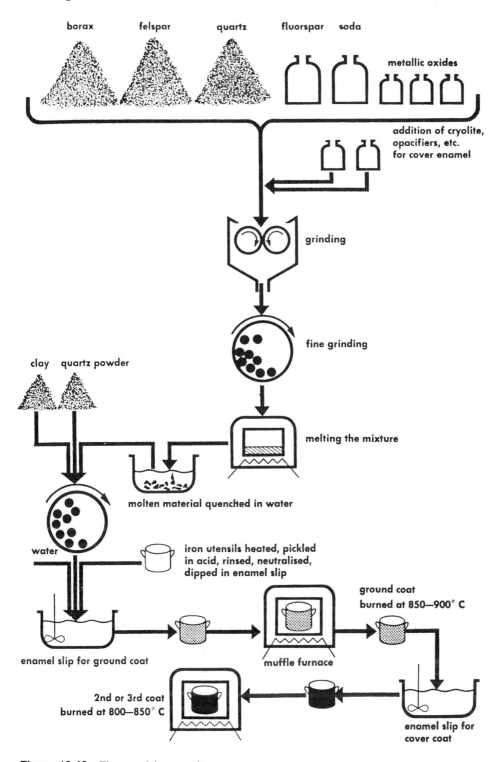

Figure 19.40 The porcelain enamel process.

As an enamel coat is always more brittle than the underlying metal, the enamel will crack or spall if the object is deformed or roughly knocked.

From the chemical point of view, enamel is a melted mixture of silicates, borates, and flourides of the metals sodium, potassium, lead, and aluminum. Color effects are produced, for example, by the admixture of various oxides to the melt (oxides of iron, chromium, cadmium, cobalt, nickel, gold, uranium, and antimony).

20

Manual Assembly

John P. Tanner   *Tanner and Associates, Orlando, Florida*

and

Jack M. Walker   *Consultant, Manufacturing Engineering, Merritt Island, Florida*

20.0   INTRODUCTION TO MANUAL ASSEMBLY

In today's complex manufacturing world, it is sometimes difficult to remember what the real purpose of manufacturing a product is—and what the total elements of the process consist of. If we assume that whatever we manufacture, we will insist on good quality products, on time delivery, and complete customer satisfaction, then we should then concentrate on the most economical method to achieve these results. We should make our manufacturing decisions based on cost. Of course, cost is not a simple thing to determine. There are a lot of different ways of looking at cost.

For a method of arriving at the lowest cost while maintaining quality, delivery, and happy customers, perhaps we should examine a one-person business operation. A few hundred years ago, there were a lot of them—and even today, there are more than most of us realize. The nation's small businesses, those with one to 10 employees, grew in numbers during recession-plagued 1991, resisting the downturn experienced by medium and larger companies, according to a U.S. Census Bureau report. The number of small businesses increased up to 1% per year between 1987 and 1991. Larger businesses increased up to 3% per year through 1990, then declined in 1991; those with 10 to 100 employees were down 0.2% in 1991, while those with more than 100 fell 1.7%. Businesses with more than 100 employees are generally concentrated in the manufacturing sector, which as recently as 1970 accounted for 35% of the workforce. By 1991, manufacturing workers made up less that 20% of the workforce.

In 1991 there were more than 4 million establishments in the United States with fewer than 10 employees, about 1½ million with 10–99 employees, and 134,000 companies with more than 100 employees (a total of 6,199,339 establishments). See Figure 20.1 for a breakdown of the manufacturing industries.

U.S. MANUFACTURING INDUSTRIES

(X 100)	1963	1967	1972	1977	1982	1987	1990	1991
Total Establishments	312	311	321	360	358	369		
UNDER 20 EMPLOYEES	207	199	203	237	230	238		
20 - 99 "	70	74	76	78	84	86		
100 - 249 "	18	20	21	22	21	22		
250 - 999 "	10	11	11	12	11	11		
1000 & OVER "	3	3	3	2	2	2		
(X MILLION)								
TOTAL EMPLOYEES	17	19.3	19	19.6	19.1	18.9	18.8	18.1
PRODUCTION WORKERS	17.3	14.4	13.5	13.7	12.4	12.2	12.1	11.5

PERCENTAGE OF PRODUCTION WORKERS IN ASSEMBLY TASKS : (1967)

MOTOR VEHICLES	45.6
AIRCRAFT	25.6
TELEPHONE, ETC.	58.9
FARM MACHINERY	20.1
HOME REFRIG. & FREEZERS	32.0
OFFICE EQUIPMENT	35.9
HOME COOKING EQUIP.	38.1
MOTORCYCLES, BBICYCLES, ETC.	26.3

CENSUS OF MANUFACTURERS
U.S. BUREAU OF CENSUS

Figure 20.1   Statistical breakdown of U.S. manufacturing industry.

Now let's get back to our one-person "factory." Upon receipt of an order, the owner makes each of the parts, assembles them, and does the finish painting, packing, and delivery. The difference between his total income and the amount of money he spent during the month is his "salary" or profit. Of course, he may be paying rent on the building, buying raw materials and supplies, and even making payments on his equipment and machines. OK, let's deduct these, and now we have his profit. Whoops—he probably pays for heating, lighting, insurance of some type—and taxes—and the remainder was his profit. We can see that even in the one-person "factory," real cost is not so easy to determine.

If business improves, one person may not be able to do everything by working at a faster pace, or working longer hours—and at some point the owner will have take some action in order to continue on-time delivery, etc. He might decide to buy the parts, and just perform the assembly operation (or vice versa). Another option might be to add helpers and continue to perform all the operations in-house. In most cases, a growing company will probably elect to continue to perform the assembly function, in order to have better control of quality, finish, delivery time, etc. It will still be "his" product as far as his customers are concerned, which will keep them satisfied and give him the opportunity to add additional sales. As the company continues to grow, the owner might reconsider his make-or-buy decisions, and perhaps add equipment to fabricate critical parts in-house. In the assembly area, the first step might be to add automated screwdrivers, nut runners, riveters, spot-welding heads, and perhaps pick-and-place mechanisms. To move parts from the fabrication or receiving department to the assembly stations, some type of transfer device might be a logical improvement. The same applies for the transfer of parts and assemblies down the assembly line. There are an infinite number of options, including redesigning the product to reduce the number of parts required to be fabricated and thereby simplifying the assembly process. (See Chapters 1 and 2.)

In the end, the assembly process may well become the key to his continuing success. Starting with a one-man assembly operation, the small firm has now grown to a multi-employee company, probably using many of the same techniques that were successful in the beginning of the one-man shop. With more orders, and probably more diverse products, it is decision time again. This is the subject of this chapter.

There are as many factors influencing the assembly process decisions as there are products, customers, and factory managers. Geoffery Boothroyd, in *Assembly Automation and Product Design*, quotes Henry Ford's principles of assembly as follows:

> First, place the tools and then the men in the sequence of the operations so that each part shall travel the least distance whilst in the process of finishing.
>
> Second, use work slides or some other form of carrier so that when a workman completes his operation he drops the part always in the same place which must always be the most convenient place to his hand and if possible have gravity carry the part to the next workman.
>
> Third, use sliding assembly lines by which parts to be assembled are delivered at convenient intervals, spaced to make it easier to work on them.

Assembly operations can be performed manually, automatically, or integrated in some manner using a combination of systems. If manual assembly is employed, an operator can adapt to changing conditions such as those brought about by part variation, mislocation, and product model mix. An operator can compensate for these changing conditions and, as a result, may not require elaborate tools and fixtures to perform the assembly tasks. However, operator error and fatigue can result in quality problems.

When production volumes are high enough, some assembly operations can be performed automatically with special-purpose machines. These automatic assembly machines consist of workstations grouped along some type of transfer system for part conveyance. Each station performs one task with the aid of dedicated station equipment, jigs, and fixtures. Part variation, misalignment, and product mix are not readily adapted to, since sensors cannot always be employed efficiently or economically to guide or monitor the assembly process. Therefore, part variations and slight misalignments can result in jamming, incomplete operations, and excessive machine downtime. However, automation can still be justified when the production volumes are high, product life is long, and assembly tasks are simple. For an assembly operation to be performed successfully on a repetitive basis, it is absolutely essential that part variation and location be minimized and consistency in dimensions and location be maximized. To achieve this in a mass production environment requires elaborate and costly tooling, fixtures, and the employment of expensive production controls. Therefore, many assembly operations are performed manually to resolve some of the problems in mating parts with variations or mislocations, which may result in increased assembly costs and lower productivity. Chapter 21 discusses the application of automation to the assembly process.

20.1 ASSEMBLY WORK INSTRUCTIONS

"Assembly processing" is another way of saying "assembly methods." Assembly methods sheets, or work instructions, must describe clearly what is to be done, in what sequence, and with what tools and materials. Assembly methods sheets should minimize operator learning time. Assembly methods sheets must be economical to prepare, reproduce, distribute, and change. Assembly process planning should include an assembly process summary or process routing, detailed work instructions for each operation called out in the summary, an operations parts list for each operation, process sketches or visual aids, and a work place layout for each operation. The work instructions should call out all tools necessary to perform the operation, and there should be a standard time on the process summary for each operation called out, broken down to the level of setup and run times.

In process planning for fabrication, whether for machining or forming, the skilled machinist or sheet metal mechanic could work to what amounts to an outline process routing supplemented by the engineering drawing of the piece part. As a skilled worker, he can set up the machine and perform the work with a minimum of written work instructions. Such is not the case with assembly operations. The work must be totally and carefully planned by the manufacturing engineer, and complete work instructions prepared. These are the two extremes. In most manufacturing plants today, process planning will fall somewhere in between. See Chapter 9.

If the plant is a high-volume producer of a single product line, then detailed assembly work instructions may be unnecessary. Once operators are trained to perform a short-cycle assembly operation, little else is needed except possibly some clear, concise visual aids showing the critical details of the operation in pictorial, or exploded view, form. However, the manufacturing engineer must plan such production down to the most detailed level. He or she must prepare a layout of the assembly lines, show each and every workstation in plan view, plan the assembly tools required, and write a complete description of the work performed at each station on the line. The ME must establish standard times, decide where visual aids are needed and prepare them, and then fine-tune or balance the line, assist in training the operators, and finally shakedown or debug the line.

All of the above documentation is necessary when assembly lines are initially established or set up, and to train the operators. Once the line is flowing smoothly and the operators are trained, there will be less and less reliance on written work instructions and even visual aids. This initial planning documentation should always be available for ready reference, and should be kept up to date by the manufacturing engineer.

If the company manufactures a variety of different product lines, in medium to high volumes, and/or sets up and produces to a job-order-type system, assembly process documentation that is complete and to the greatest level of detail is especially important. It is a proven fact that good assembly process planning and documentation significantly reduces operator learning (and relearning) time. This is especially important when the production run is relatively short. It also teaches the correct methods to operators and thereby reduces costs of assembly labor. The assembly process documentation package is essential to the operation of an ongoing production-control and time-keeping system. The assembly process routing provides the steps or sequences that materials, parts, assemblies, and work in process must follow to build the product. It provides the time standards for each operation, and the assembly parts list for each operation provides the information needed by production control to pull and kit material for production.

In a small plant, where production runs may be small to nonexistent, assembly process planning with only minimum documentation is required and can be justified for the reasons mentioned earlier. Even in the case where no formal production control system exists and the production supervisor draws material from the stockroom in one batch issue for the entire job, pictorial visual aids, workstation layouts, a tool list, and an assembly process routing should be provided.

20.2 ASSEMBLY OPERATION SEQUENCES

In assembly process planning, operation sequences usually parallel the indented parts list or engineering tree chart, since that should represent how the product goes together or is assembled. This initial assembly process sequence plan should define an assembly operation for each major and minor subassembly, and for the final assembly. See Figure 20.2. It should be emphasized that this is an initial breakdown, and normally will be followed by a more thorough analysis of the steps required to assemble the various subassemblies and the final assembly. This detailed analysis is normally done in the preproduction planning phase in the form of an operation process chart. Figure 20.3 shows an example of an operation process chart for a Coast Guard radio receiver.

The assembly process may include soldering, wiring, press fitting, brazing, shrink fitting, welding, adhesive bonding, riveting, and mechanical fastening. Within each of these assembly processes a series of sequences is required to accomplish the process, without regard to the product configuration, material, or quantity to be produced, or the rate of production. For example, many of the steps in creating a circuit card assembly, a wire harness, the frame of a truck, or in the installation of fittings on a sailboat are

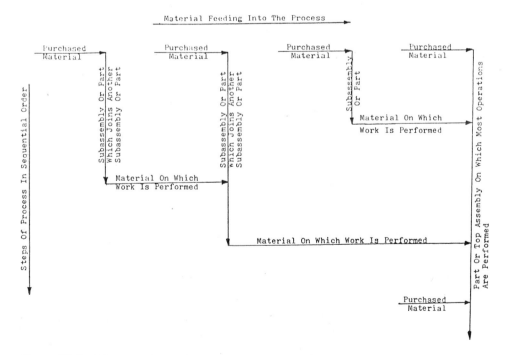

Figure 20.2 Principles of operation process chart construction.

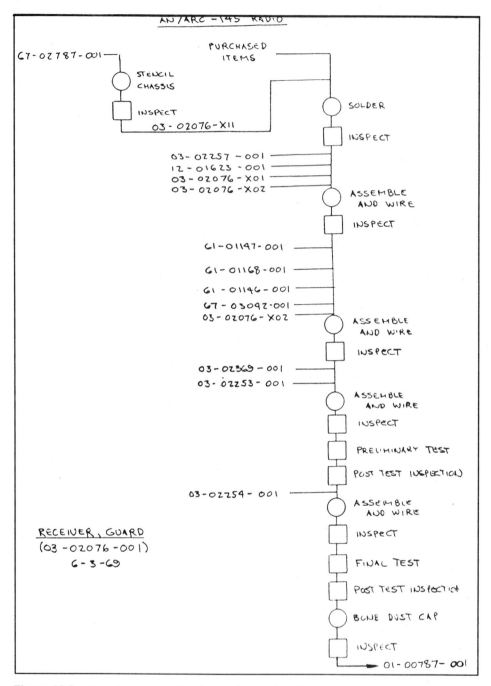

Figure 20.3 Operation process chart for a radio guard receiver.

essentially the same. The detailed instructions for the sequence should spell out the differences peculiar to the product at hand.

Often in assembly work, standard sequences or operations are possible for any product where these processes of assembly are used. The result is a considerable saving in manufacturing engineering time, and in the elapsed time required to prepare and release an assembly process plan to the shop. Such standard processes enable preprinted process planning documentation, which may only require a part number and quantity to be entered before it is ready for release. A good example of this is in the manufacture of cables for electrical or electronic equipment, where diagrams are preprinted of the various electrical connector pin configurations, requiring only that the manufacturing engineer sketch in the wires terminated to the pins for the specific cable application. Computer-aided process planning (CAPP), as discussed in Chapter 9, covers the use of a database to accomplish this task.

20.2.1   Routings, Work Instructions, and Visual Aids

Assembly process routings, such as the one shown in Figure 20.4, list the operations in the sequence in which they must occur to assemble the item or product called out in the heading. In addition to listing the operations in their proper sequence, it lists the standard times for each operation, the performing department, and the latest revision level of the process instruction sheets. The issue or revision level of the process information establishes configuration control of the product on the shop floor, since normally the assembly department does not work to engineering drawings. The importance of this cannot be emphasized enough. It is the responsibility of the manufacturing engineer always to have the latest revisions to the engineering drawings incorporated into the assembly process documentation, especially when the job is active on the production floor. In many companies the inspection department uses the process documentation to perform in-process inspections of the product. This is especially true where detailed process work instructions are used, and the process documentation is also used for shop configuration control.

As indicated earlier, the assembly process summary or routing is used by production control to move material or work in process to the next operation or sequence. A copy of the assembly process summary travels with each batch of parts and material, and in effect becomes a routing sheet or shop traveler. For this to happen, the sequences must be stamped off, either by inspection or by the operator as they are completed. If line production is involved, such a routing or traveler is unnecessary, as the progression on the assembly line is the routing followed by the assemblies. The process routing is an especially valuable tool in the job shop, where the shop is layed out by function or process, and does not follow the product flow.

Assembly work instructions are the heart of the manufacturing engineering documentation package, and explain how the product is to be assembled in production. The assembly work instructions should be available at the operator workstation, preferably on an easy-to-see holder mounted on the workbench. The assembly process routing lists all of the operations for assembly, in the sequence that they must be performed; the assembly work instructions or assembly methods sheets for each listed operation explain in detail exactly how to perform the operations. Figure 20.5 is an example of an assembly methods sheet for the wiring of a connector which becomes part of a wiring harness in a marine short-wave radio receiver.

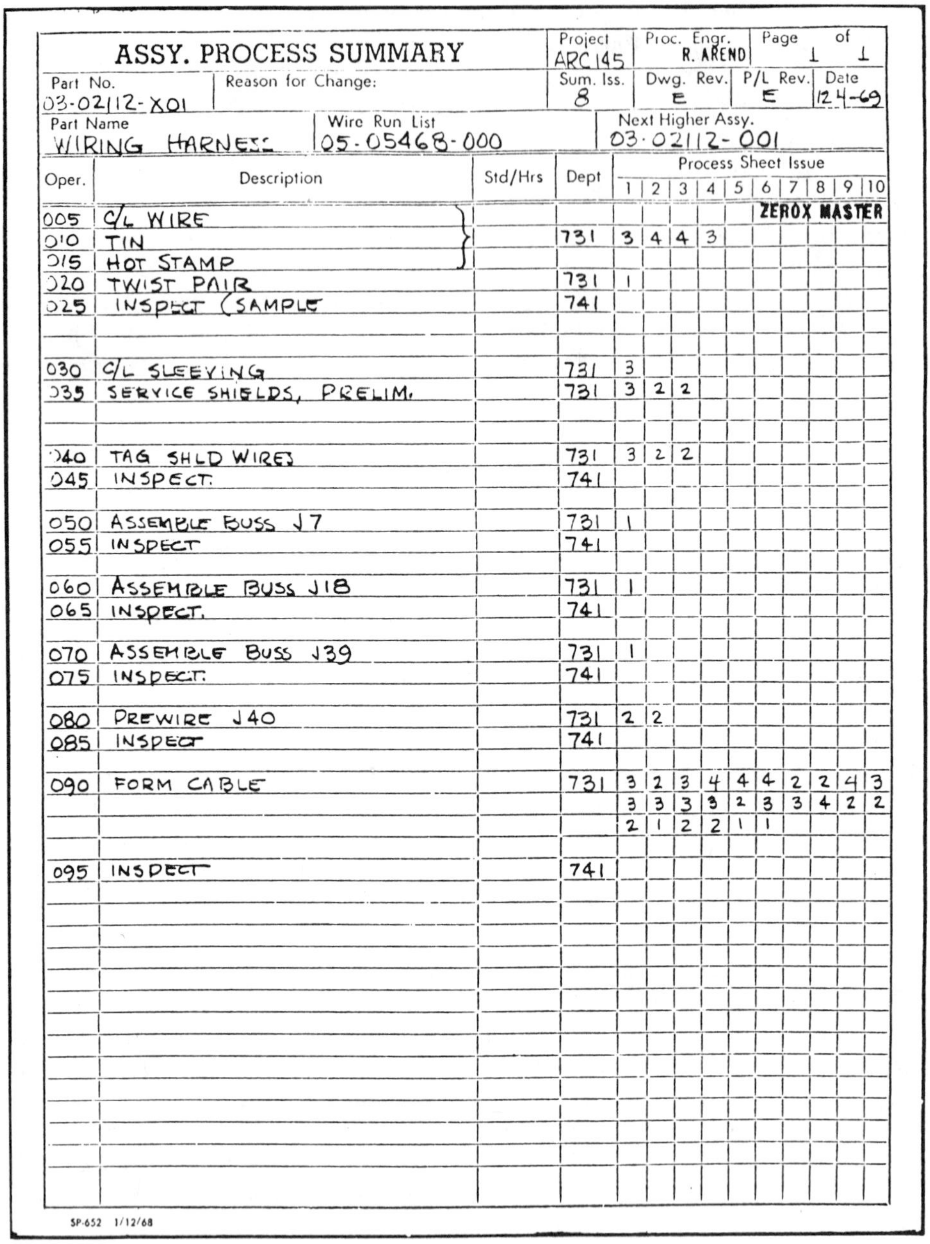

ASSY. PROCESS SUMMARY		Project ARC 145	Proc. Engr. R. AREND	Page 1 of 1
Part No. 03-02112-X01	Reason for Change:	Sum. Iss. 8	Dwg. Rev. E	P/L Rev. E	Date 12-4-69
Part Name WIRING HARNESS	Wire Run List 05-05468-000	Next Higher Assy. 03-02112-001		

Oper.	Description	Std/Hrs	Dept	Process Sheet Issue 1 2 3 4 5 6 7 8 9 10
005	C/L WIRE			ZEROX MASTER
010	TIN		731	3 4 4 3
015	HOT STAMP			
020	TWIST PAIR		731	1
025	INSPECT (SAMPLE)		741	
030	C/L SLEEVING		731	3
035	SERVICE SHIELDS, PRELIM.		731	3 2 2
040	TAG SHLD WIRES		731	3 2 2
045	INSPECT.		741	
050	ASSEMBLE BUSS J7		731	1
055	INSPECT		741	
060	ASSEMBLE BUSS J18		731	1
065	INSPECT.		741	
070	ASSEMBLE BUSS J39		731	1
075	INSPECT.		741	
080	PREWIRE J40		731	2 2
085	INSPECT		741	
090	FORM CABLE		731	3 2 3 4 4 4 2 2 4 3 / 3 3 3 3 2 3 3 4 2 2 / 2 1 2 2 1 1
095	INSPECT		741	

SP-652 1/12/68

Figure 20.4 Assembly process summary for wiring harness subassembly.

An assembly operation parts list should be included with the assembly work instructions or assembly methods sheets for each operation. This tells production control and the operator what parts and materials are required to perform the assembly operation for one unit. Figure 20.6 shows one version of such a parts list. In addition, there must also be a list of standard and special design assembly tools needed to perform the operation. In this

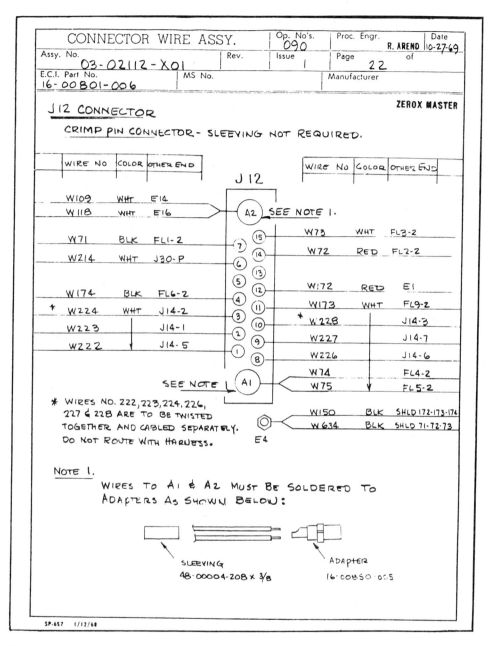

Figure 20.5   Assembly methods sheet for wiring a connector.

Seq.	E.C.I. Part Number	Description	Remarks	Qty.

ASSY. OPER. PARTS LIST — Op. No's. 005 — Operation Description: ASSEMBLE AND WIRE

Assy. No. 03-02076-001 — Rev. — Next Assy./Oper. 015 — Proc. Engr. J. WOOD — Date 7-28-69

Assy. Name RECEIVER, GUARD — Issue 1 — Page 1 of 5

ZEROX MASTER

Seq.	E.C.I. Part Number	Description	Remarks	Qty.
		TOOLS: ASSEMBLY VISE		
		W SOLDER IRON — TIP		
		INSERTION TOOL	IT-1-16-00852-004	
1	OP002	CHASSIS ASSY	POSITION IN VISE	1
2	16-00852-004	JACK (J1, J2, J3)	INSERT JACKS INTO	3
			CHASSIS USING TOOL	
			IT-1-16-00852-004	
			REF. SKETCH "A"	
3	03-02257-001	AMPL, RF (A3)	APPLY LOCTITE TO	1
4	23-01442-202	SCREW	SCREW THREADS,	2
	22-00065-005	LOCTITE, BRN	POSITION AND ASSEMBLE	A/R
			A3 TO CHASSIS	
			REF. SKETCH "A"	
5	31-00008-030	BUSSWIRE (1")	CONNECT LEADS	1
6	10-00275-044	RESISTOR (R1)	TO TERMINALS	1
7	11-00392-001	CAPACITOR (C1)	REF. SKETCH "B"	1
8	09-00415-011	COIL (L1)		1

ZEROX MASTER

SP-463 1/12/68

Figure 20.6 Assembly operations parts list.

example it is included in the operation parts list. It can also appear as a separate call-out on the work instruction or methods sheets.

An extremely important part of the assembly process documentation is visual aids. Visual aids can be anything from an actual mockup of the product to a black and white or color photograph, to a three-dimensional isometric or exploded-view drawing, to a simple sketch, or to a tracing lifted directly from the engineering drawing. Figure 20.7 shows a light table being used by a manufacturing engineer to trace parts of the engineering drawing in order to make up a visual aid. If regular office copy machines are used, visual aids can be constructed using cut-and-paste methods as shown in Figure 20.8. Illustrations may also be done by graphic artists or illustrators as shown in Figure 20.9. It should be kept in mind that all process documentation costs money, and consideration should be given to the length of the production run, anticipated changes during production, and what is really needed to instruct a particular group of assembly operators. One other very important consideration is that a visual aid supplemented by minimal notes and instructions is far superior to lengthy written work instructions. As explained earlier, visual aids that highlight key assembly details are all that are used in many companies.

20.3 WORKSTATION AND LINE LAYOUT

Workstation layouts are important from the standpoint of assembly operator methods. They tell the assembly supervisor how to set up and configure the individual workstations for optimum productivity and flow of work. Workstation layouts are usually in the form of a plan view of the workstation, and show where tooling and fixturing should be placed, where parts bins should be placed, work instructions, tote pans for staging completed work and for placing incoming work, and any other information pertinent to the operation and the set-up of the workstation. Figure 20.10 shows a workstation layout used in a small job order shop. Figure 20.11 shows how this layout might look set up in the shop.

Line layouts are used where progressive assembly lines are to be used to build the

Figure 20.7 Manufacturing engineer using a light table to make a tracing from an engineering drawing in order to prepare a visual aid.

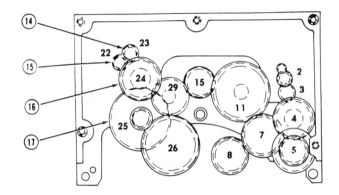

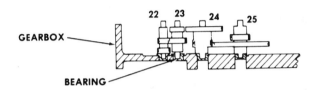

SEQ	PART NUMBER	DESCRIPTION	QTY	REQUIRED
14	39-00286-001	GEAR 23	1	POSITION TO HOUSING PER SKETCH
15	39-00287-001	GEAR 22	1	POSITION TO HOUSING PER SKETCH
16	39-00276-001	GEAR 24	1	POSITION TO HOUSING PER SKETCH
17	39-00275-001	GEAR 25	1	POSITION TO HOUSING PER SKETCH
18	—	—	—	SET HOUSING ASIDE AND POSITION COVER TO WORK AREA

Figure 20.8 Inexpensive visual aid prepared by cut-and-paste method.

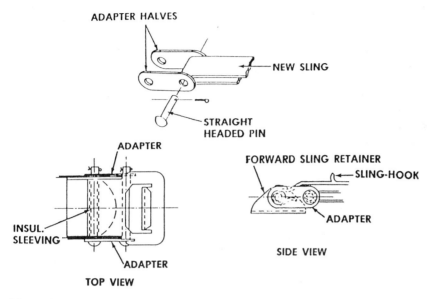

Figure 20.9 Visual aid made using formal artwork.

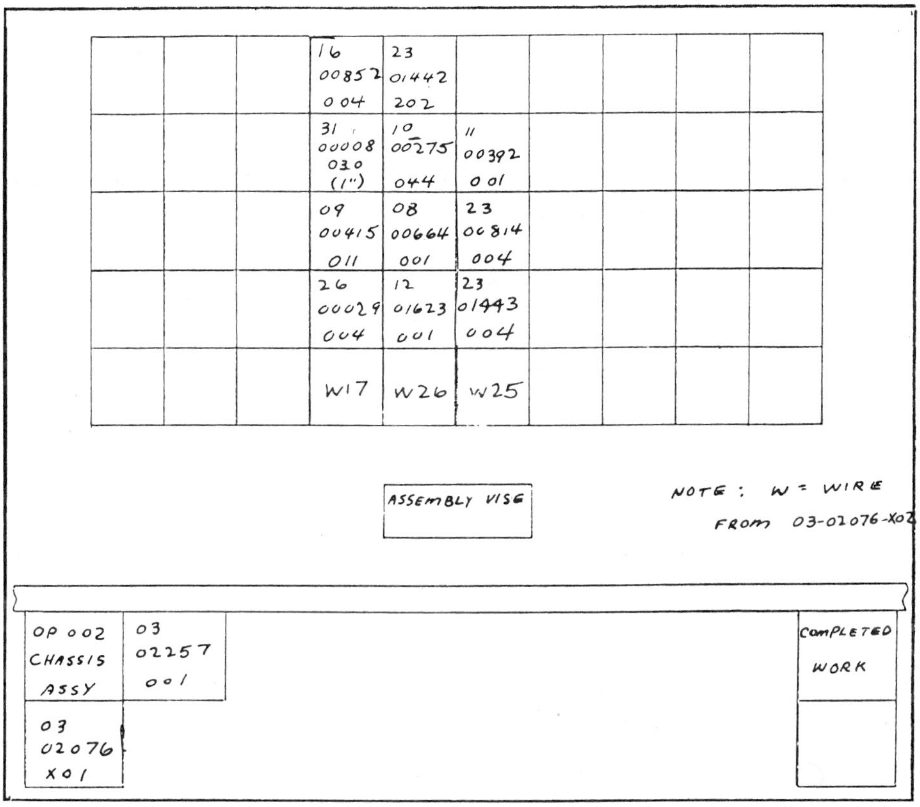

Figure 20.10 Assembly workstation layout in diagram form with parts bin setup diagram.

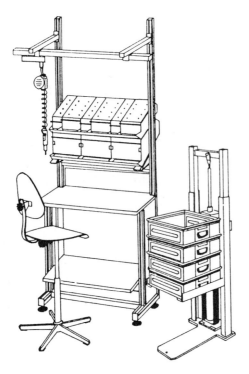

Figure 20.11   Assembly workbench arrangement as it would appear on the shop floor.

product, and the plant layout drawings do not show which workstation goes where. Again, these are used by supervision in setting up the line to conform to the assembly process flow, and to ensure optimum methods and work flow. See Figure 20.12 for an example.

20.4   MANUFACTURING METHODS ANALYSIS

In developing the manufacturing process, whether for fabrication, machining, forming, finishing, or assembly, the manufacturing engineer must specify the most economical methods for the job and for the work at hand. In order to do this he must understand and

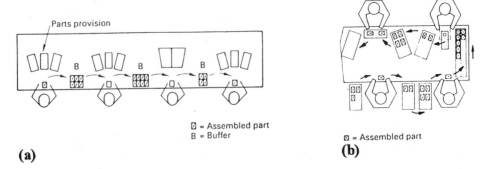

Figure 20.12   Manual line assembly with manual transfer of the workpieces (a) in a line arrangement, and (b) in rectangular form. (Courtesy IPA Stuttgart.)

be able to apply the fundamental techniques of methods analysis, motion economy, and work simplification. The presumption of good methods, and the ability of the manufacturing engineer in methods analysis is so basic, that not to be proficient in this art, is tantamount to being incompetent as a manufacturing engineer. The purpose and intent of this subchapter is to provide the basic information needed by the manufacturing engineer to gain a degree of proficiency in manufacturing methods analysis.

Even the best and most thorough process planning will sometimes overlook details, or specify methods that can be improved upon later, after the product is in production. Part of the job of the manufacturing engineer is to be alert for these opportunities to improve the process and the flow.

20.4.1 Work Simplification

Work simplification can be defined as the organized application of common sense to find easier and simpler ways of doing work. Work simplification provides a systematic, common-sense approach to make work easier, and at the same time to lower costs. The basic premise of work simplification is that there is always a better way to do any task. The work simplification pattern includes five basic steps:

1. Selection of a job to be improved
2. Recording of the job details
3. Analysis of the job details
4. Development of the improvements
5. Installation of the improvements

Selecting the job to be improved requires careful consideration and study. Efforts expended for improvement should be made first where the returns will be the greatest. Priority should be given to bottlenecks, choke points, trouble spots, jobs that require excessive amounts of time, or where generally unsatisfactory conditions exist. The following list provides assistance in making this selection:

1. Greatest cost: work that involves the greatest expenditure of funds, labor hours or use of equipment
2. Greatest workload: the largest volume of work being performed by the activity
3. Number of persons assigned: work that requires large numbers of people to perform similar tasks
4. Walking: jobs that require a lot of walking around
5. Bottlenecks: work not flowing smoothly
6. Schedules not met: failure to meet deadlines, resulting in work backlogs or overtime
7. Excessive waste: work that results in wasted materials, in scrap and in rework
8. Excessive fatigue: work that requires great physical effort or that is being done with frequent rest periods
9. Unsafe or unpleasant work practices: work that results in numerous accidents, or is undesirable because of extreme conditions such as dust, noise, fumes, vapors, or extremes of temperature

Every job is made up of three parts:

1. *Make ready*: the time and effort put into the setup, or getting ready to work.
2. *Do*: the actual work accomplished.
3. *Put away*: the time and effort put into cleaning up after the do part of the job.

For example, if we have a carpenter making a wooden box, we might expect the job to break down as follows:

1. *Make ready*: open bin, pick up nails, close bin, pick up hammer, lay nails on the bench.
2. *Do*: hammer nails.
3. *Put away*: put hammer aside, pick up box, put box aside.

Anything that reduces the time required for the "make ready" and "put away" parts of the job reduces the nonproductive time associated with the job.

Recording of the job details can best be accomplished through the use of process chart techniques, specifically the flow process chart and the flow diagram. The flow process chart is a device for recording each step of a job in a compact manner, as a means of better understanding it and improving it. The chart represents graphically the separate steps of the events that occur during the performance of work, or during a series of actions. The process chart may be used to record the flow within a unit, a section, a department, or between departments. The flow process chart has no bounds.

No matter how complicated or intricate the series of operations may be, a flow process chart can be constructed if you take one step at a time. The flow process chart, however, like other methods of graphic representation, may need to be modified to meet the requirements of a particular situation. For example, it may show in sequence the total activity of a production operator, or it may show in sequence the steps that the worker, part, or material goes through. The chart could be either the operator type or the material type, and the two types should not be combined.

A careful study and analysis of such a chart, giving a graphic picture of every step in the process is almost certain to suggest improvements. It is not uncommon to find that some operations can be eliminated or that a part of an operation can be eliminated, that one operation can be combined with another, that better routes for the parts can be found, more economical machines used, delays between operations eliminated, and other improvements made, all of which go to produce a better produce at a lower cost.

To make a flow process chart requires careful adherence to the rules outlined below:

1. State the activity being studied. Make certain you are really naming the activity you have chosen to study.
2. Choose the subject to follow. Decide on a person or a material, and follow him or it through the entire process. When you have picked a subject, stick with it.
3. Pick a starting and ending point. This is to make certain you will cover all the steps you wish to cover, but no more or less.
4. Write a brief description of each detail. Step by step, no matter how short or temporary, describe each detail.
5. Apply the symbols. The description determines each symbol. Draw a connecting line between each of the proper symbols.
6. Black in the "do" operation. Shade in the symbols for those operations that you decide are the "do" operations. This will help you later in your analysis, when you begin challenging.
7. Enter distances. Whenever there is a transportation, enter the distance traveled.
8. Enter time if required. Many times this is not necessary. However, if it will help, note the time required or elapsed.
9. Summarize. Add up all of the facts and put them in the summary block. The

summary should indicate the total number of operations, transportations, inspections, delays, storages, and distance that is traveled.

Figure 20.13 shows a flow process chart that is completely filled in, with circled numbers used to illustrate each of the rules listed above. The chart symbols and their meaning are as follows:

O   *Operation.* An operation occurs when an object is intentionally changed in any of its physical or chemical characteristics, is assembled or disassembled from another object, or is arranged or prepared for another operation, transportation, inspection, or storage. An operation also occurs when information is given or received, or when planning or calculating takes place.

⇒   *Transportation.* A transportation occurs when a person moves from one workplace to another or when an object is moved, except when such movements are part of the operation or are caused by the operator at the workstation during an operation or an inspection.

☐   *Inspection.* An inspection occurs when an object is examined for identification or is verified for quality or quantity, or any of its characteristics.

D   *Delay.* A delay occurs to an object when conditions, except those that intentionally change the physical or chemical characteristics of the object, do not permit or require immediate performance of the next planned action.

∇   *Storage.* A storage occurs when an object is kept and protected against unauthorized removal.

It is sometimes helpful to supplement a flow process chart with a flow diagram. A flow diagram is simply a layout of the area involved in the job being studied, over which you indicate by a line the path of the object or person followed in the flow process chart. It is often desirable to indicate the action taking place by using the same symbols as on the chart. They may, if desired, be keyed to each other by item numbers. Figure 20.14 shows a flow diagram to accompany the flow process chart in Figure 20.13.

The third step of the work simplification pattern involves questioning every part, aspect, or detail of the job. Here we examine each operation and ask some very pointed questions. Don't be satisfied until you have asked all possible questions and received the related "why" answers. The first thing to do is to question the entire job being studied. Why is it done? Is the job really necessary? If the answer comes back that it is, then question each "Do" operation. If you can eliminate a "do" operation, you also eliminate the "make ready" and "put away" that go with it. The "do" operations are those that add value to the product or process being studied.

There is always a better way, and our task is to find it.

Why? The overriding question, it establishes the reason for the job. The answer defines and justifies the purpose of the job.

What? What is done? What are the steps? What does each step do? What makes the step necessary?

Where? Where should this step be done? Can it be done more easily? Can it be done using less time, energy, or transportation, by changing the location of employees or equipment?

When? When should this step be done? Is it done in the right place in the sequence? Can the job be simplified by moving this step ahead or back?

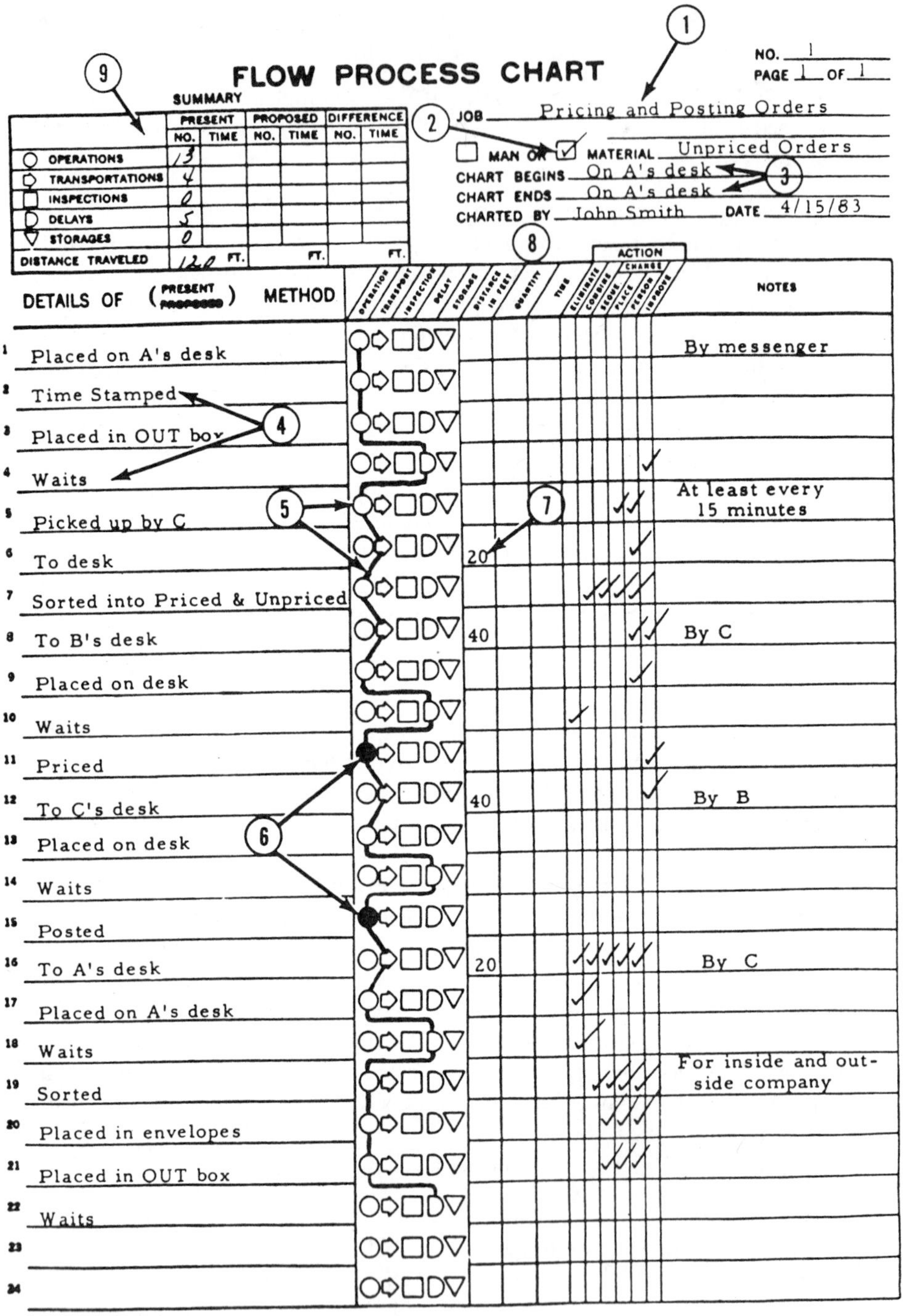

FLOW PROCESS CHART

NO. 1
PAGE 1 OF 1

SUMMARY

	PRESENT		PROPOSED		DIFFERENCE	
	NO.	TIME	NO.	TIME	NO.	TIME
◯ OPERATIONS	13					
⇨ TRANSPORTATIONS	4					
☐ INSPECTIONS	0					
D DELAYS	5					
▽ STORAGES	0					
DISTANCE TRAVELED	120 FT.		FT.		FT.	

JOB ___Pricing and Posting Orders___

☐ MAN OR ☑ MATERIAL ___Unpriced Orders___
CHART BEGINS ___On A's desk___
CHART ENDS ___On A's desk___
CHARTED BY ___John Smith___ DATE ___4/15/83___

#	DETAILS OF (PRESENT / PROPOSED) METHOD	Symbols	Distance in Feet	Quantity	Time	ACTION (Eliminate / Combine / Sequence / Place / Person / Improve)	NOTES
1	Placed on A's desk	◯⇨☐D▽					By messenger
2	Time Stamped	◯⇨☐D▽					
3	Placed in OUT box	◯⇨☐D▽					
4	Waits	◯⇨☐D▽				✓	
5	Picked up by C	◯⇨☐D▽				✓✓	At least every 15 minutes
6	To desk	◯⇨☐D▽	20			✓	
7	Sorted into Priced & Unpriced	◯⇨☐D▽				✓✓✓✓	
8	To B's desk	◯⇨☐D▽	40			✓✓	By C
9	Placed on desk	◯⇨☐D▽				✓	
10	Waits	◯⇨☐D▽			✓		
11	Priced	●⇨☐D▽				✓	
12	To C's desk	◯⇨☐D▽	40			✓	By B
13	Placed on desk	◯⇨☐D▽					
14	Waits	◯⇨☐D▽					
15	Posted	●⇨☐D▽					
16	To A's desk	◯⇨☐D▽	20			✓✓✓✓	By C
17	Placed on A's desk	◯⇨☐D▽					
18	Waits	◯⇨☐D▽				✓	
19	Sorted	◯⇨☐D▽				✓✓✓✓	For inside and out-side company
20	Placed in envelopes	◯⇨☐D▽				✓✓✓	
21	Placed in OUT box	◯⇨☐D▽				✓✓✓	
22	Waits	◯⇨☐D▽					
23		◯⇨☐D▽					
24		◯⇨☐D▽					

Figure 20.13 Flow process chart example.

Operations: Pricing and Posting Orders

Subject Charted: Unpriced Orders

PRESENT METHOD LAYOUT

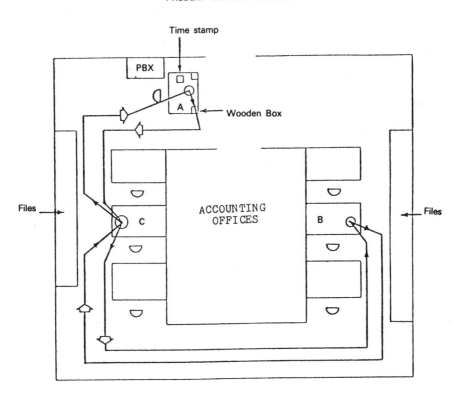

Clerk A — Receptionist and PBX Operator

Clerk B — Pricing Clerk

Clerk C — Posting Clerk

Figure 20.14 Flow diagram example.

Who? Who should do the job? Is the right person handling it? Would it be more logical to give the job to someone else?

How? How is the job being done? Can it be made easier? Can the job be done better with different equipment, or a different layout?

This question and answer approach will suggest improved methods:

Why and *what* lead to *elimination*.

Where, *when*, and *who* lead to *combine* or *change sequence*.

How leads to *simplicity*.

Careful consideration of the possibilities presented in looking for ways to eliminate, combine, change sequence, and simplify in the questioning approach brings us finally to

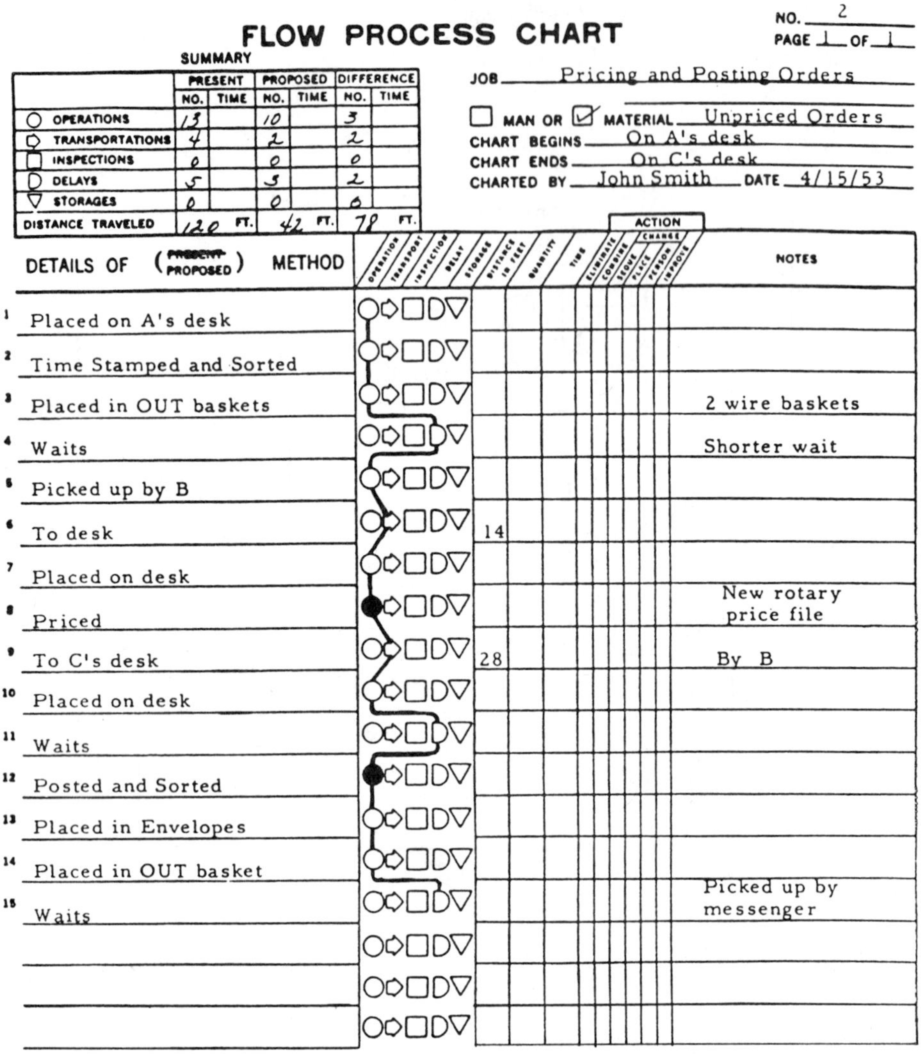

FLOW PROCESS CHART

NO. ___2___
PAGE _1_ OF _1_

SUMMARY

	PRESENT		PROPOSED		DIFFERENCE	
	NO.	TIME	NO.	TIME	NO.	TIME
○ OPERATIONS	13		10		3	
▷ TRANSPORTATIONS	4		2		2	
☐ INSPECTIONS	0		0		0	
D DELAYS	5		3		2	
▽ STORAGES	0		0		0	
DISTANCE TRAVELED	120 FT.		42 FT.		78 FT.	

JOB ___Pricing and Posting Orders___

☐ MAN OR ☑ MATERIAL ___Unpriced Orders___
CHART BEGINS ___On A's desk___
CHART ENDS ___On C's desk___
CHARTED BY ___John Smith___ DATE __4/15/53__

ACTION / CHANGE

DETAILS OF (~~PRESENT~~ / PROPOSED) METHOD

	DETAILS OF METHOD		DISTANCE IN FEET	NOTES
1	Placed on A's desk	○▷☐D▽		
2	Time Stamped and Sorted	○▷☐D▽		
3	Placed in OUT baskets	○▷☐D▽		2 wire baskets
4	Waits	○▷☐D▽		Shorter wait
5	Picked up by B	○▷☐D▽		
6	To desk	○▷☐D▽	14	
7	Placed on desk	○▷☐D▽		
8	Priced	●▷☐D▽		New rotary price file
9	To C's desk	○▷☐D▽	28	By B
10	Placed on desk	○▷☐D▽		
11	Waits	○▷☐D▽		
12	Posted and Sorted	●▷☐D▽		
13	Placed in Envelopes	○▷☐D▽		
14	Placed in OUT basket	○▷☐D▽		Picked up by messenger
15	Waits	○▷☐D▽		
		○▷☐D▽		
		○▷☐D▽		
		○▷☐D▽		

Figure 20.15 Flow process chart showing improved method plan.

the better method and provides us with the answer to "How should the job be done?" The simplest way is the best way. Figure 20.15 shows a proposed improvement flow process chart for the pricing and posting of orders in Figure 20.14.

20.5 PRINCIPLES OF MOTION ECONOMY

It is the purpose of the following discussion to explore the rules or principles of motion economy which have been and are now being used successfully in manufacturing methods studies. These principles form a basis, code, or body of rules which, if applied correctly, make it possible to greatly increase the output of manual factory labor with a minimum of fatigue. These principles will be examined under the subdivisions of operator tasks, the workplace, and as applied to tools and equipment.

20.5.1 Operator Tasks

The principles of motion economy as related to the tasks of the operator are as follows

1. The two hands should begin as well as complete their motions at the same time.
2. The two hands should not be idle at the same time except during rest periods.
3. Motions of the arms should be made in opposite and symmetrical directions and should be made simultaneously.

These three principles are closely related and should be considered together. It seems natural for most people to work productively with one hand while holding the object being worked on with the other hand. This is extremely undesirable, and should be avoided. The two hands should work together, each beginning a motion and completing a motion at the same time. Motions of the two hands should be simultaneous and symmetrical.

Many kinds of work can be accomplished better using both hands than by using one hand. For most manufacturing assembly operations, it is advantageous to arrange similar work on the left- and right-hand sides of the workplace, thus enabling the left and right hands to move together, each performing the same motions. The symmetrical movements of the arms tend to balance each other, reducing the shock and jar on the body and enabling the operator to perform the task with less mental and physical effort. There is apparently less body strain when the hands move symmetrically than when they make nonsymmetrical motions, because of balance.

The fourth principle of motion economy states that hand and body motions should be confined to the lowest classification with which it is possible to perform the work satisfactorily.

20.5.2 Classes of Hand Motions

The five general classes of hand motions emphasize that material and tools should be located as close as possible to the point of use. The motions of the hands should be as short as the work permits. In the listing of classifications shown below, the one requiring the least amount of time and effort is shown first:

1. Finger motions
2. Motions involving fingers, and wrist
3. Motions involving fingers, wrist, and forearm
4. Motions involving fingers, wrist, forearm, and upper arm
5. Motions involving fingers, wrist, forearm, upper arm, and shoulder (causes posture change)

It should be pointed out that finger motions have been found to be less accurate, slower, and more fatiguing than motions of the forearm. Evidence seems to indicate that the forearm is the most desirable member for performing light work. In highly repetitive work, motions about the wrist and elbow are superior to those of the fingers or shoulders. (Reference Subchapter 5.1 on ergonomics.)

The fifth principle of motion economy states that momentum should be employed to assist the worker wherever possible, and it should be reduced to a minimum if it must be overcome by muscular effort. The momentum of an object is defined as its mass times its velocity. In the factory environment, the total weight moved by the operator may consist of the weight of the material moved, the weight of the tools moved, and the weight of the part of the body moved. It should be a real possibility to employ momentum to advantage when a forcible blow or stroke is required. The motions of the worker should be so arranged that the blow is delivered when it reaches its greatest momentum.

The sixth principle of motion economy states that smooth, continuous, curved motions of the hands are preferable to straight-line motions involving sudden and sharp changes in direction. Abrupt changes in direction are not only time consuming but also fatiguing to the operator.

The seventh principle of motion economy states that ballistic motions are faster, easier, and more accurate than restricted or controlled movements. Ballistic movements are fast, easy motions caused by a single contraction of a positive muscle group, with no antagonistic muscle group contracting to oppose it. A ballistic stroke may be terminated by the contraction of opposing muscles, by an obstacle, or by dissipation of the momentum of the movement, as in swinging a sledge hammer. Ballistic movements are preferable to restricted or controlled movements, and should be used whenever possible.

The eighth principle of motion economy states that work should be arranged to permit an easy and natural rhythm wherever possible. Rhythm is essential to the smooth and automatic performance of any operation. Rhythm, as in a regular sequence of uniform motions, aids the operator in performing work. A uniform, easy, and even rate of work are aided by proper arrangement of the workplace, tools, and materials. Proper motion sequences help the operator to establish a rhythm which helps make the work a series of automatic motions, where the work is performed without mental effort.

The ninth principle of motion economy states that eye fixations should be as few and as close together as possible. Where visual perception is required, it is desirable to arrange the task so that the eyes can direct the work effectively. The workplace should be laid out so that the eye fixations are as few and as close together as possible.

The Workplace

The first principle of motion economy related to the workplace states that there should be a definite and fixed place for all tools and materials. The operator should always have tools and materials in the same location, and finished parts and assembled units should be placed in fixed positions or locations. For example, in the assembly of mechanical hardware, the hand should move without mental direction to the bin containing flat washers, then to the bin containing lock washers, then to the bin containing bolts, and finally to the bin containing hex nuts. There should be no thinking required on the part of the operator to do any of this.

The second principle of motion economy related to the workplace states that tools, materials, and controls should be located close to the point of use. In the horizontal plane, there is a definite and somewhat limited area which the worker can use with a normal expenditure of effort. This includes a normal working area for the right hand and one for the left hand for each working separately, and another for both hands working together. Figure 20.16 shows this and the dimensions of normal and maximum working areas in the horizontal and vertical planes. Both the standing and sitting positions are included. It also shows normal bench work surface heights, which can have a significant adverse effect if they are not correct.

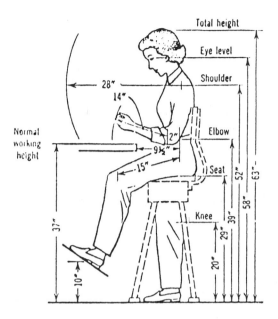

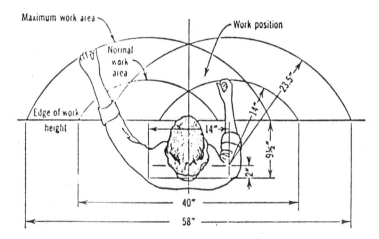

Figure 20.16 Normal and maximum working areas and heights.

Figure 20.17 shows in greater detail the areas of easiest reach for the left and right hands, for both hands working together, and the area in which small objects can be most easily picked up.

The third principle of motion economy related to the workplace states that gravity-feed bins and containers should be used to deliver the material close to the point of use. This can sometimes be accomplished by using parts bins with sloping bottoms which feed parts by gravity to the front of the bin, eliminating the need for the assembly operator to reach down into the bin to grasp parts.

The fourth principle of motion economy related to the workplace states that drop deliveries should be used wherever possible. This requires configuring the workplace, for

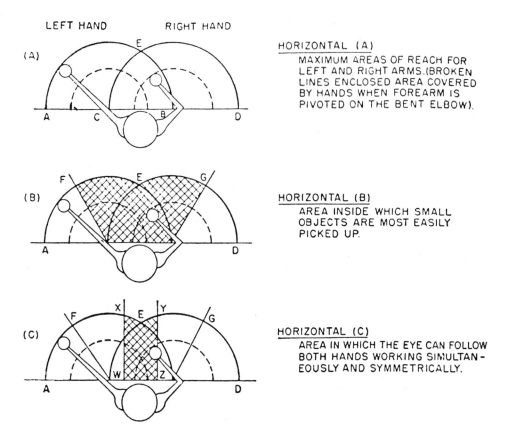

Figure 20.17 Areas of easiest reach for each hand, and for both hands working together.

example, so that finished units may be disposed of by releasing them in the position in which they are completed, delivering them to their next destination by gravity. Besides the savings in time, this frees the two hands so that they may begin the next cycle immediately without breaking the rhythm.

Other principles of motion economy related to the workplace include:

6. Materials and tools should be located to permit the best sequence of motions.
7. Provision should be made for adequate lighting.
8. The height of the workplace and chair should be arranged so that alternate sitting and standing at work are easily possible.
9. A chair of the type and height to permit good posture should be provided for the operator.

20.5.4 Tools and Equipment

Principles of motion economy as related to the design of tools and equipment include the following

1. The hands should be relieved of all work that can be done more effectively by a jig, fixture, or foot-operated device.

2. Two or more tools should be combined wherever possible.

3. Tools and materials should be prepositioned whenever possible.

4. Where each finger performs some specific movement, the load should be distributed in accordance with the inherent finger capacities (arrangement of typewriter keys).

5. Levers, crossbars, and hand wheels should be located in such positions that the operator can operate them with the least change in body position, and with the greatest mechanical advantage.

20.6 STANDARD MANUFACTURING PROCESSES

In addition to operator work instructions, process routings, visual aids, and operation parts lists, a key ingredient of the documentation package that should be provided by manufacturing engineering includes standard manufacturing processes. Standard manufacturing processes include workmanship standards, equipment operating procedures, and standard repair procedures. These standard processes are common to all products manufactured in a plant, and therefore are either referenced in the product work instructions, as in "Assemble per standard manufacturing process 367," or the appropriate text and/or illustrations are copied directly into the product instructions, thereby eliminating the requirement for the production operator to leave the work position and look up information in a separate book or document. Obviously, the latter method is preferred.

20.6.1 Workmanship Standards

Workmanship standards tell the operator what is acceptable work or practice, and what is not. These standards can be in the form line drawings supplemented with narrative text, color photographs of acceptable and unacceptable work supplemented by narrative text, or actual models or prototype units, known as standards. Figure 20.18 shows an example

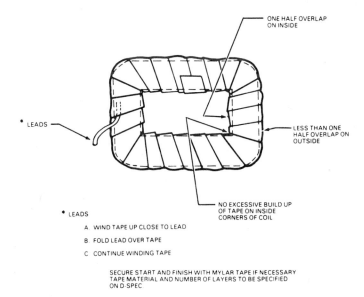

Figure 20.18 Workmanship standard for taping iron core transformers.

of a line-drawing workmanship standard for taping the core of an iron-core transformer in the magnetic components industry. It should be noted that instructions are superimposed where needed, pointing out the important points to watch for when tape-wrapping the core. Also, it should be noted that this workmanship standard tells only the requirements for making an acceptable tape wrap, and that it is necessary to refer to a product specification to determine the tape material and number of layers required.

20.6.2 Equipment Procedures

Equipment procedures are start-up, operating, and shut-down instructions for a machine or piece of production equipment. They should be posted at or near the machine or equipment, and are usually prepared by the manufacturing engineer from the manual that is prepared by the maker of the machine or equipment. These procedures are especially valuable where process variables controlled by machine or equipment settings can be critical. Equipment procedures ensure the proper training of new operators, and serve as reminders to the experienced operator. Equipment procedures should be prepared for everything from vapor degreasers to heat-treat ovens.

20.6.3 Standard Repairs

Standard repair procedures provide ready-made work instructions to be used at any time certain types of repair work must be done to product parts, subassemblies, or assemblies. These procedures or instructions have been approved in advance by inspection or quality control, and if necessary by design engineering and the customer. They not only tell manufacturing how to repair certain kinds of product defects, they eliminate the need for formal rejection by inspection before the repair procedure can be implemented.

For example, the standard repair for a mislocated or design-changed hole in a metal part might read as follows

1. The maximum number of holes to be plugged or welded in any one part by this procedure is 20% of the total holes in the part, or eight (8) holes, whichever is less.
2. The preferred method for repair shall be welding, except where heat would cause distortion.
3. In applications where surface heat due to welding will cause distortion, a press fit plug, or a formed flat head rivet are acceptable alternatives.
4. The holes to be filled by weld material shall be prepared by chamfering both sides of the hole. The chamfer shall be sufficient to insure weld penetration.
5. Surfaces to be welded shall be cleaned in accordance with established welding practices for the specific type of material involved.
6. The holes shall be completely filled with weld material.
7. Welding shall be accomplished by a certified welder.
8. The surfaces shall be ground or otherwise made flush to eliminate weld build-up.
9. Holes to be filled by plugging shall be prepared by the incorporation of a chamfer or lead on the side to accept the plug.
10. The plug shall be of such diameter that a press fit is ensured.
11. After plug insertion, the surfaces shall be ground or otherwise made flush, to eliminate excess material.
12. The plug shall be staked to the part per approved practices.

13. Holes to be plugged with a rivet shall be countersunk on both sides to accept a flat head rivet.
14. The rivet shall be formed and ground, or otherwise made flush with the mating services.
15. Rivets shall not be loose after forming.

20.7 SPECIAL MANUFACTURING INSTRUCTIONS

Frequently it becomes necessary for the manufacturing engineer to look beyond the operations in his own plant, and to make certain that material and parts received from outside suppliers comes into the plant and the manufacturing operations ready to become part of the product being produced. In the majority of cases, if the vendor meets the requirements of the purchase order and the engineering drawings (if applicable), the parts or material will be ready for processing when they arrive at the receiving dock. In some instances, however, it may be required that the vendor only partially complete a part or subassembly, or that the vendor produce a certain part on the low side of the drawing tolerance if the parts are to fit together in assembly.

To ensure that this information is correctly transmitted to the vendor, the manufacturing engineer may prepare special manufacturing instructions which are called out on the purchase order along with engineering drawings and specifications. The special manufacturing instructions along with the drawings and specifications then become the acceptance criteria for the parts or material when it goes through receiving inspection.

Special manufacturing instructions are prepared by the engineer at the time the make-or-buy determination is made and before the detailed internal work instruction documentation is prepared. These instructions can be in the form of simple hand-written information given to the buyer to incorporate directly into the purchase order, or a formal document which becomes an attachment to the purchase order. The important point to remember is that the manufacturing engineer needs to review the items to be purchased, and make certain that if these items are made exactly as the drawings and specifications say, they can enter directly into production without any problems or unplanned processing steps.

20.8 BIBLIOGRAPHY

Boothroyd, Geoffery, *Assembly Automation and Product Design*, Marcel Dekker, New York, 1992.
Tanner, John P., *Manufacturing Engineering*, Marcel Dekker, New York, 1991.

21

Assembly Automation

Jack M. Walker   *Consultant, Manufacturing Engineering, Merritt Island, Florida*

with

Vijay S. Sheth   *McDonnell Douglas Corporation, Titusville, Florida*

21.0 INTRODUCTION TO ASSEMBLY AUTOMATION

In this space age, when society regards everything as possible, there are many misconceptions about automation. Automated technology has been prescribed as the ultimate cure-all for all technical and social problems. The disinclination of workers to perform dull, repetitive tasks, considerations of worker health and safety, dramatic increases in labor rates and fringe benefits, and the increased ability of tool engineers to devise sophisticated machinery have combined to create a dramatic increase in automated equipment during the past several decades. Computer-controlled machine tools now make complex machined parts, electronic components are installed automatically in circuit boards, machines inspect detail parts on a 100% basis, and parts are positioned, assembled, and checked out automatically in the creation of assemblies—untouched by human hands. Robots of all descriptions have become part of the industrial scene.

It is true that most businesses automate primarily to reduce costs and thereby improve their competitive position in the market. However, the *real* objective of this investment is to *make* money, not just save money. When compared to manual assembly operations, the benefits usually derived from automated assembly include:

Reduced unit costs
Consistent high quality
Elimination of hazardous manual operations
Increased production standby capacity

The aerospace industry, because of its low production quantities and low-rate production, has been relatively untouched by automation except in the areas of complex machined parts and in the assembly of some types of electronic devices, when the needs for precision and accuracy justify the cost and time of automation. However, the production quantities of some of the more complex weapons are now adequate to consider automation, especially where a number of individual parts are identical on each final assembly. Examples familiar to the author are the Dragon rocket motors and firing circuit boards, all of which have multiple uses on each Dragon missile, designed and built by McDonnell Douglas.

Semiautomated operations are those in which the worker plays a substantial role in the activity. The worker's role exceeds that of supplying the automated equipment with parts or materials, or removing finished parts from the work area. To date the preponderance of factory operations have actually been semiautomated rather than fully automated, because the worker–machine combination is often the most efficient and effective in involved tasks.

21.1 ASSEMBLY MACHINES IN THE FACTORY

When they see the term *automated factory,* many people think of a Japanese factory containing hundreds of robots. But Japan is not alone in this field; many U.S. corporations are making similarly impressive strides in boosting factory productivity and efficiency. And robotics, while critical to some manufacturers, represents only a small part of the overall automation scene. Of major importance are the computer and communications networks that manage and analyze the information relevant to factory operation. In addition, every manufacturing facility has its own unique needs. The fabrication and assembly equipment and the processes that can serve them best can vary considerably.

Although lower product unit costs (and increased overall company earnings) are the prime drivers in selecting an automated process, the manufacturing process may be automated in areas where the work is dangerous or monotonous for people, or where machines can substantially improve the quality of the products or increase the production rate. Moreover, the products themselves should be designed to facilitate automated manufacturing, and changes in designs should be easily communicated to the shop floor. Many existing designs cannot be effectively automated if they were designed for manual assembly. (See Chapters 1 and 2 for more discussion.)

Equipment for the "factory of the future" will reduce the need for some of the skilled trades as we know them today. The motive on our part will not be to cope with the shortage of skilled trades, although that already exists in some companies and in some geographical areas. It will not be to solve the problem of getting young individuals to go into relatively low-paying jobs in manufacturing, when they can start in other areas for more money. The motive will be the old one: productivity and bottom-line cost.

Automated technology can achieve greater productivity, better quality, reduced costs and increased profits—if it is properly applied. However, machinery installations or business systems, may operate as designed but may not solve the ultimate problem they were created to overcome (see Subchapter 21.3).

Small details are the stumbling blocks to efficient automatic processing. This follows from the need for precise control of the movement of the various parts going into the assembly. There is no single, simple technique for obtaining control.

21.2 BASIC AUTOMATION CONCEPTS

One key rule is that part design must be compatible with the needs of automatic feeding. Parts are usually introduced to an assembly machine as bulk components. They are placed in a hopper and tracked to a loading station. Whether the hopper is vibratory, rotary, or oscillatory, it relies on gravity and/or friction for part movement. Some sort of gating or orientation device allows only those parts in the proper attitude or position to enter the track (see Subchapter 21.8).

Efficiency is affected by the system used; nonvibratory feeders, for example, are limited in the number of orientations they can perform. Part geometry is a critical factor. Soft parts may tangle in the hopper. Bowl driving forces may distort parts to the point where orienting in the bowl is impossible. Distortion of parts due to stocking and handling can also cause serious difficulties.

Parting line flash from cast and molded parts is an example of a problem that never appears on a part drawing. The sensitivity of the part to moisture, static electricity, and residual magnetism may not become apparent until mechanical handling is attempted. Sometimes a particular surface may be declared critical and must be protected for subsequent operations. This situation may preclude or restrict use of some of the automatic feeding methods.

Incomplete molded or broken parts reduce feeding efficiency. Foreign material in the hopper is certain to alter the level of performance. This includes not only the presence of material due to machine environment, but contamination of the parts from previous processes.

Turning to the main part, or body, that will receive the oriented parts, several questions arise. For example, the body must be strong enough to withstand assembly tooling forces. Typically, forces for assembly are not as high as for machining, but pressing, sizing, or machining operations might be required on the assembly machine. High production rates may increase these forces.

The main part must be placed accurately on the assembly machine. This calls for fixturing, flat locating surfaces, precision locating holes, etc. Parts are most efficiently placed in the main body with simple, short, straight-line movement. Grasping a part may be necessary. This requires some type of actuating force with its attendant timing and control elements. If spearing or grasping the part is impractical, a vacuum force may be used in transferring parts. Magnetic force is used occasionally in handling parts, but this is not recommended, because the attraction of metal chips and dirt degrades equipment performance rapidly. Also, residual magnetism often cannot be tolerated in the final assembly. The author experienced problems with the Dragon rocket motor assembly machine due to the cast retainers occasionally becoming magnetized due to improper heat treatment of the stainless steel. We had to de-gauss them all to achieve good hopper feed.

21.3 TYPES OF AUTOMATED ASSEMBLY MACHINES

21.3.1 Standard Machine Bases

There are essentially four types of standard base assembly machines:

Dial indexing machines
In-line machines
Floating work platform machines
Continuous motion machines

To meet specific assembly automation needs, we add custom tooling. For maximum cost effectiveness, we can choose from a stock of standard operating stations such as those for feeding, orienting, inspecting, and acceptance/rejection testing. Standard base machines and operating stations are outgrowths of the industry's experience in designing, building, and using machines for high-volume production programs.

Dial Indexing Machines

The dial indexing automated assembly machines incorporate a mechanical drive that rotates a circular dial table, or base plate, and indexes with a positive cam action. A circular, nonrotating table simultaneously raises and lowers a reciprocating upper tooling plate, usually mounted in the center of the larger rotating base plate. Assembly nests are installed around the outer edge of the dial table. Parts feeding, assembly, and inspection stations are installed around or above the assembly nests or on the upper tooling plate.

These machines offer the following advantages:

Greater machine accessibility and minimum floor space. The basic circular layout of dial-type machines is inherently more compact. High machine accessibility increases operator efficiency and simplifies maintenance.

Greater adaptability to a variety of operations. The dial types of automated assembly machines, containing central indexing mechanisms and reciprocating tooling plates, offer simplified rotary and up-and-down tooling motions for high adaptability to many automated assembly operations.

Figure 21.1 shows an elementary rotary indexing machine. The dial type described above has the added feature of a reciprocating tool table mounted above the indexing table.

In-Line Machine

In-line automated assembly machines feature a rectangular chassis housing an indexing mechanism driving an endless transfer chain. Nests that hold and transport the product during the various assembly operations are fastened to the transfer mechanism. The parts feeders, workstations, and inspection stations are then arranged along the work flow. Parts are fed into the assembly nests as required, and work and inspection operations are performed in sequence along the length of the machine until the product is completed.

These machines have the following advantages:

Unlimited number of workstations.

Efficient operator loading. The rectangular configurations permit machines to be placed side by side with an aisle in between. The operator can efficiently monitor all stations from the central aisle.

Work can be performed from two or three directions simultaneously.

Figure 21.2 shows an in-line indexing machine.

Floating Work Platform Machines

In floating work platform machines, parts flow into a manifold where they are located, assembled, and inspected. This system uses divergent flow channels for tandem and/or parallel operations to achieve line balancing and consists of two major elements—a parts transporting element and a modular assembly element.

The parts transporting element moves the floating work platforms sequentially to the various modular assembly elements. Each modular assembly element consists of an independently powered unit containing one or more workstations. Use of a simple transporting band permits flexibility within the system. Modular assembly elements can

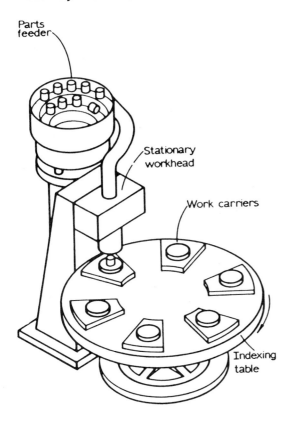

Parts feeder

Stationary workhead

Work carriers

Indexing table

Figure 21.1 Rotary indexing machine. (From Geoffery Boothroyd, *Assembly Automation and Product Design*, Marcel Dekker, New York, 1992.)

be placed in remote areas such as cubicles, barricaded hazard rooms, holding or curing rooms, or storage banks and can be returned to the main system by the parts transportation element for further processing.

Continuous band motion also permits the routing of parts onto a constant-motion machine for nonstop performance, as well as for routing to in-line or dial-type machines.

The quality of each assembly may be verified by inspection probes placed in tandem following the work performance at each of the workstations. Stations may be used for assembly function testing.

This type of machine has the following advantages:

Banks of parts may be accumulated between the workstations to cope with short station stoppages.
Work can be removed from the system, performed at a hand station, and returned to the machine.

Continuous-Motion Machines

Continuous-motion automated assembly machines provide for nonstop performance of operations. Such systems may be capable of up to 1200 assembly operations per minute. Parts are swept from a conveyor belt, oriented, and fed into the machine. Following

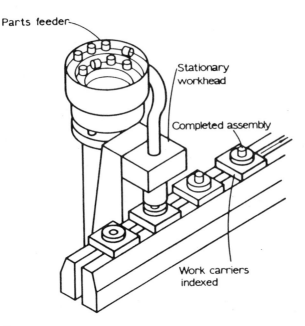

Parts feeder

Stationary
workhead

Completed assembly

Work carriers
indexed

Figure 21.2 In-line indexing machine. (From Geoffery Boothroyd, *Assembly Automation and Product Design*, Marcel Dekker, New York, 1992.)

assembly, inspection, and function testing, the assemblies are oriented and returned to the conveyor belt.

This type of machine has the advantage of higher production rates that can be achieved with other types of bases. The free-transfer machine shown in Figure 21.3 shows a "buffer" position between the two workheads. The buffer parts could be shuttled off-line to another work position, and then back to the main feed line, in either system.

In general, the higher the production rate, the lower the per-unit cost of a product. Assembly automation systems are designed to fit production rates to specific needs. Good machine design considers more than the production rate; it also considers the overall production capability, including such factors as minimum maintenance, system efficiency over years of continuous operation, minimum training required for operators, and production of a consistently high-quality product. Flexibility is important in every assembly automation system. Modularized workstations, idle stations, and standardized motions make systems adaptable to product changes with minimum downtime.

21.3.2 Robots

A *robot* can be broadly defined as a machine which copies the function of a human being in one respect or another. Industrial robots are generally equipment with a single "arm," and they are used to perform assembly-line operations and other repetitive tasks such as feeding parts into another machine.

The assembly machines we described in the previous paragraphs are designed to handle large numbers of standard workpieces. These devices could be regarded as the forerunners of the modern industrial robot, although, unlike most robots, they are controlled by the machine to which they are attached and may not be readily used to

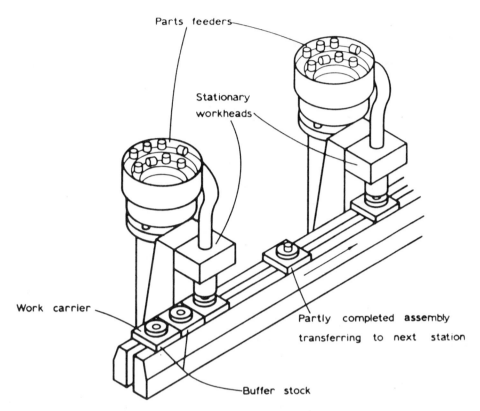

Figure 21.3 In-line free-transfer machine. (From Geoffery Boothroyd, *Assembly Automation and Product Design*, Marcel Dekker, New York, 1992.)

perform another function. The cost of reconfiguring the grippers or end effectors plus the reprogramming cost may make the robot less "flexible" in assignment than is sometimes imagined.

The real differences may be in the "eye of the beholder"—or in the mind of the reporter or author. The term "robot" comes from the Czech work *robota*, meaning work, and was first used in a play called "Rossum's Universal Robots" written in 1920 by the Czech author Karel Capec. I personally find that the term "robot" to be rather imprecise, and one that should normally be replaced by a more descriptive term. Perhaps the best definition is by example. A robot consists of three basic assemblies:

1. Motion system
2. Controller system
3. Heads and work tools

21.4 MOTION SYSTEMS

The motion systems start with the basic two-axis linear-rotary models as sketched in Figure 21.4. These high-speed, servo-controlled robot systems are designed for a variety of material handling and pick-and-place operations. The robot consists of a high-speed linear table, rotational table, arm, and work tool to provide high-speed positioning over a large

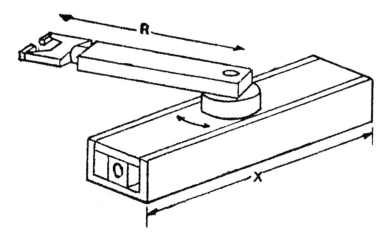

Figure 21.4 Two-axis motion system (X, R).

work area. An optional Z-axis could be added (between the arm and θ axis) to permit raising or lowering parts from one height to another.

Providing three axes of motion (X, Y, Z) the robots shown in Figure 21.5 can perform various functions within a large volume. High positional accuracy of the X–Y–Z tables make this robot ideally suited for intricate tasks, such as stuffing components onto PC boards. The Z-axis capability permits components to be selected and positioned at various vertical levels.

The basic X–Y–Z rotary model is shown in Figure 21.6. These robots, featuring X–Y–Z–θ motion plus a special work turret, offer exceptional versatility for robotic applications. The X–Y–Z–θ motion permits work to be performed on a three-dimensional workpiece from the top as well as two sides. The turret may contain an electro-optical sensor for position location and auto-centering. For performing multiple tasks, other turret positions may include a variety of tools, such as a screwdriver, Allen wrench, and grippers.

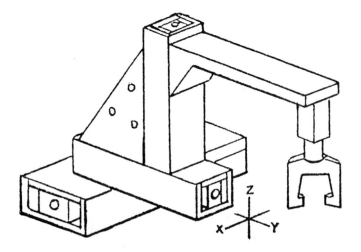

Figure 21.5 Three-axis motion system (X, Y, Z).

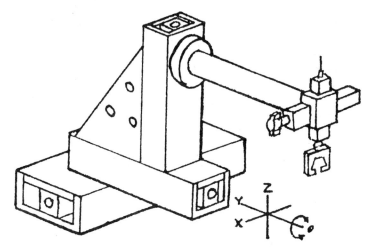

Figure 21.6   Four-axis motion system *(X, Y, Z, θ)*.

Extended-work-area robots, featuring five axes of motion $(X, Y, Z, \theta_1, \theta_2)$, provide ultimate flexibility in servicing a 360° around the robot in a large volume. When employing the same turret as the four-axis models, these robots can be programmed to provide a single robot assembly station with multiple task capability. The robot may also be placed to service an assembly line on either side of it. See Figure 21.7 for a schematic.

Robots (and assembly machines) today are normally designed by assembling standard components and assemblies to fit specific needs. The *X–Y* tables can position parts to ±0.001 in., at speeds as low as 0.001 in./s or as high as 60. in./s with accelerations to 4g.

The controls may be a programmable positioning controller consisting of a microprocessor-based CNC system with built-in hardware calculators. This permits simple manual programming via a keyboard. Program entry may also be from floppy disk, or direct from

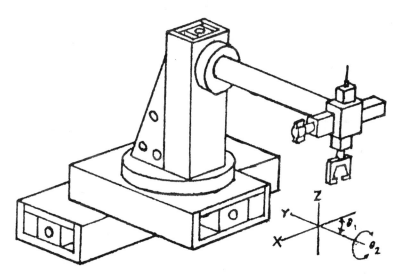

Figure 21.7   Five-axis motion system $(X, Y, Z, \theta_1, \theta_2)$.

a central or local area network (LAN) computer. Simultaneous interpolation up to six axes linear and three axes circular are becoming standard. The controller will accept positional feedback from optical encoders, interferometers, inductsyns and resolvers. Transducer outputs, such as pressure, torque, and power levels, may also be input. Other input/output functions available are usable in tuning operations, such as tuning a coil to its peak, or adjusting a potentiometer to a desired output level. Here the work tool motion is controlled by a feedback signal to perform the adjustment to the desired device output signal. Most robots today also have a "teach pendant," which can be moved manually through the desired positions to program the controller or computer. The program generated by a teach pendant can then be printed out, and minor reprogramming steps can be added to "smooth" the motions.

Figure 21.8 shows an unmanned robotic cell in a flexible manufacturing system, built around robots that handle parts and material, perform assembly tasks, operate tools, and perform other manufacturing operations. This cell is built around a Cincinnati Milicron T^3 material handling robot. The robot controller, the material handling system and its controller, the fine-resolution system and its controller, and the automatic riveter and its controller receive signals through a multiplexer from a central DEC PDP-11/34 computer.

In designing a robotic system for manufacturing, the flexible system concept described above should be the base on which the largely unmanned robotic system should be built. Figure 21.9 shows the floor layout for an application developed by the National Institute of Standards and Technology in Gaithersburg, Maryland. A deburring cell has been

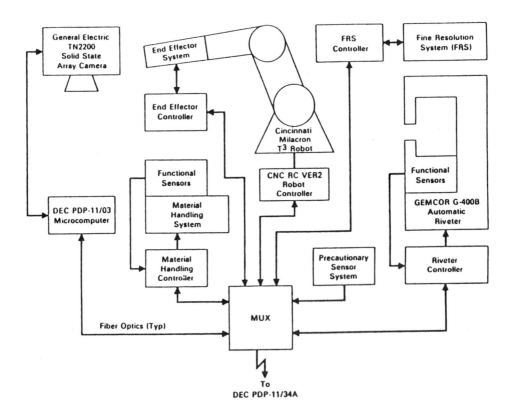

Figure 21.8   Manufacturing cell schematic. (Courtesy McDonnell Douglas Corp.)

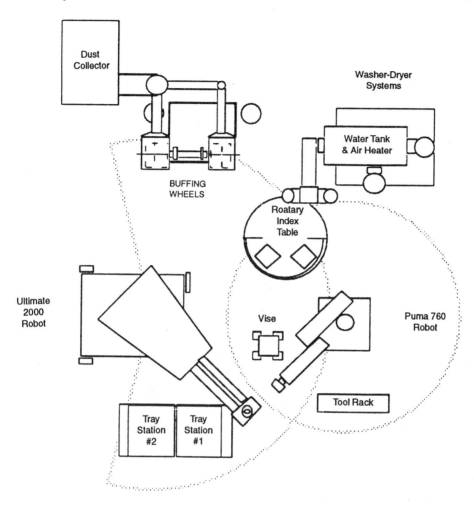

Figure 21.9 Cleaning and deburring workstation utilizing robots. (Courtesy U.S. Department of Commerce.)

developed which demonstrates the deburring of parts based on their description and a graphically developed process plan. This cell, a cleaning and deburring workstation, consists of a workstation controller, two robots, various quick-change deburring tools, a rotary vise for part fixturing, and a part transfer station.

The working head may include a turret, similar to a lathe. The variety of mechanical grippers, vacuum grippers, screwdrivers, nut runners, etc., is almost endless. The type most similar to a human hand, with multiple linkages and force feedback servos, is often pictured in the press as "standard," but in fact it is very expensive and more commonly seen as a laboratory experiment than a factory working tool.

21.4.1 Inspection and Testing During Automated Assembly

Assurance that parts are within design tolerance before they are assembled, that parts are assembled correctly, and that the product works reliably, help make assembly automation less expensive and more reliable than hand assembly. The goal is to build quality into the product, rather than just inspecting defects out. Experience has shown that product quality improves through automation. Not only does the per-unit cost decrease, the percentage of defective units also decreases with automation, reducing total costs even more. Generally, a machine will turn out a good part every time—or reject a faulty assembly—or stop!

A machine can include unique memory systems for product quality inspection and function testing. The assembly automation system knows design tolerances and assembly processes. It can make certain that the product works the way it was designed, by rejecting defective units.

A typical assembly automation system with inspection capability performs the following operations:

It checks the presence and position of internal parts in an assembly.

Defective units are automatically rejected in salvable form.

Probe stations sense the condition of the product at critical stages in the assembly process. If required, component dimensions can be checked 100% immediately prior to or following assembly.

In the event of missing, surplus, or improperly oriented components or assemblies, the machine will either reject the defective part for salvage or stop until it receives corrective action by the operator. A light on the control panel indicates the nest location of any problems.

The machines are also capable of testing products as a final acceptance/rejection decision. A simple example, which was installed at McDonnell Douglas by the author, torqued two parts together and then measured the total height of the finished assembly. This was a difficult design to work with, since the correct torque and the correct height were both considered critical to the performance of the product. Had redesign been possible at the time, one of the criteria should have been dominant. One of the most complex functional test examples was observed at the Austin, Texas, plant of IBM, where laptop computers are assembled. The completed computers are transferred to the test area via a powered transfer line which has "slack" built into the system to act as a buffer prior to test. The following operations are then performed:

Open the computer lid (containing the screen).

Remove the cardboard "floppy disk" which is put in place for protection during shipping.

Pick up and insert a test diskette, and close the top.

Plug a test cable into the back of the machine.

Transfer the unit to the powered coveyor that carries the computer through the test process.

The computer then proceeds through a complex multilevel module which runs a continuous functional test while applying heat and vibration for 24 hours.

As the computer leaves the test module, it removes the the test diskette, reinserts the cardboard shipping diskette, unplugs the test cable, and closes the top.

If the test shows an anomaly, a description of the test failure is printed out, and also the instructions for which component should be repaired or replaced. The failed computer is then shunted to a rework and repair station, where a mechanic accomplishes the required repairs and reenters it into the test cycle.

The "good" computers (which are nearly all of them) continue on down the line for packaging, etc.

21.4.2 Machine/Operator Relationships

Because the repetitive assembly operations are performed automatically, operator fatigue errors are practically nonexistent. The assembly automation system monitors its own function through its memory system, so that personnel can be quickly and easily trained as operators. Regardless of the operator's experience, system efficiency and product quality remain at a high level.

Built-in sensors detect and forewarn the operator to replenish the supply of parts at each of the automated feeding stations, thus assuring uninterrupted operation.

Techniques have been developed to assure safe machine operation and provide operator protection. These techniques, including shields, barricades, grounding conductors, etc., protect against personnel hazards and machine damage.

21.5 JUSTIFYING AUTOMATION

Calculating automation implementation payback is a subject that continues to generate controversy—and, more often than not, confusion—throughout the manufacturing world. Moreover, the issue of "How is management going to justify the vast expenditure of automation against the total needs of the business?" is still one of the hottest around.

The National Association of Electrical Manufacturers (NEMA) recently reported the results of a survey on automation. Financial payback was classified as very important by 91% of the respondents. What's more, 78% of another survey felt that "Most businesses in the United States will remain so tied to traditional investment criteria that they will be unable to realistically evaluate the potential of computer-aided manufacturing options." In other words, businesspeople are using formulas which include only traditional benefits to run the numbers, and as a result they are procrastinating—they are not putting these technologies to work.

In the automation systems business we have been developing new benefit considerations—new manufacturing economics—that go beyond the limited return-on-investment evaluations of the past to take into account the total impact of automation on the entire business. This is because automation technologies do not change the factory floor; they change the entire business in new and important ways.

The point is that business people should be skeptical of conventional wisdom in this area, because old guidelines often do not work today. Automation investments represent strategic decisions with far-reaching competitive implications, and should be evaluated as such. However, there still are financial calculations that can be made to validate the appropriateness of the investment.

So far we have talked about financial justification with words such as traditional, limited, and conventional. It's interesting that under the "enlightened" concept of financial justification—which is outlined below—we are learning there really is no "black magic" involved—no "voodoo" economics!

In fact, we still use many of the evaluation techniques that have been around for years. We still try to estimate a net present value and a payback period—these are basic decision criteria. In addition, discounted rate of return—some people call it internal rate of return—is still an important measurement. The only difference now is that we have to expand our horizons and think more globally. This means that we must project and incorporate, in our calculations, benefits that accrue business-wide—benefits that previously would not be considered at all. It also means that the long-term, macro view is the field of measurement, taking into account the benefits of automation over the long haul.

21.5.1 Benefits

Let's talk first about business-wide benefits. Up to now, whenever we evaluated the advantages of a plant or equipment investment, we identified potential savings in such terms as lower direct material costs, direct labor reductions, or improved individual machine utilization. Now, even though these benefits can—and still do—come from automation investments, we find that they pale by comparison to savings being generated in other areas. For example, we can substantially reduce indirect manufacturing expenses for such things as fork trucks, fork truck drivers, material handlers, and employees in packing and shipping. Further, automation frequently reduces the number of machines required in a factory, and a direct result of this is a reduction in maintenance and perishable tooling costs. At the McDonnell Douglas Missile Assembly Plant, we saw a reduction in the overhead areas of operating supplies and hazardous waste disposal by reducing the "touch labor" in the SMAW missile encasement department with the implementation of our semiautomated braiding system. This is real savings. (See example in Chapter 18.)

Fewer machines also means that less floor space is required. This, in turn, drives down such related expenses as heat, light, taxes, and insurance.

One of the strongest leverage factors we count on today is quality costs. The positive impact that automation can have on quality will most assuredly affect the costs associated with scrap, rework, warranty claims, quality inspectors, and after-sales service. The "ounce of prevention" that improved quality builds into our products can therefore lead to across-the-board reductions in these areas—well worth many "pounds of cure."

Up to now, we have been talking about lowering costs and expenses that appear on the operating statement. It is also important to *leverage* the balance sheet. One of the largest single items on any balance sheet is inventory. Therefore, to the extent that inventory can be reduced, cash is made available for other purposes.

We have learned that automation in general—and computer-integrated manufacturing systems in particular—reduce plantwide manufacturing cycles. These reductions usually free up large amounts of cash — cash that previously was tied up in inventory or work in process. This, in effect, becomes the "down payment" for automation.

The payoffs we have talked about so far are quite tangible and easily measured. We call these the "ripple effects" of automation. Ripple effects are "bonus points," if you will: effects such as higher productivity from increased worker satisfaction, higher market share from improved quality, higher market share from faster response time to shifts in marketplace demand. While these advantages are difficult to measure, they are even more difficult to forecast.

All these benefits are more than theoretical: They are empirical. They are results that really can be and have been achieved by many companies. GE experienced this initially at one of its own facilities, the Locomotive Products Division in Erie, Pennsylvania. GE implemented a large, multiphased project over several years which covered a broad spectrum of automation technologies. The benefits from this project were realized across several areas. Impressive savings were generated in direct material and direct labor costs, while equally important results were obtained in increased capacity, higher product quality, and reduced cash lockup in inventory. These benefits not only leveraged earnings, but established the foundation for survival in tougher times—the smaller, more competitive world market the locomotive division is dealing with today.

Let's explore some of the disciplines for using this information in the justification process. Before we start, a rule: We cannot demand large short-term paybacks. The payback may come in Phase 3 of a a three-phase project. History has proven that synergy really does come into play here: When all the phases are complete, the total system benefit is larger than the sum of its individual parts.

The payback computation requires two sets of data. These take the form of a 5- or 10-year pro-forma cash flow for the business. The pro-forma should be compiled assuming: (1) the automation alternative is undertaken, and (2) a "do nothing" or "as is" scenario, which requires a candid answer to a very tough question: "What will happen to my competitive position if we don't automate?" Often, the truth will not reflect the status quo. Instead, it may include a possible erosion of market share because of the likelihood that one's competitors will not stand still. In this case, the competitor's lower cost structure and higher quality product will allow it to gain market share.

Once the cash flow has been determined for these two alternative scenarios, the cash flow difference between the two should be used in the automation payoff calculations. For example, a corporation is evaluating whether or not to proceed with a $3 million automation project (see Table 21.1). For starters, we have assumed that the status quo will prevail if the project is not adopted. However, if the project is undertaken, the income will increase $200,000 each year—and cash flow will increase $2 million in the first year because of inventory reduction, and half a million each year thereafter. The key now is to plug the difference in cash flow into the financial calculations [see Table 21.1 (a)].

The 10-year rate of return on the investment is 24%. If we had mistakenly considered only the operating cost efficiencies and overlooked the inventory reduction, the return would have come out a meager 2%—with the result being project disapproval.

Consider another case. Let's assume a 5% erosion of market share in a $100 million market if the project is not undertaken. The resulting decrease in sales, net income, and cash flow now magnifies the benefit of automation and generates a discounted rate of return of 62% (see Table 21.1). So the lesson to be learned is that inventory reductions and share gains or losses usually have dramatic leverage on the payoff to be earned [see Table 21.1 (b)].

With regard to the "ripple effects" we discussed a above, we should deal with these "softer" measurements by means of a *sensitivity analisis*. Returning to our example, let's change it slightly by assuming no $2 million inventory reduction. The discounted rate of return, falls to 11% [see Table 21.1 (c)]. Suppose management has set a hurdle rate of 20%. That means the discounted rate of return must be at 20% before the project can be approved. We can easily determine that an additional $200,000 of cash would be needed each year if we wanted a 20% return—and this, in turn, translates into one additional point of market share. Management can then make a "gut feel" decision as to whether or not this gain in share is feasible—and whether or not to proceed A word of caution—do not let undisciplined enthusiasm influence good business decisions.

Table 21.1 Ten-Year Pro-Forma Cash Flow Analysis of Operating Results of a Product Line with and without a $3 Million Investment in Automation. (a) Considers traditional and nontraditional benefits, with no loss of market share without automation. (b) Considers traditional and nontraditional benefits, and assumes a 5% loss of market share without automation. (c) Considers traditional benefits only, and no loss in market share without automation.

Ten Year Pro Forma	
Operating Results (Millions)	
No Automation	$3.0 Automation
Sales/yr. $25.0	$25.0
Net Income/yr. $ 2.8	$ 3.0
Cash Flow/yr. $ 3.0	$ 5.0 (1st yr.)
	$ 3.5 (yr. 2–10)

(a)

TRADITIONAL BENEFITS

Direct Material
Direct Labor
Machine Utilization

NONTRADITIONAL BENEFITS

Less Inventory
Indirect Cost Reductions
Fewer Machines
Less Floor Space
Lower Costs from Higher Quality

Ten Year Pro Forma	
Operating Results (Millions)	
No Automation	$3.0 Automation
Sales/yr. $20.0	$25.0
Net Income/yr. $ 2.0	$ 3.0
Cash Flow/yr. $ 3.0	$ 5.0 (1st yr.)
	$ 3.5 (yr. 2–10)

(Project DCRR 62%)
(b)

Sensitivity Analysis	
Operating Results (Millions)	
No Automation	$3.0 Automation
Sales/yr. $25.0	$25.0
Net Income/yr. $ 2.8	$ 3.0
Cash Flow/yr. $ 3.0	$ 3.5

(Project DCRR 11%)
(c)

21.6 SOFTWARE INTERFACES TO ASSEMBLY AUTOMATION

Manufacturing automation needs to solve many technical problems before it is accepted by the majority of manufacturers. The federal government has put our tax dollars to work on clearing these barriers.

Chief among the efforts at the Automated Manufacturing Research Facilities (AMRF) is an attempt to link disparate computers, robots, and machine tools in a seamless network. At stake, says Howard M. Bloom, chief of the Factory Automation Systems Division, which contributes personnel to the AMRF, is the acceptance of computer-integrated manufacturing (CIM). "What we do here will have a dramatic effect on how manufacturers choose systems and modernize their factories," he says.

If CIM made all the elements in a production facility, in effect, "plug compatible," even the smallest factories and machine shops in the United States could build the factory of the future little by little. As yet, only a handful of large American corporations with substantial resources have started automating their production facilities. Studies commissioned by the National Bureau of Standards (NBS), where AMRF is housed, have shown that about 90% of the discrete-parts producers, those who make parts in small batches, are organizations with less than 50 employees. At the same time, they are responsible for about three-fourths of the trade in manufactured goods, and many of them subcontract from major corporations.

You can buy a computer from IBM and another vendor's operating system because IBM has provided specifications for interfacing. You can also buy languages from different vendors because well-defined interfaces allow compilers to match the operating system. The situation is different with factory systems. You cannot readily buy software or a factory system with well-defined interfaces. Even when you can, the interface may be a unique, internal interface that will not allow you to connect equipment from another vendor. To modernize factories today, you often replace everything. Moreover, factory interfaces call for a lot of wiring, and interfaces consume 30–50% of expenditures for automation.

AMRF is developing software that will "translate" data so it can be understood by robots, machine tools, and sensors anywhere in the facility. To ensure practical results, AMRF has chosen to work with many commonly available commercial computers and software packages. These include nine kinds of processors: the IBM 4341; Digital Equipment Corp.'s VAX 11/780; Hewlett-Packard Co.'s 9000/9920/9836; Sun Microsystems's line of Sun workstations; the 4404 AI Workstation from Tektronix; Symbolics LISP Machine; and Iris from Silicon Graphics. They also include nine computer languages in various versions: C, Pascal, LISP, Prolog, Fortran, BASIC, Assembly, Forth, and Praxis.

To control the machine tools, robots, sensors, and other equipment, AMRF personnel have designed control system software that runs on IBM PC-compatible micros and micros based on Motorola 68000 microprocessors. For robot control, the software uses Forth and LISP, because they are transportable, easy to use, and efficient as interpreters and operating languages, says Bloom. C is used for programming the Integrated Manufacturing Database Administration System (IMDAS), which AMRF developed.

21.7 DESIGN FOR AUTOMATED PRODUCTION

The human operator's hands have infinite control and flexibility to compensate for the many variations encountered in manufacturing processes. The automatic assembly machine can produce quality and quantity of parts only under strict, inflexible parameters. The ability to adapt automation to fabrication or assembly depends largely on the detail design of the individual elements and the complexity—or lack of it—in their assembly into the end product. The objective of this subchapter is to present some of the basic principles of piece-part design that will enable products to be assembled automatically with a minimum of difficulty.

Parts are usually introduced to the assembly machine as bulk components. They are placed in a hopper and tracked to a loading station. Whether the hopper is vibratory, rotary, or oscillatory, it relies on gravity and/or friction for part movement. Some sort of gating or orientation device allows only those parts in the proper attitude or position to enter the track.

Efficiency is affected by the system used; nonvibratory feeders, for example, are limited in the number of orientations they can perform. Part geometry is a critical factor.

Soft parts may tangle in the hopper. Bowl driving forces may distort parts to the point where orienting in the bowl is impossible. Distortion of parts due to stacking and handling can also cause serious difficulty. Parting line flash from cast and molded parts is an example of a problem that never appears on a print. The sensitivity of the part to moisture, static electricity, and residual magnetism may not become apparent until mechanical handling is attempted. Sometimes a particular surface may be declared critical and must be protected for subsequent operations. This situation may preclude or restrict use of automatic feeding methods. Use of symmetry in part design makes orientation either unnecessary (ball bearings, etc.) or very simple (plain rods, disks, etc.). Asymmetrical parts can sometimes be made symmetrical, and the added manufacturing cost is often insignificant when compared to orientation and sensing costs involved during assembly (Figure 21.10).

Parts can be designed with distinct polar properties. Asymmetrical parts can be oriented by geometry and/or weight. Obvious orientation features are exploited in vibratory feeders, as shown in Figure 21.11. The example shows drilling a hole in one end of the part to make it off-balance to aid in orientation in the feeder, and the addition of a flat to aid in sensing the hole position. Proper application of gravitational and frictional forces will move parts from the bowl onto the rising spiral track in random orientation. Improperly oriented parts passing sensing stations on the track are forced off the track and back into the feeder bowl, leaving only parts in the desired orientation moving into the pickup station.

Vibrating action in the feeder provides the forces needed to orient the piece parts. Track design features which detect shape or weight differences in parts to be rejected from the track are shown in Figure 21.12. If rejected by gravity, they fall off the track because of the weight distribution. If rejected by shape, removal is accomplished with cutouts in the track cams along the track surface, or sensing feeders. Parts are then rejected by air jets or mechanical actions.

The following techniques simplify, improve, and in some cases make feeder orientation possible.

Minimize the number of different orientations. Vibratory feed rates depend on successful orientation. Sensing features may be repeated on a part to improve the probability of correct orientation (Figure 21.13). For instance, if the track feed rate is 600 pieces/min and the orientation probability is only 1 in 10, we are effectively limited to 60 pieces/min for

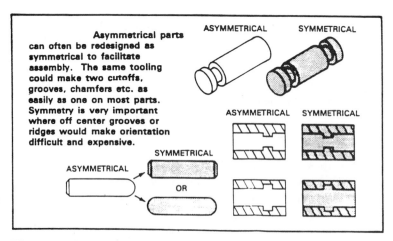

Figure 21.10 Examples of symmetrical versus asymmetrical parts.

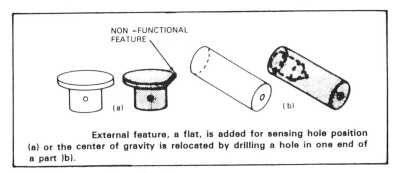

Figure 21.11 Addition of nonfunctional features to aid in feeding.

assembly. The example part, a cube, is modified to increase orienting probability from 1 in 24 to 1 in 6, which will increase feeder output by a factor of 4.

Avoid parts that tangle. Springs are probably the best example. For instance, a simple open-coil spring may be redesigned with close-wound coils at both ends and in the center. This design often permits handling by conventional hoppers and feed systems. Springs may also be fabricated in the automatic assembly machine to avoid this problem. Slotted parts may also be redesigned to minimize tangling, as shown in Figure 21.14.

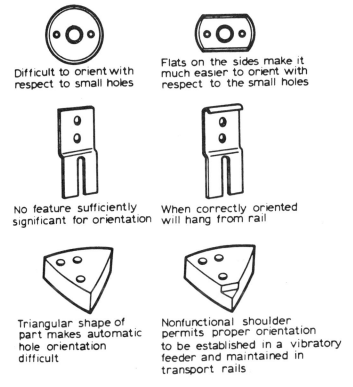

Figure 21.12 Examples of features that aid in feeding. (From Boothroyd, *Assembly Automation and Product Design,* Marcel Dekker, NY, 1992.)

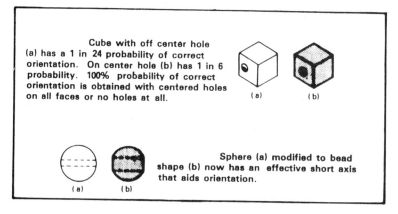

Cube with off center hole (a) has a 1 in 24 probability of correct orientation. On center hole (b) has 1 in 6 probability. 100% probability of correct orientation is obtained with centered holes on all faces or no holes at all.

(a)   (b)

(a)   (b)

Sphere (a) modified to bead shape (b) now has an effective short axis that aids orientation.

Figure 21.13   Additional features that aid in part orientation.

Avoid using parts that nest, shingle, or climb. Matching tapered surfaces will cause parts to lock or nest. Internal ribs, projections, or increasing the angle well above the locking angle will avoid the problem. Avoid shingling/climbing parts. Parts clinging onto other parts will often cause jamming in feeder tracks. If anticipated, this potential malfunction in the feeder can be avoided by redesign. Parts tend to climb over thin or shallow beveled contact surfaces. See Figure 21.15.

Determine critical tolerances. It is essential to determine dimensions that are critical for proper parts feeding and yet may be noncritical for piece-part function if the parts feeding operation is to remain trouble-free.

Assume for now that parts are oriented and coming down the feeder track; the job is now half done. There are, however, additional important design considerations if parts are to be successfully integrated in a trouble-free assembly operation.

Design parts for easy assembly. Design parts with guides for proper location during assembly. Clearances should be carefully considered, and the liberal use of chamfers and radii (Figure 21.16), will allow the parts to fall together in place unassisted. Sufficient clearance should be allowed for feeder fingers or other tooling needed to handle the parts.

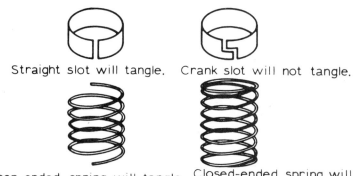

Straight slot will tangle.   Crank slot will not tangle.

Open-ended spring will tangle.   Closed-ended spring will tangle only under pressure.

Figure 21.14   Examples of part tangling. (From Tanner, *Manufacturing Engineering,* Marcel Dekker, NY, 1991.)

difficult to feed - parts overlap

easy to feed

Figure 21.15 Examples of parts climbing or shingling. (From Boothroyd, *Assembly Automation and Product Design*, Marcel Dekker, NY, 1992.)

We have not yet reached a point where we need a decision process for "automate or conventional build," comparable to "make or buy," but it is coming. Meanwhile, whenever the designer encounters a part or an assembly whose repetitive requirements to make or assemble exceeds 200,000 (project plan quantity times items per end product), he or she should immediately confer with manufacturing engineering to determine any anticipated plans to automate. When the quantity exceeds 1 million, it is almost certain that automation must be used.

Data and books on automation are available in manufacturing and elsewhere. These are illustrated and should be reviewed and studied by designers who encounter parts and assemblies which can or should be automated.

When parts pose unusually difficult handling and orientation problems, experimental mockup stations should be built to verify parts handling processes prior to building the assembly machine. The product designer, the manufacturing engineer, and the tool designer should work as a team through this development state.

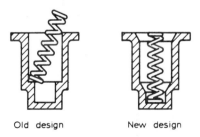

Old design   New design

Figure 21.16 Design parts for ease of assembly. (From Boothroyd, *Assembly Automation and Product Design*, Marcel Dekker, NY, 1992.)

While each part or assembly presents a special problem in automation, some general recommendations and requirements apply.

1. Parts should have at least one axis of symmetry, with ability to fabricate, feed, and assembly in respect to the axis. A sphere is, of course, ideal.
2. Parts should be suitable for sorting and alignment for feeding on equipment such as Syntron or Vibron unit.
3. Dimensioning should be such that inspection gauging can occur from one reference point or axis and one plane of reference.
4. Parts being assembled automatically must be designed to engage with or snap on to mating parts so that they will not shake loose during subsequent assembly sequences. It is impractical to balance parts on one another when using automated equipment.
5. Units being torqued together should have a shaped region such as a square or hexagon to permit grasping and torquing.
6. Messy materials, such as lubricants and adhesives, create machine maintenance and clean-up problems unless the parts receiving them are designed to accept leak-free feed orifices with "no drip," point-of-application cutoff.
7. Semiautomated operations should always be considered whenever the discriminatory capabilities of a human being will result in a significant simplification of the automation equipment.
8. In designing for automatic assembly, the designer should ask the following questions:

    Can parts be made symmetrical to avoid orientation problems?

    Do symmetrical parts have clearly defined polarity features?

    Are the number of significant part orientations minimized?

    Will parts tangle?

    Will parts nest or interlock, thereby causing problems?

    Will the part design cause shingling?

    Are critical dimensions and tolerances clearly defined?

    Do the parts lend themselves to easy location and assembly?

    Does the design have a datum surface for accurate parts location during assembly?

    Have all unnecessary handling requirements, such as turning over parts and/or assemblies, been avoided?

    Does the assembly have components that are buried or difficult to reach and position?

    Has the design been simplified and standardized as much as possible?

    Have excessive burrs and flash been eliminated?

    Can difficult-to-handle parts be assembled in an automatic system?

21.7.1 Machine/Operator Relationships

Because the repetitive assembly operations are performed automatically, operator fatigue errors are practically nonexistent. The assembly automation system monitors its own function through its memory system, so personnel can be quickly and easily trained as operators. Regardless of the operator's experience, system efficiency and product quality remain at a high level.

Built-in sensors detect and forewarn the operator to replenish the supply of parts at each of the automated feeding stations, thus assuring uninterrupted operation. Techniques have been developed to assure safe machine operation and provide operator protection.

These techniques—including shields, barricades, grounding conductors, etc.—protect against personnel hazards and machine damage.

We will find that we need a little psychology when our goal is to get the best performance possible from an automatic assembly machine. No, not on the machine—on the people who must interface with the machine to really make it work. One of the psychological problems in getting superior performance can show up early in the game. Companies which isolate line and operating personnel from the equipment prior to its delivery will be confronted by an extended learning curve in bringing the machine on line. Among the shop people there will be an explicit or implicit attitude that "Tooling bought the machine, let them make it run."

What's the solution? Management must get line supervision, machine operators, and maintenance personnel involved in the final construction and debugging of the machine. It is not expensive. When operating, an assembly machine may return $100 or more per hour; one day's operation in the future will more than cover these costs.

Another psychological problem in the introduction of automatic or semiautomatic machines is created by negative or skeptical attitudes on the part of management. If operators believe that top management questions the technical feasibility or operating capability of the machine, this will be reflected in the attitudes of line personnel. The result will be a substantial delay in realizing the potential of the equipment.

The solution lies in management understanding the debugging phase of machine use. In trying to debug and refine an assembly machine, one is attempting to pull himself up by his bootstraps. The first plateau of productivity is reached relatively easily. Each subsequent plateau generally is a lesser gain and is harder to realize.

For example, in the beginning, during rough debugging, very little attention is paid to hopper inadequacies other than absolute failure. As machine development continues, hopper performances which might be adequate in off-and-on machine operation may be marginal under full production conditions. The hoppers might have been developed using preproduction samples, or without full exposure to the possible variations that occur during production. Effort will be required to bring hopper performance up to par.

To realize the full potential of automated equipment, it is important not to show disappointment or dissatisfaction when problems become apparent. If the fundamental machine design is sound, these problems will act as stepping stones to final productivity. They should be discussed and considered as advances in debugging, not as setbacks.

We cannot overemphasize the importance of positive attitudes in the ultimate success of the system: We must express these positive attitudes to those charged with operating the equipment.

There is another hurdle to superior performance that has a psychological twist. The general evidence of the problem is that the user becomes dissatisfied months after a machine is put into production. Production goals are not being met; supplementary hand lines must be maintained; and financial management is up in arms. The irate user wants everything else dropped, and maintenance men, engineers, managers—anybody—to fix "a machine that just doesn't work." Usually, enough production has been run at the builder's plant and the user's plant that the machine truly was production-worthy at some point. A general complaint about machine productivity, rather than specific complaints, is typically evidence of either under analysis or over analysis of problems.

For example, some people cannot control the urge to play with adjustments to see if the machine can be made to work better. It is not at all uncommon to find every conceivable screw, nut, knob, turnbuckle, and compliant link completely out of its normal position. Or, when something goes wrong, instead of analyzing the problem, someone immediately turns a screw or adjusts a locknut. No one ever remembers where they were originally set, and things go from bad to worse. All that is required to solve the problem is to return the machine to its original state.

The overanalysis syndrome stems from an inability of machine operators, machine maintenance personnel, or operating supervisors to accept the limitations inherent in any unique, special-purpose machine. There are limits to the tolerances that can be imposed on component parts. There are limits to the amount of machine development that is practical before turning a machine over for production use. What is to be expected from such equipment is that the net production realized is within a range that permits reasonable justification for the purchase of the equipment. Some users appear to expect and aim for perfect performance. Every time the machine stops, three or four people stand around discussing the situation, with no one attempting to return the machine to operation.

Good assembly machines are always self-inspecting. They inspect for the proper action of their own elements, and for the presence and position of the component parts of the assembly being processed. Upon detecting some malfunction, they either shut themselves down or switch to an alternate control mode. In any reasonably designed machine, there are a few random self-clearing failures, but most failures to assemble a product indicate a problem in the machine action, or jams caused by foreign material, or improperly toleranced parts in the part feeders. Regardless of the control mode, someone must correct the deficiencies and return the machine to operation.

There comes a point where continued analysis of the cause of such failures is counter productive. We must realize that such problem occur—regularly. We must judge when the machine has reached a mature state of development—and turn from debugging to running the machine and attempting to obtain the maximum production results.

The problem is real, and the solution is a simple change in outlook. In several instances, assembly machine servicemen report they have completely cured such a problem by offering operators a cash gift. In return, the operators were to stop diagnosing the problems of that machine and attempt to operate it to meet a specific production goal. In every single instance, this change in orientation and motivation resulted in immediate production capacities that satisfied the customer.

It is interesting to note that most of the start-up problems discussed in this subchapter actually occurred at McDonnell Douglas when we introduced a propellant blanking machine, a weight and skive machine, a propellant grain heating and rolling station, an insertion machine for propellant scrolls into igniter cups, and rocket motor final assembly machines on the Dragon missile high-rate production program. We learned a lot at that time, and after a very successful implementation program, produced over 3 million live rocket motors with very high reliability and at our projected cost. The major change that solved most of our start-up problems was to decrease the machine rate from a 4-sec cycle to a 6-sec cycle, which still met our production rates and allowed machine maintenance and repair as needed. All machines have a speed where everything works best. Any speed slower or faster than this "natural" rate causes problems in feed, synchronization, or something. We found that the operators learned the sound of the machine in operation, and were the first ones to suspect an impending failure. One important fact came out of our experience with machine design and maintenance. We had one machine that contained cams and gears, electrical circuits, electronics, fluidic sensors, and computer controls. When the machine ran perfectly, there was no problem. However, when a minor "glitch" occured—who do you call for diagnostics and for repair? The machine had too many

different systems that required one or two engineers to make the diagnosis. The maintenance mechanics were experts in only one or more of the systems, but not all. Our next
machines had fewer systems, and our supervisor of maintenence was a graduate engineer.

21.8 AUTOMATED MATERIAL HANDLING

Vijay S. Sheth

21.8.0 Introduction to Automated Material Handling

Automation of a manual process or machine operation has been regarded as a stepping
stone to gradually automating the entire factory operation. This is a common approach,
since it replaces the direct labor quite visibly, and is entirely measureable. However, automation of the material handling function is another matter. Since it is a "non-value-added"
function, not quite as visible as a significant cost contributor to the product it supports,
many companies have refrained from seeking innovative approaches to automation of the
material handling function. It also quite often requires much higher capital outlays.

With the modern concepts of just-in-time, zero inventory balance, continuous-flow
manufacturing, and materials requirements planning (MRP), more emphasis is now placed
on eliminating the cost of material handling, or automating it to provide low-cost,
consistent, reliable material flow to the production areas. Essentially, an automatic material
handling system does just that—positions the parts or equipment correctly, and at the right
time for the next operation.

21.8.1 Characteristics of Material Handling Automation

Automated equipment is designed and built to fill a specific requirement. It is necessary
to define the need and coverage of the project. Following are some of the characteristics
that need consideration in order to define the automated material handling equipment:

  Definition of the beginning and end points served
  Product configuration and physical changes occurring during the process
  Production rate and fluctuation
  Quality and remedial activities in process
  Process variation
  Safety of workers around the equipment
  Coordination with support functions and overall control desired
  Avoidance of product damage and hazards due to breakdown in automated system
  Progressive time-phased plan for a large, capital-intensive installation
  Alternative plans in case of prolonged breakdown of handling equipment

Advantages and Disadvantages of Automation

Automation provides many advantages over manual handling of material. Though not all
are cost related, some of the intangible advantages provide valuable information in making
decisions. The advantages are:

  Reduction of safety and health concerns of material handlers
  Avoidance of production downtime with on-time deliveries
  Reduced cost of the handling function
  Avoidance of damage to material and equipment transported

Savings in floor space by continuously flowing material, rather than manual batch transfer

Increased productivity, with automated transport lines acting as pacing devices for progressive assembly operations

Obviously, there are also deterrents to automation. Some of them are:

High installation cost

Reduced flexibility in schedule

Obsolescence and costly reconfiguration possible if process is altered

Major breakdown in system can cause shutdown of entire plant

21.8.2 Approaches to Material Handling Automation

There are two ways of approaching automation of the material handling function in a production operation. One method is to automate portions of the plant progressively, with a plan to interconnect them in the final phase. The other method is the complete installation of automation equipment in the entire plant at one time. Both approaches are widely used, based on the type of project and the commitment of management.

Islands of Automation

An approach to total factory materials handling systems is through a progressive implementation of stand-alone processes, a few at a time. These processes can involve the use of numerical control (NC) or computer numerical control (CNC) machines, automatic tool changers, and specialized conveyors to serve a few select machines and operations. In this case, the general material flow is converted into individual manufacturing cells with some degree of automation. There are two important points in consideration of this approach:

1. Before implementing any such system, a completed overall plan is required to make sure that the systems already implemented are compatible with future islands of automation.
2. Since the overall objective is an effective materials handling system for the entire plant, there should be a cohesiveness among all the subsystems that leads to total integration.

Many companies fail to realize these requirements in the rush to implement. While they are able to justify the subsystems installed, they fail to realize the long-term potential in terms of profit, quality, and longevity possible by integrating the entire system. Sometimes it becomes difficult to justify the investment for equipment to integrate subsystems already installed.

Islands of automation exist in plants with many separate processing departments. Figure 21.17 shows an example of this concept. A system can also be easily expanded to include prior operations, or operations between existing islands. Product configuration and its development through each stage of operation plays an important part in determining the installation of this system.

The progressive installation of this system requires consideration of the automation of only one process at a time. This approach is more easily understood by management and financial executives than the installation of a complex, plant-wide materials handling system that carries a much higher price tag. Another advantage is that the total materials handling project can be spread over several years, thus causing less anxiety in lean times and reducing the up-front financial burden.

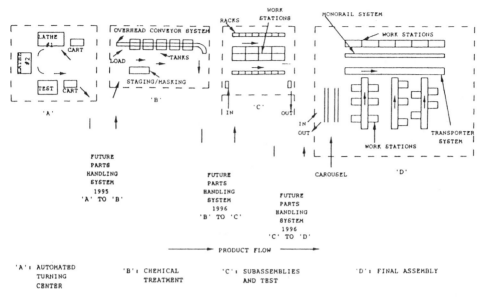

Figure 21.17 Islands of automation. Departments A, B, C, and D are considered islands for product flow inside the departments. They are to be interconnected by a materials handling system. The advantage is in spreading out investment costs over the years. However, individual systems must be compatible with the interconnecting systems of conveyance and batch processing requirements. (From Vijay Seth, *Facilities Planning and Materials Handling,* Marcel Dekker, New York, 1995.)

The progressive automation technique is useful in the following conditions:

Product design is not fully matured.

Process improvements are likely to occur in the near future due to technological advances—in either the product, or in factory systems and equipment.

Automation is difficult to justify as a whole, but phases can be justified independently.

Installations are conceived as an interim arrangement until final plans can be implemented.

Total System Installation

Many large, established corporations have considered it worthwhile to replace outdated facilities with new automated plants for products that have demand. The automobile and appliance industries fall into this category, where production is fairly continuous for many years. Recent examples are the GM Saturn and IBM PC and Proprinter automated facilities, which have replaced the manual material handling functions because of the cost, quality, and production rate advantages. This approach is sound and profitable provided the products and the processes are established. It also requires heavy dependence on the automation supplier for quality and field support. The use of computer and electronics in linking information flow has helped tremendously in simplifying automation processes.

21.8.3 Storage Equipment for Material Handling Automation

In the last two decades, more emphasis has been placed on the carrying cost of inventory and handling systems for storage equipment, since these are the two highest contributors to the cost of warehousing, and storage in general. The final objective, since both of these are "non-value-added" costs, is to totally eliminate them. Many new systems and equipment choices have been introduced. They have been further enhanced by the computer technology now available to provide accuracy, flexibility, and efficiency. Some of this equipment, now widely used in industry, is described below.

Automated Storage and Retrieval Systems

An automated storage and retrieval system (AS/RS) has two objectives. The first is to faithfully support warehousing efforts. An AS/RS provides materials handling requirements with maximum efficiency, offering a unique and cost-effective way of storing and handling materials. The second objective is to provide data in real time to the cost collection and accounting function within the plant—or sometimes to the "home office," since parts inventory and issues are tracked in digital form. For example, material issued has a contract or product code assigned along with the date and quantity. This can be converted to a dollar figure and charged to the finished product cost or contract. On some contracts, this may then be billed to the government or other customer as work in process (WIP). Analysis of such data also is used to update pricing data for bidding new contracts, since some of the items have common usage, or are allocated as part of the overhead cost.

Quite often a minicomputer is used for AS/RS operation to feed these data into the mainframe cost accounting computers. This means that networking involving a language conversion is involved. If all elements in the computer system are using one of the newer fourth-generation languages (4GL), and are compatible with open computer architecture, the problems are minimized. However, trying to tie in with older computers and languages can sometimes create serious problems in accuracy and cause delays in the retreival, delivery, and reorder of material.

Figure 21.18 shows an AS/RS system layout that is widely used. It consists of a

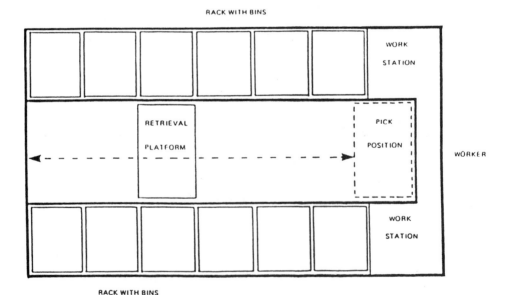

Figure 21.18 Typical AS/RS layout for high output and improved inventory control. (From Eastman, 1987.)

vertical stacking robot with a platform that travels up and down the aisle on a rail with an extended arm for safe, smooth, and precise storage and retrieval of inventory loads. On both sides of the aisle are racks full of bins, each assigned a particular location value. At a computer command, the robot travels down the aisle, adjusts to a specific height, and retreives a bin from the rack. It then moves up the aisle and places the load at the operator's workstation or conveyor. Next, it moves to an adjacent station, picks up the previously delivered bin, and deposits it onto the rack at its proper location. It begins the next cycle by picking another bin for the operator while he is still getting material out of the bin that was just delivered. The computer system tracks all possible storage locations and tells the robot where to store each load. It makes the necessary updates to the inventory record every time the bin is served. The system can be linked to a higher-level host computer for material planning and purchase order placement. Benefits of this system are:

Real-time inventory accuracy (often 100%)
Reduction in space requirements (as high as 80% reduction)
Faster order picking (up to 4 times the manual method)
Consolidation (makes off-site warehouses seem part of a central system)
Improvement of human factors (eliminates forklifts, climbing, bending, lifting, etc.)
Increased pilferage control (controlled access to material by designated employees)
Error-free operation (eliminates order picking errors)

Proper planning is required to install an AS/RS. Since most units are custom made, including the computer software applications, it is necessary to consider the following points:

Make a complete list of parts to be stored and ensure that all aspects of the AS/RS are sized correctly.
Make sure that the throughput of the AS/RS will meet your requirements.
Be specific about AS/RS computer and software requirements, if it is to be interfaced with the network system.
Discuss performance, drawbacks, and potential problems with the users of a similar operation. It is a major investment and downtime can paralyze the entire plant operation.
Plan item placement carefully. Frequently used categories should be stored in the front to reduce travel time. The number of picks per hour is greatly dependent on travel time back and forth from the aisle to the workstation and up and down the height of the racks. In most cases, the operator will be idle waiting for the next load.
Load-test the system before implementing it to make sure that it is assembled properly and can support the weight of the inventory stored.
Provide for a contingency plan in case the computer system breaks down or a load causes a jam-up. Two/three-day breaks in the operation have been experienced with computer breakdowns. This may cripple an entire manufacturing operation if the system is based on small batch sizes with short cycle times.
Installation and start up is a major task. It is necessary to decide on its location with the understanding that a future move for the AS/RS will be costly and cause production downtime.

While the AS/RS is used to store and retrieve loads on pallets and in containers, there is another version which is commonly known as the miniload system for handling and storing small parts in pans, drawers, and tote boxes. This system can efficiently automate small parts storage and handling of raw materials, finished products, kitting, and work-in-process materials. The working principle is the same, but it often requires the workstation layout to suit a particular application based on whether it is a kitting operation, production requirement, or warehousing operation.

Another variation of AS/RS is the application for high-density storage, in which the robot's extendable arm is replaced with a shuttle car. This car travels on tracks underneath the loads supported on the rack. With this arrangement, it is possible to serve unit loads on the racks deeper than one unit, thereby reducing the number of aisles and robots required to serve the system. This system provides maximum building cube utilization and maximum storage density.

Carousel Storage and Retrieval System

The carousel storage and retrievel system (CS/RS) has been in use for years in the garment industry. We have seen a moving garment conveyor application in dry cleaning shops, where a clerk activates a power conveyor supporting the garments to bring the order to the front counter. It is only in the last few years that this principle has been extended to industrial facilities for controlling inventory and handling costs. The addition of a programmable controller is useful in most cases.

Carousels are either horizontal or vertical. A horizontal carousel moves the product horizontally for use at floor level. In a vertical carousel, the product moves in a vertical loop and often can be serviced from more that one floor level.

A horizontal carousel is a closed-loop storage system comprised of a long, narrow framework supporting vertically placed storage baskets, which rotate on a horizontal plane. The baskets are connected by a common power drive system so all the baskets move at the same time as seen in Figure 21.19. The operator at the workstation activates the control to bring the desired basket to the workstation. Various applications of the horizontal carousels are shown in Figure 21.20. Many types of products may be stored with this system, including hardware, electrical parts, appliances, subassemblies, etc. It can be used in warehouse operations, burn-in for electrical equipment, assembly-line operations, kitting, and work-in-process storage among others.

Variations of the horizontal carousel include a twin-bin system, which allows an operator to pick parts from the sides of two adjacent bin columns instead of from only the front of the bin. A slightly different version is the rotary rack, in which each of the tiers rotates independently of the others. In this way, two tiers can be simultaneously rotated to a desired location for assembly-line operation. This system can be further enhanced by automating the storage and pick-up functions with the use of a fixed-location robot with an extending arm. This concept is useful in supporting parts requirements for a flexible assembly workstation.

A vertical carousel, also know as a vertical storage and retrieval system (VS/RS), operates much the same as the horizontal system, but in vertical rotation as shown in Figure 21.21. It has the following features:

Allows full use of the building cube
Provides flexibility of storage
Reduces downtime, delays, and damage to products
Allows for small batch sizes
Reduces setup time (in counter-type set up, such as tool cribs)

Figure 21.22 shows vertical carousel layouts for a variety of situations.

Figure 21.19 Horizontal carousel for random storage with computerized inventory and in-process control. (From Vijay Sheth, *Facilities Planning and Materials Handling*, Marcel Dekker, New York, 1995.)

21.8.4 Transportation Equipment

Transportation equipment for materials automation consists mainly of transporters, automatic guided vehicle systems, and power and free conveyor systems.

Transporters

One commonly used derivative of a floor-supported belt or roller conveyor is a transporter. It is a system comprised of sections of interconnected conveyors with diversion gates leading to various assembly work stations. Figure 21.23 shows a typical configuration of a transporter.

An effective manufacturing cell arrangement can be installed by combining a transporter and a horizontal carousel. The dispatcher loads the transporter to deliver a totebox from the carousel and signals the transporter to deliver the totebox to the available workstation. The power conveyor moves the load until the power-activated arm diverts the load to the desired station. After completing the operation, the workstation operator

Order Picking

Shipping objectives are met by freeing operators to do more work in less time for unsurpassed productivity!

Buffer Storage

Provides instant adjustment to peaks and valleys in production processes.

Burn-in and Test

ACCUTRIEVE™ provides efficient and effective burn-in testing for electronics manufacturing.

Records Storage

Combines the space-saving advantages of high-density archival storage with fast, easy access.

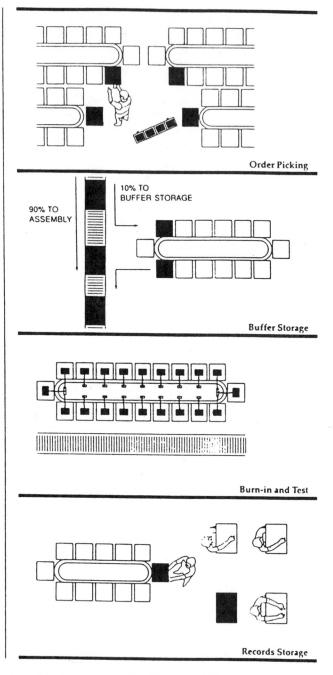

Figure 21.20 Various applications of horizontal carousels. (Courtesy of Richards-Wilcox Manufacturing Company.)

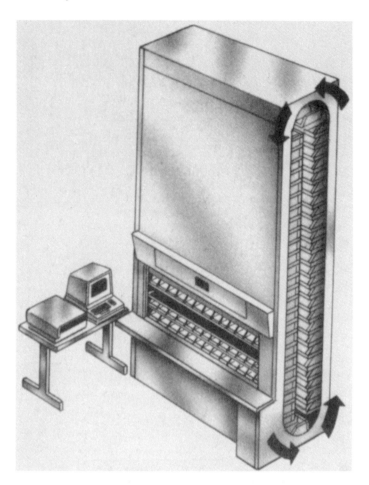

Figure 21.21　Vertical carousel. (From Vijay Sheth, *Facilities Planning and Materials Handling*, Marcel Dekker, New York, 1995.)

transfers the totebox to the dispatcher through the bottom return conveyor, which is traveling in the opposite direction. The entire system is computer controlled to also keep track of the inventory count, work order status, and the storage information on the carousel. The advantages of this system are:

Eliminates the line balance problem, since the dispatcher controls the work-in-process inventory for every station on the line.
Minimizes the supervision job of keeping workers busy, since they always have work in front of them. The employee's task at each station may vary, depending on the schedule needs and experience of the individual operators.
Flexibility exists to change product lines and to perform various product operations, without having to rebalance the line or reconfigure the conveyor.
Schedule changes can be accommodated by adding more workstations for the same operations.
Manual operations are a simple interface, and the system can also be used as a buffer.

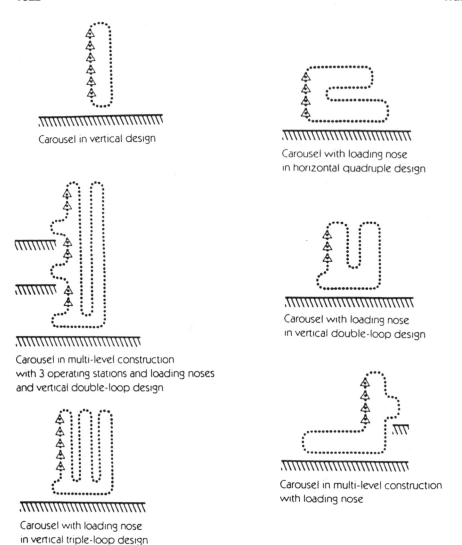

Carousel in vertical design

Carousel with loading nose
in horizontal quadruple design

Carousel in multi-level construction
with 3 operating stations and loading noses
and vertical double-loop design

Carousel with loading nose
in vertical double-loop design

Carousel with loading nose
in vertical triple-loop design

Carousel in multi-level construction
with loading nose

Figure 21.22 Vertical carousels can be configured to take advantage of tight spaces where operator or handling equipment access is difficult. (Courtesy of Baumann, Inc.)

Automated Guided Vehicle System

The automated guided vehicle system (AGVS), much like a conveyor, brings work to the operator in a driverless vehicle as shown in Figure 21.24. However, there are distinct differences between the two systems. Load sizes are usually bigger for the AGVS, and frequency of service is quite different. It is relatively easy to change the path of the vehicle with the improved track devices available. A unique use of the AGVS is by Volvo in Sweden, which uses the AGVS as a progressive assembly-line conveyor for its automobiles, with flexibility in the line and workstation layout.

AGVS have seen overwhelming growth in the last decade, with the increasing use of computer and microprocessor technology in material handling equipment. Earlier models, developed over 30 years ago, had limited applications as to load capacity and the length of the path traveled. With the development of the AS/RS, the carousel, and the emphasis

Figure 21.23 The transporter provides flexibility in tasks performed at workstations. It eliminates line balancing requirements since workstations are set independent of one another for tasks performed. (Courtesy of Speedways Conveyors, Inc.)

on flexible manufacturing systems, the role of the AGVS has been crucial in bridging so-called islands of automation. An AGVS offers many benefits not limited to just the factory floor. Other uses in the factory include such mundane tasks as delivering the mail.

One system consists of a fixed path laid out on the plant floor by cutting a groove in the concrete with a diamond saw 1/8 in. wide by 1/2 in. deep along the center of the entire path. A continuous insulated metal conductor is laid into the groove and covered with epoxy. Both ends of the conductor are connected to an electrical device to generate low-level electrical signals. The AGV is equipped with sensors that pick up the electrical signals to keep it on the path within a range of 1/2 in. The vehicle is usually propelled

Figure 21.24   Automated guided vehicle system. (Courtesy of Automated Systems Division, Bell and Howell.)

by a rechargable battery. Vehicles are also equipped with computer and microprocessor controls to identify the location and duration of their stops. Automatic load and unload robots at each station may also be installed. Other systems track a strip of metalized tape or a metallic paint stripe on the floor.

Storage AGV, unlike stackers, are capable of stacking parts at different heights as well as transporting materials to various locations in the plant. The very-narrow-aisle version (5 ft) is gaining popularity in industrial warehouse operation because of its excellent load-handling capability of medium-sized loads, making it possible to raise the storage height of the warehouse bins.

Power and Free Conveyors

One version of an industrial plant conveyor is a dual-track conveyor system referred to as a power and free conveyor and shown in Figure 21.25. The system consists of two parallel tracks, one housing the power chain and the other containing free carriers that support the load. Latch mechanisms are attached to the power chain at a predetermined spacing that engage the carriers to propel them through the free track. A carrier can be stopped by a stationary cam stop mechanism which disengages the latching mechanism.

Power and free conveyors are available in overhead and inverted styles. In general, with power and free conveyors, the product is supported and transported in three different ways: on the floor, flush with the floor, or by the overhead system. For an overhead system, the load is suspended from the overhead conveyor wirth hangars. This allows freedom of operation around the product from the top down. The conveyor can also be elevated so the floor area under the conveyor can be utilized for other productive work. This type is often required in the case of equipment assembly lines using manual labor or equipment and robots.

Figure 21.26 shows a typical layout of a power and free conveyor system. The manufacturing engineer should be able to lay out the system within specific requirements so the conveyor manufacturer can design a system to suit the customer's needs.

Benefits of power and free conveyors include the following:

They provide on-line accumulation at a desired location without disrupting the flow of other carriers in the system.

Unlike the conventional conveyor, the product on a power and free system rides on a free track. This provides an extra opportunity to enhance the performance by adding switches, line-full sensing devices, lifting and dipping possibilities, and automatic inventory counts.

Unlike the conventional conveyor, where the speed has to be synchronized with operations of the same time duration for the progressive assembly line, the power

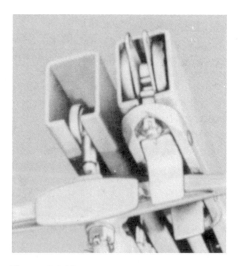

Figure 21.25 Power and free conveyor showing two-track system. One track supports a power chain with pushers and another supports trolley for product movement. (Courtesy of Richards-Wilcox Manufacturing Company.)

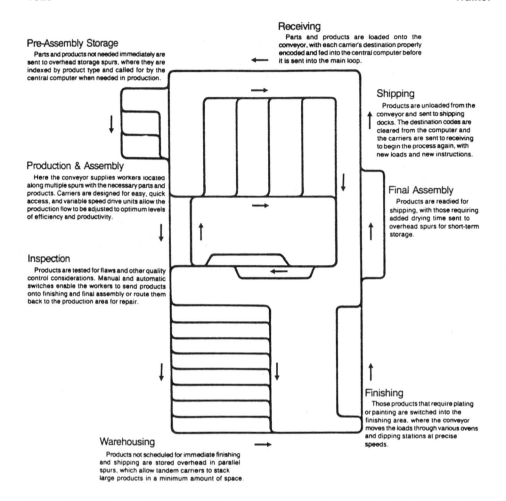

Receiving
Parts and products are loaded onto the conveyor, with each carrier's destination properly encoded and fed into the central computer before it is sent into the main loop.

Pre-Assembly Storage
Parts and products not needed immediately are sent to overhead storage spurs, where they are indexed by product type and called for by the central computer when needed in production.

Shipping
Products are unloaded from the conveyor and sent to shipping docks. The destination codes are cleared from the computer and the carriers are sent to receiving to begin the process again, with new loads and new instructions.

Production & Assembly
Here the conveyor supplies workers located along multiple spurs with the necessary parts and products. Carriers are designed for easy, quick access, and variable speed drive units allow the production flow to be adjusted to optimum levels of efficiency and productivity.

Final Assembly
Products are readied for shipping, with those requiring added drying time sent to overhead spurs for short-term storage.

Inspection
Products are tested for flaws and other quality control considerations. Manual and automatic switches enable the workers to send products onto finishing and final assembly or route them back to the production area for repair.

Finishing
Those products that require plating or painting are switched into the finishing area, where the conveyor moves the loads through various ovens and dipping stations at precise speeds.

Warehousing
Products not scheduled for immediate finishing and shipping are stored overhead in parallel spurs, which allow tandem carriers to stack large products in a minimum amount of space.

Figure 21.26 Layout of power and free conveyor system for a product going through various activities in a plant with varying output. (Courtesy of Richards-Wilcox Manufacturing Company.)

and free system provides flexibility to: provide accumulation sections, bypass and diversion, switches, empty carrier storage, and varying throughput requirements.

Support functions such as testing, inspection, or rework can be tied into the same conveyor with free tracks.

Future expansion is possible by add-on sections and carriers rather than installing a totally independent system.

21.9 BIBLIOGRAPHY

Boothroyd, Geoffrey, *Assembly Automation and Product Design*, Marcel Dekker, New York, 1992.
Sheth, Vijay, *Facilities Planning and Material Handling*, Marcel Dekker, New York, 1995.
Tanner, John P., *Manufacturing Engineering*, Marcel Dekker, New York, 1991.

22

Electronics Assembly

Michael Pecht   *CALCE Electronics Packaging Research Center, University of Maryland, College Park, Maryland*

Denise Burkus Harris   *Westinghouse Corporation, Baltimore, Maryland*

Shrikar Bhagath   *Delco Electronics Corporation, Kokomo, Indiana*

22.0 INTRODUCTION TO ELECTRONICS ASSEMBLY

Electronic assemblies consist of any number of electronic and nonelectronic devices (screws, heat sinks, carrier plates, etc.) grouped together to perform a function. Each assembly is a self-contained production item, built within a particular facility in which it follows its own process flow. For example, an insertion-mounted printed wiring board (PWB) can be built in the same facility as a surface-mounted PWB; however, different equipment and processes are used. Both boards can be populated by the same pick-and-place machine, but a wave solder system is commonly used for insertion boards, while surface-mount boards use solder paste or thick solder plate reflowed in a vapor-phase or infrared (IR) furnace.

This chapter discusses the basic subassemblies commonly used within deliverable electronic units and the assembly processing associated with each. It describes how these subassemblies are grouped together to form new assemblies, which are in turn the building blocks for larger assemblies, and the integration of these assemblies to form the final system.

22.0.1 System Breakdown

A given system may have many levels of assembly. The final system, or top assembly, can contain many subassemblies; each of these can contain its own subassemblies, and so on. Most subassemblies are manufactured as deliverable, self-contained units, built at various facilities. The system house obtains all of these subassemblies and integrates them into the final system. A crucial first step is creating what is variously called a system breakdown, system parts list, or ultimately an indented bill of material.

The list is started with the final system at the top and the major subassemblies as branches below. These subassemblies are branched out further with their own subassemblies and components. A subassembly is denoted by indenting the list at each new level.

In the example in Figure 22.1, each indentation indicates a lower level of assembly, or a subassembly within the given heading.* The quantities given are for one assembly. For example, there are 32 transmitter modules in the radar. Each transmitter module has one module housing, one cover, one PWB assembly, and so on. Thus, you would order (or build) 32 module housings, 32 covers, 32 PWB assemblies, and so on for each complete assembly. AR stands for "as required," a term used for materials purchased in bulk quantities. REF stands for "reference."

22.1 TYPICAL PACKAGE ARCHITECTURE

Single-chip packages can be broadly classified into plastic packages, ceramic packages, and metal packages, depending on the kind of casing used. Figure 22.2 shows one of the simplest component packages, encapsulated in plastic. Figures 22.3a–c show schematics of additional packages. A ceramic package typically has a substrate, to which the die is fixed using a die-attach. The die has conductor pads on its periphery to which wirebonds (or other interconnections) are attached. The other end of the wire bonds are attached to the package leads which form the basic input/output of the package. Lead seals are used to fill the gap between the leads and the package case. The lid of the package is affixed to the case with a lid seal. Table 22.1 gives some of the commonly used materials for the package elements mentioned above.

22.2 ELEMENTARY SUBASSEMBLIES

22.2.1 Die Assemblies

Die assemblies are the simplest of subassemblies, and consist of a bare die, such as a transistor or diode, mounted on a tab with a conductive surface. Figure 22.3 shows examples of typical die assemblies. Die assemblies are mounted on a substrate that is protectively overcoated (in commercial applications) or encased in a hermetically sealed package or hybrid (in military applications). Building your own assemblies has several advantages:

Die assemblies need be only slightly bigger than the dies themselves, making them much smaller than a discretely packaged or "canned" diode or transistor. While direct mounting on the higher assembly's substrate (a common practice with hybrid assemblies) would further decrease the size needed, mounting the dies first on tabs allows functional testing before committing them to the more expensive higher assembly.

Pretested die assemblies can be used for low-yield dies, such as application-specific integrated circuits (ASICs); high-power dies, such as field-effect transistors (FETs); or when it is essential to match pairs of dies for functional reasons.

The tab allows easier rework, since die assemblies supported by tabs lift up in one piece. To replace the die, the substrate assembly must be heated to reflow the solder or soften the epoxy with which the die is mounted. The die is then extracted from the

*This figure is typical of a breakdown used early during the design phase of a product, so that the designer can keep track of the materials first selected for the design. It would have part numbers and other information assigned prior to appearing on a finished drawing or CAD file.

Final System: Transmitter for a Radar
    20 #6 Screws
    1 Rack
    1 Cover
    1 Power supply module
        REF 234553 Test Spec
        1 Module housing
        1 Cover
        1 Power supply PWB assembly
            1 PWB
            25 3μ capacitors
            10 0.1μ capacitors
            1 Transformer
            1 Power hybrid
              1 Case
              1 Cover
              1 Solder preform
              2 FET assemblies
                1 FET
                1 BeO tab, thin-film metallized
                AR 0.010-in.-diameter Al wire
              1 Logic substrate assembly
                1 Substrate
                4 2N2222 transistors
                2 Rectifier chips (ICs)
                3 0.01μ capacitors
                AR conductive epoxy
                AR 0.001-in.diameter Au wire
          6 LM139 DIPs
          8 Com-04 cer paks
  1 Matrix plate
      32 8-pin connectors
      1 10-pin connector
      15 ft $\frac{1}{4}$-in. tubing
      33 Cables
  32 Transmitter modules
      1 Module housing
      1 Cover
      1 T/R PWB assembly
          1 PWB
          15 2μ capacitors
          1 Gate array
             1 Gate array die
             1 84-pin leaded chip carrier
          1 Stripline
          1 Combiner
          4 Regulator hybrids
             1 Case
             1 Cover
             1 Epoxy preform
             1 Substrate assembly
               1 Substrate
               2 10-kΩ resistor
               3 1N3600 diodes
               3 0.033μ capacitors
               AR conductive epoxy
               AR 0.001-in.-diameter Au wire
      1 Epoxy preform
      1 Cold plate
  64 Cables

Figure 22.1   Example of systems parts list or Xmas tree.

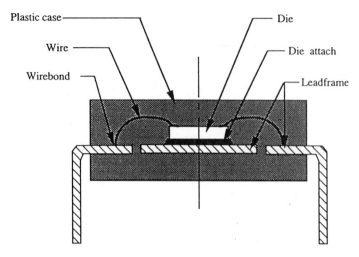

Figure 22.2 Cross section of a typical plastic-encapsulated single-chip package.

substrate. Since most dies are only 0.01-in.-thick silicon, they tend to come off in pieces. These pieces must be carefully pried off one at a time. Extreme care must be taken not to damage the neighboring components.

Die Assembly Materials

The tabs used for die assemblies are metallized ceramic or metal plated to allow for proper mounting and wirebonding. Metal tabs give the back plane of the die assembly the same electrical function as the back plane of the die. Thus, for example, the tab on which a transistor is mounted becomes the collector, allowing the back-plane input or output (I/O) to be accessed by connection to the back plane of the die assembly or the top of the tab. The most common metal tabs, Kovar and molybdenum, have high thermal conductivities and coefficients of thermal expansion (CTEs) between those of the silicon dies and the ceramic substrates on which they are mounted. The metal tabs are plated, first with 0.0001 to 0.0003 in. of nickel to provide a corrosion barrier metal, and then with 0.000050 to 0.000150 in. of gold to allow for wirebonding and solderability. If aluminum wirebonding is used, the minimum recommended thickness of gold is 0.0001 in.

Ceramic tabs are employed when the back plane of the die assembly should be inactive. The surface metallization of the tab becomes the collector when a transistor is mounted on a ceramic tab. Ceramic tabs are metallized with thick-film gold alloys or thin-film metallization. If 96% alumina is used, the ceramic substrate is laser-drilled to create a perforated array of the desired tab dimensions, called a snapstrate. The thick film is printed and fired onto the snapstrate, creating an array of bonding pads within the perforations. When snapstrate is snapped apart, individual tabs result. The back side of the snapstrate may be left as bare ceramic if the die assembly is to be dielectric epoxy-mounted, or it may be metallized to allow for solder mounting. If thin-film metallization is desired, a 99% alumina substrate is used, with a base metal, such as chromium, titanium, tungsten, or copper, deposited on it. The substrate is then nickel-gold plated like metal tabs. After plating, the substrates are diamond-sawed into tabs of the desired dimensions. If beryllium–oxygen tabs are required to provide high thermal conductivity, they are metallized using thin-film methods, since thick film does not provide good adhesion to the beryllium–oxygen.

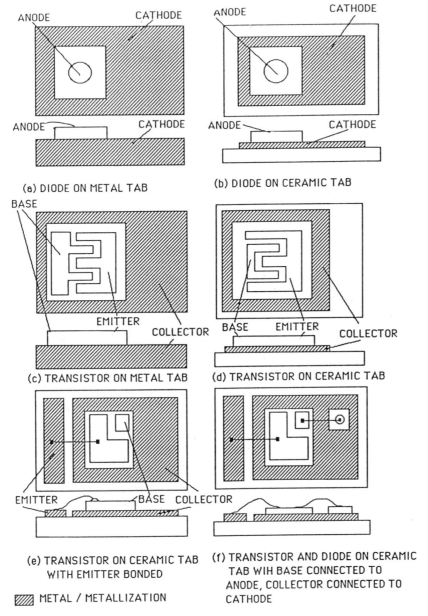

(a) DIODE ON METAL TAB

(b) DIODE ON CERAMIC TAB

(c) TRANSISTOR ON METAL TAB

(d) TRANSISTOR ON CERAMIC TAB

(e) TRANSISTOR ON CERAMIC TAB WITH EMITTER BONDED

(f) TRANSISTOR AND DIODE ON CERAMIC TAB WIH BASE CONNECTED TO ANODE, COLLECTOR CONNECTED TO CATHODE

▨ METAL / METALLIZATION

Figure 22.3 Die assembly examples.

Die Assembly Processing

Die Mounting

The dies are mounted to the tabs using either solder or eutectic mounting. Figure 22.4 shows a typical phase diagram for a solder. The x axis of the diagram gives the composition of the alloy; the y axis is temperature. The liquidus lines indicate the temperature at which a given composition will completely melt, or be in a liquid phase. The solidus lines indicate

Table 22.1 Commonly Used Materials for Various Package Elements

Package element	Materials
Chip carriers and substrates	Ceramic: alumina, beryllia, silicon carbide, aluminum nitride, silicon nitride, forsterite
	Metal: Molybdenum, copper–tungsten, tungsten, Kovar, stainless steel
Semiconductor die	Silicon, germanium, silicon carbide, gallium arsenide, cadmium sulfide, indium antimonide, lead sulfide, II–V and II–VI compounds
Die attach and adhesive	Silicone, polyurethane, acrylic, epoxy Novolac, epoxy phenolic, epoxy polyamide, solder
Wires for bonding	Gold and aluminum
Leads	Kovar
Lead seals	Glass
Lids	Same as case/chip carrier materials
Lid seal	Welding, brazing, and solder materials
Encapsulants	Silicones, epoxies, polyimides, polyamides, polyurethanes, fluorocarbons, acrylics, diallyl phthlate, polyvinyl chloride, parylene

the temperature at which a given composition will crystallize, or become completely solid. The areas between the liquidus and solidus lines represent the plastic zones of a given composition, or the zone in which both liquid and solid phases are present. In other words, as the temperature is increased for a given composition, that composition starts to melt when the temperature matches that of the solidus and has completely melted when the

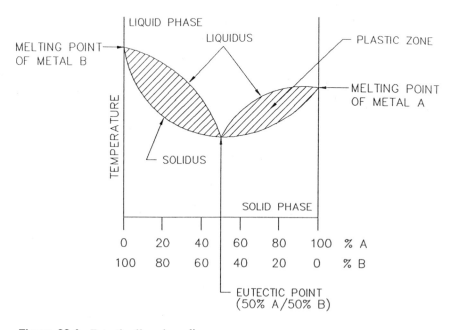

Figure 22.4 Eutectic alloy phase diagram.

temperature matches that of the liquidus. The eutectic is the point at which the solidus and liquidus intersect. A eutectic alloy is a composition with the lowest possible melting point, going from solid to liquid at one temperature (that is, a eutectic alloy has no plastic zone). Eutectic mounting is used for dies that use a eutectic alloy for attachment. This method is very labor-intensive and has a high potential for damaging the die during the rubbing, or "scrubbing," process.

In eutectic mounting without a solder preform, a silicon die is placed on the gold-plated tab. The tab is placed on the hot plate of a die-bonding station. Dry nitrogen is blown over the die-mounting area (alternatively, the entire process is performed in an environmental chamber filled with dry nitrogen). The temperatures of the tab and die are brought to approximately 400–420°C (about 20–40°C above the gold–silicon eutectic point). The die is rubbed gently back and forth into the gold plating until the gold and silicon melt together to form the eutectic alloy. At this point, the die is eutectically bonded to the tab.

Eutectic mounting with a preform, the most common die-mounting method, uses a solder preform of the desired eutectic alloy. An alloy is chosen depending on (1) the metallization of the substrate and the die backplane; (2) the desired reflow temperature; and (3) the formation of intermetallics, which can affect the long-term reliability of the unit. For example, 80% gold–20% tin has a eutectic point of 280°C. This temperature is high enough to surface mount the substrate with eutectic lead–tin (63% lead–37% tin, which reflows at 183°C) without reflowing the die mount, but it is lower than the eutectic point of some other commonly used gold eutectics; this allows limiting the exposure of the die to extreme temperature. This alloy also has good wetting properties on both the gold metallization commonly found on the substrate and the diffused gold on the die back plane. However, gold–tin intermetallics readily formed at relatively low temperatures are very brittle and can lead to mechanical fracture failures. Silver eutectics are not brittle, but silver can migrate, leading to small changes in resistivity over time. Silver is also notoriously susceptible to corrosion problems. Table 22.2 lists some commonly used solder alloys.

After an alloy is selected, a preform is punched out with a "cookie cutter" die, or simply cut out of the alloy foil with scissors. The preform is placed between the die and the mounting surface, and the die is mounted using the scrubbing method discussed earlier.

For high-volume production, furnace mounting or fluxless solder reflow is used. The die and preform are placed on the tab or substrate and sent through a furnace filled with forming gas (approximately 5% hydrogen and 95% nitrogen). Forming gas provides a reducing atmosphere, which prevents corrosion and contamination while allowing wetting to occur. Industrial furnaces normally have three to nine zones; each is programmed to maintain a specified gas flow and temperature. The parts travel on a belt at a programmed speed through the furnace. The combination of gas flow, belt speed, and zone temperatures yields a profile of the temperature the part will reach. Attaching a thermocouple to a number of parts and running them through the furnace determines the profile. A typical profile is shown in Figure 22.5.

Interconnection

Interconnection to the die assembly is most commonly accomplished with wirebonds. The standard wire is 0.0007- or 0.001-in.-diameter gold wire. When better current-carrying capacity is required, 0.005- to 0.010-in.-diameter aluminum wire is used to avoid using costly, larger-diameter gold wires. Table 22.3 lists the current-carrying capabilities and fusing limits of some standard-size wires.

Gold wirebonds are made using thermocompression or thermosonic bonding. In both

Table 22.2   Solder Alloys

Temperature (°C)			Elements (%)								
Solidus	Liquidus	Eutectic	Pb	Sn	Ag	Ln	Sb	Bi	Cd	Au	Zn
		93		42.0		44.0			14.0		
		95	32.0	15.5				52.5			
		117		48.0		52.0					
118	125			50.0		50.0					
		125	43.5					56.5			
		125	9.6	15.0		70.0			5.4		
		138				42.0		58.0			
		139		43.0				57.0			
		143			3.0	97.0					
		145	30.6	51.2					18.2		
155	149		15.0		5.0	80.0					
		146	32.0	50.0					18.0		
		162	18.0	70.0		12.0					
144	163		43.0	43.0				14.0			
156	165		25.0			75.0					
160	174		30.0			70.0					
		177		67.8					32.2		
		179	36.1	62.5	1.4						
		183	38.1	61.9							
183	186		35.0	65.0							
179	189		36.0	62.0	2.0						
183	190		40.0	60.0							
183	192		30.0	70.0							
183	195		25.0	75.0							
183	195		42.0	58.0							
		198		91.0							9.0

cases, the wire is fed through a capillary. Heat is applied to melt the wire and form a ball of material at its end. This ball is thermocompressed into place on the die, forming a thermosonic ball bond, and heat and pressure are applied while the wire is vibrated at ultrasonic frequencies. The wire is pulled through the capillary and looped over to the bond

Table 22.2 (Continued)

Temperature (°C)			Elements (%)								
Solidus	Liquidus	Eutectic	Pb	Sn	Ag	Ln	Sb	Bi	Cd	Au	Zn
183	202		20.0	80.0							
183	203		45.0	55.0							
180	209		50.0			50.0					
183	209		15.0	85.0							
183	214		50.0	50.0							
183	215		10.0	90.0							
		215	85.0		15.0						
183	218		48.0	52.0							
		221		96.5	3.5						
183	224		5.0	95.0							
183	225		55.0	45.0							
179	227		35.5	61.5	3.0						
221	229			96.0	4.0						
183	238		60.0	40.0							
233	240			95.0			5.0				
183	242		62.0	38.0							
221	245			95.0	5.0						
179	246		36.0	60.0	4.0						
		246	89.5				10.5				
183	247		65.0	35.0							
		248	82.6						17.4		
		252	88.9				11.1				
183	258		70.0	30.0							
250	264		75.0			25.0					
183	268		75.0	25.0							
		280		20.0						80.0	
183	280		80.0	20.0							
225	290		85.0	15.0							
268	299		88.0	10.0	2.0						

pad on the substrate. The capillary is used to press or scrub the wire into the gold-metallized bonding pad on the substrate, forming the "stitch." The wire is then broken, leaving a "tail." Figure 22.6a shows the steps in forming a ball bond.

Thermal bonding methods cannot be used to form aluminum wirebonds. When gold

Table 22.2 Continued

| Temperature (°C) | | | Elements (%) | | | | | | | | |
Solidus	Liquidus	Eutectic	Pb	Sn	Ag	Ln	Sb	Bi	Cd	Au	Zn
296	301		93.5	5.0	1.5						
268	302		90.0	10.0							
		304	97.5		2.5						
		309	97.5	1.0	1.5						
301	314		95.0	5.0							
		318	99.5								0.5
316	322		98.0	2.0							
304	365		94.5		5.5						

Courtesy of the International Society for Hybrid Microelectronics.

(substrate metallization) and aluminum (wire) are placed together and sufficient heat is applied, gold–aluminum intermetallics can form; these brittle intermetallics can greatly weaken the strength of the wirebond. Thus, ultrasonic bonding techniques are used for aluminum wirebonds. Ultrasonic techniques are similar to thermosonic methods but are performed at room temperature. Because no heat is applied, no ball is formed; both ends of the wirebond are stiched. These stitches, also called wedges, produce wedge bonds. Figure 22.6b depicts the formation of a wedge bond.

Military and space applications specify that bonds be tested. In "pull testing," the common military standard test, the wires are pulled until they fail. The type of failure is documented (ball lift, break in wire loop, break at stitch, etc.) and the force at which the failure occurred is noted. For a nondestructive pull test, each wire is pulled to a specified

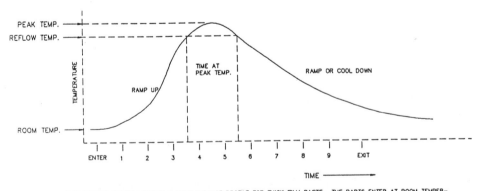

THIS EXAMPLE SHOWS A TYPICAL 9 ZONE FURANCE PROFILE FOR THICK FILM PASTE. THE PARTS ENTER AT ROOM TEMPER-
ATURE. THE FIRST 3 ZONES ARE PROGRAMMED SO THAT THEY WILL RAMP UP THE TEMPERATURE QUICKLY ENOUGH FOR
THE PASTE TO REACH ITS REFLOW TEMPERATURE BEFORE THE ACTIVE SOLVENTS TOTALLY VOLATILIZE. ZONE 4, SET AT THE
PEAK TEMPERATURE (APPROXIMATELY 30–50 C HIGHER THAN THE REFLOW TEMPERATURE) GUARANTEES THAT THE PASTE
WILL REACH AND MAINTAIN THE REFLOW TEMPERATUIRE FOR THE PROPER DURATION TO ACCOMPLISH TOTAL REFLOW, WITHOUT
OVER FIRING. THE REMAINING ZONES ARE PROGRAMMED TO COOL DOWN THE PARTS FAST ENOUGH TO ESTABLISH PROPER
GRAIN GROWTH, BUT SLOW ENOUGH TO PREVENT HIGH THERMAL STRESSING. THE PARTS EXIT THE FURANCE AT A SLIGHTLY
ELEVATED TEMPERATURE, BUT COOL ENOUGH THAT HITTING ROOM TEMPERATURE AIR WILL NOT THERMALLY STRESS THEM.

Figure 22.5 A typical furnace profile.

Table 22.3 Wirebond Fusing Currents

Wire diameter (in.)	Nonfusing current	Fusing current	Maximum wire length (in.)
0.0007 (gold)	160 mA	—	0.070
0.001 (gold)	250 mA	1 A	0.090
0.005 (aluminum)	3 A	5–7 A	0.400
0.010 (aluminum)	10 A	17 A	0.500

force. If the wirebond does not break or fail, the unit passes the test. Figure 22.7 shows a wirebond being pull-tested.

Applications

Military and space applications typically require maximum reductions in weight, area, and volume, along with the highest possible reliability. They also often require the use of state-of-the-art or leading-edge technology. For example, a radar for a fighter jet must have the most advanced capabilities, yet be small and lightweight enough to fit into the nose of the aircraft. Use of discrete packages is very uncommon because of the extra volume and weight they impose. Despite their higher cost, bare-die assemblies (i.e., hybrids) are commonly used to meet the size limitations of the system.

In addition to increased packing densities, military applications generally have higher power densities. These often require that certain transistors be tested at high temperatures prior to being mounted within the sealed unit, or that these transistors be mounted on a high-spreading tab (e.g., a metallized beryllium–oxygen tab) to prevent thermal stressing of the dies during high-temperature operations and thermal cycling tests. Furthermore, the advanced functionality of the military system often requires the use of a new die that is still immature in its design. The yields of such units are usually low, so it is necessary to test them functionally prior to placing them in the hybrid. Because of the die's active backplane, this testing can be accomplished only when the die is mounted to a conductive surface.

Die assemblies are not commonly used in commercial applications, which generally do not need the restricted sizes that merit the extra expense of additional subassemblies. Therefore, mature, high-production-volume, discretely packaged dies are commonly used commercially. These devices are very inexpensive (transistors can be as little as $0.30 each when bought in large quantities) and have high yield rates because of their simplicity and maturity; they have been produced for so long in such high volumes that the processes used to produce them and the products themselves have been fully debugged and optimized.

Die assemblies are used in commercial products with high-temperature environments; for example, car ignitions require the use of FET transistors. But discretely packaged FETs, with the extra interfaces of the discrete package, its mounting material, and its board material, do not dissipate heat enough to maintain the proper operating junction temperature. Therefore, a die assembly is used that consists of the necessary FET, eutectically mounted to a highly thermally conductive tab, such as molybdenum. This assembly is then encapsulated with a high-temperature-protective overcoat.

Another commercial application of die assemblies is the electronic sensing devices used in oil drilling. These miniature electronic devices, located at the tip of the drill, sense the conditions of the drill site—for example, pressure, temperature, and viscosity—then transmit this information through the drill rod to the surface, where it is analyzed to

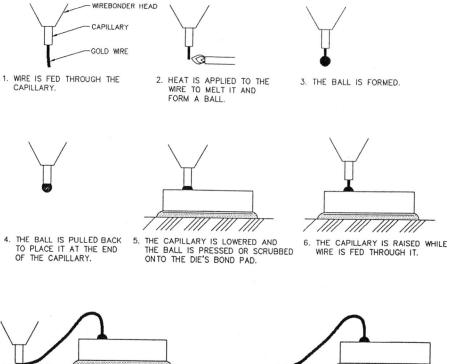

1. WIRE IS FED THROUGH THE CAPILLARY.

2. HEAT IS APPLIED TO THE WIRE TO MELT IT AND FORM A BALL.

3. THE BALL IS FORMED.

4. THE BALL IS PULLED BACK TO PLACE IT AT THE END OF THE CAPILLARY.

5. THE CAPILLARY IS LOWERED AND THE BALL IS PRESSED OR SCRUBBED ONTO THE DIE'S BOND PAD.

6. THE CAPILLARY IS RAISED WHILE WIRE IS FED THROUGH IT.

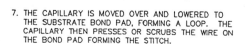

7. THE CAPILLARY IS MOVED OVER AND LOWERED TO THE SUBSTRATE BOND PAD, FORMING A LOOP. THE CAPILLARY THEN PRESSES OR SCRUBS THE WIRE ON THE BOND PAD FORMING THE STITCH.

8. THE CAPILLARY RAISES BREAKING THE WIRE OFF AT THE STITCH. THE WIREBOND IS COMPLETED.

(a)

Figure 22.6 Examples of the formation of (a) a ball bond and (b) a wedge bond.

determine how close the drill is to the oil deposit. The environment in this case can exceed 200°C, due to the heat of the earth and the friction experienced by the drill.

22.2.2 Capacitor Banks

Capacitors are commonly used in electronic systems to regulate current and to store energy for the electronic assembly. Capacitors can constitute the bulk of the parts list for a power supply; power supplies, in turn, are needed to provide and regulate the necessary power for virtually all systems. In high-power applications, higher farad values are needed to ensure device functionality; in most cases, several capacitors must be placed in parallel to obtain the necessary capacitance.

Calculating Capacitance

For capacitors in series, the total capacitance (C_T) is the reciprocal of the sum of the reciprocal capacitances of the individual capacitors:

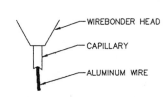

1. WIRE IS FED THROUGH THE CAPILLARY.

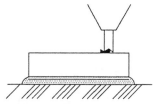

2. THE CAPILARY IS LOWERED AND THE WIRE IS ULTRASONICALLY SCRUBBED ONTO THE DIE'S BOND PAD.

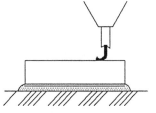

3. THE CAPILLARY IS RAISED WHILE WIRE IS FED THROUGH IT.

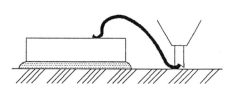

4. THE CAPILARY IS MOVED OVER AND LOWERED TO THE SUBSTRATE BOND PAD, FORMING A LOOP. THE CAPILLARY THEN ULTRASONICALLY SCRUBS THE WIRE ON THE BOND PAD FORMING THE STITCH.

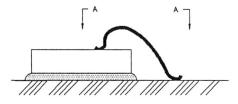

5. THE CAPILLARY RAISES BREAKING THE WIRE OFF AT THE STITCH FORMING A TAIL. THE WIREBOND IS COMPLETED.

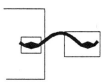

VIEW A–A

(b)

$$C_T = \frac{1}{1/C_1 + 1/C_2 + 1/C_3 + \cdots 1/C_n}$$

For capacitors in parallel, the total capacitance is the sum of the individual capacitances,

$$= C_T(5F) = \quad = C_1(1F) = C_2(1F) = C_3(1F) = C_4(2F)$$

where C_T is the total capacitance; C_1, C_2, C_3, . . . , C_n are the values of the individual

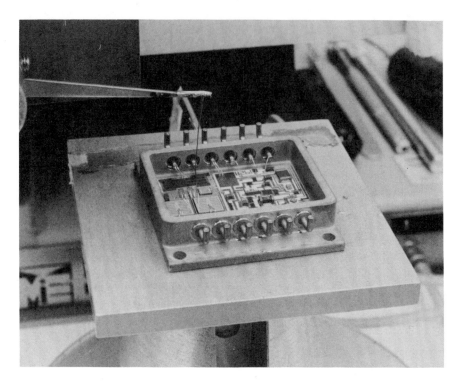

Figure 22.7   Wirebond pull test. (Courtesy of Westinghouse Electronics Systems Group.)

capacitors; F is farads, the standard measure of capacitance; and the standard symbol for a capacitor is

$$\mathrel{\substack{| \\ = \\ |}}$$

Producing Cap Banks

Since placing multiple capacitors flat on a board to be hooked up in parallel can take a great deal of board real estate, a capacitor (cap) bank subassembly may be the best alternative. To produce a cap bank, the capacitors are placed on end, side by side. The bodies of the caps are bonded together with dielectric epoxy to prevent shorting the end terminations of any one cap. The end terminations of one side of the bank are bonded together with conductive epoxy or a low-temperature solder (e.g., 60/40 SnPb or eutectic SnPb). Some applications require electrical isolation between end terminations of different caps.

Example

Figure 22.8 shows an example of a cap bank. All the negative terminations are grounded to the floor of the module housing with conductive epoxy. To achieve electrical isolation between the positive terminations, small pieces of Kapton tape are placed between each cap with dielectric epoxy. If the caps were ceramic with gold-plated terminations, the

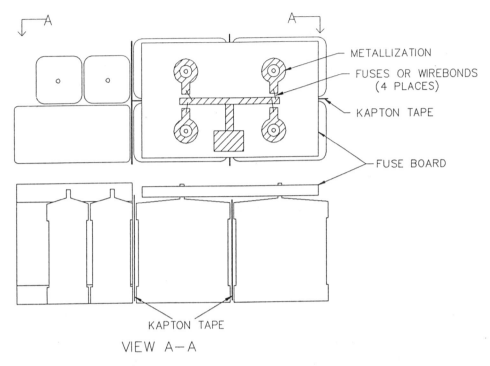

Figure 22.8 Capacitor bank with interconnecting board and fuses.

connections to the positive terminations could be accomplished with wirebonds or by attaching a gold ribbon with conductive epoxy. If the caps were tinned or had solder-coated terminations, a jumper wire could be hand-soldered to the appropriate termination. In Figure 22.8, a small ceramic board with a metallized pattern is placed over the caps; tolerancing has been calculated to allow the cap leads to fit into the drilled holes of the board. Since the metallized pattern is only on the top surface of the board, the ceramic board material in direct contact with the caps acts as an insulator. The leads are connected to the ring of metal around the holes with solder or conductive epoxy. Attached to these rings are the bonding pads needed for making the connections to the appropriate I/Os on the associated substrate. Fuse wires can also be added to this design, as needed.

22.2.3 Microwave and Radio-Frequency Subassemblies

Microwave functionality is governed as much by mechanical design as by electrical design. The placement, orientation, and proximity of components to one another can greatly influence device electromagnetic fields, and thus affect system functionality at operating frequencies associated with microwave applications. At radio frequencies (RFs), signal speeds are very fast. The signal or current travels along the metal surface; if the path length is too long, it can slow down the signal. Thus, when signal speed is the top priority, the signal path must be shortened by reducing the resistance. Resistance can be reduced by using a very electrically conductive material for the signal path, by using highly conductive die-attach materials, or by shortening the physical length of the paths. In some extreme cases, the only configuration that will meet the operating rise times required by the design is to stack the components on top of each other. In other cases, electromagnetic interference (EMI) shielding is provided by metal barriers added to the die assembly or integrated into the module housing.

Just as component placement affects electromagnetic fields, so does the wirebond configuration. Wirebonds are formed with a loop between the connection points that acts like the loop in an inductor coil. This loop, with its imposed inductance and capacitance, can change the functionality of the circuit design. Consequently, ribbon bonds are commonly used in microwave applications. A ribbon bond is not looped, and thus does not add unwanted inductance to the connection. The potential inductance can be further limited by tacking the ribbon down along the connection path, either by welding it at multiple locations or by attaching it with conductive epoxy. The ribbon is connected to the die and the substrate metallization with a welding tip split in half; one tip welds at each end of the width of the ribbon. Power is applied and the ribbon is welded across its width to the surface metallization.

Microwave subassemblies are highly customized and unique to the electrical requirements of the device in which they will be utilized. Figure 22.9 gives some examples of microwave subassemblies.

22.2.4 Summary

Elementary subassemblies are used when area restrictions are critical or when special pretesting is required. Common applications are in military and space equipment, specifically avionics or satellite applications, where weight and size are of utmost importance. Elementary subassemblies are typically composed of simple discrete functions; a die assembly is usually only a diode or transistor, and a cap bank is several caps assembled together to form one subassembly that is functionally equivalent to one large farad-valued capacitor. The need for elementary subassemblies is very design-dependent: If the design requires spreading heat under a high-power FET, a die assembly with a beryllium–oxygen tab might solve the problem; if fast rise times are required, stacking components may be the answer. These subassemblies can be built or purchased beforehand as individual products, or built as an integral part of a higher assembly, such as a hybrid.

In high-voltage applications, cross-talk and arcing potentials are critical. To prevent such problems, adequate spacing is designed into the assembly. If the real estate for such spacing is not available, spacing between two tracks can be increased by breaking up the substrate into different subassemblies. Since the current or charge travels on the surface, the path of potential arcing must travel down the side of one substrate, across the floor of the case, and up the side of another substrate. Thus, without an increase in surface area, the spacing between two high-voltage potentials is increased threefold.

22.3 CHIP CARRIER ASSEMBLIES

Chip carriers are plastic or ceramic packages that carry a single chip or die. The assembly processes used to make chip carriers are very similar to those used to make dual in-line packages (DIPs) for plastic chip carriers or ceramic packages (cerpaks) for ceramic chip carriers.

22.3.1 Plastic Chip Carrier Fabrication and Design

A plastic chip carrier package is fabricated by first punching or etching a leadframe pattern into sheet metal, typically Kovar plated with nickel and gold. The die is then epoxied to the center of the leadframe and wirebonded accordingly. The leadframe is encapsulated,

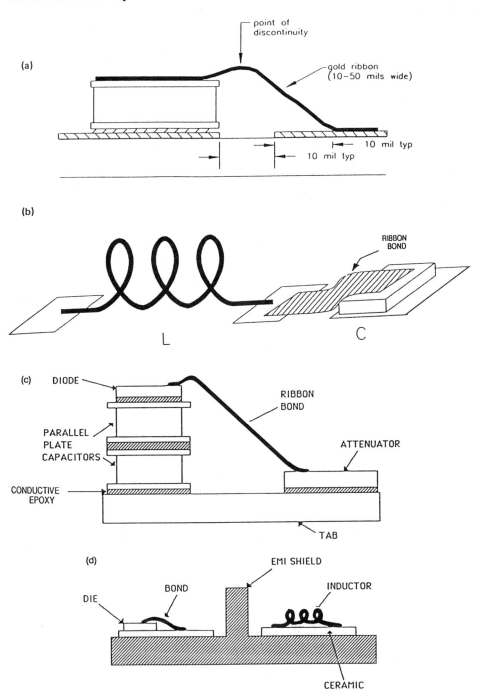

Figure 22.9 Examples of microwave subassemblies.

with die and wirebonds, within a plastic mold (plastic chip carriers are always leaded). The leads extending out of the molded body can have a gull-wing or DIP configuration, or be shaped like the letter J (J-leaded chip carriers). There are several molding techniques available for the encapsulation of plastic packages. The most widely used are transfer molding, injection molding, and reaction-injection molding.

Fabrication Using the Transfer Molding Process

In transfer molding, a plunger pushes the molding compound into a mold after the compound has been preheated to a set temperature, typically 90–95°C, for between 20 and 40 sec. The compound is then inserted into a transfer pot whose cylindrical cavity is maintained at around 170–175°C. The transfer plunger is depressed, which pushes the softened compound into the mold. The viscous molding compound flows over the chips, wirebonds, and leadframes and encapsulates the device.

There are basically two types of molds: the cavity chase mold (split-mold design with ejector pins, runners, gates, and vents) and the aperture plate mold. The latter, specifically designed for encapsulating microelectronic packages, is assembled from a series of stacked plates. The sides of the package body are formed with cutouts in the two aperture plates and the leadframes are loaded between these two plates. The runner system is in the plate above the aperture plates. On the top surface of the molded body is the surface finish plate. In a standard press without microprocessor control, the transfer pressure is controlled by throttling the higher packing pressure through a speed-control valve. Once the mold fills, the transfer pressure builds up to the packing pressure, since the flow through the throttling valve slows to a stop (Manzione, 1990).

Immediately following the molding process, the package is postcured in large ovens for 4–8 hr at 170–180°C to enable the molding compound to undergo complete conversion. The packages are then deflashed using pneumatic means or solvent deflashers (flash is the molding compound that inadvertently flows onto the leadframe). The leads are trimmed and formed (bent into shape), then solder-dipped and plated to facilitate solder attachment to the circuit board.

Fabrication Using the Injection Molding Process

The injection molding process was developed for thermoplastic materials. In this process, plastic, in the form of small pellets, falls from a feed hopper into a screw that forces the material forward through several heating zones. As the plastic is transported, it is heated to a highly viscous liquid state, arriving at the front zone fully melted, de-aired, dried, and compressed up to 20,000 lb/in.^{2}. The flights of the screw vary in each zone. When enough melt has accumulated, the screw is forced forward hydraulically as a piston, pushing the melt through a nozzle and down runners to a mold that is cooled, usually with chilled water, to freeze the plastic as quickly as possible. The removal of heat from the plastic is the time-limiting factor, which determines the overall cycle time, from seconds to minutes. The thermoplastic molding materials are not altered substantially by this process, so sprues, runners, and scrap parts can be reground and blended with virgin material for reuse (Harper, 1991).

Injection-molded parts are extremely low in cost because of the automated nature of the process and the very low cycle time. However, the molds are extremely complex, must withstand very high pressures, and are very costly. Production must be considerable to justify the mold cost and setup expense.

Reaction Injection Molding Process

Reaction injection molding is a type of injection molding used to produce thermoset shapes, usually of large sizes. Polyurethane and polyester resins are used. Reactive liquid components are prepared separately, then pumped into a mixing head, where they are thoroughly combined. The mixed liquid resins are then forced into a heated mold that may also contain fiber-mat reinforcement. The part cures like a compression-molded part. Advantages are low cost and large parts. Disadvantages are the limited choice of materials, the rather low-performance physical properties, and the difficulty of controlling the reaction process (Harper, 1991).

22.3.2 Ceramic Chip Carrier Fabrication and Design

Ceramic chip carriers are made of co-fired ceramic with tungsten metallization. The ceramic body is available in two styles, slam or flatpack, with a cavity for the die. The slam chip carrier can be single or multilayered. The die mount pad is in the center, surrounded by the wirebond pads. The bond pads via down to the buried layer and/or fan out toward the edge of the chip carrier, where they either via up to the top of the castellation (single layer) or via down to the solder-mount pad/castellation (multilayer). On the surface, surrounding the wirebond pads, is a ring frame of metallization used to solder-mount a dome-shaped lid (multilayer) or a domed lid glass (single layer).

Multilayered ceramic chip carriers are cofired ceramic layers with windows cut out in the top layers. When these layers are fired together, the windows form a cavity. The bottom layers of the chip carrier form the bottom of the package. The next layers have small windows that form the lower cavity, where the die is mounted. The third set of layers, with a slightly larger window, form the wirebonding ledge within the chip carrier. The top layer has a metallization ring surrounding the window, which is used for solder-sealing the lid onto the chip carrier. Figure 22.10 shows multilayered chip carriers.

Chip carriers are available in a cavity-up or cavity-down configuration. In the former, the cavity faces up from the surface mount pads on the bottom of the package. That is, the lid goes on the top of the package and the bottom is mounted downward to the PWB. In a cavity-down package, the lid is on the bottom of the package and is placed down on the PWB. A cavity-up package is used for traditional mounting configurations in which both routing and thermal paths are directed down through the PWB. A cavity-down package also allows the routing to be directed down to the PWB, but the thermal path is directed toward the top of the package, away from the PWB. This thermal design is sometimes used in convection-cooled systems.

Ceramic chip carriers are available in a variety of lead configurations. Leadless chip carriers have castellations. When the chip carriers are fabricated, the green tape used to make them is larger than the desired package, even considering the tape shrinkage during firing. Holes are punched into all the tape layers around the outer perimeter of the carrier, and the edges of these holes are metallized with tungsten. After the layers are laminated together, the prefired stack-up is cut to size. The outer rim of green tape is cut off through the holes, leaving evenly spaced semicircular columns around the outer perimeter of the carrier. These metallized half-circles form the castellations. The castellations provide a column of metal that allows the solder to flow up, forming the solder fillet during the surface-mounting process. Figure 22.11 shows a leadless chip carrier mounted to a board. Note the solder fillets that have formed in the castellations. Figure 22.12 shows a leaded

Figure 22.10 Examples of multilayered chip carriers. (Courtesy of Westinghouse Electronics Systems Group.)

chip carrier, also mounted to a board, on which the leads have been top-brazed to the package to give the standard gull-wing leaded style. The leads have been formed to step down to the board, providing a shock absorber. As the board temperature fluctuates, the chip carrier expands at a different rate than the board. The bent lead will give, absorbing the thermal stress between the carrier and the board.

Leaded chip carriers are fabricated like leadless ones, with the additional step of brazing on leads after the exposed tungsten metallization is plated. The leads, Kovar with nickel and gold plating, are attached using a high-temperature braze, such as a eutectic copper–silver. The leads can be brazed to the top of the package in a gull-wing configuration; to the bottom of the package; to the side of the package, for insertion mounting; or to the side of the package in a J-lead form for surface mounting.

22.3.3 Pin-Grid Array Packaging

Pin-grid array (PGA) packages are high-lead-count chip carriers. To accommodate the large number of leads without increasing the perimeter of the chip carrier, the I/Os are designed as an array of pads on the bottom of the package, instead of in a single row along the package perimeter. The leads are pedestals brazed onto the pad array, yielding a set of pins for insertion mounting on the PWB. Likewise, to accommodate the high I/O count, dual ledges for wirebonding are typically used inside a PGA. PGAs are available in both cavity-up and cavity-down configurations.

Figure 22.11 Leadless chip carrier (LCC) mounted to a PWB. (Courtesy of Westinghouse Electronics Systems Group.)

22.4 HYBRID MICROELECTRONICS ASSEMBLIES

22.4.1 Definition of a Hybrid

A hybrid microelectronics device (hybrid) with a customized electronic function consists of two or more electronic components mounted and interconnected via a substrate. The substrate is made of a dielectric material patterned with metallized signals or tracks.

Hybrids are known by many names: multicircuit packages, multichip packages (MCP), multicircuit hybrid packages (MHP), and power hybrid packages (PHP) (for hybrids with power densities greater than 10 W/in.^{2}). Some microwave hybrids are also known as integrated microwave assemblies (IMAs), but IMA is the name used for any microwave assembly; all microwave hybrids are IMAs, but not all IMAs are hybrids. The essential characteristic of a hybrid is that it contains multiple components.

22.4.2 Hybrid Design

Unlike PWBs, in which all components are individually packaged and then mounted to the board, hybrids use bare dies packaged together in a single hermetic case. A given schematic, with a parts list of needed components, can be built with individually packaged and leaded components mounted on an organic PWB, which provides the necessary routing among components. The same device can be built using bare dies, chip components (e.g., capacitors), and thick-film resistors, all mounted to a multilayer ceramic substrate

Figure 22.12 Gull-wing chip carrier mounted to a PWB. (Courtesy of Westinghouse Electronics Systems Group.)

that provides the internal routing; the substrate assembly, mounted inside a hermetically sealed metal package, forms the hybrid.

A hybrid can be as simple as two diodes in one chip carrier, or as complicated as a multiple-channel amplifier, regulator, or analog-to-digital (A/D) converter. Hybrids can be digital, analog, or a combination of both.

In the most common hybrid design, bare dies and chip components are mounted on a multilayer ceramic substrate. A hybrid substrate is typically 96% alumina with thick-film metallization; however, green tape, co-fired ceramic, and thin-film multilayering using polyamide as the dielectric may also be used, depending on the particular requirements of the design. Since bare dies are mounted, the substrate assembly must be packaged and/or sealed to protect them. The substrate assembly is either mounted inside a leaded case or encapsulated within a protective overcoat, as shown in Figure 22.13.

The advantages of hybrids are a vast reduction in volume, area, and weight; better thermal management; increased functional densities; increased frequency capabilities; and improved electrical performance. The disadvantage is increased cost.

Eliminating individual component packaging is the major factor in size reduction, allowing bare die-mounting directly onto the substrate. Components can be placed much closer, and interconnections between components are more direct. Instead of wirebonding the bare die to the package's internal leads and channeling the signal through the feedthrough to the external leads (which then have to be soldered to the PWB, which in turn routes the signal to the solder joint of another packaged die, etc.), the die on a hybrid

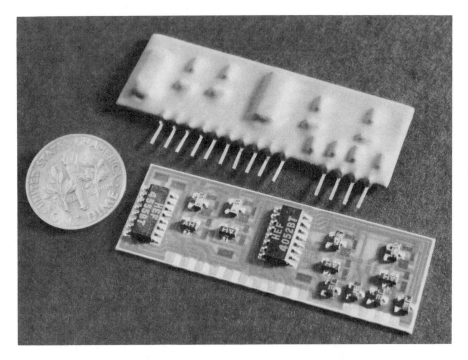

Figure 22.13 Encapsulated hybrid. (Courtesy of Westinghouse Electronics Systems Group.)

can be wirebonded to the substrate, which connects directly to the wirebond pad of the other die. Thus, electrical performance and frequencies are improved.

Thermal interfaces are also eliminated. On an organic PWB, the heat must travel from the die through the attaching media, the die package, the package leads, the solder, and finally through the PWB to a heat sink. In a hybrid, the heat travels from the die through the attaching material to the substrate and through the hybrid case to the heat sink. Not only are there fewer thermal interfaces, the thermal conductivities of the hybrid materials are higher and can dissipate more heat. Thus, thermal management is improved.

The disadvantage of using a hybrid is the increase in cost and turnaround time from electrical design to finished product. Organic PWBs use polyimide with copper metallization. Hybrids use ceramics with gold, silver, or gold alloy metallization, all of which are higher in cost.

Dies, whether discrete ones such as diodes and transistors, or integrated circuits (ICs), can be purchased in two forms: prepackaged or bare. Packaged dies come in many varieties: leadless chip carriers (LCCs) for surface mounting, DIPs for insertion mounting, plastic J-leaded or gull-winged chip carriers for surface mounting, or "canned," dies such as the three-leaded metal can packages commonly used for transistors. All packaged dies are designed for mounting on PWBs. Suppliers package the dies and then test and sort them. For example, in the case of a diode, all the bare dies are probed while still in wafer form; this "go–no go" test indicates only that the dies work like diodes or not at all. After being packaged, the diodes that passed the test are tested under full-power operating conditions. They are then sorted into bins—for example, there might be three different bins for a particular diode: one for diodes that operate only under 3 V, one for 3–7 V, and one for 7–12 V. These units can also be fully screened according to military standards, including burn-in to weed out any infant failures. Since packaged dies are in greater demand than bare dies and are manufactured in very large quantities using automated techniques, their cost is only pennies more than that of bare dies. In some cases, bare dies cost even more than their packaged counterparts because of increased handling difficulties, electrostatic discharge (ESD) damage, and decreased yields (since bare dies cannot be completely tested and screened prior to packaging). For some ICs, especially those used in microwave applications, packaging increases yield, due to the EMI and ESD protection it offers. Bare dies are more susceptible to damage than their packaged counterparts; their surface metallization can be scratched or contaminated, damaged by ESD, or chipped by tweezers, or they can simply be lost because of their tiny size (a transistor can be as small as 0.011×0.007 in.). Packaged dies have high postmounted yields, typically over 98%, while bare dies have lower yields because of their vulnerability to handling or ESD during mounting. Bare die yields can be as low as 60%, depending on the complexity, sensitivity, and process control used in their mounting, but, under proper conditions, a bare die yield of 90–97% can be expected.

The yield of a PWB assembly is normally quite high because all elements and components used in the assembly are fully tested prior to commitment to the PWB. Hybrids, however, have lower yields. Because they use several bare dies, the overall yield problem of the end product is compounded.

Hybrids can also be expensive in terms of design time. As with PWBs, layout and routing are done by computer. After being checked, the information from the design station is downloaded into a laser artwork generator (LAG), which draws out the layers on a film such as Mylar. This artwork is used to pattern the metal on the PWB. For hybrids, this artwork must first be reduced and then converted into screens and stencils to pattern the metallization onto the substrate. After board fabrication, the PWB can be assembled by mounting the components and reflowing the solder. For hybrids, components must be wirebonded before the substrate assembly can be packaged. Thus, hybrids are more labor-intensive than PWBs.

The lead time needed to order and receive components is also typically longer for hybrids. Since hybrids are used only when size limitations are extreme, they are typically not a high-volume production item. Bare dies and hybrid chip components are therefore not in as great demand as surface-mount components, and costs and delivery times are correspondingly greater.

Although extra manufacturing steps, component availability and lead times, materials and component costs, turnaround times, and yield factors can all lead to higher costs, for hybrids these cost impacts can be reduced through several procedures:

Die placement and interconnection can be automated (this requires some capital expenditure).

Full functional testing of dies can be conducted prior to mounting (requires sophisticated universal die-testing equipment that can cost approximately $2 million).

Proper precautions can be employed to reduce possible ESD damage.

Process control and operator training can be used to reduce handling damages.

Rework procedures can be used to increase final yields.

Design guidelines and checklists can be developed to reduce design cycle time.

Design engineers can be educated about the processing capabilities of the facility.

Manufacturers can be kept aware of designers' needs.

Design and manufacturing engineers can be updated on system trends and design demands.

The overall tasks of design and fabrication can be carefully planned. (Defining piece parts early and using expeditious purchasing can eliminate bottlenecks. Good planning can also keep the electrical design from going to layout prior to simulation and debugging.)

Concurrent engineering, involving all engineering fields from design through production can be utilized. (System, electrical, and mechanical designers must work with manufacturing, reliability, and production engineers to design a producible product; engineers must work with purchasing and management to coordinate efficient design and building efforts.) (See Chapter 1.)

22.4.3 Hybrid Processing

Component Mounting

The most common method of mounting components on the substrate is to attach them with epoxy. In the 1960s and 1970s, dielectric (i.e., nonconductive) epoxies were used. However, with advances in polymer engineering, epoxy adhesion was greatly improved, as were techniques for filling epoxies with metals to make them electrically conductive. To obtain a low enough resistance to allow for proper conduction, the epoxies must be 70–80% gold or silver. Because of cost, silver is the more commonly used material. Other metals are either not conductive enough, too expensive, too corrosive, or not conducive to filling.

In recent years, the use of silver-filled epoxies in military applications has been questioned. The silver in epoxies can oxidize and reduce conductivity over time; silver can also migrate over lengthy periods, collecting in one area of the epoxy joint, leaving behind an insulating layer. Data showing gradual resistance changes in the epoxy support these two concerns. However, recent studies have shown that if the epoxy is properly processed and is contained in a hermetic environment away from potential contaminants and oxidizing agents, such as air and moisture, resistance changes are insignificant. Silver migration still occurs, but its effect is negligible.

Proper Storage

The organics within the epoxies can spoil if not stored properly. Premixed epoxies are kept frozen or refrigerated. The shelf life of frozen epoxy is 1 year; of refrigerated epoxy, 6 months; and of epoxy stored at room temperature, 1–3 months. When the epoxy is to be used, it is taken out of the freezer and allowed to thaw before the container is opened; if the container is opened before the material reaches room temperature, humidity in the air can condense in the material and contaminate it. The material has to be protected by using only clean stainless steel utensils to mix or apply it; wood "tongue depressors" can leave fibers that can react with the epoxy organics. Old material left on the screen after printing should never be mixed back in with the fresh epoxy; this material should be gathered in its own container and recertified prior to use on deliverable products, or it should be set aside for engineering or prototyping use only.

Conductivity

If the device is mounted with conductive epoxy, the epoxy acts as the electrical connection; it should only be used under end terminations to avoid shorting. In high-current or microwave applications, in which the resistivity of the conductive epoxy or the minute changes of resistance over time do not meet the electrical performance criteria of the design, the capacitors should be mounted with dielectric epoxy under the entire length of the capacitor body, and the end terminations should be wirebonded to provide the electrical connection. While the same component mounting techniques can be used for both dielectric and conductive epoxies, using dielectric epoxy dictates the need for additional interconnections, usually in the form of wirebonds.

Applying Adhesives

Epoxy can be applied as either a film or a paste. Some epoxy films use a fiberglass mesh support in the prepreg or B stage to laminate the layers of an organic PWB. Epoxy films can also be formed by printing a layer of epoxy on a nonstick, inert surface (e.g., Mylar tape) and partially curing it, leaving it flexible but no longer fluid. Film adhesives can be purchased in approximately 12×12 in. sheets or can be preformed by cutting or punching. Film adhesives are from 0.003 to 0.012 in. thick. Most film adhesives require the application of pressure during cure.

Film adhesives are used for mounting large-area devices (that is, mounting LCCs approximately 0.300 in. square to larger die assemblies or chip carriers with higher I/O counts), for mounting substrates to the case floor, for mounting the case to the PWB, or for attaching the cold plate to the module housing. Preforms are not practical for mounting small dies or chip components—pastes are used for this purpose. Epoxy pastes should be hand-applied only when building prototypes and then using a stainless steel spatula or dental picks. If the facility uses automatic dispensing, the epoxy paste is placed in the dispenser gun and pressure is used to push it out. With a manual dispenser, the pressure is controlled through a trigger or foot pedal. More advanced equipment applies a set pressure pulse automatically, giving uniform amounts of epoxy. Some automatic dispensers can be programmed to deposit epoxy in predetermined locations, much as pick-and-place machines are programmed to place components.

The most common method of epoxy application is printing the epoxy, just as thick-film paste is screen-printed, using a larger mesh (typically 80 count) and a thicker emulsion (0.002–0.010 in.). The larger mesh accommodates the epoxy's larger particle size and higher viscosity. The thicker emulsion is used to regulate the thickness of the epoxy deposit. Stencil printing can also be used, as in solder paste printing.

Soldering

Other component mounting methods use solders. The die can be scrubbed down using eutectic alloy formation between the die and the substrate metallization or eutectic solder alloy preforms. Furnace mounting, using eutectic alloy preforms, is another process used with hybrids.

Chip components can also be mounted using solder paste or solder wire. The solders used for these applications are normally lead–tin alloys, rather than the gold alloys used for die attachment. They reflow at lower temperatures and are not necessarily eutectic, though they can be. To get good uniform adhesion, or wetting, of the solder to the surface metal, oxides must be removed. This can be accomplished in two ways. In the first method, the solder is reflowed in a reducing atmosphere in a furnace filled with forming gas (a mixture of inert nitrogen and approximately 5% hydrogen). The nitrogen prevents the introduction of any new oxides and the hydrogen, at the reflow temperature, reduces the oxides that have formed on the solder. This method is used for high-temperature or eutectic solders.

The other method utilizes flux to increase wetting. The flux chemically reduces any oxides on the metal surface or in the solder, and can be applied to the surface or component prior to mounting. For example, flux can be printed on a pretinned or thick-solder plated board, providing a tacky surface that temporarily holds components in place prior to solder reflow. Pretinned components can also be dipped into flux before placing them on the board surface, but this labor-intensive method is not frequently used. Solder pastes and most solder wires have flux in them. Solder pastes suspend particles of the solder alloy in a solution containing flux, emulsifiers, and volatile solvents, which burn off during reflow. Solder wires have a rosin core that contains the flux.

Solder pastes or wires are reflowed by applying heat locally with a soldering wand (soldering gun) or hot-air gun. Solder paste is applied to the substrate by hand or by screening; the component is placed in the solder and the solder is reflowed.

The solder paste can be reflowed locally, by submitting the entire substrate to the necessary heating cycle. The substrate is placed on top of a hot plate or on a reflow system that employs a belt that travels the length of the reflow equipment through zones heated conductively (by coils under the belt) and/or convectively (by hot air or nitrogen blown down onto the belt).

Furnace reflow can involve fluxless solder reflow, as in eutectic die attach, or fluxed solder. In the case of fluxed solder, the reflow is performed in a nitrogen atmosphere; a reducing atmosphere is not needed because the flux provides the reducing or wetting agents. Another way to provide the energy needed to reflow the solder exposes the substrate in the furnace to infrared radiation (IR furnace).

Finally, in the vapor-phase method, the substrate is lowered into a chamber filled with Freon vapor; the boiling point of the Freon determines the temperature of the vapor. As the part is lowered into the vapor, the solder reflows as the vapor condenses on the surface of the substrate and transfers its latent condensation energy to the solder. (Figure 22.14 shows the conceptual construction of typical vapor-phase systems.)

When using these methods, the substrates must be defluxed or cleaned prior to

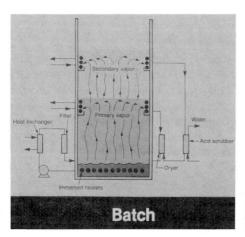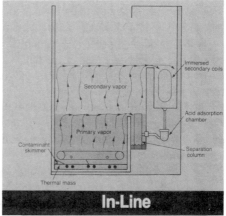

Figure 22.14 Conceptual outline of vapor-phase reflow. (Courtesy of Westinghouse Electronics Systems Group.)

mounting any bare dies. It is best to deflux the unit before it has a chance to cool down fully, or the flux may bake onto the board and penetrate the surface. The board must be cleaned while it is still warm and the flux is still tacky to minimize flux residue which can outgas later and damage the bare die. Flux damages bare dies, so they cannot be mounted using flux methods.

Dies are passivated (given a protective overcoat of a glass or an oxide) only at the wafer level. When the wafer is sawn into individual dies, the passivation is also cut. Thus, only the die surface, not its sides, is protected. Flux and flux fumes can deposit on the die, migrate under the passivation at the sides, then attack the oxides and metallization, causing loss of functionality over time.

Interconnection

Wirebonding and Ribbonbonding

The most common interconnections in a hybrid are wirebonds. In high-power and/or microwave applications, a single wirebond connection does not always meet thermal or electrical requirements. In high-power FETs, the current, and thus the heat, are dissipated through the gate connection; a single wirebond used for this connection, even if it has a large diameter (0.005 in.), might not be able to handle the thermal transfer. When the heat travels from the bond pad surface of the gate to the much smaller surface area of the wirebond, the power is bottlenecked and cannot transfer fast enough, resulting in a burned-out gate (this power loss should not be confused with the power loss in the silicon due to FET switching). Consequently, multiple wirebonds are often used, although the wire itself may be rated for the carried current. Sometimes it is better to use three or four smaller-diameter gold wirebonds instead of a single larger-diameter aluminum wirebond; the increased surface area of multiple bonds, coupled with the higher thermal conductivity of the gold wire, can greatly decrease the potential for a thermal failure under high-power conditions.

There are several different kinds of wirebonds, including the wedge bond, the ball bond, and the stitch bond (Harper, 1970). Wedge bonds are made with a wedge or chisel-shaped tool that applies pressure to the lead wire on a preheated bonding pad. Difficulties with wedge bonding include imprecise temperature control, poor wire quality, inadequately mounted silicon chips, and a poorly finished bonding tool.

Ball bonding is a process in which a small ball is formed on the end of the wire by severing the wire with a flame; the ball is then deformed under pressure against the pad area on the silicon chip. The number of steps in this bonding operation is small and the strength of the bond obtained is strong. Aluminum wire cannot be used because of its inability to form a ball when melted with a flame. However, gold wire, which is an excellent electric conductor and is more ductile than aluminum, can be used. A disadvantage of ball bonding is that a relatively large bond pad is required. Figure 22.6a shows the steps in forming a ball bond.

Stitch bonding combines some of the advantages of both wedge and ball bonding. The wire is fed through the bonding capillary, but the bonding area is smaller than for ball bonds, and no hydrogen flame is required. Both gold and aluminum wires can be bonded at a high rate. Figure 22.6b depicts the formation of a wedge bond.

Gold wire is typically bonded using thermocompression; aluminum wire is typically bonded using an ultrasonic process. Thermocompression wirebonding depends on heat and pressure. In general, the bonding equipment includes a microscope, a heated stage, and a heated wedge or capillary that applies pressure to the wire at the interface with the bonding surface. In addition, a wire-feed mechanism is required, as is some mechanism for manipulation and control. Three primary parameters affecting thermocompression bonding are force, temperature, and time. These parameters are interdependent and are affected by other conditions and factors. Minor changes in these variables can cause significant differences in bond characteristics. Low bonding temperature is desirable to avoid degradation of wire bonds due to gold–aluminum interactions. Low pressures avoid fracturing or otherwise damaging the silicon beneath the bond. The bonding tool used in the process may be of tungsten carbide, titanium carbide, sapphire, or ceramic (Schafft, 1972).

In a ball bond the weakest link occurs in the annealed wire leading to the bond. In a stitch or wedge bond, it occurs in the region of the wire in which the cross section has been reduced by the bonding tool.

Ultrasonic wirebonding also involves heat and pressure, but heat is supplied by ultrasonic energy rather than by a heated stage or capillaries. Pressure is also used, but is incidental to the effect of the ultrasonic energy. Three primary factors affecting ultrasonic wirebonds are force, time, and ultrasonic power. The ultrasonic power available for making the bond is dependent on the power setting of the oscillator power supply and the frequency adjustment of the tool. The force used is generally of the order of tens of grams and is large enough to hold the wire in place without slipping and to channel the ultrasonic energy into the bonding site without causing deformation of the wire. High power and short bonding time are usually preferred to avoid metal fatigue and to prevent the initiation of internal cracks. Lower power nevertheless gives a good surface finish and a large pull strength.

The third bonding method is a combination of ultrasonic and thermocompression wirebonding. In ultrasonic ball bonders, the ultrasonic heat is identical to that in ordinary ultrasonic bonders, but a straight-wire capillary is used to feed the wire, as on the thermocompression bonder. Also included is the flame-off device necessary to form the ball on the gold wire (Pecht, 1991).

In ribbon bonding, a gold ribbon is split-tip welded to the die and substrate. Ribbons range from 0.005 to 0.050 in. wide and from 0.001 to 0.005 in. thick. The cross-sectional area of a ribbon can offer more current-carrying capacity than a wirebond, and the larger surface area allows power transfer from the die to the ribbon. An example comparison of cross-sectional and bonding surface area is given below.

	Cross-sectional area $(10^{-5}$ in.$^2)$	Bond surface $(10^{-5}$ in.$^2)$
0.005-in.-diameter wire	1.9	2.0
0.010×0.002 in. ribbon	2.0	5.0

As can be seen in this tabulation, a ribbon gives 2.5 times the heat-transferring area for approximately the same cross-sectional area. This factor can be further improved by tacking down the ribbon at multiple places on the pad site.

Tape Automated Bonding

Another interconnection method used in hybrids is tape automated bonding (TAB). With TAB, copper metallization is built up on top of a sheet or "tape" of Kapton. The copper is patterned using photolithography or etching. If the design requires multilayering, another layer of Kapton is placed on top of the patterned copper and the etching is repeated for the second layer. The copper between the layers of Kapton is sealed as protection against corrosion; the exposed copper, including the leads, is nickel- and gold-plated to generate a corrosion barrier and a bondable surface, respectively.

Figure 22.15 shows a close-up of tape that has been patterned and plated. Bumped tab leads are shown in Figure 22.16. The die and the substrate or chip carrier are connected to bumps of copper and/or plating at the ends of the leads by a process similar to resistance welding. After the die has been bonded, it can be probed and tested while still in tape form. Figure 22.17 shows a tape-automated-bonded die. Note that the leads fan out to larger probing pads, which enable the die to be functionally tested. These probe pads are cut off when the die and its bonds are punched out of the tape. After removal from the tape, the leads must be formed, or bent, to be mounted and bonded to the chip carrier or substrates.

TAB has several advantages: The bonding area is much larger than that of a ball or wedge bond; the lead itself provides a larger cross-sectional area for carrying the current and power; and the copper also gives one of the highest possible thermal conductivities and current capacities. The disadvantages of TAB are the time and cost of designing and fabricating the tape and the capital expense of the bonding equipment. Each die must have its own tape, patterned for its bonding configuration, and each TAB design requires its own equipment program and setup. Consequently, TAB has typically been limited to high-volume production applications.

Flip-Chip Bonding

Flip-chip technology for interconnection of dies is gaining popularity owing to the fact that the number of input/output connections per chip achieved using flip-chip bonding is much greater than can be achieved using wirebonding or TAB.

The flip-chip uses a solder bump to connect the bond pad to the substrate (Figure 22.18). The flip-chip assembly consists of a ball-limiting metallurgy (BLM) on the chip bond pads, the solder bump, and a top surface metallurgy (TSM) on the substrate bond

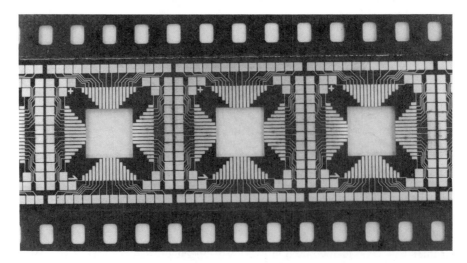

Figure 22.15 Patterned tape for TAB. (Courtesy of Westinghouse Electronics Systems Group.)

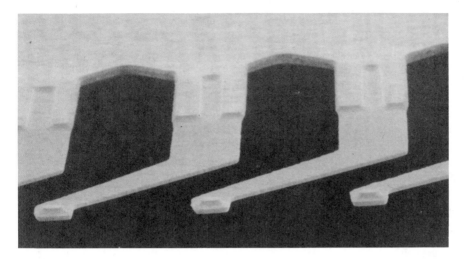

Figure 22.16 Bumped TAB leads. (Courtesy of Westinghouse Electronics Systems Group.)

Figure 22.17 Tape-automated-bonded die. (Courtesy of Westinghouse Electronics Systems Group.)

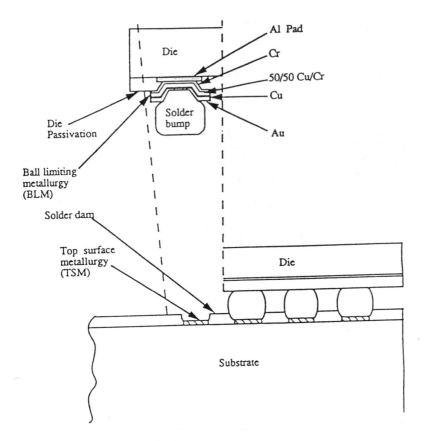

Figure 22.18 Example of a flip-chip interconnect.

pads. BLM is a multilayered structure, with an adhesive layer, a barrier layer, and a bonding layer. Common materials for the BLM structure include chromium and titanium for the adhesion layer; copper, palladium, platinum, and nickel for the barrier layer; and gold for the bonding layer. Common materials for TSM structures include nickel for the lower layer and gold for the upper bonding layer. Common materials for the solder bump include 95 Pb/5 Sn, and 50 Pb/50 In solders. Solder bump diameters typically range from 4 to 10 mils. Typical solder bump heights range from 2 to 8 mils. Minimum I/O pitches achieved using flip-chip are in the neighborhood of 10 mils. The maximum number of I/Os bonded using flip-chip interconnects are in the neighborhood of 700 or greater.

Substrate Mounting

A substrate assembly can be mounted mechanically, with epoxy, or with solder.

Mechanical Mounting

Mechanical mounting requires the engineer to attach the substrate assembly to a carrier plate that supports screws used to mount the substrate to the case. The screw support is usually a flange that extends out from under the substrate. Because ceramic cannot withstand the torque or stress a screw would impose on it, the ceramic is first attached to a metal carrier plate, and then the carrier is screwed into the case. The ceramic is formed or machined to provide either a rounded notch out of a corner or a hole in the center for the screw. This method is not commonly used for military hybrids, since a screw that penetrates the case floor would result in a loss of hermeticity. Furthermore, a floor thick enough to support the screws would counteract the weight and volume reduction offered by hybridization. In commercial applications, substrates are mounted to carrier plates, protectively overcoated, and then mechanically mounted within the systems because this method provides for quick and easy replacement of the substrate assembly; the associated cost is increased weight and volume, with decreased hermeticity.

Epoxy Mounting

Epoxy mounting can be done with epoxy paste, but this requires extensive process control of printing and curing. If the printed deposit is not smooth and uniform, air pockets or voids can form when the substrate is mounted, while curing epoxy outgasses the volatile solvents used to liquefy it. As outgassing occurs under the larger area of a substrate, pockets of gas can accumulate, forming voids that decrease adhesion and heat-transferring area. If a void is located directly under a "hot" component, the thermal resistance is greatly increased, and thermal failure can occur. Epoxy preforms, which resolve these problems, are therefore the norm. The preform can be cut or punched to the desired size and shape and comes in a very uniform thickness. The preform is placed in the case with the substrate on top of it and pressure is applied by placing a weight (a few hundred grams) on top of the substrate, if component spacing allows, or by clamping down the ends of the substrate. The unit is placed in a vacuum or nitrogen-purged oven to cure it; vacuum ovens are recommended for their ability to extract outgassed material. The door to the oven must not be opened at any time during the cure; this would greatly affect the profile, or epoxy cure schedule. If the epoxy does not reach cure temperature in the right amount of time, complete molecular interlocking cannot be obtained. If the ramp up to the cure temperature is prolonged because heat escapes from the oven, the organics that promote the interlocking can volatilize before the process is complete, reducing the adhesive strength of the epoxy.

Solder Mounting

Another way to mount substrates is with a solder preform, using a low-temperature solder, such as a lead–tin composition. The case is typically gold-plated to allow solder wetting. The back side of the substrate is also metallized. If the substrate is thick-film alumina, the metallization can be a thick-film paste of platinum–gold, platinum–palladium–gold, or silver. Gold paste is not used because of the tendency of gold to leach off the ceramic surface into the solder, forming intermetallics; this mass transfer accelerates at elevated temperatures (upper operating conditions) until all of the gold has diffused into the solder and none is left to maintain adhesion to the ceramic. The addition of platinum or palladium to the gold alloy will prevent substantial leaching. If the substrate is beryllium–oxygen, thin-film metallization must be used.

The actual process flow for solder mounting of a substrate varies for different companies and applications. One approach is to reflow the solder on the floor of the case with flux and a hot plate. Then the case is defluxed, typically in a degreaser. The substrate backplane is burnished to roughen the surface and increase surface tension and wetting. The substrate is then placed on the solder and scrubbed into place while reflowing the solder on a hot plate under a dry nitrogen flow or in an environmental glove box, as shown in Figure 22.19. The substrate can also be reflowed in place by clamping the preform and substrate into the case and sending the unit through a forming gas furnace. This method

Figure 22.19 Substrate mounting in a glove box. (Courtesy of Westinghouses Electronic Systems Group.)

cannot be used if thick-film resistors are on the substrate being mounted; exposing these thick-film resistors to a reducing atmosphere can cause changes of up to 600% in the resistor value because it reduces the oxides comprising the resistor paste.

After the substrate is in place, the I/Os are wirebonded to the case leads. A final internal visual is performed and full functional electrical tests are conducted. Figure 22.20 shows examples of delidded hybrid assemblies.

Biomedical Hybrids

Biomedical hybrids are usually not mounted to any carrier plate or case; they are implanted, obviating the need for a carrier plate. Mounting within a metal case would increase the volume and produce package corners, which could irritate the unit's surroundings. Although metal cases provide a hermetic seal against air and moisture, they would not protect the unit from the hostile environment of body fluids (acids, enzymes, and other organics). Biomedical hybrids must be protectively overcoated with a polymeric material that will be inert in its environment, yet withstand the chemical hostility of its surroundings.

Biomedical companies have found that be eliminating all contaminants prior to and during the application of the protective overcoat and by carefully controlling the curing process, complete and total adhesion can be obtained. The entire surface is coated; no voids or pockets are formed along the substrate assembly surface. Although the coating will absorb moisture from its surroundings, the moisture has no place to go; it penetrates to a certain depth of the coating and stops since there is no void to which the moisture can

Figure 22.20 Hybrid assemblies. (Courtesy of Westinghouse Electronics Systems Group.)

mass-transfer and condense. The moisture remains trapped in the overcoat material; once the surface is saturated, it absorbs no more moisture (Troyk, 1987).

Thermal stress in a biomedical hybrid is not as great as in military and most commercial applications; although the temperature is slightly elevated (98.6°F), it is fairly constant. The maximum terperature gradient that a biomedical device would experience is only a few degrees. Thus, thermal management is not a crucial issue for these devices.

Package Sealing

Once the hybrid passes precap inspection and electrical testing, it is sealed. The most common sealing methods are welding and solder sealing.

Welding

In resistance welding, an electrode rolls over the edges of the lid while current and pressure are applied. The current heats the lid and case because of the resistance of the case material, so this method works only on cases and lids made of materials with high resistivities, such as Kovar or stainless steel. If the case itself is made of a copper alloy for thermal and grounding purposes, then a Kovar or stainless steel seal ring can be brazed to the top of the case side walls to provide the necessary resistance.

Another welding method uses a laser to hermetically seal the lid and case together. Laser welding can be done on a wide variety of materials, including aluminum–silicon alloys, ferrous alloys, and nickel alloys. Figure 22.21 shows the different lid configurations used for soldering and welding.

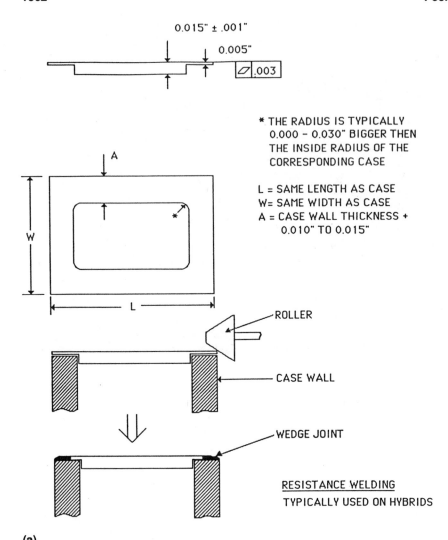

0.015" ± .001"

0.005"

.003

* THE RADIUS IS TYPICALLY
0.000 – 0.030" BIGGER THEN
THE INSIDE RADIUS OF THE
CORRESPONDING CASE

L = SAME LENGTH AS CASE
W= SAME WIDTH AS CASE
A = CASE WALL THICKNESS +
   0.010" TO 0.015"

ROLLER

CASE WALL

WEDGE JOINT

RESISTANCE WELDING
TYPICALLY USED ON HYBRIDS

A

W

L

(a)

Figure 22.21   Examples of (a) sealing lids and (b) solder sealing for hybrid packages.

Lid Deflection

Lid deflection must be taken into account when designing an electronic package, whether for a hybrid or for a module. The metal must be thin enough to allow the lid and seal ring surface to reach welding temperature without the entire case being heated. Usually, the lids used for resistance welding have a outer perimeter thickness of 0.005 in., while the center metal is 0.010 to 0.015 in. thick, as shown in Figure 22.21. This added center thickness gives the lid more rigidity. However, it still allows lid deflection to occur during operation and testing conditions.

Example

In military applications, hybrids must pass various environmental tests, including leak testing, centrifuging, and vibrational testing. These tests impose stresses that can cause lid

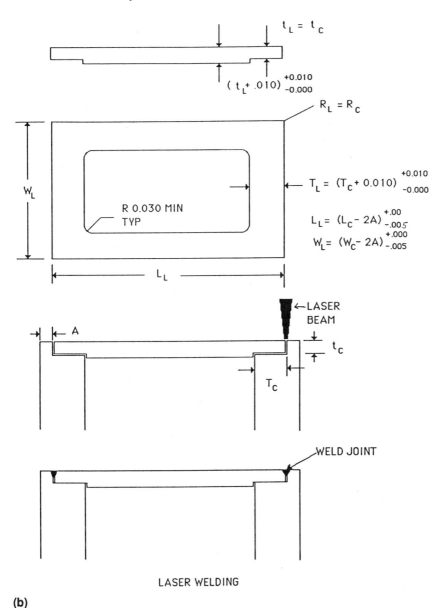

LASER WELDING

(b)

deflection. For example, when a unit is leak-tested, the sealed package is exposed to a differential pressure of 2 atm (29.4 psi). If a standard 1 in. × 2 in. hybrid is leak-tested, the total deflection for a uniformly distributed load can be expressed as

$$y = k_1 \frac{\omega r^4}{E t^3}$$

where y is the total deflection, E is Young's modulus of the lid material, r is the width of the lid, R is the length of the lid, t is the lid thickness, ω is a uniformly distributed load in psi, and k_1 is a coefficient dependent on the ratio of R to r and the way the lid is supported (Avallone and Baumeister, 1987).

A standard 1 in. $\times$ 2 in. hybrid would have a 1 in. $\times$ 2 in. $\times$ 0.015 in. Kovar lid. Thus $r = 1$ in., $R = 2$ in., $t = 0.015$ in., and $E = 20 \times 10^6$ psi (see Table 22.4). This lid is welded or attached to the case, implying that the lid is fixed on all edges.

Using a fixed support model and given that $R/r = 2$, $k_1 = 0.0277$ (Avallone and Baumeister, 1987). Substituting the given information into the equation yields the following:

$$y = 0.0277 \frac{29.4 \times 1^4}{20,000,000 \times 0.015^3}$$

$$= 0.012 \text{ in.}$$

Checking this number against deflection measured during leak testing, the calculated value is much smaller than the measured value of 0.030 in. Further investigation indicates that the case walls are not totally rigid; they can flex at the floor of the case during leak testing. This suggests that the lid is not totally fixed.

Recalculating for a simply supported model changes k_1 from 0.0277 to 0.1106 (Avallone and Baumeister, 1987), which in turn gives a deflection of 0.048 in. This does indeed indicate that the actual lid deflection (0.030 in.) is between that of a fixed lid (0.012

Table 22.4 Typical Properties of Metallic and Ceramic Materials Commonly Used in Frenchtown Ceramic-Metal Brazed Assemblies

Property				(Metallic materials)				
	Steel	Stainless steel 304	Kovar	Alloy 42	Nickel	Molybdenum	OFHC (copper)	WCu10
Composition by weight %	Fe 99+	Fe 67 Cr 20 Ni 10	Fe 54 Co 17 Ni 29	Fe 58 Fe 42	Ni 99.0 min.	MO 99.9	CU 99.96	Cu 10 W 90
Specific gravity (g/cm^3)	8.7	7.8	8.4	8.2	8.8	10.2	8.9	17.2
Hardness Rockwell	50	90	70-85	60-80	50-80	55-60	10-45	30
(B, C, or 45N scale)	B	B	B	B	B	B	B	C
Tensile strength (x 10^3 psi)	55	85	70	60	60	170	30	125
Compressive strength (x 10^3 psi)	—	—	—	—	—	—	—	—
Young's modulus of elasticity (x 10^6 psi)	28	28	20	20	30	50	17	37
Coefficient of linear thermal expansion (x 10^{-7}/°C, 20-500°C)	140	190	53	73	150	56	180	60
Thermal conductivity (cal cm/cm^2 see°C) 20°C	0.14	0.33	0.040	0.036	0.204	0.380	0.940	0.50
100°C	0.16	0.39	0.043	0.041	0.200	0.330	0.940	0.49
Specific heat (cal/g °C)	—	0.12	0.104	0.120	0.130	0.060	0.092	0.039
Electrical conductivity	—	—	—	—	—	—	100	40

Property	(Ceramic materials)		
	Alumina FA-94	Alumina 2082	Beryllia thermal
Composition by weight %	Al$_2$O$_3$ 94	Al$_2$O$_3$ 94	BeO[b] 99.5
Specific gravity (g/cm^3)	3.63	3.60	2.80
Hardness Rockwell	80	80	60
(B, C, or 45N scale)	45N	45N	45N
Tensile strength (x 10^3 psi)	35	33	22
Compressive strength (x 10^3 psi)	410	410	225
Young's modulus of elasticity (x 10^6 psi)	45	45	21
Coefficient of linear thermal expansion (x 10^{-7}/°C, 20-500°C)	71	70	73
Thermal conductivity (cal cm/cm^2 see°C) 20°C	0.05	0.05	0.55
100°C	0.04	0.04	0.40
Specific heat (cal/g °C)	0.19	0.19	0.25
Electrical conductivity	—	—	—

[a] These materials are most often used by Frenchtown Ceramics to produce ceramic-metal brazed assemblies. Other materials are available.
[b] Registered trademark, Brush Wellman.
Courtesy of Frenchtown Ceramics, Inc., Frenchtown, N.J.

in.) and that of a simply supported lid (0.048 in.). For future calculations of this hybrid's deflections, a simply supported model should be used to ensure that the calculated value always reflects the worst-case situation.

This example demonstrates that lid deflection must be accounted for when designing a hybrid. To ensure that the lid does not come into contact with any components, potentially creating shorts or damaging wirebonds, the bottom of the lid must be 0.048 in. above the highest component or wirebond in the hybrid.

Standoff Posts

If the additional height needed to allow for lid deflection is not available because of system height limitations, the design must be modified in another way to compensate for lid deflection. One approach is to use a solder seal or laser weld so the lid can be thicker and experience less deflection. Another possibility is to reinforce the lid by adding ridges or dimples. The most common solution, however, is to incorporate standoff posts, or spacers, in the assembly. This can be done by mounting posts on the surface of the substrate assembly that support the lid and limit its deflection. Care must be taken when designing these posts. They must support the lid without deforming it (pointed posts could dent or even puncture the lid during leak testing). They must also account for stack-up tolerances to adequately limit deflection without exerting a force on the lid during sealing.

Another necessity in standoff post design is accounting for all the forces and resulting deflections to which the case will be exposed. To continue the earlier example, a 1 in. $\times$ 2 in. hybrid tested per MIL-H-38534 will be exposed to a 2-atm pressure differential during leak testing, $5000 \times g$ during centrifuging, and vibration at $200 \times g$ rms during particle impact noise detection (PIND) testing. The previous example indicated a worst-case deflection of 0.048 in. Similar calculations yield expected worst-case deflections of 0.037 in. and 0.0015 in. for centrifuging and PIND, respectively. These calculations yield a maximum expected deflection of 0.048 in. during leak testing. Since the lid will vibrate 0.0015 in. during PIND, the posts must be designed to prevent the lid from contacting the components. This design will involve a complex structural analysis to determine what number of posts at what height and placement will accommodate the vibration potential. The posts must be short enough to remain below the vibration range, so that the lid will not tap against the posts during PIND and cause a noise detection failure.

22.4.4 Hybrid Packaging

Once the substrate assembly, mounting, and wirebonding are completed, the substrate is ready to be mounted in the package or case. Typical hybrid cases are metal with glass or ceramic feedthroughs, or ceramic with brazed-on leads.

Metal Packaging

Metal cases are fabricated in three ways. The first is two-piece construction with a Kovar plate brazed to a window frame of Kovar. Both pieces are nickel- and gold-plated prior to brazing. The window frame is typically machined out of a solid piece of metal. A second technique for fabricating a metal package is to machine a "bathtub" out of a solid piece of metal. A third fabrication method molds the metal in the shape of a bathtub. However the side walls of a gull-winged leaded case are formed—by brazing, machining, or molding—the side walls must have holes for feedthroughs drilled prior to plating. For insertion-mounted hybrids, the holes are drilled in the bottom or floor of the case along the inside perimeter.

Glass Seals

To fabricate a matched seal, a donut of glass or a glass bead is inserted into the unplated holes in the window frame or case; an unplated Kovar lead is placed in the center of the donut. The case is sent, with glass beads and leads in place, through a furnace profiled to allow the glass to melt. The glass reflows and fuses with the oxide on the surface of the Kovar. A mild etch or cleaning method is used to prepare the remaining Kovar surface for plating. The glass used in this process must have a coefficient of thermal expansion (CTE) close, or equal, to that of the Kovar. Figure 22.22 shows a side view of a matched glass seal.

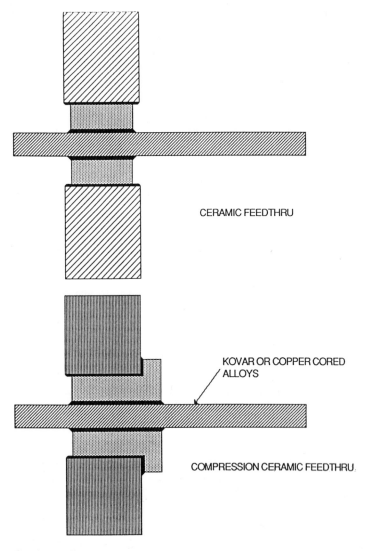

Figure 22.22 Various feedthrough configurations.

In a compression glass seal, another common seal, the glass is reflowed to a Kovar eyelet or ferrule. The lead and ferrule are plated as usual. Finished feedthroughs are brazed into the side walls of the case. Figure 22.23 shows the machined case, the feedthroughs, the solder preforms, the fixturing used to braze the feedthroughs into the case, and a finished metal case with compression feedthroughs. The fixturing used for the brazing is typically graphite, which can withstand the brazing temperatures without losing its shape, and will not adhere to the brazing material. The case wall material can be Kovar or, in power applications, a copper alloy. The ferrule and the glass are put in compression because the expansion rate of the surrounding metal is greater than that of the glass. Glasses typically used in feedthroughs have a design tensile strength of only 1000 psi; tensile strengths of approximately 2000 psi may be obtained with certain glass compositions. The strength can be further increased to 3000 psi by annealing the glass or to 20,000 psi by tempering it (Stamps, 1990). Although these strengthening methods cannot be used for feedthroughs, the strength of the glass can be greatly increased by placing it in compression. Actual compression strength measurements are difficult to obtain; the glass usually exhibits tensile failure before the compression strength is gauged, because of slight bending or torquing during the measurement. Theoretically, the compression strength is in the 100 ksi range. While such a case configuration involves more processing, the glass seals are much stronger.

Glass-to-metal cases have been used in the industry for decades. They are readily available, use proven technology, are relatively inexpensive, and provide hermeticity. However, glass forms a thin, brittle, uncompressed meniscus during the reflow process (see

Figure 22.23 Compression feedthrough case components and fixturing. (Courtesy of Westinghouse Electronics Systems Group.)

Figure 22.22), so the strength of the glass within the meniscus is low. Glass-to-metal seals are plagued by meniscus cracking due to thermal and mechanical stresses applied during temperature cycling and chip-out caused by lead forming or centrifugal forces. When the glass chips out in the meniscus area, the lead base metal, typically unplated Kovar, is exposed and becomes a corrosion site. If the chip-out occurs inside a hermetically sealed package, corrosion forms much more slowly; however, on the exterior of the case, a chip-out will eventually corrode. Cracks can lead to chip-outs and, in the matched glass configuration, can propagate, theoretically producing a loss of hermeticity. The military has established visual inspection criteria specifying cracks and chip-outs as rejectable.

Some methods for repairing glass chip-outs have been tried. The most common is to cover the chipped-out area with an epoxy or overcoat. One argument against this practice is that the epoxy or overcoat would not be sealed within a hermetic atmosphere and would absorb moisture that would eventually migrate and reach the Kovar, leading to corrosion or dendritic growth. A second argument is that the epoxy would make it impossible to see the extent of the chip-outs, cracks, or other possible problems, such as foreign material, embedded material in the glass, or glass overrun. The military has thus rejected this repair process.

Another method sometimes attempted is to brush-plate the exposed base metal with gold. Unfortunately, the gold by itself, without a nickel underplating, does not provide a corrosion barrier; thus this method only postpones the development of a corrosion site. Furthermore, the gold in direct contact with the current-carrying Kovar lead sets up an electrolytic reaction that acts like a tiny battery and can result in even more extensive corrosion.

Ceramic Feedthroughs

In an effort to eliminate cracked-glass yield problems, U.S. Department of Defense contractors incorporate ceramic feedthroughs. Ceramic and metal packages with ceramic feedthroughs have been used in space applications for many years. Ceramic is much stronger than glass; 95% alumina, for example, has a tensile strength of 30,000 psi and a compression strength of 300,000 psi, making a ceramic feedthrough an order of magnitude stronger than a glass feedthrough (Stamps, 1990). Furthermore, it does not have a meniscus. Ceramic feedthroughs are formed by metallizing the inside hole and outer ring of a ceramic donut, then brazing in the lead. These feedthroughs can then be brazed into the case wall at lower temperatures.

Figure 22.22 depicts the differences between glass and ceramic feedthroughs. Ceramic feedthroughs offer greater strength and the absence of a meniscus, with its associated cracking and chip-out problems. Because of the increased strength of the material, less metal must surround the feedthrough to ensure the package's mechanical integrity. With glass seals, MIL-H-38534 requires that there be at least 0.040 in. of metal between the glass and the seal surface, to protect the glass from stress during resistance welding. No standard for ceramic is given in MIL-H-38534; however, ceramic feedthrough packages used for space applications typically measure under 0.040 in. from the ceramic to the seal surface. Of course, any package used for military or space applications must be qualified for very stringent environments and testing levels.

Although ceramic is stronger than glass and better able to withstand the stresses imposed during processing, handling, operation, and testing, ceramic feedthrough packages have disadvantages. They are more expensive than glass feedthrough packages, in which the feedthrough and lead are inserted into the case in one step, in a furnace programmed to reflow the glass. Ceramic feedthroughs, in contrast, are much more labor-intensive. The ceramic must be formed into donuts by machining a sheet of fired ceramic the same thickness as the desired donut height by laser drilling or ultrasonic milling.

As the laser cuts through the ceramic, it melts the material, or slag, splattering it onto the surrounding surface. Slag forms burrs with very smooth, glasslike surfaces to which metallization will not adhere properly. To prepare the ceramic donut for metallization, the surfaces must be cleaned of slag. One cleaning method is sandblasting, in which abrasive slurry is forced through the holes under high pressure. An etching process is sometimes used, but this can lead to surface pitting. Another method, developed by Laserage, a laser machining company, involves heat-treating the laser-machined ceramic, which leaves the machined surfaces with a finish similar to that of fired ceramic. Once the ceramic is prepared, it must be metallized both inside the hole and on the outer perimeter. This can be done by painting on a thick-film metal paste and firing it to the ceramic.

Another method of forming ceramic donuts is to mold them into shape while the ceramic is still green, or nonfired. Before firing, a refractory metal is applied to the donut hole and perimeter, then the unit is cofired. The metal must be nickel-plated to provide a corrosion barrier, and the nickel must in turn be plated with a solderable or brazable metal, typically gold. Once the ceramic donut is formed and metallized, leads or pins must be brazed into the donut holes, and the feedthrough must be brazed or soldered into the metal package. In some cases, both brazing steps are performed together, but this requires special tooling to hold the leads and donuts in place during the process.

Not only are ceramic feedthrough cases more labor-intensive than using glass cases, their manufacture also requires higher temperatures. Because ceramic firing and brazing temperatures are much higher than glass reflow temperatures, more sophisticated furnaces are required. The additional labor, higher material cost, and elaborate equipment needed all make ceramic feedthrough cases more expensive than their glass counterparts.

Ceramic also has different electrical properties than the standard glass used in electronic packaging, which may affect the package design, depending on the electrical requirements of the unit. For example, ceramic may enhance high-voltage packages because its porosity keeps voltage from traveling across the ceramic insulator and shorting to the case. However, in analog or microwave applications, ceramic feedthroughs may be undesirable because of their higher dielectric constant. The dielectric constant of the insulating material will determine the impedance of the feedthrough. The impedance of a coaxial cable or a feedthrough is

$$Z = \frac{60}{\sqrt{\varepsilon}} \ln \frac{b}{a}$$

where Z is the characteristic impedance in ohms (Ω), ε is the dielectric constant of the insulating material (i.e., the glass or ceramic), b is the radius of the insulating material, and a is the radius of the lead (Leahy, 1989).

The dielectric constant of the glass used for feedthroughs is approximately 5; dielectric constants of ceramics are higher [e.g., 96–99% alumina ceramics have dielectric constants of approximately 9–10 (Leahy, 1989)]. If a glass feedthrough were converted to ceramic with no change in dimensions, the impedance would be decreased.

Ceramic Packaging

Co-fired ceramic is a frequently used alternative case material. Although they are manufactured the same way as ceramic chip carriers, co-fired cases are larger and usually come with top-brazed leads on two sides (full-winged lead configuration).

Ceramic cases can have an interior-mounted substrate, or the base of the case can act as a multilayered substrate (an integral substrate package, an example of which is shown in Figure 22.24). In medium- to high-volume applications (e.g., orders of approximately 1000 or more cases), integral substrate packaging can greatly reduce manufacturing costs. Once the electrical design is fixed and the vendor has tooled up to make the case, these cases can be produced less expensively than metal cases.

In addition to reducing the cost, an integral substrate case eliminates the need to fabricate a substrate, which lowers the hybrid's material and labor costs. The integral substrate also eliminates the need to mount the substrate in a case and wirebond it to the package leads; interconnection is internal to the multiple layers of the case, further reducing labor costs.

Other advantages of ceramic packaging are improved yields and reliability. Ceramic

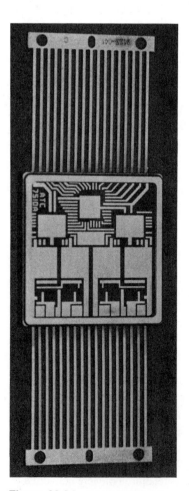

Figure 22.24   Integral substrate ceramic hybrid case. (Courtesy of Kyocera America.)

cases eliminate yield problems associated with glass feedthrough cracking and chip-outs. The ceramic package is appealing for space and military applications because of its lower weight and noncorrosiveness. As with ceramic feedthroughs, conversion from metal to all-ceramic cases is not always practical. Again, in analog or microwave devices, the refractory metals used to fabricate co-fired ceramic packages may not carry current fast enough, so the unit cannot operate at the necessary frequencies. Furthermore, the dielectric constant of the ceramic package may affect the electrical performance of a microwave device, causing an impedance mismatch. Microwave packaging design is an example of the critical need for concurrent mechanical, electrical, and material engineering.

Customization of Electronic Packaging

Many designs dictate specific requirements that cannot be met with standard metal or ceramic cases. High-power hybrids might utilize a case with a heat-spreading beryllium–oxygen bottom and a weldable Kovar window frame; others might have a metal base, ceramic side walls, special grounding requirements, and so on. Ceramic and metal combinations are frequently used. Just as the hybrid itself has a customized electronic function, hybrid packaging often requires customized cases to meet specifications.

The ceramics most commonly used in production today are 92–99% alumina and 99% beryllia. Common packaging metals include Kovar, copper alloys, molybdenum, copper- or nickel-clad molybdenum, copper-clad Invar, copper tungsten, and aluminum. These materials can be combined through special techniques and advanced materials to meet packaging requirements. Table 22.4 lists the properties of commonly used industrial ceramics and metals.

Special compliant adhesives can absorb the mismatch in thermal expansion between the metals and the ceramic. Other methods of attaching metals to ceramic have been developed, such as direct bonding methods.

Research is being conducted with an increasing number of materials that offer even higher thermal conductivities, increased strength, better-matched coefficient of thermal expansion (CTE), lighter weight, and so forth. Today's leading-edge materials include aluminum nitride, silicon carbide, A40 (silicon–aluminum material), and other metal-matrix and composite materials.

22.5 PRINTED WIRING BOARD ASSEMBLIES

22.5.1 Introduction to Printed Wiring Board Assemblies

There are various kinds of printed wiring boards (PWBs) or printed circuit boards. The oldest and most common are organic boards. Ceramic boards, offering better thermal properties with closer CTE matching to the components, have been around for more than 20 years, but recently have been used increasingly in high-power applications, such as standard electronic module (SEM) power supplies. The combination of polyamide and integrated circuit technology has produced high-density surface-mount boards with finer line definition of less than 0.001 in. Another board fabrication method directly bonds copper foil to ceramic; directly bonded copper enables vast improvements in power dissipation.

PWBs can use mixed technology; they can have both insertion-mounted components, plugged into the PWB, and surface-mounted components on top of the board. However, a single component must be either inserted or surface-mounted.

All PWBs have similar assembly capabilities, but the type of board will limit the type of assembly. An organic board with through-hole intralayer routing can be assembled either by insertion or surface mounting, depending on the components and the layout design. Ceramic boards are only surface-mounted. However, ceramic boards can easily be made double-sided by metallizing both sides or by sandwiching two boards around a carrier plate or heat sink.

Direct-bond copper offers strong, lightweight boards for high-power applications, but has limited layering and line definition. Another option, using thin-film polyamide boards, is capable of supporting bare-die applications.

22.5.2 Organic PWB Assemblies

Introduction to Organic PWB Assemblies

Organic boards are composite structures; for example, in epoxy glass boards, glass fibers are suspended in an epoxy resin. Polyamide boards are composed of glass fabric, with polyamide as the dielectric material. These are the most common industry combinations. The individual layers of organic boards are metallized with copper, have patterns etched on them, and then are laminated together. Interconnection between the layers is accomplished with through-holes, blind vias, or buried vias. The method of intralayer connection and the design of the mounting pads determine the type of assembly used. If only insertion-mounted components are used, then all connections from the components to the board are through-holes; the mounting pads are actually holes in the board. Routing signals from one component to another can be accomplished by through-hole, blind via, or buried via. Since the assembled board will have holes with leads extending through them, insertion boards typically do not have components mounted on both sides. Any components mounted on the bottom of the board are typically small, hermetic chip components, mounted with epoxy to hold them in place as they travel through the solder wave.

Boards routed with buried vias have limited reworkability and lower yields, but they offer increased surface-mounting area and fewer routing layers. These surface-mount designs can also be routed to allow double-sided mounting. Surface-mount boards must be laid out to include mounting pads of the right size, component tolerances, and proper solder fillets. These boards often have both leadless and leaded components; the latter require surface-mount lead forming.

Electrical parameters must also be taken into account during board layout. Components with a high voltage potential between them should be separated to prevent arcing. When an arc occurs, the voltage travels across the board material and can burn a trail, or carbon track, in the polymer on the board surface. If the arcing itself does not short out the circuit or damage the components, the carbon track left behind can cause shorts later on. Carbon tracking can be avoided by putting additional space between all tracks and components, but size limitations usually preclude this solution. The more common prevention method is to give the board assemblies a protective overcoat that reduces the available surface path for the voltage.

Boards can also be designed to accommodate both inserted and surface-mounted components together. The layout of mixed-technology boards greatly influences the assembly processing. If all inserted components are mounted on one side, low-profile, hermetic, surface-mounted chip components can be tacked down, or epoxied, to the bottom side and reflow can be accomplished with wave solder. Therefore, mixed-technology processing is similar to that for insertion mounting. If inserted and surface-mounted components are placed on the same side, the processing is similar to surface mounting.

Heat Sink Attachment

For low-power commercial applications, components are mounted directly on the PWB. For high-power or military applications, the first step in the assembly process is to attach the PWB to a heat sink or carrier plate. With an insertion-mounted board, the carrier is either an edge support or a thermal ladder. The edge support, which can be attached to the bottom or top of the PWB or wrapped around its sides, braces the edge of the PWB, leaving the center free for device placement. A thermal ladder is a heat sink attached to the top of the PWB with windows or slots cut in it; the component (i.e., a plastic DIP, ceramic DIP, or cerpak) straddles the ladder and the leads go through the slots into the PWB, as shown in Figure 22.25. This design provides a thermal path directly from the ceramic component to the heat sink, rather than through the leads.

Surface-mount boards typically use some edge supports, whether leadless or with a configuration that permits both leads and package body to be mounted on the plane of the board. Straddling a thermal ladder is difficult with these designs.

Heat sinks can also be attached to the bottom using a B-stage epoxy for organic surface-mount PWBs. The PWB, epoxy film, and carrier plate are cured together under controlled pressure and temperature. In some cases the PWB assembly is double-sided, sandwiching two boards over the heat sink; this "cored" board doubles the surface-mountable area with a minimal increase in board assembly height or volume. Although this design forces a thermal path through the PWB, the components are in contact with the surface and assist with primary cooling, as depicted in Figure 22.25. Thermal management of this design is enhanced by placing vias under the packages, which are not only used for routing, but also act as thermal vias, to provide a more efficient and direct path from component to heat sink.

Heat sink or core materials offer improved thermal conductivity over the PWB material. These materials include aluminum, Kovar, copper-clad Invar, copper-clad nickel, graphite composites, copper-clad molybdenum, and metal-matrix composites, such as silicon–aluminum materials.

Aluminum has high thermal conductivity and is lightweight, less corrosive than most metals, inexpensive, easy to machine, and readily available. Boards mounted to aluminum heat sinks are often used for SEM modules. Copper has many of the same properties as aluminum, but it is highly corrosive and heavy. Both aluminum and copper have excessive CTEs compared with those of either organic or surface-mount components. Copper-clad Invar offers the superior thermal and electrical conductivity of copper, while an Invar center gives the composite material a more desirable CTE. Copper-clad Invar has a lower CTE than copper, but a higher one than Invar. Bonding the board to the core material gives the assembly a CTE approaching that of the leadless ceramic chip carriers (LCCCs). However, copper-clad Invar is many times heavier and more expensive than aluminum, and is available only from Texas Instruments or its sales representatives.

Graphite fiber composites offer high thermal conductivity, approaching that of copper-clad material, yet are lightweight like aluminum. However, graphite is difficult to machine. Composites, like graphite and metal-matrix heat sinks, can have thermal conductivity and CTE tailored to the assembly. Since the direction or orientation of the fibers in a graphite composite directs the thermal path, resistance along the path between the components and the heat sink can be reduced by orienting the fibers in the same direction as the path. Likewise, controlling the composition of the composite controls, to a degree, conductivity and CTE. For example, the thermal conductivity of an aluminum–silicon

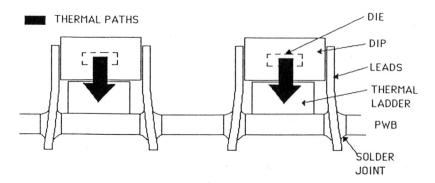

(a) INSERTION MOUNT ORGANIC PWB WITH THERMAL LADDER. HEAT IS DISSIPATED FROM THE DIE THROUGH THE DIP PACKAGE FLOOR TO THE HEAT SINK.

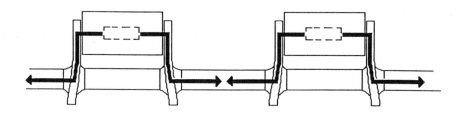

(b) INSERTION MOUNTED ORGANIC PWB. SINCE THE DIP STANDS OFF THE PWB, THE HEAT MUST BE TRANSFERRED FROM THE DIE, THROUGH THE DIP PACKAGE, THROUGH THE LEADS, THROUGH THE SOLDER, TO THE PWB, OUT TO AN EDGE SUPPORT HEAT SINK.

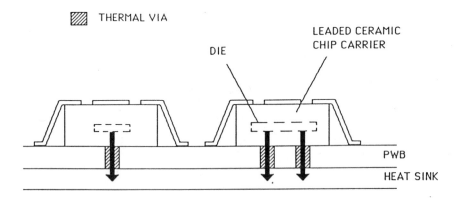

(c) LEADED SURFACE MOUNTED ORGANIC PWB WITH CAVITY UP CHIP CARRIERS

Figure 22.25 Thermal paths.

composite can be increased by using more aluminum—a better conductor—in the material, at the cost of an increase in CTE. As the amount of aluminum is increased, both the conductivity and CTE move toward those of aluminum. Thus, composite materials must be designed both metallurgically and mechanically to optimize heat sink design. Research and development of new composite materials is ongoing.

Most heat sinks can be attached with epoxy films, usually thermally conductive B-stage or prepreg sheets, or with other film adhesives. Material and adhesive choices depend on the requirements of the system or PWB assembly. If cost and weight are more critical than thermal management—that is, if the assembly will not be subjected to severe or rapid temperature changes—aluminum attached with standard B-stage is appropriate. If thermal management or power dissipation is critical but there are no restrictions on weight or outgassing, then copper Invar copper attached with a flexible silicon adhesive to absorb the CTE mismatch might be preferred. If thermal management and weight are both critical, the answer might be a graphite heat sink with silicone adhesive.

When military specifications are followed, the components of the PWB determine heat sink and adhesive design. For example, a PWB assembly mounted inside a hermetic module along with nonhermetic or bare-chip devices could not use a silicon adhesive because of military outgassing limitations. The heat sink would not only have to dissipate any heat exceeding the maximum junction temperature, it would also need a CTE closely matching that of the PWB. In this case, the more expensive developmental composite materials would have to be considered, and a less flexible film adhesive meeting outgassing limitations would have to be used. The components would have to be designed or packaged to withstand their CTE mismatch with the PWB, either by being small enough that the mismatch would have little effect, or by incorporating leads formed to absorb the stresses caused by the CTE differential.

Insertion Mounting

On an insertion-mounted board, all components and subassemblies must be leaded, and the leads must be configured or mechanically formed for insertion into the mounting holes of the board. On pin-grid arrays, the leads are already in the correct configuration for insertion, perpendicular to the bottom of the package. Dual in-line packages (DIPs) are also designed for insertion, with the leads along the side of the package, pointing down. The military standard configuration specifies leads at a 6° angle from perpendicular.

Components such as resistors and capacitors are leaded round cylinders, with leads formed into a right angle for insertion by pick-and-place equipment. This machine sends a robotic arm to the bin, tape, or feeder to select the component, picks it up by vacuum or tweezer grabbing, transports it to the lead former, where it is pushed into a troughlike fixture that bends the leads as they are forced against its side walls. Lead forming can also be done by two robotic arms that grab the leads and turn to bend them appropriately. The holding arm then carries the lead-formed part to the board, where it inserts or mounts it in a preprogrammed location, as shown in Figure 22.26. Of course, there are many other ways of accomplishing the lead forming. This is done within tight tolerances and within seconds. DIPs are normally supplied in tubes or feeders for the placement equipment, with leads already formed. Chip carriers and hybrids normally have preformed leads created by arms that grab the leads and force them into the necessary L shape while the package is tightly cradled. In high-volume applications, the feeders form the leads. Figure 22.27 shows an example of lead forming. After the components have been inserted, automatic lead trimmers cut and flatten the leads coming out of the bottom of the board (see Figure 22.28).

Figure 22.26 Automatic component placement. (Courtesy of Westinghouse Electronics Systems Group.)

To prepare the boards for component placement, the PWB is plated with a tin flash that protects the copper metallization against corrosion, or a thick plate of solder. Once all the components are inserted, the board is sent on a belt that pulls it across a wave of flux foam and then a wave or waterfall of liquid solder. The bottom of the board floats over these waves. The flux, activated by temperature, deposits flux over the entire board, removes oxides on the surfaces to be soldered, and allows solder to wet these surfaces and wick up the leads and into the holes, forming the solder joints. Usually, components are mounted only on the top side of the board; components mounted on the bottom must be small enough to ride in the waves without compromising surface contact, and they must be attached to the bottom with an epoxy so they will not fall off before the solder fillet interconnections are formed. After soldering, the boards are cleaned and inspected prior to final electrical testing. Figure 22.29 shows an example of an insertion-mounted PWB assembly.

Surface Mounting

Advantages

Surface-mount components can be either leaded or leadless, such as chip capacitors, chip resistors, or LCCs. Surface-mount organic boards can be routed with any combination of through-holes, blind vias, or buried vias; PWBs can have heat sinks attached directly to the back side of the board, since there is no need to leave this area clear for lead protrusions.

The advantages of surface-mount technology over insertion mounting are reductions in routing layers, in the assembly's profile, and in board area. Components can be mounted

Figure 22.27 Lead forming (Courtesy of Westinghouse Electronics Systems Group.)

on pads on the board's surface, obviating the need for through-holes for component mounting. In addition, board routing can be accomplished with blind or buried vias, which also eliminate the need for through-holes and extra layering to allow for routing around the holes. Surface-mounted components can also be reworked with an extractor, which heats only the component to be replaced and avoids heating and reflowing the entire assembly. Figure 22.30 shows a component being removed with an extractor.

Surface-mount components usually sit lower than their inserted counterparts and sit directly on the board surface, not slightly above, as shown in Figure 22.25. They also offer

(a) COMPONENT AND BOARD

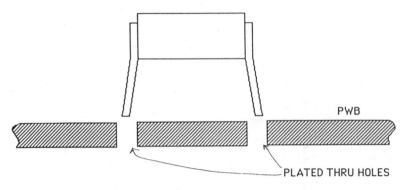

PWB

PLATED THRU HOLES

(b) COMPONENT INSERTED IN A BOARD WITH LEADS CUT AND
   CLENCHED

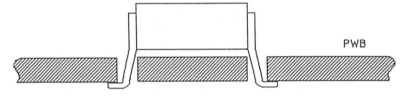

PWB

(c) AFTER WAVE SOLDER

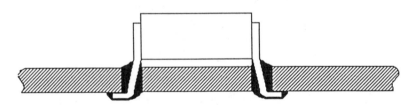

Figure 22.28   Insertion mounting and lead trimming.

great reductions in board area compared with comparable DIPs or cerpaks, a decrease further enhanced if leadless chip carriers are employed. For example, a 20-pin DIP has 0.100-in. center-to-center lead spacing or pitch, making it at least 1.1 in. long by approximately 0.250 in. wide; the area of the component itself is 0.275 in.^{2}. The footprint for this DIP would add an additional 0.150 in. to each side, so its required mounting area would be 0.605 in.^{2}. The body of the DIP would be approximately 0.250 in. thick and would sit approximately 0.050 in. off the board, making its profile 0.300 in. overall. In contrast, a 20-pin chip carrier has a body 0.300 in. square, with a height as low as 0.060 in. and I/Os along the perimeter on all four sides. If the chip carrier is leaded, the footprint would extend 0.150 in. out from the body of the carrier, making the overall mounting area 0.600^2 or 0.360 in.^{2}. The footprint of an LCC would extend only 0.050 in. out from its body, or require a mounting area of only 0.160 in.^{2}. Whether leaded or leadless, the chip

Figure 22.29 Insertion-mounted organic PWB assembly. (Courtesy of Westinghouse Electronics Systems Group.)

carrier would sit only 0.005 in. above the board. The leaded device would have an epoxy preform under it, 0.003 to 0.005 in. in height, while the LCC would have a 0.005-in. solder joint standoff. The magnitude of the area reductions increases with pin count.

With system trends toward increased complexity, decreased size, and cost effectiveness, surface-mount designs offer great reductions in height, area, and system volume and cost no more than inserted components. For example, a system requiring 30 PWB assemblies uses 10 large-scale integrated (LSI) circuits with I/O counts of 60 to 68 for each assembly. Assuming that their footprints account for 60% of the needed board area (the other 40% is required for spacing, small discrete components, and connector attachment), an insertion PWB version would require a minimum area of 64.5 in.^{2}, and each board would be approximately 8 in. × 8 in. They would also require center-to-center spacings of 0.450 in. (0.300 in. component height, plus 0.040 in. board thickness, plus 0.010 in. clearance for bottom-side lead extension, plus 0.050 in. spacing between boards). This same design, using surface-mount technology, would require boards of approximately 6 in. × 4 in. or 5 in. square, with center-to-center spacings of 0.165 in. Thus, an insertion-mount system of 30 such assemblies would require a minimum volume of 14 in. × 8.5 in. × 8.5 in., or 7.024 ft^3, while the same system in a leaded surface-mount design would require a minimum volume of only 5.1 in. × 6.5 in. × 4.5 in., or 1.036 ft^3. This conversion would result in a 6.8:1 volume reduction. Figure 22.31 shows a leaded-component, surface-mounted PWB assembly.

This reduction in volume at comparable cost can be further improved by converting to a leadless surface-mount design. Using the previous example, leadless board assemblies would require the same spacing of 0.165 in., but would only have to be 4 in. square. Thus, the system would need only 0.717 ft^3, an overall reduction of 9.8:1. The general rule of

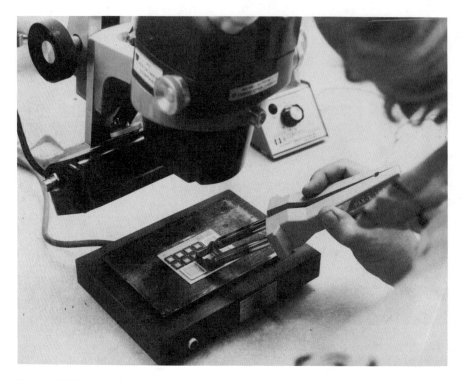

Figure 22.30 Extractor rework equipment. (Courtesy of Westinghouse Electronics Systems Group.)

thumb to estimate the area reductions for conversion from insertion mounting to leadless surface mounting is 8:1.

The price of leadless components is also comparable to or less than that of their leaded counterparts. A leaded chip carrier is slightly more expensive than the leadless version, since it requires brazing the leads onto the leadless version; leadless components also allow larger placement tolerancing due to their ability to self-align.

Surface-Mounted LCCCs on Organic PWBs

The feasibility of surface mounting leadless ceramic chip carriers on organic PWBs has been debated over the years. While such designs can solve many sizing and spacing problems without requiring leads to provide strain relief for the CTE mismatch between the PWBs and the carriers, initial attempts at LCCC mounting to organic PWBs were discouraging. Mounting small LCCCs, approximately 0.400 in. square with 28 pins or less, to organic PWBs had some limited success. However, as the LCCCs increase in size, the effect of the CTE mismatch also increases; if the PWB assemblies are subjected to extreme temperature changes, as required in military applications, the thermal stress imposed by the mismatch can be great enough to cause catastrophic solder joint failures. Consequently, only smaller LCCCs have been surface-mounted to organic PWBs in military devices.

Cored Boards

Attempts have been made to improve the CTE match between LCCCs and PWBs by sandwiching the PWBs around a heat sink material, or core, and laminating them with

Figure 22.31 Leaded surface-mounted organic PWB. (Courtesy of Westinghouse Electronics Systems Group.)

B-stage or film epoxy, using heat and pressure. The resulting cored board has a composite CTE between those of the PWB material and the core material; this composite CTE more closely matches the LCCC's, but the cored PWB dimensions are slightly changed. The core material, with a higher CTE than the PWB, expands more than the PWB during lamination heating. As the cored board cools, the core contracts more than the PWB normally would, causing some shrinkage of the PWB. The artwork layer for the solder mask must be modified, as a one-to-one image of the solder pads will be slightly larger than the pads on the board. The adjustment will vary with the core material, the dimensions of the board, and the CTEs of the board and the core material. An aluminum or copper-clad Invar-cored board has to be adjusted to approximately 96–98% of its original size.

Components to Boards: Solder Interconnection

Typically, eutectic Pb–Sn solders with RMA (rosin mildly active) flux or 60 Pb–40 Sn alloys are used to connect components to boards. These solders can withstand common commercial and military operating temperature ranges, yet reflow at substantially lower temperatures (approximately 183°C) than most die-attach solder alloys. This eliminates the potential for die-attach reflow during the surface-mount reflow process.

Solder Deposition

Solder is applied in one of two ways. The first is to print solder paste onto a tin or solder-flashed board (the copper-metallized boards have flashes or coatings to prevent corrosion). Solder paste is normally wet-printed 0.009 to 0.011 in. thick, yielding reflowed solder joint heights of 0.005 to 0.006 in. that provide stress relief during temperature cycling. Any additional print thickness would make processing more difficult and cause solder balling and bridging (see Figure 22.32), while only slightly increasing stress relief benefits.

Solder pastes can be screen- or stencil-printed. An 80-mesh screen with 9 mils of emulsion will produce the desired thickness, but paste coverage on the mounting pads will be nonuniform. Screen printing can leave too much solder on the screen, and the wire mesh leaves peaks and valleys in the paste surface after printing. This uneven paste distribution can result in solder bridging between two adjacent pads with excess paste and in starved solder joints with too little paste. In extreme cases, LCCCs will not sit parallel to the board after reflow or may be skewed against the footprint.

Self-alignment occurs during solder reflow, provided the LCCC has not been bonded to the board with a thermally conductive adhesive for heat-transfer enhancement. As the solder melts, its surface tension rises, increasing wetting (or adhesion) to the PWB mounting pads and the LCCC castellations. These increases in surface tension cause the solder joint to pull the LCCC into alignment over its footprint. Self-alignment allows the process control placement window to open, improving manufacturability. However, uneven paste distribution leans to nonuniform surface tensions and inadequate self-alignment, leaving some larger and heavier LCCCs skewed on the board.

Excessive solder paste can slump during prebake or reflow, spreading beyond the footprint metallization. During reflow, only the solder wetting the surface is pulled by surface tension onto the metallized pad; the surface tension within the remaining, nonadhering solder causes it to form a ball. Solder balls, which come in various sizes, can potentially roll around and lodge between two electrically uncommon tracks, causing a short or initiating a dendritic growth site. Dendritic growth occurs when moisture, condensing on the PWB, becomes polarized by the electromagnetic fields. The polarized particles, or ions, will start to attract each other and form a chain, which grows between conductive surfaces that attract the ions. If the ion chain grows enough to bridge the conductive pads, it can carry enough low current to cause intermittent or complete shorting.

Given these problems, screen-printed boards require substantial rework by a trained operator who manually removes or adds solder to individual joints and takes out any solder balls. This process is very labor-intensive, so an alternative method is recommended: replacing the wire mesh screen with a brass stencil. The stencil thickness should be the

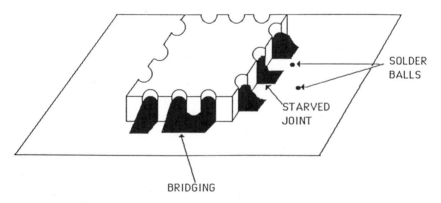

Figure 22.32   Solder joint defects.

desired wet print thickness—that is, a 0.010-in. stencil will yield a uniform 0.010-in.-thick wet print. Since a more uniform wet print yields more uniform solder joints, a stencil gives better print resolution, which makes alignment easier and more accurate and decreases the formation of solder balls and bridges.

To further reduce solder balling and bridging, the solder paste is subjected to a prebake prior to reflow. The prebake volatilizes some of the paste solvents that provide a printable viscosity, making the paste tacky so it can set up. The prebake must be controlled, and time limitations between printing, baking, and reflow must be observed, to assure that only viscosity-lowering solvents are removed, not flux and other reactive solvents needed for proper reflow and wetting. Best results are obtained if all soldering steps occur within 4 hr. Some solder pastes—for example, Cermalloy—use organic wetting agents that cannot be exposed to elevated temperature prior to reflow. These need to be reflowed immediately following printing.

The second deposit method is to thick-plate solder onto the board. When plating the solder to the desired thickness, all top-layer metal that is not to be plated is first masked. This can be done in a variety of ways. To apply a permanent mask, a photosensitive polymer is deposited over the entire surface. The coated surface is exposed to a specific frequency of ultraviolet light through a photomask; the mask can expose either the areas to be soldered or the areas to be masked, depending on whether a positive or negative photoresist is used. The board is sent through a developer that dissolves the unwanted polymer, leaving the desired areas masked. Another way to apply a permanent solder mask is to cover the areas to be masked with a film cut into the desired shapes and sizes. With either method, the solder mask must be capable of withstanding the plating and soldering processes.

A temporary mask, another alternative, can be printed on with a stencil or screen. The masking material is a dielectric that the solder will not wet, and it must be able to withstand the soldering temperature and exposure to the flux. It must also be soluble after reflow in a degreaser or other simple cleaning process. Temporary masking has been accomplished by printing on dielectric or glass thick-film paste, which can be printed on the board but left unfired. After the mask is printed and dried, the solder paste can be printed. The mask can withstand soldering temperatures without curing. Although these materials typically contain oxides that are attacked by the flux, only the surface of the deposit is affected since they are thick-film printed. They can be washed off with an isopropyl alcohol (IPA) rinse. One added advantage of a temporary mask is that as it is washed away, so is the flux left on its surface. Disadvantages are that this method can be used only when the solder is printed on the boards as a paste, and any unsoldered top metallization must be protected from corrosion. Thus, the copper must be tin- or nickel-plated prior to masking, or the top conductor layer must contain only solder pads and/or tracks that can be solder-coated, such as digital tracks or signal lines. However, a solder-coated track may significantly slow the signal, a problem at RF and microwave frequencies, at which the speed of conductor-surface signals is highly critical. Solder alloys typically have higher resistivities than the copper metallization of the board, so even though the solder coating increases the cross-sectional area, it does not necessarily improve signal speeds for RF designs.

The best design for thick-plate boards is to lay them out so that only the mounting pads are on the top layer and all signal tracks are buried, as required for military designs. This eliminates the need for any solder masking, but such a layout may require additional layering.

Solder Reflow

The most common reflow methods are vapor-phase and infrared (IR) reflow. Belt reflow is sometimes used for prototype or small-volume products, but is not practical for production applications.

Both vapor-phase and IR furnace reflow require product-specific equipment programming. The thermal mass and configuration of the product determine what program will provide a temperature profile that gives the product consistently correct ramp-up, holds it at the proper peak temperature for the right interval, and cools it to allow solder-joint formation without thermal overstress. If the product has a large thermal mass (e.g., several large ceramic chip carriers), it will take more time and energy to heat the components and their solder joints. Small units may need to enter the vapor furnace faster, while larger devices may require a slower entry and longer vapor dwell time or higher furnace temperature.

The first step, therefore, in optimizing a vapor-phase reflow process is to develop a profile that provides steady temperature ramp-up rapidly enough to avoid volatilizing reactants before the melting point of the solder and complete reflow are achieved, but slowly enough for the flux to work and self-alignment to occur. The profile must also keep the part in the vapor at peak temperature long enough for complete reflow, but not long enough to allow board degradation, overformation of intermetallics, bake-on flux or discoloration, or delamination of polyimide or epoxy glass boards. Some intermetallics must form to allow the metals in the solder to combine and adhere to the metals on the board surface, but if all the surface metals diffuse into intermetallics, adhesion is compromised; dwelling at the peak temperature too long can result in overformation of intermetallics. The profile also must enable rapid cooling to provide proper grain growth without thermally overstressing the boards, by slowly lowering the parts into the vapor, letting them sit there for the correct duration, then pulling them out of the chamber rapidly. After the boards cool to approximately 90°C, they are immediately sent through a degreaser for cleaning and defluxing before the flux has a chance to harden onto the board.

Adjusting the vapor-phase profile depends on the component with the largest thermal mass. The vapor hits all surfaces simultaneously, and the heat is transferred from the vapor to the component. Thus, the temperature at the surface of the component and its solder joints is independent of surface area or location. The rate of heat transfer into the solder joints depends on the rate of transfer into the component, which in turn depends on the thermal mass. Therefore, if the vapor phase is programmed to allow for complete reflow of the component with the largest thermal mass, it will reflow all other components.

Even with an optimized printing process and vapor-phase profile, an unforeseen problem may arise. As the vapor contacts the PWB, it transfers its latent heat and condenses. The condensed liquid accumulates on the board surface until eventually it runs off the sides, accumulating before the solder fully melts under large LCCCs that reflow slowly; this keeps the liquid solder's surface tension from getting high enough to hold the LCCCs in place. In other words, the condensed vapor lifts the LCCCs, washing away the solder paste and floating the LCCCs off the board. To remedy this problem, the board is placed on a slight angle—about 10°—during the vapor phase; this angle is just great enough to allow the condensation to run off without permitting gravity to move any of the LCCCs during reflow.

The solvents used in vapor-phase reflow create additional problems. The Freon used can be toxic if inhaled in large doses, and will react chemically with the organics (e.g., flux) in the solder paste. The flux and other chemical by-products accumulate over time and can alter the vapor temperature or introduce contaminants on the PWBs; consequently, lot-to-lot process control can be difficult. Solvents, which are very costly, must be replaced periodically, and require special safety precautions, along with costly waste removal. They also represent a danger to the ozone layer.

Reflow in multiple-zone furnaces, although successful for small assemblies such as solder-sealing chip carriers or solder-chip mounting or sealing of hybrids, has not been very successful for solder-mounting components onto PWBs. This is due not to the limitations of the equipment but to the limitations of the modes of heat transfer. In a standard multiple-zone furnace, heat is transferred by convection and conduction and is dependent on surface area. The increase in temperature of a given component and its board depends on its area, its mass, its specific heat, the area of the overall assembly, and the component's location on the assembly. The outer edges of the unit are the first to heat up. As the heat continues to transfer into the unit, a temperature gradient will form as the board under the component acts as a heat sink and conductively pulls heat away from the component. Thus, components at the center of the board will lag thermally behind those along the edges. In some cases, the board layout can be designed to compensate for this phenomenon; the larger components can be placed on the perimeter of the board, with increasingly smaller devices placed closer to the center. However, this approach typically works only for smaller PWB assemblies, because significant gradients still form. Furthermore, this placement is usually not electrically conducive and causes unnecessary layering or routing difficulties.

Furnace reflow, which uses radiation as the heat-transfer mode, has been made a viable reflow process. In an IR furnace, in addition to convective and conductive heat transfer, energy is transferred into the components and their solder joints by exposing all surfaces simultaneously to IR radiation as the primary mode of heat transfer. There are still thermal gradients and heat transfer is still dependent on component area and location, but to a much lesser degree. With proper equipment programming, most PWB assemblies can be re-flowed using an IR furnace. The advantages of IR furnaces are improved long-term process control and operating costs and elimination of the use of expensive and dangerous solvents.

In both vapor-phase and IR reflow, once an optimized profile window is established for one product, it cannot be used for subsequent products. The program controlling the descent speed, dwell time, and accent speed (vapor-phase) or the belt speed, energy levels, or zone settings (furnace), must be adjusted for each product. A product of higher thermal mass requires lower processing speeds and higher energy settings to reach the same profile as a device with lower thermal mass.

Whether to use single or double reflow was an issue during the early stages of surface-mount technology process development. Double reflow involves stenciling on the solder paste, reflowing it, mounting the components, and reflowing the solder again; both components and boards are thus pretinned prior to assembly. One theory held that reflowing the solder twice reduced voiding—presumed to weaken solder joints or to cause cracks to form and propagate—by forcing the gases trapped in the solder to outgas. While double reflow heightened thermal stressing of the assemblies, increased intermetallics that embrittled solder joints, and was more expensive, the gains of double reflow were assumed to cancel out the risks. Still another theory posited that voiding acted as a relief stress and that crack propagation would stop, not start, at a void.

Comparing the voiding after one reflow and after two, using X-rays and microsection-ing, determined that there is no correlation between the number of reflows and the amount of voiding, nor between voiding and solder-joint failures. A second reflow has mixed effects on voiding. Solder joints not located under LCCCs can completely outgas small voids near the surface during the second reflow, which can thus decrease voiding. However, multiple voids, especially if they are located under LCCCs, can join together to form large single voids after the second reflow. In both cases, the number of voids decreases; however, in the latter case, the percent of voiding remains the same. The larger voids formed after a second reflow are usually directly under the LCCCs, where the solder is thinnest and the thermal stress due to CTE mismatch is greatest. Voiding directly under a self-aligned LCCC, or an LCCC with solder-mounted thermal pads under the dies, greatly hinders thermal power dissipation from the dies inside the LCCCs through the solder, into the PWB, and finally to the heat sink or core of the assembly. This thermal solder voiding issue can be addressed by mounting the body of the LCCC with a thermally conductive adhesive that then becomes the primary thermal path. It has also been determined that the amount and location of voids within solder joints is totally random and is not a function of the size or location of the LCCCs.

Studies in which both single- and double-reflowed boards were subjected to over 1000 temperature cycles of –55 to +125°C have shown no correlation between solder joint failures—typically defined as specified percent increases in resistance—and voiding within the joints. Failure analysis, through microsectioning of the board assemblies and/or delidding of the LCCCs, showed that the decreases in resistance were not due to solder joint voiding, cracks, or even wirebond failures; rather, the stress concentration was located within the plated through-holes. In organic boards cored to aluminum, copper-clad Invar, and graphite, the CTE mismatch between the LCCCs and the PWB is overshadowed by the mismatch between the epoxy glass or polyamide and the core material; consequently, the blind vias and plated through-holes of the PWB fail before the solder joints. Redesigning cored PWBs to reduce the thermal stress between the PWB and the heat sink can reduce thermal failures in through-holes and vias. This can be accomplished by using a core material with a CTE closer to that of the organic PWB and by laminating the PWB to the core with a flexible adhesive capable of absorbing the thermal stress. However, this once again compromises solder joint reliability. The solder responds to the applied strains and resulting stresses with time-dependent plastic deformation, leading to solder-joint failures due to accumulating fatigue damage.

Special Mounting Techniques

In production applications, components are typically mounted with pick-and-place equipment. If solder paste is applied by printing onto the boards, then placement follows immediately. The tackiness of the paste holds the parts in place until the solder is reflowed and self-alignment or the surface tension of the liquid solder takes over, just as inserting the leads in an insertion-mount board aligns the components and holds them in place until the solder joints are formed. Single-sided solder-paste PWBs thus do not need additional placement or mounting techniques. For high heat dissipation and improved conductive cooling, thermally conductive adhesives are used between the components and the boards. Solder paste can mount pretinned or nonpretinned components; the flux in the paste allows it to wick up and wet the component castellations or terminations, so pretinning is not absolutely necessary. Pretinning is still recommended, however, to ensure complete and uniform wetting to all I/Os. The paste's initial tackiness and its later liquid surface tension do not overcome gravity as well as a wave-soldered, inserted, and clenched component. In addition, the solder paste must be processed within the time limitations of its solvents.

Thus, a solder-paste surface-mounted board does require extra process control and handling.

Parts on thick-solder-plated boards must be in place with an epoxy or a tacking flux prior to reflow. Epoxy preforms can be manually or automatically placed under larger chip carrier or hybrid components. Partial curing enables the epoxy to hold the components in place while reflow forms the solder joints and finishes the curing of the epoxy. The same technique can be used with manually applied or printed epoxy pastes. Tacking involves printing a viscous flux over the thick-plated solder pads. The flux serves two purposes: It provides a tacky surface that holds the components in place, and it provides the cleaning/reducing agents needed to rid the solder surfaces of any oxides that may have formed during storage.

Double-sided mounting also requires special processing. The parts mounted on the bottom of the boards must be held in place both prior to and during reflow. The tackiness of the solder paste or the printed flux in thick-solder plating may hold the parts in place, but if they are too heavy, the surface tension of the liquid solder cannot secure them until the solder joint is formed. Therefore, bottom-mounted components must be epoxied in place.

22.5.3 Ceramic PWB Assembly

Ceramic PWBs can have co-fired, thick-film, or thin-film metallization. All interlayer connection or routing is accomplished with buried vias, or with windows left out of a printed dielectric layer or punched out of the green tape. All components are surface-mounted. Figure 22.33 shows a ceramic surface-mount assembly that may have through-holes. In a co-fired substrate, these are stacked, metallized vias that produce a finished hole

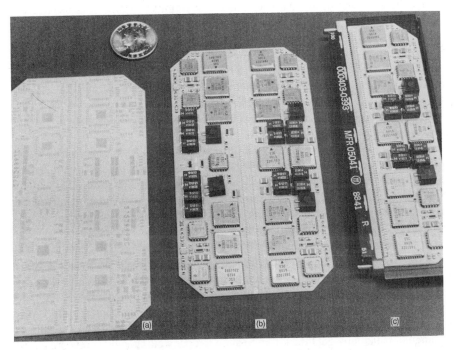

Figure 22.33 Ceramic PWB assembly. (Courtesy of Westinghouse Electronics Systems Group.)

through the substrate similar to an organic PWB plated through-hole. In a thick-film substrate, a hole is machined—punched out in the green stage, or laser-drilled or sonic-milled out of the fired ceramic. This hole is then cleaned, and metallized by coating the walls with a thick-film conductor paste. The result is a plated through-hole that can be used to route a signal from one side of the board to the other, or to increase the thermal path at a specific location on the substrate.

A patterned or metallized substrate, the starting point for any ceramic PWB assembly, can be mounted to a heat sink or left unmounted. A ceramic PWB is a rigid flat board that maintains its shape. The camber of such boards is typically 0.002 to 0.003 in./in., maintained approximately throughout processing and operation. These boards will not warp during solder processing as organic boards will. By definition, the ceramic board's CTE matches that of the ceramic components mounted to it. However, although there is no CTE mismatch, thermal stress is still generated by temperature gradients formed because of the difference in the rate of heating due to the variant thermal masses of the components and the board. Nonetheless, the need for ceramic chip carriers (CCCs) to be leaded to allow for stress relief is greatly reduced. Leadless chip carriers can be mounted to ceramic PWBs with minimal effect on the long-term reliability of the assembly. Furthermore, the ceramic material from which the boards are fabricated offers much better thermal conductivity than its organic PWB counterparts. Use of more advanced ceramic materials, such as AlN, and incorporation of thermal through-holes in the substrates, can further enhance the thermal performance of a ceramic PWB.

Ceramic PWBs are particularly valuable in military and space applications. They offer reductions in thermal stress, reductions in volume (they can incorporate LCCCs in their design), improved thermal management (due to their superior conductivity and thermal via capability), and improved long-term reliability (due to a CTE matched with that of their components, the reduction of thermal stress, and their rigidity). However, as usual, these improved properties typically mean higher costs and longer design times.

Heat Sink Attachment

Although ceramic boards are rigid and can stand on their own, heat sinks are commonly attached for thermal reasons. The heat sink increases the ability of the assembly to dissipate heat. It can also act as a carrier plate, providing a flanged support that can be readily bolted or inserted into a chassis. In standard electronic modules (SEMs), the module housing is the heat sink/carrier plate for ceramic PWBs.

The materials commonly used are Kovar, tungsten–copper alloys, composites such as graphic fiber plates or silicon–aluminum alloys, and molybdenum. All of these materials offer thermal conductivities at least as good as those of the ceramics, and CTEs closely similar to those of ceramics. The CTE match, thermal environmental requirements, outgassing restrictions, and linear size of the PWB dictate the attachment method.

Ceramic PWBs can be attached to their heat sinks with epoxy or solder. The attachment methods used for ceramic PWBs are the same as those for ceramic hybrid substrates. The mismatch in CTE is more critical for a ceramic PWB because it is larger than a hybrid substrate. A flexible, stress-absorbing adhesive, such as silicon, is typically used, provided that no bare dies are mounted on the ceramic PWB.

The back of the ceramic PWB is metallized depending on the attachment method. For example, if the PWB is to be mounted to the heat sink with solder, its back is metallized with a platinum–palladium–gold thick film to allow proper wetting, while preventing total leaching of the metallization into the solder. If the PWB is a thin-film or co-fired ceramic, the base metal is nickel- and gold-plated. The gold plating provides a good wetting surface, while the nickel acts as a barrier metal providing corrosion protection and preventing total leaching. Ceramic PWBs can also be bonded directly to a heat sink during the conductive patterning step.

Surface Mounting

As discussed, ceramic PWBs are not insertion-mounted; all components are surface mounted. These components can have leads formed for surface mounting, but this is not necessary and lessens the area and volume reduction benefits of using a surface-mount design.

Leadless components can be used with high reliability because their CTEs are very close to those of the ceramic PWBs. Leadless components include ceramic chip carriers, ceramic chip capacitors, and ceramic thin-film and/or thick-film resistor chips. Resistors can easily be incorporated into the layout of the board, making it a printed circuit board or PCB, not just a PWB.

The most common mounting method for ceramic boards is to print solder paste on them, mount the components, and reflow the boards in a vapor-phase or IR furnace. Thick-solder plating of the mounting pads can be done to a thin-film or cofired board; however, solder-paste printing is still commonly used.

22.5.4 Connector Attachment

In commercial applications, connectors need not be mounted to the PWB assembly. Instead, the PWB assembly or card has I/O pads along one edge, allowing the card to be inserted or plugged into an end-card connector, typically an integral part of the back plane or motherboard of the system. Therefore, the PWB itself acts as the connector for the PWB assembly.

In some commercial and most military applications, connectors are usually mated pairs; half of the pair is attached to the PWB assembly, while the mating half is mounted to the system backplane or motherboard. The connector can be attached before, during, or after component mounting, though attachment before mounting is not commonly done because it can interfere with solder-paste printing and component placement by preventing the PWB from lying flat on the printer stage and the placement equipment. Since assembling a PWB with connector in place requires special fixturing and handling, the connector is usually attached at the same time or after component mounting.

Connectors are typically placed along the edge of the PWB. Some can be mounted and reflowed at the same time as the components, a common practice with thick-plated organic PWB assemblies. The connector can be snapped onto the edge of the assembly and aligned with rivets at each end. The assembly is run, with connector and components in place, through the wave solder (insertion mount boards) or through a vapor-phase or IR furnace (surface-mount boards). To decrease the possibility of dislodging components while snapping on the connector, the connector is attached after component mounting and solder reflow. The I/O pads to which the connector pins are attached are pretinned by thick-plating them while plating or wave-soldering the component mounting pads or by printing solder paste on them and letting the paste reflow during component mounting and reflow. After the components are mounted and reflowed and the boards are cleaned, the connector is positioned with the spring-loaded pins over the pretinned pads. Locally the connector pads are reflowed by exposing the area to a hot-air gun (shielding the rest of the board with a metal plate), or by placing a hot bar over the connector pads.

Another method of connector attachment uses Raychem's Solder Kwik product, which works with either pretinned or plain connector pads. After component mounting and reflow, the connector is positioned and the Solder Kwik preform is placed next to the pads, aligning them with the capillaries through which the solder will flow. A hot bar is placed over the reservoir of solder. The solder melts and flows through the capillaries of the Solder Kwik preform onto the connector pads, where it wets the pads while wicking up over the pins.

In addition to edge connectors, flex cables are sometimes mounted to the PWB assembly in microwave or power modules (SEMs). Since the PWB is typically mounted to the floor (or central web) of the module, it cannot have a connector snapped over its edge, so a flex cable is surface-mounted to the board. The other end of the flex usually has a plug-in connector inserted into the module wall to provide interconnection from the module to the system.

22.6 SYSTEM INTEGRATION

Components, such as bare dies and capacitors, are the building blocks of many electronic assemblies. Subassemblies, including these components, chip carriers, PGAs, and hybrid assemblies, are the building blocks of PWB assemblies. In turn, PWBs are the building blocks for final systems. The PWB assemblies can be components inserted into the motherboard or backplane of the system (see Figure 22.34), or can be mounted inside SEMs, which in turn are plugged into the system chassis. How these building blocks participate in the final system will dictate their design.

A typical packaging hierarchy of an electronic system is shown in Figure 22.35. The lowest level of packaging, or the *zeroth level*, is generally considered to be the semicon-

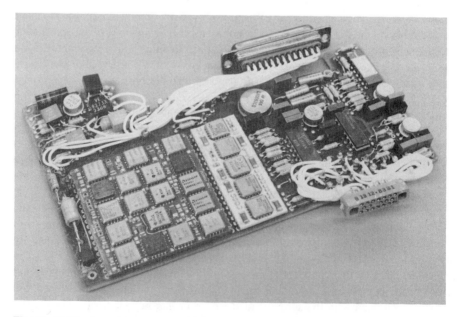

Figure 22.34 Insertion-mount motherboard assembly with two connectors, jumper wires, one ceramic daughterboard assembly, one organic surface-mount assembly, and discrete insertion-mount components. (Courtesy of Westinghouse Electronics Systems Group.)

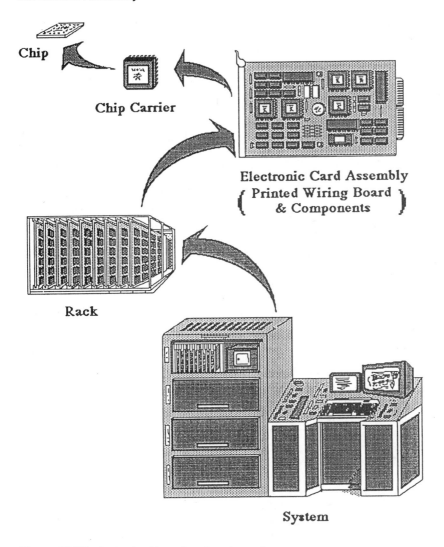

Chip

Chip Carrier

Electronic Card Assembly
{ Printed Wiring Board }
{ & Components }

Rack

System

Figure 22.35 Packaging hierarchy of an electronic system.

ductor chip, although discrete passive devices such as resistors and capacitors may also be included. The packaging of a chip or a set of chips in a functional and protective chip carrier is referred to as the *first level* of packaging. Chip carriers can range from single-chip (monolithic) carriers to very sophisticated multichip modules containing hundreds of chips and devices. The *second level* of packaging is often referred to as the electrical circuit assembly (ECA). At this level, the individual chip carriers are mounted on a common base, usually a printed wiring board. The *third level* of packaging typically involves the interconnection of circuit boards and power supplies to a physical interface, such as a chassis, control, and/or electromechanical device or system. The third level of packaging may also involve the connection of several boards within a supporting or protective structure such as a cabinet. Several such cabinets are joined together to form the *fourth level* of packaging.

22.7 REFERENCES

Avallone, E. A., and T. Baumeister III, *Marks' Standard Handbook for Mechanical Engineers*, 9th ed., McGraw-Hill, New York, 1987, p. 5–54.

Harper, C. A. (ed.), *Handbook of Materials and Processes for Electronics,* McGraw-Hill, New York, 1970.

Harper, Charles A., *Electronic Packaging and Interconnection Handbook*, McGraw-Hill, New York, 1991.

Leahy, K., Microwave hybrid design tutorial, Presented at Sect. 1-3, Capital Chapter of International Society for Hybrid Microelectronics, Twentieth Annual Symp., May 1989.

Manzione, L. T., *Plastic Packaging of Microelectronic Devices*, Van Nostrand Reinhold, New York, 1990.

Pecht, M. (ed.), *Handbook of Electronic Package Design,* Marcel Dekker, New York, 1991.

Schafft, H. A., Testing and fabrication of wire bonds electrical connections—A comprehensive survey, *Natl. Bur. Std. (U.S.) Tech. Note 726*, pp. 80, 106–109, 1972.

Stamps, Jerry, Sr., Mechanical Engineer, Structural Analysis Group, Westinghouse Electronics Systems Group, Personal interview, Baltimore, MD, February 27, 1990.

Troyk, P., Encapsulants as packaging for implanted electronics, Presented at Capital Chapter of International Society for Hybrid Microelectronics Symp., May 1987.

22.8 SUGGESTED READING

Clark, R. H., *Handbook of Printed Circuit Manufacturing,* Van Nostrand Reinhold, New York, 1985.

Coombs, C. F., *Printed Circuits Handbook*, McGraw-Hill, New York, 1990.

Dally, J. W., *Electronic Packaging—A Mechanical Engineering Perspective,* New York, 1989.

Dostal, C. A. (sr. ed.), *Electronic Materials Handbook, Vol. 1, Packaging*, ASM International, Materials Park, OH, 1989.

Hunadi, R., et al., New ultra-high purity, electrically conductive epoxy die attach adhesive for advanced microelectronic applications, *Proc. 1985 Int. Symp. on Microelectronics*, Anaheim, CA, November 1985.

Johnson, R. R., Multichip modules: Next-generation packages, *IEEE Spectrum* 27(3) (1990).

Jowett, C. E., *Reliable Electronic Assembly Production*, Tab Books, Blue Ridge Summit, PA, 1971.

Kear, F. W., *Printed Circuit Assembly Manufacturing*, Marcel Dekker, New York, 1987.

Oscilowski, A., and D. L. Sorrells, Use of thermogravimetric analysis (TGA) in predicting outgassing characteristics of electrically conductive adhesives, *Proc. Technical Conference—IEPS, Fourth Annual International Electronics Packaging Society*, Baltimore, MD, October 1984.

Pandiri, S. M., Behavior of silver flakes in conductive epoxy adhesives, *Adhesives Age* 30(11):31–35 (1987).

Ruska, W. S., *Microelectronic Processing—An Introduction to the Manufacture of Integrated Circuits,* McGraw-Hill, New York, 1987.

Sorrells, D. L., et al., Selection and cure optimization of conductive adhesives for use in AuSn sealed microelectronic packages, *Proc. 1984 Int. Symp. on Microelectronics*, Dallas, TX, September 1984.

Tummala, R. R., and E. J. Rymaszewski (eds.), *Microelectronics Packaging Handbook*, Van Nostrand Reinhold, New York, 1989.

About the Contributors

Frank Altmayer President, Scientific Control Laboratories, Chicago, Illinois. Mr. Altmayer began his career as a laboratory technician at SCL, eventually earning his B.S. (Honors) in chemical engineering and M.S. in metallurgy from the Illinois Institute of Technology while working full-time, and advanced to the presidency of the company before becoming its owner in 1986. He continues to head the staff operations of consulting on metal finishing and related problems on a worldwide basis, designing plating shops and waste treatment systems, and providing customer service/contact. He is an American electroplaters' and Surface Finishers' Society (AESF) Certified Electroplater-Finisher (CEF) and has been serving the metal finishing industry for over 25 years. The SCL client list includes over 1900 companies nationwide. Mr. Altmayer was chosen by the World Environmental Center to accompany U.S. EPA representatives on a pollution-control mission to Indonesia. He has taught numerous college-level courses at Triton College and the Illinois Institute of Technology, and for the AESF. Mr. Altmayer has written over 75 articles on metal finishing and waste treatment, and he is the author of a monthly column on metal finishing problems and regulatory compliance for *Plating and Surface Finishing* magazine. A complete list of his professional affiliations, achievements, and awards is too lengthy for inclusion here.

Shrikar Bhagath Project Engineer, Reliability, Parts and Materials Engineering Department, Delco Electronics Corporation, Kokomo, Indiana. He received an M.S. in Mechanical Engineering from the University of Maryland at College Park, where he conducted research at the CALCE Electronics Packaging Research Center. His research interests focused on the third level of electronic packaging and he developed a physics-based mechanostochastic model for the determination of connector contract interface reliability. His interests also span the areas of multichip modules, electronic package reliability and physics-of-failure issues.

Geoffrey Boothroyd President, Boothroyd Dewhurst, Inc., Wakefield, Rhode Island, and developer of the Design for Assembly (DFA) process while at the University of Massachusetts. Dr. Boothroyd formed Boothroyd Dewhurst, Inc., with Peter Dewhurst to develop personal computer software for DFA. Major supporters of the system include Ford, IBM, DEC, Xerox, and AMP. The system has expanded to become Design for Manufacture and Assembly (DFMA)®. Dr. Boothroyd is currently a Professor of Industrial Engineering and Manufacturing Engineering at the University of Rhode Island, in addition to being involved with DFMA. He is the author or coauthor of more than 100 journal articles and several books in the field, including *Assembly Automation and Product Design* (Marcel Dekker, Inc.). He is a Fellow of the Society of Manufacturing Engineers and a member of the National Academy of Engineering, among other professional societies. He received his Ph.D. and D.Sc. from the University of London, England.

Robert S. Busk President, International Magnesium Consultants, Inc., Hilton Head, South Carolina. Mr. Busk received a B.S. degree from Colgate and a D.Eng. degree from Yale. He is a member of the International Magnesium Association (IMA), a Fellow of the American Society for Metals (ASM), and Chairman of Mg subcommittees in the American Society for Testing and Materials (ASTM), and is involved in several other related associations. He authored *Magnesium Product Design* (Marcel Dekker, Inc.) and contributed chapters on magnesium to several other published books. He retired after a lifetime of experience in magnesium with metallurgical research and wrought products, and as Assistant Director of Research at Dow Chemical. His broad list of international clients now includes Dow Chemical, Dow Corning, Brasmag (Brazil), the International Magnesium Association, the Massachusetts Institute of Technology, Norsk Hydro, and Honjo Chemical (Japan).

Greg Chandler Manager, Manufacturing Engineering, Hubbell Premise Wiring, Inc., Marion, North Carolina. Mr. Chandler's company manufactures telecommunications modular products and accessories including voice and data connectors, cross-connect products, adapters, baluns, and workstation outlets for commercial and industrial applications worldwide. He is experienced in production of products at rates of 300 per year to over 1 million per year. He is responsible for the introduction and adoption of computer simulation of factory flow to discover bottlenecks and for establishing tooling requirements at McDonnell Douglas, where he held a variety of industrial engineering positions including estimating, shop floor support, and strategic planning. He has a B.S. in industrial engineering from Auburn University, and an M.B.A./A from Embry-Riddle Aeronautical University.

Stephen C. Cimorelli Manufacturing Systems Analyst, Square D Company, Asheville, North Carolina. Mr. Cimorelli holds a B.S. in industrial engineering form the University of Central Florida, is certified in production and inventory control (CPIM) from the American Production and Inventory Control Society (APICS), and is a registered Professional Engineer (PE). At NASA he developed software systems in resource planning; at McDonnell Douglas he was an industrial engineer, computer systems analyst, and manager of production scheduling. In the commercial industries, Mr. Cimorelli was manager of manufacturing systems at Learjet, and he is currently involved in high-rate manufacturing—improving the forecasting and production planning processes, including electronic Kanban, to reduction of lot sizes and cycle times.

Denise Burkus Harris Senior Electronics Packaging Engineer, Mechanical Design and Developmental Engineering Department, Westinghouse Corporation, Baltimore, Maryland. Ms. Harris received her B.S. and M.S. in chemical engineering from the University of Notre Dame. She worked at Texas Instruments in the areas of photolithography engineering and new processes and product development in the Hybrid Microelectronics Lab. She later worked at Magnavox in hybrid layout and packaging design. Ms. Harris has been an active member for 10 years in the International Society of Hybrid Microelectronics (ISHM) and was past president of the Capital Chapter. She wrote a chapter on electronic assemblies for *Handbook of Electronic Package Design,* edited by Michael Pecht (Marcel Dekker, Inc.).

Alexander Houtzeel Founder and Chairman, Houtzeel Manufacturing Systems Software, Waltham, Massachusetts, and President, Organization for Industrial Research. Born and raised in the Netherlands, Mr. Houtzeel received his graduate degree in manufacturing from the Delft Institute of Technology. He also earned a graduate degree in nuclear engineering from the National Institute of Nuclear Science in Saclay, France. He was employed in the nuclear engineering field in the U.S. and France for 12 years. Since 1972, Mr. Houtzeel has been heavily involved in computer-aided manufacturing in the United States, and has pioneered in the fields of computer-aided group technology and process planning. He has written numerous papers on CAD/CAM for the Numerical Control Society and the Society of Manufacturing Engineers.

Robert L. Lints Quality Consultant, Quality Assurance Systems, St. Louis, Missouri. Mr. Lints lectures and instructs on the various aspects of the Quality discipline. He has more than 30 years' experience in inspection, quality engineering, quality planning, and reliability, in both manager- and director-level positions. His company affiliations include Curtiss-Wright, Marquardt, Hughes, Sikorski, and McDonnell Douglas. His B.S. degree was from St. Louis University (Parks College), and he conducted graduate work in statistics at Rutgers. He is a registered Professional Engineer (Quality), a senior member of the American Society for Quality Control (ASQC), and a Certified Baldrige Internal Examiner.

Dr. John F. Maguire Program Director, Staff Scientist, Materials and Structures Division, Southwest Research Institute, San Antonio, Texas. His current position involves the development of a program in advanced materials characterization and processing, with an emphasis on polymers and organic polymeric composites. He received his B.S. (Honors) in chemistry and Ph.D. in physical chemistry from the University of Ulster. Dr. Maguire has organized and chaired sessions at conferences, presented papers nationally and internationally, and delivered presentations to the Society for the Advancement of Materials and Process Engineering, The British Association for the Advancement of Science, the American Chemical Society, and the American Institute of Chemical Engineers. He was elected Fellow of Royal Society of Chemistry and has received the SwRI Scientific Achievement Award. His industry experience includes work on the Lear FanJet, and work as a Staff Engineer for General Electric.

Timothy L. Murphy Group Manager, Human Resources, McDonnell Douglas Corporation, Titusville, Florida. Mr. Murphy is in charge of industrial safety, industrial hygiene, first aid, and workers' compensation. He has 25 years' experience in the issues that affect workers today, as well as relations with the local community. He is a Certified Fire Instructor (CFI) and a Certified Commander for Hazardous Spills (CCHS).

Clyde S. Mutter Manager of tooling, CMS Defense Systems, Titusville, Florida, and former manager of tooling for McDonnell Douglas. Mr. Mutter started his career as a designer for Emerson Electric. After receiving his associate's degree in tool design at University of Southern Illinois, he became a tool designer for McDonnell Aircraft. He is experienced in the design, fabrication, and operation of all types of tooling, including computer-controlled special-purpose machines and equipment for handling hazardous materials at low to very high production rates. One of his projects was the first "pop-top" aluminum can lid machine; another was a machine for the assembly (and eventually disassembly) of rocket motors on 3–5 second cycles. Mr. Mutter has also worked as an ME, laying out production lines and establishing tooling concepts for the production of Tomahawk cruise missiles.

Dr. Michael Pecht Professor and Director, CALCE Electronics Packaging Research Center (EPRC), University of Maryland, College Park, Maryland. The CALCE EPRC, sponsored by over 35 organizations, conducts reliability research to support the development of competitive electronic products in a timely manner. Dr. Pecht has a B.S. in acoustics, an M.S. in electrical engineering, and an M.S. and a Ph.D. in engineering mechanics from the University of Wisconsin. He is a Professional Engineer, an IEEE Fellow, an ASME Fellow, and a Westinghouse Fellow. He is the chief editor of the *IEEE Transactions on Reliability,* and an associate editor for the *SAE Reliability, Maintainability and Supportability Journal,* and the *International Microelectronics Journal,* and he is on the advisory board of *IEEE Spectrum* and the *Journal of Electronics Manufacturing.* Dr. Pecht serves on the board of advisors for various companies and consults for the U.S. Government, providing expertise in strategic planning in the area of electronics packaging. He is the author of *Handbook of Electronic Package Design* and *Placement and Routing of Electronic Modules* (both titles Marcel Dekker, Inc.).

Robert E. Persson   Senior Corrosion Engineer, EG&G, Cape Canaveral, Florida. Mr. Persson attended St. Louis University and received a B.A. in Chemistry and Mathematics from Coe College. He has 22 years of corrosion control and coating experience and is a Level 3 Certified Coating Inspector from the National Association of Corrosion Engineers (NACE). He has responsibility for assessing protective coating requirements based on the current and projected condition of all buildings and equipment at the Kennedy Space Center. Mr. Persson implemented closed-loop recycling of abrasive cleaning and blasting media systems, and environmental regulatory compliance. His other positions in manufacturing engineering have been as process planner and manufacturing control supervisor for McDonnell Douglas; superintendent of layout planning, estimating, and quality control for Welco Industries; and project analyst for Dow Chemical, using computerized scheduling for work in process.

Allen E. Plogstedt[†]   Former Staff Director, MDC Fellow, McDonnell Douglas Corporation, Titusville, Florida. Mr. Plogstedt earned B.S. in infrared technology at the University of Michigan, followed by an M.S.E.E. from the University of Cincinnati. He did additional work in modulation theory at MIT, and was involved in developing the first 3–5µ forward-looking IR imaging system. He published several papers on the use of optics with computers. In addition to his various accomplishments in the research and advanced development fields, he gained valuable experience working with electrical power distribution systems and the production of commercial electronic devices at high production rates. He recently contributed to the utilization of computers in the manufacturing industries by providing real-time control of configuration, work instructions, material control, cost collection, and production analysis.

Marc Plogstedt   Co-founder and Project Executive for themed projects, ITEC Production, Orlando, Florida. Mr. Plogstedt's project experience stems from nine years working with the Walt Disney Company. His interaction with Walt Disney Imagineering allowed him to develop his skills over many diverse projects and locations, including new attractions at Tokyo Disneyland, EPCOT Center, and Disneyland, and enhancements and rehabs for attractions at the Magic Kingdom at Walt Disney World. After leaving Disney to start ITEC Productions, he applied his diverse background to domestic projects such as Mickey's Birthdayland for Disney, and Kongfrontation, Ghostbusters, Earthquake, and JAWS for Universal Studios. Internationally, he directed ITEC's work on projects such as Sanrio Puroland, the Darinkai Water Expo, and the highly successful KIA Motors exhibit for the Taejon Korea Expo. His knowledge of networked programmable controllers, minicomputers, and digital playback audio equipment make him unique in his industry. His design of the first attraction to use only solid-state memory, and of the industry's first compact disc system is now the standard, having replaced previous analog tape machines. Mr. Plogstedt attended the University of Central Florida.

Paul R. Riedel   Principal Industrial Hygienist, DynaCorp, Westminister, Colorado. Mr. Riedel is currently involved with nuclear waste storage and cleanup at the Rocky Flats environmental technology site. He is a member of the American Board of Industrial Hygiene, a Certified Industrial Hygienist (CIH), and a Certified Hazardous Materials Manager (CHMM). He has 16 years of experience in the industrial hygiene field since receiving his B.S. from the University of Florida.

Lawrence J. Rhoades   President, Extrude Hone Corporation, Irwin, Pennsylvania. Mr. Rhoades is one of the leaders in the field of nontraditional machining, finishing, and measurement. He holds

---

*Deceased.

numerous patents and has authored a variety of works in the field. After graduating from Brown University, where he studied economics and mechanical engineering, he received an M.B.A. from Northwestern University. Mr. Rhoades is a member of the Advisory Committee of the Manufacturing Systems Engineering Program at the University of Pittsburgh, chair of the Board of Directors of the Southwestern Pennsylvania Industrial Resource Center, and a member of the Board of Concurrent Technologies Corporation, which operates Centers of Excellence in key manufacturing technologies for the U.S. Department of Defense, as well as being active in many other important associations.

Thomas J. Rose Technical Director, Advanced Processing Technology/Applied Polymer Technology (AvPro) Inc., Norman, Oklahoma. His company provides control systems, consultation for processing, and training in the field of advanced composites. Chemical and physical testing is used to develop process cycles and control feedback parameters. Mr. Rose made numerous contributions to the materials and processing sciences for advanced composite materials, after receiving his B.S. in chemistry from the University of Minnesota. Materials developed under his supervision include graphite fabrics, chopped mats, Fiberite 900 series, and other matrix and prepreg materials. His technical projects include impact damage, dimensional stability for satellite structures, and ultrahigh-modulus fiber flow effects. He has authored industry specifications including the Society of Automotive Engineers' recommended practice for dielectric cure monitoring. The control technology that he has helped develop has been applied to the Lear Fan 2100, the Starship, MD11, F22, Airbus, B1 repair, Space Shuttle, and others. He also has other experience working at Lear Fan, Lear Fan Ltd., Ford Aerospace, Carsonite, Inc., and Fiberite Corporation. Mr. Rose has contributed to several patents and made numerous presentations for SAMPE and SME.

Vijay S. Sheth Principal Manufacturing Engineer, McDonnell Douglas Corporation, Titusville, Florida. Mr. Sheth has a B.S. in mechanical engineering from Birla Engineering Institute, Anand, India; and an M.S. in industrial engineering from the University of Tennessee. His experience includes all the major functions of a working manufacturing engineer, including methods, processes, tooling, materials handling, and equipment selection—in mechanical and electrical assembly, composites, welding, sheet metal, tooling, and OSHA requirements. He was manager of the industrial engineering department for a textile machinery plant and of the facilities department for a Fortune 500 container plant, and is the author of *Facilities Planning and Materials Handling* (Marcel Dekker, Inc.).

John P. Tanner President, Tanner and Associates, Orlando, Florida. Mr. Tanner also serves as Adjunct Professor of Manufacturing Engineering at the University of Central Florida and is active as a manufacturing consultant. He was previously Director of Manufacturing at Applied Devices Corporation and Director of Industrial and Production Engineering at the ECI Division of E-Systems. While manager of Manufacturing Engineering and Planning for McDonnell Douglas Corporation, he planned production lines, tooling, and processes for their high-rate low-cost missile plant. He holds a B.B.A. in Economics, a B.S. in Industrial Engineering, and an M.B.A. in management. He is the author of *Manufacturing Engineering* (Marcel Dekker, Inc.) and numerous papers and articles published in leading technical and trade magazines.

Dr. V. M. Torbilo Research Fellow, Mechanical Engineering Department, Ben-Gurion University of the Negev, Beer-Sheva, Israel. Dr. Torbilo received his M.S. in engineering from the Zaporojie Engineering Institute, followed by his Ph.D. and D.Sc. in mechanical engineering from the Moscow Automotive Institute in the former Soviet Union. He was a professor in finish machining at the Perm Polytechnic Institute and Research Fellow in the Faculty of Manufacturing Engineering at the Moscow Automotive Institute. He has more than 30 years experience in shop supervision and product and process design in the metal-working industries. His research includes surface integrity and process optimization in the burnishing of metals, and cemented carbide and silicon nitride cutting tool wear. His broad teaching, research, and design work has resulted in 24 patents, 5 books, and numerous professional technical papers.

Jeffrey W. Vincoli   President, J. W. Vincoli and Associates, Titusville, Florida. Mr. Vincoli's company offers consulting services and training in safety, health, and environmental issues. He has a B.S., M.S., and M.B.A. in aeronautical science and management, as well as experience in the safety and environmental areas. He is a Certified Safety Professional (CSP), a Certified Hazard Control Manager (CHCM), a Registered Environmental Professional (REP), and a WSO-Certified Safety Specialist (WSO-CSS). He was a member of the first delegation of U.S. Safety and Health Professionals to visit the former Soviet Union and led similar delegations to the Pacific Rim countries. Mr. Vincoli has authored six books in his field, the most recent being *Risk Management for Hazardous Chemicals.*

Jack M. Walker   Director of Factory Operations (retired), McDonnell Douglas Corporation, Florida Missile Production. Currently consulting with small manufacturers in the southeastern United States. With more than 30 years of experience in most of the manufacturing engineering, tooling, and other manufacturing disciplines, Mr. Walker has been Director of Advanced Programs, Manager of Co-production Programs in Europe, and Branch Manager of Production Design. He was instrumental in installing the first robotic operations, the first automatic assembly machines, and the automatic composites line at MDC. He earned a B.S. degree in engineering from St. Louis University and has done graduate work in both industrial engineering at Washington University and business administration at Stetson University. He is a senior member of SME (Composites Section), a professional member of AIAA, past president of the Brevard County Manufacturers Association, and a member of the University of Florida Engineering Advisory Council. Mr. Walker has taught college-level courses in manufacturing and design, presented papers to technical societies, and written the handbooks *Production Design* and *Introduction to Composites.* He holds the patent for a pyrotechnic-operated electrical device.

William L. Walker   Head of Facilities Engineering and Operations, National High Magnetic Field Laboratory, Florida State University, Tallahassee, Florida. Mr. Walker is responsible for facility design, arrangement, and maintenance, including preparation of analyses, cost trades, and specifications for procurement of equipment and machines. He has worked with several A & E firms in the design and construction of airports, hospitals, and other major projects. He was superintendent and manager of a construction firm, and chief designer and department head of one of Florida's largest consulting engineering companies. He has designed and patented drawings for sports equipment, and consulted in composites design for Kidder, Ski Supreme, and O'Brien sports equipment manufacturers. His formal education started in aerospace engineering at Georgia Tech and continued at the University of Florida, where he received a B.S. in architectural design as he progressed to Project Architect, Field Architect, Program Manager, and other positions. His Masters thesis in Architectural Design from Florida A&M University emphasized programming, defining the detailed needs of the ultimate user as the driver for design and arrangement. He is currently involved in the program and preliminary design for FSU's Technology Transfer Facility, where private industry works with research scientists to develop new products and build state-of-the-art prototype and preproduction hardware—using the MAGLAB's shops and support. He is also consulting with Florida A&M University and FSU for their Composites R & D laboratory, and was heavily involved with the overall preparation of this Handbook.

Don Weed   Staff Scientist, Southwest Research Institute, San Antonio, Texas. Mr. Weed's 30 years of experience in the fabrication and bonding of composite materials includes development of the tube-fabrication technology at General Dynamics and work with the IIT Research Institute. He has valuable experience with graphite, boron, Kevlar, and glass-epoxy composites, and in developing specialized forming techniques for making specimens and structural members. Mr. Weed contributed to the first application of the boron/epoxy tape laying machine. He has also worked on the development of articulated armor vests employing armor plate, on processing techniques for automobile leaf springs, and on a study of moisture effects on the mechanical performance of graphite/epoxy materials.

Bruce Wendle   Senior Manufacturing Engineer, Advanced Input Devices, Coeur d'Alene, Idaho. Mr. Wendle has a B.S. in chemical engineering from University of Idaho. He has worked with Boeing, Mobay Chemical, E. I. du Pont de Nemours, and others. He is experienced in tooling, manufacturing engineering, and product design. Mr. Wendle is a member of the Society of Plastic Engineers, and was a honored service member in 1990. He has published several books on plastics, including *What Every Engineer Should Know About Developing Plastics Products* (Marcel Dekker, Inc.).

Kjell Zandin   Senior Vice President, International Business Development and Director, H. P. Maynard & Company, Inc., Pittsburgh, Pennsylvania. A native of Gothenburg, Sweden, Mr. Zandin joined Maynard Sweden as a project engineer. He developed the new concept of work measurement, which subsequently became the Maynard Operation Sequence Technique, MOST® Systems. After introducing manual MOST® Systems to the United States, he directed the development of MOST® Computer Systems, which is considered to be the state of the art in work measurement. Mr. Zandin holds a degree in mechanical engineering from Chalmers University of Technology in Gothenburg. He has written numerous articles on work measurement, work management, and work participation published by technical and management publications in the United States and Sweden. He has spoken at international conferences and seminars arranged by industrial engineering societies and associations. He was recipient of the first Technical Innovation in Industrial Engineering Award presented by the IIE, and later the Royal Charter Award from the Institution of Production Engineers in Great Britain. Mr. Zandin authored *MOST® Work Measurement Systems* (Marcel Dekker, Inc.).

Index